246 Solved Structural Engineering Problems

Third Edition

C. Dale Buckner, PhD, PE

PROFESSIONAL PUBLICATIONS, INC.

How to Locate Errata and Other Updates for This Book

At Professional Publications, we do our best to bring you error-free books. But when errors do occur, we want to make sure that you know about them so they cause as little confusion as possible.

A current list of known errata and other updates for this book is available on the PPI website at **www.ppi2pass.com**. From the website home page, click on "Errata." We update the errata page as often as necessary, so check in regularly. You will also find instructions for submitting suspected errata. We are grateful to every reader who takes the time to help us improve the quality of our books by pointing out an error.

246 SOLVED STRUCTURAL ENGINEERING PROBLEMS
Third Edition

Current printing of this edition: 1

Printing History

edition number	printing number	update
1	4	Minor corrections.
2	1	New edition. Copyright update.
3	1	New edition. Copyright update.

Printed in the United States of America

Professional Publications, Inc.
1250 Fifth Avenue, Belmont, CA 94002
(650) 593-9119
www.ppi2pass.com

Library of Congress Cataloging-in-Publication Data
Buckner, C. Dale.
246 solved structural engineering problems / C. Dale Buckner.--3rd ed.
p. cm.
Includes bibliographical references and index.
ISBN: 1-59126-003-5
1. Structural engineering--Problems, exercises, etc. I. Title: Two hundred forty-six solved structural engineering problems. II. Title: Two hundred and forty-six solved structural engineering problems. III. Title.

TA640.4.B83 2003
624.1'7'076--dc22

2003063952

Table of Contents

Preface

Significant changes have occurred in structural design criteria since the first printing of *246 Solved Structural Engineering Problems*, which was based on the 1988 *Uniform Building Code*. Design specifications for steel, concrete, timber, and masonry structures have been revised several times to reflect new research findings and to correct deficiencies. The *National Design Specification for Wood Construction* was revised extensively in 1991 to better represent member and connector strengths and to update values of allowable stresses. As a first step toward a model building code for the United States, the 1994 edition of the *Uniform Building Code* was completely reformatted by moving topics from one chapter to another and by reorganizing and renumbering sections. Further changes were made in the 1997 edition by expressing earthquake loading on the basis of ultimate rather than working load values and incorporating new design specifications. The building code requirements for reinforced concrete, ACI 318-02, have been revised to provide more realistic limits on reinforcement. While none of these changes substantively affect the solutions presented in previous editions, it became necessary to revise the problems to reflect the latest design expressions, allowable stresses, and code citations.

Perhaps the greatest impetus for changes in design criteria was the behavior of structures during the Northridge earthquake of 1994. Severe damage to welded steel moment frames during the earthquake invalidated the prescriptive design rules for steel moment frame connections that were included in earlier codes. Currently, there are many research projects underway to verify connection details that should ensure ductile behavior of steel moment frames in future earthquakes. While I have attempted to follow the 1997 *Uniform Building Code* (UBC-97) for problems related to seismic design, the lack of specific criteria in UBC-97 for steel moment frames requires alternatives. In revising the problems for steel moment frames for this edition (four problems), I have followed interim guidelines issued by the Federal Emergency Management Agency (FEMA). These criteria represent the state-of-practice as of spring 1997. Refinements to these criteria may occur as research progresses.

Different building codes are used in various regions of the United States. For simplicity, only the 1997 UBC is cited in the problems appearing in this book. The differences between codes are found primarily in the specified loadings. Thus, readers who are accustomed to other building codes should be able to interpret the problems without difficulty.

To provide a reasonably sized publication, it was necessary to limit the solutions in this book to an average of one page per problem. Within this limit, the essential steps to every solution are presented. Textbook-type information, such as derivations and theory, could not be included. Wherever possible, I have used a standard method and have cited an appropriate reference for the procedure.

C. Dale Buckner, PhD, PE

Acknowledgments

Many organizations and individuals have contributed to this book. The problems appearing in this book are reprinted or reconstructed from California Structural Engineering Examinations between 1968 and 1983, with permission by the Board of Registration for Professional Engineers and Land Surveyors. These examinations are copyrighted by the Board of Registration for Professional Engineers and Land Surveyors. Complete examinations may be available from the Board at a nominal cost. The solutions provided in this book are the work of the author and do not represent the Board's official solutions to the examination questions.

Figures accompanying the problem statements were prepared by Carol L. Irvine and Rhonda A. Jones for the *Structural Engineering Practice Problem Manual*, a precursor to this book, published in 1985. About 120 solutions were contributed by users of that book and many of these were used to check and revise the solutions in this book.

Many helpful suggestions were obtained from my colleagues and from reviews. I am especially grateful to Vijaya K.A. Gopu, of Louisiana State University, for his advice and generosity in providing notes that proved useful in solving several problems. Readers of the earlier printings have been helpful in reporting errors, which I have endeavored to correct.

I am indebted to the staff at Professional Publications, Inc., especially Jessica Holden, Cathy Schrott, and Aline Magee, for their patience and assistance during the preparation of this edition. Finally, I express appreciation to my wife, Jackie, and son, Brandon, for the help and encouragement they provided throughout the course of this work.

How To Use This Book

246 Solved Structural Engineering Problems provides a comprehensive set of practice problems, with solutions, for those studying for registration exams in structural or civil engineering. The problems are arranged by chapters in the broad categories of structural analysis, seismic analysis, foundation design, structural steel, structural concrete, timber, and masonry.

Within each chapter, problems are arranged in order of increasing complexity. As a rough guide, the first quarter of the problems in each chapter are basic, relatively short problems that should require about 30 to 45 minutes to solve. The middle half of the chapter contains more difficult problems, which should require about 1 hour to solve. The last quarter of the problems in each chapter are more in-depth and require $1^1/_2$ to 2 hours to solve. These latter problems are similar in scope to those on the NCEES Structural II Examination.

This book also contains problems with questions that require specific short answers, rather than calculations and sketches. These questions cover a broad range of topics and are especially helpful in preparing for the NCEES Structural I Examination's multiple-choice questions.

Any method of linear structural theory can be used to solve the analysis problems in the Structural chapter. I have used the method that I considered best suited to a particular problem. Some engineers might prefer an alternative method. Moment distribution, for example, is a popular choice for manual solutions to statically indeterminate problems. I usually avoided moment distribution because the tabular format was not well suited to the layout and space limitations of this book.

Many of the design problems in this book require assumptions that lead to unique solutions. My solutions to such problems are intended only to serve as a basis of comparison for someone studying to take the NCEES structural engineering examinations. The solutions should not be interpreted as recommended design procedures for similar situations.

Problems in the Steel chapter are solved using either the AISC allowable stress design (AISC 1989) or the Load and Resistance Factor Design (LRFD) method (AISC 1994). Allowable Stress Design (ASD) is the more widely used method in current practice. LRFD is relatively new, but is gaining acceptance as LRFD-based textbooks become available and universities begin to emphasize LRFD. I believe that LRFD is the superior approach and that it will gradually replace ASD (just as strength design replaced working stress design for reinforced concrete). In the meantime, illustrative examples are needed for both methods. Each method is used to work approximately one-half of the steel problems. The method used is indicated at the beginning of each solution.

Most of the timber design problems are solved using criteria from the *AITC Timber Construction Manual*, third edition. Essentially the same design criteria are published in the *National Design Specification* (NDS) *for Wood Construction* (NFAP 1991). The *AITC Manual* (AITCM) contains additional design aids and data that are not included in the NDS. The AITCM is referenced in the belief that it is simpler for a reader to consult only one reference, if needed, rather than several. Some timber problems also require plywood diaphragm shear capacities, which are tabulated in both the 1997 UBC and in a bulletin published by the American Plywood Association. For problems requiring diaphragm shear capacity, both the AITCM and the 1997 UBC are usually referenced.

Problems in the Concrete chapter are solved using the strength design method of ACI 318-02. For the Masonry problems, the allowable stress method of the 1997 UBC is used for all solutions.

Codes, Handbooks, and References

American Concrete Institute Committee 318. *Building Code Requirements for Reinforced Concrete and Commentary* (ACI 318-02). Detroit: American Concrete Institute, 2002.

American Concrete Institute Committee 439. *Mechanical Connections of Reinforcing Bars* (ACI 439.3R-2). Detroit: American Concrete Institute, 1994.

American Institute of Steel Construction. *Engineering for Steel Construction*. Chicago: American Institute of Steel Construction, 1984.

American Institute of Steel Construction. *Manual of Steel Construction—Load and Resistance Factor Design*. Chicago: American Institute of Steel Construction, 1994.

American Institute of Steel Construction. *Manual of Steel Construction—Allowable Stress Design*, 9th ed. Chicago: American Institute of Steel Construction, 1989.

American Institute of Timber Construction. *Timber Construction Manual*, 3rd ed. New York: John Wiley and Sons, 1985.

American Plywood Association. *Construction Guide—Diaphragms*. Tacoma, WA: American Plywood Association, 1989.

Applied Technology Council. *Guidelines for the Design of Horizontal Wood Diaphragms*, Report No. ATC-7. Berkeley, CA: Applied Technology Council, 1981.

Becker, R., F. Naeim, and E.J. Teal. "Seismic Design Practice for Steel Buildings." *Steel Tips* (1988). Walnut Creek, CA: Steel Committee of California.

Blodgett, O.W. *Design of Welded Structures*. Cleveland: The James F. Lincoln Welding Foundation, 1966.

Bresler, B., T.Y. Lin, and J.B. Scalzi. *Design of Steel Structures*, 2nd ed. New York: John Wiley and Sons, 1968.

Darwin, D. *Design of Steel and Composite Beams with Web Openings*. Chicago: American Institute of Steel Construction, 1989.

FEMA-267, *Interim Guidelines Advisory No. 1: Evaluation, Repair, Modification and Design of Welded Steel Moment Frame Structures*. Report No. SAC-96-03. Washington, D.C.: Federal Emergency Management Agency, 1997.

Fisher, J.M., and D.R. Buettener. *Light and Heavy Industrial Buildings*. Chicago: American Institute of Steel Construction, 1979.

International Conference of Building Officials. *Uniform Building Code*. Whittier, CA: International Conference of Building Officials, 1997.

Iwankin, N. "Ultimate Strength Considerations for Seismic Design of the Reduced Beam Section (Internal Plastic Hinge)," *AISC Engineering Journal*, 1st Quarter, 1997.

Naeim, Farzad, ed. *The Seismic Design Handbook*. New York: Van Nostrand Reinhold Company, 1989.

National Forest Products Association. *National Design Specification for Wood Construction*, 9th ed. Washington, D.C.: National Forest Products Association, 1991.

Norris, C.H., and J.B. Wilbur. *Elementary Structural Analysis*, 2nd ed. New York: McGraw-Hill Book Company, 1960.

Paulay, T., and M.J.N. Priestley. *Seismic Design of Reinforced Concrete and Masonry Buildings*. New York: Wiley, 1992.

Paz, Mario. *Structural Dynamics—Theory and Computation*, 2nd ed. New York: Van Nostrand Reinhold Company, 1988.

Portland Cement Association. *Design and Control of Concrete Mixtures*, 13th ed. Skokie, IL: Portland Cement Association, 1988.

Prestressed Concrete Institute. *PCI Design Handbook*, 5th ed. Chicago: Prestressed Concrete Institute, 1999.

Rice, P.F., and E.S. Hoffman. *Structural Design Guide to the ACI Building Code*. New York: Van Nostrand Reinhold Company, 1972.

Roark, R.J., and W.C. Young. *Formulas for Stress and Strain*, 5th ed. New York: McGraw-Hill Book Company, 1975.

Salmon, C.G., and J.E. Johnson. *Steel Structures—Design and Behavior*, 3rd ed. New York: Harper and Row, 1990.

Scalzi, J.B., W. Podolny, Jr., and W.C. Teng. *Design Fundamentals of Cable Roof Structures*. Pittsburgh: United States Steel Corporation, 1969.

Schnable, Harry, Jr. *Tiebacks in Foundation Engineering and Construction*. New York: McGraw-Hill Book Company, 1982.

Structural Engineers Association of California. *Recommended Lateral Force Requirements and Tentative Commentary*. San Francisco: Seismology Committee Structural Engineers Association of California, 1996.

Taranath, Bungale S. *Structural Analysis and Design of Tall Buildings*. New York: McGraw Hill Book Company, 1988.

Winter, G., L.C. Urquhart, C.E. O'Rourke, and A.H. Nilson. *Design of Concrete Structures*, 7th ed. New York: McGraw-Hill Book Company, 1964.

1 STRUCTURAL

STRUCTURAL-1

REQUIRED

Draw the bending moment diagrams for the frames with the loadings shown in figures S-1(a) and (b) and indicate the magnitude of the maximum bending moments.

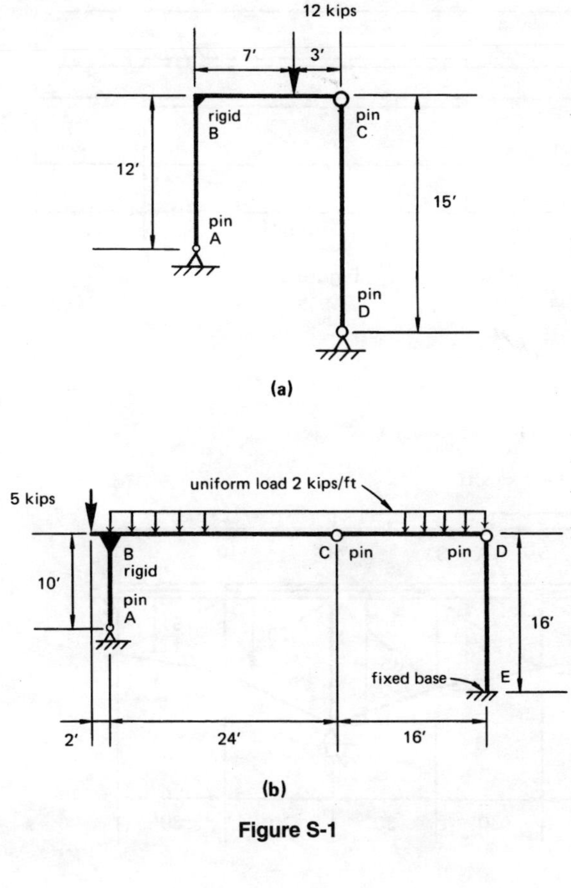

Figure S-1

STRUCTURAL-2

The roof truss shown is one of a series spaced 12 feet on center and loaded uniformly on the top chord by joists spaced 16 inches on center. The roof dead load (including effective truss weight) is 20 psf on a horizontal projection. Live load is per UBC requirements for non-snow load area. The upper chord follows a curve described by: $Y = X(1 - X/48)$, with origin of coordinates at left support.

DESIGN CRITERIA

- All members are pin-connected.
- The truss is statically determined for external reactions.

REQUIRED

Determine the internal design force in members A, B, and C for the loading shown in figure S-2.

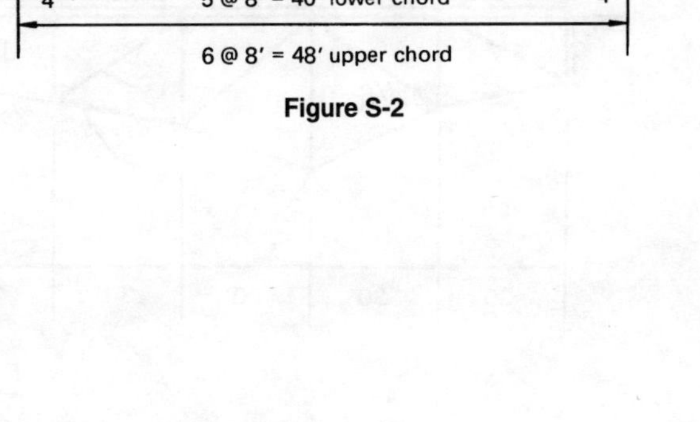

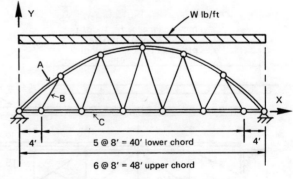

Figure S-2

STRUCTURAL-3

REQUIRED

Given the cantilevered beam shown in figure S-3, find the maximum shear stress at point O.

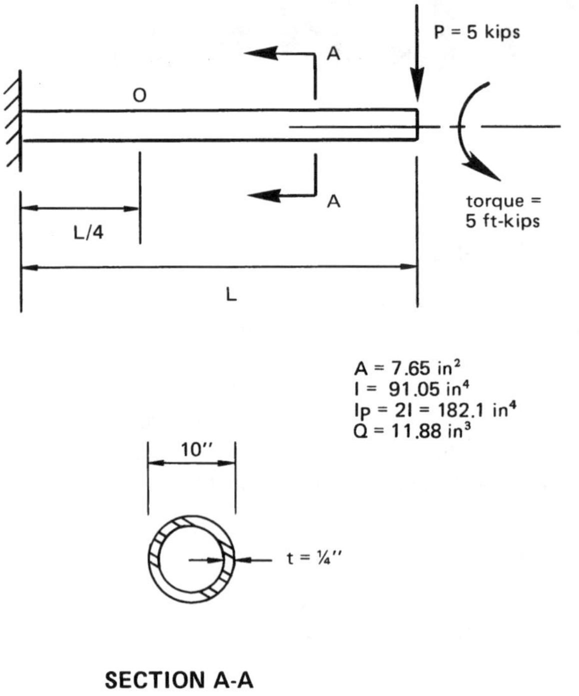

Figure S-3

STRUCTURAL-4

A steel bridge truss is shown in figure S-4.

REQUIRED

1. Construct influence line for axial force of members A, B, and C. Show coordinates.
2. Calculate maximum and minimum axial forces on members A, B, and C by using influence lines in part (a).

STRUCTURAL-5

One of the beams on a project must carry extra load, and can be reinforced through the addition of cover plates.

DESIGN CRITERIA

- The total design load, including the dead load of the beam, is 10,000 pounds applied at the center of the span.
- The resultant values of moments of inertia (I) are shown in figure S-5.

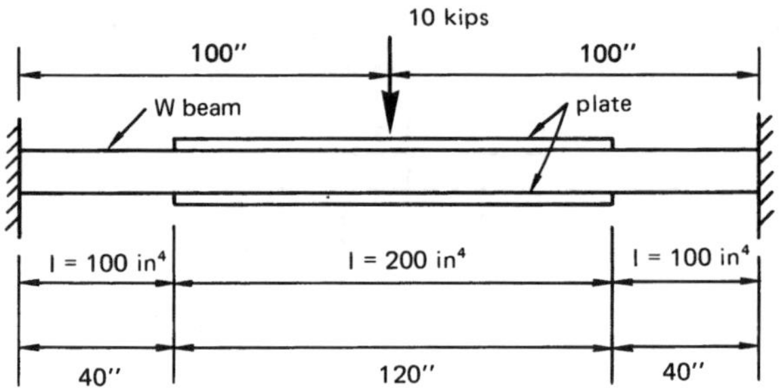

Figure S-5

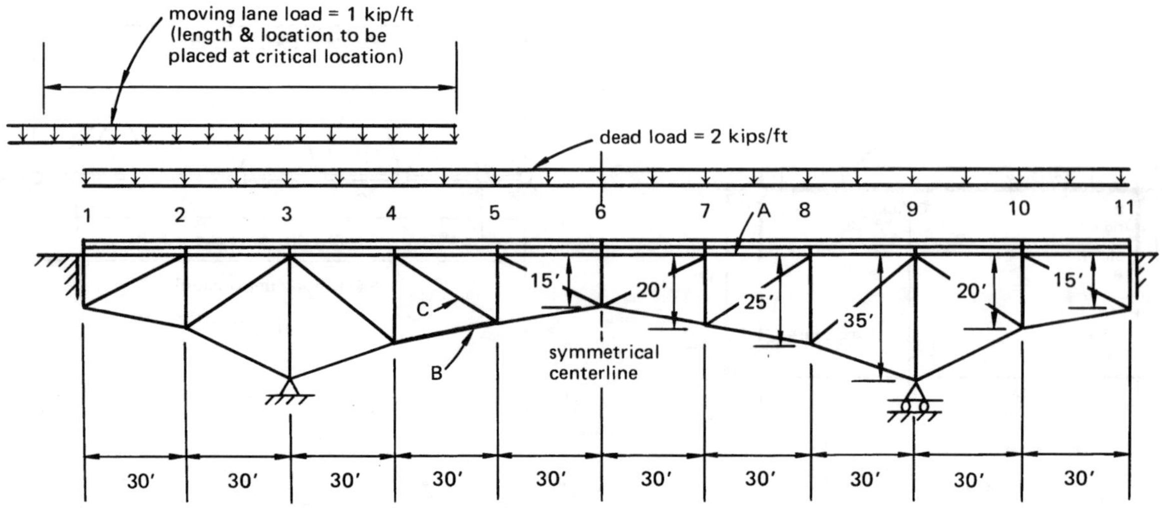

Figure S-4

STRUCTURAL-6

A rectangular opening must be framed as indicated in figure S-6 to support a temporary load.

DESIGN CRITERIA

- The members shown are 10-inch by 12-inch wood beams (net) with strong axis vertical.
- Assume no movement of supports at the edge of the opening.

REQUIRED

Compute the maximum bending stress in each beam.

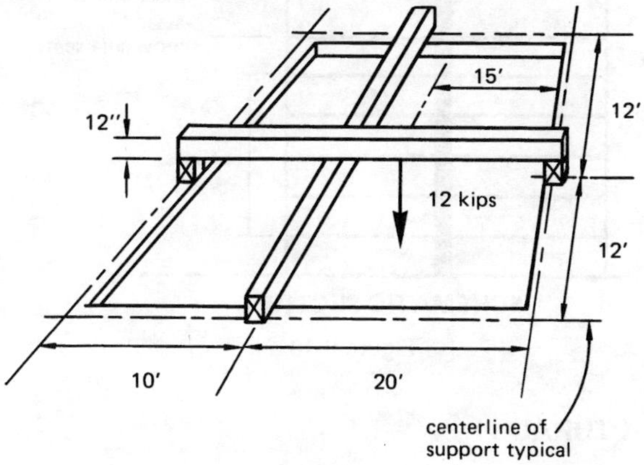

Figure S-6

STRUCTURAL-7

An existing king post truss is made from a 4 x 14 (nominal) wood beam, a 1 sq. in. (net) steel tie rod and a 4 x 6 (nominal) strut as shown in figure S-7.

DESIGN CRITERIA

- Assume beam is laterally supported full length in horizontal direction.
- Strut is laterally stayed at each end.
- $E_s = 30 \times 10^6$ psi
- $f_s = 22{,}000$ psi
- $E_w = 1.6 \times 10^6$ psi
- $f_b = 1500$ psi
- $f_c = 1250$ psi

REQUIRED

(a) Determine the allowable load (P) which can be applied to the truss at centerline without overstressing any one of three members. Which member governs allowable load (P)?
(b) Compute deflection (Δ) of truss at centerline of span due to load (P).
(c) Compute stress in each member.

Lumber

4 × 14	4 × 6
$A = 46.4$ in^2	$A = 19.3$ in^2
$S = 102.4$ in^3	$S = 17.7$ in^3
$I = 678.5$ in^4	$I = 48.5$ in^4

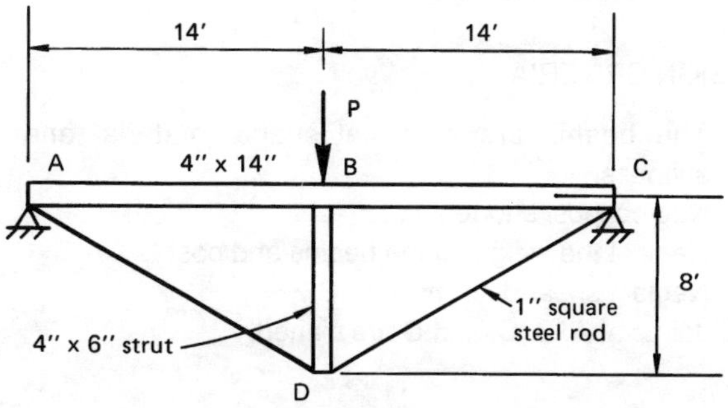

Figure S-7

STRUCTURAL-8

A section of crane rail girder is supported on piles. The piles at A and C are rigid supports. The pile at B was discovered to yield as load was applied. Tests showed that pile B acts like a spring with a constant K = 600 kips per inch of vertical displacement. A wheel load P of 100 kips is placed at B.

DESIGN CRITERIA

- Assume that the force in pile B is zero under dead load of the girder.
- The crane girder is 2.5 ft. square, regular weight concrete.
- $E_c = 3.0 \times 10^6$ psi
- Use the moment of inertia of the gross concrete section for this problem.

REQUIRED

Determine the moment and deflection in the girder at point B due to the wheel load.

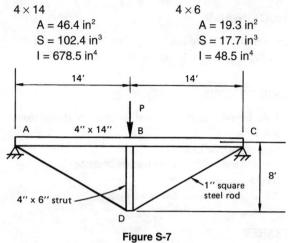

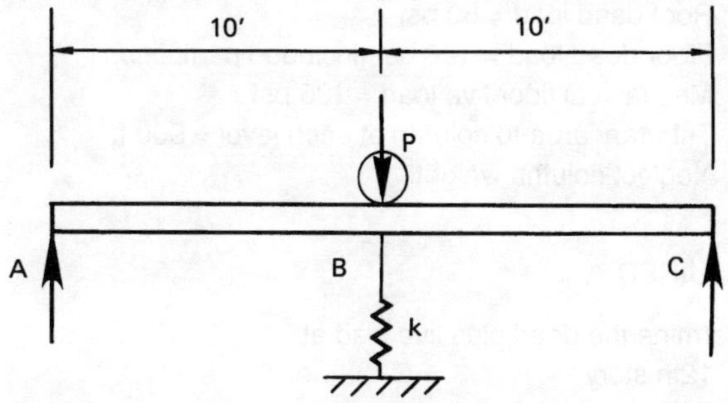

Figure S-8

STRUCTURAL-9

Beam AB and CD are joined together by a post and are supporting a load of 4.5 kips, as shown in figure S-9.

DESIGN CRITERIA

- The beams are of equal shape, material and stiffness.
- Neglect post shortening.
- Neglect the weight of the beams and post.
- Neglect shear deformation ($G = \infty$).
- Joints at A, B, C, and D are pinned.

REQUIRED

Find values of reactions at A, B, C, and D.

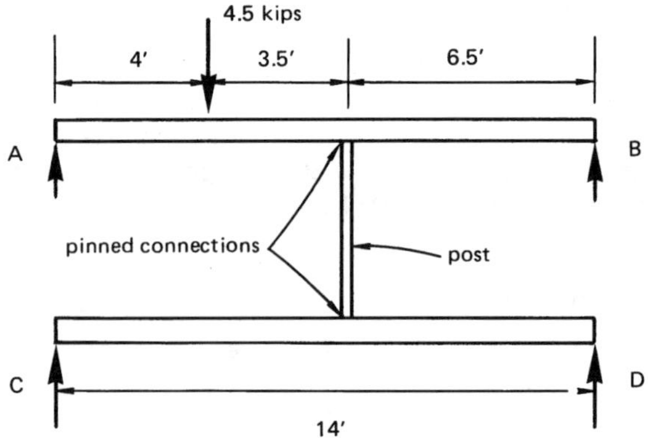

Figure S-9

STRUCTURAL-10

A schematic section which indicates the floor occupancy of an 18-story multi-use building is shown in figure S-10.

DESIGN CRITERIA

- Roof dead load = 80 psf
- Floor dead load = 100 psf (included partitions)
- Mechanical floor live load = 125 psf
- Tributary area to column at each level = 900 ft^2
- Neglect column weight.

REQUIRED

Determine the dead plus live load at:
(a) 12th story
(b) 6th story
(c) 1st story

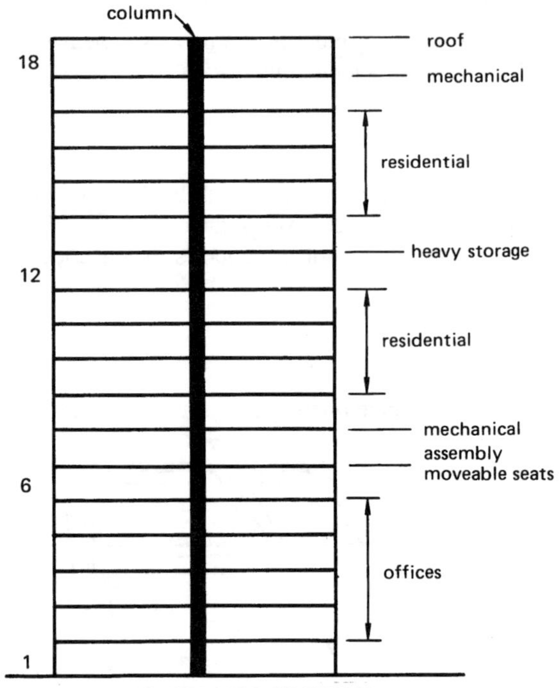

SCHEMATIC SECTION

Figure S-10

STRUCTURAL-11

A frame is loaded as shown in figure S-11.

DESIGN CRITERIA

- $E = 29 \times 10^6$ psi

REQUIRED

(a) Determine the moments at each joint.
(b) Sketch the moment diagram.
(c) Sketch the deflected shape.
(d) Determine the maximum deflection in span A-B.

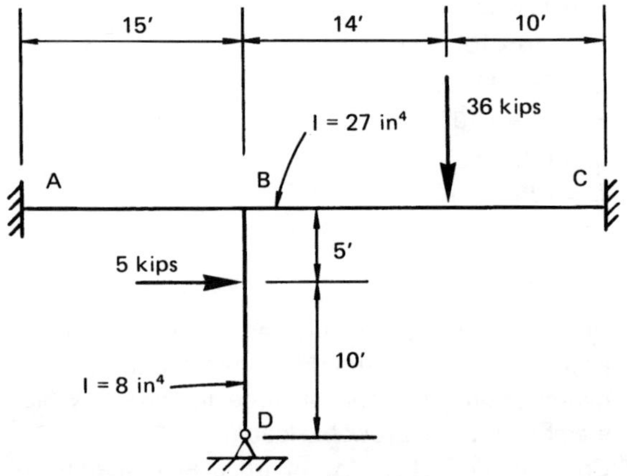

Figure S-11

STRUCTURAL-12

An existing W 16 x 67 steel beam (see figure S-12) was found to have settled 1/4 inch at support B with respect to support A and 1/8 inch at support C with respect to support A. Support A did not settle.

DESIGN CRITERIA

- The original design loads shown in figure S-12 are valid.
- The compression flange is adequately stayed so that buckling will be prevented under the load shown.
- The beam complies with ASTM A36 steel requirements.
- The beam is continuous from A to C.

REQUIRED

Determine whether the beam is over-stressed. Show all necessary calculations.

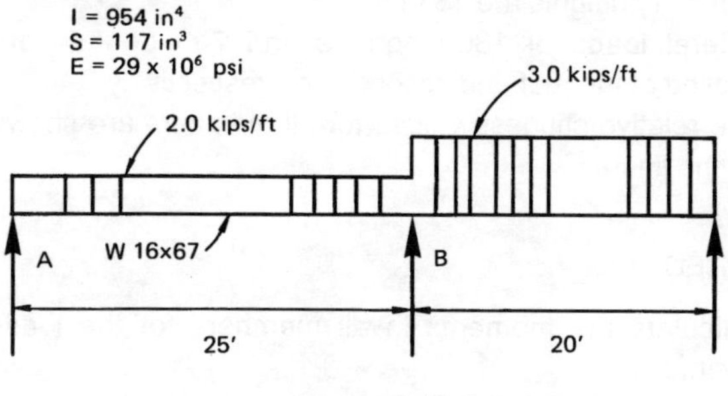

Figure S-12

STRUCTURAL-13

A rigid frame is shown in figure S-13. The right support has settled 6 inches.

DESIGN CRITERIA

- $I_1 = 500 \text{ in}^4$ (at AB)
- $I_2 = 1600 \text{ in}^4$ (at BC)
- $I_3 = 600 \text{ in}^4$ (at CD)
- $E = 29 \times 10^6$ psi

REQUIRED

(a) Find moments at joints B and C due to the settlement.
(b) Draw the moment diagram.
(c) Find the vertical and horizontal forces at joints A and D. Draw the moment diagram.

NOTE: Use following units: moments (ft-kips) and forces (kips). Round to nearest whole number.

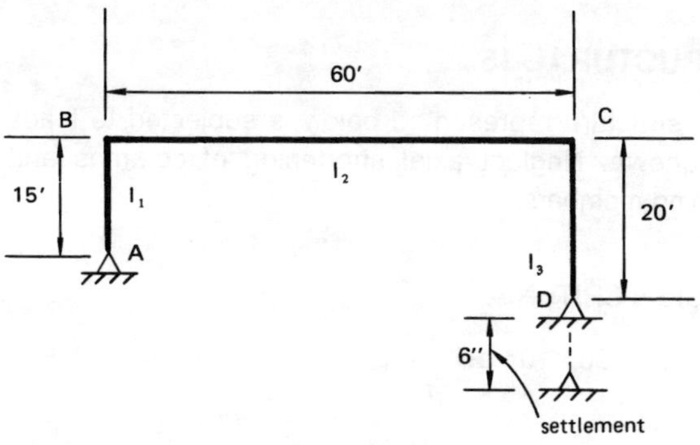

Figure S-13

STRUCTURAL-14

All members in figure S-14 are timber except diagonal DG which is a steel rod. All connections are assumed pinned.

DESIGN CRITERIA

- $E_W = 1.6 \times 10^6$ psi
- $E_S = 29 \times 10^6$ psi

REQUIRED

Determine for the loading shown (Show calculations):
(a) External reactions.
(b) Internal member forces (shear, moment, axial)
(c) Horizontal deflection of point C (neglect shear distortions within the members). What makes the greater contribution to the overall deflection: axial or bending loads?

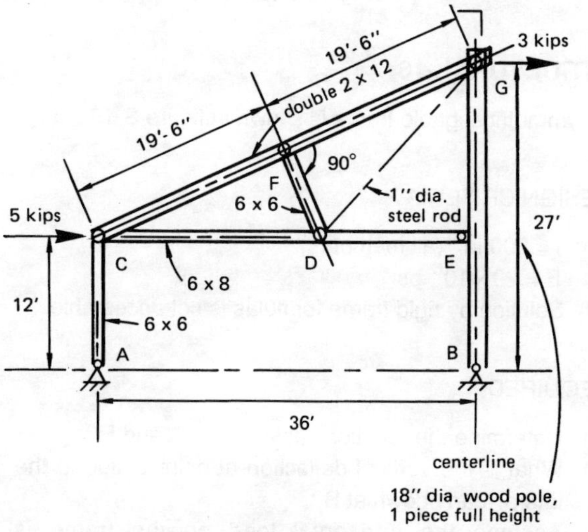

Figure S-14

STRUCTURAL-15

The structure represented below is subjected to loads as shown. Neglect axial shortening of columns and sloping members.

DESIGN CRITERIA

- $E_s = 30 \times 10^6$ psi

REQUIRED

Show all calculations.

(a) Load in member BD.
(b) Deflected position of B.

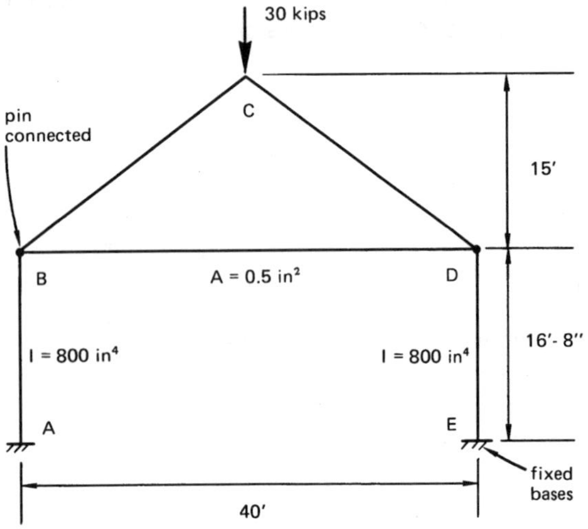

Figure S-15

STRUCTURAL-16

A symmetrical gable frame is shown in figure S-16.

DESIGN CRITERIA

- $I = 300$ in^4 (all members)
- $E = 29 \times 10^6$ psi
- Solution by rigid frame formulas is not acceptable.

REQUIRED

(a) Determine the reactions at supports A and E.
(b) What is the vertical deflection at point C due to the lateral load shown at B?
(c) Assume the horizontal load on the frame is removed. Apply a 5 kip vertical load at C. What is the horizontal deflection at B?

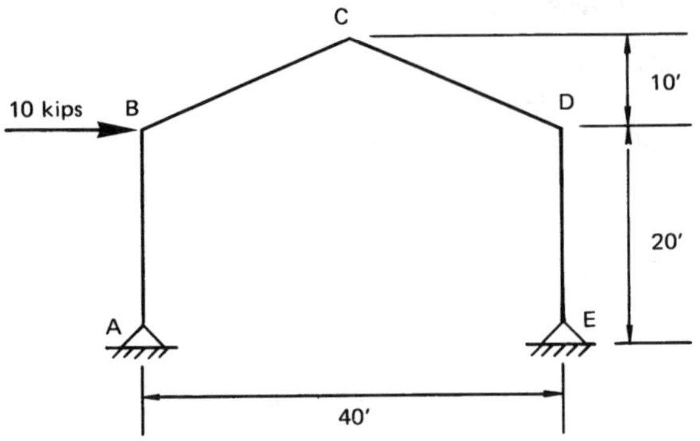

Figure S-16

STRUCTURAL-17

A two-story building frame is shown in figure S-17.

DESIGN CRITERIA

- The beams span 26 feet.
- The story heights are 13 feet.
- Lateral loads of 1300 pounds and 700 pounds are applied at the roof and second floor, respectively.
- The relative stiffness values for all members are shown on the figure.

REQUIRED

(a) Calculate the moments in all members for the loads given.
(b) Draw the moment diagram and show all critical values.
(c) Draw the deflected shape of the frame and identify the points of contraflexure (location can be closely approximated).

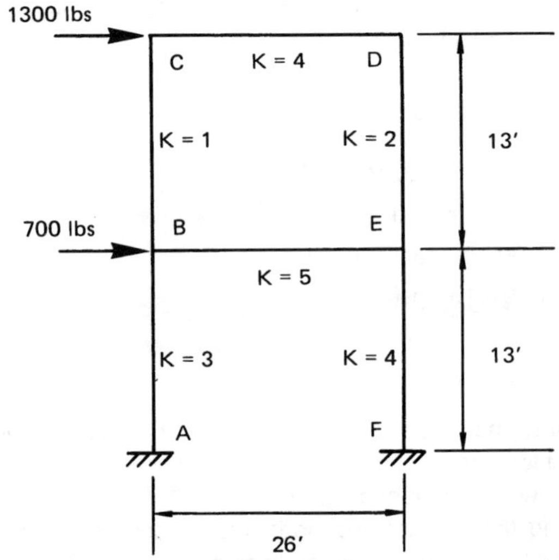

Figure S-17

STRUCTURAL-18

A two bay rigid shown in figure S-18. Assume the girders are infinity stiff.

REQUIRED

What are the base shears at points A, E, and G?

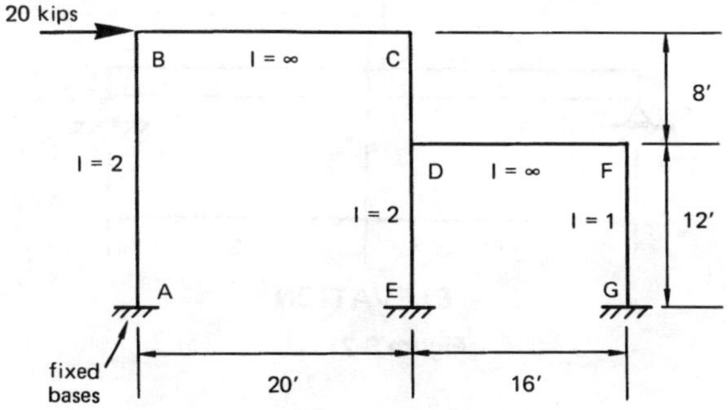

Figure S-18

STRUCTURAL-19

A structural steel plane rigid frame is shown in figure S-19.

DESIGN CRITERIA

- The centers of gravity of story weights are at the respective roof and floor levels.
- Lateral forces and corresponding total deflections are shown on the figure.
- Stresses in steel members remain within the elastic range.

REQUIRED

(a) Compute the fundamental period of the frame using the appropriate UBC formula.
(b) Assume the lateral forces are doubled. Compute the fundamental period of the frame.
(c) What is the significance of the results when (a) is compared to (b)?

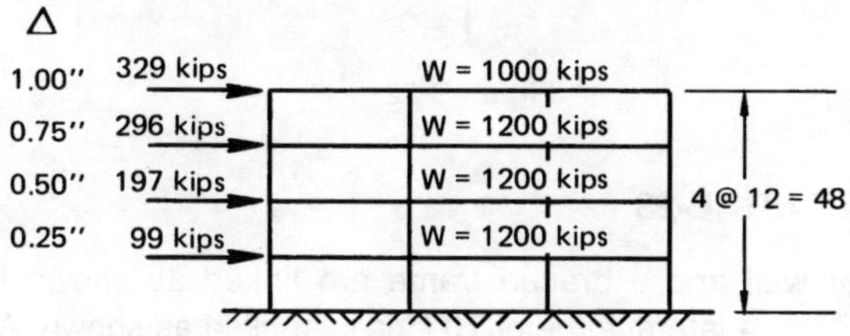

Figure S-19

STRUCTURAL-20

For the evaluation of moment, shears, and axial loads in high rise buildings there are three approximate methods of analysis in common use as follows:

1. Portal method
2. Cantilever method
3. Moment distribution method (Hardy Cross)

The frame shown in figure S-20(a) was analyzed by each of these methods for overturning forces at the third floor. Figures S-20(b), (c), and (d) show the results of this analysis for the overturning forces only - not necessarily in the same order as listed above.

REQUIRED

(a) For each of the three figures S-20(b), (c), and (d) identify the method that was used to obtain the results shown.
(b) List the principal assumptions that are used as the basis for each of the three methods.
(c) For what type and configuration of structural frame would the portal method be chosen rather than the cantilever method? Explain.

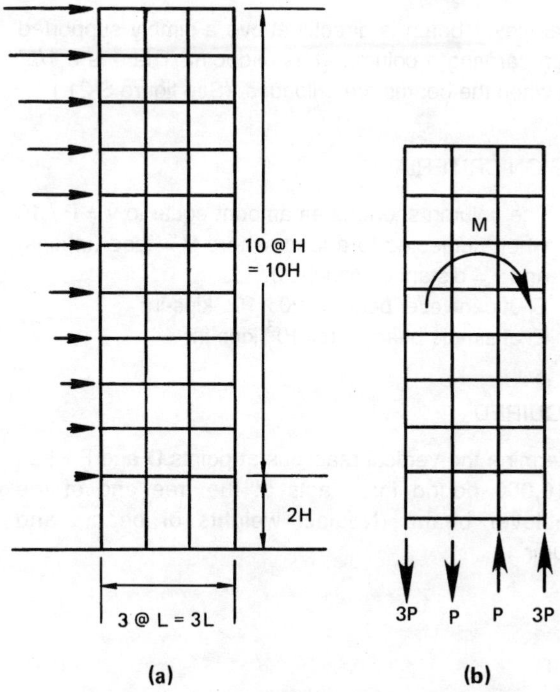

Figure S-20

(continued on next page)

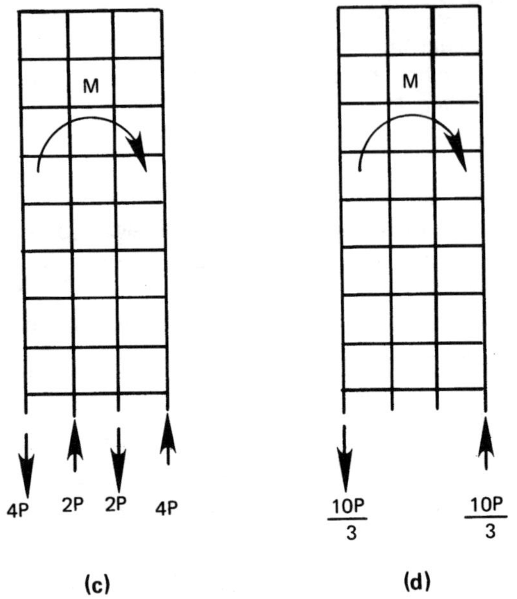

Figure S-20 (cont.)

STRUCTURAL-21

A cantilever beam is directly above a simply supported beam carrying a column at its midpoint. There is a 1/2" gap when the beams are unloaded. (See figure S-21.)

DESIGN CRITERIA

- The column shortens an amount equal to y = P / 10 when subjected to a load, where P = load (in kips) and y = distance (in inches).
- EI of cantilever beam = 20×10^5 kips-in^2
- EI of simple beam = 5×10^5 kips-in^2

REQUIRED

Determine the vertical reactions at points D and F when a 10,000 pound force acts at the free end of the cantilever beam. (Neglect weights of beams and column.)

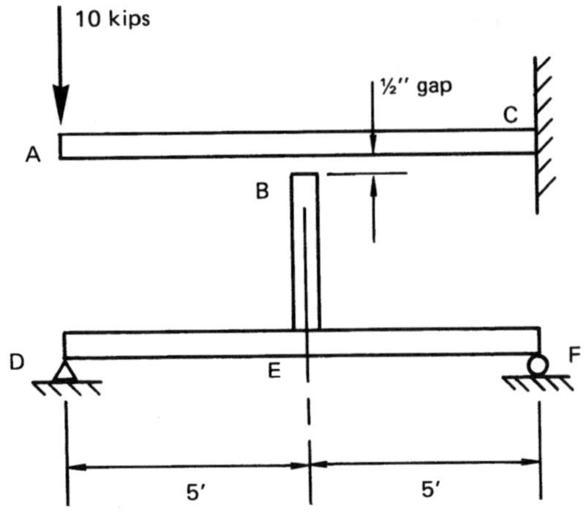

Figure S-21

STRUCTURAL-22

A beam rests on a rigid support at one end and on a spring at the other end as shown in figure S-22. (Neglect beam weight and damping effect.)

DESIGN CRITERIA

- $E = 1.5 \times 10^6$ psi
- $K = 200$ lb/in

REQUIRED

(a) Determine the maximum stress in the beam caused by the 90 pound weight falling through 4 inches.

(b) Determine the maximum stress in the beam caused by the 90 pound weight falling through 4 inches if the spring at the right end is replaced by a firm support.

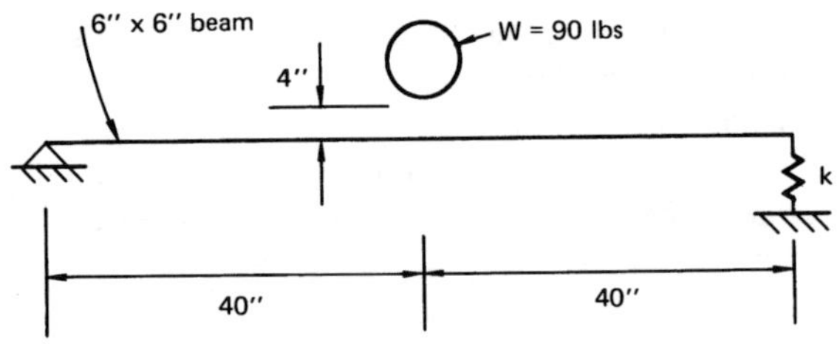

Figure S-22

STRUCTURAL-23

A shear wall and a braced frame are linked as shown in figure S-23. A lateral force of 70 kips is applied as shown. All connections may be assumed adequate.

DESIGN CRITERIA

- $f'_c = 3$ ksi
- $G = .4\ E$
- A501 pipe
- $F_y = 36$ ksi
- $E_s = 29 \times 10^6$ psi

REQUIRED

Determine the following:

(a) Force resisted by the shear wall.
(b) Force resisted by the link.
(c) Force resisted by the braced frame.
(d) Deflection of shear wall.
(e) Deflection of braced frame.

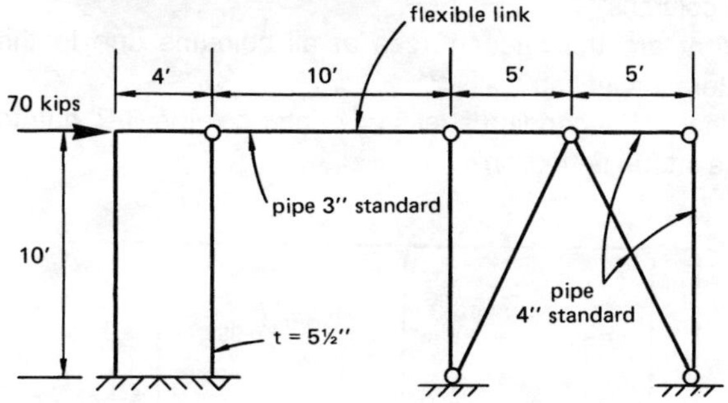

Figure S-23

STRUCTURAL-24

The floor and roof plans of a two-story hospital building with shear walls at the ends are shown in figure S-24. Each acts as a diaphragm.

DESIGN CRITERIA

- Roof dead load = 50 psf
- Roof live load = 20 psf
- Floor dead load = 70 psf
- Floor live load = 50 psf
- Ignore weight of walls.
- Chords take all bending stresses and are located six inches in from all diaphragm edges.
- Assume concrete takes all shear forces.
- Seismic zone 3
- $f'_c = 3000$ psi
- $G = \infty$ (infinity)
- Soil profile type S_D
- Near-fault factors of unity

REQUIRED

(a) Determine total seismic load to the roof, and to the second floor in the design of the diaphragms.
(b) Determine the second floor shear forces at section A-A due to North-South lateral loads of 200 pounds per foot along line C and along line D.
(c) What would be the effect on the shear force calculated if the shear modulus (G) were less than ∞ (infinity)? Why?

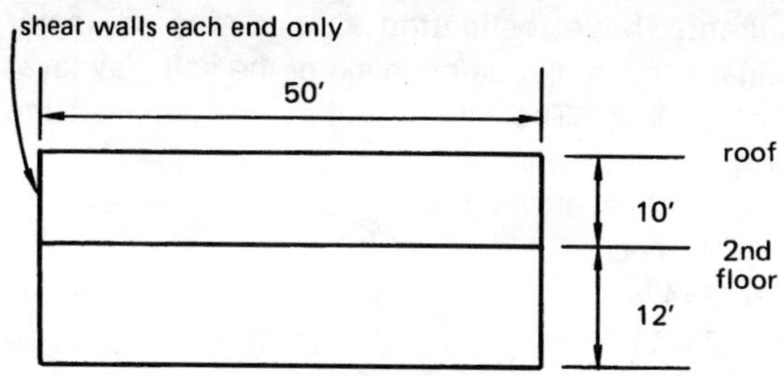

ELEVATION
typical for end walls

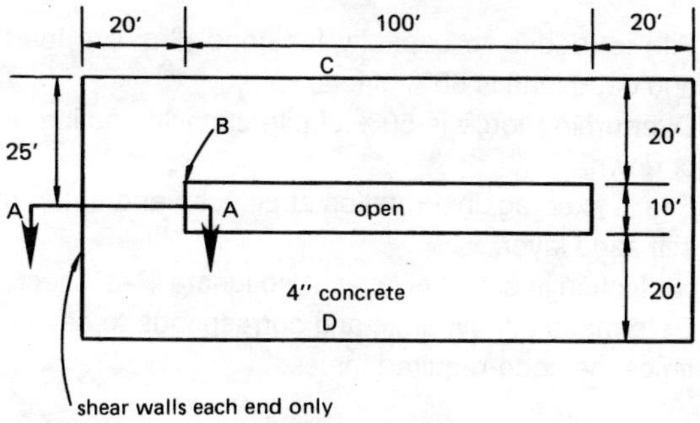

SECOND FLOOR AND ROOF PLAN

Figure S-24

STRUCTURAL-25

The parts of this question relate to a field investigation to be made during the construction phase of some jobs in your area.

REQUIRED

(a) A multi-story steel framed building is supported on a pile and grade beam foundation system. The piles are in place and the contractor is getting ready to pour the grade beams. List six items you should check and verify during a field investigation.
(b) A subterranean parking structure under a building uses concrete columns and flat slab design. The concrete columns are poured and the contractor is ready to pour the flat slab. List six items you should check and verify during a field investigation.
(c) A one-story type V building is supported on spread footings with a slab on grade. The contractor has finished all rough carpentry and structural sheathing. He has asked for final framing inspection before covering up the structural elements. List six items you should check and verify during a field investigation.

STRUCTURAL-26

A 120 ton capacity, 16 inch square precast, prestressed (pretensioned) pile is used to support a portion of a building.

The pile passes through a layer of soft clay and then enters stiff sand 15 feet below the pile cap. Soils consultants have estimated that during a major earthquake the lateral deformation of the soft clay layer will be 0.3 inch. The pile manufacturer supplied the following specifications:

f'_c = 5000 psi at 28 days
W = 145 pcf
I = 5344 in^4
A = 254 in^2
S = 668 in^3
effective prestress = 765 psi

DESIGN CRITERIA

- Pile is loaded to capacity for dead plus live load, and dead load is 80% of total load.
- Overturning force is 50% of pile capacity, acting up or down.
- Pile is fixed against rotation at pile cap and at top of stiff sand layer.
- No friction in soft clay layer. Also ignore P–Δ effect.
- Deformation is dynamic and corresponds to several times the code-required forces.

REQUIRED

(a) Calculate stress with overturning, up or down.
(b) Can pile withstand the deformation? Give reasons.
(c) Would the addition of mild steel reinforcing help the pile withstand the deformation? State reason. (No calculation required for this part.)

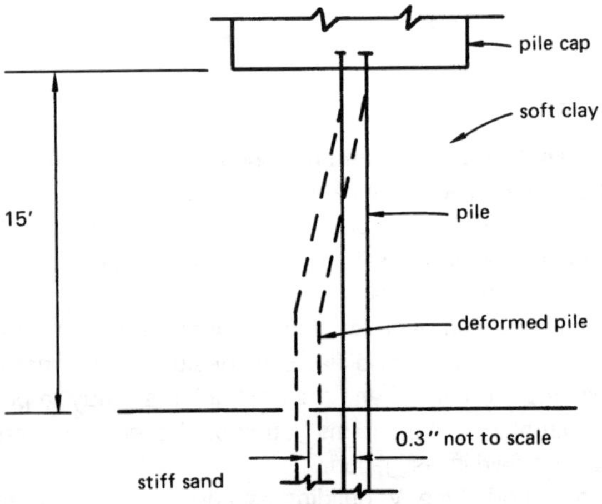

ELEVATION

Figure S-26

STRUCTURAL-27

The frame shown in figure S-27 is fixed at the base, and pinned at the top of the columns.

DESIGN CRITERIA

- Change in temperature of girder only, $\Delta T = 60°$ F
- Temperature coefficient $D_F = 6 \times 10^{-6}$ ft / ft–° F
- $E = 30 \times 10^6$ psi
- Neglect girder shortening induced from column shear.
- Neglect effect of temperature change on columns.

REQUIRED

(a) What is the magnitude and direction of deflection at top of columns?
(b) What are the shear forces at all columns due to the lateral deflection?
(c) What is the bending stress on center column B-B' due to the lateral deflection?

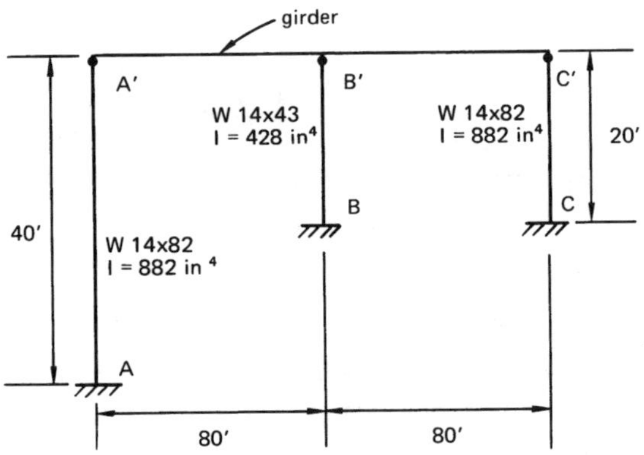

FRAME ELEVATION

columns are fixed at base, and pinned at top

Figure S-27

STRUCTURAL-28

Figure S-28 represents a Vierendeel roof truss which has limited clearance of 6 inches above installed equipment.

DESIGN CRITERIA

- ASTM A36 steel
- $E = 29 \times 10^6$ psi
- I = 200 in^4 typical all members
- Two 20 kip concentrated loads

REQUIRED

(a) Determine if the given clearance is adequate.
(b) Would you recommend any changes? Comment on the deflection of this roof member.

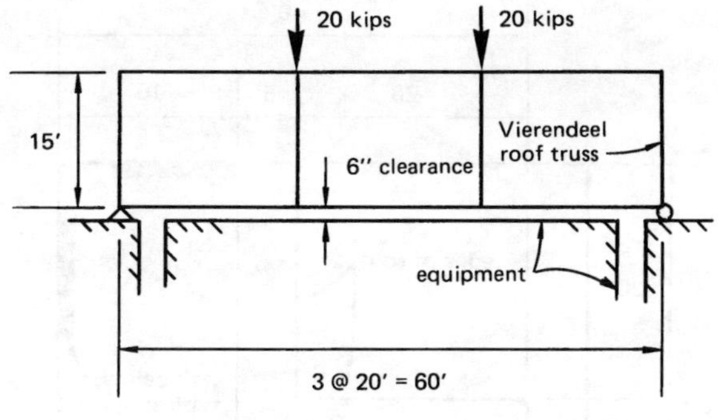

Figure S-28

STRUCTURAL-29

REQUIRED

Find the fixed-end moments, stiffness, and carry-over factors for the girder shown in figure S-29. Neglect weight of beam. Show all calculations.

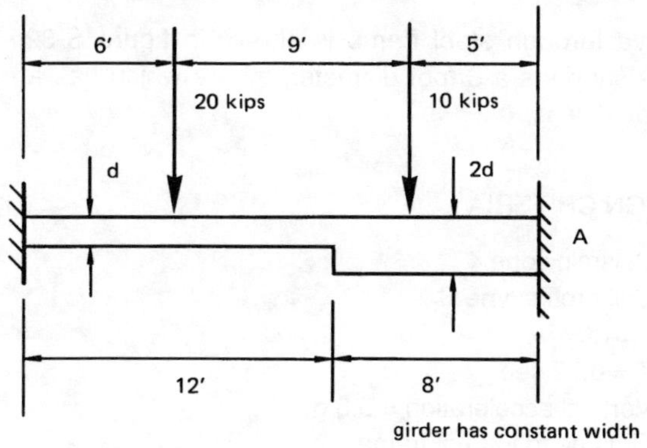

Figure S-29

STRUCTURAL-30

DESIGN CRITERIA

- See each figure.

REQUIRED

Calculate the relative rigidities of the shear resisting elements in figure S-30. (Assume the rigidity of the shear element in (a) is unity, for comparison.)

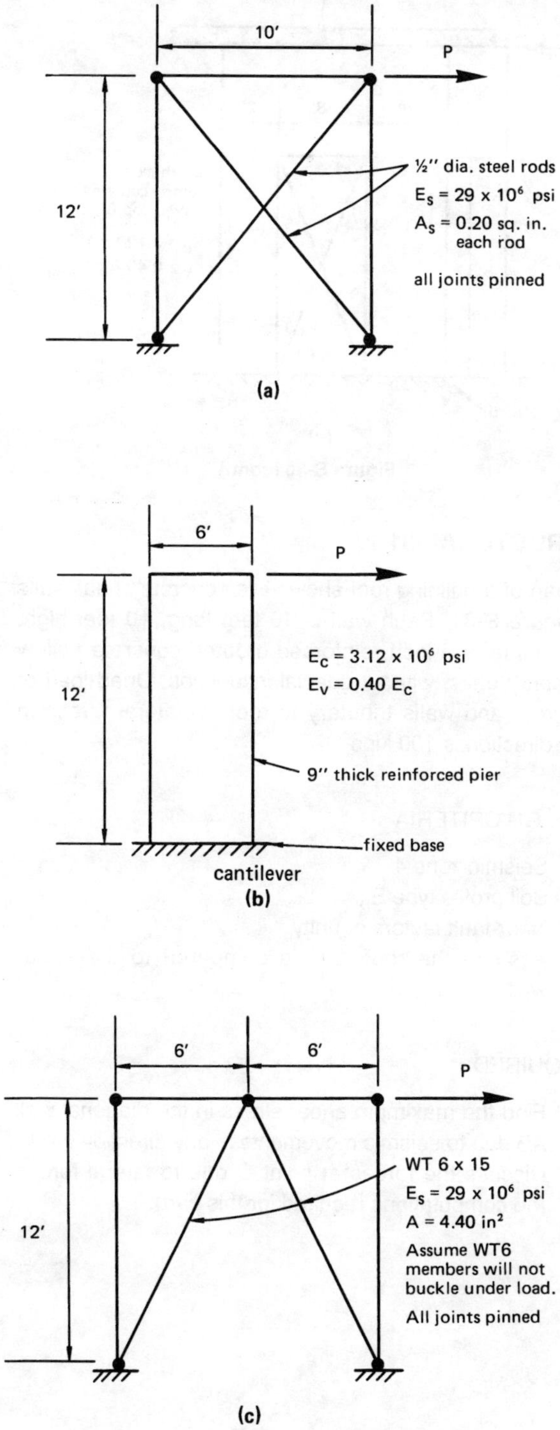

Figure S-30
(continued on next page)

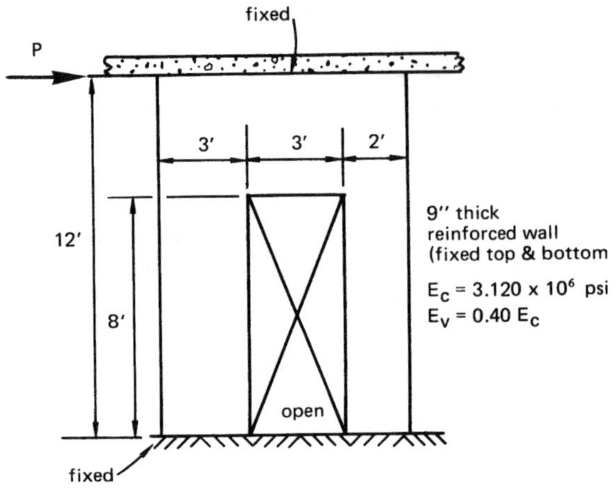

(d)

Figure S-30 (cont.)

STRUCTURAL-31

A plan of a building roof shows five concrete shear walls in figure S-31. Each wall is 10 feet long, 10 feet high, and made of 8-inch reinforced grouted concrete hollow masonry units, without special inspection. Dead load of the roof and walls tributary to roof for lateral forces in any direction is 100 kips.

DESIGN CRITERIA

- Seismic zone 4
- Soil profile type S_D
- Near-fault factors of unity
- Assume the roof is rigid compared to the shear walls.

REQUIRED

(a) Find the maximum shear stress in the diagonal wall AB due to seismic movements in any direction.

(b) Discuss the forces at point C due to lateral forces (no computations required for this part).

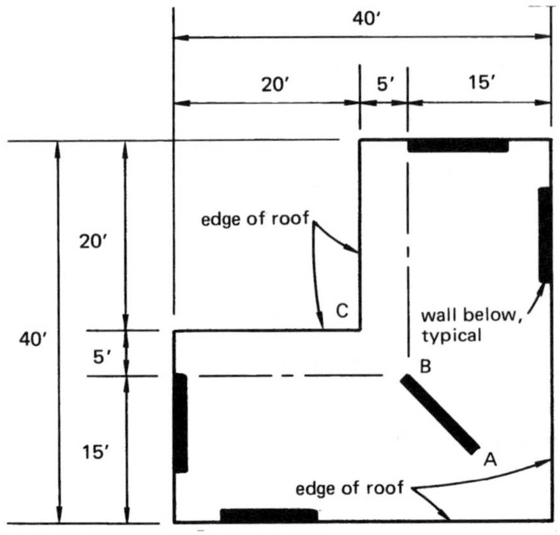

ROOF PLAN

Figure S-31

STRUCTURAL-32

A drive through steel frame is shown in figure S-32. The frame supports a 6 foot diameter sphere which has a dead load of 40 kips.

DESIGN CRITERIA

- Seismic zone 4
- Soil profile type S_D
- I = 1.0
- T = 0.79 sec
- Vertical acceleration = 0.3 g
- Omit dead load of frame.
- The frame and sphere are braced adequately in the perpendicular direction.

REQUIRED

Draw the moment diagram for the dead load plus seismic lateral load condition. Show values of moments (neglect wind load).

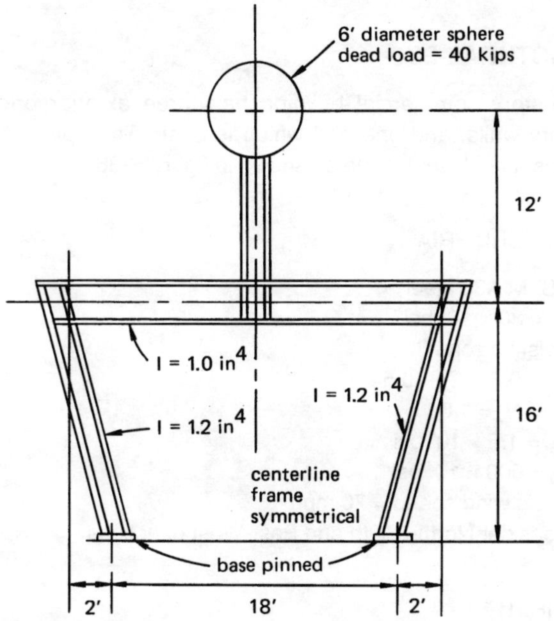

Figure S-32

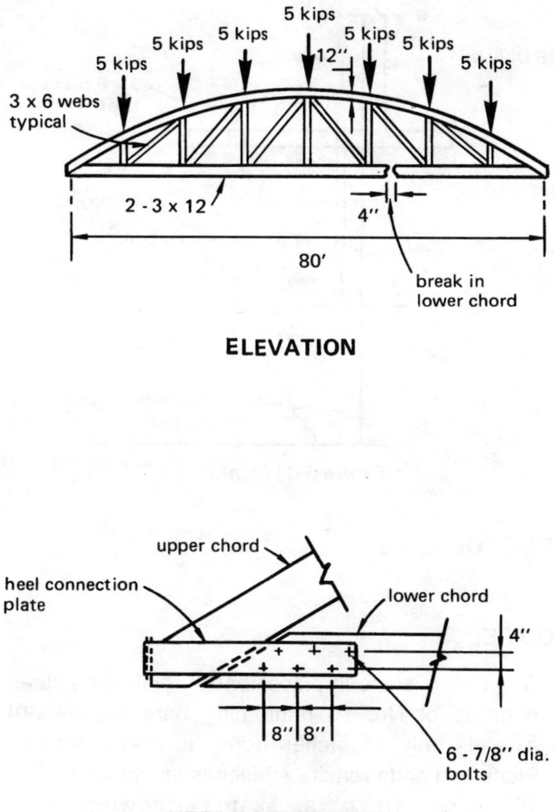

Figure S-33

STRUCTURAL-33

You have been requested to prepare a design for the repair of a wood bowstring truss which has developed a failure in the lower chord. The truss is shown in figure S-33.

DESIGN CRITERIA

- Truss height-to-span ratio is 1:7.
- The break in the lower chord is to be repaired by use of new tension rods between the heel connections.
- The truss is supported temporarily until repairs can be completed.
- No field welding is permitted.

REQUIRED

(a) Determine the size and number of new tension rods required.

(b) Sketch a heel connection showing all pertinent information. (Do not induce eccentricity into the existing heel connection due to the new rods.)

(c) Sketch a section through lower chord which shows the relationship of new rods with respect to the lower chord.

(d) Describe the repair procedure you would pursue.

(e) What additional consideration would you give to the actual break point in the lower chord?

STRUCTURAL-34

A field inspection revealed that the anchor bolt for a hold-down is mislocated as shown in figure S-34.

REQUIRED

Design a welded angle connection to the 4 × 6. Use minimum bolt distances. The connection must fit in the wall. Additional drilled anchors or bars not permitted. Provide sketch of your design. Show calculations.

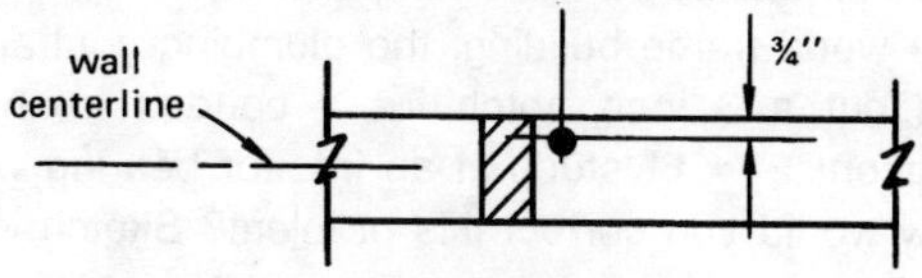

Figure S-34
(continued on next page)

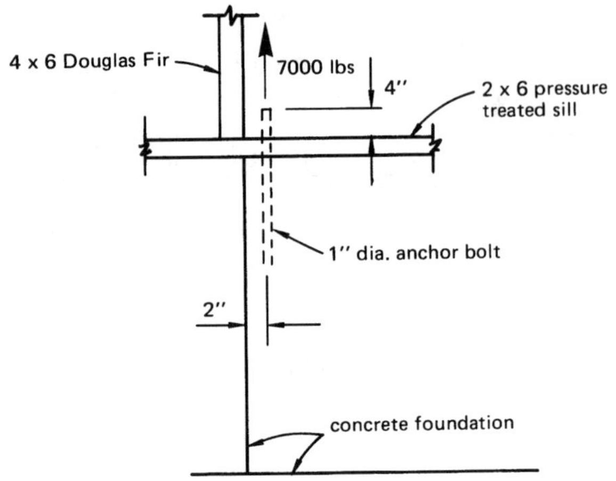

Figure S-34 (cont.)

STRUCTURAL-35

REQUIRED

(a) Due to a mistake in placement of reinforcing steel, a group of No. 10 reinforcing bars (Grade 60) extends only 18 inches from the previous pour. Plans and code require 48 inches lap splice for No. 10 bars. Draw a sketch showing your recommended solution to develop the bar in tension.

(b) A No. 7 reinforcing dowel (Grade 40) for a jamb bar was inadvertently omitted from a foundation pour (12" wide by 24" deep continuous footing) for an 8 inch thick concrete block wall. f'_c = 3000 psi. Knowing that the block wall has not been installed, how would you correct this problem? Describe the procedure.

(c) It is required that non-destructive testing be performed on a major welded frame structure. List two types of non-destructive testing other than visual that you would recommend and for what types of weld each would be applicable.

(d) In a wood frame building, the plumbing contractor has cut a 3 inch notch for a conduit in three adjacent 2" × 6" studs at an interior bearing wall. How would you correct this problem? Sketch your solution.

(e) List the items for an inspection checklist for preparing the excavation for a belled caisson prior to pouring concrete.

(f) An 8 3/4" × 18" glulam beam with twelve 1 1/2" laminations spans 22 feet. A void (no glue) is detected between the seventh and eighth laminations from the bottom. The void starts five feet from one and is three feet long. The beam is in place and the roof installation is complete. Describe two solutions to repair this defect.

STRUCTURAL-36

A one-story commercial building has three exterior brick masonry walls, and one end which is open. The open end contains a rigid steel frame as shown in figure S-36.

DESIGN CRITERIA

- ASTM A36 steel
- $E_s = 30 \times 10^6$ psi
- Seismic zone 3
- $C_a = 0.36$; $C_v = 0.54$
- $N_a = N_v = 1.0$
- $E_m = 1.5 \times 10^6$ psi
- $E_v = 600 \times 10^3$ psi
- Basic wind speed = 70 mph
- Consider North-South and East-West directions

REQUIRED

Determine the values of the lateral forces that must be carried by the three exterior masonry walls, and by the rigid steel frame.

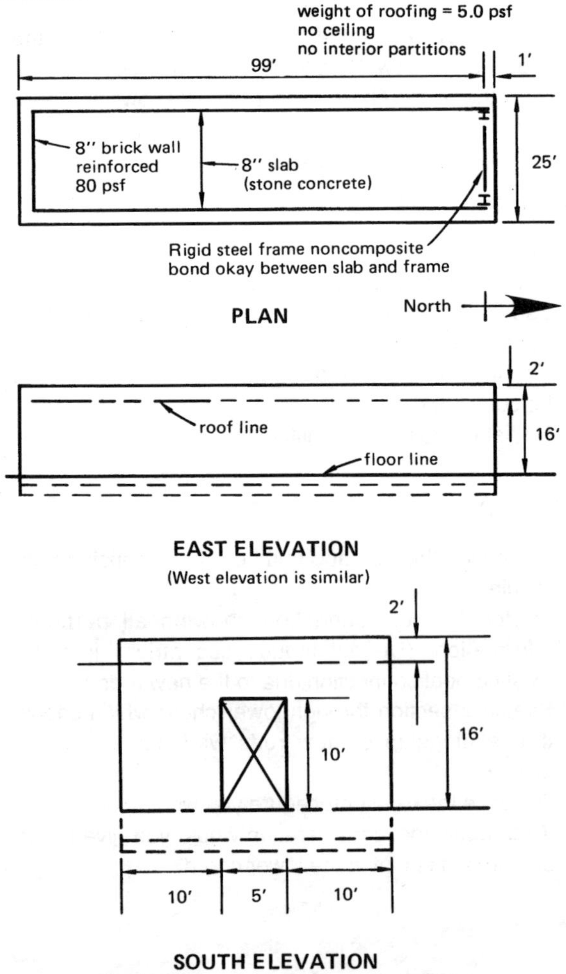

Figure S-36 (continued on next page)

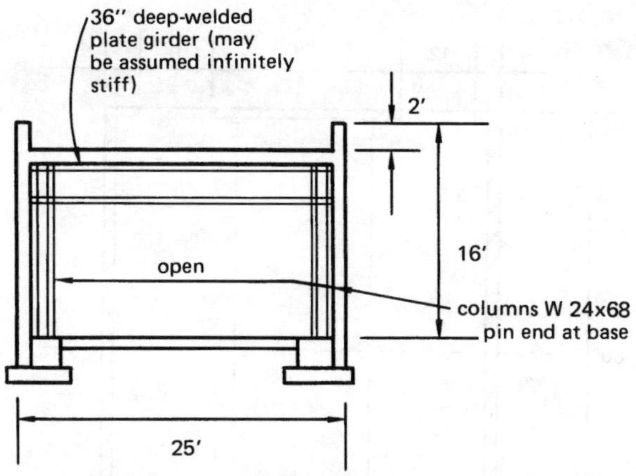

NORTH ELEVATION

Figure S-36 (cont.)

STRUCTURAL-37

Figure S-37 represents a framing condition which has been utilized in a 100 unit subdivision. It is assumed that the construction was in accordance with the details as shown. Plans were reviewed and approved prior to construction.

DESIGN CRITERIA

- Adequate drains and waterproofing have been provided for the masonry wall.
- No connection exists between the 2 × 6 brace and the web members of the trusses.
- The trusses span 22 feet, and are supported by masonry walls which do not retain earth.
- Horizontal steel in the masonry wall is of no concern.
- The plan dimensions of the garage are 22 feet by 22 feet.

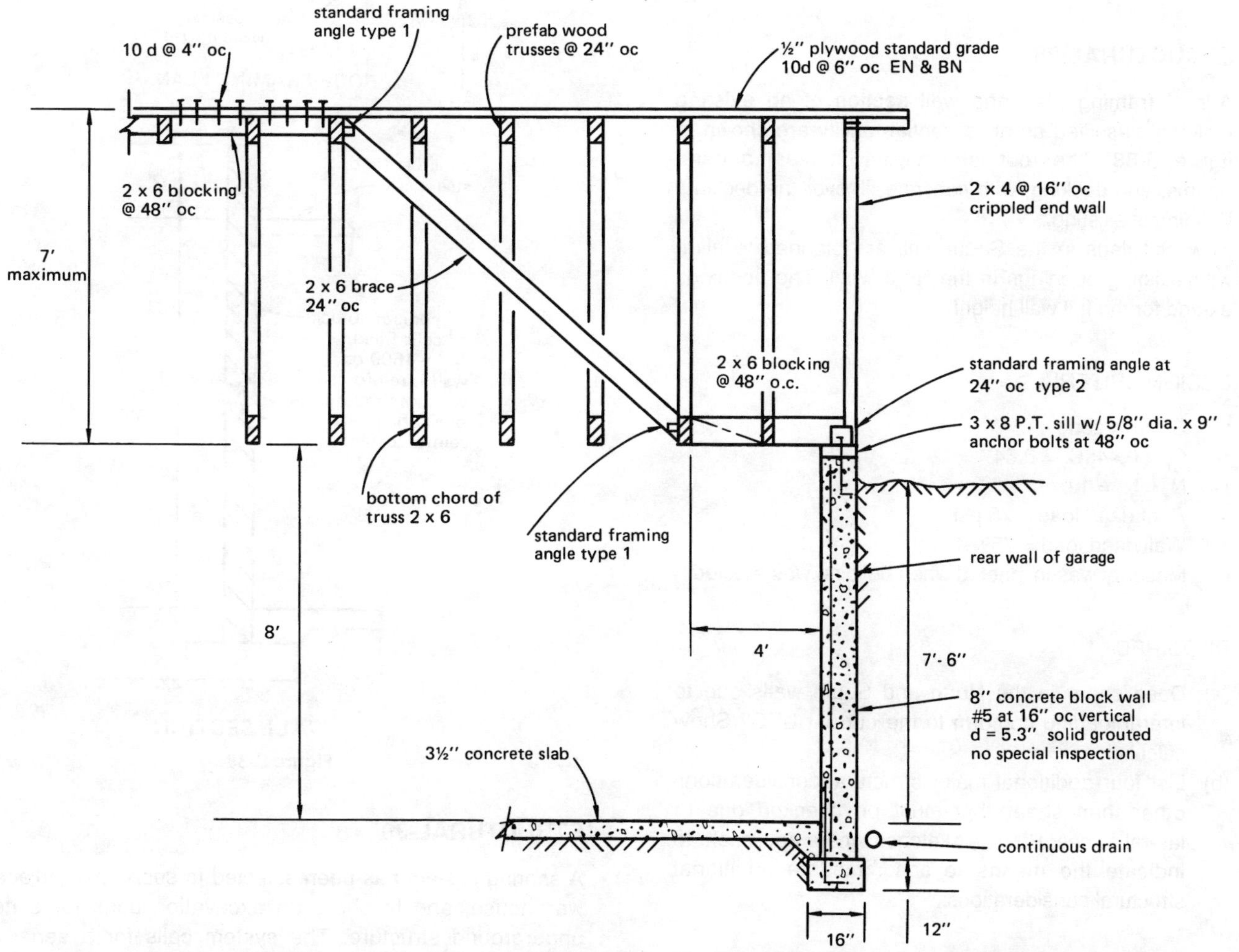

Figure S-37

REQUIRED

(a) Prepare a written critique of the details shown in figure S-37. Include substantiating calculations and assumptions.

(b) If you find that the details prove to be inadequate, then prepare recommendations for remedial repair. Include calculations, assumptions, and suitable sketches which show your intentions as to the recommended repairs. (Freehand sketches are okay.) Identify specifically any details which you deem inadequate.

Note:

standard framing angles	permissible capacity for normal duration loading
type 1	260 lbs
type 2	1000 lbs

STRUCTURAL-38

A roof framing plan and wall section of an existing building classified as an essential facility are shown in figure S-38. The roof is framed with steel girders, beams, and decking, with concrete fill over the decking. Building is existing.

New openings in the South wall are planned to align with existing openings in the North wall. The openings extend for the full wall height.

DESIGN CRITERIA

- Seismic zone 4
- $C_a = 0.44$; $C_v = 0.64$
- $N_a = N_v = 1.0$
- Roof dead load = 75 psf
- Wall dead load = 75 psf
- Masonry was inspected when building was erected.

REQUIRED

(a) Does shear in the North and South walls due to lateral loading conform to the current UBC? Show calculations.

(b) List four additional major structural considerations other than shear that must be checked due to lateral forces. Provide sketches where applicable to indicate the means to accommodate additional structural considerations.

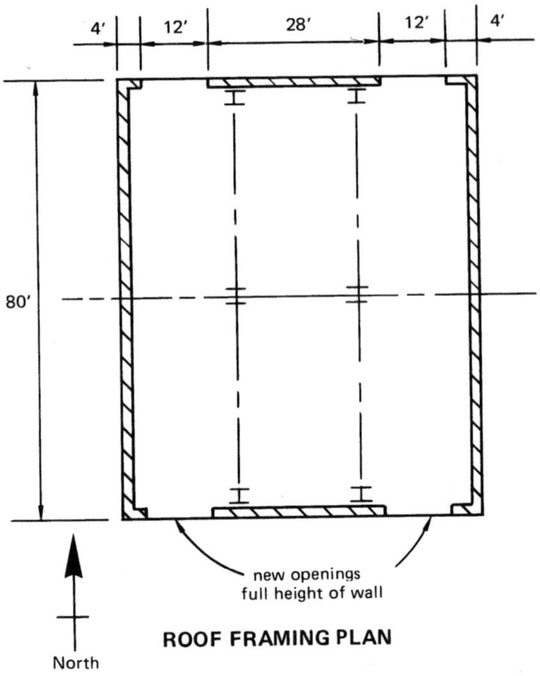

ROOF FRAMING PLAN

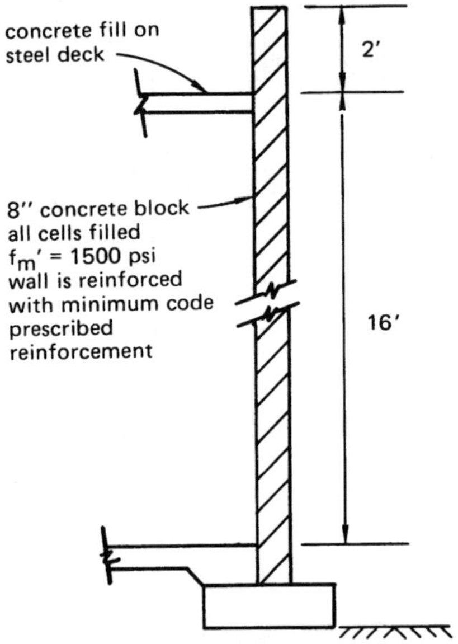

WALL SECTION

Figure S-38

STRUCTURAL-39

A shoring system has been selected to support an adjacent warehouse, and to shore an excavation bank for a new underground structure. The system calls for a series of soldier beams dropped into predrilled holes at 8 feet on center with timber lagging placed between. Two tiebacks are placed to resist lateral pressure. A typical section is shown in figure S-39.

DESIGN CRITERIA

- Assume tie rods provide rigid support.
- $N = Q_f - 120D_f$ (net pressure)
- See figure S-39 for other criteria.

REQUIRED

(a) Show all lateral pressures acting on soldier beams. Calculate critical values.

(b) Calculate forces at A, B, and C.

(c) Check adequacy to toe length assumed.

(d) Calculate grouted anchor length of tiebacks A and B. Draw a sketch to show how length is measured.

(e) Describe briefly the installation procedure of this system (step by step).

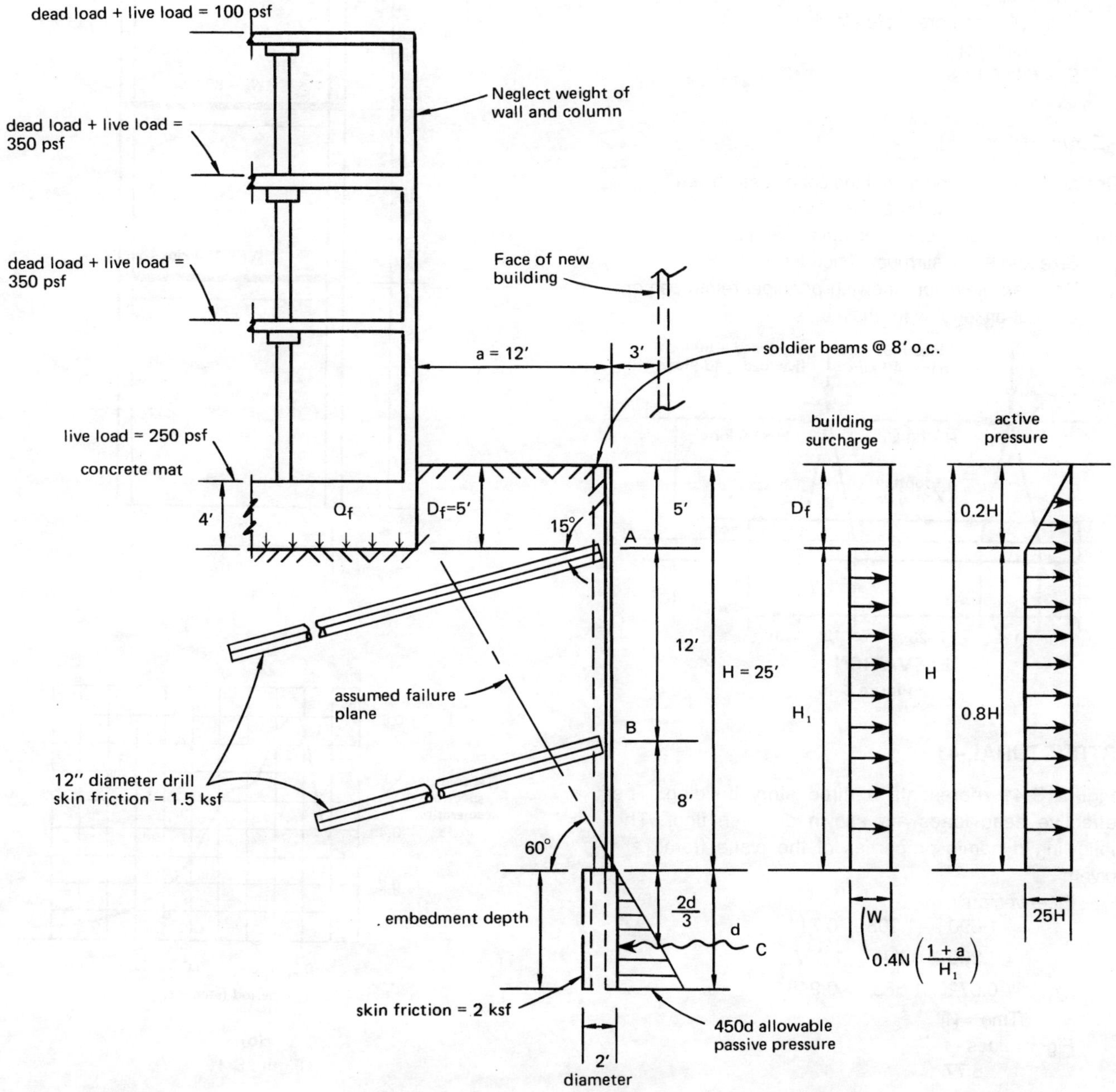

Figure S-39

STRUCTURAL-40

A large industrial plant has two interior A-frames which are supported on a 5 foot deep footing. The A-frames support vertical loads, and resist seismic forces, as shown in figure S-40.

DESIGN CRITERIA

- The seismic loads are resisted at the footing centerline.
- The footing is infinitely stiff.
- The soil is compressible elastically.
- $F'_c = 3000$ psi
- $F_y = 60{,}000$ psi

REQUIRED

Design the foundation to suit the conditions shown.

(a) Determine the soil pressure diagram.
(b) Size and place reinforcement top and bottom.
(c) Size and place stirrups - if needed.
(d) Draw an elevation showing principal reinforcement and cut-off lengths for main bars.

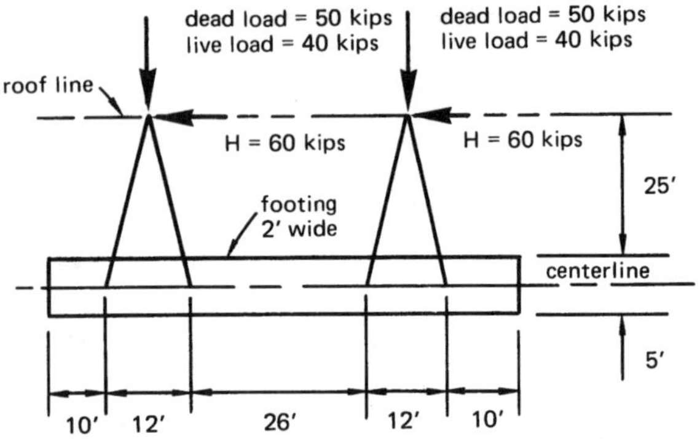

ELEVATION

Figure S-40

STRUCTURAL-41

Figure S-41 represents a three story building. The effective dead loads are shown on each floor. The following dynamic properties of the plane frame are given:

Eigenvectors

$$\phi = \begin{matrix} 1.690 & -1.208 & -0.714 \\ 1.220 & 0.704 & 1.697 \\ 0.572 & 1.385 & -0.948 \end{matrix}$$

$$\phi^T m \phi = [1]$$

Eigenvalues

$$W = \begin{matrix} 8.77 \\ 25.18 \\ 48.13 \end{matrix} \text{ radians/second}$$

REQUIRED

(a) Compute the participation factors.
(b) How can you check that your participation factors are correct?
(c) Calculate the displacements of each floor based on the spectra given in figure S-41.
(d) Calculate the interstory drift for each floor.

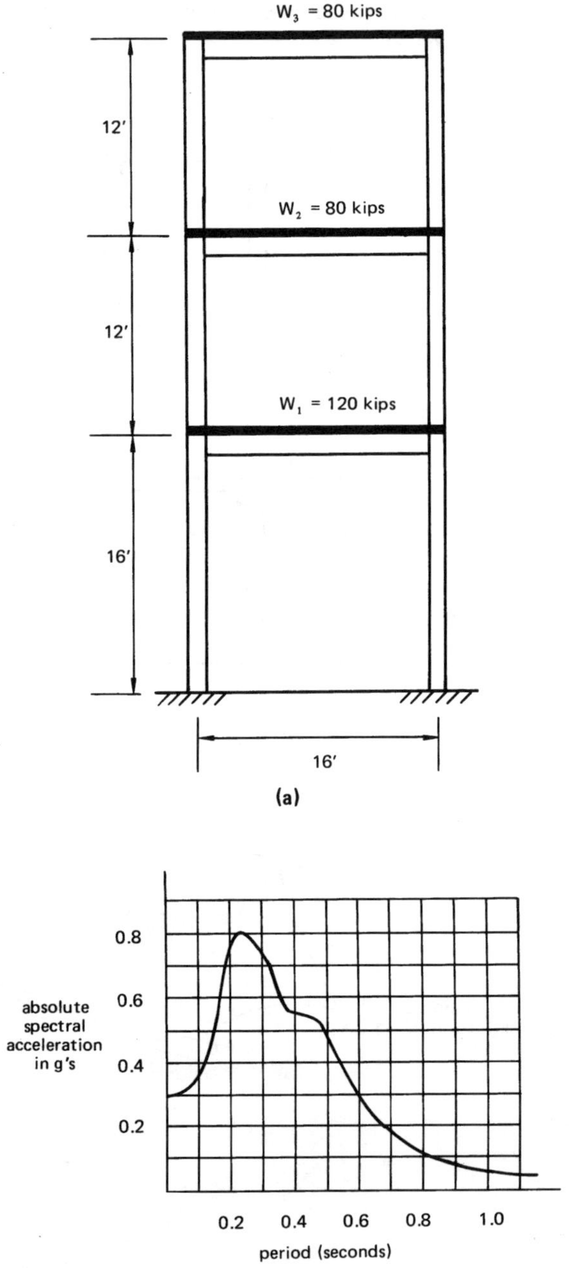

(b)

Figure S-41

STRUCTURAL-42

For the steel rigid frame shown in figure S-42, consider the structure to have girders infinitely stiff compared to the columns, masses concentrated at the plane of the girders, 7% damping and neglect axial strains in columns.

REQUIRED

(a) Calculate the natural frequencies and characteristic shapes of the structure.

(b) Determine the spectral acceleration and velocity of the first mode.

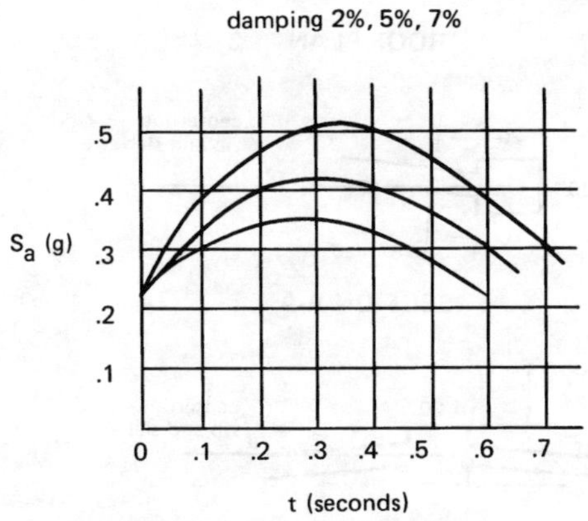

acceleration response spectra

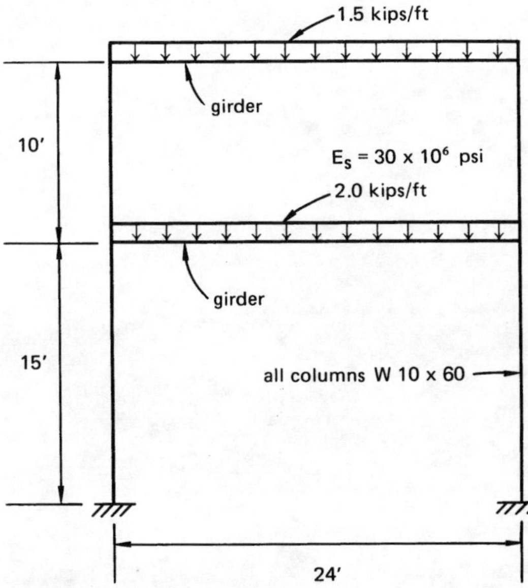

Figure S-42

STRUCTURAL-43

The plan and typical profile shown in figure S-43 shows a single cable supported roof structure. The member sizes for the outer rings and inner rings, as well as the plan dimensions, are shown on the sketch.

DESIGN CRITERIA

- Neglect the lateral force system for this problem.

REQUIRED

(a) What is the horizontal component of cable tension at point A on the cable? (See profile view.) Explain.

(b) Verify the adequacy of the tension ring indicated for combined stresses. (Show calculations.)

(c) Verify the adequacy of the compression ring indicated for combined stresses. (Show calculations.)

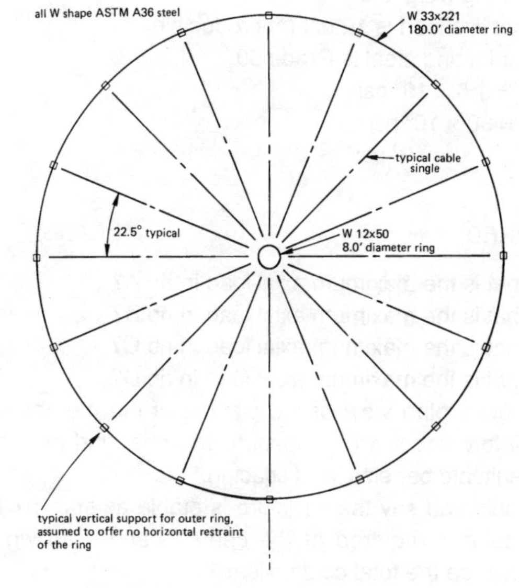

PLAN

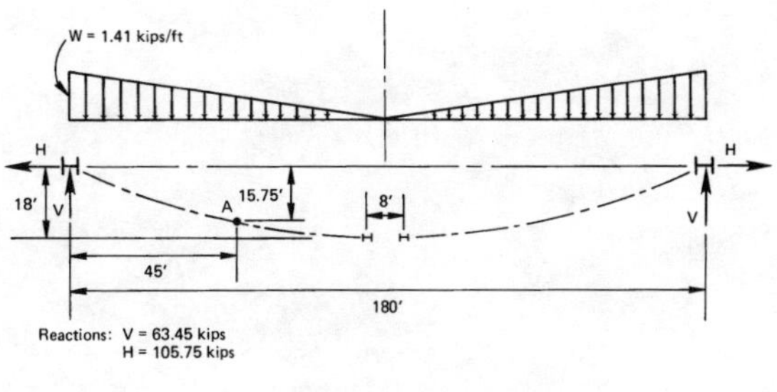

TYPICAL PROFILE

Figure S-43

STRUCTURAL-44

A roof structure is composed of a four quadrant hyperbolic paraboloid which joins the straight ribs (edge members) as shown in figure S-44.

DESIGN CRITERIA

- Seismic zone 3
- Wind pressure - map area 20
- The slab, or membrane, is 3 inches thick.
- All ribs are 20 inches by 20 inches in cross section. (See sections A-A and B-B of figure S-3.)
- All concrete is lightweight (110 pcf air dry with $f'_c = 3000$ psi)
- Vertical support is provided by the abutments at the four corners.
- A perfect hinge condition may be assumed at the juncture of the ribs to the various abutments.
- Ceiling weighs 3 psf.
- Roofing weighs 3 psf.
- Roof live load is 12 psf (not reducible).
- Reinforcing steel is Grade 60.
- $E_m = 1.5 \times 10^6$ psi
- $E_s = 30 \times 10^6$ psi
- $E_v = 600 \times 10^3$ psi

REQUIRED

(a) What is the maximum axial load in rib A?
(b) What is the maximum axial load in rib B?
(c) What is the maximum axial load in rib C?
(d) What is the maximum axial load in rib D?
(e) Show a plan view of the placing of the membrane reinforcement with respect to the ribs, and provide a suitable bar size, and spacing.
(f) Would you say the structure is stable as shown? Is a column required at the center - and if so what would be the total column load?

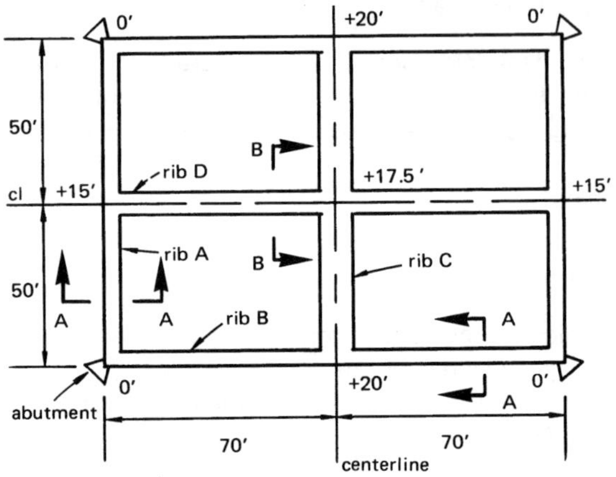

ROOF PLAN

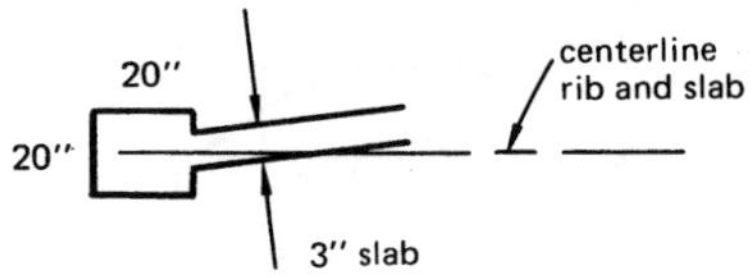

SECTION A-A

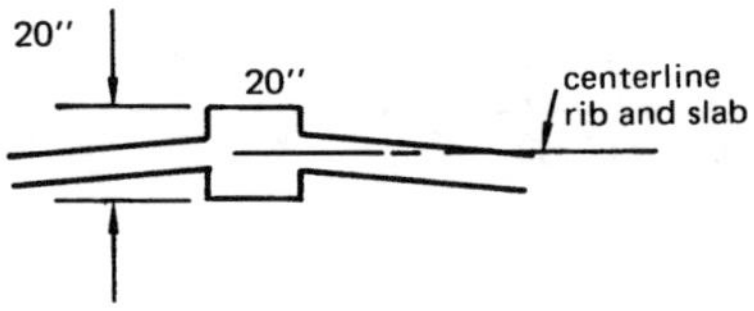

SECTION B-B
Figure S-44

2 CONCRETE

CONCRETE-1

A beam is shown in figure C-1 which is made up of lightweight concrete. The beam is not part of a ductile moment-resisting frame. No redistribution of the negative moment is necessary.

DESIGN CRITERIA

- Concrete uses lightweight aggregates.
- Splitting tensile strength = 0.274 ksi
- f'_c = 3 ksi
- f_y = 40 ksi
- The distance from face of concrete to the center of gravity (CGS) is 2 1/2 inches.
- Identify any design aids you may have used.

At A: dead load: M = +140 ft-kips (tension at bottom)
V = 0
live load: M = +70 ft-kips
V = 9 kips
At B: dead load: M = −200 ft-kips (tension at top)
V = 36 kips
live load: M = −100 ft-kips
V = 18 kips

REQUIRED

(a) Determine the minimum areas of flexural reinforcement required at points A and B.
(b) Determine the minimum areas of stirrup reinforcement required at points A and B (in^2/ft).

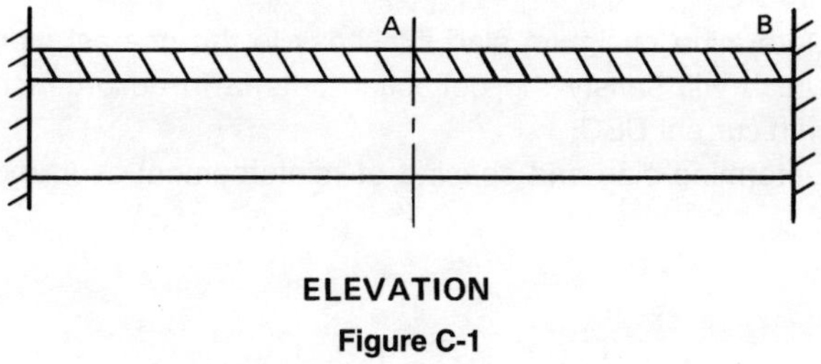

ELEVATION

Figure C-1

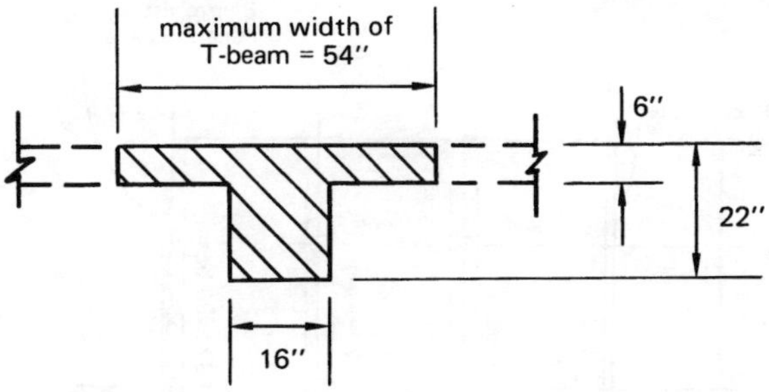

SECTION

Figure C-1 (cont.)

CONCRETE-2

REQUIRED

(a) List and discuss two possible effects on concrete beams if shoring is removed too soon. What measures should you take to ensure proper scheduling of shore removal?
(b) What is the major cause of plastic shrinkage cracks in slabs? List five measures or precautions which can be taken to minimize or prevent plastic shrinkage cracking.
(c) The following types of admixtures are used in concrete mixes: air entraining agents, retarders, accelerators, and water reducers.
    (1) Discuss the principal benefits of each type.
    (2) List the precautions which should be taken when using each type of admixture.

CONCRETE-3

A part plan and a cross section of a typical floor for a multistory building are shown in figure C-3. The columns are 24 inches square and the beams are shown in section A-A. Assume for this problem that the beams and columns are adequate.

DESIGN CRITERIA

- Suggested code: current ACI
- f'_c = 4000 psi stone aggregate concrete
- f_y = 40 ksi reinforcing steel
- Live load = 250 psf (do not reduce)

REQUIRED

(a) What slab thickness would you establish for your design? Show calculations to support your development.
(b) Determine the maximum slab moments.
(c) Design and sketch the slab reinforcement at the critical line. Indicate the size of bars, spacing, clearances, placing, and cut-off points. (Do not use truss bars.)

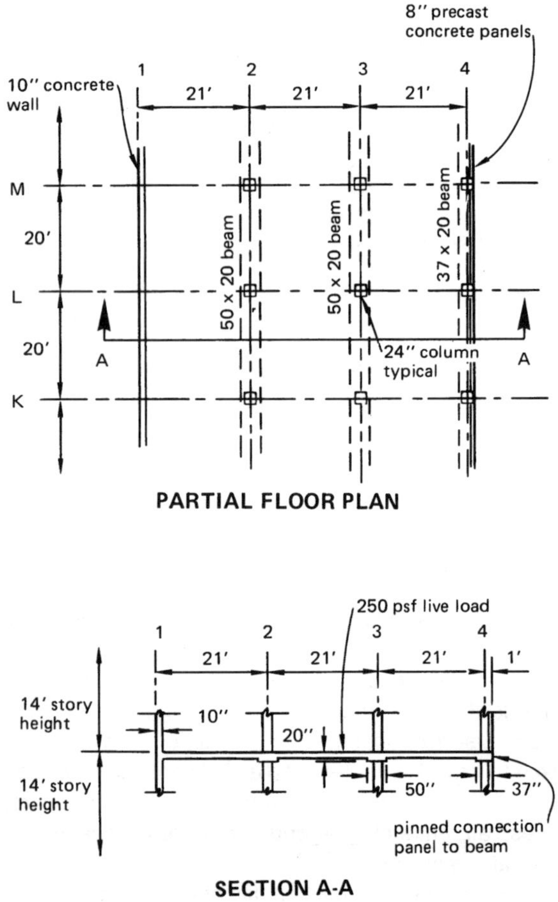

Figure C-3

CONCRETE-4

A reinforced concrete pile as shown in figure C-4 is subjected to bending about the X-X axis. Axial load equals zero.

DESIGN CRITERIA

- f'_c = 4000 psi at 28 days hard rock concrete
- ASTM A615 grade 60 longitudinal reinforcing steel

REQUIRED

(a) Verify the location of the neutral axis as shown.
(b) Determine the ultimate moment capacity about the X-X axis of the pile in accordance with the ultimate strength provisions of the current UBC.

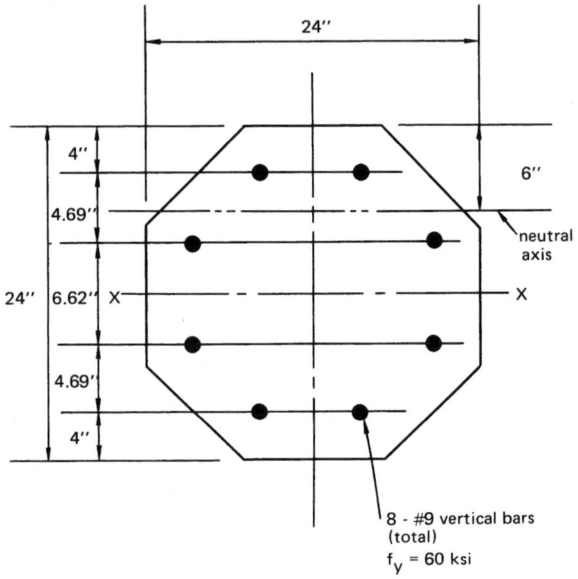

Figure C-4

CONCRETE-5

A one-way cast-in-place concrete slab has been proposed to span over a garage area as shown in figure C-5.

DESIGN CRITERIA

- Design load: partitions = 20 psf
  live load = 80 psf (short term)
- f'_c = 3000 psi, w = 150 pcf concrete
- Grade 40 reinforcing steel
- Neglect continuity between slab and block walls.
- Maximum deflection = L/480
- Assume partitions will be installed while slab is still shored.

REQUIRED

(a) Determine minimum slab thickness to the nearest inch which will satisfy the deflection criteria in accordance with current UBC.
(b) Determine size and spacing of reinforcement at A and B.

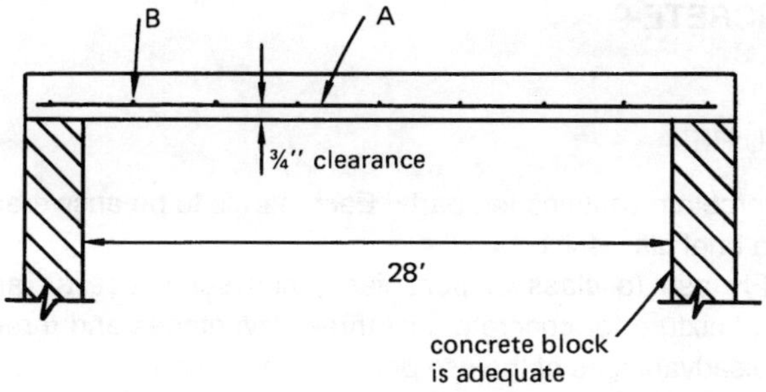

Figure C-5

CONCRETE-6

An existing 16-inch by 36-inch concrete beam shown in figure C-6 supports a uniform dead load of 1500 pounds per lineal foot including weight of the beam.

DESIGN CRITERIA

- f'_c = 3000 psi normal weight concrete
- f_y = 40,000 psi
- d = 33.5 inches
- Neglect any torsion consideration.

REQUIRED

Based upon the shear capacity of the beam as indicated by figure C-6 and the criteria above, how much uniform live load can the beam support?

CONCRETE-7

A short concrete column is shown in cross section in figure C-7. It is 21 inches wide x 17 inches deep with four #9 bars and two #8 bars placed as shown. The column carries combined axial and bending loads about the Y-Y axis.

DESIGN CRITERIA

- P_u = 700 kips
- M_u = 330 ft-kips
- Do not use column tables.
- f'_c = 4500 psi
- f_y = 60,000 psi

REQUIRED

(a) Plot the ultimate strength interaction diagram for axial compression and bending moment.

(b) Check adequacy of column for P_u = 700 kips and M_u = 330 ft-kips

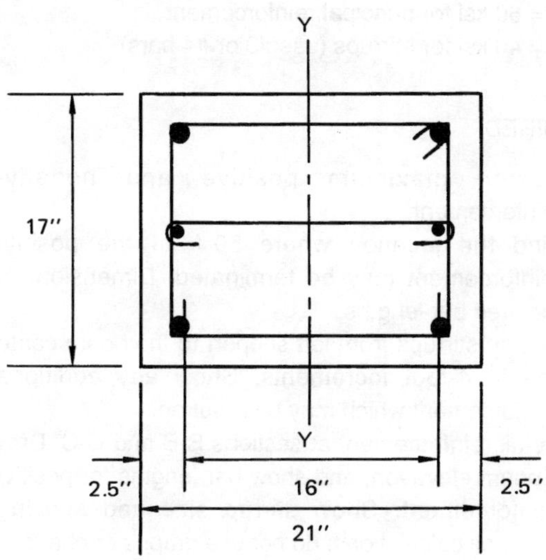

Figure C-7

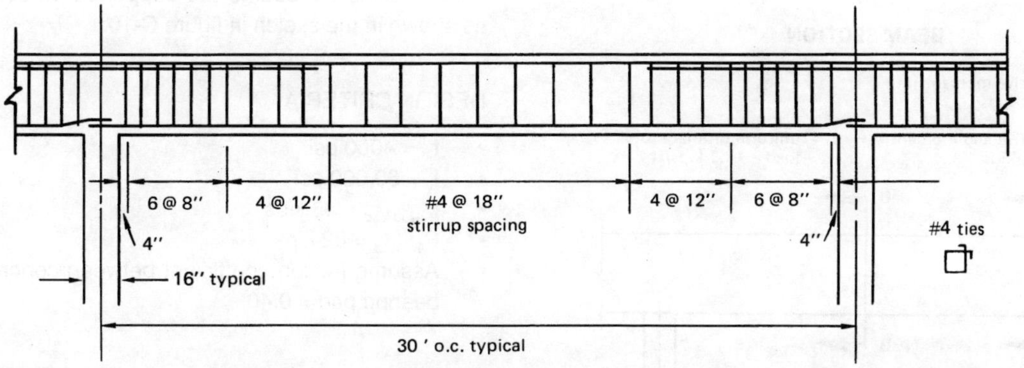

Figure C-6

CONCRETE-8

Shown in figure C-8 is a typical interior span of a cast-in-place one-way slab and beam system. The beam span is 36 feet. Beams are spaced 22 feet on center. The typical beam depth is 36 inches with a web width of 18 inches. Beam loading and end moment are given as ultimate loads and moments. All principal reinforcement is to be #8 bars. Neglect column width and any unbalanced loading.

DESIGN CRITERIA

- f'_c = 4000 psi hard rock aggregate
- f_y = 60 ksi for principal reinforcement
- f_y = 40 ksi for stirrups (use #3 or #4 bars)

REQUIRED

(a) Design maximum positive and negative reinforcement.

(b) Find the location where 50% of the positive reinforcement may be terminated. Dimension the required bar lengths.

(c) Design stirrups from left support to the beam center line in 4-foot increments. Show any additional reinforcement which may be required.

(d) Detail reinforcement at sections B-B and C-C. Draw a beam elevation, and show bar lengths for positive reinforcement. Show stirrup size and spacing. Show all calculations; do not use graphs or charts.

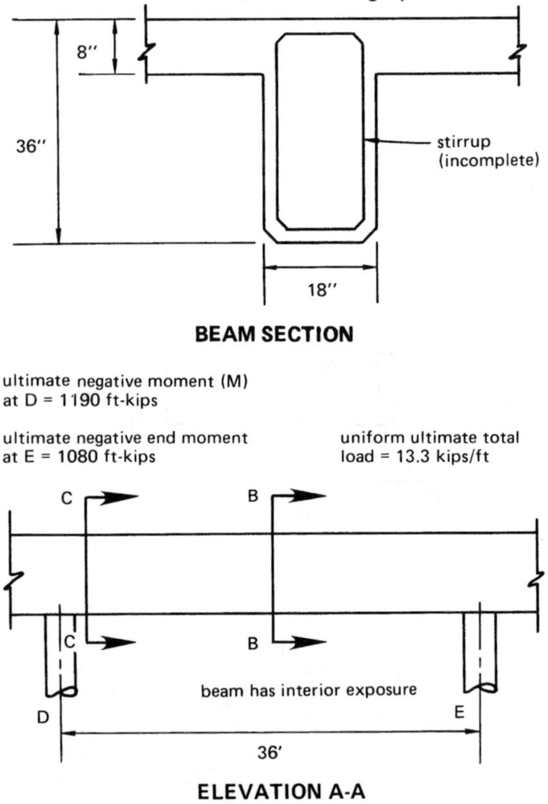

Figure C-8

CONCRETE-9

REQUIRED

This problem contains ten parts. Each part is to be answered with a brief discussion.

(a) Fly ash (a class of pozzolans) has some use as an admixture for concrete. List three advantages and three disadvantages of the use of fly ash in concrete.

(b) When welding reinforcing steel, what is meant by (1) preheat, (2) carbon-equivalent, and (3) heat-affected zone?

(c) A full-welded tension butt splice of reinforcing steel must develop what capacity of the bar? What minimum capacity would a weld of grade 60 #14 reinforcing bar be required to develop?

(d) What were the two main purposes for developing reinforcing steel identified as ASTM A706?

(e) List three types of mechanical splices currently in use for reinforcing bars.

(f) Epoxy adhesives are used in concrete work. List two advantages and two disadvantages when using this group of materials.

(g) Identify three methods used to cure concrete.

(h) Why is it necessary to check both the jacking force and the tendon elongation during stressing procedures in prestressed concrete construction?

(i) Concrete deformations due to creep and shrinkage may be two or more times greater than the elastic deformations. Thus shrinkage must be given proper considerations. List four factors which influence shrinkage of concrete.

(j) The harmful effects of hot weather on freshly poured concrete may be minimized by a number of practical procedures in the construction operation. List three of those specific procedures.

CONCRETE-10

Precast concrete beams are supported on concrete corbels as shown in the sketch in figure C-10.

DESIGN CRITERIA

- f'_c = 4000 psi
- f_y = 60,000 psi
- $P_{dead\ load}$ = 48 kips
- $P_{live\ load}$ = 32 kips
- Assume friction coefficient between concrete beam and bearing pad = 0.40.

REQUIRED

(a) Design the corbels using shear friction concept.

(b) Draw a sketch large enough to clearly indicate all reinforcement and dimensions for your design.

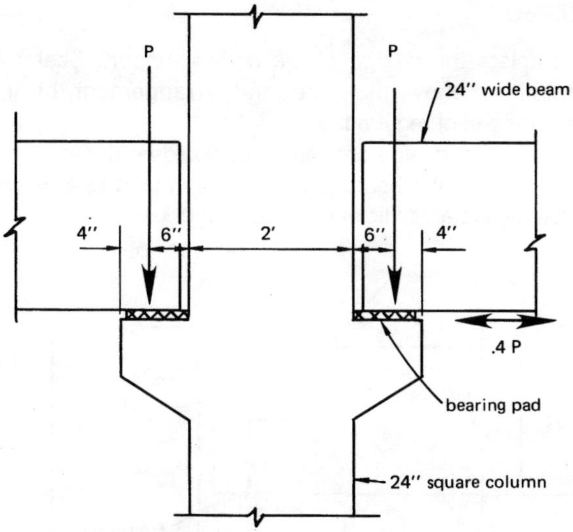

Figure C-10

CONCRETE-11

An owner has an existing 2-story office building to be converted to a warehouse. You are to determine the maximum allowable live load that the second floor will safely support. A typical interior bay is shown in figure C-11.

DESIGN CRITERIA

- f'_c = 3500 psi (from core samples) hardrock concrete
- f_y = 40 ksi
- Slab reinforcement (3/4 inches clearance)
  - #4 at 12 inches on center, bottom
  - #4 at 8 inches on center, top at supports
- beam reinforcement (1 1/2 inches clearance to stirrups)
  - 4 - #9 bottom
  - 4 - #10 top to supports
  - 8 - #3 stirrups at 8 inches on center each end (starting 4 inches from support)

REQUIRED

(a) What is the maximum unit live load the second floor slab and beam will support?

(b) List some other items that should be checked for which no information has been given.

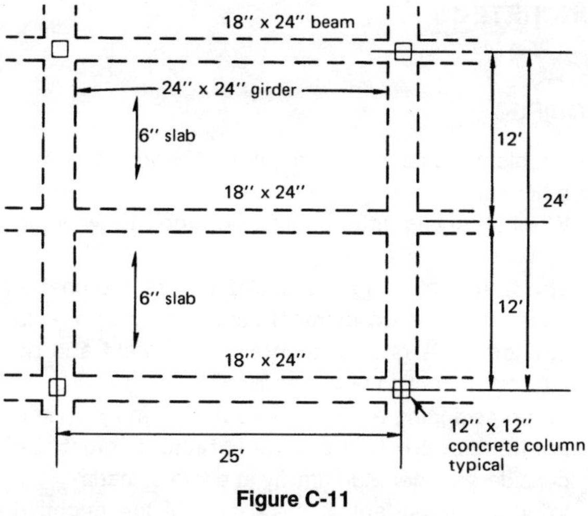

Figure C-11

CONCRETE-12

A column cross section in figure C-12.

DESIGN CRITERIA

- f'_c = 5 ksi
- f_y = 60 ksi
- (Kl_u) = 18'-0"
- C_m = 1.0

REQUIRED

(a) Draw the interaction diagram for the column cross section shown. Locate at least three critical points.

(b) Determine the maximum allowable factored total load moment about the X-X axis for the column when a factored axial total load of 100 kips is applied. (Assume ratio of factored dead load moment to total load moment is 0.6.)

Note: Column design tables are not to be used. Show all calculations.

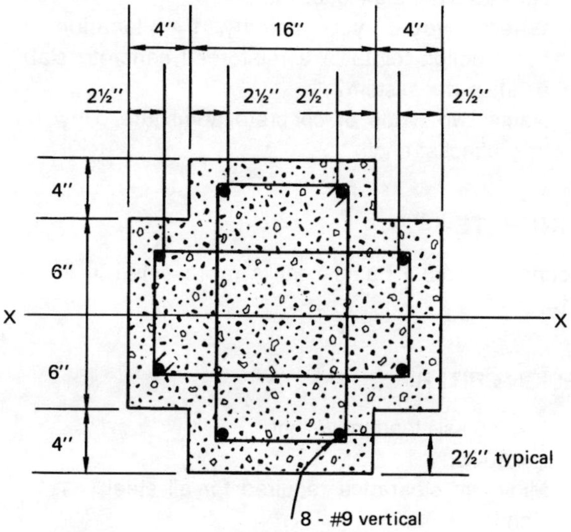

Figure C-12

CONCRETE-13

REQUIRED

This problem contains several parts. Answer each part with a brief discussion.

(a) What analysis must be made prior to welding reinforcing steel and why?
(b) The web reinforcing in a ductile flexural member is based on ultimate moment capacities with a yield strength 25% greater than specified yield and no capacity reduction factor ϕ. Why?
(c) When designing an exterior beam-column joint in a posttensioned frame, what factors must be considered other than gravity and lateral loads?
(d) Why is it important to place part of the negative reinforcement in the flange area of a T-beam?
(e) What are the effects of excessive vibration on concrete?
(f) What type of electrodes are preferable when welding reinforcing steel?
(g) What considerations should you give to the connection of masonry or concrete walls to framed slabs in a posttensioned concrete structure and why?
(h) In a multi-story reinforced concrete construction:
    (1) When and where would you permit reshoring to be placed?
    (2) What determines the amount of reshores you would require?
    (3) When would you permit reshoring to be removed?
    (4) What physical property of concrete other than compressive strength is important when considering form removal and why?
(i) Why is the embedment, or splice length, longer for top bars in reinforced concrete?
(j) What are three factors which affect deflection of a flat slab other than loads?
(k) Where would you specify the location of construction joints in a reinforced concrete slab-beam-girder system?
(l) Name two types of concrete admixture. How do they affect strength?

CONCRETE-14

A concrete corbel beam seat is represented in figure C-14.

DESIGN CRITERIA

- f'_c = 3.0 ksi standard weight
- f_y = 60 ksi
- Minimum clearance required for all steel = 1 1/2 inches

REQUIRED

(a) Complete the design of the corbel support. Sketch a detail to show the size and arrangement of the reinforcement required.
(b) Discuss the design changes that would be necessary in the corbel if the bearing pad were changed to a friction type having a coefficient of friction of 0.3.

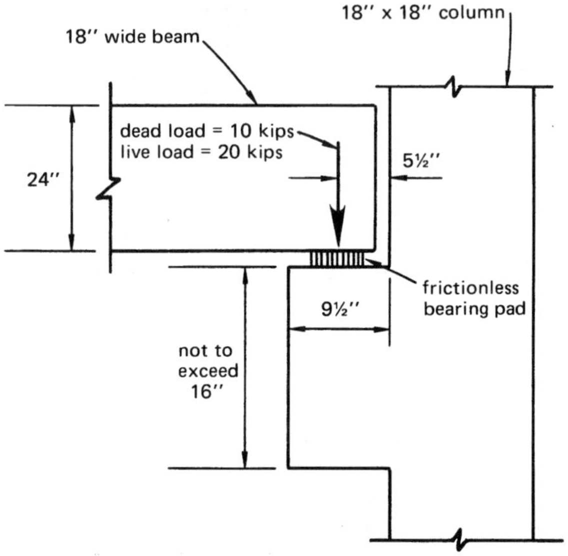

Figure C-14

CONCRETE-15

Figure C-15 shows a pile-supported footing used to span the load from a 20-inch square reinforced concrete column over an underground utility tunnel.

DESIGN CRITERIA

- f'_c = 3000 psi concrete
- f_y = 60,000 psi steel
- pile service loads: dead = 22.0 kips
  live = 10.0 kips
- Current UBC

REQUIRED

(a) Determine the minimum uniform thickness of footing required assuming the head of piles is to be embedded 3 inches into footing.
(b) Design all required footing reinforcement. (Show calculations and assumptions.)
(c) Draw detail of footing showing all dimensions and reinforcing steel. Use approximate scale 3/8" = 1'-0".

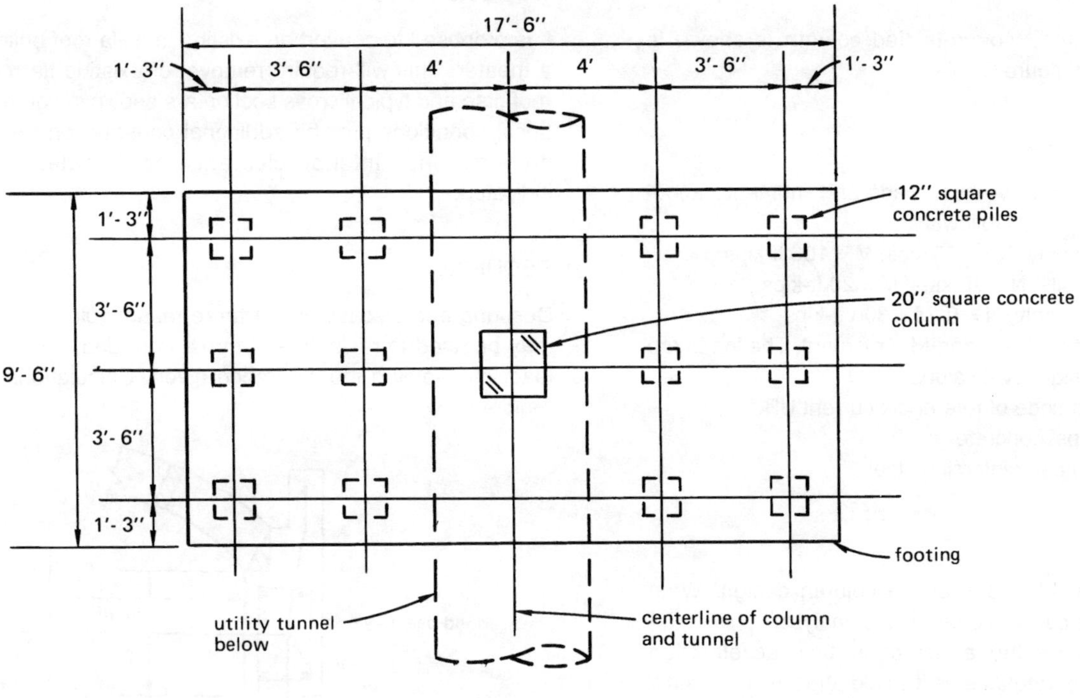

Figure C-15

CONCRETE-16

REQUIRED

(a) You have been retained to investigate a concrete slab that is exhibiting greater than anticipated deflections. The slab was placed Nov. 26, 1972, and stripped 5 days later when the laboratory-cured cylinders reached design strength. Job records are available.
   (1) List six items other than design details you would check, and explain why.
   (2) What three design details might have affected deflection of the slab?

(b) What is the philosophy of ductile concrete as applied to seismic design?

(c) Why isn't mild steel, such as A615 grade 40, used for prestressing?

(d) New codes restrict the tack welding of crossing rebars for assembly of reinforcement except under special conditions. Why?

(e) List at least six items that might be included in a good cold weather concreting procedure.

(f) (1) What are two major benefits of air entrainment in concrete?
   (2) What is one disadvantage?
   (3) What is considered a normal range of air entrainment?

(g) What are the readily identifiable defects of poorly consolidated concrete?

(h) What two principal alloying elements used in the manufacture of steel rebar affect electrode selection and preheat for welding?

(j) (1) What is the primary function of calcium chloride when used in a concrete mix?
   (2) What are two disadvantages of its use?

(k) What is the most important single factor affecting compressive strength for a well-cured concrete of a particular age?

(l) What harmful effect on concrete is caused by high carbon dioxide concentrations? Assume freshly placed concrete.

(m) A prestressed beam has been installed in a building when you discover the effective stress in the tendons is 30% less than you specified. The beam was designed for zero tension in the concrete at service loads (pretensioned).
   (1) What would be the loss in ultimate capacity of the beam?
   (2) Name two other features of the beam that should be evaluated and how they have been affected by loss in prestress.

CONCRETE-17

A short reinforced concrete tied column is shown in cross section in figure C-17.

DESIGN CRITERIA

The column carries vertical loads and moments about the center line X-X as follows:

- Dead load only, N = 150 kips, M = 100 ft-kips
- Live load only, N = 70 kips, M = 20 ft-kips
- Earthquake only, N = 0, M = 300 ft-kips

where N = vertical load, and M = moment. The loads are unfactored, design-level values.

- Suggested code of reference: current UBC
- f'_c = 4000 psi concrete
- f_y = 50,000 psi reinforcing steel

REQUIRED

(a) Verify the adequacy of the column design. What comments can you offer on this analysis?

(b) Comment on the effect of vertical acceleration during an earthquake as it would affect this column.

Note: Table values are not acceptable. Calculations are required.

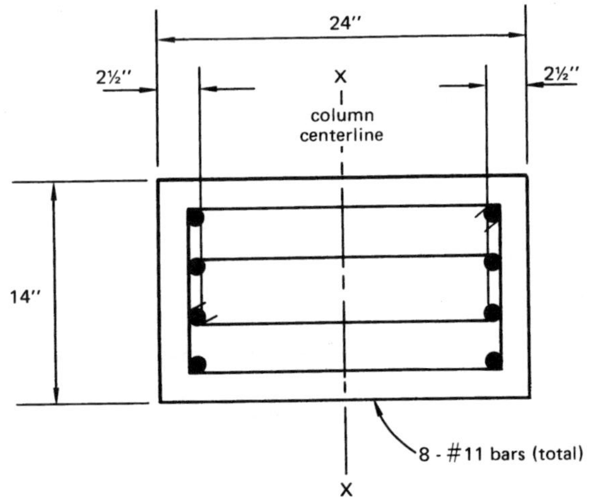

TYPICAL COLUMN SECTION

Figure C-17

CONCRETE-18

It is proposed to convert an existing lamella roof building into a theater. This will require removal of existing tie rods. The roof plan and typical cross section are shown in figure C-18. Local conditions prohibit additional construction beyond the property line. Interior clearance for theater usage is indicated.

REQUIRED

Describe and discuss briefly three reasonable methods that may be used to resist lateral thrust from dead load plus live load, maintaining the clearance given. Calculations are not required.

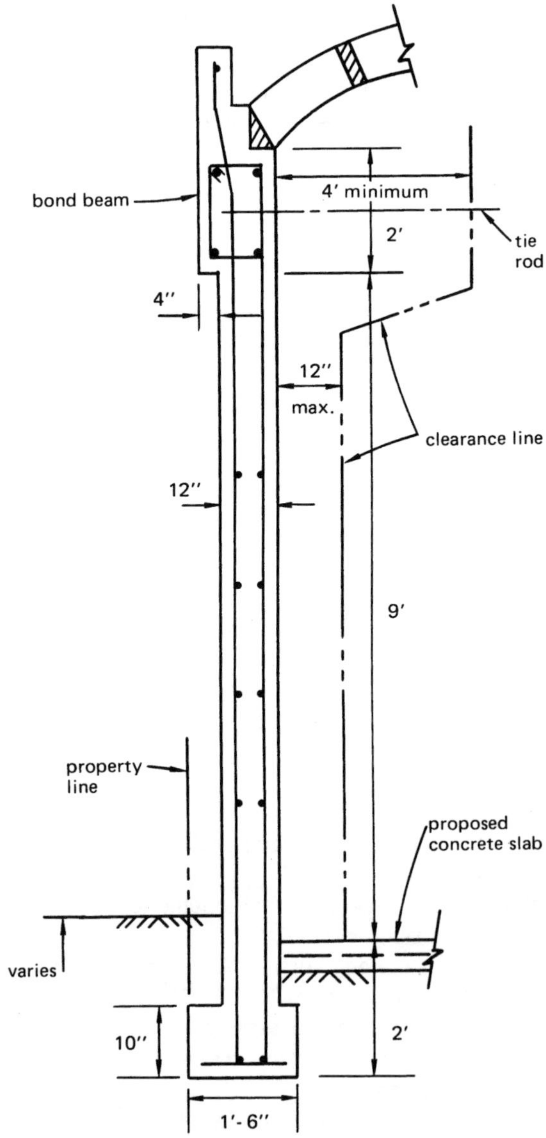

WALL SECTION

Figure C-18

(continued on next page)

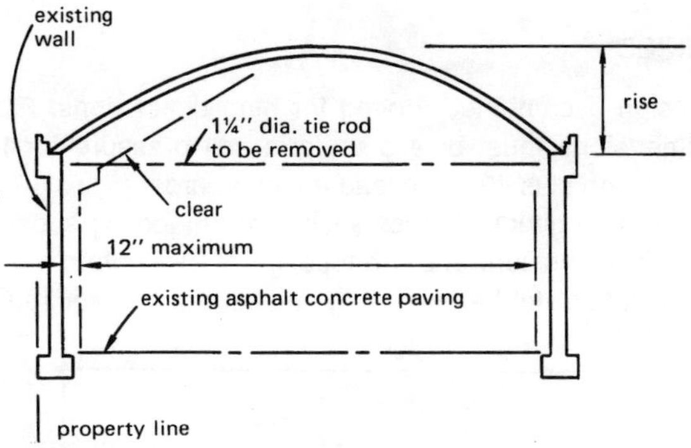

SECTION A-A

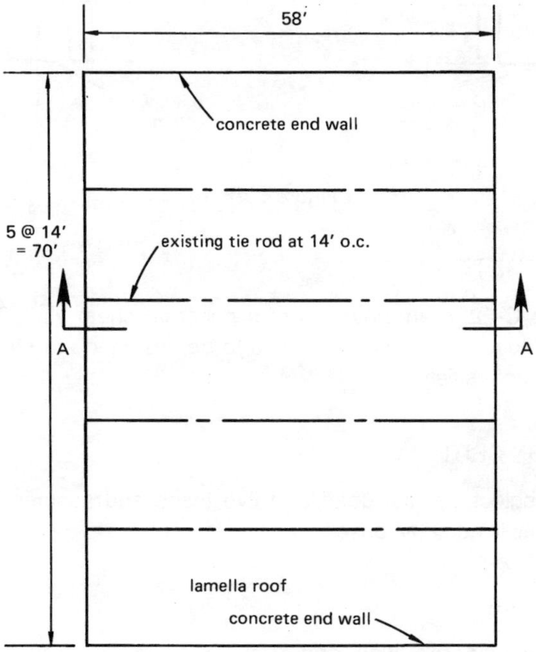

ROOF PLAN

span 58', rise 9'- 8", horizontal thrust 870 lb/ft

Figure C-18 (cont.)

CONCRETE-19

A pretensioned cored slab is shown in figure C-19.

DESIGN CRITERIA

- f'_c of plank and topping = 4 ksi normal weight concrete
- f_u = 270 ksi prestressing steel
- Ten 3/8-inch-diameter 7-wire strands reinforcement
- Area of one 3/8-inch-diameter strand = 0.085 square inches

REQUIRED

Determine the ultimate moment capacity. Show all calculations.

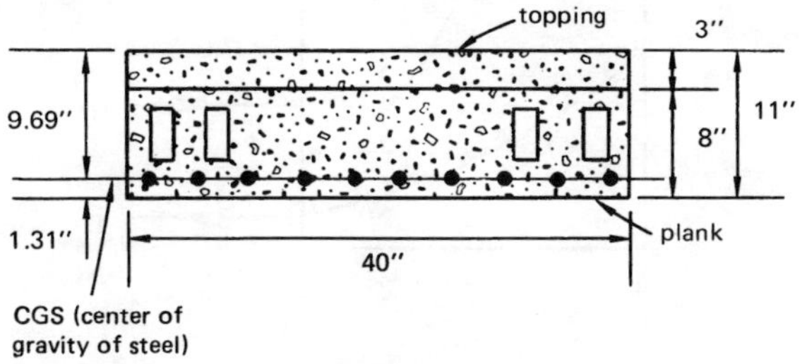

Figure C-19

CONCRETE-20

REQUIRED

(a) Write a brief description for the basic difference between the internal resisting couple of a reinforced concrete beam and a prestressed concrete beam.

(b) Sketch the shape of the concrete stress block due to location of the compression force C for the prestressed beams shown in figures C-20 (a) through (d).

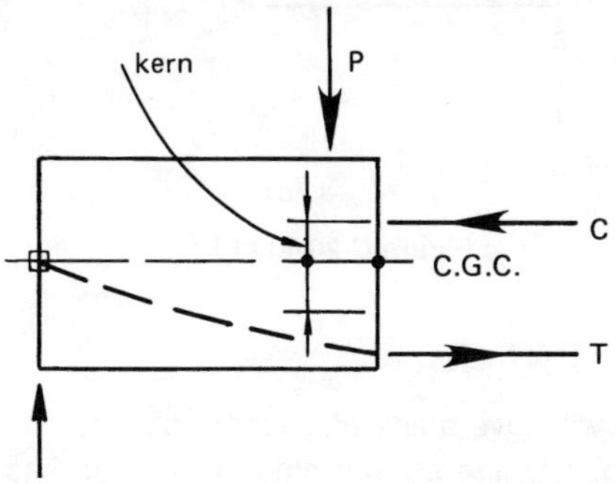

(a)

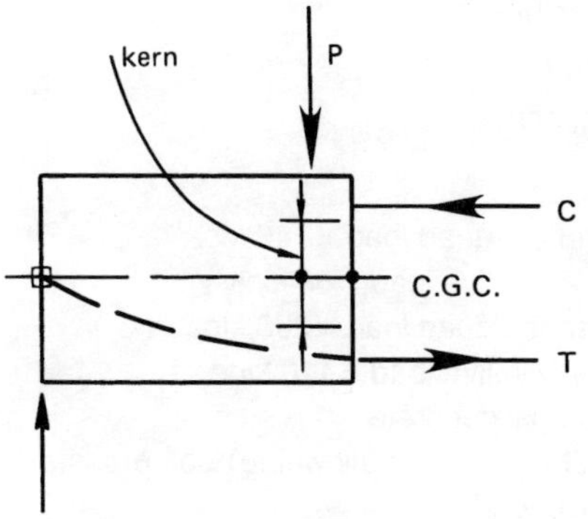

(b)

Figure C-20

(continued on next page)

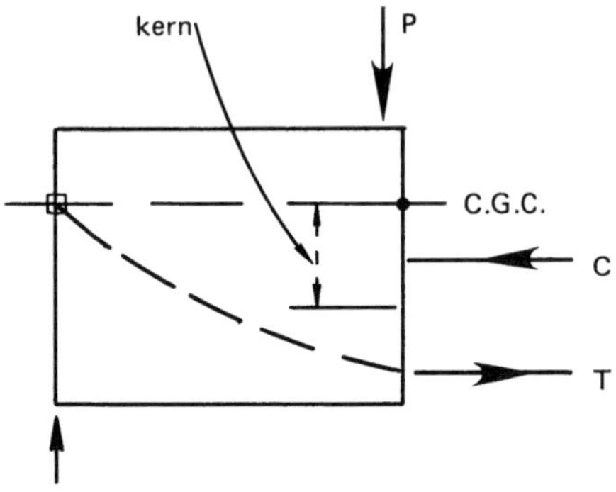

(c)

NOTE: C.G.C. is the location of the center of gravity for the concrete section.

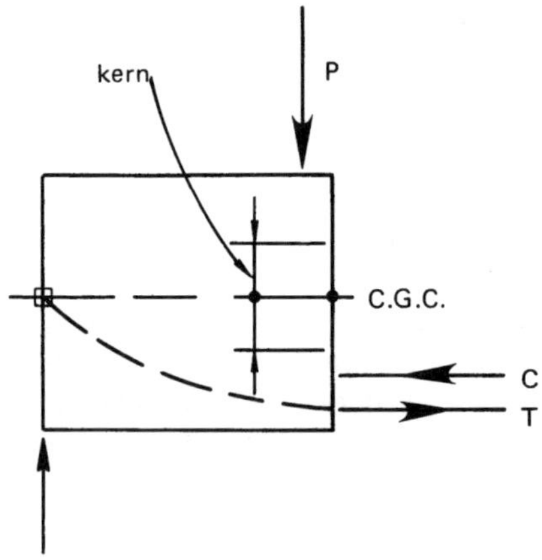

(d)

Figure C-20 (cont.)

CONCRETE-21

A building will have a line of exterior columns bordering on the property line as indicated in the plan shown in figure C-21. A continuous footing parallel to the property line is not feasible.

DESIGN CRITERIA

- Column loads:
  - exterior dead load = 140 kips
    - live load = 50 kips
  - interior dead load = 200 kips
    - live load = 110 kips
- f'_c = 3000 psi concrete
- 4000 psf (maximum allowable) soil pressure, dead load + live load

REQUIRED

(a) Design a combined footing for plan dimensions. Final dimensions must give a uniform soil pressure for full dead load plus 40% live load (no overburden).

(b) Draw a longitudinal cross section of the footing showing depth dimension, and the location of all reinforcement. Calculations for bar sizes and number are not required.

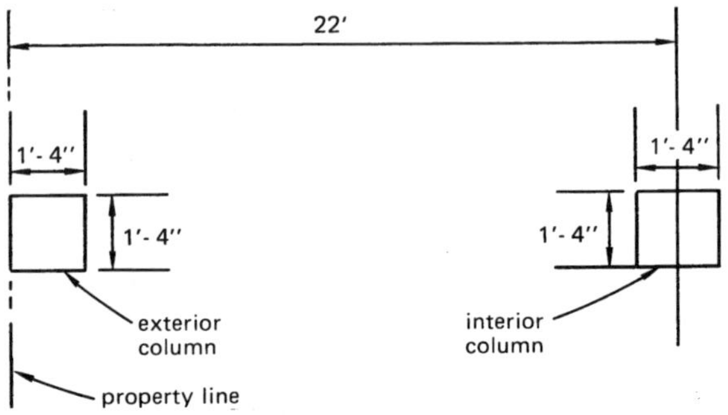

Figure C-21

CONCRETE-22

Figure C-22 is an elevation of a concrete shear wall. The lateral loads for which the wall is to be designed are shown on the right side.

DESIGN CRITERIA

- Neglect vertical dead and live loads and assume the wall fixed at the base.

REQUIRED

(a) Analyze the wall and draw a series of free-body diagrams which show the forces acting on each resisting element of the wall.

(b) Draw an elevation of the wall and show all additional boundary and trim reinforcement needed to resist lateral forces.

NOTE: It is not necessary to completely detail the reinforcement. Show the amount of reinforcement needed, approximate lengths, and placing. Do not show typical wall reinforcement.

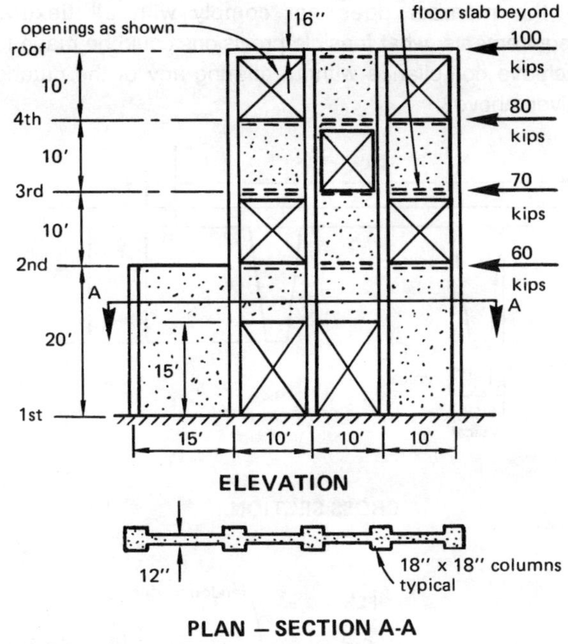

Figure C-22

CONCRETE-23

A part plan of a concrete waffle floor slab is shown in figure C-23. A typical column panel for an interior column is shown. The columns are 24 inches square and are spaced 36 feet on center each way. Use tributary area for loads. Do not provide for unbalanced shears. At one of the interior columns a 36-inch by 18-inch hole is necessary to pass duct work and pipes.

DESIGN CRITERIA

- f'_c = 3000 psi hard rock aggregate
- f_y = 40 ksi
- Dead load = 90 psf (averaged)
- Live load = 100 psf

REQUIRED

(a) What are the maximum unit shear stresses at the critical sections? Three are expected: on the face of the solid shear head (joist to solid panel), at a distance d from the face of the column, and at a distance d/2 from the face of the column.

(b) Design the reinforcing for shear where required. Assume that longitudinal bars in joists are not effective, and that no shear reinforcement is provided in the solid concrete panel.

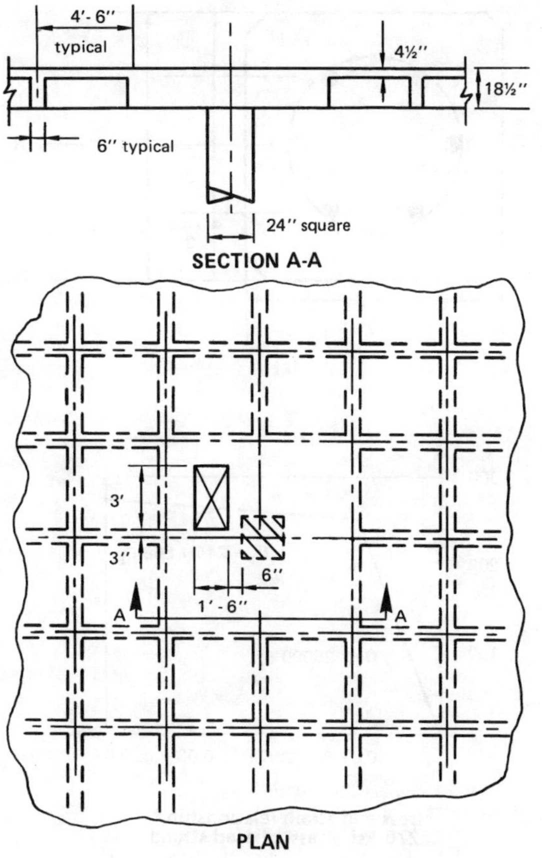

Figure C-23

CONCRETE-24

A prestressed concrete pile is shown in cross section in figure C-24.

DESIGN CRITERIA

- f'_c = 5000 psi @ 28 days concrete
- f_{pu} = 270,000 psi prestress strands
- A_s = 0.115 square inches
- A_c = 144 square inches
- $E_s = 27 \times 10^6$ psi
- The fabricator shall use 0.75 f_{pu} as jacking stress.
- Assume 10% loss in prestress.

REQUIRED

(a) Calculate the allowable service load for the pile under axial load only.

(b) Due to the soil condition, the pile has to resist both axial load and bending moment during earthquake.

$P_{dead + live + seismic}$ = 100 kips at service load

load factor = 1.4

ϕ = 0.7 for both axial and bending

Determine the maximum allowable bending moment about the X-X axis.

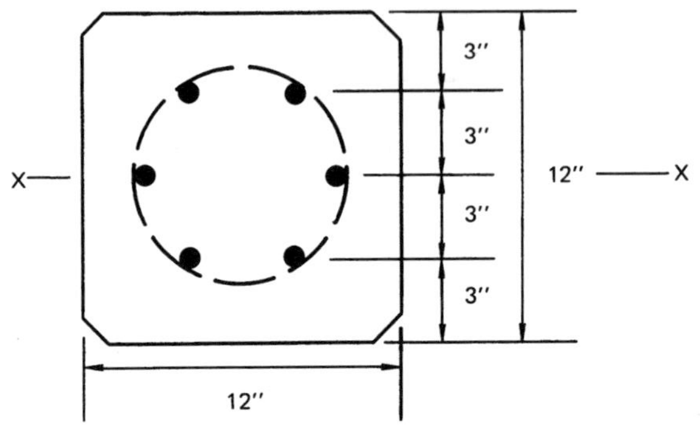

(a)

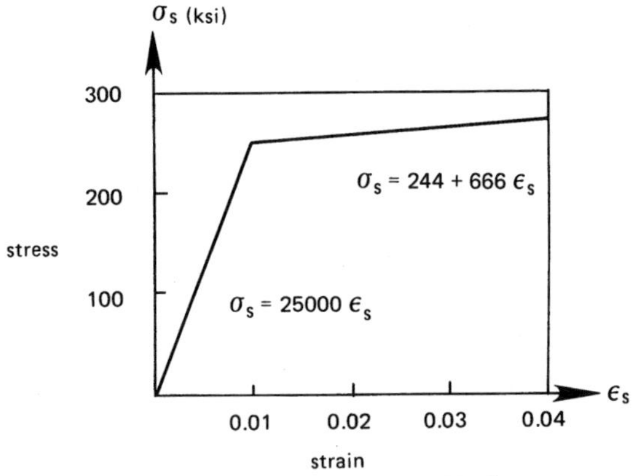

stress and strain relationship for 270 ksi stress-relieved strand

(b)

Figure C-24

CONCRETE-25

A double-tee beam shown in figure C-25 is to be used as a simple span beam having an overhang of 8 feet at one end. The member is prestressed with bonded tendons. At the overhang support, the tendons are located 2.50 inches from the top of the member. Design requirements shall be in conformance with the current UBC.

DESIGN CRITERIA

- $A = 187.5$ in^2
- $I = 4256$ in^4
- $Y_t = 5.17$ in
- $Y_b = 10.83$ in
- Area of tendons $= 0.58$ in$^2 = A_{ps}$
- $f_{pu} = 270.0$ ksi
- $f'_c = 5000$ psi normal weight
- Seismic zone 3

REQUIRED

At the overhang support:

(a) Determine the adequacy of the member from the standpoint of flexural strength requirements.

(b) If the member does not comply with all flexural requirements, what feasible provisions could be made to achieve compliance without altering any of the criteria given above?

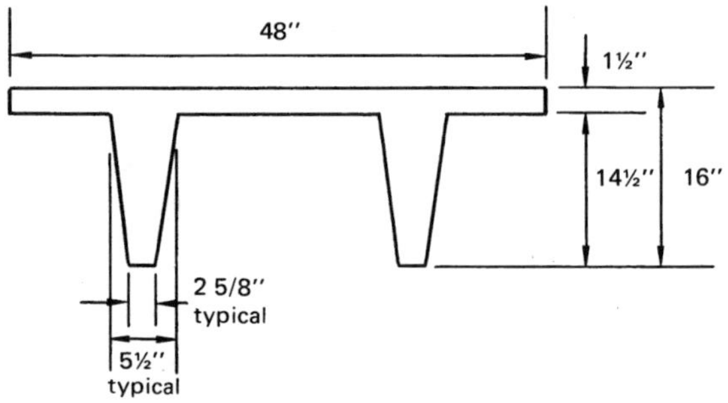

CROSS SECTION

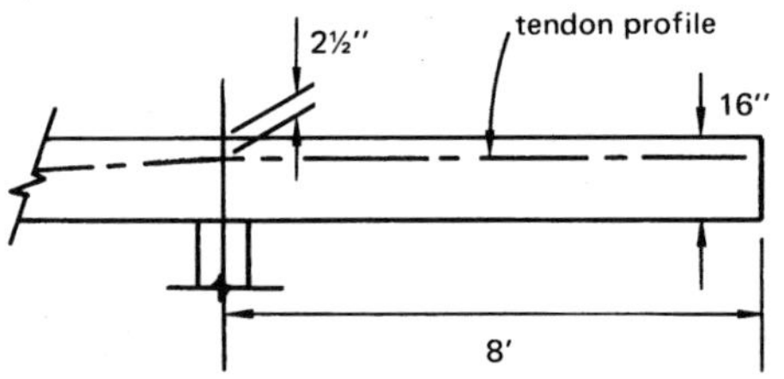

PARTIAL ELEVATION

Figure C-25

CONCRETE-26

REQUIRED

This problem contains several parts. Each part may be answered with a brief discussion. Answer all parts.

(a) In reinforced concrete moment-resisting frame structures, what seismic hazard to columns with high-strength reinforcement may over-strength girder reinforcement present?

HINT: The actual yield strength of rebars having a minimum specified yield strength of $f_y = 40$ ksi may be 60 to 70 ksi.

(b) How is ductility provided in reinforced concrete columns in addition to providing adequate main reinforcing steel?

(c) Why does the ordinary grade 60 rebar present more of a welding problem than does ASTM A36 structural steel?

(d) Why are columns usually poured and allowed to harden before the floor system is poured in reinforced concrete high-rise construction?

(e) What is the primary purpose of a column spiral?

(f) Does excessive air entrainment (above 10%) have a harmful effect in fresh concrete mixes? If so, what is it?

(g) Name at least three precautions that should be observed when a concrete slab is to be poured on a hot, windy day.

(h) When would you specify a retarder admixture for concrete?
(i) If calcium chloride is allowed as an admixture, what three precautions should be observed?
(j) A concrete mix is found to be so harsh that it cannot be pumped readily. Name two adjustments to the mix that may be used without reducing strength.
(k) When handling fresh concrete the final shrinkage may be adversely affected by at least two factors. Describe each.
(l) A horizontal construction joint is needed in a highly stressed concrete shear wall. What treatment would you require for the concrete contact surface?
(m) What are three key factors which affect deflection of a flat slab other than loads?
(n) If grade 40 bars have a minimum yield of 40 ksi, what is the requirement for maximum yield?
(o) What is creep? Are creep limits normally specified? if so, how?

CONCRETE-27

A cast-in-place reinforced concrete building is being designed with exterior bearing and shear walls. A typical cross section is shown in figure C-27. Dimensions and member sizes are shown. Minimum wall reinforcement satisfies the requirements for lateral force design. Assume that the foundation provides fixity to exterior walls at the first floor line. Assume interior column stiffness is negligible.

DESIGN CRITERIA

- f'_c = 3000 psi hard rock concrete
- ASTM A615 grade 40 reinforcing steel
- f_y = 40,000 psi
- Floor slab design live load = 150 psf
- Floor slab dead load including ceiling = 120 psf

REQUIRED

(a) Determine the critical second floor slab loading for design of joint A. For this loading condition draw a complete moment diagram of the second floor. Indicate maximum moments, and locate points of inflection.
(b) Check adequacy of the reinforcement at joint A.
(c) Determine the size, spacing, and required length of additional reinforcement needed at joint A.

NOTE: Do not consider the slab to be pin-connected at joint A.

(d) Draw joint A to scale 1" = 1'-0". Show all required reinforcement, lengths of bars, and clearances.

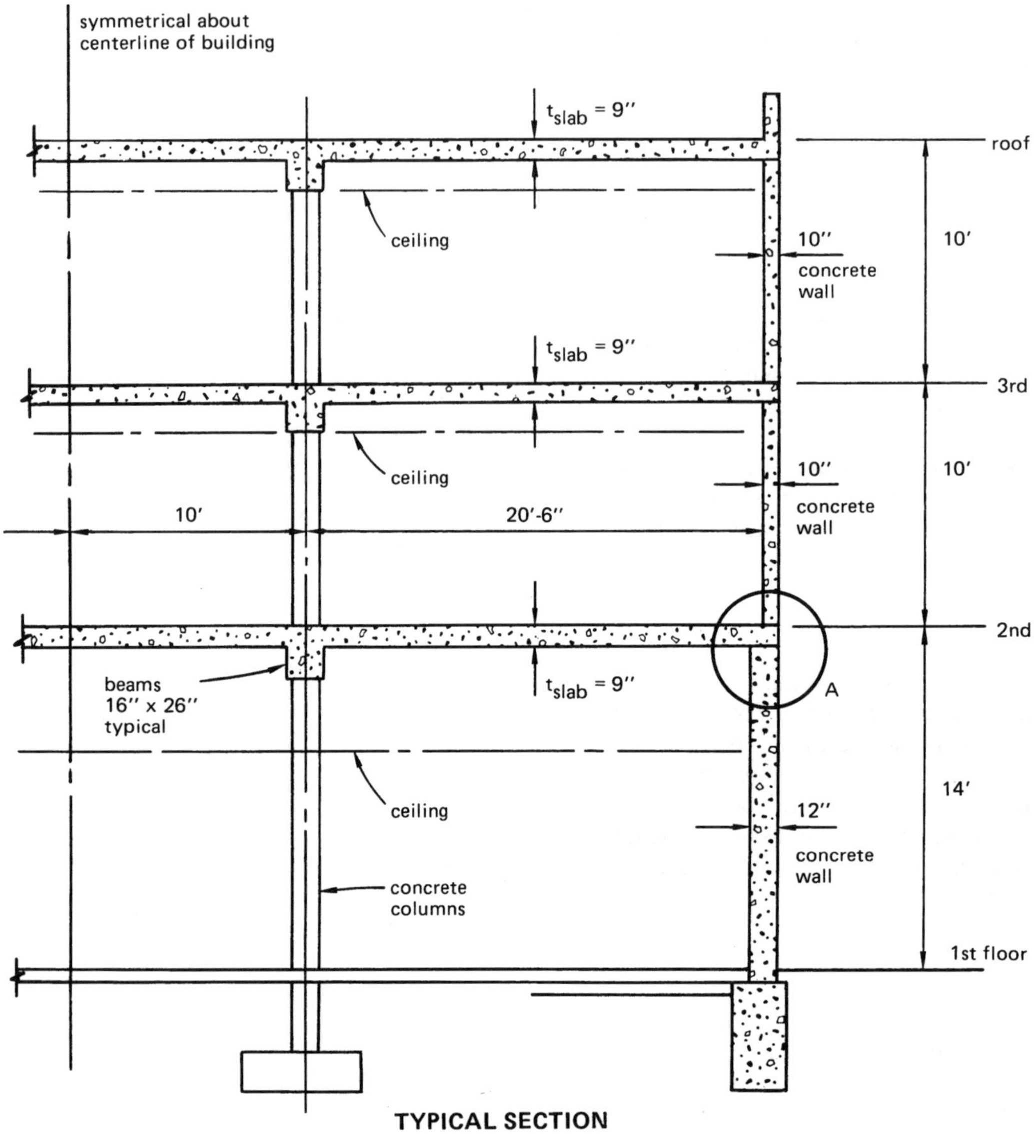

TYPICAL SECTION

Figure C-27
(continued on next page)

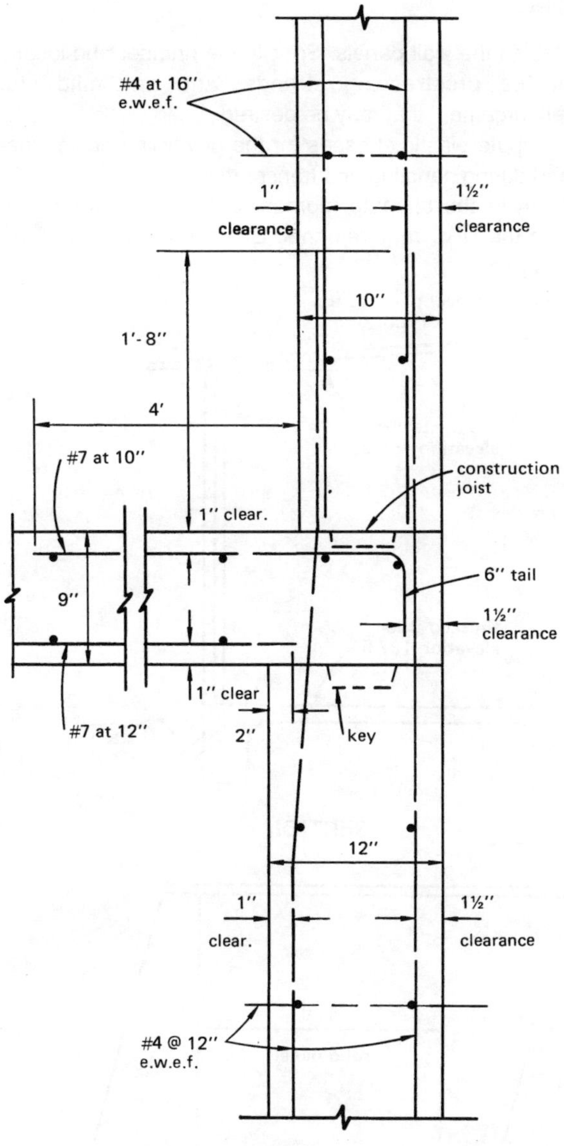

JOINT A

Figure C-27 (cont.)

CONCRETE-28

Figure C-28 indicates a canopy frame with loads as shown. The footing is capable of providing partial fixity. The top support can resist a maximum horizontal load of 3 kips.

DESIGN CRITERIA

- f'_c = 3000 psi, 150 pcf stone concrete
- f_y = 40,000 psi reinforcing steel
- 13 inches typical width of members
- Neglect axial compression.

REQUIRED

Design the frame for the condition shown, and show the size and placing of all reinforced steel. (Footing reinforcement not required.)

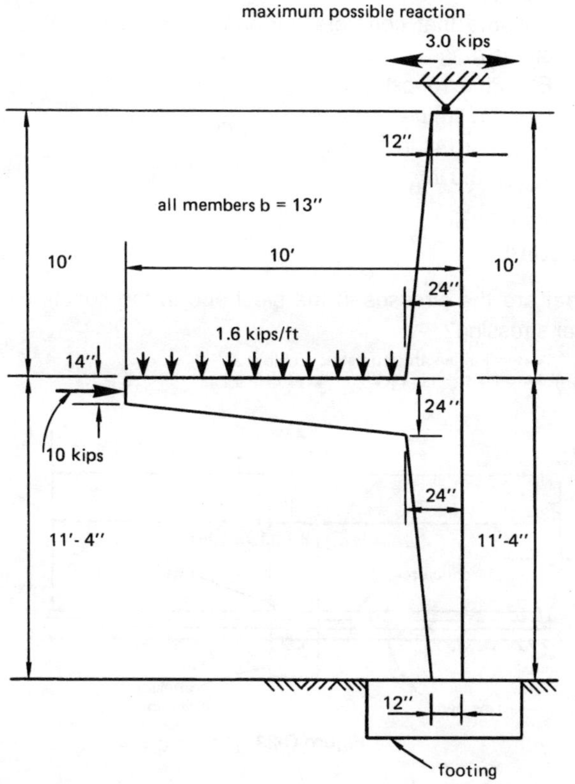

Figure C-28

CONCRETE-29

Two concrete members are post-tensioned concentrically as shown in figure C-29. The prestress wire is a single cable extending between exterior anchorages and pulled from one end. The interior anchorages are fixed, but they move toward each other under the prestress load.

After stressing, the distance between the interior anchorages has been reduced by 0.25 inches. The total extension of the prestress cable at the exterior anchorage was 24 inches before stressing and 26.25 inches after stressing. This difference in length is due to the movement of interior supports, compressive strain in the concrete, and tensile strain in the prestress cable.

DESIGN CRITERIA

- Assume no friction between concrete members and the bed.
- Assume that concrete performs elastically in this stress range.
- $E_s = 30 \times 10^6$ psi
- $E_c = 4 \times 10^6$ psi
- $A_{concrete} = 144$ in^2
- $A_{steel} = 3.0$ in^2

REQUIRED

What are the stresses in the steel and in the concrete after stressing?

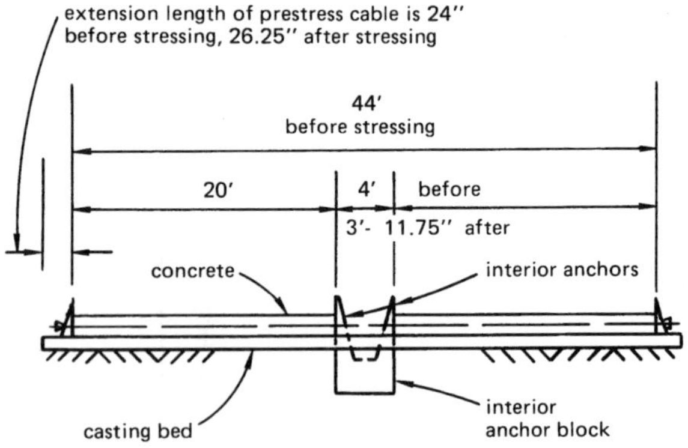

Figure C-29

CONCRETE-30

The drawing in figure C-30 shows a part elevation of a one-story building which has precast prestressed concrete wall panels. A section through a typical wall panel is also shown. The wall panels are tensioned concentrically with 3/8-inch diameter seven-wire stress-relieved strands which have an ultimate strength of 250,000 psi, and a cross sectional area of 0.080 square inches each.

DESIGN CRITERIA

- Suggested code: current UBC
- Seismic zone 3: soil profile type S_D
- Design wind pressure = 15 psf
- $f'_c = 4000$ psi concrete, 150 pcf
- Loss of prestress = 35,000 psi
- No tension is allowed in concrete at design load.
- Allowable concrete tension = $3\sqrt{f'_c}$ during handling and transportation
- Impact factor = 50% during handling and transporting
- Pickup and support points 1'-8" from top and bottom of panel
- $f_y = 40,000$ psi mild steel reinforcement

REQUIRED

(a) Design the wall panels. Specify the number and location of the prestressing strands, and any mild steel reinforcement that may be desired.
(b) Compute elastic stresses for the governing design load, and during handling and transporting.
(c) Analyze the ultimate moment capacity for compliance with the UBC requirements. Explain the results of your analysis.

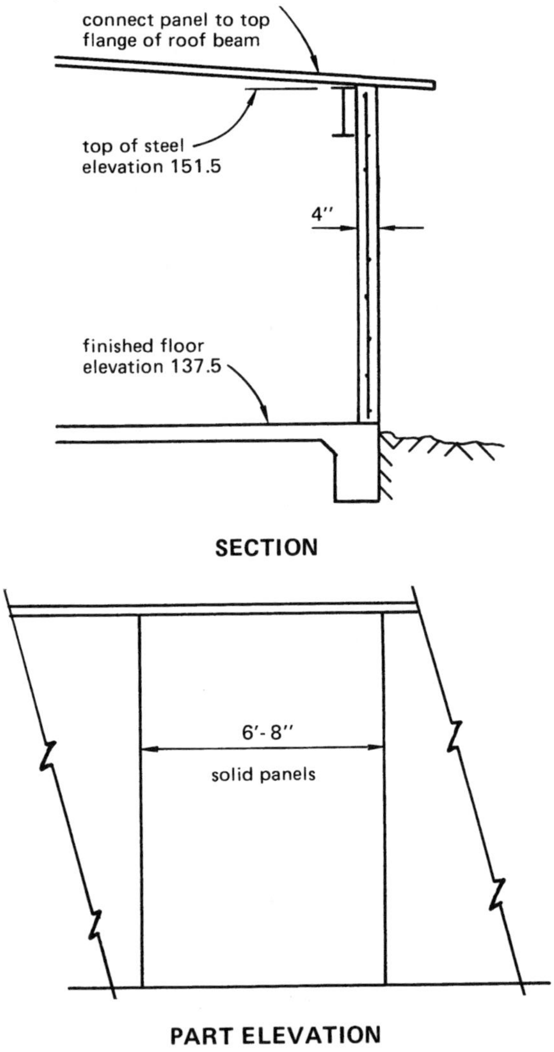

Figure C-30

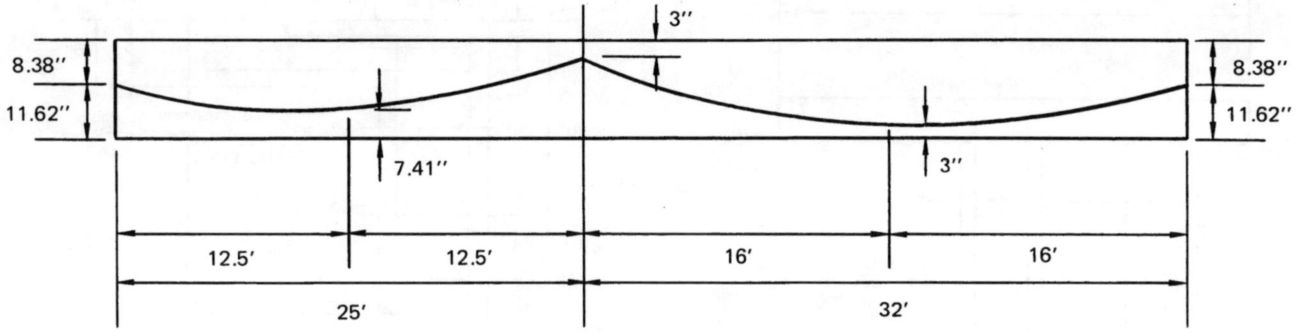

Figure C-31

CONCRETE-31

The two-span continuous prestressed concrete beam shown in figure C-31 is posttensioned with a cable of 10 strands having a total area of 1.53 square inches.

DESIGN CRITERIA

- Effective prestressing force is 248 kips.
- Ultimate tensile strength of steel = 270,000 psi
- f'_c = 5500 psi
- The tendon trajectory can be assumed to be a simple parabola in each span.

REQUIRED

(a) What is the magnitude of the secondary moment due to prestressing at the center support?
(b) What is the secondary reaction at each support?
(c) What are the stresses in the extreme top and bottom fibers at the center support due to prestressing alone?

CONCRETE-32

The drawing in figure C-32 shows a precast prestressed girder with a depth of 42 inches. The deck slab is poured in place to form a T-section as shown. No other dead loads are involved.

DESIGN CRITERIA

- 65 foot span
- 6 foot slab width
- 125 psf live load
- f'_c = 5000 psi hard rock concrete
- Prestress steel:

f'_s ultimate	260,000 psi	
f_{st} transfer	182,000 psi	(0.7 f'_s)
losses	35,000	
f_{sw} working	147,000	
f_{si} initial	167,000	

REQUIRED

(a) Calculate the prestress force required. (Do not select a prestressing system.)
(b) Calculate the stresses at top and bottom at transfer. (Temporary tensile stresses in excess of $6\sqrt{f'_c}$ may be resisted by mild steel reinforcement.)
(c) Calculate top and bottom stresses (girder) at dead load and live load.
(d) In the design of the joint at the girder support, what are the expected effects of shrinkage, creep, and temperature on the assumptions for fixity?

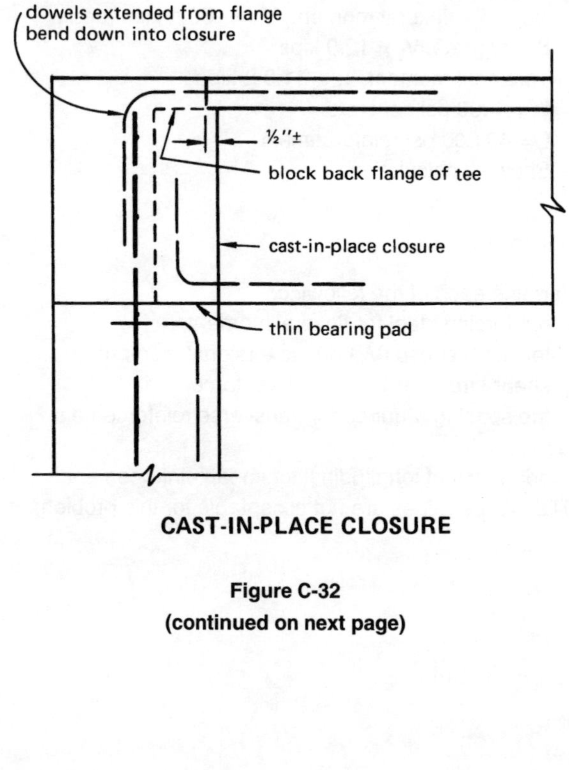

CAST-IN-PLACE CLOSURE

Figure C-32

(continued on next page)

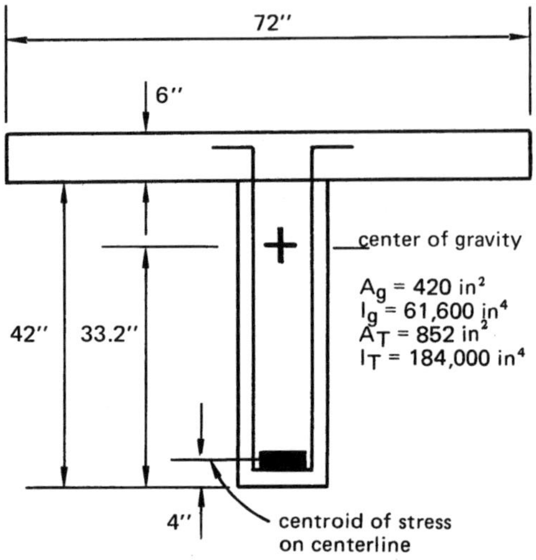

Figure C-32 (cont.)

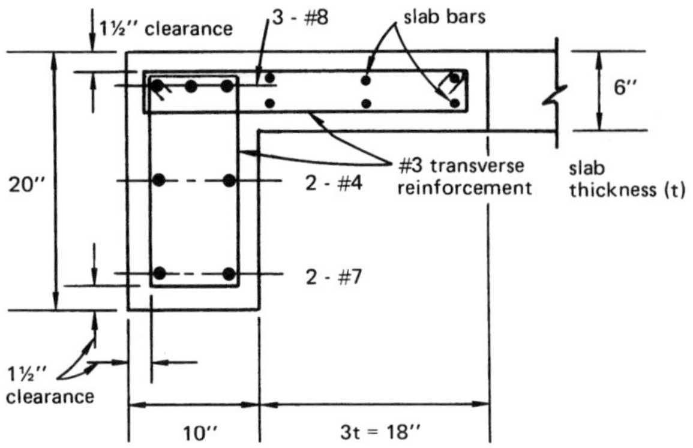

Figure C-33

CONCRETE-33

A reinforced concrete 10-inch by 20-inch spandrel beam is joined integrally with a 6-inch slab as shown in figure C-33.

DESIGN CRITERIA

- Negative flexural moment, M_u = 100 ft-kips (beam)
- Shear force, V_u = 12.0 kips
- Torsional moment, T_u = 11.0 ft-kips
- f'_c = 4000 psi concrete
- f_y = 40,000 psi reinforcement
- Show all calculations.

REQUIRED

Determine each of the following:

(a) reinforcing steel for flexural moment,

(b) torsional stress (V_{tu}) due to torsional moment,

(c) shear stress (V_u) due to shear force,

(d) the spacing required for transverse reinforcement, and

(e) adequacy of longitudinal torsional reinforcement.

NOTE: Table values are not acceptable for this problem.

CONCRETE-34

The diagram in figure C-34 shows service loads for a combined footing which cannot extend beyond the property line.

DESIGN CRITERIA

- f'_c = 2500 psi
- f_y = 40,000 psi
- Maximum allowable soil pressure dead load + live load = 3000 psf
- Total depth of footing is 28 inches.
- Show all calculations.

REQUIRED

(a) Design a rectangular footing for the column loads shown and the allowable soil pressure indicated. Proportion footing for total load.

(b) Determine the size and number of reinforcing bars and prepare a drawing to show location, size, and number of rebars. (Do not calculate, or show, intermediate cut-offs for rebar.)

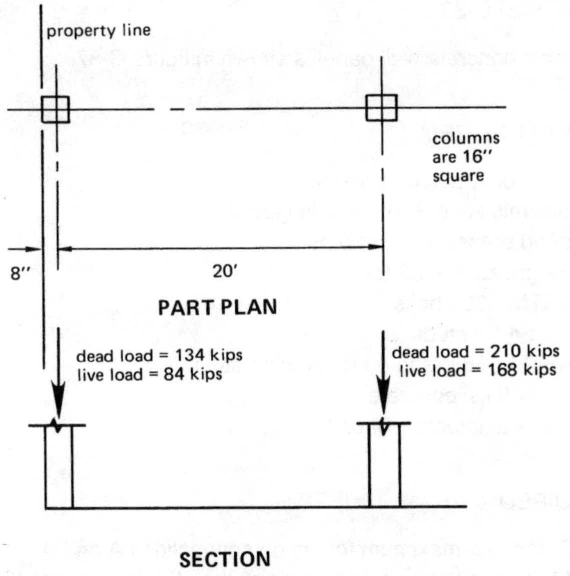

Figure C-34

CONCRETE-35

Two identical one-story buildings are composed of a continuous posttensioned roof slab that is 6 inches thick supported on concrete bearing walls. One building has bonded tendons, the other has unbonded tendons. The roof slab is prestressed with tendons placed along parabolic profiles. The effective stress in the tendons is 15.0 kips per foot of beam width. The tendons have an area of 0.11 square inches per foot of width. Figure C-35 shows the building cross section and the tendon profile.

DESIGN CRITERIA

- $f'_c = 4500$ psi
- $f_{pu} = 270.0$ ksi
- $M_{cf} = 6.14$ kip-ft/ft
- $P_p = \dfrac{0.11}{12 \times 4.5} = 0.00204$

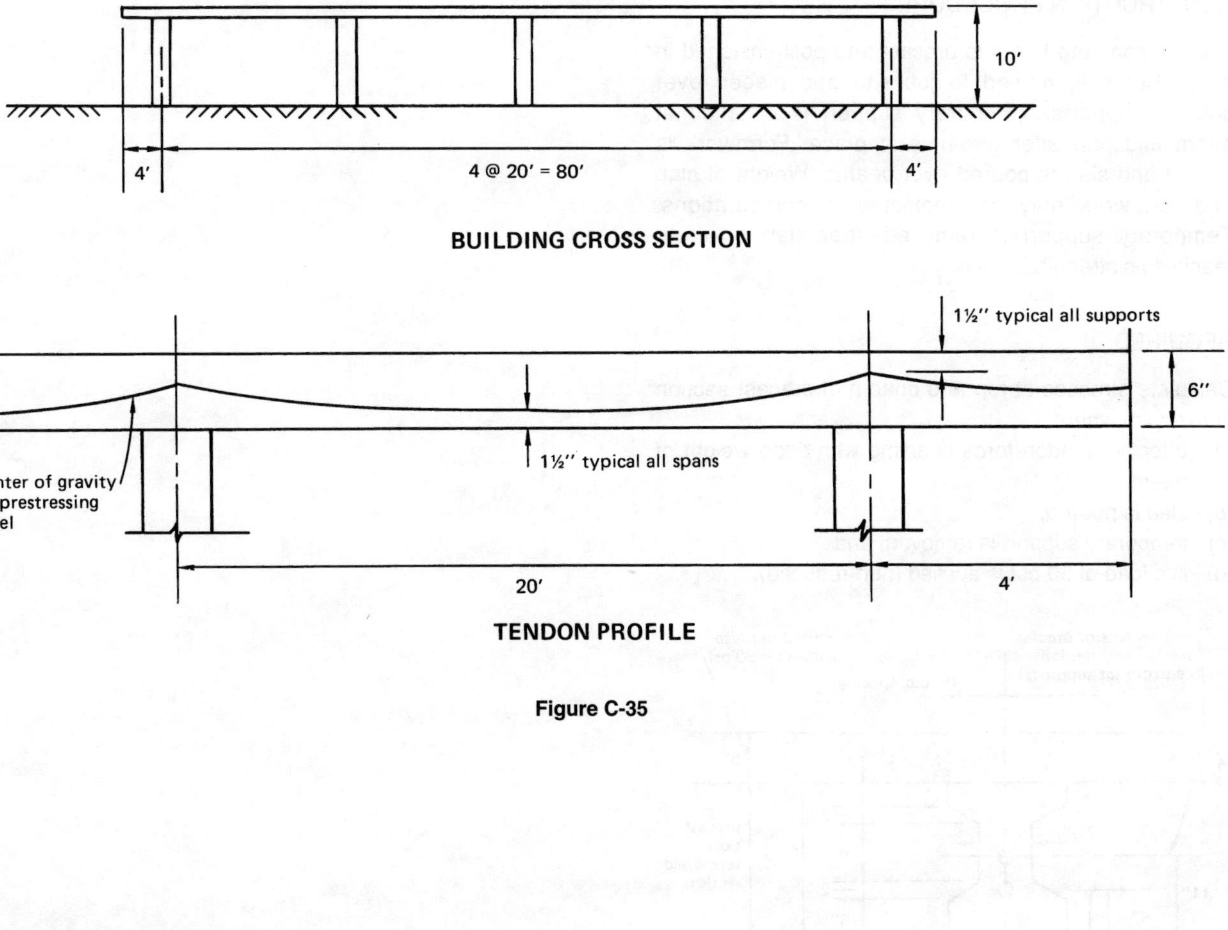

Figure C-35

CONCRETE-36

Figure C-36 depicts a precast concrete girder with a cast-in-place slab which is intended to work compositely with the girder.

DESIGN CRITERIA

- Beam span 40'-10"
- Dead load of beam = 210 lb/ft
- Precast concrete beam:
  - $A = 265\ in^2$
  - $S_T = 1200\ in^3$
  - $S_B = 1250\ in^3$
- Composite section:
  - $S_T = 13{,}500\ in^3$
  - $S_B = 2350\ in^3$
- No continuity is assumed at the columns.
- Tendon force = 200 kips effective (after all losses have occurred).
- Beam and slab have the same value of modulus of elasticity, E.

CONSTRUCTION PROCEDURE

Precast concrete beam is precast and posttensioned in shop. Beam is hauled to job site and placed over column supports. Temporary support is installed at beam midspan after girder is in place. Formwork is placed and slab is poured over beams. Weight of slab and formwork may be neglected in computations. Temporary support is removed after slab concrete reaches its strength.

REQUIRED

Calculate stresses at top and bottom of precast section at midspan when:

(a) effective tendon force is acting with dead weight of beam,
(b) slab is poured,
(c) temporary support is removed, and
(d) live load of 50 psf is applied (non-reduced).

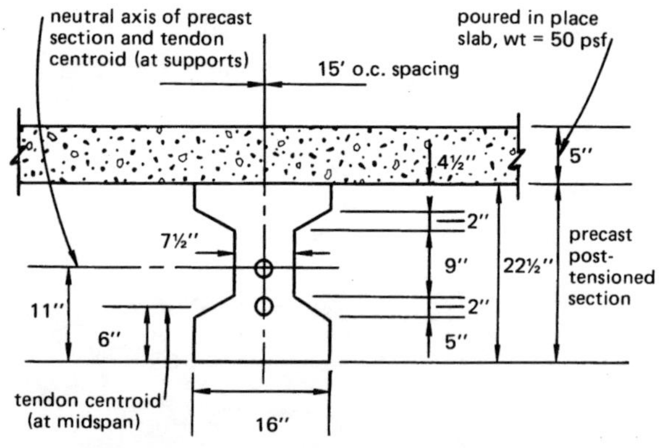

Figure C-36

CONCRETE-37

A precast concrete wall panel is shown in figure C-37.

DESIGN CRITERIA

- Type of structure: hospital
- Seismic zone 3; soil profile type S_D
- Wind pressure area 20 psf
- Height zone < 99 feet
- ASTM A325 bolts
- f_y = 54 ksi studs
- ASTM A36 grade 40 reinforcing steel
- f'_c = 4.0 ksi concrete
- Near-fault factors of unity

REQUIRED

(a) Determine maximum forces on connections A and B.
(b) At connection A determine the thickness of the connection angle and the number and size of studs that will support the load and fit the connection plate.
(c) Determine ultimate pull-out capacity of the bolt in connection B.

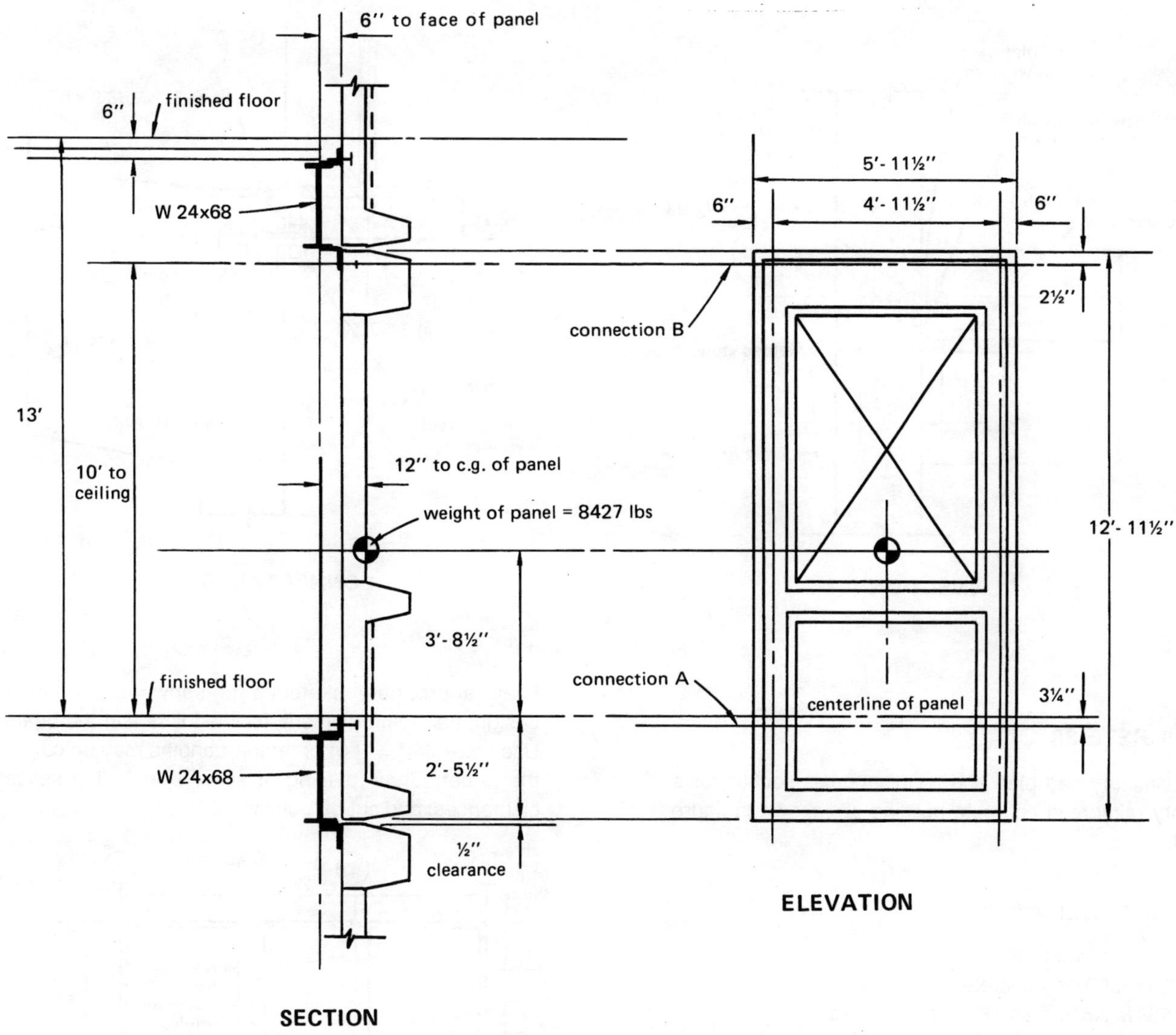

Figure C-37
(continued on next page)

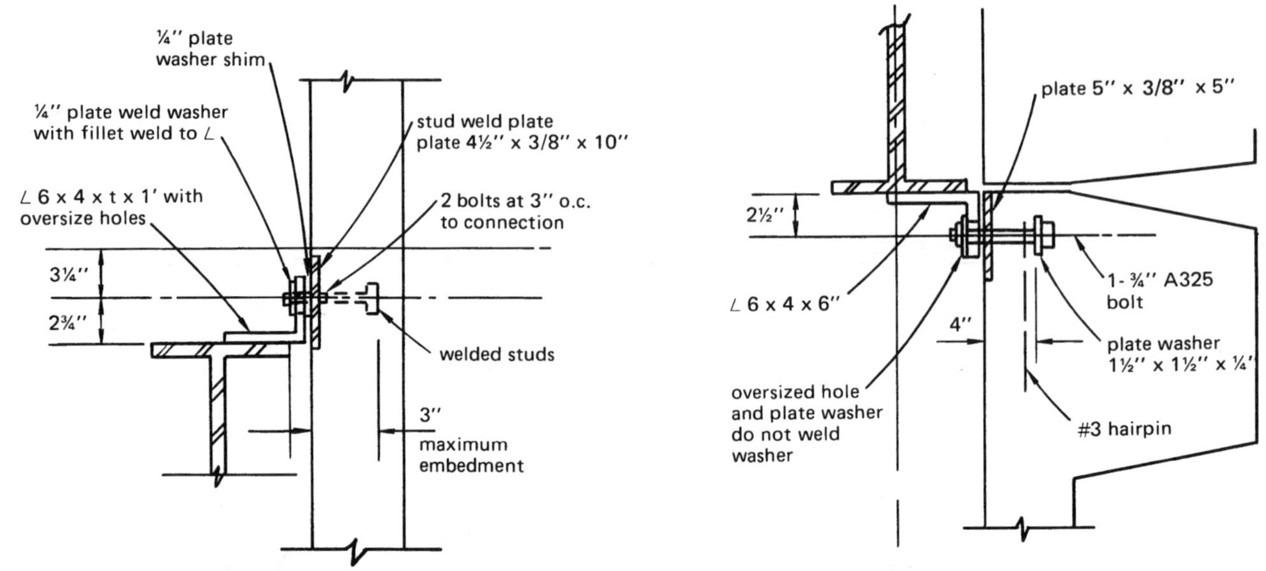

CONNECTION A **CONNECTION B**

Figure C-37 (cont.)

CONCRETE-38

A preliminary part plan and a typical cross section for a 2-story reinforced concrete building are shown in figure C-38.

DESIGN CRITERIA

- Code: current UBC and ACI
- f'_c = 3000 psi concrete
- ASTM A615 grade 60 reinforcement
- Design live load = 100 psf (may be reduced)
- Partitions = 20 psf
- Ceiling and mechanical = 10 psf
- Columns = 18 inches square, maximum

In addition to these general criteria, the following will apply to the second floor framing system:

- 4-inch-thick concrete slab
- 20-inch-wide by 14-inch-deep joist pans
- Joists are oriented to span East-West as indicated.
- Beams on lines 2 and 3 are limited to a depth of 22 inches. (18 inches + 4-inch slab)

REQUIRED

(a) Consider the vertical loads only, and prepare a design for the second floor slab, and for the second floor joists. Assume a pin-centered support at exterior walls.

(b) Design the second floor beams on lines 2 and 3.

(c) Draw appropriate sketches to show your complete design, and reinforcement, for the joints and the beams. Use scale 3/4" = 1'-0" or larger. Lengths may be cut to fit the paper. Show placing of all bars including splices, clearances, and cut-off points.

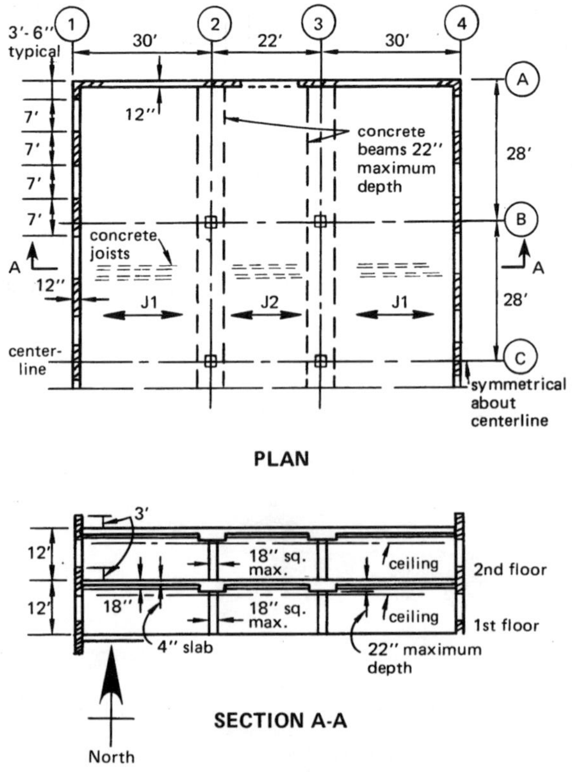

SECTION A-A

Figure C-38

CONCRETE-39

A concrete column is shown in figure C-39 which extends from a cast-in-place pile cap at the bottom to a concrete girder at the top.

DESIGN CRITERIA

- Ductile frame
- f'_c = 4.0 ksi normal weight concrete
- f_y = 60 ksi reinforcing steel
- Column P_u = 164.30 kips
- M_u = 11,000 ft-kips at top of column
- Spiral reinforcement: #5 bars, f_y = 60 ksi

REQUIRED

Determine the length of embedment of the vertical reinforcement into the girder and the pile cap. How far should the spiral extend at top and bottom? What spacing should be used for the spiral?

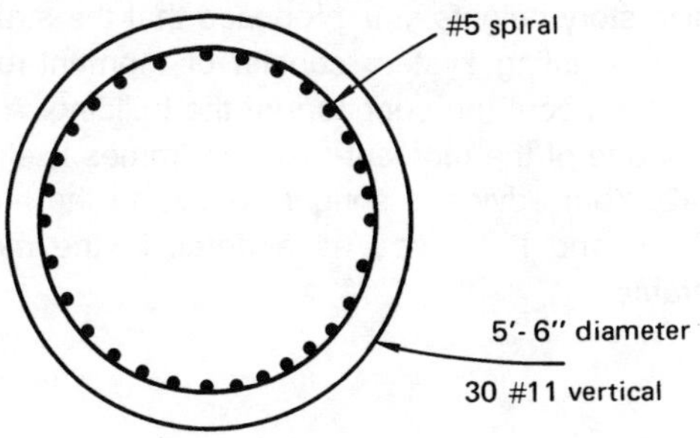

SECTION A-A

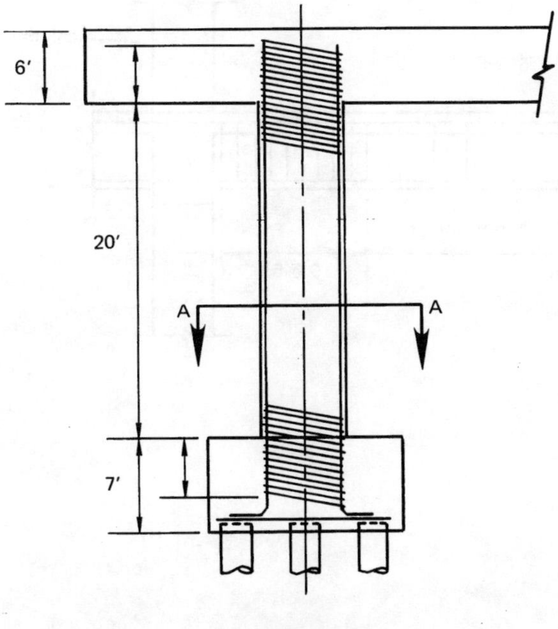

ELEVATION

Figure C-39

CONCRETE-40

The shear wall in elevation A of figure C-40 is part of a 2-story building. The building has a dual bracing system consisting of a ductile moment-resisting space fram and shear walls.

DESIGN CRITERIA

- f'_c = 3000 psi hardrock concrete
- A615 grade 60 reinforcing steel
- Seismic zone 3

REQUIRED

(a) Calculate the maximum shear stress in the wall and determine the wall reinforcement.

(b) Determine the minimum dimension *t* for square vertical boundary reinforcing. Sketch a cross section showing reinforcing. Use scale of 3/4" = 1'-0". Assume adequate lateral support at floor lines perpendicular to the pane of the wall.

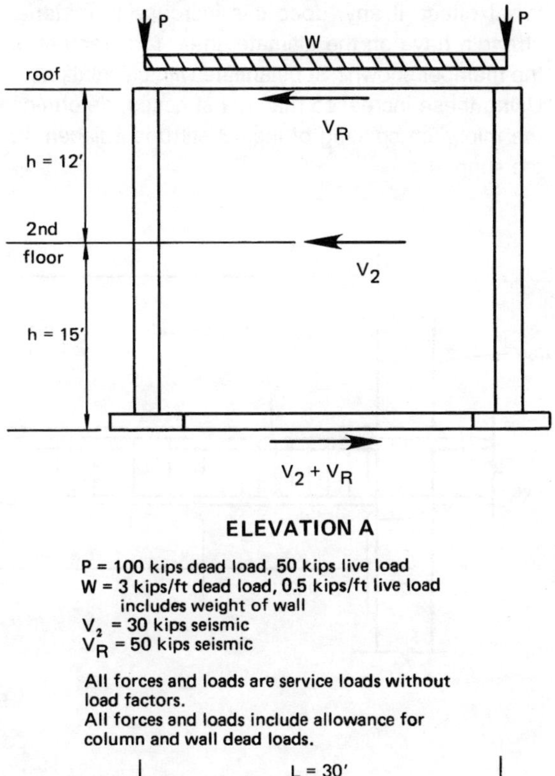

ELEVATION A

P = 100 kips dead load, 50 kips live load
W = 3 kips/ft dead load, 0.5 kips/ft live load includes weight of wall
V_2 = 30 kips seismic
V_R = 50 kips seismic

All forces and loads are service loads without load factors.
All forces and loads include allowance for column and wall dead loads.

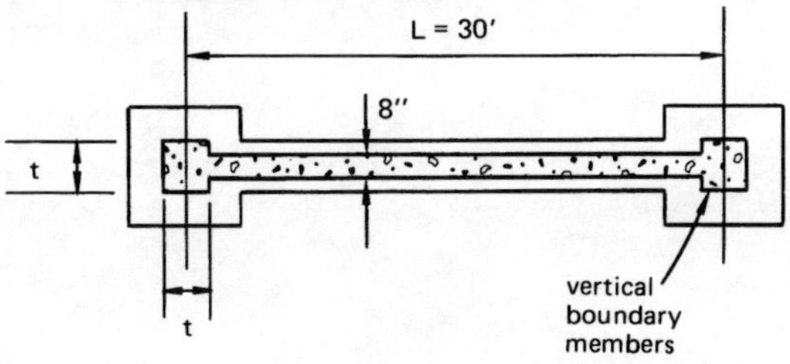

PLAN

Figure C-40

CONCRETE-41

The reinforced concrete beam shown in figure C-41 is part of a reinforced concrete ductile moment-resisting space frame in a building more than 160 feet in height in seismic zone 3. The frames have been designed to resist the seismic forces in accordance with the current UBC for the structure. There are no shear walls.

DESIGN CRITERIA

Original design:

- f'_c = 3000 psi
- ASTM A615 grade 40 reinforcing steel for longitudinal reinforcement and stirrups

Actual testing of material during construction:

- Longitudianl reinforcement has actual yield strength of 72,000 psi.
- f'_c = 3400 psi at 28 days
- Stirrups were not tested.

REQUIRED

(a) What effect, if any, does this increase in material strength have on the ultimate shear requirement of the member shown? Substantiate with calculations.

(b) Using these increased material stresses, determine the minimum spacing of the #3 stirrups adjacent to the support.

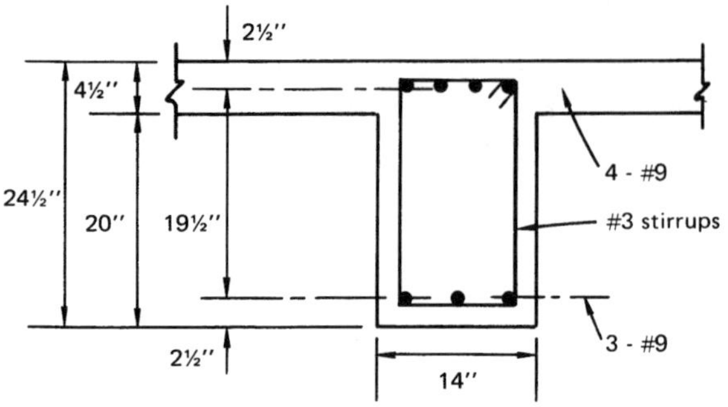

SECTION

Figure C-41 (cont.)

CONCRETE-42

A reinforced concrete high-rise is being planned on a site 6 miles from an active earthquake fault. A consulting seismologist reports that this fault has the capacity to generate a magnitude 7.5 earthquake (Richter).

To minimize story heights, it is proposed that the structure's lateral force-resisting system consist of moment-resisting frames located about the periphery of the building. A typical elevation of one of the moment-resisting frames is shown in figure C-42. Your advice is sought relative to the merits of details D, E, and F for use as a detail in the moment-resisting frame.

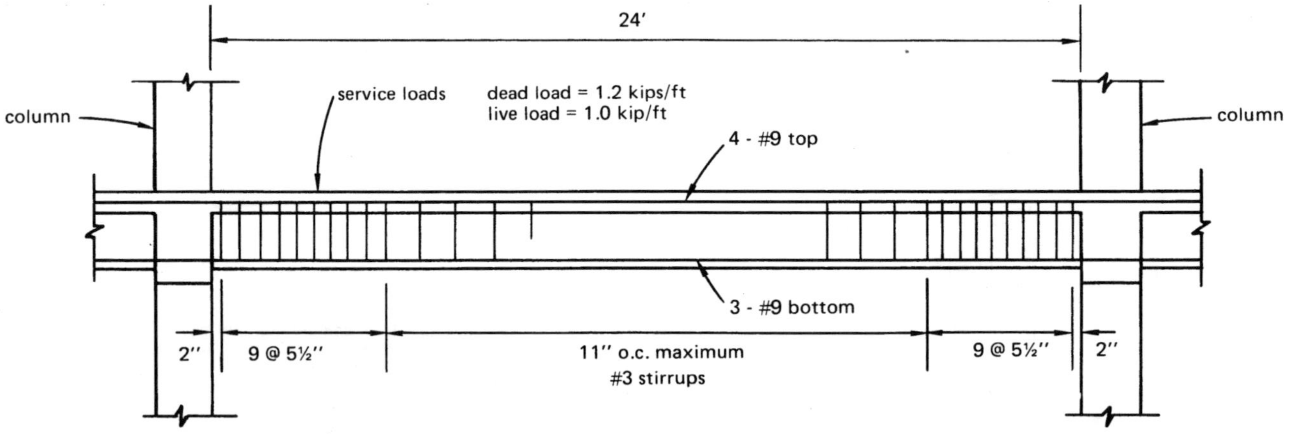

ELEVATION

Figure C-41

REQUIRED

Answer the following questions. No calculations or codes are required to complete this problem.

(a) Discuss the relative merits of details D, E, and F under severe seismic overloads associated with a possible magnitude 7.5 earthquake.
(b) Which detail is preferred? Why?
(c) What are the pitfalls to be avoided when proportioning concrete dimensions and amounts of principal reinforcement in the detail selected?

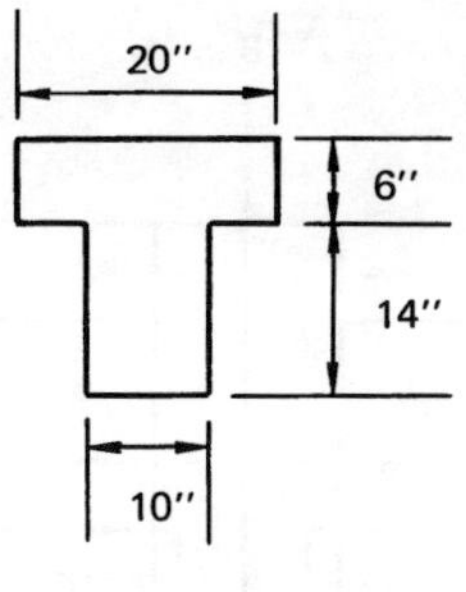

TYPICAL SECTION

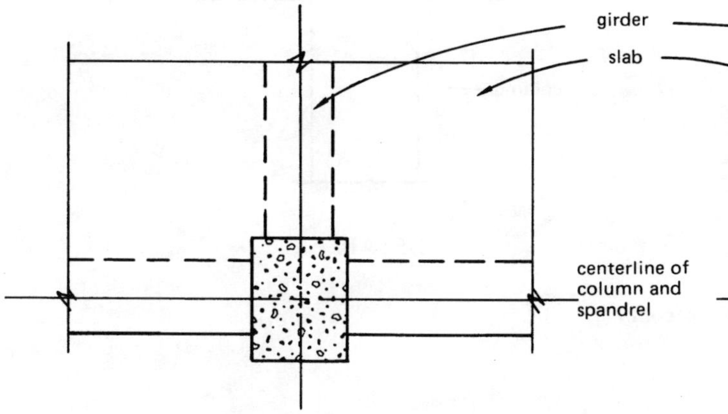

SECTION A-A

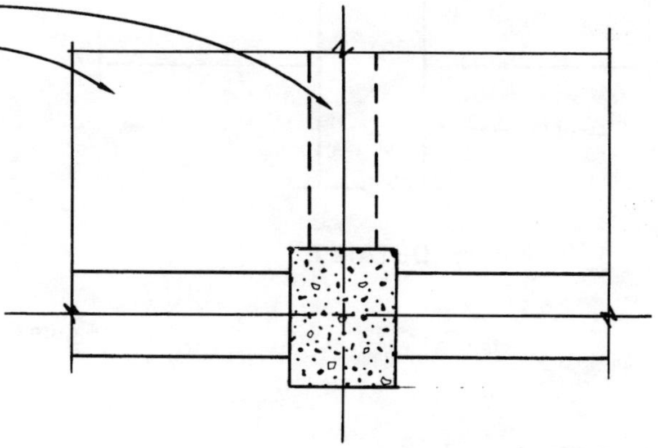

SECTION B-B

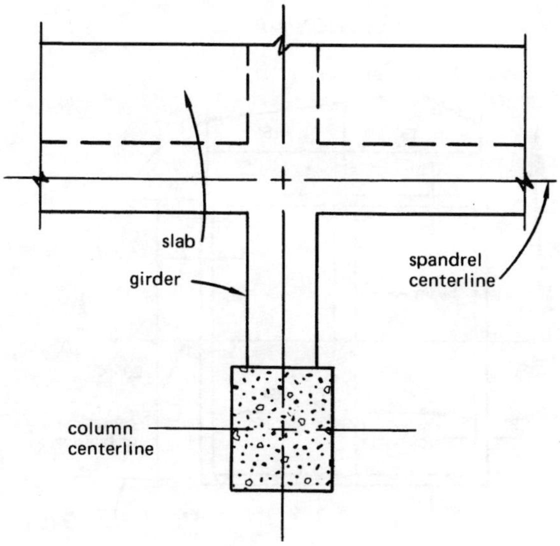

SECTION C-C

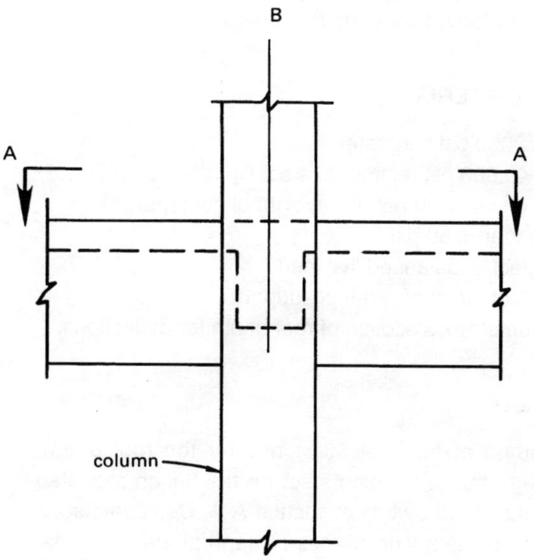

DETAIL D

Figure C-42
(continued on next page)

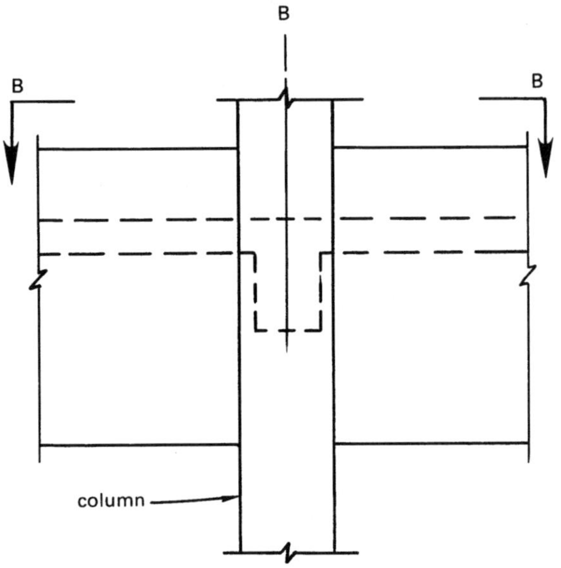

DETAIL E

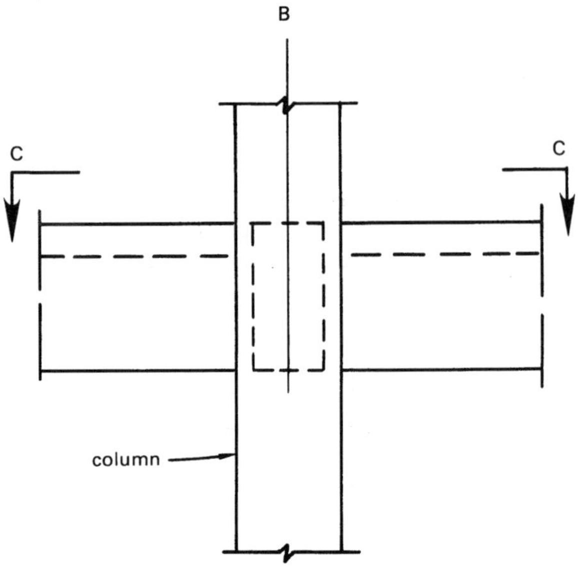

DETAIL F

Figure C-42 (cont.)

CONCRETE-43

A roof plan and a cross section through a one-story building are shown in figure C-43. The 8-inch roof slab is supported along the building centerline by a wide, but shallow, reinforced concrete roof beam.

DESIGN CRITERIA

- f'_c = 3000 psi concrete
- f_y = 40,000 psi reinforced steel
- Dead load = 20 psf plus weight of concrete
- Live load = 50 psf
- Neglect unbalanced live load.
- Neglect effect of width of support.
- Assume gross section of roof beam for deflections.

REQUIRED

(a) Assume a depth of 30 inches for the roof beam. Design the reinforcing steel for the 8-inch roof slab (as indicated by S1) at section A-A. Use continuous reinforcement with splices at appropriate locations. Draw a section to show your design. Do not detail the wall steel.

(b) Assume a depth of 15 inches for the roof beam. Prepare a design similar to that called for in part (a).

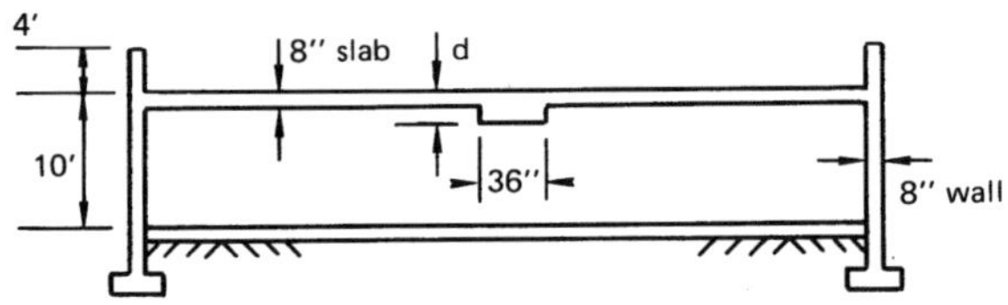

SECTION A-A

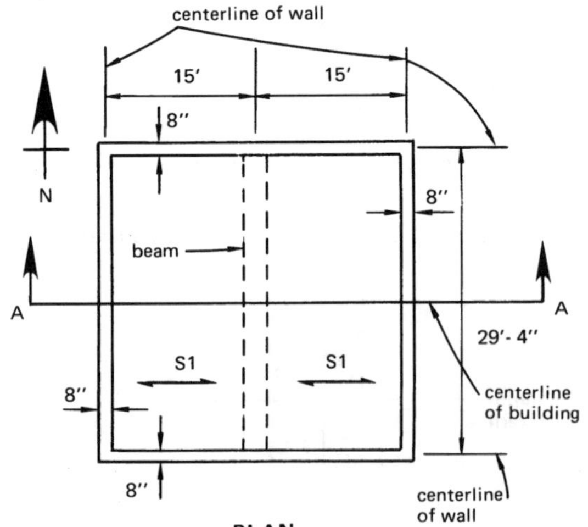

PLAN

Figure C-43

CONCRETE-44

An existing concrete beam A shown in figure C-44 is to be used to support a new concrete floor.

DESIGN CRITERIA

Existing concrete properties:

- $F'_c = 2000$ psi
- $F_y = 40{,}000$ psi
- u = 240 psi (bond)
- $E = 3.1 \times 10^6$ psi
- Measured deflection = 3/8 inch
- New concrete: $f'_c = 3000$ psi (normal weight)
- Office occupancy

REQUIRED

(a) Design a composite girder for the new condition. Assume shoring is provided.

(b) Sketch an elevation and typical cross section of the composite girder showing details of the new design.

(c) List any features of the composite design not in compliance with the code.

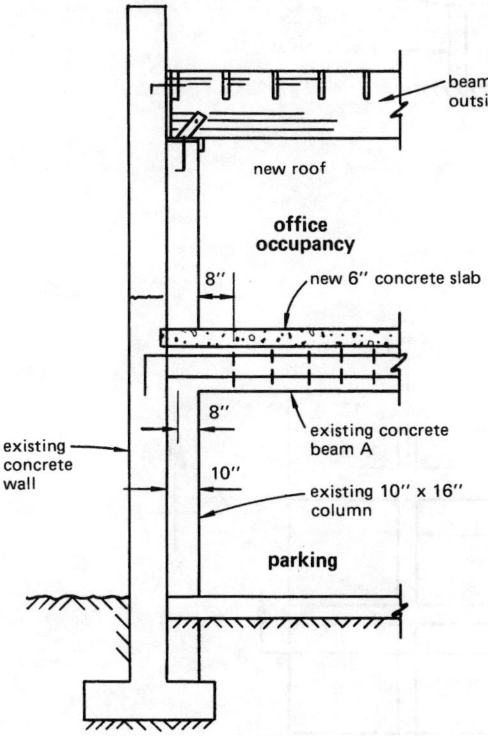

SECTION A-A

Figure C-44

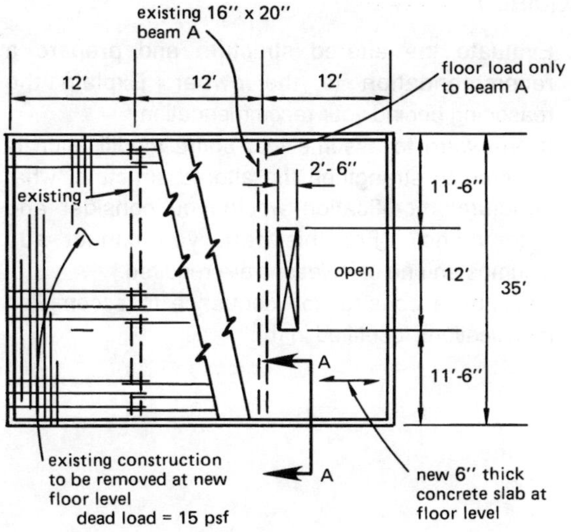

FLOOR PLAN

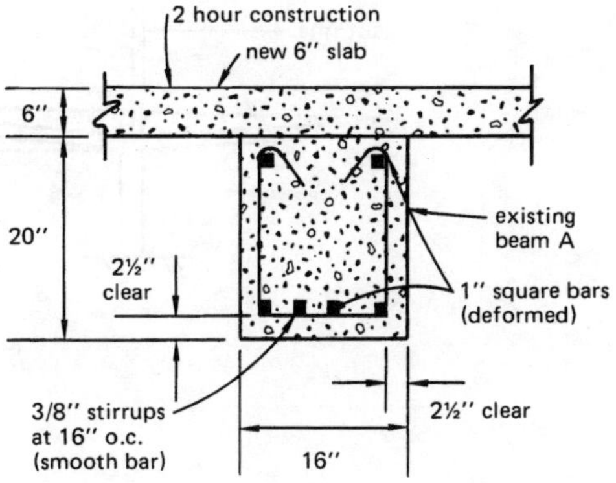

TYPICAL SECTION

Figure C-44 (cont.)

CONCRETE-45

The plan of a cantilevered canopy is shown in figure C-45, along with the adjacent part of the roof structure. The owner proposes to remove the cantilevered canopy and to cut it off at the face of the girder shown on line 3. The cantilever to the east of line 4 is to remain.

DESIGN CRITERIA

- $f'_c = 3000$ psi lightweight concrete
- Dead load of concrete = 112 pcf
- Grade 40 reinforcing steel
- Assume no roofing: follow UBC.
- Structure was first erected in 1960.
- Disregard lateral forces.
- The present structure (before modification) is adequate for dead and live loads.
- Assume the central beam is loaded uniformly.

REQUIRED

(a) Evaluate the altered structure and prepare a recommendation to the owner. Explain the reasoning behind your recommendation.

(b) If one were to assume that some modification is required to strengthen the altered structure, what structural modification would you consider and recommend? (For this part, you cannot add columns and no calculations are required.)

(c) Describe a construction sequence to accomplish modifications identified in (b).

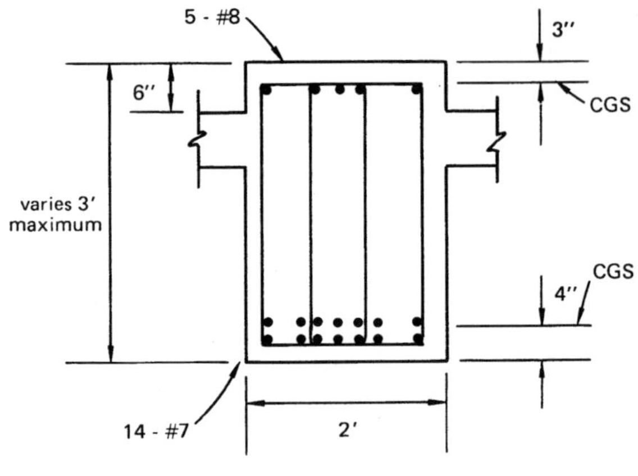

SECTION B-B

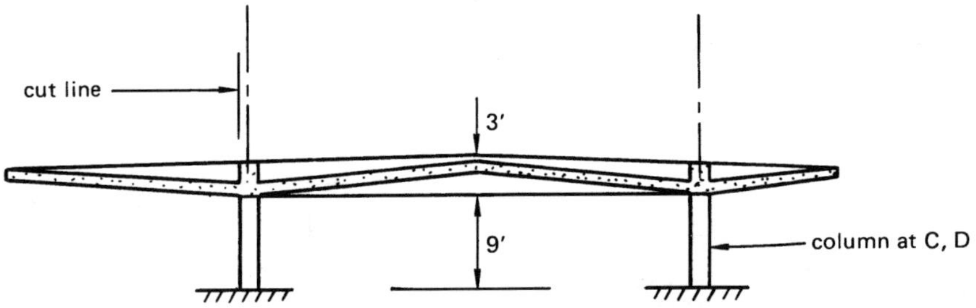

SECTION A-A

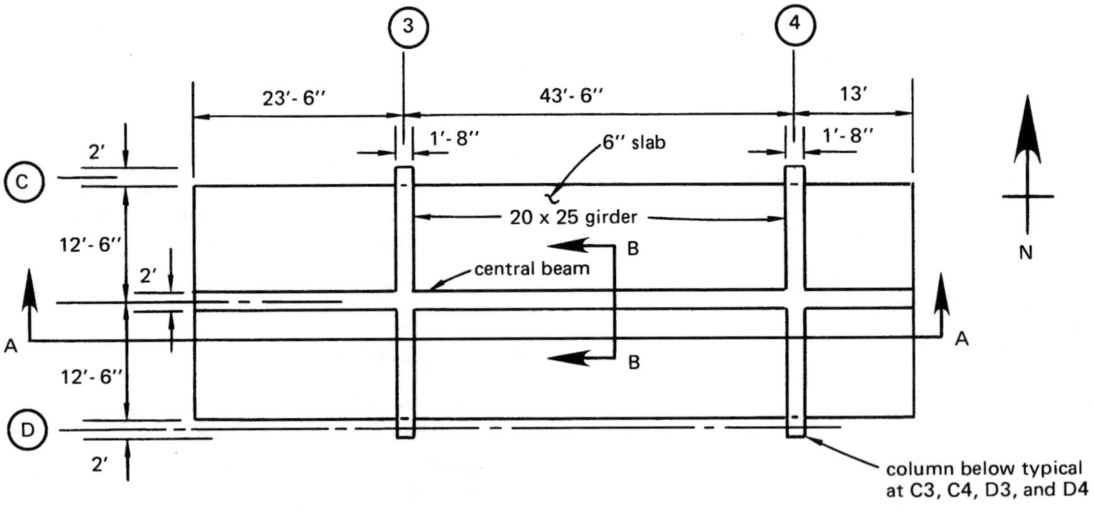

CANOPY ROOF PLAN

Figure C-45

CONCRETE-46

Figure C-46 shows an idealization of a typical design condition. For simplification, the reinforcement and details of the column are not given, and the strength of the column is assumed as accurately represented by the interaction curve (figure C-469(b)). The system is assumed to be laterally braced, and column strength reduction factors are to be neglected. Anchorage of beam reinforcement in the column is adequate. Assume the column and beam are laterally held against buckling. Neglect the effect of the two #4 bars acting as compression steel.

DESIGN CRITERIA

- f'_c = 3000 psi hard rock aggregate concrete
- f_y = 40,000 psi reinforcing steel

REQUIRED

Answer each of the following questions, supporting your answers with appropriate calculations.

(a) What is the safe working load P on the beam? Assume the column is adequate.

(b) At what load P will the tension steel in the beam yield? Assume the column is adequate.

(c) What is the ultimate load P for the beam? Assume the column is adequate.

(d) Define ductility for the beam section and calculate the ductility ratio, μ, of the beam.

(e) Describe the expected mode of failure of the beam.

(f) Based on the interaction curve given in figure C-46(b), is the column adequate? Support your conclusion. Would you investigate secondary stresses in the column? Explain.

(g) If the top reinforcement in the beam were increased from two #8 bars to two #11 bars, describe the expected mode of failure of the beam. Assume the column is adequate.

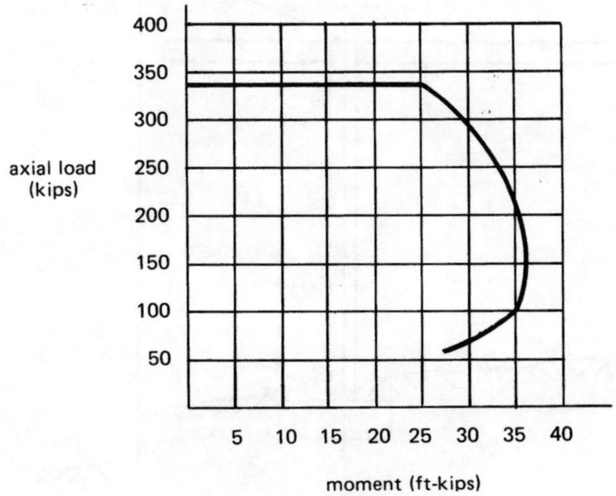

interaction curve for column (ultimate capacity)

Figure C-46

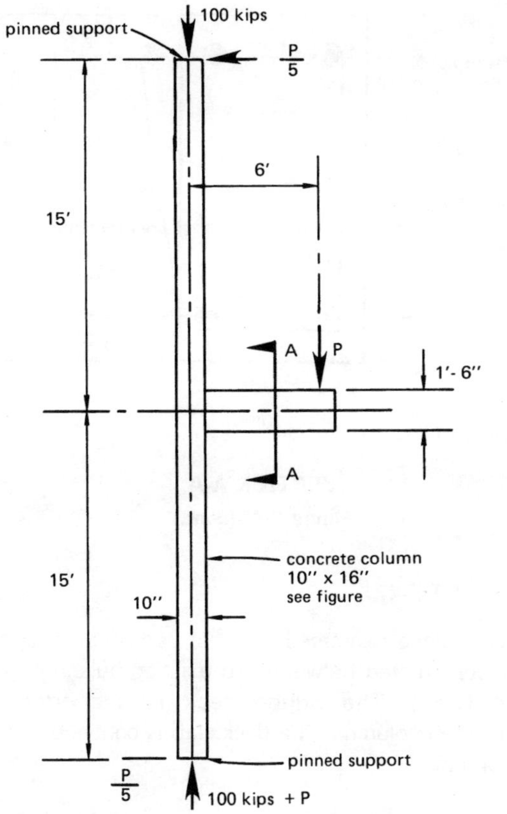

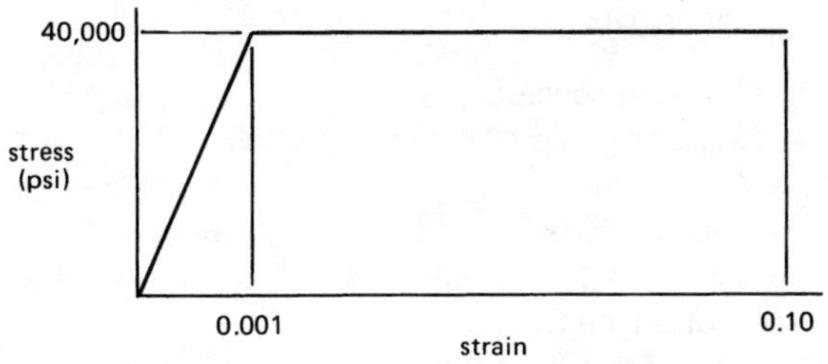

idealized stress-strain curve for reinforcing steel

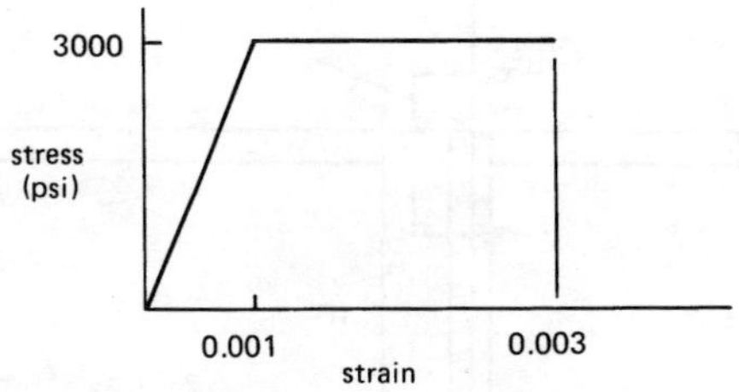

idealized stress-strain curve for concrete

Figure C-46

(continued on next page)

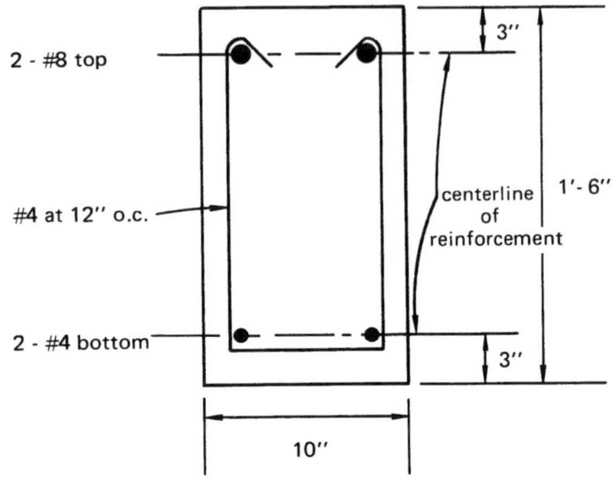

SECTION A-A

Figure C-46 (cont.)

CONCRETE-47

A cast-in-place reinforced concrete pedestrian bridge is to be constructed between two existing buildings (see figure C-47). The bridge deck is supported by cantilevered columns. The deck slab is continuous over all supports.

DESIGN CRITERIA

- f'_c = 3000 psi concrete
- ASTM A615 grade 40 reinforcing steel
- f_y = 40,000 psi
- Live load = 100 psf
- Seismic zone 3
- Soil profile type S_D
- Near-fault factors = 1.0

REQUIRED

(a) Design all reinforcing steel for the pedestrian bridge deck slab for full dead load + live load on all spans. Provide 1 1/2 inches clear to top reinforcement and 1 inch clear to bottom reinforcement.

(b) Draw a detail showing all slab reinforcement.

(c) Determine the deflection of the cantilevered slab for dead load on the entire deck considering long term effects.

(d) Design the reinforcing steel for columns using 13-inch square columns. Do not combine live load with dead load under lateral loading. Consider lateral loading perpendicular to bridge. Use V = .2W for lateral loads.

(e) Detail column reinforcement.

(f) Determine maximum column deflection under lateral load. Neglect footing rotation.

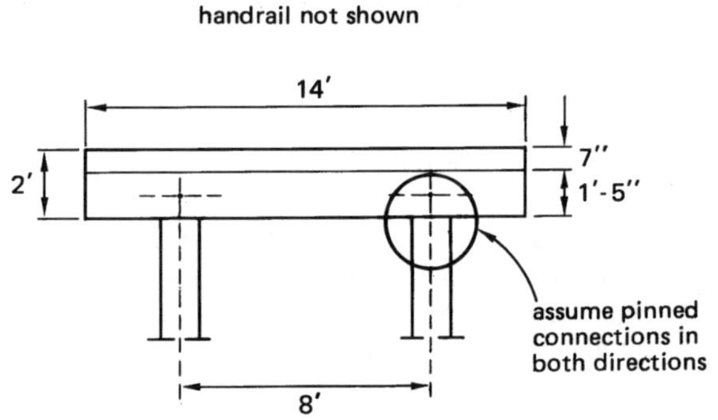

SECTION A-A

Figure C-47

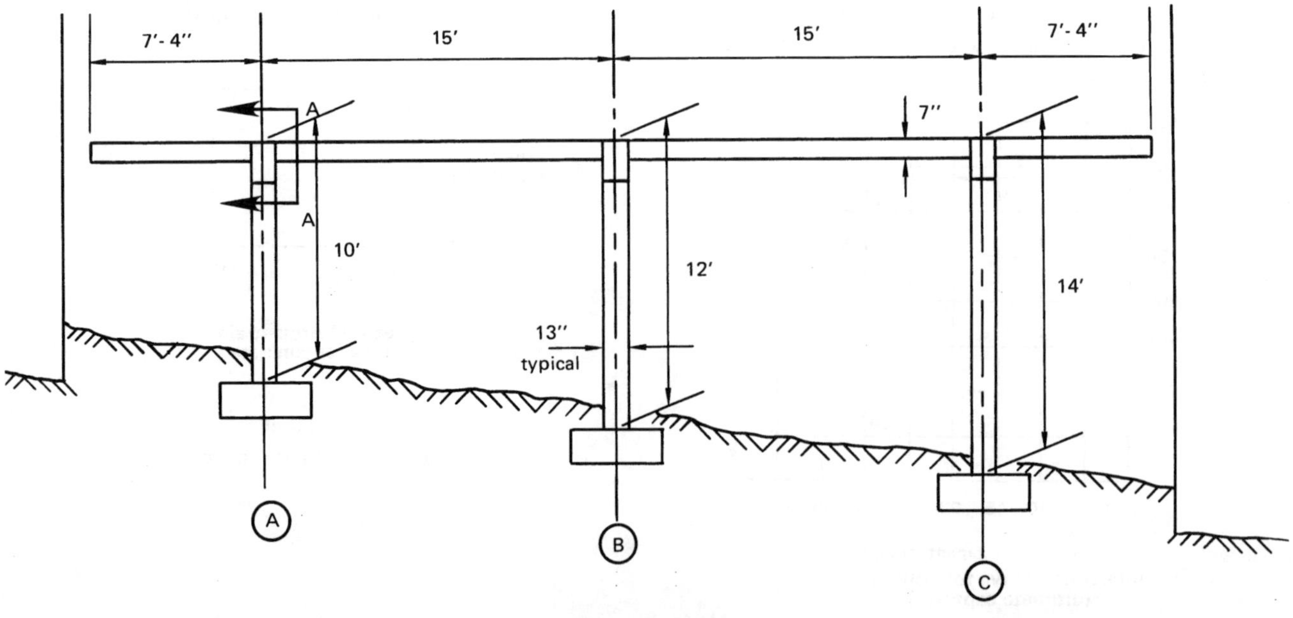

Figure C-47 (cont.)

CONCRETE-48

A one-story structure supported by grade beams and a caisson type foundation is shown in figure C-48.

DESIGN CRITERIA

- f'_c = 3000 psi, 150 pcf concrete
- Grade 40 reinforcing steel
- Roof dead load = 150 psf
- Roof live load = 12 psf
- Base shear coefficient = 0.186
- Torsion may be neglected.
- All lateral forces are resisted by shear walls.
- Foundation support is provided only by caissons which are assumed pinned at base of walls.

REQUIRED

(a) Design the end wall of the structure including the grade beam.

(b) Draw an end wall elevation and show all reinforcement. Show bar sizes and placement with approximate lengths. Do not calculate cut-offs. Show all calculations and any assumptions made.

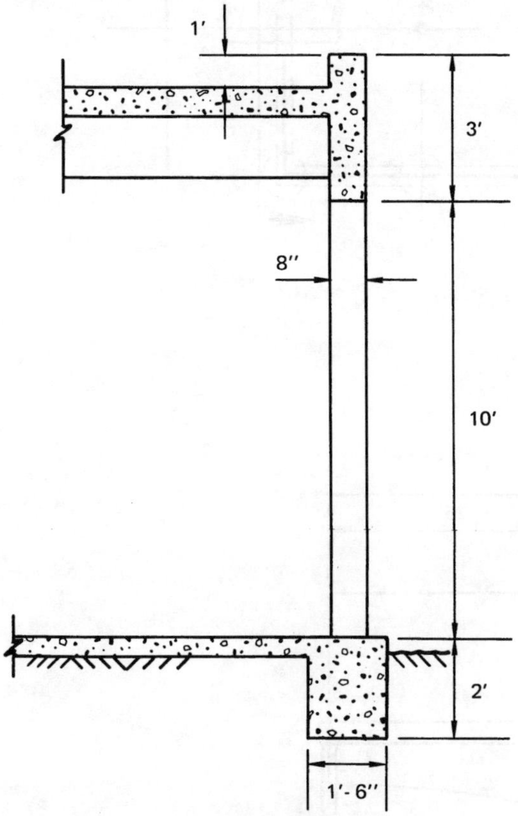

Figure C-48

(continued on next page)

CONCRETE-49

Figure C-49 illustrates a ductile moment-resisting space frame for a building in seismic zone 3.

DESIGN CRITERIA

- f'_c = 3000 psi hardrock concrete
- ASTM A615 grade 60 reinforcing steel

REQUIRED

(a) Determine the vertical web reinforcement required for the maximum ultimate design shear at the end of the beams.

(b) Design the special transverse reinforcement for the beam column connection at joint A. (Assume plastic hinges form in the beams.) Draw a sketch showing how the joint would look. Use approximate scale 3/4" = 1'-0".

(c) Write a short commentary on ductile moment-resisting space frames. Explain why they are required to be ductile and discuss four principles to be followed in the design of ductile concrete.

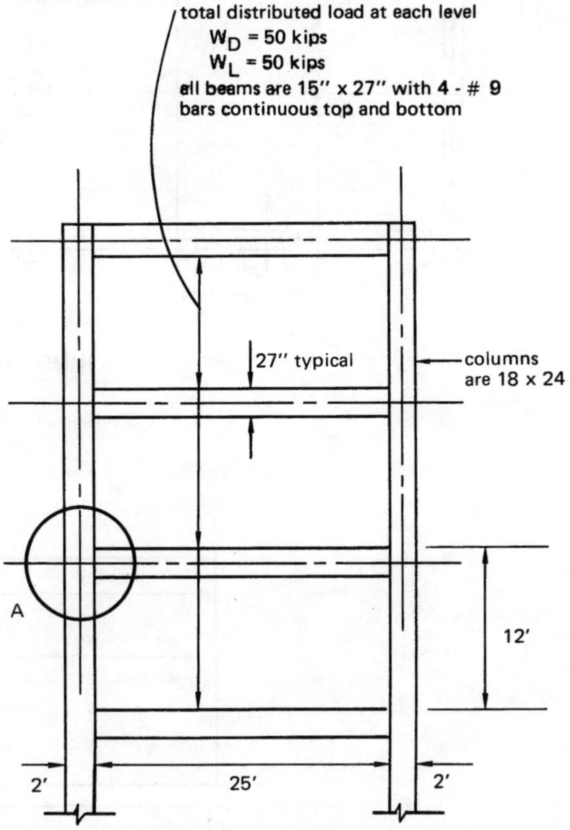

Figure C-49

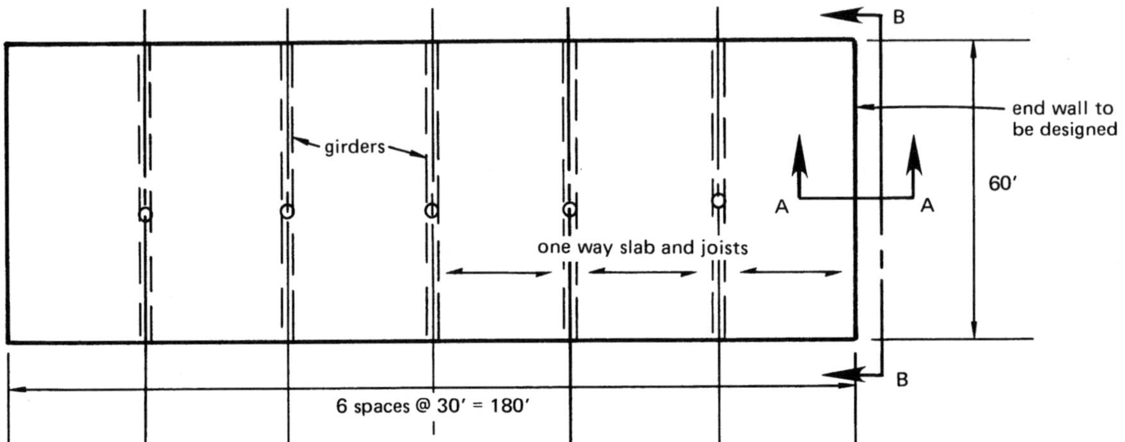

ROOF PLAN

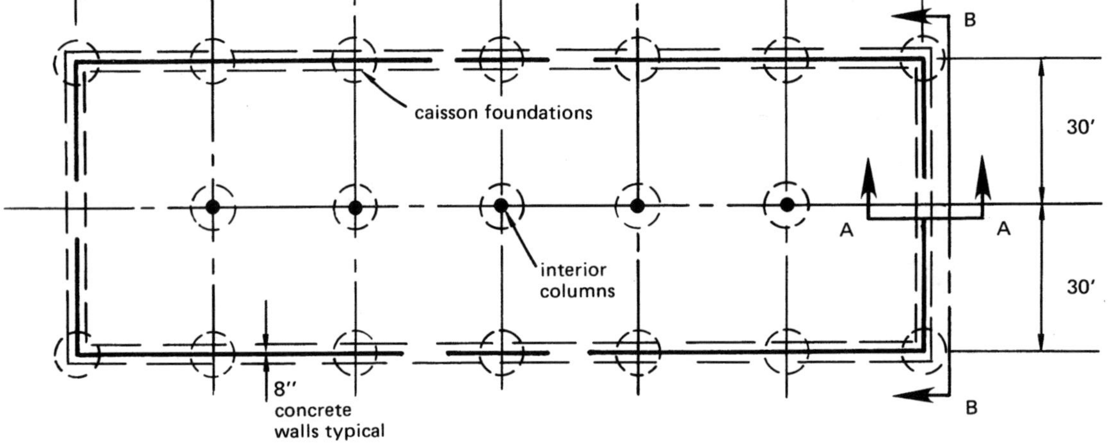

FOUNDATION PLAN

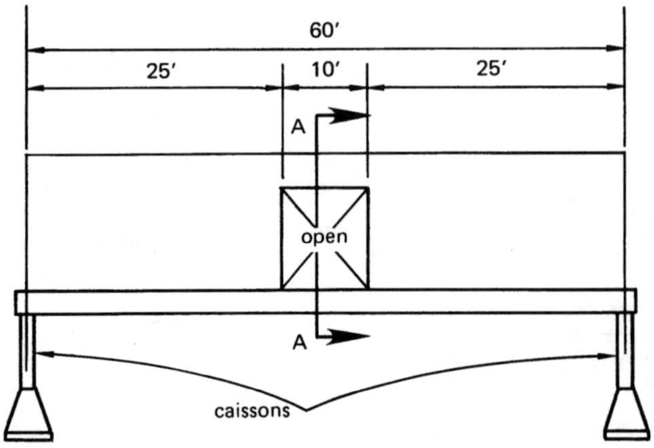

SECTION B-B

Figure C-48 (cont.)

CONCRETE-50

A 3-story concrete building has been analyzed for code prescribed lateral forces in the North-South direction. The analysis indicated the total deflection of each level relative to the base as follows:

roof = 0.35" second floor = 0.16"
thrid floor = 0.25" first floor = 0.0" (base)

A partial framing plan and typical section through the building are shown in figure C-50. The story heights are shown on the figure.

The interior vertical load-carrying frames are not considered as part of the lateral load-resisting system; these consist of 18-inch by 18-inch tied columns with four #9 vertical bars and a 6-inch cast-in-place slab-and-girder system (girders are 18 inches wide by 22 inches deep measured from bottom of slab).

DESIGN CRITERIA

- f'_c = 4000 psi standard weight concrete columns
- $E = 3.64 \times 10^6$ psi (standard)
- $E = 2.57 \times 10^6$ psi (lightweight)
- 4000 psi lightweight concrete, 115 pcf, for slab and girders
- ASTM A615 grade 60 reinforcing steel
- The base provides full fixity for each column.
- The ratio of $\dfrac{M_{dead}}{M_{total}} = 0.70 = \beta_d$
- Cracked moment of inertia of the girder I_{cr} = 6300 in^4. Use this value for the entire length of girder (both positive and negative regions).
- Inflection points occur at midspan of girders for the lateral loading (North-South displacement).

REQUIRED

(a) Determine the resulting moments to the interior columns. Neglect dead load and live load considerations. Conform to the requirements of current UBC.

(b) What is the reason for the requirement of the section of the UBC used for moments to interior columns?

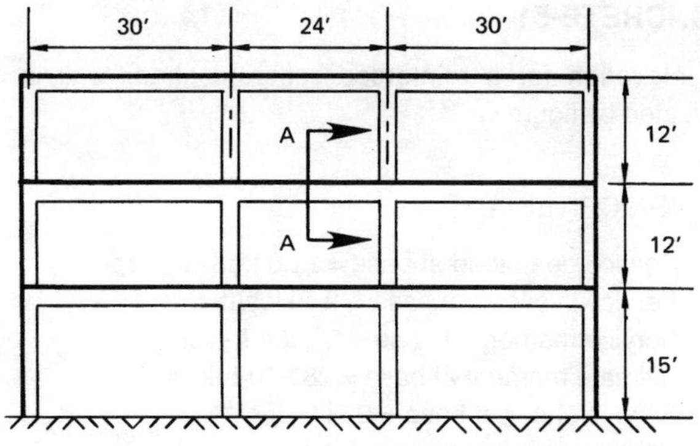

TRANSVERSE SECTION THROUGH BUILDING

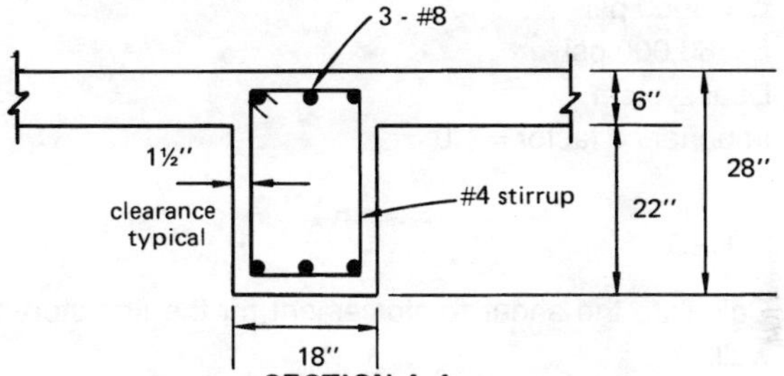

SECTION A-A

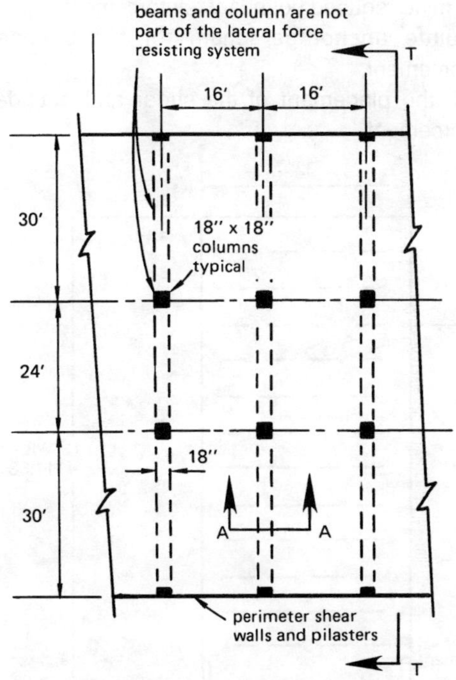

PARTIAL FRAMING PLAN

Figure C-50

CONCRETE-51

An elevation for a reinforced concrete shear wall is indicated by figure C-51.

DESIGN CRITERIA

- Service dead load at base = 21.0 kips/ft
- Service live load at base = 2.7 kips/ft
- Service moment at base = 20,264 ft-kips
- Ultimate moment at base = 28,370 ft-kips
- Service shear at base = 228.0 kips
- Ultimate shear at base = 456 kips
- Seismic zone 4
- f'_c = 3000 psi
- f_y = 60,000 psi
- Dual system
- Importance factor = 1.0

REQUIRED

(a) Calculate the shear reinforcement for the first story wall.
(b) Calculate maximum service stresses at top of footing.
(c) Determine required boundary reinforcement.
(d) Calculate anchorage length for the boundary reinforcement.
(e) Detail the placement of the shear and boundary reinforcement.

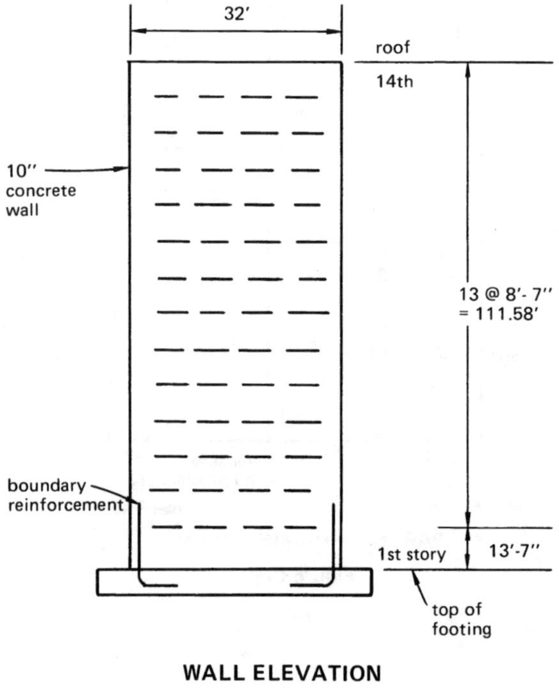

Figure C-51

CONCRETE-52

A pedestrian bridge across an expressway provides a separation for school children. At each end of the span a concrete stair is provided as shown in figure C-52. The stair is a concrete slab, 6 feet wide, supported by a single cantilevered beam. The stair beam is cantilevered out from a 12-inch concrete wall which is adequate for the stair load. The design live load is 100 psf (not reducible). Lateral loads and torsional loads may be neglected.

DESIGN CRITERIA

- f'_c = 4000 psi hard rock aggregate
- f_y = 60 ksi reinforcing steel

REQUIRED

(a) Design the beam using an 18-inch wide stem. The centerline of the beam is coincident with the centerline of the 12-inch wall. Show calculations for the steel required, including stirrups, and check for shear, etc.
(b) Draw an elevation of the beam, and show the placing of all reinforcing steel necessary to complete the design. This includes bar sizes, cut-offs, spacings, anchorage, and clearances. Wall reinforcement is not required.

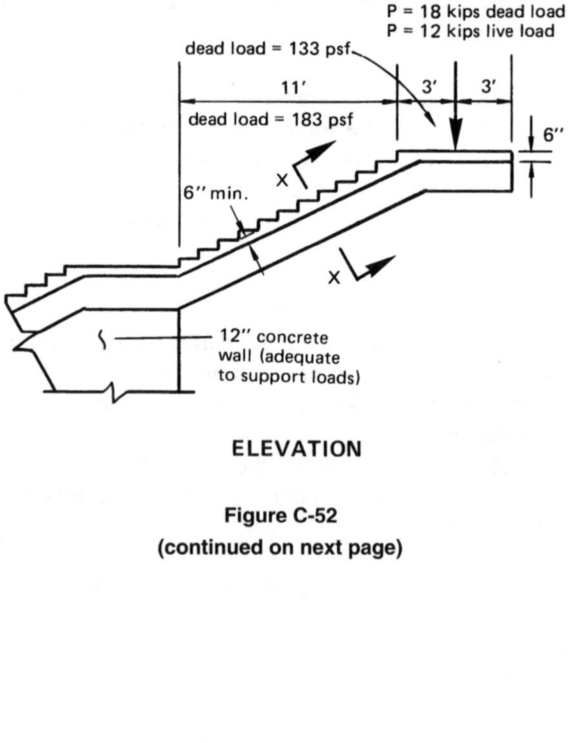

Figure C-52
(continued on next page)

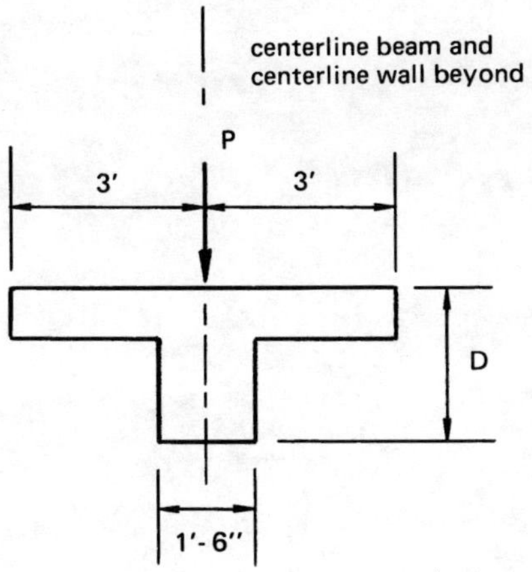

SECTION X-X

Figure C-52 (cont.)

3 STEEL

STEEL-1

Fig. ST-1 shows one story of a column tier for a moment-resisting space frame for a building with no other bracing system. The girder and column sizes given are the result of a preliminary design and the axial load and moments are the result of a computer analysis.

DESIGN CRITERIA

- ASTM A36 steel
- Neglect axial shortening of column.
- Moments given are center-to-center of members and are for dead load plus live load plus wind.
- Do not investigate beam-to-column connections.

REQUIRED

Determine the adequacy of the column due to the forces given. Show all calculations.

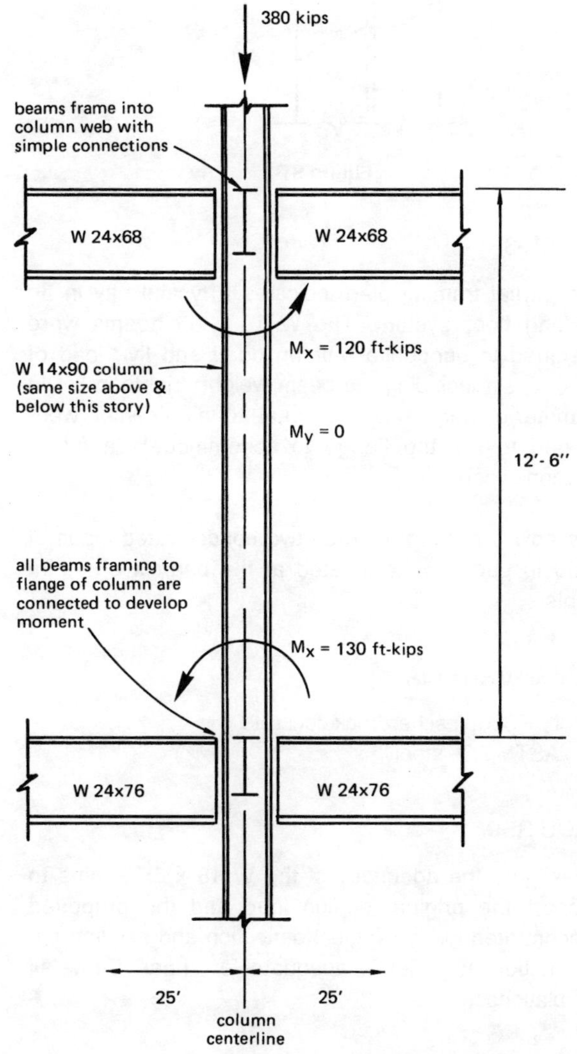

Figure ST-1

STEEL-2

The steel fabricator has submitted the following detail for a steel beam-to-column connection. Your calculations show the beam to have a reaction of 16.0 kips.

DESIGN CRITERIA

- ASTM A36 steel
- E70XX welding electrodes
- 7/8 in diameter A325-X bolts

REQUIRED

Evaluate the connection and determine if the detail is acceptable. Show all calculations used to develop your determination.

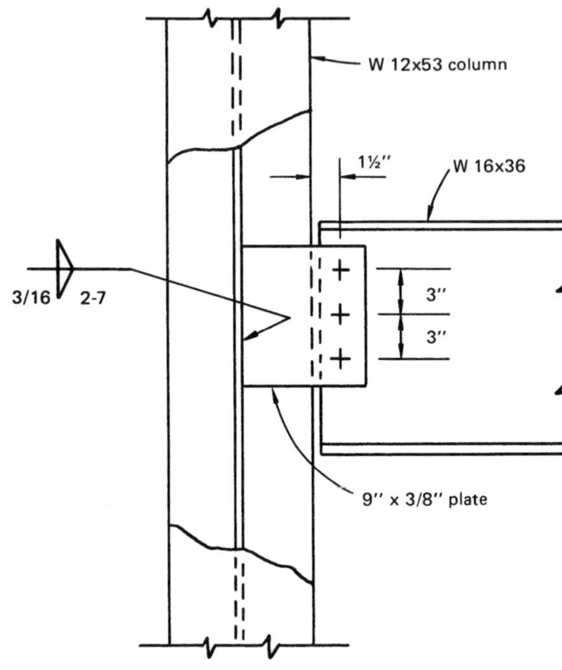

Figure ST-2

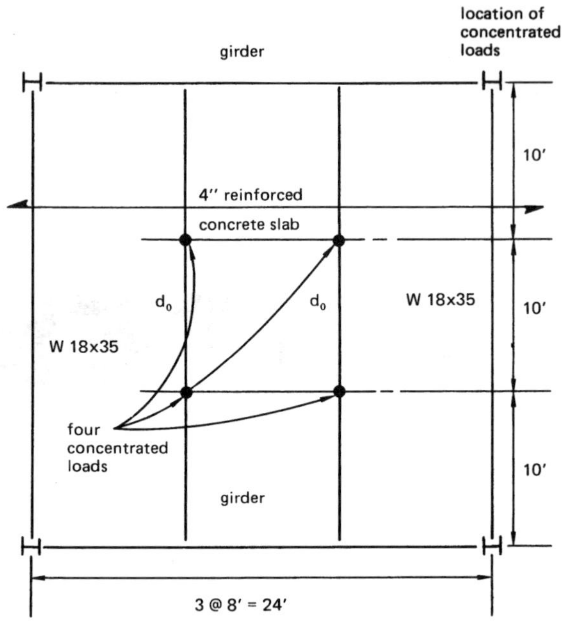

Figure ST-3

STEEL-3

The partial framing plan indicates a typical bay in an existing floor system. The W 18 x 35 beams were designed to support a uniform dead and live load of 1000 kips/ft including the beam weight. Eighteen 3/4 in diameter by 3 in studs, spaced at 20 in on center, were welded to the top flange to accommodate a future additional load.

It is now proposed to place two concentrated loads of 3000 lb each, to be located at the beam's one-third points.

DESIGN CRITERIA

- f'_c = 3000 psi hardrock concrete
- ASTM A36 steel

REQUIRED

Investigate the adequacy of the W 18 x 35 beams to support the original design load and the proposed concentrated loads. Neglect deflection and assume pin connection at girder is adequate for shear. Show all calculations.

STEEL-4

A bracket for a beam-to-column connection is shown in Fig. ST-4.

DESIGN CRITERIA

- P = 20 kips
- ASTM A36 steel plate and column
- E70XX welding electrodes
- Minimum yield point = 57 ksi
- Tensile strength range = 70 ksi minimum
- Follow AISC, current edition.

REQUIRED

(a) Determine the size of fillet welds for static loads (see *a* on figure).

(b) Determine the size of fillet welds for 100% reversible stresses for over 2,000,000 cycles. Show all calculations.

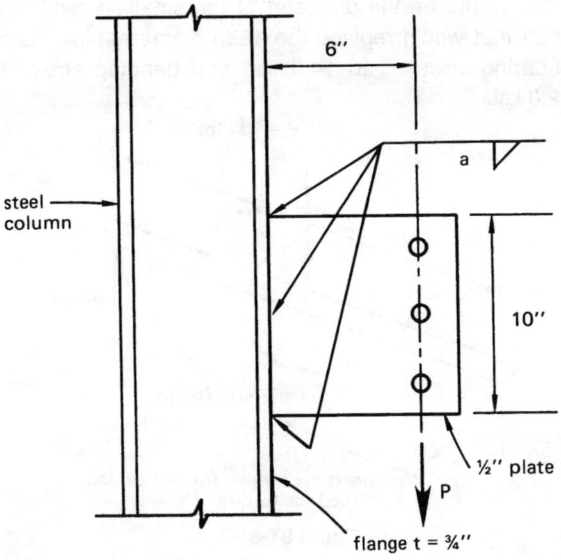

Figure ST-4

STEEL-5

REQUIRED

(a) When and why would you specify the use of high-strength friction bolts?
(b) What is the difference in installation between high-strength bolts in a friction-type connection and in a bearing-type connection?
(c) The W 12 x 50 beams in Fig. ST-5 are loaded in such a way that a moment connection equal in capacity to that of the W 12x beams has to be provided at the built-up girder. Draw a moment connection to approximately 1 in = 1 ft 0 in scale. Describe briefly the advantages of your solution. No calculations are required.

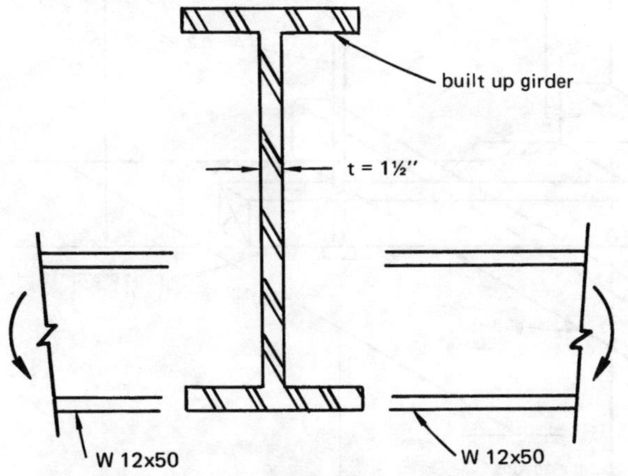

Figure ST-5

STEEL-6

REQUIRED

(a) When you arrive on the job site you find that the contractor has used a cutting torch to enlarge the bolt holes in a beam because of erection difficulties. (See Fig. ST-6(a).) Steel is ASTM A36. Identify three possible corrective measures to correct the deficiency.
(b) A beam-to-column connection is shown in Fig. ST-6(b). Assume the tab plate and welds are adequate. Determine if the design for a beam reaction of 82 kips dead load plus live load is adequate. Steel is ASTM A36.
(c) High strength steel bolts such as ASTM A325-F can be installed by various methods. Identify three.

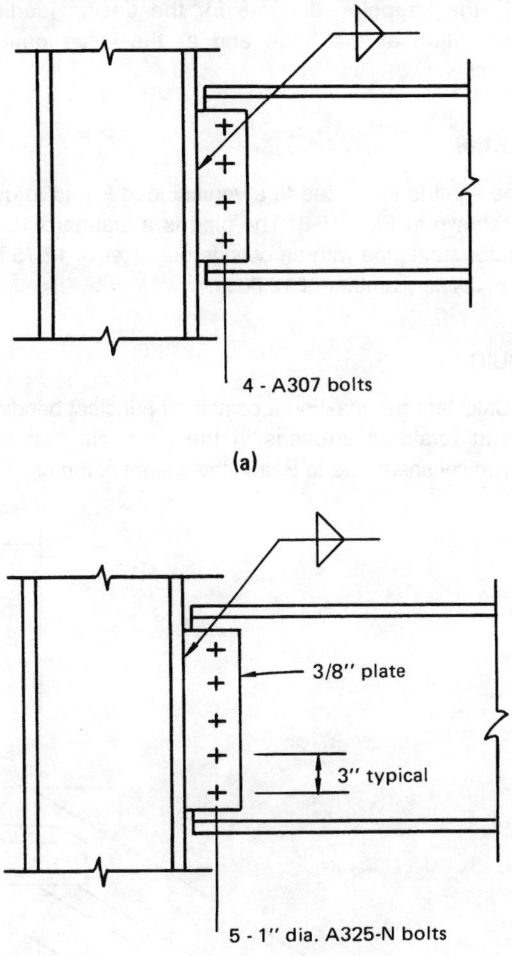

Figure ST-6

STEEL-7

A flag pole 50 ft tall is located in a region corresponding to a basic wind speed of 80 mph. The flag will generate a 250 lb force at the top of the pole. The flag pole is made up of 10 ft lengths of standard steel pipe columns with a 6 in diameter at the bottom and stepping down as it goes up to 5 in, 4 in, and 3 1/2 in diameters with a top 10 ft length of 3 in diameter schedule 40 pipe.

Assume the manufacturer will provide a totally fixed base connection, and that the connection at each joint is adequate. Show all calculations.

REQUIRED

(a) Draw a loading diagram.
(b) Is the flagpole adequate for the design loading condition at the base and at the other critical sections?

STEEL-8

A pipe shaft is subjected to a vertical load P and torque T as shown in Fig. ST-8. The pipe is a standard 12 in diameter steel pipe with an outside diameter of 12.75 in, and an inside diameter of 12.00 in.

REQUIRED

(a) Calculate the maximum combined principal bending and torsional stresses in the pipe. Neglect the vertical shear due to P, and the weight of the pipe.
(b) What would be the diameter of the smallest size steel shaft that would replace the 12 in diameter pipe? Limit shearing stresses to 14.5 ksi, and bending stress to 22.0 ksi.

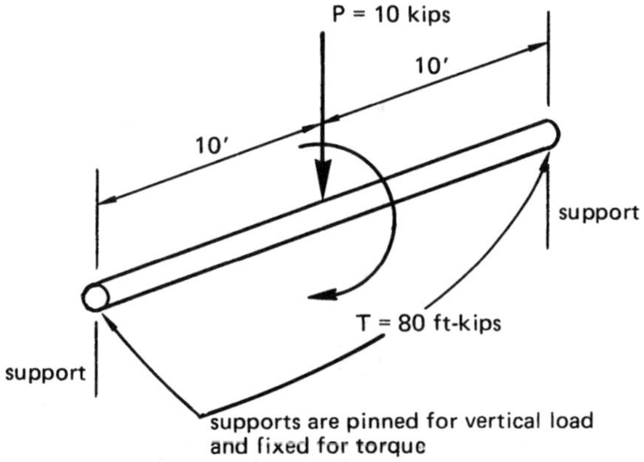

Figure ST-8

STEEL-9

The system shown in Fig. ST-9 supports a concentrated load of 14 kips. The load is located at the center of an infinitely stiff cross beam.

DESIGN CRITERIA

- Neglect the dead loads of beams, girder, and hanger.
- The cross beam is located at the centers of span of the girders.
- All members are steel, E = 30 x 10^6 psi.

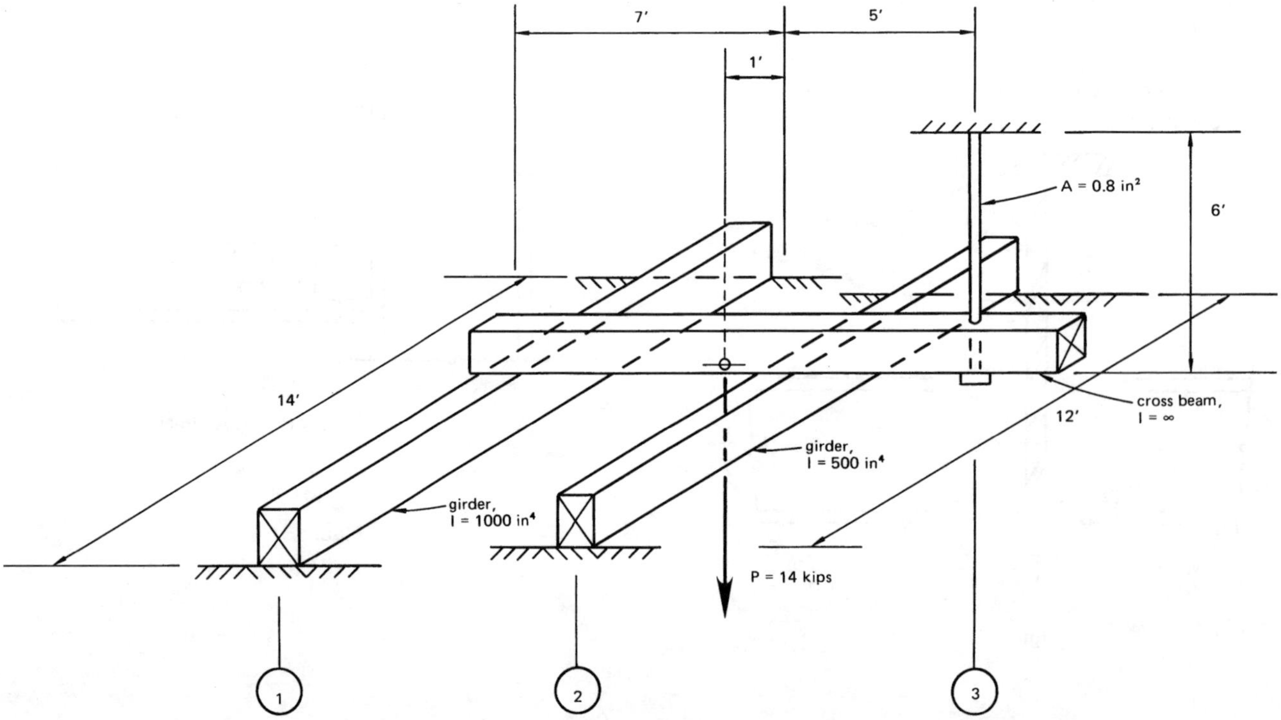

Figure ST-9

REQUIRED

(a) Calculate the load supported by each girder and the hanger. Show all work.
(b) Calculate the deflection of each girder and the hanger. Show all work.

STEEL-10

A machine shop monorail is supported by three steel cable hangers. Due to variations in the roof structural framing, the three hangers are varying in length. The cable lengths, the position of each, and the cross sectional area of each are shown in Fig. ST-10. The beam may be assumed infinitely stiff.

DESIGN CRITERIA

- The condition for design has a total load of 15.0 kips placed 7 ft from the right-hand cable hanger.
- The dead load of the beam is assumed included.

REQUIRED

Calculate the unit stresses in each of the three hangers due to the vertical load of 15.0 kips.

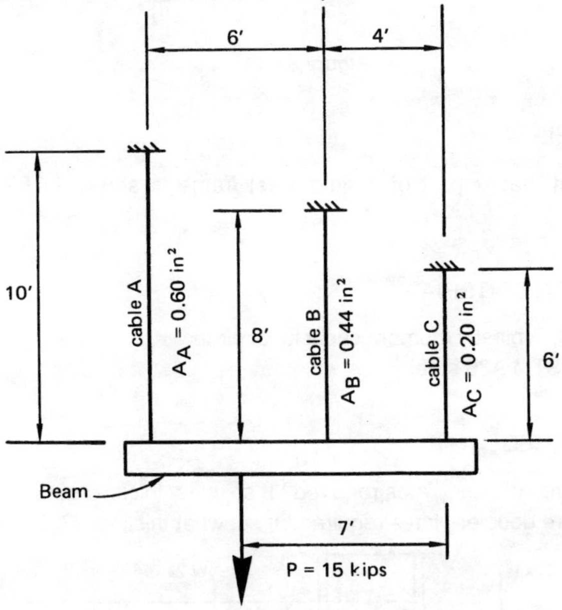

Figure ST-10

STEEL-11

The detail in Fig. ST-11 shows the construction stage at which a local job was stopped. At this point it was determined that the steel beam will be 35% overstressed in bending when fully loaded. Remedial action is needed now. The beam cannot be replaced, and new columns may not be installed.

The beam spans 30 ft, and was designed for a uniformly distributed load.

REQUIRED

(a) Show all calculations, and draw a complete detail to an approximate scale of 1 1/2 in = 1 ft 0 in, for your remedial solution. Calculate the maximum load (in kips per foot) that the reinforced beam can safely support.
(b) Describe briefly your instructions to the contractor regarding any action he must undertake prior to proceeding with your remedial solution.

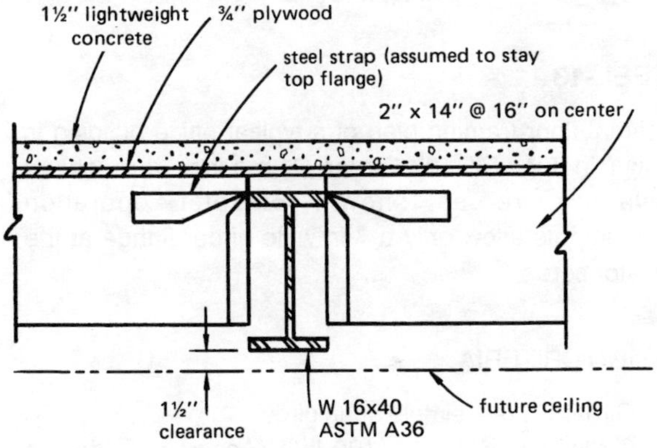

Figure ST-11

STEEL-12

The figure shown in Fig. ST-12 represents a column splice detail. The wide flange sections that are to be spliced are shown in their relative positions.

DESIGN CRITERIA

- E70XX welding electrodes
- A325-F (1 in maximum) bolts
- ASTM A36 steel (F_b = 24 ksi)
- The column splice must be developed for 50% of moment capacity and 50% of shear capacity.

REQUIRED

(a) What is the size of the partial penetration bevel weld shown as weld A?
(b) Determine the web plates (sizes and thickness) and the bolts to complete the detail. Bolts act in double shear. Draw a detail to approximate scale 1 in = 1 ft 0 in showing placing of bolts including spacing and edge distances.

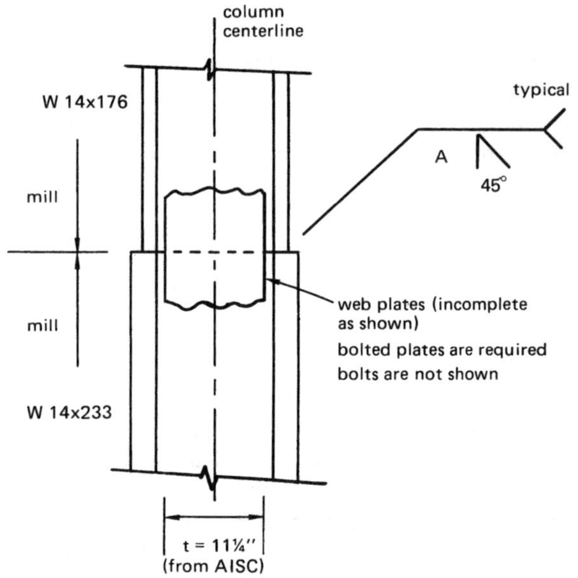

Figure ST-12

STEEL-13

A partial floor framing plan of a typical office building is shown in Fig. ST-13. Girder G1 passes through an elevator core as shown. Elevator operation requirements allow only a 7 in wide girder flange at the elevator core.

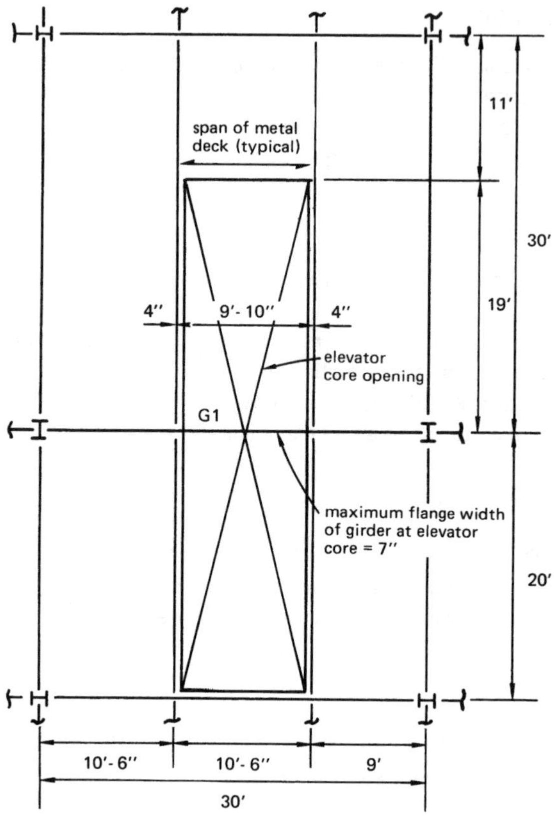

Figure ST-13

DESIGN CRITERIA

- Girder G1 is a simple span girder.
- Girder self-weight = 100 lb/ft (Assume for design load.)
- Maximum depth of girder is 25 in.
- A built-up girder is not an acceptable solution.
- No lateral ties can be placed within the elevator core.
- Live load reduction allowable per UBC
- ASTM A36 steel
- Dead loads:
  3 1/4 in lightweight fill on 3 in steel deck = 48 psf
  Partitions = 20 psf
  Ceiling, mechanical, and miscellaneous = 10 psf
  Beam framing = 5 psf
  Live loads = 80 psf

REQUIRED

Design girder G1. Use rolled section or modified rolled section to meet the design requirements. Explain how you would modify a rolled section to satisfy the requirements.

STEEL-14

A joint that is part of a rigid steel frame is shown in Fig. ST-14.

DESIGN CRITERIA

- M_p = plastic moment capacity of girder
- ASTM A36 steel

REQUIRED

(a) Are stiffener plates required? If so, what thickness?
(b) Are doubler plates required? If so, what thickness?

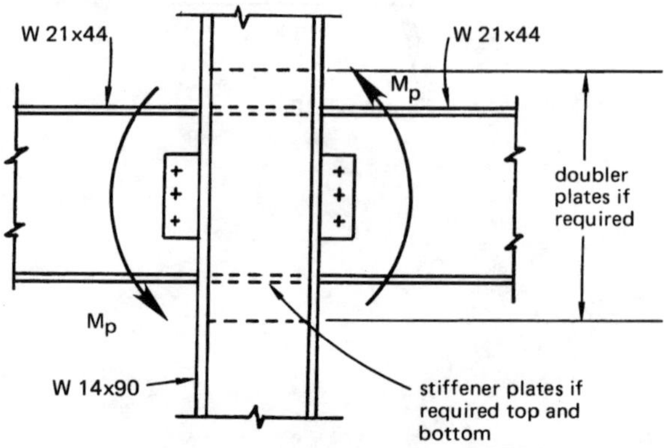

Figure ST-14

STEEL-15

A steel column with an externally applied moment at mid-height is shown in Fig. ST-15.

DESIGN CRITERIA

- ASTM A36 steel
- Axial load = 80 kips
- Moment about X-X axis = 95 ft-kips
- Moment about Y-Y axis = 75 ft-kips
- No stress increases are permitted.
- Show all calculations.

REQUIRED

What is the lightest 14 in wide flange permissible for this column?

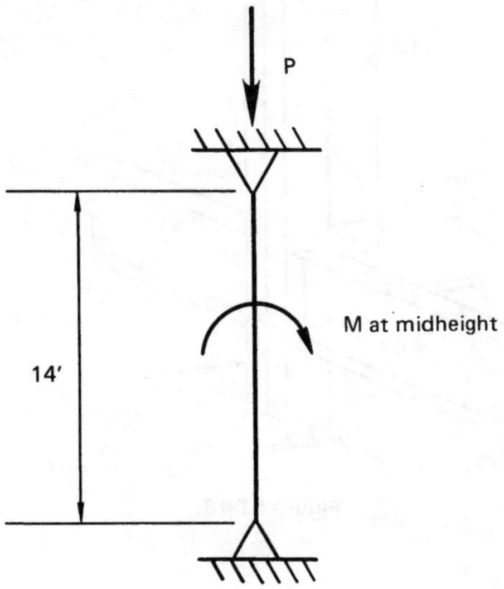

Figure ST-15

STEEL-16

A 60 ft high steel tower will be used in a forested area for field surveillance. The loads acting on the tower include a total dead load of 50 kips at the 60 ft platform level, and a 20 kip seismic force. Both of these loads are indicated on the tower elevation in Fig. ST-16.

DESIGN CRITERIA

- All column bases are hinged.
- Dimensions are shown to centerline of members.
- The tower is supported on a concrete foundation which may be deemed adequate for all design loads.
- ASTM A36 structural steel pipe columns
- Unfinished A307 bolts
- f'_c = 3000 psi concrete

REQUIRED

(a) Calculate the design loads, and specify a size for column A. Use schedule 40 standard steel pipe.

(b) Draw a detail for the connection at joint B. Include connection to foundation. No field welding is permitted.

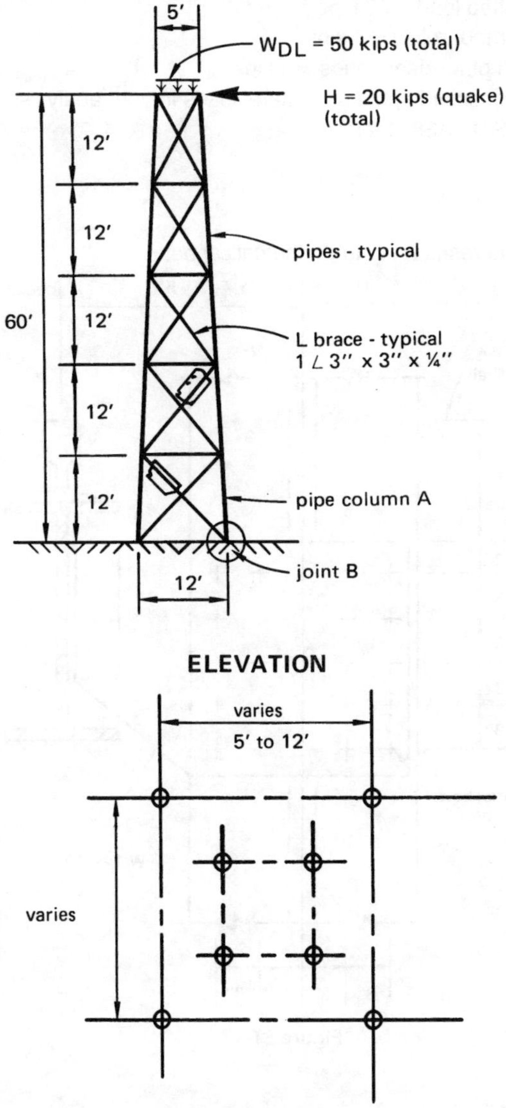

Figure ST-16

STEEL-17

A supporting bracket for a traveling crane runway has been designed as shown in Fig. ST-17, using friction-type high strength bolts (ASTM A325). Gravity and lateral forces are carried by the bracket. Longitudinal forces are carried elsewhere.

DESIGN CRITERIA

Loads acting on the bracket are:

- Lifted load = 50 kips
- Crane trolley = 4 kips
- All other dead loads = 2 kips
- Consider impact and lateral loads in your analysis.
- ASTM A36 steel

REQUIRED

Find the resultant load on the critical bolt.

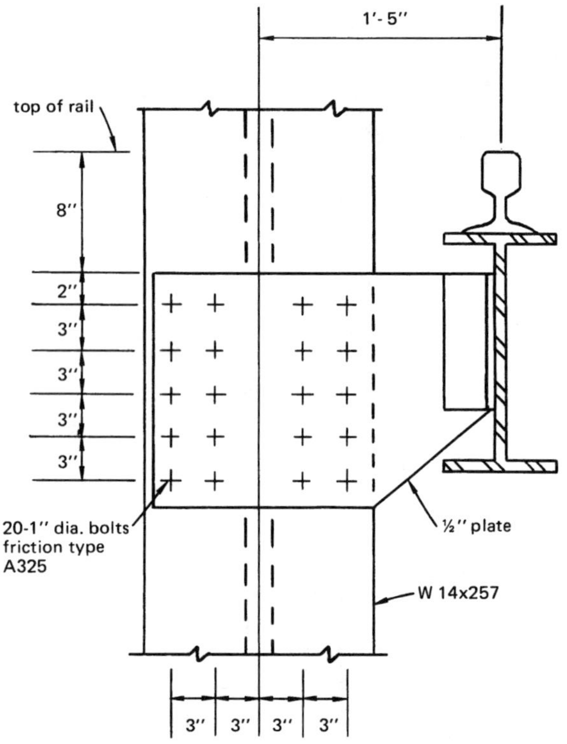

Figure ST-17

STEEL-18

The steel frames in Fig. ST-18 are to provide lateral stability in each direction for a building.

DESIGN CRITERIA

- Use ASTM A36 steel.
- Use loads and moments shown. Do not increase allowable stresses by 1.33, or reduce loads and moments by 0.75.
- Assume beam connections to column web and flanges sufficient to develop beams.
- Assume column in tier above and below to be same as column in question.
- Do not design joists.

REQUIRED

Design the minimum weight W 14 series steel column to meet specifications listed above.

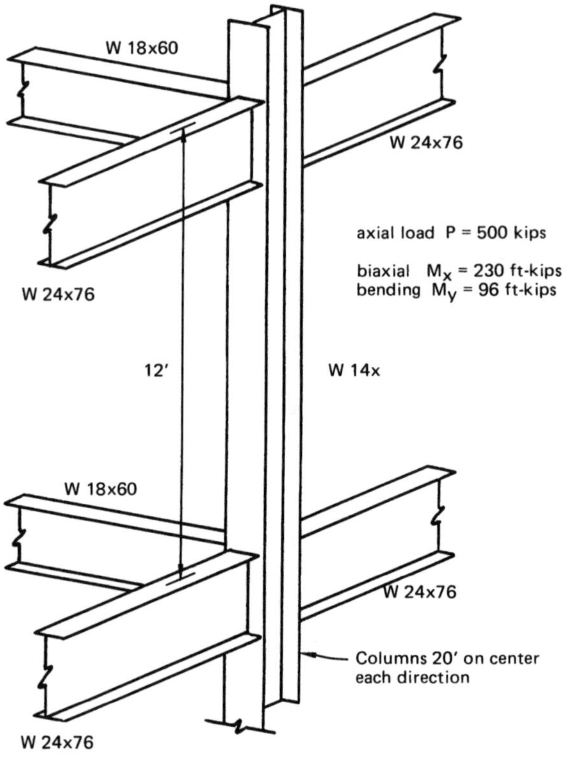

Figure ST-18

STEEL-19

The drawing in Fig. ST-19 shows a bridge that is supported by a runway girder spanning simply 22 ft between columns.

DESIGN CRITERIA

- The girder is a rolled W-shape with a standard channel cap.
- The crane rolls on an 85 lb ASCE rail fastened to the top flange.
- ASTM A36 steel
- Consider impact and lateral loads in your analysis.

REQUIRED

(a) Design the runway girder in the most economical proportions, taking into account both vertical and horizontal loading. Only the top flange of W-shape may be combined with channel to resist horizontal loading.

(b) Design welds connecting channel to W-shape. Give type and length.

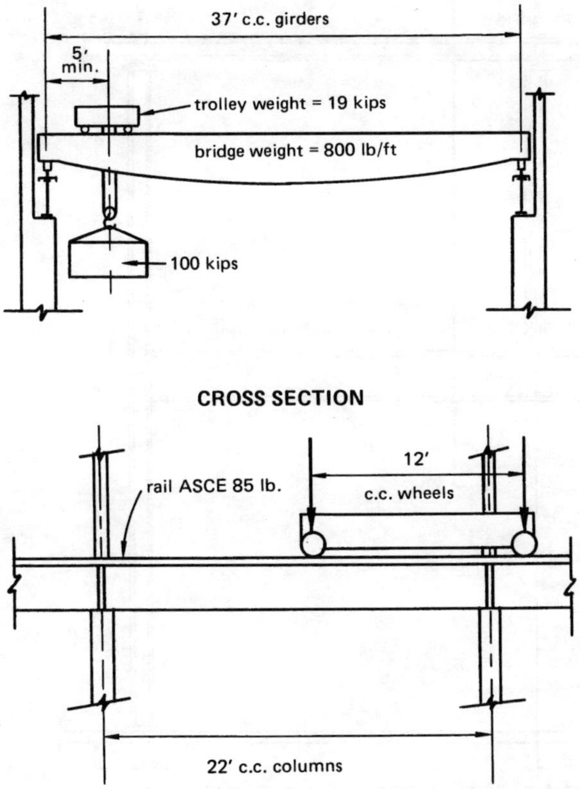

GIRDER ELEVATION

Figure ST-19

STEEL-20

A large clear floor area must be provided as shown in Fig. ST-20. Column 2 is to be supported on a beam that spans columns 1 and 3.

DESIGN CRITERIA

- ASTM A36 steel
- ASTM A307 bolts
- Sidesway is prevented in both directions at the roof.
- Lateral support normal to the beam is provided by the struts shown on section X-X.

REQUIRED

(a) What is the lightest W30 that may be used for beam A? Assume lateral support is provided at mid-span only.

(b) Draw a detail at approximate scale 1 in = 1 ft 0 in of an all bolted unstiffened beam seat connection at B. Show supporting data.

(c) What is the lightest W8 that may be used for the column on line 1?

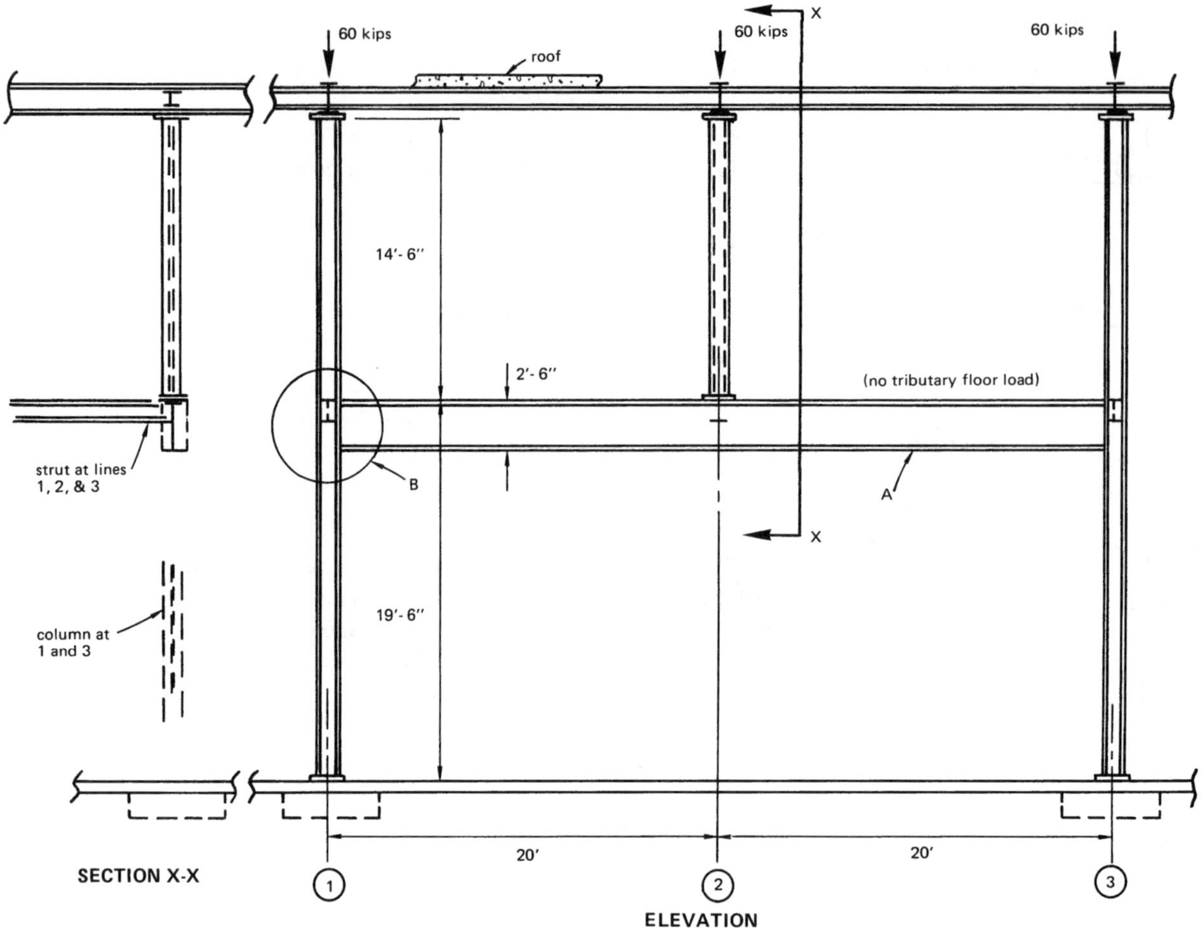

Figure ST-20

STEEL-21

A 2-story braced frame for a storage building receives vertical and lateral forces from both roof and second floors. See Fig. ST-21.

DESIGN CRITERIA

- ASTM A36 rolled sections, plates, and milled steel rods
- ASTM A325-F high-strength bolts
- E70XX welding electrodes
- Braced frame members consist of columns, horizontal struts, rod diagonals, and base plates.
- Assume pinned end conditions at first floor and roof.
- Columns are continuous from first floor to roof.
- Maximum concrete bearing capacity is 1000 psi for vertical loads, with 1/3 increase for load combinations that include lateral forces.
- Horizontal struts have pinned ends.
- Use current edition of AISC.

REQUIRED

(a) Design the lightest W 8x rolled section column *AC* that will satisfy the loading requirements.

(b) Design the smallest diameter threaded rod *CE*, and the smallest diameter unthreaded rod *BF*.

(c) Check adequacy of horizontal struts, *CD* and *BE*.

(d) For the load condition shown, design the most economical base plate size and thickness at point *A* (no base plate stiffeners). See Fig. ST-21(d).

(e) Check adequacy of strut connection at point *D*. See Fig. ST-21(e).

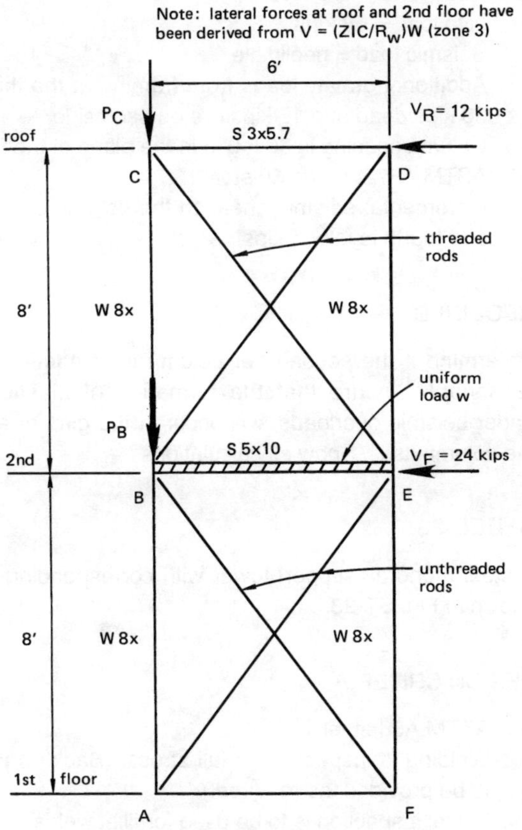

(a) ELEVATION OF BRACED FRAME

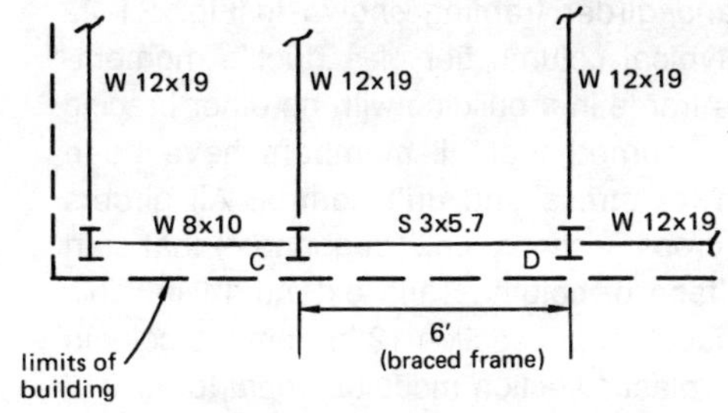

(b) PARTIAL ROOF PLAN

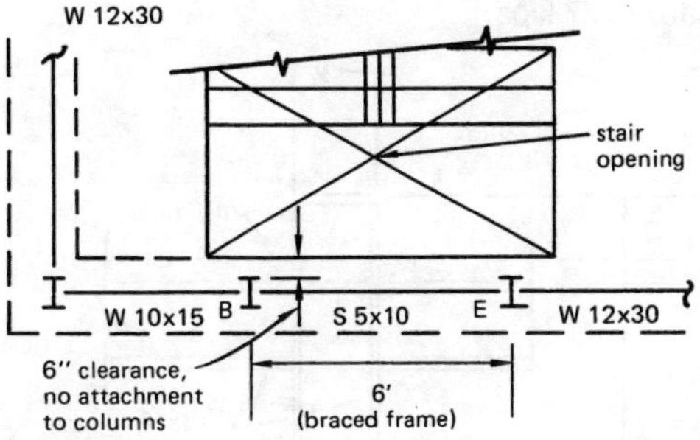

(c) PARTIAL SECOND FLOOR PLAN

load	vertical	
	dead load	live load
P_C	34 kips	5 kips
P_B	40 kips	10 kips
w	0.3 kips/ft	0

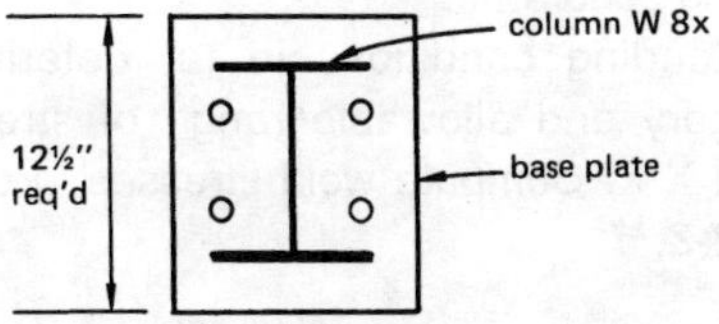

(d) PLAN AT A

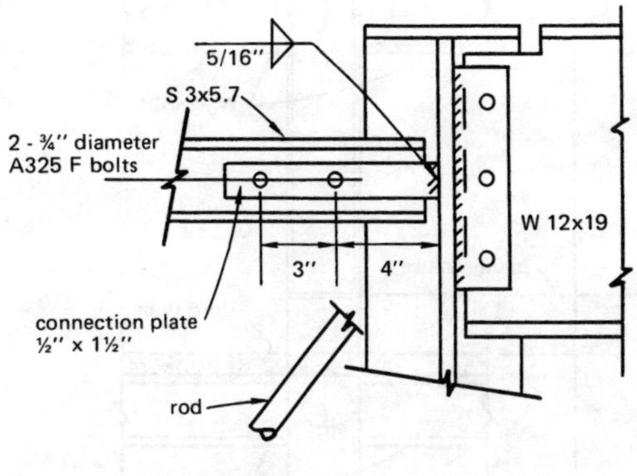

(e) ELEVATION AT D

Figure ST-21

STEEL-22

The column and girder framing shown in Fig. ST-22 represents a typical column tier of a ductile moment-resisting space frame in a building with no other bracing system. It is assumed that all members have been properly sized for stress and drift control. All girders have reduced cross sections (i.e., "dogbones") that start 5 in from the face of columns and extend 14 in. This results in a reduced beam section 12 in from the column face that has a plastic section modulus equal to 60% of the unreduced section.

DESIGN CRITERIA

Axial load in column at fourth story level:

- Dead load = 177 kips

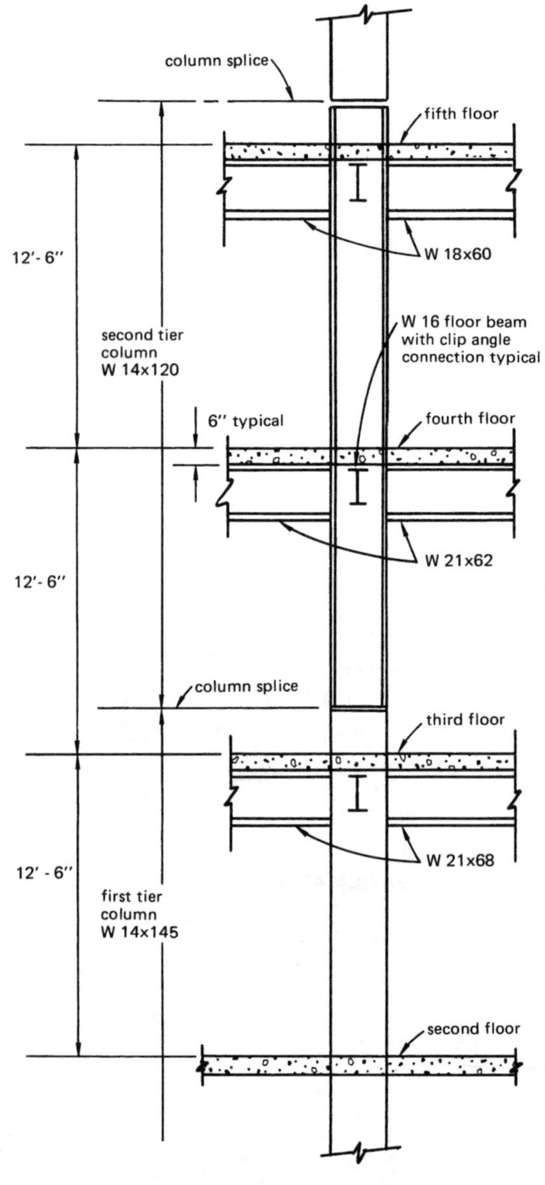

ELEVATION

Figure ST-22

Axial load in column at fourth story level:

- Dead load = 177 kips
- Live load = 68 kips
- Seismic load = negligible
- Additional gravity loads from framing at the third floor = 30 kips dead and 15 kips live on each side
- Column spacing is 30 ft 0 in in the plane of the frame
- ASTM A572, grade 50 steel
- Incremental seismic shear to the column between third and fourth level = 5 kips

REQUIRED

Determine if the second tier column is of adequate size to reasonably ensure that the formation of plastic hinging, under seismic overloads, will occur in the girders and not in the column itself. Show all calculations.

STEEL-23

A steel monorail support tower with corresponding details is shown in Fig. ST-23.

DESIGN CRITERIA

- ASTM A36 steel
- Welding is inspected for full stress. (Back-up plates are to be provided, as required.)
- Visual inspection is to be used for fillet welds.
- Ultrasonic inspection is to be used for groove welds.
- Neglect girder weight for all computations.
- Operation period is for a 25-year life, 300 cycles per day, and 300 days per year.
- Fatigue design per AISC current edition.
- Variable load P = 50 kips (Neglect all lateral loads.)

REQUIRED

(a) Compute the stress cycles and determine the fatigue loading conditions.

(b) For loading conditions in (a) determine the stress category and allowable range of stress for metal at detail X-X. Compute weld stresses in details X-X, Y-Y, and Z-Z.

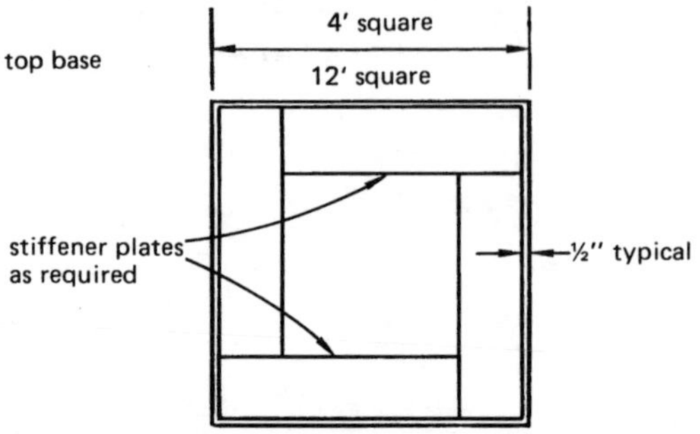

TYPICAL PLAN SECTION A-A

Figure ST-23

(continued on next page)

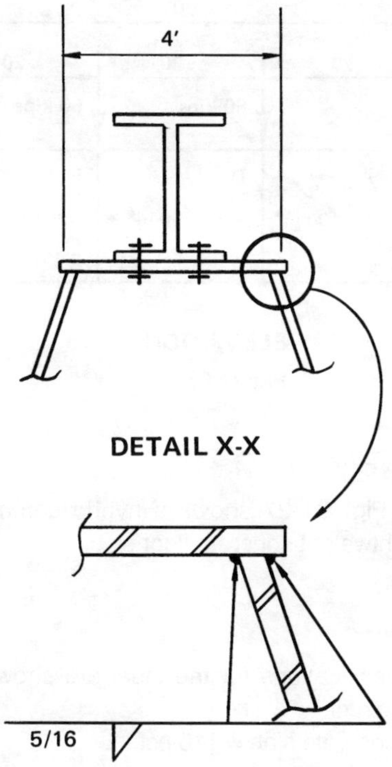

DETAIL X-X

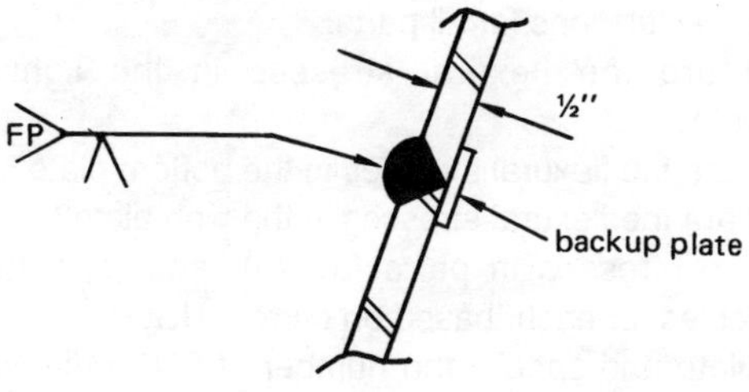

SPLICE DETAIL Y-Y

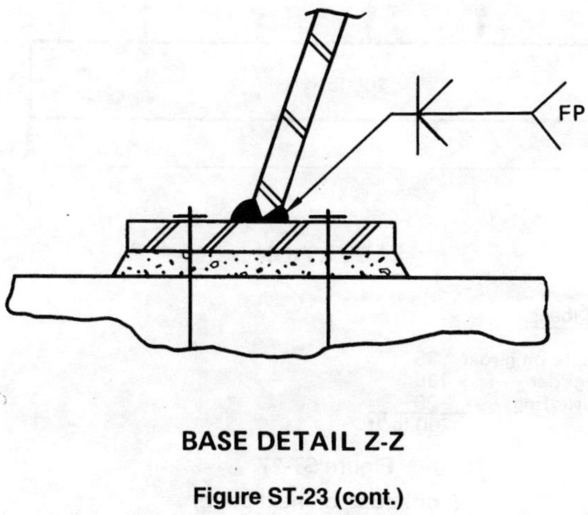

BASE DETAIL Z-Z

Figure ST-23 (cont.)

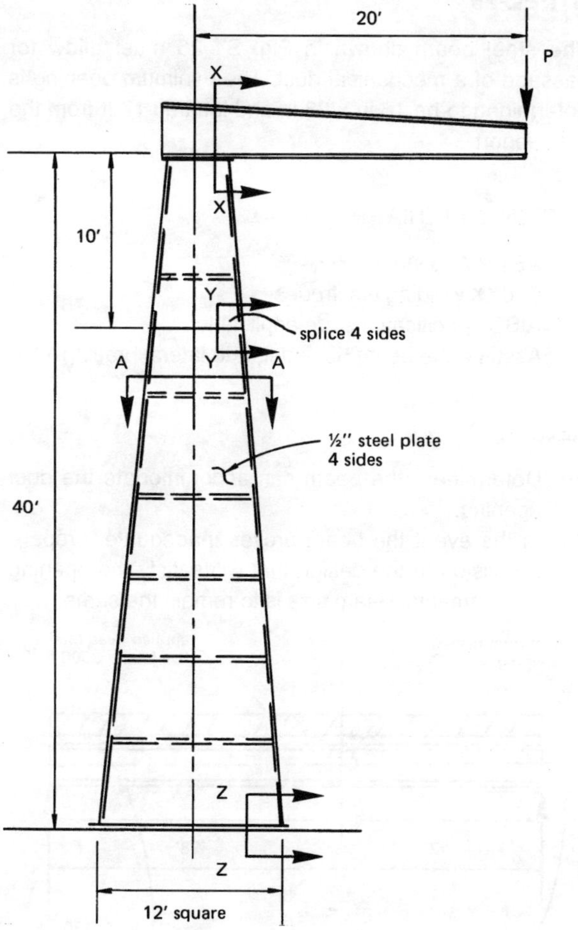

ELEVATION

Figure ST-23 (cont.)

STEEL-24

A steel rigid frame that carries a uniform load of 2 kips/ft (including the dead load of the steel) is shown in Fig. ST-24.

REQUIRED

Determine the values of the plastic moments for the beams and columns using plastic design.

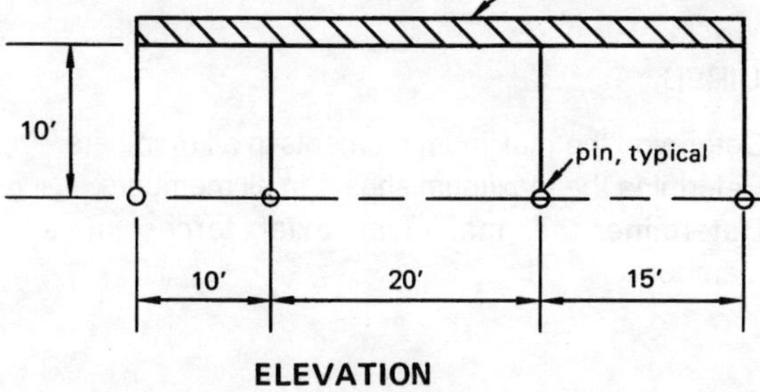

ELEVATION

Figure ST-24

STEEL-25

The steel beam shown in Fig. ST-25 must allow for passage of a mechanical duct. The minimum opening is determined to be 16 in x 28 in and located 17 ft from the left support.

DESIGN CRITERIA

- ASTM A36 steel
- E70XX welding electrodes
- AISC specifications are applicable.
- Assume the beam has adequate lateral bracing.

REQUIRED

(a) Determine if the beam can accommodate the duct opening.
(b) In the event the beam proves inadequate, prepare a revision to the design that will satisfy the opening requirements. Beam size is to remain the same.

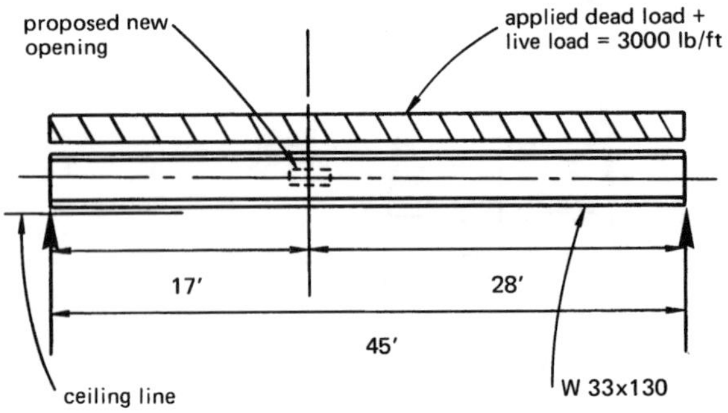

Figure ST-25

STEEL-26

A steel-framed Vierendeel girder (see Fig. ST-26) is 10 ft deep, and it spans 60 ft.

DESIGN CRITERIA

- The principal loads are located at the one-third panel points, and are 50 kips each.
- The weight of the girder may be neglected.

REQUIRED

(a) Determine the maximum moments in all members.
(b) Determine the maximum shears in all members.
(c) Determine the maximum axial forces in all members.

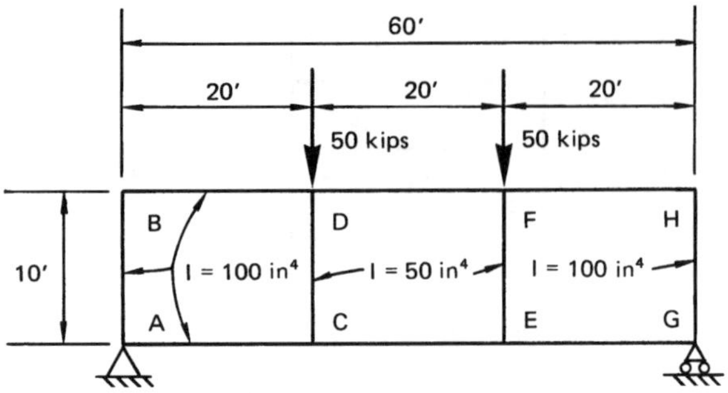

Figure ST-26

STEEL-27

The sketch in Fig. ST-27 shows a hybrid composite steel girder with a lightweight concrete floor slab.

DESIGN CRITERIA

- Various specifications for the steel are shown on each part of the girder.
- Weight of concrete slab = 110 pcf
- f'_c = 4000 psi lightweight concrete

REQUIRED

Show your calculations for all parts.

(a) What are the flexural stresses in the lightweight concrete?
(b) What are the flexural stresses in the bottom plate?
(c) What are the flexural stresses in the web plate?
(d) Are the stresses in parts (a), (b), and (c) within the allowables for each, based on current UBC?
(e) Calculate and specify the number of 3/4 in diameter by 3 in studs required.
(f) Comment on the shoring for the girder. Assume that 55% of the moment is applied when the concrete is placed. Does the girder require shoring?

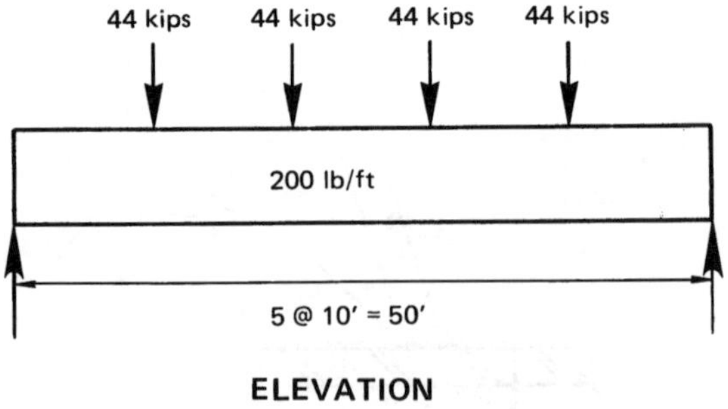

Dead loads

concrete on girder	35
steel girder	136
fire proofing	29
	200 lb/ft

Figure ST-27

(continued on next page)

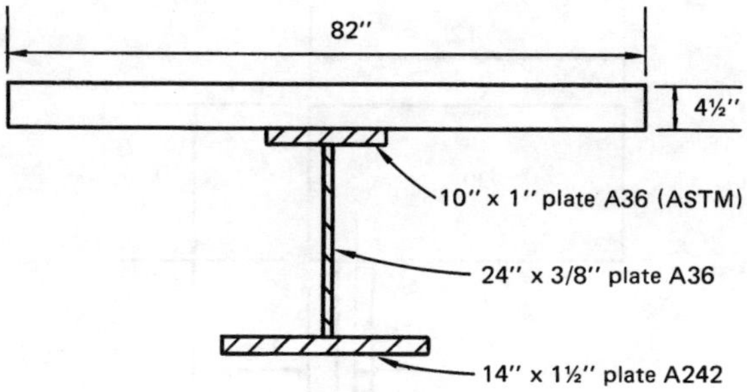

SECTION

Figure ST-27 (cont.)

STEEL-28

An existing steel beam must be modified to provide an opening as indicated in Fig. ST-28. The beam is a W 24 x 84 with a 30 kip load at midspan.

DESIGN CRITERIA

- Assume that the compression flange is stayed against buckling.
- Neglect the weight of the beam.
- ASTM A36 steel

REQUIRED

(a) Certain reinforcement plates are shown. Verify the adequacy of the plates.
(b) Determine the length of plates that will avoid overstressing the web of the beam.
(c) Calculate and show the welding required (E70 electrodes).
(d) Describe a construction sequence to accomplish the desired changes.

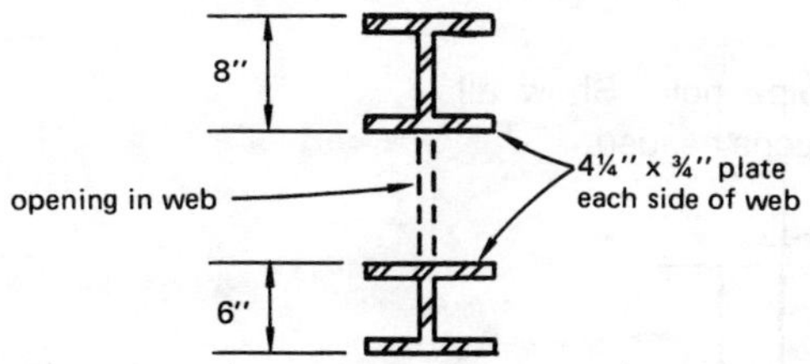

SECTION B-B

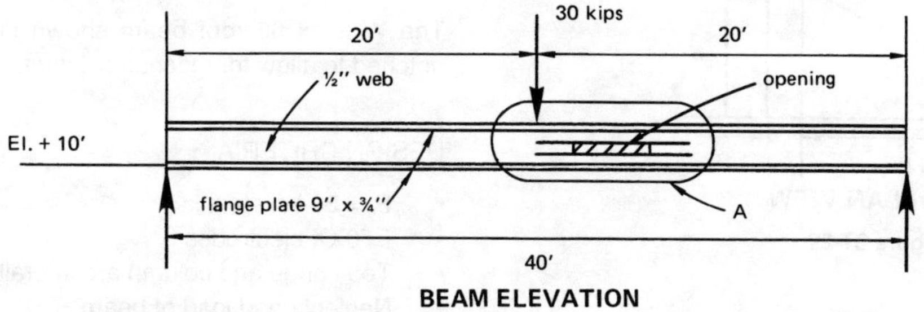

BEAM ELEVATION

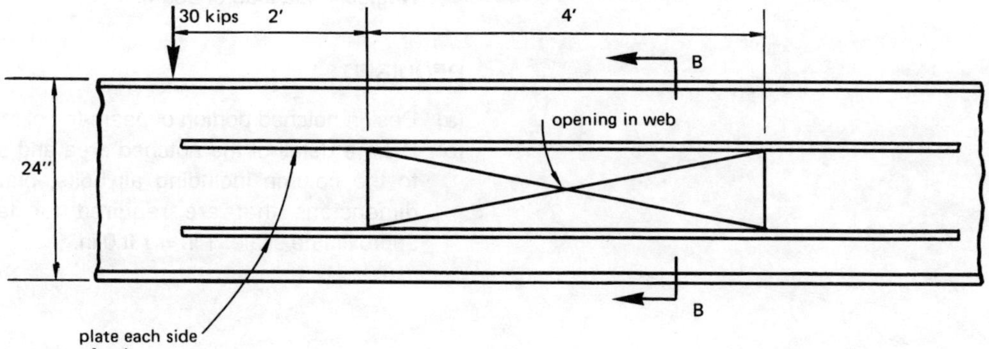

DETAIL A

Figure ST-28

STEEL-29

Fig. ST-29 shows a sign mounted on a steel pole.

DESIGN CRITERIA

- Weight of sign and sign frame = 4800 pounds
- Steel pipe column: F_y = 36 ksi (1.33 increase for lateral permitted)
- Wind load: 0 ft to 30 ft = 15 psf
  30 ft to 50 ft = 20 psf
  50 ft to 99 ft = 25 psf
- Seismic factor: 0.30 g (working load)
- Dead load deflection at C is limited to 1.0 in maximum.
- Assume sign frame is infinitely rigid and fully developed to top of pole at A. Pole is fully fixed at base B.
- The foundation may be assumed adequate.

REQUIRED

Design a constant section steel pipe pole. Show all calculations and any design assumptions needed.

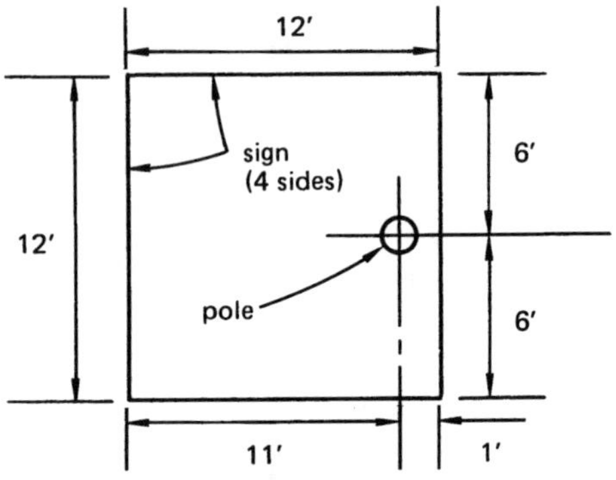

TOP PLAN VIEW

Figure ST-29

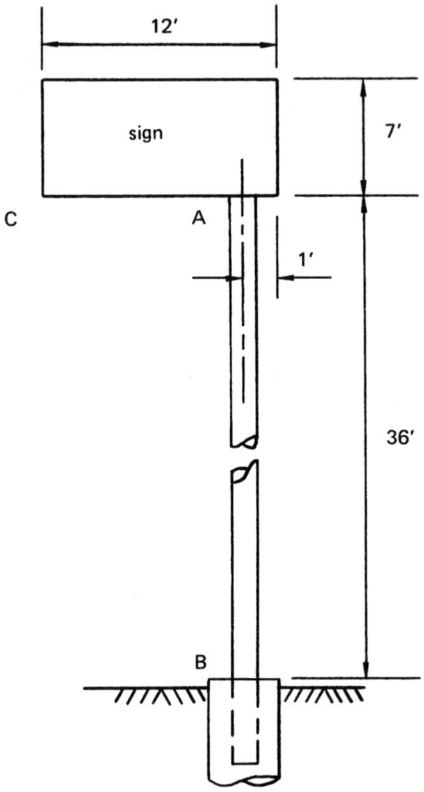

Figure ST-29 (cont.)

STEEL-30

The W 24 x 62 roof beam shown in Fig. ST-30 has been notched to allow for mechanical duct.

DESIGN CRITERIA

- F_y = 36 ksi steel
- E70XX electrodes
- Top flange and column are laterally braced.
- Neglect dead load of beam.

REQUIRED

(a) Design notched portion of beam-to-column connection.
(b) Draw a detail of the notched area and show connection to the column including all bolts, plates, welds, and dimensions that are required for fabrication. Use approximate scale: 1 in = 1 ft 0 in.

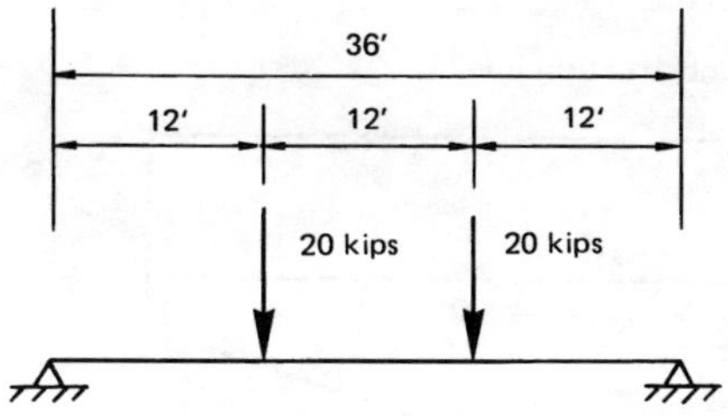

BEAM LOADING

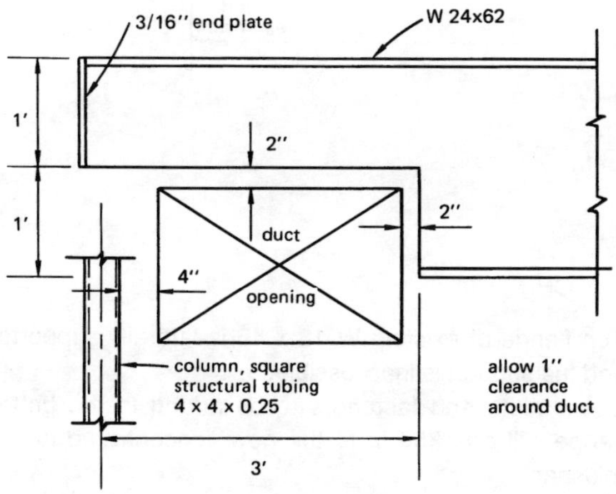

Figure ST-30

STEEL-31

A traveling bridge crane is shown in Fig. ST-31. The capacity of the lift hook is 40 tons.

DESIGN CRITERIA

- Span of runway beam = 24 ft (simple span)
- Weight of trolley = 20 kips (dead load)
- Wheel load = 53 kips (maximum static, includes bridge dead load)
- Design vertical load = wheel load + rail + 25% hook load + runway beam (impact is 25%)
- Transverse load = 20% of hook load + trolley

REQUIRED

(a) Determine the maximum flexural stress in the bottom flange of the runway beam.

(b) Determine the maximum flexural stress in the top flange of the runway beam.

STEEL-32

An elastically designed 2-story steel rigid frame is shown in Fig. ST-32.

DESIGN CRITERIA

- $E = 29 \times 10^6$ psi
- $I_g = 1500$ in^4
- $I_c = 1000$ in^4
- Assume points of inflection are at midpoints of members.

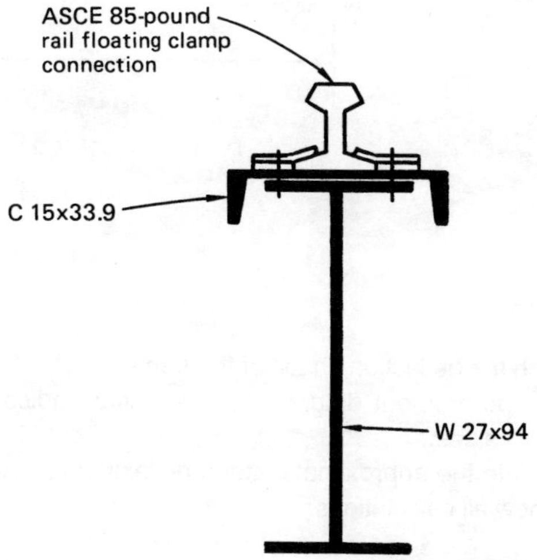

RUNWAY BEAM

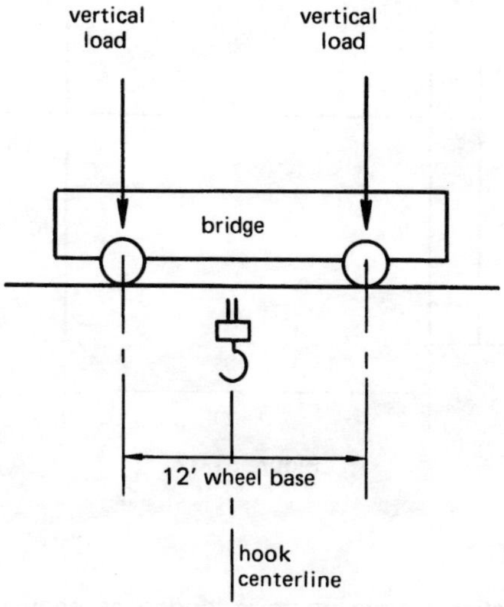

SECTION A-A

Figure ST-31

(continued on next page)

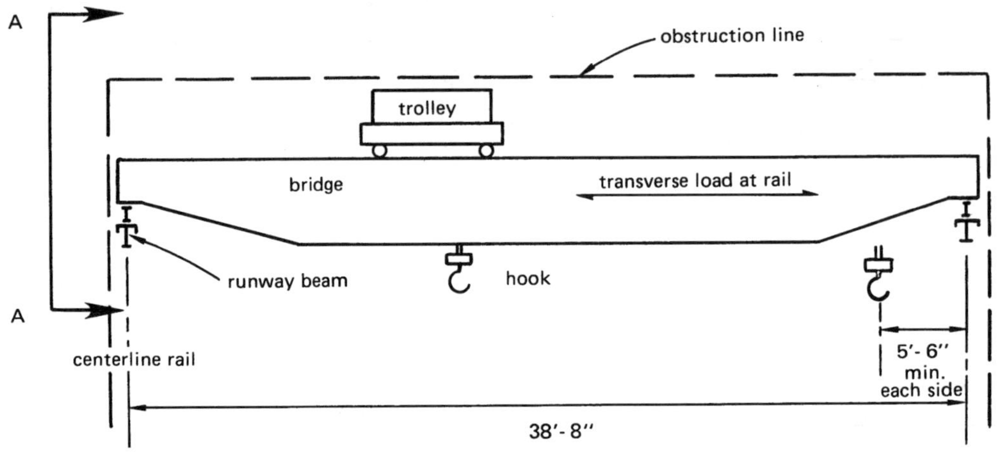

Figure ST-31 (cont.)

REQUIRED

(a) Sketch the deflection shape of the frame.
(b) Draw the moment diagram for the frame. Indicate critical values.
(c) Calculate the approximate lateral deflection of point A. Show all calculations.

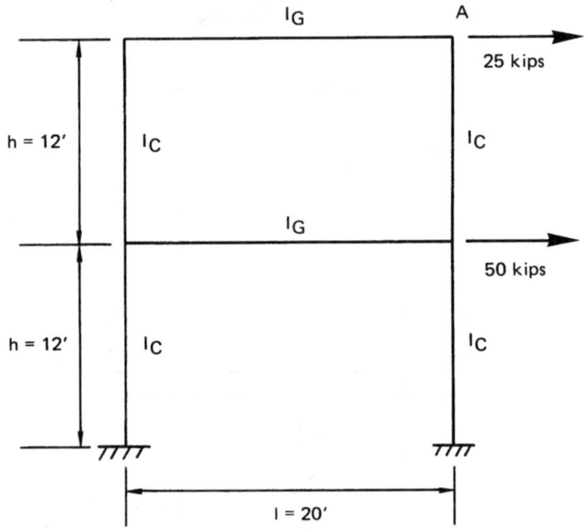

Figure ST-32

STEEL-33

An existing simple beam is loaded as shown in Fig. ST-33.

DESIGN CRITERIA

- Top flange of existing W 18 x 46 is laterally supported, and the support is inaccessible.
- Cover plate and fasteners to be added to the bottom flange will add 350 lb to the new concentrated load at midspan.
- Beam and cover plate: F_y = 36 ksi
- Cover plate is fastened with 5/8 in diameter high tension (A325-F) bolts in two rows, 3 1/2 in gage.
- Existing beam will be shored and existing total load deflection will be essentially eliminated prior to beam modification.

REQUIRED

(a) Determine the minimum thickness (to the nearest 1/8 in) and length of the tension cover plate 7 1/2 in wide attached to the bottom flange of the W 18 x 46 to resist a concentrated load of 5150 lb at midspan. (Indicate cover plate thus: 7 1/2 in x length x thickness.)
(b) Calculate top and bottom stress of modified girder at midspan.
(c) Sketch cover plate and indicate cover plate length and spacing for pairs of 5/8 in diameter tension bolts (A325-F).

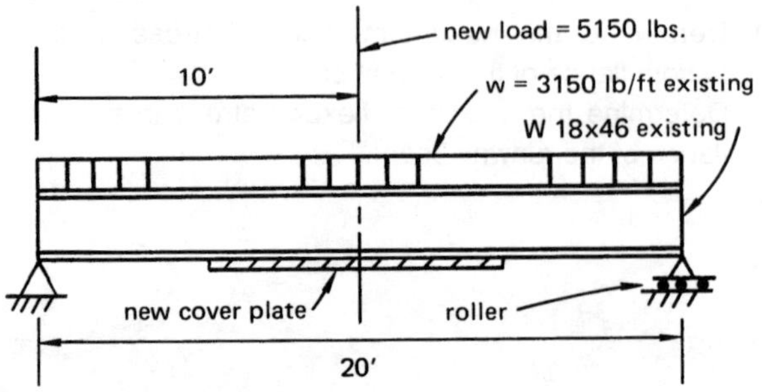

Figure ST-33

STEEL-34

An existing steel beam, which is continuous over one support, supports a uniform load along its full length. The load and spans are shown by Fig. ST-34. Certain modifications require the support of B to be relocated to D. Two procedures are proposed.

1) Snug the new support at D tight to the existing beam soffit, and then remove the support at B.
2) At point B, release restraints by support B against upward movement of the beam. Jack up point D to the same elevation as A and C. Install the support at D, and then remove the support at B.

DESIGN CRITERIA

- ASTM A36 steel
- Top flange of beam is continuously supported.

REQUIRED

(a) Determine the final reaction at D.
(b) Verify the adequacy of the existing beam section.
(c) Determine the minimum column size for a new support at D. The unsupported column height is 10 ft 0 in.
(d) Design and draw a detail to approximate scale 1 in = 1 ft 0 in for a welded beam-to-column connection at D.

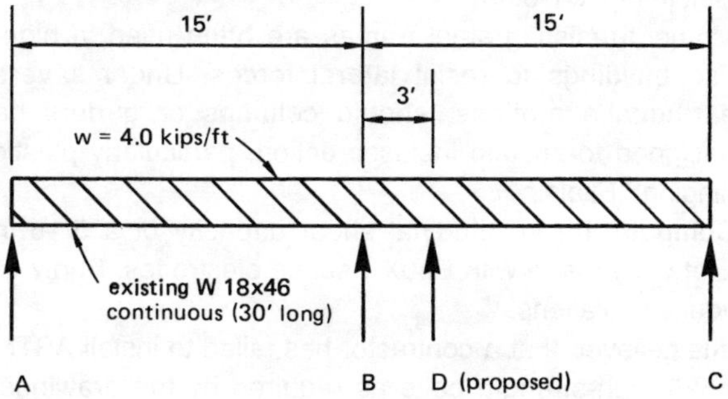

Figure ST-34

STEEL-35

A simple supported beam is shown in Fig. ST-35.

DESIGN CRITERIA

- Uniform load is 1140 lb/ft (includes beam weight).
- Beam top flange is fully stayed.
- $F_b = 22.0$ ksi

REQUIRED

(a) What is the minimum flange width (nearest 1/4 in in width) at the critical depth?
(b) What is the location of the critical depth?
(c) What is the bending stress at critical depth?
(d) What is the deflection of the beam at midspan due to total load?

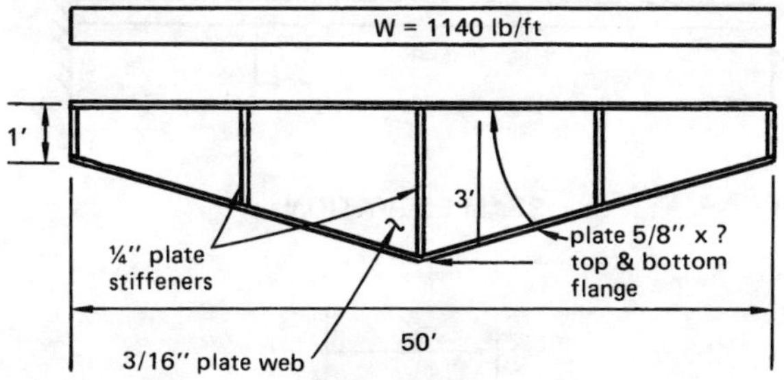

Figure ST-35

STEEL-36

A rectangular steel tube spans between two rigid supports as shown in Fig. ST-36. A load of 12 kips is supported by an outrigger at mid-span.

DESIGN CRITERIA

- Assume steel supports.
- Use E70 welding rods.
- Assume support arms and connections are adequate.

REQUIRED

(a) Calculate the maximum torque and shear forces in the beam, and determine the maximum combined torque and shear stress. Indicate the location of the maximum combined stress.
(b) Calculate the maximum combined total shear and moment stress at supports A and C. Show the minimum required welding.
(c) What are the relative merits of each of the following steel sections with respect to torsional resistance? (No calculations are required.)
    (1) Rectangular tube sections
    (2) Pipe sections
    (3) Wide-flange sections
    (4) Channel sections
    (5) Angle sections

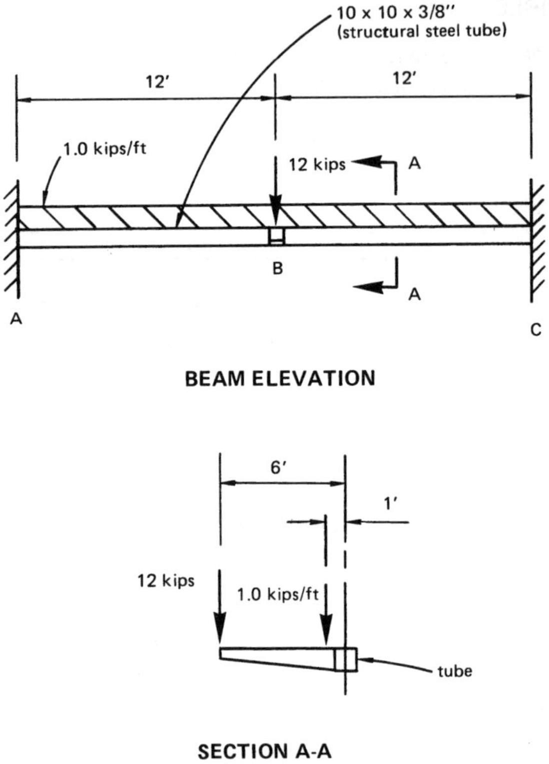

Figure ST-36

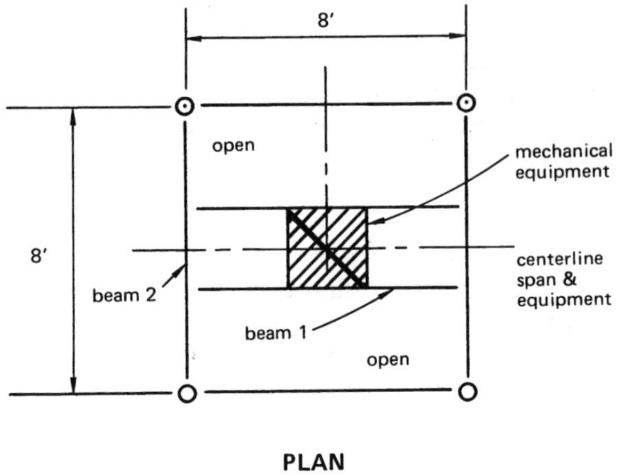

Figure ST-37

STEEL-37

The mechanical equipment shown in Fig. ST-37 has moving parts that could subject the structure to an annoying vibration.

DESIGN CRITERIA

- Equipment weight = 300 pounds
- 1725 rpm motor
- ASTM A36 steel
- Use 10 in wide flange series beams.

REQUIRED

(a) Select the lightest weight size for beams 1 and 2 to support the gravity load.
(b) In what ways can the beams be modified or changed to eliminate the vibration problem?
(c) What change would you recommend, if any, to minimize (or solve) the vibration problem? Show calculations.

STEEL-38

REQUIRED

Provide brief answers to the following questions. Indicate any assumptions used in your solutions.

(a) In high-rise design, what is the P-Δ effect?
(b) What is a prequalified welded joint?
(c) What is delamination and where might it occur in the joint in Fig. ST-38(a)?
(d) Moment-resisting steel frames are often used in high-rise buildings to resist lateral forces. Under severe earthquake motions, should columns or girders be designed to sustain inelastic action, particularly plastic hinging? Explain.
(e) Compute the longitudinal shear capacity of a 5/16 in fillet weld made with E80XX series electrodes. Show all your calculations.
(f) It is believed that a contractor has failed to install ASTM A325 high-strength bolts as required by the drawings. Heads of already-installed bolts have marks as shown in Fig. ST-38(b). Are these the required high-strength bolts? Explain.
(g) Using the AISC code, compute the ultimate capacity of each of the two double-angle connections shown in Fig. ST-38(c). Which would be expected to fail first under inelastic seismic overloads? Why?

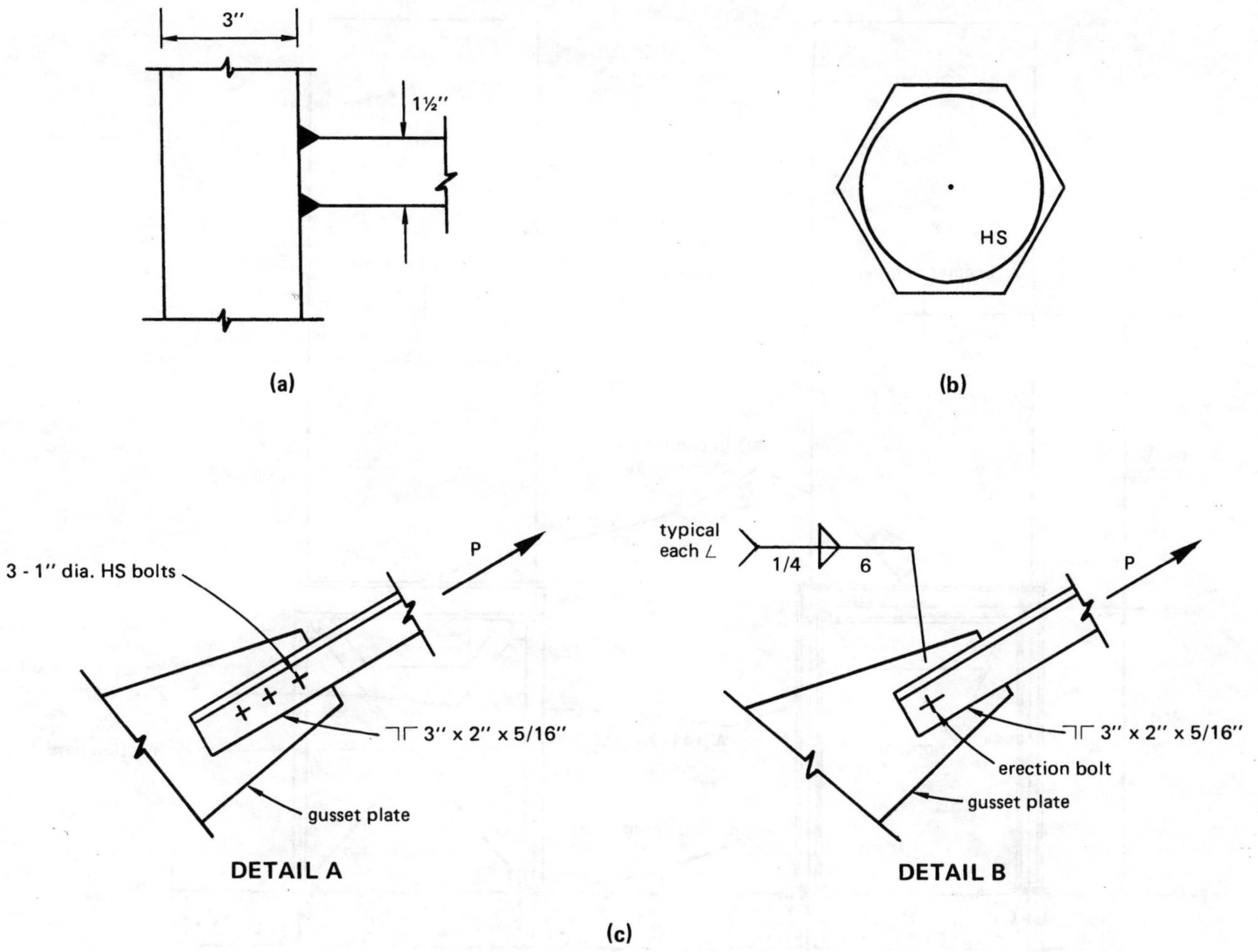

Figure ST-38

STEEL-39

A steel storage bin used for agricultural purposes is supported by a steel frame. Fig. ST-39 shows the bin and the sizes of the structural members. The bin is square.

DESIGN CRITERIA

- Seismic zone 4
- Soil profile type S_D
- Near-fault factors = 1.0
- I = 1.0
- ASTM A36 steel (E = 29 x 10^6 psi)
- Total dead load of bin plus contents = 400 kips
- The dead load of the steel frame may be neglected.
- Assume concrete foundation is satisfactory.
- Provide calculations to support your work.
- Assume girder connections to be satisfactory.

REQUIRED

(a) Determine the design forces and reactions at the supporting frame for each direction.

(b) Verify the adequacy for the member sizes shown in Fig. ST-39. Assume all connections are satisfactory.

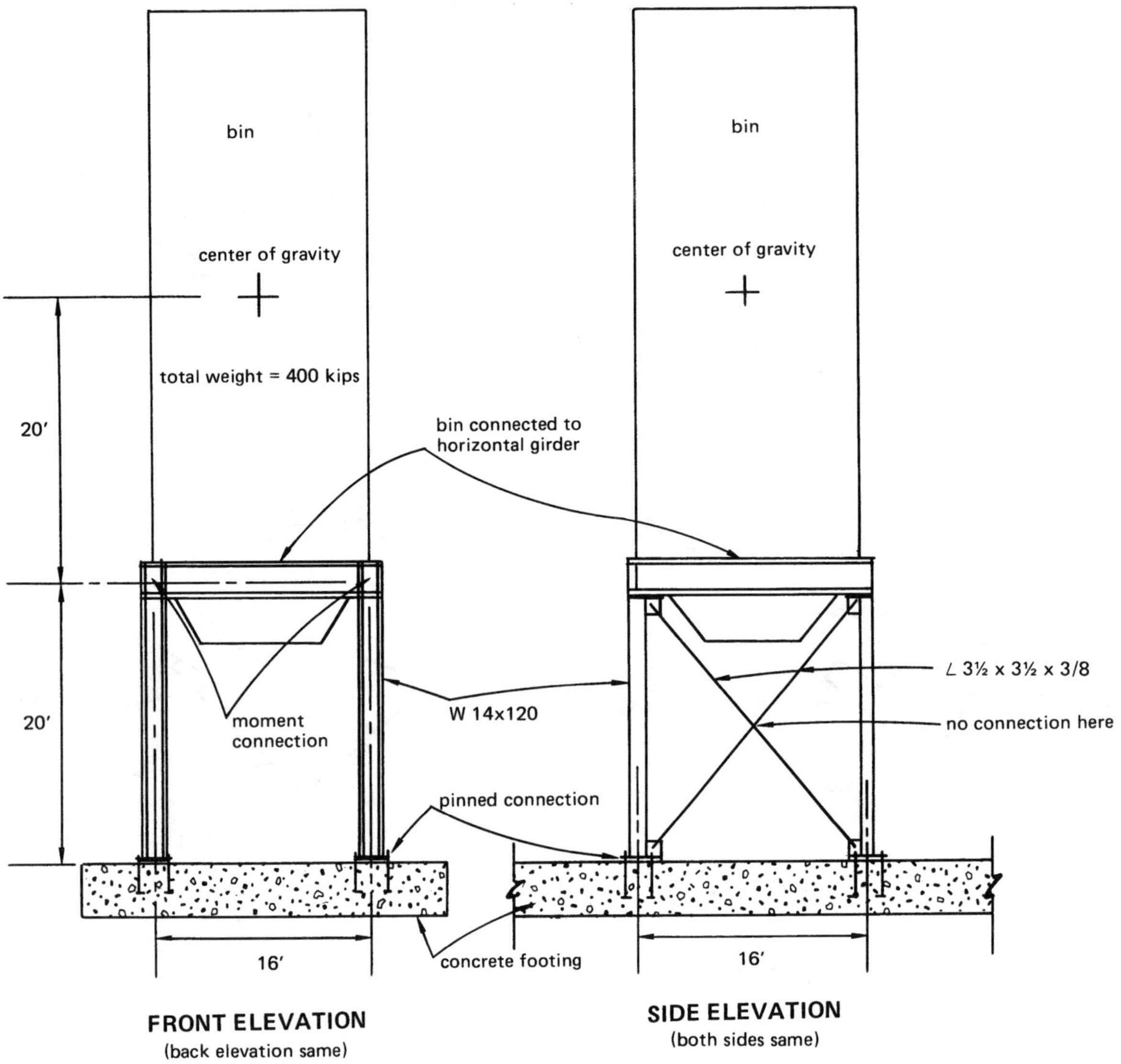

Figure ST-39

STEEL-40

It is necessary to investigate the possibility of upgrading the live load capacity of the existing floor (shown in Fig. ST-40) from 50 psf to 75 psf (reducible due to tributary area). It had been found that the concrete floor slab is adequate for the increased live load.

DESIGN CRITERIA

- AISC, current edition
- F'_c = 3000 psi
- n = 9
- F_y = 36 ksi for beams and columns
- Dead load of floor = 50 psf + weight of beams and girders
- No shoring is allowed, and existing live load is removed while work is being done.

REQUIRED

(a) Utilizing the existing steel beams and concrete slab to provide full composite action, find
  (1) the maximum unit stress at bottom of W 14 x 26,
  (2) the maximum unit stress at top of concrete slab, and
  (3) the number and approximate spacing of 5/8 in diameter x 3 in long headed studs (allowable horizontal shear per stud = 8 kips).

(b) Check adequacy of the beam end connection, including block shear, and if found inadequate, state your solution. Show all calculations to support your findings and design.

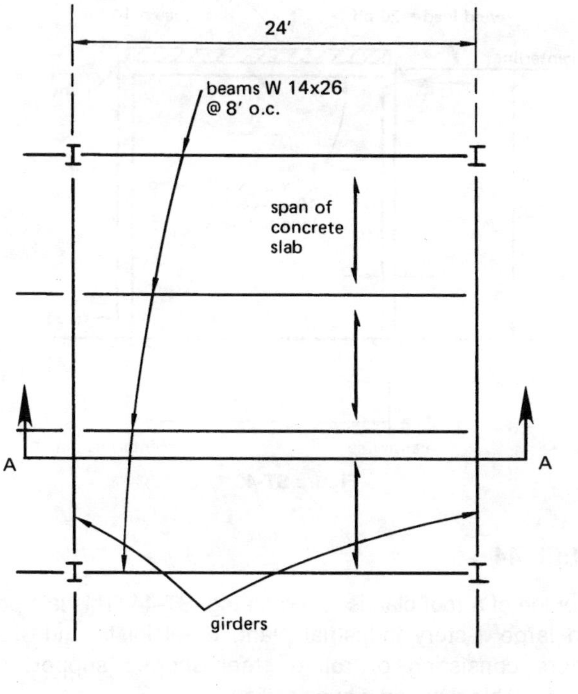

FLOOR FRAMING PLAN PARTIAL

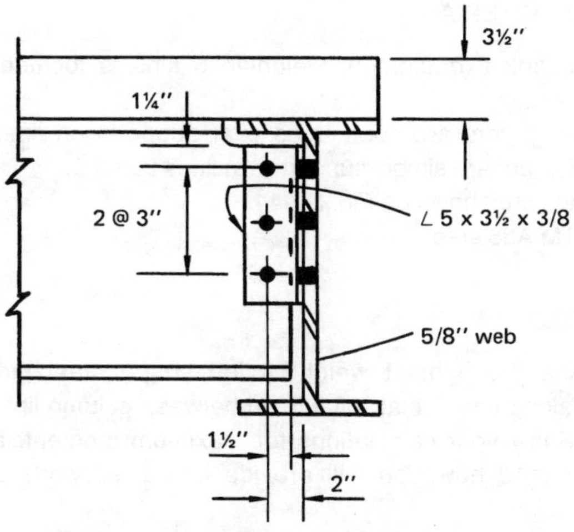

SECTION A-A

Figure ST-40

STEEL-41

A rigid steel frame was designed for the loading condition shown in Fig. ST-41. After the frame had been completed, and had been in service for short period, it was discovered that the right support was settling at a more rapid rate than was the left support. A differential settlement of 1/2 in was observed.

DESIGN CRITERIA

- Neglect the weight of the frame members.
- Assume the frame is stable through adequate lateral support in the plane perpendicular to the frame.
- All members W 21 x 62
- $I = 1330 \text{ in}^4$

REQUIRED

(a) What is the moment at joint B before settlement occurs?
(b) What is the moment at joint B after settlement occurs?

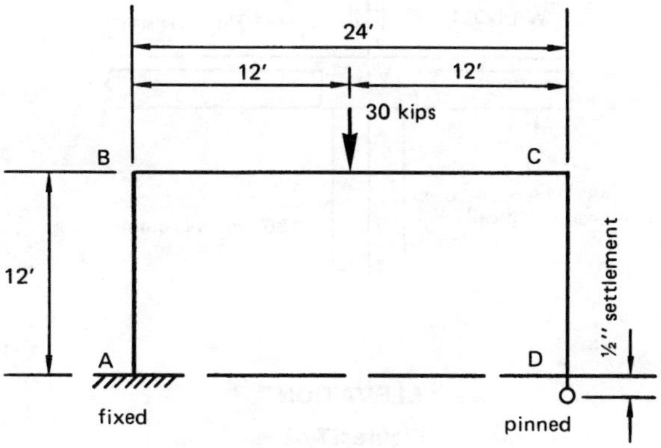

Figure ST-41

STEEL-42

A high rise steel frame will be erected on top of reinforced concrete deck. All work is new. Full continuity from the steel column through the column base to the concrete below is desired. The design loads are shown in Fig. ST-42.

DESIGN CRITERIA

- $f'_c = 3000$ psi concrete
- ASTM A36 steel

REQUIRED

Design a square steel base plate and complete base connection to satisfy the given condition. Show the plate thickness, stiffeners (if any), required welding (or other connection method), and size and placing of anchor bolts. Draw a detail of your base connection showing all the necessary information for others to complete the installation.

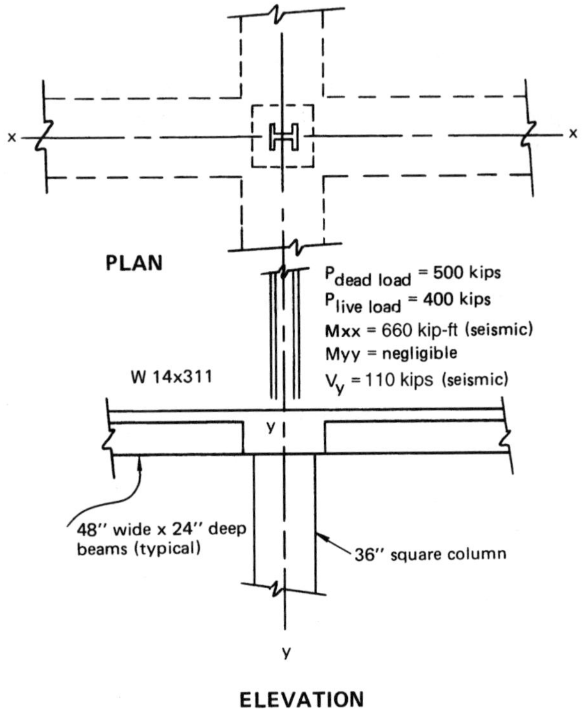

Figure ST-42

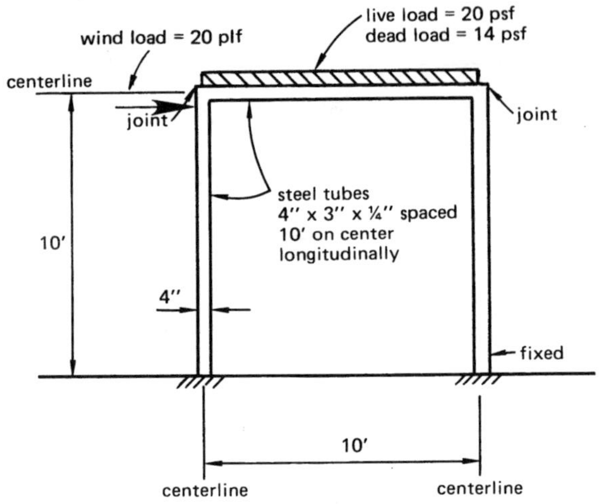

Figure ST-43

STEEL-43

The drawing in Fig. ST-43 shows a cross section of a long covered walk between two hotel units. The bent shown is made up of steel tubing as noted.

DESIGN CRITERIA

- Bents are spaced 10 ft on center longitudinally.
- Column legs are considered to be fixed at the base.
- Neglect the weight of the tubes.
- ASTM A500, Gr. B (F_y = 46 ksi)

REQUIRED

(a) Investigate and verify the adequacy of the tube sizes that are given. Consider both the vertical and lateral loads. Show your calculations for maximum moments, shears, and any other critical design values, for the beam member, and the columns.

(b) Draw a detail of the joint at the upper right-hand corner of the bent. Show the welding necessary, and any other considerations, to provide a completely fabricated job ready for final painting.

STEEL-44

A portion of a roof plan is shown in Fig. ST-44. This is a part of a large 1-story industrial plant. Steel joists and steel girders consisting of rolled steel shapes support an incombustible roof and a hung ceiling.

DESIGN CRITERIA

- Mechanical equipment weighing 6 kips is located at point A.
- Roof girders are continuous over supports on lines 3 and 4, but are simply supported at lines 2 and 5.
- Loads are shown on Fig. ST-44.
- ASTM A36 steel

REQUIRED

Determine the lightest weight rolled shape steel girder section along line C that will extend between column lines 2 and 5. Show your calculations for maximum moment, and buckling, and how you will provide lateral support. Use plastic design.

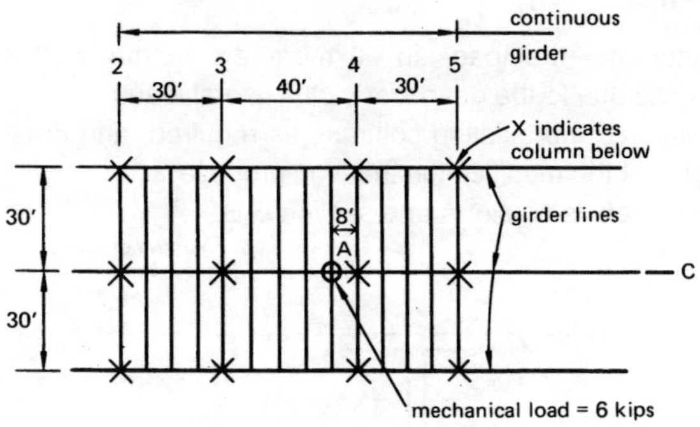

PARTIAL ROOF PLAN

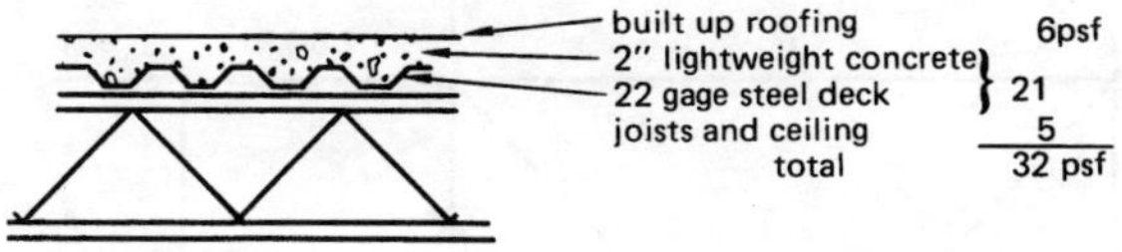

TYPICAL SECTION

Figure ST-44

STEEL-45

The sketch in Fig. ST-45 shows a trussed girder that is isolated as part of a multi-storied structure.

DESIGN CRITERIA

- ASTM A36 steel
- Use current AISC specifications.
- Design loads are shown on the detail for connection A.
- The columns are box sections that will be shop fabricated (complete between the splices, including connections). The usual procedure is to first fabricate the box section, cut it accurately to length, and then provide any necessary connection details in the shop so that field connections can be made readily. Field connections can be welded or bolted.

REQUIRED

Prepare a design for connection A using the design forces shown. Provide the computations to support your design. State any assumptions that you may make, including reasons for them.

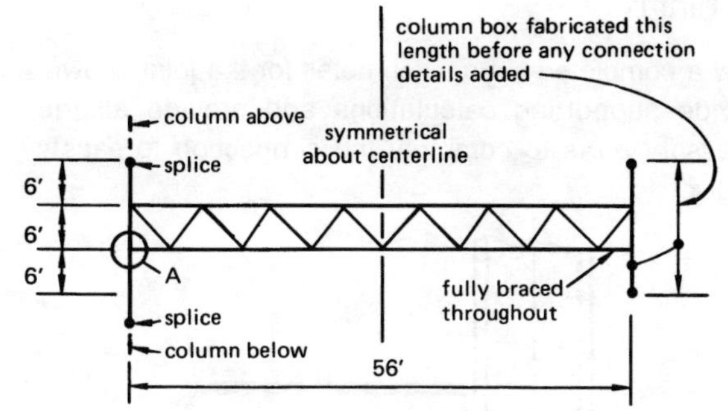

ELEVATION

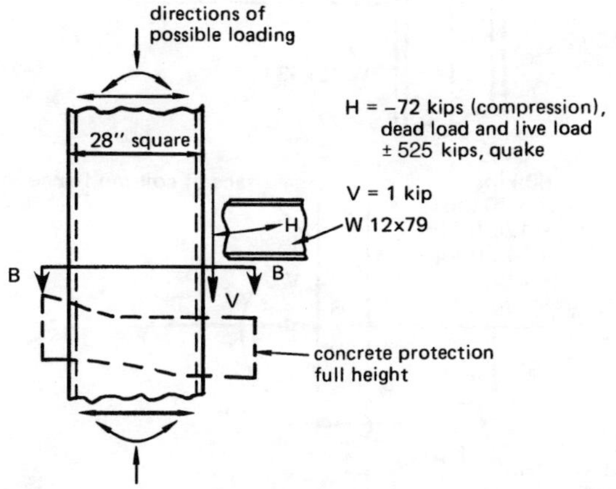

CONNECTION A

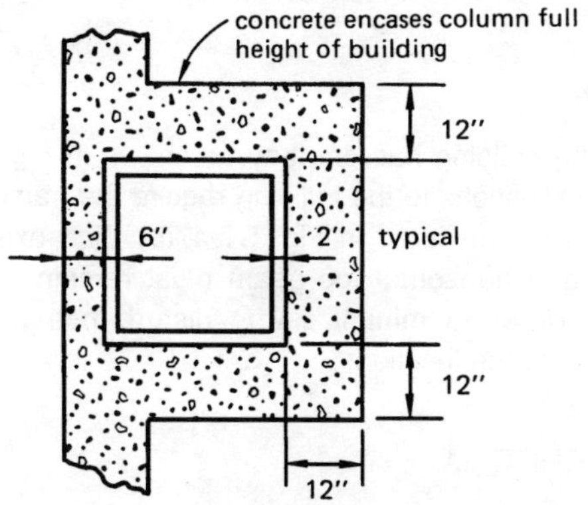

SECTION B-B

Figure ST-45

STEEL-46

The connection detail of a steel beam to a steel column is incompletely shown in Fig. ST-46. The design forces and moments are given as shown.

DESIGN CRITERIA

- ASTM A36 steel
- Use current AISC specifications.

REQUIRED

Draw a completed connection detail for the joint shown. Provide supporting calculations and provide all the necessary parts to complete the connection to satisfy the given loads.

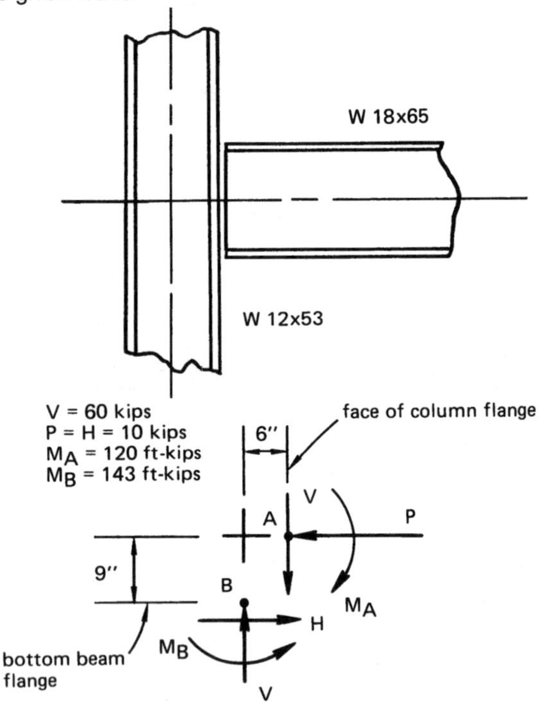

Figure ST-46

STEEL-47

An existing building has one bay as shown in Fig. ST-47(a). Modifications to the building require the frames to be altered as indicated in Fig. ST-47(b). The existing columns and horizontal top beam must remain in the frame in order to minimize any disturbance to the structure or to the function.

DESIGN CRITERIA

- Column bases and all joints are assumed pinned.
- Columns are continuous full height, and web members are pinned at the column connection.
- Dimensions are given to centerline of members.
- Columns are braced perpendicular to the plane of the frame between the floor and the new lower chord.
- Neglect the weight of the frames.
- ASTM A36 steel
- Use current UBC code.

REQUIRED

(a) Determine the loads in all members of the modified frames due to the dead loads plus lateral loads.
(b) Reinforce the existing columns as required, and draw a detail of the modification that you propose.
(c) Draw details of joint A and section B-B.

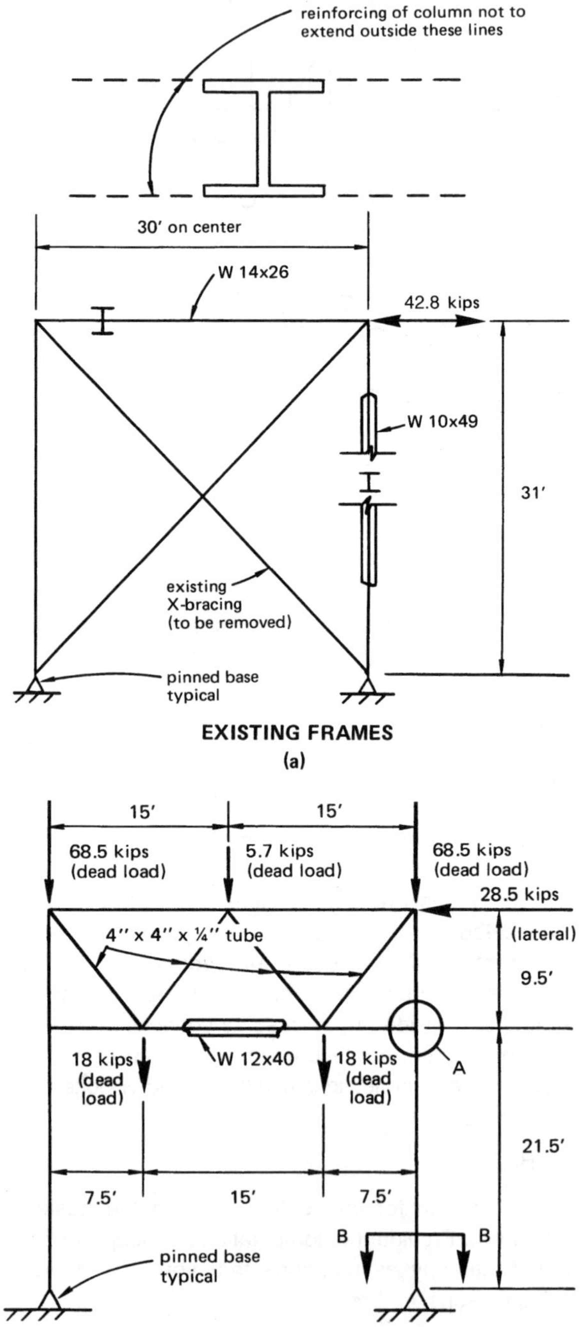

Figure ST-47

STEEL-48

A truss joint is shown in Fig. ST-48.

DESIGN CRITERIA

- ASTM A36 steel
- ASTM A325 friction-type bolts

REQUIRED

Design and detail this connection using bolts to connect the angles to the gusset and welds to connect the gusset to the bottom chord. Show by computation that you have accounted for the transfer of all the forces, and that the stresses in the critical sections are within the allowables.

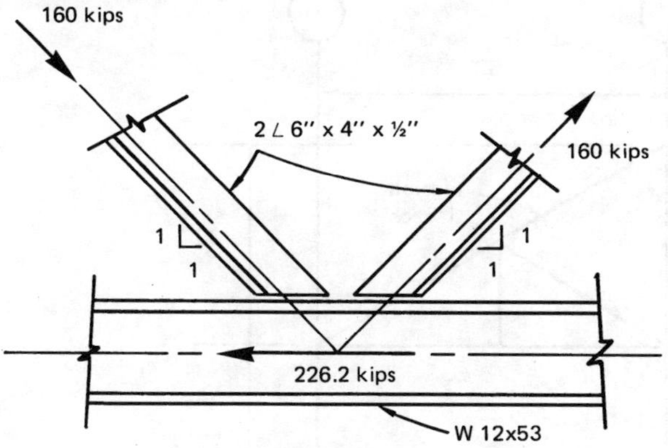

Figure ST-48

STEEL-49

A water storage tank is shown in Fig. ST-49 with a cross-braced four-leg tower.

DESIGN CRITERIA

- Seismic zone 3
- Soil profile type S_D
- Near-fault factors of unity
- Tower is to be designed to current UBC code and AISC specifications.
- Neglect weight of steel frame.
- Assume frame to be identical at north, south, east, and west elevations.
- Assume full height columns (no splices).
- Assume pinned ends at diagonals, struts, and columns.

REQUIRED

(a) Find critical axial loads in columns.
(b) Find maximum bending moments in columns.
(c) Find maximum axial forces in upper and lower struts, and diagonals (assume compression diagonals not acting).
(d) Design tower legs using standard pipe sizes (standard weight) ASTM A501: f_u = 58,000 psi, f_y = 36,000 psi.
(e) Discuss effect on column design if the detail attaching column to footing provides full fixity. A sketch may be used. Calculations are not necessary.

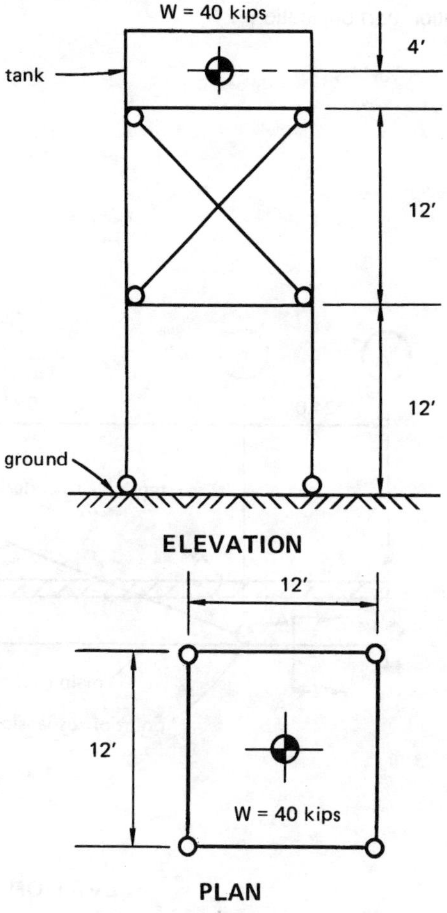

Figure ST-49

STEEL-50

The line diagram in Fig. ST-50 depicts a cantilever-type roof construction used as an aircraft hangar. The main roof girder supports open web steel joists at 8 ft 4 in on center. (See detail A-A.) The indicated uniform dead load includes an allowance for the weight of the main roof girder ABF.

DESIGN CRITERIA

- ASTM A36 steel
- Consider all joists as pinned.

	dead load	live load
P	9.5 kips	---
W	0.62 kips/ft	0.63 kips/ft

REQUIRED

(a) Design the main roof girder using the most economical W36 section.
(b) Complete the details of section A-A. Add any necessary items and state your reason for each addition or modification.

STEEL-51

The joint shown in Fig. ST-51 occurs in a ductile moment-resisting space frame.

DESIGN CRITERIA

- ASTM A913 / Grade 50
- E70XX electrodes
- Column W 14 x 120
- Girders W 21 x 62 (spans 28 ft center-to-center)
- Gravity load only:
  - V_g = 39.2 kips
  - M_g = 187 ft-kips
  - $V_{c1} = V_{c2} = 0$
- Dead-to-total gravity load ratio = 0.6

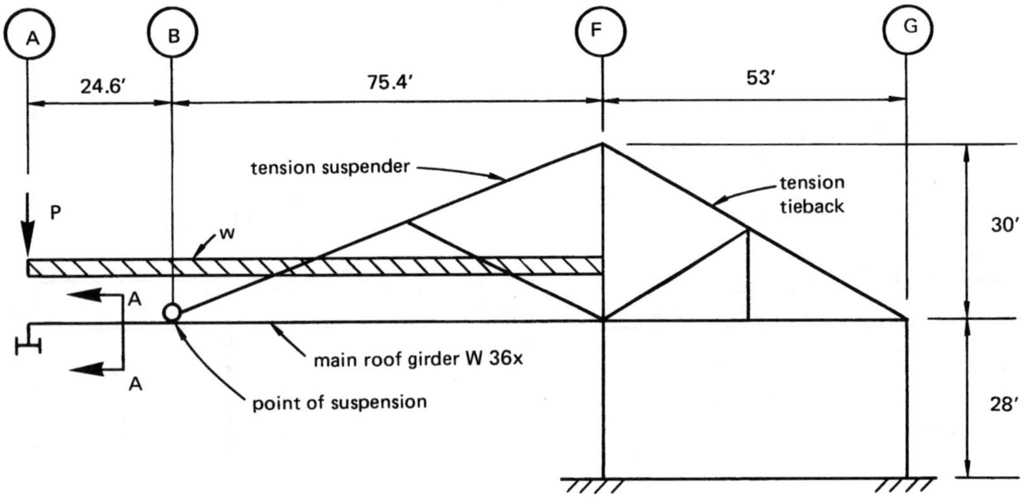

ELEVATION – TYPICAL CANTILEVER BENT

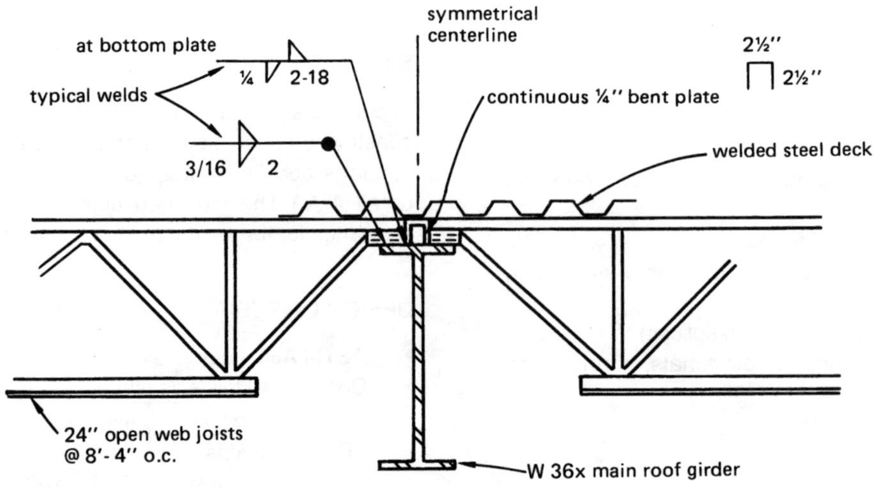

SECTION A-A

Figure ST-50

- Seismic load only:
  $V_g = \pm 9.2$ kips
  $M_g = \pm 139$ ft-kips
  $V_{c1} = \pm 20.3$ kips
  $V_{c2} = \pm 24.0$ kips
- Column stress (15 ksi above; 18 ksi below)

REQUIRED

Design and draw the complete joint detail using all welded construction. Show all weld symbols (weld sizes not required), and indicate all detailing material required. Use approximate scale: 1 in = 1 ft 0 in.

NOTE: Assume points of inflection in the columns occur 5.5 ft above and below the girder.

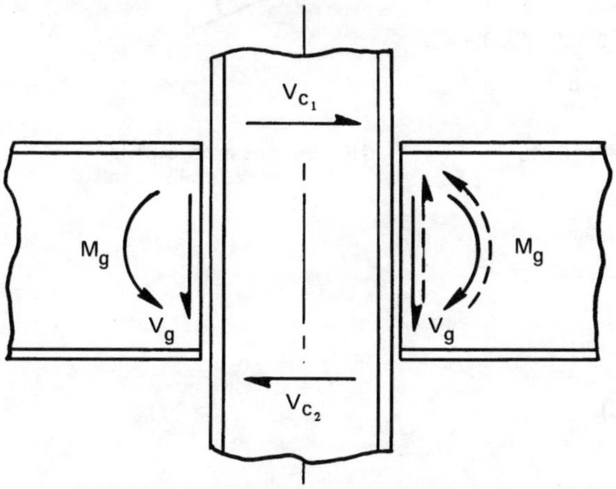

Figure ST-51

STEEL-52

Fig. ST-52 shows a truss joint from an all welded truss whose partial elevation is shown.

DESIGN CRITERIA

- ASTM A36 steel
- E70XX welding electrodes

REQUIRED

(a) Complete the design of the joint using welding throughout, transfer all the forces given into the W 10 x 45. Show all calculations.

(b) Draw a complete detail to approximate 1 in = 1 ft 0 in scale showing the amount and size of all welds and component detail material.

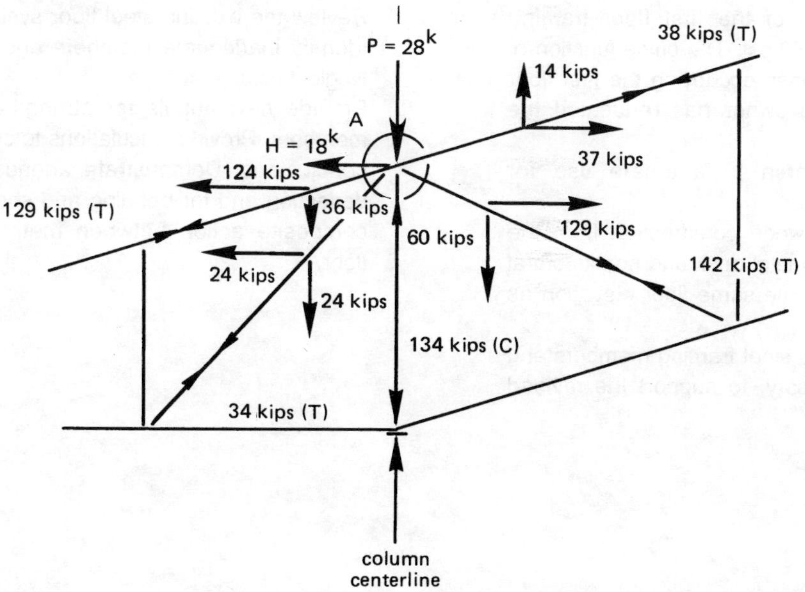

PARTIAL ELEVATION

Figure ST-52

(continued on next page)

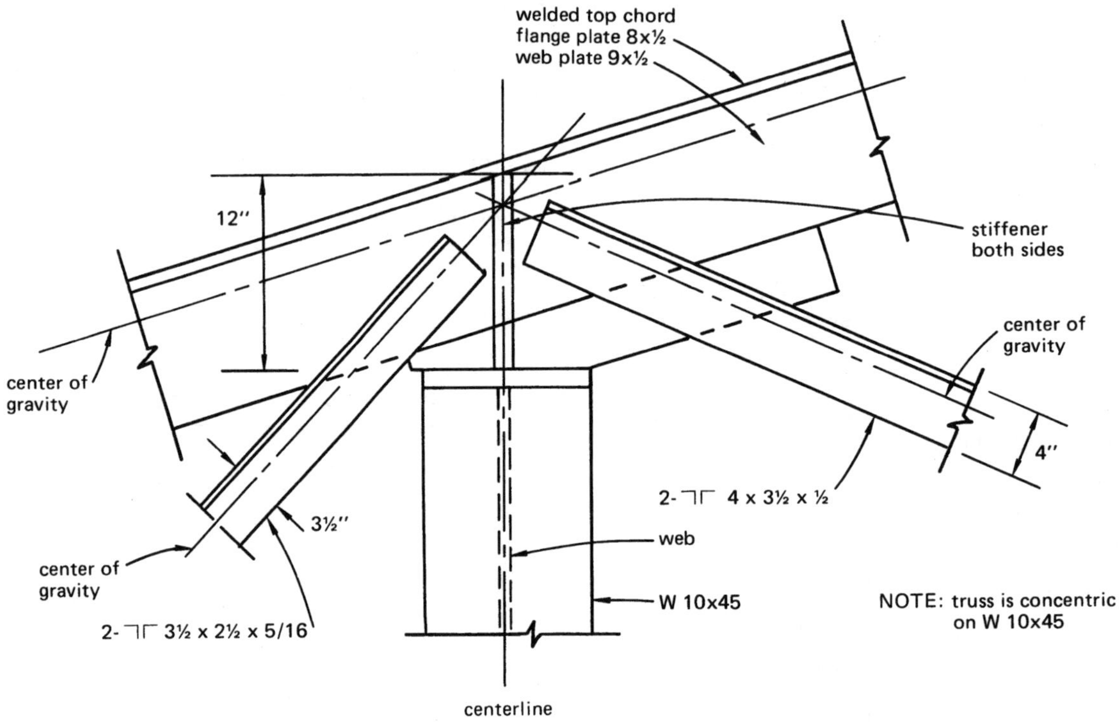

DETAIL A

Figure ST-52 (cont.)

STEEL-53

Fig. ST-53 shows a plan of the first floor framing designed for a live load of 50 psf. The office function of a local printing plant has been occupying the first floor area shown. The building owner has requested the following changes.

(1) Convert the office area to alternate use for composing and linotype.
(2) Remove the existing wood construction from the floor and replace with metal deck and nonstructural concrete fill to provide the same floor elevation as at present.
(3) Strengthen the existing steel framing members and connections, if necessary, to support the revised design.

DESIGN CRITERIA

- ASTM A36 steel
- ASTM A307 bolts
- E70XX welding electrodes
- Floor dead load = 50 psf (partitions included)
- For floor live load, use current UBC code.
- Show all calculations.

REQUIRED

(a) Review the existing steel floor system and connections. Identify inadequate members and connections, if any. Neglect column stresses.
(b) Provide new details for strengthening of inadequate members. Provide calculations to support your work and conclusions. Demonstrate adequacy of members for deflection and for bending and shear stresses. Neglect composite action between metal deck and concrete floor.

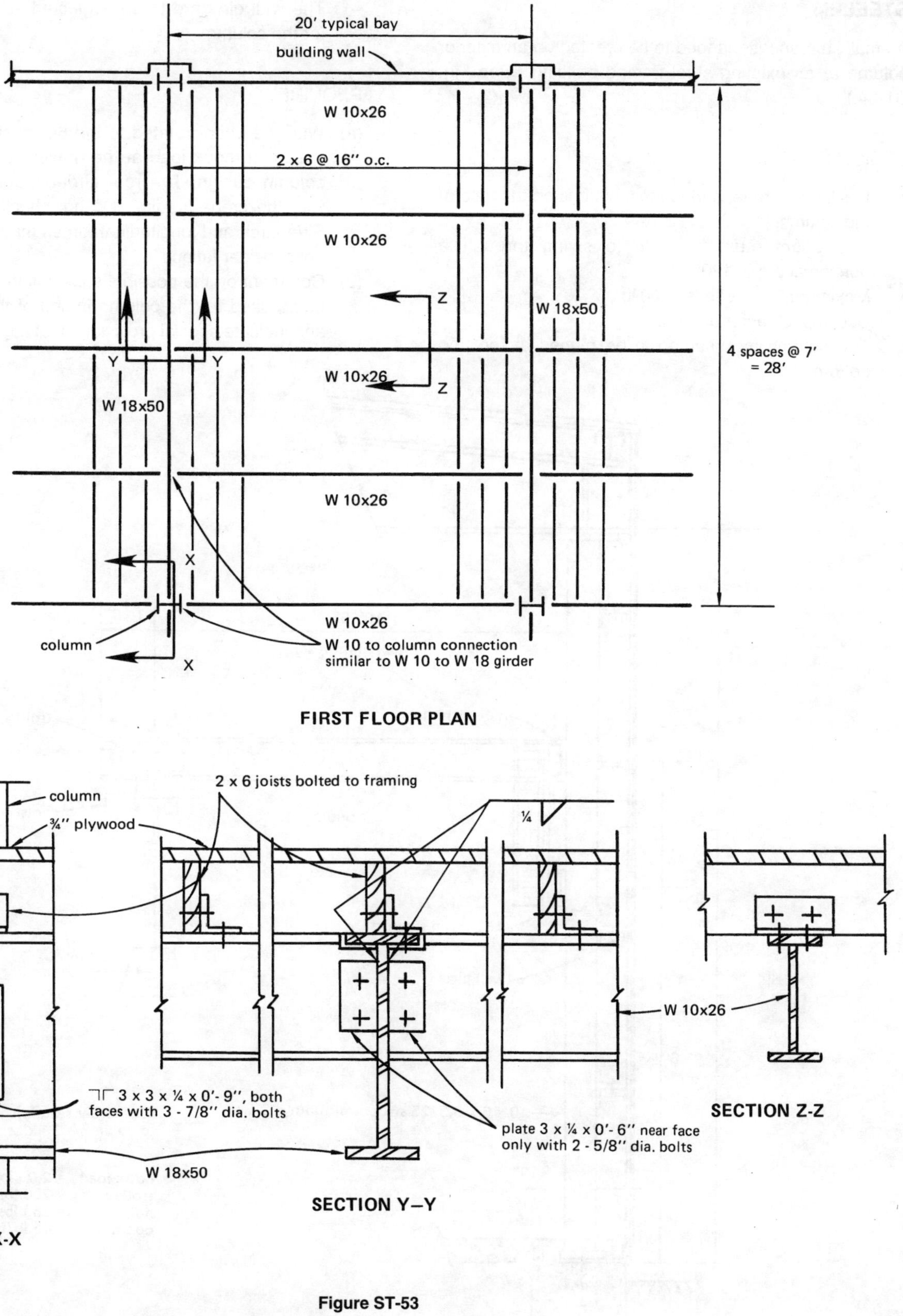
20′ typical bay
building wall
W 10x26
2 x 6 @ 16″ o.c.
W 10x26
Z
W 18x50
Y
Y
W 10x26
Z
4 spaces @ 7′ = 28′
W 18x50
W 10x26
X
W 10x26
column
W 10 to column connection similar to W 10 to W 18 girder
X
FIRST FLOOR PLAN
column
¾″ plywood
2 x 6 joists bolted to framing
¼
⌐⌐ 3 x 3 x ¼ x 0′- 9″, both faces with 3 - 7/8″ dia. bolts
W 18x50
W 10x26
plate 3 x ¼ x 0′- 6″ near face only with 2 - 5/8″ dia. bolts
SECTION X-X
SECTION Y–Y
SECTION Z-Z

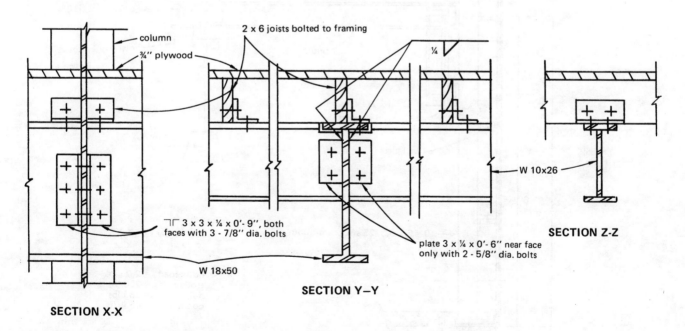

Figure ST-53

STEEL-54

A small jib crane is planned to be erected on an exterior column at an existing steel framed building. (See Fig. ST-54.)

DESIGN CRITERIA

- For the jib crane, surface mounting is on the face of the column.
- The boom must be able to swing through a horizontal arc of 120°.
- Maximum lifted load is 2000 lb.
- Use impact factor of 25%.
- The column is assumed to be pinned at top and bottom.
- The wall girts are to be neglected as providing bracing for the column.

REQUIRED

(a) With the boom rotated to the 60° position and lifting its maximum hook load at the maximum reach, check the column design. The roof girder reaction is concurrent with the crane loading. Do not check the connections. Side lurch and longitudinal forces may be neglected. Do not consider fatigue.

(b) Comment on the possible substitution of a rolled section being used for the column in lieu of the indicated tube, and indicate special provisions that might be necessary.

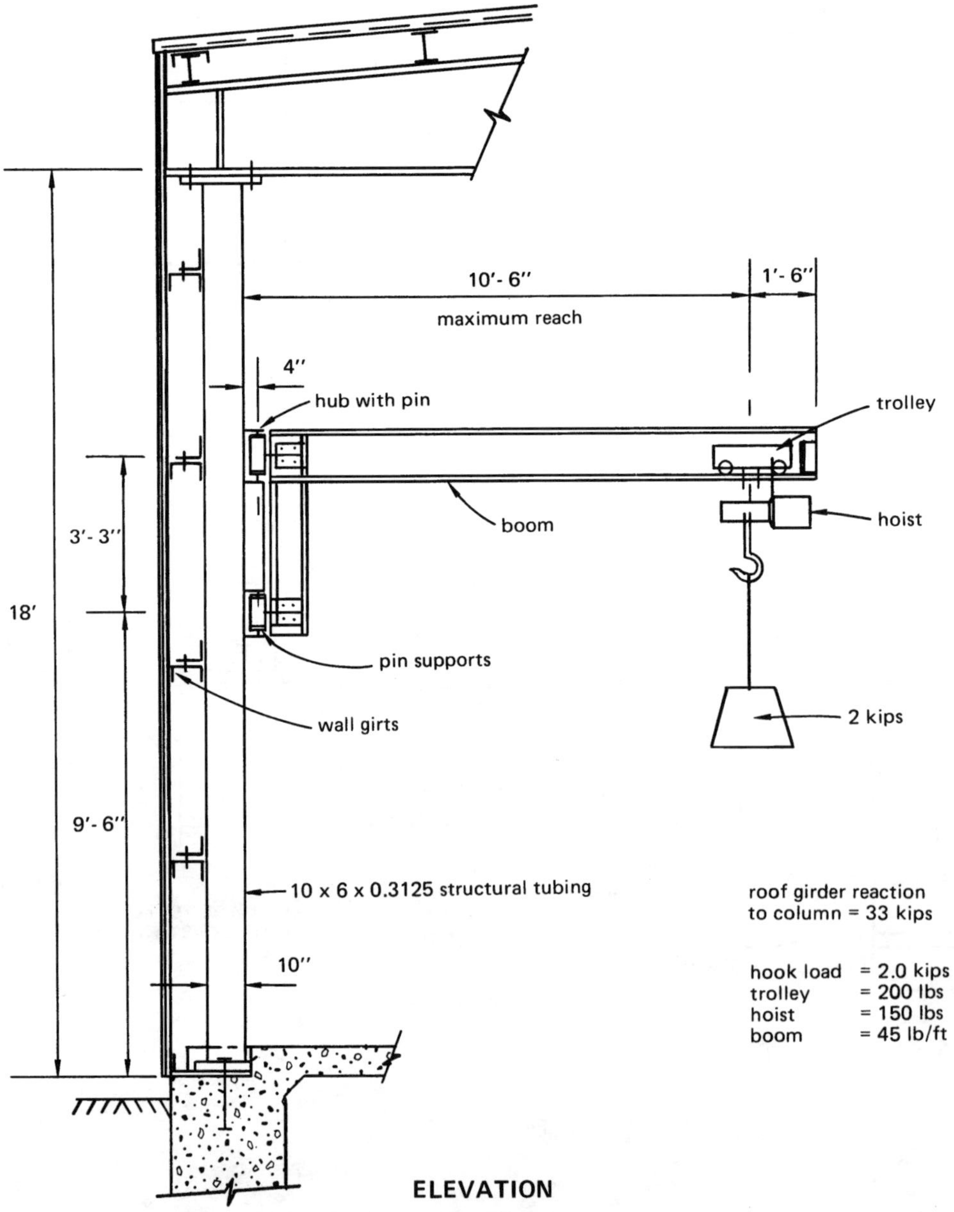

Figure ST-54

(continued on next page)

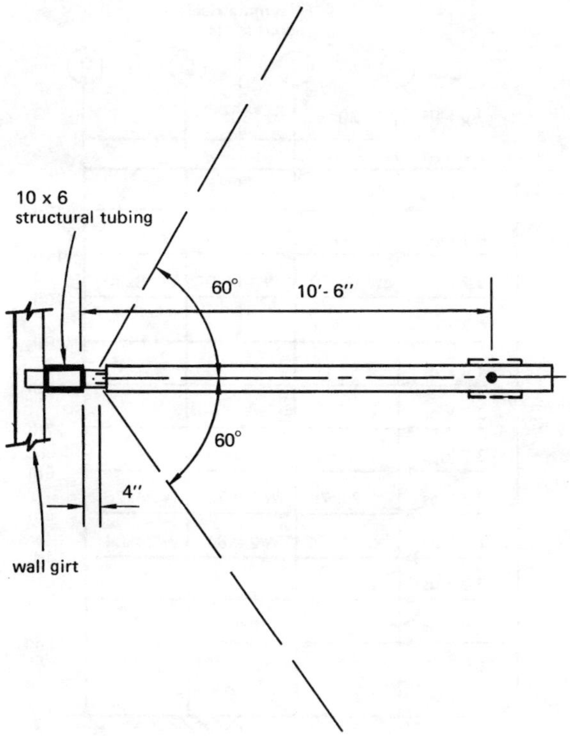

PLAN

Figure ST-54 (cont.)

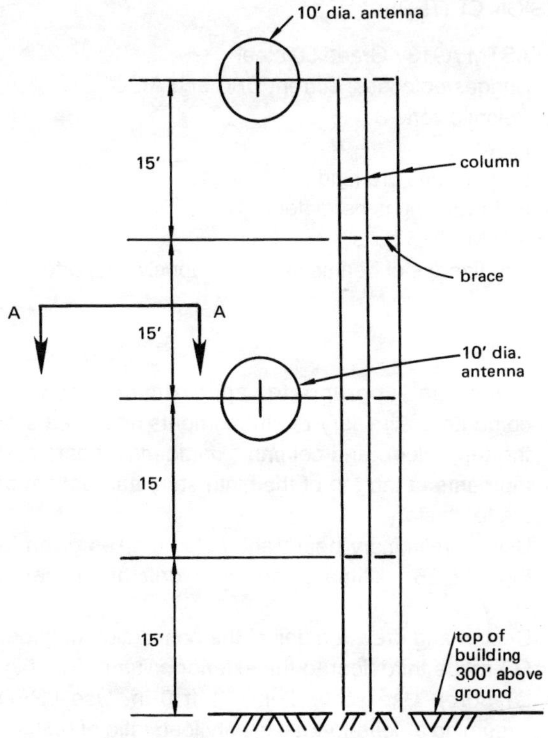

ELEVATION OF TOWER

STEEL-55

A tower for a microwave antenna is placed on top of a high-rise building. See elevation in Fig. ST-55. Two 10 ft diameter antennae are shown. The axis of each antenna must not rotate more than 1° about the vertical tower axis during wind loading.

DESIGN CRITERIA

- Basic wind speed = 70 mph
- Exposure C
- Shape factor of antenna = 1.0
- Assume the building does not move.
- Assume horizontal brace members at 15 ft centers are rigid for bending and torsion.
- Assume antenna brace is rigid.
- Assume wind effect on tower does not contribute to rotation.
- Assume shear and axial deformations are insignificant.

REQUIRED

Design a W-shaped column for the tower. All four columns are to be the same size. Show calculations.

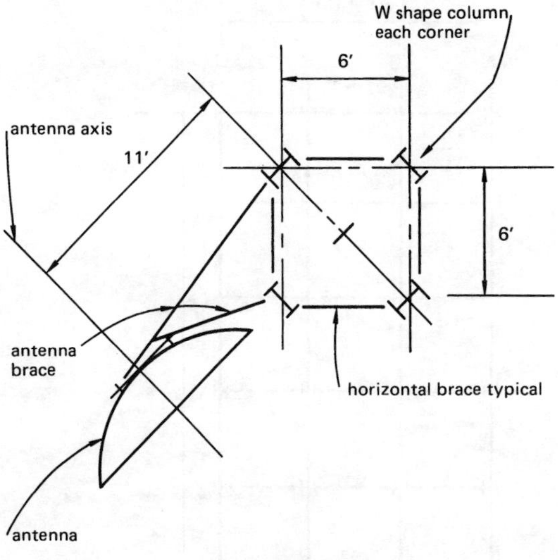

SECTION A-A

Figure ST-55

STEEL-56

A 12-story office building 90 ft by 150 ft in plan dimension is to be designed using a structural steel space frame to resist 100% of the lateral forces. A seismic analysis has been made and the F_x forces tributary to the typical interior frame are shown in Fig. ST-56. Forces shown are the working stress values.

DESIGN CRITERIA

- ASTM A913 / Grade 50 steel
- Suggested codes: current UBC and AISC
- Seismic zone 3
- I = 1.0
- Diaphragms are rigid
- Field welding is permitted
- ASTM A325 bolts
- Top flanges of beams are continuously supported.

REQUIRED

(a) Using an appropriate approximate method, compute preliminary beam moments and shears at the tenth floor, and column axial loads, shears and moments at the top of the ninth story and bottom of the tenth story.

(b) Using preliminary beam and column sizes given in Fig. ST-56, estimate the story drift at the tenth story.

(c) Design and draw a detail of the connection of girder G3 at the third floor to the exterior column. See Fig. ST-56(c). Use scale 1 in = 1 ft 0 in. Use LRFD assuming a dead-to-total gravity load ratio of 0.6.

NOTE: For part (a) present your answer in the form of free body and moment diagrams. Show all calculations.

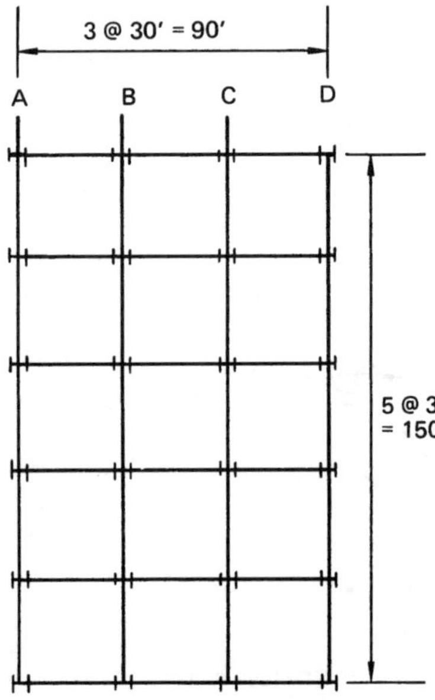

TYPICAL FLOOR PLAN

Figure ST-56

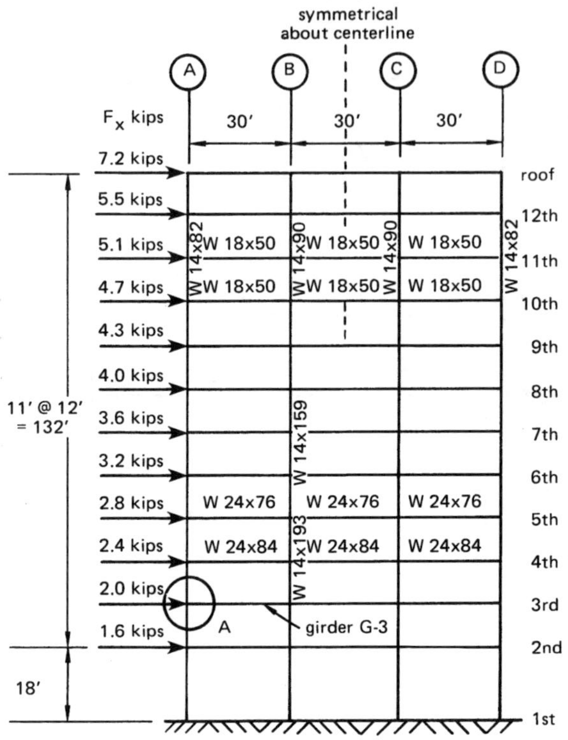

TYPICAL W-FRAME ELEVATION

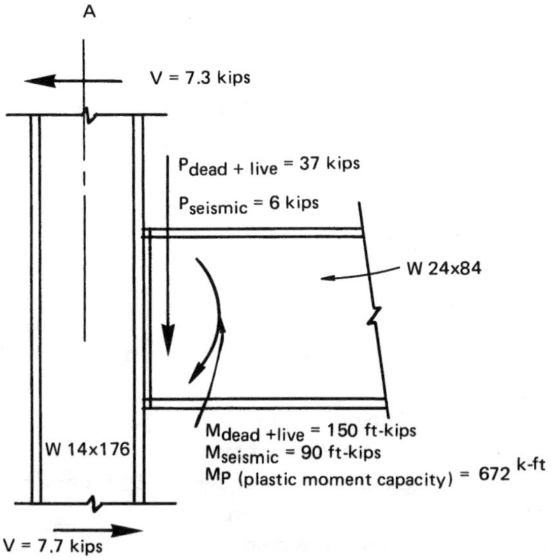

Figure ST-56 (cont.)

STEEL-57

A large vertical sign is shown in Fig. ST-57. The sign is cantilevered from the base, and the wind loading is perpendicular to the 16 ft wide sign face.

DESIGN CRITERIA

- Seismic zone 4
- Soil profile type S_D
- Near-fault factors of unity
- The sign is located in an 80 mph wind zone; exposure condition C
- ASTM A36 steel
- Design parallel to the face of the sign not required.
- State any assumptions made.

REQUIRED

(a) Assume that wind controls the design, and that allowable stresses may not be increased the usual one-third. Determine the size, length, and cut-off points for added steel plates for wind and dead load of the sign.

(b) Determine the period of the sign, and the seismic force. Now that you have determined the seismic force, which would control the design?

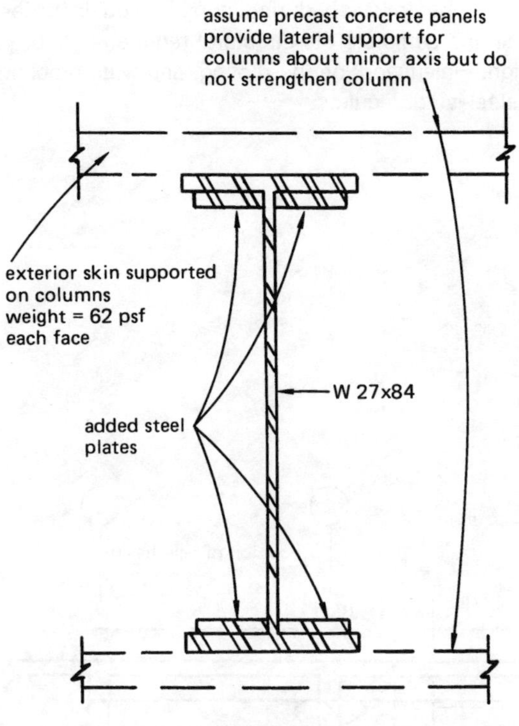

SECTION A

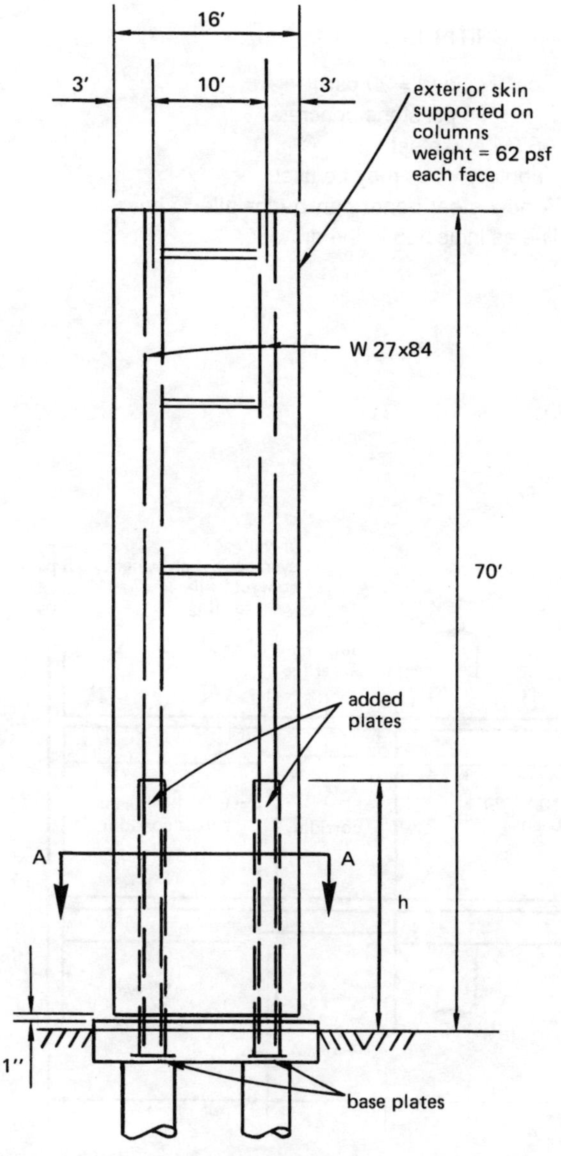

ELEVATION

Figure ST-57

STEEL-58

The sketch in Fig. ST-58 shows the end of one of the wings of a 2-story concrete building that is being used as a hospital. The column on line 2 must be removed in order to permit the installation of a new operating room. The existing ceiling line must be maintained, and the ceiling space cannot be used for the new structure.

The 8 in exterior wall on line 1 can support a new load, and the columns on line 3 can also support a new load. The deflection of the existing roof cannot be permitted to increase over the condition as it now exists.

DESIGN CRITERIA

- Roof live load = 20 psf
- f'_c = 3000 psi stone concrete
- ASTM A36 steel
- Tributary loads may be used.
- A new steel beam can be installed above the roof line as indicated in the drawing.

REQUIRED

(a) Design a new steel beam to support the existing roof in the region where the existing column will be removed. Show your calculations for moment, shear, and buckling. Size the beam, and provide any other necessary appurtenances.

(b) Design a suitable connection detail that will support the existing roof at line 2 after the new beam is installed and the column removed. Draw a detail of your design showing all the necessary structural elements.

(c) Describe in outline form a series and a sequence of steps to be followed during construction in order to install and complete the structural requirements of your design. Finish work on the roofing, and waterproofing of the steel is not required.

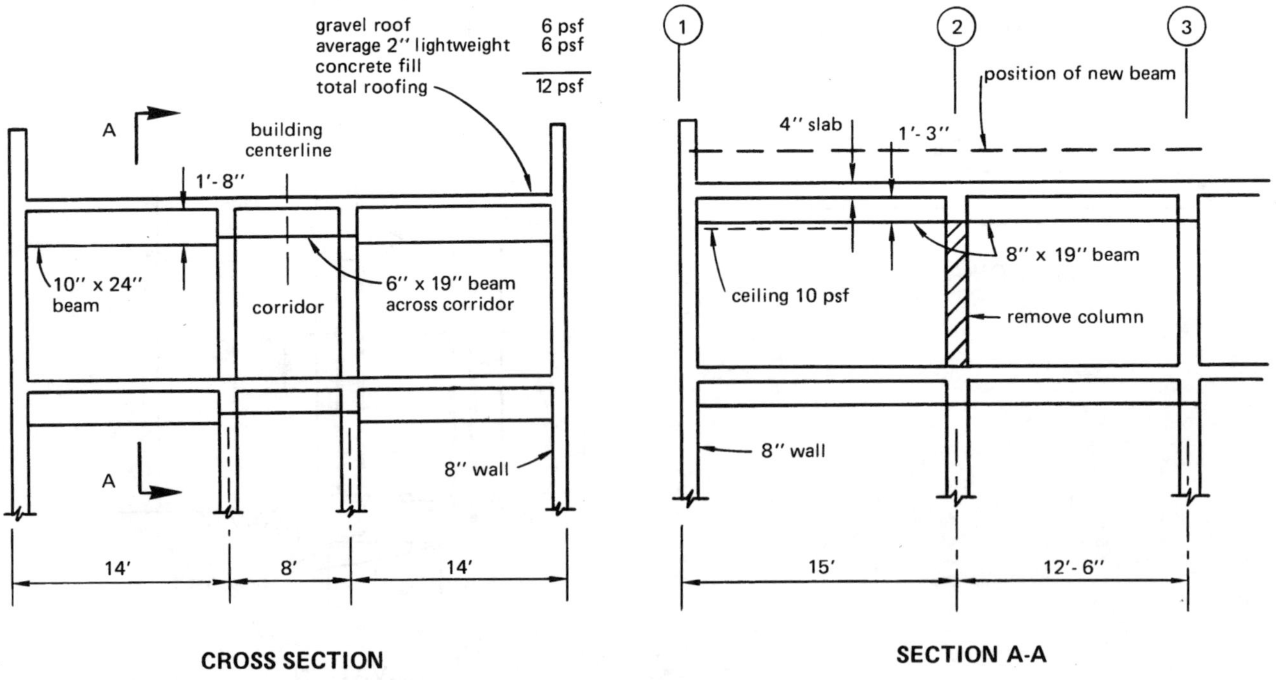

Figure ST-58

STEEL-59

A roof system is composed of a 4 in lightweight concrete slab supported on tapered steel girders as shown in Fig. ST-59. The slab and the girder are designed to act as a composite section and the girders are shored.

Prior to removing the shoring it was determined that the ridge beams are not designed to resist the upward component of the compressive stress in the concrete flange of the composite girder at the ridge.

DESIGN CRITERIA

- Girder moment at midspan = 2200 ft-kips
- Composite section I = 63,700 in^4 (Refer to section A-A for location of neutral axis.)
- ASTM A36 steel
- f'_c = 3000 psi lightweight concrete (W = 110 pcf, n = 14)
- ASTM A-307 existing bolts
- Welding was performed under continuous inspection with E70 electrodes.
- The concrete slab is sufficiently anchored to ridge beams and to top flange of tapered girders to transfer the upward component of the compressive stress in the concrete.
- Assume compressive stress in concrete is uniformly distributed within the effective width of concrete flange shown in section A-A.

REQUIRED

(a) Calculate the upward component of the compressive stress in the concrete. Neglect the weight of the concrete slab and all other gravity loads.

(b) Develop a plan to strengthen the ridge beams for load calculated in part (a). Draw details to show your design to an appropriate scale.

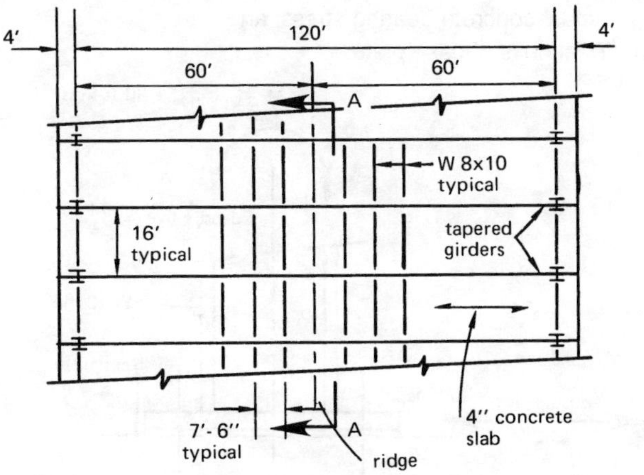

PARTIAL ROOF PLAN

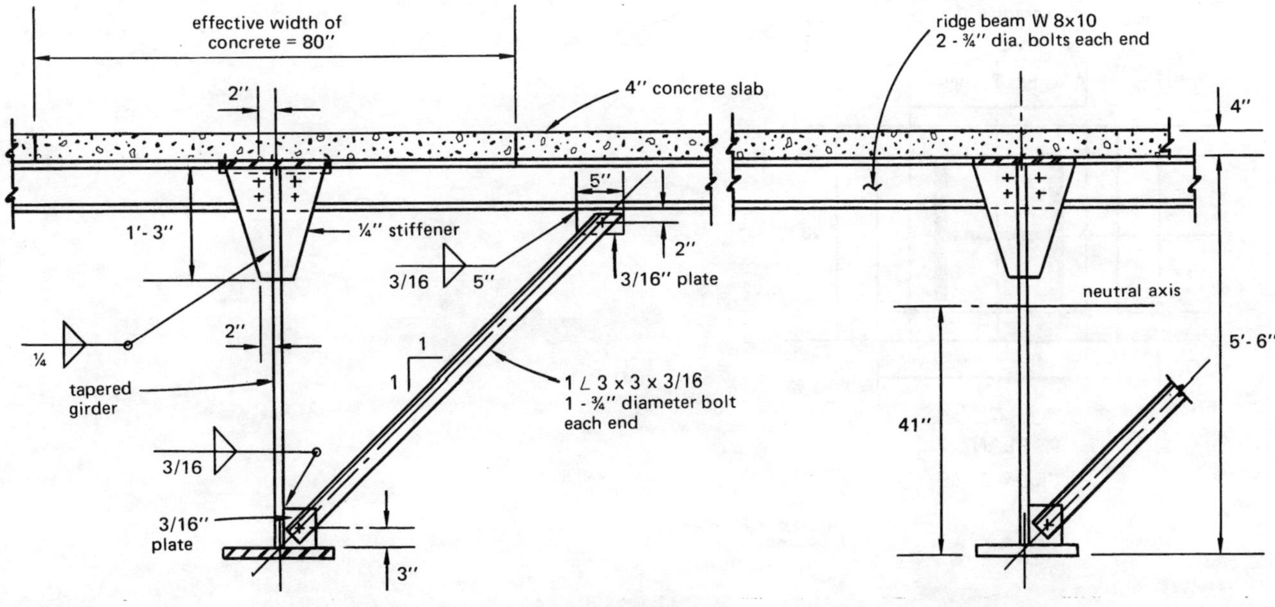

SECTION A-A
(existing condition)

Figure ST-59

STEEL-60

Fig. ST-60 shows a column base. Assume base plate adequately welded to column and anchor bolts fully developed in concrete.

DESIGN CRITERIA

- ASTM A36 steel
- ASTM A307 anchor bolts, UNC threads
- Allowable concrete bearing stress = 1125 psi

REQUIRED

Show all calculations for the following:
(a) Size of anchor bolts,
(b) Actual concrete bearing stress, and
(c) Thickness of base plate.

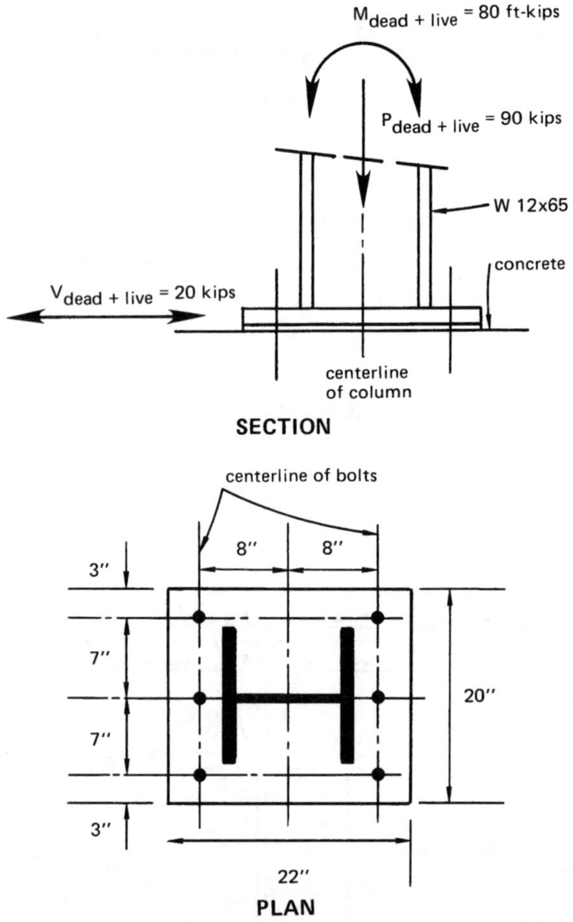

Figure ST-60

4 SEISMIC

SEISMIC-1

A 12-story building which is square in plan has a series of moment-resisting space frames around the perimeter of the building. The floor loads and the frame dimensions are shown in Fig. EQ-1.

DESIGN CRITERIA

- Seismic zone 3
- Importance factor, I = 1.0
- Soil profile type S_d

REQUIRED

(a) Calculate the story shear at each floor.

(b) Calculate the overturning moment at the first floor due to seismic loads only.

(c) Calculate the overturning moment at the fourth floor due to seismic loads only.

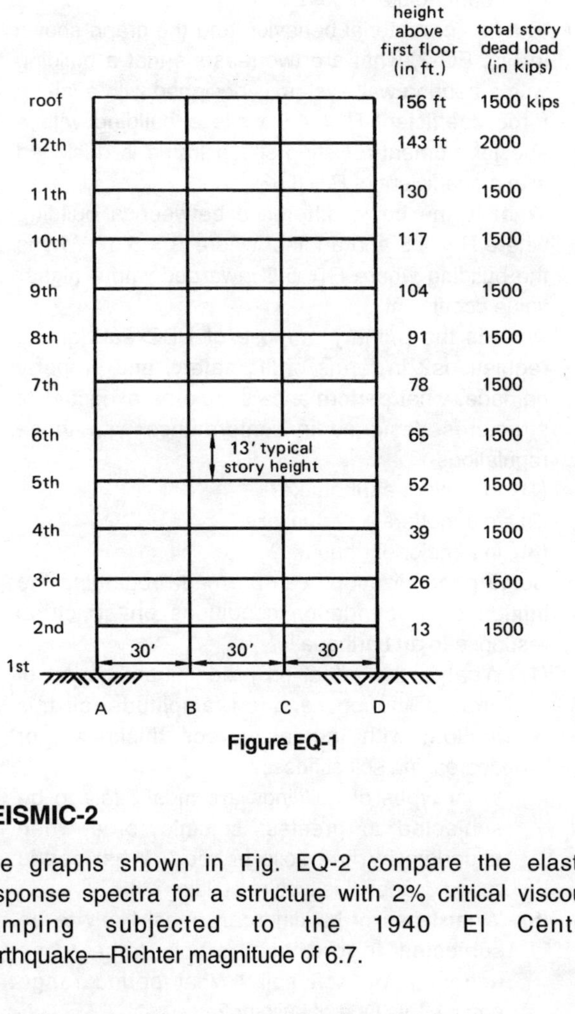

Figure EQ-1

SEISMIC-2

The graphs shown in Fig. EQ-2 compare the elastic response spectra for a structure with 2% critical viscous damping subjected to the 1940 El Centro earthquake—Richter magnitude of 6.7.

DESIGN CRITERIA

- Seismic zone 3
- Seismic design requirements for a moment-resisting frame building.

REQUIRED

(a) Explain what is meant by "elastic response spectra."
(b) In terms of material behavior, what is the single most important factor that reconciles the apparent discrepancy between earthquake force level and structural design level as illustrated by the graph in figure EQ-2?
(c) What are two other possible factors that might significantly modify the response of any given structure to an earthquake?
(d) What is the fundamental requirement for the survival of moment-resisting frame structures during major earthquakes if they are designed for UBC earthquake forces?
(e) In terms of material behavior, and the graph shown in Fig. EQ-2, what are two reasons that a building with a bearing wall system is designed with a lateral force coefficient R = 4.5 while a building with a special moment-resisting space frame is designed with a coefficient of R = 8.5?
(f) What is the basic difference between a building where R = 4.5 and another where R = 5.5? Why is the building where R = 5.5 rewarded with a higher value coefficient?
(g) What is the primary purpose of UBC earthquake regulations? In terms of life safety, and property damage, what performance should be expected of structures designed in conformance with these regulations
    (1) in a minor earthquake?
    (2) in a moderate earthquake?
    (3) in a major earthquake?
(h) Some generalizations can be made regarding the influence of foundation conditions on structural response to an earthquake.
    (1) What is the effect on predominant period of ground motion, and on amplitude of this motion, with increasing soil thickness, or decreasing soil stiffness?
    (2) What types of buildings are most likely to be subjected to greatest seismic force when underlain by firm soils or rock? What period range defines this type of building?
    (3) What types of buildings are most likely to be subjected to greatest seismic force when underlain by soft soil? What period range defines this type of building?

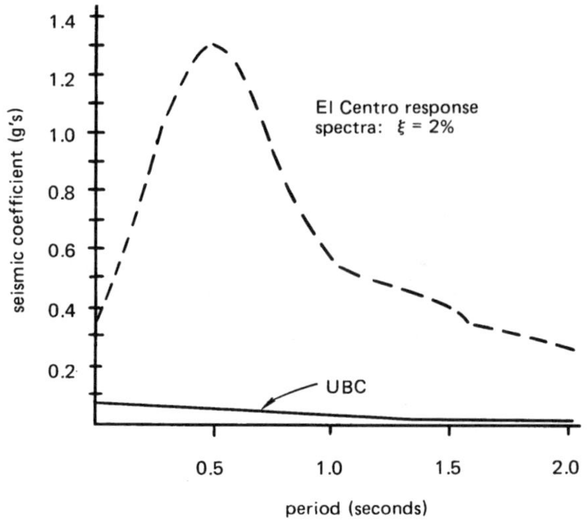

Figure EQ-2

SEISMIC-3

A typical floor framing plan for a 3-story building is shown in Fig. EQ-3. The floors are reinforced concrete slabs supported on steel beams, and the exterior walls are nonbearing filler walls.

Braced bents are provided at lines 1, 5, and D. The chevron bracing indicated is not symmetrical due to the position of the window openings. A 3-story rigid frame occurs on line A, which will be detailed as a special moment resisting frame. The relative rigidity (R) is shown on the plan.

The total weight (W) at each level, including framing and tributary walls, is shown on section X-X. The center of mass may be considered as 38.0 feet east of line A.

DESIGN CRITERIA

- ASTM A36 steel
- Seismic zone 3; soil profile type S_d
- Wind = 20 psf
- ASTM A325 bearing type bolts with threads excluded from shear plane
- No field welding permitted
- Columns are not stayed in weak direction.
- Neglect the effect of story drift.
- Omit vertical loads on braces.
- Torsional effect may be neglected, except as indicated for part (b).
- Use elastic analysis (no plastic design).

REQUIRED

Show calculations for all parts.
(a) What are the base shear, and the overturning moment in the north-south direction? What is your reason for the selection of the system quality factor (R) you have used?

(b) What is the applied horizontal moment at the second floor?

(c) Design double angle braces B1 and B2 as indicated on section X-X.

(d) Design and draw a detail of the brace to girder connection marked Y on section X-X. Assume the girder to be a W16. Use scale 3/4 in = 1 ft 0 in.

(e) Design column D4 at the first story for combined vertical and lateral loads:

vertical dead load = 240 kips
live load = 35 kips

(f) At column A4 in the first story the moment and axial load have been determined as:

M = 165 ft-kips
P = 140 kips

The frame is pinned at the base (K_x = 2.0). Verify the adequacy of a W 10 x 100 for combined vertical and lateral loads.

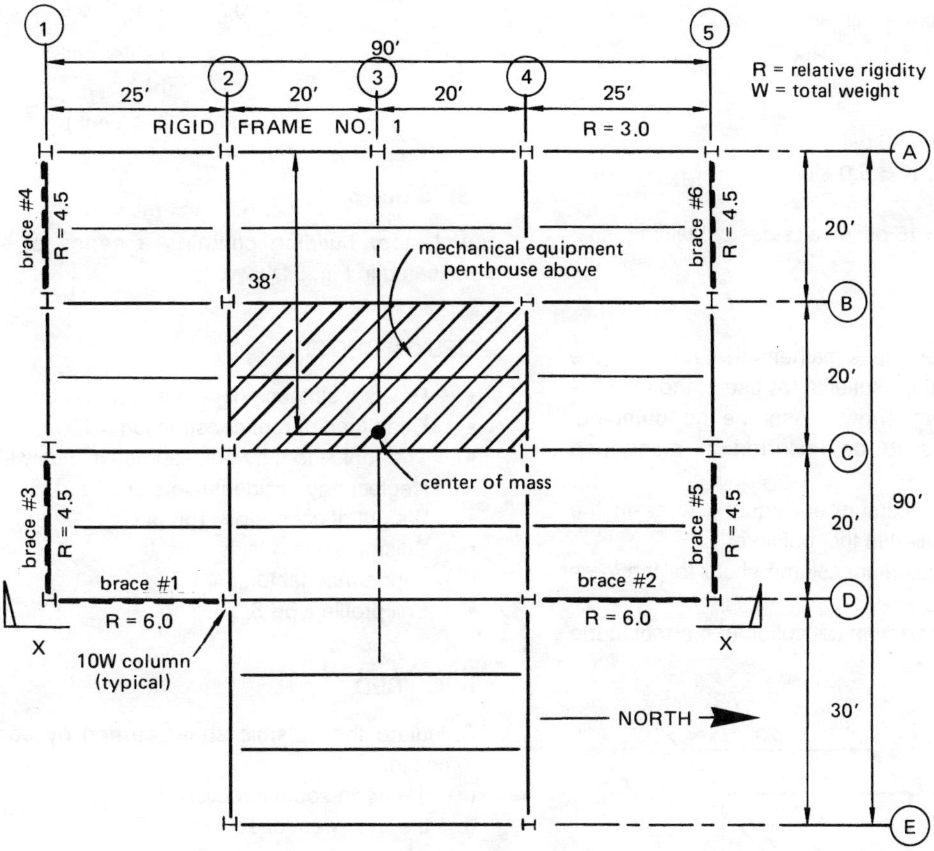

TYPICAL FLOOR FRAMING PLAN

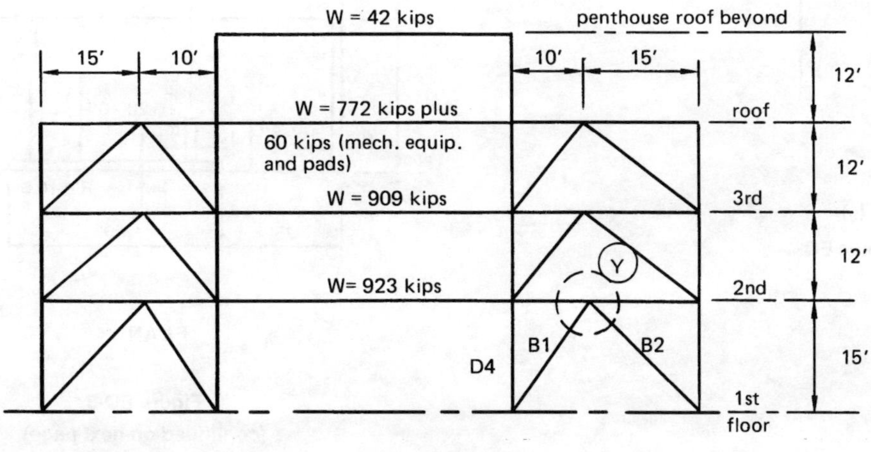

SECTION X-X

Figure EQ-3

SEISMIC-4

A symmetrical 1-story structure is shown in Fig. EQ-4(a). The roof girders have an infinite stiffness relative to the columns. Dimensions are shown on the figure.

The structure will be designed for earthquake ground motions that might be expected to occur at the site. The ground motions that may be expected are represented by the elastic response spectrum shown in Fig. EQ-4(b). Assume that the structure will remain elastic.

DESIGN CRITERIA

- Total roof weight = 40 psf
- Column stiffness K = 5.0 kip/in each way, for each column
- Weight of walls = 15 psf on all sides of building

REQUIRED

(a) Develop a lumped mass mathematical model of the structure that will be suitable for use in the analysis for the spectrum shown. Assume no damping. Explain what is to be included in mass and stiffness.

(b) Determine the maximum earthquake force in the X-X direction applied to the roof level.

(c) Determine the maximum column shear for the force in part (d).

(d) Determine the maximum deflection of the roof in the X-X direction.

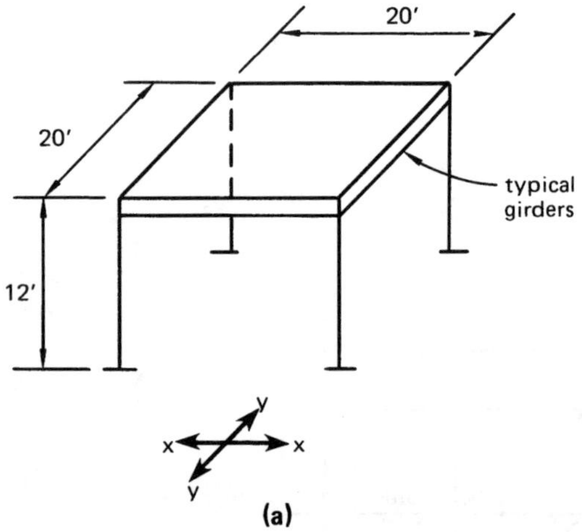

(a)

Figure EQ-4

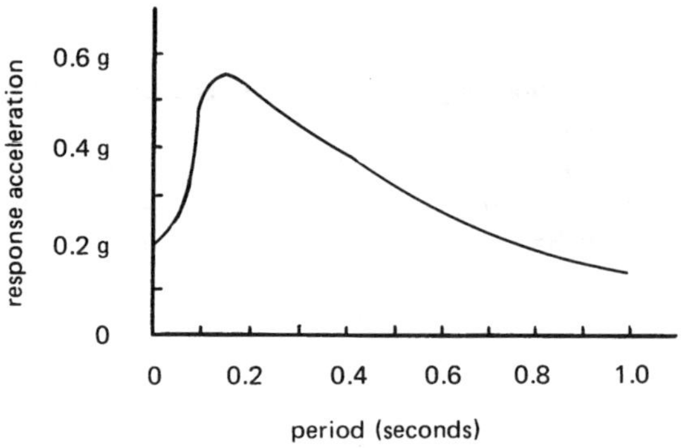

(b)

Figure EQ-4 (cont.)

SEISMIC-5

A 1-story building contains a series of shear walls as indicated in Fig. EQ-5.

DESIGN CRITERIA

- Relative stiffness (R) of each wall is given.
- Dead load of roof construction = 100 psf.
- The roof slab may be assumed infinitely stiff.
- Neglect any accidental torsion.
- Weight of concrete = 150 pcf
- Seismic zone 3
- Importance factor, I = 1.0
- Soil profile type S_d

REQUIRED

Calculate the seismic shear carried by each wall for the forces in:

(a) the north-south direction

(b) the east-west direction

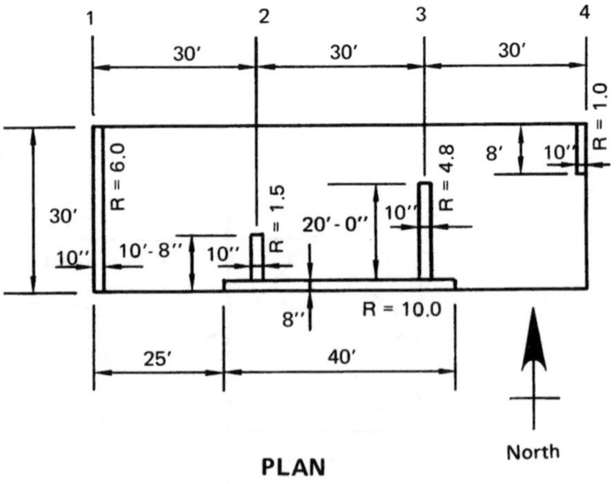

Figure EQ-5

(continued on next page)

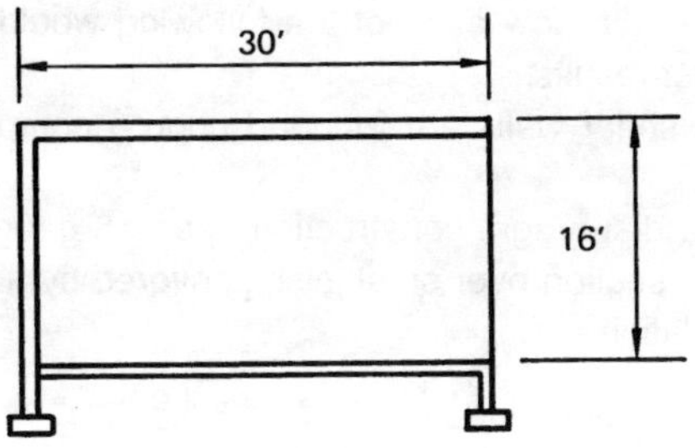

SECTION

Figure EQ-5 (cont.)

SEISMIC-6

A 2-story concrete building with shear walls at three sides and a concrete rigid frame at the fourth side (line J) is shown in Fig. EQ-6.

DESIGN CRITERIA

- Assume rigid diaphragms at the second floor and roof levels.
- Neglect any frame-action resistance involving the slabs and columns at all other lines.

REQUIRED

(a) Calculate the maximum lateral forces V_R and V_2 that result on the concrete frame due to the lateral loads shown in section A-A. Note that the center of mass coincides with the centerline of the building for both directions. Neglect accidental torsion.

(b) Using the Portal Method find the moment and shear stresses in the first story column J3. Assume that the foundation at line J is sufficiently rigid to fix the column bases.

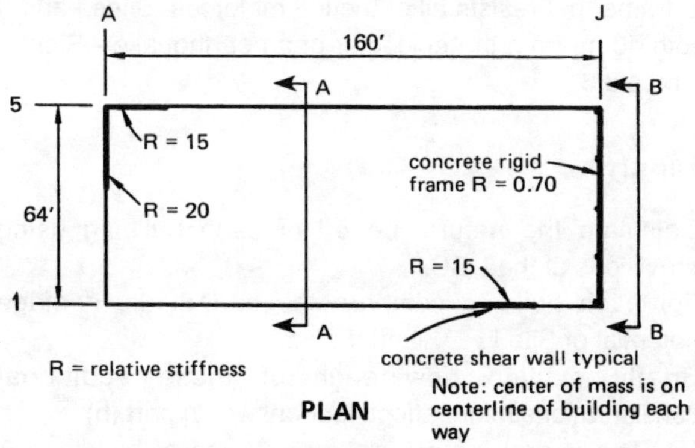

PLAN

Figure EQ-6

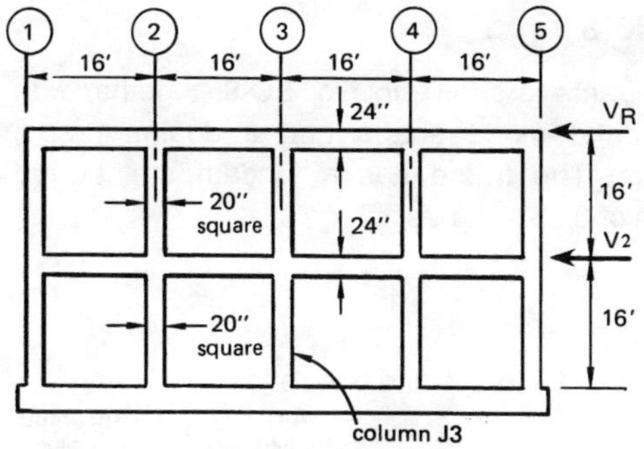

FRAME ELEVATION B-B

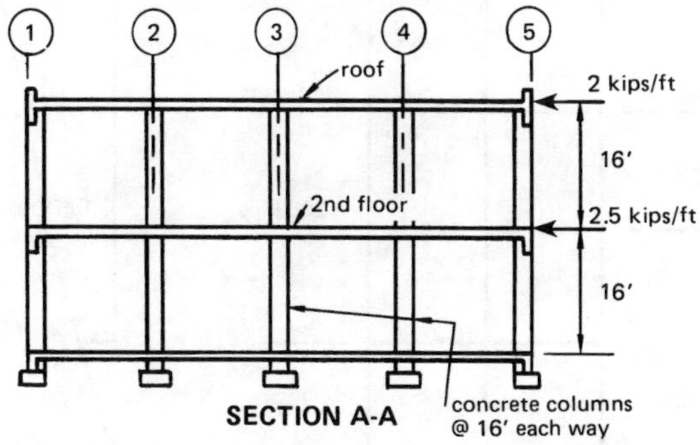

SECTION A-A

Figure EQ-6 (cont.)

SEISMIC-7

REQUIRED

(a) Describe briefly the structural system that is identified as a bearing wall system. What is the primary distinguishing characteristic of a bearing wall system? What value is assigned as the horizontal force R in a bearing wall system structure?

(b) What is the basis of various *zones* with respect to earthquakes and earthquake loads? How many zones are recognized in the United States? How are they identified by number and damage?

(c) Earthquakes are measured by various scales. Name two of these scales and briefly describe the basis for the scale, range of various factors and general meaning of a given scale reading.

(d) What is the minimum requirement that a structural engineer must consider in connection with horizontal torsional moments and story shear of a building? How are negative torsional shears considered?

(e) Explain the term *critical damping*. How is the amount of damping related to the critical damping? In the normal range of accepted values, what are the maxima and minima of damping factors for building? What would be a normal range for steel buildings?

(f) What is meant by the term *confined concrete*?

(g) In a typical framed structure the period for the first mode of vibration is 1.5 seconds. How would the period be expected to change for the second mode? How would the period be expected to change for the third mode?

SEISMIC-8

A steel frame is shown in Fig. EQ-8, together with its fundamental mode shape and a design response spectrum. The frame has a fundamental period of vibration of 0.35 seconds.

REQUIRED

Determine story forces and base shear.

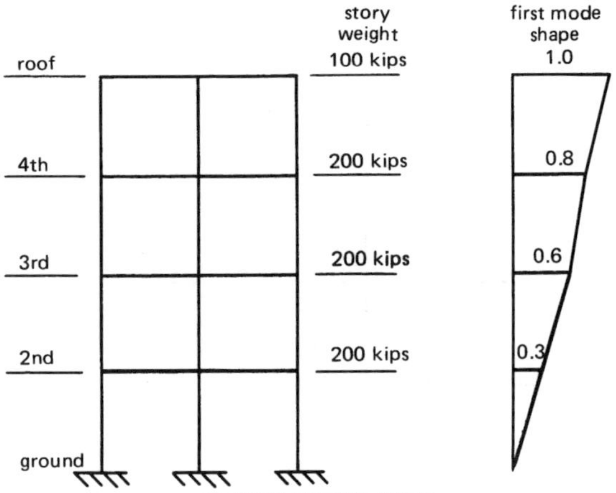

FRAME ELEVATION

(a)

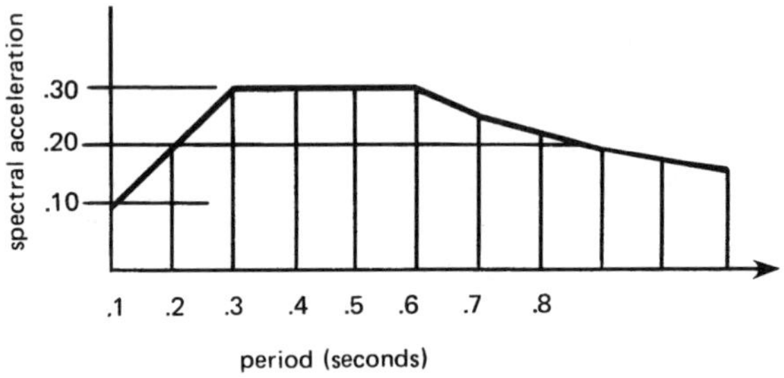

DESIGN SPECTRUM

(b)

Figure EQ-8

SEISMIC-9

A preliminary analysis is required of the 1-story building shown in Fig. EQ-9. It is braced in the north-south direction by shear walls A, B, and C whose relative rigidities when founded on rock are indicated. Bracing in the east-west direction is not shown. An earthquake force of 0.5 kip/ft is applied to the roof diaphragm. (On plan shown, R = relative rigidity.)

REQUIRED

(a) What is the shear (in kips) to be resisted by each wall if the roof diaphragm construction is

  (1) A 6 in reinforced concrete slab spanning concrete beams?

  (2) 1/2 in plywood nailed to wood joists?

(b) Briefly explain how each of the following would tend to affect your results.

  (1) The shear walls are founded upon a compressible soil.

  (2) Roof diaphragm construction is an 18 gauge metal deck section over steel joists, covered by fiberglass insulation.

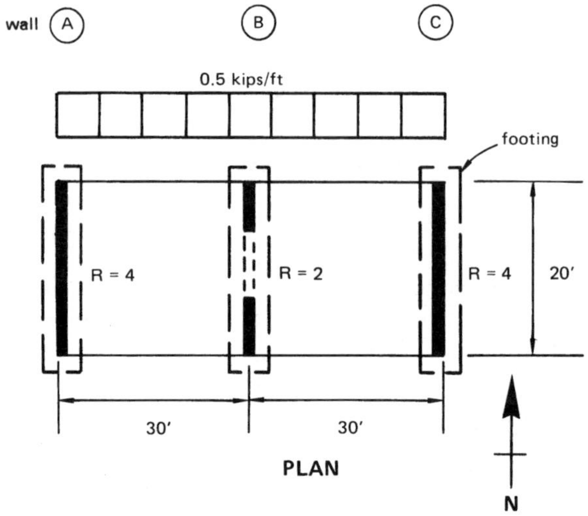

PLAN

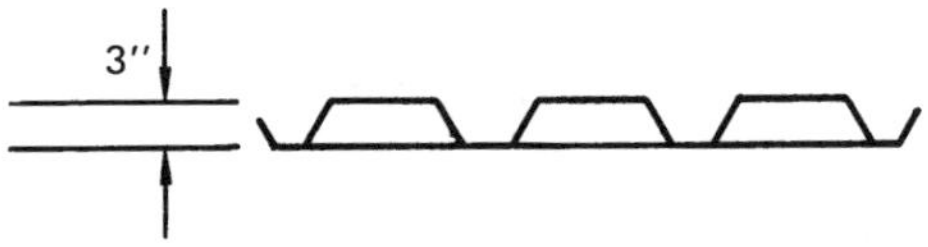

METAL DECK SECTION

Figure EQ-9

SEISMIC-10

Two square symmetrical buildings A and B (see Fig. EQ-10) have been designed and constructed in conformance with the current UBC. Lateral forces in building A are resisted by shear walls. Building B has a special moment-resisting space frame that resists all of the lateral forces. Sites I and II are both 40 mi from the epicenter of an earthquake—Richter magnitude 8.3.

REQUIRED

(a) Calculate the natural period of each building using provisions of the UBC.

(b) For each building compare the earthquake damage potential of Site I vs. Site II. Explain.

(c) Briefly explain how each of these additional considerations might affect your answer in part (b).

  (1) The lateral force resisting systems of buildings A and B are interchanged.

  (2) Building B has been previously subjected to a magnitude 5.2 earthquake. Broken glass was replaced and plaster cracks were patched. (NOTE: Assume building B as described in the first paragraph.)

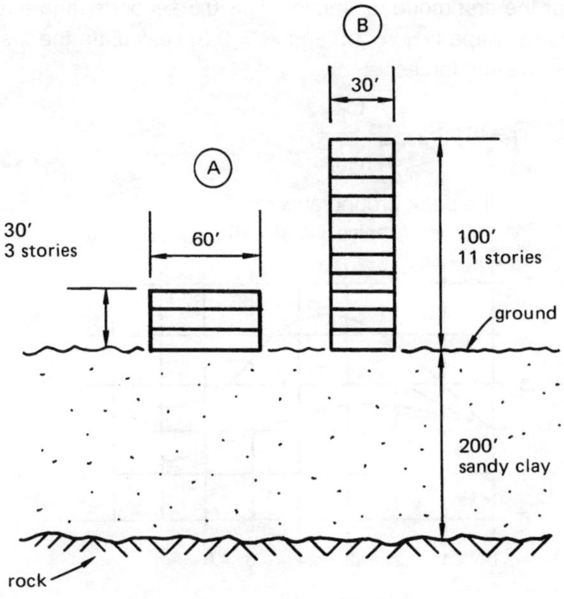

SITE I

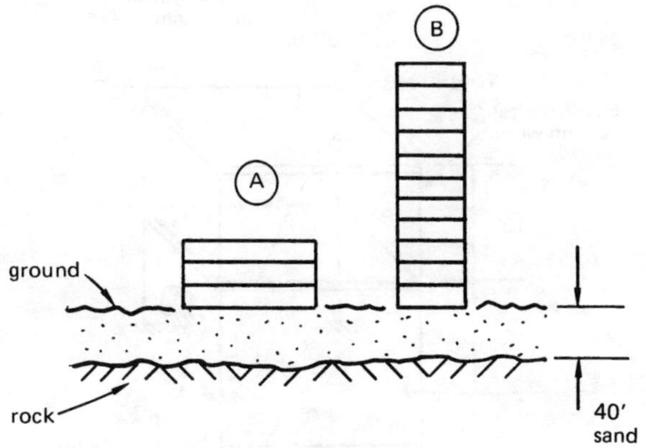

SITE II

Figure EQ-10

SEISMIC-11

A uniformly distributed seismic force of 52 kips is transmitted through a 6 in concrete diaphragm to shear walls of equal rigidity at the east and west ends as shown in Fig. EQ-11.

DESIGN CRITERIA

- Seismic zone 3
- $f'_c = 3000$ psi
- $f_y = 40{,}000$ psi

REQUIRED

For the north-south force only, use an approximate analysis to

(a) calculate the maximum diaphragm shear in psi.

(b) develop the steel reinforcement around the opening. Show the area of steel required and how it is to be placed around the opening. Provision for vertical loads is assumed as adequate for the problem. Show the steel required at the north-south edges of the diaphragm.

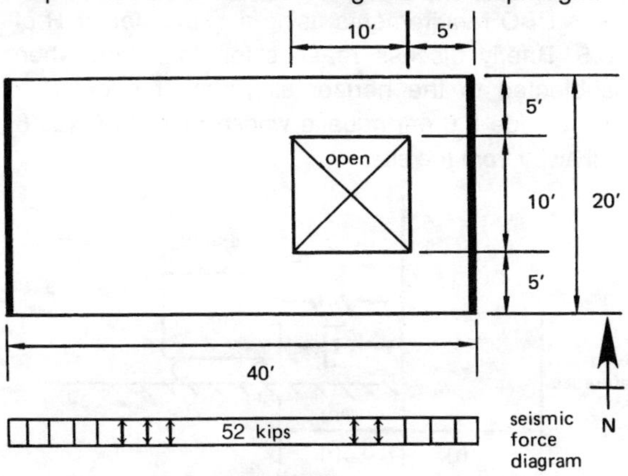

PLAN

Figure EQ-11

SEISMIC-12

REQUIRED

(a) What is an acceleration response spectrum?

(b) What is liquification? How might liquification affect buildings?

(c) What is Tsunami?

(d) A single degree of freedom system is shown in Fig. EQ-12(a). Write the equation of motion for this system when it is subjected to the earthquake ground acceleration Ug(t).

(e) How would a shear wall structure behave differently than a frame structure with respect to nonstructural seismic damage?

(f) A building in California is known to have performed in accordance with the primary purpose of the UBC earthquake regulations while being subjected to a given earthquake of magnitude 7.0. Briefly discuss three reasons why its performance in a later earthquake might not be acceptable.

(g) A 12-story building 120 ft in height has a substantially complete vertical load-carrying space frame with no calculated lateral resistance capability. All forces are resisted by concrete shear walls. What system quality factor (R) would you use under the following circumstances?

(1) Where frame columns are structural steel and are embedded in the shear walls.

(2) Where the frame columns are conventional reinforced concrete (nonductile) and are monolithic with the shear walls.

(3) Where the frame columns are ductile reinforced concrete and are monolithic with the shear walls.

Briefly explain your choice of R for each condition.

(h) The exterior wall elevation in Fig. EQ-12(b) of the symmetrical reinforced concrete building drawn shows all of its vertical and lateral force-resisting elements. The building was designed in accordance with UBC requirements using a quality factor R of 8.5. Briefly discuss reasons for its failure when subjected to the horizontal ground motion of a magnitude 6.6 earthquake whose epicenter was 6 mi away from the site.

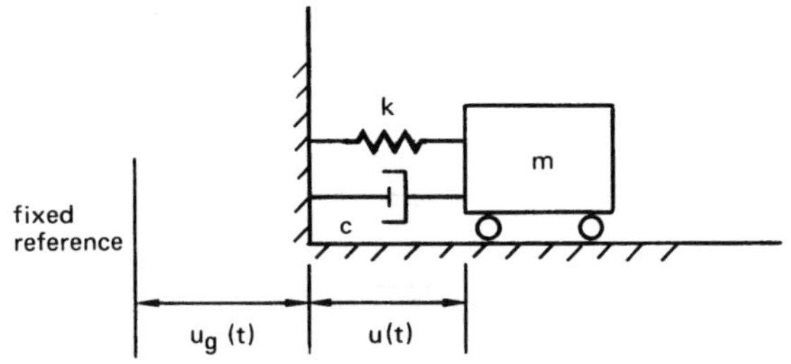

(a)

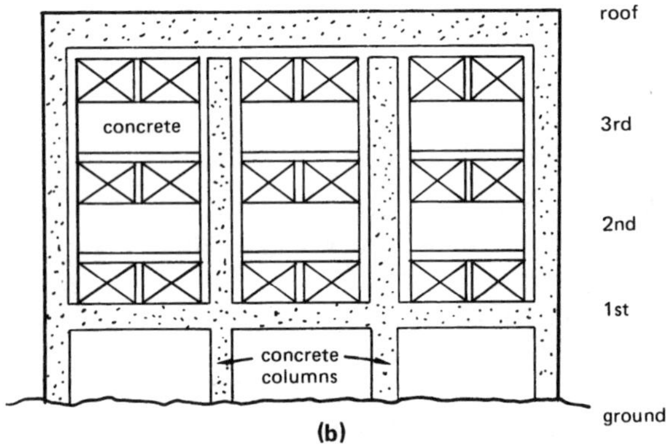

(b)

Figure EQ-12

SEISMIC-13

A 2-story steel frame structure is shown in Fig. EQ-13. Consider the structure to have 7% damping. Each story is known to deflect 0.1 in under a 10 kip story shear.

REQUIRED

For X-direction response:

(a) Determine the mathematical model for dynamic analysis, and summarize this in a sketch indicating story masses and stiffness.

(b) Sketch the approximate first mode shape, and calculate the fundamental period of vibration of the mathematical model. Do not use UBC equations to calculate period, but you may use recognized approximate methods such as the Rayleigh method.

Hint: $$T_1 \cong 2\pi \sqrt{\frac{\sum w\Delta^2}{g\sum f\Delta}}$$

(c) For the first mode, assuming T_1 = 0.5 seconds, and that mode shape is ϕ_2 = 1.0 and ϕ_1 = 0.67, calculate the first mode story forces.

Hint: $$F_i = m_i\phi_i S_a \frac{\sum m_i\phi_i}{\sum m_i\phi_i^2}$$

(d) What is the peak ground acceleration?

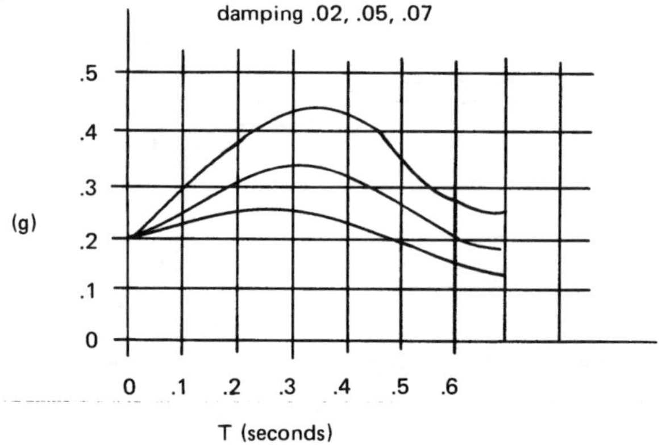

acceleration response spectra

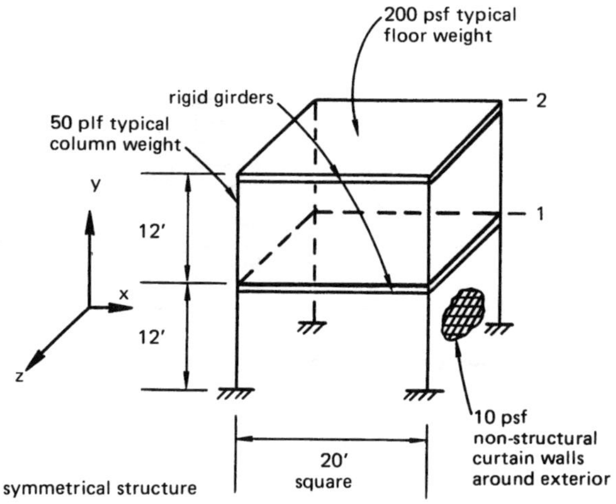

Figure EQ-13

SEISMIC-14

The unfactored seismic shear load resisted by the 10 in wall shown in Fig. EQ-14 is 105 kips. The structure does not have a 100% moment-resisting space frame.

DESIGN CRITERIA

- Assume footings are adequate to fix the base of the wall.
- f'_c = 3000 psi
- f_y = 40,000 psi

REQUIRED

(a) Determine the maximum ultimate shear stress in the wall.

(b) Determine the reinforcement required for element 1 due to the bending produced by the shear force in (a).

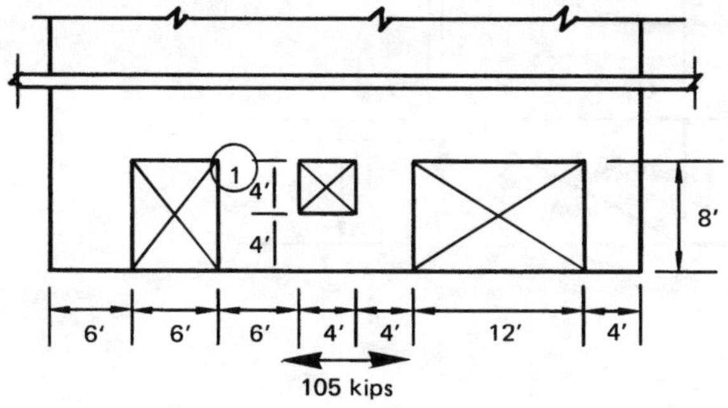

ELEVATION

Figure EQ-14

SEISMIC-15

A mezzanine floor used for the support of mechanical equipment is shown in Fig. EQ-15. The equipment will be anchored to the mezzanine floor to resist seismic forces. In order to properly evaluate the seismic forces to which the equipment will be subjected, it is necessary to determine the period of the supporting structure and to know the period of the equipment.

The mezzanine's horizontal, lateral force-resisting system in the north-south direction is the indicated horizontal truss system. The fundamental period of vibration of the mezzanine can be assumed to be that of the horizontal truss. Walls 5 and 6 are very rigid relative to the truss. The floor deck does not afford any diaphragm action. The lateral system for the east-west direction is not involved in this problem.

DESIGN CRITERIA

- The total contributory weight (dead load plus equipment) is 60 psf uniformly distributed over the entire mezzanine.
- The truss is of steel, $E = 29 \times 10^6$ psi.
- The lateral forces on the truss may be applied as concentrated forces at the panel points.
- All joints are pinned.
- The dimensions and notations given in the unit-load stress diagrams will be used in the problem solution.

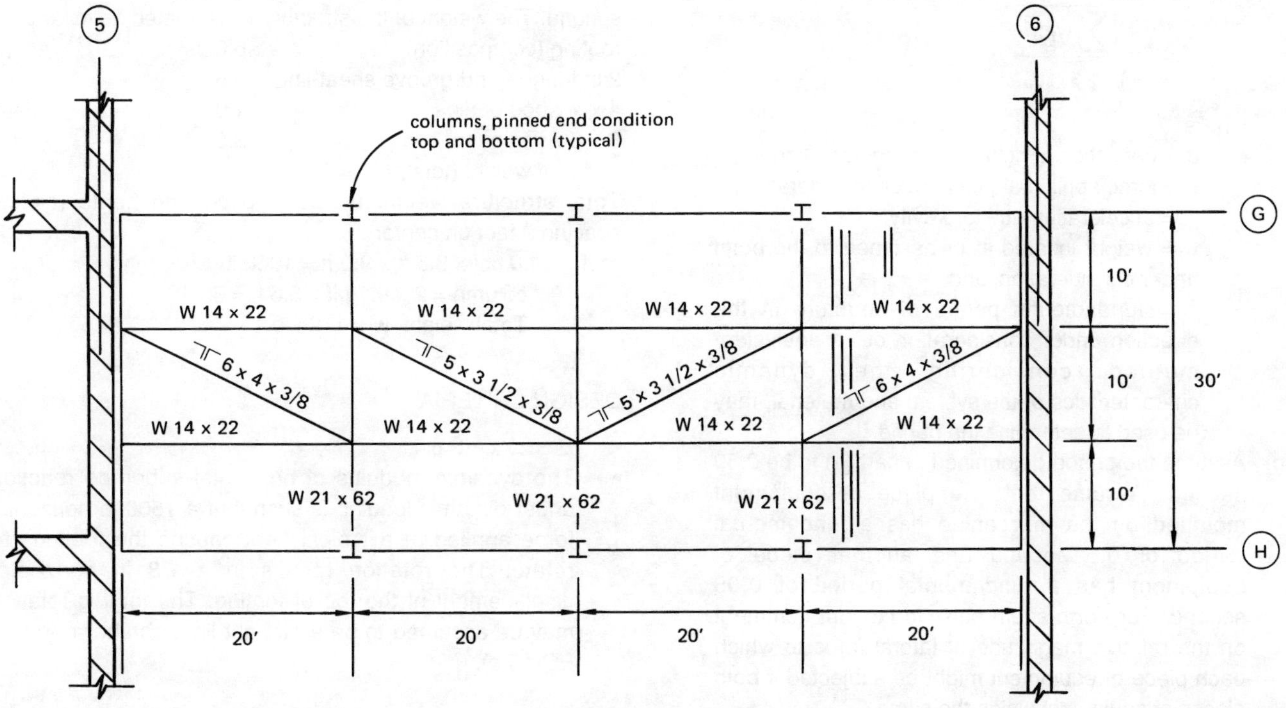

MEZZANINE PLAN

Figure EQ-15
(continued on next page)

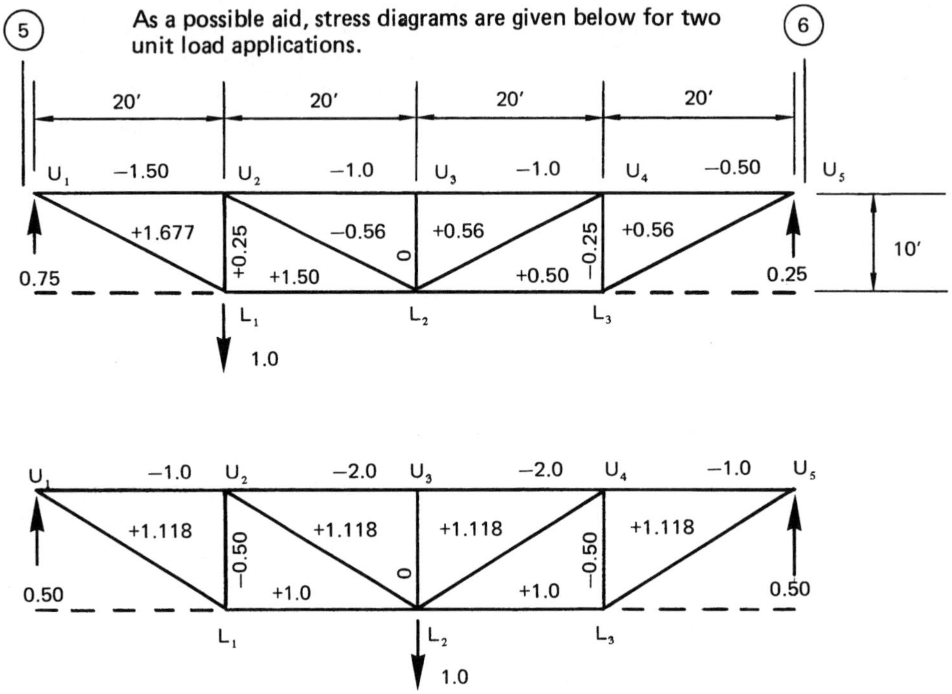

Figure EQ-15 (cont.)

(a) Determine the fundamental period of vibration of the horizontal truss in the north-south direction. Use the Rayleigh method, with

$$T = 2\pi\sqrt{\frac{\sum w_i d_i^2}{g\sum f_i d_i}}$$

where:

d_i = deflection at point under consideration

f_i = force applied at point under consideration,

g = acceleration due to gravity

w_i = weight located at or assigned to the point under consideration, and

T = fundamental period of structure in the direction under consideration or an equivalent method considering the dynamic characteristics of the system and material, may be used to determine the period.

(b) Assume the period determined in part (a) to be 0.50 second. Assume that one piece of equipment mounted on the mezzanine has a fundamental period of 0.5 second and another piece of equipment has a fundamental period of 0.05 second. For north-south seismic motion, comment on the relative magnitude of lateral force to which each piece of equipment might be subjected if both pieces of equipment weigh the same.

SEISMIC-16

The small arcade structure shown in Fig. EQ-16(a) suffered damage in the San Fernando earthquake. Based on a simple, single degree of freedom, lumped mass model (see Fig. EQ-16(b)) with complete fixity at the base, the fundamental period of vibration was calculated as 0.31 second. The weight of the structure is estimated as follows:

roofing (composition)	6.0 psf
2 in tongue-and-groove sheathing	4.5
4 x 4 wood purlins	1.0
6 x 8 wood beam	2.5
unit roof weight (total)	14.0 psf

Total structure weight tributary to column bents (typical spacing 9 feet on center).

roof = 14.0 psf x 8.5 ft x 9.0 ft = 1060 lb

column = 2 x 9.1 plf x 3.6 ft = 65 lb

Total weight, w = 1125 lb

DESIGN CRITERIA

- $E = 29 \times 10^6$ psi
- The dynamic modulus of horizontal subgrade reaction under dynamic loading is such that a 1500 lb horizontal force applied at a typical bent causes the footings to rotate. The rotation results in a 1/8 in horizontal displacement of the top of footing. The footing rotation may be assumed to be a straight line, as shown in Fig. EQ-16(c).

REQUIRED

Taking into account the indicated foundation flexibility, what is the revised period of the simple lumped mass structure?

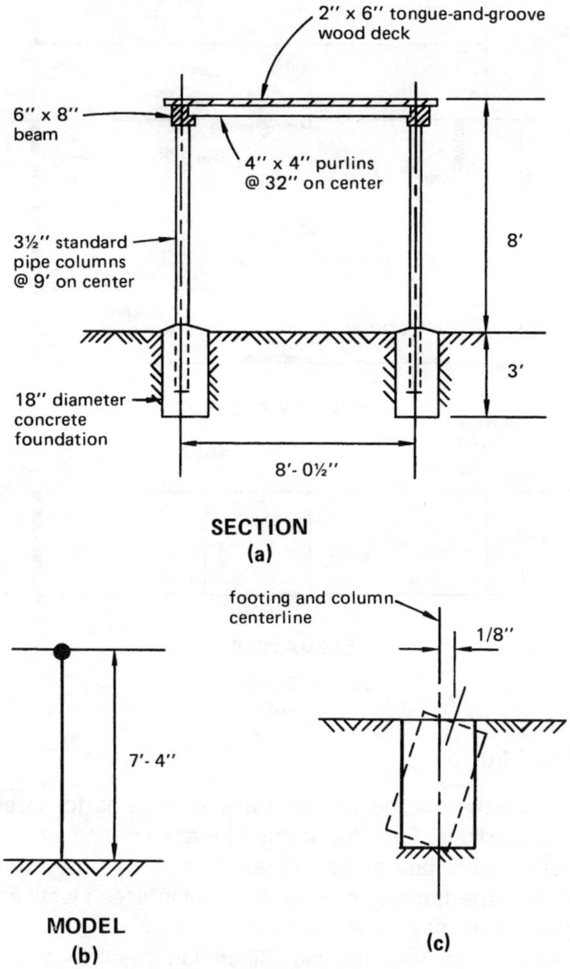

Figure EQ-16

SEISMIC-17

It is considered prudent to rely on dynamic analysis to evaluate the behavior of certain types of structures under seismic loadings.

REQUIRED

(a) List two types of structure that should be subjected to such analysis.

(b) What is meant by the damping characteristics of a structure?

(c) Briefly discuss the anticipated response of the following structures founded on the types of underlying soil condition described:
 (1) a tall building founded on soft soils, and
 (2) a low shear wall building founded on stiff soils.

(d) Which of the two buildings above do you think will be more significantly affected by a distant earthquake? Explain what the effect might be and why.

SEISMIC-18

REQUIRED

(a) (1) In a 10-story office building what change in the stiffness (if any) might be expected during extended high intensity ground motion? What effect accrues to the period (T)?
 (2) Show by diagram a typical code seismic load distributions on a building of 10 stories. Show a typical shear diagram for the same building.
 (3) Define critical damping.

(b) (1) Assume an optional choice between a 20-story building and a 3-story building on the same site. Why is it that the 20-story building may be designed for a lower base shear factor (assuming that the basic frame for each is similar)?
 (2) For the joints in a lateral load-resisting concrete frame why would special care be exercised to confine the concrete?
 (3) For a given earthquake motion what is an acceleration response spectrum?

(c) (1) What is the approximate proportion of inelastic strain to elastic strain if ASTM A36 steel fails in tension at 20% elongation?
 (2) A member in a building (vertical load only) was erected with ASTM A325 bearing type bolts loaded in shear. At the time of erection, the impact wrench used was miscalibrated and the applied torque was 50% of the torque specified. What is the resulting effect on the design strength of the connection? Explain.
 (3) Identify three of the most important factors that influence the magnitude of the available ductility in a reinforced concrete frame.

(d) (1) During the San Fernando earthquake in 1971 many engineered buildings were subjected to a base shear in excess of the code design requirement. Many of these buildings showed only limited distress. Give three reasons why.
 (2) The second story of a 4-story rigid steel frame is infilled with an unreinforced tile wall design. (See Fig. EQ-18(a).) The infill part was not considered in the lateral load design. Describe two important effects on the behavior of the building when subjected to earthquake motion.

(e) (1) What are two of the major factors that determine the dynamic characteristics of a building?
 (2) A building site may have a deep soft soil foundation condition. What might be the effect of this condition on the input ground motion to a tall flexible building and thus on the seismic response?
 (3) Fig. EQ-18(b) shows a proposed shear wall arrangement. Do you see anything wrong with this design? What provision in the Uniform Code would relate to this particular wall layout?

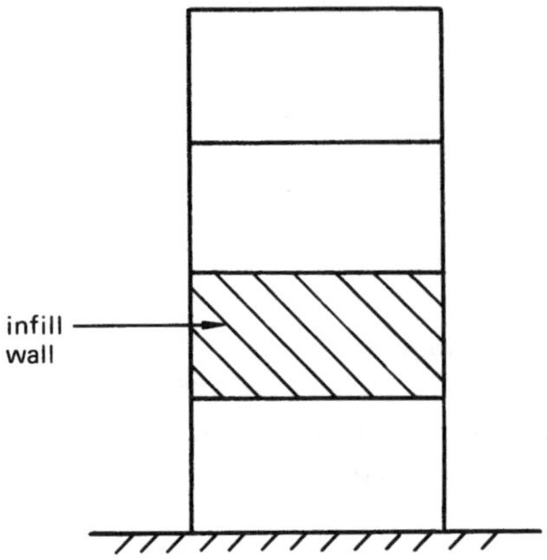

(a)

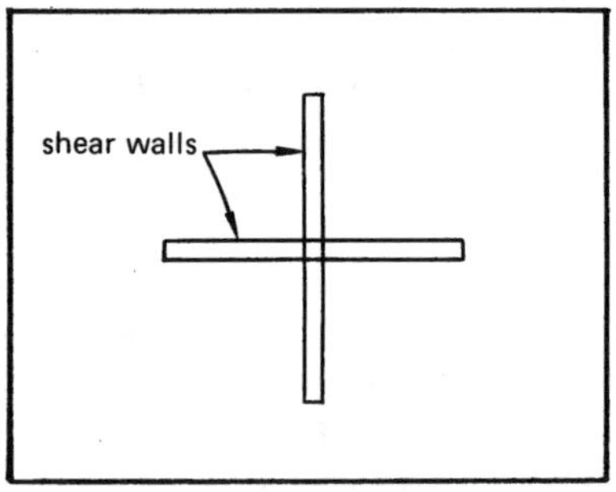

(b)

Figure EQ-18

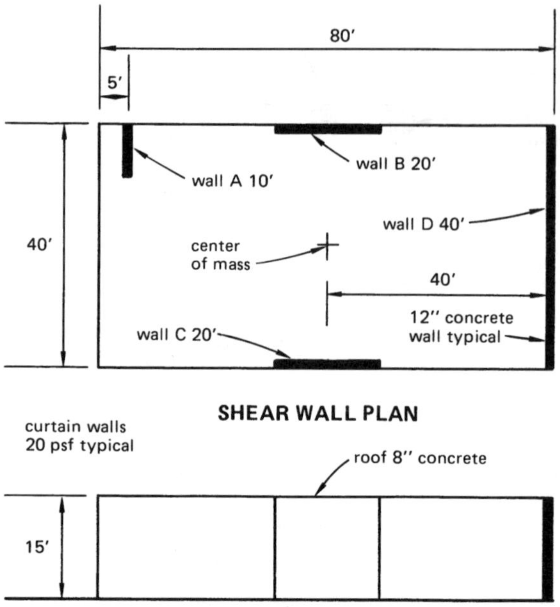

Figure EQ-19

SEISMIC-19

The building shown in Fig. EQ-19 is to be analyzed for an earthquake with motion in the north-south direction.

DESIGN CRITERIA

- All walls shown are bearing walls that serve as shear walls (12 in thick concrete).
- The roof is equivalent to an 8 in concrete slab.
- The center of mass is located on the plan.
- Assume regular weight concrete.
- Base shear coefficient = 0.1

REQUIRED

(a) Determine the force to wall A resulting from a north-south earthquake. Consider shear deflection only in walls.

(b) How would the distribution of lateral load be changed if wall A were made a masonry pier with a stiffness half that of concrete? No calculations are required for part (b).

(c) What effect does the torsional analysis have on wall D (refer to part (a))? No calculations are required for (c).

SEISMIC-20

Fig. EQ-20 depicts the various framing systems for three different buildings. Each has a lateral force-resisting system for north-south lateral forces as shown by the three elevations. The frames occur at the column lines identified. All other frames may be neglected.

All three buildings have the same floor plan dimensions, and all three have rigid floor diaphragms.

DESIGN CRITERIA

Building A

- Columns are fixed at the base.
- Tributary weight to roof = 800 kips
- Tributary weight to each floor = 1000 kips
- All columns meet the ductile moment-resisting space frame requirements.
- $E_c = 4 \times 10^6$ psi

Building B

- Steel frames are designed to meet ductile moment-resisting frame requirements.

Building C

- Steel frames are designed to meet ductile moment-resisting frame requirements.
- Shear wall meets the earthquake-resisting concrete shear wall requirements.

REQUIRED

(a) For each building, draw the first mode shape.
(b) Assume a strong motion earthquake that has a maximum ground acceleration of 0.4g with strong shaking for 30 sec, and explain the comparative performance of each structural system.
(c) Select design system quality factor (R) values for each building. Give reasons for your selection.
(d) What is the period for building A? Neglect axial stiffness of columns. Do not use the formulas for period in UBC.

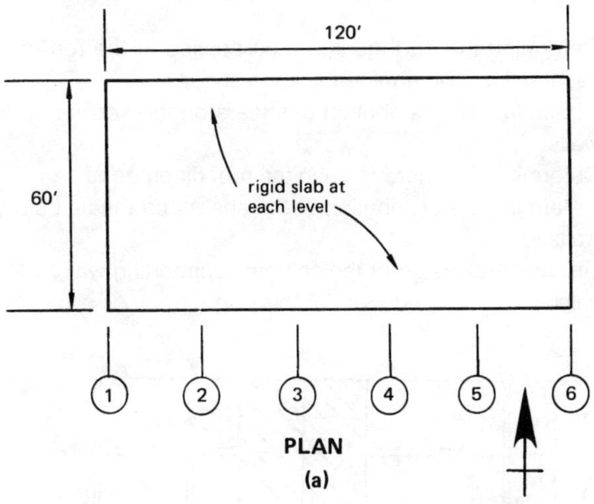

PLAN
(a)

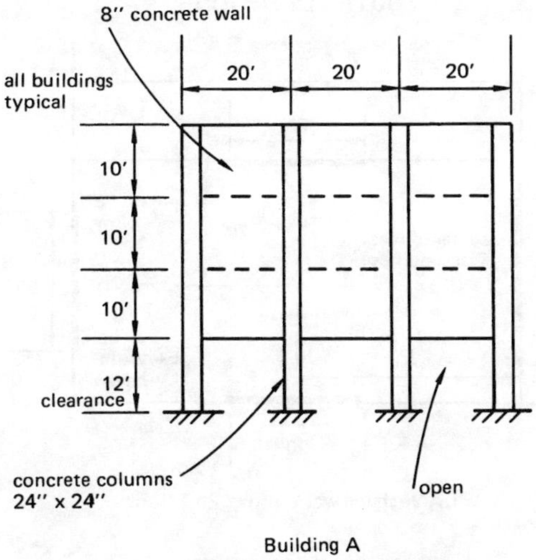

Building A
ELEVATION at lines 1, 6
(b)

Figure EQ-20

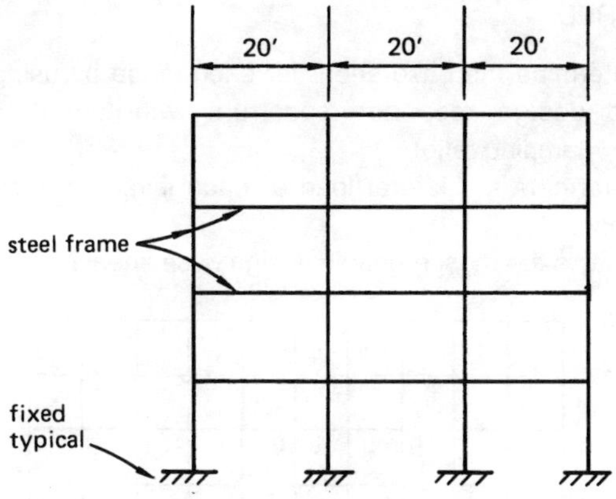

Building B
ELEVATION at lines 1, 2, 3, 4, 5, 6
(c)

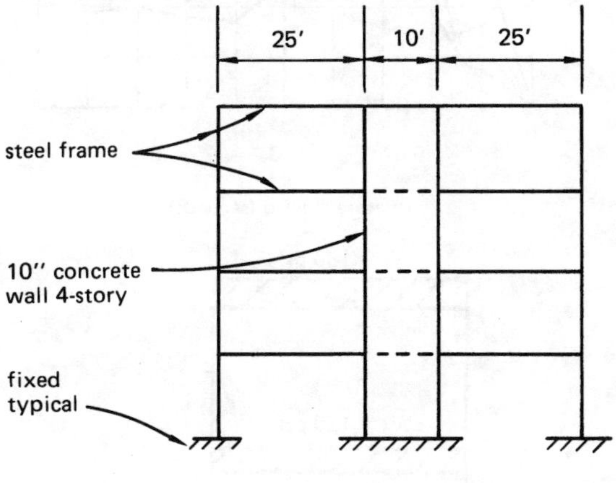

Building C
ELEVATION at lines 1, 6
(d)

Figure EQ-20 (cont.)

SEISMIC-21

Fig. EQ-21 represents a 3-story building with plan dimensions of 100 ft by 100 ft. The effective dead load on each floor is shown on the figure.

DESIGN CRITERIA

- Assume the following matrices.

$$\text{mode shape matrix } \phi = \begin{bmatrix} 2.860 & -0.657 & 0.378 \\ 1.959 & 0.725 & -1.610 \\ 1.000 & 1.000 & 1.000 \end{bmatrix}$$

$$\text{modal frequency matrix} = \begin{bmatrix} 15.1 \\ 38.5 \\ 61.7 \end{bmatrix} \text{ rad/sec}$$

REQUIRED

(a) Determine the base shear for each mode by using the design response spectrum with $\xi = 0.05$ (ξ = damping ratio).
(b) Determine the lateral load at each level for each mode.
(c) What is the most probable design base shear?

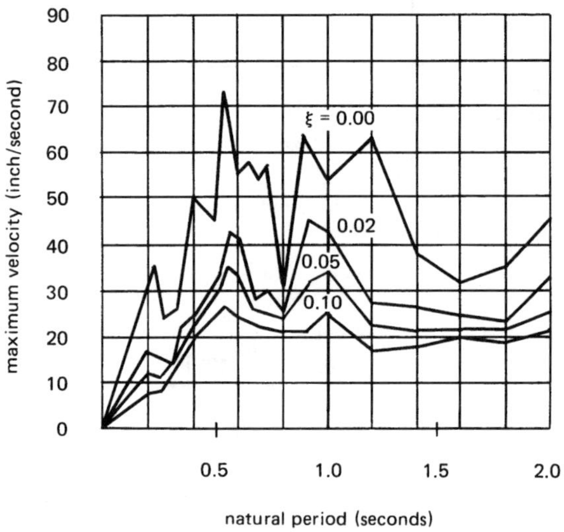

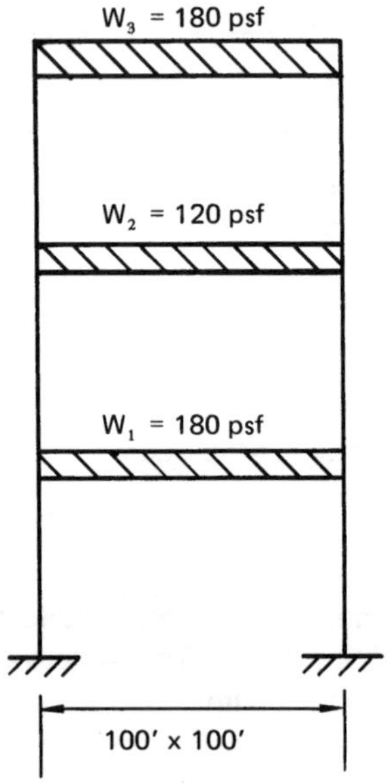

Figure EQ-21

SEISMIC-22

A 2-story building shown in Fig. EQ-22 has a plywood roof and a concrete slab on the second floor. The shear walls are all 8 in concrete, and also act as bearing walls.

DESIGN CRITERIA

- Roof dead load = 25 psf
- Second floor dead load = 120 psf
- These dead loads include nonstructural elements.
- Consider earthquake in the north-south direction.
- Neglect flexural deflection in the shear walls.
- Seismic zone 3
- Importance factor, I = 1.0
- Soil profile type S_A

REQUIRED

(a) Determine the earthquake force applied at the roof and second floor. Determine the base shear.
(b) Determine the distribution of shears on the second story walls.
(c) Determine the chord force in the roof diaphragm.
(d) Determine the distribution of the shears on the first story walls.
(e) Discuss the design of the columns supporting wall D. No calculations are required for this part.

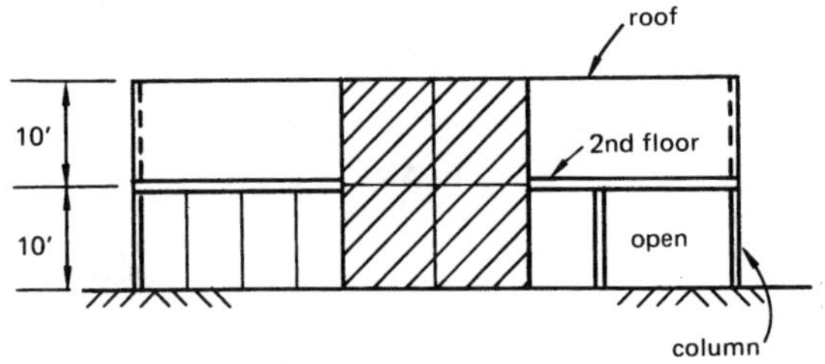

SOUTH ELEVATION

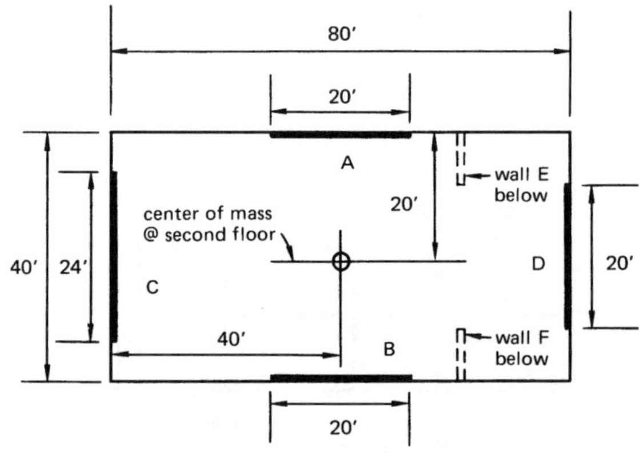

PLAN- shear walls above 2nd floor

Figure EQ-22
(continued on next page)

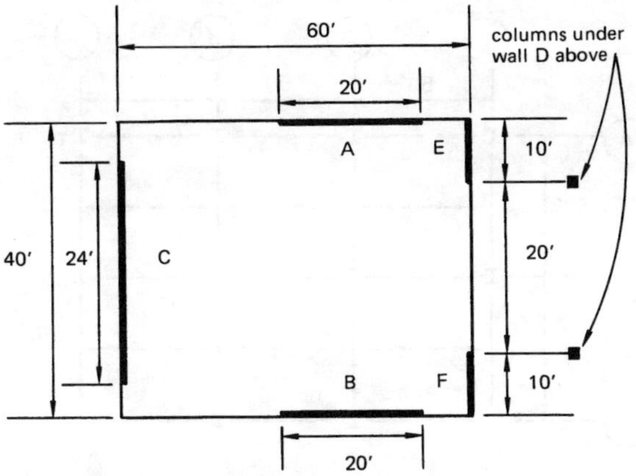

PLAN - shear walls at 1st floor

Figure EQ-22 (cont.)

SEISMIC-23

The parameters for the design of a new 1-story office building include the following.

The north, east, and west exterior walls of the structure are to be plywood shear panels. The south side is to be glass except for a single 4 ft long steel-braced frame located as shown by the roof plan in Fig. EQ-23.

DESIGN CRITERIA

- Seismic zone 4
- Soil profile type S_D
- Near-fault factors of unity
- No determination of characteristic site period
- f'_c = 3000 psi concrete
- Grade 60 reinforcing steel
- ASTM A36 steel tubes and all other structural steel
- E70 series welding electrodes
- The tributary weight W to roof diaphragm is 250 kips.
- Neglect the gravity loads for the design of the frame and its connections.
- The center of mass is located at the center of the structure.
- All indicated member sizes and reinforcement are to be assumed as adequate.

REQUIRED

(a) Determine the seismic force in the glulam beams at joints A and B.

(b) Prepare calculations for the complete detailing of the joints A and C. Assume the force in the glulam beam due to seismic to be 14.5 kips at joint A, and the total force delivered by the diaphragm to the braced frame to be 19.3 kips.

(c) Draw details of joints A and C to an approximate scale of 1 in = 1 ft 0 in. Show all relevant information including bolts, plates, sizes, welds, dimensions, etc.

(d) Provide a general comment on the available lateral force-resisting system. Would you recommend a change in the system?

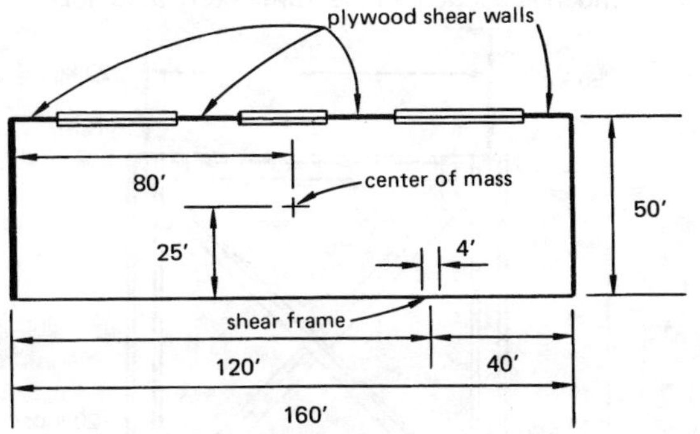

ROOF PLAN

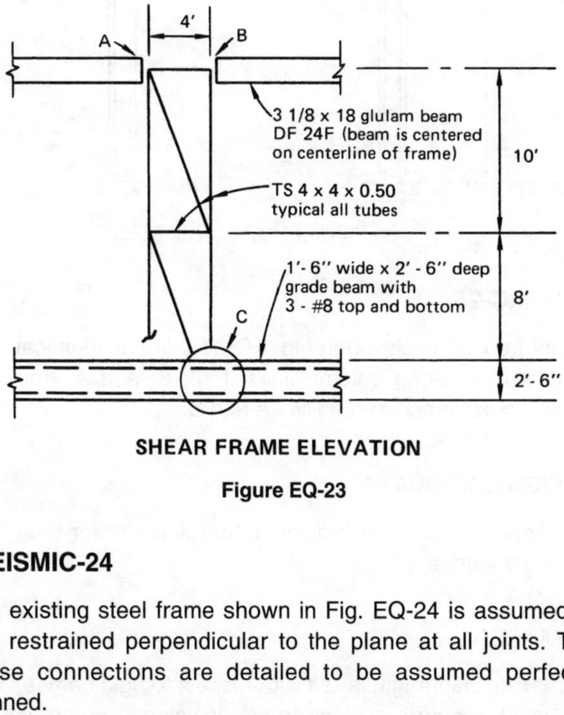

SHEAR FRAME ELEVATION

Figure EQ-23

SEISMIC-24

An existing steel frame shown in Fig. EQ-24 is assumed to be restrained perpendicular to the plane at all joints. The base connections are detailed to be assumed perfectly pinned.

DESIGN CRITERIA

- Seismic zone 4
- Tributary dead loads at joints A, B, C, and D are shown, the weight of the frame included.
- f_y = 50 ksi steel
- Characteristic site period is not available.
- Occupancy importance factor = 1.0

REQUIRED

(a) Calculate forces in the frame due to horizontal seismic loads, neglecting axial deformations.
(b) Are the W 10 x 45 columns adequate? Show calculations.
(c) How much live load can the frame carry at joint D?

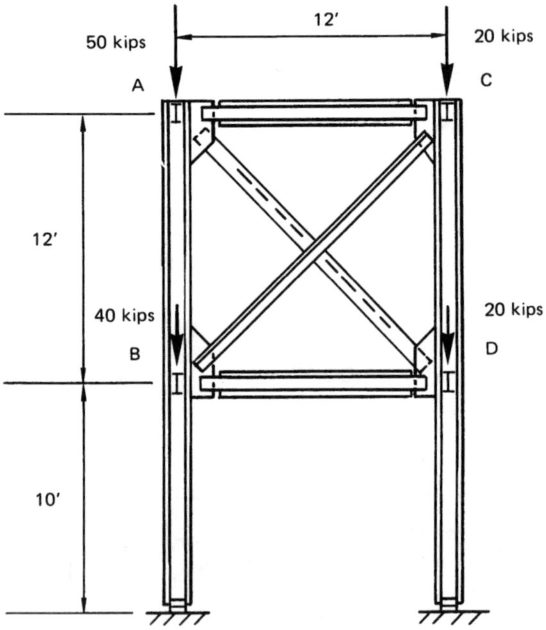

Figure EQ-24

SEISMIC-25

A roof framing is shown in Fig. EQ-25 that has identical rigid frames along column lines 1, 2, 3, and 4, and vertical cross bracing along line A and C.

DESIGN CRITERIA

- Assume the cross bracing is 100 times stiffer than rigid frames.

REQUIRED

(a) If roof diaphragm is stiff with respect to rigid frames, what proportion of north-south seismic shear is taken by the frame on line 4?
(b) If roof diaphragm is flexible with respect to rigid frames, what proportion of north-south shear is taken by the frame on line 4?
(c) If the frame on line 4 is modified so that its lateral stiffness is increased by a factor of 6, what proportion of north-south seismic shear is taken by the frame on line 4? (Assume the roof diaphragm is rigid.)
(d) Discuss roof diaphragm stiffness with respect to frame stiffness. What determines whether a diaphragm should be considered stiff or flexible?

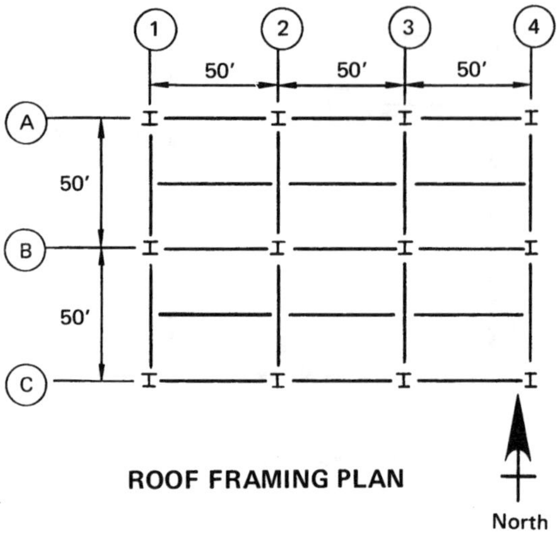

Figure EQ-25

SEISMIC-26

A 9 in thick two-way flat slab system is shown by the plan in Fig. EQ-26.

DESIGN CRITERIA

- Seismic zone 4
- Shear wall braced building
- All shear walls are 8 in thick and 10 ft high.
- Columns are 14 in in diameter and 10 ft high.
- Assume all walls are fixed at bottom and pinned at top.
- Columns are not part of the lateral bracing system.
- Assume columns are fixed at top and bottom.
- Gravity moments in column from the frame are 38 ft-kips top and 19 ft-kips bottom.
- Shear walls $\Delta = P/Et\left[4(h/d)^3 + 3(h/d)\right]$
- Diaphragm $\Delta = \dfrac{WL}{6.4Et}\left[(L/b)^3 + 2.13(L/b)\right]$

REQUIRED

Determine the moments at top and bottom of column from seismic loading for column at line 2-D only.

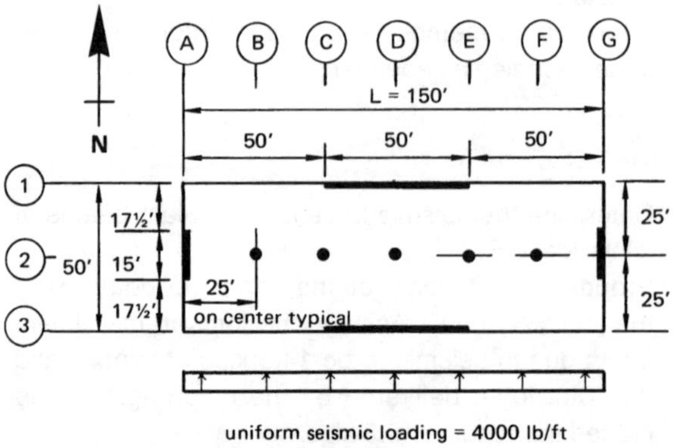

Figure EQ-26

SEISMIC-27

The wall shown in Fig. EQ-27 is an interior shear wall of 8 in thick (nominal) concrete block; all cells are fully grouted. Seismic load from a flexible roof diaphragm, applied at the top of the wall was determined to be 30 kips.

DESIGN CRITERIA

- ASTM A615 grade 40 reinforcing steel
- $f'_m = 1500$ psi
- $n = 40$
- No special inspection

REQUIRED

(a) Determine the lateral load in piers 1, 3, and 4 due to the given load. Neglect weight of wall for seismic effects.
(b) Determine the maximum anchorage load from the drag strut to wall. Assume pin ends and axial deformation to be negligible.
(c) Determine the axial load in pier 5. Assume that the seismic load at the top of wall B is 25 kips. Neglect weight of wall for seismic effects.

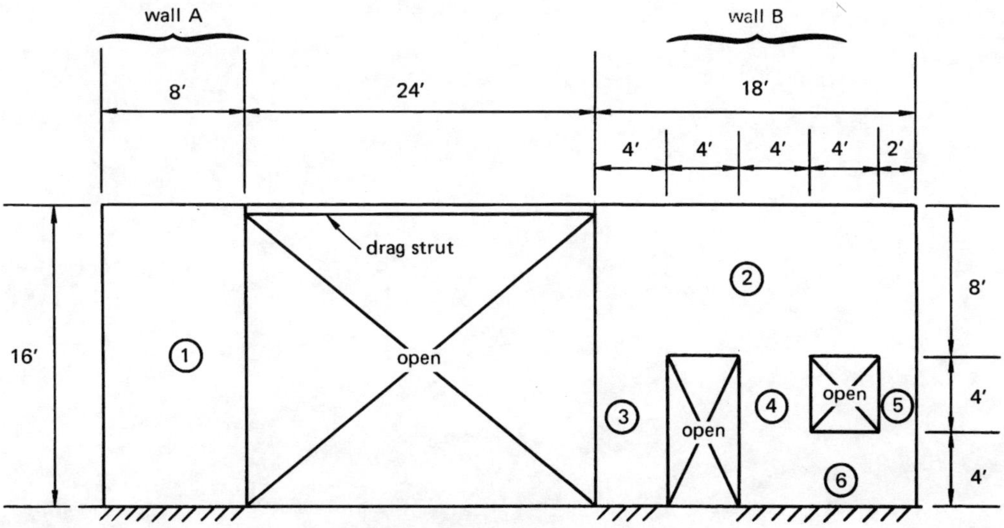

SHEAR WALL ELEVATION

Figure EQ-27

5 FOUNDATIONS

FOUNDATIONS-1

A pile cap and a system of batter piles support a horizontal force of 138 kips which is delivered to the cap by a tension rod 3 1/4 in in diameter. See Fig. F-1.

DESIGN CRITERIA

- The cap ties together two batter tension piles and two batter compression piles.
- All piles are 12 in square.
- The forces are transmitted to the soil through skin friction on the pile with values as follows:
  - 0 to 15 ft below bottom of cap = 500 psf
  - 15 to 60 ft below bottom of cap = 900 psf
- All lateral loads must be resisted by the piles.
- Neglect the dead load of the pile cap.
- f'_c = 3000 psi
- f_y = 60,000 psi

REQUIRED

(a) Calculate the load carried by the tension piles and show a design for a 12 in square precast tension pile.

(b) Calculate the necessary pile penetration measured from bottom of pile cap to resist the forces in tension.

(c) Draw a section through the pile to show placing of reinforcement. Identify and locate for proper positioning all bars including ties. Show clearances.

(d) Draw a detail in elevation to show the method of attachment of cap to pile to transfer the forces in tension adequately.

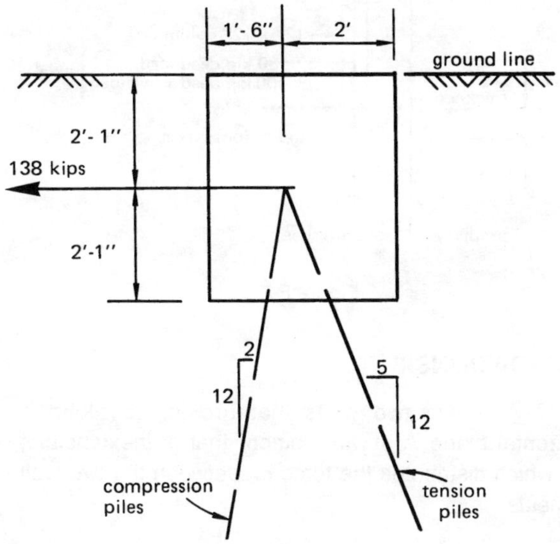

Figure F-1

FOUNDATIONS-2

Fig. F-2 shows a footing location for a 12 in x 12 in column. The design loads are shown on the drawing and include vertical loads as well as an overturning moment. The actual footing cap is not shown.

DESIGN CRITERIA

- The soil report recommends 24 in diameter drilled piers with allowable design loads of 35 kips and 50 kips for dead load and live load respectively.
- The minimum spacing between piers is recommended as 5 ft.
- The edge distance from the centerline of any pier to the edge of the cap is 2 ft.

REQUIRED

(a) Calculate the individual loadings to each pier. Position each pier for uniform loading under dead load only.

(b) Draw a plan view of the pier cap (to scale) showing the placing of piers for minimum cap dimensions.

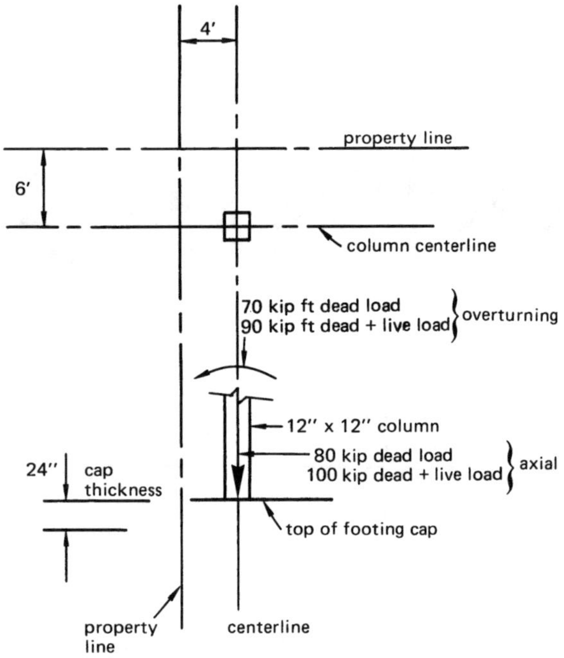

Figure F-2

FOUNDATIONS-3

Fig. F-3 shows two walls that are joined along a horizontal plane A by an element that is inextensible, and which distributes the force F_Q between the two wall elements.

DESIGN CRITERIA

- The walls are 12 in thick reinforced concrete that rest upon footings embedded in soil that will compress 1/100 in for each 1000 psf increase in soil pressure.
- The vertical loads W_1 and W_2 are sufficient to maintain compression throughout the length of both footings.
- E_c = 3000 ksi
- G = 1200 ksi

REQUIRED

(a) What portion of F_Q is resisted by each wall? Consider flexural and shearing deformations of the wall above the footings as well as rotation of the footings due to soil compression. Neglect any deformation of the footings themselves.

(b) What proportion of the force F_Q would each wall resist if flexure and shear in concrete were neglected and only soil compression were considered?

(c) What proportion of the force F_Q would each wall resist if only flexure and shear in concrete were considered and rotation due to soil yield were neglected?

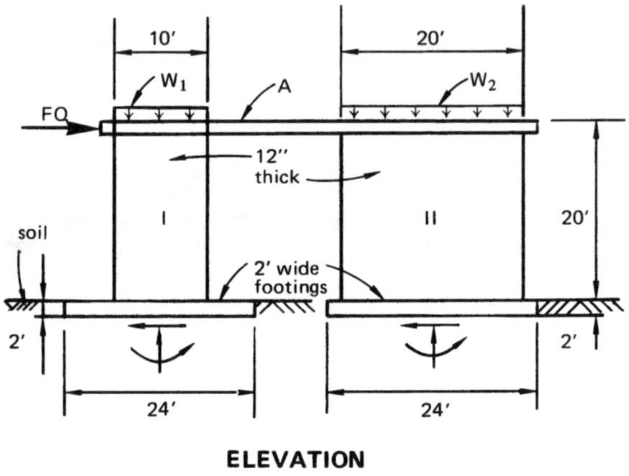

Figure F-3

FOUNDATIONS-4

You have been provided with a soil investigation report that shows a bearing value of 1800 lb/ft^2, a passive pressure equivalent to a fluid weighing 300 lb/ft^3, and a friction factor of 0.4. See Fig. F-4.

REQUIRED

Show all calculations for the design of the footing shown and draw a detail to an approximate scale of 1 in = 1 ft 0 in showing all reinforcing bars and dimensions for your solution.

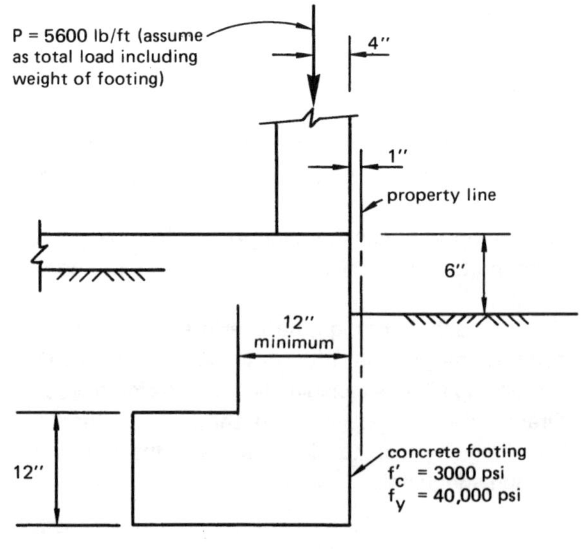

Figure F-4

FOUNDATIONS-5

A pole type retaining wall with a dead man anchor is shown in Fig. F-5.

DESIGN CRITERIA

- Pole spacing is 10 ft on center.
- The dead man anchor may be assumed as adequate.
- The wall, poles, and footing may be assumed as adequate for the applied loads.
- Equivalent fluid pressure against wall for slope shown = 40 pcf
- Passive resistance = 350 pcf against foundation.

REQUIRED

(a) Draw a load diagram showing the forces acting on one pole and footing.

(b) Find depth *d* (to nearest 6 in of embedment) of footing for a height *h* of 9 ft.

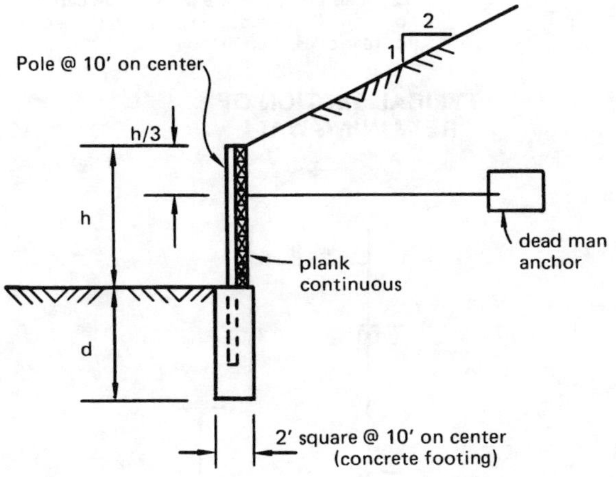

Figure F-5

FOUNDATIONS-6

Fig. F-6 shows the foundation for a reinforced concrete shear wall.

DESIGN CRITERIA

- Vertical loads given are ultimate loads.
- Roof live load is not included.
- Assume seismic shear is resisted at floor line.
- Unfactored overturning moment is about bottom of foundation.

REQUIRED

(a) Solve for the maximum soil pressure for the loading system shown.

(b) Draw the soil pressure diagrams.

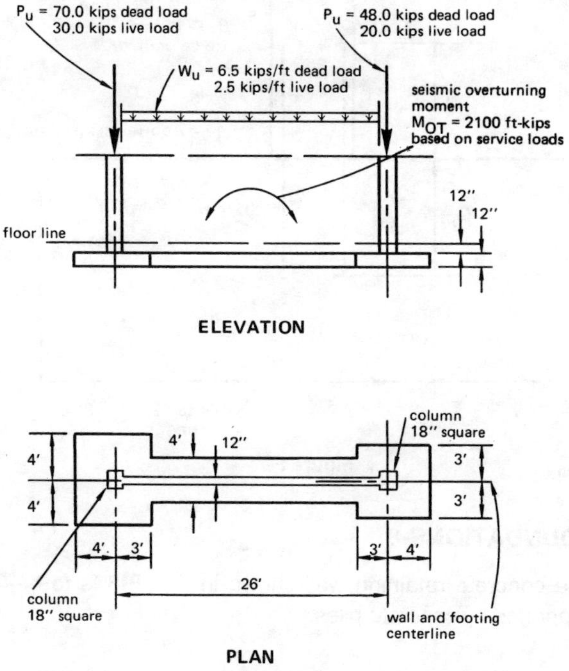

Figure F-6

FOUNDATIONS-7

Due to an error during construction, four #7 dowels were placed in a pile cap in lieu of the dowels specified on the structural plans. The concrete column has not yet been poured. The column is designed to be 16 in x 16 in with eight #11 bars to carry an ultimate axial load of 800 kips. See Fig. F-7.

DESIGN CRITERIA

- f'_c = 3 ksi pile cap
- f'_c = 4 ksi column
- f_y = 60 ksi

REQUIRED

Determine if four #7 dowels are adequate. Show all calculations. If you determine they are not, what recommendations would you make to correct the present field condition? Show all calculations.

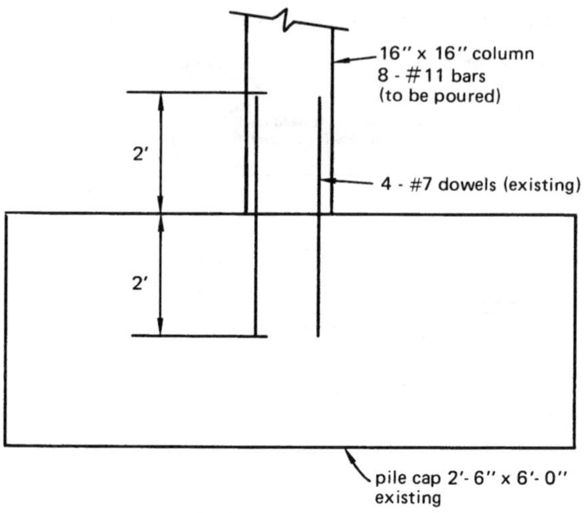

Figure F-7

FOUNDATIONS-8

The concrete retaining wall shown in Fig. F-8 is to be supported by concrete piles.

DESIGN CRITERIA

- Equivalent fluid pressure = 36 psf
- Concrete weight = 150 pcf
- Soil weight = 100 pcf
- No surcharge, free draining soil
- All loads acting on retaining wall to be resisted by the piles. The piles take axial loads only.
- Do not consider passive pressure on key. Ignore friction on wall face.

REQUIRED

Show all calculations for the following.

(a) Find the axial loads on the piles.

(b) If f'_c = 3 ksi and f_y = 40 ksi, what is the maximum spacing for #5 vertical bars at the bottom of the stem? Use spacing to nearest 1/2 in.

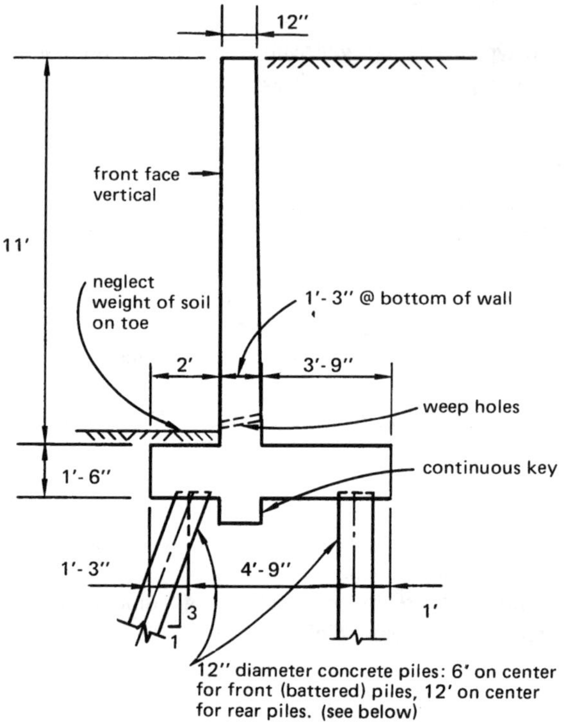

TYPICAL SECTION OF RETAINING WALL

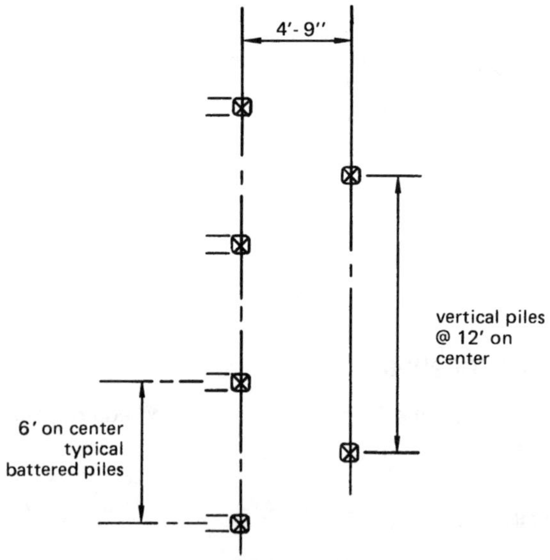

PILE LAYOUT PLAN

Figure F-8

FOUNDATIONS-9

A pre-engineered prefabricated building is shown in Fig. F-9. Steel rigid frames span north-south and simple span purlins running east-west support the roof. Walls at lines A, B, 1, and 2 are nonbearing and nonshear elements. There is no wall at line C. The building houses fire-fighting equipment.

DESIGN CRITERIA

- Southern California job site in seismic zone 4
- Snow loading is not required.
- Roof dead load including framing = 4 psf
- Wall dead load including framing = 3 psf
- Undisturbed clayey silt, level site
- 3000 psi at 28 days concrete
- Wind zone 20 = 15 psf
- Rigid frames are pin-ended at anchorage to floor slab.
- Framing, diagonal bracing, and anchor bolts are adequate for all loads.
- Rigid frame base outward thrust is tied through the floor slab.
- Footing depth is limited to 12 in below grade due to ground water problems.

REQUIRED

(a) Design a square spread footing for frame base at line B at line 1 only. Determine size of footing and reinforcement requirements.

(b) Draw a sketch of a section of footing showing dimensions and schedule and location of reinforcing.

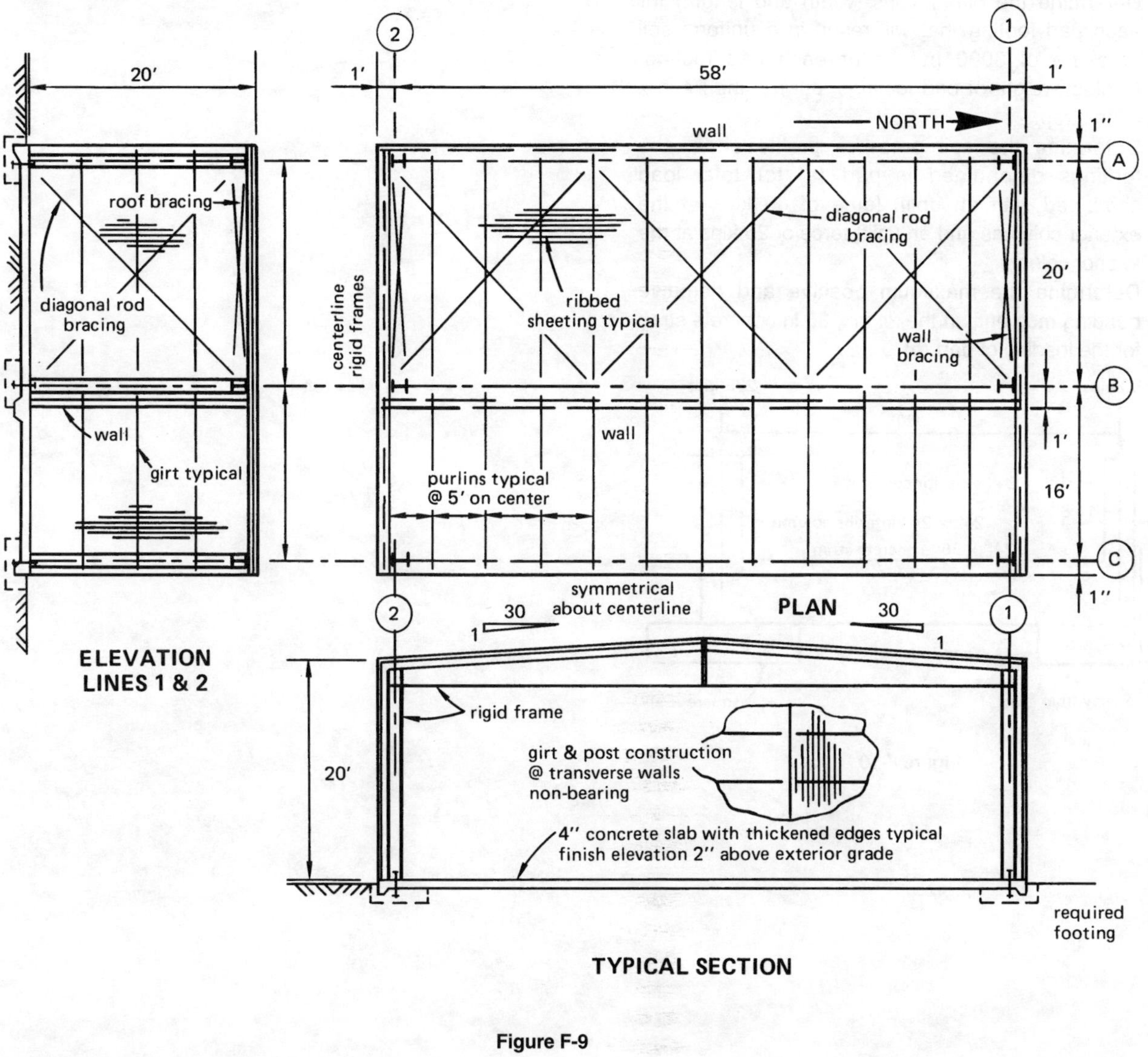

Figure F-9

FOUNDATIONS-10

An exterior 24 in x 18 in column with a total vertical load of 150 kips and an interior 24 in x 24 in column with a total load of 225 kips are to be supported by a pad footing at each column connected by a 24 in x 36 in concrete strap. (See Fig. F-10.)

DESIGN CRITERIA

- Assume the 24 in x 36 in concrete strap is placed such that it does not bear directly on the soil.
- Concrete weighs 150 pcf.

REQUIRED

(a) Determine the dimensions (width and length) for each pad footing that will result in a uniform soil pressure of 3000 psf under each pad footing. Neglect weight of pad footings, but not the 24 in x 36 in strap.

(b) Determine the soil pressure profile under the footings determined in part (a) for total load combined with an uplift force of 80 kips at the exterior columns and an uplift force of 25 kips at the interior column.

(c) Determine the maximum positive and negative bending moments in the 24 in x 36 in concrete strap for the loading of part (b).

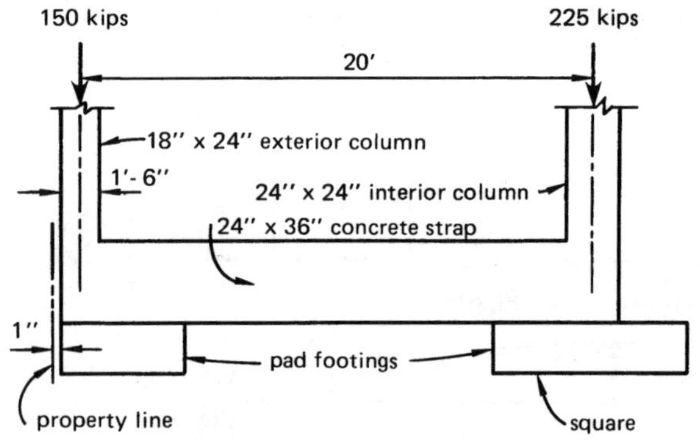

Figure F-10

6 TIMBER

TIMBER-1

A portion of a wood framing system is shown in Fig. T-1.

DESIGN CRITERIA

- No. 1 Douglas fir-larch (north) floor joists and ledgers
- Girder: any wood material that suits the system
- Take required values from current UBC.

REQUIRED

(a) Design the floor joists for the minimum board footage using standard dimension Douglas fir. Give size and spacing and show your calculations.

(b) Calculate a size for the main wood girder. Assume a simple span. No detail is necessary. Calculate all the principal design considerations that may be critical.

(c) Provide a suitable wall ledger. Give the size and show anchorage for vertical and lateral loads. Draw a detail of the ledger showing all the principal elements that will be required to complete the installation.

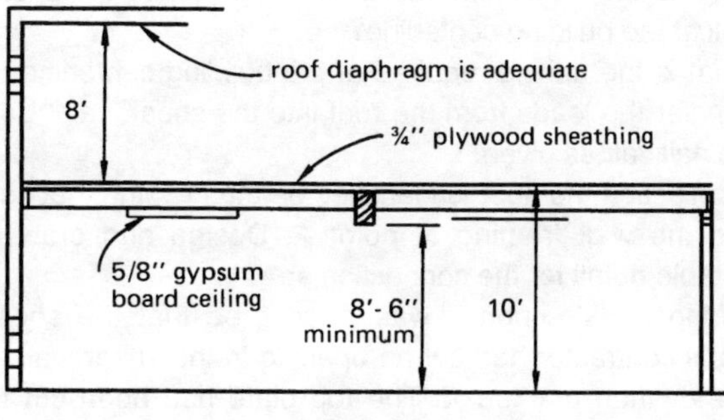

SECTION

Figure T-1

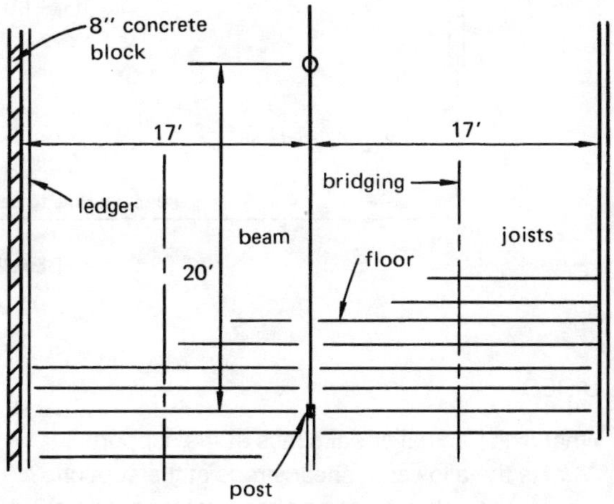

PLAN

Figure T-1 (cont.)

TIMBER-2

A 1-story wood frame building has a roof supported with glued laminated timber beams that are 9 in wide, spaced at 16 ft on center, and which span 80 ft to centerlines of wall supports. Fig. T-2 shows the principal dimensions of the beams.

DESIGN CRITERIA

- Roof dead load = 18 psf (includes dead load of beam)
- Roof live load = 12 psf (not reducible)
- Douglas fir 22F combination
- Beam is deemed satisfactory for deflection and camber.

Wood values are as follows:

- F_b = 2200 psi
- F_t = 1600 psi
- F_c = 1500 psi
- F_v = 165 psi
- $F_{c\perp}$ = 450 psi
- E = 1.8 x 10^6 psi (dry conditions of use)

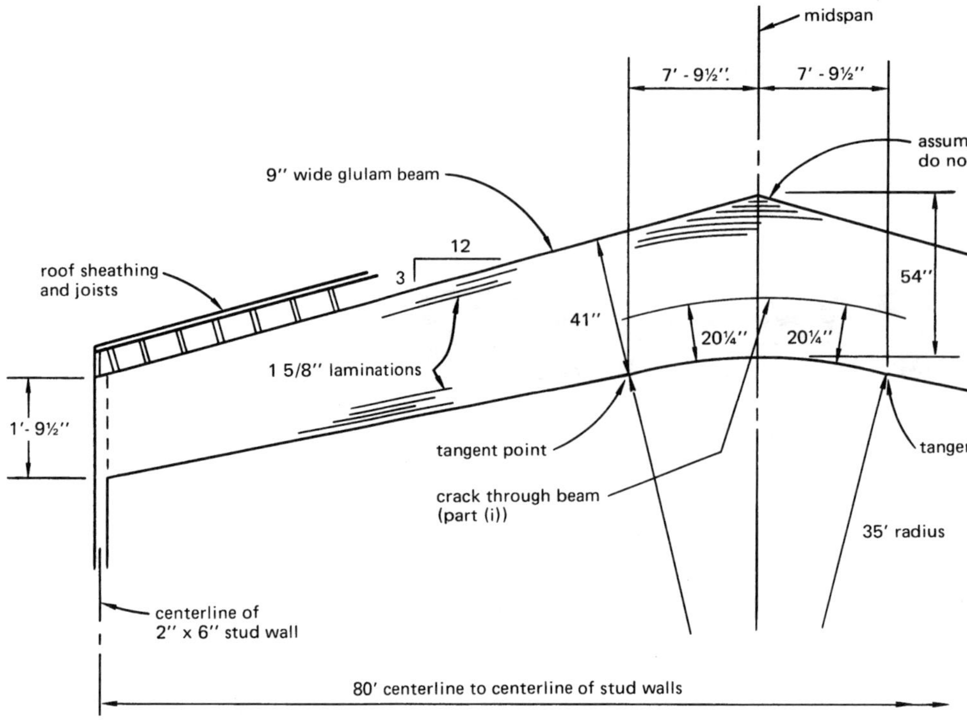

BEAM ELEVATION

Figure T-2

REQUIRED

(a) What is the actual shear stress at the supports?
(b) What is the allowable shear stress at the supports?
(c) What is the actual bending stress at tangent point?
(d) What is the actual bending stress at midspan?
(e) What is the allowable bending stress at tangent point?
(f) What is the allowable bending stress at midspan?
(g) What is the actual radial stress?
(h) What is the allowable radial stress?
(i) After the building was completed and it had been in use for one year, a crack appeared at the curved portion of the beam, as indicated on the figure. The crack was in the lamination and not in the glued joint. The crack went through the 9 in thickness of the beam. How would you repair the beam so it will continue to carry the original design loads? Support your conclusions with calculations and a sketch to show what you would use, what size, and how placed in order to complete your repairs.

TIMBER-3

A 1-story wood frame building is 40 ft wide x 80 ft long. It has 8 ft long shear walls at each end, and a 12 ft long shear wall at the center of the building. Fig. T-3 shows the details of the 12 ft shear wall at the building centerline.

DESIGN CRITERIA

- Seismic zone 3
- Exterior walls weigh 20 psf.
- Ca = 0.36
- Importance factor = 1.0
- 80 mph basic wind speed; exposure condition C
- The roof and the shear walls are sheathed with grade C-C exterior plywood.
- Nail and bolt values may be taken from the current UBC.
- Douglas fir-larch framing

REQUIRED

(a) What is the maximum lateral force carried by the shear wall at the building centerline?
(b) What is the nailing required at the building centerline to transfer the loads from the roof into the shear wall? Use the nail values given.
(c) A collection member is identified on the drawing that ties into the wall framing at point A. Design and draw a suitable detail for the connection at A.
(d) During construction it was discovered that the sheet metal contractor had cut an opening in the shear wall at the location marked X. The top plate had been cut to pass an 18 in x 18 in duct. The top of the duct was set at 4 in below the soffit of the roof plywood. To avoid delay at the job site, it is necessary to redesign a collection member around the duct opening for the transfer of lateral forces from the roof to the shear wall.

Design an installation that will act to transfer the lateral loads around the duct opening.

(e) Draw a suitable detail that shows your design and intent to accomplish the tie.

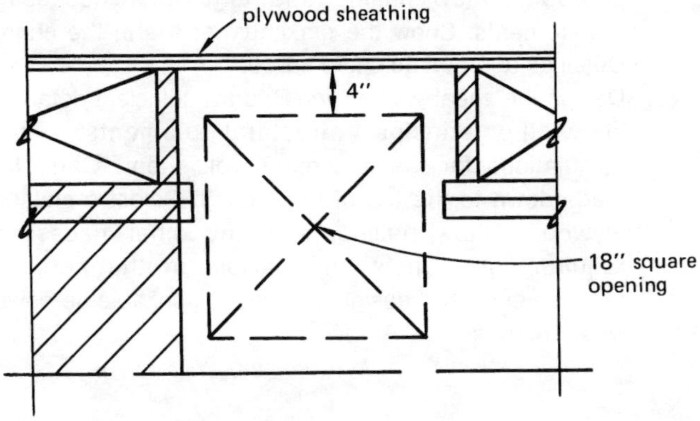

POINT X

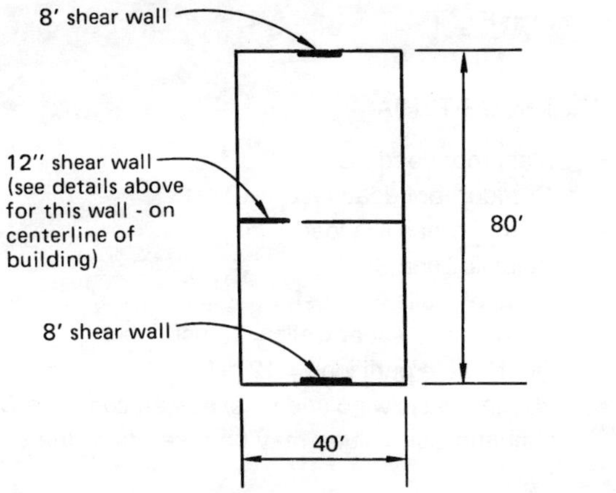

KEY PLAN

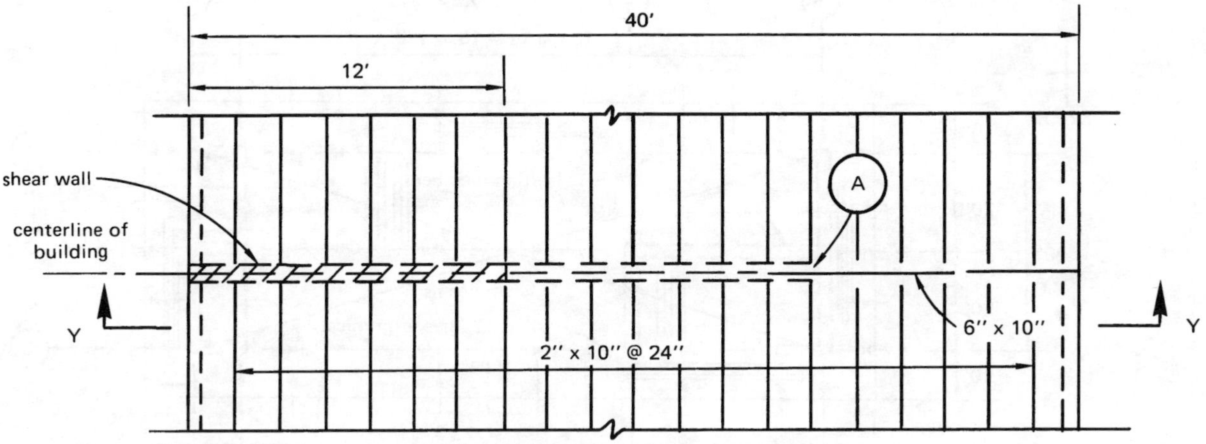

PART PLAN

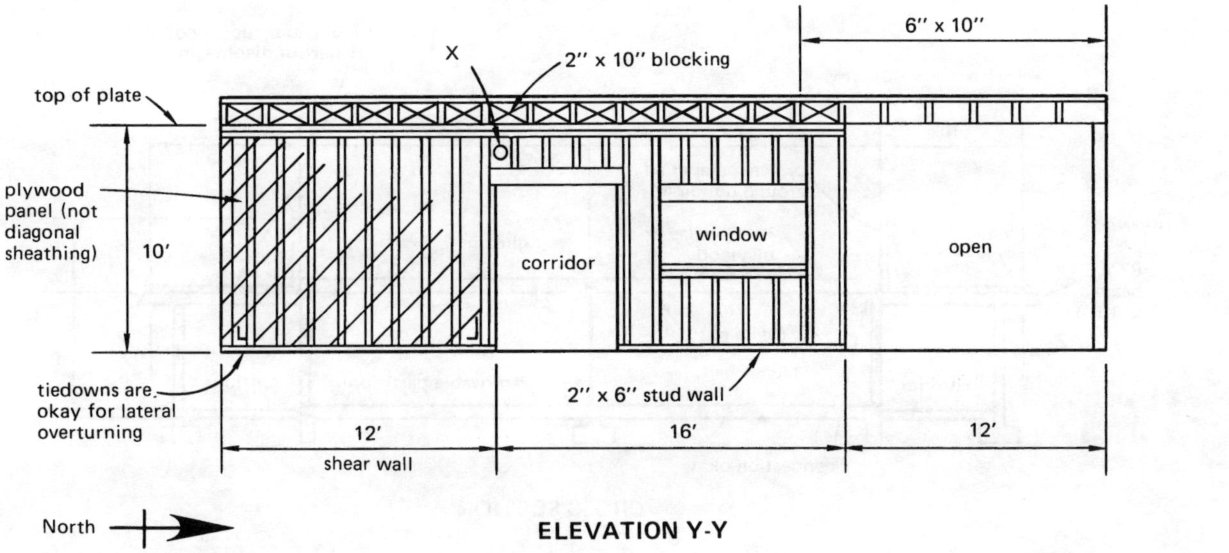

ELEVATION Y-Y

Figure T-3

TIMBER-4

A plan and cross section of a 1-story frame building are shown in Fig. T-4.

DESIGN CRITERIA

- Main roof dead load = 16 psf
- Corridor roof dead load = 12 psf
- All roof areas live load = 20 psf
- Seismic zone 3
- Parapet walls = 10 psf
- Exterior and shear walls = 12 psf
- Removable partitions = 12 psf
- 80 mph basic wind speed; exposure condition C
- Nail and bolt values may be taken from the current UBC.
- Table T-4 shows nail values for plywood diaphragms.

REQUIRED

(a) Design the horizontal roof diaphragm for lateral forces in the north-south direction. Show calculations for maximum bending and shear stresses. Select a suitable plywood. Show nailing and any other necessary requirements. Show the maximum stress in the chord. Detail an appropriate chord splice.

(b) Design the shear wall on line B; draw sufficient details to show the principal structural elements. Show calculations for the maximum wall shear. Carry the loads down to the top of the footing. Select a suitable plywood. Show nailing, and any other necessary requirements. Show an analysis of the tie-down requirements, and design and detail a suitable tie-down, where required.

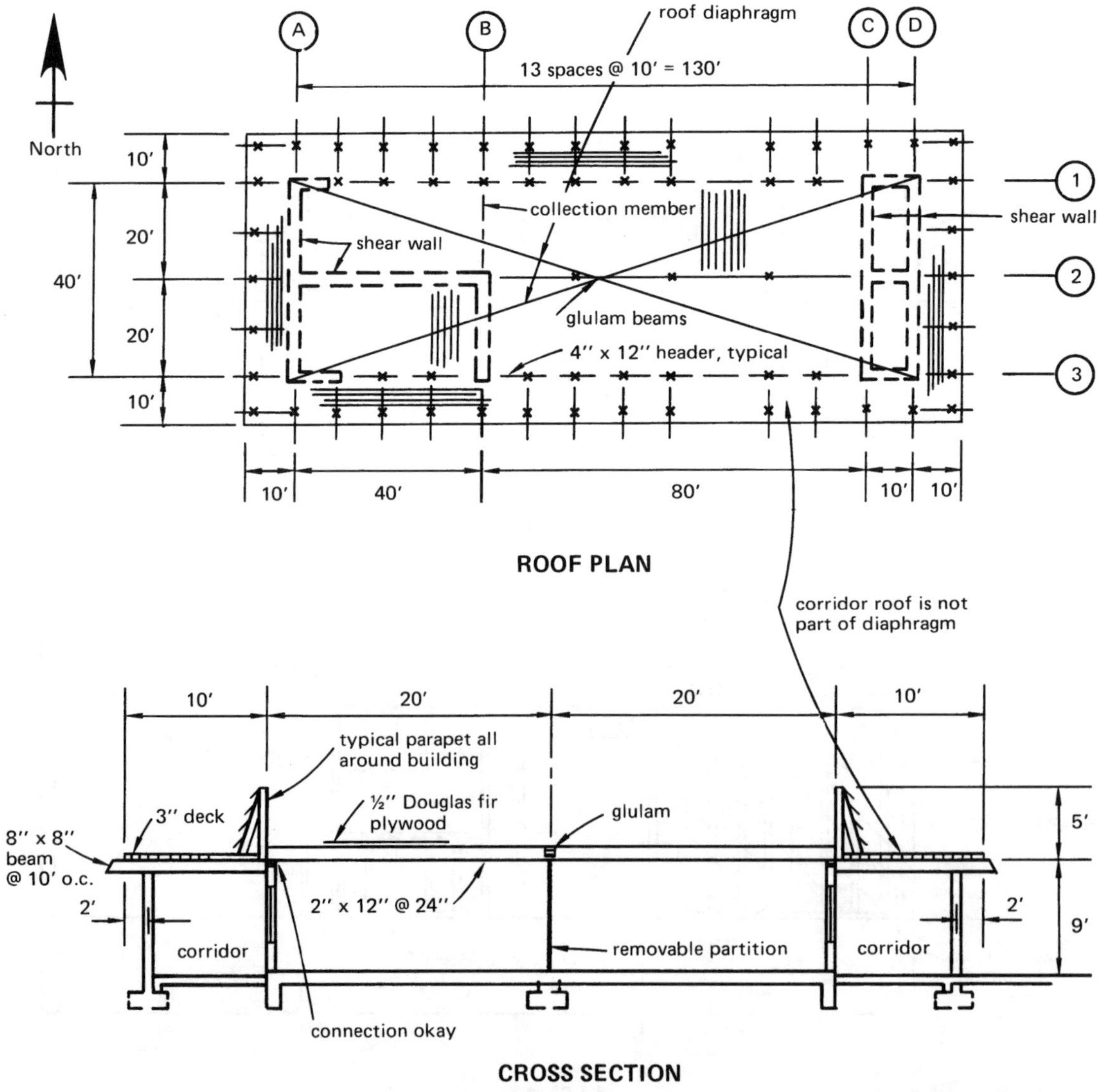

Figure T-4

Table T-4									
Plywood diaphragms - Allowable nail shear (lb / ft)									
Common nail size				6d		8d		10d	
Minimum nail penetration in framing (inches)				1 1/4		1 1/2		1 5/8	
Nominal plywood thickness (inches)				5/16		3/8		1/2	
				wind	seismic	wind	seismic	wind	seismic
Width of framing lumber (nominal)				2	3	2	3	2	3
Blocked									
nail spacing at diaphragm boundaries and continuous panel edges	6	nail spacing at other plywood panel edges	6	188	210	270	300	318	360
	4		6	250	280	360	400	425	480
	2 1/2		4	375	420	530	600	640	720
	2		3	420	475	600	675	730	820
Unblocked (nails spaced 6" maximum at supported edge) Load perpendicular to unblocked edges and continuous panel joints				167	187	240	267	283	320
All other configurations				125	140	180	200	212	240

TIMBER-5

The drawing in Fig. T-5 represents a three-hinged arch with dimensions as shown and with loads as stated.

DESIGN CRITERIA

- Arch spacing = 20 ft on center
- Arch material: glued laminated timber (24F combination)
- Arch width = 7 in net
- Current UBC is suggested.
- Laminations are 3/4 in.
- Assume duration of full load is 7 days.
- Compute live load on tributary area of one-half arch span.
- Dead load = 320 lb/ft (includes dead load of arch)

REQUIRED

(a) Determine the magnitude and direction of the reactions at the base and crown for each of the following conditions:
   (1) Dead load plus live load on full span
   (2) Dead load only
   (3) Wind left only (neglect uplift)
   (4) Wind right only (neglect uplift)
   (5) Life load left only
   (6) Live load right only

(b) At point X find the bending moment, axial load, and shear on the cross section for:
   (1) Dead load plus live load on full span
   (2) Dead load and wind left

(c) Using the following given loads find the required depth of glulam section at X for dead load plus live load.
   Moment = 100 ft-kips
   Axial load = 18.4 kips
   Shear = 2.0 kips
   Consider allowable increase or reduction in fiber stress as indicated in UBC.

(d) Which load condition controls the design at the crown?

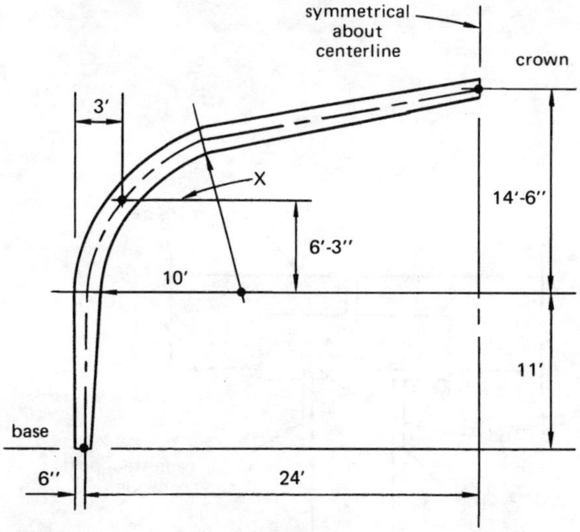

Figure T-5

TIMBER-6

A roof plan of a 1-story building and two connection details are shown in Fig. T-6.

REQUIRED

(a) Determine if the connections are adequate for lateral loads. Show calculations.

(b) How would you revise or improve the details shown? (Is revision necessary?) Explain.

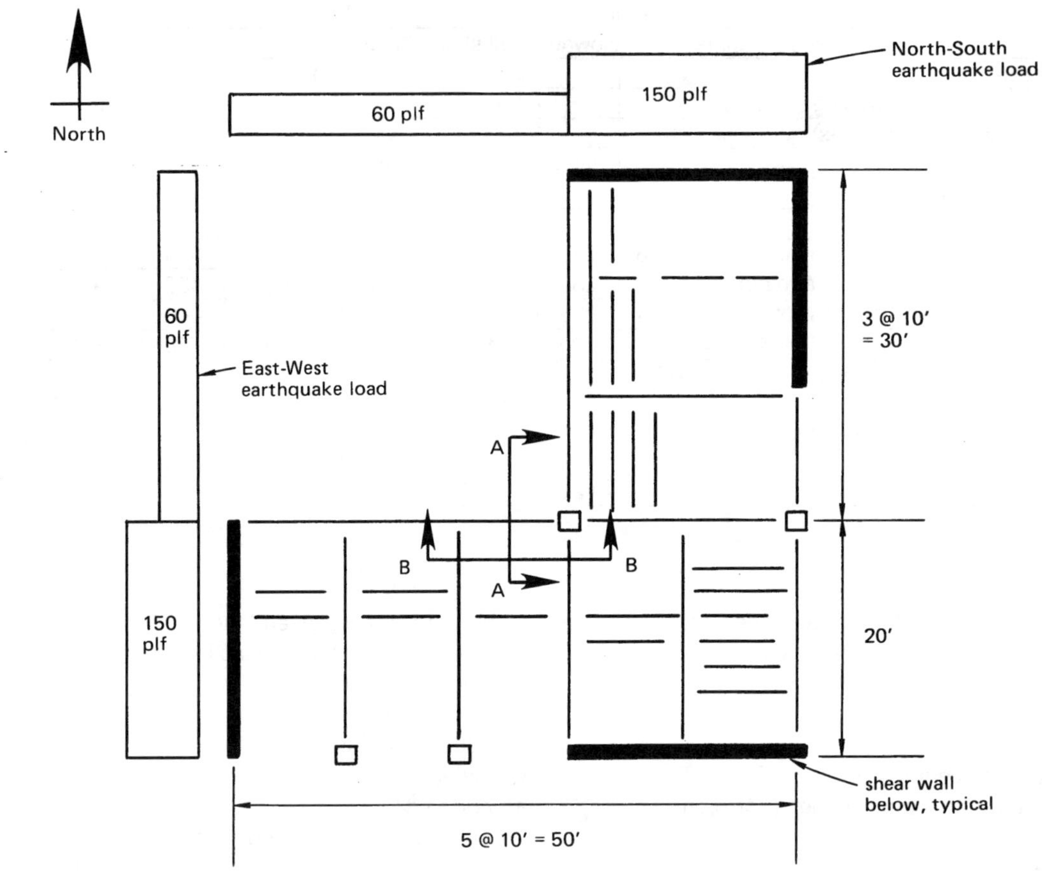

PLAN

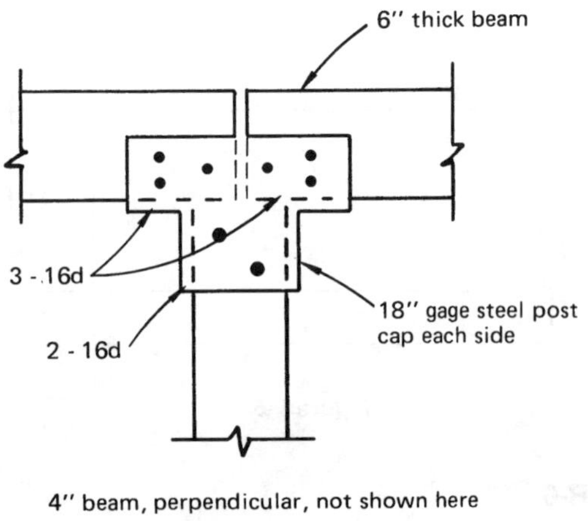

SECTION A-A

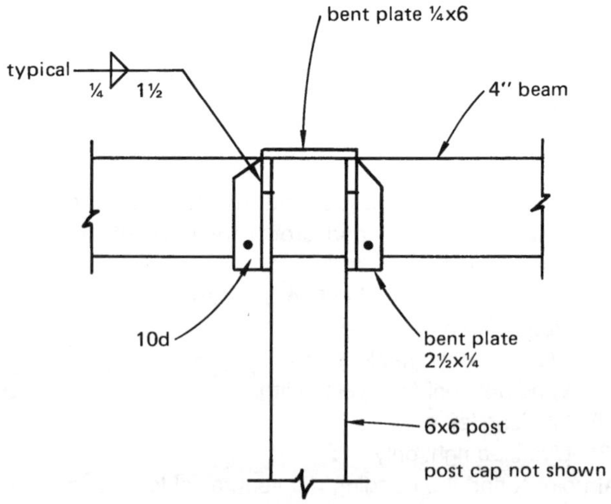

SECTION B-B

Figure T-6

TIMBER-7

An existing warehouse roof is to be investigated with a prospect to add a new opening for a skylight. See Fig. T-7.

DESIGN CRITERIA

- 24F glulam beam (dry condition), 5 1/8 in x 30 in
- Roof dead load = 10 psf
- Assume all existing connections are adequate.
- Skylight dead load = 10 psf (uniformly supported by the beams)

REQUIRED

(a) What is the bending stress in the 5 1/8 x 30 in glulam beam girder?
(b) Does the 5 1/8 x 30 glulam beam girder work for the new skylight? Show calculations.

TIMBER-8

A part elevation of the side wall of a 1-story wood frame building is shown by Fig. T-8. The total load and dead load reactions from roof beams are indicated.

DESIGN CRITERIA

- Design should conform to current UBC.
- Seismic zone 3
- All wood is Douglas fir-larch.
- Show supporting calculations for all work.

REQUIRED

(a) Design lintel A using glued laminated Douglas fir 22F combination. (Select size and check critical factors that influence choice of member, etc.)
(b) Design member B and its connections. Limit deflection to 1/240.
(c) Design member C and its connections.
(d) Design member D and its connections.
(e) Design the plywood shear panel. Use 1/2 in plywood. Show nailing, blocking, sill bolts, allowable shear value, etc. (Neglect dead load of wall.)
(f) Design the connection of member E to the concrete foundation.
(g) Draw a sketch of the plywood shear panel and show all the necessary information.

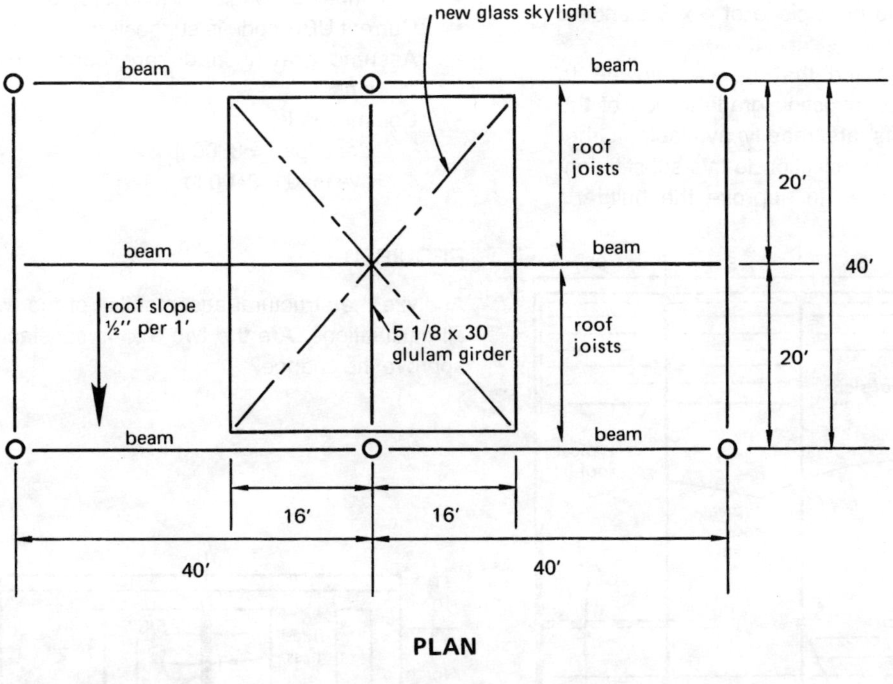

Figure T-7

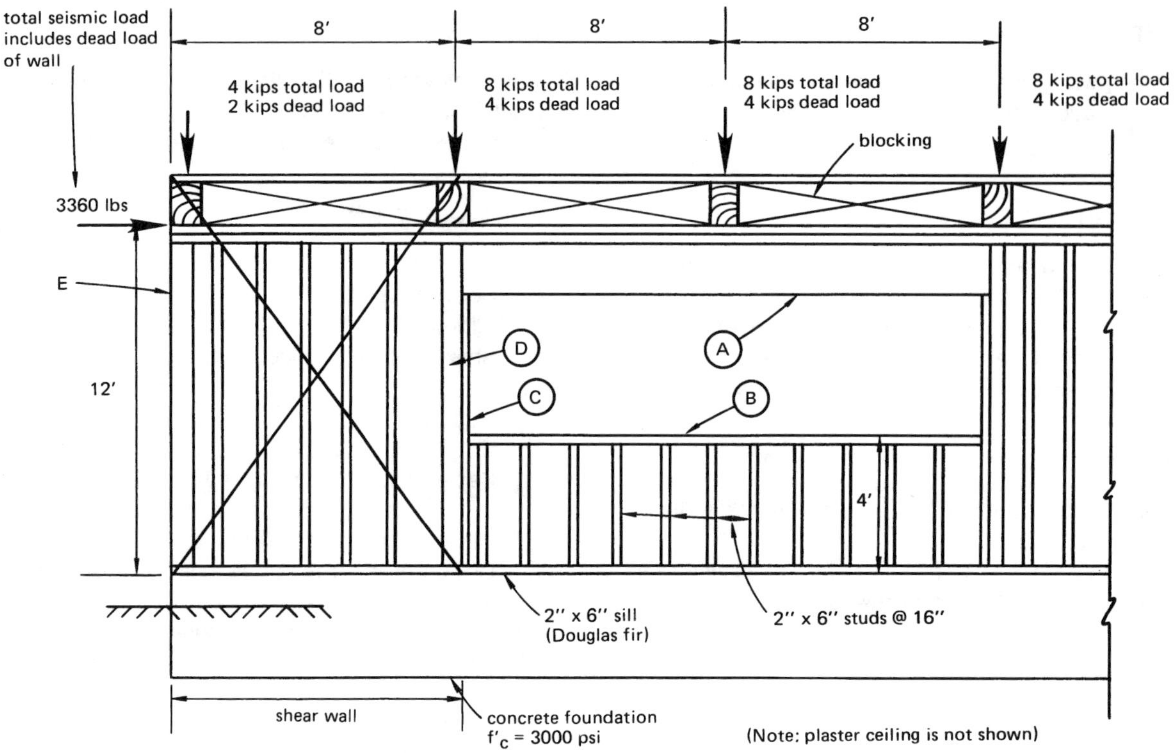

PART WALL ELEVATION

Figure T-8

TIMBER-9

The roof framing plan and the front elevation of a 1-story building are shown in Fig. T-9. The original plans showed the mullion A to be a piece of 4 x 6 standard grade lumber.

The builder has requested that he be allowed to substitute two 3 x 4s construction grade in lieu of the 4 x 6 since the 3 x 4s are readily available. Other interested parties have no objection to this substitution, and the engineer is asked to approve the builder's request.

DESIGN CRITERIA

- All lumber is Douglas fir (coast region).
- Current UBC code is suggested.
- Assume gravity loads are concentrically applied to mullions.
- Column loads:
  - dead load = 3600 lb
  - live load = 2400 lb

REQUIRED

Analyze the structural acceptability of the two 3 x 4s. Show all calculations. Are the two 3 x 4s satisfactory? Would you approve the change?

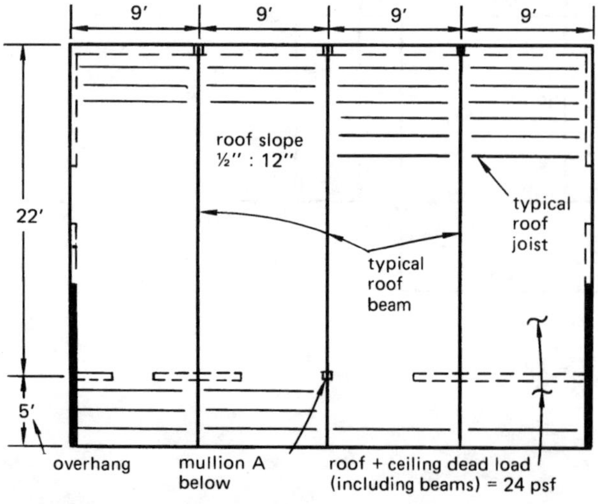

ROOF FRAMING PLAN

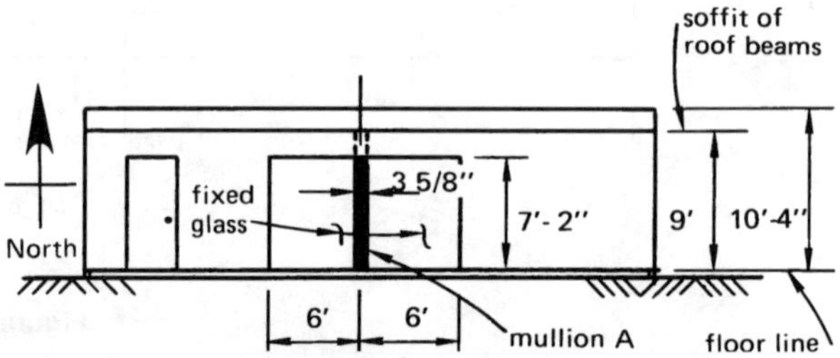

SOUTH ELEVATION

Figure T-9

TIMBER-10

A second floor balcony is shown in plan and elevation in Fig. T-10. The balcony is supported by a building wall and glued laminated beams with wood posts at 10 ft on center.

DESIGN CRITERIA

- Balcony dead load = 20 psf (includes rails and plaster ceiling for first floor below)
- Balcony live load = 100 psf
- Floor joists are spaced at 16 in on center.
- Glued laminated beams are supported by wood posts at 10 ft on center except at corner, where beams are cantilevered.
- Assume adequate bracing for bottom of beams.
- Beams are connected at cantilever ends adequate for distribution of vertical loads. Assume beams are connected together prior to application of any dead and live load.
- Douglas fir-larch joists
- $E = 1.8 \times 10^6$ psi
- 24F Douglas fir glulam beams
- $F_b = 2400$ psi
- $F_v = 165$ psi
- $E = 1.8 \times 10^6$ psi

REQUIRED

Determine:

(a) Size of floor joists,

(b) load to be transmitted through connection at ends of cantilever beams, and

(c) size of beams with cantilever spans for deflection and strength requirements.

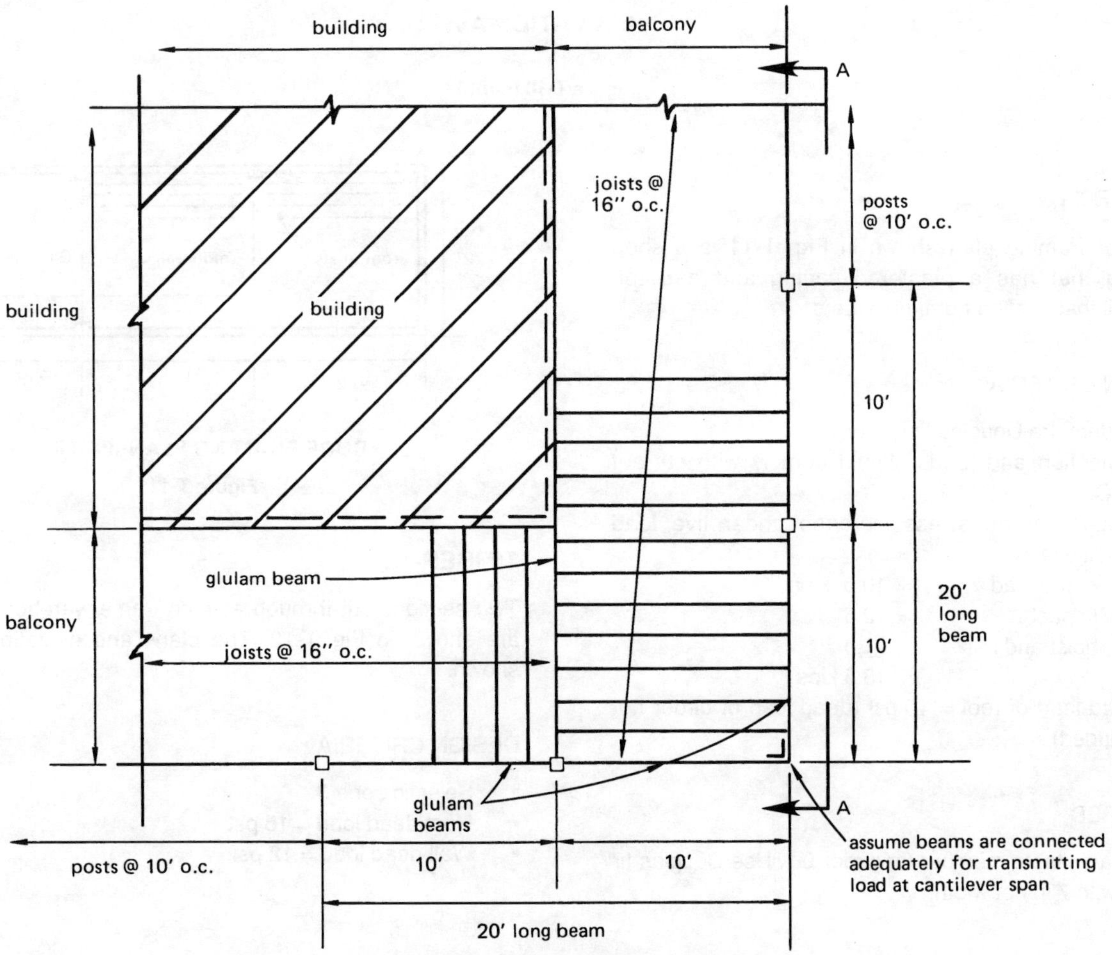

SECOND FLOOR BALCONY PLAN

Figure T-10

(continued on next page)

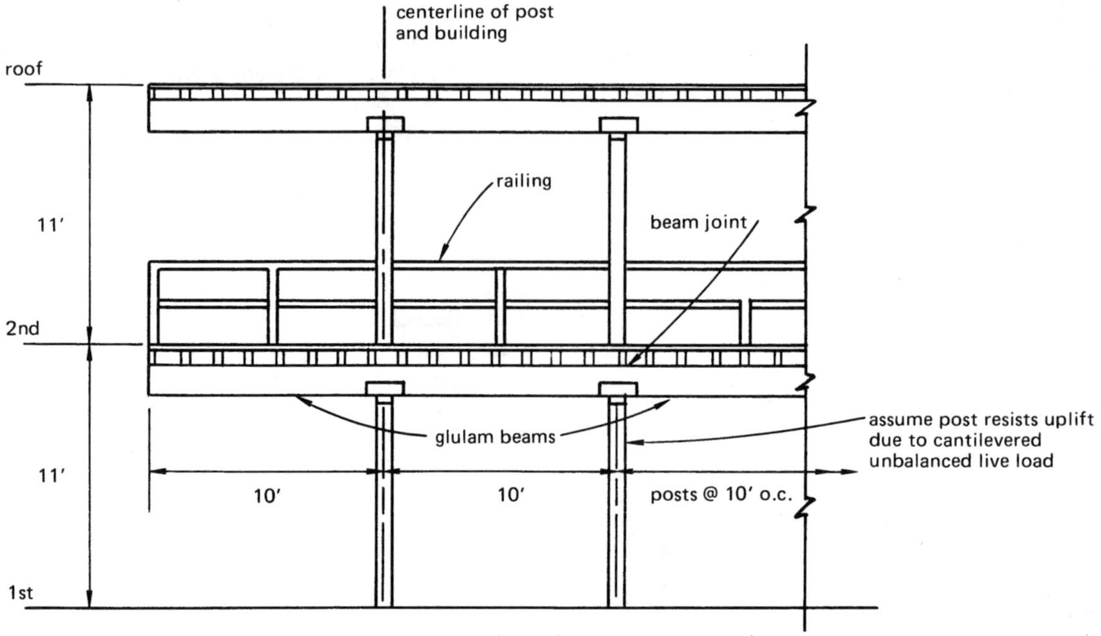

ELEVATION A-A

Figure T-10 (cont.)

TIMBER-11

The roof framing plan shown in Fig. T-11 is a shop building that has a plastered ceiling and a 5 ton monorail that is hung from the girders.

DESIGN CRITERIA

- Girders are Douglas fir glulams.
- Deflection and loads should comply with current UBC.
- Design for no stress increase under live load conditions:

crane load	=	10.0 kips
impact	=	2.5
hoist and rail	=	1.0
		13.5 kips

- Dead load of roof = 16 psf (dead load of girder not included)

REQUIRED

Design a suitable member for girder G1. Use Douglas fir glulam with 7 in net width.

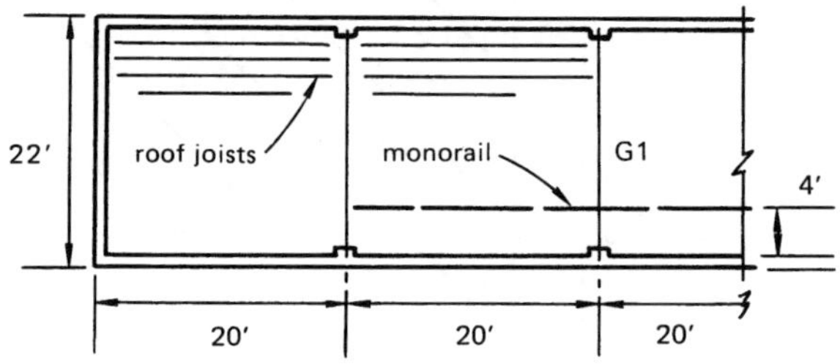

ROOF FRAMING PLAN (PART)

Figure T-11

TIMBER-12

Two sections cut through a wood frame warehouse building are shown in Fig. T-12. The plans and elevation are also shown.

DESIGN CRITERIA

- Seismic zone 3
- Roof dead load = 15 psf
- Wall dead load = 12 psf

REQUIRED

(a) If a lateral load-resisting element (rigid frame or shear wall) is not provided between lines 2 and 3, what structural element(s) would be required to maintain diaphragm integrity? Give at least two.

(b) Assume the lateral load on the roof in the north-south direction is 190 lb/ft, and no rigid frame or shear wall is to be used below the flat roof at lines 2 and 3. Describe all the necessary structural elements required to make the structure work from a lateral load standpoint. Determine the loads the resisting elements should be designed for.

(c) What is your opinion of a stepped diaphragm? Observe section A-A. There is no shear wall or rigid frame located where the step in the diaphragm occurs.

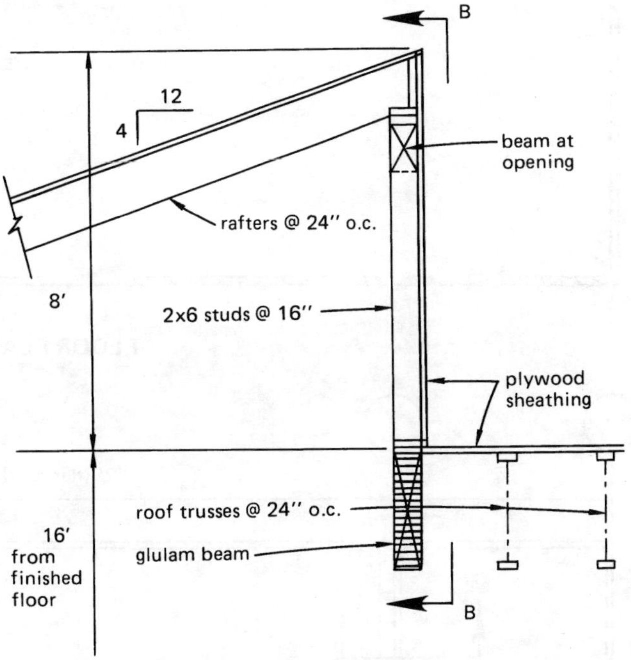

SECTION A-A

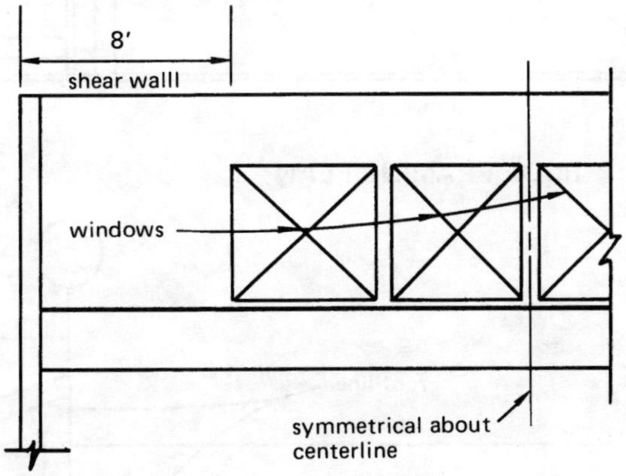

SECTION B-B

Figure T-12
(continued on next page)

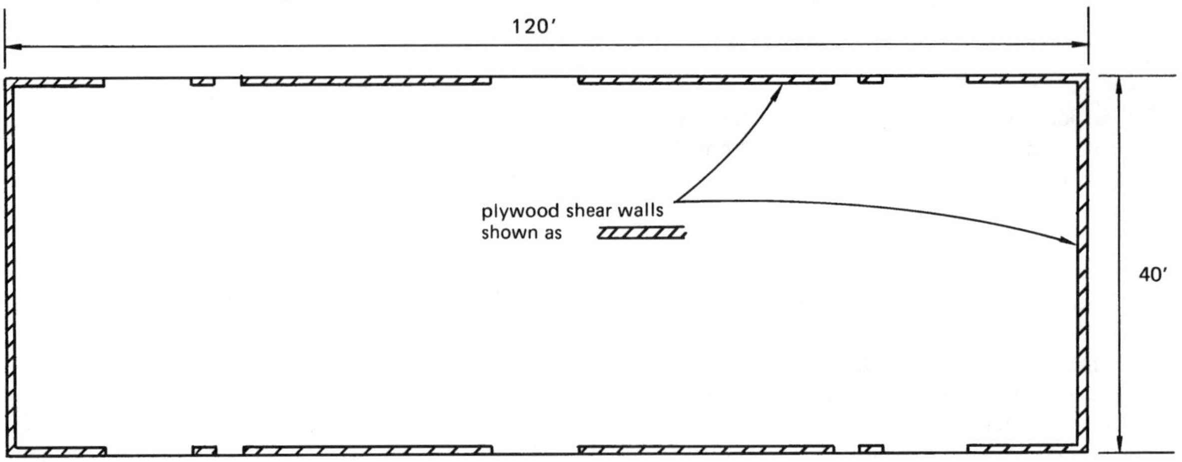

FLOOR PLAN

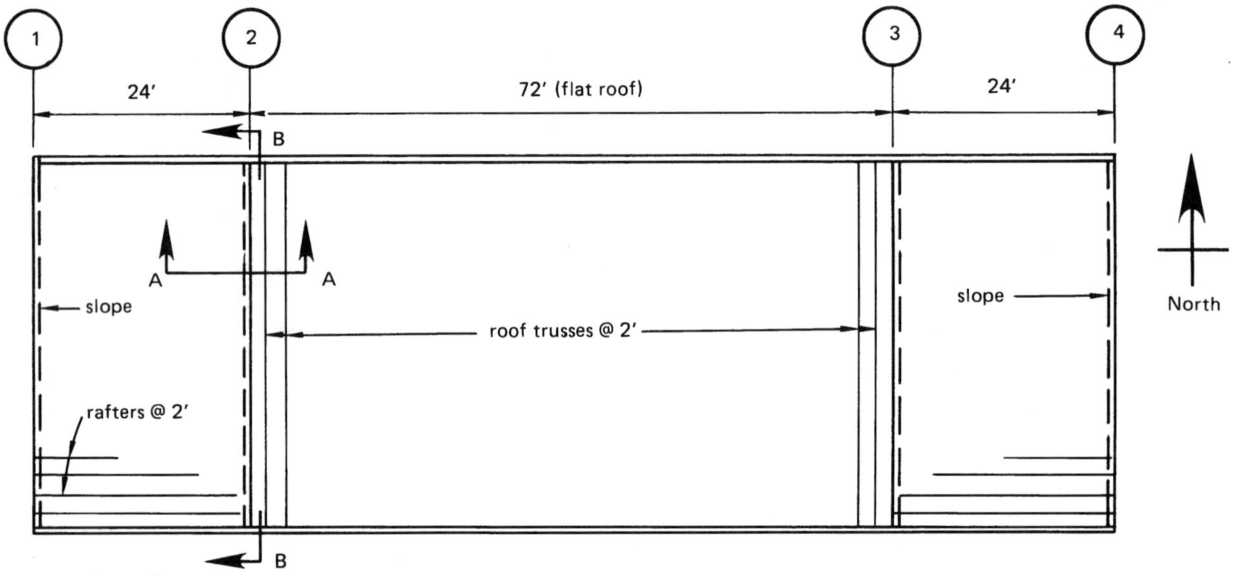

ROOF FRAMING PLAN

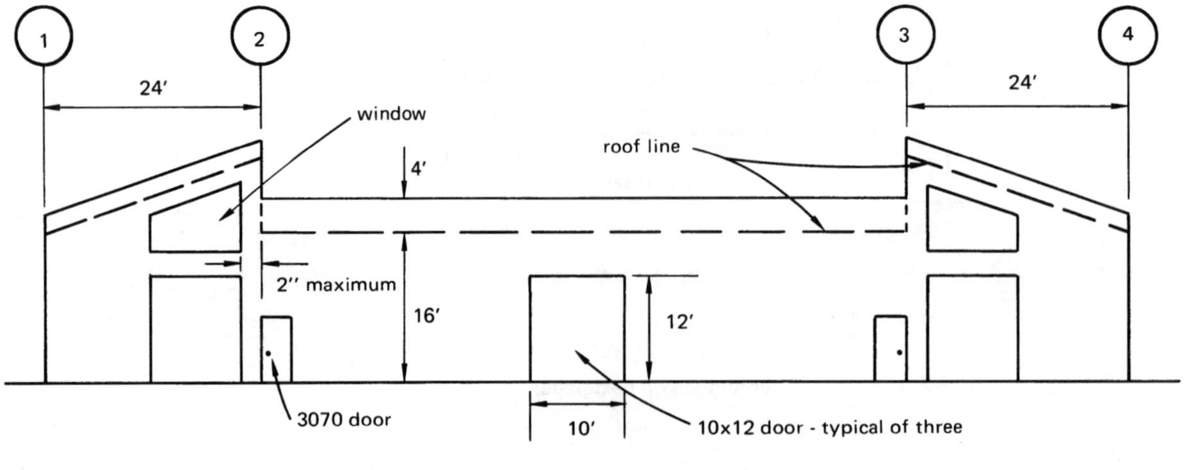

NORTH AND SOUTH ELEVATION

Figure T-12 (cont.)

TIMBER-13

The 1-story building represented by the roof plan shown in Fig. T-13 is to be analyzed for lateral forces. Concrete columns on line A are fixed against rotation at the level of the slab on the grade floor. The concrete columns on line D and all walls are free to rotate at the floor for forces perpendicular to the wall.

DESIGN CRITERIA

- Seismic zone 3
- Do not consider wind.

REQUIRED

(a) Determine the required plywood diaphragm nailing. Check the chord.

(b) Are the columns adequate if the plywood diaphragm is assumed to deflect as shown? Use the deflection formula given here as

$$\Delta = \frac{5vL^3}{8\ EAb} + \frac{vL}{4Gt} + 0.094\ Le_n$$

Δ = deflection (in)
v = shear (lb/ft)
L = length (ft)
b = width (ft)
A = area of chord (in^2)
E = elastic modulus of chords (psi)
G = 110,000 psi
t = plywood thickness (in)
e_n = nail deformation (see graph G)

(For this particular problem do not consider any resisting forces that may be offered by the columns (as they deflect) to the diaphragm.)

(c) Section C-C shows the method proposed to connect the roof to the precast wall. Demonstrate by computation the adequacy of this proposed method. What comments do you have regarding this detail, based on your own experience or observations?

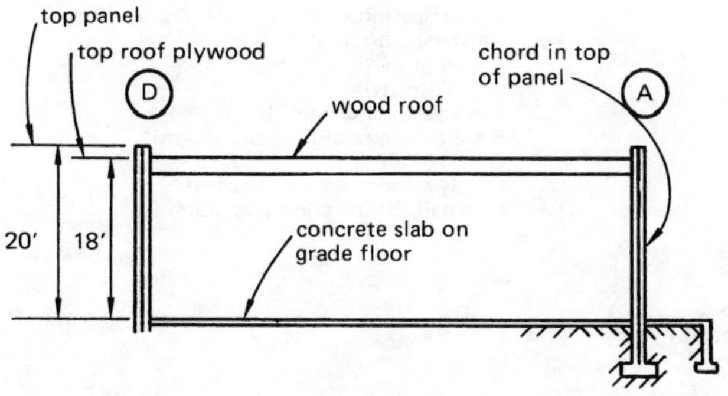

SECTION A-A

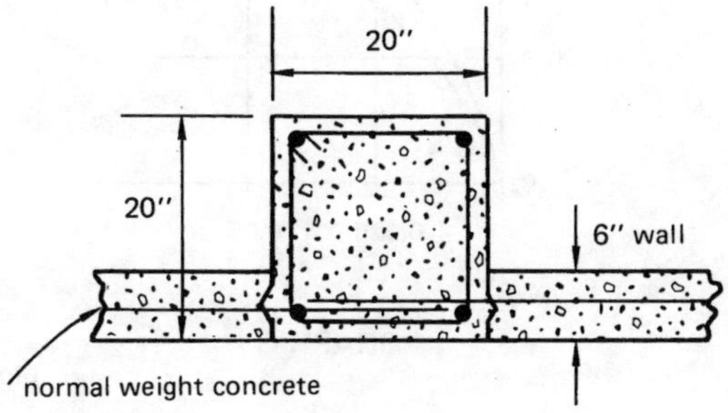

DETAIL B

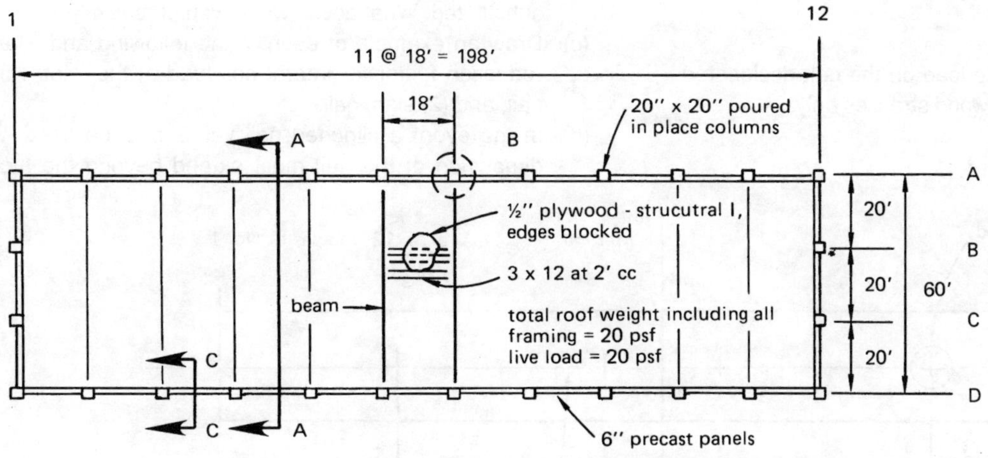

ROOF FRAMING

Figure T-13
(continued on next page)

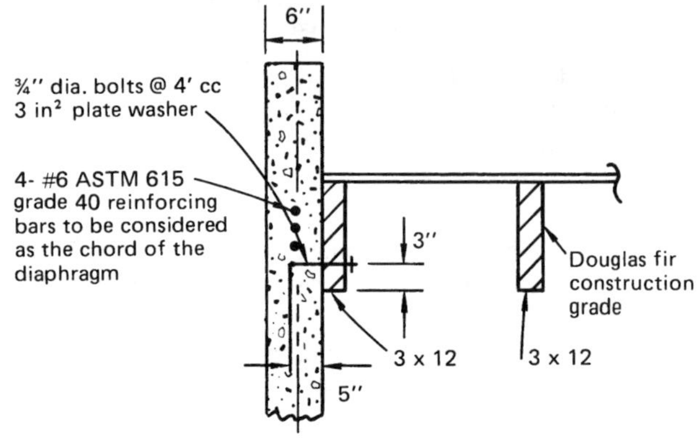

SECTION C-C

DIAPHRAGM DEFLECTION DATA

column data $\left\{ \begin{array}{l} M_{all} = 70 \text{ ft-kips (including 1/3 increase)} \\ EI = 8.0 \times 10^{10} \text{ in}^2\text{-lbs} \end{array} \right.$

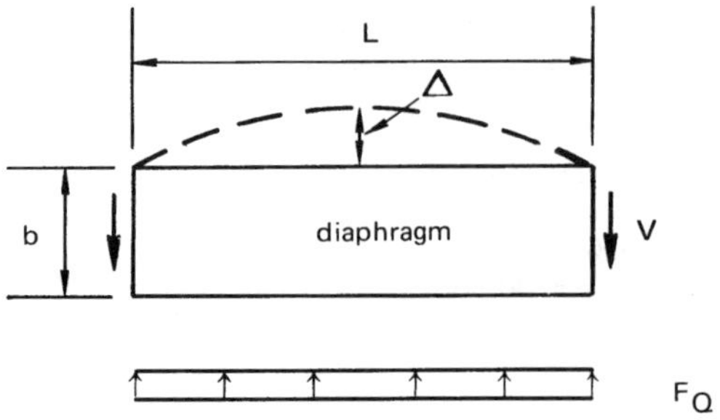

KEY PLAN

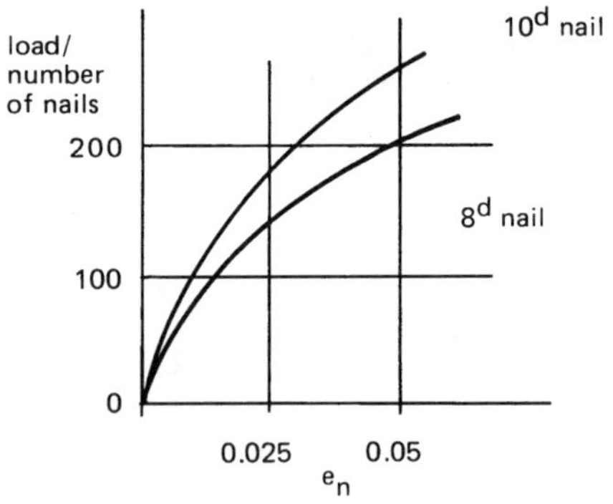

GRAPH G

$$\Delta = \frac{5\,vL^3}{8\,EAb} + \frac{vL}{4Gt} + 0.094\,Le_n$$

Δ = deflection (inches)
v = shear (lb/ft)
L = length (ft)
b = width (ft)
A = area of chord (in²)
E = elastic modulus of chords (psi)
G = 110,000 (psi)
t = plywood thickness (inches)
e_n = nail deformation (see graph G)

Figure T-13 (cont.)

TIMBER-14

REQUIRED

(a) Calculate the allowable load on the bolt designated in Fig. T-14. Consider wood stresses only.

(b) On a design drawing, a specification is shown that calls for structural I plywood for a roof diaphragm. Following a visit to the job site, you find that CDX plywood has been substituted. What action would you pursue?

(c) Draw an example of each of the following and give the reduction factor for lateral nail load at the joint: (1) toe nail, and (2) slant nail.

(d) In the event a clinched nail value is to be used, what dimension of the nail must extend beyond the face of

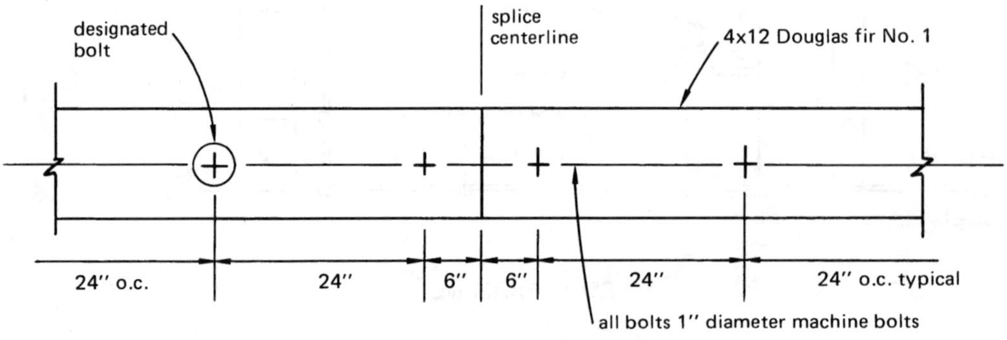

Figure T-14

the material?

TIMBER-15

A 1-story wood frame structure is represented in Fig. T-15.

DESIGN CRITERIA

- Roof dead load = 25 psf (Use for all calculations.)
- Roof live load = 250 psf snow load (Assume two-month duration.)
- Roof joists are Douglas fir-larch (coast region).
- F_b = 1500 psi roof joists
- Glulam beams are Douglas fir.
- F_b = 2600 psi glulam beams
- Compression parallel to grain = 1500 psi
- Horizontal shear = 165 psi
- Compression perpendicular to grain = 450 psi
- Roof deck, walls, and foundations are assumed adequate.

REQUIRED

(a) Verify the adequacy of the typical framing shown in Fig. T-15. Show all calculations.

(b) Indicate the revisions, if any, you would recommend to improve the design shown.

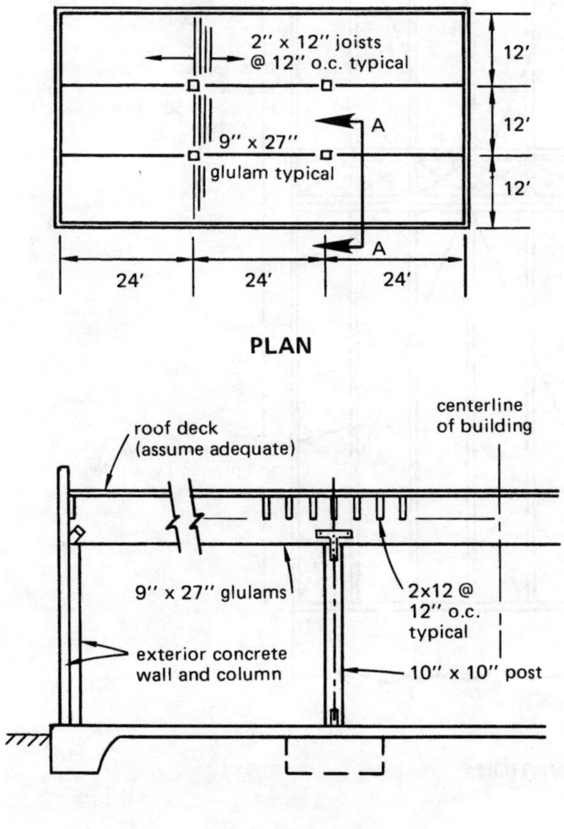

SECTION A-A

Figure T-15

TIMBER-16

The detail in Fig. T-16 represents a hold-down anchor for a shear wall in a wood frame structure.

DESIGN CRITERIA

- Douglas fir-larch
- ASTM A36 steel
- ASTM A307 bolts
- F_b = 1500 psi lumber
- f'_c = 2500 psi concrete
- Neglect bearing of angle on wood and tension of bolts through wood.
- No increase for metal side plates.

REQUIRED

(a) Design the anchor assembly for an uplift force of 4000 lb.

(b) Draw the completed detail and show all member sizes, dimensions, etc., to approximate scale 1 in = 1 ft 0 in.

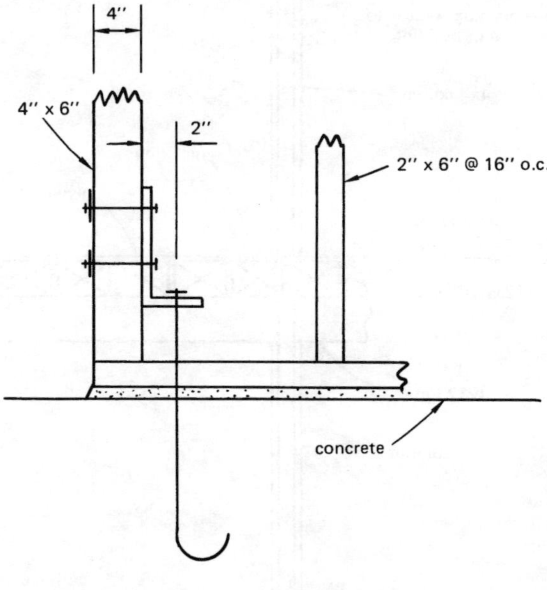

Figure T-16

TIMBER-17

Fig. T-17 shows a portion of a wood-framed building wall. The plywood-sheathed wall panel must resist all lateral loads shown on the diagram.

REQUIRED

Show (approximately to scale) all horizontal and vertical ties that are required to transfer the given lateral loads to the resisting elements. Show by calculations the load for which each tie must be designed. Do not detail the ties. Indicate all ties necessary to carry indicated loads into foundation.

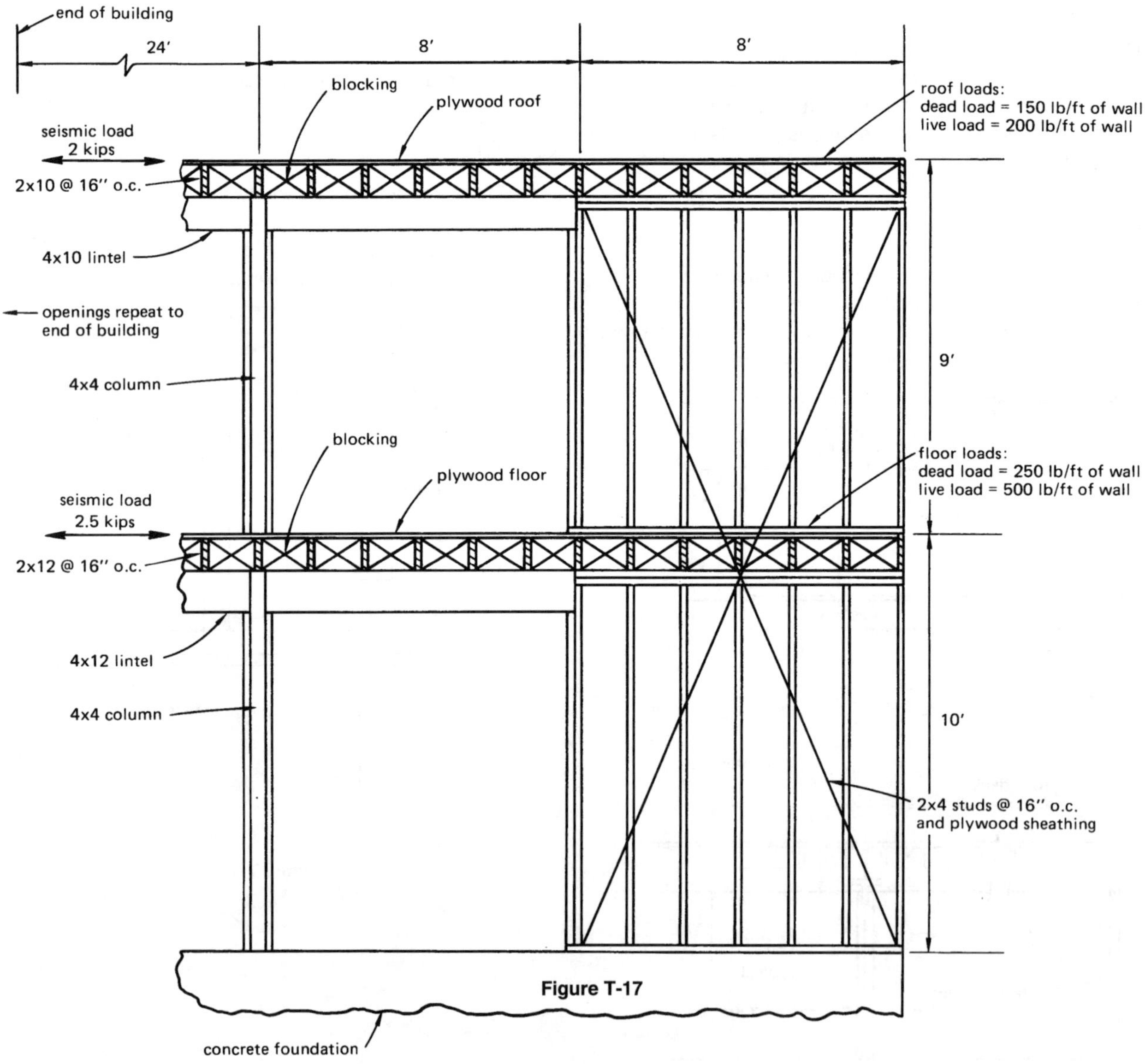

Figure T-17

PARTIAL WALL ELEVATION

TIMBER-18

The existing wood truss has been subjected to a temporary overload from the ceiling, and the lower chord is split through one row of bolts as shown in Fig. T-18. Assume the original design was adequate.

REQUIRED

(a) Describe a method whereby the damage can be repaired.
(b) Describe a construction sequence to be followed during operations. Show any calculations and additional comments that may be applicable.

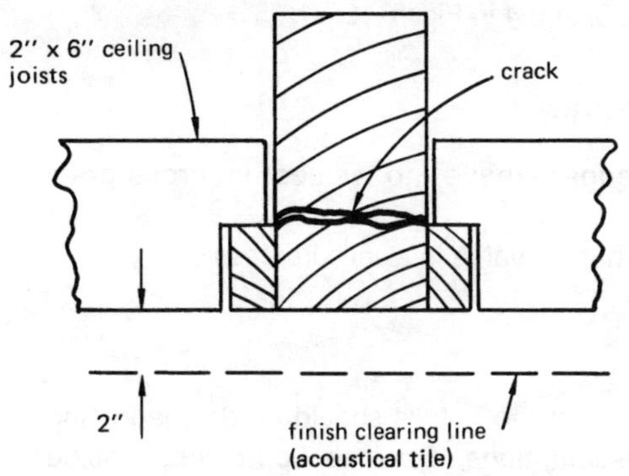

SECTION THRU CHORD

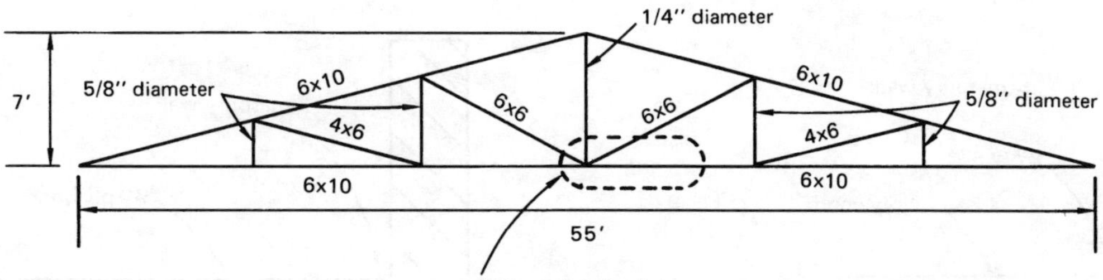

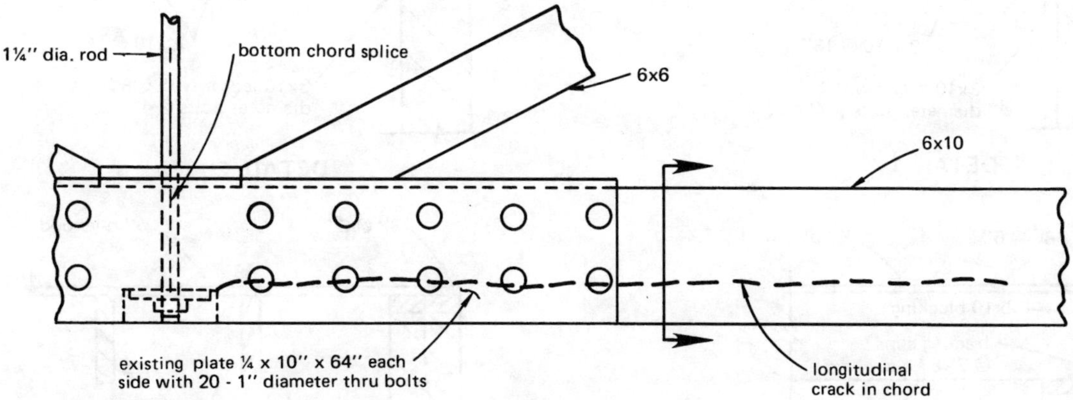

Figure T-18

TIMBER-19

A building with wood roof framing and masonry walls has roof-to-wall details that were constructed as shown in details 1, 2, 3, and 4 in Fig. T-19.

DESIGN CRITERIA

- Wood ledgers have no values in cross-grain bending.
- Plywood has no value in axial compression.

REQUIRED

Show how each of the details should be detailed using the stated assumptions. Draw to approximate scale 1 in = 1 ft 0 in.

TIMBER-20

Two roof beams for a 1-story wood frame building are shown in Fig. T-20.

DESIGN CRITERIA

- Beam A has a span of 18 ft plus a cantilevered end.
- Beam B has a simple span of 24 ft.
- Both beams bear on wood posts at the ends which do not provide moment resistance.
- Assume the beams are stayed against sidesway at their top edges and at all supports.
- The building will have a suspended ceiling, so appearance is not a factor.
- Use the current UBC.
- Assume no axial force in either beam A or B.

REQUIRED

(a) Design beams A and B using combination C glulams.
(b) Provide complete design calculations for a support connection at D for beam B.
(c) Draw a complete detail of connection at D to approximate scale 1 in = 1 ft 0 in.

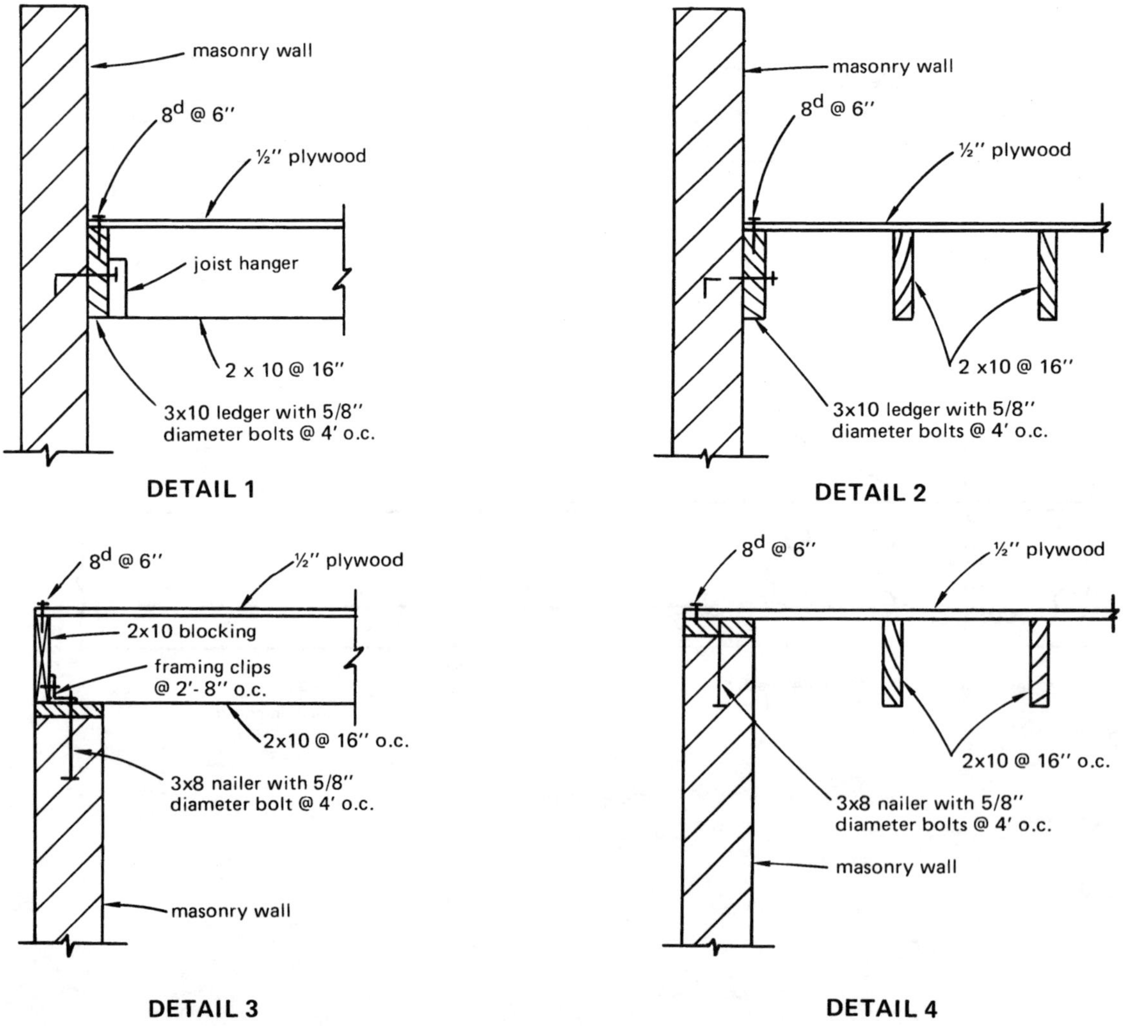

Figure T-19

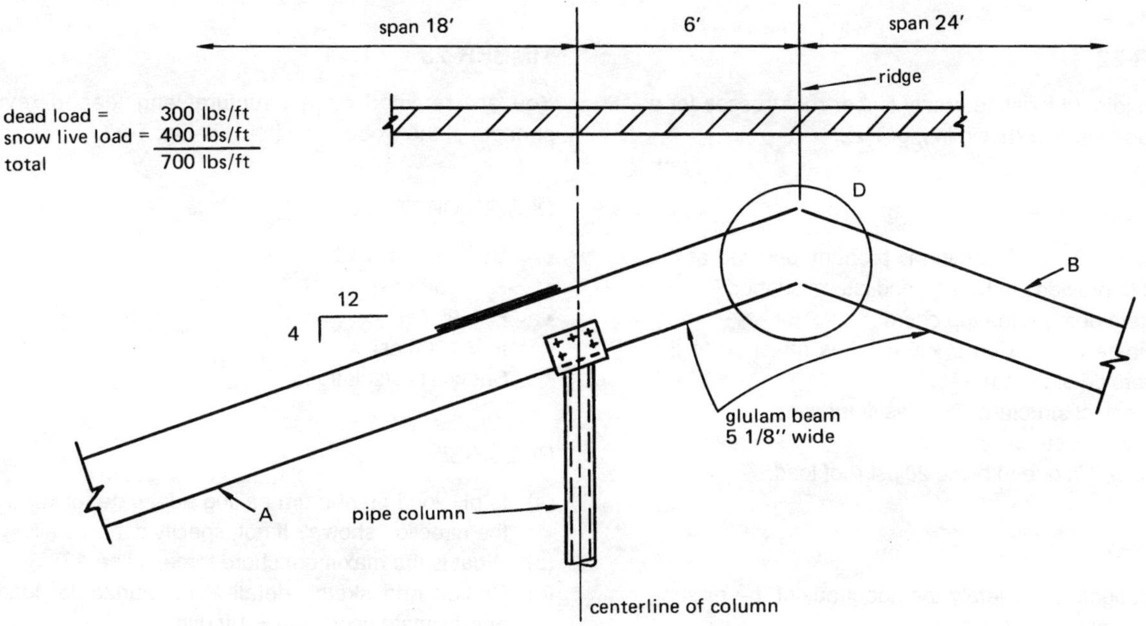

Figure T-20

TIMBER-21

A typical header beam for a first story window wall of a 2-story office building is shown in Fig. T-21.

DESIGN CRITERIA

- Assume the columns are adequate and the top edge of the beams is laterally stayed.
- ASTM A36 steel

REQUIRED

Show complete calculations where applicable.

(a) Check the adequacy of the beams using 22F combination glulams (Douglas fir, dry condition of use).

(b) Design a support connection at A.

(c) Draw a complete detail of the connection at A to the scale of 1 in = 1 ft 0 in.

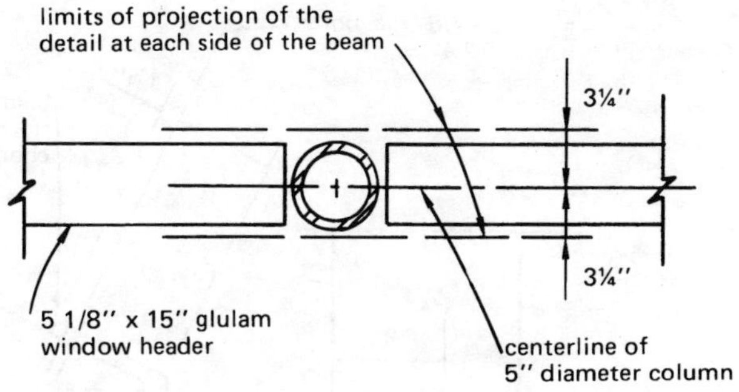

SECTION X-X

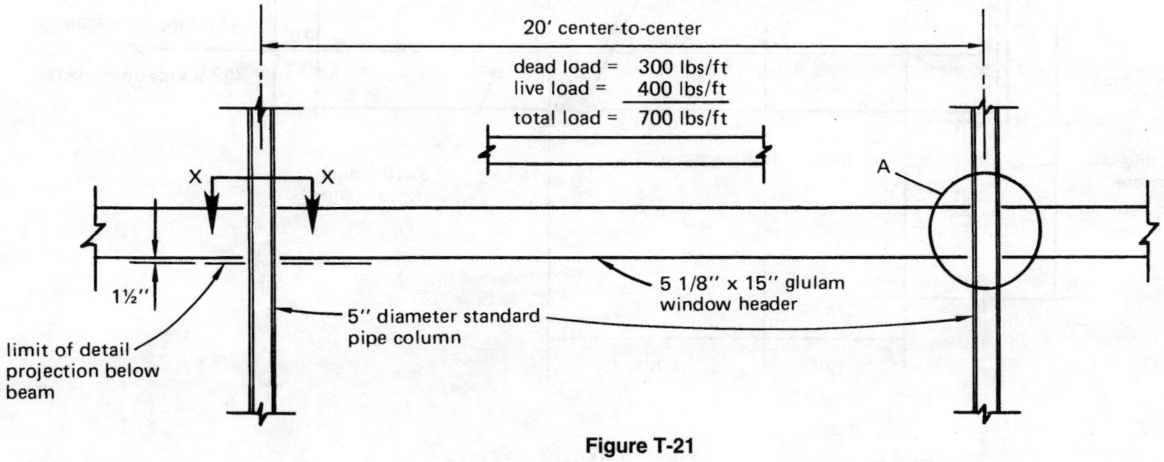

Figure T-21

TIMBER-22

The heel joint of existing heavy timber roof trusses for a public auditorium is shown in Fig. T-22.

DESIGN CRITERIA

- Assume the roof framing is properly detailed at the wall to provide anchorage and lateral support.
- Rafters bear on the top chord.
- A plaster ceiling on 2 x 6 joists is attached to the underside of the trusses.
- Use select structural Douglas fir lumber.
- ASTM A36 steel
- Live load is due to basic 20 psf roof load.

REQUIRED

(a) Investigate completely the adequacy of the existing heel joint.

(b) Comment on the condition of the joint briefly. Should the auditorium be closed temporarily for strengthening? (The facility is used daily, and closing would cause considerable inconvenience and loss of revenue for the owner.)

TIMBER-23

You are retained as a structural engineer to review the strength of the wood roof shown in Fig. T-23.

DESIGN CRITERIA

- Use current UBC.
- $F_b = 1500$ psi
- $F_v = 95$ psi
- $F_c = 1200$ psi
- Show all calculations.

REQUIRED

(a) Is plywood diaphragm nailing adequate for wind force in the direction shown? If not, specify correct nailing.

(b) What is the maximum chord force at line A?

(c) Design and sketch detail P for horizontal loads. Use approximate scale 1 in = 1 ft 0 in.

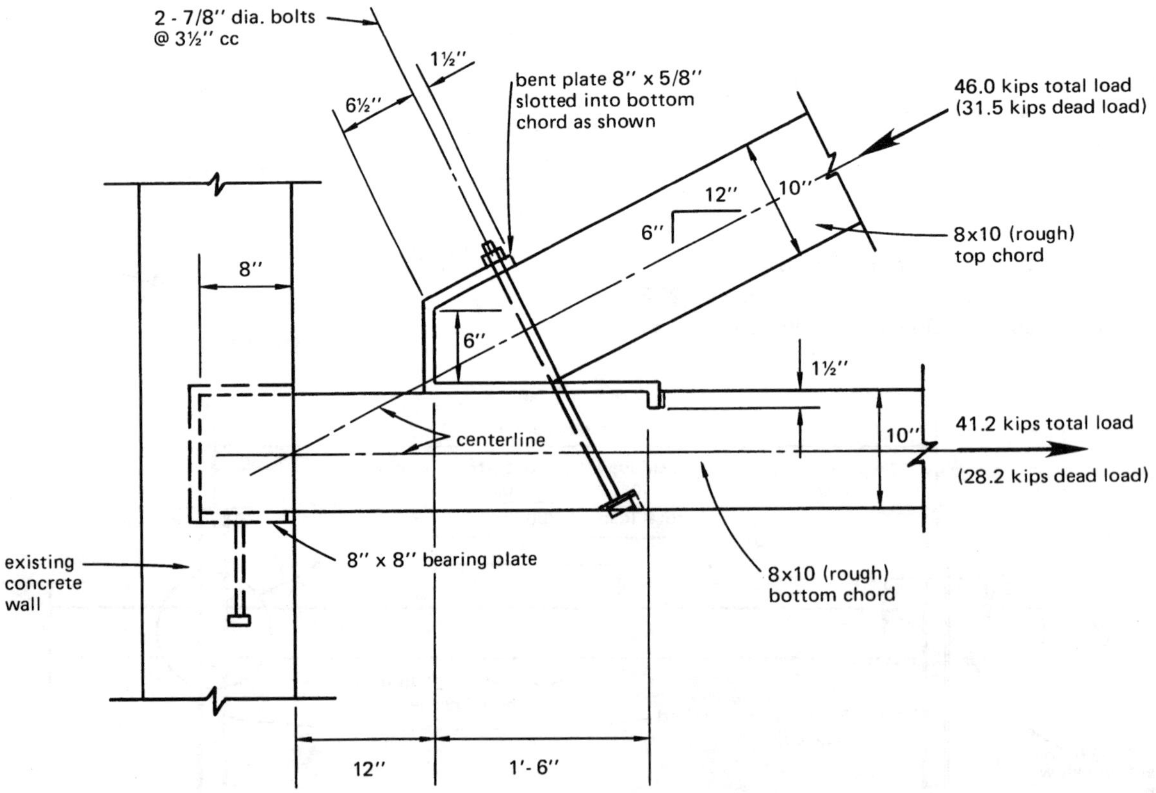

Figure T-22

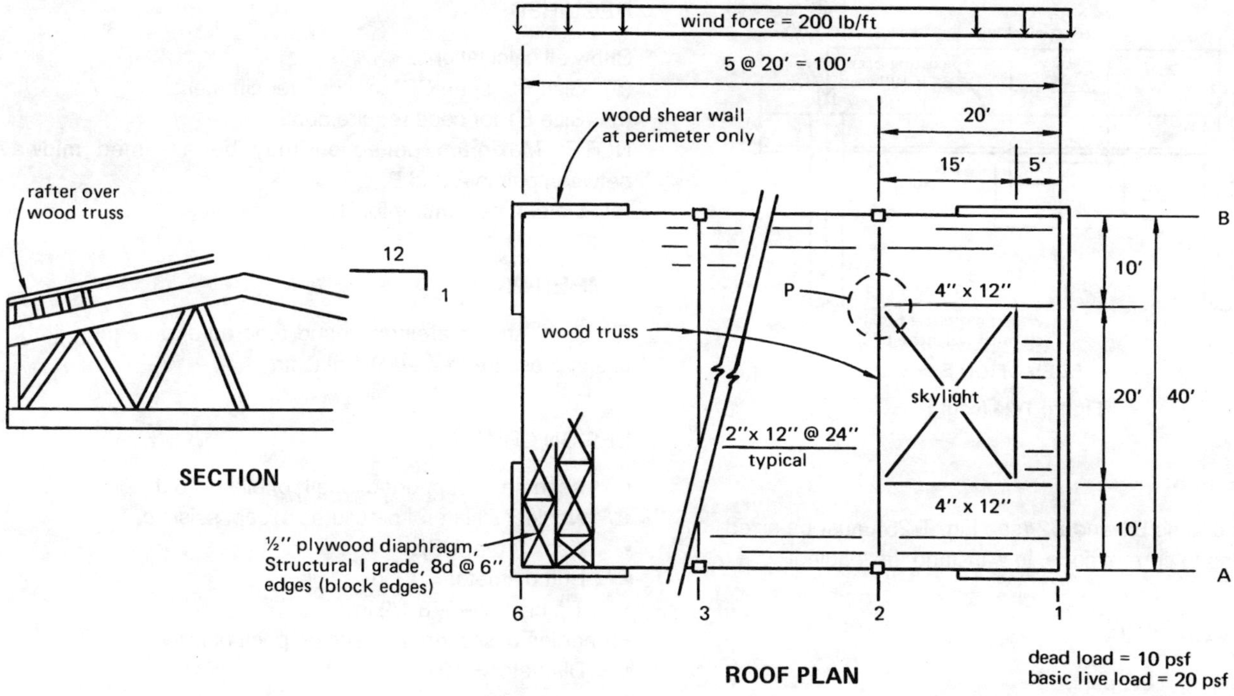

Figure T-23

TIMBER-24

The wood bents shown in Fig. T-24 support a 6 x 18 beam along which runs a monorail. It is desired to remove that portion noted and install a new wood brace as shown.

DESIGN CRITERIA

For sawn lumber, allowable stresses,

- $F_b = 1350$ psi
- $F_c = 925$ psi
- $F_{c\perp} = 385$ psi
- $F_v = 85$ psi
- $E = 1.6 \times 10^6$ psi

For nontapered wood pile, allowable stresses,

- $F_b = 1800$ psi
- $E = 1.1 \times 10^6$ psi
- $F_c = 1200$ psi maximum or $0.225\ E/(l/d)^2$

REQUIRED

Show all calculations for all parts of problem.

(a) Will the remaining portion of the 6 x 16 beam be adequate for the design load?

(b) Will the remaining 14 in diameter nontapered wood pole be adequate for the design load?

(c) Assuming that the previous connection will not interfere with the new connection, design and sketch to an approximate scale the joint noted as 1 on the sketch.

(d) Determine the minimum size of the 4x wood brace which may safely be used to support the design load.

(e) Design and sketch to an approximate scale the joint noted as 2 on the sketch.

NOTE: Neglect impact loading, lateral loading, and deflection in your solution.

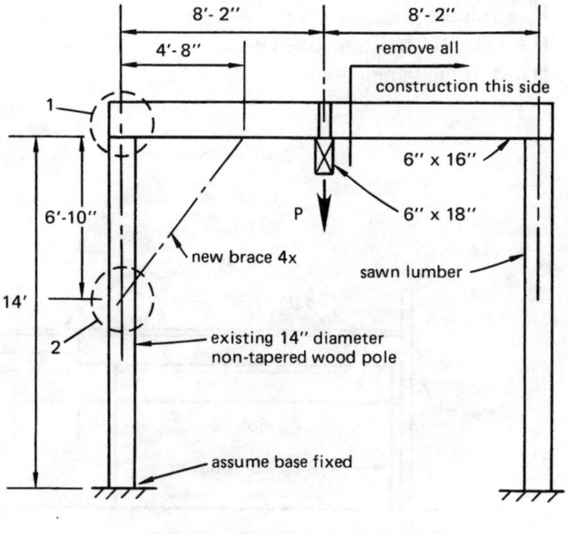

Figure T-24

(continued on next page)

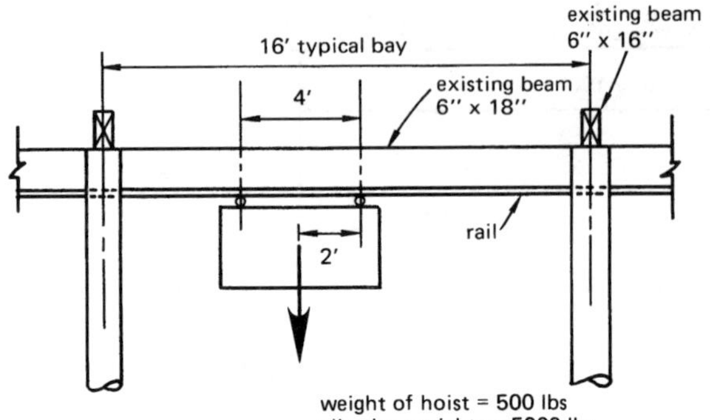

ELEVATION E

Figure T-24 (cont.)

TIMBER-25

Glulam beams B1 and B2 (see Fig. T-25) must be sized according to current code, to withstand the loading.

DESIGN CRITERIA

- Roofing = 6.0 psf
- Sheathing = 2.0 psf
- Purlins = 2.0 psf
- Plaster ceiling = 10.0 psf
- Dead load (total) = 20.0 psf
- Live load = 20.0 psf (reducible per code)
- Joists framing perpendicular to glued laminated beams adequately stay beams at points B and C.
- Use current UBC.
- Douglas fir
- F_b = 2200 psi
- F_v = 165 psi
- E = 1.8 x 10^6 psi, seasoned wood
- Use 1 1/2 in laminations

REQUIRED

Show all calculations.

(a) Size B2 (span CD) for code requirements.

(b) Size B1 for code requirements.

NOTE: Maximum deflection may be assumed midway between points A and B.

(c) Determine camber for B2.

TIMBER-26

Fig. T-26 shows a single round tapered-end bearing timber pile of pressure-treated Douglas fir.

DESIGN CRITERIA

- Assume unsupported length of pile is 16 ft.
- Neglect all lateral pressures except seismic.
- Use current UBC.
- Butt diameter = 12 in
- Tip diameter – 8 1/2 in

Properties of section at assumed point of fixity:

- Diameter = 10 in
- Area = 78.5 in^2
- I = 490 in^4
- E = 1.5 x 10^6 psi
- $Fc = \dfrac{3.619E}{(1/r)^2}$

REQUIRED

Investigate maximum stresses in pile for given loads only. Is pile adequate? Show all calculations.

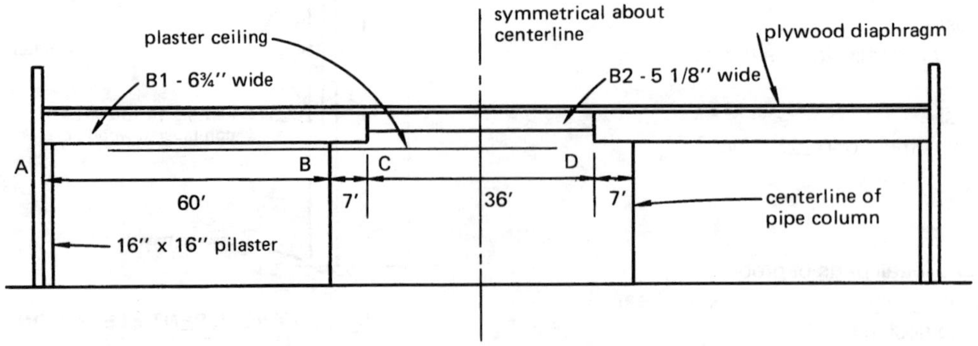

TYPICAL SECTION THROUGH BUILDING

(beams on 16″ centers)

Figure T-25

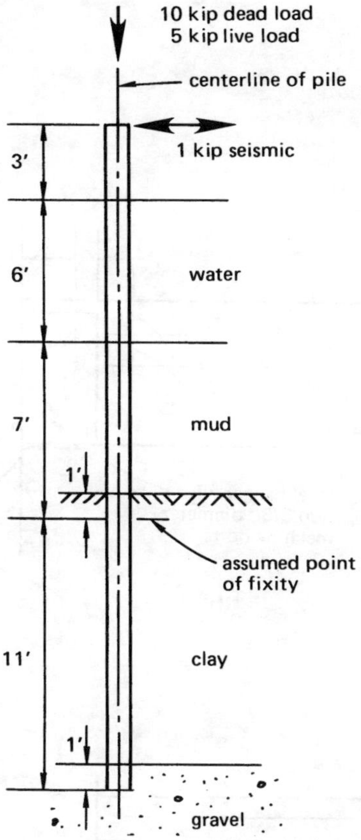

Figure T-26

TIMBER-27

Fig. T-27 shows a partial elevation of a 2-story wood frame building with a frame door opening. The plywood sheathed wall must resist all the lateral loads shown. Assume 75% of dead loads to resist uplift.

DESIGN CRITERIA

- Douglas fir structural I grade plywood, 1/2 in
- Douglas fir construction grade framing
- F_b = 1500 psi framing

REQUIRED

(a) Give nailing of entire plywood wall sheathing.

(b) Design all boundary members below the second floor and describe where located. Show all calculations for parts (a) and (b).

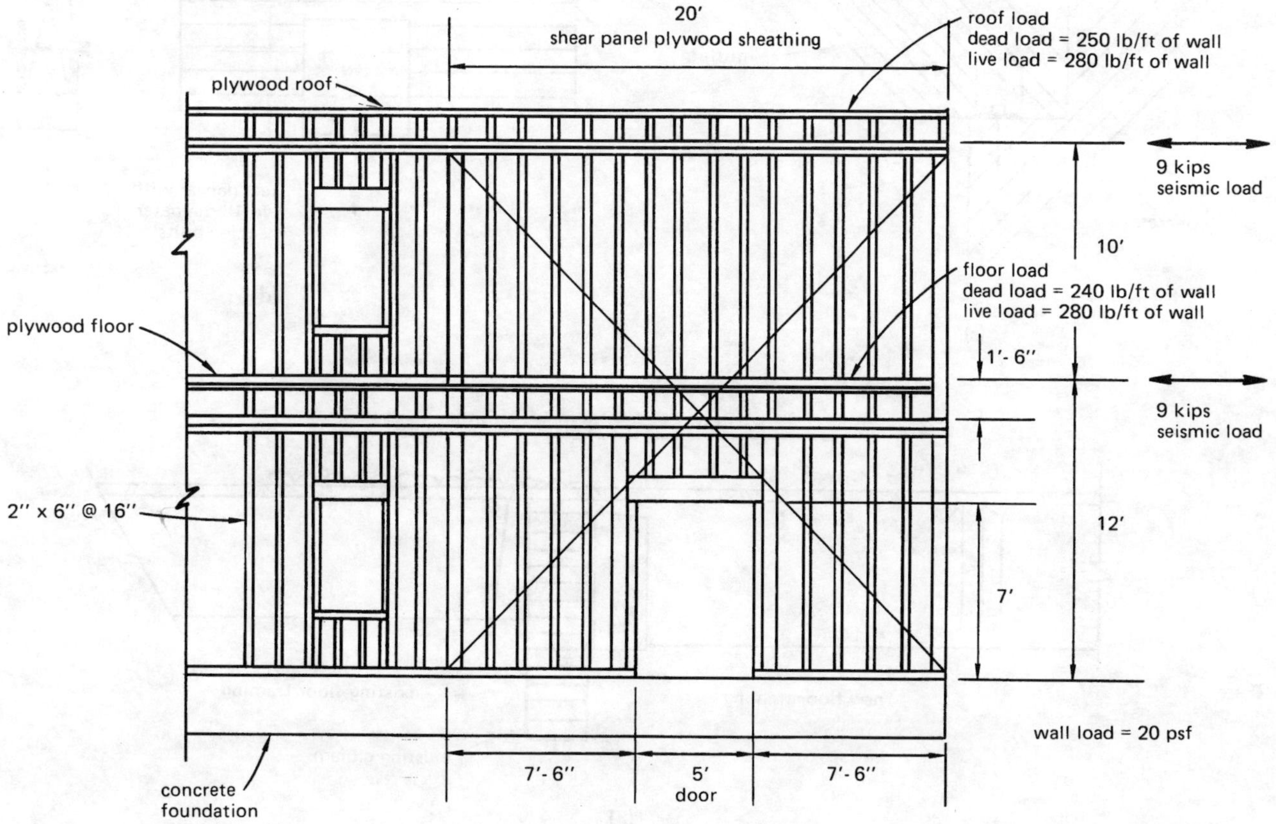

ELEVATION LOOKING NORTH

Figure T-27

TIMBER-28

During an inspection of a building under construction, you observe the details shown in Fig. T-28. Comment on the suitability of each and identify any potential problems.

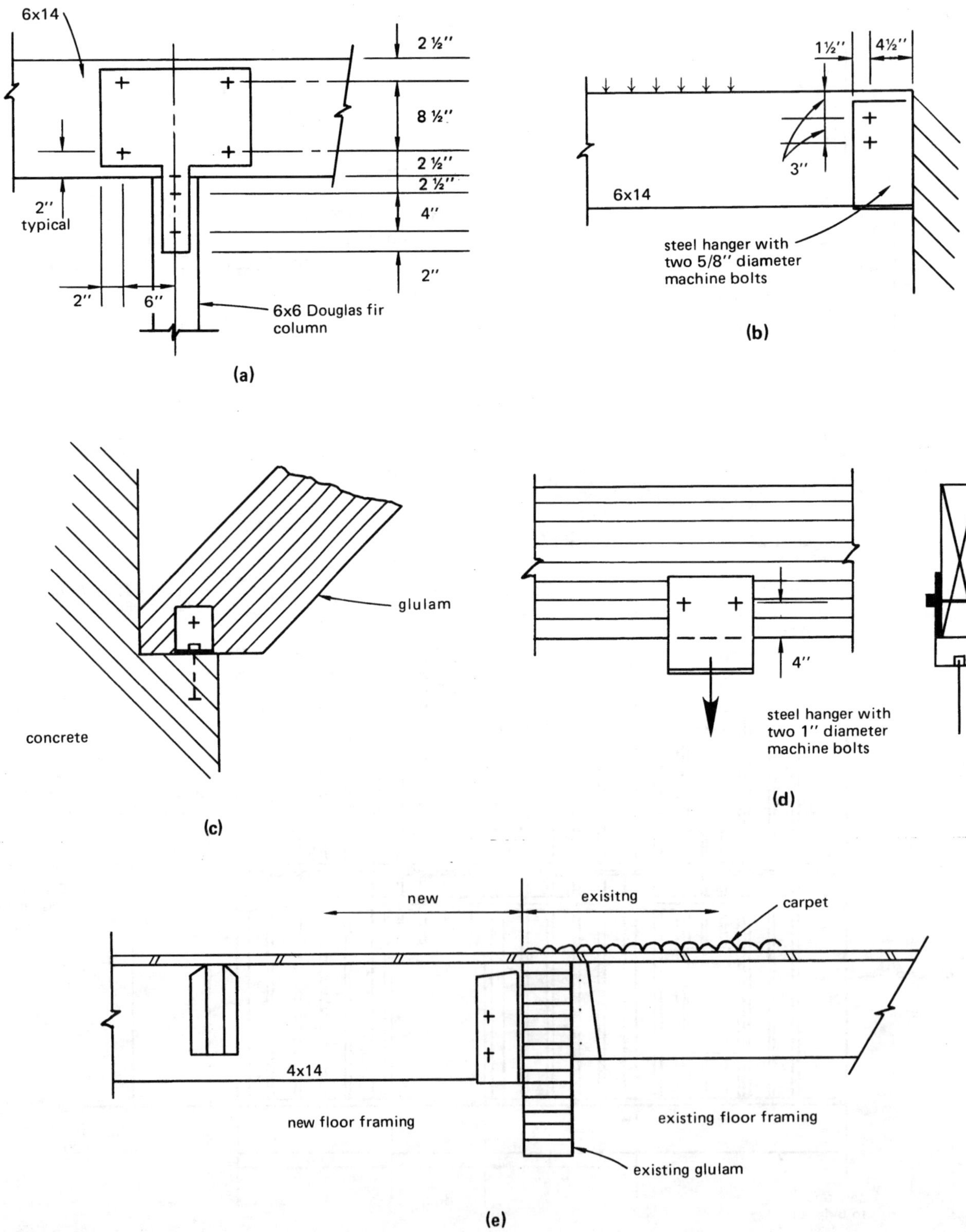

Figure T-28

TIMBER-29

A single story wood frame warehouse building is illustrated in Fig. T-29. The roof framing plan and a typical cross section are indicated.

DESIGN CRITERIA

- Current UBC is suggested.
- 80 mph basic wind speed; exposure C
- Seismic zone 4
- Douglas fir-larch No. 1 wood
- Douglas fir-western larch combination 22F glulam beams
- 1 1/2 in laminations
- Beam width = 5 1/8 in
- 1/2 in plywood structural I roof sheathing
- Built-up composition roofing = 7 psf
- Roof is level (no slope).
- 5/8 inch gypsum board ceiling = 3 psf
- Exterior walls: 2 x 6 studs with 7/8 in exterior cement plaster and interior surface of 5/8 in gypsum board. Exterior sides of walls are sheathed with 3/8 in structural I plywood. Studs are to extend to top of parapet with no continuous plate at the roof level. Assume the weight of exterior wood frame walls to be 15 psf. The west exterior wall is of reinforced concrete block construction that weighs 80 psf.
- Soil profile type S_D
- Near-fault factors = 1.0

REQUIRED

(a) Design roof joists. Neglect ponding. Allowable stresses under live load plus dead load may be increased by 25%.

(b) Design glulam beams and specify camber. Allowable stresses under live load plus dead load may be increased by 25%.

(c) Calculate lateral loads to roof diaphragms (w_1 and w_2) for the north-south directions, and (w_3 and w_4) for the east-west direction.

(d) For north-south loads:

  (1) Calculate maximum shear and cord stresses in roof diaphragm.

  (2) Specify plywood nailing for maximum shear.

  (3) Design and detail a diaphragm chord at the north edge (line A) of the diaphragm for maximum chord stress.

NOTE: Use an approximate scale of 3/4 in = 1 ft 0 in for all details.

(e) Provide design calculations and draw section B-B showing the roof-to-wall connection at the end wall 3. Provide for all lateral and gravity loads and also develop a continuous diaphragm chord that will taken an axial chord stress of 1500 lb.

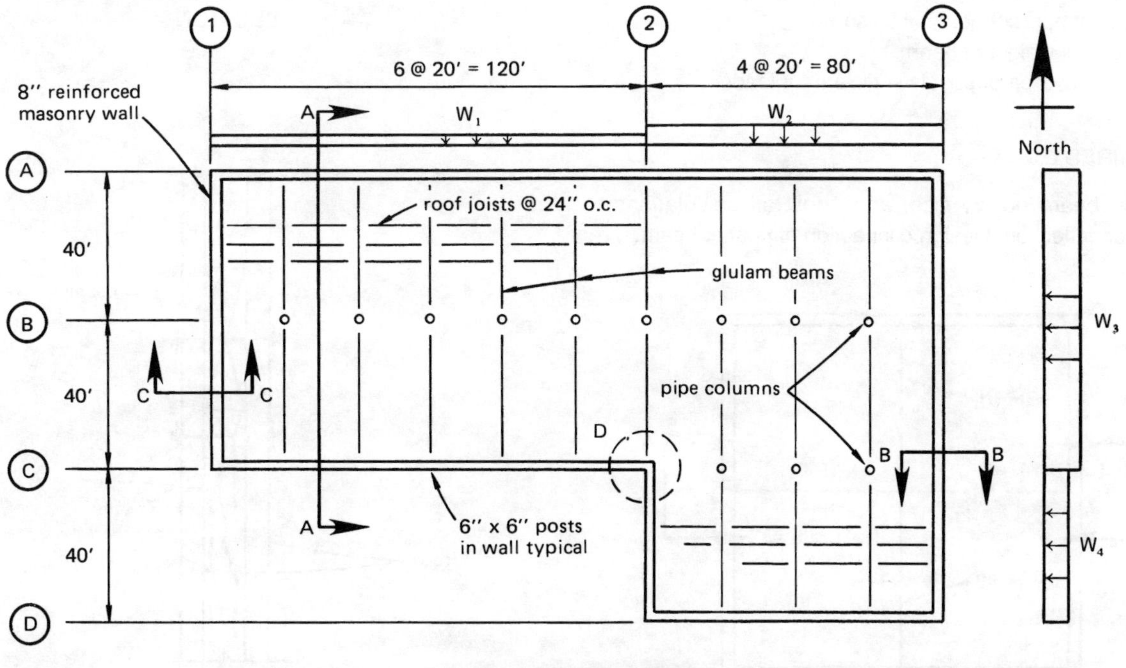

ROOF FRAMING PLAN

Figure T-29
(continued on next page)

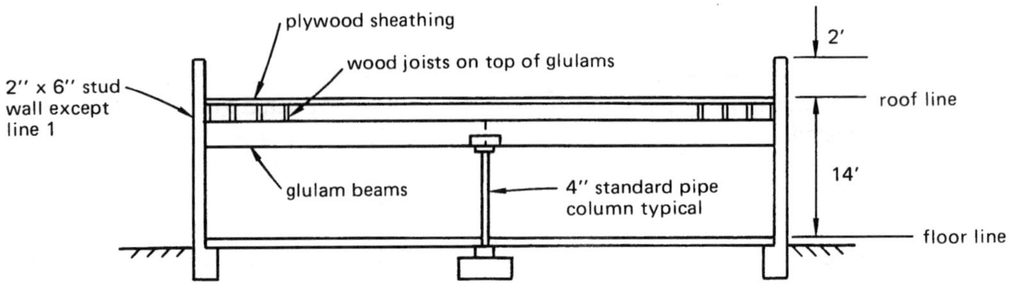

SECTION A-A

Figure T-29 (cont.)

TIMBER-30

You are commissioned to locate a beam 3 ft below the load it is supporting, as shown in the plan and section in Fig. T-30.

DESIGN CRITERIA

- Douglas fir glued laminated beams
- $F_b = 2000$ psi
- $F_v = 165$ psi
- $E = 1.8 \times 10^6$ psi
- 1 1/2 in laminations
- No increase in allowable stresses should be made for loads of short duration.
- Width of beam no. 3 = 3 1/8 in
- Span of beam no. 3 = 20 ft
- Load is at midpoint.
- Load from post = 3100 lb
- End support of beam is as shown.
- Neglect weight of beam.
- Bearing area of post is sufficient for load.

REQUIRED

Design beam no. 3, size, and show all calculations. Neglect deflection. Design connection of post to beam.

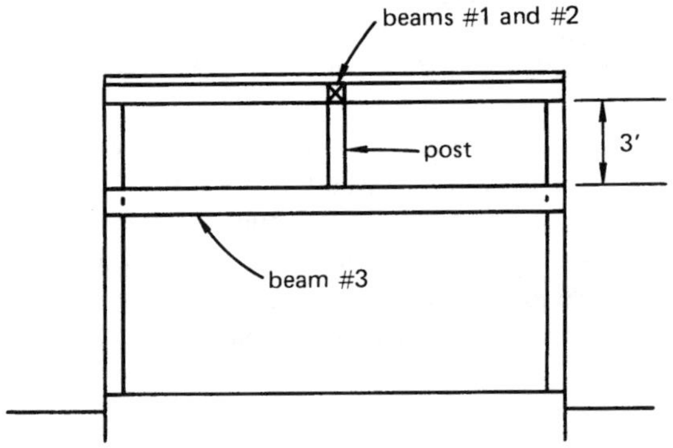

SECTION

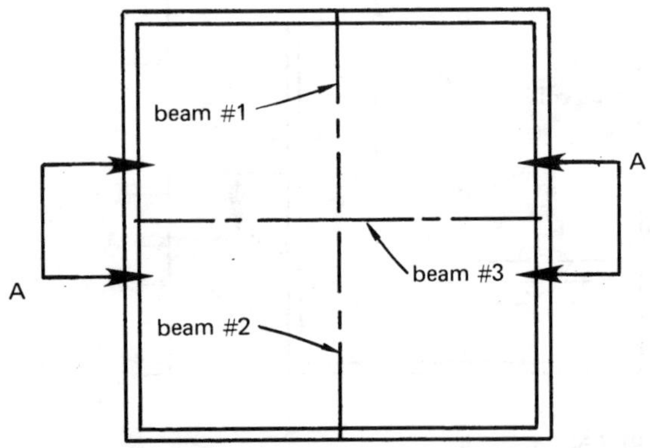

PLAN

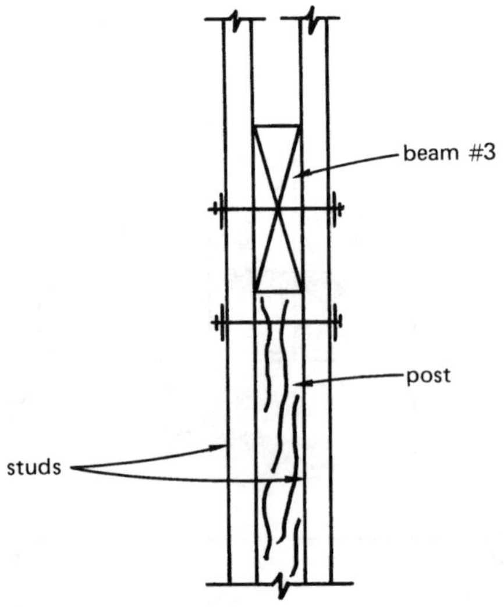

DETAIL A

Figure T-30

TIMBER-31

Several bowstring roof trusses over a supermarket are exhibiting failure by splitting along the bolts of the bottom chord heel and splice connections. See Fig. T-31. The trusses have been temporarily shored to prevent collapse. The owner desires to repair the trusses at minimum cost, and considers intermediate supports unacceptable.

DESIGN CRITERIA

- Current UBC with 25% allowable increase in stresses for short-term live load
- Bottom chord: 8 x 8 Douglas fir-larch select structural, cut as shown in sketch
- ASTM A36 steel plates
- ASTM A307 bolts

REQUIRED

(a) Check the design of the bottom chord. List two possible reasons why the failures may have occurred.

(b) Design and provide sketches of the system of repair that you would propose. A system that only replaces the damaged timbers with new timbers will not be considered acceptable.

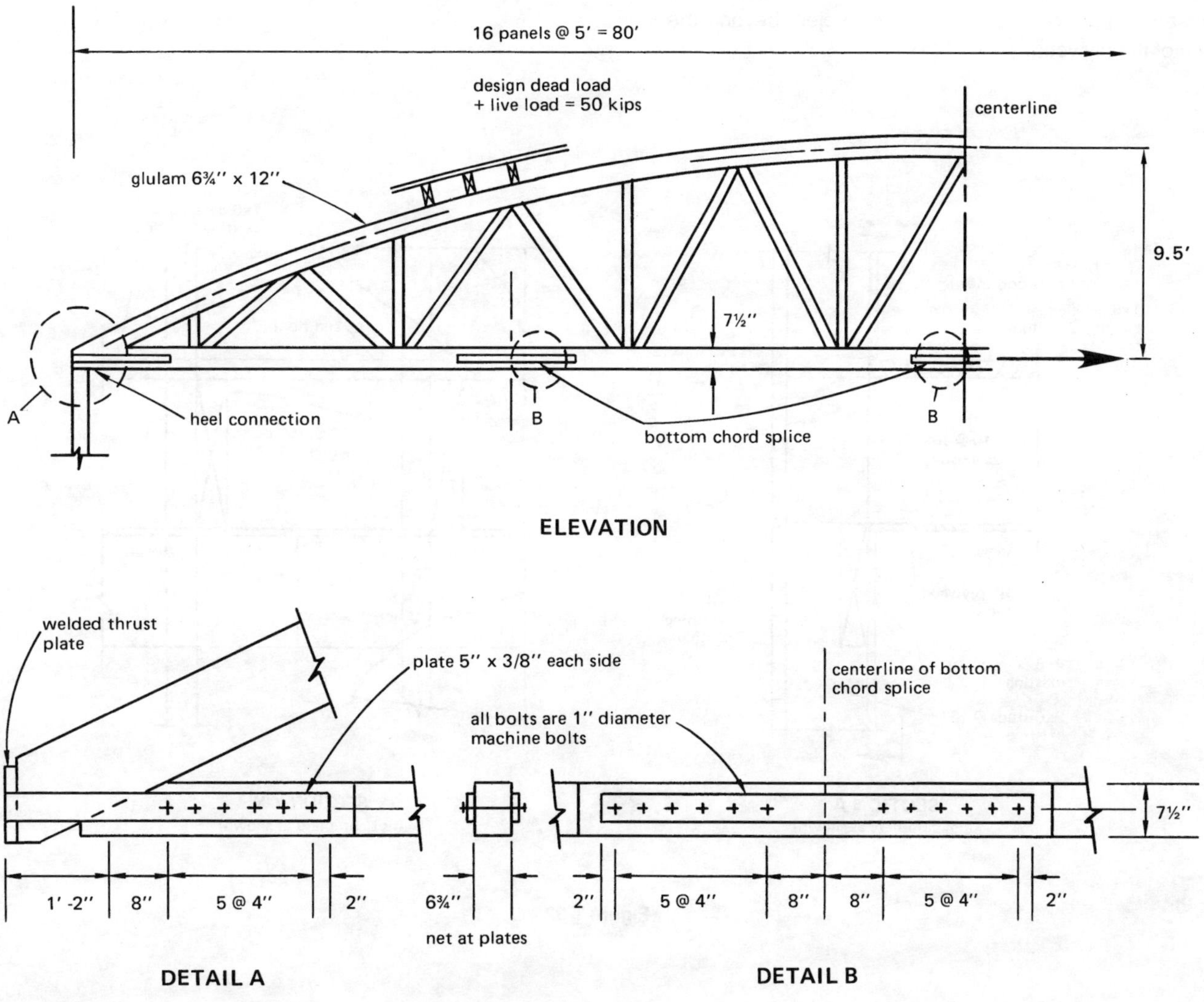

Figure T-31

TIMBER-32

The sections shown in Fig. T-32 are taken at the second floor of an existing building. New 1/2 in plywood is to be added at the inside face of studs, and at the bottom of floor joists, which will act as vertical and horizontal diaphragms, respectively.

DESIGN CRITERIA

- Wall shear = 400 lb/ft above second floor
- Diaphragm shear = 200 lb/ft at second floor

REQUIRED

(a) Reproduce section A and complete the needed details. Show nailing and shear flow across second floor to wall sheathing below the second floor.

(b) Draw a detail for the existing 2 x 8 shown in section X-X for a chord force of 1000 lb.

NOTE: For your details, use scale of 1 in = 1 ft 0 in. To preserve appearances, nothing can project beyond the face of the plywood.

TIMBER-33

Shown in Fig. T-33 is a typical wood roof truss at 4 ft spacing. Bottom chord is raised above its usual location to satisfy the architectural requirements.

DESIGN CRITERIA

- Dead load = 14 psf projected upon horizontal plane
- Live load = 16 psf projected upon horizontal plane
- Douglas fir-larch select structural
- Use 25% increase in allowable stress for roof live loads.
- Assume supports offer no horizontal restraint.
- Current UBC is suggested.

REQUIRED

(a) Check adequacy of member ABC.

(b) Design and draw a detail of joint B.

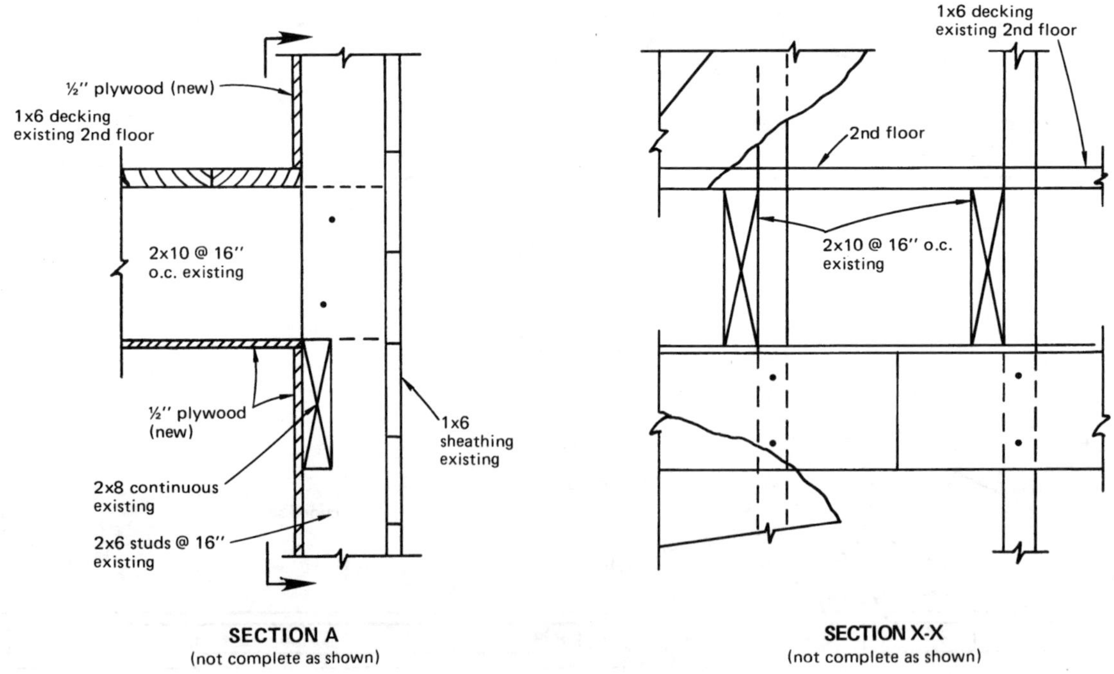

Figure T-32

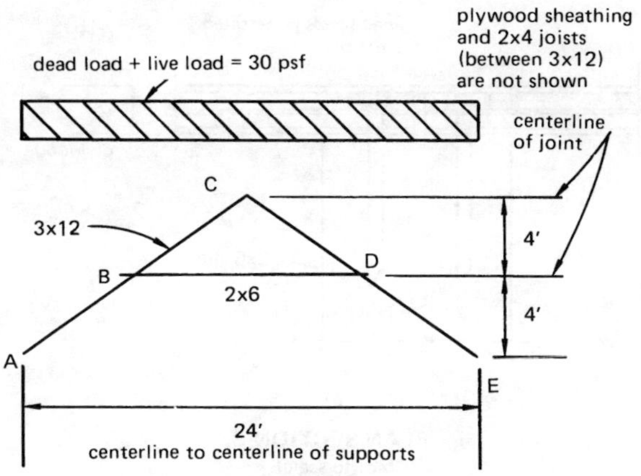

Figure T-33

TIMBER-34

Given in Fig. T-34 is a pyramid-shaped wood roof. The space enclosed by the pyramid is clear and without any structural members.

DESIGN CRITERIA

- Dead load = 16 psf projected upon a horizontal plane
- Live load = 16 psf projected upon a horizontal plane
- Douglas fir-larch no. 1 lumber
- Use 25% increase in allowable stress for roof live load.
- Neglect unbalanced live load condition.
- Current UBC is suggested.
- Design for vertical loads only.

REQUIRED

(a) Find forces acting at joint A.
(b) Design and sketch joint A.

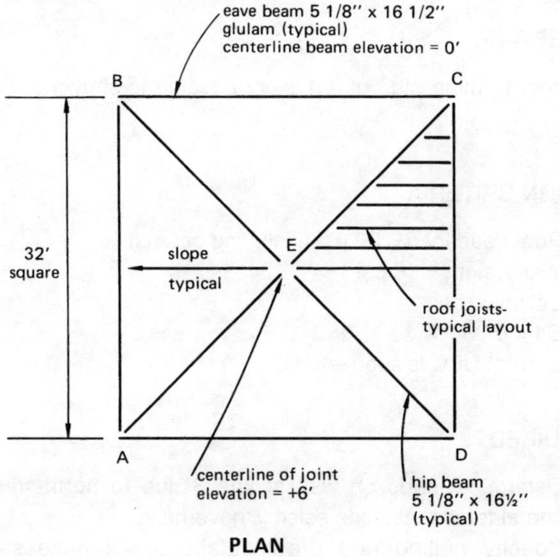

PLAN

Figure T-34

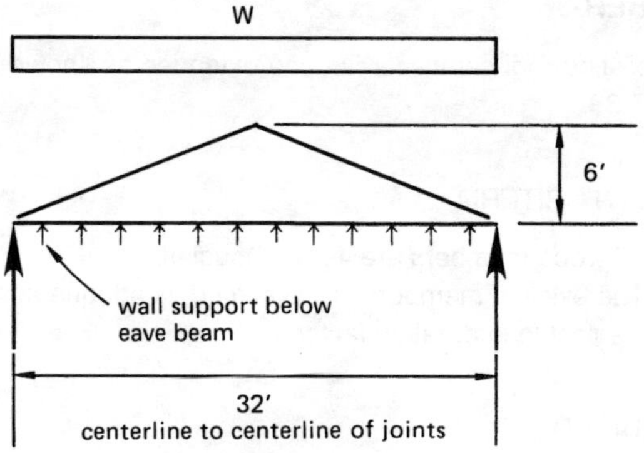

ELEVATION

Figure T-34 (cont.)

TIMBER-35

An elevation of glulam roof beam is shown in Fig. T-35.

DESIGN CRITERIA

- F_b = 2400 psi glulam
- 5 1/8 in x 18 in glulam beam
- 1 1/2 in laminations
- E = 1.8 x 10^6 psi
- S = 276.75 in^3
- A = 92.25 in^2
- Radial tension reinforcement available: 3/4 in lag screws
  - 5240 lb allowable tension
  - 528 lb/in of penetration allowable withdrawal
- No increases are allowed for load duration.
- There is no need to check deflection.

REQUIRED

(a) What is the allowable bending stress?
(b) What is the maximum spacing of the top flange bracing to permit loading as shown in the figure?
(c) Check radial tension (f_t allowable = 15 psi).
(d) How would you provide reinforcement for radial tension? Use a convenient scale.

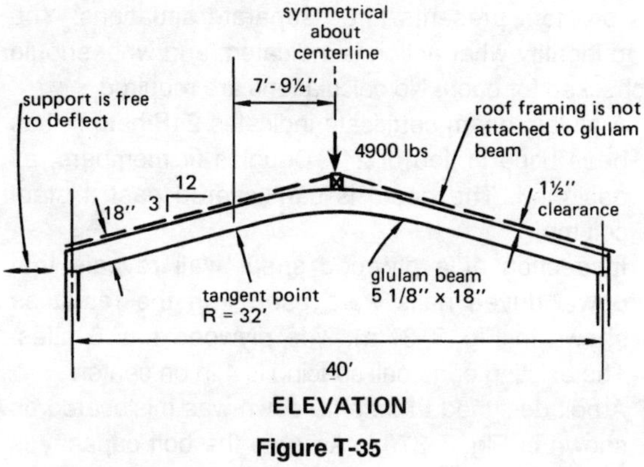

ELEVATION

Figure T-35

TIMBER-36

An existing roof cantilever is to be extended as shown in Fig. T-36.

DESIGN CRITERIA

- All wood members are 4 x 12 Douglas fir.
- The existing members are assumed as adequate.
- Neglect load duration factor.

REQUIRED

What is the maximum load P that the joint connection can support?

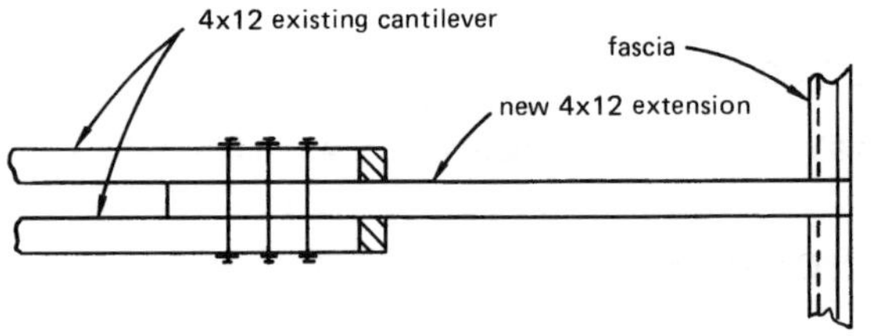

PLAN

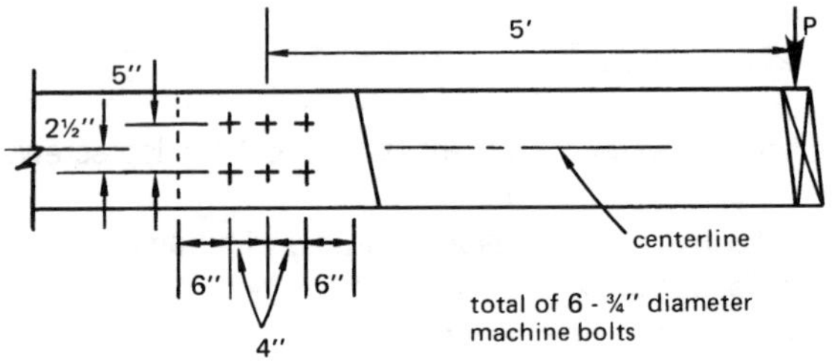

ELEVATION

Figure T-36

TIMBER-37

REQUIRED

This problem presents three separate situations. You are to identify what action is indicated, and what should be checked for each. No calculations are required.

(a) A glulam beam certificate indicates 24F hem fir has been used in lieu of 24F Douglas fir members, as designed. The beam is cantilevered past a steel column.
(b) Inspection of a plywood shear wall reveals that power-driven nails were used with the result as shown in Fig. T-37(a). The plywood has 5 plies. The existing edge nail spacing is 4 in on center.
(c) A bolt designed to be a tie-down was mislocated as shown in Fig. T-37(b). Assume the bolt capacity is adequate at the position shown.

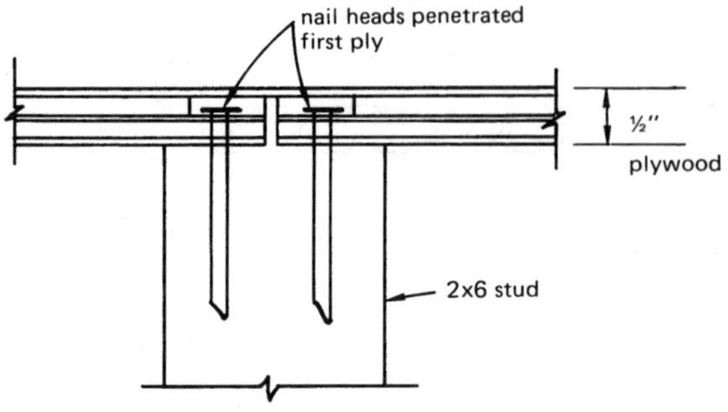

(a)

PLAN SECTION B

(not to scale)

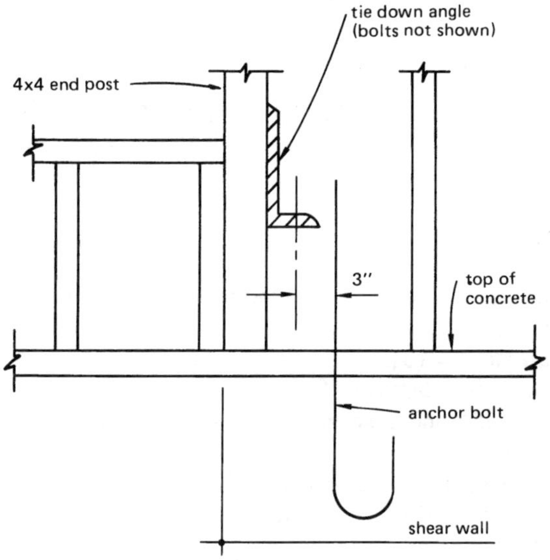

(b)

ELEVATION C

(not to scale)

Figure T-37

TIMBER-38

The roof framing plan of a 1-story building is shown in Fig. T-38.

DESIGN CRITERIA

- Roof dead load = 20 psf (including beams)
- Wall weight = 16 psf
- Seismic zone 3
- Ca = 0.33
- Current UBC is suggested.

REQUIRED

(a) Determine loads on wall at line 3 due to north-south lateral forces. Assume seismic governs.
(b) Specify nailing and draw details of all necessary connections to the shear wall on line 3 at roof level.

Design and detail a typical section of wall on line 3 between lines A and B for north-south lateral forces. Use structural I plywood.

(c) Calculate forces between wall on line 3 and footing. Design and detail wall-to-wall footing connections for these forces. Do not design footing reinforcement. Assume soil is adequate for pressures imposed.

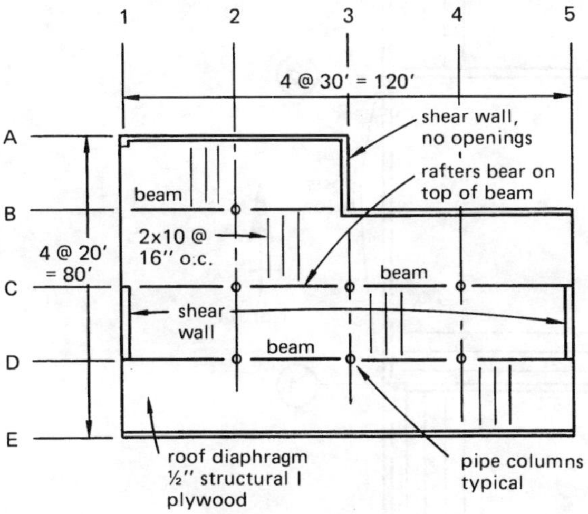

ROOF FRAMING PLAN

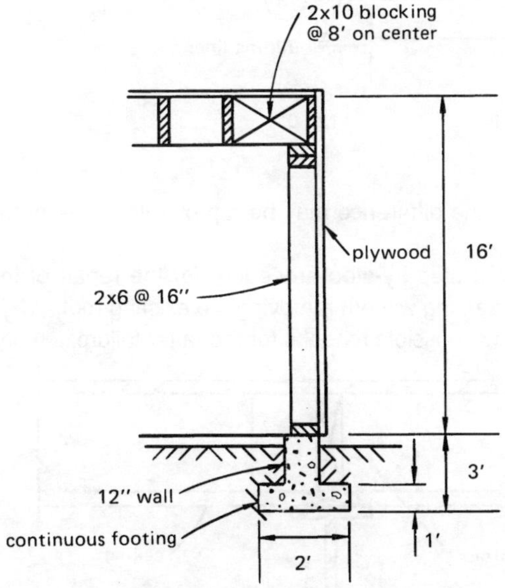

WALL SECTION
typical shear wall

Figure T-38

TIMBER-39

A 1-story wood frame building with center roof portion 4 ft above the building roof is shown in plan in Fig. T-39.

DESIGN CRITERIA

- All roofs are level.
- Allowable stresses and design loads are as per current UBC.
- Seismic zone 3
- Ca = 0.33
- High roof elevation = + 16 ft 0 in
- Low roof elevation = + 12 ft 0 in
- High roof and low roof dead load = 20 psf (including beams)
- High roof and low roof walls dead load = 20 psf
- Elevations are to bottom of joists.
- Roof and wall framing is plywood sheathed.
- Roof joists bear on top of glued laminated beams and to plates of wall studs.
- Posts do not project above low roof.

REQUIRED

(a) Compute and draw the load, sheer, and moment diagrams at low roof due to the seismic lateral forces.

(b) Compute maximum chord forces at the exterior walls of the low roof.

(c) Provide sketch of details required at the corners of the opening for the tension forces due to the lateral forces in both directions. Calculations are required.

(d) Provide sketch of details for the chord splices at the low roof exterior walls for the maximum chord forces. Calculations are required.

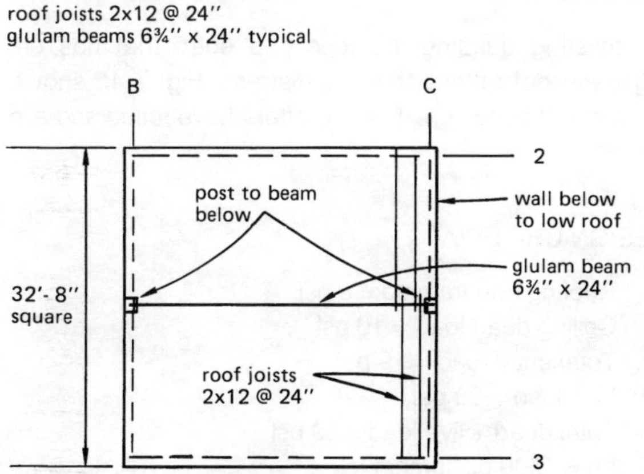

HIGH ROOF FRAMING PLAN

Figure T-39

(continued on next page)

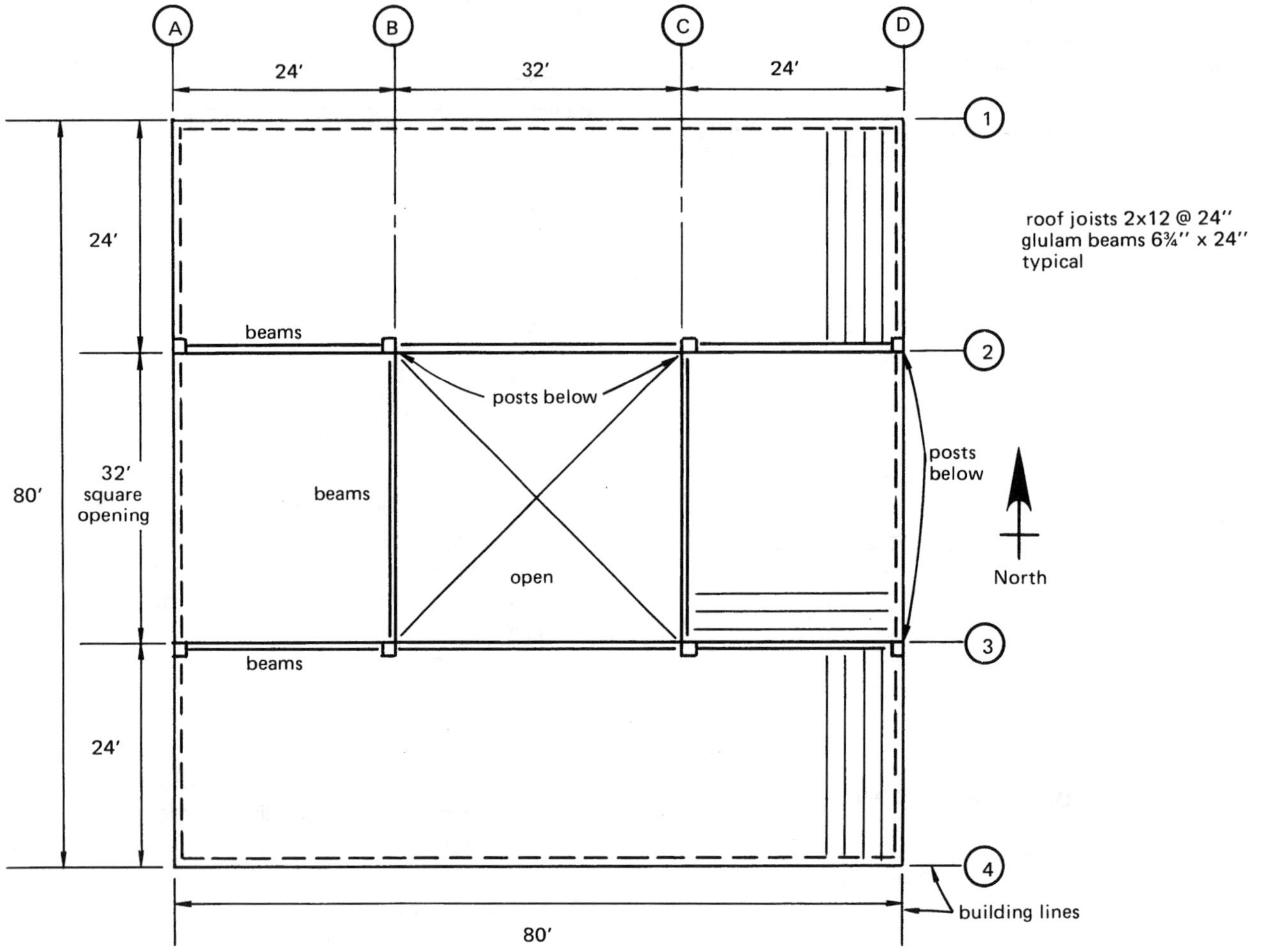

LOW ROOF FRAMING PLAN +12′- 0″

Figure T-39 (cont.)

TIMBER-40

An existing building includes one span that has six adjacent roof rafters showing distress. Fig. T-40 shows the present condition. The six rafters have failed and are sagging.

DESIGN CRITERIA

- Roofing and framing = 9 psf
- Ceiling dead load = 10 psf
- Total dead load = 19 psf
- Live load = 20 psf
- Total dead + live load = 39 psf
- Fb = 1500 psi lumber
- Fv = 95 psi
- E = 1.8 x 10^6 psi lumber
- Use present lumber dimensions

REQUIRED

(a) Use the design loads given and check the bending stress in the rafters. Assume dry condition of use. Compare the computed stress to the allowable. Do you think the difference may be a probable cause of the failure?

(b) Describe a step-by-step procedure for the repair of the existing framing without removing the existing roof.

(c) Identify two possible reasons for the rafter failure.

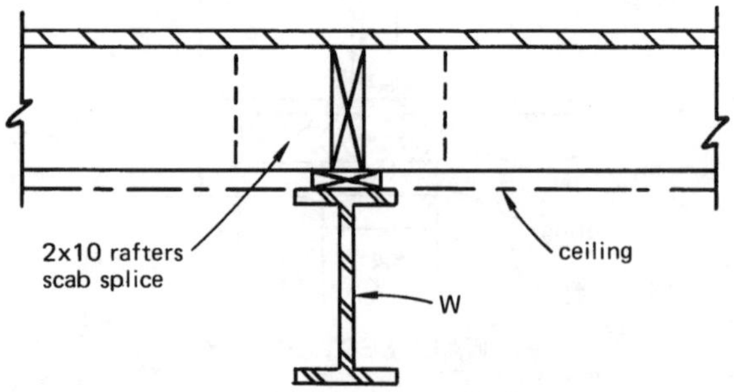

DETAIL A

Figure T-40

(continued on next page)

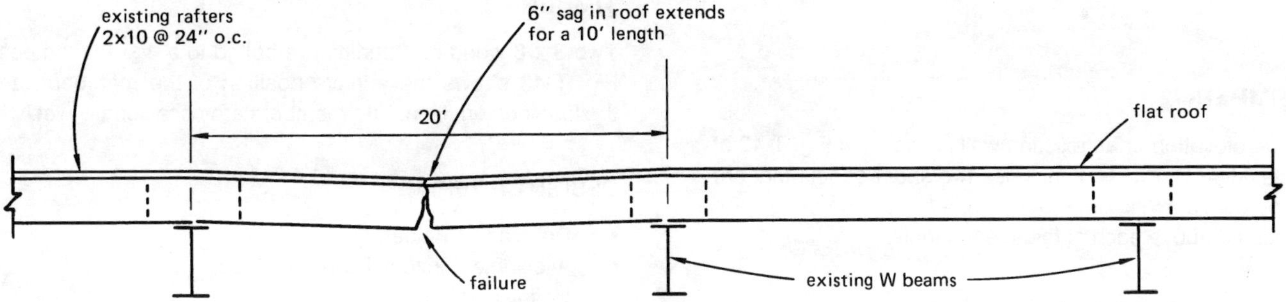

SECTION AT ROOF

Figure T-40 (cont.)

TIMBER-41

The framing that supports the ceiling of an existing building is shown in Fig. T-41. It is necessary to add mechanical equipment in the attic space above the ceiling framing. The equipment loads are shown in elevation B.

DESIGN CRITERIA

- Additional framing must be limited to the space of 5 1/2 in over the top of the existing 4 x 8s.
- The existing 4 x 8s and the ceiling joists cannot be removed during construction.
- The equipment will be installed through an existing skylight in the roof.
- Douglas fir no. 1 lumber (existing)
- Ceiling dead load = 10 psf (includes framing)

REQUIRED

Design a method to support the new equipment utilizing the existing ceiling framing. Show calculations for new or modified supporting members. Limit total deflection of beam to 1/240.

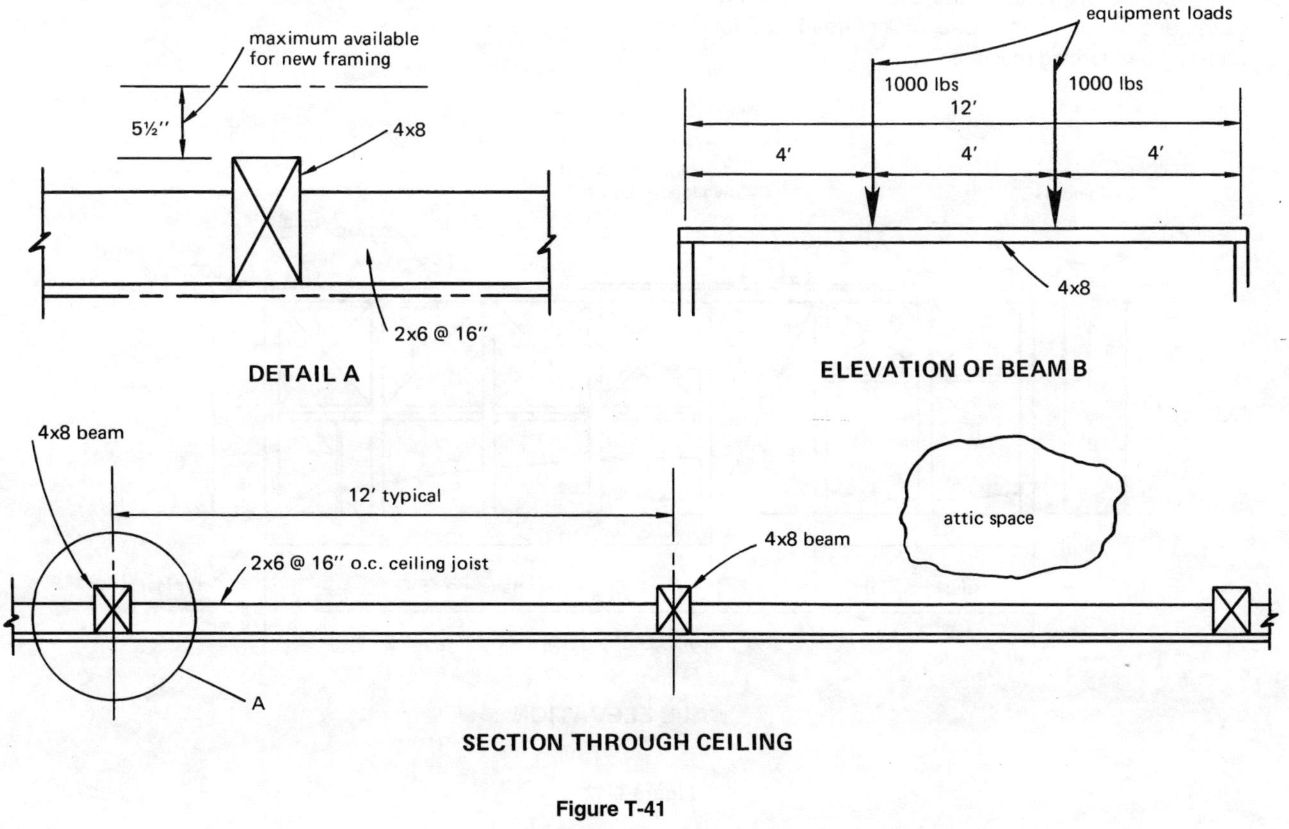

Figure T-41

TIMBER-42

The elevation of an existing wall is shown in Fig. T-42 at elevation C. The wall includes two shear panels that are separated by a door opening. Window openings are located above each of the shear panels.

DESIGN CRITERIA

- F_b = 1200 psi lumber
- F_v = 85 psi
- $F_{c\perp}$ = 355 psi
- F_c = 1000 psi
- Structural I plywood
- Assume that the 2 x 6 and 4 x 6 top plates are adequate for strut action.
- Assume that the end posts are adequately tied to the window sills.

REQUIRED

(a) Compute lateral force on each shear panel. Determine the nailing requirements for the plywood.
(b) Check the stress in the 6 x 6 posts for maximum load condition. Are they adequate?
(c) Design connections at A and B. Prepare a sketch of each to approximate scale 1 in = 1 ft 0 in.
(d) Compute the uplift due to the overturning moment at the end of each panel. Use 75% of dead load for resisting overturning moment.

TIMBER-43

Two 3 x 8 wood roof purlins are bolted to a 4 x 4 wood post. Fig. T-43 shows the relative position of the members, and the loading condition. The members are occasionally wet.

DESIGN CRITERIA

- Douglas fir lumber
- Allowable horizontal shear = 95 psi
- Unfinished bolts

REQUIRED

(a) Determine the number and size of bolts for the load indicated. Show all calculations.
(b) Check the stresses at the connection. Show all calculations.
(c) Draw a detail of the complete connection to approximate scale 1 in = 1 ft 0 in.

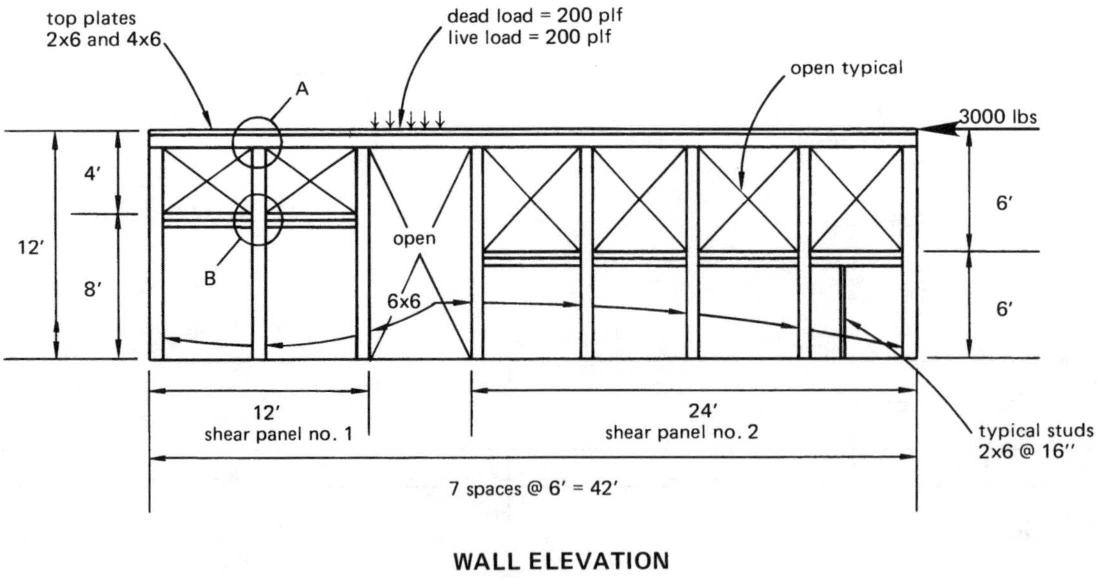

Figure T-42

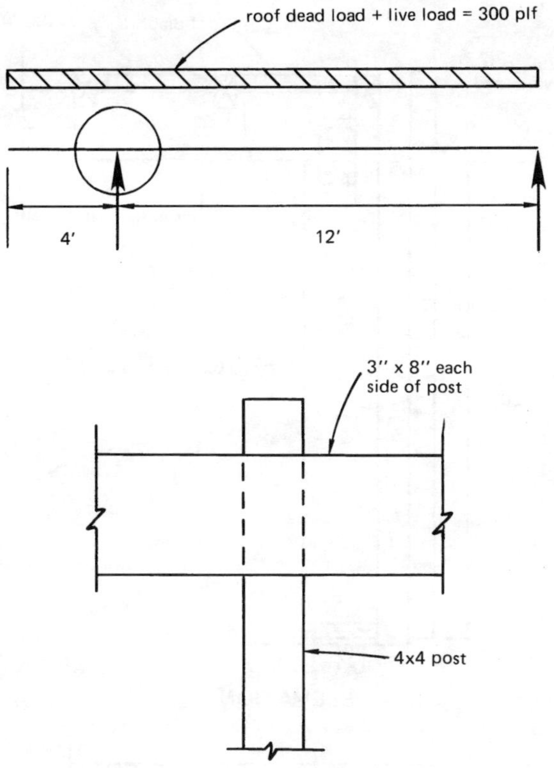

Figure T-43

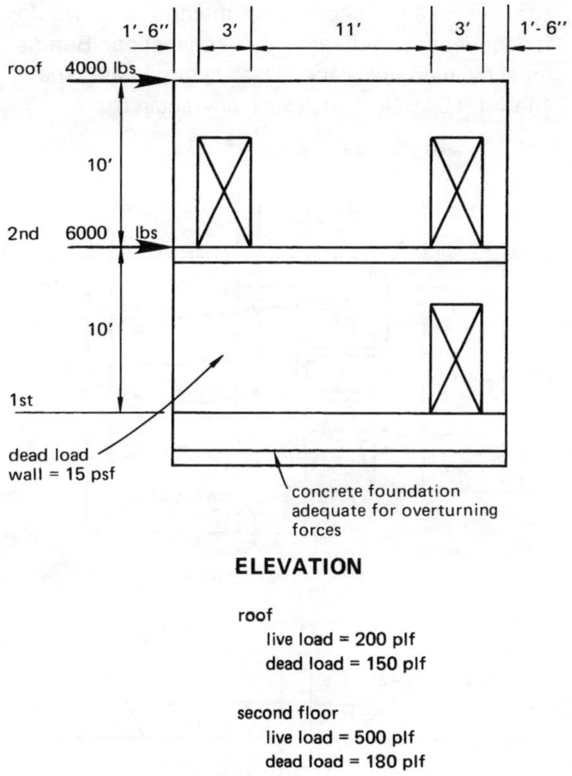

Figure T-44

TIMBER-44

The drawing in Fig. T-44 shows an elevation of the end wall of a 2-story wood frame building. The end wall is sheathed plywood. The doors open out to a second story balcony. The rough door openings are 3 ft wide x 6 ft high.

DESIGN CRITERIA

- Structural I plywood
- Assume loads are seismic

REQUIRED

(a) Compute the shear in the plywood wall sheathing in the first and second story. Determine the size and spacing of nails.

(b) Draw an elevation in your workbook and show the location of tie-down anchors. Show the load each anchor must resist. Use 75% of dead load for resisting overturning moment.

(c) Sketch the various types of tie-down anchors that may be used. It is not necessary to fully design and detail each one.

TIMBER-45

You are to investigate the glued laminated beams which have failed and which are connected as shown in Fig. T-45.

DESIGN CRITERIA

- Stress grade combination for beams A and B is 24F Douglas fir-larch, glued laminated beam.
- Beam A is simply supported with a span of 35 ft.
- Reaction from beam B to connection at midpoint of beam A = 16,000 lb (roof total load)
- Neglect the effect of the eccentricity of beam B on one side of beam A.
- Existing connection has been in place for two years.
- Present moisture content was measured at 9%.
- Job records show that beams had been exposed to extreme wet conditions before final protection.
- All bolts are 1 in diameter machine bolts.

REQUIRED

(a) Describe briefly the probable mode of failure in beams A and B. Show all necessary calculation that will confirm your findings.

(b) Draw a sketch showing a method you would recommend to reinforce the connection. Beams must remain in place. Use approximate scale 1 in = 1 ft 0 in. No calculations are required.

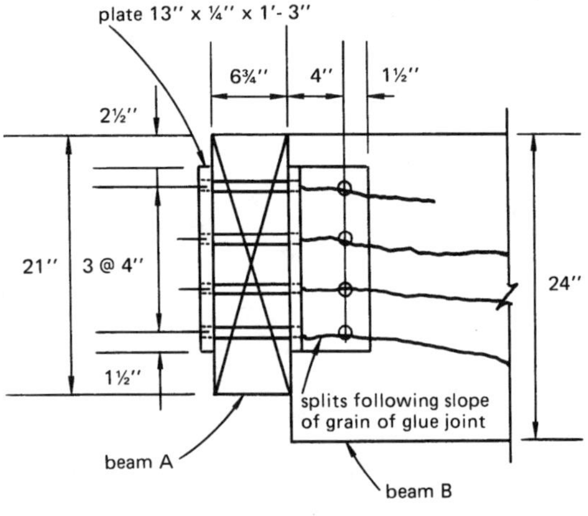

SECTION

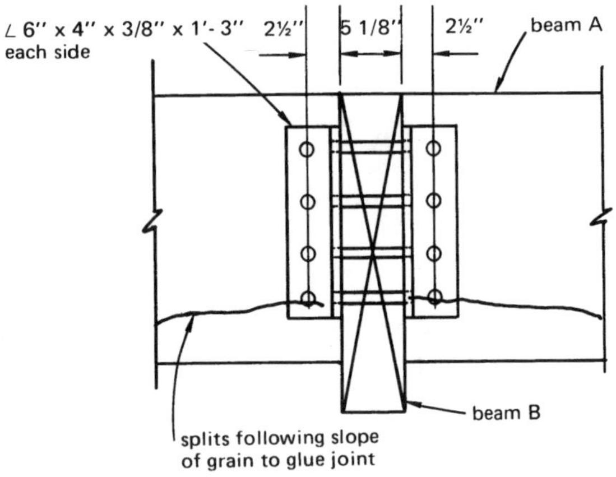

ELEVATION

Figure T-45

TIMBER-46

Fig. T-46 shows a glulam roof beam-to-wood column connection. All wood is Douglas fir-larch no. 1. A design condition for wind load is shown. The port frame shown is assumed to be symmetrical.

REQUIRED

Determine the minimum size of bolts. All bolts must be the same size.

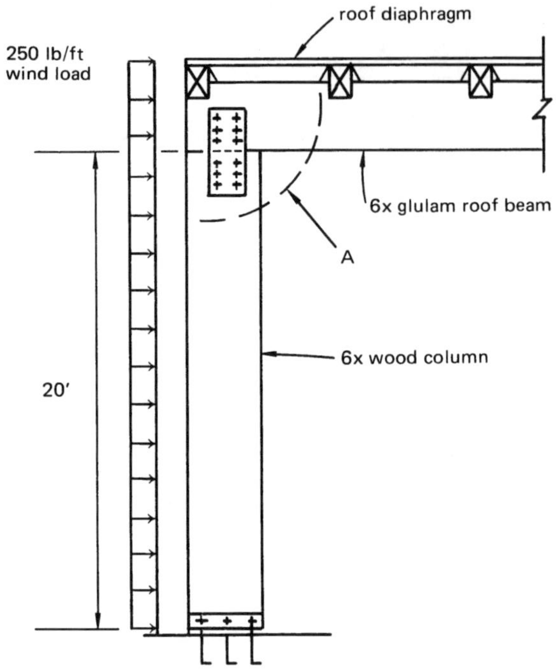

ELEVATION

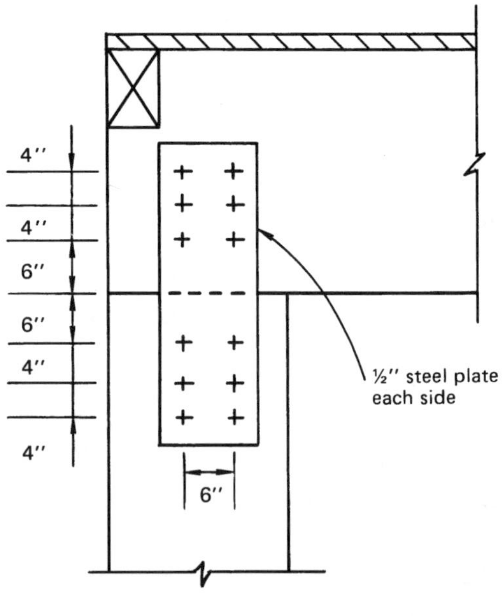

DETAIL A

Figure T-46

TIMBER-47

The design for a 2-span glued laminated beam installation is shown in Fig. T-47. The bay spacing is 18 ft. Loading is from floor purlins connected across the top, spaced 8 ft from the wall end support 1. Lateral bracing of the bottom of cantilever beam A is provided at the interior column 2.

DESIGN CRITERIA

- 24F Douglas fir-larch glulam beam (dry use)
- Floor occupancy = office (F-2) with partitions
- Total floor dead load = 30 psf (Neglect weight of glulam beam.)

REQUIRED

(a) Investigate design of beams A and B for compliance with UBC.
(b) What changes or additions, if any, would you require? Use $P_{dead\ load}$ = 4.3 kips, $P_{live\ load}$ = 6.8 kips.
(c) Specify camber requirement for beam B. Use purlin load P of (b), above.
(d) Show camber diagram shape appropriate for fabrication of beam A. (No magnitudes are required.)
(e) Draw a detail for connection of beam B to A (joint 3). Use scale 1 in = 1 ft 0 in.

TIMBER-48

You are commissioned to make an inspection of a building that was designed by a member of your firm. Fig. T-48 depicts a central shear wall constructed as shown, with details shown as well. Everything agrees with the drawings.

REQUIRED

Would you approve the construction as a result of your inspection of the given conditions? List any deficiencies you suspect. What do you when you return to the office?

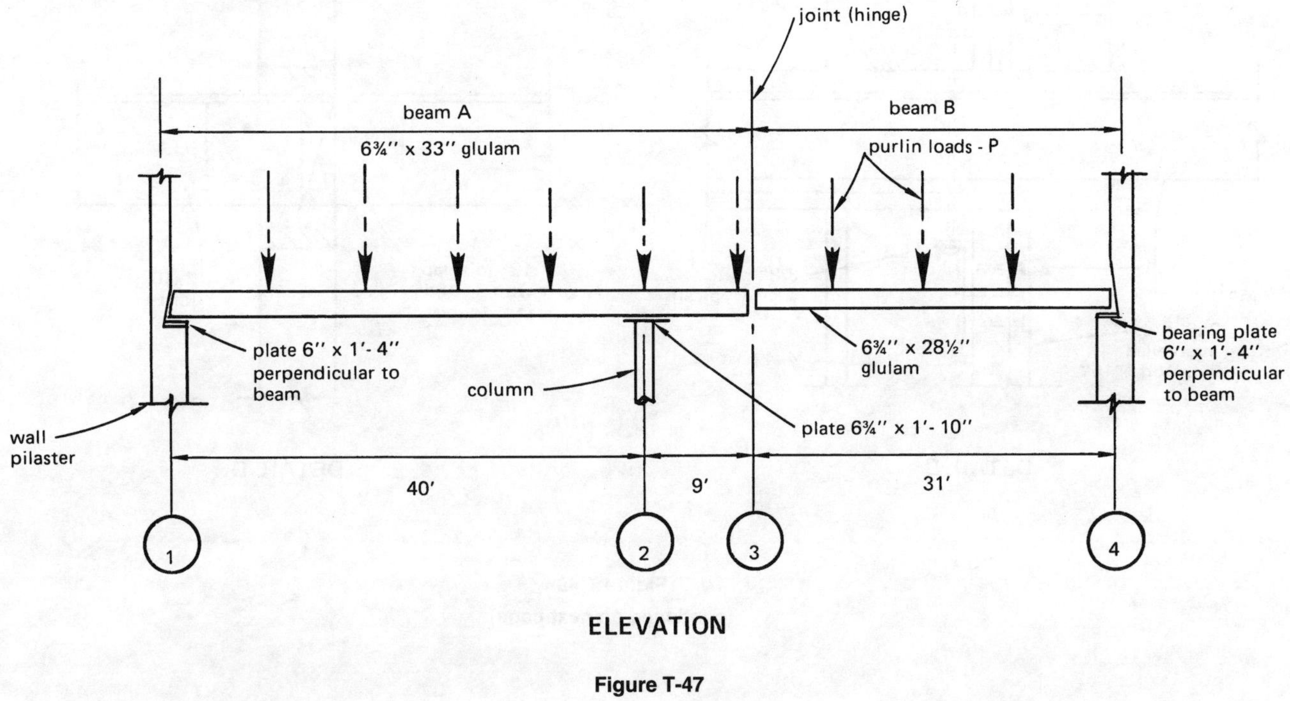

Figure T-47

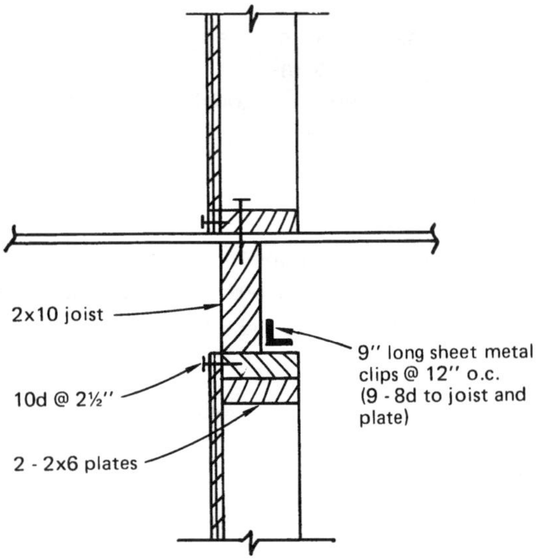

SECTION A (Typical)

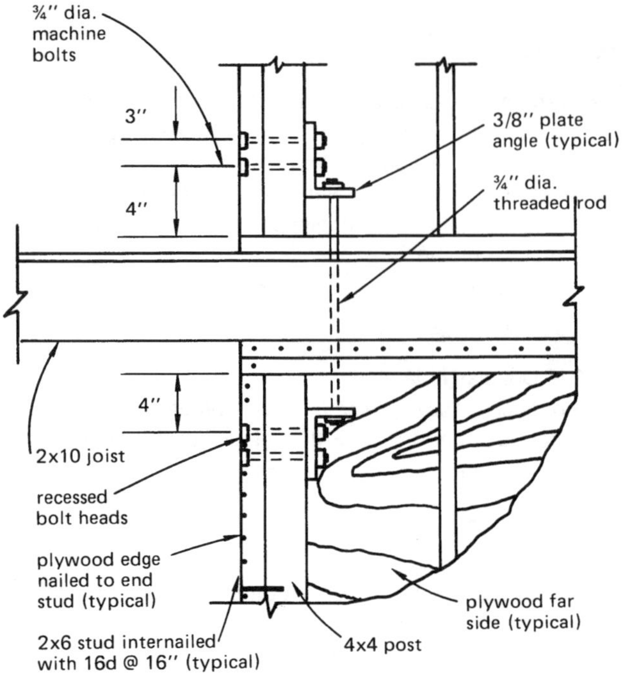

DETAIL C

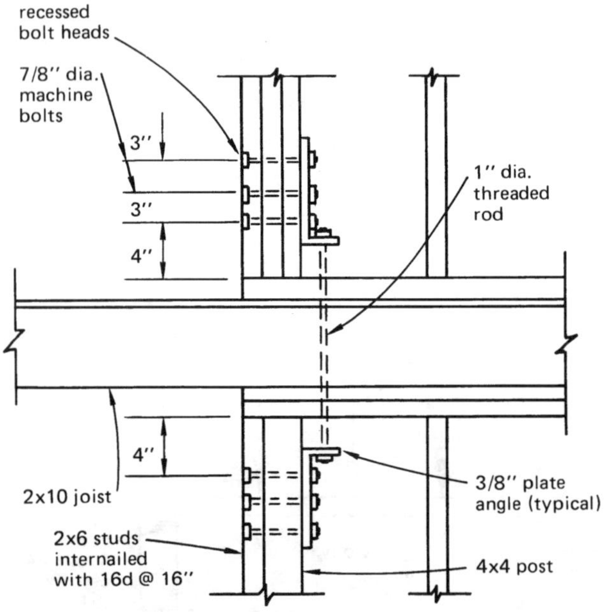

DETAIL B

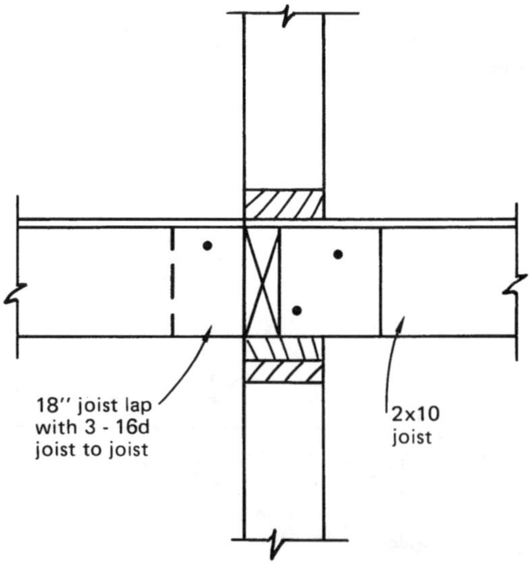

DETAIL D

Figure T-48
(continued on next page)

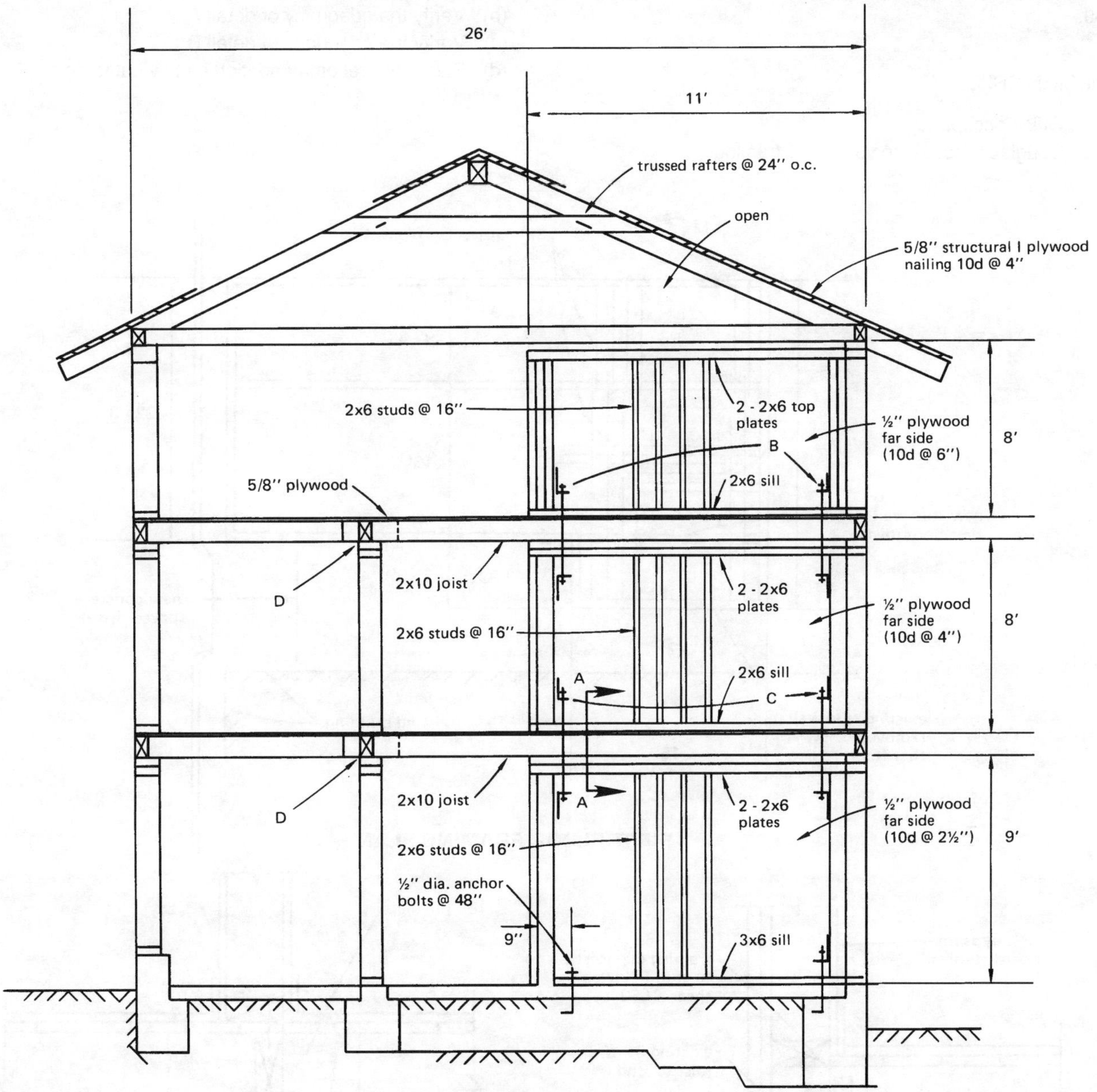

ELEVATION

Figure T-48 (cont.)

TIMBER-49

In order to provide for a driveway, a contractor has relocated an existing bearing wall as indicated in Fig. T-49.

DESIGN CRITERIA

- Office occupancy
- Douglas fir (coast region) no. 1 framing

REQUIRED

(a) Verify the adequacy of the 3 x 10 floor joists to suit the revised condition.
(b) Verify the adequacy of detail A.
(c) Verify the adequacy of detail B.
(d) State your recommendation as to what to do.

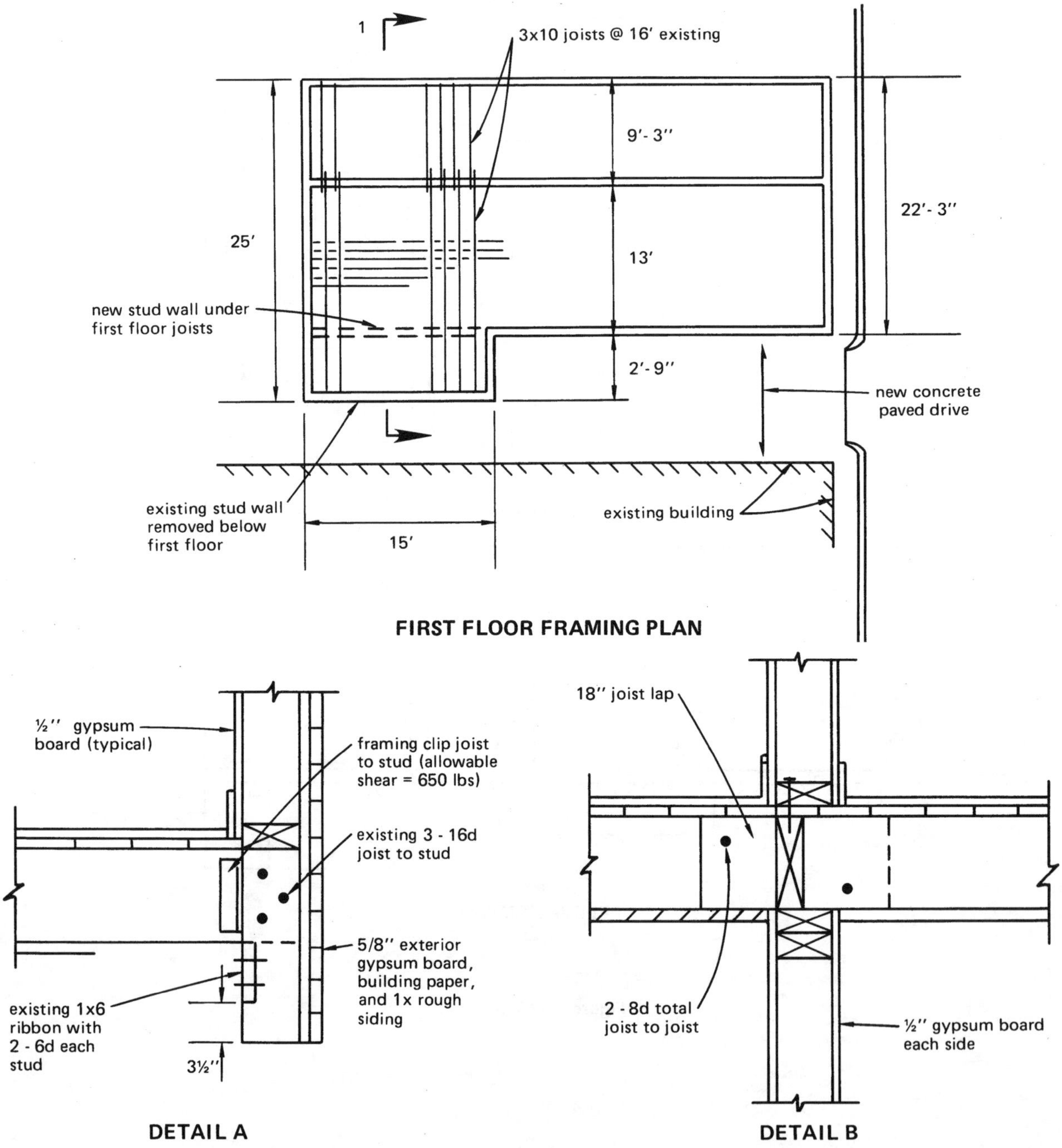

Figure T-49
(continued on next page)

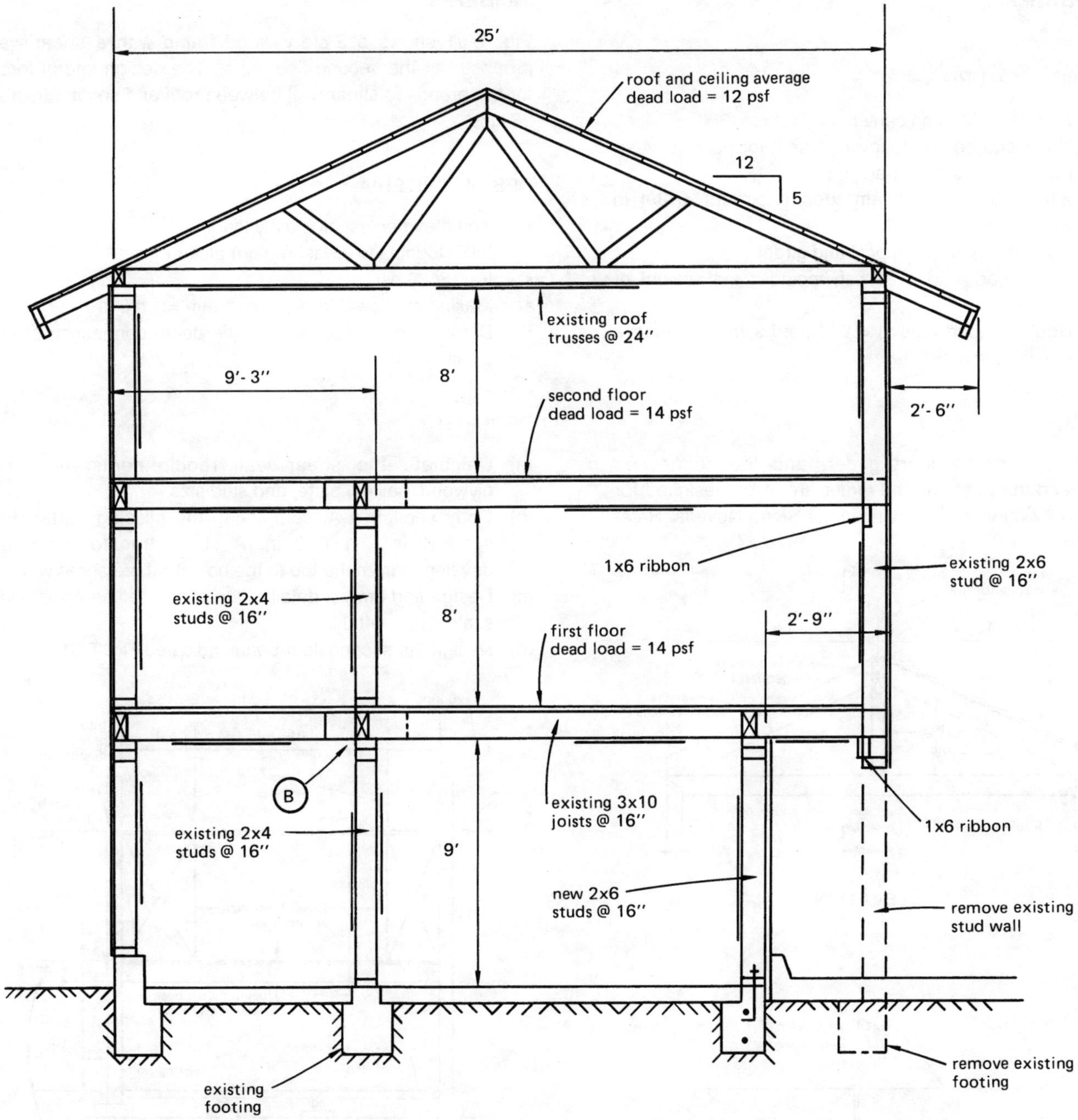

ELEVATION 1

Figure T-49 (cont.)

TIMBER-50

DESIGN CRITERIA

- Rafters at 24 in on center
- No. 1 and better Douglas fir-larch for post between ridge beam and support girder
- 24F Douglas fir glulam support girders, 6 3/4 in wide
- Connection between post and girder
- Roof dead load = 20 psf (includes self weight of framing)
- Roof live load (snow) = 240 psf basic (less than 2 months duration)

REQUIRED

Rafters, post, support girder, and the connection between the post and the girder are to be designed for the roof system of a building in the Sierra Nevada. Refer to Fig. T-50.

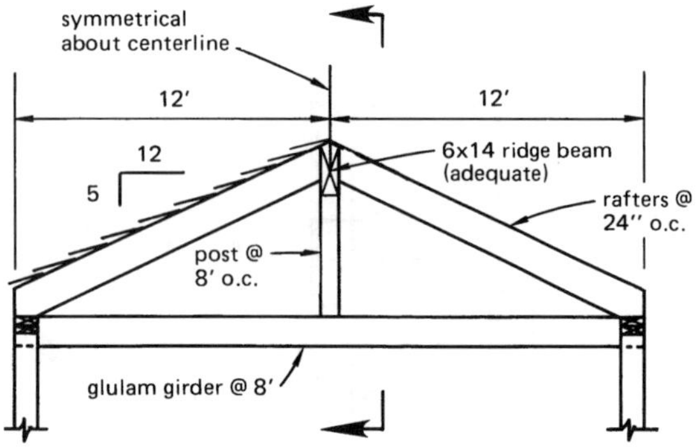

ELEVATION

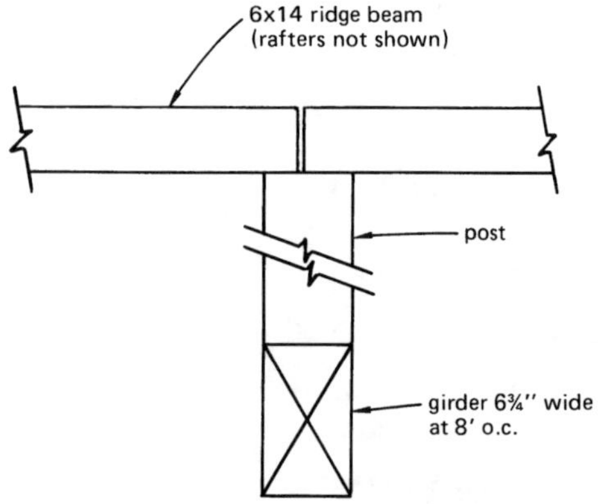

SECTION

Figure T-50

TIMBER-51

Fig. T-51 shows a 2-story wood frame with a shear wall proposed at the second floor level. The design lateral force for the proposed shear wall between roof and second floor is 26 kips.

DESIGN CRITERIA

- You may neglect all gravity loads.
- 24F Douglas fir (coast region) glulam
- F_b = 2400 psi
- Maximum allowable depth of beam = 1 ft 6 in
- Details are not required for the beam connection at the existing wall.

REQUIRED

(a) Calculate the shear wall requirements including plywood, nailing, plate, and stud size.
(b) Copy section A-A, completing the missing parts. Use scale 1 in = 1 ft 0 in. A shear transfer must be developed from the top to the bottom of the shear wall.
(c) Design and draw a detail of the shear wall tie-down. Use scale 1 in = 1 ft 0 in.
(d) Design the second floor beam indicated Fig. T-51.

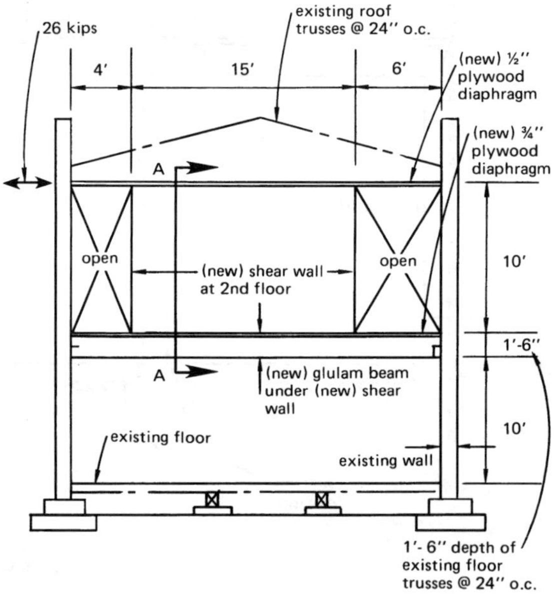

BUILDING SECTION

Figure T-51
(continued on next page)

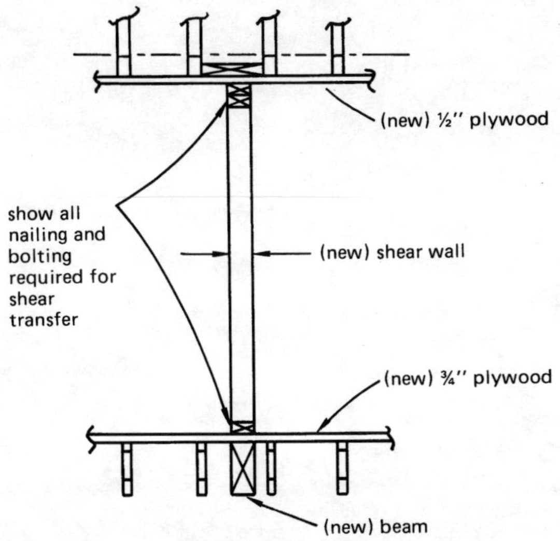

SECTION A-A

Figure T-51 (cont.)

7 MASONRY

MASONRY-1

A cantilevered masonry beam projects out from the wall of a masonry building as shown in Fig. M-1. The load on the beam is 500 lb per linear ft (including the dead load of the beam).

The wall is constructed of 8 in (nominal) hollow unit concrete block masonry. Special inspection is not provided. The wall and the beam had been placed and grouted up to within 10 in of the top when it was discovered that the *A* bars were still lying around and had not been placed as shown on the design drawings.

You have been called in to advise on any possible expedient measure that might save the work done thus far.

DESIGN CRITERIA

- Suggested code: current UBC
- Grade 60 reinforcing steel
- $E_s/E_m = 44$

REQUIRED

What are your recommendations for corrective measures? Show calculations to support any suggested corrective measures. (A column at the free end is not an acceptable solution.)

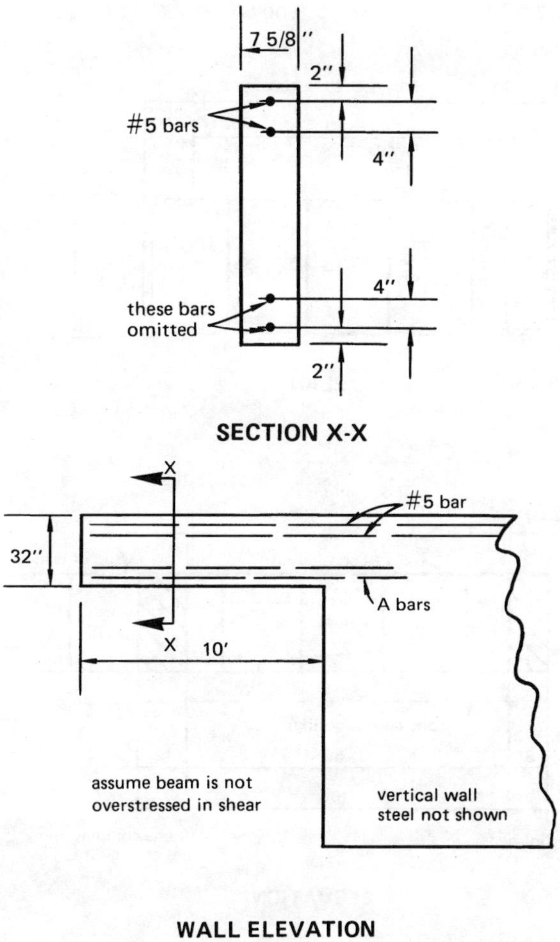

Figure M-1

MASONRY-2

Fig. M-2 depicts an equipment storage garage with rigid frames, masonry walls, and a continuous steel eave strut connected to the masonry walls as shown and noted. The measured dimension between walls at 100°F is 80 ft.

DESIGN CRITERIA

- Neglect the rigidity of the roof diaphragm.
- Assume masonry walls are fixed at the footing.
- Assume the eave strut is adequately braced laterally and is rigidly attached to the masonry walls.
- Masonry walls:

  $E_m = 1.5 \times 10^6$ psi

  $E_v = 600 \times 10^3$ psi

REQUIRED

For a temperature drop of 80°F determine the change in force in the eave strut and the deflections at the top of 8 ft and 12 ft masonry walls (shown on the figure as Δ_1 and Δ_2).

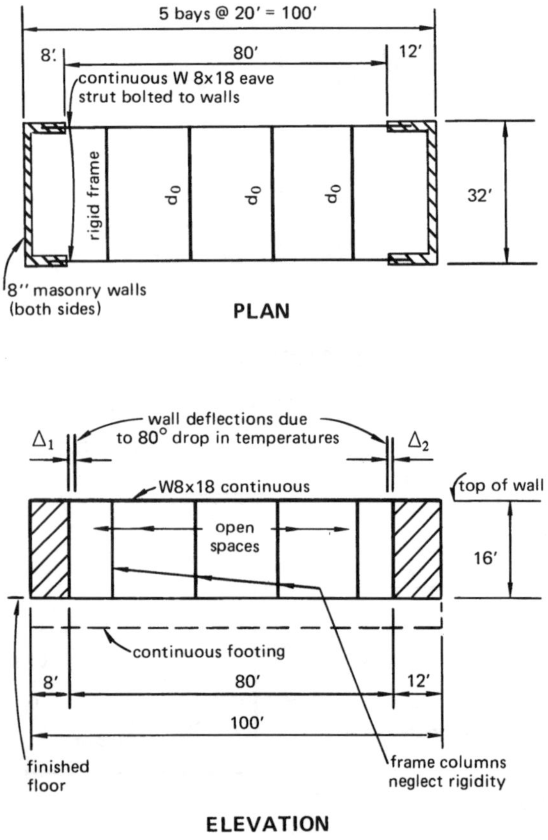

Figure M-2

MASONRY-3

The north wall of a building contains a number of masonry piers. A portion of the wall elevation is shown in Fig. M-3. The seismic forces indicated were determined from the UBC formula.

DESIGN CRITERIA

- Neglect dead and live loads acting on the wall.
- The grade beam is 1 ft 4 in wide.
- Masonry units are 8 in CMU solid grouted hollow concrete block.
- f'_m = 1500 psi special inspection
- Grade 60 reinforcement
- f'_c = 3000 psi standard weight concrete
- $V = \pm 20$ kips
- $M = \pm 40$ kip-ft

REQUIRED

(a) Design a typical wall and typical grade beam.

(b) Provide detailed sketches of sections A-A and B-B.

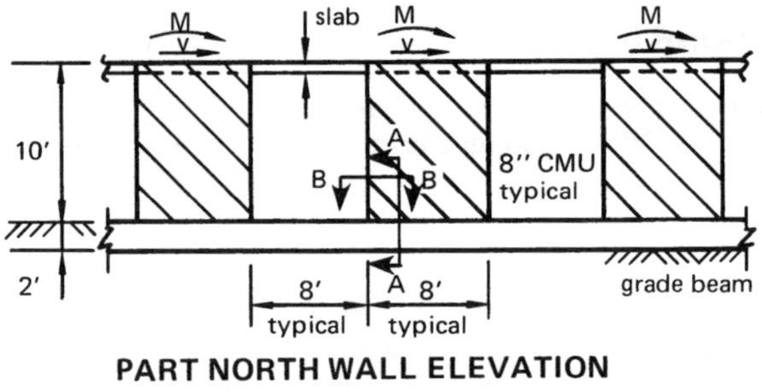

Figure M-3

MASONRY-4

A retaining wall is shown in Fig. M-4. It is to be constructed with concrete masonry units (CMU) in conformance with the current UBC.

DESIGN CRITERIA

- Equivalent fluid pressure = 30 pcf
- The resultant of the soil pressure is to be in the middle half of the footing.
- Soil bearing = 2500 psf maximum
- Coefficient of friction = 0.4
- Soil weight = 90 pcf
- 8 in (CMU) by 12 in (CMU) by 8 in high
- f'_m = 1500 psi
- Special inspection to be employed
- Grade 60 reinforcement
- f'_c = 2000 psi normal weight concrete

REQUIRED

(a) Design the masonry cantilever retaining wall and size the concrete footing. Do not design the footing reinforcement.

(b) Provide a sketch detail at approximately 3/4 in = 1 ft 0 in indicating all dimensions, reinforcement, and any other pertinent data. Show reinforcement for wall as designed. Indicate the placement of footing reinforcement, but do not design steel for footing.

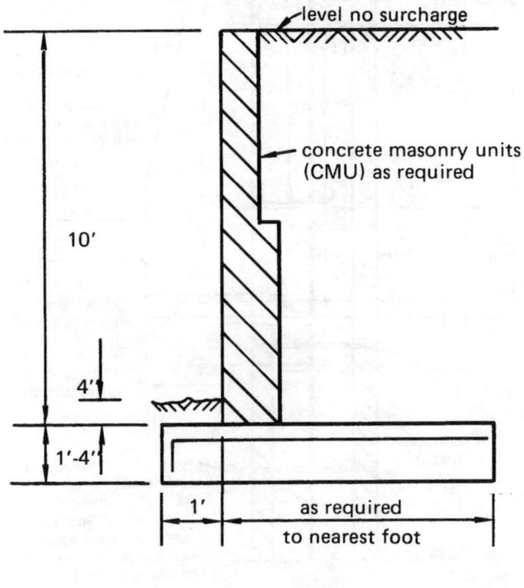

Figure M-4

MASONRY-5

A reinforced hollow concrete block masonry wall is shown in Fig. M-5. The wall is stayed in both directions by the roof system. Assume the wall is pinned at the floor line; special inspection is not feasible.

DESIGN CRITERIA

- f'_m = 1500 psi
- Wind = 15 psf
- Seismic zone 3
- Ca = 0.24
- Importance factor = 1.0
- Concrete block units are 16 in long with two cells per unit grouted solid.
- f_y = 40,000 psi

REQUIRED

(a) Determine the critical stresses in masonry and steel for the loads indicated, with the steel placed as shown in Fig. M-5.

NOTE: $k = \sqrt{2pn + (pn)^2} - pn$

(b) Following through with the conclusions developed in part (a), what are your recommendations to bring the design of the wall up to the requirements of the building code?

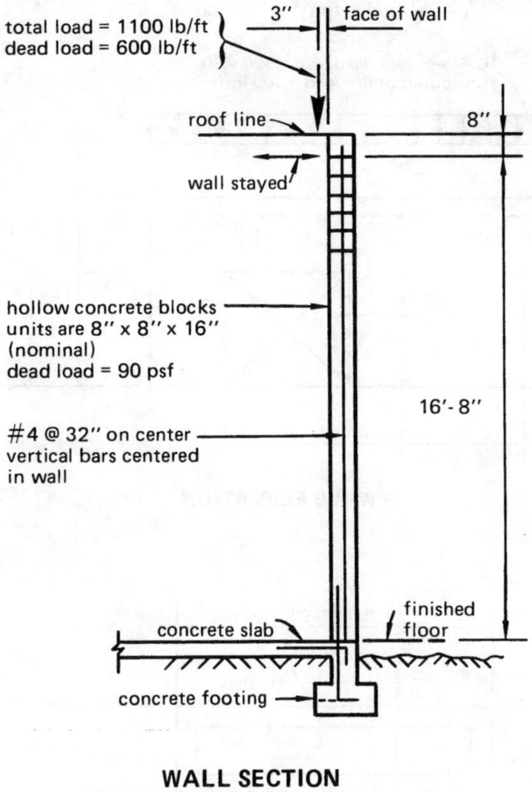

Figure M-5

MASONRY-6

The uniform tributary roof load on a fully grouted exterior bearing brick masonry wall is shown on the elevation in Fig. M-6. The brick wall weighs 135 psf, and it is 13 1/2 in thick.

DESIGN CRITERIA

- f'_m = 2500 psi
- f_y = 60,000 psi
- Seismic zone 4
- Ca = 0.32
- Importance factor = 1.0
- Continuous inspection
- Neglect eccentricity.

REQUIRED

(a) Calculate reinforcement at pier A-A to satisfy combined vertical and seismic forces perpendicular to wall.

(b) Sketch a plan detail showing vertical and horizontal reinforcing steel. Use approximate scale 1 in = 1 ft 0 in.

NOTE: $K = \sqrt{2pn + (pn)^2} - pn$

$$k = \sqrt{2pn + (pn)^2} - pn$$

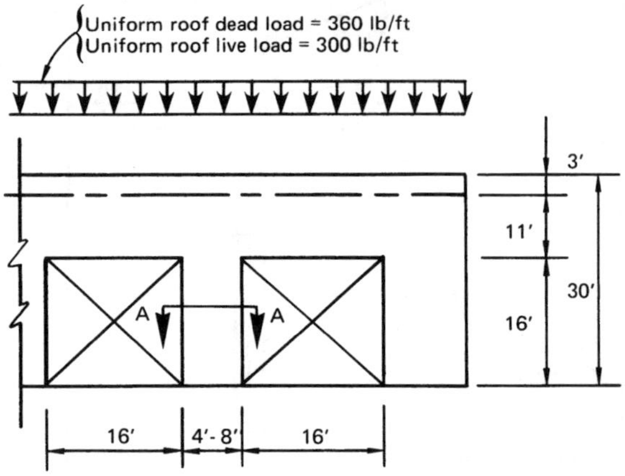

WALL ELEVATION

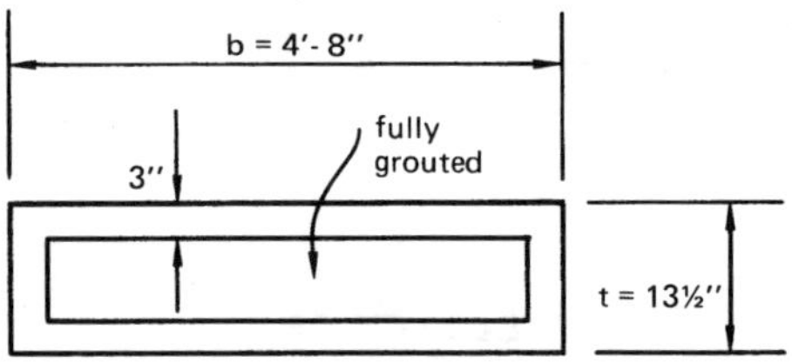

PIER A-A

Figure M-6

MASONRY-7

A reinforced concrete masonry wall supports a wood roof beam with the detail shown in Fig. M-7.

DESIGN CRITERIA

- Special inspection: all cells grouted
- Wall dead load = 92 psf
- f'_m = 1500 psi
- Seismic zone 4

REQUIRED

(a) Calculate the allowable beam reaction *P*. (Do not consider anchor bolt shear value.)

(b) Investigate the wall design at the connection. Assume top of wall elevation is 12 ft; there are no other superimposed loads.

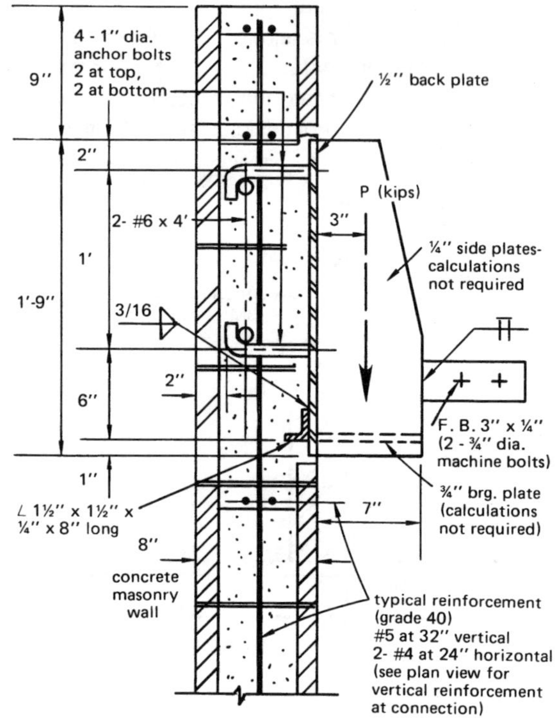

SECTION

(no scale)

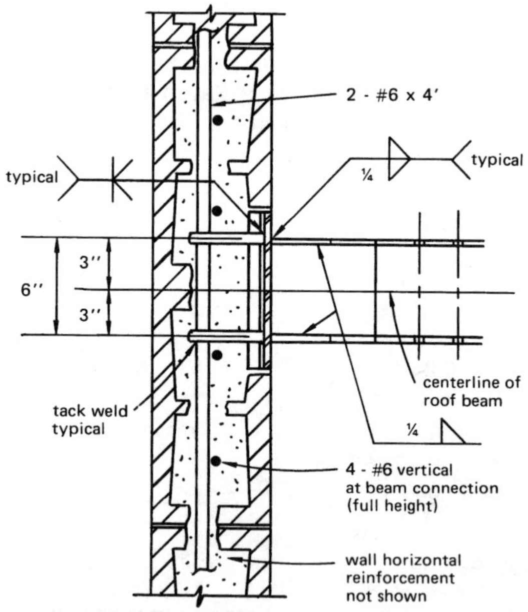

PLAN VIEW

(no scale)

Figure M-7

MASONRY-8

The drawing in Fig. M-8 shows a basement wall of regular weight hollow concrete block with all cells filled. Weight of wall is 85 psf.

DESIGN CRITERIA

- Assume the connection of slab to wall is adequate.
- f'_m = 1400 psi no inspection

REQUIRED

(a) Determine the reinforcement to be placed in center of wall for the loads shown.
(b) Check combined stresses. Show all calculations.

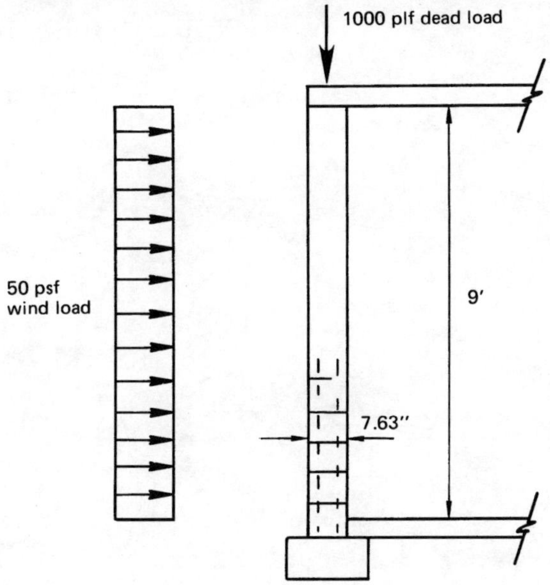

Figure M-8

8 SOLUTIONS

Structural-1

Required: V- and M- diagrams

a)

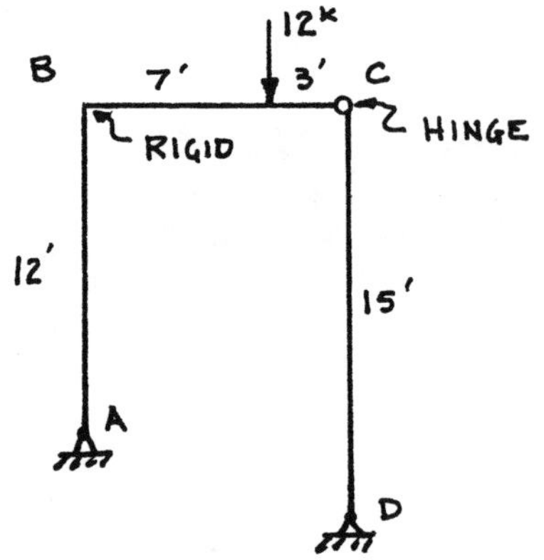

Member CD is a two-force member. Therefore, there is no shear or moment in CD.

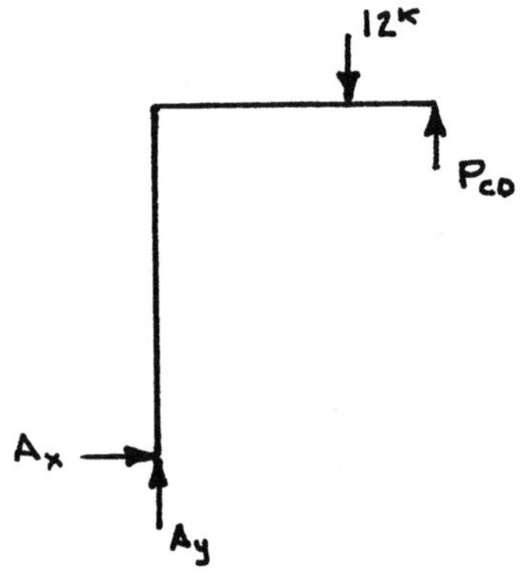

$\Sigma M_A = 0$:

$10P_{CD} - 12(7) = 0$

$\therefore \quad P_{CD} = 8.4^k \uparrow$

$\Sigma F_y = 0$:

$\therefore \quad$ **$\underline{A_y = 12 - 8.4 = 3.6^k \uparrow}$**

$\Sigma F_x = 0$: $\underline{A_x = 0}$

$\therefore \quad$ No shear or moment in member AB

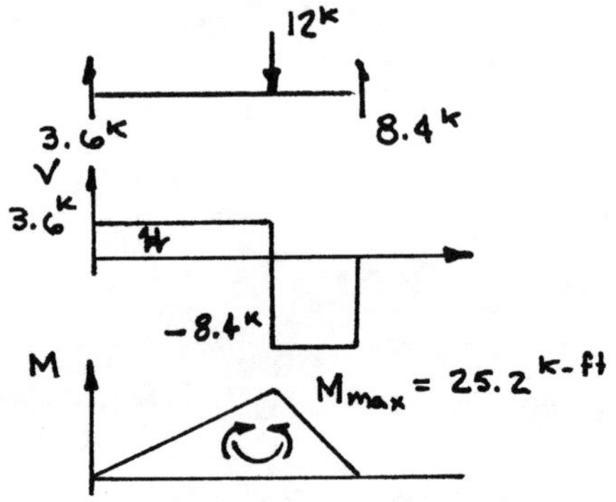

b) Free Body Diagrams:

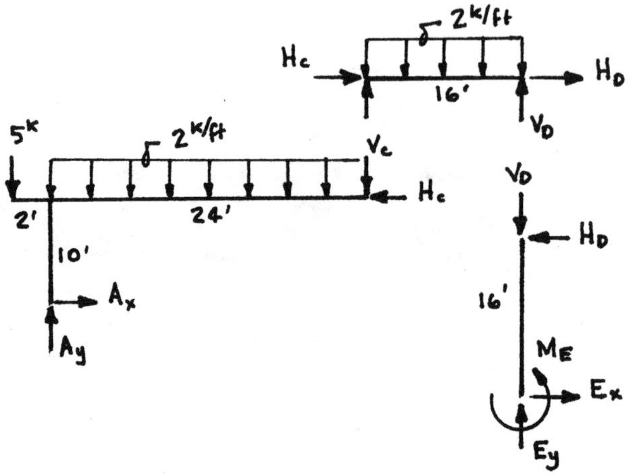

Free body CD:

$\Sigma M_C = 0$: $\underline{V_D = 16^k \uparrow}$

$\Sigma F_y = 0$: $\underline{V_C = 16^k \uparrow}$

$\Sigma F_x = 0$: $\underline{H_C = -H_D}$

Free body ABC (Note $V_C = 16^k \downarrow$):

$\Sigma M_A = 0$: $5(2) - 2(24)(12) - 16(24) + H_C(10) = 0$

$\therefore \quad H_C = 95^k \leftarrow$

$\Sigma F_x = 0$: $\underline{H_A = 95^k \rightarrow}$

$\Sigma F_y = 0$: $-5 - 48 - 16 + A_y = 0$

$\underline{A_y = 69^k \uparrow}$

Going back to free body CD: **$\underline{H_D = 95^k \leftarrow}$**

From free body DE:

$\Sigma F_x = 0$: $\underline{E_x = 95^k \leftarrow}$

$\Sigma F_y = 0$: $\underline{E_y = 16^k \uparrow}$

$\Sigma M_E = 0$: $M_E = 95(16) = 1520^{k\text{-ft}}$ (c.w.)

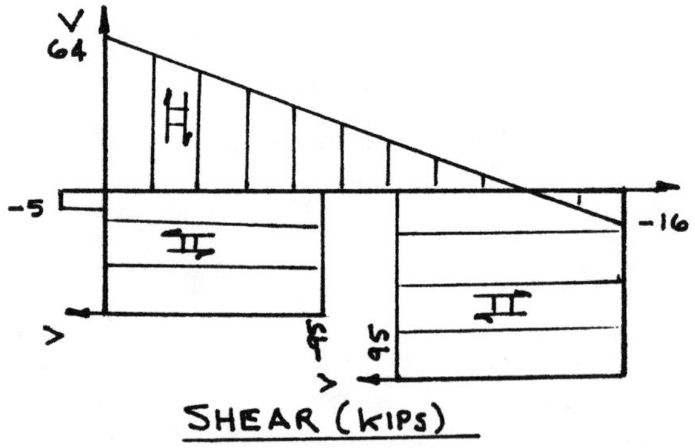

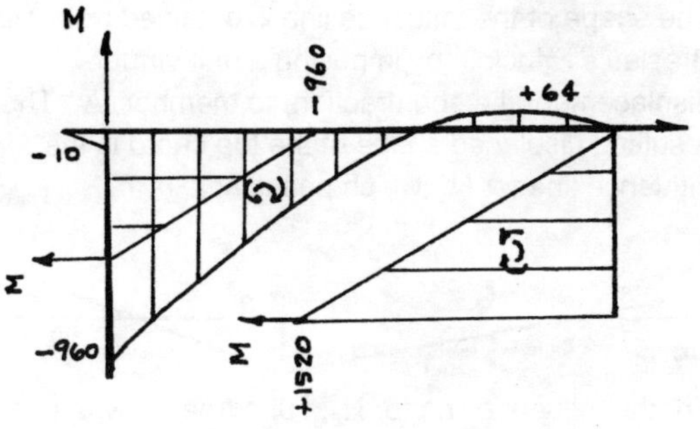

MOMENT (KIP-FT)

Structural-2

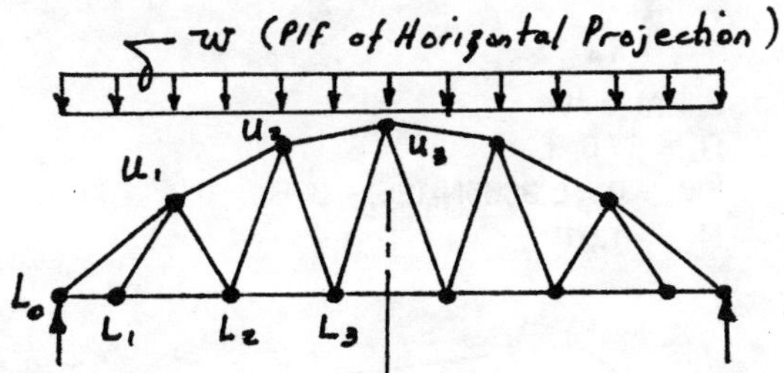

JT	X	Y
L_0	0	0
L_1	4	0
L_2	12	0
L_3	20	0
U_1	8	6.67
U_2	16	10.67
U_3	24	12.00

From Table No. 16-C (UBC 1997), Case 2 for an arch with a rise of at least one-eight the span but less than three-eighths of the span applies. Thus, using Method 2 of Table 16-C, apply a uniform live load of 16 lbf/ft^2 and reduce the live load at a rate of 0.06, with a maximum reduction of 25 percent.

(1 / 8) Span < rise = 12.0' < (3 / 8) span ∴ by Method 2:

live = 16 psf; reducible; r = 0.06

$$R \le \begin{cases} 0.06[12' \times 48' - 150 \text{ sq ft}] = 26\% \\ 0.23(1 + 20/16) = 52\% \\ 25\% \longleftarrow \text{Controls} \end{cases}$$

Top chord panel loads (trusses spaced 12' o.c.; 8' horizontal between joints)

$\therefore \quad P = (20 + (1 - 0.25) \times 16) \times 12 \times 8$

$= 3070^{\#} \downarrow$

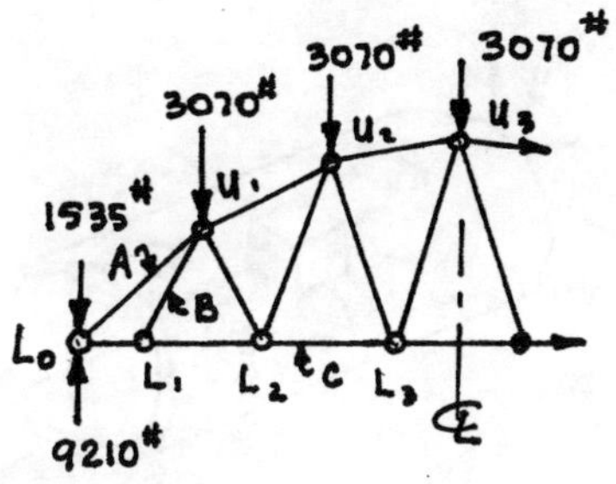

Find the design force in members 'A','B', and 'C':

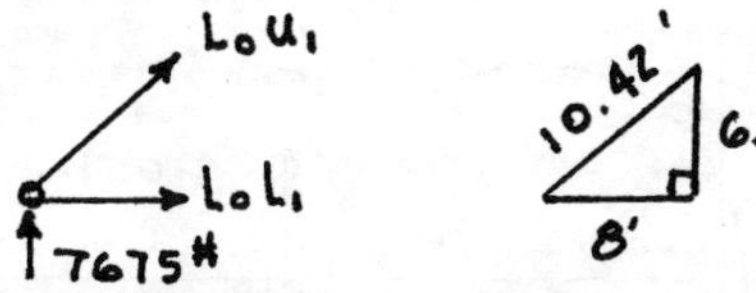

$\Sigma F_y = 0$: $(6.67 / 10.42)\ L_0U_1 + 7675 = 0$

∴ $\underline{L_0U_1 = -12{,}000^{\#} = 12^{k}\ \text{(comp)}}$

Note: since the upper chord supports joists spaced 16" o.c., there is bending plus axial compression in member 'A':

$M_{max} = [(20 + 0.75 \times 16) \times 12]\ 8^2 / 8 = 3072^{\text{lb-ft}}$

For member 'B' (i.e., L_1U_1):

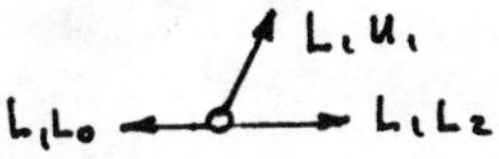

$\Sigma F_y = 0$: $\underline{L_1U_1 = 0}$

For member 'C' (i.e., L_2L_3):

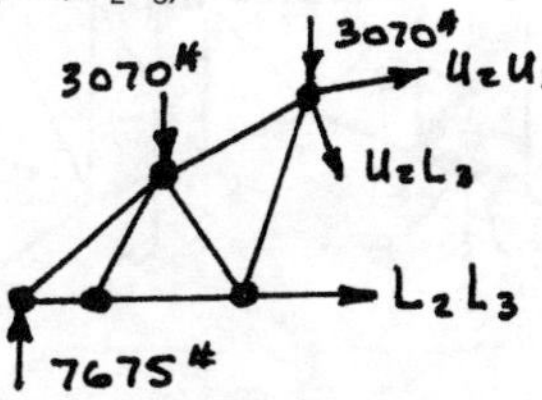

$\Sigma M_{u2} = 0$: $L_2L_3\ (10.67) + 3070(8.0) - 7675(16) = 0$

∴ $\underline{L_2L_3 = +9200\ \text{lb} = 9.2^{k}\ \text{(Tension)}}$

Structural-3

The maximum shear stress at every cross section occurs at the neutral axis on the side where stress due to P is additive to stress due to T:

$\tau_{max} = PQ / Ib + Tr / J$

$= 5 \times 11.88 / (91.05 \times 2 \times 0.25) + 5 \times 12 \times 5 / 182.1$

$\tau_{max} = 2.96^{\text{ksi}}$

The above value of τ_{max} would be used to check against a code specified allowable shear stress. It is not necessarily the absolute maximum shear stress in the member. Since the value of L is not given, we cannot calculate the normal stress due to bending in the member. If we consider point 'O' to be the top fiber of the beam at L/4 from the support, the normal stress is $\sigma_x = Mc / I$; where $M = P(3L / 4)$; $C = 5"$; $I = 91.05\ \text{in}^4$.

The shear stress at this point is caused only by torsion:

$\tau_0 = Tr / J = 5 \times 12 \times 5 / 182.1 = 1.65^{\text{ksi}}$

$\therefore \tau_{xz} = \tau_{zx} = 1.65^{\text{ksi}}$

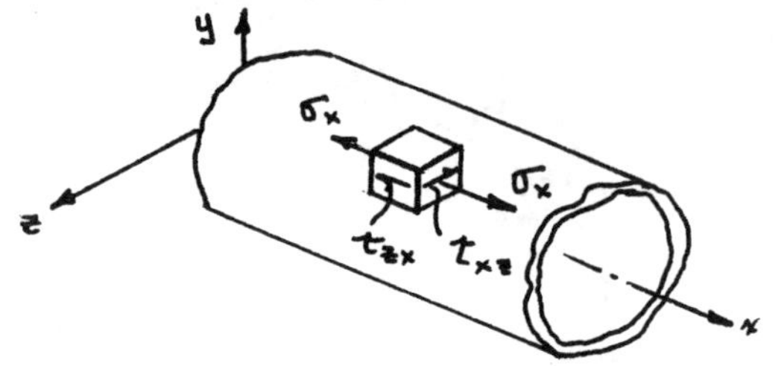

From Mohr's circle for the biaxial state of stress at the top of beam:

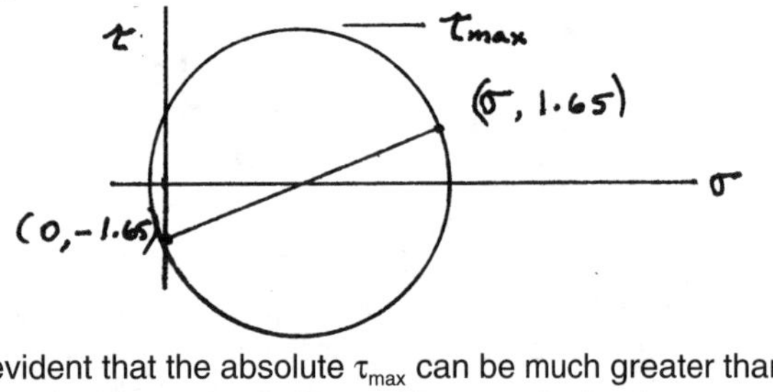

It is evident that the absolute τ_{max} can be much greater than on a vertical face, depending on the magnitude of σ.

Structural-4

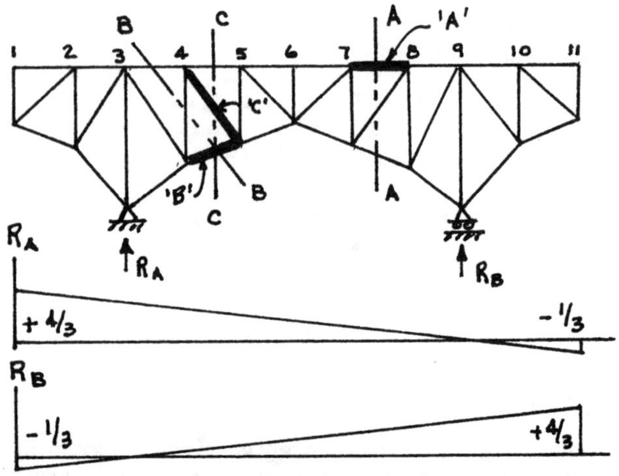

The influence lines for vertical reactions at "A" and "B" are shown above.

a) Construct influence lines for axial forces in members "A," "B," and "C".

For member "A, " cut a free body diagram "A-A":

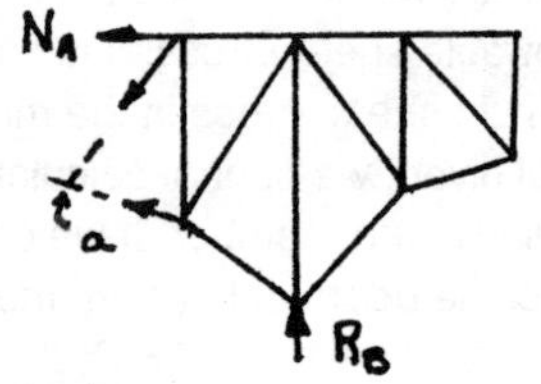

For $1^k \downarrow$ at joint "7":

$R_B = 2/3^k \uparrow$

$\Sigma M_a = 0 = 20N_a + 60R_B$

$\therefore \quad N_a = -3R_B = -2.0^k$

The shape of the influence line is obtained from Mueller-Breslau's Principle by imposing a unit virtual displacement (i.e., contraction) to member "A." The resulting displaced shape of the top chord is the influence line for N_a, which peaks at joint 7.

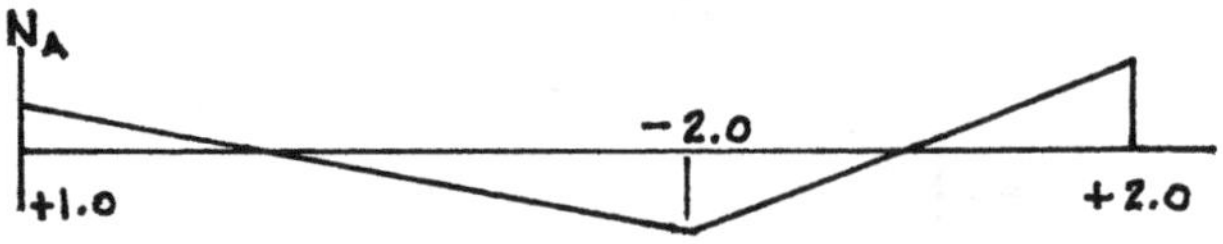

For the influence line for N_B, cut a free body diagram through "B-B":

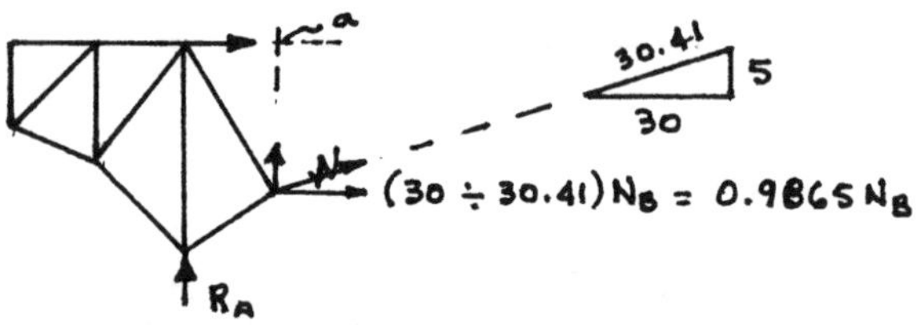

For $1^k \downarrow$ at joint 4:

$R_A = 5/6^k \uparrow$

$\Sigma M_a = 0 = 0.9865N_B(25) - 30R_A = 0$

$\therefore \quad N_B = +1.01^k$

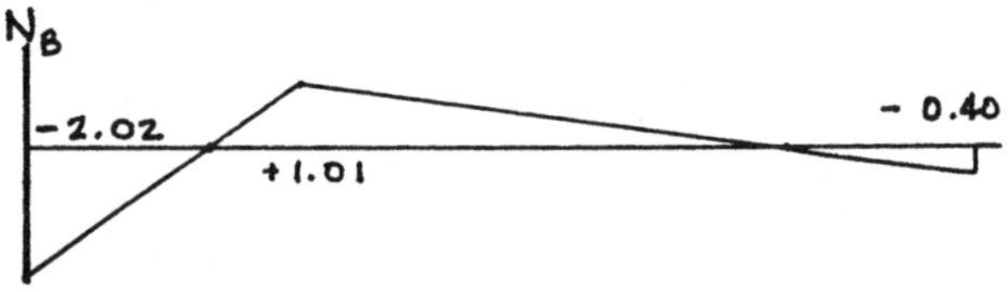

For N_C, cut free body diagram through "C-C":

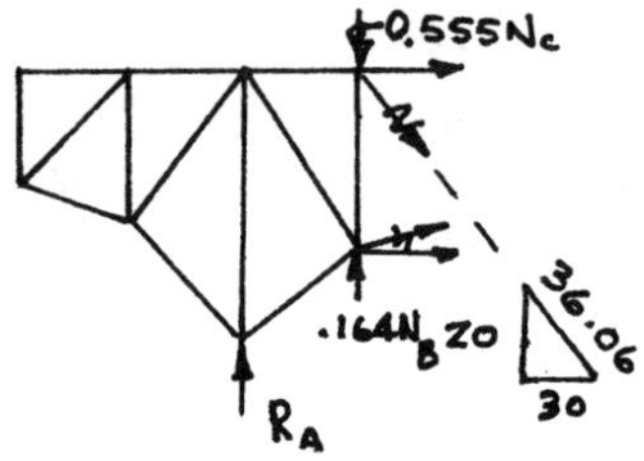

For $1^k \downarrow$ at 4:

$R_A = 5/6^k \uparrow$

$N_B = 1.01^k$ (T)

$\Sigma F_y = 0 = R_A + 0.164N_B - 0.555N_C - 1 = 0$

$\therefore \quad N_C = 0$

For $1^k \downarrow$ at joint 5: $R_A = 2/3^k$; $N_B = 0.81^k$

$\Sigma F_y = 0$: $N_C = +1.44^k$

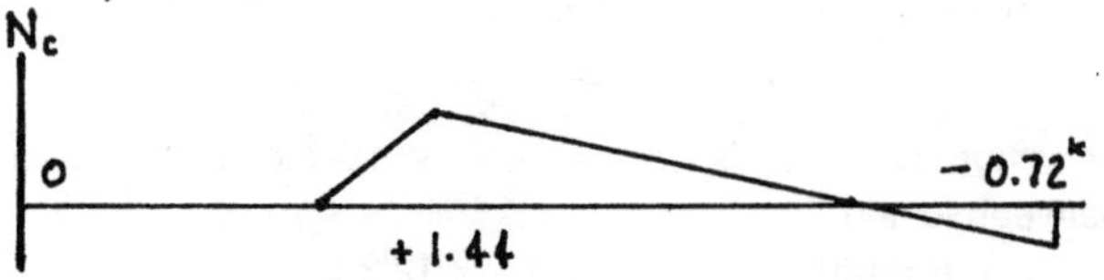

b) Calculate the maximum and minimum axial forces in members "A," "B" and "C" under uniformly distributed dead load of $2^{k/f} \downarrow$and lane load of $1^{k/f} \downarrow$ (uniform load times average ordinate). For member "A":

$N_{A,D} = 2(0.5)\{1.0(60) + 2(60) - 2(180)\}$

$= -180^k$ (comp)

$N_{A,L1} = 1(0.5)\{(-2)180\} = -180^k$ (comp)

$N_{A,L2} = 1(0.5)\{(3)60\} = +90^k$ (ten)

$\therefore \quad -360^k \le N_A \le -90^k$

For member "B":

$N_{B,D} = 2(0.5)\{(-2.02 - 0.4)60 + 1.01(180)\}$
$= 36.6^k$ (ten)

$N_{B,L1} = 1(0.5)\{(-2.02 - 0.4)60\} = -72.6^k$ (comp)

$N_{B,L2} = 1(0.5)\{(1.01)180\} = +90.9^k$ (ten)

$\therefore$ **$-36^k \le N_A \le 127.5^k$**

For member "C":

$N_{C,D} = 2(0.5)\{1.44(150) - 0.72(60)\}$
$= 172.8^k$ (ten)

$N_{C,L1} = 1(0.5)\{(1.44)150\} = 108^k$ (ten)

$N_{C,L2} = 1(0.5)\{(-0.72)60\} = -21.6^k$ (comp)

$\therefore$ **$151.2 \le N_C \le 280.8^k$**

In the expressions above, the subscript "D" denotes dead load; "L1" corresponds to the critical compression due to lane load; ""L2" corresponds to the maximum lane load tension.

Structural-5

The column analogy is by far the simplest method for solving this problem (Blodgett 1967).

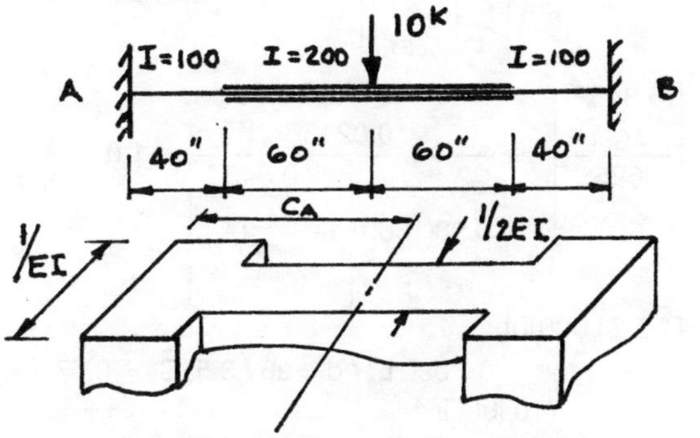

From symmetry of beam and loading, we need only the area of the analogous column:

$C_A = C_B = 100$ in

$A = 200 / (2EI) + 2(40) / (2EI)$
$= 140 / EI$

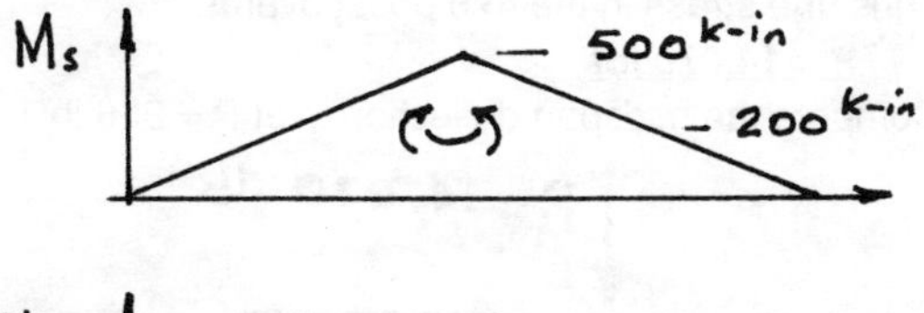

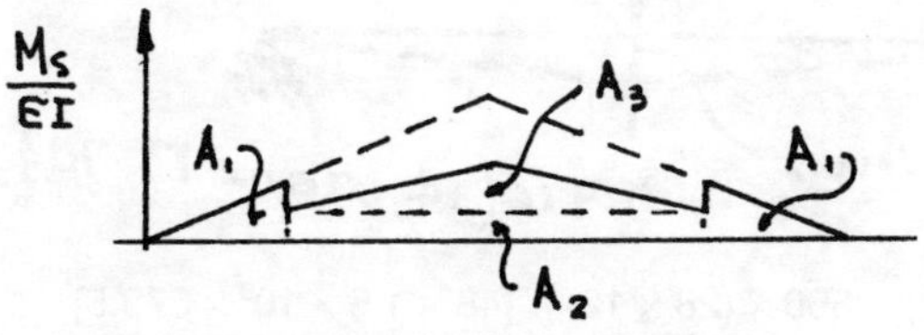

$A_1 = 0.5(40)(200) / EI = 4000 / EI$

$A_2 = 120(100) / EI = 12{,}000 / EI$

$A_3 = \{0.5(250 - 100) / EI\}(60)(2) = 9000 / EI$

$\therefore \quad P = 2A_1 + A_2 + A_3 = 29{,}000 / EI$

By symmetry, $FEM_{AB} = -FEM_{BA}$

$FEM_{AB} = P / A = (29{,}000 / EI) / (140 / EI)$

$FEM_{AB} = 207^{k\text{-}in}$ (c.c.w.)

Note: an alternative approach to this problem is the method of consistent displacements, which is best applied by treating the fixed end moments as the redundants.

Structural-6

Compute the maximum bending stresses in the beams of the statically indeterminate structure. Analyze by the method of consistent displacements. Let P = vertical force between beams at the intersections be the redundant. Given: 10" x 12" wood beams; EI = constant. Release P and solve for displacements in terms of EI:

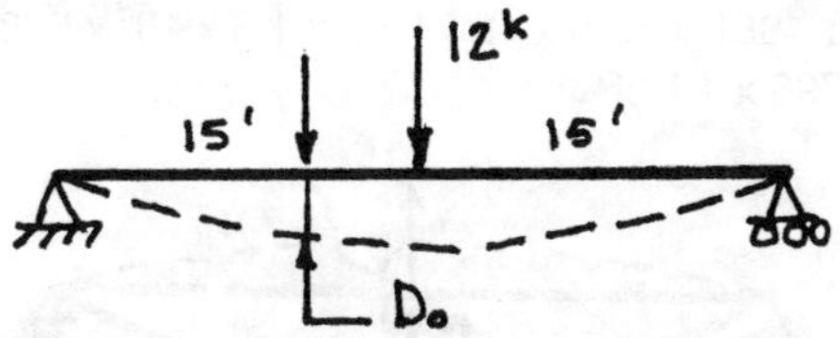

$D_o = 12(10)\{(3)(30)^2 - 4(10)^2\} / (48EI) = 5750 / EI$

Apply equal and opposite unit forces at points of intersection in the released structure:

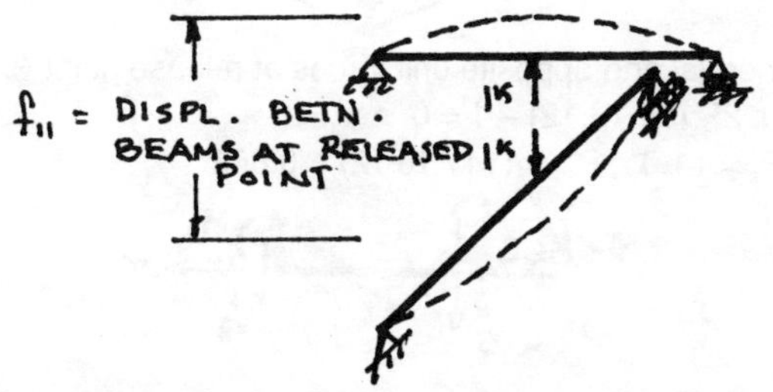

$f_{11} = 1(24)^3 / (48EI) + 1(10)^2(20)^2 / (90EI) = 732 / EI$

For consistent displacement:

$P(732 / EI) - 5750 / EI = 0$

$\therefore \quad P = 7.86^k$

Thus, for the beam spanning 24':

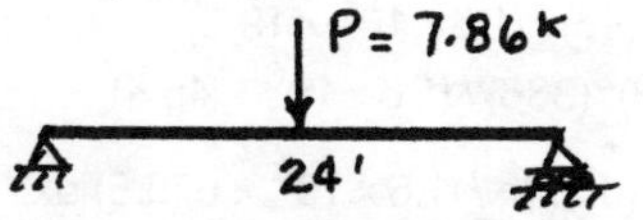

$M_{max} = 7.86(24) / 4 = 47.2^{k\text{-}ft}$

For the 30' long beam:

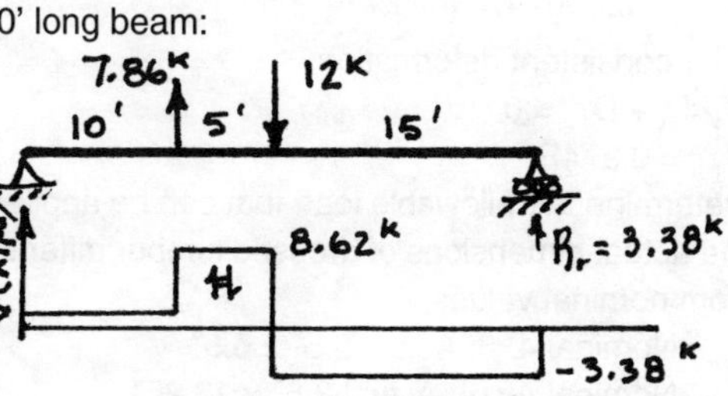

$V = 0$ at $L / 2 = 15'$

$\therefore \quad M_{max} = 3.38(15) = 50.7^{k\text{-}ft}$

Thus, the maximum bending stresses (based on the 10" x 12" actual size) are

$S = 10(12)^2 / 6 = 240 \text{ in}^3$

$\therefore$ for the 24' long beam:

$f_b = 47.2(12) / 240 = 2.36^{ksi} = 2360$ psi

for the 30' long beam:

$f_{b,max} = 50.7(12) / 240 = 2.54^{ksi} = 2540$ psi

Structural-7

The structure is one degree statically indeterminate. Analyze by consistent deformation taking the axial force in the post as the redundant force.

Release the redundant = P_{4x6}:

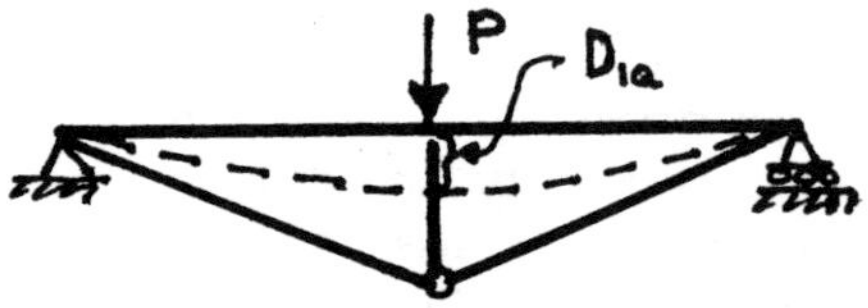

$D_{1Q} = PL^3 / 48EI = P(28 \times 12)^3 / (48 \times 1.6 \times 10^6 \times 678.5)$
$= (7.2796 \times 10^{-4})P$

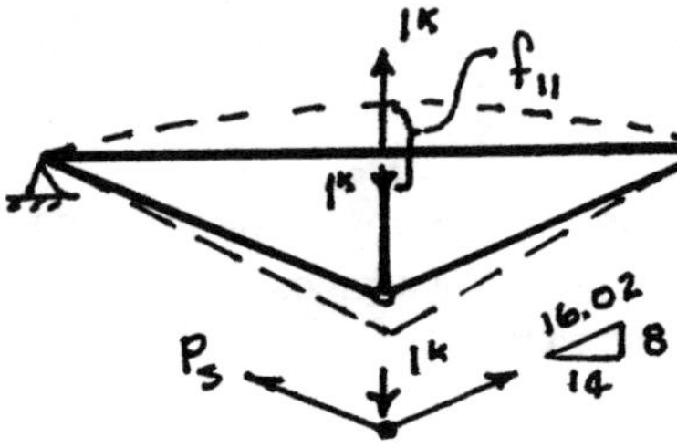

Apply equal and opposite unit forces at release point

$\Sigma F_y = 0$: $2P_s(8 / 16.12) - 1 = 0$

$\therefore \quad P_s = +1.0^k$; $P_{sh} = (14 / 16.12) = 0.868^k$

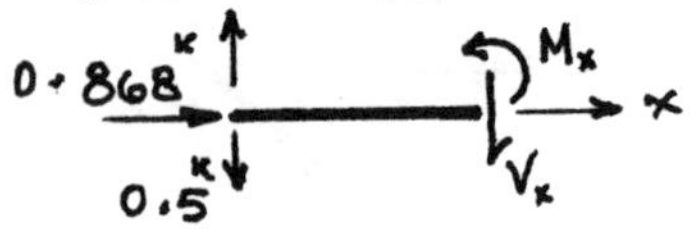

$\Sigma M_x = 0$: $M_x = -0.5x$; $0 \le x < 14'$

By virtual work:

$$(1) f_{11} = \sum (pPL/AE) + \int_L \frac{mMdx}{EI})_i$$

$$f_{11} = 2(1)(1)(193.4) / (30 \times 10^6 \times 1.0)$$
$$+ (-1)(-1)(96) / (1.6 \times 10^6 \times 19.3)$$
$$+ (-0.868)^2(336) / (1.6 \times 10^6 \times 46.4)$$
$$+ 2\int_0^{168} [(0.5x)^2 / (1.6 \times 10^6 \times 678.5)]dx$$

$\therefore f_{11} = 0.0007474$ in / k

Thus, for consistent deformation,

$P_{4x6}f_{11} + D_{1Q} = 0$

$\therefore \quad P_{4x6} = 0.974P$

a) Determine the allowable load that can be applied. Note: the actual dimensions of dressed lumber differ slightly from nominal values:

Nominal 4×6: Actual 3.5" × 5.5"

Nominal 4×14: Actual 3.5" × 13.25"

1. The top chord, which is subjected to combined axial and bending, appears as follows

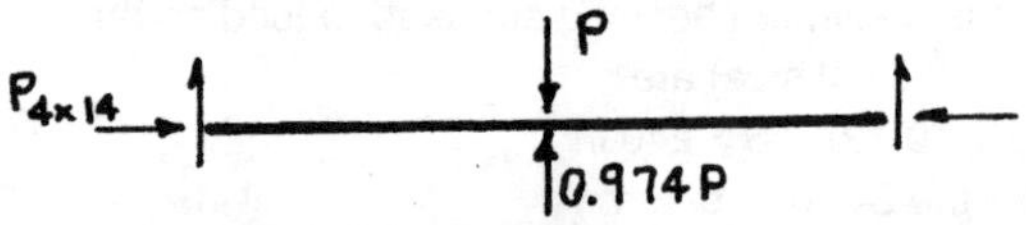

$P_{4x14} = 0.868(0.974P) = 0.845P$

The top chord is braced continuously about its weak axis

$\therefore \quad L_e / d = (28 \times 12 / 13.25) = 25.3$

$$F_{CE} = \frac{0.3E'}{(L_e / d)^2} = \frac{0.3(1,600,000 \text{ lbf / in}^2)}{(25.3)^2} = 750 \frac{\text{lbf}}{\text{in}^2}$$

$$C_p = \frac{1 + F_{CE} / F_c^*}{2c} - \sqrt{\left[\frac{1 + F_{CE} / F_c^*}{2c}\right]^2 - \frac{F_{CE} / F_c^*}{c}}$$

$$= \frac{1 + (750 / 1250)^2}{1.6} - \sqrt{\left[\frac{1 + (750 / 1250)}{1.6}\right]^2 - \frac{(750 / 1250)}{0.8}}$$

$= 0.5$

$F_c' = C_pF_c^* = 0.5(1250) = 625 \text{ lbf / in}^2$

The allowable bending stress is

$C_F = (12 / d)^{1/9} = 0.98$

$M_{max} = (P - 0.974P)28 / 4 = 0.182P$

$\therefore \quad f_b = (0.182P) \times 12 / 102.4 = 0.0213P$

For combined axial plus bending stresses

$$\left[\frac{f_c}{F'_c}\right]^2 + \frac{f_{b1}}{F'_{b1}\left[1 - \left(\frac{f_c}{F_{cE1}}\right)\right]} \le 1$$

$$\left[\frac{\frac{0.845P}{46.4}}{625}\right]^2 + \frac{0.0213P}{0.98(1500)\left[1 - \frac{\frac{0.845P}{46.4}}{750}\right]} \le 1.0$$

$P \le 21,000$ lbf

2. For the 4×6 Post, $L_e / d = 96 / 3.5$, $C_P = 0.57$

   $F_c' = 710 \text{ lbf / in}^2$

   $f_c = (0.974P / 19.3) \le 710$

   $\therefore P \le 14,070$ lbf

3. For the steel rods: $F_s = 22,000$ psi

   $P / 1.0 \le 0.974 \times 22,000$

   $P \le 21,400$ lbf

Thus, the stress in the 4×6 post governs:

<u>P = 14,070 lbf</u>

b) Compute the midspan deflection: Net P = 366 lbf↓

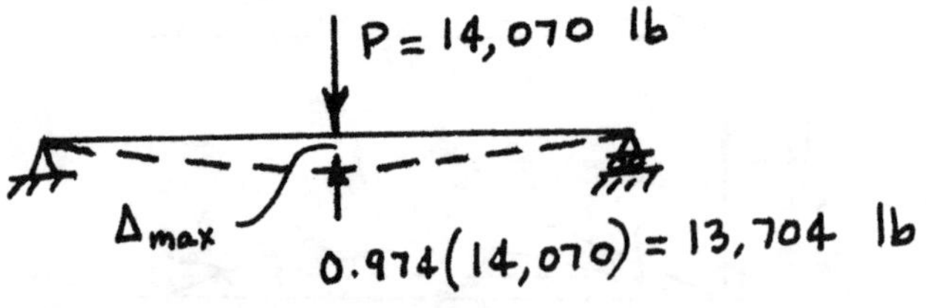

$\Delta_{max} = 366 \times (28 \times 12)^3 / [48 \times 1.6 \times 10^6 \times 678.5]$

Δ_{max} = 0.27 in ↓

c) The calculated member stresses are:

1) Top chord (combined axial plus bending):

   $f = [(0.182 \times 12 / 102.4) + (0.845 / 46.4)] \times 14,070$

   f = 556 lbf / in²

2) Post (governs):

   f = 730 lbf / in²

3) Steel rods:

$f = 0.974 \times 14{,}070 / 1.0 = 13{,}700$ lbf / in^2

Structural-8

Based on the gross area of concrete:

$E_c I_g = (3 \times 10^3 \text{ ksi})30 \times 30^3 / 12$
$= 2.025 \times 10^8$ k-in^2

Take the force in the pile as the redundant and release the structure:

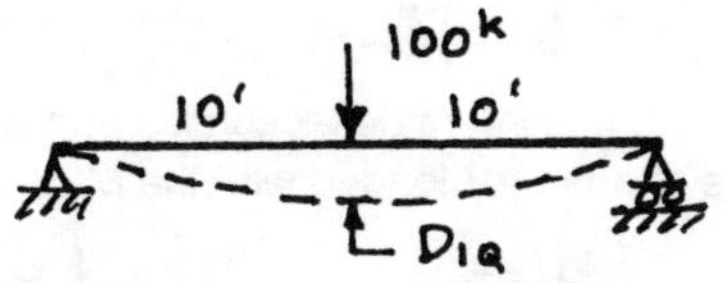

$D_{IQ} = 100(240)^3 / (48EI_g) = 0.1422" \downarrow$

Apply equal and opposite 1^k at release point:

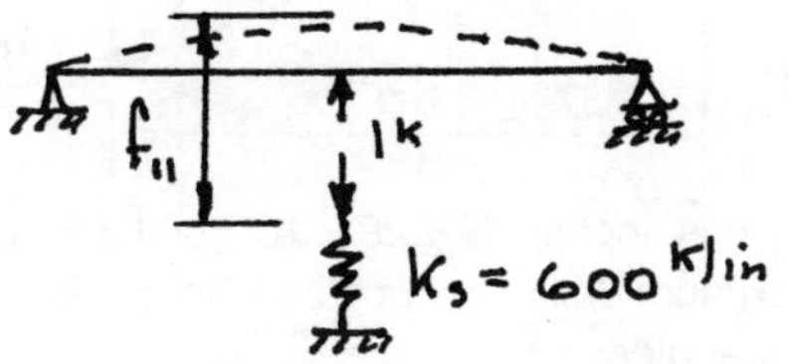

$f_{11} = 1(240)^3 / (48\,EI_g) + 1 / 600 = 0.003089\ ^{in}/_{kip}$

For consistent displacement:

$f_{11} P_B = D_{IQ}$

$\therefore \quad P_B = 0.1422 / 0.003089 = 46.0^k$

Thus,

$M_B = (100 - 46) \times 20 / 4 = 270^{k\text{-ft}}$

$\Delta_B = (100 - 46) \times 240^3 / (48EI_g)$

$= 0.08$ in $\downarrow$

Note: The above values of M_B and Δ_B are due only to the wheel load, which was required by the problem statement.

Structural-9

Solve for the reactions in the structure, which is one degree statically indeterminate. Analyze by consistent displacements. Choose the axial force in the post as the redundant force. Release:

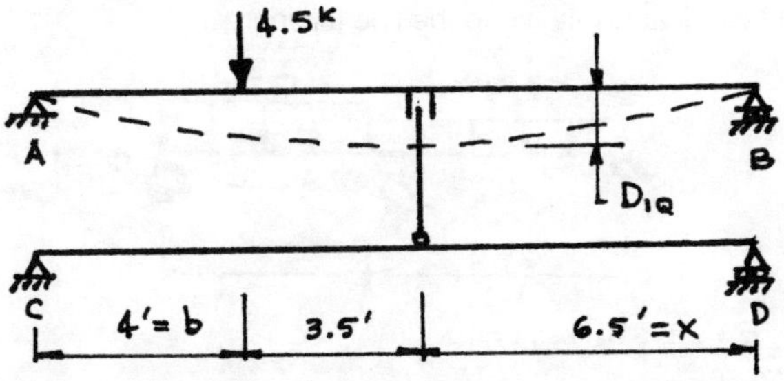

$D_{IQ} = (Pbx / 6EIL(L^2 - b^2 - x^2)$

EI = constant in beams; EA = ∞ in post

$D_{IQ} = [(4.5 \times 4 \times 6.5) / (6 \times 14EI)](14^2 - 4^2 - 6.5^2)$

$D_{IQ} = 191.86 / EI$

Apply equal and opposite 1^k forces to beam and post at the released point:

$f_{11} = 2 \times (1 \times 6.5^2 \times 7.5^2) / (3 \times 14EI) = 113.17 / EI$

For consistent displacement, if P is the axial force in the post:

$Pf_{11} = D_{IQ}$

$\therefore \quad P = 191.86 / 113.17 = 1.7^k$ (comp)

Thus:

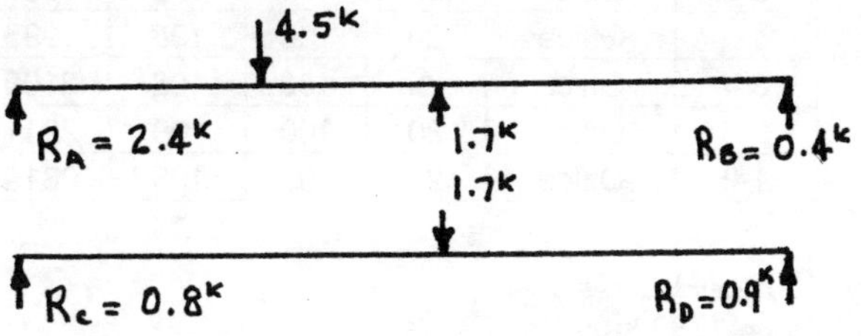

$R_A = 2.4$ kips
$R_B = 0.4$ kips
$R_C = 0.8$ kips
$R_D = 0.9$ kips

Structural-10

Determine column load at the 12th, 6th, and 1st stories. Follow the Uniform Building Code (UBC 1997):

- Roof live load (Method 2: $w_L = 20$ PSF; $r = 0.08$; Tributary Area = 900 ft^2)

$$R \le \begin{cases} r(A - 150) = 0.08(900 - 150) = 60\% \\ 23(1 + 80/20) = 115\% \\ 40\% \leftarrow \text{Governs} \end{cases}$$

$\therefore \quad w_{Lr} = (1 - 0.4)20 = 12$ PSF

- The live load for the mechanical equipment room exceeds 100lbf/ft^2 and cannot be reduced
- In residential areas , live loads of 40 lbf/ft^2 can be reduced at the rate of 0.08 percent of the tributary area that exceeds 150 ft^2

$$R \le \begin{cases} 0.08(A - 150) = 60\% \\ 23(1 + 100/40) = 81\% \\ 40\% \text{ below one level; } 60\% > \text{one} \end{cases}$$

$\therefore \quad w_{Lr} = (1 - 0.40)40 = 24$ PSF one level

or $\quad w_{Lr} = (1 - 0.60)30 = 16$ PSF all others

- for the office areas: $w_L = 50$ PSF; $w_{Lr} = 0.4 \times 50$ PSF
- for the assembly areas (moveable seats): $w_L = 100$ PSF; cannot be reduced
- for heavy storage areas: $w_L = 250$ PSF; UBC Sec. 1607.5 permits a 20% reduction for columns
  $\therefore w_{Lr} = 0.8 \times 250 = 200$ psf

Level	Category	w_{Lr} (Psf)	w_d (Psf)	ΔP (kips)	P (kips)
18-R	Roof	12	80	83	83
17-18	Mech.	125	100	203	286
16-17	Resid.	24	100	112	398
15-16	Resid.	16	100	104	502
14-15	Resid.	16	100	104	606
13-14	Resid.	16	100	104	710
12-13	H. Stor.	200	100	270	980
11-12	Resid.	16	100	104	1084
10-11	Resid.	16	100	104	1188
9-10	Resid.	16	100	104	1292
8-9	Resid.	16	100	104	1396

7-8	Mech.	125	100	203	1599
6-7	Assem.	100	100	180	1799
5-6	Office	20	100	108	1887
4-5	Office	20	100	108	1995
3-4	Office	20	100	108	2103
2-3	Office	20	100	108	2211
1-2	Office	20	100	108	2319

(a) **$P_{1\text{-}2}$ = 2319 kips**
(b) **$P_{6\text{-}7}$ = 1799 kips**
(c) **$P_{12\text{-}13}$ = 980 kips**

Structural-11

(a) Analyze the frame by moment distribution.

$K_{AB} = 4E(27) / 15 = K_{BA} = 7.2E$
$K_{BC} = 4E(27) / 24 = K_{CB} = 4.5E$
$K_{BD} = 3E(8) / 15 = 1.6E$
$\Sigma K_B = 13.3E$; $DF_{AB} = 0$; $DF_{BA} = 0.541$; $DF_{BD} = 0.120$;
$DF_{BC} = 0.338$; $DF_{CB} = 0$;
$CO_{BD} = 0$; All other carryover factors = 0.5
$FEM_{BC} = -36 \times 14 \times 10^2 / 24^2 = -87.5^{\text{k-ft}}$
$FEM_{CB} = 36 \times 14^2 \times 10 / 24^2 = 122.5^{\text{k-ft}}$
$FEM_{BD} = (5 / 15^2)[(5 \times 10^2) + (0.5 \times 5^2 \times 10)] = 13.9^{\text{k-ft}}$

0		.541	.120	.338		0
			13.9	-87.5		122.5
19.9	c.o. ←	39.8	8.8	24.9	c.o. →	12.5
19.9		39.8	22.7	-62.6		135.0

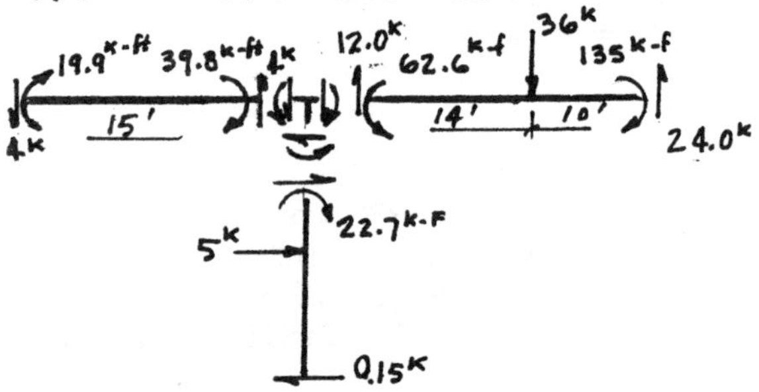

b) Moment diagram: +M

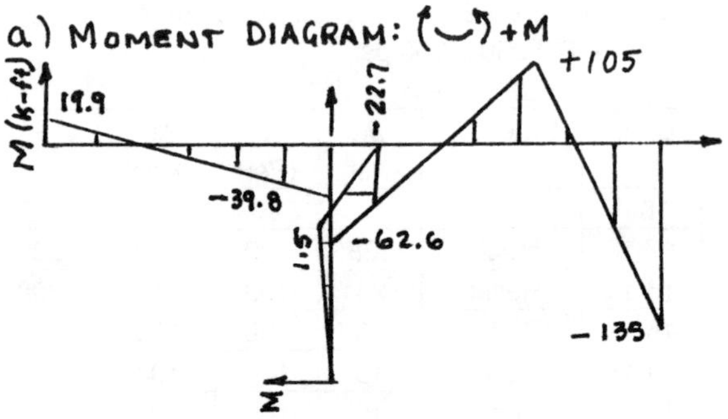

c) Deflected shape

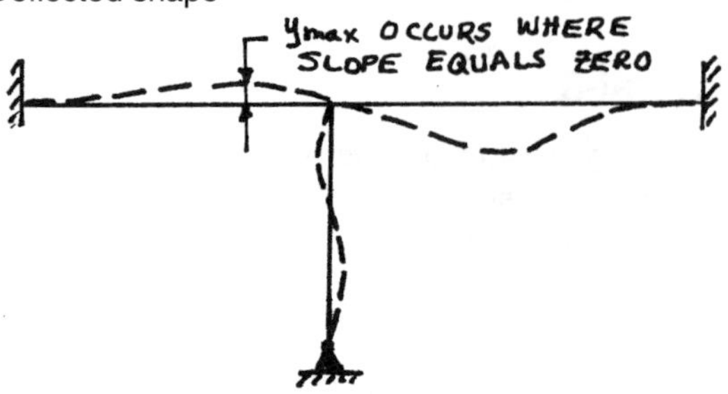

d) Calculate the maximum deflection in span AB. Take the conjugate of span AB loaded with the M/EI diagram.

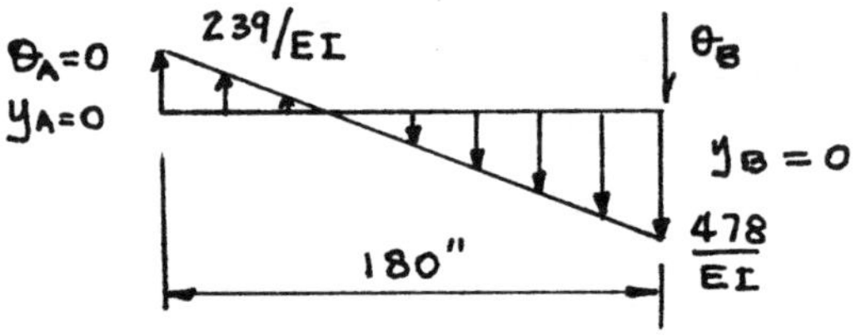

Work in kips, inches: $E = 29 \times 10^3$ ksi; $I = 27$ in^4
The maximum deflection occurs where θ_x is zero. By similar triangles:
$(^X/_2) / 239 = 180 / (239 + 478)$ ∴ $X = 120$ in

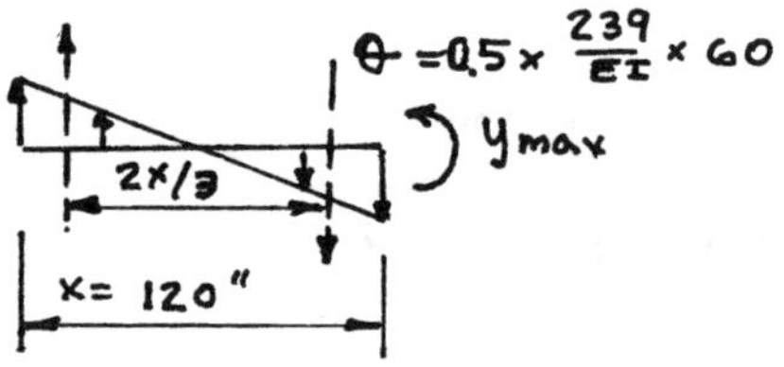

$y_{max} = 0.5(239 / EI)60(^2/_3)(120)$
y_{max} = 0.73 in ↑

Structural-12

Analyze the beam by the method of consistent displacement, taking the vertical reaction at B as the redundant force. Given: Δ_{BV} = 0.25" ↓ and Δ_{CV} = 0.125" ↓, which is equivalent to a differential settlement equal to 0.25" – 0.125(25 / 45) = 0.181" ↓ at B relative to the line joining AC.

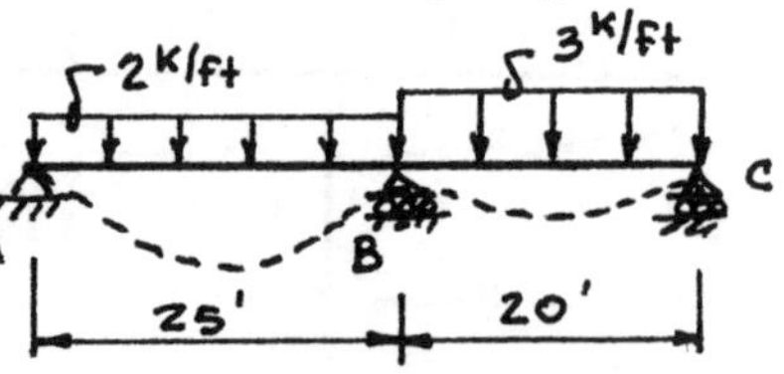

$E = 29 \times 10^3$ ksi; $I = 954$ in^4
Release:

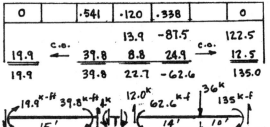

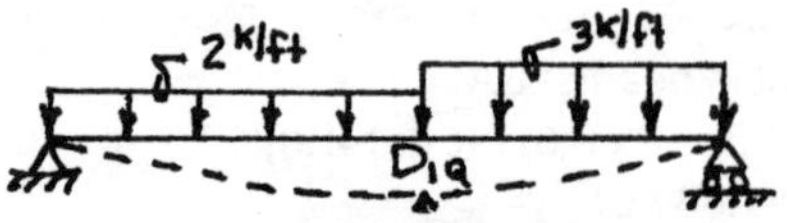

Compute the deflection at the release point

$D_{1Q} = (2 \times 25^2 \times 20)[(4 \times 25 \times 45) - (3 \times 25^2)]$
$\times [1728 / (24 \times 29{,}000 \times 954 \times 45)]$
$+ (3 \times 20^2 \times 25)[(4 \times 20 \times 45) - (3 \times 20^2)]$
$\times [1728 / (24 \times 29{,}000 \times 954 \times 45)] = 7.956$ in ↓

The flexibility influence coefficient is

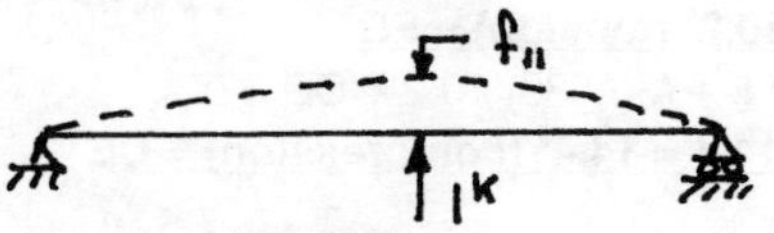

$f_{11} = 1 \times 25^2 \times 20^2 \times 1728/[3EI \times 45]$

$= 0.1157\ ^{in}/_{kip} \uparrow$

For consistent displacement:

$R_B f_{11} - D_{1Q} = -0.181$

<u>$R_B = (7.956 - 0.181) / 0.1157 = 67.2^k \uparrow$</u>

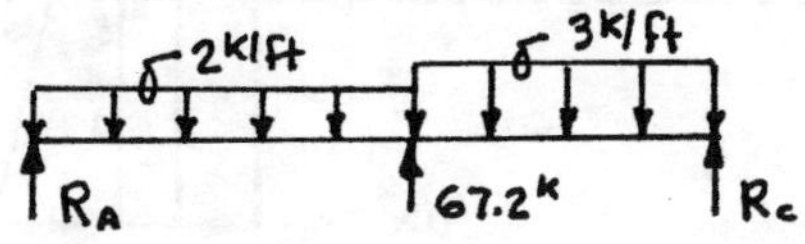

$\Sigma M_A = 0 = 2 \times 25^2 / 2 + 3 \times 20 \times 35 - 67.2 \times 25 - 45R_c$

$R_c = 23.2^k$

$\Sigma F_y = 0 = R_A - 2 \times 25 + 67.2 - 3 \times 20 + 23.2$

$R_A = 19.6^k$

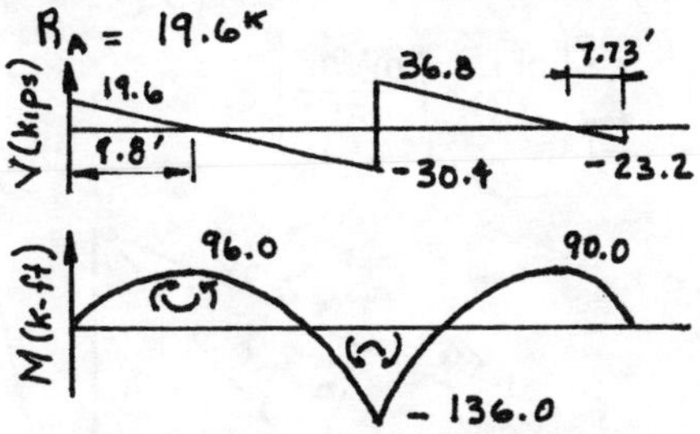

Check shear and flexure using allowable stress design (AISC 1989). Given that the compression flange is braced laterally; the W16×67 of A36 steel is compact $\therefore$ $F_b = 24^{ksi}$. Thus,

$f_b = M / S = 136 \times 12 / 117 = 13.9^{ksi} < F_b$

$f_v = V / dt_w = 36.7 / (16.33 \times 0.395) < F_v = 0.4F_y$

$\therefore$ **<u>The W16×67 is acceptable</u>**

Structural-13

a) Analyze the frame by the method of consistent displacements taking the horizontal reaction at 'D' as the redundant.

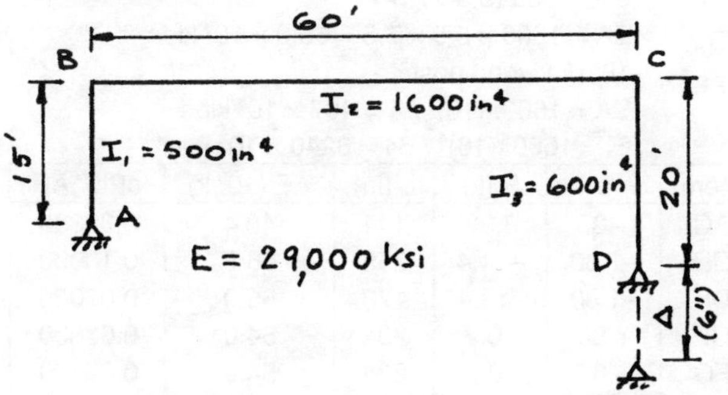

The structure is a one degree statically indeterminate frame and experiences a 6" vertical settlement at 'D'. Release:

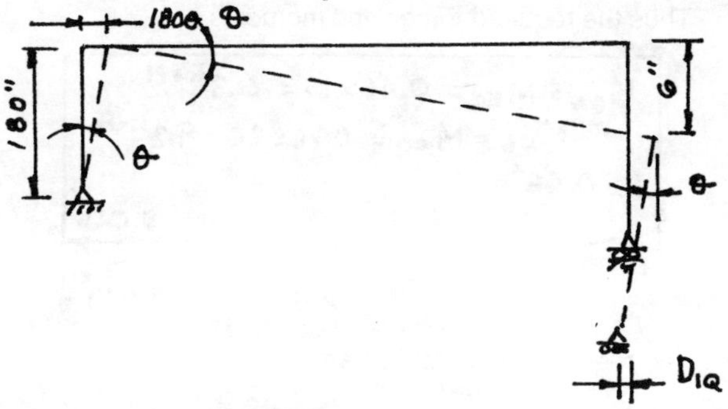

The released statically determinate structure experiences a rigid body rotation about 'A':

$\theta = 6" / (60 \times 12)" = 0.00833$ rad

$\therefore$ $D_{IQ} = 180"\theta - 240"\theta = 0.5" \leftarrow$

From linear elastic theory, small displacements are calculated with respect to the original, undeformed geometry. Thus,

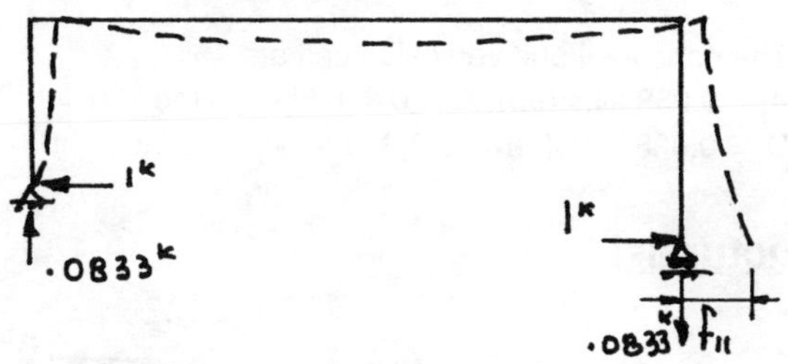

By the Dummy Load Method:

$$f_{11} = \sum_{i=1}^{m} \int_{L_i} \left(\frac{mMdx}{EI}\right)_i$$

Mem	Internal	M = m
AB	$0 \le x \le 180$	$+1x$
BC	$0 \le x \le 720$	$+180 + 0.0833x$
DC	$0 \le x \le 240$	$-1x$

$$Ef_{11} = \int_0^{180} \frac{x^2}{500}dx + \int_0^{720} \frac{(180 + 0.0833x)^2}{1600}dx + \int_0^{240} \frac{(-1x)^2}{600}dx$$

$$Ef_{11} = \left.\frac{x^3}{1500}\right|_0^{180} + \left(\frac{32,400x}{1600} + \frac{30x^2}{3200} + \frac{0.00694x^3}{4800}\right)\Bigg|_0^{720} + \left.\frac{x^3}{1800}\right|_0^{240}$$

$\therefore$ $f_{11} = 31,548 / E = 1.0879\ ^{in}/_{kip}$

For consistent displacements:

$f_{11}D_x + D_{IQ} = 0$

$\therefore$ $D_x = 0.5 / 1.0879 = 0.46^k \rightarrow$

From statics:

$\Sigma F_x = 0$: $A_x = 0.46^k \leftarrow$

$\Sigma M_A = 0$: $D_y = 0.038^k \downarrow$

$\Sigma F_y = 0$: $A_y = 0.038^k \uparrow$

Thus the required forces and moments are:

$$M_{BA} = M_{BC} = 0.46 \times 15 = 6.9^{k\text{-}ft}$$
$$M_{CB} = M_{CD} = 0.46 \times 20 = 9.2^{k\text{-}ft}$$

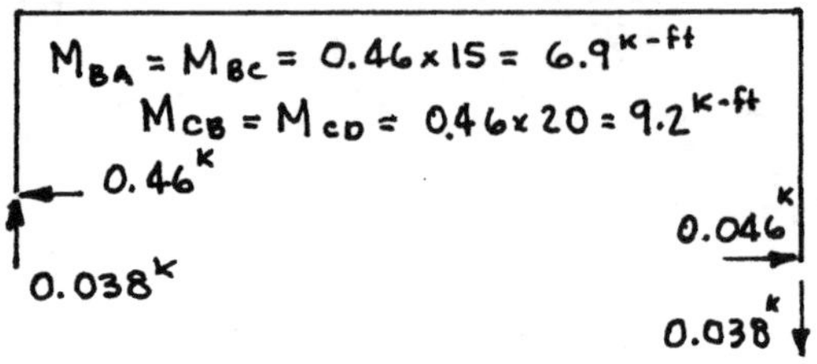

b) The required moment diagram is:

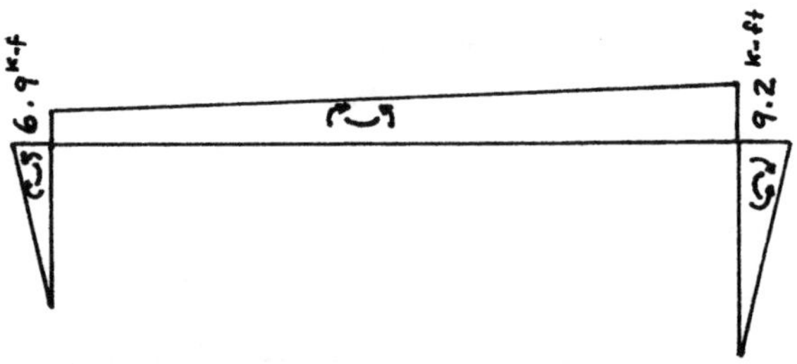

c) The horizontal and vertical forces are
$A_y = 0.038$ kips (up); $A_x = 0.046$ kips (to the left)
$D_y = 0.038$ kips (down); $D_x = 0.46$ kips (to the right)

Structural-14

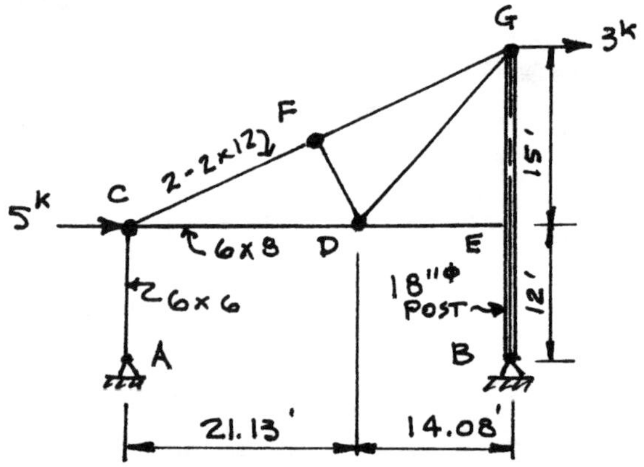

a) External reactions
$\Sigma M_A = 0$: $5(12) + 3(27) - B_y(36) = 0$
$\therefore B_y = 3.92^k \uparrow$
$\Sigma F_y = 0$: $A_y + 3.92^k = 0$
$\therefore A_y = 3.92^k \downarrow$
$\Sigma F_x = 0$: $B_x + 5 + 3 = 0$
$\therefore B_x = 8^k \leftarrow$

b) Internal forces: Member "BEG" is subjected to shear, bending, and axial force; all others are two-force members, which resist only axial force. By inspection, member FD has zero force, from which it follows that DG is also a zero force member.
Taking free body of joint C:

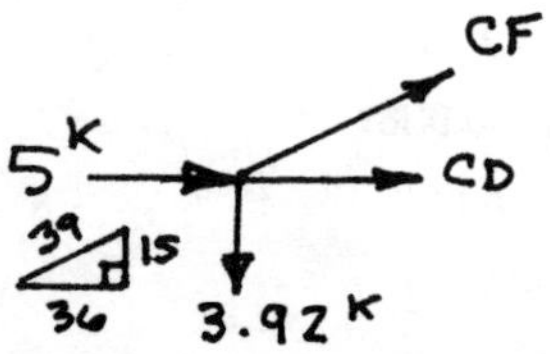

$\Sigma F_y = 0$: $(15/39)\,CF - 3.92 = 0$
CF = +10.2^k (tension) = FG
$\Sigma F_x = 0$: $5 + (36/39)\,10.2 + CD = 0$
CD = –14.4 = 14.4^k (compression) = DE

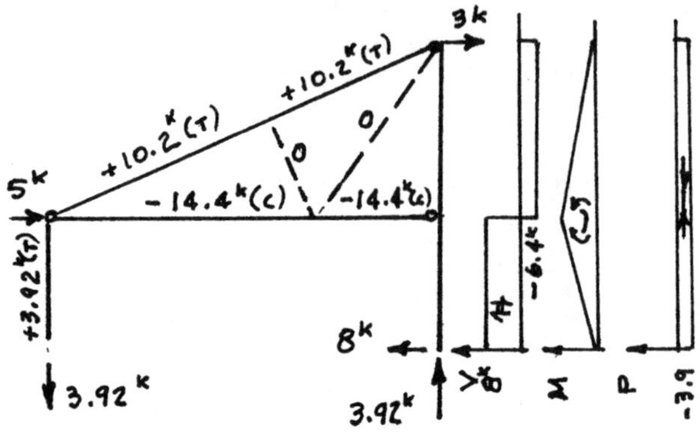

c) Compute the horizontal deflection at 'C'. Use the virtual work (i.e., dummy load) method: apply unit force at 'C'.

$$(1)\Delta_{CH} = \sum_{i=1}^{mem}\left(\frac{pPL}{Ae} + \int_L \frac{mMdx}{EI}\right)_i$$

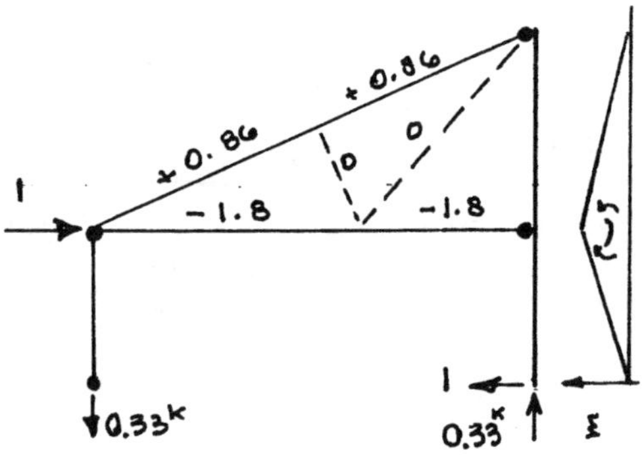

Calculate properties of solid sawn sections using dressed dimensions (E = 1600 ksi):
- 6×6 wood (5.5"×5.5")
  EA = 1600 × 5.5^2 = 48.4 × 10^3 kips
- 2–2×12 wood (1.5"×11.25")
  EA = 1600 × 2 × 1.5 × 11.25 = 54.0 × 10^3 kips
- 6×8 wood (5.5"×7.5")
  EA = 1600 × 5.5 × 7.5 = 66.0 × 10^3 kips
- 18" dia wood post
  EA = 1600π(18)^2 / 4 = 407 × 10^3 kips
  EI = 1600π(18)^4 / 64 = 8240 × 10^3 k-in^2

Mem	p	P (k)	L (in)	AE (10^3 k)	pPL / AE
AC	+.33	3.92	144	48.4	0.00385
CD	–1.80	–14.4	255	66.0	0.10000
DE	–1.80	–14.4	178	66.0	0.07000
CF	+.86	10.2	234	54.0	0.03800
FG	+.86	10.2	234	54.0	0.03800
BE	–.33	–3.92	144	407	0.00050
EG	–.33	–3.92	180	407	0.00060

Contribution due to axial: Σ = 0.25100
Contribution due to bending:

$$\sum\int\frac{mMdx}{EI}=\int_0^{144}[(x)(8x)/EI]dx$$
$$+\int_0^{180}[(0.8x)(6.4x)/EI]dx$$
$$=8x^3/(3EI)\Big|_0^{144}+5.12x^3/(3EI)\Big|_0^{180}$$
$$=2.174''\rightarrow$$

Thus, $\Delta_{CH}=2.174+0.251=2.424''\rightarrow$

(d) **<u>90% of the deflection is due to bending</u>**

Structural-15

a) Solve for the load in member BD by consistent deformation. The frame is two degree statically indeterminate. Choose M_E and the axial force in BD, P_{BD}, as the redundant components.
Release:

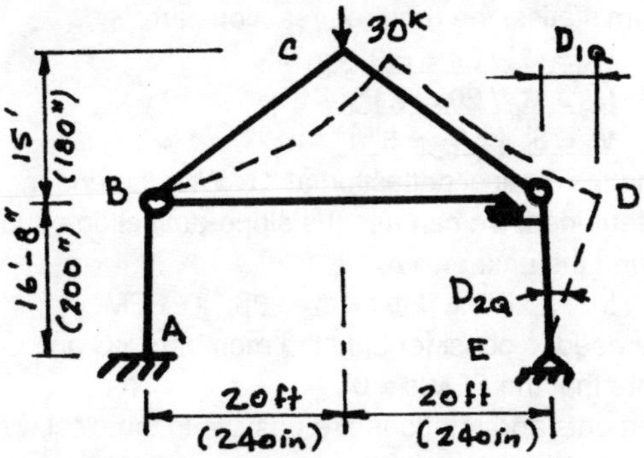

Consider axial deformation only in member BD: $AE=(0.5)(30\times10^3)^k$; bending deformation only in columns and rafters: $EI=(800)(30\times10^3)$ k-in^2. For the $30^k\downarrow$ at joint C of the <u>released</u> structure, member DE is a two-force member; BD has zero force; AB carries axial force only; and, by symmetry, the bending moments in CD are mirror images of those in BC.

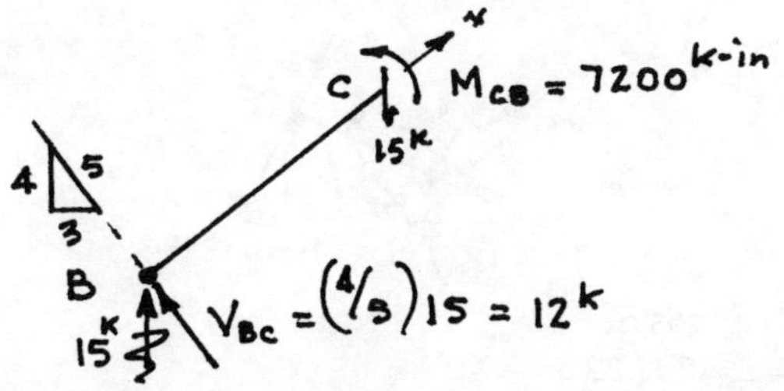

$\therefore\quad M_x=12x;\ 0\le x\le300''$

Apply unit equal and opposite horizontal forces at joint D to the released structure at each release point:

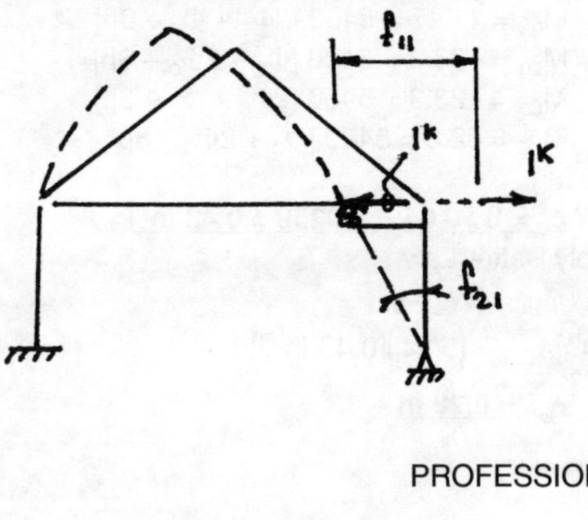

This creates an axial force of $+1^k$ in BD

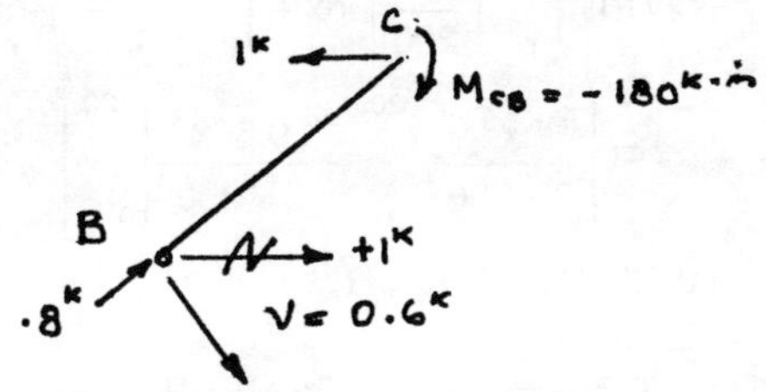

$M_x=-0.6x;\ 0\le x\le300''$

Apply a unit couple at E

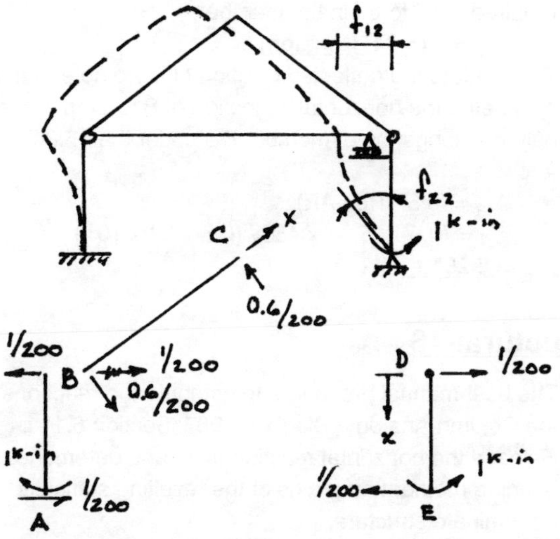

Member DE: $M_x=-x/200;\ 0\le x\le200''$
Member AB: $M_x=x/200;\ 0\le x\le200''$
Member BC: $M_x=-0.6x/200;\ 0\le x\le300''$

Calculate the displacements in the released structure:

$$(1)\ D_{IQ}=2\int_0^{300}(mM/EI)dx$$

$$D_{IQ}=2\int_0^{300}[(0.6x)(12x)/EI]dx$$

$$=14.4x^3/3EI\Big|_0^{300}=5.40\text{ in (relative displacement)}$$

From geometry (small deformation)

$$D_{2Q}=D_{1Q}/200=0.027^{\text{rad}}\ \text{(c.w.)}$$

The flexibility influence coefficients are

$$(1)\ f_{11}=\sum\left[\frac{pPL}{AE}+\int_L\frac{mMdx}{EI}\right]$$
$$=(1)(1)(480)/(0.5\times30\times10^3)$$
$$+2\int_0^{300}\left[(0.6x)^2/EI\right]dx$$
$$=0.032+0.72x^3/3EI\Big|_0^{300}$$
$$=0.302\text{ in (relative displacement)}$$

$$(1)\ f_{21}=f_{12}=2\int_0^{300}\left[(0.6x)\left(\frac{-0.6x}{200}\right)/EI\right]dx$$
$$=-0.72x^3/600EI\Big|_0^{300}$$
$$=0.00135^{\text{rad}}/_k\ \text{(c.c.w.)}$$

$$(1)\ f_{22} = (2/EI)\left[\int_0^{200}\left(\frac{x}{200}\right)^2 dx + \int_0^{300}\left(\frac{0.6x}{200}\right)^2 dx\right]$$

$$= 2/EI\left[\left.\frac{x^3}{3\times 200^2}\right|_0^{200} + \left.\frac{0.36x^3}{3\times 200^2}\right|_0^{300}\right]$$

$$= 0.0000123^{rad}/_{k-in}\ \text{(c.c.w.)}$$

For consistent deformation:

$5.40 - 0.302\ P_{BD} - 0.00135\ M_E = 0$

$-0.027 + 0.00135P_{BD} + 0.0000123M_E = 0$

Eliminating M_E and solving for P_{BD} we obtain the required axial force in the member:

P_{BD} = +15.84^k (tension)

b) To calculate the deflected position of B we note that, by symmetry, the horizontal deflection of B is simply one half the elongation of member BD under the 15.84^k force:

$\therefore\ \Delta_{BH} = (0.5)\ (PL / AE)$

$= (0.5)\ (15.84 \times 480 / (0.5 \times 30 \times 10^3))$

Δ_{BH} = 0.25" ←

Structural-16

a) The best manual procedure to compute the reactions is the Column Analogy (Blodgett 1967; Section 6.1). Release the horizontal reaction at 'E' and determine the bending moment diagrams of the resulting statically determinate structure:

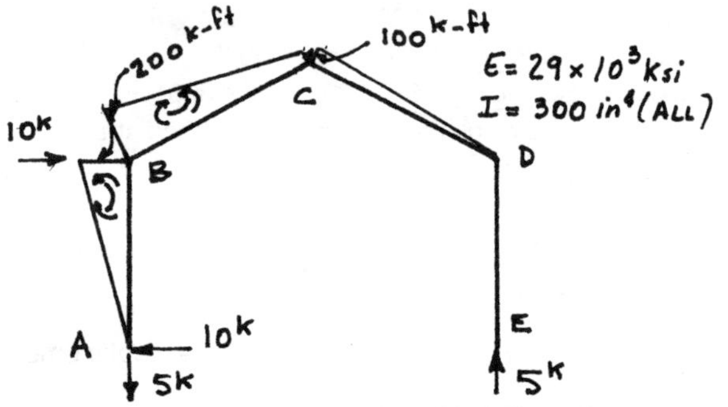

Apply the M / EI of the statically determinate structure as "axial loads" on the analogous column:

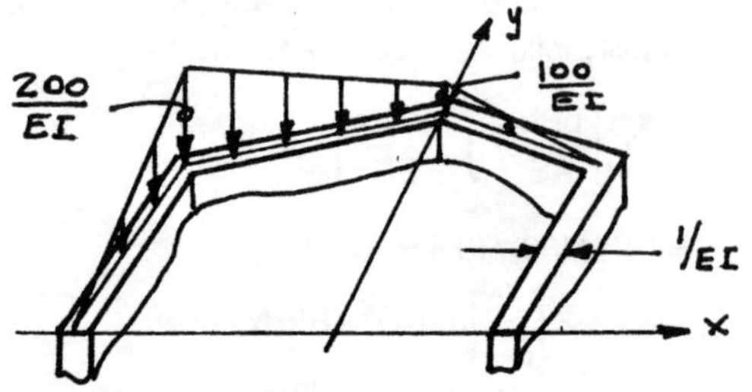

Note: at hinged bases 1 / EI → ∞ ; therefore, the area and moment of inertia about the y-axis → ∞ and the x-axis is at the bottom of the column

$I_x = 2[20^3 / (EI \times 3) + (1 / 12EI)\ 22.36 \times 10^2$

$+ (22.36 / EI) \times 25^2] = 33{,}656 / EI$

We need only compute the moment of the M / EI load about the x-axis (other "stress" components are zero because both I_y and area → ∞).

A_i	P_i	Y_i	$M_{xi} = P_iY_i$
1	0.5(20)200 / EI	40 / 3	26,667 / EI
2	0.5(22.36)200 / EI	23.33	52,099 / EI
3	0.5(22.36)100 / EI	26.67	29,851 / EI
4	0.5(22.36)100 / EI	26.67	29,851 / EI
ΣM_{xi} =			(138,467) / EI

(Note: slight discrepancy due to roundoff)

The final end moments are obtained by superimposing upon the moments in the released structure the "stresses" in the analogous column. Thus,

$M_A = 0$

$M_B = +200 - (138{,}467 / EI)\ 20 / (33{,}656 / EI)$

$= 117.7^{k\text{-}ft}$ c.c.w.

$M_C = +100 - (138{,}467 / EI)\ 30 / (33656 / EI)$

$= -23.3 = 23.3^{k\text{-}ft}$ c.w.

$M_D = 0 - (138{,}467 / EI)\ 20 / (33{,}656 / EI)$

$= -82.3 = 82.3^{k\text{-}ft}$ c.w.

From statics, the required reactions are:

$H_A = M_B / 20 = 5.9^k$ ←

$H_E = M_D / 20 = 4.1^k$ ←

$V_A = 5^k$ ↓ ; $V_E = 5^k$ ↑

b) Find the vertical deflection at 'C'. With the end moments determined, we can use the slope-deflection equations to find the unknown Δ_{CV}.

$M_{NF} = 2EI / L\ [2\Phi_N + \Phi_F - 3\beta_{NF}] \pm FEM_{NF}$

We need to consider only the members BC and CD (note that the FEM's ≡ 0).

Moments and rotations are positive in the clockwise sense at member ends.

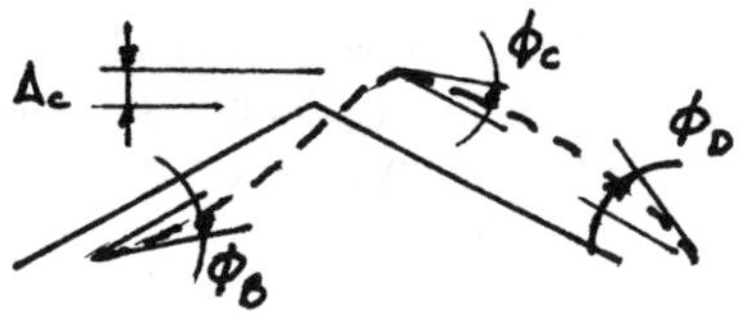

In terms of Δ_C:

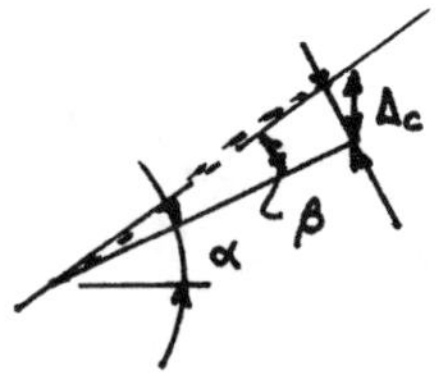

$\Delta = \Delta_C \cos\alpha$

$= \Delta_C\ (20 / 22.36)$

$\beta = \Delta / L = \Delta_C\ (20 / 22.36^2)$

Let $K' = 2EI / L = 2 \times 29{,}000 \times 300 / (22.36 \times 12)$

$= 64{,}850^{k\text{-}in} = 5400^{k\text{-}ft}/_{Rad}$

$M_{BC} = 117.7 = 5400\ [2\Phi_B + \Phi_C + 3\beta]$

$M_{CB} = +23.3 = 5400\ [\Phi_B + 2\Phi_C + 3\beta]$

$M_{CD} = -23.3 = 5400\ [2\Phi_C + \Phi_D - 3\beta]$

$M_{DC} = 82.3 = 5400\ [\Phi_C + 2\Phi_D - 3\beta]$

Solving:

$\Delta_C = \beta / 0.04 = 0.0359 = 0.43$ in ↑

c) Apply Betti's Law:

$(W_{ext})_{I\rightarrow II} = (W_{ext})_{II\rightarrow I}$

$\left(1\vec{0}^k\right)\Delta_{BH} = \left(5^k \downarrow\right)\left(0.43 \text{ in} \uparrow\right)$

$\therefore$ **Δ_{BH} = 0.22 in ←**

Structural-17

a) Find the moments in all the members. Use the slope-deflection method (note: moment distribution is a good alternative method, which avoids solving a system of linear, simultaneous equations).

The slope-deflection equation (Norris, et. al. 1991) is:

$$M_{NF} = {}^{2EI}/_{L}\ (2\Phi_N + \Phi_F - 3\beta_{NF}) \pm FEM_{NF}$$

For our problem, there are six kinematic unknowns: Φ_B; Φ_C; Φ_D; Φ_E; $\beta_{AB} = \beta_{FE} = \beta_1$; $\beta_{BC} = \beta_{ED} = \beta_2$; all FEM = 0; K = 2EI / L. Thus:

$$M_{AB} = 3[\Phi_B - 3\beta_1]$$
$$M_{BA} = 3[2\Phi_B - 3\beta_1]$$
$$M_{BC} = 1[2\Phi_B + \Phi_C - 3\beta_2]$$
$$M_{CB} = 1[\Phi_B + 2\Phi_C - 3\beta_2]$$
$$M_{BE} = 5[2\Phi_B + \Phi_E]$$
$$M_{EB} = 5[\Phi_B + 2\Phi_E]$$
$$M_{DC} = 5[2\Phi_D + \Phi_C]$$
$$M_{CD} = 5[2\Phi_C + \Phi_D]$$
$$M_{DE} = 2[2\Phi_D + \Phi_E - 3\beta_2]$$
$$M_{ED} = 2[\Phi_D + 2\Phi_E - 3\beta_2]$$
$$M_{EF} = 4[2\Phi_E - 3\beta_1]$$
$$M_{FE} = 4[\Phi_E - 3\beta_1]$$

Six equilibrium equations:

1. $\Sigma M_B = 0;\ M_{BA} + M_{BC} + M_{BE} = 0$

$\therefore\ 18\Phi_B + \Phi_C + 5\Phi_E - 9\beta_1 - 3\beta_2 = 0$

2. $\Sigma M_C = 0;\ M_{CB} + M_{CD} = 0$

$\therefore\ \Phi_B + 12\Phi_C + 5\Phi_D - 3\beta_2 = 0$

3. $\Sigma M_D = 0;\ M_{DC} + M_{DE} = 0$

$\therefore\ 5\Phi_C + 14\Phi_D + 2\Phi_E - 6\beta_2 = 0$

4. $\Sigma M_E = 0;\ M_{ED} + M_{EB} + M_{EF} = 0$

$\therefore\ 5\Phi_B + 2\Phi_D + 22\Phi_E - 12\beta_1 - 6\beta_2 = 0$

5. $\Sigma F_h = 0$ for top story

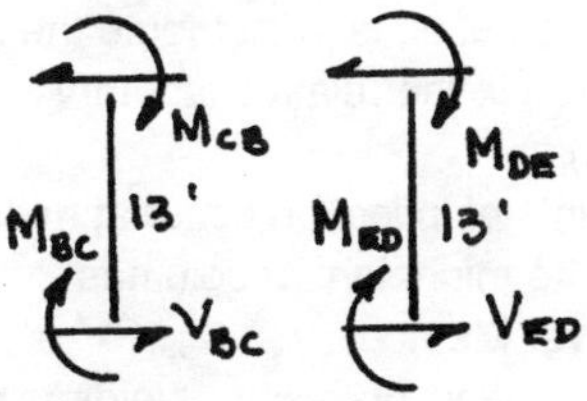

$$V_{BC} = (M_{BC} + M_{CB}) / 13$$
$$V_{ED} = (M_{ED} + M_{DE}) / 13$$
$$\therefore\ V_{BC} = [2\Phi_B + \Phi_C - 3\beta_2 + \Phi_B + 2\Phi_C - 3\beta_2] / 13$$
$$V_{ED} = [2\Phi_D + 4\Phi_E - 6\beta_2 + 4\Phi_D + 2\Phi_E - 6\beta_2] / 13$$
$$\Sigma F_h = 0\text{: } V_{BC} + V_{ED} + 1.3 = 0$$
$$\therefore\ 3\Phi_B + 3\Phi_C + 6\Phi_D + 6\Phi_E - 18\beta_2 = -1.3(13)$$

6. $\Sigma F_h = 0$ for entire frame

$$V_{AB} = [M_{AB} + M_{BA}] / 13$$
$$\therefore\ V_{AB} = [3\Phi_B - 9\beta_1 + 6\Phi_B - 9\beta_1] / 13$$
$$V_{FE} = [4\Phi_E - 12\beta_1 + 8\Phi_E - 12\beta_1] / 13$$
$$\Sigma F_h = 0\text{: } V_{AB} + V_{FE} + 1.3 + 0.7 = 0$$
$$\therefore\ 9\Phi_B + 12\Phi_E - 42\beta_1 = -2.0(13)$$

The equation of equilibrium in matrix form: [K] {D} = {Q}

$$\begin{bmatrix} 18 & 1 & 0 & 5 & -9 & -3 \\ 1 & 12 & 5 & 0 & 0 & -3 \\ 0 & 5 & 14 & 2 & 0 & -6 \\ 5 & 0 & 2 & 22 & -12 & -6 \\ 3 & 3 & 6 & 6 & 0 & -18 \\ 9 & 0 & 0 & 12 & -42 & 0 \end{bmatrix} \begin{Bmatrix} \phi_B \\ \phi_C \\ \phi_D \\ \phi_E \\ \beta_1 \\ \beta_2 \end{Bmatrix} = \begin{Bmatrix} 0 \\ 0 \\ 0 \\ 0 \\ -16.9 \\ -26.0 \end{Bmatrix}$$

Solving the system of equations:

$\{D\}^T$ = (0.4957, 0.1258, 0.4688, 0.7531, 0.9404, 1.450)

Substituting into the slope-deflection equations yields the final end moments:

$M_{AB} = -6.98^{K\text{-}F}$; $M_{BA} = -5.49^{K\text{-}F}$
$M_{BC} = -3.23^{K\text{-}F}$; $M_{CB} = -3.60^{K\text{-}F}$
$M_{BE} = +8.72^{K\text{-}F}$; $M_{EB} = +10.01^{K\text{-}F}$
$M_{CD} = +3.60^{K\text{-}F}$; $M_{DC} = +5.32^{K\text{-}F}$
$M_{DE} = -5.32^{K\text{-}F}$; $M_{ED} = -4.75^{K\text{-}F}$
$M_{EF} = -5.26^{K\text{-}F}$; $M_{FE} = -8.27^{K\text{-}F}$

(b) The moment diagrams and deflected shape are

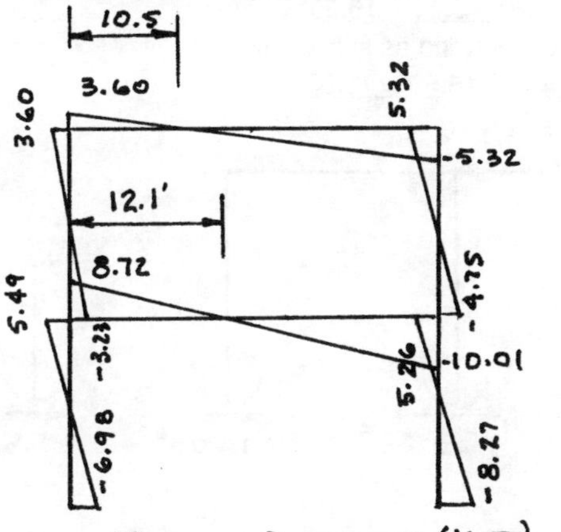

MOMENT DIAGRAMS (K-FT)

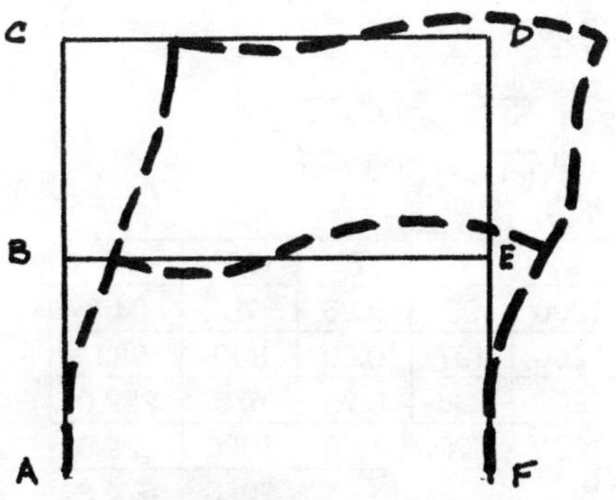

DEFLECTED SHAPE

Structural-18

Find the base shears in the frame. The girders are infinitely stiff; therefore, consider bending deformation only:

$\Delta = 12EI / h^3$

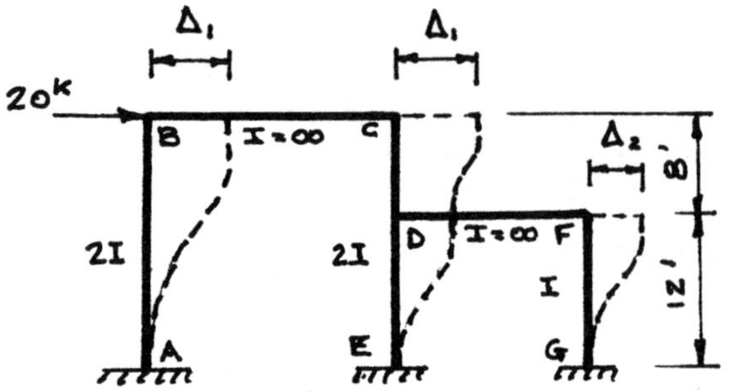

Let K_1 = translational stiffness of portal EDFG:

$K_1 = 12E[2I + I] / (12^3) = 36EI / 1728$

K_e = translational stiffness of column CD in series with portal EDFG:

$1 / K_e = 1 / [12E(2I) / 8^3] + 1 / [36EI / 1728]$

∴ $K_e = 36EI / 2496$

Now, distribute 20^k between columns AB and CD:

$V_{AB} = \{[12E(2I) / 20^3] / [(24EI / 8000) + (36EI / 2496)]\} \times 20 = \mathbf{3.44^k}$

∴ $\underline{\mathbf{V_{CD} = 20 - 3.44 = 16.56^k}}$

Column DE is twice as stiff as GF

∴ $\underline{\mathbf{V_{ED} = (^2/_3)\ 16.56 = 11.03^k}}$

$\underline{\mathbf{V_{FG} = 5.53^k}}$

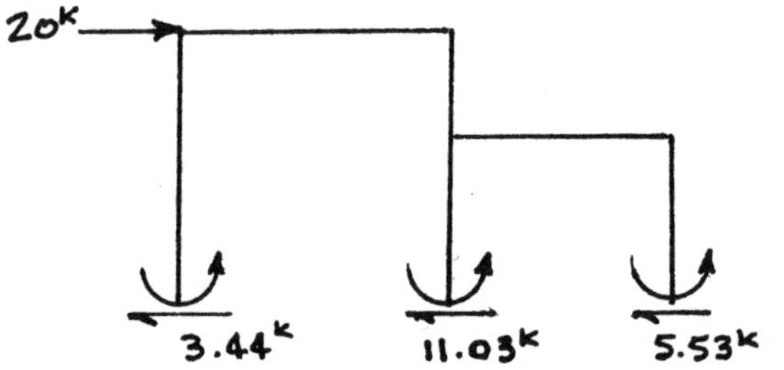

Structural-19

a) Calculate the fundamental period using equation 30-10 (UBC 1997)

$$T = 2\pi\sqrt{\left(\sum_{i=1}^{n} w_i\delta_i^2\right) / \left(g\sum_{i=1}^{n} f_i\delta_i\right)}$$

In tabular form:

i	w_i	f_i	δ_i	$w_i\delta^2_i$	$f_i\delta_i$
1	1200	99	0.25	75	24.8
2	1200	197	0.50	300	98.5
3	1200	296	0.75	675	222.0
4	1000	329	1.00	1000	329.0
Σ =				2050	674.3

Thus,

$$T = 2\pi\sqrt{\frac{2050^{k-in^2}}{\left(32.2 \times 12^{in/sec^2}\right)674.3^{k-in}}}$$

$\underline{\mathbf{T = 0.56\ sec}}$

b) Double the applied lateral forces and recalculate T. Note that the structure is linearly elastic; therefore, doubling the f_i's in Equation 30-10 will double each corresponding δ_i. Thus,

$$T = 2\pi\sqrt{\sum_{i=1}^{n} w_i(2\delta_i)^2 / \left[g\sum_{i=1}^{n}(2f_i)(2\delta_i)\right]}$$

$$T = 2\pi\sqrt{4\sum_{i=1}^{n} w_i\delta_i^2 / \left[4g\sum_{i=1}^{n} f_i\delta_i\right]}$$

which is identical to the equation used to calculate T in part a ∴ $\underline{\mathbf{T = 0.56\ sec}}$

c) The significance of the result in part b compared to the result in part a is that the magnitude of the lateral forces used to approximate the fundamental frequency by Equation 30-10 is not important. Any rationally distributed system of lateral forces can be used, along with the corresponding set of elastic deflections, to obtain a reasonable estimate of the fundamental period.

Structural-20

a) The analysis method used to obtain the results shown:
Fig (b) – Results were obtained by the cantilever method
Fig (c) – Results are consistent with moment distribution
Fig (d) – Results obtained by portal analysis

b) The principal assumptions are:

1) Portal Method:
- The lateral force is distributed to columns on the basis that interior columns carry twice as much shear as exterior columns (a refinement that is sometimes used is to distribute the lateral force in proportion to the relative tributary area of each column)
- The points of inflection occur at midspan of girders and at the midheight of columns

2) Cantilever Method
- Points of inflection occur at midspan of girders and midheight of columns
- Axial forces in columns resist the overturning moments due to lateral forces, and vary linearly as the distance of each column from the centroid of column areas at a particular level (this is analogous to the variation in normal stress due to bending in a cantilevered beam by linear elastic principles).

3) Moment Distribution
- Linear elastic behavior (ie, small displacements; material obeys Hooke's Law)
- Only bending deformation is considered (neglects axial deformation and shear)

c) Of the two approximate methods, the portal method is better suited to buildings with low height-to-width ratios. The cantilever method is better suited to buildings with large height-to-width ratios, where sway due to column axial deformation is relatively important.

Structural-21

Calculate the required reactions in the structure. First, calculate the force required to close the 0.5 in gap:

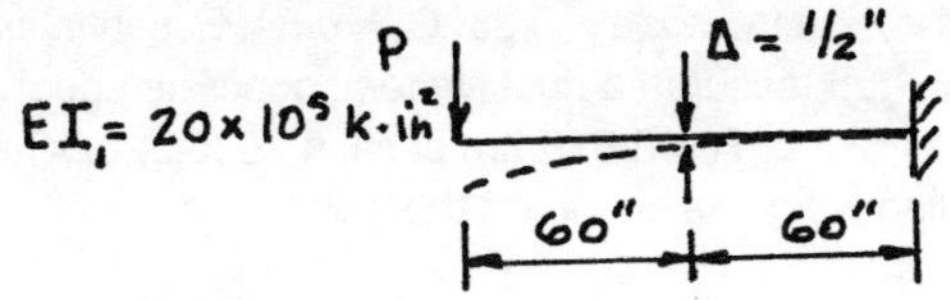

$EI_1 = 20 \times 10^5$ k-in^2

$\Delta_x = (P / 6EI)[2L^3 - 3L^2x + x^3]$

$\therefore\ 0.5 = (P / 6EI_1)[2 \times 120^3 - 3 \times 120^2 \times 60 + 60^3]$

$P = 5.56^k$

Now, the remaining 10 – 5.56 = 4.44^k acts on a structure that is one degree statically indeterminate. Choose the force in the strut EB as redundant and release the structure.

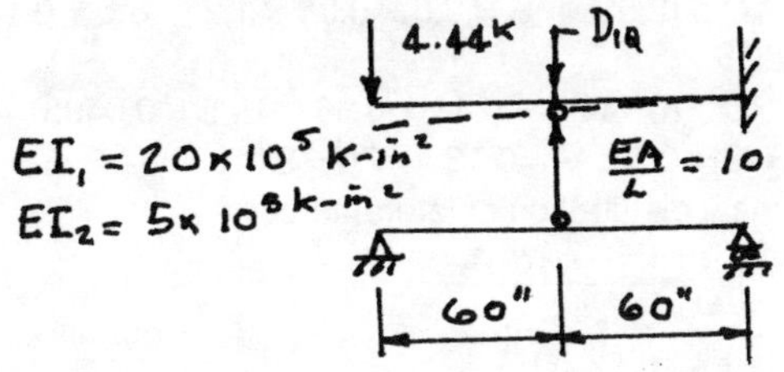

$EI_1 = 20 \times 10^5$ k-in^2; $EI_2 = 5 \times 10^5$ k-in^2

$D_{1Q} = (4.44 / 5.56) \times 0.5 = 0.40"$

Calculate the flexibility influence coefficient

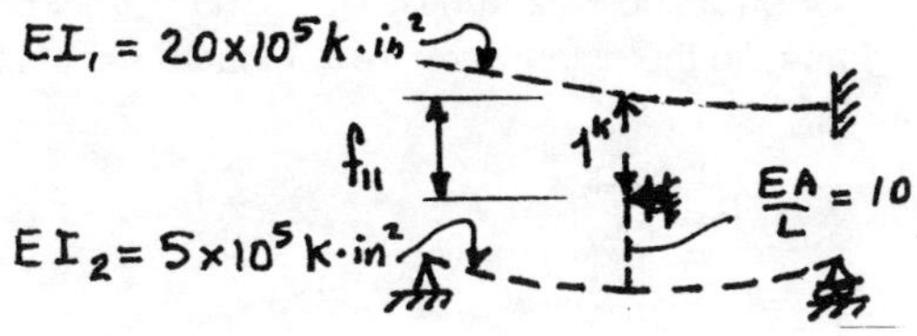

$f_{11} = [(1)(60^3) / (3EI_1)] + (1 / 10) + [1(120)^3 / (48EI_2)]$

$= 0.208\ ^{in}/_{kip}$

For consistent displacement:

$P_{BE}f_{11} - D_{1Q} = 0$

$P_{BE} = 0.40 / 0.208 = 1.92^k$ (comp)

The 1.92^k reacts at the midspan of the simple beam

$\therefore\ V_E = V_F = 1.92 / 2 = 0.96^k \uparrow$

Structural-22

a) To find the maximum stress, we need to find the force exerted on the beam when the ball comes to rest.

Given: 6" × 6" beam; $E = 1.5 \times 10^6$ psi; $I = 6 \times 6^3 / 12 = 108$ in^4; $S = 36$ in^3

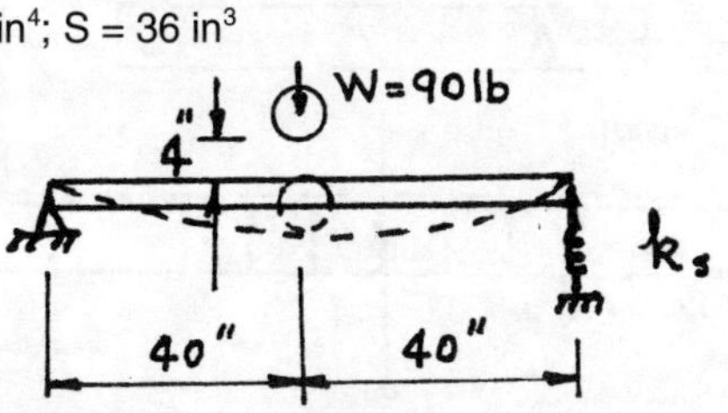

Let:

Δ = Midspan deflection when 90 lb weight comes to rest

F = Force between beam and weight

Apply conservation of energy:

$U_{int} = E_p = 90(4 + \Delta)$

Internal energy consists of the strain energy due to bending of beam plus compression of the spring:

$$U_{int} = \frac{1}{2}\left(\int_L \frac{M^2dx}{EI}\right) + \frac{1}{2}\left(\frac{(F/2)^2}{k_s}\right)$$

$$= \frac{1}{2}\left[2\int_0^{40}\left((Fx/2)^2 / EI\right)dx + \frac{F^2}{4 \times 200}\right]$$

$$= 0.0006579F^2$$

The midspan deflection is the superposition of the beam bending plus the displacement due to spring compression

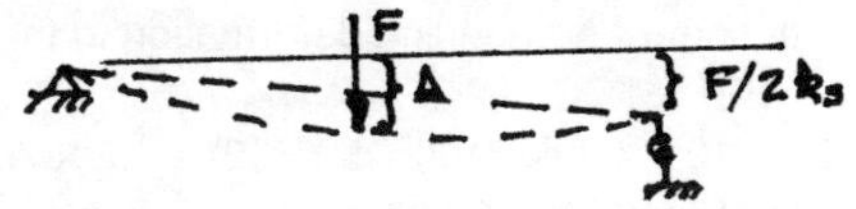

$\Delta = FL^3 / 48EI + 0.5(F / 2k_s)$

$= F(80^3 / 48EI + 1 / (4 \times 200))$

$= 0.0013158F$

$\therefore\ 0.0006579F^2 = 90(4 + 0.0013158F)$

$F^2 - 180F - 547{,}196 = 0$

F = 835 lb

The maximum bending stress is

$M_{max} = 835 \times 80 / 4 = 16{,}700$ lb-in

$\sigma_{max} = M_{max} / S = 16{,}700 / 36 = 464$ psi

b) Repeat the problem if the right support is rigid, i.e., $k_s \rightarrow \infty$:

$\therefore\ U_{int} = 3.292 \times 10^{-5}F^2$

$\Delta = FL^3 / 48EI = 6.584 \times 10^{-5}F$

$3.292 \times 10^{-5}F^2 = 90(4 + 6.584 \times 10^{-5}F)$

F = 3398 lb

$M_{max} = 6790$ lb-in

$\sigma_{max} = 1888$ Psi (four-fold increase)

Structural-23

Analyze the frame below:

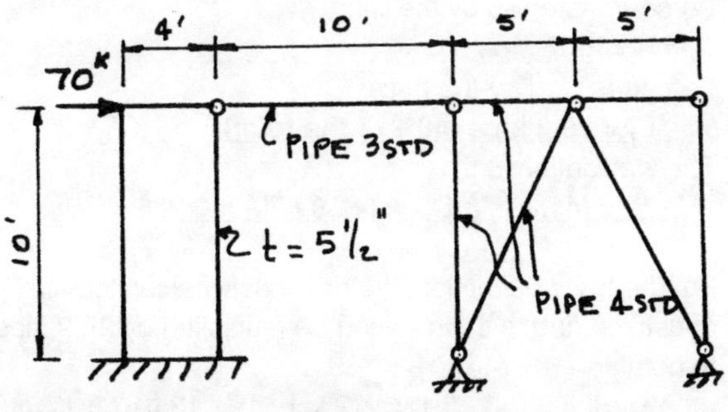

For the concrete: $f''_c = 3000$ psi; $E_c = 3100$ ksi; $G = 0.4E_c = 1240$ ksi

For steel: $E_s = 29{,}000$ksi

Pipe 3 std: $A = 2.23$ in^2; $I = 3.02$ in^4

Pipe 4 Std: $A = 3.17\ \text{in}^2$

Check the Euler buckling strength of the 10 ft long Pipe 3 Std; pinned-pinned end conditions $\therefore$ KL = L:

$$P_E = \pi^2 EI / L^2 = \pi^2 \times 29{,}000 \times 3.02 / 120^2 = 60\ \text{kips}$$

Thus, the Pipe 3 Std can transfer a maximum of 60 kips, which should be less than the amount transferred between the shearwall and braced bent. Let the force in this member be the unknown and release the indeterminate structure:

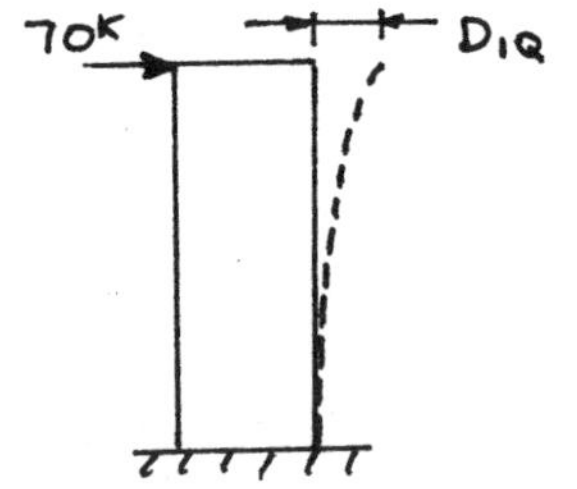

Properties of the 5-1/2" x 48" wall: $A = 5.5 \times 48 = 264\ \text{in}^2$; $I = 5.5 \times 48^3 / 12 = 50{,}700\ \text{in}^4$

Consider both shear and bending deformation in the wall:

$$\Delta = Ph^3 / 3EI + 1.2Ph / AG$$

$\therefore$ $D_{1Q} = 70 \times 120^3 / (3 \times 3100 \times 50{,}700) + 1.2 \times 70 \times 120 / (1240 \times 264) = 0.288$ in

Now, apply equal and opposite unit forces at the released points:

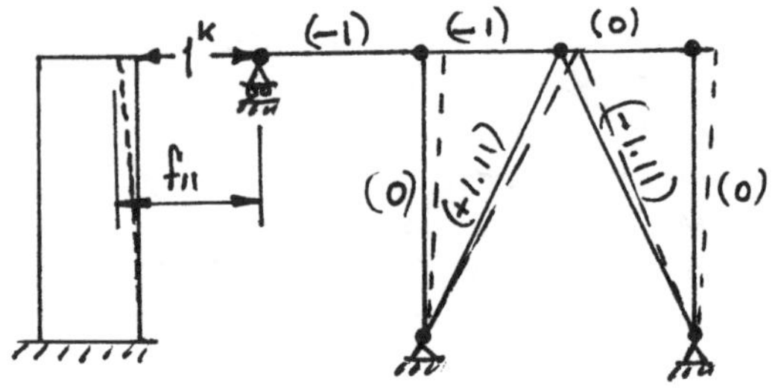

$f_{11} = 1 \times 0.288 / 70 + 1 \times 120 / (2.23 \times 29{,}000) + 1 \times 60 / (3.17 \times 29{,}000) + 2 \times 1.11^2 \times 134.1 / (3.17 \times 29{,}000)$

$= 0.0102$ in / kip

For consistent displacement:

$$Pf_{11} = D_{1Q}$$

$\therefore$ $P = 0.288 / 0.0102 = 28.2$ kips

Thus, the answers are:

a) Force resisted by the wall:

$F_W = 70 - 28.2 = 41.8$ kips (60% of the total)

b) Force transferred by the link:

P = 28.2 kips

c) Force resisted by the truss:

$F_T = 28.2$ kips (40% of the total)

d) The wall deflection:

$\Delta_W = (41.8 / 70)D_{1Q} = (41.8 / 70) \times 0.288 = 0.17$ in $\rightarrow$

e) For the truss deflection, the previously calculated deflection due to 1 kip $\rightarrow$ can be multiplied by 28.2 kips to obtain

$\Delta_T = 28.2[(60 / 3.17) + (2 \times 1.11^2 \times 134.1 / 3.17)] / 29{,}000$

$\Delta_T = 0.120$ in $\rightarrow$

Structural-24

a) Seismic zone 3 – compute force to diaphragm using UBC-97. Assume soil profile type S_D; near-source factors of unity $\therefore$ Ca = 0.36; $C_v = 0.54$; Essential facility $\therefore$ I = 1.25; assume a dual system, concrete shearwalls with SMRF $\therefore$ R = 8.5. From section 1630.2, UBC-97, building weight is based on the given 70 psf on the 1st floor; 50 psf at the roof

$W_1 = 0.07(140 \times 50 - 10 \times 100) = 420^k$

$W_2 = 0.05(140 \times 50 - 10 \times 100) = 300^k$

$W = W_1 + W_2 = 720^k$

$V = CW$

$T = C_t(h_n)^{3/4} = 0.02(22)^{3/4} = 0.2$ sec

$C = (C_vI / RT) = 0.54 \times 1.25 / (8.5 \times 0.2) = 0.40$

But

$C \le (2.5C_aI / R = 2.5 \times 0.36 \times 1.25 / 8.5 = 0.13)$

and

$C \ge (0.11C_aI = 0.11 \times 0.36 \times 1.25 = 0.0495)$

$\therefore$ $C = 0.13$; $V = 0.13 \times 720 = 94^k$.

Vertical distribution of lateral force:

$$F_x = \frac{(V - F_t)w_x h_x}{\sum_{i=1}^{n} w_i h_i}$$

Since T = 0.2 sec < 0.7 sec, $F_t = 0$

$F_1 = 94 \times 420 \times 12 / (420 \times 12 + 300 \times 22) = 40.7^k$

$F_2 = 94 \times 300 \times 22 / (420 \times 12 + 300 \times 22) = 53.3^k$

The forces to the diaphragm are computed per section 1633.2.9

$$F_{px} = \frac{F_t + \sum_{i=x}^{n} F_i}{\sum_{i=x}^{n} w_i} w_{px} \ge 0.5C_aIw_{px}$$

$F_{p1} = (40.7 + 53.3)/(420 + 300)]420 = 54.8^k$

$\ge 0.5 \times 0.36 \times 1.25 \times 420$ **$= 94.5^k$ (governs)**

$F_{p2} = (40.7)/(300)]300 = 53.3^k$

$\ge 0.5 \times 0.36 \times 1.25 \times 300$ **$= 67.5^k$ (governs)**

b)

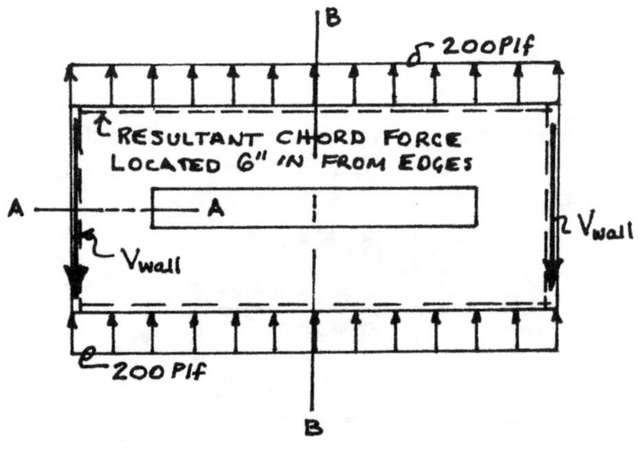

SECOND FLOOR PLAN

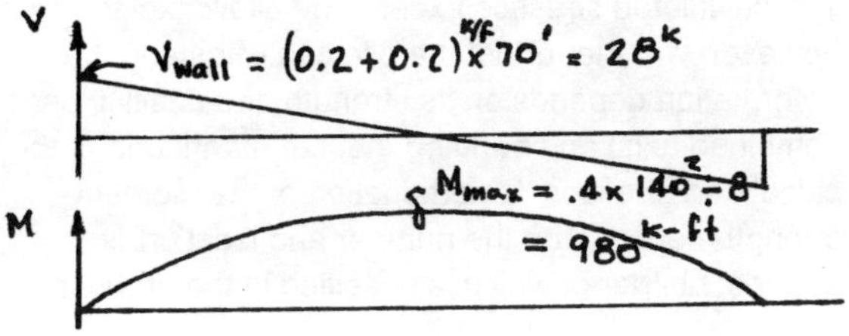

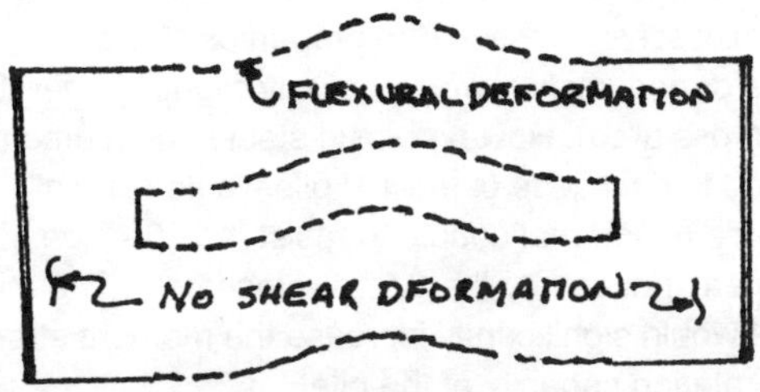

If G is infinite, the diaphragm will deform under lateral forces in the manner shown. Each 20' × 100' region will deform in flexure with a point of inflection approximately at 1/4 the 100' span. Thus,

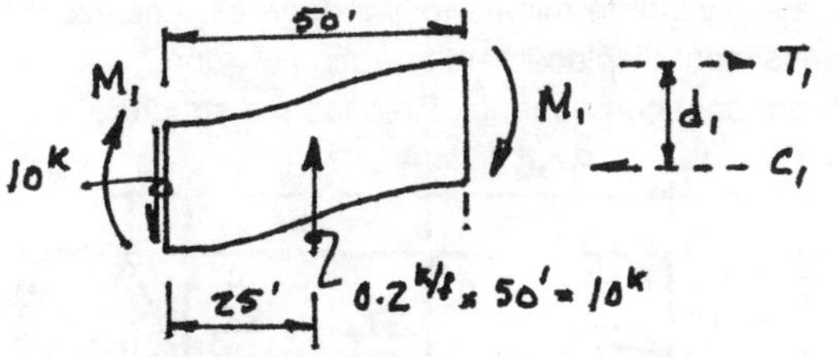

$\Sigma M_{PC} = 0$: $10^k(25') - 2M_1 = 0 \therefore M_1 = 125^{k\text{-}ft}$

Chord forces are located 6" in from edges

$\therefore d = 20 - 2 \times 0.5 = 19'$

$C_1 = T_1 = M_1 / d = 125 / 19 = 6.6^k$

Now, consider freebody of one-half the diaphragm

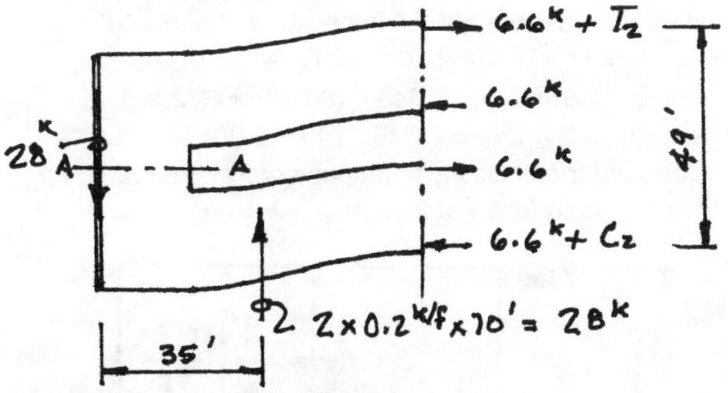

$\Sigma M = 0$: $28^k(35) - 2 \times 125 - T_z(49) = 0$

$\therefore \quad T_z = 14.9^k = C_z$

Freebody through A-A:

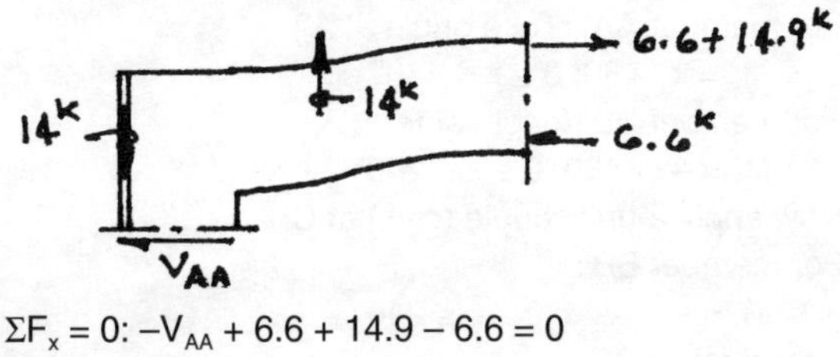

$\Sigma F_x = 0$: $-V_{AA} + 6.6 + 14.9 - 6.6 = 0$

$\underline{V_{AA} = 14.9^k}$

c) If G is finite, the segment of the diaphragm will experience shear deformation. The 20' × 100' panels deform more nearly as a simply supported rather than fixed end beam (i.e.  rather than ). In this case, the T_1-C_1 couple will involve larger forces, the $C_2 - T_2$ would have smaller forces ; hence, $V_{AA} = T_2$ decreases.

Structural-25

a) A partial checklist of inspection items for a grade beam-on-pile foundation:
   1) Piles positioned within specified tolerances and cut off at the proper elevations.
   2) Pile uplift anchorage devices in place, if applicable.
   3) Proper surface condition on rebars (i.e., mud, rust, form oil, etc).
   4) Proper size and position of grade beam reinforcement.
   5) Rebars securely tied and supported to maintain alignment, clearances and cover.
   6) Sleeves, inserts, blockouts, bulkheads – with supplemental reinforcement , if required.

b) Inspection checklist for a flat slab system prior to pour:
   1) Proper size and position of reinforcement.
   2) Placement of reinforcement for slab-to-column moment transfer (Re: Equation 13-1, ACI 318).
   3) Placement of shearheads, if required (Re: Section 11.11.4, ACI 318).
   4) Size and depth of drop panels.
   5) Openings, and supplementary reinforcement at openings, consistent with design drawings.
   6) Surface condition of rebars; rebars securely tied and supported.

c) Inspection checklist for type V, single story wood construction:
   1) Proper size, grade and spacing of joists, rafters, studs, etc.
   2) Joist bridging properly installed.
   3) Cuts, notches and penetrations for electrical, plumbing and HVAC properly located or reinforced, if necessary.
   4) Proper nail size and spacing, especially in shearwalls and diaphragms.
   5) Diaphragm blocking, if required, properly located and installed.
   6) Connections and anchorages properly installed and tightened.

d) Inspection checklist for a single story steel frame (rigid joints made with high strength bolts):
   1) All pieces erected and aligned.
   2) All bolts installed and tightened.
   3) Solid seating of connected parts.
   4) Grout, if required, beneath column bases and beam seats.
   5) Fire protective or other coatings applied.

6) Evidences of cuts or alterations in the structural steel made to facilitate erection or to accommodate the work of other trades.

Structural-26

Given:

Pile capacity = 120^{ton} = 240 kips = P_g;

$P_d = 0.8\ P_g$ = 192 kips;

$P_L = 0.2\ P_g$ = 48 kips;

$P_o = \pm\ 0.5\ P_g = \pm$ 120 kips

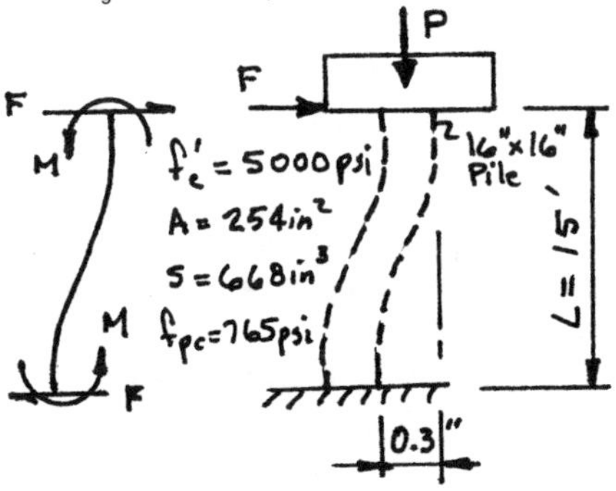

Find the moment induced in the pile by the 0.3-in lateral translation. Neglect the interaction between the pile and soft clay:

$E_c = 57{,}000\sqrt{f'_c} = 57{,}000\ (5000)^{1/2}$

$= 4.03 \times 10^6$ psi = 4030 ksi

$I = 5344\ in^4$

$M = 6EI\Delta / L^2$

$= 6 \times 4030 \times 5344 \times 0.3" / (180)^2$

$M = 1200^{k\text{-}in} = 100^{k\text{-}ft}$

a) Calculate stresses – follow conventional method of computing stresses using properties of the gross, uncracked section. Take $^1/_3$ increase in allowable stress for gravity plus seismic. The allowable compression stress per section 1808.5.3, UBC-97:

$F_c = {}^4/_3\ (0.33\ f'_c - 0.27f_{pc})$

$= {}^4/_3\ (0.33 \times 5000 - 0.27 \times 765)$

$= 1925$ psi

Allowable tension stress:

$F_t = {}^4/_3\ (6\sqrt{f'_c}) = {}^4/_3\ (6\sqrt{5000})$

$= 565$ Psi

Maximum compression stress:

$f_c = (P_g / A) + (P_o / A) + (M / S)$

$= (240{,}000 + 120{,}000) / 254 + 1{,}200{,}000 / 668$

$f_c = 3214$ Psi $>> F_c$

Note that the effective prestress is treated as a reduction to the allowable stress rather than as a component of the compression stress.

Critical tension stress:

$f_t = (P_d / A) - (P_o / A) + f_{pc} - M / S$

$= (192{,}000 - 120{,}000) / 254 + 765 - 1{,}200{,}000 / 668$

$f_t = -748$ psi = 748 psi (Tension) $> F_t$

Thus, the calculated stresses exceed the allowable stresses in concrete tension and compression.

b) The calculated stresses exceed the allowable stresses; however, whether or not the pile can withstand the deformation depends on its strength and ductility under combined axial and bending, not the magnitude of the calculated stresses. Determination of the moment strength depends on the number and location of strands, neither of which is specified in the problem statement.

c) The addition of mild steel would not do much to alleviate the excessive stresses (the properties of the transformed section would be only slightly larger than for the gross area). However, mild steel reinforcement added to each side of the 16" pile would certainly provide an internal couple to resist the $100^{k\text{-}ft}$ bending moment induced by the 0.3" displacement. The mild steel would significantly increase the moment strength and rotation capacity of the pile.

Structural-27

a) Solve for the reactions in the structure. The structure is statically indeterminate to two degrees. Analyze by consistent displacements treating M_A and M_C as redundant components. Release the structure

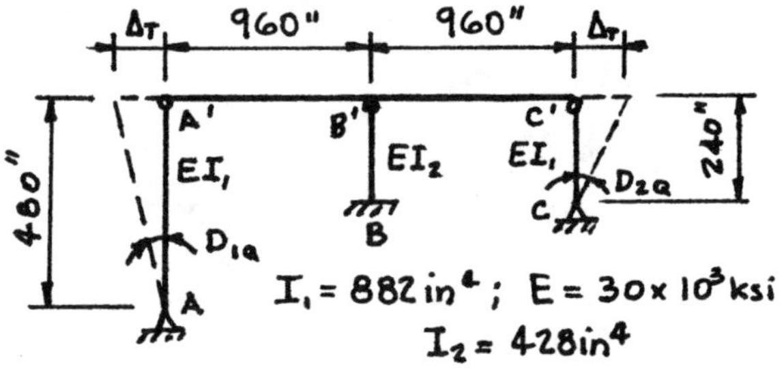

Compute the rotations at joints A & C of the released structure due to a 60° F temperature change:

$\Delta_T = D_F\Delta TL = 6 \times 10^{-6} \times 60 \times 960 = 0.346"$

$\therefore\quad D_{IQ} = \Delta_T / 480 = 0.346 / 480 = 0.00072$ rad

$D_{2Q} = \Delta_T / 240 = 0.346 / 240 = 0.00144$ rad

Compute the flexibility influence coefficients. First, apply a unit couple (c.w.) at joint A:

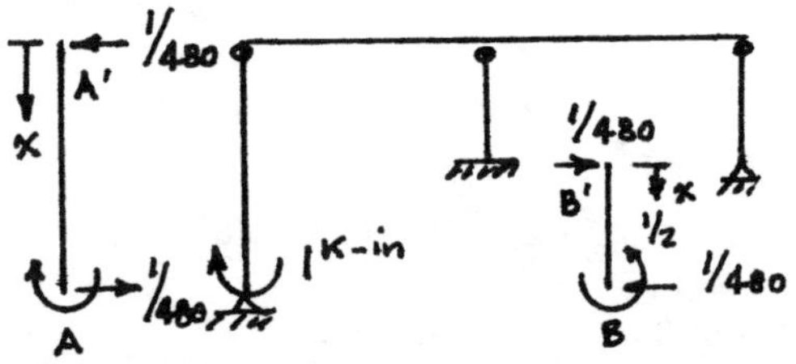

For member AA' (Take origin at A'):

$M_x = x / 480;\ 0 \le x \le 480"$

For member BB' (origin at B'):

$M_x = -x / 480;\ 0 \le x \le 240"$

Now apply a unit couple (c.w.) at C:

For member BB':

$M_x = -x / 240;\ 0 \le x \le 240"$

For member CC':

$M_x = x / 240;\ 0 \le x \le 240"$

Using the dummy load method:

$$(1)\ f_{11} = \sum \int_L (mM / EI)\,dx$$
$$= \int_0^{480} \left[(x/480)^2 / EI_1\right]dx + \int_0^{240}\left[(x/480)^2 / EI_2\right]dx$$
$$= x^3 / \left(480^2 \times 3EI_1\right)\Big|_0^{480} + x^3 / \left(480^2 \times 3EI_2\right)\Big|_0^{240}$$
$$= 0.0000076 \text{ rad / k-in}$$

$$(1)\ f_{21} = \int_0^{240} \left[(x/480)(-x/240)/EI_2\right]dx$$
$$= -x^3 / \left(480 \times 240 \times 3EI_2\right)\Big|_0^{240}$$
$$= -0.0000031 \text{ rad / k-in}$$

By Maxwell's Law: $f_{12} = f_{21}$

$$(1)\ f_{22} = \int_0^{240} \left[(x/240)^2 / EI_1\right]dx + \int_0^{240}\left[(x/240)^2 / EI_2\right]dx$$
$$= 0.0000093 \text{ rad / k - in}$$

For consistent displacements, rotations at A and C must be zero. Thus,

$$0.0000076M_A + 0.0000031M_C = 0.00072$$
$$0.0000031M_A + 0.0000093M_C = -0.00144$$

Solving simultaneously:

<u>$M_C = -216^{\text{K-in}} = 216^{\text{K-in}}$ (c.c.w.)</u>

<u>$M_A = +183^{\text{K-in}} = 183^{\text{K-in}}$ (c.w.)</u>

b) The column shears in AA' and BB' are found by summing moments about the bases. Thus,

<u>$V_{AA'} = 183 / 480 = 0.381^K \rightarrow$</u>

<u>$V_{CC'} = 216 / 240 = 0.900^K \leftarrow$</u>

<u>$\Sigma F_x = 0$: $V_{BB'} = 0.900 - 0.381 = 0.519^K \rightarrow$</u>

With the column shears known, the deflections at the top of columns are simply $Vh^3 / 3EI$

$\therefore$ <u>$\Delta_{B'} = 0.381 \times 480^3 / (3EI_1) = 0.531" \leftarrow$</u>

<u>$\Delta_{C'} = 0.900 \times 240^3 / (3EI_1) = 0.157" \rightarrow$</u>

Note a simple check on the solution can be made by equating the elongation of the girder to the sum of $\Delta_{B'}$ and $\Delta_{C'}$.

$$\Delta L = 6 \times 10^{-6} \times 60 \times 2 \times 960 = 0.69"$$
$$\cong \Delta_{B'} + \Delta_{C'}$$

c) The maximum bending stress in Column BB' occurs at the base

$M_B = 0.519 \times 240 = 124.6^{\text{K-in}}$

For a W14×43: $S_x = 62.7 \text{ in}^3$

$\therefore$ <u>$f_b = 124.6 / 62.7 = 2.0$ ksi</u>

Structural-28

Analyze the truss below. Since the problem asks for midspan deflection, the slope-deflection method will provide the answer directly.

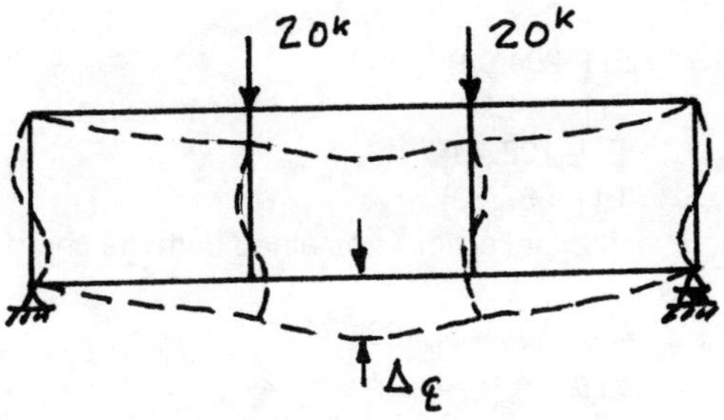

Considering only bending deformation, and taking advantage of symmetry, the midspan deflection of the vierendeel truss can be obtained by analysis of a simpler model:

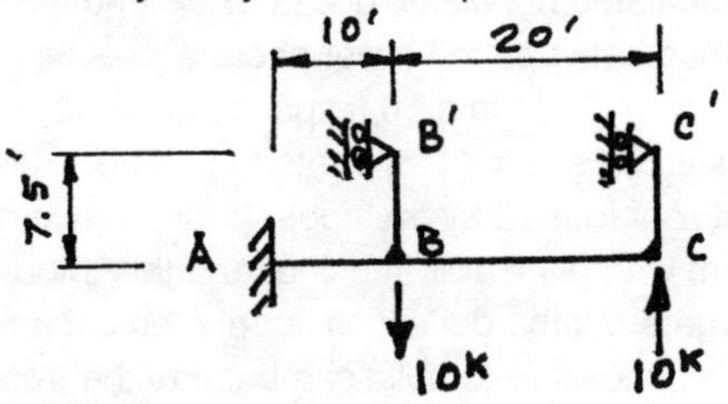

$EI = \text{Constant} = 29{,}000 \times 200$

$= 5.8 \times 10^6 \text{ k-in}^2 = 40{,}278 \text{ k-ft}^2$

By slope-deflection (Φ_B, Φ_C, β_{AB}, and β_{BC} are unknowns):

$$M_{AB} = 2EI / 10\ [\Phi_B - 3\beta_{AB}]$$
$$M_{BA} = 2EI / 10\ [2\Phi_B - 3\beta_{AB}]$$
$$M_{BB'} = 3EI / 7.5\ [\Phi_B]$$
$$M_{BC} = 2EI / 20\ [2\Phi_B + \Phi_C - 3\beta_{BC}]$$
$$M_{CB} = 2EI / 20\ [\Phi_B + 2\Phi_C - 3\beta_{BC}]$$
$$M_{CC'} = 3EI / 7.5\ [\Phi_C]$$

For equilibrium:

$\Sigma M_B = 0$: $M_{BA} + M_{BB'} + M_{BC} = 0$

$\therefore$ $EI\ [\Phi_B + 0.1\Phi_C - 0.6\beta_{AB} - 0.3\beta_{BC}] = 0$

$\Sigma M_C = 0$: $M_{CB} + M_{CC'} = 0$

$\therefore$ $EI\ [0.1\Phi_B + 0.6\Phi_C - 0.3\beta_{BC}] = 0$

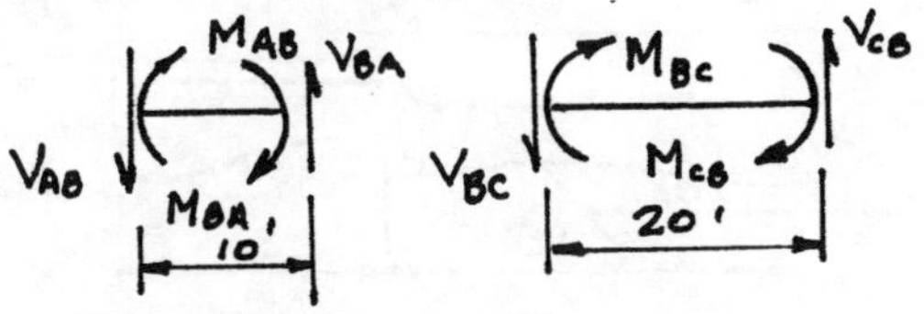

$$V_{AB} = [M_{AB} + M_{BA}] / 10$$
$$V_{BC} = [M_{BC} + M_{CB}] / 20$$

Taking the entire structure as a free body:

$\Sigma F_y = 0$: $V_{AB} - 10 + 10 = 0$

$\therefore$ $V_{AB} = 0 = [M_{AB} + M_{BA}] / 10$

$\therefore$ $EI\ [0.6\Phi_B - 1.2\beta_{AB}] = 0$

Taking the portion BCC' as a freebody

$\Sigma F_y = 0$: $V_{BC} - 10 = 0$

$\therefore$ $EI\ [0.3\Phi_B + 0.3\Phi_C - 0.6\beta_{BC}] = 200$

The set of equilibrium equations in matrix form:

$$EI\begin{bmatrix} 1.0 & 0.1 & -0.6 & -0.3 \\ 0.1 & 0.6 & 0 & -0.3 \\ 0.6 & 0 & -1.2 & 0 \\ 0.3 & 0.3 & 0 & -0.6 \end{bmatrix}\begin{Bmatrix} \Phi_B \\ \Phi_C \\ \beta_{AB} \\ \beta_{BC} \end{Bmatrix} = \begin{Bmatrix} 0 \\ 0 \\ 0 \\ 200 \end{Bmatrix}$$

Solving:

$\Phi_B = 1 / EI\ [-204.08]$

$\Phi_C = 1 / EI\ [-244.90]$

$\beta_{AB} = 1 / EI\ [-102.40]$

$\beta_{BC} = 1 / EI\ [-557.83]$

a) The midspan deflection is obtained from the chord rotations:

$\Delta_{centerline} = \beta_{AB}(10') + \beta_{BC}(20')$

$= [102.4 \times 10 + 557.8 \times 20] / EI$

$\Delta_{centerline} = 0.302\ \text{ft} = 3.62\ \text{in} \downarrow$

Thus, the midspan deflection is less than the specified clearance; therefore, the clearance is adequate.

b) The calculated deflection does not exceed the 6 in clearance between the lower chord and equipment. A deflection of 3.62" in a 60 ft span is generally not excessive for a roof (it amounts to L / 200). However, the flat roof formed by the upper chord may create a problem with rainwater ponding, and this should be investigated. If the deflection is deemed excessive, it may be necessary to build camber into the truss or to stiffen it by increasing the moment of inertia of its members.

Structural-29

The simplest approach to determine the required properties of the beam is by column analogy (Blodgett 1966).

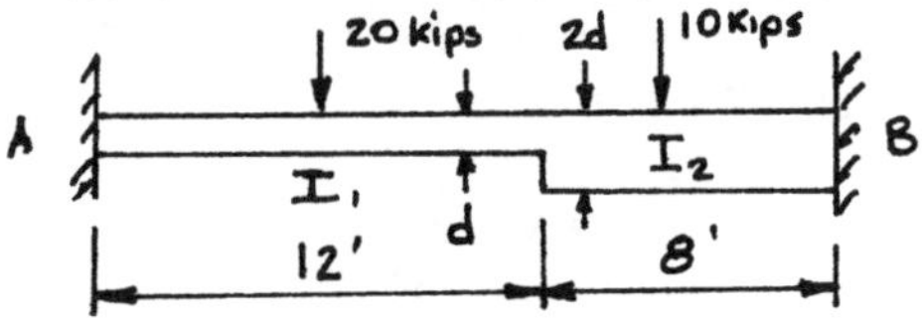

$I_1 = bd^3 / 12 = I$; $I_2 = b(2d)^3 / 12 = 8I$

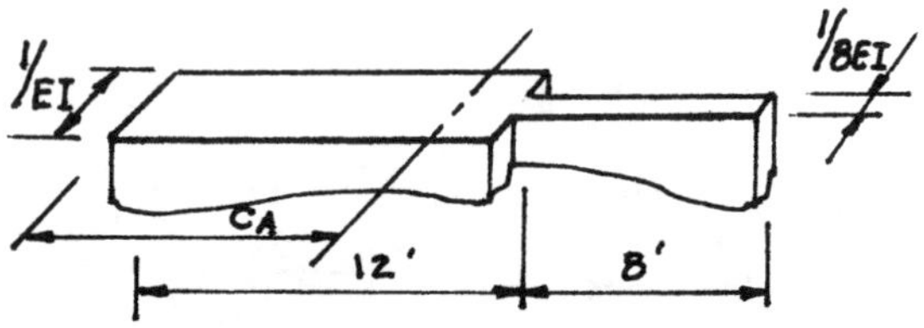

$A_C = 12 / EI + 8 / 8EI = 13 / EI$

$C_A = [(12 / EI)6 + 1 / EI(12 + 4)] / (13 / EI) = 6.77'$

$C_B = 20 - 6.77 = 13.23'$

$I_C = (12^3 / 12EI) + [(12 / EI)(0.77)^2] + [(1 / 8EI)8^3 / 12] + [(1 / EI)(16 - 6.77)^2] = 241.6 / EI$

a) Compute the stiffness and carry-over factors. Apply 1 rad angle change to the column at end A. The combined "stress" in the analogous column at end A is the stiffness:

$K_{AB} = \sigma_A = 1 / A_C + Mc_{AB} / I_C$

$= 1 / (13 / EI) + 1(6.77)(6.77) / (241.6 / EI)$

$\therefore$ **$K_{AB} = 0.267EI = 5.33EI / L$**

$C_{AB}K_{AB} = 1 / (13 / EI) - 1(6.77)(13.23) / (241.6 / EI)$

$C_{AB}K_{AB} = -0.294EI = -5.87EI / L$

Thus,

$C_{AB} = -5.87 / 5.33 = -1.10$

Now, apply the 1^{rad} to end B:

$K_{BA} = 1(13 / EI) + 1(13.23)^2 / (241.6 / EI)$

$K_{BA} = 0.801EI = 16.02\ EI / L$

By Betti's Law: $C_{BA}K_{BA} = C_{AB}K_{AB} = -0.29EI$

Thus,

$C_{BA} = -0.367$

b) Fixed end moments – calculate static moment for a simple beam

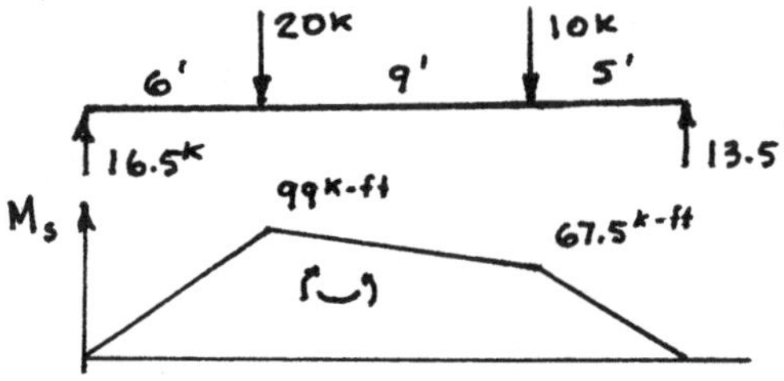

Load the analogous column with M_S / EI

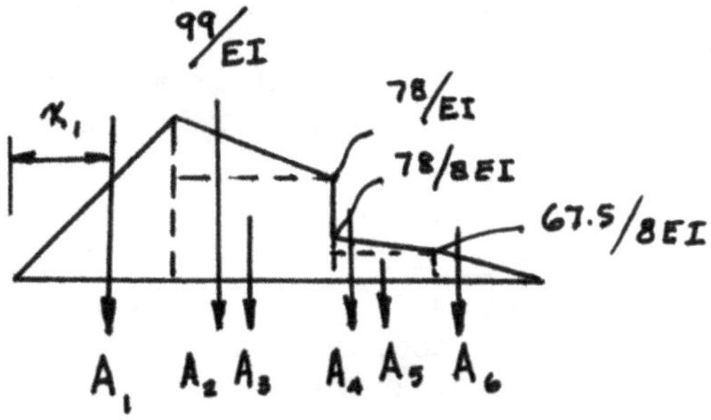

Area	x_i (ft)	EIA_i	$EI(x_iA_i)$
1	4	297	1188
2	8	63	504
3	9	468	4212
4	13	2	26
5	13.5	25.3	341.4
6	16.7	21.1	351.3
$\Sigma =$		876	6622.7

$\therefore\ \bar{x} = 6622.7 / 876 = 7.56'$

$P = 876 / EI$

$M = (876 / EI)(7.56 - 6.77) = 692 / EI$

Thus,

$FEM_{AB} = \sigma_A = (876 / EI) / (13 / EI) - (692 / EI) \times [6.77 / (241.6 / EI)]$

$FEM_{AB} = 48.0^{k\text{-}ft}$ (c.c.w.)

Since EI cancels in each term

$FEM_{BA} = 876 / 13 + 692 \times 13.23 / 241.6 = 105.2^{k\text{-}ft}$ (c.w.)

Structural-30

Calculate the relative rigidities of the shear resisting elements below (express relative to the structure in part a).

a) for the braced frame:

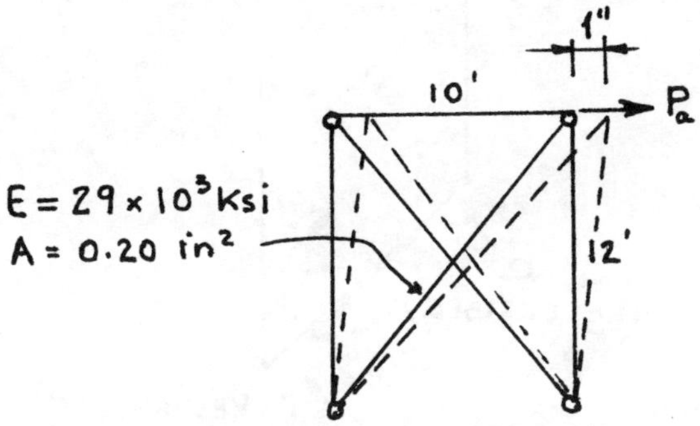

Assume AE = ∞ in column and beam; E = 29 × 10^3 ksi; A = 0.20 in^2 for braces (i.e., only the braces contribute to horizontal deflection). Also, assume rods are not pretensioned and that they are too slender to resist compression.

By method of real work:

$(0.5)P_a(1") = \Sigma(0.5)P_i^2L_i / A_iE$

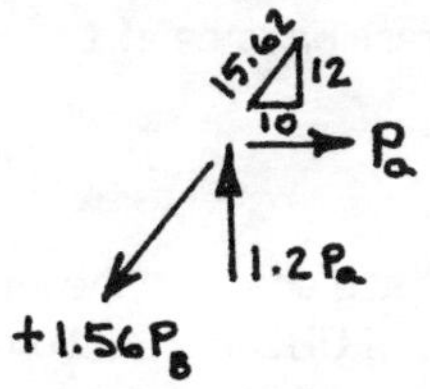

$\therefore$ $P_a = 29 \times 10^3 \times 0.20 / (1.56^2 \times 15.62 \times 12)$

<u>P_a = 12.7 k/in</u>

b) For the cantilevered wall – consider both shear and bending deformation:

$\Delta = 1" = P_bh^3 / 3E_cI + 1.2P_bh / E_mA$

$1" = P_b\,(144^3 / (3 \times 3120 \times 9 \times 72^3 / 12) + 1.2 \times 144 / (0.4 \times 3120 \times 9 \times 72))$

$\therefore$ <u>P_b = 739 k/in</u>

c) For the K-braced truss – again, assume EA of horizontal and vertical members infinite relative to braces:

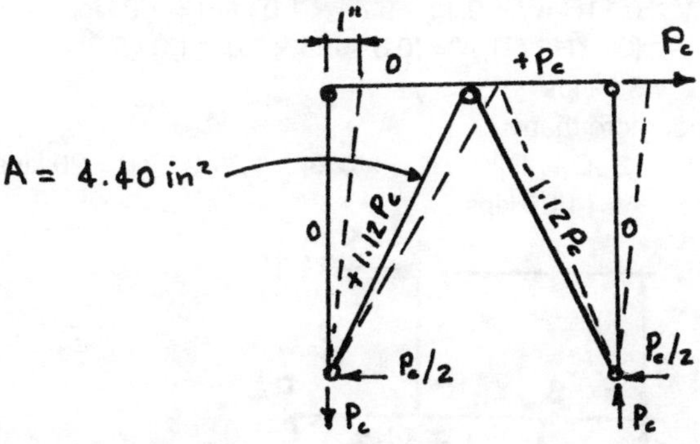

$$\frac{1}{2}P_C(1") = \sum_{i=1}^{m}\left(\frac{1}{2}\right)P_i^2L_i / A_iE$$

$1" = 2 \times (1.12^2 \times 13.42 \times 12 / (29 \times 10^3 \times 4.4))P_c$

$\therefore$ <u>P_c = 316 k/in</u>

d) For the pierced shearwall:

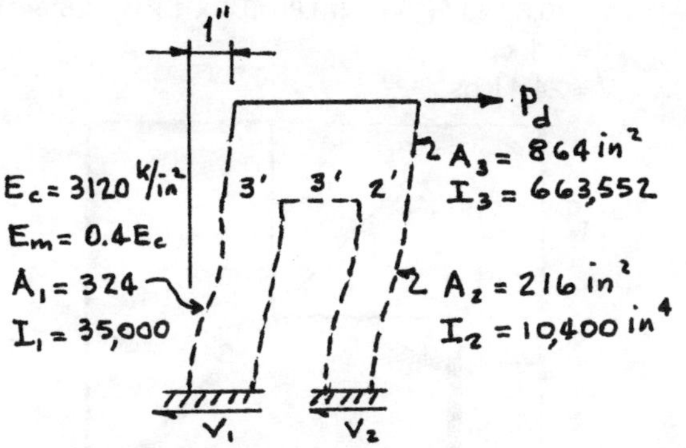

Proportion the force P_d to the lower walls on the basis of their translational stiffness, assuming zero rotation at top and bottom. Thus,

$K_1 = 1 / [h^3 / 12E_cI_1 + 1.2h / E_mA_1]$

$= 1 / [96^3 / (12E_c \times 35{,}000) + (1.2 \times 96) / (0.4E_c \times 324)]$

$= E_c / 2.995 = 0.3339E_c$

$K_2 = 1 / [96^3 / (12E_c \times 10{,}400) + (1.2 \times 96) / (0.4E_c \times 216)] = 0.1188E_c$

$\therefore$ $V_1 = [0.3339 / (0.3339 + 0.1188)]\,P_d$

$= 0.738\,P_d$

$\therefore$ $\Delta = 1" = P_d[48^3 / (3 \times 3120 \times 663{,}552) + 1.2 \times 48 / (0.4 \times 3120 \times 864) + 0.738(96^3 / (3120 \times 12 \times 35{,}000) + 1.2 \times 96 / (0.4 \times 3120 \times 324))]$

<u>P_d = 1170 k/in</u>

Thus, relative rigidities:

a) <u>**1.0**</u>
b) <u>**739 / 12.7 = 58**</u>
c) <u>**316 / 12.7 = 25**</u>
d) <u>**1170 / 12.7 = 92**</u>

Structural-31

a) Determine the maximum seismic force in the wall "AB" shown below. Follow provisions of UBC-97.

The building has a type 2 plan irregularity (re-entrant corners); however, the static force procedure is permitted (1629.8.3, UBC-97) provided that the diaphragm is designed and detailed per section 1633.2.9, items 6 and 7. For structures in zone 4 on soil profile type S_D, with near-fault factors of unity, Tables 16-Q and 16-R give C_a = 0.44 and C_v = 0.64. For a bearing wall system with masonry shearwalls, Table 16-N gives R = 4.5 and Ω_o = 2.8. The given dead load is 100 kips and the importance factor is unity.

The period can be estimated using UBC Eqn. 30-8

$T = C_T(h_n)^{3/4} = 0.02 \times (10)^{3/4} = 0.1$ sec

For structures with periods less than about 0.5 sec (i.e., stiff structures), the base shear is controlled by UBC Eqn 30-5

$V \le (2.5C_aI / R)W = (2.5 \times 0.44 \times 1 / 4.5) \times 100$

$= 24.4$ kips

Eqn. 30-7 imposes a lower bound on the base shear

$V \geq (0.8ZN_vI / R)W = [(0.8 \times 0.4 \times 1 \times 1) / 4.5] \times 100$

$= 7.1$ kips

$\therefore \quad V = 24.4$ kips

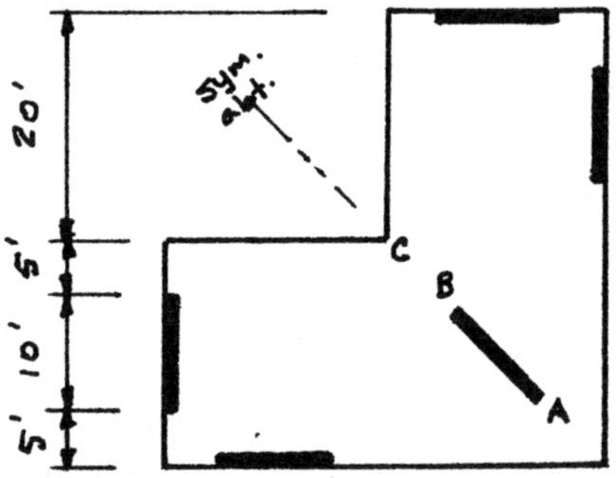

ROOF PLAN

The center of rigidity of the wall system coincides with the line of action of diagonal wall AB. Thus, only the lateral force affects the shear in wall AB (torsion has no effect). The maximum shear occurs when the lateral force acts diagonally at 45°. The rigidity of each 10' long wall (8" grouted concrete block) depends on both shear and flexural deformation:

$\Delta = R[L^3 / 3EI + 1.2L / GA]$

where G = 0.4E; L = 10'; Δ = 1':

$I = 0.63 \times 10^3 / 12 = 52.5$ ft^4

$A = 0.63 \times 10 = 6.3$ ft^2

$\therefore \quad R = 1 / [10^3 / (3E \times 52.5) + 1.2 \times 10 / (0.4E \times 6.3)]$

$= 0.09E$

When the side wall translates diagonally an amount Δ at 45°, the movement in the direction of shear resistance is $\Delta\sin45°$:

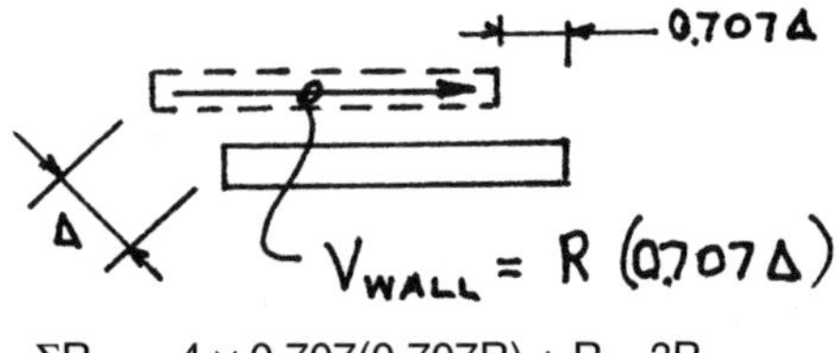

$\therefore \quad \Sigma R_{45°} = 4 \times 0.707(0.707R) + R = 3R$

Therefore, the shear force in wall BC is

$\underline{V_{AB} = VR / \Sigma R_{45°} = V / 3 = 24.4 / 3 = 8.13 \text{ kips}}$

The required nominal shear stress is

$\underline{v_{AB} = V_{AB} / BL = 8.13 / (7.6 \times 120) = 0.0089 \text{ ksi} = 89 \text{ psi}}$

b) The force acting at 'C' can be inferred by cutting a free body of the roof diaphragm along the diagonal through CBA.

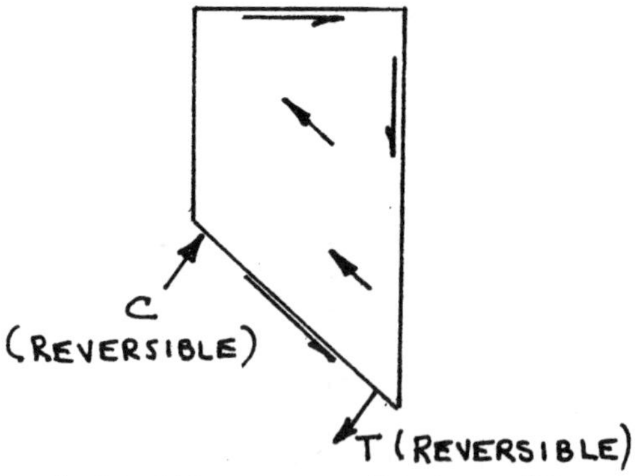

When the lateral force acts at 45°, there will be no shear at 'C'. However, there will be moment produced by the lateral force and side wall shear forces, which must be resisted by a T-C couple as shown above. The reentrant corner at 'C' creates a severe stress concentration at that point. Eccentricity of the lateral force (a minimum 5% of building dimension) will further increase the tension or compression force at 'C'.

Structural-32

Compute the design base shear on the structure using the static force procedure of UBC-97. Section 1630.11 requires consideration of the vertical acceleration only on horizontal cantilevers $\therefore$ the given 0.3g acceleration is not a factor in this problem. Per problem statement, the dead load of the frame can be neglected. From Table 16-P, the appropriate R-factor is 2.2. Using the given values T = 0.79 sec; $C_a = 0.44$; $C_v = 0.64$; $I = 1.0$; $N_v = N_a = 1.0$; $Z = 0.4$. Thus, from section 1630.2

$V = [C_vI / (RT)]\, W = [0.64 \times 1.0 / (2.2 \times 0.79)]40$

$= 14.7$ kips

The base shear must be at least

$V \geq 0.11C_aIW = 0.11 \times 0.44 \times 1.0 \times 40 = 1.9$ kips

$V \geq [0.8ZN_vI / R]W = [0.8 \times 0.4 \times 1.0 \times 1.0 / 2.2]40$

$= 5.8$ kips

but not more than

$V \leq [2.5C_aI / R]W = [2.5 \times 0.44 \times 1.0 / 2.2]40 = 20$ kips

$\therefore \quad V = 14.7$ kips

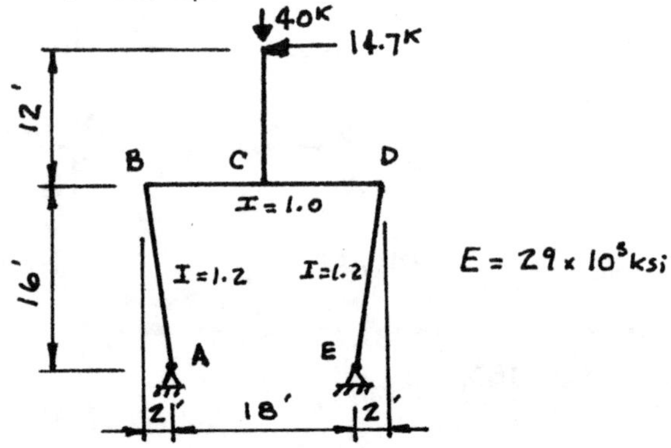

Analyze the frame for the combined loading shown above using the method of consistent displacements. Let the horizontal reaction at point "E", E_x, be the unknown and release this component:

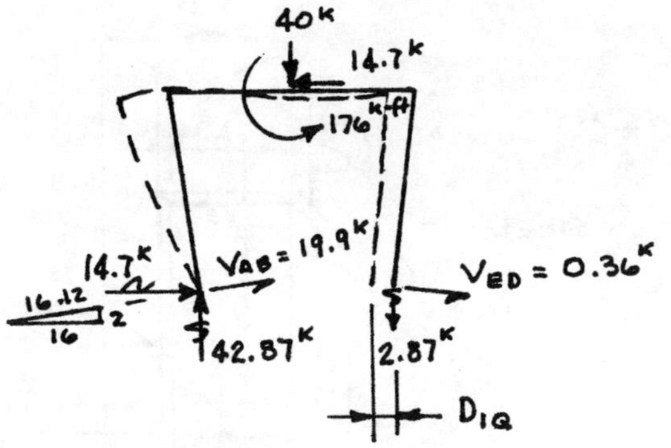

Apply a dummy unit load at E →

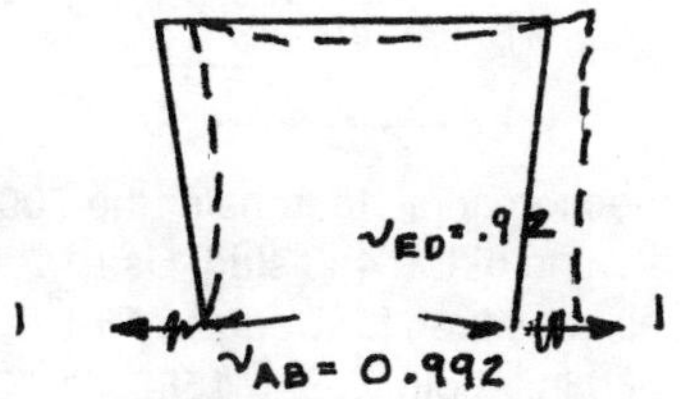

By the Dummy Load Method:

Mem	Internal	M	m
AB	$0 \le x \le 193.5$	$-19.9x$	$0.992x$
BC	$0 \le x \le 132$	$-3851 + 42.87x$	192
DC	$0 \le x \le 132$	$-310 + 2.87x$	192
ED	$0 \le x \le 193.5$	$-0.36x$	$-0.992x$

$$EID_{1Q} = \int_0^{193.5} (-19.9x)\left(\frac{0.992x}{1.2}\right)dx$$
$$+ \int_0^{132} (-310 + 2.87x)(192)dx$$
$$+ \int_0^{132} (-3851 + 42.87x)(192)dx$$
$$+ \int_0^{193.5} (-0.36x)\left(\frac{-0.992x}{1.2}\right)dx$$
$$D_{1Q} = -67{,}990{,}990 / EI$$
$$EIf_{11} = 2\int_0^{193.5} [(0.992x)^2 / 1.2]dx + \int_0^{264} (192)^2 dx$$
$$f_{11} = 13{,}692{,}979 / EI$$

Thus, for consistent displacement

$$E_x f_{11} + D_{1Q} = 0$$

∴ $E_x = 67{,}990{,}900 / 13{,}692{,}979 = 4.97$ kips →

From statics:

$A_x = 9.73$ kips →; $A_y = 42.87$ kips ↑; $E_y = 2.87$ kips ↓

Thus, $V_{AB} = 14.98$ kips; $V_{ED} = 5.28$ kips

The required moment diagrams are:

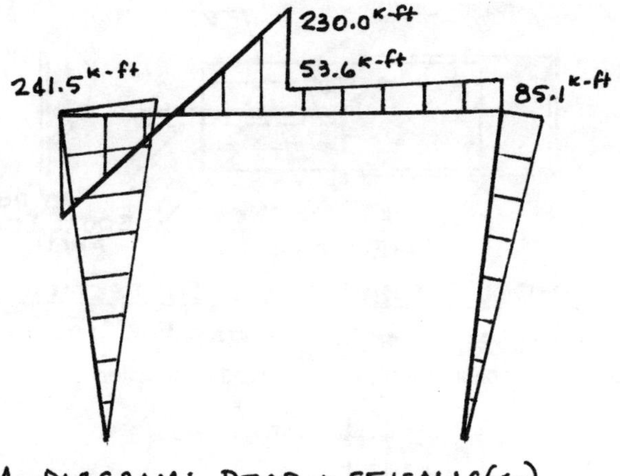

M-DIAGRAM: DEAD + SEISMIC(←)

Structural-33

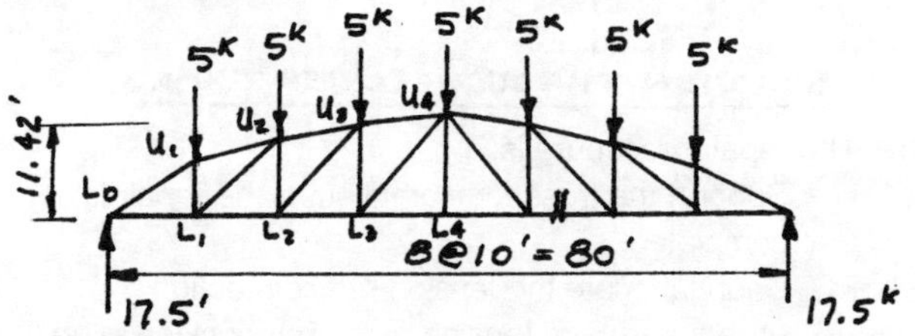

a) Provide tensioned rods to repair the fractured lower chord. For for a bowstring truss under uniform loads, the lower chord axial force is constant from panel-to-panel. The tension can be calculated most easily at midspan where the rise = 80(1 / 7) = 11.42 ft. Cutting through U_4U_5-to-L_3L_4 and summing moments about U_4

$\Sigma M_{U4} = 0$: $17.5(40) - 5(10 + 20 + 30) - T(11.42) = 0$

∴ **T = 35.0 kips**

Size the tension rods assuming 2 rods, one on each side; A-36 steel; use allowable stress design (AISC 1989):

$F_T = 0.33F_u = 0.33 \times 58 = 19.1$ ksi

$A_g \ge (35 / 2) / 19.1 = 0.91$ in$^2 = \pi D^2 / 4$

$D \ge (4 \times 0.91 / \pi)^{1/2} = 1.08$"

∴ **Use 2-1$^1/_8$" diameter rods (one each side)**

b) Since the tension in every panel of the truss lower chord is the same, the existing heel connection should be adequate to handle the 35 kip tension. The design problem is to provide a steel plate of the same size as the existing side plates, then punch the new plates for the $^7/_8$" diameter bolts to match the gage and pitch of the existing bolts.

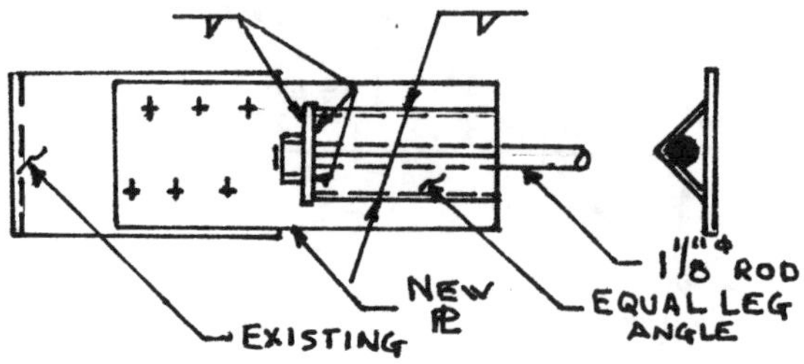

b) HEEL & TENSION ROD CONNECTION
(TRUSS MEMBERS OMITTED FOR CLARITY)

c) The section through the repaired connection is

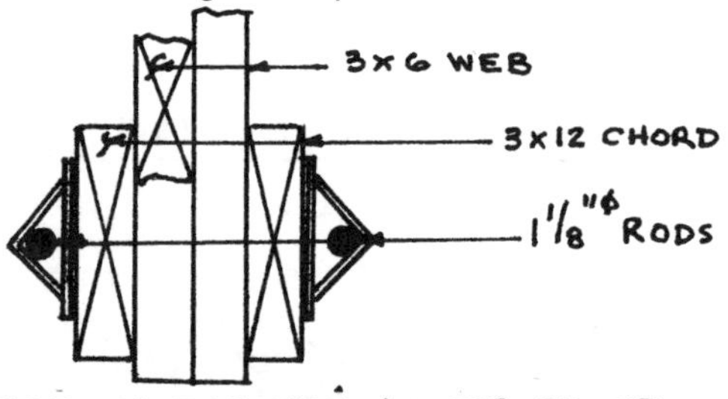

SECTION THROUGH LOWER CHORD

d) The repair procedure is:
   i Support the truss temporarily until the repair is completed.
   ii Shop fabricate the tension rod connection (4 required) and tension rods. Turnbuckles to be provided to field splice the rods and to permit tensioning of rods after installation.
   iii Stabilize the lower chord in the exterior panels and remove 6 – 7/8" diameter bolts between the lower chord and heel connection plates. Position the new connection plates (one each side) at the heel plate and connect it using 7/8" dia bolts.
   iv Install the tension rods and turnbuckles.
   v Establish and maintain the alignment of lower chord at fracture points. Tension the rods by tightening the turnbuckles and closing the gap in the lower chord at fracture point. Remove temporary supports as tension is applied.

e) The tension rods resist all tensile force of the lower chord. There is no need to restore tensile capacity at the fracture point (at least not for the given, uniform loading). It is reasonable to insert a spacer block between the 2 – 3 × 12 members at the break point to stabilize the chord at this point.

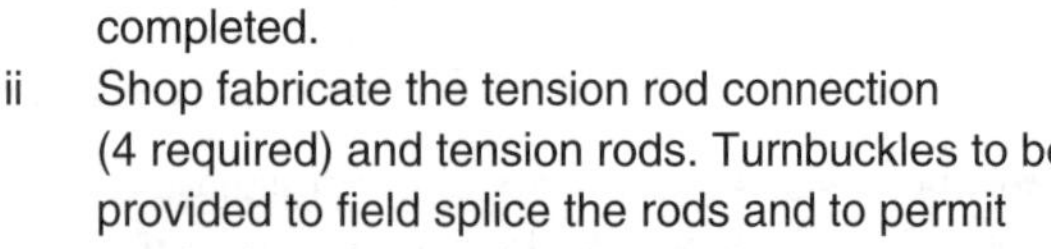

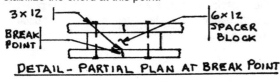

DETAIL - PARTIAL PLAN AT BREAK POINT

Structural-34

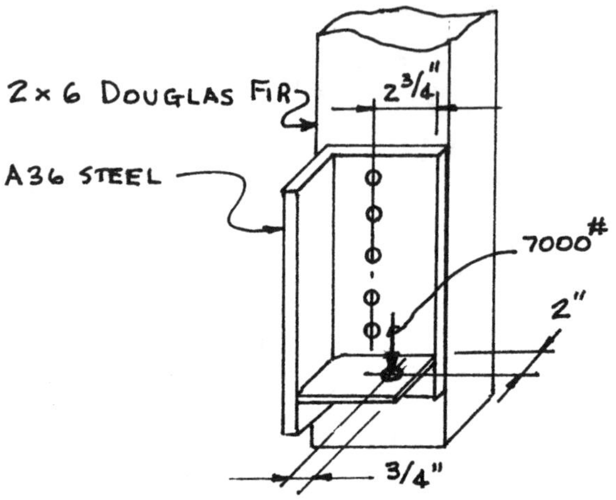

Design the connection angle to transfer the 7000# eccentric force to the centerline of the 4×6 stud. Use 1/2-in diameter bolts and the design values for Douglas Fir-Larch.

$Z_{parallel}$ = (1580 lbf / bolt) C_dC_g = 1580 × 1.33 × 0.75
= 1580 lbf / bolt

$Z_{perpendicular}$ = (650 lbf / bolt)C_dC_g = 650 × 1.33 × 0.75
= 650 lbf / bolt

The above design values are adjusted by the load duration factor C_d = 1.33 for wind or seismic, and by the group action factor $C_g \cong 0.75$ obtained by preliminary design assuming 7 bolts in a row. The problem statement stipulates that the minimum bolt spacing = 4d = 4×0.75 = 3". (Note: the eccentric force creates a large component of bolt force perpendicular to grain, which would be alleviated by increasing the spacing between bolts). For 7 bolts at 3" o.c.:

$r_{parallel} = 7000 / 7 = 1000$ # / bolt ↑

$\Sigma y_i^2 = 2(3^2 + 6^2 + 9^2) = 252$ in^2

$M_t = 7000^{\#} (2.75 - 0.75)" = 14{,}000^{\#\text{-in}}$

∴ $r_{perpendicular} = M_t y / \Sigma y_i^2 = 14{,}000 \times 9 / 252$
= 500 # / bolt ← or →

$r = (r_{parallel}^2 + r_{perpendicular}^2)^{1/2} = (1000^2 + 500^2)^{1/2}$

r = 1120 # / bolt

$\theta = \tan^{-1}(r_{perpendicular} / r_{parallel}) = \tan^{-1}(500 / 1000) = 26°$

Thus, the critical force on the extreme fasteners is 1120# at 26° to the grain.

Check the allowable load per Hankinson's Equation

$N = Z_{parallel}Z_{perdendicular} / (Z_{parallel}\sin^2\theta + Z_{perpendicular}\cos^2\theta)$

$= (1580)(650) / [1580 \times \sin^2(26°) + 650 \times \cos^2(26°)]$

= 1240 lbf / bolt

Thus, N > r ∴ **Use 7 – 3/4" diameter bolts.** The 7 bolts will create a clamping force between the plate and stud when installed. Assuming the plate is 5 1/2" × 21":

$S = 5.5 \times 21^2 / 6 = 404$ in^3

$M_O = 7{,}000 \times 2 = 14{,}000$ $^{\#\text{-in}}$

$f_b = M_O / S = 14{,}000 / 404 = 35$ psi

The seven bolts must produce contact pressure > 35 psi when installed. Thus,

$T_{bolt} \geq (35 \times 5.5 \times 21 / 7) = 580$ $^{\#}$

For Douglas Fir: $F_{c\text{-perpendicular}}$ = 335 psi

∴ $A_{wash} \geq 580 / 335 = 1.72$ in^2

Need either a plate or circular washers with $A \geq 1.72$ in^2 behind nuts of 3/4" diameter bolt.
Design the angle using A36 steel and E70xx electrodes (AISC 1989)

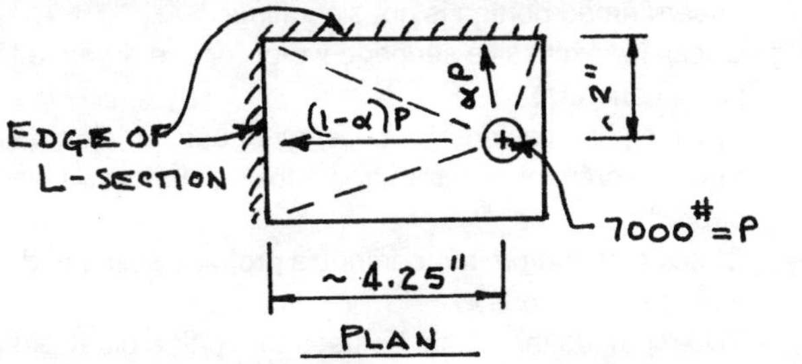

Transfer 7000# into the angle. Let α = % transferred the short way; constant EI each way; $\Delta = Pa^3 / 3EI$

$\alpha(7000)(2)^3 / 3EI = (1 - \alpha)(7000)(4.25)^3 / 3EI$

$\therefore$ $\alpha = 0.9$ (virtually all the force transfers the shorter direction. Try PL 5 x $^5/_8$; 5" long

$M_f = 0.9 \times 7000 \times (2 - 0.625) / 5$

$= 1.8\ ^{k\text{-}in}/_{in}$

$f_b = M_f / S = 1.8 / (0.625^2 / 6) = 26.9$ ksi

$F_b = {}^4/_3\,(0.75F_y) = 36$ ksi $> f_b \therefore$ O.K.

Use PL $^5/_8$"; Provide full penetration weld to long leg of angle; Use minimum size ($^1/_4$") fillet weld to outstanding leg.

Detail of connection angle:
(Note: the L5 $\times$ 3$^1/_2$ $\times$ $^5/_8$ is the minimum size that works. AISC indicates that this is not readily available $\therefore$ may need to substitute a heavier L5 $\times$ 3$^1/_2$ $\times$ $^3/_4$)

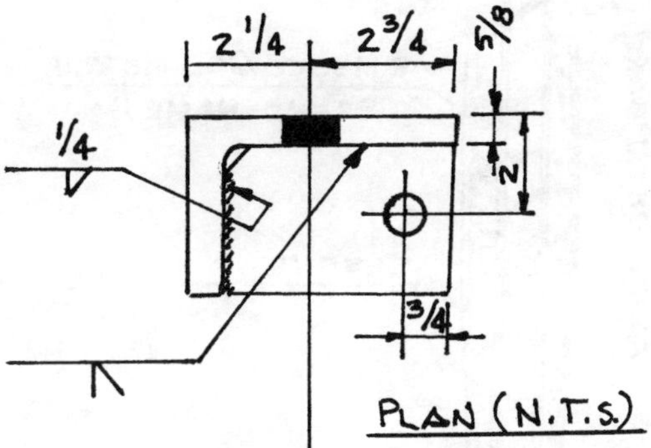

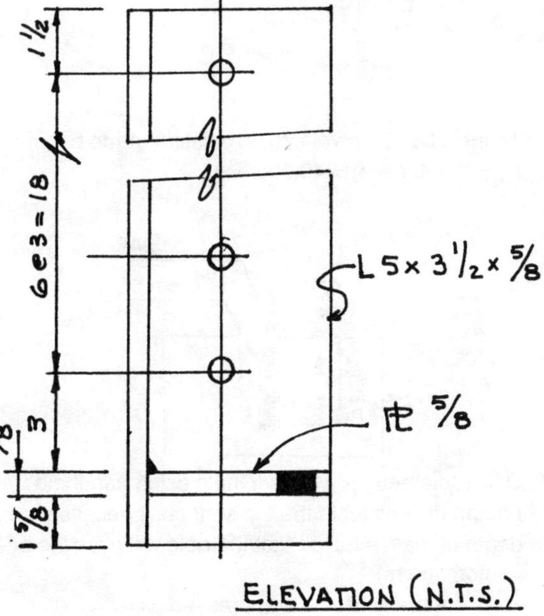

Structural-35

a) There are several methods that can be used to splice a #10 grade 60 rebar with only an 18" projection (ACI 439 1983). Special devices and tools are required and the decision depends on whether the splice is for tension or compression, access, clearance between bars, and cost. One possibility, that will develop the bar in either tension or compression, is the mortar filled sleeve (ACI 318R3):

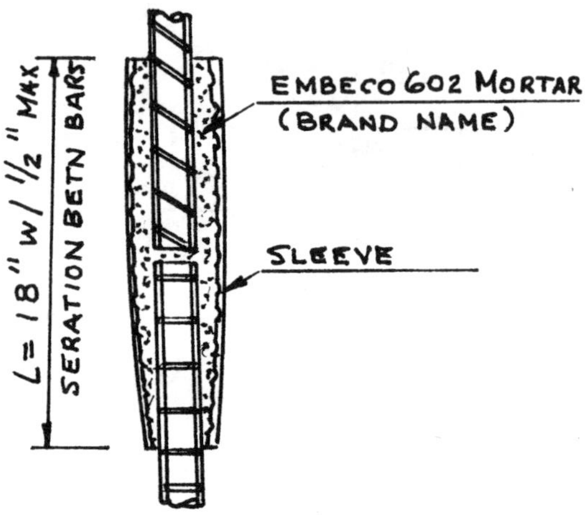

b) To install a #7 dowel into an existing grade beam (f'_c = 3,000 psi; f_y = 40,000 psi):

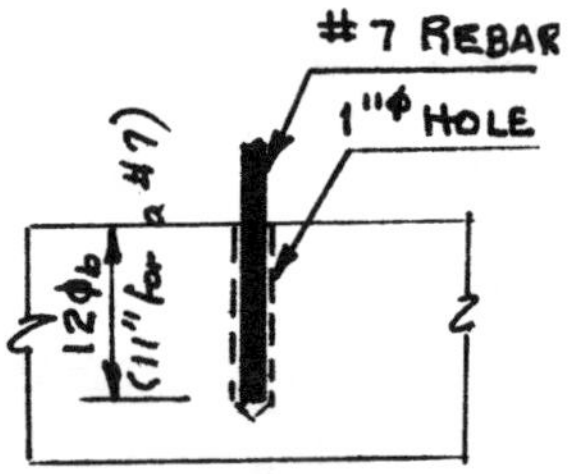

- Drill 1" diameter × 11" deep hole using handheld pneumatic drill (chip the top surface, if necessary, to the depth of main steel so that the hole will miss the beam's reinforcement)
- Blow out the hole with compressed air
- Fill the hole with high modulus epoxy gel using a caulking gun with polyethylene tube extension
- Insert the dowel, working up and down to ensure complete embedment
- Maintain alignment in center of hole until the epoxy sets.

c) Weld inspection methods include visual, dye penetrant, magnetic particle, radiographic and ultrasonic (AISC 1994). A brief description of two of the non-visual methods follows:

- Dye Penetrant – used to detect surface cracks on fillet and groove welds. Dye is applied to the weld surface; excess is removed by wiping ; a developer is used to indicate where dye penetrated the surface cracks.
- Radiographic – used to detect porosity, voids, cracks, etc., in groove welds by film exposure using x-rays.

d) If it is permissible to increase the wall thickness, 2×4's can be scabbed the full length to the 2×6 studs. This will fur the wall an extra $1^1/_2$", but it will restore the area and stiffness loss to the notches.

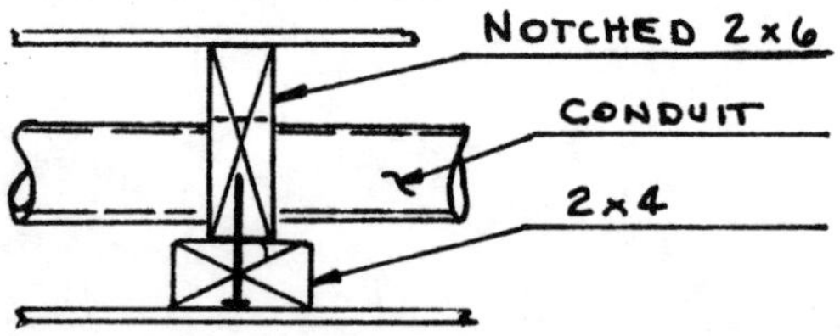

e) Partial checklist for inspection prior to pouring a belled caisson:

- Verify location; vertical alignment and diameter of shafts; diameter of underream bottom
- Ensure that loose soil is removed adequately and that underreamed portion is not sloughing
- Check for excessive seepage water (or use a tremie to place concrete).
- Provide shaft casing, if necessary, or use an "elephant trunk" to prevent concrete from striking shaft walls when poured
- Check that shaft reinforcement is properly sized, tied, free of oil or debris

f) Two methods that might be used to repair a glue void in a laminated timber beam (Note: void occurs near the neutral axis at a location near midspan; thus horizontal shear strength needs to be restored):

- Epoxy injection – an epoxy gel is used to seal the surface and to surround injection ports. A low viscosity epoxy is pressure injected through ports near one end of void and caused to flow to open ports at the far end. The ports are then plugged and the epoxy allowed to set. The surface is refinished after epoxy cures.
- Lag bolts designed to resist the shear flow over the length of the void can be installed vertically into countersunk holes in the beam soffit (assuming the net section at the location of lag bolts is sufficient to resist flexure). After bolts are installed, wood plugs can be glued in to hide the bolt head.

Structural-36

Compute the design lateral forces on the structure considering both wind and seismic forces in the east-west and north-south directions. Follow provisions of UBC-97.
For lateral forces in the east-west direction:

- Wind loads are computed for the design wind speed of 70 mph assuming exposure condition "C" (Re: Table 16-G, C_e = 1.06); importance factor = I_w = 1.0; q_s = 12.6 psf. The appropriate pressure coefficients from Table 16-H are

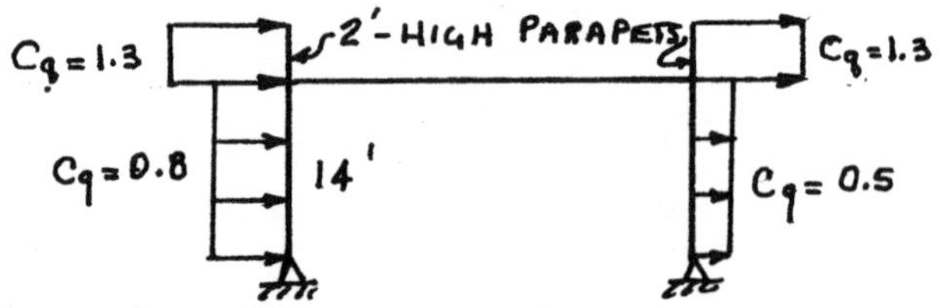

Thus,

$$p = C_e C_q q_s I_w = 1.06 \times 12.6 \times 1 \times C_q$$

The equivalent lateral force per lineal foot at the roof level is found by summing forces about the base

$$w = [(2 \times 1.3 \times 1.06 \times 12.6\text{) psf} \times 2' \times 15' + ((0.8 + 0.5) \times 1.06 \times 12.6) \text{ psf} \times 14' \times 7'] / 14' = 196 \text{ plf}$$

- For seismic forces, the structure has torsional plan irregularity; however, the static force procedure of 1630.2 is still applicable (single story building < 65' height). Compute the effective weight at the roof level on a per lineal foot basis

$w_r = (5 + 8 \times 12.5)$ psf $\times (25 - 2 \times 0.67)' + 2 \times (2 + 14 / 2)' \times 80$ psf $= 3925$ plf

For a bearing wall system with reinforced masonry walls, Table 16-N gives R = 4.5. The period can be estimated using Eqn. 30-8 as

$T = C_t(h_n)^{3/4} = 0.02 \times (14)^{3/4} = 0.14$ sec

Thus, from section 1630.2, using the given values $C_a = 0.36$, $C_v = 0.54$ and $N_a = N_v = 1.0$

$v = [C_vI / (RT)]\,W = [0.54 \times 1.0 / (4.5 \times 0.14)]3925$
$= 3365$ plf

The base shear must be at least

$v \geq 0.11C_aIW = 0.11 \times 0.36 \times 1.0 \times 3925 = 155$ plf

and

$v \geq [0.8ZN_vI / R]W = [0.8 \times 0.3 \times 1.0 \times 1.0 / 4.5]3925$
$= 210$ plf

but not more than

$v \leq [2.5C_aI / R]W = [2.5 \times 0.36 \times 1.0 / 4.5]3925$
$= 785$ plf $\leftarrow$ governs

Thus, seismic governs the lateral loading in the east-west direction

$\therefore \quad v = 785$ plf

$V = vL = 785 \times 100 = 78.5$ kips

The required wall shears require calculation of the translational stiffness of wall and frame elements

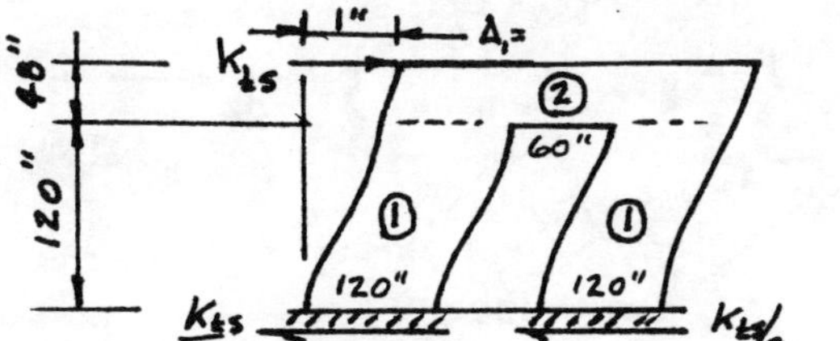

For the south wall: 8" masonry; $E_m = 1.5 \times 10^6$ psi; $E_v = 600 \times 10^6$ psi. Consider both shear and flexural deformation in the piers ($A_1 = 8 \times 120 = 960$ in^2; $I = 8 \times 120^3/12 = 1.15 \times 10^6$ in^4). Consider only shear deformation in wall above ($A_2 = 8 \times 300 = 2400$ in^2).

$\Delta = 1" = (K_{ts} / 2)(1.2 \times 120) / (960 \times 600)$
$+ (K_{ts} / 2)(120^3) / (12 \times 1.5 \times 10^3 \times 1.15 \times 10^6)$
$+ (K_{ts})(1.2 \times 48) / (2400 \times 600)$

$\therefore \quad K_{ts} = 4830$ k / in

For the steel rigid frame, the bases are pinned and the girder is rigid. Consider only the bending deformation in the two columns ($E = 30 \times 10^3$ ksi; $I = 1830$ in^4)

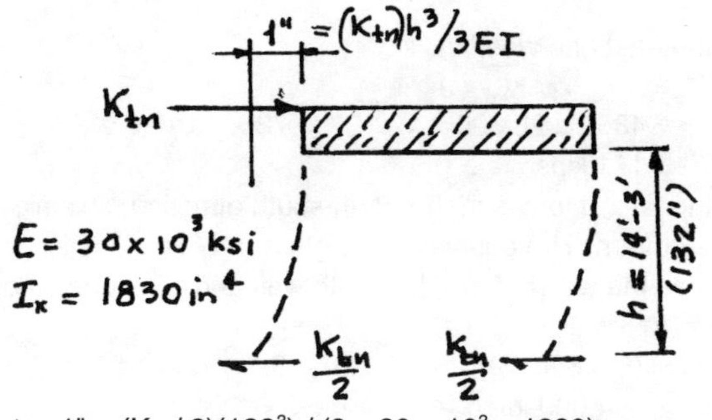

$\Delta = 1" = (K_{tn} / 2)(132^3) / (3 \times 30 \times 10^3 \times 1830)$

$\therefore \quad K_{tn} = 143$ k / in

For the east and west side walls, only shear deformation is significant

$K_{te} = 8 \times (98.7 \times 12) \times 600 / (1.2 \times 14 \times 12)$
$= 28{,}300$ k / in

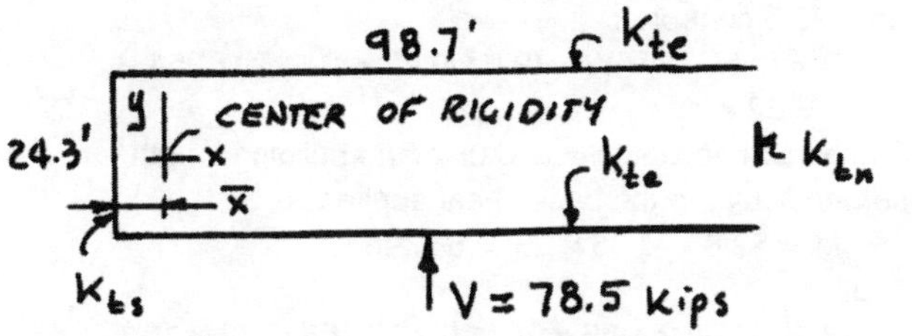

The center of rigidity is found using centerline dimensions

$\bar{x} = (143 \times 98.7) / (143 + 4830) = 2.84'$

The location of $\bar{y}$ is found by symmetry to be at the midpoint of the 24.3' direction.

The distance from mass center to center of rigidity is very large, and the accidental eccentricity limit, $A_x = 3.0$, of UBC 1630.7 will apply

$e_{accidental} = 3 \times 0.05 \times 100' = 15'$

$\therefore \quad M_t = 78.5(50 - 0.33 - 2.84 + 15) = 4850$ k-ft

$J = \Sigma(K_t x^2 + K_t y^2) = 4830 \times 2.84^2 + 143 \times (99 - 0.33 - 2.84)^2 + 2 \times 28{,}300 \times 12.17^2$

$J = (9.73 \times 10^6)$ k-ft^2/in

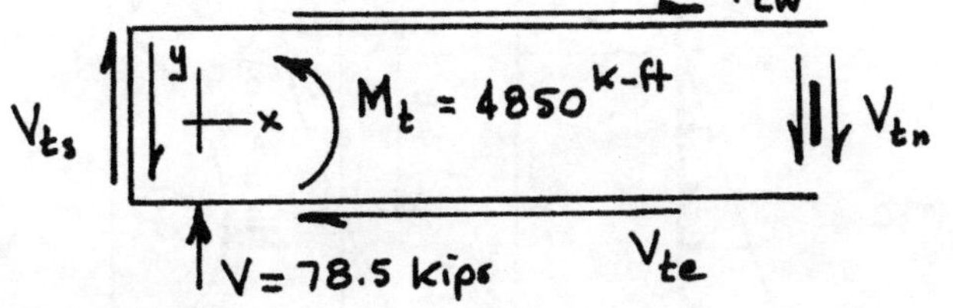

Lateral force is resisted by the north and south walls only

$V_{vs} = 78.5 \times 4830 / (4830 + 143) = 76.2$ kips $\downarrow$

$V_{vn} = V - V_{vs} = 78.5 - 76.2 = 2.3$ kips $\downarrow$

The torsional moment is resisted by all elements

$V_{tn} = M_tK_{tn}y / J = 4850 \times 143 \times 95.8 / (9.73 \times 10^6)$
$= 6.8$ kips $\downarrow$

$V_{ts} = -V_{tn} = 6.8$ kips $\uparrow$

Since only the increase in wall shear due to torsion is considered in design (Re: UBC 1630.7)

$\therefore \quad V_s = V_{vs} = 76.2$ kips $\downarrow$

$V_n = V_{vn} + V_{tn} = 2.3 + 6.8 = 9.1$ kips $\downarrow$

For the east and west walls

$V_e = -V_w = M_t K_{tn} y / J$

$= 4850 \times 28{,}300 \times 12.15 / (9.73 \times 10^6)$

$= 171$ kips

For the lateral forces in the north-south direction, seismic loads govern. The equivalent weight at roof level, which includes the weight from the south wall and excludes other wall is

$W = (5 + 8(12.5))(25 - 2(0.67))(99) + (2 + 14 / 7)(25)(80)$

$= 264{,}000$ lbf

$= 264$ kips

Use UBC Eqn 30-9 to estimate the period

$T = 0.1 / \sqrt{A_C}$

$A_c = \Sigma A_e[0.2 + (D_e / h)^2]$; $D_e / h < 0.9 \leftarrow$ governs

$A_c = 2 \times 0.67' \times 100'[0.2 + 0.9^2] = 135$ ft^2

$\therefore$ $T = 0.1 / (135)^{1/2} = 0.009$ sec

Eqn. 30-5 controls

$V = (2.5C_a I / R)W = (2.5 \times 0.36 \times 1 / 4.5) \times 264$

$= 52.8$ kips

A minimum eccentricity of 0.05 of the building length perpendicular to the base shear applies

$M_t = 52.8 \times (0.05 \times 25) = 66$ k-ft

Thus,

$V_e = 52.8 / 2 + 66 \times 28{,}300 \times 12.15 / (9.73 \times 10^6)$

$= 28.7$ kips

For the south wall

$V_s = 66 \times 2.84 \times 4380 / (9.73 \times 10^6) = 0.08$ kips

and for the rigid frame

$V_n = -V_s = 0.08$ kips

Structural-37

a) Critique the proposed construction details. Assume a well drained, granular backfill for which $k_O = 0.3$; $\gamma_s = 110$ Pcf. $\therefore$ equivalent fluid weight = 33 pcf. The wall spans vertically between the sill plate and the slab centerline approximately 8.2'.

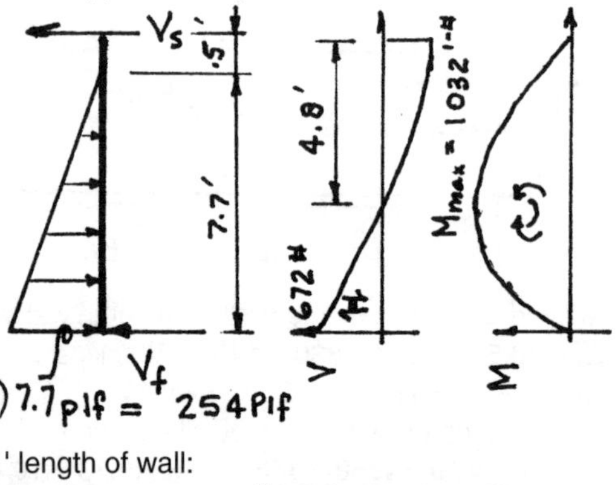

For 1' length of wall:

$V_s = (0.5 \times 254 \times 7.7^2 / 3) / 8.2 = 306$ plf

$V_f = 0.5 \times 254 \times 7.7 - 306 = 672$ plf

Check the strength of reinforced masonry wall (UBC 1997):

For #5 @ 16" o.c.: $A_s = 0.31 / 1.33 = 0.23$ in^2 / ft; f'_m taken as 1500 psi; b = 12"; d = 5.3"

$n = E_s / E_m = 29 \times 10^6 / (750 \times 1500) = 26$

$np = 26 \times 0.23 / (12 \times 5.3) = 0.094$

$k = (0.094^2 + 2 \times 0.094)^{1/2} - 0.094 = 0.35$

$j = 1.00 - 0.35 / 3 = 0.88$

$f_b = (M / bd^2)(2 / jk) = [(1032 \times 12) / (12 \times 5.3^2)][2 / (0.35 \times 0.88)]$

$= 238$ psi $< 0.5 \times (0.33 \times 1500) = 250$ psi

(Note: the allowable stresses are for construction without special inspection, Chapter 21, UBC-97)

$f_s = M / A_s jd = 1032 \times 12 / (0.23 \times 0.88 \times 5.3)$

$= 11{,}500$ psi (O.K. for all grades)

$f_v = V / bjd = 672 / (12 \times 0.88 \times 5.3) = 12$ psi $<$ $(0.5)(1500)^{1/2} = 20$ psi

$\therefore$ **<u>The Wall strength is adequate</u>**

Check the transfer of 306 plf to sill plates using $^5/_8$" dia. anchor bolts at 4' o.c.:

$V = 306 \times 4 = 1223$ $^\#/_{bolt}$

From Table No. A-21-B, UBC, the required spacing of 5/8" anchor bolts is 24' o.c. for a backfill height of 7 ft. **Thus, anchor bolts are not adequate** to transfer force from wall-to-sill. The type 2 framing angles at 2' o.c. are O.K. (I assume here that there are 2 angles – one each side – of the 2×6 blocking). Several deficiencies occur in the transfer of the lateral force into the roof diaphragm. For example, consider the 2×6 brace at the 7' maximum height:

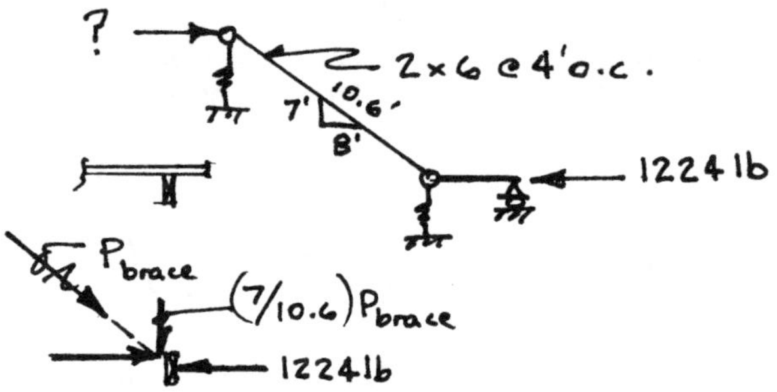

$\Sigma F_x = 0$: $1224 - (8 / 10.6)P_{brace} = 0$

$P_{brace} = 1630$ lb

1. The 2×6 brace has a weak axis unsupported length = 10.6' = 127". Thus, $L_e / d = 127 / 1.5 = 85 > 50$. The 2×6 cannot resist the design load!
2. The type 1 framing angles connecting the brace to the truss chords are not adequate to transfer $(7 / 10.6) \times 1630 = 1080$ lb.
3. The vertical component of force in the brace imposes loads on the roof trusses that are probably excessive (e.g., 1080 lb down on the lower chord ; 1080 lb up on the upper chords where the braces are maximum height).

b) **<u>Remedial action is needed.</u>** One possibility is to create a diaphragm in the plane of the lower chords of the roof trusses.

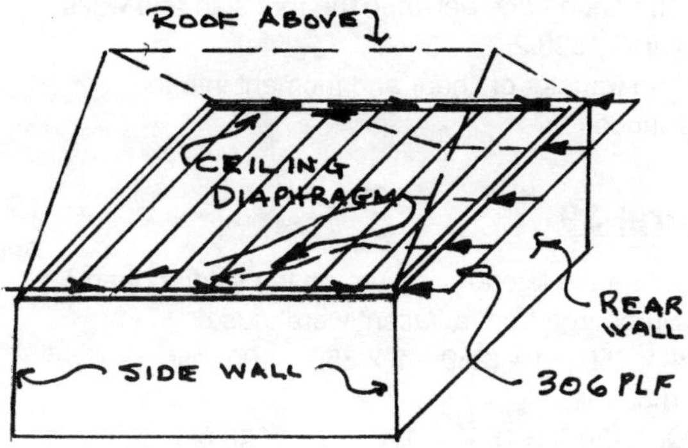

We would need to provide additional strength to transfer the 306 plf into the diaphragm (to supplement the $^5/_8$" dia anchor bolts). One method is to install a cleat on the 3×8 sill, which would transfer the force into the sill:

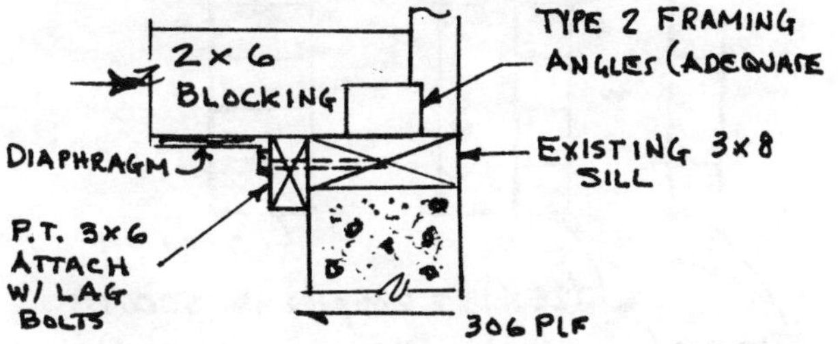

Collector and anchorage elements along the side walls should be investigated and strengthened, if necessary. Front and rear walls plus sills are adequate as chords. The shear transfer through the ceiling diaphragm to the side walls is:

$$V = (306 \times 22 / 2) / 22 = 153 \text{ plf}$$

Taking the allowable shear for plywood diaphragms from Table 23-II-H, UBC (Note: table values are for wind or seismic – adjust by 0.75×0.9 for sustained loads):

$$V_{au} = 0.75 \times 0.9 \times 215 = 145 \text{ plf – close enough}$$

Requires $^{15}/_{32}$" min. structural I plywood; 10-d nails 6" max. at supported ends.

Structural-38

a) Check the shear in North-South walls (UBC 1997). Assume a rigid roof diaphragm and simple connections in the steel frame. From Table 16-N, for a building frame system with masonry shearwalls, R = 5.5. The building is classified as essential ∴ the importance factor = I = 1.25. For a 75 psf wall continuous around the perimeter and 75 psf roof dead load

$$W = 75(80 - 2 \times 0.67)(60 - 2 \times 0.67)$$
$$+ 75(79.3 + 59.3) \times 2 \times (18 \times 9 / 16)$$
$$= 556{,}000 \text{ lb} = 556 \text{ kips}$$
$$T = C_t h_n^{3/4} = 0.02 \times 16^{3/4} = 0.16 \text{ sec}$$

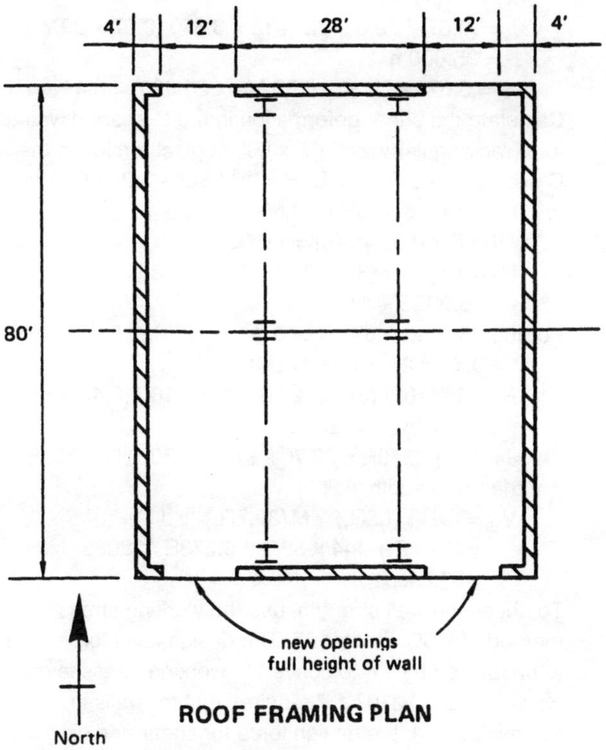

ROOF FRAMING PLAN

Thus, from UBC 1630.2, using the given values $C_a = 0.44$, $C_v = 0.64$ and $N_a = N_v = 1.0$

$$V = [C_v I / (RT)] W = [0.64 \times 1.0 / (5.5 \times 0.16)]556$$
$$= 404 \text{ kips}$$

The base shear must be at least

$$V \geq 0.11 C_a I W = 0.11 \times 0.44 \times 1.0 \times 556 = 27 \text{ kips}$$

and

$$V \geq [0.8 Z N_v I / R] W = [0.8 \times 0.4 \times 1.0 \times 1.0 / 5.5]556$$
$$= 32 \text{ kips}$$

but not more than

$$V \leq [2.5 C_a I / R] W = [2.5 \times 0.44 \times 1.0 / 5.5]556$$
$$= 111 \text{ kips} \leftarrow \text{governs}$$

The maximum shear in the North-South walls occurs during East-West ground motion. Centers of mass and rigidity coincide; however, UBC requires a minimum eccentricity

$$e_{min} = 0.05 \times 80 = 4 \text{ ft}$$

$$\therefore \quad M_t = Ve = 111 \times 4 = 444 \text{ k-ft}$$

For the 80' long flanged walls (use centerline dimensions). Locate the shear center, x_0 (Roark 1975; table 14)

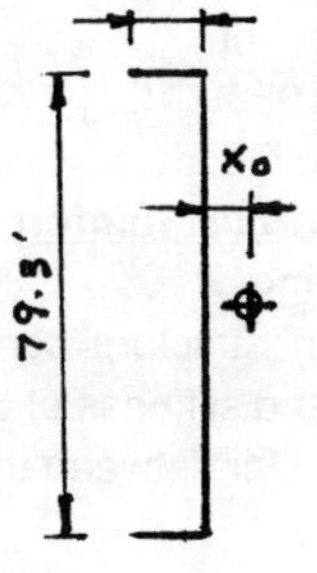

$x_0 = b^2h^2t / 4I_x$

$I_x = (0.67 \times 79.3^3 / 12) + (2 \times 3.67 \times 0.67 \times 39.7^2)$
$= 35{,}600\ ft^4$

$x_0 = 3.67^2 \times 79.3^2 \times 0.67 / (4 \times 35{,}600) = 0.41' - nil$

Calculate the shear deformation in the C-shaped wall as for a rectangular wall 0.67' x 80'; neglect torsion in the C-section; take $G = 0.4E$; $R = [h^3 / 3EI + 1.2h / GA]^{-1}$

$\therefore\ R_y = 1 / [16^3 / (3E \times 35{,}600) + 1.2 \times 16 / (0.4E \times 0.67 \times 80)] = 1.07E$

$R_x = 1 / [16^3 / (3E \times 0.67 \times 3.33^3 / 12)]$
$= 0.002E - nil$

For the 28' long walls: $R_y = 0$

$I = 0.67 \times 28^3 / 12 = 1226\ ft^4$

$R_x = 1 / [16^3 / (3E \times 1226) + 1.2 \times 16 / (0.4E \times 0.67 \times 28)] = 0.273E$

$\therefore\ J = 2 \times [0.273E \times 39.7^2 + 1.07E \times 30.1^2] = 2800E$

For the north-south wall

$V_{28} = (VR_{x28} / \Sigma R_x) \pm M_t(39.7R_{x28}) / J$
$= 111 / 2 + 444 \times 39.7 \times 0.273E / 2800E$
$= 57.2\ kips$

To check the wall strength, use the working stress method of UBC chapter 27. The design level forces are to be divided by 1.4 to convert to working stress level. However, UBC 2107.1.7 requires that for walls in seismic zone 4, the design force for shear and diagonal tension be increased by 50%. Therefore, the shear to check in the north-south wall is

$V = (57.2 / 1.4) \times 1.5 = 61.2\ kips$

The nominal shear stress is

$v = V/bjd = 61.2 / (7.63 \times 7/8 \times 22 \times 12) = 0.035\ ksi$
$= 35\ psi$

Assuming that masonry resists the shear:

$M / Vd = 61.2 \times 16 / (61.2 \times 7 / 8 \times 22) = 0.83 < 1.0$

From UBC Eqn. 7-19

$F_v \le 4 / 3 \times 1 / 3 \times (4 - M / Vd) \sqrt{f'_m}$
$= 4 / 3 \times 1 / 3 \times (4 - 0.83)(1500)^{1/2} = 55\ psi$

$F_v \le 4 / 3(80 - 45M / Vd) = 4 / 3(80 - 45 \times 0.83)$
$= 57\ psi$

Thus, $F_v = 55\ psi > v = 35\ psi$, the wall is adequate in shear. Check flexural strength using the minimum vertical wall steel (UBC 2106.1.12.4) and neglecting gravity load

$A_v = (0.002 - 0.0007)A_e = 0.0013 \times 7.63" \times 12" / ft$
$= 0.12\ in^2 / ft$

$M_o = Vh = 61.2 \times 16 = 980\ k\text{-}ft$

Approximate the resistance by assuming that the steel smeared over a length of 10 ft acts at its centroid ($jd \cong 22 - 5 = 17$ ft). Thus,

$f_s = M_o / jdA_s \cong 980 \times 12 / [(17 \times 12) \times (10 \times 0.12)]$
$= 48\ ksi$ – No Good

Thus, minimum vertical reinforcement is inadequate to resist overturning.

b) Four other important structural considerations:

1. The strength and stiffness of the roof diaphragm
2. The shear transfer between the walls and diaphragm
3. The anchorage between the roof slab and walls (UBC 1633.2.8)
4. The transfer of shear and moment into the foundation

Structural-39

a) Determine the lateral pressure on the soldier beams.

Q_f = Surcharge from adjacent warehouse
$= 4' \times 150\ pcf + (250 + 2 \times 350 + 100)\ psf$
$= 1650\ psf$

$N = Q_f - 120D_f = 1650 - 120 \times 5 = 1050\ psf$

Compute the lateral pressure on soldier beams given that beams are spaced 8' o.c.; a = 12'; H = 25'; $H_1 = 0.8H = 20'$

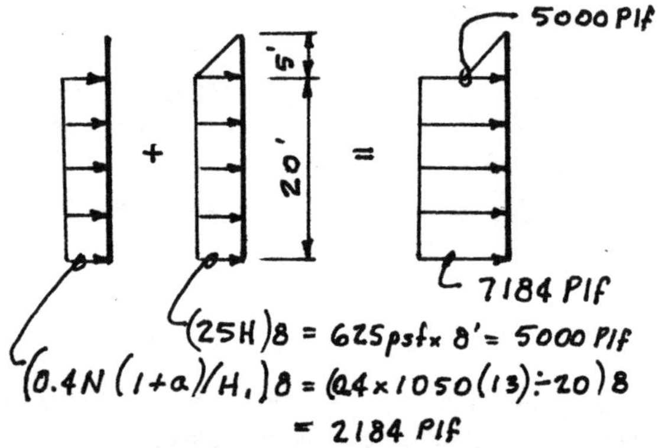

Make an approximate analysis of the tie-back forces in order to obtain a trial value for toe length. The toe will have relatively short embedment ; therefore, the restraint against rotation is neglected. The resultant lateral force (per 8' of wall length) is

$F_x = 0.5 \times 5000 \times 5 + 7184 \times 20 = 156{,}000\ lb$
$= 156^k$

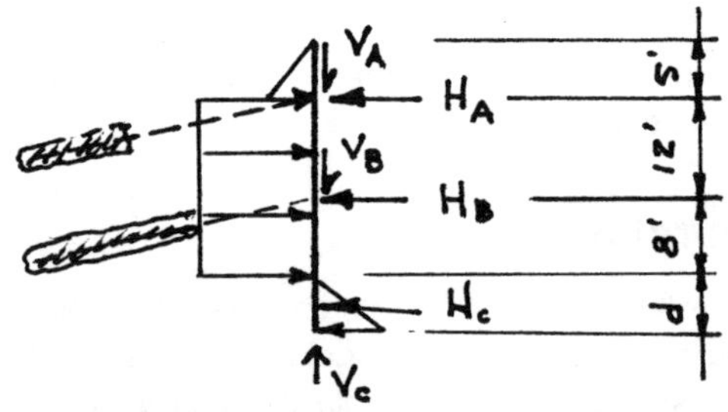

b) Calculate the forces at the supports

$H_A \cong 12.5^k + 0.4 \times 7.184 \times 12 = 47^k \leftarrow$

$H_B \cong 7.184(0.6 \times 12 + 0.75 \times 8) = 95^k \leftarrow$

$\therefore\ H_C \cong 156 - 47 - 95 = 14^k \leftarrow$

Thus, the toe length required to develop 14^k by passive soil pressure is

$0.5(0.45d)d = 14\ \therefore\ d^2 = 62\ ft^2$

$d \cong 8'$

The vertical component of tendon force, V_A and V_B, is to be resisted by skin friction along the 2' diameter toe. Let f = skin friction (ksf) $\le 2^{ksf}$

The tendon slopes 15° with respect to the horizontal. Thus,

$V_A = (47^k / \cos 15°) \sin 15° = 12.6^k$

$V_B = (95^k / \cos 15°) \sin 15° = 25.5^k$

$V_C = 12.6 + 25.5 = 38.1^k = \pi 2f(8)$

$\therefore$ $f = 0.76^{ksf} < 2^{ksf}$ $\therefore$ O.K.

Now, a more refined analysis of tieback will be made using the trial toe length of 8'. Do a linearly elastic analysis treating the horizontal reaction at B, H_B, as redundant.

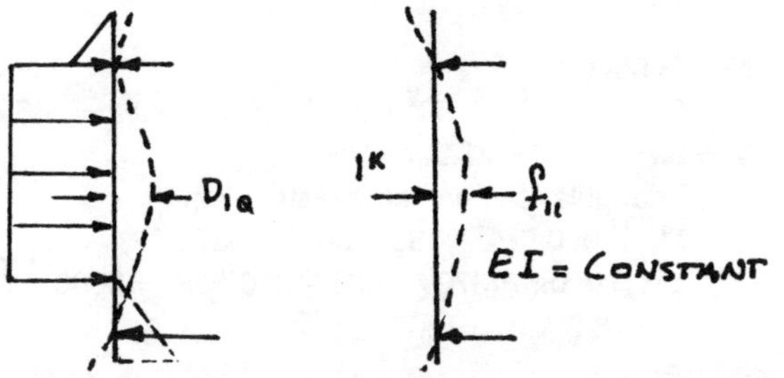

Using appropriate formulas to compute deflections at point 'B' in the released structure (Roark 1976 ; table 3):

$D_{IQ} = 5.789 \times 10^{10} / EI \rightarrow$

$f_{11} = 0.0580 \times 10^{10} / EI \rightarrow$

For consistent displacement:

$H_B f_{11} = -D_{IQ}$

$\therefore$ **$\underline{H_B = -5.789 / 0.0580 = 99.8^k \leftarrow}$**

From $\Sigma M_c = 0$:

$\underline{H_A = 47.8^k \leftarrow}$

From $\Sigma F_x = 0$:

$\underline{H_C = 8.4^k \leftarrow}$

Given the uncertainties in lateral force distribution and support conditions, the above analysis is accurate enough.

c) To verify adequacy of the 8' long toe length:

$V_C = ((98.8 + 47.8) / \cos 15°)\sin 15°$

$= 39.5^k$

$\therefore$ $f = 39.5 / (2\pi \times 8) = 0.79^{ksf} < 2^{ksf}$ $\therefore$ O.K.

Passive pressure on 8' toe

$f_p = 8.4 / (0.5 \times 8) = 2.1^{ksf} < 0.45 \times 8^{ksf}$

$\therefore$ **<u>8' toe length is O.K.</u>**

d) Compute the grouted anchor lengths (skin friction on 12" dia. drilled shaft = 1.5 ksf)

$T_A = H_A / \cos 15° = 47.8 / \cos 15° = 49.5$ kips

$\therefore$ $1.5 \times \pi \times 1' \times L_A = 49.5$ kips

$\underline{L_A = 10.5'}$

$T_B = 99.8 / \cos 15° = 103.3$ kips

$\underline{L_B = 103.3 / (1.5 \times \pi \times 1) = 22'}$

Grouted portion extends beyond failure plane, which makes a 60° angle from the horizontal. Thus, total tendon lengths are:

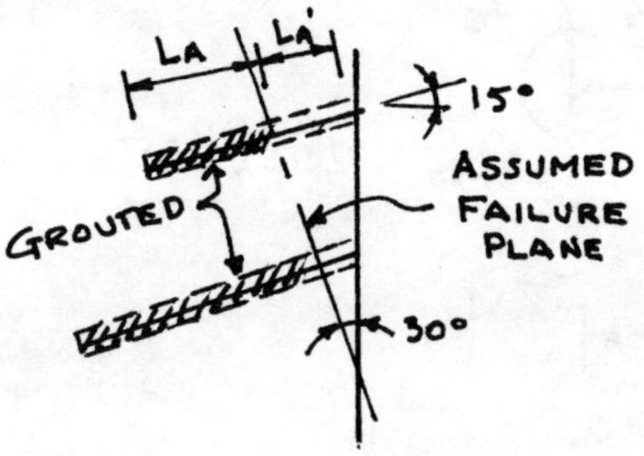

$L_A' / \sin 30° = 20' / \sin 75° \rightarrow L_A' = 10.4'$

$\therefore$ **$\underline{L_A + L_A' = 10.5' + 10.4' = 20.9' \text{ say } 21'}$**

$L_B' = 8 \sin 30° / \sin 75° = 4.1'$

$\underline{L_B + L_B' = 22 + 4.1 = 26.1'}$

e) Installation procedure (Schnabel 1982):

- Drill (case, if necessary) 2 ft dia. vertical shafts at soldier beam locations
- Pour concrete in lower (i.e., toe) portion of shaft; insert and align soldier beam.
- Excavate to region below 'A', installing lagging planks as work progresses.
- Drill 1' dia. shaft inclined 15° between soldier beams at elevation 'A'.
- Grout tendon 'A' (usually by pressure injection rather than by gravity flow).
- Install wales horizontally between soldier beams to transfer tendon tieback forces to soldier beams (usually two C-sections back-to-back][to form wales).
- Tension tendons to their proof load (e.g., 80% of ultimate) then anchor the tendons at their working load.

Structural-40

a) Determine the soil pressure (note that the given loads are working loads)

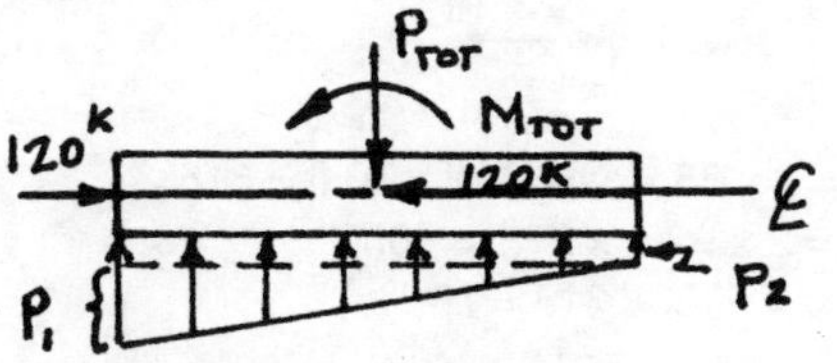

The footing weights are

$W = 2' \times 5' \times 0.15^{kcf} = 1.5$ klf

$P_{TOT} = 2(50 + 40) + 70 \times 1.5 = 285^k$

$M_{TOT} = 2 \times 60(25 + 2.5) = 3300^{K\text{-}F}$

$\bar{x} = 3300 / 285 = 11.6' < (70 / 6)'$

$\therefore$ compression throughout

$\Sigma M = 0 : 0.5(70)P_1(70 / 6) = 3300$

$P_1 = 8.08$ klf

$\Sigma F_y = 0 : 0.5(8.08)70 + 70P_2 = 285$

$P_2 = 0.03$ klf (practically nil)

Thus, the maximum soil pressure under 2' wide footing is

$\underline{f_p = (8.08 + 0.03) / 2 = 4.06^{ksf}}$

b) Design the footing using UBC-97. Check the loading conditions of UBC 1612.2

$$U = \begin{cases} 1.4D + 1.7L \\ 1.1(1.2D + 0.5L + E) \\ 0.9D + 1.1(E) \end{cases}$$

The design-level seismic force is 1.4 times the working value = $1.4 \times 60 = 84$ kips.

- For the gravity loading condition

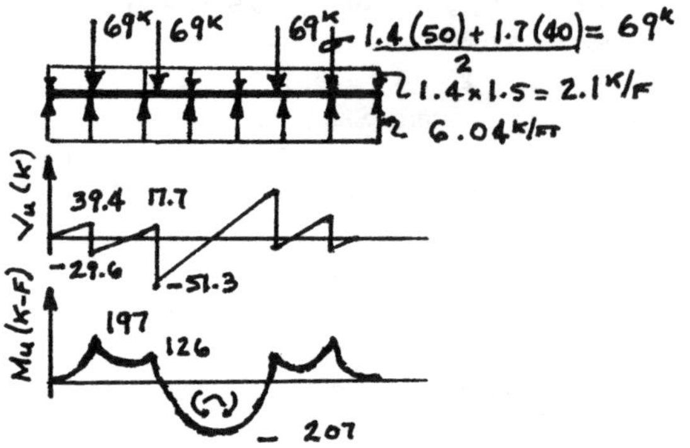

- For U = 1.1(1.2D + 0.5L + E)

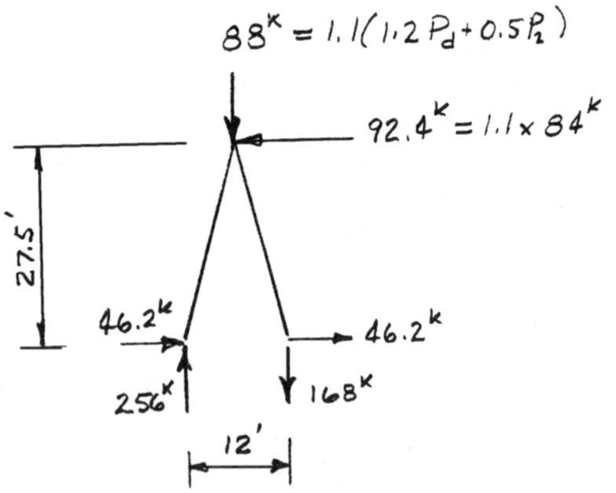

P_{TOT} = 2 × 88 kips + 2 klf × 70 ft = 316^K
M_{TOT} = 2 × 92.4 × 27.5 = $5080^{K\text{-}F}$
$\bar{x}$ = 5080 / 316 = 16' > $L/6$
∴ Triangular soil pressure

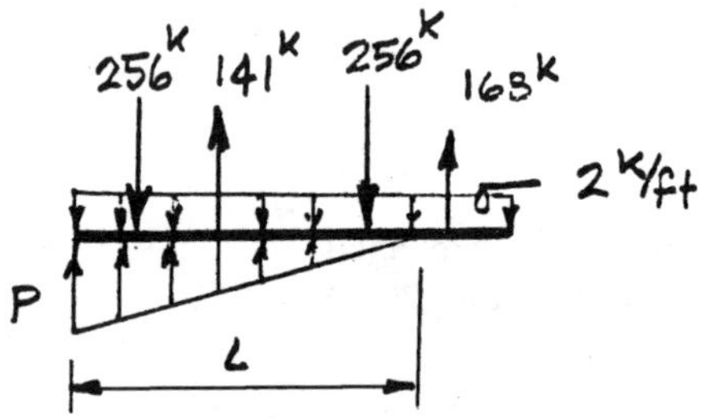

$\Sigma F_y = 0$: 0.5PL – 316 = 0 → L = 632 / P
$\Sigma M = 0$: 0.5P(632 / P)(35 – 632 / 3P) = 5080
∴ P = 11.1 $^K/_{ft}$; L = 632/P = 56.9'

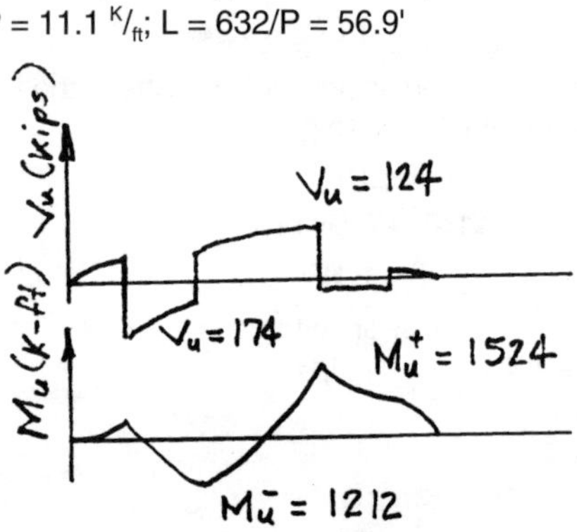

- For U = 0.9D + 1.1(E)

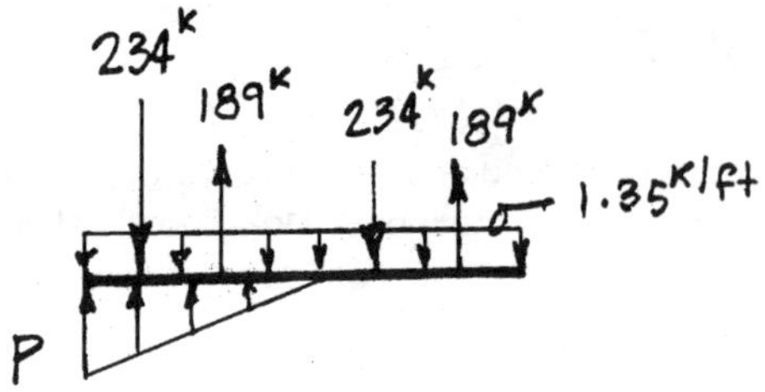

M_{TOT} = 5080 $^{k\text{-}ft}$
$P_{TOT} = 2 \times 0.9 \times 50^k + 0.9 \times 1.5$ klf × 70 ft = 185 kips
$\bar{x}$ = 5080 / 185 = 27.5 ft > $L/6$
∴ Triangular soil pressure distribution
$\Sigma F_y = 0$: 0.5PL – 185 = 0 → L = 370 / P
$\Sigma M = 0$: 0.5P(370 / P)(35 – 370 / 3P) = 5080
∴ P = 16.3$^k/_{ft}$; L = 22.7'

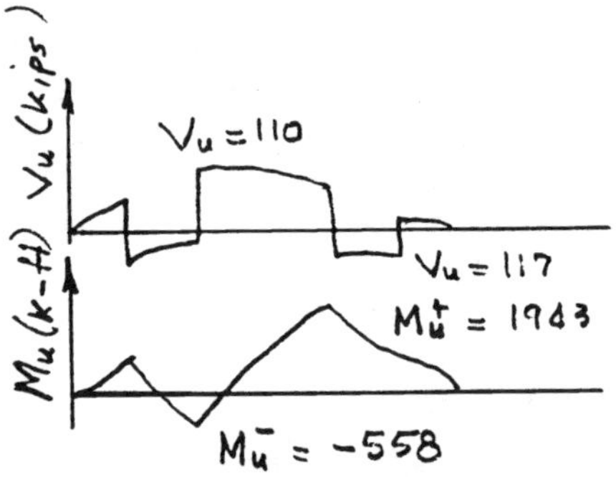

Note that bending moments caused by seismic load combinations are several times greater than those caused by gravity alone. Design reinforcement using section 1921, UBC-97: f'_c = 3000 psi ; grade 60 rebars (f_y = 60,000 psi); b = 24"; h = 60"; d ≅ h – 4 = 56"
$A_{s,\,min} = (200 / f_y)bd = (200 / 60{,}000) \times 24 \times 56 = 4.50$ in^2
$A_{s,\,max} = 0.025\, bd = 0.025 \times 24 \times 56 = 33.60$ in^2
Governing moments: $M_u^+ = 1943^{K\text{-}ft}$
∴ $A_s^+ = 8.33$ in^2; say **7# 10 in the Bottom**
For $M_u^- = 1524^{K\text{-}ft}$; $A_s^- = 6.36$ in^2 say **5 # 10 in the Top**
Calculate the shear corresponding to probable moment strength (Φ = 1.0; $f_s = 1.25 f_y$):
For 5 # 10 top: a = 5 × 1.27 × 1.25 × 60 / (0.85 × 3 × 24) = 7.78"
∴ M_{pr} = 1.25 × 60 × 5 × 1.27(56 – 7.78 / 2) / 12 = $2068^{K\text{-}F}$
For 7#10 bottom: a = 10.89"; $M_{pr} = 2853^{K\text{-}FT}$

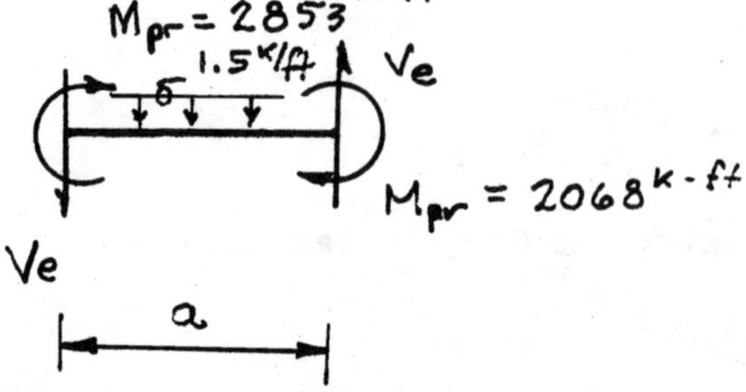

$V_e = (2068 + 2853) / a + V_g$

Use #6 double hoops: $A_v = 2 \times 0.44 = .88\ \text{in}^2$

The minimum spacing of the transverse reinforcement between plastic hinges (UBC 1921.3.3.2.) is

$s \le (\,^d/_4 = 14"; 8d_b = 8 \times {}^{10}/_8 = 10"; 24d_h = 24 \times {}^6/_8 = 18; 12") \rightarrow 10"$ o.c.

Between the legs of the A-frames: a = 12' and

$V_g = 6 \times 1.5 = 9$ kips

$\therefore\quad V_e = (2068 + 2853) / 12 + 9 = 419$ kips

The shear is predominately seismic $\therefore V_c = 0;\ \Phi = 0.6$

Stirrup spacing: $S \le A_v f_y d / V_s$

$s \le 0.88 \times 60 \times 56 / (419 / 0.6) = 4"$ o.c.

Between the A-frames (i.e., the central zone)

$V_c = (2068 + 2853) / 26 + 1.5 \times 13 = 208$ kips

$S \le 0.88 \times 60 \times 56 / (208 / 0.6) = 8"$ o.c.

Outside the A-frames requires hoop steel for confinement only $\therefore$ s = 10" o.c.

Thus,

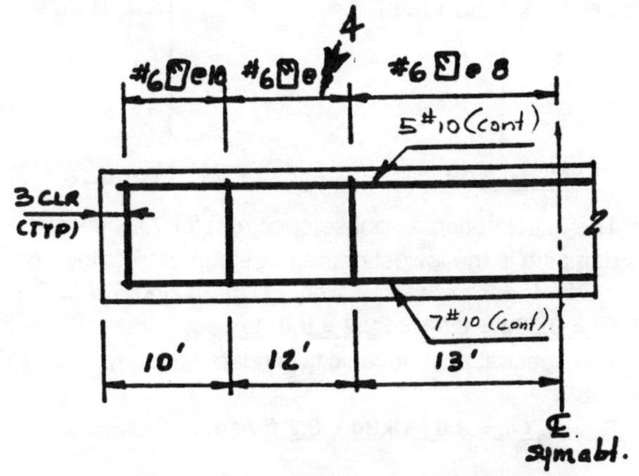

PARTIAL ELEVATION (N.T.S.)

Structural-41

a) Calculate the participation factors (SEAOC 1996).

$M_1 = W_3 / g = 80 / 32.2 = 2.48\ \text{k-sec}^2/\text{ft}$

$M_2 = W_2 / g = 2.48\ \text{k-sec}^2/\text{ft}$

$M_3 = W_1 / g = 120 / 32.2 = 3.74\ \text{k-sec}^2/\text{ft}$

$M = M_1 + M_2 + M_3 = 8.70\ \text{k-sec}^2/\text{ft}$

For the first mode: $\omega_1 = 8.77\ ^{rad}/_{sec}$; the normalized eigenvector is

$\{\Phi_1\}^T = (1.69, 1.22, 0.572)$

$$M_1 = \sum_{i=1}^{3} \phi_{i1}^2 M_i = 1.69^2 \times 2.48 + 1.22^2 \times 2.48 + 0.572^2 \times 3.74 = 12.0\ \text{k-sec}^2/\text{ft}$$

$$P_1 = \sum_{i=1}^{3} \phi_{i1} M_i \Big/ \sum_{i=1}^{3} \phi_{i1}^2 M_i = [1.69 \times 2.48 + 1.22 \times 2.48 + 0.572 \times 3.74] / 12.0 = \underline{0.780}$$

Similarly, for modes 2 and 3:

$\{\Phi_2\}^T = (-1.208, 0.704, 1.385)$

$M_2 = 12.0\ \text{k-sec}^2/\text{ft}$

$P_2 = [-1.208 \times 2.48 + 0.704 \times 2.48 + 1.385 \times 3.74] / 12.0 = \underline{0.327}$

$\{\Phi_3\}^T = (-0.714, 1.697, -0.948)$

$M_3 = 11.76\ \text{k-sec}^2/\text{ft}$

$P_3 = [-0.714 \times 2.48 + 1.697 \times 2.48 - 0.948 \times 3.74] / 11.76 = \underline{-0.094}$

b) As a check, the sum of the participation factors should be unity:

$P_1 + P_2 + P_3 = 0.780 + 0.327 - 0.094 \cong 1$

c) Calculate the displacement at each floor level based on the given response spectrum.

For Mode 1: $f = \omega / 2\pi$; $T = 1/f$

$\omega_1 = 8.77$ rad / sec; $f_1 = 1.4$ cps; $T_1 = 0.7$ sec

$s_{a1} = 0.15g = 0.15 \times 32.2 = 4.8\ \text{ft/sec}^2$

$s_{d1} = s_{a1} / \omega_1^2 = 4.8 / 8.77^2 = \underline{0.062\ \text{ft}}$

For Mode 2:

$\omega_2 = 25.18\ ^{rad}/_{sec}$; $f_2 = 4.0$ cps; $T_2 = 0.25$ sec

$s_{a2} = 0.8g = 25.8\ \text{ft/sec}^2$

$s_{d2} = 25.8 / 25.18^2 = 0.041$ ft

For Mode 3:

$\omega_3 = 48.13\ ^{rad}/_{sec}$; $f_3 = 7.7$ cps; $T_3 = 0.13$ sec

$s_{a3} = 0.4g = 12.9\ \text{ft/sec}^2$

$s_{d3} = 12.9 / 48.13^2 = 0.0056$ ft

The displacement at each level: $x_{ind} = \Phi_{in} P_i S_{dn}$

$\therefore\quad x_{11d} = (1.690)(0.780)(0.062) = 0.0817$ ft

$x_{21d} = (1.220)(0.780)(0.062) = 0.0590$ ft

$x_{31d} = (0.572)(0.780)(0.062) = 0.0277$ ft

$x_{12d} = (-1.208)(0.327)(0.041) = -0.0162$ ft

$x_{22d} = (0.704)(0.327)(0.041) = 0.0094$ ft

$x_{32d} = (1.385)(0.327)(0.041) = 0.0186$ ft

$x_{13d} = (-0.714)(-0.094)(0.0056) = 0.0004$ ft

$x_{23d} = (1.697)(-0.094)(0.0056) = -0.0009$ ft

$x_{33d} = (-0.948)(-0.094)(0.0056) = 0.0005$ ft

Combine the modal maxima using the square-root-of-sum-of-squares method:

$$x_i = \left[\sum_{j=1}^{3} x_{ijd}^2\right]^{\frac{1}{2}}$$

$\therefore\quad x_1 = [0.0817^2 + (-0.0162)^2 + 0.0004^2]^{1/2}$

$x_1 = 0.083$ ft = 1.0 in

$x_2 = [0.0590^2 + 0.0094^2 + (-0.0009)^2]^{1/2}$

$x_2 = 0.0597$ ft = 0.72 in

$x_3 = [0.0277^2 + 0.0186^2 + 0.0005^2]^{1/2}$

$x_3 = 0.033$ ft = 0.40 in

d) The interstory drifts are

Base-to-Level 1: $D_{01} = 0.40$ in

Level 1-to-Level 2: $D_{12} = 0.72 - 0.40 = 0.32$ in

Level 2-to-Level 3: $D_{23} = 1.0 - 0.72 = 0.28$ in

Structural-42

a) Determine the natural frequencies and mode shapes for the two-story steel frame below. Idealize as a shear building (i.e., infinitely rigid girders; masses lumped at story levels; bending deformation only).

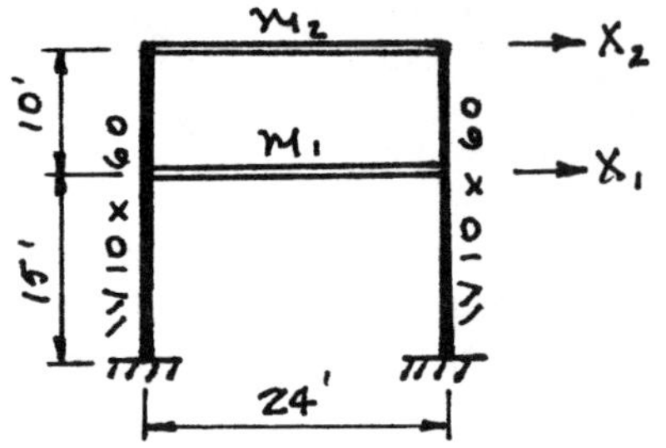

W10 x 60:

$I_x = 341\ \text{in}^4$

$E = 30 \times 10^6$ psi

$W_1 = 2000\ ^{lb}/_{ft} \times 24' + 60\ ^{lb}/_{ft}(7.5 + 5)^1 \times 2$

$= 49{,}500$ lb

$\therefore\quad m_1 = 49{,}500 / (32.2 \times 12) = 128\ \text{lb sec}^2 / \text{in}$

$W_2 = 1500 \times 24 + 2 \times 60 \times 5 = 36{,}600$ lb

$m_2 = 36{,}600 / (32.2 \times 12) = 95\ \text{lb sec}^2 / \text{in}$

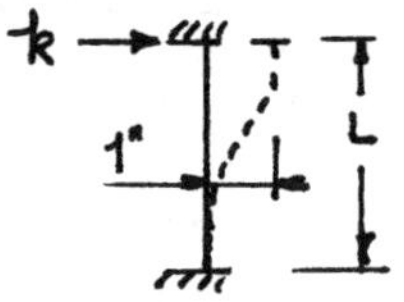

$k_1 = 12EI / L^3$

$= 12E \times 341 / (180)^3$

$= 21{,}050$ lb / in

$k_2 = 12EI / L^3$

$= 12E \times 341 / (120)^3$

$= 71{,}040$ lb / in

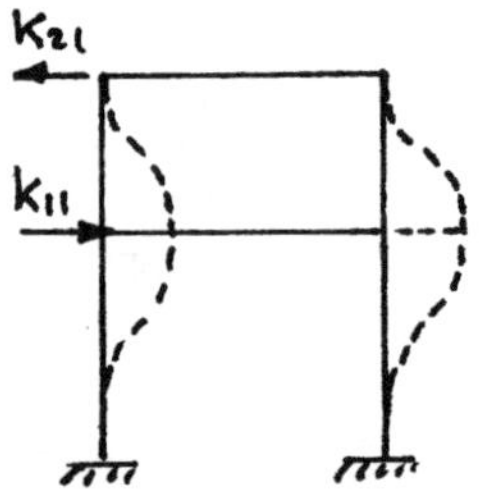

$k_{11} = (k_1 + k_2)2$

$= 184{,}183$ lb / in

$k_{21} = -k_2$

$= -71{,}040$ lb / in

$k_{12} = k_{21} = -71{,}040$ lb / in

$k_{22} = k_2 = 71{,}040$ lb / in

Damping has negligible effect on natural frequencies and mode shapes. The equations of motion can be written (Paz 1988):

$$[M]\,\{\ddot{x}\} + \{k\}\,\{x\} = \{0\}$$

The solution for the lateral translation, x_i, will be of the form $x_i = a_i \sin(\omega t - \alpha)$. Substituting and simplifying,

$$[[K] - \omega^2[M]]\,\{a\} = \{0\}$$

For a nontrivial solution, the determinant of the coefficient matrix is set to zero.

$$\left|[[K] - \omega^2[M]]\right| = 0$$

Thus, $\begin{vmatrix} (184{,}183 - 128\omega^2) & -142{,}080 \\ -142{,}080 & (142{,}080 - 95\omega^2) \end{vmatrix} = 0$

Expanding the determinant yields the characteristic equation:

$$(184{,}183 - 128\omega^2)(142{,}080 - 95\omega^2) - (142{,}080)^2 = 0$$

$$\therefore\quad \omega^4 - 2934.2\omega^2 + 4.919 \times 10^5 = 0$$

Solving for ω^2:

$\omega^2 = (178.6;\ 2755)$

$\therefore\quad \underline{\omega_1 = 13.34\ \text{rad / sec} = 2.1\ \text{cps}}$

$\underline{\omega_2 = 52.50\ \text{rad / sec} = 8.4\ \text{cps}}$

To obtain the mode shapes, let $a_1 = 1.0$, substitute ω_1 and ω_2 and solve. Thus,

<u>For ω_1 = 13.34 rad / sec; a_{21} = 1.136</u>

<u>For ω_2 = 52.50 rad / sec; a_{22} = –0.842</u>

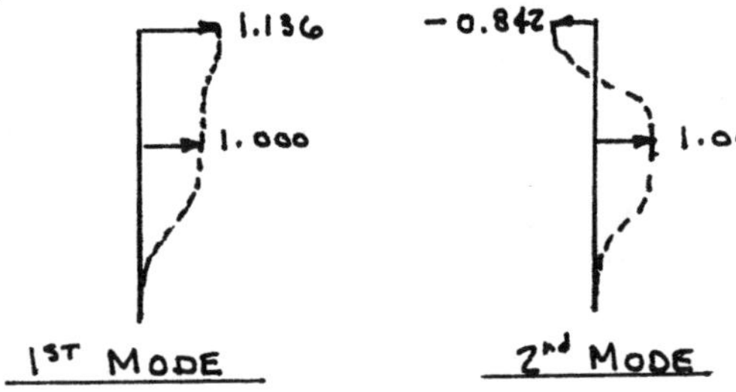

b) The acceleration response spectrum for 7% critical damping is the lowest of the three curves. For the first mode, $f_1 = 2.1$ cps; $T_1 = 1 / f_1 = 0.49$ - say 0.5 sec. Thus,

<u>$S_a = 0.28g = 0.28 \times 32.2 = 9.0\ \text{ft / sec}^2$</u>

The spectral pseudovelocity is related directly to S_a (Paz 1988)

<u>$S_v = S_a / \omega = 9.0 / 13.34 = 0.7$ ft / sec.</u>

Structural-43

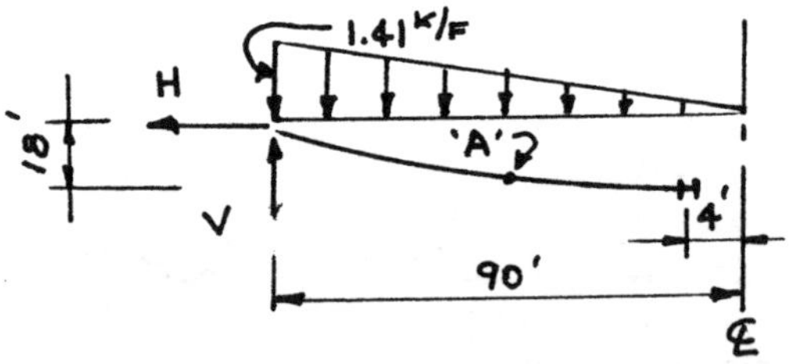

a) Horizontal component of cable tension at point "A". Consider typical strand as FBD:

$$\sum F_y = 0{:}\ V - 0.5 \times 1.41 \times 90 = 0$$

$$\therefore\quad V = 63.45^k \uparrow$$

$$\sum M_e = 0{:}\ 63.45(90) - 63.45(60) - H(18) = 0$$

$\therefore\quad$ **<u>$H = 105.75^k \leftarrow$</u>**

Now, cut FBD through "A"

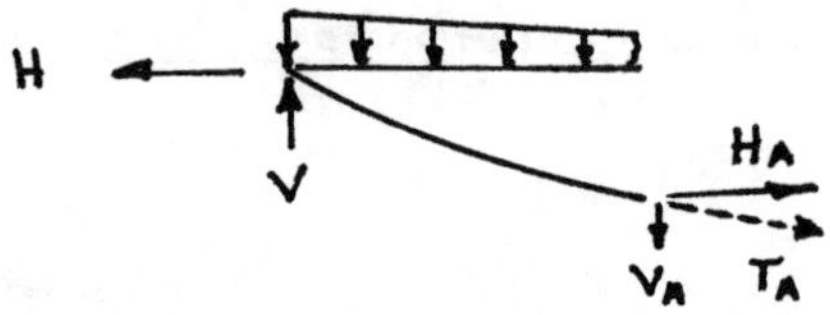

Clearly, at any point "A",

$\sum F_x = 0$ requires $H_a = -H$;

$\therefore$ **$H_a = 105.75^k \rightarrow$**

b) Verify W12×50 of A36 steel for tension ring: $A_g = 14.7$ in^2; $S_x = 64.7$ in^3 (ASD AISC 1989)

Maximum ring tension occurs midway between cable connection points (Roark 1976; table 17, case 7)

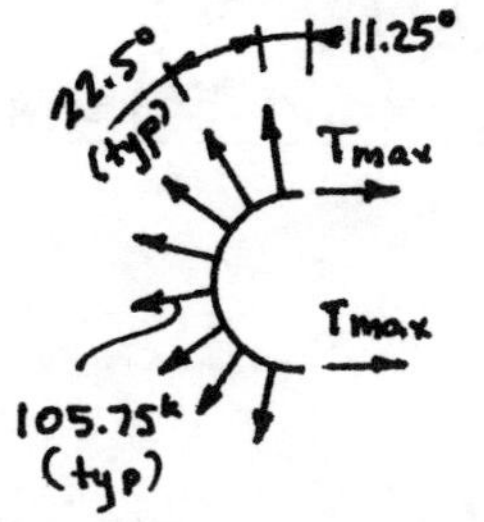

$T_{max} = 105.75(\cos 11.25° + \cos 33.75° + \cos 56.25° + \cos 78.75°)$

$\therefore$ $T_{max} = 271.0^k$

$M^+ = WR / 2(1 / \sin\theta - 1 / \theta)$

$\theta = 11.25° = 0.19635$ rad

$M^+ = 105.75 \times 4(1 / \sin 11.25° - 1 / 0.19635) / 2$
$= 7.0^{k\text{-}f}$

$M^- = WR / 2(1 / \theta - 1 / \tan\theta)$
$= 105.75 \times 4(1 / 0.19635 - 1 / \tan 11.25°) / 2$
$= 13.9^{k\text{-}f}$

Negligable weak axis bending. Assume that cable-to-ring connection will be detailed such that full area of W12×50 resists tension. Use allowable stress design (AISC 1989)

Combined tension and bending:

$f_t / F_t + f_{bx} / F_{bx} \le 1.0$

$(271 / 14.7) / 22 + (13.9 \times 12 / 64.7) / 24$
$= 0.84 + 0.11 = 0.95 < 1.0$

$\therefore$ **W12×50 is o.k.**

c) Check W33×221 compression

Ring: $A_g = 65.0$ in^2; $r_x = 14.1$ in; $I_x = 12{,}800$ in^4;

$S_x = 757$ in^3; $d / t_x = 43.8 \approx 257 / \sqrt{F_y}$ $\therefore$ o.k.

$C_{max} = T_{max} = 271.0^k$

$M_x^+ = WR / 2(1 / \sin\theta - 1 / \theta)$
$= 105.75 \times 90(1 / \sin 11.25° - 1 / 0.19635) / 2$
$= 156.0^{k\text{-}f}$

$M_x^- = 105.75 \times 90(1 / 0.19635 - 1 / \tan 11.25°) / 2$
$= 313^{k\text{-}f}$

Ring buckling: Calculate the equivalent radial pressure and use available solution (Roark 1976; table 34, case 8)

$P' = 3EI / R^3 = 3 \times 29{,}000 \times 12{,}800 / (90 \times 12)^3$
$= 0.88$ k / in

$\therefore$ $P' = p'R = 0.88 \times (90 \times 12) = 950^k$

Equivalent effective length from Euler equation:

$P_e = \pi^2 EI / (KL)^2 = P' = 950^k$

$\therefore$ $kL_x = [(950)^{-1}\pi^2 \times 29{,}000 \times 12{,}800]^{1/2}$
$= 1960$ in $= 163$ ft

$\therefore$ $(kL / r)_x = 1960 / 14.1 = 139 > C_c$

$\therefore$ $F_a = 7.73$ ksi $= F_{ex}'$

Axial only: $F_a = 271 / 65.0 = 4.17$ ksi

$f_a / F_a = 4.17 / 7.73 = 0.54 > 0.15$ $\therefore$ check interaction with moment magnification

Bending only: Compact section $\therefore$ $F_{bx} = 24$ ksi

$f_{bx} = 313 \times 12 / 757 = 5.0$ ksi; $C_{mx} = 1.0$

Combined axial and bending:

$f_a / F_a + C_{mx}F_{bx} / (1 - f_a / F_{ex}')F_{bx} \le 1.0$

$0.54 + 1.0(5.0) / [(1 - 0.59)(24)] = 0.99 < 1.0$

$f_a / .6F_y + f_{bx} \le 1.0$

$4.17 / 22 + 5.0 / 24 = 0.40 < 1.0$

$\therefore$ **W33×221 is o.k.**

Note: in the above solution, the flexural stresses are computed by the simple flexure formula $f_b = M / S$, which neglects curvature in the rings. This is consistent with an accepted design reference (Scalzi, et. al. 1969).

Structural-44

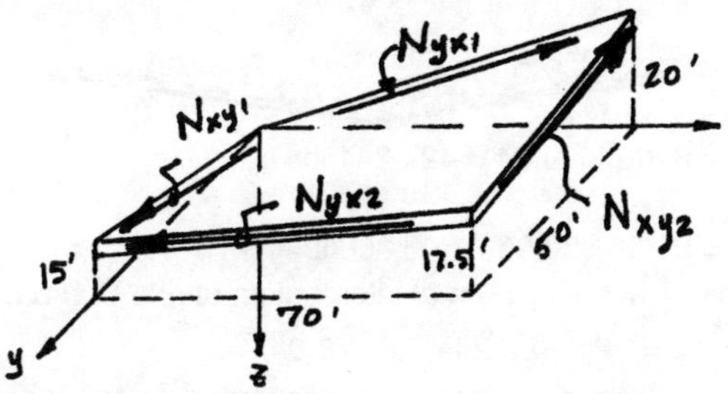

$L_{y1} = (15^2 + 50^2)^{1/2} = 52.2$ ft; $C_{xy1} = 15$ ft

$L_{x1} = (20^2 + 70^2)^{1/2} = 72.8$ ft; $C_{yx1} = 20$ ft

$L_{y2} = (2.5^2 + 70^2)^{1/2} = 70.04$ ft; $C_{xy2} = -2.5$ ft

$L_{y2} = (2.5^2 + 50^2)^{1/2} = 50.06$ ft; $C_{xy2} = +2.5$ ft

Follow usual practice of expressing gravity force as an equivalent uniformly distributed load in psf of horizontal projection (include the rib weight):

$W_{rib} = (110 / 144) \times 20 \times 17 = 260$ plf

$W_d = 3"$ lt. wt. conc. + ceiling + roof + rib
$= (3 / 12) \times 110 + 3 + 3 + [260 \times 740 / (1000 \times 140)]$
$= 47.2$ psf

service dead + live over the horizontal projection:

$P_e = 47.2 \times (72.8 + 70.04)(52.2 + 50.06) / (4 \times 70 \times 50)$
$+ 12 = 61$ psf

equilibrium of the membrane:

$\sum F_x = 0$:

$N_{yx1}(72.8)(70 / 72.8) - N_{yx2}(70.04)(70 / 70.04) = 0$

$\therefore$ $N_{yx1} = N_{yx2}$

Similarly, $\sum F_y = 0$:

$N_{xy1} = N_{xy2}$

$\sum M_{z\text{-axis}} = 0$: $N_{yx2}(70.04)(70 / 70.04) \times 50$
$+ N_{xy2}(50.06)(50 / 50.06) \times 70 = 0$

$\therefore$ $N_{yx2} = N_{xy2} = N_{xy}$

Thus,

$\sum F_y = 0$:

$61 \times 50 \times 70 - N_{xy}[(52.2 \times (15 / 52.2) + 72.8 \times (20 / 72.8)$
$+ 50.06 \times (2.5 / 50.06) - 70.04 \times (2.5 / 70.04)] = 0$

$\therefore$ **$N_{xy} = 6100$ plf**

For Rib A:

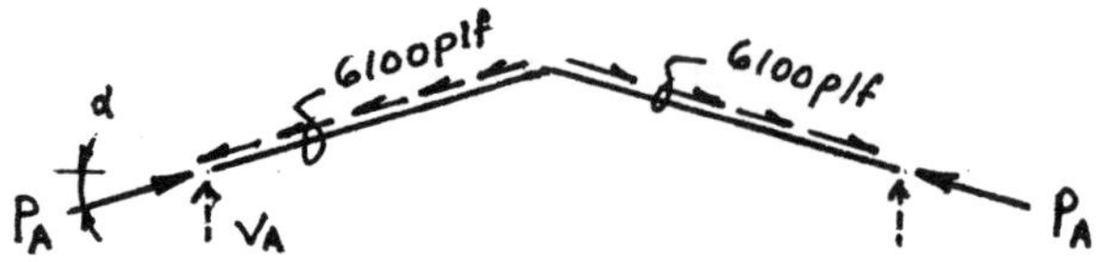

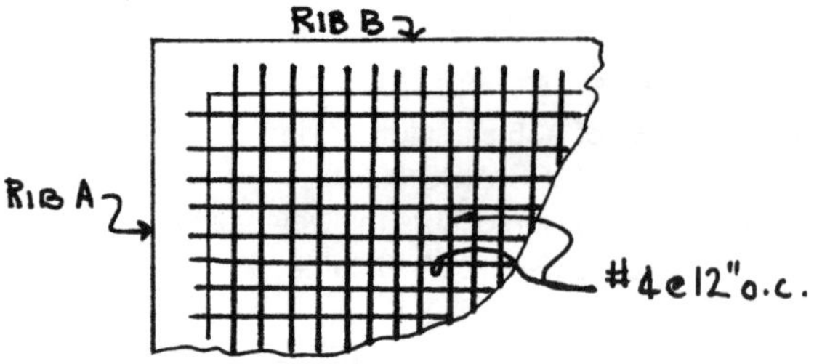

PARTIAL ROOF FRAMING PLAN

a) $P_a = 6100 \times 52.2 = 318{,}420$ lb

$V_a = P_a \sin\alpha = 318{,}420(15 / 52.2) = 91.5^k \uparrow$

Rib B:

b) $P_b = 6100 \times 72.8 = 444{,}080$ lb

$V_b = 444{,}080 \times 20 / 72.8 = 122.0^k \uparrow$

Total vertical reaction at corner:

$V_{tot} = V_a + V_b = 213.5^k$

Check vertical equilibrium of the entire roof:

f) $\sum F_z = 61 \times 10 \times 140 - 4 \times 213{,}500 = 0$

Thus, the vertical forces are transferred to the corners; there is no need for a center column.

Rib D:

$P_d = 6100 \times 70.04 = 427{,}244$ lb

Rib C:

$P_c = 6100 \times 50.06 = 305{,}366$ lb

Check vertical equilibrium at intersection of ribs C and D:

$$\sum F_z = -2 \times 427{,}244(2.5 / 70.04) + 2 \times 305{,}366(2.5 / 50.06) = 0$$

$\therefore$ **Equilibrium is satisfied; the roof is stable.**

e) Membrane reinforcement–follow section 19.4 ACI code (ACI 318). Use the alternate design method with $f_s = 24{,}000$ psi (grade 60 rebars):

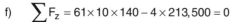

Principal stresses at 45° to x, y axes with same magnitude

$f_c = 6100 / (3 \times 12) = 169$ psi – ok

$A_s \geq 6100 / 24{,}000 = 0.25\ \text{in}^2/\text{ft}$

$A_{s,\,min} = .0018 \times 12 \times 3 = 0.065\ \text{in}^2/\text{ft}$

Check ductility (section 19.4.4):

$N_{max,\,c} = 0.85 \times 3 \times 12 \times 3 = 91.8^k$

$N_{max,\,t} = 0.25 \times 60{,}000 = 15.0^k \ll n_t$

Steel will yield prior to crushing of concrete.

The most efficient pattern to place reinforcement is at 45° with respect to ribs (i.e., parallel to principal stresses). However, since the principal stresses are relatively low (169 psi), a simpler pattern running parallel to ribs will satisfy stress and minimum steel requirements:

$A_s = 0.25\ \text{in}^2/\text{ft} = A_{st}(\cos^2 45° + \sin^2 45°)$

$\therefore$ **Use #4 @ 9" o.c. e.w.**

Concrete-1

Assume that the given shears and moments are the design (i.e., factored) values. Also, assume that the shear specified at point "B" has been reduced, if applicable, to the value at d-distance from the face of support.

Given: $f'_c = 3$ ksi; $f_y = 40$ ksi; $d = 22 - 2.5 = 19.5$ in; lightweight concrete; $f_{ct} = 0.274$ ksi

a) Design flexural steel at midspan: $b = b_{eff} = 54$ in; Use the strength design method (ACI 318); $b_w = 16$ in; $h_s = 6$ in. Assume, initially, a rectangular beam:

$$\varphi M_n = \varphi \rho b d^2 f_y\{1 - 0.59\rho(f_y / f'_c)\} \geq M_u$$

$$M_u = 1.4\ M_d + 1.7 M_l = 140 + 70 = 210^{k\text{-}ft}$$

$$0.9 \times 54 \times 19.5^2 \times 40\rho\{1 - 0.59\rho(40 / 3)\} = 210 \times 12$$

$\therefore$ $\rho = 0.0035$

$$A_s = 0.0035 \times 54 \times 19.5 = 3.69\ \text{in}^2$$

Check the assumption of a rectangular beam:

$$a = A_s f_y / (0.85 f'_c b) = 3.69 \times 40 / (0.85 \times 3 \times 54)$$
$$= 1.07\ \text{in} < h_s$$

$\therefore$ the assumption is valid – check $A_{s,min}$:

$$A_{s,min} = (200 / f_y) b_w d = (200 / 40{,}000) \times 16 \times 19.5$$
$$= 1.56\ \text{in}^2 < 3.69\ \text{in}^2$$

$\therefore$ **Use $A_s = 3.69$ in^2; say 3 # 10**

Design flexural steel at the face of support. Here, the bottom is in compression $\therefore$ $b = b_w = 16$ in.

$$M_u = 200 + 100 = 300^{k\text{-}ft}$$

$$0.9 \times 16 \times 19.5^2 \times 40\rho\{1 - 0.59\rho(40 / 3)\} = 300 \times 12$$

$$\rho = 0.0193 > \rho_{min} = 0.005; < \rho_{max}$$

$\therefore$ Use $A_s = 0.0193 \times 16 \times 19.5 = 6.03$ in^2

Say 8 # 8; 4 in stem and 2 in each flange

b) Determine the stirrup requirements on an in^2 / ft basis (Re: Chapter 11, ACI 318):

–For the midspan (i.e., Point "A"): $V_u = 9^k$

$$fct / 6.7 = 274 / 6.7 = 41\ \text{psi} < \sqrt{f_c'}$$

$$\therefore V_c = 2(fct / 6.7) b_w d = 2 \times 41 \times 16 \times 19.5$$
$$= 25{,}600\ \text{lb} = 25.6^k$$

Stirrups are not required at point "A"

– At Point "B": $V_u = 36 + 18 = 54^k$

$$\phi V_c = 0.85 \times 25.6 = 21.8^k \prec V_u$$

$$V_s = V_u / \phi - V_c = 54 / 0.85 - 25.6 = 37.9^k$$

For the design basis, let s = 12"

$$A_v = V_s(s) / f_y d = 37.9 \times 12 / (40 \times 19.5)$$
$$= 0.58^{sq.in} / _{ft}$$

Minimum stirrups (Re: Eqn 11-4, ACI 318):

$$A_v \geq 50 b_w(s) / f_y = 50 \times 16 \times 12 / 40 = 0.24^{sq.in} / _{ft}$$

$\therefore$ **Strength governs, use $A_v = 0.58\ ^{sq.in}/_{ft}$ with spacing not to exceed d / 2 = 9" o.c. (e.g., # 3 U-stirrups at 4-1/2" o.c.)**

Concrete-2

a) Two possible effects of early shore removal in reinforced concrete are: 1) Increased deflection (E_c lower than intended) and increased crack width (larger than intended curvature). To establish a schedule for shore removal, it is necessary to set a minimum compressive strength at which the structure is satisfactorily self supporting. Proper curing (moisture retention and temperature control) must be maintained to ensure adequate strength gain. Field-cured test specimens provide a more dependable gauge of in-place strength than does the usual practice of specifying age at removal.

b) Plastic shrinkage cracking is caused by rapid evaporation of water near the slab surface. Cracking is particularly severe during hot weather, low relative humidity and windy conditions. Some methods to alleviate the problem are (PCA 1988):

1. Moisten subgrade & forms
2. Moisten aggregates that are dry
3. Erect temporary wind breaks
4. Erect temporary sunshades
5. Protect surface with temporary covering (e.g., plastic sheet)
6. Protect surface immediately after finishing (e.g., fog spray, wet burlap or sawdust, or other curing method.

c) Admixtures for concrete (PCA 1988):

1. Air entrainment is used primarily to increase the resistance of hardened concrete to cycles of freeze-thaw. The entrainment also improves workability. The primary disadvantage of an air-entraining admixture is that the compressive strength of the mix is lowered. Precautions include testing air content of fresh concrete to ensure the proper air content is obtained through batching and mixing.
2. Retarders are used to delay the set of fresh concrete. This admixture might be needed to compensate for hot weather or to avoid cold joints in pours of massive elements. The disadvantage to the use of retarders is that strength gain is slowed, which might be cause for delays in subsequent construction phases. Also, the amount of shrinkage in the hardened concrete is less predictable.
3. Accelerators (usually calcium chloride) are used to accelerate strength gain. Precaution in using calcium chloride is that it should be added to the mix water, dissolved, then added to the batch. This is to prevent high local concentrations that might occur if flakes are added dry to the batch. Primary disadvantage to calcium chloride is that the CL^- ion speeds corrosion of metal in the hardened concrete.

4. Water reducers (superplasticizers) are added to improve workability without sacrificing strength. For normal strength ranges (3000-to-5000 psi), the admixture is generally used to produce high slump, cohesive concrete that is flowable. In the normal slump range (1-to-3 in), the water reducer is usually added to produce high strength concrete. Precautions to use include awareness that the effect of the admixture is short lived (typically about 30 min). Primary disadvantage is cost.

Concrete-3

a) Determine the required slab thickness for the floor system below.

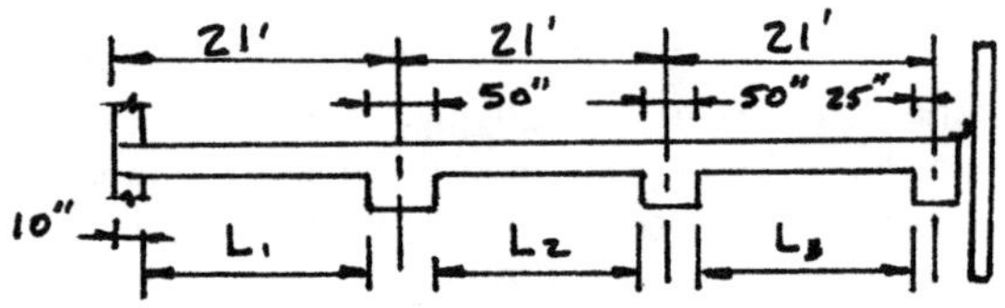

Use the strength design method (ACI 318); $f'_c = 4$ ksi; $f_y = 40$ ksi. The slab is integral with the 10" wall at the left end; integral with the spandrel beam at the right end. For Section 8.3, ACI 318, the clearspans are:

$L_1 = 21 - (10/12) - (25/12) = 18.1'$

$L_2 = 21 - (50/12) = 16.8'$

$L_3 = 21 - (25/12) - (25/12) = 16.8'$

Assume that fire rating does not constrain slab thickness. For serviceability, Table 9.5(a) requires

$$h \geq \begin{cases} 18.1 \times 12/28 = 7.75\text{" (both ends cont.)} \\ 16.8 \times 12/24 = 8.4\text{" (one end cont.)} \end{cases}$$

Note: A footnote to Table 9.5(a) permits a slightly thinner slab because the steel is grade 40 rather than grade 60. However, the problem will be worked using an 8.5" slab.

Try an 8.5" slab – check shear:

$w_u = 1.4w_d + 1.7w_l = 1.4 \times 8.5 \times 12.5 + 1.7 \times 250 = 574$ psf

Design on a per foot width basis $\therefore$ b = 1' = 12";

$w_u = 574$plf = 0.574 klf

Allowing 3/4" cover and 1/2" additional to centroid:

d = 8.5 – 1.25 = 7.25".

$$V_u \geq \begin{cases} w_u L_1/2 - w_u d = 0.58(18.1/2 - 7.25/12) \\ \quad = 4.9^k \\ 1.15 w_u L_3/2 - w_u d = 0.58(1.15 \times 16.8/2 \\ \quad - 7.25/12) = 5.3^k \leftarrow \text{governs} \end{cases}$$

$$\phi V_c = 2\phi\sqrt{f'_c}\, b_w d = 2 \times \sqrt{4000} \times 12 \times 7.25$$
$$= 9{,}350\# = 9.4^k$$

$\therefore$ shear strength is more than adequate – **use a solid 8.5" one-way slab.**

b) Maximum slab moments (8.3. ACI 318):

$M^-_{u1L} = w_u l_n^2/16 = 0.58 \times 18.1^2/16 = 11.9^{k\text{-}ft}/_{ft}$

$M^+_{u1} = w_u l_n^2/14 = 0.58 \times 18.1^2/14 = 13.6\ ^{k\text{-}ft}/_{ft}$

$M^-_{u1R} = w_u l_n^2/10 = 0.58 \times \{(18.1 + 16.8)/2\}^2/10 = 17.7\ ^{k\text{-}ft}/_{ft}$

$M^+_{u2} = w_u l_n^2/16 = 0.58 \times 16.8^2/16 = 10.2\ ^{k\text{-}ft}/_{ft}$

$M^-_{u3L} = w_u l_n^2/10 = 0.58 \times 16.8^2/10 = 16.4\ ^{k\text{-}ft}/_{ft}$

$M^+_{u3} = w_u l_n^2/11 = 0.58 \times 16.8^2/11 = 14.9\ ^{k\text{-}ft}/_{ft}$

$M^-_{u1R} = w_u l_n^2/24 = 0.58 \times 16.8^2/24 = 6.8\ ^{k\text{-}ft}/_{ft}$

c) Design of slab reinforcement:

$A_{s,min} = 0.002bh = 0.002 \times 12 \times 8.5 = 0.20^{sq.in}/_{ft}$

For flexural steel ($f_y = 40$ ksi; $f'_c = 4$ ksi)

$\varphi M_n = \varphi\rho b d^2 f_y\{1 - 0.59\rho(f_y/f'_c)\} \geq M_u$

For $M_u = 11.9\ ^{k\text{-}ft}/_{ft}$:

$0.9 \times 12 \times 7.25^2 \times 40\rho\{1 - 0.59\rho(40/4)\} = 11.9 \times 12$

$\therefore$ $\rho = 0.007$

$\therefore$ $A_s = 0.007 \times 12 \times 7.25 = 0.59^{sq.in}/_{ft}$

Use # 5 @ 6" o.c.

Similarly, for $M_u = 13.6\ ^{k\text{-}ft}/_{ft}$:

$A_s = 0.68^{sq.in}/_{ft}$; **Use # 5 @ 5.5" o.c.**

$M_u = 17.7\ ^{k\text{-}ft}/_{ft}$:

$A_s = 0.90^{sq.in}/_{ft}$; **Use # 6 @ 6" o.c.**

$M_u = 10.2\ ^{k\text{-}ft}/_{ft}$:

$A_s = 0.50^{sq.in}/_{ft}$; **Use # 5 @ 7.5" o.c.**

$M_u = 16.4\ ^{k\text{-}ft}/_{ft}$:

$A_s = 0.83^{sq.in}/_{ft}$; **Use # 6 @ 6" o.c.**

$M_u = 14.9\ ^{k\text{-}ft}/_{ft}$:

$A_s = 0.75^{sq.in}/_{ft}$; **Use # 5 @ 5" o.c.**

$M_u = 6.8\ ^{k\text{-}ft}/_{ft}$:

$A_s = 0.34^{sq.in}/_{ft}$; **Use # 4 @ 6" o.c.**

Locations for bar cutoffs (Winter, et.al., 1963, Graph 11):

For negative steel at the left end of span 1:
$L = 0.15l_n = 0.15 \times 18.1 = 2.7'$

For negative steel at the right end of span 1:
$L = 0.3l_n = 0.3 \times 18.1 = 5.5'$

For negative steel on interior spans: $L = 0.22l_n = 0.22 \times 16.8 = 3.7'$

For negative steel at the left end of span 3: $L = 0.3l_n = 0.3 \times 16.8 = 5.0'$

For negative steel at the right end of span 3:
$L = 0.1l_n = 0.1 \times 16.8 = 1.7'$

Positive steel: Extend alternate bars 6" into supports; terminate other bars at points of inflection. Thus for the exterior spans:

$L' = [M^+_u \times 8/w_u]^{1/2} = [14.9 \times 8/0.58]^{1/2} = 14.5'$

For the interior span:

$L' = [10.2 \times 8/0.58]^{1/2} = 12.0'$

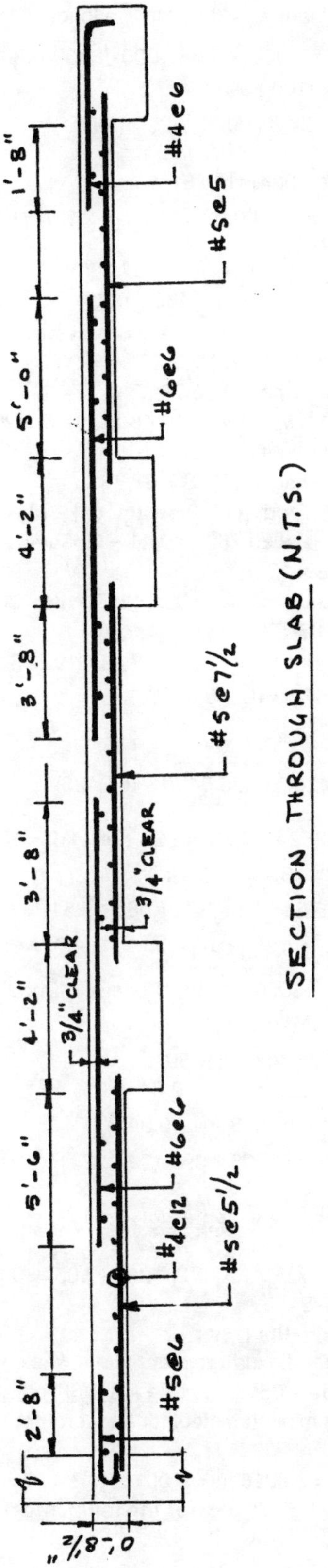

Concrete-4

a) Verify that the neutral axis is 6" below the top edge.

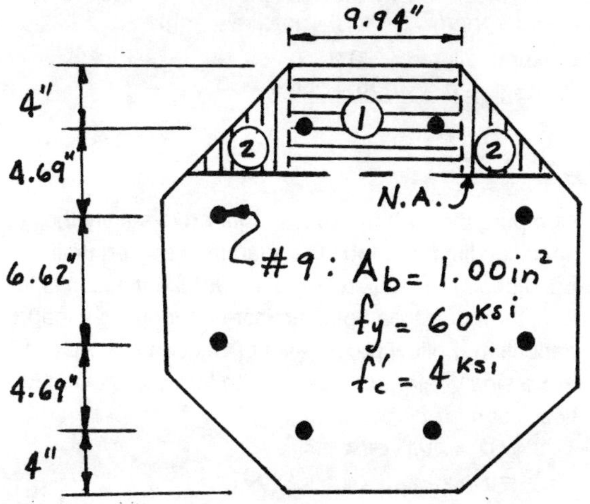

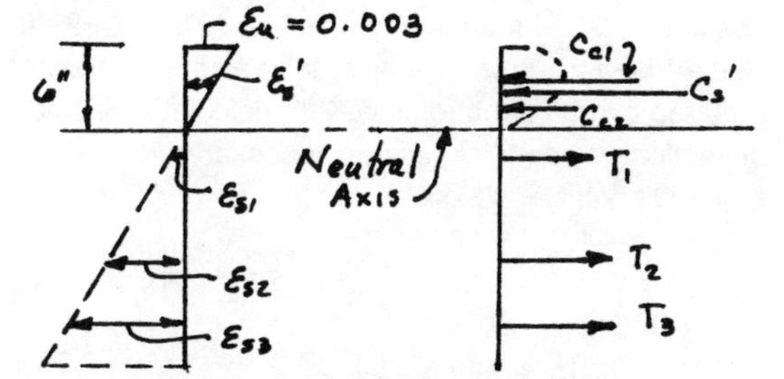

Strains vary linearly. Therefore, using similar triangles

$\varepsilon'_s / (6.0 - 4.0) = 0.003 / 6.0$

$\varepsilon'_s = 0.001 < \varepsilon_y = f_y / E_s = 60 / 29{,}000 = 0.00207$

$\therefore \quad f'_s = E_s\varepsilon'_s = 29{,}000 \times 0.001 = 29$ ksi

Similarly,

$\varepsilon_{s1} / (8.69 - 6.0) = 0.003/6.0$

$\varepsilon_{s1} = 0.00135 < \varepsilon_y$; $f'_{s1} = 29{,}000 \times 0.00135 = 39$ ksi

$\varepsilon_{s2} / (15.31 - 6.0) = 0.003 / 60$

$\varepsilon_{s2} = 0.0046 > \varepsilon_y$; $f'_{s2} = f_y = 60$ ksi

$\varepsilon_{s3} > \varepsilon_{s2} \therefore f'_{s3} = f_y = 60$ ksi

Break the compression zone into component rectangles and triangles:

$C_{c1} = 0.85f'_c b(\beta_1 c) = 0.85 \times 4.0 \times 9.94 \times 0.85 \times 6.0$
$= 172^k$

$C_{c2} = (0.85 \times 4.0 \times 0.5 \times 5.1 \times 0.85 \times 6.0) \times 2 = 88^k$

The equilibrium requirement for the proper position of the neutral axis is that the summation of forces along the longitudinal axis must be zero

$\therefore \quad \Sigma F_{long} = 0 = C_{c1} + C_{c2} - T_1 - T_2 - T_3 = 0$

$172 + 88 - 2.00\,(29 + 60 + 60) = 0$

Therefore, equilibrium is satisfied and the **given position of the neutral axis is confirmed**. (Note: the solution above ignores the 2 # 8 in the compression zone. A more precise solution that accounts for the area of concrete occupied by the 2 # 8 would require that the neutral axis be slightly lower than the given 6".)

b) Compute the moment strength of the section, φM_n:

$M_n = 172(20 - 0.85 \times 6.0 / 2) + 88(20 - (2/3) \times 0.85 \times 6.0) + 29 \times 2 \times (20 - 4) - 78(20 - 8.69) - 120(20 - 15.31) = 3944^{k\text{-}in} = 329^{k\text{-}ft}$

Flexural failure: $\varphi = 0.9$

$\therefore$ **$\varphi M_n = 0.9 \times 329 = 296^{k\text{-}ft}$**

Concrete-5

a) Compute the slab thickness required to satisfy the specified limit, L / 480. The system is shored until partitions are installed. Thus, the total deflection is the long term dead load plus an instantaneous live load deflection. Follow procedures of Section 9.5, ACI 318. For a singly reinforced beam with loads sustained 5 years or more:

$\lambda = \xi / (1 + 50\rho') = \xi = 2.0$

$\Delta_{max} = (1+\lambda)\Delta_{id} + \Delta_{il} \le L / 480$

L ≤ Clearspan + h ≤ centerline-to-centerline of bearing; assume a 12" wide masonry wall (h ≥ 12") ∴ L =29 ft. Use a trial-and-error approach: take an initial estimate of required overall depth of, say, 20"; calculate the total deflection; revise thickness as necessary. For h = 20"; $f'_c = 3000$ psi; $f_y = 40$ ksi:

$d \approx h - 1.25" = 18.75"$

$w_u = 1.4w_d + 1.7w_l = 1.4(20 \times 12.5 + 20) + 1.7 \times 80 = 515 \text{ plf} = 0.515 \text{ klf}$

$M_u = 0.515 \times 29^2 / 8 = 54.1^{k\text{-}ft}$

$\varphi M_n = \varphi\rho b d^2 f_y\{1 - 0.59\rho(f_y / f'_c)\} \ge M_u$

$0.9 \times 12 \times 18.75^2 \times 40\rho\{1 - 0.59\rho(40/3)\} = 54.1 \times 12$

$\therefore\ \rho = 0.0044;\ A_s = 0.0044 \times 12 \times 18.75 = 0.99^{sq.in}/_{ft}$

Compute I_e using Eqn 9-7, ACI 318:

$I_g = bh^3/12 = 12 \times 20^3 / 12 = 8000 \text{in}^4$

$\rho n = 0.0044 \times 10 = 0.044$

$k = \sqrt{(\rho n)^2 + 2(\rho n)} - \rho n$

$= \sqrt{(0.044)^2 + 2(0.044)} - (0.044) = 0.26$

$\therefore kd = 0.26 \times 18.75 = 4.9";\ (d - kd) = 14.0"$

$I_{cr} = b(kd)^3/3 + nA_s(d - kd)^2$

$= 12 \times 4.9^3 / 3 + 10 \times 0.99 \times 14.0^2 = 2410 \text{in}^2$

$M_{cr} = 7.5\sqrt{f'_c} I_g / (h/2)$

$= [7.5\sqrt{3000}(8000)/(20/2)]/12000$

$= 27.4^{k-ft}$

The total service load moment (i.e. live plus dead load) is

$M_a = wL^2/8 = 0.35 \times 29^2 / 8 = 36.8^{k-ft}$

$M_{cr}/M_a = 27.4/36.8 = 0.74$

$I_e = (M_{cr}/M_a)^3 I_g + [1 - (M_{cr}/M_a)^3] I_{cr} \le I_g$

$= 0.74^3 \times 8000 + [1 - 0.74^3] \times 2410 = 4675 \text{ in}^4$

$w = w_d + w_l = 20 \times 12.5 + 20 + 80 = 350 \text{ plf} = 0.35 \text{ klf}$

For dead load, $w_d = 0.27$ klf; live load, $w_l = 0.08$ klf

$E_c = 57{,}000\sqrt{f'_c} = 57{,}000\sqrt{3000} = 3.1 \times 10^6 \text{ psi} = 3100 \text{ ksi}$

$\Delta_d = (5/384)0.27 \times 29 \times (29 \times 12)^3 / (3100 \times 4675) = 0.30"$

$\Delta = (5/384)wL^4 / EI$

$\Delta_l = (5/384)0.08 \times 29 \times (29 \times 12)^3 / (3100 \times 4675) = 0.09"$

$\Delta_{max} = (1+\lambda)\Delta_d + \Delta_l = 3(0.30) + 0.09" = 0.99" \succ L/480 = 0.73"$

Therefore, a 20" slab is inadequate. Need a stiffer slab – try $h = 20(0.99/0.73)^{1/3} = 22"$; d = 20.75. Repeat the calculations with d = 20.75"

$w_u = 1.4w_d + 1.7w_l = 1.4(22 \times 12.5 + 20) + 1.7 \times 80 = 550 \text{ plf}$

$M_u = 0.550 \times 29^2 / 8 = 57.8^{k\text{-}ft}$

$\varphi M_n = \varphi\rho b d^2 f_y\{1 - 0.59\rho(f_y / f'_c)\} \ge M_u$

$0.9 \times 12 \times 20.75^2 \times 40\rho\{1 - 0.59\rho(40/3)\} = 57.8 \times 12$

$\therefore\ \rho = 0.004257;\ A_s = 0.004257 \times 12 \times 20.75 = 1.06^{sq.in}/_{ft}$

$I_g = bh^3/12 = 12 \times 22^3/12 = 10{,}648 \text{ in}^4$

$\rho n = 0.00426 \times 10 = 0.0426$

$k = \sqrt{(\rho n)^2 + 2(\rho n)} - \rho n$

$= \sqrt{(0.0426)^2 + 2(0.0426} - (0.0426)$

$= 0.252$

$\therefore kd = 0.252 \times 20.75 = 5.24";\ (d - kd) = 15.51"$

$I_{cr} = b(kd)^3/3 + nA_s(d - kd)^2$

$= 12 \times 5.24^3/3 + 10 \times 1.06 \times 15.51^2 = 3125 \text{in}^2$

$M_{cr} = 7.5\sqrt{f'_c} I_g / (h/2)$

$= [7.5\sqrt{3000}(10648)/(22/2)]/12000$

$= 33.1^{k-ft}$

$M_a = 0.375 \times 29^2/8 = 39.4^{k-ft}$

$M_{cr}/M_a = 33.1/39.4 = 0.84$

$I_e = 0.84^3 \times 10{,}648 + [1 - 0.84^3] \times 3125 = 7580 \text{in}^4$

$\Delta_d = (5/384)0.295 \times 29 \times (29 \times 12)^3 / (3100 \times 7580) = 0.20"$

$\Delta_l = (5/384)0.08 \times 29 \times (29 \times 12)^3 / (3100 \times 7580) = 0.05"$

$\Delta_{max} = (1+\lambda)\Delta_d + \Delta_l = 3(0.20) + 0.05" = 0.65" \prec L/480 = 0.73"$

$\therefore$ **Use 22" thick slab**

b) The required flexural reinforcement was computed above to be $1.06^{sq.in}/_{ft}$ **∴ use # 8 @ 9" o.c. for the main steel**. The minimum steel for temperature and shrinkage is

$A_{s,min} = 0.0018bh = 0.0018 \times 12 \times 22 = 0.53^{sq.in}/_{ft}$

$\therefore$ **use # 5 @ 7" o.c. for the temp. steel.**

Concrete-6

Use the strength design method (ACI 318) to find the service live load capacity, w_l. Given f'_c = 3 ksi; f_y = 40 ksi.; normal weight concrete; d = 33.5"; b_w = 16"; w_d = 1.5 klf. Since the longitudinal steel is not given, use the simple equation for V_c.

$$V_c = 2\sqrt{f'_c}\,b_w d = 2\sqrt{3000} \times 16 \times 33.5 = 58{,}700 \text{ lb}$$
$$= 58.7^k$$
$$V_s = A_v f_y d / s$$

For # 4 U-stirrups, $A_v = 2 \times 0.20 = 0.40$ in^2

For s = 8" o.c.:

$$V_s = 0.40 \times 40 \times 33.5 / 8 = 67^k$$

For s = 12" o.c.:

$$V_s = 0.40 \times 40 \times 33.5 / 12 = 44.6^k$$

For s = 18" o.c.:

$$V_s = 0.40 \times 40 \times 33.5 / 18 = 29.8^k$$

Strength requirement:

$$V_u \le \phi(V_s + V_c);\ \phi = 0.85$$

From statics (simply supported, uniformly loaded beam):

$$V_u = w_u(L/2 - x) = w_u(15 - x)$$

The critical shear where s = 8" o.c. occurs at d-distance from the face of support

x = (8 + 33.5) / 12 = 3.46' from centerline of support

∴ $w_u \le 0.85(58.7 + 67) / (15 - 3.46) = 9.26$ klf

At the location where stirrup spacing changes to 12" o.c., x = (8 + 4 + 6 × 8) / 12 = 5'

$$w_u \le 0.85(58.7 + 44.6) / (15 - 5) = 8.8 \text{ klf}$$

At the location where stirrup spacing changes to 18" o.c., x = 5' + 4' = 9'

$$w_u \le 0.85(58.7 + 29.8) / (15 - 9) = 12.5 \text{ klf}$$

Note: in the strict sense, the given 18" stirrup spacing exceeds the current ACI limit of d / 2 = 16.75" and the stirrups are not fully effective in this region. In this case, V_s should be taken as 0 and the shear strength would be $\varphi V_c/2 = 25^k$. On this basis

$$w_u \le 25 / (15 - 9) = 4.2 \text{ klf}$$
$$w_u = 1.4w_d + 1.7w_l \le w_u = 4.2 \text{ klf} = 1.4(1.5) + 1.7w_l$$

∴ **$w_l \le (4.2 - 2.1) / 1.7 = 1.2$ klf**

Concrete-7

a) Plot the interaction diagram for the given column.

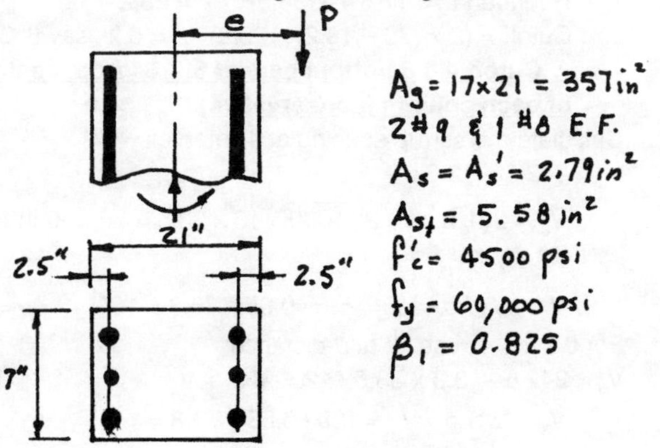

- For the concentric load case (M = 0):

$$P_o = 0.85f'_c(A_g - A_{st}) + f_y A_{st}$$
$$= 0.85 \times 4.5(357 - 5.58) + 60 \times 5.58$$

$P_o = 1679^k$

- For pure bending (P = 0):
  Iterate to find the neutral axis position: Trial 1.
  Assume $f'_s = 0$

  a = 60 × 2.79 / (0.85 × 4.5 × 17) = 2.57"

  x = a / β_1 = 2.57 / 0.825 = 3.12"

  ε'_s = [(3.12 – 2.5) / 3.12]0.003 = 0.0006

  $f'_s = E_s\varepsilon'_s$ = 29,000 × 0.0006 = 17.4 ksi

Trial 2. Assume f'_s = 10 ksi $C_c + C'_s = T$

  a = 2.79[60 – (10 – 0.85 × 4.5)] / (0.85 × 4.5 × 17) = 2.31"

  x = a / β_1 = 2.31/ 0.825 = 2.80"

  ε'_s = [(2.80 – 2.5) / 3.12]0.003 = 0.00032

  $f'_s = E_s\varepsilon'_s$ = 29,000 × 0.00032 = 9.4 ksi ∴ close enough: f'_s = 10 ksi

  $M_n = C_c(d - a/2) + C'_s(d - d')$ = 0.85 × 4.5 × 0.825 × 2.80 × 17(18.5 – 1.16) + 2.79(10 – 0.85 × 4.5)(18.5 – 2.5)

∴ **$M_n = 2878^{k\text{-in}} = 239^{k\text{-ft}}$**

- Balanced conditions: $\varepsilon_y = f_y / E_s$ = 60 / 29,000 = 0.00207

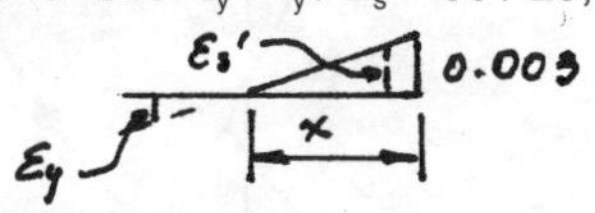

By similar triangles

  x = (0.003 / 0.00507)18.5 = 10.93"

  ε'_s = ((10.93 – 2.50) / 10.93)0.003 = 0.0023

∴ $\varepsilon'_s > \varepsilon_y$ ∴ $f'_s = f_y$ = 60 ksi

  T = 60 × 2.79 = 167.4^k

  $C_c = 0.85f'_c b(\beta_1 x)$ = 0.85 × 4.5 × 17 × 0.825 × 10.93 = 586^k

  $C'_s = (f'_s - 0.85 \times f'_c)A'_s$ = (60 – 0.85 × 4.5)(2.79) = 156.7^k

∴ **$P_b = C_c + C'_s - T$ = 586 + 156.7 – 167.4 = 575^k**

  $M_b = C_c(h/2 - a/2) + C'_s(h/2 - d') + T(h/2 - 2.5")$

  = 586(10.5 – 0.825 × 10.93 / 2) + 156.7(10.5 – 2.5) + 167.4(10.5 – 2.5)

  $M_b = 6103^{k\text{-in}} = 509^{k\text{-ft}}$

- Find some intermediate points to define the shape of the interaction curve:

1. The simplest point is where f_s = 0; x = 21 – 2.5 = 18.5"; $\varepsilon'_s > \varepsilon_y$ ∴ $f'_s = f_y$ = 60 ksi

   C_c = 0.85 × 4.5 × 17 × 0.825 × 18.5 = 992^k

   C'_s = (60 – 0.85 × 4.5)2.79 = 156.7^k

   P = 992 + 156.7 = 1149^k

   M = 992(10.5 – 0.825 × 18.5 / 2) + 156.7(10.5 – 2.5)

   = 4099$^{k\text{-in}}$ = 342$^{k\text{-ft}}$

2. Take x = 8" (tension controls)

   ε'_s = ((8 – 2.50) / 8)0.003 = 0.00207

∴ $\varepsilon'_s > \varepsilon_y$ ∴ $f'_s = f_y$ = 60 ksi; C'_s = 156.7^k

   T = 60 × 2.79 = 167.4^k

   C_c = 0.85 × 4.5 × 17 × 0.825 × 8 = 429^k

   P = 429 + 156.7 – 167.4 = 418^k

M = 429(10.5 – 0.825 × 8 / 2) + 156.7 × 8 + 167.4 × 8
= 5682$^{k\text{-}in}$ = 473$^{k\text{-}ft}$

For design values, apply $\Phi = 0.7$ for compression failure; $\Phi = 0.9$ for flexure; Φ varies linearly from 0.7-to-0.9 as P varies from $0.1f'_cA_g = 0.1 \times 4.5 \times 357 = 160^k$

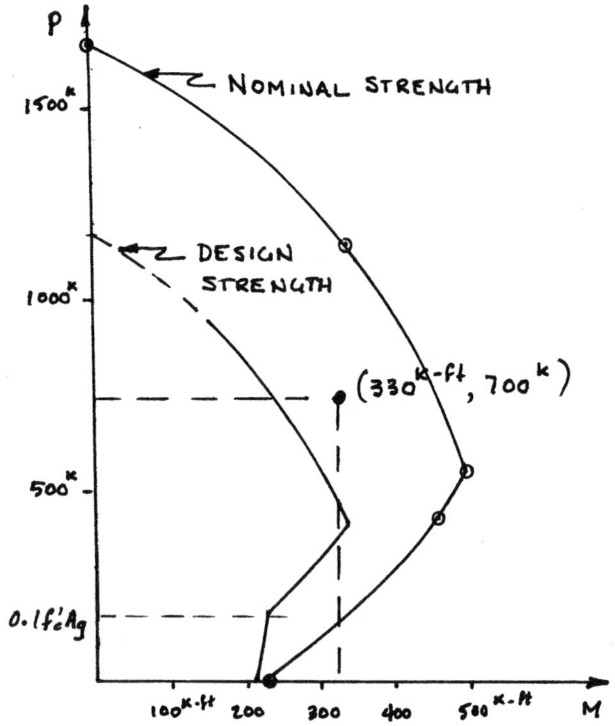

b) Check adequacy of the column for the given loading. The point $M_u = 330^{k\text{-}ft}$; $P_u = 700^k$ falls outside the design interaction diagram for the column; therefore, **the column is inadequate for this loading.**

Concrete-8

The design moments and loadings are given; $f'_c = 4$ ksi; $f_y = 60$ ksi for longitudinal steel; $f_y = 40$ ksi for stirriups. Use # 8 for main steel; no redistribution of moments or reduction of moments to the face of support. Also, it will be assumed that the given shears and moments occur under the same pattern of live loading.

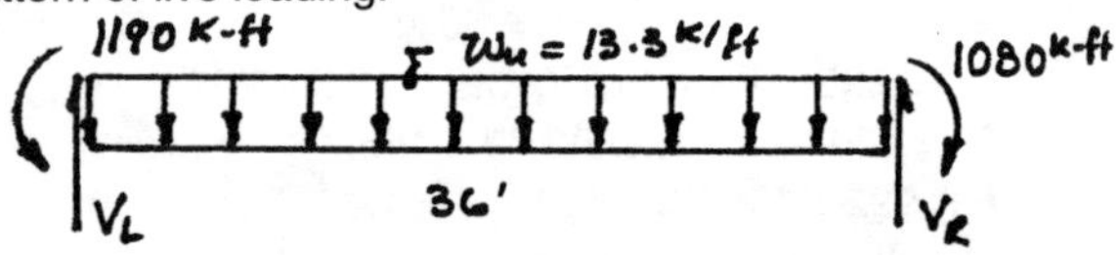

$V_L = 13.3 \times 36 / 2 + (1190 - 1080) / 36 = 242.5^k$

$V = V_L - w_ux = 242.5 - 13.3x$; $V = 0$ @ $x = 18.2'$

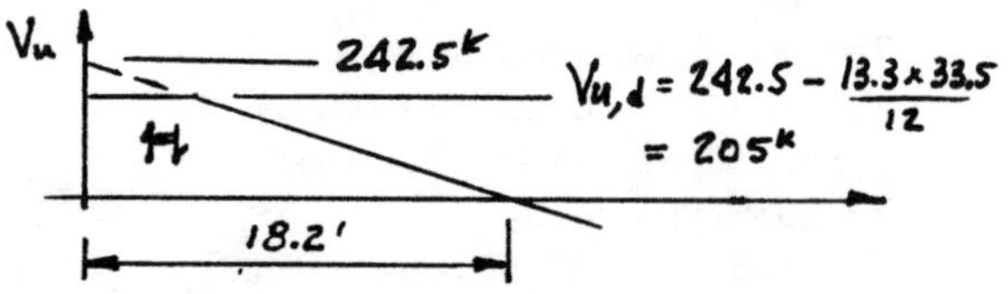

$M_u^+ = -1190 + 0.5 \times 242.5 \times 18.2 = 1017^{k\text{-}ft}$

$$b_{eff} \le \begin{cases} L/4 = 36 \times 12/4 = 108" \\ \text{beam spacing} = 22 \times 12 = 264" \\ b_w + 16h_s = 18 + 16 \times 8 = 146" \end{cases}$$

$\therefore\ b_{eff} = 108"$

a) Design the positive and negative flexural reinforcement:

- Positive reinforcement: $M_u^+ = 1017^{k\text{-}ft}$; Tee beam; Use the strength design method

  $d^+ = 36 - 2.5 = 33.5"$

  $\phi M_n = \phi\rho bd^2f_y\{1 - 0.59\rho(f_y / f'_c)\} \ge M_u$

  $0.9 \times 108 \times 33.5^2 \times 60\rho\{1 - 0.59\rho(60 / 4)\} = 1017 \times 12$

  $\therefore\ \rho = 0.0019$

  $\therefore\ A_s = 0.019 \times 108 \times 33.5 = 6.86^{sq.in}/_{ft}$

  Check the neutral axis depth:

  $x = a / (\beta_1) = 6.86 \times 60 / [0.85(0.85 \times 4 \times 108)] = 1.12" < h_s \therefore$ O.K.

  $A_{s,min} = (200 / 60000)b_wd = 0.0033 \times 18 \times 33.5 = 2.01$ in^2.

  Use 6.86 / 0.79; **say 9 # 8 bottom**

- Negative reinforcement: $M_u^- = 1190^{k\text{-}ft}$; $b = b_w = 18"$

  $0.9 \times 18 \times 33.5^2 \times 60\rho\{1 - 0.59\rho(60 / 4)\} = 1190 \times 12$

  $\therefore\ \rho = 0.0151 < 0.75\rho_b = 0.0214$

  $\therefore\ A_s = 0.0151 \times 18 \times 33.5 = 9.10^{sq.in}/_{ft}$; **say 12 # 8 top left**

- For $M_u^- = 1080^{-ft}$; $b = b_w = 18"$

  $0.9 \times 18 \times 33.5^2 \times 60\rho\{1 - 0.59\rho(60 / 4)\} = 1080 \times 12$

  $\therefore\ \rho = 0.0137 < 0.75\rho_b = 0.0214$

  $\therefore\ A_s = 0.0137 \times 18 \times 33.5 = 8.26\ ^{sq.in}/_{ft}$; **say 11 # 8 top right**

b) Find the locations where 50% of A_s^+ can terminate:

$A_s = 0.5 \times 6.86 = 3.43$ in^2

$a = 3.46 \times 60 / (0.85 \times 4 \times 108) = 0.56"$

$$\phi M_n = \phi A_sf_y(d - a/2) = 0.9 \times 3.43 \times 60(33.5 - 0.56/2)/12 = 512^{k-ft}$$

$w_uL'^2/8 = 512^{k-ft}$

$L' = [8 \times 512 / 13.3]^{1/2} = 17.6'$

The distance L' centers about the point of M_{max}; i.e., 18.2' from the left support. However, bars must extend beyond the theoretical cutoff point at least d = 33.5" or $12d_b$ = 12" (ACI 318, Section 12.10). Thus,

Required L' = 17.6 + 2(33.5) / 12 = 23.2'

$\therefore$ Cutoff = (2 × (36 – 18.2) – 23.2) / 2 = 6.2'; say 6'-3"

$\therefore$ Cutoff 4 # 8 bottom bars at 6' – 3" from center of each column (conservative)

c) Calculate the stirrup spacing at 4' intervals:

$V_u = 242.5 - 13.3x$

$$V_c = 2\sqrt{f'_c}b_wd = 2\sqrt{4000} \times 18 \times 33.5 = 76{,}300 \text{ lb} = 76.3^k$$

$$V_s = V_u/\phi - V_c = 205/0.85 - 76.3 = 165^k$$

For $0 \le x \le 4'$, use V_u at d-distance:

$V_u = 242.5 - 13.3 \times 33.5 / 12 = 205^k > V_c$

$$V_s = V_u/\phi - V_c = 189/0.85 - 76.3 = 146^k$$

Try a spacing of approximately 6" o.c.

$$A_V = V_S(s) / f_y d = 165 \times 6 / (40 \times 33.5) = 0.74^{\text{sq.in}} /_{\text{ft}}$$

$\therefore$ Use # 4 double stirrups ($A_v = 4 \times 0.20 = 0.80$ in^2)

Actual required spacing over this region is (0.74 / 0.80)6 = 5.6"; **say # 4 double stirrups at 5-1/2" o.c.**

For $4' \le x \le 8'$: $V_u = 242.5 - 13.3 \times 4 = 189^k > V_c$

$$V_S = V_u / \phi - V_C = 189 / 0.85 - 76.3 = 146^k$$

$$s = A_V f_y d / V_S = 0.80 \times 40 \times 33.5 / 146 = 7.3,$$

say 7" o.c.

For $8' \le x \le 12'$:: $V_u = 242.5 - 13.3 \times 8 = 136.1^k > V_c$

$$V_S = V_u / \phi - V_C = 136.1 / 0.85 - 76.3 = 84^k$$

$$s = A_V f_y d / V_S = 0.80 \times 40 \times 33.5 / 84 = 12.7,$$

say 12" o.c.

For $12' \le x \le 16'$: $V_u = 242.5 - 13.3 \times 12 = 83^k > V_c$

$$V_S = 83 / 0.85 - 76.3 = 21^k$$

switch to a single leg stirrup: $A_v = 0.40$ in^2

$$s = A_V f_y d / V_S = 0.40 \times 40 \times 33.5 / 21 = 25".$$

Check the minimum spacing in this region:

$$s \le A_V f_y / 50 b_W = 0.40 \times 40{,}000 / (50 \times 18) = 17.8"$$

$$s \le d / 2 = 33.5 / 2 = 16.7, \text{ say } 16" \text{ o.c.}$$

Therefore, minimum spacing governs, **use single leg # 4 @ 16" o.c.. See the elevation below for summary of stirrup spacings**

d) Detail reinforcement – Re: 10.6.6, ACI 318, distribute the top flexural reinforcement over a flange width of 1/10 the span = 3.6'; say, 3'-6". Intermediate web reinforcement is not required (10.6.7, ACI 318).

For structural integrity (7.13.2.3, ACI-318), splice two of the bottom bars over the column with a class A splice (Re: 12.15; 12.2): $l_d = 60{,}000 \times 1" / (20 \times \sqrt{4000}) = 47"$

– Re: 12.11.3, at the point of inflection:

$$x \cong [36 - (8 \times 1017 / 13.3)]^{1/2} / 2 = 5.6'$$

$$V_u = 242.4 - 13.3 \times 5.6 = 168^k$$

$$M_n = 512 / 0.9 = 569^{k\text{-ft}}$$

$$l_d \le (M_n / V_u) + l_a;\ l_a = d = 33.5"$$

$$l_d \le 569 / 168 + 33.5 / 12 = 6.17'$$

$\therefore$ O.K., the # 8 bars have adequate embedment beyond the points of inflection.

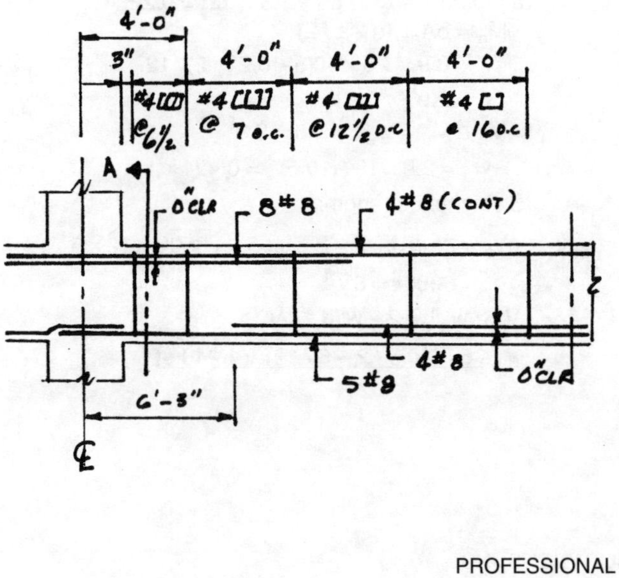

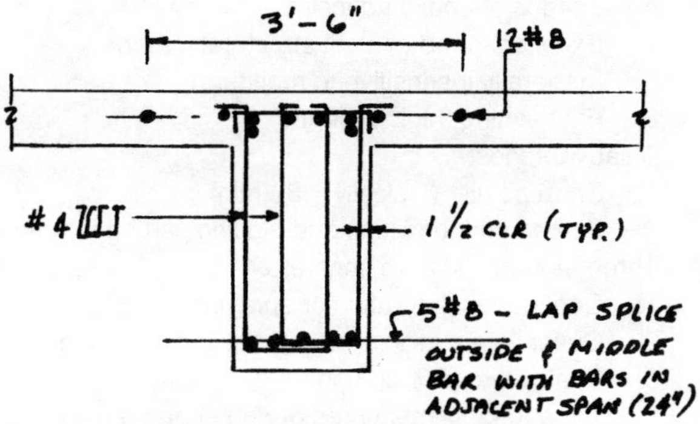

SECTION - A (N.T.S.)

(SLAB REINF. OMITTED FOR CLARITY)

Concrete-9

a) Some advantages of fly ash:

1. Improves workability (lowers water demand, reduces segregation, more pumpable, etc.)
2. Lowers heat of hydration
3. Improves permeability (ie, lowers it) and durability (eg, resistance to sulfate attack)

Some disadvantages:

1. Slower strength gain
2. Longer curing period required
3. Extra material to store and handle in batching

b) 1) Preheat is the localized application of heat prior to welding (minimum of 3" to each side of weld). 2) To ascertain the weldability of steel rebars, the chemical composition based on mill analysis must be known. Elements in the steel (both impurities and alloy elements) are expressed in percentages. The percentage of the various elements are multiplied by factors and summed to obtain a quantity called the carbon equivalent. The carbon equivalent must be less than 0.55 in order for the rebar to be welded without preheat. 3) The heat affected zone is the base metal adjacent to a weld over which temperature during welding becomes high enough that too rapid cooling (ie, quenching) might cause hardening and embrittlement.

c) The splice must develop 125 percent of the minimum specified yield strength. For example, the splice for a #14 grade 60 bar would need to develop

$$T = 1.25 f_y A_b = 1.25 \times 60 \times 2.25 = 169^k.$$

d) Re: 3.5.3.1, ACI 318 - ASTM A706 covers low-alloy bars intended for special applications where welding, bending or both are of importance. Better quality control of chemical composition helps to ensure that yield strength does not deviate excessively from the minimum specified. This makes A706 the preferred reinforcement when seismic loadings govern (re: 21.2.5.1, ACI 318).

e) Three types of mechanical splices:

1. Threaded couplers
2. Wedge-sleeve
3. Thermite welds

f) Advantages of epoxy adhesive:
   1. Excellent bond to most structural materials (generally insensitive to moisture)
   2. Fast setting, high strength

   Disadvantages:
   1. Short pot life (typically < 30 min)
   2. Fumes during mixing and placing can be irritating

g) Three methods to cure concrete:
   1. Continuous saturation or spraying
   2. Cover with plastic sheet or wet sand or burlap
   3. Seal with curing compound

h) If only the force is measured during strand tensioning, it is possible that extraordinary friction loss occurs at some point, which prevents adequate stress over a significant length of strand. If only the elongation is measured, it is possible that a strand slipped or fractured and goes undetected.

i) Factors that influence shrinkage:
   1. Time
   2. Environment (especially the relative humidity)
   3. Shape of element (surface/volume)
   4. Curing condition (moist vs. accelerated)

j) Some hot weather concrete practices:
   1. Cool aggregates and mix water
   2. Moisten forms
   3. Erect temporary sunshades

Concrete-10

Use the strength design method (ACI 318) to design the corbel. Given: $f'_c = 4$ ksi; $f_y = 60$ ksi; follow prescriptive criteria in Section 11.9.

$a = 6"$; $a / d \le 1.0 \therefore d \ge 6"$

To simplify formwork, let b = column width = 24". Use trial and error procedure to find h,

Try: $h = 13"$; $d = 13 - 0.75 - 0.25 = 12"$

$P_u = 1.4P_d + 1.7P_l = 1.4 \times 48 + 1.7 \times 32 = 122^k$

$N_u = \mu_s P_u = 0.4 \times 122 = 49^k \succ 0.2P_u$

$M_u = P_u a + N_u(h - d) = 122 \times 6 + 49(13 - 12) = 781^{k\text{-in}}$

$A_f \ge M_u / (\phi f_y 0.95d) = 781 / (0.85 \times 60 \times 0.95 \times 12)$
$= 1.34 \text{ in}^2$

Check the approximation:

$a = 1.34 \times 60 / (.85 \times 4 \times 24) = 0.99"$

$d = 12 - 0.99 / 2 = 11.5" \cong 0.95d \therefore$ O.K.

Additional steel needed to resist the direct tension

$A_n \ge N_u / (\phi f_y) = 49 / (0.85 \times 60) = 0.96 \text{ in}^2$

$\therefore \quad A_s = A_v + A_n = 1.34 + 0.96 = 2.30 \text{ in}^2$

Check minimum steel

$A_{s,min} = 0.04(f'_c / f_y)bd = 0.04 \times (4 / 60) \times 24 \times 12$
$= 0.77 \text{ in}^2 \therefore$ O.K.

Check shear transfer:

$V_u = P_u \le \phi V_n = \phi A_{vf} f_y \mu$

$A_{vf} \ge V_u / \phi f_y \mu = 122 / (0.85 \times 60 \times 1.4) = 1.71 \text{ in}^2$

Re: Section 11.9.3.2.1 (note that $0.2f'_c = 800$ psi)

$\therefore \quad V_n \le 0.2f'_c bd = 0.2 \times 4 \times 24 \times 12 = 230^k$ is O.K.

The primary tension steel (11.9.3.5.):

$A_s \ge (2A_{vf} / 3 + A_n) = (2 / 3) \times 1.71 + 0.96 = 2.10 \text{ in}^2$

$\therefore \quad A_s = 2.30 \text{ in}^2$ governs; **use 3 # 8 top**

Stirrups are required to furnish the additional shear strength (11.9.4)

$A_h \ge 0.5(A_s - A_n) = 0.5(2.30 - 0.96) = 0.67 \text{ in}^2$

Use 2 # 4 closed stirrups: $A_h = 0.80 \text{ in}^2$; detail as shown below.

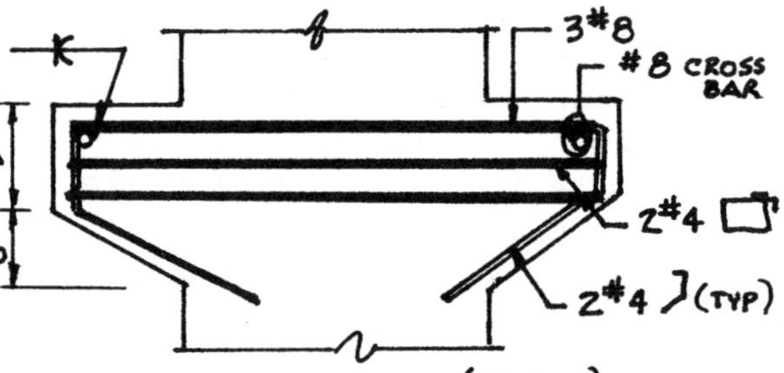

ELEVATION (N.T.S.)
(COLUMN STEEL OMITTED FOR CLARITY)

Concrete-11

a) Compute the service live load capacity based on the strength of the one-way slab and beam. Use the strength design method (ACI 318); $f'_c = 3.5$ ksi; $f_y = 40$ ksi. For the typical interior spans: $M^+_u = w_u l_n^2 / 16$; $M^-_u = w_u l_n^2 / 11$; $V_u = w_u l_n / 2$.

For the 6" slab: $d = 6 - 0.75 - 0.25 = 5"$; unit width basis $\therefore b = 12"$

- In the negative moment region, $A_s^- = 0.30^{\text{sq.in}}/_{\text{ft}}$;

  $l_n = 12 - 1.5 = 10.5'$

  $a = 0.30 \times 40 / (0.85 \times 3.5 \times 12) = 0.34"$

  $\phi M_n = \phi A_s f_y(d - a / 2)$
  $= 0.9 \times .30 \times 40(5 - 0.34 / 2) / 12$
  $= 4.35^{k-ft}$

  $\phi M_n \ge M_u = w_u l_n^2 / 11 = 4.35^{k-ft}$

  $\therefore \ w_u = 4.35 \times 11 / (10.5)^2 = 0.434$ ksf

- For the positive moment region: $A_s^+ = 0.20^{\text{sq.in}}/_{\text{ft}}$;

  $l_n = 12 - 1.5 = 10.5'$

  $a = 0.20 \times 40 / (0.85 \times 3.5 \times 12) = 0.22"$

  $\phi M_n = \phi A_s f_y(d - a / 2)$
  $= 0.9 \times .20 \times 40(5 - 0.22 / 2) / 12$
  $= 2.9^{k-ft}$

  $\phi M_n \ge M_u = w_u l_n^2 / 16 = 2.9^{k-ft}$

  $\therefore \ w_u = 2.9 \times 16 / (10.5)^2 = 0.421$ ksf

- Slab shear strength:

  $\phi V_c = 2\phi\sqrt{f'_c}\, b_w d = 2 \times 0.85 \times \sqrt{3500} \times 12 \times 5$
  $= 6000\# = 6^{k/ft}$

  $V_u \ge w_u l_n / 2 - w_u d \le \phi V_c$

$\therefore$ **$w_u \le 6 / (10.5 / 2 - 5 / 12) = 1.24$ ksf**

For the 18" x 24" beam:

- Negative bending: $d^- = 24 - 1.5 - 0.375 - 10/16$
  $= 21.5" \cong d^+$; $l_n = 25 - 2 = 23'$; $b = b_w = 18"$;
  $A_s^- = 4 \times 1.27 = 5.08$ in^2
  $a = 5.08 \times 40 / (0.85 \times 3.5 \times 18) = 3.8"$
  $$\phi M_n = 0.9 \times 5.08 \times 40(21.5 - 3.8/2)/12 = 299^{k-ft}$$
  $$\phi M_n \geq M_u = w_u l_n^2 / 11 = 299^{k-ft}$$
  $$\therefore\ w_u = (299 \times 11/(23)^2)/\text{spacing} = 6.21 \text{ klf}/12'$$
  $$= 0.518 \text{ ksf}$$
- Positive moment: $A_s^+ = 4 \times 1.00 = 4.00$ in^2; tee beam
  $$b_{eff} \leq \begin{cases} L/4 = 23 \times 12/4 = 69" \\ \text{beam spacing} = 12 \times 12 = 144" \\ b_w + 16h_s = 18 + 16 \times 6 = 114" \end{cases}$$
  $$\therefore\ b_{eff} = 69"$$
  $a = 4.00 \times 40 / (0.85 \times 4 \times 69) = 0.78"$
  $$\phi M_n = 0.9 \times 4.00 \times 40(21.5 - 0.78/2)/12 = 253^{k-ft}$$
  $$\phi M_n \geq M_u = w_u l_n^2 / 16 = 253^{k-ft}$$
  $$\therefore\ w_u = (253 \times 16/(23)^2)/\text{spacing} = 7.66 \text{ klf}/12'$$
  $$= 0.638 \text{ ksf}$$
- Shear: # 3 stirrups @ 8" o.c.; $A_v = 0.22$ in^2; $s = 8"$
  $$V_s = A_v f_y d / s = 0.22 \times 40 \times 21.5/8 = 23.7^k$$
  $$V_c = 2\sqrt{f'_c}\, b_w d = 2\sqrt{3500} \times 18 \times 21.5 = 45{,}800 \text{ lb}$$
  $$= 45.8^k$$
  $$\phi V_n = \phi(V_c + V_s) = 0.85(45.8^k + 23.7) = 59.0^k$$
  $$V_u = w_u l_n / 2 - w_u d \leq \phi V_n = 59.0^k$$
  $$w_u = [59.0/(23/2 - 21.5/12)]/12 = 0.506 \text{ ksf}$$

Check shear capacity in the central region where there are no stirrups (note: current ACI 318 limits shear on the unreinforced web to $\Phi V_c / 2$

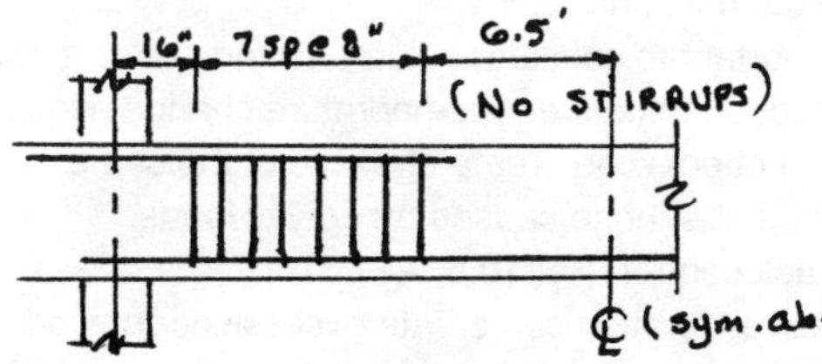

$$V_u = w'_u (6.5)' \leq \phi V_c / 2 = 0.85 \times 45.8 / 2 = 19.5^k$$
$$w_u = (19.5/6.5)/12 = 0.25 \text{ ksf}$$

Thus, the floor system strength is governed by shear capacity in the central region. The design load capacity is

$w_u = 1.4w_d + 1.7w_l = 0.25$ ksf

Assume that normal weight concrete is the only dead load

$w_d = 6" \times 12.5^{psf}/_{in} + (150/144) \times (18)"(24 - 6)"/12'$
$= 104$ psf

$\therefore$ $w_l = (250 - 1.4 \times 103)/1.7 = 62$ psf

Shear on unreinforced beam web governs: $w_l \leq 62$ psf.

b) Other considerations: to establish true capacity, it is necessary to check capacity of girders in shear and flexure; capacity of column and foundation; edge and corner bay locations; and special locations (e.g., around stairwells). Also, in each case, the appropriate building code would likely permit a live load reduction for elements supporting large tributary areas, which could lead to a specified live load that is larger than calculated above. Serviceability considerations (e.g., live load deflection, cracking) might also limit the acceptable loading. Finally, any lateral force effects that might combine with the gravity loads would have to be considered.

Concrete-12

a) Plot the interaction diagram for the given column; $f'_c = 5000$ psi; $\beta_1 = 0.8$; $f_y = 60{,}000$ psi.

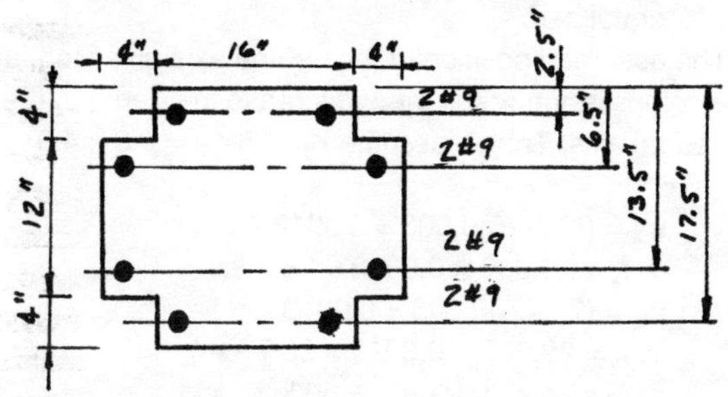

$A_{st} = 8 \times 1.00 = 8.00$ in^2; $A_g = 16 \times 20 + 8 \times 12 = 416$ in^2

- For the concentric load case (M = 0):
  $$P_o = 0.85f'_c(A_g - A_{st}) + f_y A_{st}$$
  $$= 0.85 \times 5(416 - 8) + 60 \times 8$$
  $P_o = 2214^k$
- Balanced conditions: $\varepsilon_y = f_y / E_s = 60/29{,}000 = 0.00207$; section is symmetrical $\therefore$ plastic centroid coincides with geometric centroid.

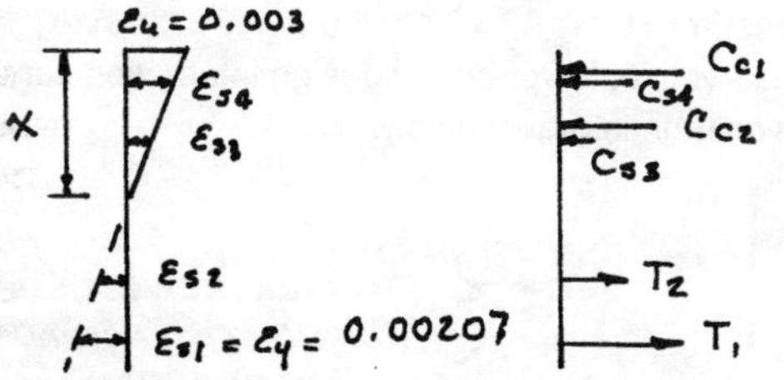

By similar triangles

$x = (0.003/0.00507)17.5 = 10.3"$
$\varepsilon_{s4} = ((10.3 - 2.50)/10.3)0.003 = 0.0023$
$\varepsilon_{s4} > \varepsilon_y \therefore f'_{s4} = f_y = 60$ ksi
$\varepsilon_{s3} = ((10.3 - 6.50)/10.3)0.003 = 0.0011$
$\varepsilon_{s3} < \varepsilon_y \therefore f_{s3} = 29{,}000 \times 0.0011 = 32.1$ ksi
$\varepsilon_{s2} = ((13.5 - 10.3)/[17.5 - 10.3])0.00207 = 0.00092 < \varepsilon_y$

$\therefore$ $f_{s2} = 29{,}000 \times 0.0092 = 26.8$ ksi
$C_{c1} = 0.85 \times 5 \times 16 \times 4 = 272^k$
$C_{c2} = 0.85 \times 5 \times 24 \times (8.27 - 4) = 436^k$
$C_{s3} = (60 - 0.85 \times 5) \times 2.0 = 112^k$
$C_{s4} = (32.1 - 0.85 \times 5) \times 2.0 = 56^k$

$T_2 = 26.8 \times 2.0 = 53.6^k$

$T_1 = 60 \times 2.0 = 120.0^k$

$\therefore \quad P_b = C_{c1} + C_{c2} + C_{s3} + C_{s4} - T_1 - T_2$

$= 272 + 436 + 112 + 56 - 53.6 - 120 = 702^k$

$M_b = P_b e_b = 272(10 - 2) + 436(6 - 4.27 / 2)$

$+ 112(10 - 2.5) + 56(10 - 6.5) + 53.6(13.5 - 10)$

$+ 120(17.5 - 10) = 5980^{k\text{-}in} = 498^{k\text{-}ft}$

- For pure bending (P = 0):

  Iterate to find neutral axis position: Trial 1. Assume $x = 5"$; $a = 0.8 \times 5 = 4"$

  $\varepsilon_{s4} = ((5 - 2.50) / 5)0.003 = 0.0015 < \varepsilon_y$

  $\therefore f_s = 29{,}000 \times 0.0015 = 43.5$ ksi

  $\varepsilon_{s3} = (1.5 / 5)0.003 = 0.0009 < \varepsilon_y$

  $\therefore f_s = 29{,}000 \times 0.0009 = 26.1$ ksi

  $\varepsilon_{s2} = (8.5 / 5\,)0.003 = 0.0051 > \varepsilon_y \therefore f_{s2} = 60$ ksi

  $\varepsilon_{s1} > \varepsilon_{s2} > \varepsilon_y \therefore f_{s1} = 60$ ksi

  $\Sigma F_{long} = 0$: $0.85 \times 5 \times 16 \times 4 + 2.0\,(43.5 - 4.25)$

  $- 26.1 \times 2.0 - 60 \times 2.0 - 60 \times 2.0 = 58.3^k$ – not equal zero

The assumed position of the neutral axis gives too large a compression resultant ∴ decrease the depth to the neutral axis. Trial 2: assume $x = 4.5"$; $a = 0.8 \times 4.5 = 3.6"$

$\varepsilon_{s4} = ((2 / 4.5)0.003 = 0.0013 < \varepsilon_y$

$\therefore f_s = 29{,}000 \times 0.0013 = 38.7$ ksi

$\varepsilon_{s3} = (2 / 4.5)0.003 = 0.0013 < \varepsilon_y$

$\therefore f_s = 29{,}000 \times 0.0013 = 38.7$ ksi

$\varepsilon_{s1} > \varepsilon_{s2} > \varepsilon_y \therefore f_{s1} = f_{s2} = 60$ ksi

$\Sigma F_{long} = 0$: $0.85 \times 5 \times 16 \times 3.6 + 2.0\,(38.7 - 4.25)$

$- 38.7 \times 2.0 - 60 \times 2.0 - 60 \times 2.0 \cong 0$

∴ The neutral axis is 4.5" from top edge

$M_n = 0.85 \times 5 \times 16 \times 3.6(17.5 - 3.6 / 2)$

$+ 2.0(38.7 - 4.25)(17.5 - 2.5) - 2.0 \times 38.7 \times (17.5 - 6.5)$

$- 2.0 \times 60(17.5 - 13.5) = 3545^{k\text{-}in} = 296^{k\text{-}f}$

For design values, apply Φ= 0.7 for compression failure; $\Phi = 0.9$ for flexure; Φ varies linearly from 0.7-to-0.9 as P varies from $0.1f'_c A_g = 0.1 \times 5.0 \times 416 = 208^k$. A conservative, lower bound design interaction curve for this column is drawn below

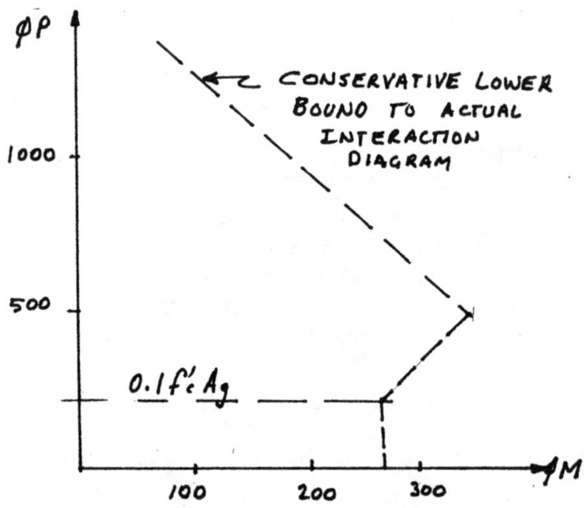

b) Compute the maximum value of M_{ux} that can be applied simultaneously with an applied $P_u = 100^k$. Reading from the interaction diagram for $\Phi P_n = 100^k$, $\Phi M_n \cong 260^{k\text{-}ft}$. Check slenderness effect using the approximate method of section 10.10, ACI 318. Given: $kL_u = 18' = 216"$; $C_{mx} = 1.0$; $\beta_d = 0.6$. Assume that the column is part of a non-sway frame (Section 10.12 applies). from Eqn 10-13:

$EI = 0.4E_c I_g / (1 + \beta_d)$; $E_c = 57000(5000)^{1/2}$

$= 4.03 \times 10^3$ ksi

$I_g = (16 \times 20^3 + 8 \times 12^3) / 12 = 11{,}820$ in^4

$\therefore \quad EI = 0.4 \times 4.03 \times 10^3 \times 11{,}820 / (1 + 0.6)$

$= 11.9 \times 10^6$ k-in^2

$P_c = \pi^2 EI / (kL_u)^2 = (3.14)^2 \times 11.9 \times 10^6 / 216^2$

$= 2515^k$

$\delta_{ns} = C_m / [1 - P_u / (0.75 \times P_c)]$

$= 1.0 / [1 - 100 / (0.75 \times 2515)] = 1.06$

$\therefore$ **$\underline{M_{ux} = 260 / 1.06 = 245^{k\text{-}ft}}$**

Concrete-13

a) A mill test analysis of the chemical composition must be available. Elements in the steel are converted to carbon equivalents for the purpose of determining weld electrode and preheat.

b) Under seismic lateral forces, plastic hinges should form in the girders near the column faces. To ensure ductile behavior, it is important that shear and development strengths exceed flexural strength. Since the yield strength of rebars is generally higher than the minimum specified yield stress, the flexural capacity of the member is proportionally higher. This gives rise to higher shears in the girder than if the bars yielded at or below the minimum f_y.

c) It is important to consider volume change effects caused by creep, shrinkage and temperature change. Restraint to these effects can create stresses of the same magnitude as those caused by gravity loads.

d) For crack control (Re: 10.6, ACI 318)

e) Excessive vibration can create excessive formwork pressure, and can cause segregation of the mortar and coarse aggregate.

f) Weld electrodes for reinforcing steel must meet AWS-ASTM specifications and have a tensile strength of at least 1.25 times the minimum specified yield of the bar to be welded. Low hydrogen, flux coated electrodes are usually employed for shielded metal arc welding (e.g., Exx15 or Exx16).

g) It may be necessary to isolate the joint between slab and wall so that unintentional restraint to the posttensioning is not created. Two problems might occur if the unintentional bond Is not broken: 1) If the wall is strong and rigid enough, it can prevent the intended precompresssion of the slab, and 2) more often, the wall is rigid but relatively weak and is damaged as it receives accidental loading from the prestress.

h) Reshoring is used:
   1. After removal of formwork and before the floor system is capable of supporting loads to be placed upon it. In multistory construction, the loads are primarily the weight of forms and wet concrete for the floors to be cast above. Usually, several floors adjacent to the one to be constructed are reshored so that the weight is distributed safely among the several floors.
   2. The amount of reshoring depends on the construction load and the strength and relative stiffness of the supporting floors below.
   3. Shores can be removed after the floor above gains sufficient strength so that its formwork can be removed and it becomes self-supporting.
   4. An important property, other than compressive strength, is the modulus of elasticity. This property determines the rigidity of the floors, which then determines the number of floors to be reshored.

I) In fresh concrete, excess mix water and air "bleed" upward and accumulate under the horizontal rebar. This weakens the concrete and reduces the bond.

j) Three factors that affect deflection in the flat slab:
   1. Variation of flexural rigidity, EI
   2. Camber (specified or actual)
   3. Support conditions (i.e., continuous one or both ends)

k) Construction joints should be located near midspan, where shear transfer across the joint is relatively low.

l) Two types of admixture:
   1. Air entrainment - reduces compressive strength
   2. Accelerators - increases rate of strength gain. Thus, the accelerator has a significant effect on early strength, but only a slight (<5%) increase in eventual compressive strength.

Concrete-14

Use the strength design method (ACI 318) to design the corbel. Given: $f'_c = 3$ ksi; $f_y = 60$ ksi; follow prescriptive criteria in Section 11.9. (Note: Section 11.9.3.4 requirement that $N_u \geq 0.2V_u$ is waived in Part a to contrast the effect of the specified frictionless bearing pad to the pad in Part b; i.e., $N_u = 0$ for part a)

$b = 18"$; $a = 5.5"$; try $h = 16"$; $d = 16 - 2 = 14"$; $a / d \leq 1.0$

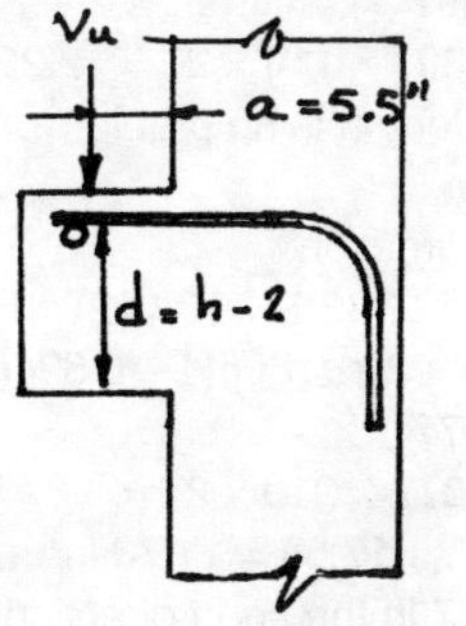

$V_u = 1.4V_d + 1.7V_l = 1.4 \times 10 + 1.7 \times 20 = 48^k$

$M_u = V_u a + N_u(h - d) = 48 \times 5.5 + 0(13 - 12) = 264^{k\text{-}in}$

Approximate the required area of flexural steel assuming $(d - a / 2) \cong 0.95d$

$A_f \geq M_u / (\phi f_y 0.95d) = 264 / (0.85 \times 60 \times 0.95 \times 14)$
$= 0.39 \text{ in}^2$

Check the approximation:

$a = 0.39 \times 60 / (.85 \times 3 \times 18) = 0.50"$

$d = 14 - 0.50 / 2 = 13.75" > 0.95d$ ∴ O.K.

Check minimum steel

$A_{f,min} = 0.04(f'_c / f_y)bd = 0.04 \times (3 / 60) \times 18 \times 12$
$= 0.43 \text{ in}^2$ – controls

Check shear transfer:

$V_u \leq \phi V_n = \phi A_{vf} f_y \mu$

$A_{vf} \geq V_u / \phi f_y \mu = 48 / (0.85 \times 60 \times 1.4) = 0.67 \text{ in}^2$

Re: Section 11.9.3.2.1 (note that $0.2f'_c < 800$ psi)

∴ $V_n \leq 0.2f'_c bd = 0.2 \times 3 \times 18 \times 14 = 151^k$ is O.K.

The primary tension steel (11.9.3.5.):

$A_s \geq (2A_{vf} / 3 + A_n) = (2 / 3) \times 0.67 + 0 = 0.45 \text{ in}^2$

∴ $A_s = A_{f,min} = 0.43 \text{ in}^2$ governs; **use 2 # 5 top**

Stirrups are required to furnish the additional shear strength (11.9.4)

$A_h \geq 0.5(A_s - A_n) = 0.5(0.43 - 0) = 0.22 \text{ in}^2$

Use 2 # 3 closed stirrups: $A_h = 0.44 \text{ in}^2$; detail as shown below.

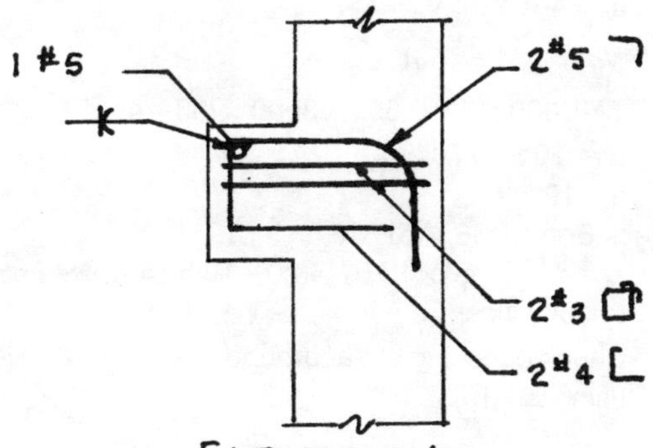

ELEVATION
(NOTE: COLUMN STEEL OMITTED FOR CLARITY)

b) Discuss the changes required if the frictionless pad is replaced by a pad with a coefficient of friction of 0.3. In this case, the moment to be resisted is supplemented by a term $0.3V_u(h - d)$, which requires additional steel to resist the moment. In addition, a direct tension force, $0.3V_u$, requires additional steel (i.e., additive to A_f required for flexure).

$N_u = 0.3V_u = 0.3 \times 48 = 14.8^k$

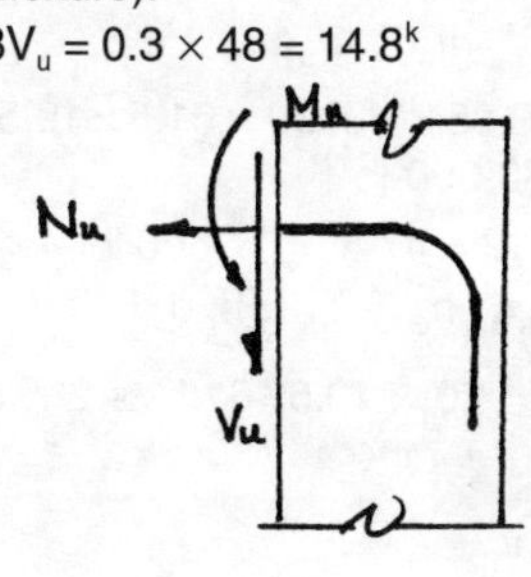

Thus,

$M_u = V_u a + N_u(h - d) = 48 \times 5.5 + 0.3 \times 48(16 - 14)$
$= 293^{k\text{-}in}$

$A_f \cong (293 / 264)0.39 = 0.43\ in^2$

$A_n \geq N_u / (\phi f_y) = 0.3 \times 48 / (0.85 \times 60) = 0.28\, in^2$

$A_s = A_v + A_n = 0.43 + 0.28 = 0.71\ in^2 > A_{s,min}$

The shear transfer requirement is the same as in Part a, $\therefore$ O.K.

The only changes required are to increase the 2 # 5 top bars to 2 # 6 and to use a # 6 instead of # 5 cross bar to anchor the tension reinforcement.

Concrete-15

Use the strength design method (ACI 318) to design the footing. Given: f'_c = 3 ksi; f_y = 60 ksi; given service loads on each of the 12, 12" x 12" piles: $P_l = 10^k$ and $P_d = 22^k$. Assume that the pile cap is rigid for purposes of distribution of loads to piles and that the weight of the cap is negligible.

Design load in 20" x 20" columns:

$P_u = 12(1.4P_d + 1.7P_l) = 12(1.4 \times 22 + 1.7 \times 10)$
$= 574^k$

a) Minimum cap thickness:

1. For punching shear around 20" x 20" column:

   $b_o = 4(h + d);\ V_u = 547^k$

   $V_u \leq \phi V_c = 4\phi\sqrt{f'_c}\, b_o d$

   $574{,}000 \leq 4 \times 0.85 \times \sqrt{3000} \times (80 + 4d)d$

   $d^2 + 20d - 770 \geq 0$

   $d \geq 19.5"$

   Check Eqn 11-36, ACI 318

   $\alpha_s d / b_o = 40 \times 19.5 / [4(20 + 19.5)] = 4.9 \succ 2$

   $\therefore$ not critical

2. Check punching shear around the 12" x 12" piles using d = 19.5":

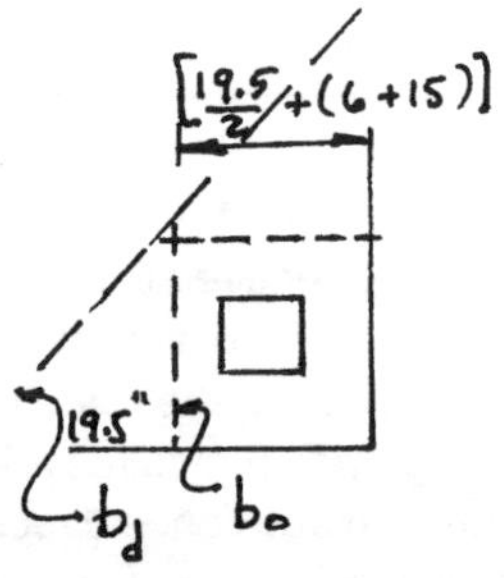

   $b_o = 2(15 + (12 + 19.5) / 2) = 61.5"$

   $V_u \leq \phi V_c = 4\phi\sqrt{f'_c}\, b_o d$

   $= 4 \times 0.85 \times \sqrt{3000} \times 61.5 \times 19.5$

   $\phi V_c = 233{,}000$ lb

   $V_u = 574/12 = 47.8^k = 47{,}800$ lb

   $\phi V_c \leq (\alpha_s d / b_o + 2)\sqrt{f'_c}\, b_o d$

   $\alpha_s d / b_o = 20 \times 19.5 / 61.5 \succ 2\ \therefore$ not critical

   $V_u \leq \phi V_c\ \therefore$ O.K.

3. Check wide beam shear across the corner diagonal: $b_d = 2[19.5 / 2 + (6 + 15)] / 0.707 = 87"$

   $\phi V_c = 2\phi\sqrt{f'_c}\, b_w d = 2 \times 0.85 \times \sqrt{3000} \times 87 \times 19.5$

   $= 158{,}000\ lb \succ\succ V_u = 47.8^k\ \therefore$ O. K.

4. Wide beam shear across the short direction: $b = 9.5 \times 12 = 114"$

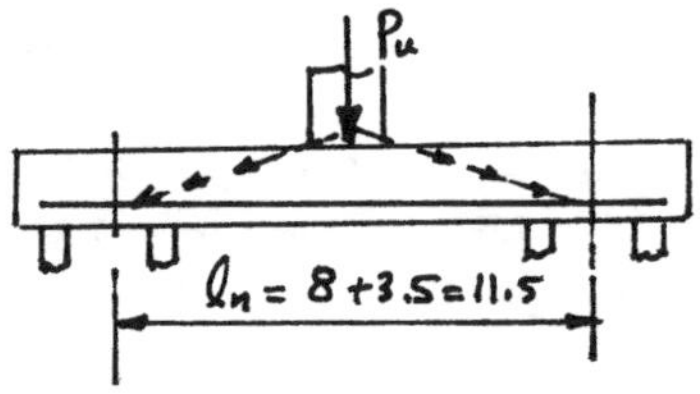

   $l_n / d = 11.5 \times 12 / 19.5 = 7 > 5\ \therefore$ slender member

   $V_u = 574 / 2 = 287^k$

   $\phi V_c = 2\phi\sqrt{f'_c}\, b_w d = 2 \times 0.85 \times \sqrt{3000} \times 114 \times 19.5$

   $= 207{,}000\ lb = 207^k \prec V_u\ \therefore$ N.G.

Wide beam shear governs – increase d so that stirrups are not required:

$\frac{d_{req}}{19.5} = \frac{287}{207};\qquad \therefore\ d_{req} \geq 27"$

Allow 3" cover plus 1" to steel centroid

$h = 27 + 3 + 1 = 31"$; **the required cap thickness is 2' – 7" thick**

b) Design the flexural reinforcement:

$A_{s,min} = 0.0018bh = 0.0018 \times 12 \times 31 = 0.67^{sq.in}/_{ft}$

For the long (17.5') direction

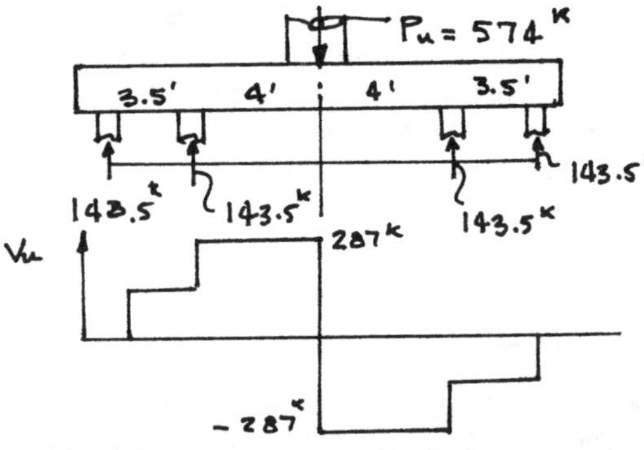

Critical bending occurs at column face:

$M_u^+ = 143.5 \times 3.5 + 287\,(4 - 10 / 12) = 1411^{k\text{-}ft}$

$\phi M_n = \phi \rho b d^2 f_y\{1 - 0.59\rho(f_y / f'_c)\} \geq M_u$

$0.9 \times 114 \times 27^2 \times 60\rho\{1 - 0.59\rho(60 / 3)\}$
$= 1411 \times 12$

$\rho = 0.00397$

$A_s = 0.00397 \times 114 \times 27 = 12.22^{sq.in}/_{ft}$

Development length is no problem $\therefore$ **Use 10 # 10 in the long direction**.

For the short direction:

$M_u^+ = (574 / 3)(3.5 - 10 / 12) = 510^{\text{-}ft}$

$0.9 \times 210 \times 27^2 \times 60\rho\{1 - 0.59\rho(60 / 3)\} = 510 \times 12$

$\rho = 0.00077$

$A_s = 0.00077 \times 210 \times 27 = 4.37^{sq.in} < A_{s,min}$

$= (0.67^{sq.in}/_{ft})(17.5') = 11.73\ in^2$

$\therefore$ Use 18 # 7 in the short direction.

c) Details are:

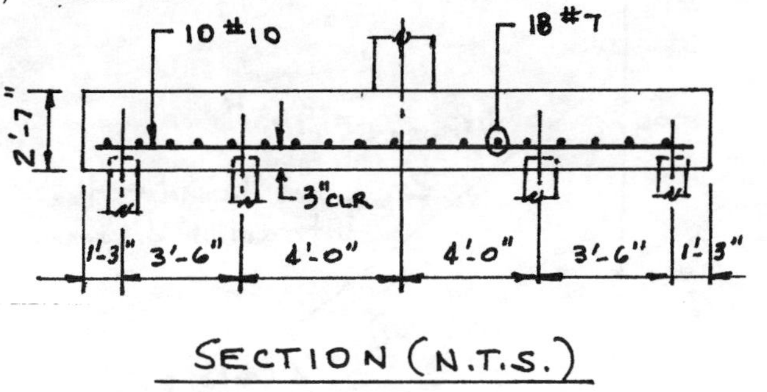

Concrete-16

a) Six items other than design details to check:
1. Curing conditions - were temperature and moisture controlled to allow slab to gain strength?
2. As-built dimensions - is the slab thickness correct?
3. Rebars - were correct bar sizes and positions used?
4. Construction loads - was the slab overloaded by temporary forces?
5. Falsework - did settlement occur while concrete was plastic?
6. In-place strength - was water added to the mix after cylinders were cast?

Design details to check:
1. Variation in EI properly computed?
2. Was assumed joint rigidity (i.e., end condition) achieved?
3. Were deformations of supporting members considered?

b) Design philosophy is that the structure have enough toughness to respond inelastically to severe ground motion.

c) Mild steel cannot be used to prestress concrete because creep and shrinkage strains in concrete practically eliminate the prestress.

d) Welding can cause local embrittlement and the notches caused by weld uncut causes stress concentrations that can significantly reduce the tensile strength.

e) Cold weather concrete practices:
1. Protect the subgrade from freezing.
2. Maintain proper concrete temperature during placement.
3. Determine curing procedures to ensure strength gain and prevent freezing
4. Consider use of accelerating admixture.
5. Maintain temperature records during construction and cure.
6. Ensure in-place strength.

f1) Two benefits of air entrainment:
1. Increase resistance to freeze-thaw deterioration.
2. Improve workability.

f2) Principal disadvantage to air entrainment is the reduction in compressive strength for a given water-cement ratio.

f3) Normal range: 3-to-7%

g) Poorly consolidated concrete will show "honeycomb" and variation in color (in extreme cases, cold joints will be evident between portions placed at different times)

h) The two alloying elements which affect weld electrode selection and preheat are manganese and nickel, both of which yield relatively high carbon equivalents.

j1) Calcium chloride is used as an accelerating admixture.

j2) Two disadvantages of the use of $CaCl_2$ are: 1) it stiffens the mix, creating need for more water to maintain workability, and 2) the Cl^- ions accelerate corrosion of reinforcement and metal embedments.

k) Water-cement ratio

l) CO_2 can form a weak carbonic acid, which attacks some compounds in the cement.

m) The 30% additional loss of prestress has practically no effect on the moment strength (ultimate). The elastic strain corresponding to effective prestress is generally small relative to ultimate strain in properly designed flexural members.

m2) The lowered effective prestress can create serviceability problems (i.e., additional deflection and cracking of tension zones).

Concrete-17

Seismic loading ∴ follow provisions of UBC-97. For loading combinations, apply Sections 1909.1 and 1612.2.1, UBC.

For gravity loading:

$$P_u = 1.4P_d + 1.7P_l = 1.4 \times 150 + 1.7 \times 70 = 329^k$$

$$M_u = 1.4M_d + 1.7M_l = 1.4 \times 100 + 1.7 \times 20 = 174^{k\text{-}ft}$$

For combined gravity plus seismic (assume that live loads fall in category of "other" in Section 1612.2.1):

$$P_u = 1.1(1.2P_d + 1.0P_E + 0.5P_l)$$
$$= 1.1(1.2 \times 150 + 1.0 \times 0 + 0.5 \times 70) = 237^k$$

$$M_u = 1.1(1.2M_d + 1.0M_E + 0.5M_l)$$
$$= 1.1(1.2 \times 100 + 1.0 \times 300 + 0.5 \times 20) = 473^{k\text{-}ft}$$

$$P_u = 0.9P_d \pm 1.1P_E = 0.9 \times 150 + 1.1 \times 0 = 135^k$$

$$M_u = 0.9M_d \pm 1.1M_E = 0.9 \times 100 + 1.1 \times 300 = 420^{k\text{-}ft}$$

Thus, there are three loading combinations that might be critical on the given column:

$$(M_u, P_u) = \begin{cases} (174, 329) \\ (473, 237) \\ (420, 135) \end{cases}$$

a) Plot the interaction diagram for the tied column; $f'_c = 4000$ psi; $\beta_1 = 0.85$; $f_y = 50$ ksi; $A_{st} = 8 \times 1.56 = 12.48$ in^2; $A_g = 14 \times 24 = 336$ in^2; d = 21.5"; d' = 2.5"

- For the concentric load case (M = 0):

$$P_o = 0.85f'_c(A_g - A_{st}) + f_yA_{st}$$
$$= 0.85 \times 4(336 - 12.48) + 50 \times 12.48$$
$$P_o = 1724^k$$

For design, we have to apply a reduction factor $\Phi = 0.7$ and a factor of 0.8 to account for accidental eccentricity. Thus,

$$\phi P_{o,max} = 0.7 \times 0.8 \times 1724 = 965^k$$

- Balanced conditions: $\varepsilon_y = f_y / E_s = 50 / 29{,}000 = 0.0017$; section is symmetrical $\therefore$ plastic centroid coincides with geometric centroid.

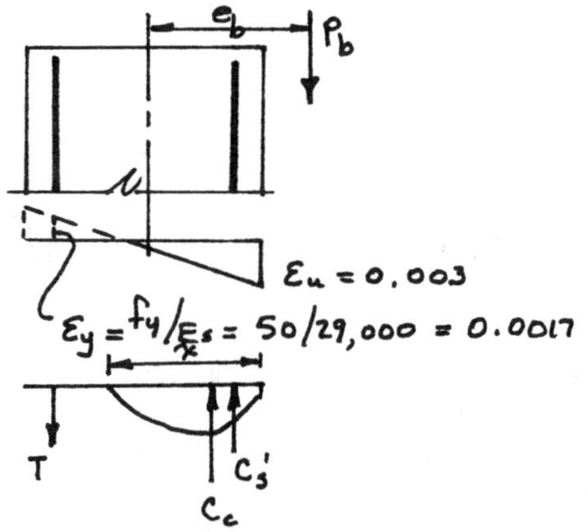

By similar triangles

$x = (0.003 / (0.003 + 0.0017)21.5 = 13.7"$

$\varepsilon'_s = ((13.7 - 2.50) / 13.7)0.003 = 0.0025$

$\varepsilon'_s > \varepsilon_y \therefore f'_s = f_y = 50$ ksi

$\varepsilon_s > \varepsilon'_s > \varepsilon_y \therefore f_s = 50$ ksi

$T = 50 \times 4 \times 1.56 = 312^k$

$C'_s = (50 - 0.85 \times 4) \times 4 \times 1.56 = 291^k$

$C_c = 0.85 \times 4 \times 14 \times 0.85 \times 13.7 = 555^k$

$\therefore$ $P_b = C_c + C'_s - T = 555 + 291 - 312 = 534^k$

$M_b = P_b e_b = 555(12 - 0.85 \times 13.7 / 2)$
$+ 291(12 - 2.5) + 312(12 - 2.5) = 9157^{k\text{-in}} = 763^{k\text{-ft}}$

For design, apply $\Phi = 0.7$: $\Phi P_b = 0.7 \times 534 = 374^k$; $\Phi M_n = 0.7 \times 763 = 534^{k\text{-ft}}$.

- For pure bending ($P = 0$): By trial-and-error, the stress in the compression side reinforcement is found to be $f'_s = 29$ ksi, with the neutral axis at 3.76" from top edge. Verify these values:

$\Sigma F_{long} = 0 = f_y A_{st} / 2 - 0.85 f'_c ba - (f'_s - 0.85 f'_c) A'_s$

$\therefore$ $a = [(50 \times 4 \times 1.56) - (29 - 0.85 \times 4) \times 4 \times 1.56]$
$/ [0.85 \times 4 \times 14] = 3.20"$

$x = a / \beta_1 = 3.20 / 0.85 = 3.76"$

$\varepsilon'_s = ((3.76 - 2.5) / 3.76)0.003 = 0.001 < \varepsilon_y$

$\therefore f_s = 29{,}000 \times 0.001 = 29$ ksi

$\Sigma F_{long} = 0$: $0.85 \times 4 \times 14 \times 3.20 + 4$
$\times 1.56(29 - 0.85 \times 4) - 50 \times 4 \times 1.56 = 0$

$\therefore$ the values of f'_s and x are correct.

$M_n = (f'_s - 0.85 f'_c) A'_s (d - d') + 0.85 f'_c ba(d - a / 2)$
$= (29 - 0.85 \times 4) \times 4 \times 1.56 \times (21.5 - 2.5) + 0.85 \times 4$
$\times 14 \times 3.2(21.5 - 3.2 / 2)$
$= 6066^{k\text{-in}} = 506^{k\text{-f}}$

Re: 1909.3.2.2, UBC-97, Φ may be increased linearly from 0.7 at $P = 0.1 f'_c A_g$ to 0.9 at $P = 0$

$$\Phi P_{0.1Ag} = 0.7 \times 0.1 \times 4 \times 336 = 94^k$$

A conservative lower bound approximation to the interaction diagram is shown in the figure below, which corresponds to the calculated design values.

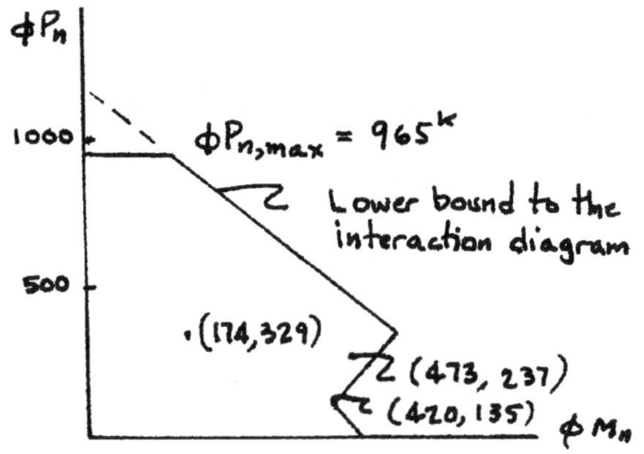

The column is governed by the loading combination corresponding to $P_u = 0.9P_d \pm 1.1P_E$, and the **design is adequate** in that the point corresponding to this set of axial force and moment falls on the design interaction curve.

b) Vertical acceleration components are not normally considered in the design of columns. However, the effect of a vertical acceleration ↓ would be to increase the likelihood of failure by reducing the design axial force on the column for the short duration loading. This is because the critical loading point is near the tension failure side of the balanced point on the interaction diagram. In this region, axial compression is beneficial in increasing the column's resisting moment strength (in a loose sense, the axial compression "prestresses" the column in this region). Conversely, acceleration ↑ would increase the moment resistance and be beneficial.

Concrete-18

Alternative 1. Provide transfer girders to span horizontally 70' between new tie rods located at end walls. The tie rods at the ends would furnish strength

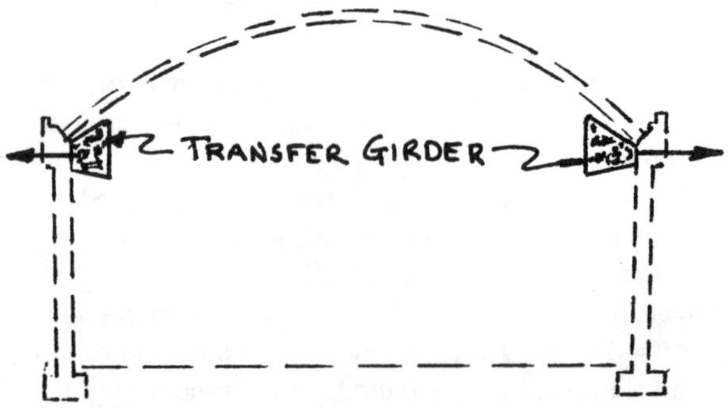

equivalent to the existing rods to be removed. Given the limited depth of girder, I suspect that a post tensioned member would be required. The post tensioning could be controlled so that the tension from rods could be transferred prior to their removal. This would minimize the horizontal deflection at the eave. Careful attention would need to be given to details so that the girder could deform independently of the main structure. Long term deformation would also require study.

Alternative 2. Construct rigid frames within the clearance envelope and install new tie rods below finish floor. The frames could be

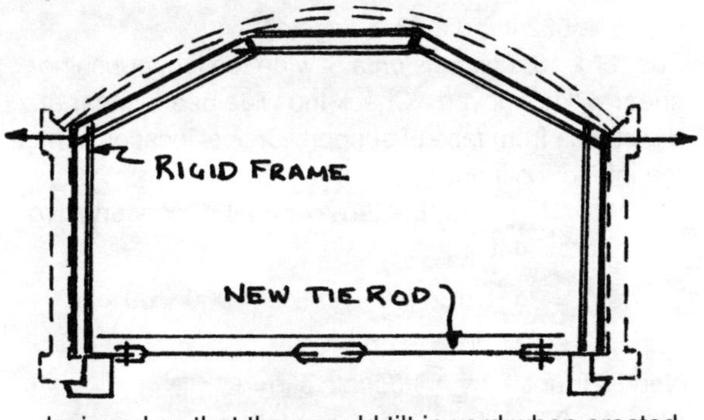

designed so that they would tilt inward when erected, then deflect outward so that they become vertical when the tension is transferred to them.

Alternative 3. A third possibility - at least from a structural standpoint - is to resist the tension using a truss or frame constructed over the existing roof. For example:

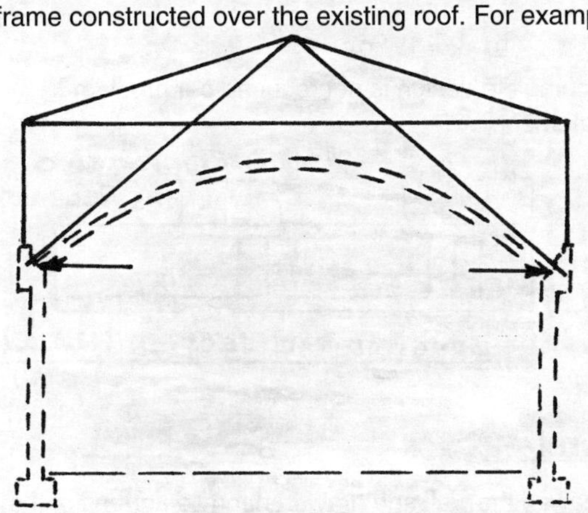

Connection details would be difficult, but are feasible. The construction sequence could be controlled to permit force transfer without excessive horizontal deflection.

Concrete-19

Use the strength design method (ACI 318) to compute the moment capacity. Given: $f'_c = 4$ ksi; $f_{pu} = 270$ ksi; $A_{ps} = 10 \times 0.085 = 0.85$ in^2; $d_p = 9.69$"; $b = 40$"; assume adequate shear transfer at the interface between the precast and cast-in-place slab (Re: 17.5, ACI 318); stress-relieved strands with $\gamma_p = 0.4$; $\beta_1 = 0.85$.

Using Eqn. 18-3.,

$\rho_p = A_{ps} / bd_p = 0.85 / (40 \times 9.69) = 0.00219$; $\omega = \omega_p = 0$

$f_{ps} = f_{pu}\{1 - \gamma_p / \beta_1[\rho_p(f_{pu} / f'_c)]\} = 270\{1 - 0.4 / 0.85[0.00219(270 / 4)]\} = 251.2$ ksi

$\omega_p = \rho_p f_{ps} / f'_c = 0.00219(251.2 / 4) = 0.137 < 0.36\beta_1 = 0.3$

$\therefore$ Ductile flexural failure

$a = A_{ps}f_{ps} / (0.85 \times b \times f'_c) = 0.85 \times 251.2 / (0.85 \times 40 \times 4) = 1.57$"

$$\phi M_n = \phi A_{ps} f_{ps}(d_p - a / 2)$$
$$= 0.9 \times 0.85 \times 251.2(9.69 - 1.57 / 2)$$
$$\phi M_n = 1711^{k-in} = 143^{k-ft}$$

Concrete-20

a) Internal resisting couple:

1. For a conventional reinforced concrete beam, the lever arm of the internal couple remains essentially constant as bending moment is increased. Equilibrium is maintained by increases in the internal forces ($C = T = f_sA_s$)
2. For a prestressed concrete beam, the internal couple increases with the bending moment by increasing the internal lever arm, with the internal T = C forces remaining essentially constant.

b) Concrete stress distribution:

case a) - Resultant C on upper kern; zero stress on bottom fiber:

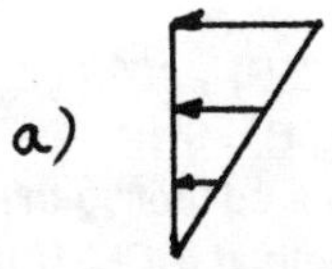

case b) - Resultant C above upper kern; tension in bottom fiber (assuming tensile stress is less than the modulus of rupture):

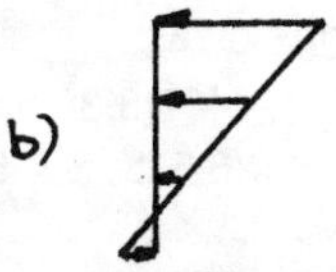

case c) - Resultant C is within the kerns and below the c.g.c.; section is in compression throughout with higher stress at bottom than at top

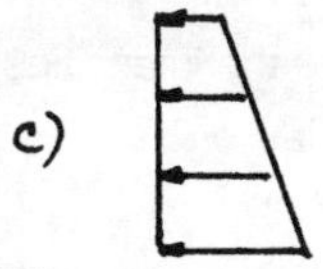

case d) - Resultant C below bottom kern; opposite condition to case b

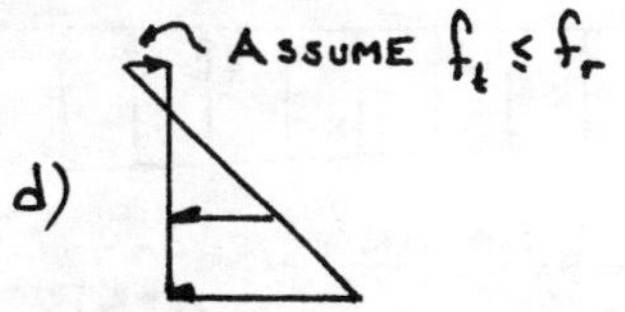

Concrete-21

a) Proportion the footing under dead plus 40% live such that the bearing pressure does not exceed an allowable 4 ksf.

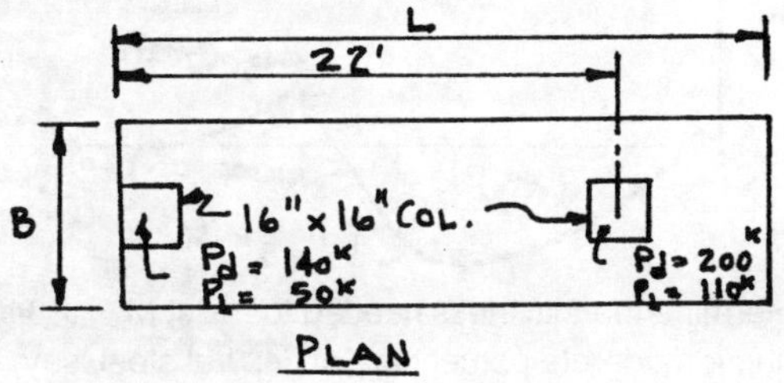

Design loads on 16" x 16" columns:

$P_{ext} = 140 + 0.4 \times 50 = 160^k$

$P_{int} = 200 + 0.4 \times 110 = 244^k$

$P_{tot} = 160 + 244 = 404^k$

Locate resultant with respect to the left end:

$\bar{x} = [160(0.67) + 244(22)] / 404 = 13.55'$

$\therefore$ $L = 2\bar{x} = 2 \times 13.55' = 27.1'$; say, $\underline{27'-3"}$

For full live plus dead load:

$P_{ext} = 140 + 50 = 190^k$

$P_{int} = 200 + 110 = 310^k$

$P_{tot} = 190 + 310 = 500^k$

$M = 190(13.55 - 0.67) - 310(22 - 13.55) = -172$
$= 172^{k\text{-ft}}$ c.w.

$A = 27.25B$; $S = (27.25^2 / 6)B = 123.8B$

$f_p = P / A + M / S$
$= 500 / (27.25B) + 172 / (123.8B) \leq 4$

$\therefore$ $B = 4.93'$; say, **B = 5'-0"**

b) Determine the required footing thickness. Use the strength design method (ACI 318); $f'_c = 3$ ksi; $f_y = 60$ ksi.

$P_{ue} = (1.4P_d + 1.7P_l) = (1.4 \times 140 + 1.7 \times 50) = 281^k$

$P_{ui} = (1.4 \times 200 + 1.7 \times 110) = 467^k$

$P_{u,tot} = 467 + 281 = 748^k$

$M_u = 281(12.95) - 467(8.38) = 275^{k\text{-ft}}$ c.w.

$A = 5 \times 27.25 = 136.3\ \text{ft}^2$; $S = 5 \times 27.25^2 / 6 = 619\ \text{ft}^2$

$f_{pu} = P_u / A \pm M_u / S$

$f_{pl} = 748 / 136.3 - 275 / 619 = 5.04$ ksf; $w_l = 5.04 \times 5$
$= 25.2$ klf

$f_{pr} = 748 / 136.6 + 275 / 619 = 5.93$ ksf; $w_r = 5.93 \times 5$
$= 29.65$ klf

Where w_l and w_r are the intensities of soil reactions (klf) at the design loading.

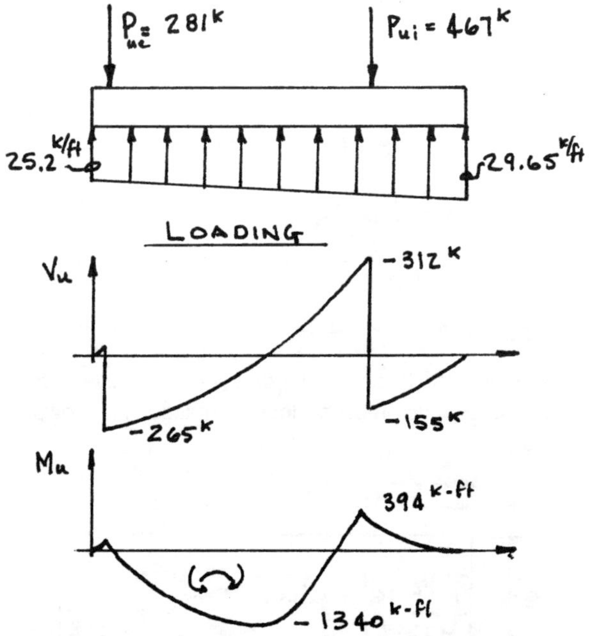

Determine the thickness needed to resist $M_u^- = 1340^{k\text{-ft}}$ using a moderate percentage of flexural steel, say, $\rho = 0.01$ ($f_y = 60$ ksi).

$$\phi M_n = \phi \rho b d^2 f_y \{1 - 0.59\rho(f_y / f'_c)\} \geq M_u$$

$0.9 \times 60 \times d^2 \times 60 \times 0.01\{1 - 0.59 \times 0.01(60 / 3)\}$
$= 1340 \times 12$

$\therefore$ $d^2 \geq 562\ \text{in}^2$; $d \geq 23.9$, say 24"

For 16" x 16" columns on a 5' wide footing, punching shear will not govern. Check the wide beam shear at d-distance from face of support. Critical location is near the interior column

$V_u = 312 - 25.2(8 + 24) / 12 = 245^k$ (conservative approximation)

$$\phi V_c = 2\phi\sqrt{f'_c}\, b_w d = 2 \times 0.85 \times \sqrt{3000} \times 60 \times 24$$
$$= 134{,}000\# = 134^k$$

Need stirrups – try # 4 double stirrup ($A_v = 4 \times 0.20 = 0.80\ \text{in}^2$)

$V_s = V_u / 0.85 - V_c = (245 - 134) / 0.85 = 130^k$

$$s \leq \begin{cases} A_v f_y d / V_s = 0.80 \times 60 \times 24 / 130 \\ \quad = 8.8,\ \text{say 9" o.c.} \\ A_v f_y / 50 b_w = 0.80 \times 60{,}000 / (50 \times 60) \\ \quad = 16"\ \text{o.c.} \\ d / 2 = 12"\ \text{o.c} \end{cases}$$

A complete design is not required per problem statement.

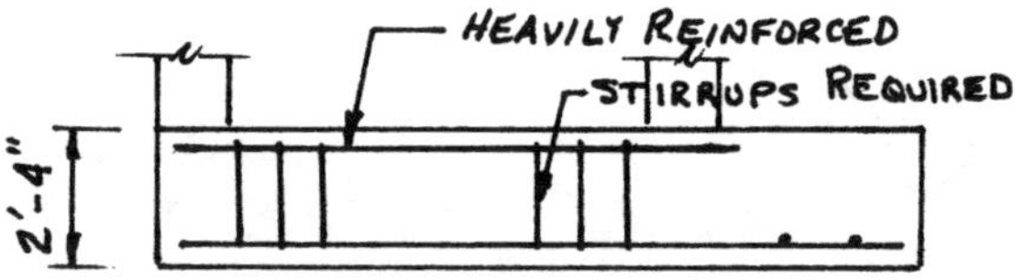

LONGITUDINAL SECTION (N.T.S.)

Concrete-22

a) Analyze the wall and draw freebody diagrams of the resisting elements. Assume that the floor slab (beyond walls) acts as a rigid diaphragm and distributes the concentrated lateral force at each level among the walls. The lateral forces are reversible; therefore, all elements act in both compression and tension. Since wall panels are significantly stiffer in compression than tension the lateral force transfers from a given level to the level below primarily through compression struts with tension ties.

- From roof to level 4:

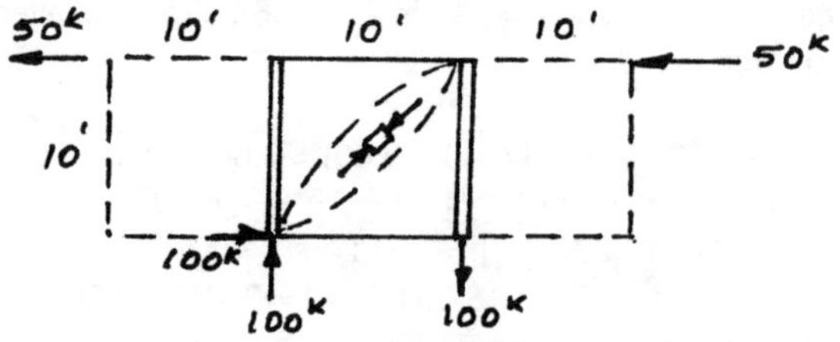

- Level 4 to level 3:

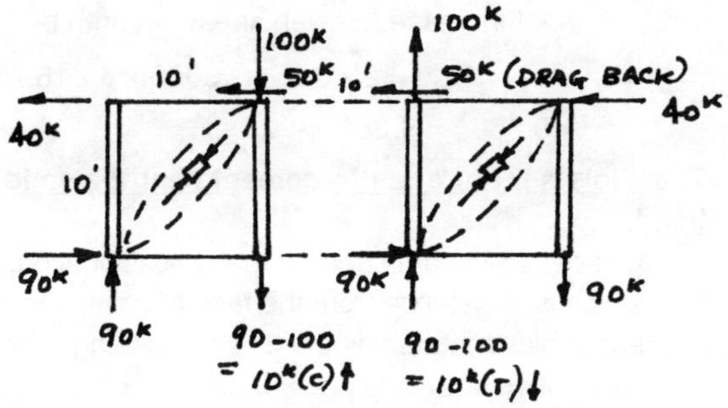

- Level 3 to level 2:

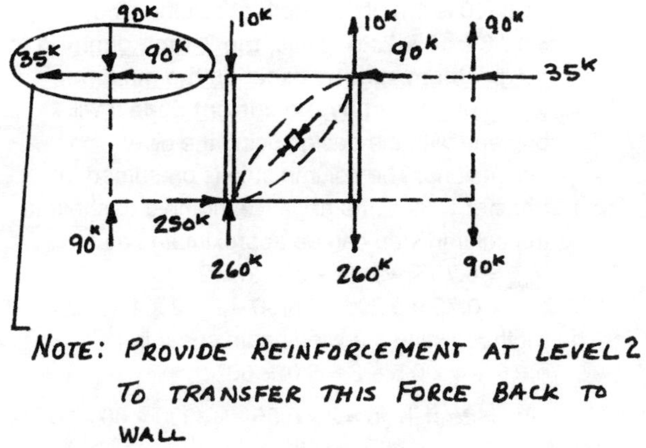

- Level 2 to level 1:

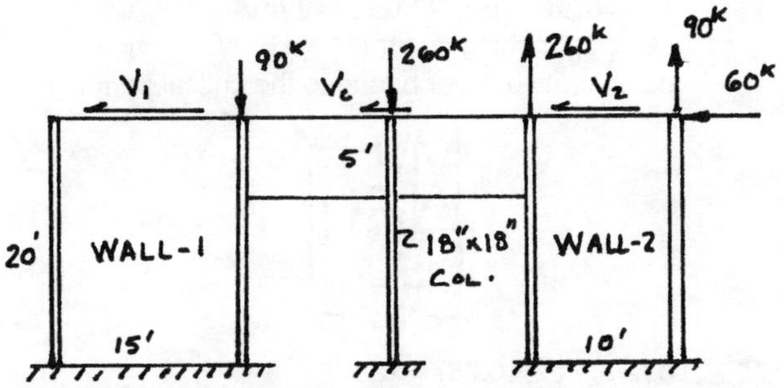

Distribute the base shear, 310^k, to the wall and column elements from ground to level 2 on the basis of their translational stiffness.

- For wall-1: Cantilevers from a fixed base; H = 20'; include both shear and flexural deformations; take $G = 0.4E_c$

  $$\frac{K_1H^3}{3E_cI_1} + \frac{1.2K_1H}{GA_1} = 1$$

  $A_1 = (15 + 1.5)\,1 = 16.5\ \text{ft}^2$

  $I_1 = 1(15 - 1.5)^3 / 12 + 2[1.5^4 / 12 + 2.25 \times 7.5^2]$
  $= 459\ \text{ft}^4$

  $$K_1 = \frac{1}{\dfrac{20^3}{3E_c(459)} + \dfrac{1.2 \times 20}{0.4E_c(16.5)}} = 0.106E_c$$

- For wall-2:

  $A_2 = (10 + 1.5)\,1 = 11.5\ \text{ft}^2$

  $I_2 = 1(10 - 1.5)^3 / 12 + 2[1.5^4 / 12 + 2.25 \times 5^2]$
  $= 165\ \text{ft}^4$

  $$K_2 = \frac{1}{\dfrac{20^3}{3E_c(165)} + \dfrac{1.2 \times 20}{0.4E_c(11.5)}} = 0.047E_c$$

- For the column: fixed top and bottom; flexural deformations only

  $$\frac{K_cH^3}{12E_cI_c} = 1; \quad \therefore\ K_c = \frac{12E_c(0.42)}{15^3} = 0.0015E_c$$

  $\Sigma K = (0.106 + 0.047 + 0.0015)E_c = 0.155E_c$

$\therefore$ $V_1 = (0.106 / 0.155) \times 310 = 213^k$

$V_2 = (0.047 / 0.155) \times 310 = 94^k$

$V_c = (0.0015 / 0.155) \times 310 = 3^k$

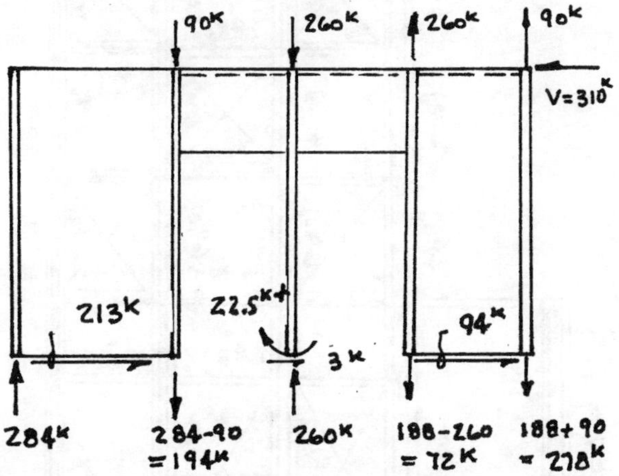

Check equilibrium of the entire structure:

$\Sigma F_x = -310 + 213 + 3 + 94 = 0$

$\Sigma F_y = 284 - 194 + 260 - 72 - 278 = 0$

$\Sigma M_{left} = 100(50) + 80(40) + 70(30) + 60(20)$
$- 194(15) + 260(25) - 72(35) - 278(45) - 22.5 \cong 0$
(residual value is practically nil compared to the overturning moment) $\therefore$ equilibrium checks

b) Determine the boundary and trim reinforcement. Gravity loads are to be neglected per problem statement. Note: UBC-97, Section 1630.8.2.1 requires that columns supporting discontinuous stiff walls be designed for the special loading of Section 1612.4. But since gravity effects are to be ignored, this check will not be included. Also, section 1921.4.4.5 applies and special detailing of confinement steel would likely be required.

Try a minimum amount of longitudinal steel for the 18" x 18" columns:

$A_{st} = \rho_g bh = 0.01 \times 18 \times 18 = 3.24\ \text{in}^2$; say, 4 # 8:
$A_{st} = 4 \times 0.79 = 3.16\ \text{in}^2$ is close enough.

$\phi P_{n,max} = \phi 0.8[0.85f'_c(A_g - A_{st}) + f_yA_{st}]$

$\phi P_{n,max} = 0.7 \times 0.8[0.85 \times 4 \times (324 - 3.16) + 60 \times 3.12] = 716^k$

Assume given loads are service values $\therefore$ apply 1.4 factor

$P_u = 1.4 \times 284 = 398^k < \phi P_{n,max}$ $\therefore$ reinforcement is governed by seismic tensile forces (in all columns since loads are reversible)

$T_{u,max} = 398^k$

$A_s \geq T_{u,max} / \phi f_y = 398 / (0.9 \times 60) = 7.37 \text{ in}^2$, **say 8 # 9**

Similarly, for

$T_u = 1.4 \times 194 = 272^k$; $A_s = 5.03 \text{ in}^2$, **say 6 # 9**

$T_u = 1.4 \times 260 = 364^k$; $A_s = 6.74 \text{ in}^2$, **say 6 # 10**

$T_u = 1.4 \times 278 = 389^k$; $A_s = 7.20 \text{ in}^2$, **say 8 # 9**

For $T < 150^k$, minimum steel governs **(4 # 8)**

For the horizontal steel

$A_4 = 1.4 \times 50 = 70^k$; $A_s = 1.30 \text{ in}^2$, say **2 # 8**

$A_3 = 1.4 \times 125 = 175^k$; $A_s = 3.24 \text{ in}^2$, say **3 # 10**

$A_2 = 1.4 \times 253 = 354^k$; $A_s = 6.55 \text{ in}^2$, **say 7 # 9**

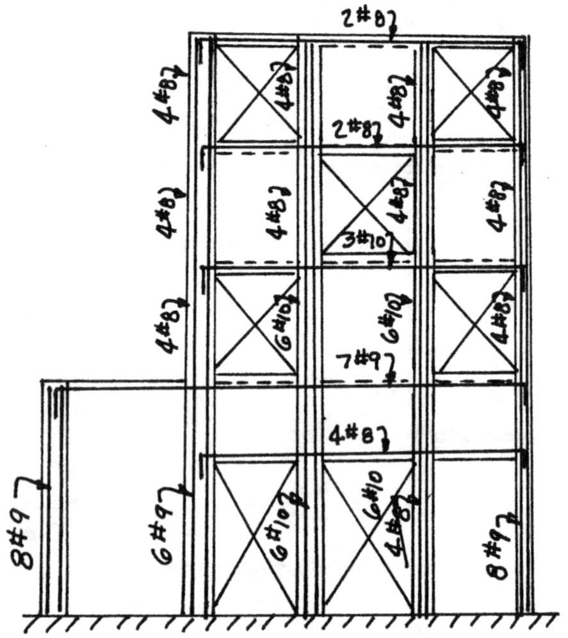

ELEVATION- BOUNDARY REINF.
(WALL & HOOP STEEL OMITTED FOR CLARITY)

Concrete-23

a) Calculate the nominal shear stresses. Use the strength design method (ACI 318); $f'_c = 3$ ksi; $f_y = 40$ ksi. A typical interior bay is 36' x 36'. Assume that the given loads are service loads and that the live load is non-reducible.

$w_u = 1.4w_d + 1.7w_l = 1.4 \times 90 + 1.7 \times 100 = 296$ psf $= 0.296$ ksf

For an overall depth given as 18.5", allow 3/4" cover and estimate the average effective depth at the crossing bar interface as d = 18.5 – 0.75 – 1 = 16.75".

- Critical shear in ribs occurs at d-distance from face of solid head

  $V_u = w_u A_{tributary}$
  $= 0.296[36 \times 36 - (6.5 + 2 \times 16.75 / 12)^2] = 358^k$

  The total shear is resisted by 12 joists

$\therefore$ $V_{u,j} = V_u / 12 = 358 / 12 = 29.8^k/_{joist}$

The joist spacing does not satisfy 8.11.3, ACI 318 $\therefore$ cannot increase the web shear strength by 10%.

$$\phi V_c = 2\phi\sqrt{f'_c}\, b_w d = 2 \times 0.85 \times \sqrt{3000} \times 6 \times 16.75 = 9{,}360^{\#} = 9{,}36^k \prec\prec V_{u,j}$$

Thus, **joists require reinforcement near the solid head.**

- Check wide beam shear at the critical location, which is d-distance from the face of column where shear plane intersects the 3'-long opening. The width of column strip is 2 × 0.25 × 36 = 18'. The maximum width of opening permitted per 13.4.2.3, ACI 319 is 1/8 of the width of column strip = 1 / 8 × 18 = 2.25'. Thus, **the 3' long opening in the problem slightly exceeds the maximum width permitted by the current code**. I will proceed with the design using the given dimensions. The column strip is presumed to transfer 75% of the negative moment $\therefore$ shear in the column strip can be approximated as

  $V_{u,cs} = 0.75 \times w_u l_2 (l_1 - c - 2d) / 2$
  $= 0.75 \times 0.296 \times 36(36 - 2 - 2 \times 1.4) / 2 = 125^k$

The width resisting the wide beam shear is

$b = 6 + 4 \times 0.5 - 3 = 5.0' = 60"$

$$\phi V_c = 2\phi\sqrt{f'_c}\, b_w d = 2 \times 0.85 \times \sqrt{3000} \times 60 \times 16.75 = 93{,}600^{\#} = 93.6^k \prec V_u$$

$\therefore$ **Shear reinforcement is required**

- Check punching shear (11.12.5, ACI 318): a conservative lower bound to the critical perimeter is

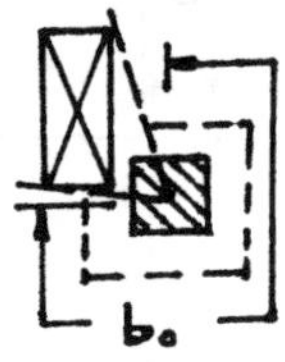

$b_o \geq 3(24 + 16.75) = 122"$

$$\phi V_c = 4\phi\sqrt{f'_c}\, b_o d = 4 \times 0.85 \times \sqrt{3000} \times 122 \times 16.75 = 381000^{\#} = 381^k$$

$V_u = w_u A_{trib} = 0.296[36 \times 36 - ((24 + 16.75) / 12)^2]$
$= 380^k < \phi V_c$

$\therefore$ **Punching shear is O.K.**

The current ACI code expresses shear strength as force rather than stress. The problem statement, however, requests answers as stresses.

Joist: $v_u = V_u / b_w d = 29{,}800 / (6 \times 16.75) = 297$ psi

Wide beam: $v_u = V_u / b_w d = 125{,}000 / (60 \times 16.75) = 124$ psi

Punching: $v_u = V_u / b_o d = 380{,}000 / (122 \times 16.75) = 186$ psi

b) Analysis shows that shear reinforcement is needed in the joist ribs. The narrow (6") width makes stirrup placement difficult (a better approach generally is to enlarge the solid head). Use a single leg # 4 stirrup: $A_v = 0.20 \text{ in}^2$

$$V_s = V_u / \phi - V_c = 29.8/0.85 - 9.36/0.85 = 24^k$$

$$s \le \begin{cases} A_v f_y d / V_s = 0.20 \times 40 \times 16.75 / 24 = 5\text{" o.c.} \\ A_v f_y / 50 b_w = 0.20 \times 40{,}000 / (50 \times 6) = 27\text{" o.c.} \\ d/2 = 8\text{" o.c} \end{cases}$$

$\therefore$ **Use # 4 @ 5" o.c.**

Since wide beam shear is excessive at the solid head, we also need stirrups in that region

$$V_s = V_u / \phi - V_c = (125 - 93.6)/0.85 = 37^k$$

try # 4 vertical leg stirrups spaced at d/2 = 8" o.c. in this region

$$A_v = V_s(s) / f_y d = 37 \times 8 / (40 \times 16.75) = 0.44 \text{ in}^2$$

Use 3 # 4 single leg stirrups across width of column strip at openings spaced 8" o.c.e.w.

Concrete-24

I assume that the given value of $0.75f_{pu}$ is the temporary jacking stress and that the specified loss of 10% "at prestress" pertains only to strand anchorage, relaxation and elastic shortening at release. I will assume an additional 12% loss to account for creep, shrinkage and additional strand relaxation before the member is placed in service. Thus, for $f_{pu} = 270$ ksi

$$f_{pe} = 0.75f_{pu}(1 - 0.22) = 0.75 \times 270(1 - 0.22) = 158 \text{ ksi}$$

$$P = f_{pe}A_{ps} = 158 \times 6 \times 0.115 = 109^k$$

$$f_{pc} = P / A = 109 / 144 = 0.757 \text{ ksi}$$

a) allowable service load under axial compression (Section 1808.5.3, UBC-97)

$$f_c = 0.33f'_c - 0.27f_{pe} = 0.33 \times 5.0 - 0.27 \times 0.757 = 1.45 \text{ ksi}$$

$\mathbf{P_{all} = f_c A_g = 1.45 \times 144 = 209^k}$

Note: The value above is the structural capacity. The capacity based on skin friction and/or end bearing would also need to be considered, but there is insufficient information to make this calculation.

b) Calculate the moment strength about the x-axis corresponding to an axial force of $P_u = 1.4 \times 100 = 140^k$.

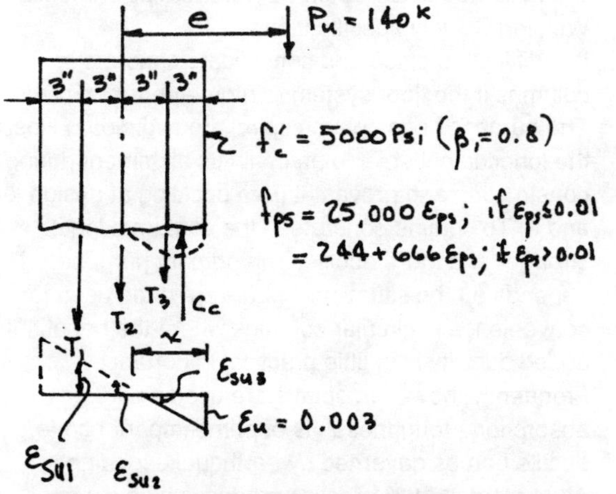

Establish the position of the neutral axis by a trial-and-error procedure to satisfy strain compatibility and equilibrium. Assume initially that the depth of the neutral axis is 6" from the compression edge.

$$\varepsilon_{ps1} = \varepsilon_{pe} + \varepsilon_{ce} + \varepsilon_{su1}$$

$$\varepsilon_{pe} = f_{pe} / E_{ps} = 158 / 25{,}000 = 0.00632$$

$$\varepsilon_{ce} = f_{ce} / E_c = 0.757 / 4030 = 0.00019$$

By similar triangles

$$\varepsilon_{su1} / 3.0 = 0.003 / 6.0; \ \therefore \ \varepsilon_{su1} = 0.0015$$

$$\varepsilon_{ps1} = 0.00632 + 0.00019 + 0.0015 = 0.00801 < 0.01$$

$$f_{ps1} = E_{ps}\varepsilon_{ps1} = 25{,}000 \times 0.00801 = 200.3 \text{ ksi}$$

$$\varepsilon_{ps2} = 0.00632 + 0.00019 = 0.00651 < 0.01$$

$$f_{ps2} = E_{ps}\varepsilon_{ps2} = 25{,}000 \times 0.00651 = 162.8 \text{ ksi}$$

$$\varepsilon_{ps3} = 0.00632 - 0.0015 = 0.00482 < 0.01$$

$$f_{ps3} = E_{ps}\varepsilon_{ps3} = 25{,}000 \times 0.00482 = 120.5 \text{ ksi}$$

$\therefore$ $T_1 = 2 \times 0.115 \times 200.3 = 46.1^k$

$T_2 = 2 \times 0.115 \times 162.8 = 37.4^k$

$T_3 = 2 \times 0.115 \times 120.5 = 27.7^k$

Neglect the chamfer ($A = 12 \times 12 = 144$ in^2)

$$\Sigma F_y = 0: P_u + T_1 + T_2 + T_3 - C_c = 0$$

$$140 + 46.1 + 37.4 + 27.7 - 0.85 \times 5 \times 12a = 0$$

$a = 4.93$"; $x = a / \beta_1 = 4.93 / 0.8 = 6.16$" --- close to the assumed value. Trial 2: try x = 6.12" (should be a value greater than the 6" originally assumed but less than the 6.16" computed based on that assumption).

$$\varepsilon_{su1} / 2.88 = 0.003 / 6.12 \ \therefore \ \varepsilon_{su1} = 0.00141$$

$$\varepsilon_{ps1} = 0.00632 + 0.00019 + 0.00141 = 0.00792 < 0.01$$

$$f_{ps1} = E_{ps}\varepsilon_{ps1} = 25{,}000 \times 0.00792 = 198 \text{ ksi}$$

$$\varepsilon_{ps2} = 0.00632 - 0.000059 = 0.00626 < 0.01$$

$$f_{ps2} = E_{ps}\varepsilon_{ps2} = 25{,}000 \times 0.00626 = 156.5 \text{ ksi}$$

$$\varepsilon_{ps3} = 0.00632 - 0.00152 = 0.00480 < 0.01$$

$$f_{ps3} = E_{ps}\varepsilon_{ps3} = 25{,}000 \times 0.00480 = 120 \text{ ksi}$$

$T_1 = 2 \times 0.115 \times 198 = 45.5^k$

$T_2 = 2 \times 0.115 \times 156.5 = 36.0^k$

$T_3 = 2 \times 0.115 \times 120 = 27.6^k$

$$\Sigma F_y = 0: 140 + 45.5 + 36.0 + 27.6 - 0.85 \times 5 \times 12a = 0$$

$a = 4.88$"; $x = a / \beta_1 = 4.88 / 0.8 = 6.11$"

$\therefore$ Converges! x = 6.12" is close enough.

$$C_c = 0.85 \times 5 \times 12 \times 4.88 = 249^k$$

Taking moments about the plastic centroid of the section and applying a capacity reduction factor, Φ, of 0.7 gives

$$\phi M_n = \phi[T_1(3) - T_3(3) + C_c(6 - a/2)]$$

$$= 0.7[45.5(3) - 27.6(3) + 249(6 - 4.88/2)]$$

$\mathbf{\Phi M_n = 658^{k\text{-}in} = 54.9^{k\text{-}ft}}$

Concrete-25

Check the double tee using UBC-97. Given: seismic zone 3; $f'_c = 5000$ psi; $f_{pu} = 270$ ksi; $A_{ps} = 0.58$ in^2 (0.29 in^2 / stem); $A = 187.7$ in^2; $I = 4256$ in^2; $y_t = 5.17$"; $y_b = 10.83$"; $e = 5.17 - 2.5 = 2.67$" (above c.g.c.)

a) Check flexural requirements at supports: assume effective prestress of

$$f_{pe} = 0.7f_{pu} - 35 = 154 \text{ ksi.}$$

$$P = f_{pe}A_{ps} = 154 \times 0.58 = 89.3^k$$

$M_{ps} = Pe = 89.3(5.17 - 2.5) = 238^{k\text{-}in}$ (top in compression)

Stresses due to prestress:

$f_{pt} = P / A + M_{ps} / S_t$

$= 89.3 / 187.5 + 238 \times 5.17 / 4256$

$= 0.765$ ksi (comp.)

$f_{pb} = P / A - M_{ps} / S_b$

$= 89.3 / 187.5 - 238 \times 10.83 / 4256$

$= -0.129$ ksi (ten.)

Cracking moments: $f_r = 7.5(f'_c)^{1/2} = 7.5(5000)^{1/2} = 530$ psi $= 0.530$ ksi. To crack the top:

$M^-_{cr} = I / y_t[0.765 + f_r] = (4256 / 5.17)[0.765 + 0.530]$

$= 1066^{k\text{-}in} = 88.8^{k\text{-}ft}$

To crack the bottom:

$M^+_{cr} = I / y_b[-0.129 + f_r]$

$= (4256 / 10.83)[-0.129 + 0.530]$

$= 157^{k\text{-}in} = 13.1^{k\text{-}ft}$

Section 1918.8.3, UBC-97, requires that the moment strength exceed $1.2M_{cr}$, unless the shear and flexural strength of the member exceeds twice the required strength. Calculate the moment strength in the negative region. The stem tapers ∴ use a trial-and-error approach to obtain the depth of the compression block. Work with one stem: $A_{ps} = 0.58 / 2 = 0.29$ in^2.

$b = 2.625 + 0.198a$; try $a = 4"$; ∴ $b = 3.4"$

Using Eqn. 18-3, with $\gamma_p = 0.4$

$\rho_p = A_{ps} / bd_p = 0.29 / (13.5 \times 3.4) = 0.00263$;

$\omega = \omega_p = 0$

$f_{ps} = f_{pu}\{1 - \gamma_p / \beta_1[\rho_p(f_{ps} / f'_c)]\}$

$= 270\{1 - 0.4 / 0.8[0.00263(270 / 5)]\} = 224$ ksi

∴ $a = A_{ps}f_{ps} / (0.85f'_c b) = 0.29 \times 224 / (0.85 \times 5 \times 3.4)$

$= 4.5"$; try again with $b = 3.5"$

$a = 0.29 \times 224 / (0.85 \times 5 \times 3.5) = 4.4"$ – close enough.

Thus,

$C_1 = 0.85 \times 5 \times 4.5 \times 2.63 = 50.3^k$

$C_2 = 0.85 \times 5 \times 4.5 \times (3.5 - 2.63) = 16.6^k$

$$\phi M_n = \phi[C_1(d_p - a / 2) + C_2(d_p - 2a / 3)]$$

$= 0.9[50.3(13.5 - 4.5 / 2) + 16.6(13.5 - 2(4.5) / 3)]$

$\Phi M_n^- = 666^{k\text{-}in} = 55.5^{k\text{-}ft}$

Thus, the moment capacity for the two stems is

$\Phi M_n^- = 2 \times 55.5^{k\text{-}ft} = 111^{k\text{-}ft} > 1.2M_{cr}^-$

Calculate ΦM_n^+: $d_p = 2.5"$; $b = b_e = 48"$

$\rho_p = A_{ps} / bd_p = 0.58 / (48 \times 2.5) = 0.0048$;

$\omega = \omega_p = 0$

$f_{ps} = f_{pu}\{1 - \gamma_p / \beta_1[\rho_p(f_{ps} / f'_c)]\}$

$= 270\{1 - 0.4 / 0.8[0.0048(270 / 5)]\} = 235$ ksi

$a = A_{ps}f_{ps} / (0.85f'_c b) = 0.58 \times 235 / (0.85 \times 5 \times 48)$

$= 0.66"$

$\Phi M_n^+ = 0.9 \times 235 \times 0.58(2.5 - 0.66 / 2) / 12 = 22^{k\text{-}ft}$

$\Phi M_n^+ > 1.2M_{cr}^+$

Check limits of 1918.8.1 (negative region):

$\omega_p = \rho_p f_{ps} / f'_c = 0.29 / (13.5 \times 3.4)(224 / 5)$

$= 0.28 < 0.36\beta_1 = 0.29$

∴ Ductile flexural failure. Check development length for $f_{ps} = 235$ ksi (1912.9.1, UBC-97); strand diameter was not given – assume 1/2" diameter $l_d = (f_{ps} - (2 / 3)f_{se})d_b = (235 - (2 / 3)154)0.5 = 66" <$ embedment $= 8 \times 12 = 96"$ ∴ o.k.

Check the calculated strength against the design bending moments. Assume 12 psf roof live load; 10 psf additional dead load

$w_d = 187.5 \times 150 / 144 + 4 \times 10 = 235$ plf

$w_l = 12 \times 4 = 48$ plf

$w_u = 1.4w_d + 1.7w_l = 1.4 \times 235 + 1.7 \times 48$

$= 411$ plf ↓

$M_u^- = w_u l^2 / 2 = 0.41 \times 8^2 / 2 = 13^{k\text{-}ft} < \Phi M_n^-$ ∴ o.k.

Section 1630.11, UBC-97, requires that in seismic zone 3 and 4, cantilevers must be designed for an upward force of $0.7C_aIW_p$. Assuming an importance factor of unity; soil profile, S_D; near-fault factor unity: $C_a = 0.36$

$w_u = 0.7 \times 0.36 \times 1 \times 235 = 59$ plf ↑

$M_u = 0.059 \times 8^2 / 2 = 1.9^{k\text{-}ft} < \Phi M_n^+$

Therefore, flexural requirements are satisfied

Concrete-26

a) During an earthquake, several cycles of inelastic deformation are expected. In a ductile frame, the design intent is usually to force plastic hinges to form in the girders rather than columns (Re: FEMA-267, Advisory No.1, 1997). If the girder reinforcement has higher than expected yield strength, the moment capacity will be higher than expected, and the hinges might form instead in the columns.

b) Ductility is achieved by using an appropriate amount of confinement steel and limiting the steel ratio.

c) The chemical composition of ASTM A 615 steel is not controlled as carefully as for ASTM A 36. ASTM sets limit only on Phosphorous for ASTM A 615; whereas, limits are established for six other elements in ASTM A36. A615 is not intended to be welded and reliable welds can be obtained only by preheat per American Welding Society specifications.

d) It is difficult to place and consolidate concrete in columns if the floor system reinforcement is in place.

e) The purposes of a column spiral are twofold: 1) Restrain the longitudinal steel to maintain its alignment during construction and prevent it from buckling at design load and 2) To confine concrete in the core, making it stronger and more ductile. The primary purpose depends on the situation. Spirals are often used for convenience in circular columns where the benefit of added ductility is of little practical importance. Frequently, however, spirals are used where energy absorption - toughness - is of prime importance (e.g., ductile frames governed by earthquake loading).

f) Air content of 10% or more produces high slump (flowable) concrete, which might tend to segregate during transport and placement.

g) The risks of placing concrete for slabs during hot, windy conditions are that the concrete may set too fast and that the surface may craze due to plastic shrinkage cracking. Some precautions that might be taken under these conditions are:
   1. Moisten forms
   2. Erect temporary sunshades and windbreakers
   3. Cool the concrete aggregates or mix
   4. Speed up the operation using more manpower and equipment.

h) A retarding admixture might be specified to offset the effect of temperature, to avoid cold joints in massive pours, and to permit special finishes for architectural concrete.

i) Some precautions when calcium chloride ($CaCl_2$) is to be added to concrete:
   1. Added to mix water rather than to batch
   2. Proper amount, by weight, is added
   3. Completely dissolve in water prior to addition to mix

j) To make a harsh mixture more workable (pumpable):
   1. Increase the amount of fines and cement paste without changes to water-cement ratio.
   2. Add mineral admixture (e.g., flyash)

k) Two factors related to handling fresh concrete that might affect shrinkage:
   1. Adding water (e.g., to make the concrete more workable)
   2. Admixtures, such as calcium chloride, which increase the demand for mix water

l) Horizontal construction joint treatment: deliberately roughen contact surface (e.g., raking fresh concrete with metal tines); remove laitance from hardened concrete (e.g., wet and blast with compressed air); moisten contact surface before placing fresh concrete.

m) Three factors other than load that affect deflection of flat slab:
   1. Camber
   2. Variation of EI over the span
   3. Support condition (i.e., continuous both ends *vs.* one end only)

n) Table 2, ASTM A615 establishes minimum yield and tensile strength for grade 40 rebar, but there is no maximum value for F_y specified.

o) Creep is time dependent strain under constant stress. Creep limits are usually established indirectly through limits on long-term deformation (i.e., deflection, camber or axial shortening.).

Concrete-27

Design on a per lineal foot basis. Use the strength design method (ACI 318); f'_c = 3 ksi; f_y = 40 ksi. Analyze the live load pattern that produces critical bending moment at joint 'A'; no live reduction; use relative stiffness to wall$_1$, (4EI / L) / K_1, taking K_1 = 100; . $I_{slab} = 12 \times 9^3 / 12 = 729$ in^4; $I_{wall1} = 12 \times 10^3 / 12 = 1000$ in^4; $I_{wall2} = 12 \times 12^3 / 12$ = 1728 in^4.

$w_u = 1.4w_d + 1.7w_l = 1.4 \times 120 + 1.7 \times 150 = 423$ plf
= 0.423 klf

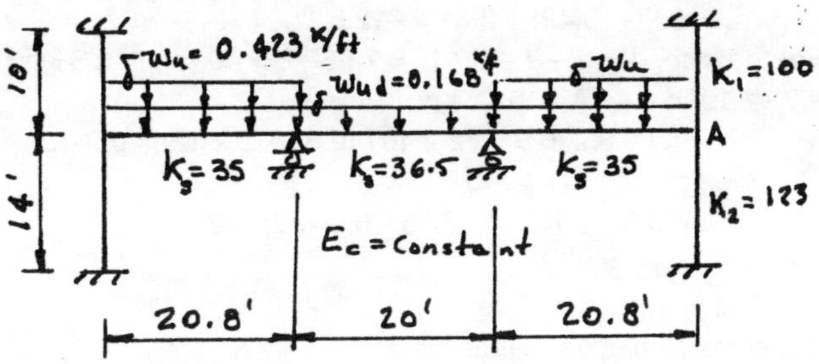

Moment distribution (units k-ft):

0.39	0.13	0.49	0.51	0.51	0.49	0.13	0.39
0.48	−15.3	15.3	−5.6	5.6	−15.3	15.3	0.48
	1.99	0.99					
	−2.62	−5.24	−5.45	−2.72			
	0.34	0.17	3.17	6.33	6.08	3.04	
	−0.82	−1.63	−1.70	−0.85	−1.19	−2.38	
	0.11	0.06	0.52	1.04	1.00	0.50	
	−0.14	−0.28	−0.29	−0.15	−0.03	−0.07	
	0.02	0.01	0.04	0.09	0.09	0.05	
	----	−0.02	−0.03	----	----	----	
Σ =	−16.42	9.35	−9.35	9.35	−9.35	16.42	

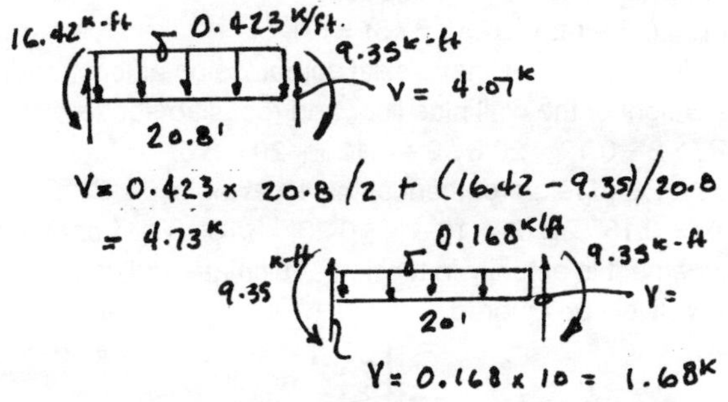

Moments caused by gravity loads may be reduced per section 8.4, ACI 318. The design is such that $\rho < 0.5\rho_b$ ∴ qualifies for a 10% reduction in negative moment. Also, take a reduction to face of support.

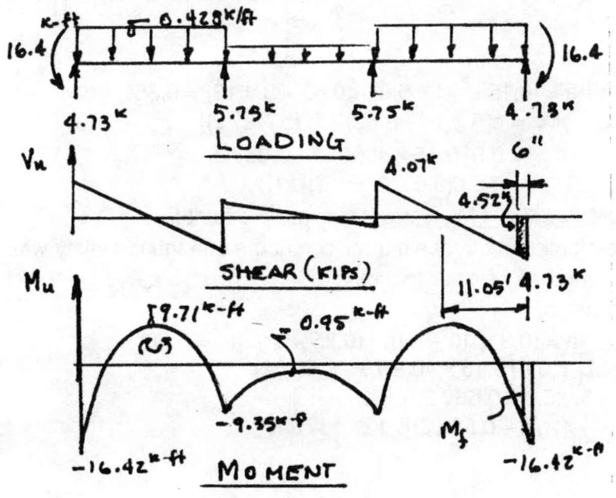

The design moment at face of support is

$M_{uf} = 0.9M_u^- + 0.5(4.73 + 4.52) \times 0.5$

$= 0.9(-16.42) + 2.31 = -12.5^{k\text{-}ft}/_{ft}$

For the slab, $d^- = 9 - 1 - 7/16 = 7.56$"; $f_y = 40$ ksi; $f'_c = 3$ ksi; # 7 @ 10" o.c. ∴ $A_s = 0.72\ in^2/_{ft}$

$a = A_s f_y / 0.85 f'_c b = 0.72 \times 40 / (0.85 \times 3 \times 12) = 0.94$"

$$\phi M_n = \phi A_s f_y (d - a/2)$$
$$= 0.9 \times 0.72 \times 40(7.56 - 0.94/2)/12$$
$$= 15.3^{k-ft}$$

∴ **$\Phi M_n \geq M_u$ and slab is o.k.**

Verify the slab shear capacity:

$$\phi V_c = 2\phi\sqrt{f'_c}\,b_w d = 2 \times 0.85 \times \sqrt{3000} \times 12 \times 7.56$$
$$= 8450^{\#} \succ\succ V_u$$

∴ shear strength is more than adequate.

Check the moment transfer to the wall:

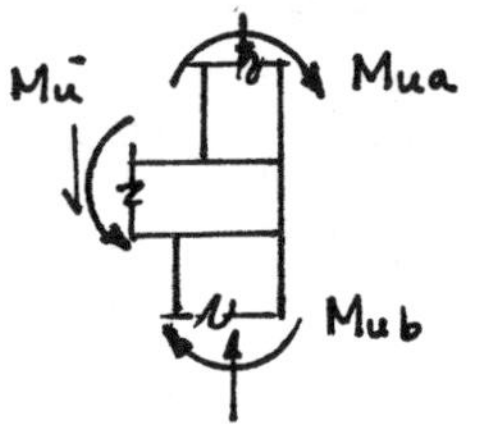

$M_{ua} = [K_a / (K_a + K_b)]M_u^- = [100 / (100 + 223)]0.9 \times 16.42$

$= 6.6^{k\text{-}ft}/_{ft}$

Reduce the moment to top of slab:

$M_u = 6.6 - (4.5 / 12)(6.6 / 5) = 6.1^{k\text{-}ft}/_{ft}$

The wall at joint 'A' will have axial compression at least equal to the weight of the wall plus floor and roof slab above

$P_d \geq 2 \times 0.12 \times 20.8 / 2 + 0.125 \times 20 = 5.0^k$

The developed reinforcement at this location is # 4 @ 16" o.c.: $A_s = 0.15^{sq.in}/_{ft}$; $d^+ = 10 - 1 - 0.25 = 8.75$". The front face reinforcement is actually in tension at ultimate and can conservatively be ignored

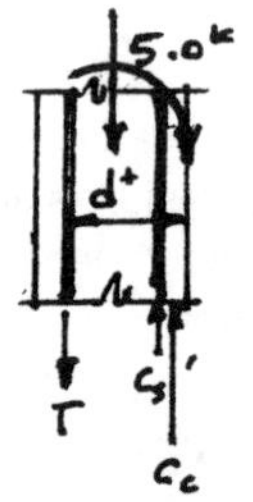

$a = [0.15 \times 40 + 5.0] / (0.85 \times 3 \times 12) = 0.36$"

$$\phi M_n = \phi[A_s f_y (d - a/2) + P_d(5 - a/2)]$$
$$= 0.9[0.15 \times 40(8.75 - 0.36/2) + 5.0(5.0 - 0.36/2)]/12$$

$\Phi M_n = 5.7^{k\text{-}ft} < M_u$ ∴ n.g. Consider the front face reinforcement to see if it will contribute enough to satisfy wall strength. By similar triangles (d' = 1.75"), $\varepsilon'_s > \varepsilon_y$ ∴ $f'_s = 40$ ksi (tension)

$a = (0.3 \times 40 + 5.0) / (0.85 \times 3 \times 12) = 0.56$"

$$\phi M_n = 0.9[0.15 \times 40(8.75 - 0.56/2) + 5.0(5.0 - 0.56/2) + 0.15 \times 40(1.75 - 0.56/2)]/12$$

$\Phi M_n = 6.2^{k\text{-}ft} > M_u$ ∴ o.k. The strength of the wall above is adequate. The strength of the wall below is obviously adequate since it has a larger A_s; larger effective depth; and larger axial force due to the additional load from the floor framing into it.

Thus, the only revision necessary to the detail provided is to extend the bottom reinforcement 6" past the face of the wall, rather than the 2" extension shown.

Concrete-28

Use the strength design method (ACI 318) to design the frame. Given: $f'_c = 3$ ksi; $f_y = 40$ ksi. Assume that the given loads are the design (i.e., factored) loads and that the member weights have been included.

- Design the horizontal member (neglect axial forces per problem statement):

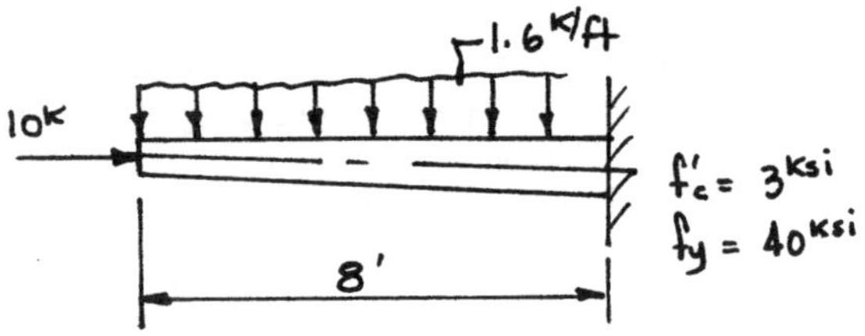

The 10^k axial force has two beneficial effects: 1) it increases the member's strength and 2) it acts to reduce the negative bending moment at the face of support. Both effects are neglected.

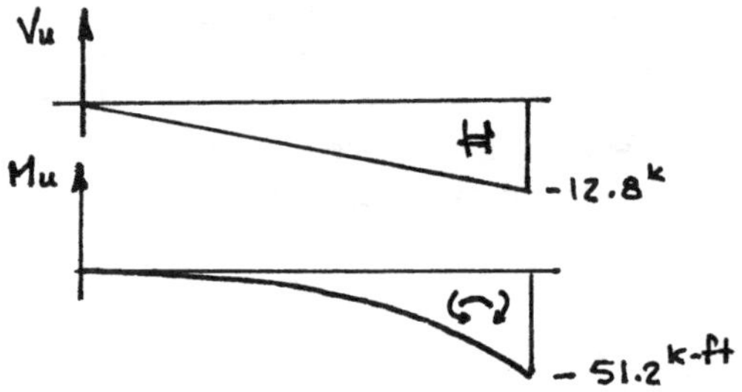

Effective depth: $d \cong h - 3 = 24 - 3 = 21$"

$$\phi M_n = \phi \rho b d^2 f_y \{1 - 0.59\rho(f_y / f'_c)\} \geq M_u$$

$0.9 \times 13 \times 21^2 \times 40\rho\{1 - 0.59\rho(40/3)\} = 51.2 \times 12$

∴ $\rho = 0.0031 < \rho_{min} = 200 / f_y = 200 / 40{,}000 = 0.005$

∴ $A_s = 0.005 \times 13 \times 21 = 1.37\ in^2$

Use 2 # 8 top; provide 2 # 5 bottom for stirrup support bars

Shear design (Chapter 11, ACI 318):

$V_u = 12.8 - w_u d = 12.8 - 1.6 \times 21 / 12 = 10^k$

$d = 21 - ((21 - 9) / 8)(21 / 12) = 18.4$"

$$\phi V_c = 2\phi\sqrt{f'_c}\,b_w d = 2 \times 0.85 \times \sqrt{3000} \times 13 \times 18.4$$
$$= 22{,}300^{\#} = 22.3^k$$

$V_u < \Phi V_c / 2$ ∴ stirrups are not required. However, use # 3 @ 12" o.c. to maintain bar alignment.

- Vertical member (Note: I assume that the 1.6 klf vertical load ends at the face of support).

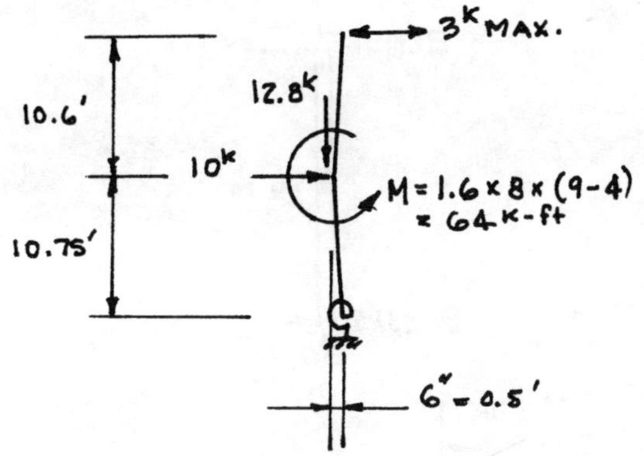

$\Sigma M_{base} = 0$: $10(10.75) - 1.6 \times 8\ (9.5 - 4) - H(21.35) = 0$

$\therefore$ $H = 1.74^k \leftarrow < 3^k$ $\therefore$ since the reaction at top is less than the given capacity, it is not necessary to develop a moment at the base for equilibrium.

The plastic centroid of the columns varies due to inclined face: $\alpha = \tan^{-1}(6 / (12 \times 10.6)) = 2.7^0$.

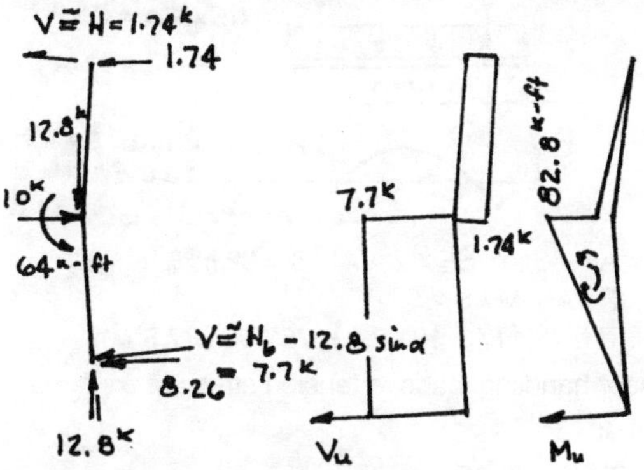

The maximum bending moment:

$$M_u = 7.7 \times 10.75 / \cos(2.7^0) = 82.8^{k\text{-}ft}$$

With d = 21"; b = 13"

$$\phi M_n = \phi \rho b d^2 f_y\{1 - 0.59\rho(f_y / f_c^{'})\} \geq M_u$$

$$0.9 \times 13 \times 21^2 \times 40\rho\{1 - 0.59\rho(40 / 3)\} = 82.8 \times 12$$

$\therefore$ $\rho = 0.0054 > \rho_{min} = 200 / f_y = 200 / 40{,}000 = 0.005$

$\therefore$ $A_s = 0.0054 \times 13 \times 21 = 1.47\ in^2$

Use 2 # 8 outside face

The shear near the joint is smaller than in the horizontal member $\therefore$ stirrups are not required – use # 3 @ 12" o.c. (adequate for ties). The inside face is in compression throughout. Provide 2 # 5 on inside face as supplemental column steel.

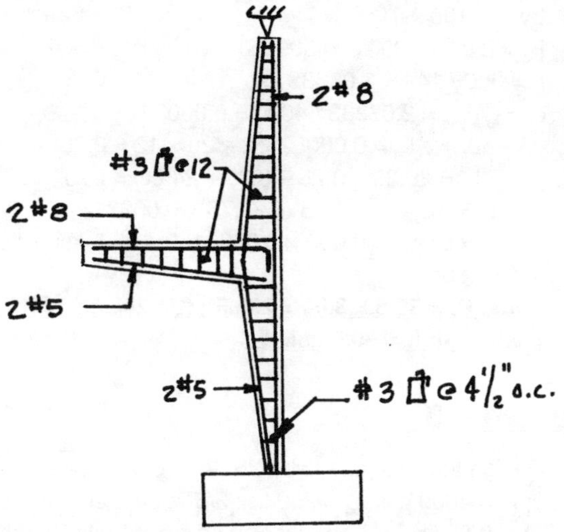

ELEVATION (N.T.S.)
(FOOTING REINFORCEMENT & DOWELS OMITTED FOR CLARITY)

Note: at the bottom, the effective depth is d = 12 – 3 = 9" and $V_u = 7.7^k$

$$\phi V_c = 2\phi\sqrt{f_c^{'}}\, b_w d = 2 \times 0.85 \times \sqrt{3000} \times 13 \times 9$$

$$= 10{,}900^{\#} = 10.9^k$$

Since $V_u > \Phi V_c / 2$ $\therefore$ need minimum stirrups # 3 @ d / 2 = 4-1/2" o.c. at the bottom.

Concrete-29

Calculate the stresses in the steel and concrete after prestressing. The 0.25" reduction in the distance between the interior anchorage points amounts to a rigid body movement of the concrete elements. Thus, the original unstressed length of strand is

$$L_s = (44 \times 12) - 0.25 = 527.75"$$

Use a trial-and-error approach to satisfy strain compatibility and equilibrium:

1. try $f_s = 50$ ksi

   $P_s = 3\ in^2 \times 50\ ksi = 150^k$

   Assume that the given $A_c = 144\ in^2$ is the net area (deduction already taken for strand duct)

   $f_c = P_s / A_c = 150 / 144 = 1.0417$ ksi.

   $\varepsilon_c = f_c / E_c = 1.0417 / 4000 = 0.00026$

   $\therefore$ $\Delta L_c = \varepsilon_c L_c = 0.00026 \times 2 \times 20 \times 12 = 0.125"$

   $\Delta L_s = 26.25 - 0.25 - 0.125 - 24.00 = 1.875"$

   $\varepsilon_s = \Delta L_s / L_s = 1.875 / 527.75 = 0.00355$

   $f_s = E_s\ \varepsilon_s = 30{,}000 \times 0.00355 = 107$ ksi

   Thus, the actual stress is somewhere between the assumed 50 ksi and the computed 107 ksi; actually closer to 107 ksi.

2. try $f_s = 100$ ksi
   $P_s = 3\text{ in}^2 \times 100\text{ ksi} = 300^k$
   $f_c = 300 / 144 = 2.08333$ ksi.
   $\varepsilon_c = f_c / E_c = 2.08333 / 4000 = 0.00052$
   $\therefore$ $\Delta L_c = \varepsilon_c L_c = 0.00052 \times 2 \times 20 \times 12 = 0.25"$
   $\Delta L_s = 26.25 - 0.25 - 0.25 - 24.00 = 1.75"$
   $\varepsilon_s = \Delta L_s / L_s = 1.75 / 527.75 = 0.003316$
   $f_s = E_s\,\varepsilon_s = 30{,}000 \times 0.0033169 = 99.5$ ksi – close enough!
   Thus, $P_s = 99.5 \times 3.00 = 298.5^k$; $f_c = 298.5 / 144 = 2.07$ ksi; $f_s = 99.5$ ksi.

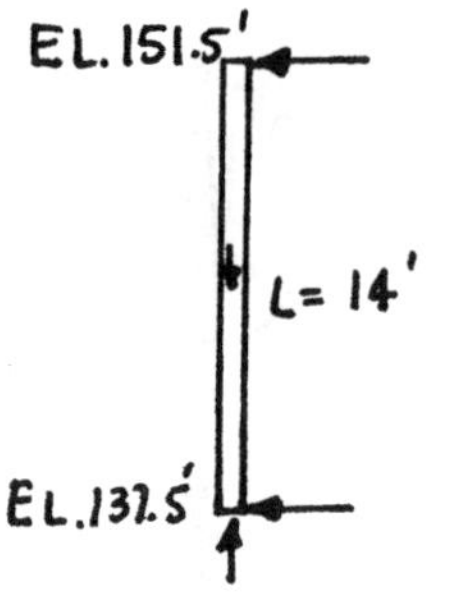

Concrete-30

a) Design the wall panels by UBC-97. Given: seismic zone 3; $f'_c = 4000$ psi; $f_{pu} = 250$ ksi; 3/8" diameter strands ($A = 0.08\text{ in}^2$ / strand). Design on a per foot of width basis: $A = 4 \times 12 = 48\text{ in}^2$; $w_p = 48 \times (150 / 144) = 50$ psf.
   Compute the seismic load per Section 1632.2: soil profile type S_D $\therefore$ from Table 16-Q, $C_a = 0.36$; importance factor $I_p = 1.0$. From Table 16-O for exterior walls at or above grade: $a_p = 1.0$; $R_p = 3.0$ (note: 1632.2 requires that the panel connectors be designed using $R_p = 1.5$; thus, twice the loading required on the panel itself).

   $$F_p = (a_p C_a I_p / R_p)[1 + 3(h_x / h_r)]w_p$$

   For this case, $h_x = h_r = 14'$

   $$F_p = (1.0 \times 0.36 \times 1.0 / 3.0)[1 + 3(14 / 14)]50 = 24\text{ psf}$$

   This value is ultimate. For checking service load stresses, the lateral seismic force is $24 / 1.4 = 17$ psf (Re: 1612.3.2), which governs lateral loading over the given wind loading of 15 psf. Bending moment in the panel under handling results in a critical value of $960^{\text{lb-ft}}$. Solve for the effective prestress required to resist both bending moments. No tension is allowed under the design load

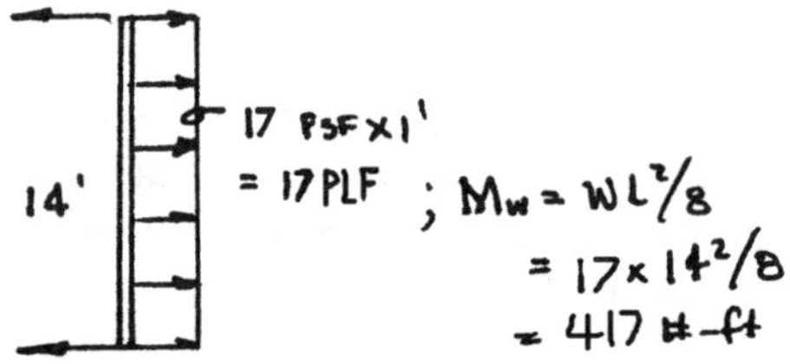

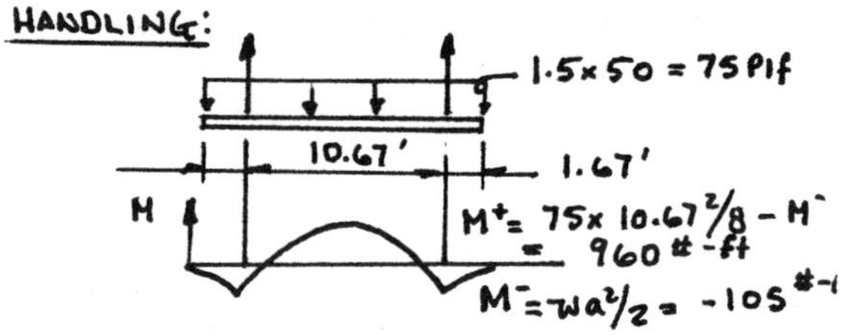

$A = 48\text{ in}^2$; $S = 12 \times 4^2 / 6 = 32\text{ in}^3/_{\text{ft}}$

$P / A - M / S \geq 0$

$\therefore$ $P \geq 48[417 \times 12 / 32] = 7500\text{ lb} = 7.5^k/_{\text{ft}}$

Under handling loads, a tension stress of $3\sqrt{f'_c}$ is permitted

$P / A - M / S + 3\sqrt{f'_c} \geq 0$

$\therefore$ $P \geq 48[(960 \times 12 / 32) - 3\sqrt{4000}\,] = 8180$ lb $= 8.18^k/_{\text{ft}} \leftarrow$ governs

For 3/8" dia. strand, $A_p = 0.08\text{ in}^2$; losses of 35 ksi are given

$f_{pe} = 0.74 f_{pu} - \text{losses} = 0.74 \times 250 - 35 = 150$ ksi

$\therefore$ $A_{ps} \geq P / f_{pe} = 8.18^k / 150^{\text{ksi}} = 0.055\ ^{\text{sq.in}}/_{\text{ft}}$

For an 8-ft wide panel:

$n \geq 8 \times 0.055\text{ in}^2 / 0.08\text{ in}^2$ / strand = 5.5; say, 6 strands

Transfer reinforcement:

$A_s \geq 0.002bh = 0.002 \times 12" \times 4" = 0.096\ ^{\text{sq.in}}/_{\text{ft}}$

Check the handling stresses across the 8' width of panel:

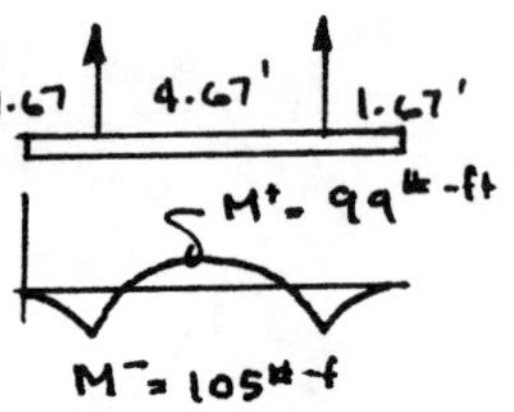

$f_b = M_{max} / S = 105 \times 12 / 32 = 39\text{ psi} < 3\sqrt{f'_c}$

$\therefore$ flexural reinforcement is not needed; use the minimum computed above, say **# 3 @ 12" o.c. in the transfer direction.**

b) Calculate the actual stresses based on the 6-3/8" diameter strands:

$P_e = 6 \times 0.08\ in^2 \times 150\ ksi = 72^k = 9^k/_{ft} = 9000^{\#}/_{ft}$

Handling: $M_{max} = 960^{\#\text{-}ft}$

$f = P / A + M / S = 9000/48 + 960 \times 12 / 32$

$= 548$ psi (comp)

$f = P / A - M / S = 9000 / 48 - 960 \times 12 / 32$

$= -173$ psi (ten)

Design lateral loading: $M_{max} = 417^{\#\text{-}ft}$

$f = P / A + M / S = 9000 / 48 + 417 \times 12 / 32$

$= 344$ psi (comp)

$f = P / A - M / S = 9000 / 48 - 417 \times 12 / 32$

$= 31$ psi (comp)

c) Check the moment capacity:

$M_u = 1.1M_E = 1.1(0.024 \times 14^2 / 8) = 0.65^{k\text{-}ft}/_{ft}$

Using Eqn. 18-3., with $\gamma_p = 0.4$; $\beta_1 = 0.85$

$\rho_p = A_{ps} / bd_p = (6 \times 0.08 / 8) / (12 \times 2) = 0.0025;$

$\omega = \omega_p = 0$

$f_{ps} = f_{pu}\{1 - \gamma_p / \beta_1[\rho_p(f_{ps} / f'_c)]\}$

$= 250\{1 - 0.4 / 0.85[0.0025(250 / 4)]\} = 232$ ksi

Check limits of 1918.8.1:

$\omega_p = \rho_p f_{ps} / f'_c = 0.0025(232 / 4) = 0.145 < 0.36\beta_1 = 0.31$

∴ Ductile flexural failure

$a = A_{ps}f_{ps} / (0.85f'_c b) = 0.06 \times 232 / (0.85 \times 4 \times 12) = 0.34"$

$\phi M_n = \phi A_{ps} f_{ps}(d_p - a / 2)$

$= 0.9 \times 0.06 \times 232(2 - 0.34 / 2)$

$\phi M_n = 22.9\ ^{k\text{-}in}/_{ft} = 1.91\ ^{k\text{-}ft}/_{ft} \succ M_u = 0.65\ ^{k\text{-}ft}/_{ft}$

∴ **The section is adequate**

Concrete-31

Properties of the gross area:

$A = 20 \times 6 + 10 \times 14 = 260\ in^2$

$$\bar{y} = \frac{120(14+3)+140(7)}{260} = 11.62"$$

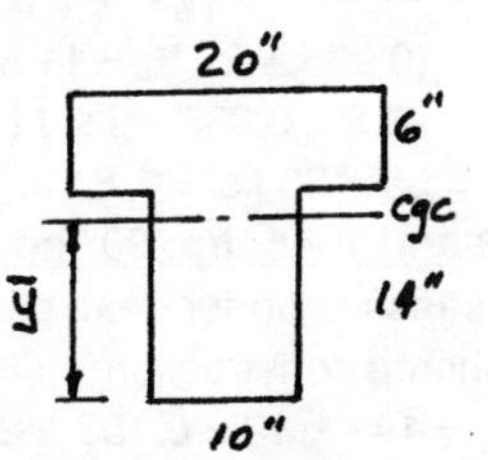

$$I = \sum_{i=1}^{n} (I_{xo} + Ad^2)_i$$

$= 20 \times 6^3 / 12 + 120(17 - 11.62)^2 + 10 \times 14^3 / 12 + 140(11.62 - 7)^2$

$= 9108\ in^4$

a) Calculate the secondary effects of prestressing. Use the equivalent load concept and the method of consistent displacements (note: other methods of linear structural analysis could also be used (e.g., moment distribution) but I believe that consistent displacements is the best approach). Release the interior support:

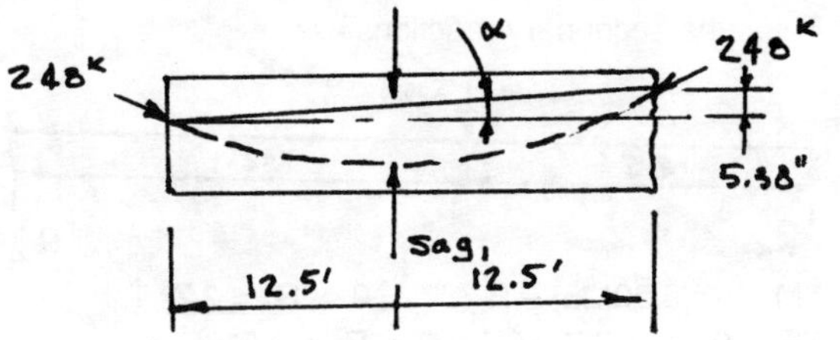

$sag_1 = (11.62" - 7.41") + 5.38" / 2 = 6.9"$

$w_{e1} = P(sag_1)(8) / L_1^2 = 248 \times 6.9 \times 8 / 300^2$

$= 0.153^k/_{in}$

$\alpha \cong \sin\alpha = 5.38 / (300) = 0.01793$ radians

Thus, the equivalent load on span 1:

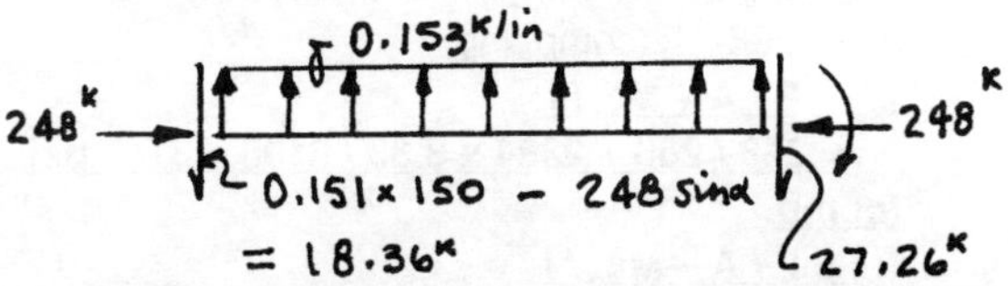

For span 2:

$sag_2 = 5.38" / 2 + 8.62" = 11.31"$

$w_{e2} = P(sag_2)(8) / L_2^2 = 248 \times 11.31 \times 8 / (12 \times 32)^2$

$= 0.153^k/_{in}$

$\alpha \cong \sin\alpha = 5.38 / (12 \times 32) = 0.01401$ radians

The equivalent load on span 2:

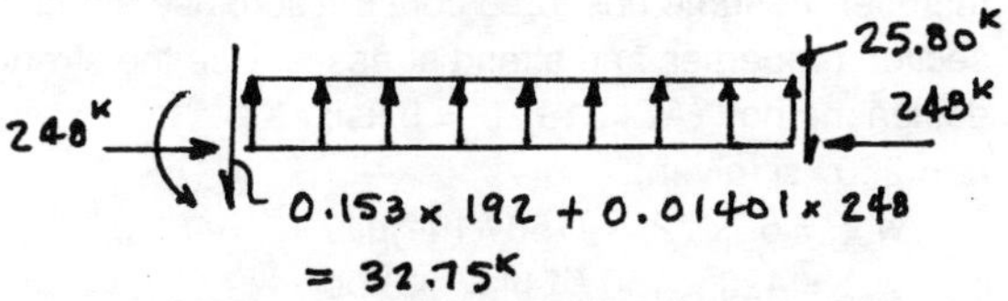

Thus, the equivalent loading on the released structure:

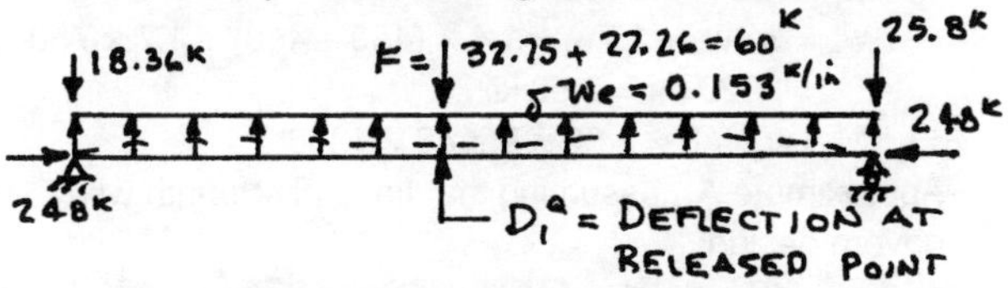

$D_1^Q = (w_1L_1 / 24EI)\{L^3 - 2LL_1^2 + L_1^3\} - FL_1^2L_2^2 / (3EIL)$

$(D_1^Q)EI = 0.153(300)(684^3 - 2 \times 684 \times 300^2 + 300^3) / 24 - 60 \times 300^2 \times 384^2 / (3 \times 684)$

$= 4.01539 \times 10^7 \uparrow$

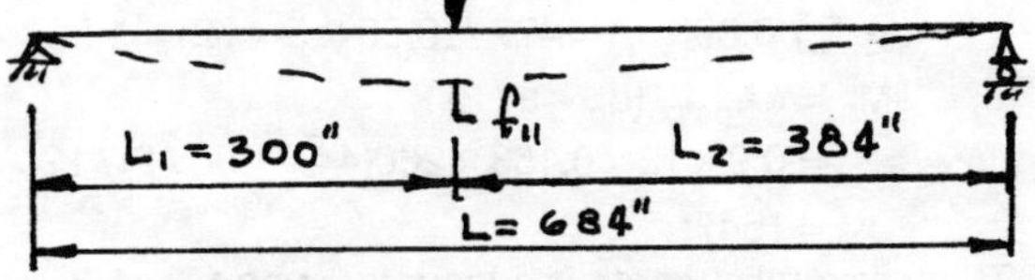

$f_{11} = (1)L_1^2L_2^2 / (3EIL) = 300^2 \times 384^2 / (3EI \times 684)$

$(f_{11})\ EI = 6.4670 \times 10^6 \downarrow$

For consistent displacements:

$R_B f_{11} + D_1^Q = 0$

$R_B(6.4670 \times 10^6) = 4.01539 \times 10^7$

∴ $R_B = 6.20^k \downarrow$

b) Thus, the secondary reactions are

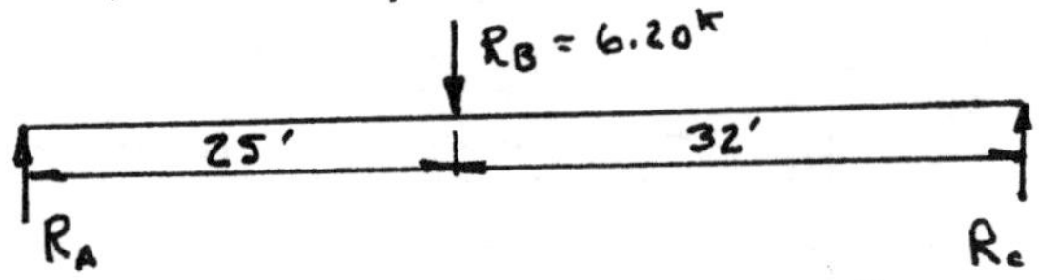

$\Sigma M_A = 0$: $6.20(25) - R_C(57) = 0$ ∴ $\underline{R_C = 2.7^k \uparrow}$

$\Sigma F_y = 0$: $R_A + 2.7 - 6.2 = 0$ ∴ $\underline{R_A = 3.5^k \uparrow}$

$\underline{(M_B)_{sec} = R_A L_1 = 3.5 \times 25 = 87.5^{k\text{-}ft}}$ (compression on top)

c) Compute the stresses at the center support due to prestress:

$M_{ps} = M_{primary} + M_{secondary} = 248(8.38 - 3) + 3.5 \times 300$
$= 2384^{k\text{-}in}$ (comp. top)

$f_{top} = P / A + Mc_t / I$
$= 248 / 260 + 2384 \times 8.38 / 9108 = 3147$ psi (comp.)

$f_{bot} = P / A - Mc_b / I$
$= 248/260 - 2384 \times 11.62 / 9108$
$= -2088$ psi (ten.)

Concrete-32

a) Calculate the required prestress force in the composite member. Assume unshored construction; use the given section properties and strand stresses; Use the strength design method (ACI 318); $f'_c = 5$ ksi; $f_{ps} = 260$ ksi (given).

$w_{slab} = 6" \times 72" \times (150 / 144)\text{plf / in}^2 = 450$ plf
$w_{p.c.} = 420 \times 150 / 144 = 440$ plf
$w_l = 125 \times 6 = 750$ plf
$w_u = 1.4w_d + 1.7w_l = 1.4 \times (450 + 440) + 1.7 \times 750$
$= 2520$ plf $= 2.52$ klf
$M = w_u l^2 / 8 = 2.52 \times 65^2 / 8 = 1330^{k\text{-}ft}$

Approximate A_{ps} assuming that flexural strength will govern design:

$d_p = 42 + 6 - 4 = 44"$
$d_p - a / 2 \cong 0.95d = 0.95 \times 44 = 41.8"$
∴ $A_{ps} = M_u / (\Phi f_{ps} 0.95 d_p) \cong 1330 \times 12 / (0.9 \times 260 \times 41.8) = 1.65 \text{ in}^2$

For 1/2" diameter strand (0.153 in² / strand):

$n \geq 1.65 / 0.153 = 10.7$ (say 12 strands)
$T_p = f_{ps} A_{ps} = 260 \times 12 \times 0.153 = 477^k$
$a = T_p / 0.85 f'_c b_e) = 477 / (0.85 \times 5 \times 72) = 1.56"$
$\phi M_n = \phi A_{ps} f_{ps} (d_p - a / 2)$
∴ $= 0.9 \times 12 \times 0.153 \times 260(44 - 1.56 / 2) / 12$
$= 1547^{k\text{-}ft}$

Thus, flexural strength is adequate – **use 12 – 1/2" diameter strands**: $A_{ps} = 1.84 \text{ in}^2$

b) Check the stresses at transfer: The critical location occurs at the end of the transfer length $\cong 50d_b \cong 24"$. Assume a straight strand pattern with $e = 21 - 4 = 17"$

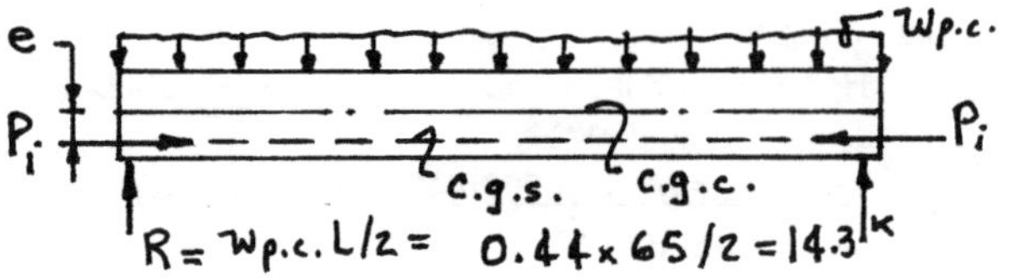

$P_i = f_{pi} A_{ps} = 167 \text{ ksi} \times 1.84 \text{ in}^2 = 307^k$

at x = 2' from the end of member:

$M_d = 0.5(14.3 + (14.3 - 2 \times 0.44)) \times 2 = 27.7^{k\text{-}ft}$ (comp. top)
$M_{ps} = 307 \times 17 / 12 = 435^{k\text{-}ft}$ (ten. top)

Stresses ($A = 420 \text{ in}^2$; $S = I / 21 = 2933 \text{ in}^3$)

$f = P_i / A \pm (M_d - M_{ps})/S$
$f_{top} = 307 / 420 + (27.7 - 435) \times 12 / 2933 = -0.935$ ksi (tension)
$f_{top} = 307 / 420 - (27.7 - 435) \times 12 / 2933 = 2.39$ ksi (comp)

The strength of concrete at transfer was not specified. ACI 318 limits the compressive stress at transfer to $0.6f'_{ci}$. This requires

$f'_{ci} \geq 2.39 / 0.6 = 3.98$ ksi $= 3980$ psi; say 4000 psi.

The computed tensile stress at transfer is limited to $6\sqrt{f'_{ci}}$, which cannot be accommodated by high enough strength. There are several options:

1. Depress the strands to reduce the end eccentricity
2. Blanket several strands to reduce the prestress near the end of member
3. Add mild steel to resist the total tensile force at the top at transfer
   I choose the last option: to add mild steel at a working stress of 30 ksi (grade 60 rebars):

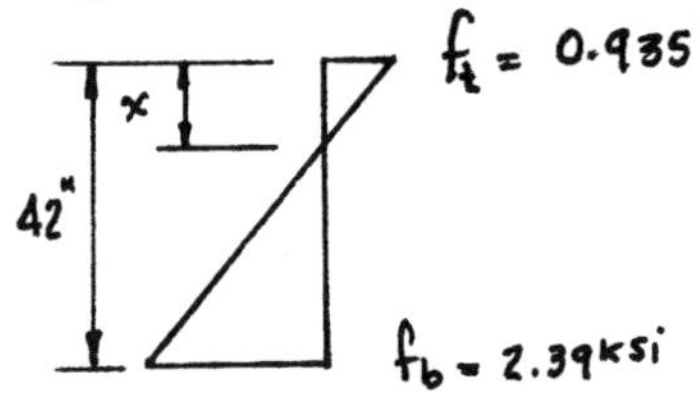

$x = 0.935 + (0.935 + 2.39) = 11.8"$
$T = 0.5 f_t x b = 0.5 \times 0.935 \text{ ksi} \times 11.8" \times 10" = 55^k$
$A_s \geq T / f_s = 55^k / 30 \text{ ksi} = 1.84 \text{ in}^2$
Use 2 # 8 and 1 # 6 top; 90° hook at end

c) Calculate the stresses under dead plus live load assuming unshored construction:

$P_e = f_{pe} A_{ps} = 147 \times 12 \times 0.153 = 270^k$

Dead plus prestress at midspan

$M_d = (w_{slab} + w_{pc})L^2 / 8 = (0.45 + 0.44)65^2 / 8 = 470^{k\text{-}ft}$ (comp.top)
$M_{ps} = P_e e = 270 \times 17 = 4590^{k\text{-}in} = 383^{k\text{-}ft}$ (ten.top)
$f_{top} = P_e / A + (M_d - M_{ps}) / S$
$= 270 / 420 + (470 - 383) \times 12 / 2933 = 1.0$ ksi
$= 1000$ psi (comp)
$f_{bot} = P_e / A - (M_d - M_{ps}) / S$
$= 270 / 420 - (470 - 383) \times 12 / 2933$
$= 0.287$ ksi $= 287$ psi (comp)

Superimpose the live load stresses acting on the composite section:

$M_L = w_L L^2 / 8 = 0.75 \times 65^2 / 8 = 396^{k\text{-}ft}$ (comp.top)

$f'_{bot} = -M_L / S_{tb} = -396 \times 12 \times 33.2 / 184{,}000$
$= -0.857$ ksi (ten)

$f'_{top} = 396 \times 12 \times (42 - 33.2) / 184{,}000$
$= 0.227$ ksi (comp)

$f_{top} = 396 \times 12 \times (48 - 33.2) / 184{,}000 = 0.382$ ksi (comp)

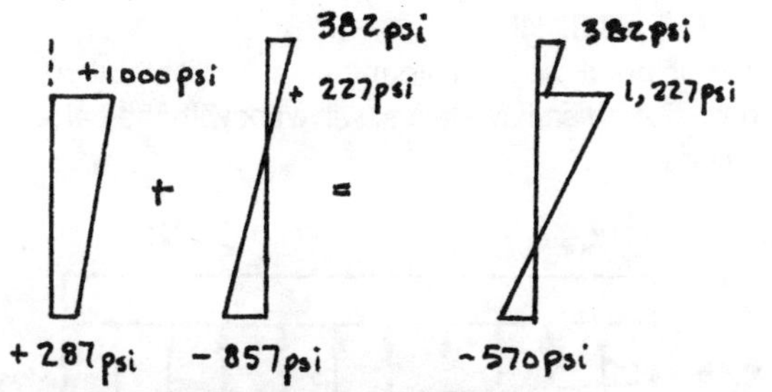

Thus, the calculated stresses are:

f_{bot} = 570 psi (tension)

$f_{top,pc}$ = 1227 psi (compression)

f_{top} = 382 psi (compression)

Note that the calculated tension stress exceeds $6\sqrt{f'_c}$ but is less than $12\sqrt{f'_c}$. This is acceptable provided that deflections are computed per section 18.4.2, ACI 318.

d) Depending on the age of the precast at the time the slab and closure are cast, the beam will experience long term deformation that will likely result in shortening and upward deflection (i.e., additional camber). This, in turn, will cause a separation between the precast member and the closure concrete. Consequently, with application of the live load, there will not be significant moment resistance at the connection. Thus, the connection is flexible, behaving essentially as a simple connection.

Concrete-33

Use the strength design method (ACI 318); $f'_c = 4$ ksi; $f_y = 40$ ksi.; all given forces and moments are the design (i.e., factored) values.

a) Determine the flexural steel required to resist the negative moment of $100^{k\text{-}ft}$; $d^- = 20 - 1.5 - 0.375 - 0.5 = 17.63"$; $b = b_w = 10"$

$$\phi M_n = \phi \rho b d^2 f_y \{1 - 0.59\rho(f_y / f'_c)\} \ge M_u$$

$$0.9 \times 10 \times 17.63^2 \times 40\rho\{1 - 0.59\rho(40/4)\} = 100 \times 12$$

$\therefore$ $\rho = 0.0117 > \rho_{min} = 200 / f_y = 0.005$

$A_s = 0.0117 \times 10 \times 17.63 = 2.07$ in^2

Note: the 3 # 8 top bars are adequate for flexure; however, current ACI 318, Section 10.6 requires that a portion of this steel be distributed into the flange of the tee-section for crack control.

b) Calculate the torsion stress. The dimensions of the equivalent hollow section used in the current code procedure are defined in the figure below.

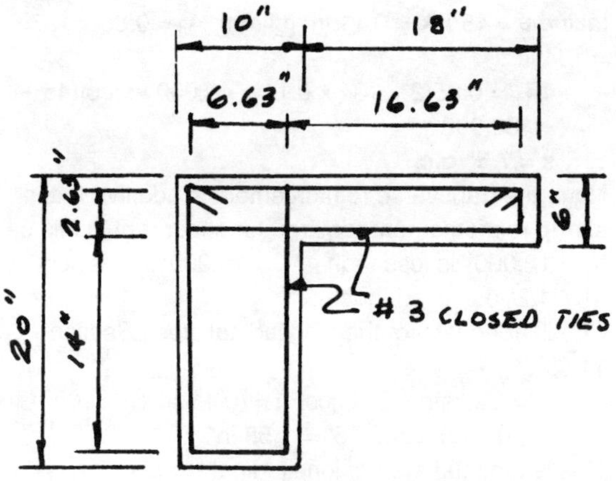

SECTION (N.T.S.)
(Longitudinal Steel Omitted for Clarity)

$p_h = 2[(16.63 + 6.63) + (18 + 2.63)] = 87.8$ in

$A_{oh} = (16.63 \times 6.63) + (18 \times 2.63) = 157.6$ in^2

$p_{cp} = 2[(10 + 20) + (6 + 18)] = 108$ in

$A_{cp} = 10 \times 20 + 6 \times 18 = 308$ in^2

The torsional shear stress can be calculated consistent with the current code approach from the expression $\tau = T / (2A_0 t)$, where A_0 = gross area of the shear flow path and can be approximated as $0.85A_{ch}$; t = wall thickness. For the given section, $t = 2 \times 1.63 = 3.26"$; $A_0 = 0.85 \times 157.6 = 134$ in^2

$\tau_u = (11{,}000^{lb\text{-}ft} \times 12^{in}/_{ft}) / (2 \times 134 \text{ in}^2 \times 3.26")$
= 151 psi

c) Vertical shear stress:

$v_u = V_u / b_w d = 12{,}000 / (10 \times 17.63) = 68$ psi

d) Check the spacing of transverse reinforcement (Section 11.6.3, ACI 318):

$$T_u = 11{,}000 \times 12 = 132{,}000^{lb-in}$$

$$> \phi\sqrt{f'_c}\left(\frac{A_{cp}^2}{p_{cp}}\right) = 0.85 \times \sqrt{4000}\left(\frac{303^2}{108}\right) = 45{,}700^{lb-in}$$

$\therefore$ Torsion must be considered. The spandrel beam is a case of compatibility torsion and the upper bound $T_u \le 4\phi\sqrt{f'_c}\left(\frac{A_{cp}^2}{p_{cp}}\right) = 183{,}000^{lb-in}$ applies; however, the given $T_u = 132{,}000^{lb\text{-}in}$ controls.

Checking Eqn. 11-18:

$$\sqrt{\left(\frac{V_u}{b_w d}\right)^2 + \left(\frac{T_u p_h}{1.7A_{oh}^2}\right)^2} \le \phi\left(\frac{V_c}{b_w d} + 8\sqrt{f'_c}\right)$$

$$\sqrt{\left(\frac{12{,}000}{10 \times 17.63}\right)^2 + \left(\frac{132{,}000 \times 87.3}{1.7 \times 157.6^2}\right)^2}$$

$$= 281 \text{ psi} \le 0.85\left(\frac{22{,}300}{10 \times 17.63} + 8\sqrt{4000}\right)$$

$$= 537 \text{ psi} - \text{o.k.}$$

From section 11.6.3.5

$\Phi T_n = (2A_o A_t f_{yv} / s) \cot\theta > T_u$

taking $\theta = 45°$; $A_t = 0.11$ in^2 (given); $A_0 = 0.85 \times 157.6$ $= 134$ in^2:

$\Phi T_n = 0.85(2 \times 134 \times 0.11 \times 40{,}000 / s)\cot(45°)$
$= 132{,}000$

$\therefore$ **s = 7.5" o.c.**

Note: the transverse reinforcement is additive to the stirrup reinforcement required for shear, but in this case, $V_u = 12{,}000^{lb}$ is less than $\Phi V_c = 18{,}900^{lb}$. Therefore, # 3 @ 7-1/2" o.c. is adequate.

e) Check adequacy of the longitudinal steel, Section 11.6.3.7

$A_l = (A_t / s)p_h(f_{yv} / f_{yl})\cot^2\theta = (0.11 / 7.5) \times 108 \times (40 / 40) \times \cot^2 45° = 1.58$ in^2

This is the total area of longitudinal steel, which is additive to the steel required for shear and flexure. This steel is to be distributed around the perimeter, p_h, at a spacing not to exceed 12" o.c. For the 3 # 8 top bars,

$A_{s,provided} = 3 \times 0.79 = 2.37$ in^2

$A_{s,required} = 2.05$ in^2

$\therefore$ $A_{s,available} = 2.37 - 2.05 = 0.32$ in^2

This excess steel combined with the # 4 longitudinal bars and the 2 # 7 bottom bars gives 0.32 + 0.40 + 1.20 = 1.92 in^2, which satisfies the longitudinal steel requirement. Thus, **longitudinal steel is o.k.**

Concrete-34

a) Design the combined footing for full dead and live load condition: $P = 596^k$

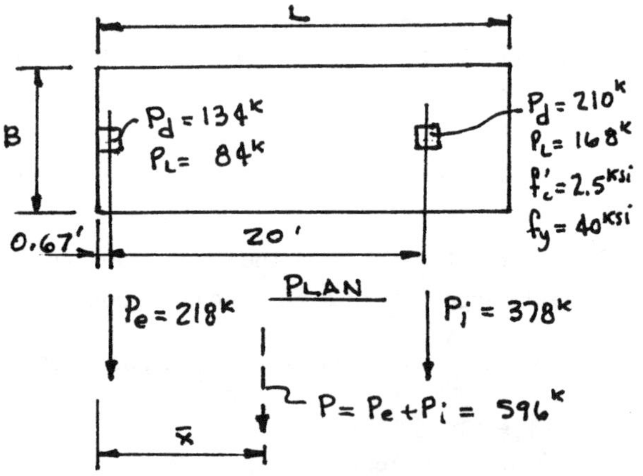

Proportion footing length to achieve uniform bearing pressure (3 ksf net pressure):

$$\bar{x} = [218^k(0.67') + 378^k(20.67')] / 596^k = 13.35'$$

$$\therefore \quad L = 2\bar{x} = 2 \times 13.35' = 26.7', \text{ say } 26'\text{-}8"$$

The width required to limit the bearing pressure to 3 ksf is

$B \geq P / f_pL = 596^k / [(3 \text{ ksf}) \times 26.7'] = 7.4'$, say 7'-6"

Thus, use a footing with plan dimensions **7'-6" x 26'-8"**

b) Design of reinforcement: Use the strength design method (ACI 318); $f'_c = 2500$ psi; $f_y = 40$ ksi.; h = 28" (given)

$P_{ue} = 1.4P_d + 1.7P_l = 1.4 \times 134^k + 1.7 \times 84^k = 330^k$

$P_{ui} = 1.4 \times 210^k + 1.7 \times 168^k = 580^k$

$M_u = 330^k(13.33' - 0.67') - 580^k(20.67' - 13.33')$
$= -79^{k\text{-}ft} = 79^{k\text{-}ft}$ c.w.

Thus, the bearing pressure is non uniform under design loads: L = 26.7'; $S = 26.7^2 / 6 = 119$ ft^2

$w_{uL} = P_u / L - M_u / S = (330 + 580) / 26.7 - 79 / 119$
$= 33.4$ klf

$w_{ur} = P_u / L + M_u / S = (330 + 580) / 26.7 + 79 / 119$
$= 34.7$ klf

This is practically a uniform pressure distribution; however, to satisfy statics, will work with the calculated values.

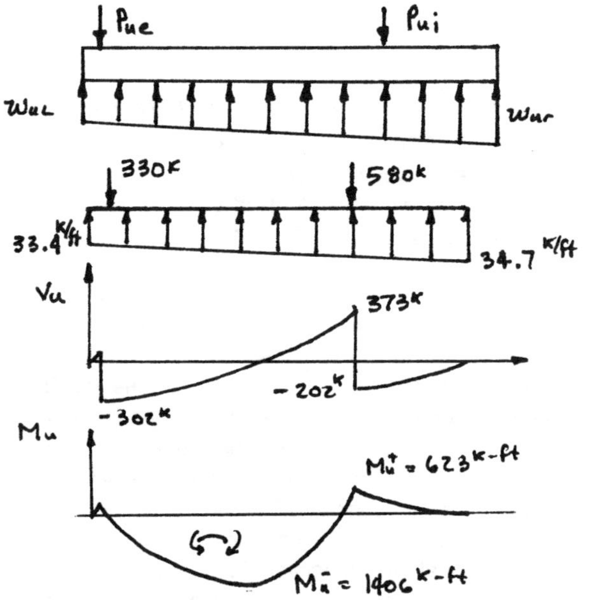

Check punching shear on the interior column: d = 28 – 4 = 24"; $b_o = 4(h + d) = 4(16 + 24) = 160"$

$V_u = 580^k - (33.9 / 7.5)^{ksf}(5.3')^2 = 451^k$

$$V_u \leq \phi V_c = 4\phi\sqrt{f'_c}\, b_o d$$
$$= 4 \times 0.85 \times \sqrt{2500} \times 24 \times 160$$
$$\phi V_c = 653{,}000 \text{ lb} \succ V_u \therefore \text{o.k.}$$

Check the punching shear on the exterior column: $b_o = (16 + 24) + 2(16 + 12) = 96"$

$V_u = 330^k - (33.4 / 7.5)^{ksf}(5.3')(3.3') = 250^k$

$$V_u \leq \phi V_c = 4\phi\sqrt{f'_c}\, b_o d = 4 \times 0.85 \times \sqrt{2500} \times 24 \times 96$$
$$\phi V_c = 392{,}0000 \text{ lb} \succ V_u \therefore \text{o.k.}$$

Check the wide beam shear across the short direction: b = 7.5 × 12 = 90"

$V_u \leq 373^k - (33.9)^{klf}(8 + 24) / 12' = 283^k$

$$\phi V_c = 2\phi\sqrt{f'_c}\, b_w d$$
$$= 2 \times 0.85 \times \sqrt{2500} \times (7.5 \times 12) \times 24$$
$$= 183{,}000 \text{ lb} \prec V_u \therefore \text{need stirrups}$$

$$V_s = V_u / \phi - V_c = (283 - 183) / 0.85 = 118^k$$

Try # 5 double stirrups: $A_v = 4 \times 0.31 = 1.24\ in^2$

$s \le A_v f_y / 50 b_w = 1.24 \times 40 / (50 \times 90) = 11"$

$s \le d/2 = 24/2 = 12"$

$s \le A_v f_y d / V_s = 1.24 \times 40{,}000 \times 24 / 118$

$= 10"$ o.c. ← governs

Flexure: $M_u^- = 1406^{k\text{-}ft}$

$\phi M_n = \phi \rho b d^2 f_y \{1 - 0.59\rho(f_y / f'_c)\} \ge M_u$

$0.9 \times 90 \times 24^2 \times 40\rho\{1 - 0.59\rho(40/2.5)\}$

$= 1406 \times 12$

$\rho = 0.00998$

∴ $A_s^- = 0.00998 \times 90 \times 24 = 21.56\ in^2$

For $M_u^+ = 623^{k\text{-}ft}$

$0.9 \times 90 \times 24^2 \times 40\rho\{1 - 0.59\rho(40/2.5)\} = 623 \times 12$

$\rho = 0.00424$

∴ $A_s^- = 0.00424 \times 90 \times 24 = 9.16\ in^2$

Minimum footing reinforcement:

$A_{s,min} = 0.002bh = 0.002 \times (12 \times 7.5) \times 28 = 5.04\ in^2$; not critical

∴ **use 17 # 10 top; 8 # 10 bottom and double # 5 stirrups @ 10" o.c.**

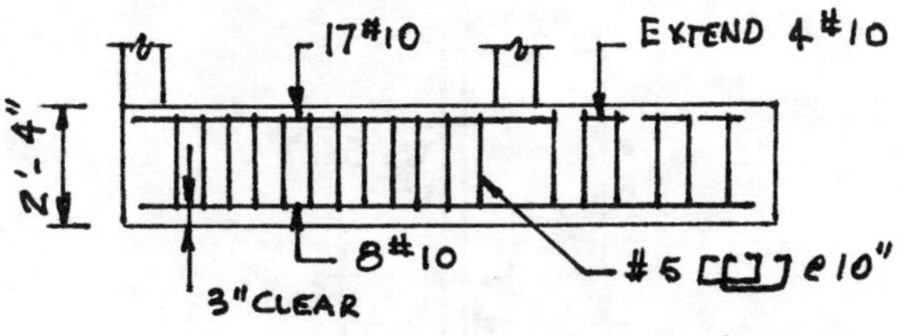

ELEVATION (N.T.S.)

Concrete-35

Given: $f'_c = 4.5$ ksi; $\beta_1 = 0.825$; $f_{pu} = 270$ ksi; $A_{ps} = 0.11\ in^2/ft$; $d_p = 6 - 1.5 = 4.5$; $f_{pe} = 15/0.11 = 136.4$ ksi; assume no mild steel. Use the strength design method (ACI 318); stress-relieved strands with $\gamma_p = 0.4$.

a) Calculate the nominal moment strength:

1. For the slab with bonded strands, $f_{pe} = 136.4$ ksi $> 0.5 f_{pu}$ ∴ use Eqn 18-3

$\rho_p = A_{ps} / bd_p = 0.11 / (12 \times 4.5) = 0.00204$;

$\omega = \omega_p = 0$

$f_{ps} = f_{pu}\{1 - \gamma_p / \beta_1[\rho_p(f_{pu} / f'_c)]\}$

$= 270\{1 - 0.4/0.825[0.00204(270/4.5)]\}$

$= 254$ ksi

$\omega_p = \rho_p f_{ps} / f'_c = 0.00204(254/4.5) = 0.115 < 0.36\beta_1$

$= 0.30$

∴ Ductile flexural failure

$a = A_{ps} f_{ps} / (0.85 \times b \times f'_c)$

$= 0.11 \times 254 / (0.85 \times 12 \times 4.5) = 0.60"$

$\phi M_n = \phi A_{ps} f_{ps}(d_p - a/2)$

$= 0.9 \times 0.11 \times 254(4.5 - 0.60/2)$

$\phi M_n = 106^{k\text{-}in}/_{ft} = 8.8^{k\text{-}ft}/_{ft}$

2. For the slab with unbonded strands – for a 4' cantilever: $L = 2a = 2 \times 48 = 96"$; $L/d = 96/4.5 = 21 < 35$, ∴ Eqn. 18-4 applies

$f_{py} = 0.85 f_{pu} = 0.85 \times 270 = 230$ ksi

$f'_c / 100\rho_p = 4.5 / (100 \times 0.00204) = 22$ ksi

∴ $f'_c / 100\rho_p + 10 < 60$ ksi

$f_{ps} = f_{se} + 10 + f'_c / 100\rho_p = 136.4 + 10 + 22 = 168.4$ ksi $< f_y$

$a = A_{ps} f_{ps} / (0.85 \times b \times f'_c)$

$= 0.11 \times 168.4 / (0.85 \times 12 \times 4.5) = 0.40"$

$\phi M_n = \phi A_{ps} f_{ps}(d_p - a/2)$

$= 0.9 \times 0.11 \times 168.4(4.5 - 0.40/2)$

$\phi M_n = 72^{k\text{-}in}/_{ft} = 6.0^{k\text{-}ft}/_{ft}$

For a load w = 950 klf ↓ on the cantilever:

$M^- = 0.95 \times 4^2 / 2 = 7.6^{k\text{-}ft}/_{ft}$

The load causes a bending moment that exceeds the design strength of the member with unbonded strands, but is less than that of the member with bonded strands. Thus, in theory, the overload would cause collapse of the cantilever with unbonded strands, which then destroys the end anchorage. Hence, a progressive collapse could occur in the adjacent spans, which lose their prestress.

Note: proper design of the cantilever requires a minimum amount of bonded mild steel per section 18.9, ACI 318. This mild steel would increase the resisting moment strength. Since no information was given regarding the mild steel, it was not included in the analysis,

b) An overload of 1200 psf ↓ causes a bending moment of $9.6^{k\text{-}ft}/_{ft}$, which exceeds the design flexural strength of both designs. In this case, failure of the cantilever would likely occur in both cases. The member with the bonded strands would experience only a localized failure, whereas, the unbonded design would again experience progressive collapse affecting a larger region of the structure.

Concrete-36

Compute the top and bottom midspan stresses under the specified loading. I assume a parabolic strand profile with a sag = 11 – 6 = 5".

a) Under the dead plus effective prestress at the midspan:

$w_{eq} = 8P_e(\text{sag}) / L^2 = 8 \times 200 \times (5/12) / 40.83^2$

$= 0.40$ klf ↑

The strand is concentric to the precast centroid at ends ∴ the equivalent load diagram is

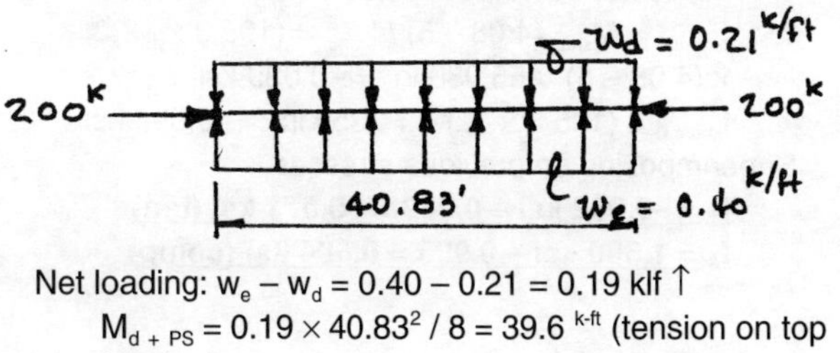

Net loading: $w_e - w_d = 0.40 - 0.21 = 0.19$ klf ↑

$M_{d+PS} = 0.19 \times 40.83^2 / 8 = 39.6^{\ k\text{-}ft}$ (tension on top fiber)

For the precast section: A = 265 in^2; S_T = 1200 in^3; S_B = 1250 in^3

$f_T = P / A - M_{d+PS} / S_T$

= 200 / 265 – (39.6 × 12) / 1200 = 0.359 ksi (comp.)

$f_B = P/A + M_{d+PS} / S_B$

= 200 / 265 + (39.6 × 12) / 1250 = 1.134 ksi (comp.)

b) With the precast section shored at midspan and subjected to the wet concrete load of 50 psf × 15' = 750 plf applied after the shore is positioned, calculate the midspan stresses and superimpose them with the stresses in the previously simply supported member.

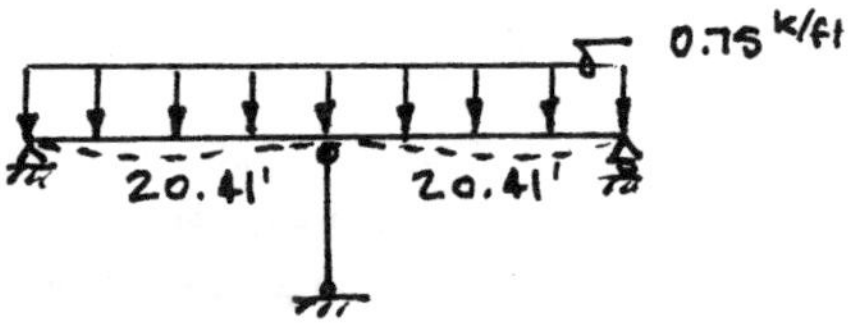

$M^- = wL'^2 / 8 = 0.75 \times 20.42^2 / 8 = 39.1^{k\text{-}ft}$ (tension top)

$f_{T,slab} = M^- / S_T = (39.1 \times 12) / 1200 = -0.391$ ksi (ten)

$f_{B,slab} = M^- / S_B = (39.1 \times 12) / 1250 = 0.375$ ksi (comp)

Superimpose:

f_T = 0.359 – 0.391 = –0.032 ksi (ten)

f_B = 1.134 + 0.375 = 1.500 ksi (comp)

c) Remove the midspan shore after the concrete hardens and compute the resulting stresses. This is equivalent to applying a concentrated force at the midspan of the now simply supported member equal to the reaction of the shore under the wet load (note: in all computations for this problem, ideal linear elastic behavior is assumed without adjustments for the effects of creep and shrinkage).

$R' = 2[wL'(5 / 8)] = 2[0.75 \times 20.42 \times (5 / 8)] = 19.1^k$

$\therefore$ $M_{shore} = R'L / 4 = 19.1 \times 40.83 / 4 = 195.3^{k\text{-}ft}$ (compression top)

The stresses caused by this bending moment are induced in the composite section. Thus, we need to calculate the stresses at the bottom and at the interface of the precast and cast-in-place slab using properties of the composite section. The neutral axis location was not given; however, it can be found from the given section moduli and dimensions (S_T = 13,500 in^3; S_B = 2350 in^3)

$I = S_T c_T = 13{,}500 c_T = S_B c_B$

But $c_B = 22.5 + 5 - c_T = 27.5 - c_T$

$\therefore$ $c_T = 64{,}625 / 15{,}850 = 4.08$"; $I_{comp} = 55{,}080$ in^4

$\therefore$ $f_{shore,T} = M_{shore}(4.08 - 5) / I_{comp} = (195.3 \times 12)^{k\text{-}in} \times (4.08 - 5)" / 55{,}080 \text{ in}^4 = -0.039$ ksi

$f_{shore,B} = (195.3 \times 12)^{k\text{-}in} / 2350 \text{ in}^3 = -0.996$ ksi

Superimposing on previous stresses

f_T = –0.032 ksi – 0.039 = –0.071 ksi (ten)

f_B = 1.500 ksi – 0.996 = 0.504 ksi (comp)

d) Apply live load of 50 psf (750 plf) to the composite section

$M_L = 0.750^{klf} \times 40.83'^2 / 8 = 156.3^{k\text{-}ft}$ (comp. top)

$f_{fL} = (-156.3 \times 12)^{k\text{-}in} \times 0.92" / 55{,}080$ = –0.031 ksi (ten)

$f_{BL} = (-156.3 \times 12)^{k\text{-}in} / 2350 \text{ in}^3 = -0.798$ ksi (ten)

Superimpose

f_T = –0.071 – 0.031 = –0.102 ksi (ten)

f_B = 0.504 ksi – 0.798 = –0.294 ksi (ten)

The stress at the top of the slab is

$f_{TT} = M_L / S_T = 156.3 \times 12 / 13{,}500 = 0.138$ ksi (comp).

Concrete-37

a) Determine the forces on connections "A" and "B"

- Gravity loads:

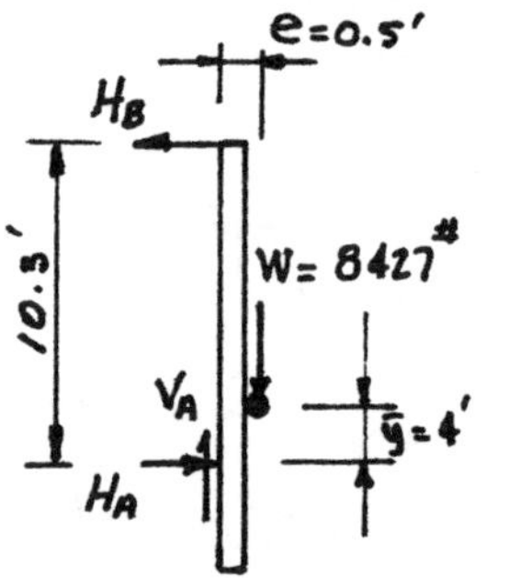

$\Sigma M_A = 0$: $10.3H_B - 0.5 \times 8.43 = 0$ $\therefore$ $H_B = 0.409^k \leftarrow$

$\Sigma F_X = 0$: $H_A = 0.409^k \rightarrow$

$\Sigma F_y = 0$: $V_A = 8.43^k \uparrow$

- Compute the seismic load (UBC-97): seismic zone 3; hospital (I_p = 1.5); soil profile type S_D; near-fault factor = 1.0; Table 16-Q, C_a = 0.36; Table 16-K, I_p = 1.5; Table 16-O, a_p = 1.0, R_p = 3.0 (use 1.0 for embedded connectors per 1633.2.4.2; use Eqn. 32-2 assuming that the panel is located near the roof such that $h_x / h_r \cong 1.0$; thus

$$F_p = \frac{a_p C_a I_p}{R_p}\left(1 + 3\frac{h_x}{h_r}\right)W_p$$

$$= \frac{1.0 \times 0.36 \times 1.5}{3.0}(1 + 3 \times 1.0)W_p$$

$F_p = 0.72W_p > 0.7C_aI_pW_p$

$\therefore$ $F_p = 0.72 \times 8427 = 6.1^k$

This value is ultimate. For working stress applications, divide by 1.4

$F_{ps} = 6.1 / 1.4 = \pm 4.4^k$

$\Sigma M_A = 0$: $4(4.4) - 10.3H_B = 0$ $\therefore$ $H_B = \pm 1.73^k$

$\Sigma F_X = 0$: $H_A - 4.4 + 1.73 = 0$ $\therefore$ $H_A = \pm 2.67^k$

- Wind (1616, UBC-97) – From Table 16-G, C_e = 1.61 ($h \le 99'$ and exposure condition "C" given); Table 16-H, C_q = 1.5 (near corners; Table 16-K, I_w = 1.15; Table 16-F, q_s = 16.4 psf

$p = C_e C_q q_s I_w = 1.61 \times 1.5 \times 16.4 \times 1.15 = 46$ psf

$\Sigma M_A = 0$: $10.3H_B - (0.046 \times 6 \times 12 \times 4) = 0$

$\therefore H_B = 1.29^k$

$\Sigma F_X = 0$: $H_A - 0.046 \times 6 \times 12 + 1.29 = 0 \therefore H_A = 2.02^k$

Wind reactions are well below those created by seismic $\therefore$ use gravity in combination with seismic reactions for design of connections.

Thus, for design of the connection body ($R_p = 3.0$) using working stresses

$\underline{H_B = 0.41^k + 1.73 = 2.14^k \leftarrow (0.409 + 3(1.73) = 5.6^k}$ <u>on fasteners)</u>

$\underline{H_A = 0.409^k + 2.67 = 3.08^k \rightarrow (0.409 + 3(2.7) = 8.5^k}$ <u>on fasteners)</u>

$\underline{V_A = 8.43^k \uparrow}$

b) Determine the thickness of an L 6 x 4 x t x 1'-0" required at connection "A". Use allowable stress design (AISC 1989) following the design procedure from the PCI Manual (PCI 1999, p.6-22)

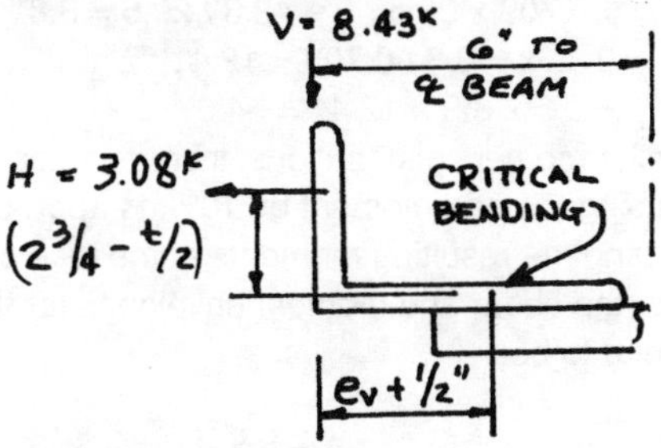

By trial-and-error: 1) try t = 0.5"

$e_v = 6 - b_f / 2 = 6.0 - 4.5 = 1.5$"

$M = V(e_v + 0.5) + H(2.75 - t / 2)$

$= 8.43(1.5 + 0.5) + 3.08(2.75 - 0.25) = 24.6^{k\text{-}in}$

$F_b = (4 / 3)(0.75F_y) = (4 / 3)(0.75 \times 36) = 36$ ksi

$S_x = bt^2 / 6 = 12 \times t^2 / 6 = 2t^2$

$f_b \leq F_b$: $24.6 / (2t^2) \leq 36$

$\therefore$ $t^2 \geq 0.344$ in^2; $t \geq 0.58$" – close to assumed t

$\therefore$ Use L 6 x 4 x 5 / 8 x 1'-0"

Determine the studs required to attach the 3/8" plate. Use strength design with $\phi = 0.85$; $U = 1.1(1.2D \pm E)$:

$V_u = 1.1(1.2 \times 8.43) = 11.1^k$

$T_u = 1.1[1.2 \times 0.41 + 1.4(3 \times 2.67)] = 12.9^k$

Provide confinement around connectors to ensure a ductile failure. Use the PCI method (Section 6.5.2, PCI 1999). Maximum stud diameter = 1/2" on a 3/8" plate; maximum six stud diameter spacing = 3" = x. Therefore, we can fit no more than 4 studs on the given plate.

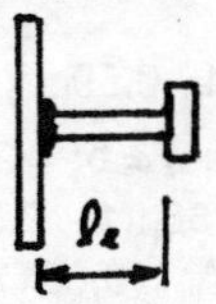

$l_e \cong 3.0 - 0.375 - 0.5 = 2.13$"

Try 4-1/2" diameter studs spaced 3" o.c. At "A", the distance from the back connector to the free edge is $1.5 + 6 = 7.5" = 15d_b$. Therefore, PCI equation 6.5.8 governs the group shear capacity

$V_c = (800A_b\lambda\sqrt{f'_c})n$

$= 800 \times 0.20 \times 1.0 \times (4000)^{1/2} \times 4 = 40{,}500$ lb

$= 40.5^k$

The tension strength of the group is given by Case 1, PCI Table 6.5.6, where x = y = 3" and $\lambda = 1.0$

$P_c = 4\lambda\sqrt{f'_c}\,(x + 2l_e)(y + 2l_e)$

$= 4 \times 1.0 \times (4000)^{1/2}(3 + 2 \times 2.13)(3 + 2 \times 2.13)$

$= 13{,}300$ lb $= 13.3^k$

Check the combined shear and tension loading on the group

$1 / \phi[(V_u / V_c)^2 + (T_u / P_c)^2] = (1 / 0.85)[(11.1 / 40.5)^2 + (12.9 / 13.3)^2] = 1.2 > 1.0$

Thus, the connection is no good. **<u>The design forces per UBC-97 are too large to resist using headed studs on the given plate.</u>**

c) Compute the pullout strength of the 3/4" diameter embedded bolt with plate washer at connection 'B".

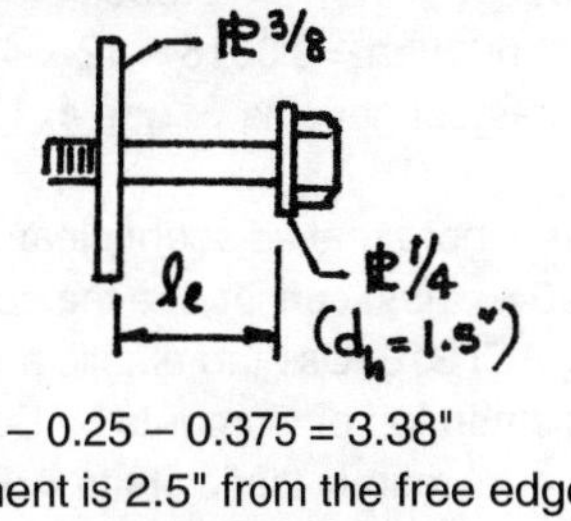

$l_e = 4.00 - 0.25 - 0.375 = 3.38$"

The embedment is 2.5" from the free edge

$l_e + d_h = 3.38 + 1.5 = 4.88$", which is greater than the distance to the free edge = 2.5" $\therefore$ reduce the pull out strength by a factor $C_{es} = d_e / l_e = 2.5 / 3.38 = 0.74$.

$A_o = \sqrt{2}l_e\pi(l_e + d_h) = 1.414 \times 3.38 \times \pi \times (4.88)$

$= 73$ in^2

$\phi P_c = C_{es}\phi A_o 2.8\lambda\sqrt{f'_c}$

$= 0.74 \times 0.85 \times 73 \times 2.8(4000)^{1/2} = 8130^\# = 8.1^k$

Check the shear strength of the concrete with a # 3 'hairpin" bent bar (Av = 0.22 in^2) serving as shear friction reinforcement at the corner (1911.7, UBC-97).

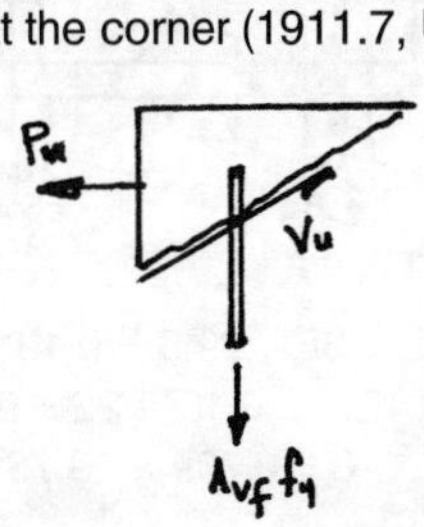

$P_u \leq \phi\mu f_y A_{vf}\cos(20)^\circ$

$= 0.85 \times 1.4 \times 40 \times 0.22\cos(20)^\circ = 9.8^k$

$\therefore$ bolt pullout strength governs: **$\underline{P_u = 8.1^k}$**

Concrete-38

a) Design the 4"one-way slab over the joist on a per lineal foot basis. Use the strength design method (ACI 318); $f'_c = 3$ ksi; $f_y = 60$ ksi; normal weight (150 pcf) concrete $\therefore$ slab weight = 50 psf; partitions plus ceiling impose additional 30 psf dead; live = 100 psf. Thus,

$w_u = 1.4w_d + 1.7w_l = 1.4 \times (50 + 30) + 1.7 \times 100$
$= 280$ plf

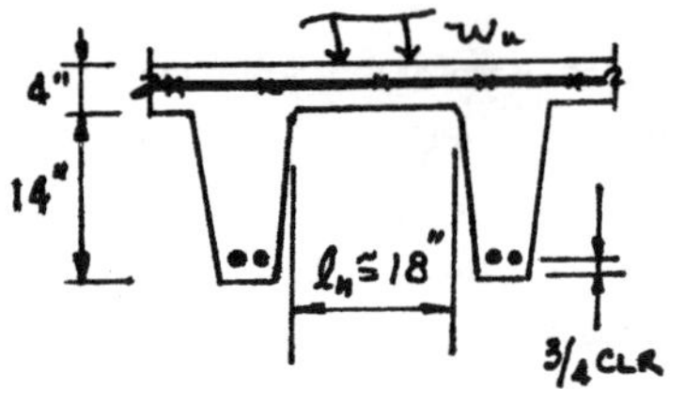

Reinforce the slab using welded wire fabric. Take $d^- = d^+ = 2$"; joists taper such that clearspan, L_n, is approximately 18". Re: 8.3, ACI 318:

$M_u^- = w_u L_n^2 / 12 = 0.28 \times 1.5^2 / 12 = 0.053^{k\text{-}ft}/_{ft}$

$\phi M_n = \phi \rho b d^2 f_y \{1 - 0.59\rho(f_y / f'_c)\} \geq M_u$

$0.9 \times 12 \times 2^2 \times 60\rho\{1 - 0.59\rho(60 / 3)\} = 0.053 \times 12$

$\rho = 0.00026$

$\therefore$ $A_s = 0.00026 \times 12 \times 2 = 0.00624^{sq.in}/_{ft}$

$A_{s,min} = 0.0018bh = 0.0018 \times 12 \times 4 = 0.0864\ ^{sq.in}/_{ft}$ – minimum steel governs $\therefore$ **use 4x12 – W2.9xW0.9 W.W.F.**

For the joists – note that adjacent clearspans differ by more than 20%; thus, cannot use the coefficients of Section 8.3, ACI 318. Use the elastic analysis per ACI 8.9 with redistribution of moments by Section 8.4. Limit the flexural steel ratio in the joists to $0.5\rho_b$ and redistribute moments by 10%. Assume girders 30" wide and reduce moments to face of support for design. Pan forms are standard module (20" wide with 5" soffit) $\therefore$ joists are spaced 25" = 2.08' o.c.

$w_{ud} = 1.4[(80 \text{ psf} \times 2.08 + (150 / 144)6 \times 14 \text{ plf}]$
$= 356 \text{ plf} = 0.356 \text{ klf}$

$w_{uL} = 1.7 \times 100 \text{ psf} \times 2.08' = 354 \text{ plf} = 0.354 \text{ klf}$

Loading 1: Maximum M_u^+ in exterior spans – analysis by moment distribution:

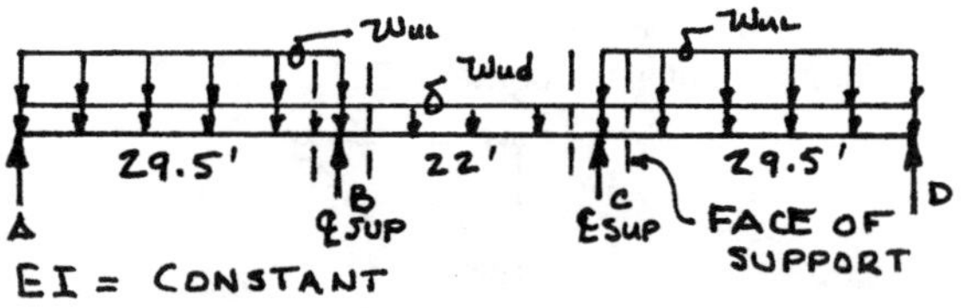

$K_{BA} = 3EI / L = 3EI / 29.5 = 0.101EI$

$K_{BC} = K_{CB} = 4EI / L = 4EI / 22 = 0.182EI$

$FEM_{BA} = wL^2 / 8 = 0.709 \times 29.75^2 / 8 = 77.1^{k\text{-}ft}$
$= -FEM_{CD}$

$FEM_{BC} = -FEM_{CB} = -wL^2 / 12 = -0.356 \times 22^2 / 12$
$= -14.3^{k\text{-}ft}$

$DF_{BA} = DF_{CD} = 0.101 / (0.101 + 0.182) = 0.36;$

$DF_{BC} = DF_{CB} = 0.64$

DF:	---	0.36	0.64	0.64	0.36	---
FEM:	0	77.1	–14.3	14.3	–77.1	0
		–22.6	–40.1	–20.1		
			26.5	53.0	29.8	
		–9.5	–17.0	–8.5		
			2.7	5.4	3.1	
		–1.0	–1.7	–0.9		
			0.3	0.6	0.3	
	---	–0.1	–0.2	---	---	---
Σ =	**0**	**43.9**	**–43.9**	**43.9**	**–43.9**	**0**

Re: Section 8.4, can increase the negative moment by 10% with corresponding decrease in positive moment

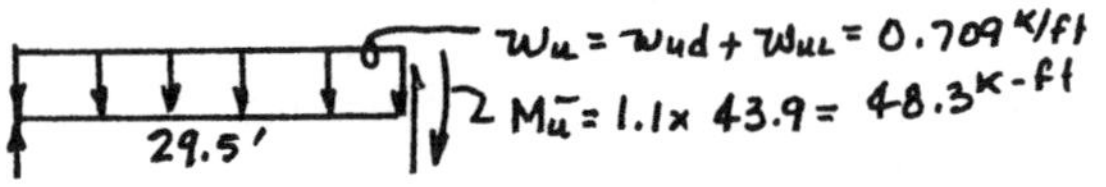

$V_{uL} = 0.709 \times 29.5 / 2 - 48.3 / 29.5 = 8.8^k$

$V = 0$ @ $x = 8.8 / 0.709 = 12.41'$

$M_u^+ = 0.5 \times 8.8 \times 12.41 = 54.6^{k\text{-}ft}$

By similar moment distributions, increasing or decreasing the end moment by 10% as appropriate, and calculating the resulting moments at the face of the girders, the shear and moment envelopes for the joists are found to be:

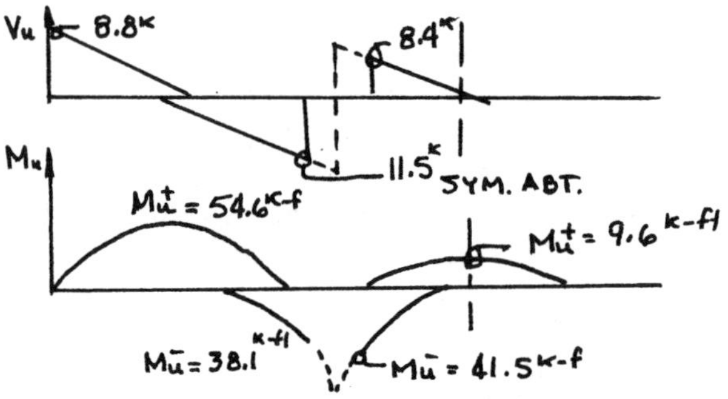

Shear design (re: 8.11.8)

$\phi V_c = 2.2\phi\sqrt{f'_c}\,b_w d = 2.2 \times 0.85 \times \sqrt{3000} \times 5 \times 17$
$= 8700$ lb.

$V_u = 11.5 - 0.709 \times 17 / 12 = 10.5^k = 10{,}700^{lb}$

V_u exceeds ΦV_c. $\therefore$ **need to use tapered end forms to increase shear capacity near supports.**

Flexural design:

$A_{s,min} = (200 / f_y)b_w d = (200 / 60{,}000) \times 5 \times 17$
$= 0.28 \text{ in}^2$

$\phi M_n = \phi \rho b d^2 f_y \{1 - 0.59\rho(f_y / f'_c)\} \geq M_u$

For $M_u^+ = 54.6^{k\text{-}ft}$: $b = b_e = 25$"; $d = 17$"

$0.9 \times 25 \times 17^2 \times 60\rho\{1 - 0.59\rho(60 / 3)\} = 54.6 \times 12$

$\rho = 0.00176$; $\therefore$ $A_s = 0.00176 \times 25 \times 17 = 0.75 \text{ in}^2$;
say **1 # 5 & 1 #6**

For $M_u^- = 41.5^{k\text{-}ft}$: $b = b_w = 9$" (near end of tapered joists); $d = 17$"

$0.9 \times 9 \times 17^2 \times 60\rho\{1 - 0.59\rho(60 / 3)\} = 41.5 \times 12$

$\rho = 0.00396$; $\therefore$ $A_s = 0.00396 \times 9 \times 17 = 0.60 \text{ in}^2$;
say **2 # 5**

For $M_u^+ = 9.6^{k\text{-}ft}$: $b = b_e = 25"$; $d = 17"$

$0.9 \times 25 \times 17^2 \times 60\rho\{1 - 0.59\rho(60/3)\} = 9.6 \times 12$

$\rho = 0.000306$; $\therefore A_s = A_{s,min} = 0.28$ in^2; say **2 # 4**

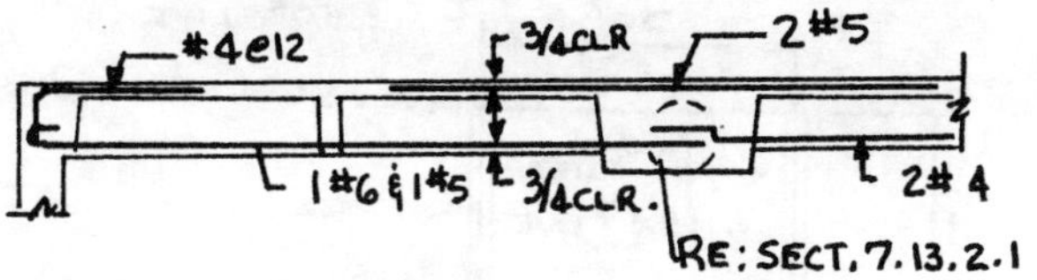

ELEVATION - TYP. JOIST (N.T.S.)

b) Girders on Lines 2 & 3: 30" x 22"; $L_e = 28 - 0.5 - 0.75 = 26.75'$; $L_i = 28 - 1.5 = 26.5'$. The girders qualify for the coefficients of Section 8.3, $d^- = d^+ \cong 22 - 3 = 19"$.

$w_u = 1.4w_d + 1.7w_l$ = reactions from joists plus additional weight from girder

$w_u = (0.709 / 2.08)^{ksf} \times (29.5 + 22)' / 2 + 1.4 \times (0.15 \times 30(22 - 4) / 144)^{klf} = 9.6$ klf

For member supported by wall at exterior, use minimum steel

$A_{s,min} = (200 / f_y)b_w d = (200 / 60{,}000) \times 30 \times 19 = 1.9$ in^2, say 4 # 6 – keep continuous across span; supplement as required over first interior support

$M^-_{u1} = w_u l_n^2 / 10 = 9.6 \times \{(26.75 + 26.5) / 2\}^2 / 10 = 680^{\ k\text{-}ft}$

$0.9 \times 30 \times 19^2 \times 60\rho\{1 - 0.59\rho(60/3)\} = 680 \times 12$

$\rho = 0.0155$; $\therefore A_s = 0.0155 \times 30 \times 19 = 8.84$ in^2; **use 8 # 9 w/ 4 # 6**

For M_u^+ in the exterior span (non-integral supports at exterior)

$M^+_{u1} = w_u l_n^2 / 11 = 9.6 \times 26.75^2 / 11 = 624^{\ k\text{-}ft}$

$0.9 \times 30 \times 19^2 \times 60\rho\{1 - 0.59\rho(60/3)\} = 624 \times 12$

$\rho = 0.0142$; $\therefore A_s = 0.0142 \times 30 \times 19 = 8.11$ in^2; **use 7 # 10**

For M_u^+ in the interior span

$M^+_{u1} = w_u l_n^2 / 16 = 9.6 \times 26.\ 5^2 / 11 = 421^{\ k\text{-}ft}$

$0.9 \times 30 \times 19^2 \times 60\rho\{1 - 0.59\rho(60/3)\} = 421 \times 12$

$\rho = 0.00908$; $\therefore A_s = 0.00908 \times 30 \times 19 = 5.18$ in^2; **use 4 # 10**

Shear design – exterior spans:

$$V_{u,ext} = 1.15 w_u L_e / 2 - w_u d = 9.6(1.15 \times 26.75 / 2 - 19/12) = 132.5^k$$

$$\phi V_c = 2\phi\sqrt{f'_c}\,b_w d = 2 \times 0.85 \times \sqrt{3000} \times 30 \times 19 = 53{,}000\# = 53^k$$

Need stirrups – try # 4 stirrup ($A_v = 2 \times 0.20 = 0.40$ in^2)

$V_s = V_u / 0.85 - V_c = (132.5 - 53) / 0.85 = 94^k$

$$s \le \begin{cases} A_v f_y d / V_s = 0.40 \times 60 \times 19 / 94 = 5" \text{ o.c.} \\ A_v f_y / 50 b_w = 0.40 \times 60{,}000 / (50 \times 30) \\ \qquad = 16" \text{ o.c.} \\ d/2 = 19/2 = 9.5" \text{ o.c} \end{cases}$$

$\therefore$ use **# 4 U-stirrups @ 5" o.c. at critical location**

For the interior span:

$V_{u,int} = w_u L_i / 2 - w_u d = 9.6(26.5 / 2 - 19/12) = 112^k > \phi V_c$

Need stirrups –

$V_s = V_u / 0.85 - V_c = (112 - 53) / 0.85 = 69^k$

$$s \le \begin{cases} A_v f_y d / V_s = 0.40 \times 60 \times 19 / 69 = 6.5" \text{ o.c.} \\ A_v f_y / 50 b_w = 0.40 \times 60{,}000 / (50 \times 30) \\ \qquad = 16" \text{ o.c.} \\ d/2 = 19/2 = 9.5" \text{ o.c} \end{cases}$$

$\therefore$ **use # 4 U-stirrups @ 6-1/2" o.c. at critical location**

Stirrups terminate where $V_u \le \phi V_c / 2 = 27^k$

$V_u = 1.15 \times 9.6 \times 26.75 / 2 - 9.6x \le 27 \ \therefore x = 12.6'$ from interior face of first interior support. At interior spans:

$V_u = 9.6 \times 26.5 / 2 - 9.6x \le 27 \ \therefore x = 10.4'$ from other faces of supports

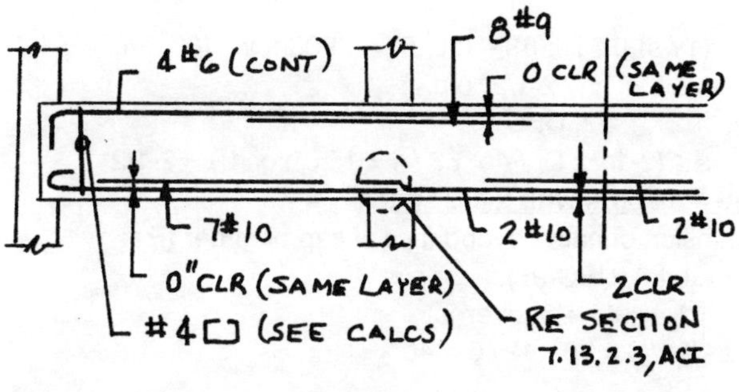

ELEVATION – GIRDER LINES "1" & "2" (N.T.S)

Note: space limitation will not permit drawing to scale specified in problem statement. Minor details (e.g., bar cutoff locations) omitted from the above drawing.

Concrete-39

Determine the required embedment length of the #11 bars into the top and bottom. Given that $P_u = 164^K << P_{bal} = 3549^k$, the column is on the tension failure side of the interaction diagram with the axial force contributing practically nothing to the flexural resistance. Furthermore, given that $V_u = 0$, the moment is constant throughout the length of the column; thus, the longitudinal steel must be developed in tension at both top and bottom. The ductile frame stipulation implies that the member must satisfy the seismic provisions of UBC-97, Chapter 1921.

Since no information is given regarding longitudinal steel in the girder or footing, the possibility of the spiral serving as joint reinforcement will not be investigated.

For the column: $f'_c = 4$ ksi; $f_y = 60$ ksi; # 11 longitudinal bars; # 5 spiral; $A_g f'_c/10 = (\pi \times 66^2 / 4) \times 4/10 = 1370^k > P_u$; I assume that the longitudinal steel will not be lap spliced.

The development length (1921.5.4.1, UBC-97) is

$$l_{dh} \ge \begin{cases} f_y d_b / 65\sqrt{f'_c} = \dfrac{60{,}000 \times 11/8}{65\sqrt{4000}} = 20" \\ 8 d_b = 8 \times 11/8 = 11" \\ 6" \end{cases}$$

$\therefore \ l_{dh} = 20"$

$\therefore \ l_d = 2.5 l_{dh} = 2.5 \times 20 = 50"$

For the straight portion at top (no reduction for "top bar" effect in a vertical bar)

Note: the above values assume anchorage within the confined core of the intersecting girder and enclosure within spirals.

Transverse reinforcement (Re: 1921.4..4.1, UBC-97)

$$\rho_s \geq \begin{cases} 0.45(A_g / A_c - 1)f'_c / f_y \\ \quad = [0.45(66/60)^2 - 1]4000/60{,}000 \\ \quad = 0.006 \\ 0.12f'_c / f_{yh} = 0.12 \times 4000 / 60{,}000 = 0.008 \end{cases}$$

$V_s > \rho_s V_c$; V_s is the volume of spiral reinforcement = $\pi D_c A_b s$; where A_b = 0.31 in^2; D_c = 66 – 3 = 63" (assumes 1-1/2" cover to the spiral)

$\therefore$ $(1/s)(0.31)\pi(63 - 0.625) = 0.008 \times \pi \times (63^2/4)$

$s \leq 2.4$, say 2-1/2" o.c.

Re: 1921.4.4.2, UBC-97

$s \leq$ (2-1/2"; $D_c/4 = 63/4 = 15.8$"; or 4") = 2-1/2"

Use # 5 spiral with 2-1/2" pitch

Extension of bars beyond face of cap or girder (Re: 1921.4.4.4, UBC-97).

$$l_o \geq \begin{cases} d_c = 63\text{"} \leftarrow \text{governs} \\ h_n / 6 = 20 \times 12 / 6 = 40\text{"} \\ 18\text{"} \end{cases}$$

I assume that the column longitudinal steel must lap splice for tension transfer into dowels from the pile cap. Thus, the confinement steel must also extend full depth into the pile cap.

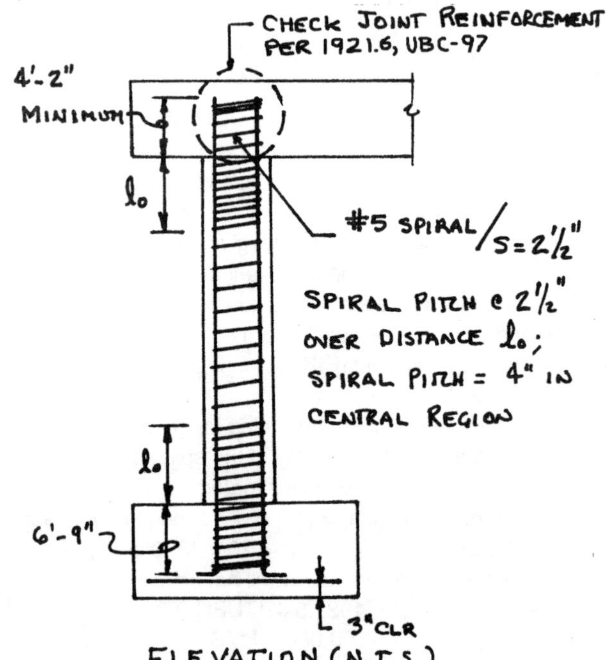

ELEVATION (N.T.S.)

Concrete-40

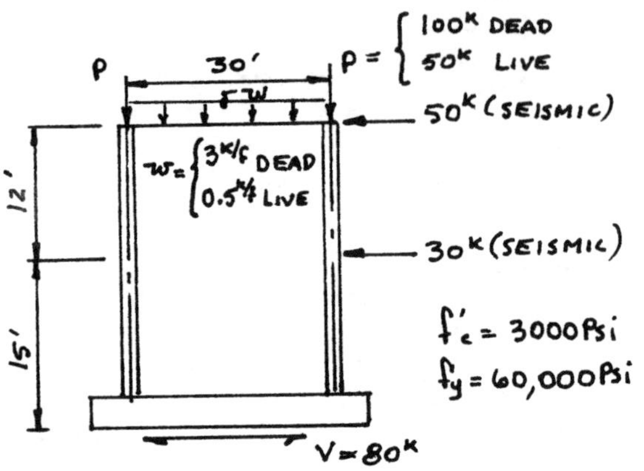

Design by UBC-97, Chapter 1921; f'_c = 3000 psi; f_y = 60,000 psi. The given loads are service; therefore increase seismic forces by 1.4 to obtain design values: $V_R = 1.4 \times 50 = 70^k$; $V_2 = 1.4 \times 30 = 42^k$; thus, $V_u = 70 + 42 = 112^k$; $M_u = 70 \times 27 + 42 \times 15 = 2520^{k\text{-ft}}$.

a) Design shear reinforcement based on the given 8" thick wall. In order to design shear reinforcement, we need the overall wall length, l_w, which depends on the boundary member dimension, t. Make a preliminary sizing of the boundary member based on the strength required to support the factored gravity loads at each end

$P_u = 1.4P_d + 1.7P_l = 1.4 \times 100 + 1.7 \times 50 = 225^k$

Problem statement requests minimum feasible size boundary member, therefore, used the maximum steel percentage (Re: 1921.4.3.1), $\rho_g = 0.06$

$\phi P_n = \phi 0.8[0.85f'_c(A_g - A_{st}) + f_y A_{st}] \geq P_u$

$0.7 \times 0.8[0.85 \times 3(A_g - 0.06A_g) + 0.06A_g \times 60] = 225$

$A_g \geq 67$ in^2; $t \geq 8.2$"; but minimum size per 1921.4.1.1 is 12" $\therefore$ use 12" x 12"; l_w = 31'.

$h_w / l_w = 27/31 = 0.87 < 2.0$ (squat wall) $\therefore$ use Eqn. 21-7

$\phi V_n = \phi[A_{cv}(\alpha_c \sqrt{f'_c} + \rho_n f_y)]$

try minimum steel $\rho_n = \rho_v = 0.0025$; $A_{cv} = 31 \times 12 \times 8$ = 2976 in^2; $\alpha_c = 3.0$ (maximum limit); $\phi = 0.6$; thus,

$\phi V_n = 0.6[2976(3 \times (3000)^{1/2} + 0.0025 \times 60{,}000)]$

$= 561{,}000 = 561^k$

A wall with minimum steel has strength well above the design shear $\therefore$ use $A_v = A_n = 0.0025 \times 12 \times 8 = 0.24$ in^2/ft. Since V_u is less than $2A_{cv}\sqrt{f'_c} = 326^k$, a single curtain of reinforcement is adequate $\therefore$ **use # 4 @ 10" o.c.e.w.**

b) Check boundary reinforcement per 1921.6.6. For the maximum axial compression loading

$P_u = 1.4P_d + 1.7P_l$

$= 1.4 \times (100 + 30 \times 3) + 1.7 \times (50 + 0.5 \times 30)$

$= 377^k$

$P_u < 0.1A_g f'_c = 0.1 \times 2976 \times 3 = 893^k$

$M_u = 42(15) + 70(27) = 2520^{k\text{-ft}}$

$M_u / (V_u l_w) = 2520 / (112 \times 31) = 0.73 < 1.0$

∴ Special boundary zone detailing is not required. Size the longitudinal reinforcement in the 12" x 12" columns to resist the concentrated factored load $P_u = 225^k$ at each end. Check the column strength using minimum longitudinal steel $A_{st} = 0.01 \times 12 \times 12 = 1.44$ in^2 (say 4 # 6: $A_{st} = 4 \times 0.44 = 1.76$ in^2)

$\phi P_{n,max} = 0.7 \times 0.8[0.85 \times 3 \times (144 - 1.76) + 60 \times 1.76] = 263^k > P_u$

Use # 3 closed hoops at 12" o.c. (special confinement steel of 1921.6.6.6 is not required). Terminate longitudinal wall steel with 90° hooks in the column core.

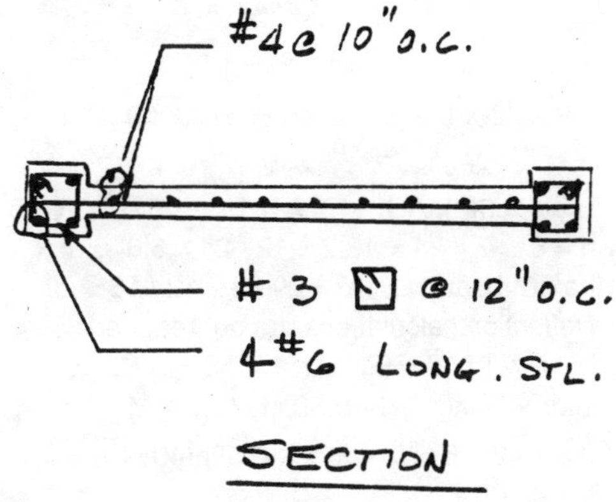

Concrete-41

a) Calculate the effect that the actual strength will have on the shear requirements. Use UBC-97. For the original design: $f'_c = 3000$ psi; $f_y = 40{,}000$ psi ($f_{ypr} = 1.25f_y = 50{,}000$ psi). The test values indicate actual strengths of $f'_c = 3400$ psi and $f_y = 72{,}000$ psi. For negative bending at the face of columns:

$d^- = 24.5 - 2.5 = 22$"; $b = b_w = 14$"; $A_s^- = 4.00$ in^2; $A_s' = 3.00$ in^2

Neglecting the compression steel

$a = A_s f_y / 0.85 f'_c b = 4.00 \times 72 / (0.85 \times 3.4 \times 14) = 7.11$"

$x = a / \beta_1 = 7.11 / 0.85 = 8.36$"

$\varepsilon'_s = ((8.36 - 2.50) / 8.36)0.003 = 0.0021$

∴ $\varepsilon'_s \cong \varepsilon_y$ ∴ $f'_s \cong f_y = 72$ ksi

Recalculate the depth assuming the compression steel is at about 50 ksi

$a = (A_s f_y - A_s' f_s) / (0.85 f'_c b) = (4.00 \times 72 - 3.00 \times 50) / (0.85 \times 3.4 \times 14) = 3.41$"

$x = a / \beta_1 = 3.41 / 0.85 = 4.01$"

$\varepsilon'_s = ((4.00 - 2.50) / 4.00)0.003 = 0.00113$

∴ $f'_s \cong 29000 \times 0.00113 = 33$ ksi – try again using an intermediate value, say 40 ksi

$a = (4.00 \times 72 - 3.00 \times 40) / (0.85 \times 3.4 \times 14) = 4.15$"

$x = a / \beta_1 = 3.41 / 0.85 = 4.88$"

$\varepsilon'_s = ((4.88 - 2.50) / 4.88)0.003 = 0.00146$

∴ $f'_s \cong 29000 \times 0.00146 = 42$ ksi – close enough

$M_{pr} = 0.85 f'_c ba(d - a / 2) + (f'_s - 0.85 f'_c) A'_s(d - d')$

$M_{pr} = 0.85 \times 3.4 \times 14 \times 4.15(22 - 4.15 / 2) + (42 - 0.85 \times 3.4) \times 3.00 \times (22 - 2.5) = 5634^{k\text{-}in} = 469^{k\text{-}ft}$

For positive bending at the face of the opposite support: $d^+ = 22$"; $b = b_e$; $A_s = 3.00$ in^2

$$b_{eff} \le \begin{cases} L/4 = 24 \times 12 / 4 = 72" \\ \text{beam spacing (na)} \\ b_w + 16h_s = 14 + 16 \times 4.5 = 86" \end{cases}$$

∴ $b_{eff} = 72$"

$a = 3.00 \times 72 / (0.85 \times 3.4 \times 72) = 1.03$"

$x = 1.03 / 0.85 = 1.22$" ∴ the top steel is in the tension zone, ignore it.

$M_{pr} = 3.00 \times 72(22 - 1.03/2) / 12 = 387^{k\text{-}ft}$

Re: section 1921.3.4.1, UBC-97, the seismic shear is computed based on the probable moments acting at faces of supports in combination with the tributary gravity load (unfactored live plus dead)

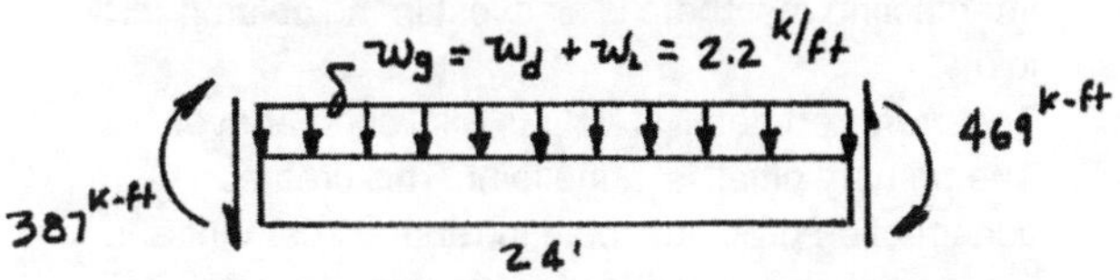

$V_e = (M_{prl} + M_{prr}) / L + wL / 2 = (387 + 469) / 24 + 2.2 \times 24 / 2 = 62.1^k$

Given that the member was designed based on the specified material properties, it should have been designed for probable moment strengths based on 1.25fy = 50 ksi. Considering that the f'_c has practically no effect on moment strength, the probable moment strengths can be estimated in direct proportion to the yield stresses. Thus,

$(V_e)_{original} \cong (50 / 72)(387 + 469) / 24 + 2.2 \times 24 / 2 = 51^k$

Thus, the effect of the higher material strength is to increase the design shear approximately 20 percent (62.1^k vs. 51^k)

b) Calculate the spacing of # 3 stirrups to resist the shear: $A_v = 0.22$ in^2; assume $f'_c = 3400$ psi and $f_y = 40{,}000$ psi for the stirrups; $b_w = 14$"; $d = 22$". Re: Section 1921.3.1 since the seismic contribution to the design shear exceeds one-half the gravity contribution, the contribution V_c is to be neglected. Thus,

$V_s = V_e / 0.85 = 62.1 / 0.85 = 73^k$

$s \le A_v f_y d / V_s = 0.22 \times 40 \times 22 / 73 = 2.6$"

In the hinge regions, closed hoops are required at minimum spacing (Re:1921.3.3)

$s \le (d / 4 = 22 / 4 = 5.5$"; $8d_{bl} = 9$"; $24d_{bs} = 9$"; 12")

∴ **Use # 3 stirrups @ 2-1/2" o.c.**

Concrete-42

a) **Detail 'D' is the preferred detail**. It has the relative disadvantage that the top layers of longitudinal steel in intersecting members conflict; however, the structural advantages outweigh this consideration.

- **Detail 'E' is bad** in several respects. The dimensions do not satisfy Section 1921.3.1.3 of UBC-97, which requires that the width-to-depth ratio must be ≤ 0.3. More important, the design philosophy for the ductile moment resisting frame is that plastic hinges should form in the beams rather than the columns. It would be difficult to satisfy this objective using relative depths of beam and column shown in 'E'. Also, the spandrel would be difficult to construct. The top position would probably have to be formed and poured separately sometime after the floor system was placed.
- **Detail 'F' is impractical** for a moment resisting exterior frame. The moment transfer from spandrel-to-columns would have to be through torsion in the girder, which would be difficult to accomplish. Also, the girder now has to transfer the additional vertical load from the spandrel beams. The benefit of confinement by intersecting members is also lost in the column-girder joint.

b) The preferred detail is "D", as explained above.

c) The primary pitfall is in the joint. This creates construction problems both in rebar placement and in placement and consolidation of concrete.

Concrete-43

Assume that the roof system is simply supported by walls. For the 8" slab with the given superimposed loads

$w_u = 1.4w_d + 1.7w_l = 1.4 \times (20 + 8 \times 12.5) + 1.7 \times 50$
$= 253$ psf

Use the strength design method (ACI 318); $f'_c = 3$ ksi; $f_y = 40$ ksi.

a) For the case with a 30" deep girder, compare the stiffness in each direction. For this analysis, use an effective flange width of $16h_f + b_w = 16 \times 8 + 36 = 164$".

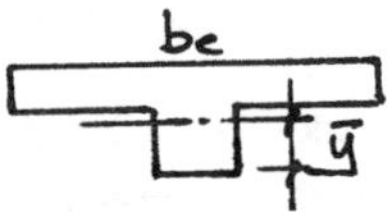

$A = 164 \times 8 + 36(30 - 8) = 2104$ in²

$$\bar{y} = \frac{164 \times 8 \times 26 + 36 \times 22 \times 11}{2104} = 20.35"$$

$I_b = 164 \times 8^3 / 12 + 164 \times 8\,(26 - 20.35)^2 + 36 \times 22^3 / 12 + 36 \times 22 \times (20.35 - 11)^2$

$I_b = 150{,}000$ in⁴

Compare to 8" thick slab of same width spanning the perpendicular direction. Let r = percentage of force (applied at intersection) carried by the beam

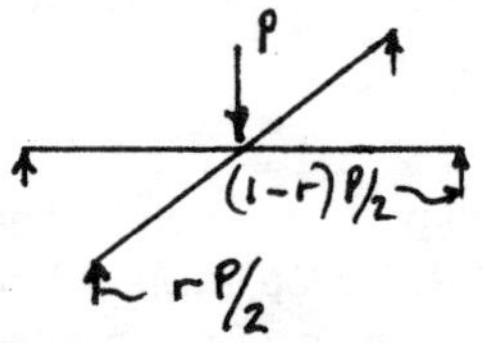

$$\frac{rP(29.33)^3}{(48E)\ 150{,}000} = \frac{(1-r)P(30)^3}{(48E)(6997)}$$

∴ r = 0.96

Thus, the beam is rigid relative to the slab. Design the slab as a one-way slab spanning from wall to beam. Per problem statement, use the moments at support centerline and neglect alternate span loading conditions. Consider a unit width of slab, b = 12"; take d = 8 – 1.5 = 6.5".

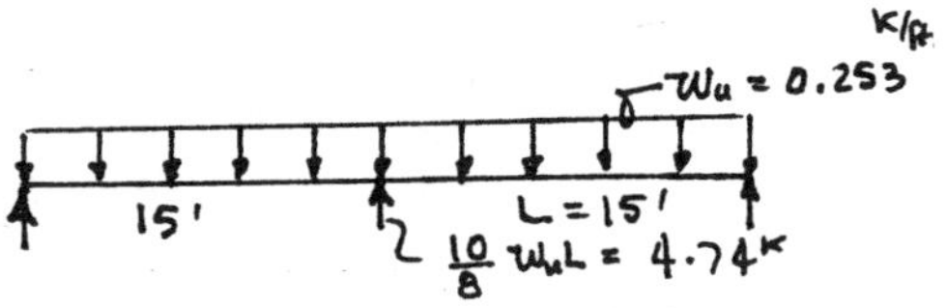

$V_u \cong 0.253(5 / 8 \times 15 - 2) = 1.86^k$

$\phi V_c = 2\phi\sqrt{f'_c}\,b_w d = 2 \times 0.85 \times \sqrt{3000} \times 6.5 \times 12$
$= 7260^{\#} = 7.3^k \succ\succ V_u$

Re: Table 9.5(a), ACI 318: $l_n = 15 - 1.5 - 0.3 = 13.2'$
$h = 8" > l_n / 24 = 13.2 \times 12 / 24 = 6.6"$ ∴ o.k.

For the 30" beam: $l_n / 16 = 29.33 \times 12 / 16 = 22"$ – o.k.

∴ Deflection calculations are not required.

$M_u^- = w_u L^2 / 8 = 0.253 \times 15^2 / 8 = 7.1^{k\text{-}ft}/_{ft}$,

$\phi M_n = \phi\rho b d^2 f_y\{1 - 0.59\rho(f_y / f'_c)\} \geq M_u$

$0.9 \times 12 \times 6.5^2 \times 40\rho\{1 - 0.59\rho(40 / 3)\} = 7.1 \times 12$

$\rho = 0.005$

∴ $A_s = 0.005 \times 12 \times 6.5 = 0.39^{sq.in}/_{ft}$

Minimum Slab reinforcement:

$A_{s,min} = 0.002bh = 0.002 \times 12 \times 8 = 0.19^{sq.in}/_{ft}$

Use # 5 @ 9" o.c.

For $M_u^+ = 9w_u L^2 / 128 = 4.0^{k\text{-}ft}/_{ft}$, minimum steel governs

Use **# 4 @ 12" o.c.**

For the beam; d = 30 – 2.5 = 27.5"

$w_u = 4.74 + 1.4 \times 36 \times 22 \times 0.15 / 144 = 5.89$ klf

$M_u = w_u L^2 / 8 = 5.89 \times 29.33^2 / 8 = 633^{k\text{-}ft}$

$b_{eff} = L / 4 = 29.33 \times 12 / 4 = 88"$

$\phi M_n = \phi\rho b d^2 f_y\{1 - 0.59\rho(f_y / f'_c)\} \geq M_u$

$0.9 \times 88 \times 27.5^2 \times 40\rho\{1 - 0.59\rho(40 / 3)\} = 633 \times 12$

$\rho = 0.00326$; $A_s = 0.00326 \times 88 \times 27.5 = 7.89$ in²

Check: $a = 7.89 \times 40 / (0.85 \times 3 \times 88) = 1.4" < h_f$ ∴ o.k.

$A_{s,min} \geq (200 / f_y)b_w d = (200 / 40{,}000)36 \times 27.5$
$= 5.0$ in² ∴ **use 8 # 9**

Check shear at d-distance from face of support

$V_u = 5.89(29.33 / 2 - (27.5 + 4) / 12) = 70.9^k$

$\phi V_c = 2\phi\sqrt{f'_c}\,b_w d = 2 \times 0.85 \times \sqrt{3000} \times 36 \times 27.5$
$= 92{,}000^{\#} = 92^k$

$\phi V_c > V_u$ ∴ use # 4 stirrups @ d / 2 ≅ 12" o.c.; discontinue stirrups where $V_u < \phi V_c / 2$ (i.e., 7' from ends)

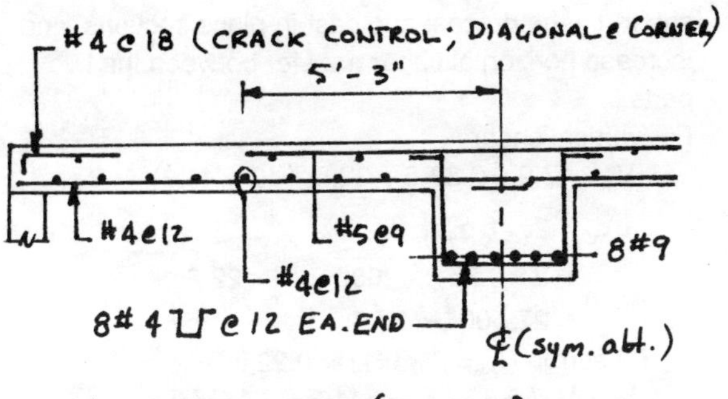

SECTION (N.T.S.)

b) Redesign for the case that the beam is 36" wide by 15" deep. For the beam strip: A = 1564 in^2; $\bar{y}$ = 9.78"; I = 19,900 in^4; I_{slab} = 6997 in^4

$$\frac{rP(29.33)^3}{(48E)19{,}900} = \frac{(1-r)P(30)^3}{(48E)(6997)}$$

∴ **r = 0.75**

In this case, the system has similar stiffness in the two directions ∴ design as a two-way beam-slab. The stiff central strip will be designed to carry 65% of the slab loading. Thus, for the 29.33' span,

$w_u = 0.65 \times 0.253 \times 30 + 1.4 \times 7 \times 36 \times 0.001$
$= 5.3$ klf

$M_o = w_u L^2 / 8 = 5.3 \times 29.33^2 / 8 = 570^{k\text{-}ft}$

Use the distribution criteria of Chapter 13, ACI 318:
$\alpha_1 = 19{,}900 / 6997 = 2.84$; $L_2 / L_1 \cong 1.0$; $\alpha L_2 / L_1 = 2.84$ ∴ 75% of the static moment is resisted by the central strip; 85% of this is to be resisted by the beam

$M_u^+ = (0.85)(0.75M_o) = 0.85 \times 0.75 \times 570 = 363^{k\text{-}ft}$
$d = 15 - 2.5 = 12.5$"; Re: 13.2.4, $b_{eff} \le b_w + 2h_s$
$= 36 + 2 \times 7 = 50$"

$$\phi M_n = \phi\rho b d^2 f_y \{1 - 0.59\rho(f_y / f'_c)\} \ge M_u$$

$0.9 \times 50 \times 12.5^2 \times 40\rho\{1 - 0.59\rho(40/3)\} = 363 \times 12$
$\rho = 0.0212 < \rho_{max} = 0.75\rho_b = 0.0278$
∴ $A_s = 0.0212 \times 50 \times 12.5 = 13.25$ in^2

Check the rectangular beam assumption:
$a = 13.25 \times 40 / (0.85 \times 3 \times 50) = 4.16" < 8"$ ∴ o.k.

Use 9 # 11 bottom

Shear – check shear at d-distance from face of support
$w_u = 0.75 \times 0.85 \times 0.253 \times 30 + 1.4 \times 7 \times 36 \times 0.001$
$= 5.2$ klf
$V_u = 5.2[29.33/2 - (12.5 + 4)/12] = 69^k$

$$\phi V_c = 2\phi\sqrt{f'_c}\, b_w d = 2 \times 0.85 \times \sqrt{3000} \times 36 \times 12.5$$
$$= 41{,}900^{\#} = 41.9^k$$

$\phi V_c < V_u$ ∴ use # 4 stirrups
$V_s = V_u / 0.85 - V_c = (69 - 41.9)/0.85 = 31.9^k$

$$s \le \begin{cases} A_v f_y d / V_s = 0.40 \times 40 \times 12.5 / 31.9 = 6.3" \text{ o.c.} \\ A_v f_y / 50 b_w = 0.40 \times 40{,}000 / (50 \times 36) \\ \quad = 16" \text{ o.c.} \\ d/2 = 12.5/2 = 6.25" \text{ o.c} \leftarrow \text{governs} \end{cases}$$

∴ use **# 4 U-stirrups @ 6" o.c. at critical location** terminate where $V_u < \phi V_c / 2 = 41.9 / 2 = 21^k$
$V_u = 5.2[29.33/2 - x] \le 21$; ∴ $x \ge 10.6'$

The slab must resist the remainder of the moment, M_o
$m_1 = 0.15 \times 0.75 \times 570 / (15 - 50/12) = 5.9^{k\text{-}ft}/_{ft}$
$m_2 = 0.25 \times 570 / 15 = 9.5^{k\text{-}ft}/_{ft}$

For simplicity, design the slab spanning parallel to resist $9.5^{k\text{-}ft}/_{ft}$ as well.

Minimum Slab reinforcement:
$A_{s,min} = 0.002bh = 0.002 \times 12 \times 8 = 0.19^{sq.in}/_{ft}$

$$\phi M_n = \phi\rho b d^2 f_y \{1 - 0.59\rho(f_y / f'_c)\} \ge M_u$$

$0.9 \times 12 \times 6.5^2 \times 40\rho\{1 - 0.59\rho(40/3)\} = 9.5 \times 12$
∴ $\rho = 0.00756$; $A_s = 0.00756 \times 12 \times 6.5 = 0.59^{sq.in}/_{ft}$

Use # 6 @ 9" o.c.

For the perpendicular direction:
$M_o = (0.25 + 0.253 \times 29.33 \times 30^2 / 8) / 29.33$
$= 7.11^{k\text{-}ft}/_{ft}$
$0.9 \times 12 \times 6.5^2 \times 40\rho\{1 - 0.59\rho(40/4)\} = 7.11 \times 12$
∴ $\rho = 0.00487$; $A_s = 0.00487 \times 12 \times 6.5 = 0.38^{sq.in}/_{ft}$

Use # 5 @ 10" o.c.

Even though simple support conditions were assumed, provide minimum top steel (# 4 @ 12" o.c.) for crack control extending nominal (e.g., 0.2L) into the span. Also, diagonal top and bottom steel is required at the corners (Re: 13.4.6, ACI 318): **use # 6 @ 9" o.c.**

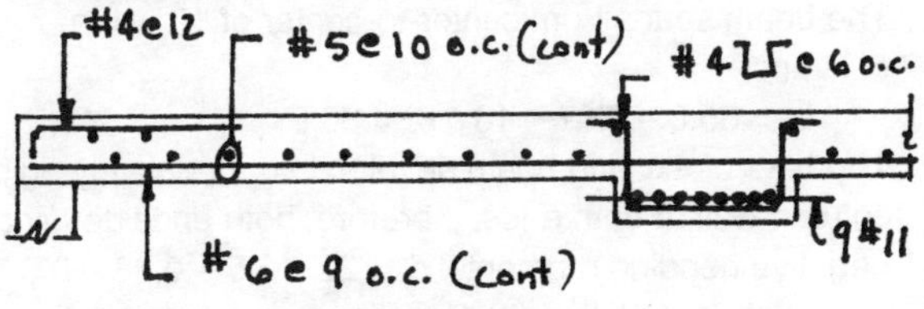

SECTION (N.T.S.)
(WALL REINF. OMITTED FOR CLARITY)

Concrete-44

a) Design a composite beam for the specified conditions. Check adequacy of the given longitudinal steel assuming that composite action can be achieved and that the shear strength of the section is adequate. Use the strength design method (ACI 318); f'_c = 3 ksi; f_y = 40 ksi. For office occupancy, Table 16-A, UBC-97, requires w_l = 50 psf. Reduce live loads using basic method of Section 1607.5.

$$R \le \begin{cases} 0.08(A - 150) = 0.08(12 \times 35 - 150) \\ \quad = 22\% \leftarrow \text{governs} \\ 23(1 + 100/50) = 70\% \\ 40\% \text{ for one level} \end{cases}$$

$w_u = 1.4w_d + 1.7w_l = 1.4[(20 + 6 \times 12.5) \times 12 + 16 \times 20 \times (150/144)] + 1.7 \times (1 - 0.22) \times 50 \times 12 = 2860$ plf
$= 2.86$ klf

The critical section is at the midspan where the opening destroys the tee-flange on one side

$$b_{eff} \le \begin{cases} b_w + L/12 = 16 + 35 \times 12/12 = 51" \\ \text{spacing plus edge} = 72 + 8 = 80" \\ b_w + 6h_s = 16 + 6 \times 6 = 52" \end{cases}$$

∴ $b_{eff} = 51"$

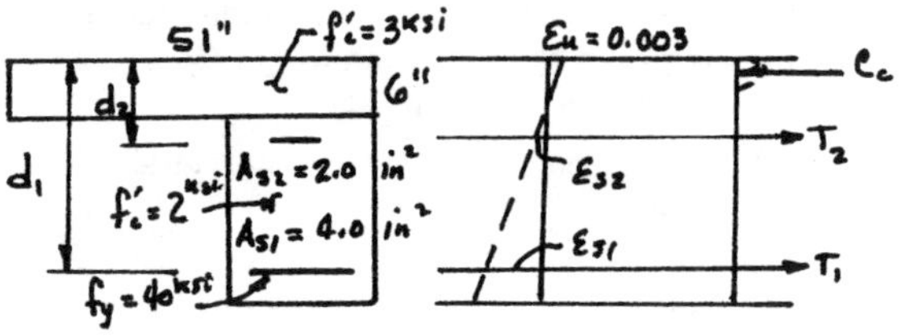

$d_1 = 26 - 2.5 - 0.38 - 0.5 = 22.6$";

$d_2 = 6 + 2.5 + 0.38 + 0.5 = 9.4$"

Assume that both layers of steel yields and that the beam behaves as a rectangular beam of width $b = b_{eff}$.

$a = (A_{s1} + A_{s2})f_y / (0.85f'_c b) = 6.00 \times 40 / (0.85 \times 3 \times 51) = 1.85" < h_s.$

Therefore, the beam behaves as rectangular. Check strains in the two layers; by similar triangles

$x = a / \beta_1 = 1.85 / 0.85 = 2.18"$

$\varepsilon_2 = ((9.4 - 2.18) / 2.18)0.003 = 0.0099 > \varepsilon_y$

$\therefore f_{s1} = f_{s2} = f_y = 40$ ksi

∴ Both layers of steel yield.

$$\begin{aligned}\phi M_n &= \phi[A_{s1}f_y(d_1 - a/2) + A_{s2}f_y(d_2 - a/2)]\\ &= 0.9[4.0 \times 40(22.6 - 1.86/2) + 2.0 \times 40(9.40 - 1.86/2)]/12\\ &= 311^{\text{k-ft}}\end{aligned}$$

The beam spans from center-to-center of 10" wide columns:

$L_n = 35.0 - 0.67 - 10/12 = 33.5'$

Top bars of existing beam develop into a relatively rigid column-wall at both ends; therefore, both ends develop negative bending moments: $d^- = 20 - 2.5 - 0.38 - 0.5 = 16.6$"; $b_w = 16$"; $f'_c = 2000$ psi (compression is in the existing, weak concrete)

$a = 2.00 \times 40 / (0.85 \times 2 \times 16) = 2.94"$

$$\begin{aligned}\phi M_n &= \phi A_s f_y (d - a/2)\\ &= 0.9 \times 2.00 \times 40(16.6 - 2.94/2)/12\\ &= 90.8^{\text{k-ft}}\end{aligned}$$

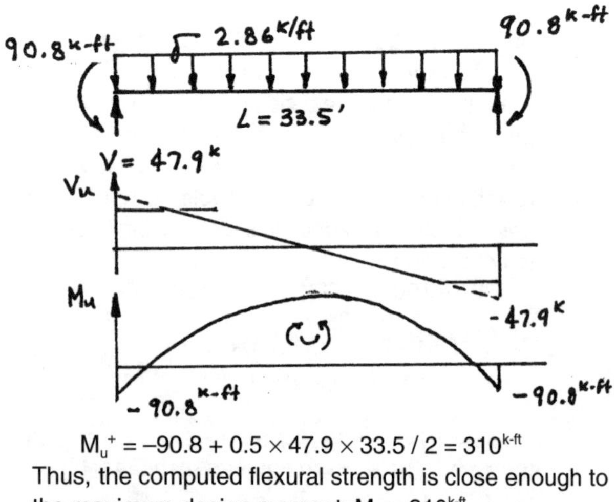

$M_u^+ = -90.8 + 0.5 \times 47.9 \times 33.5 / 2 = 310^{\text{k-ft}}$

Thus, the computed flexural strength is close enough to the maximum design moment, $M_u = 310^{\text{k-ft}}$.

Check shear capacity of the composite section. The existing smooth # 3 bars are not acceptable under current codes; therefore, disregard them. Cover equal to 2-1/2" clear to existing stirrups will permit slots 2" deep cut into beam and new # 3 stirrups installed. The new stirrups will perform multiple functions: resist vertical shear, tie the precast and cast-in-place sections, and increase horizontal shear transfer between the two parts.

For vertical shear:

$V_u = 47.9 - 2.89(5 + 22.6) / 12 = 41.3^k$

$$\begin{aligned}\phi V_c &= 2\phi\sqrt{f'_c}\, b_w d\\ &= 2 \times 0.85 \times \sqrt{2000} \times 16 \times 22.6\\ &= 27{,}500^{\#} = 27.5^k\end{aligned}$$

try # 3 stirrup ($A_v = 2 \times 0.11 = 0.22$ in²)

$V_s = V_u / 0.85 - V_c = (41.3 - 27.5) / 0.85 = 16.3^k$

$$s \le \begin{cases} A_v f_y d / V_s = 0.22 \times 40 \times 22.6 / 16.3 = 12.2" \text{ o.c.}\\ A_v f_y / 50 b_w = 0.22 \times 40{,}000 / (50 \times 16) = 11" \text{ o.c.}\\ d/2 = 22.6/2 = 11.3" \text{ o.c}\end{cases}$$

∴ **use # 3 U-stirrups @ 11" o.c.**

Check the distance to stirrup termination ($V_u < \phi V_c / 2 = 13.8^k$)

$V_u = 47.9 - 2.86x \le 13.8$; $x \ge 11.9'$ close to midspan

∴ use 11" spacing throughout.

b) The required elevation is:

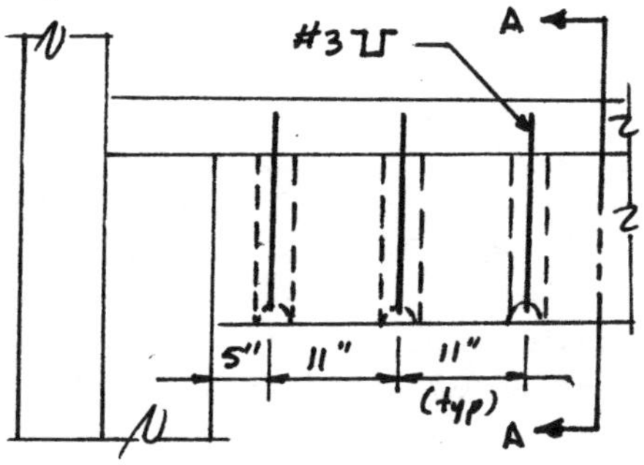

ELEVATION

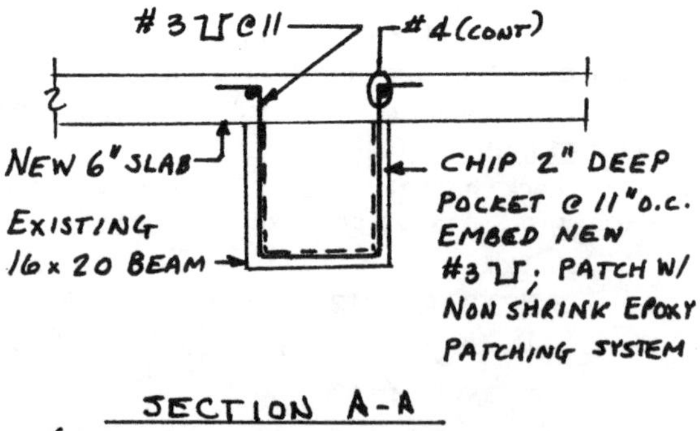

SECTION A-A

(NOTE: SLAB REINFORCEMENT AND EXISTING BEAM REINFORCEMENT OMITTED FOR CLARITY)

c) An additional required check is for horizontal shear transfer (17.5, ACI 318). Let v_{nh} = required horizontal shear strength required (psi)

$V_u \le \phi V_{nh} = \phi v_{nh} b_w d$

$47.9 \le 0.85 v_{nh} \times 16 \times 22.6$

$v_{nh} = 0.156$ ksi $= 156$ psi

The required shear transfer can be accomplished by deliberately roughening the contact surface and using the minimum ties, which are provided by the # 3 stirrups at 11" o.c.

Concrete-45

a) Evaluate the proposed structure. The column sizes are not given. Assume that moment transfer from the central beam to supports is negligible (i.e., simple supports). For roof live load, use UBC-97 Table 16-C, (based on the tributary area for the shorter cantilever $= 13 \times 2 \times 12.5 = 325\ \text{ft}^2$); therefore, $w_L = 16$ psf. Analyze for the critical pattern of live loading on the structure. Critical loading occurs with live plus dead on the 43.5' span with dead only on the cantilevers. (Note: the problem statement indicates uniformly distributed loading. Therefore, no consideration is given to possible snow drift accumulation in this solution.)

Dead load of 112 pcf; say, 115 pcf to allow for reinforcement; the roof slopes 3:22.5

∴ $w_{slab} = 6 \times (115 / 12) \times (22.7 / 22.5) \times 2 \times 12.5$
$= 1450$ plf

Add for the 24" x 36" additional beam weight on the central span:

$w_{bm} = 24 \times (36 - 6) \times 115 / 144 = 575$ plf

For the cantilevers, the w_{bm} varies linearly from 575 plf at the supports to zero at the ends. Use the strength design method (ACI 318-95); $f'_c = 3$ ksi; $f_y = 40$ ksi.

For the 23.5' cantilever:

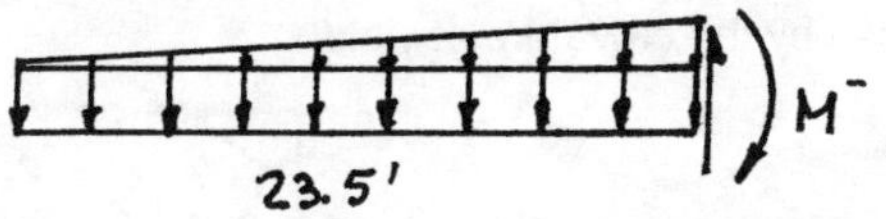

$M^- = 1.4 \times 1.45 \times 23.5^2 / 2 + 0.5 \times 1.4 \times 0.575 \times 23.5^2 / 3 = 635^{k\text{-}ft}$

For the 13' cantilever:

$M^- = 1.4 \times 1.45 \times 13^2 / 2 + 0.5 \times 1.4 \times 0.575 \times 13^2 / 3 = 194^{k\text{-}ft}$

Critical loading on the 43.5' span:

$w_u = 1.4w_d + 1.7w_l = 1.4 \times (1.45 + 0.575) + 1.7 \times 0.016 \times 2 \times 12.5 = 3.5$ klf

As originally built:

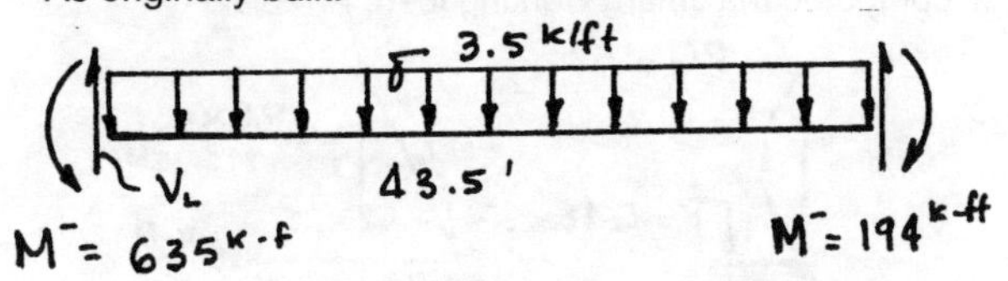

$V_L = 3.5(43.5 / 2) + (635 - 194) / 43.5 = 86.3^k$

$V = 86.3 - 3.5x = 0$; $x = 24.6'$ (point of M_{max})

∴ $M_{u,max} = -635 + 0.5(86.3)24.6 = 426^{k\text{-}ft}$

As modified (23.5' cantilever removed)

$V_L = 3.5(43.5 / 2) - 194 / 43.5 = 71.7^k$

$V = 71.7 - 3.5x$; $x = 20.5'$

$M_{u,max} = 0.5 \times 71.7 \times 20.5 = 734^{k\text{-}ft}$

Thus, there is a significant increase in bending moment caused by removal of the cantilever

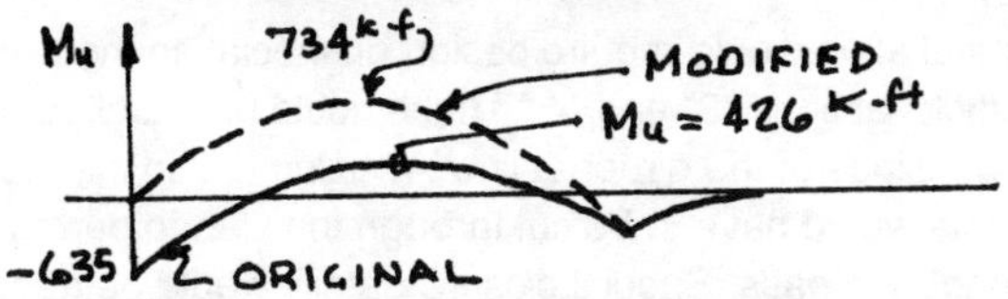

Removing the cantilever creates a 72% increase in the design moment at the critical flexural section. Deflection in the 43.5' span will also increase significantly. The reaction on the right support will increase about 10% from its original value, which might be a concern. Check the capacity of the member to handle the increased M_u.

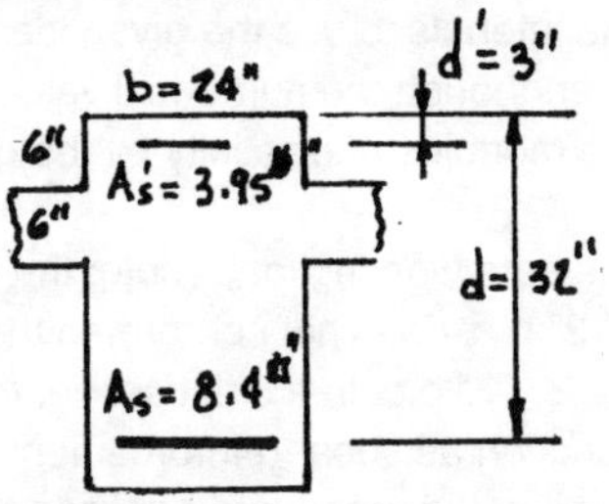

Doubly reinforced beam – solve by trial-and-error: try $f'_s = 30$ ksi

$a = (A_s f_y - A_s f_s) / (0.85 f'_c b)$
$= (8.4 \times 40 - 3.95 \times (30 - 0.85 \times 3)) / (0.85 \times 3 \times 24) = 3.72"$

$x = a / \beta_1 = 3.72 / 0.85 = 4.37"$

$\varepsilon'_s = ((4.37 - 3) / 4.37)0.003 = 0.00094 < \varepsilon_y$

∴ $f'_s \cong 29000 \times 0.00094 = 27.3$ ksi – try again using an intermediate value, say 28.5 ksi

$a = (8.4 \times 40 - 3.95 \times (28.5 - 0.85 \times 3)) / (0.85 \times 3 \times 24) = 3.81"$

$x = a / \beta_1 = 3.81 / 0.85 = 4.48"$

$\varepsilon'_s = ((4.48 - 2.50) / 4.48)0.003 = 0.00099$

$f'_s \cong 29000 \times 0.00099 = 28.8$ ksi – close enough

$M_n = \phi[C_c(d - a / 2) + C_s A'_s(d - d')]$

$M_n = 0.9[0.85 \times 3 \times 24 \times 3.81 \times (32 - 3.81 / 2) + (28.5 - 0.85 \times 3) \times 3.95 \times (32 - 3)] / 12 = 749^{k\text{-}ft}$

Thus, $\phi M_n > M_u$ ∴ **the section has adequate flexural strength to resist the increased moment.**

b) If it were necessary to either strengthen the beam or compensate for the increased deflection, a scheme that should prove feasible is to externally prestress the central span between lines "3" and "4". Anchors could be positioned such that negative eccentricity occurs at line "3" with a straight tendon sloped to zero eccentricity at line "4." The prestress force could be computed to induce an end moment at line "3" equivalent to the moment produced by the 23.5' cantilever.

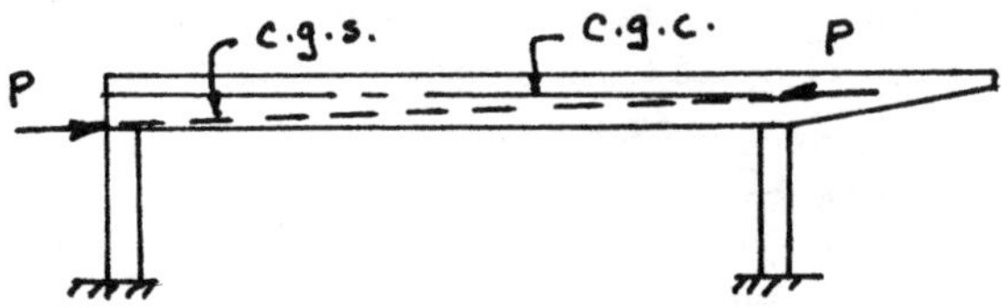

c) Installation would require design of special anchorage blocks at lines "3" and "4." These could be installed on both sides of the girder at line "4" using drilled inserts. Slots would have to be cut through the slab to permit the tendon to pass. Special closure details would be needed to patch the slots after the tendons were installed. There are several post tensioning systems that could be adapted for this type application (e.g., the Dywidag "threadbar" system).

Concrete-46

I assume that the intent is to use the given idealized stress-strain curves, even though the numerical values are not realistic (e.g., the modulus of elasticity by the given curve is 40,000 ksi).

a) Compute the safe working load (note: the given stirrup spacing of 12" o.c. does not conform to the code minimum d / 2 = 7" o.c. In a strict sense, the stirrups might not function as shear reinforcement because a diagonal crack could propagate between them. However, since this problem relates to theory rather than code criteria, I will include the stirrup capacity in the analysis),

$$V_c = 2\sqrt{f'_c}\,b_w d = 2 \times \sqrt{3000} \times 10 \times 15$$
$$= 16,400^{\#} = 16.4^k$$

$V_s = A_v f_y\, d / s = 0.40 \times 40{,}000 \times 15 / 12 = 20{,}000^{\#}$

$V_u = \phi(V_c + V_u) = 0.85(16.4 + 20) = 30.9^k$

For a load factor of 1.7, the safe load based on shear is

$P \le V_u / 1.7 = 30.9 / 1.7 = 18.2^k$

To determine the working load limit based on flexure, use working stress design with an allowable steel stress of 20,000 psi and concrete compressive strength of $0.45f'_c$.

$n = E_s / E_c = 40{,}000 / 3 = 13$

$\rho n = (A_s / bd)n = (2 \times 0.79 / (10 \times 15))13 = 0.137$

$$k = \sqrt{(\rho n)^2 + 2(\rho n)} - \rho n$$
$$= \sqrt{(0.137)^2 + 2(0.137)} - (0.137)$$
$$= 0.404$$

$\therefore\ kd = 0.404 \times 15 = 6.06"; \quad (d - kd) = 8.94"$

$$I_{cr} = b(kd)^3 / 3 + nA_s(d - kd)^2$$
$$= 10 \times 6.06^3 / 3 + 13 \times 1.58 \times 8.94^2$$
$$= 2380\,\text{in}^4$$

$f_s = (M(d - kd) / I_{cr})n \le 20$ ksi

$\therefore\ M \le 20 \times 2380 / (8.94 \times 13) = 410^{k\text{-}in} = 34.2^{k\text{-}ft}$

$f_c = M(kd) / I_{cr} \le 0.45f'_c = 0.45 \times 3\text{ ksi} = 1.35$ ksi

$\therefore\ M \le 1.35 \times 2380 / 6.06 = 530^{k\text{-}in} = 44.1^{k\text{-}ft}$

Steel governs: $M = 34.2^{k\text{-}ft} = P(6 - 5/12)'$

<u>$P \le 6.12^k$ governs</u>

b) Stress at yield = 40 ksi ∴ since behavior is linear, the result above leads directly to

<u>$P_y = 2 \times 6.12 = 12.24^k$</u>

c) Ultimate capacity – I assume that the given stress-strain relationship applies to the linear strain distribution that exist in the compression zone and that the usual assumptions of flexure theory apply (i.e., no tension in concrete, complete bond between steel and concrete, steel yields, etc.). Thus, for the trapezoidal stress distribution

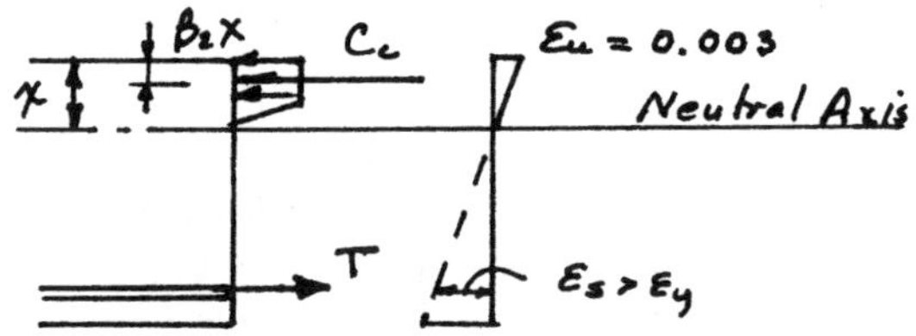

$C_c = \beta_1\beta_3 f'_c bx;\ \beta_3 = 1.0$

$\beta_1 f'_c bx = (2/3) \times f'_c b + (1/6) \times f'_c b$

$\therefore\ \beta_1 = 0.8$

The position of the resultant with respect to the top fiber is

$\beta_2 x = [(2/3) \times f'_c b(x/3) + (1/6) \times f'_c b(2/3 + 1/6)x] / [(5/6)\,f'_c bx]$

$\therefore\ \beta_2 = 0.45$

Thus,

$\Sigma F_h = 0$: $C_c = T$: $0.8 \times 1.0 \times 3 \times 10x = 1.58 \times 40$

$\therefore\ x = 2.63"; \ \beta_2 x = 1.19"$

$\Sigma M_T = 0$:

$M_n = T(d - \beta_2 x) = 1.58 \times 40(15 - 1.19) = 873^{k\text{-}in}$

$= 72.7^{k\text{-}ft}$

<u>$P_u = M_n / (6 - 5/12) = 72.7 / 5.58 = 13.0^k$</u>

d) Define the ductility ratio, μ, as the ultimate curvature, ϕ_u, divided by the curvature at yield

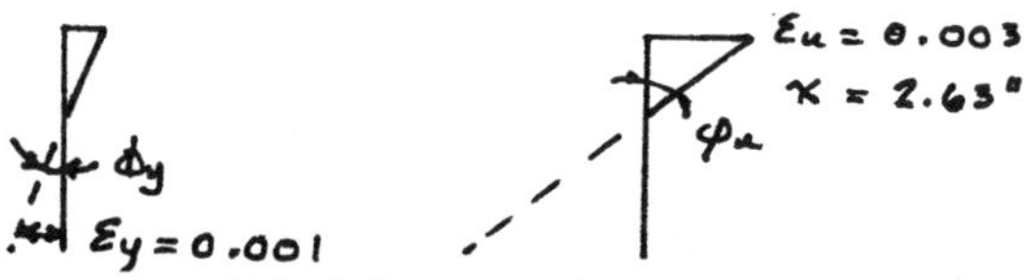

$\phi_y = \varepsilon_y / (d - kd) = 0.001 / 8.94 = 0.000112\ ^{rad}/_{in}$

$\phi_u = \varepsilon_u / x = 0.003 / 2.63 = 0.00114\ ^{rad}/_{in}$

$\therefore$ <u>$\mu = \phi_u / \phi_y = 0.00114 / 0.000112 = 10$</u>

e) The failure mode would be described as a **<u>ductile, flexural failure</u>**.

f) To determine adequacy of the column, apply the computed maximum working load, $P = 6.12^k$

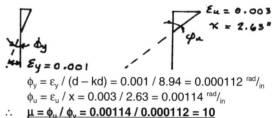

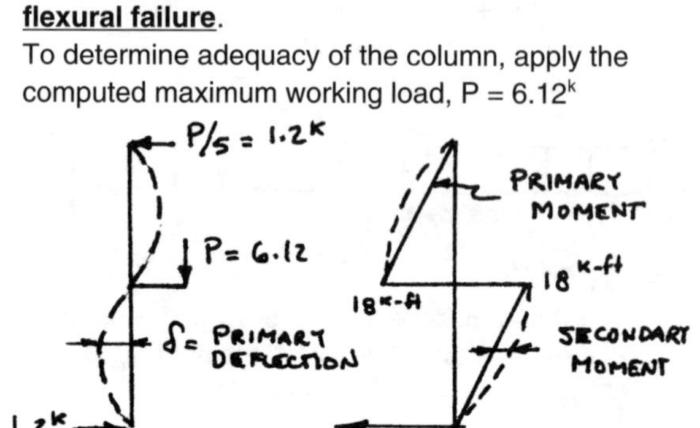

The column is subjected to critical combined axial compression plus bending: $P_u = 1.7 \times 106 = 180^k$; $M_u = 1.7 \times 18 = 30.6^{k\text{-}ft}$, therefore, check this loading against the given interaction diagram. The point (30.6, 180) is well inside the curve; therefore, column strength is sufficient.

Secondary moments, $P\delta$, will increase along the column (see diagram above), but not at the point of maximum moment. Thus, it is not likely that secondary, or slenderness, effects will become critical for this loading (I assume here that the problem statement that the column is braced against buckling is still applicable).

g) If A_s is increased to 2 # 11, $A_s = 2 \times 1.56 = 3.12$ in^2

$\Sigma F_h = 0$: $C_c - T = 0$

$\therefore$ $x = 3.12 \times 40 / (0.8 \times 3 \times 10) = 5.2''$

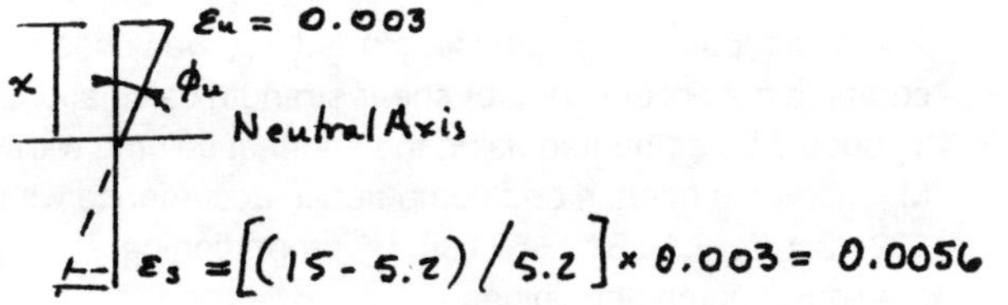

$\varepsilon_s > \varepsilon_y$ $\therefore$ still have ductile behavior ($\phi_u \cong 5\phi_y$)

$M_n = 2 \times 1.56 \times 40 \times (15 - 0.45 \times 5.2) / 12 = 132^{k\text{-}ft}$

$P_u = 132 / (6 - 5 / 12) = 23.6^k$

The ultimate load computed on the basis flexural strength is still below the computed shear strength $\therefore$ still expecting a **ductile, flexural failure**.

Concrete-47

a) Design the one-way deck slab. Use the strength design method (ACI 318); $f'_c = 3$ ksi; $f_y = 40$ ksi.;
$d^- = 7 - 1.5 - 1.25 = 5.25''$; $d^+ = 7 - 1 - 0.25 = 5.75''$.

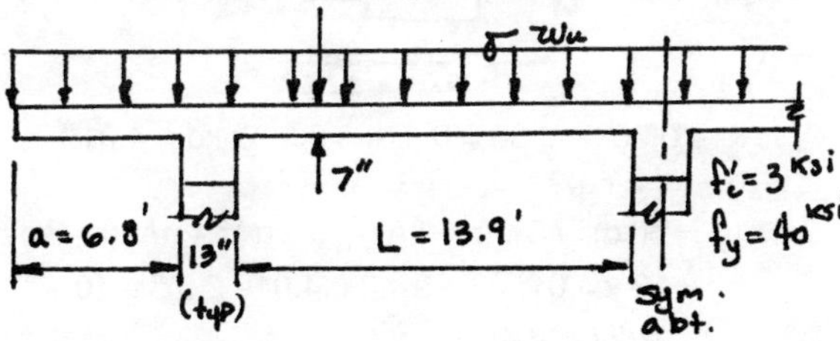

$w_u = 1.4w_d + 1.7w_l = 1.4 \times 7 \times 12.5 + 1.7 \times 100$
$= 293$ plf $= 0.293$ klf

Shear design (use the ACI coefficients, ACI 8.3):

$$V_u \geq \begin{cases} 1.15w_uL/2 - w_ud^- \\ = 293(1.15 \times 13.9/2 - 5.25/12) = 2220^\# \\ w_ua - w_ud^- = 293(6.8 - 5.25/12) = 1880^\# \end{cases}$$

$$\phi V_c = 2\phi\sqrt{f'_c}\,b_wd = 2 \times 0.85 \times \sqrt{3000} \times 12 \times 5.25$$
$$= 5870^\# \succ V_u$$

$\therefore$ shear strength is adequate. Minimum steel (temp/shrinkage)

$A_{s,min} = 0.002bh = 0.002 \times 12 \times 7 = 0.17^{sq.in}/_{ft}$, say #4 @ 14" o.c.

Flexural steel:

$M_u^- = w_ua^2/2 = 0.293 \times 6.8^2/2 = 6.7^{k\text{-}ft}/_{ft}$

$$\phi M_n = \phi\rho bd^2f_y\{1 - 0.59\rho(f_y/f'_c)\} \geq M_u$$

$0.9 \times 12 \times 5.25^2 \times 40\rho\{1 - 0.59\rho(40/3)\} = 6.7 \times 12$

$\therefore$ $\rho = 0.0072$; $A_s = 0.0072 \times 12 \times 5.25 = 0.45^{sq.in}/_{ft}$

Use # 5 @ 8" o.c. top

To satisfy section 1637-11, UBC-97, using $C_a = 0.36$ (soil profile type D; near-fault factor = 1.0) and an importance factor of 1.0, design upward loading due to vertical acceleration on cantilever is

$w_u = 0.7C_aIW = 0.7 \times 0.36 \times 1.0 \times (7 \times 12.5$ psf$)$
$= 22$ psf $\uparrow$

Provide minimum steel at least **# 4 @ 14" o.c. bottom** fully developed at face of cantilever. For the interior span

$M^+_u = w_ul_n^2/11 = 0.293 \times 13.9^2/11 = 5.09^{k\text{-}ft}$

$0.9 \times 12 \times 5.75^2 \times 40\rho\{1 - 0.59\rho(40/3)\}$
$= 5.09 \times 12$

$\rho = 0.0044$ $\therefore$ $A_s = 0.0044 \times 12 \times 5.75 = 0.31$ in$^2/_{ft}$;

use # 5 @ 12" o.c.

$M^-_u = w_ul_n^2/9 = 0.293 \times 13.9^2/9 = 6.2^{k\text{-}ft}$

$\rho = 0.00666$; $A_s = 0.00666 \times 12 \times 5.25 = 0.42$ in$^2/_{ft}$;

use # 5 @ 8" o.c.

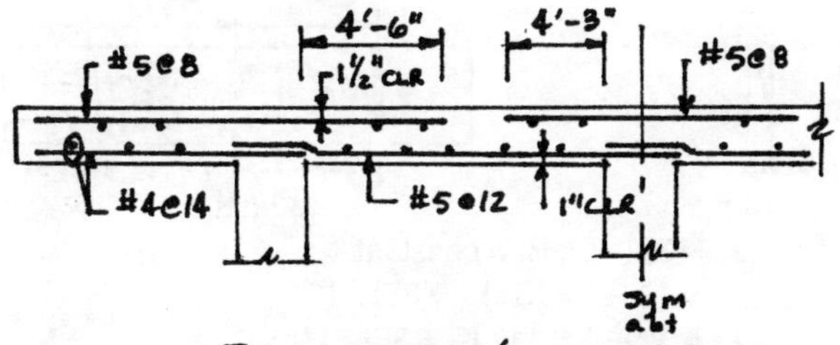

ELEVATION (N.T.S.)

b) Calculate the short and long term dead load deflection in the cantilever. Use the centerline dimensions (Re: Section 9.5, ACI 318, Eqn 9-7)

$$I_g = bh^3/12 = 12 \times 7^3/12 = 343\text{in}^4$$
$$\rho = 0.45/(12 \times 5.25) = 0.007;\ \ n = 10;$$
$$\therefore\ \rho n = 0.007 \times 10 = 0.070$$
$$k = \sqrt{(\rho n)^2 + 2(\rho n)} - \rho n$$
$$= \sqrt{(0.07)^2 + 2(0.07)} - (0.07) = 0.31$$
$$\therefore kd = 0.31 \times 5.25 = 1.65'';\ \ (d - kd) = 3.60''$$
$$I_{cr} = b(kd)^3/3 + nA_s(d - kd)^2$$
$$= 12 \times 1.65^3/3 + 10 \times 0.45 \times 3.60^2 = 76\text{in}^4$$

Service load moment (sustained dead only):

$w = w_d = 7 \times 12.5 = 87.5$ plf $= 0.0875$ klf

$$M_{cr} = 7.5\sqrt{f'_c}I_g/(h/2)$$
$$= [7.5\sqrt{3000}(343)/(7/2)]/12000 = 3.35^{k-ft}$$
$$M_a = wL^2/8 = 0.0875 \times 7.3^2/2 = 2.33^{k-ft}$$
$$M_{cr}/M_a = 3.35/2.33 = 1.43;\ \text{thus, } I_e = I_g = 343\text{in}^4$$

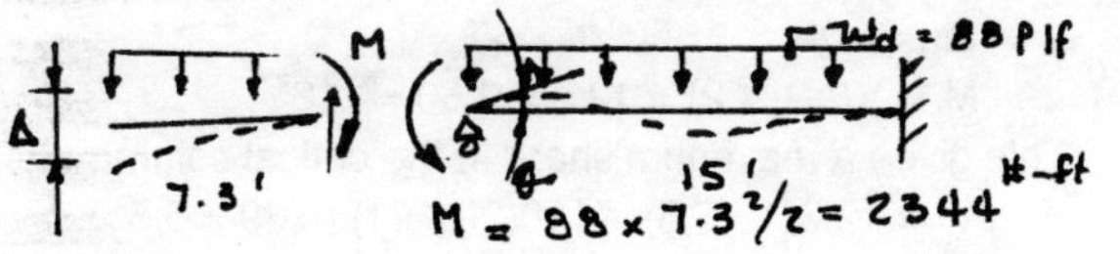

$EI = 3{,}100{,}000 \times 343 = 1.07 \times 10^9$ lb-in^2

$\Delta = wL^4 / (48EI) + \theta a$

From Roark, 1975, Table 3, the rotation at the support is

$\theta = wL^3 / 48EI - ML / 4EI$

$\theta = [(87.5 \times 15)(15 \times 12)^2 / 48 - 2344 \times 12 \times (15 \times 12) / 4] / 1.07 \times 10^9$

$\theta = 0.000346$ rad (counterclockwise on the left end)

$$\Delta_i = [0.0875 \times 7.3 \times (7.3 \times 12)^3] / (48 \times 1.07 \times 10^9) + (0.000346 \times 7.3 \times 12) = 0.039" \downarrow$$

$$\Delta_{lt} = (1+\lambda)\Delta_d + \Delta_l = 3(0.039) = 0.12" \downarrow$$

For a singly reinforced beam with loads sustained 5 years or more: $\lambda = 2.0$

c) Column Design – given that the (service) lateral load is 0.2W

W = 7" slab + 1/2 col. wt.

$= (87.5 \times 14 \times 44.7) + 176 \times 2 \times (5 + 6 + 7)$

$W = 61{,}400^{\#} = 61.4^k$

$\therefore \quad V = 0.2W = 12.3^k$

Consider columns to be pinned at top and fixed at base (per problem statement).

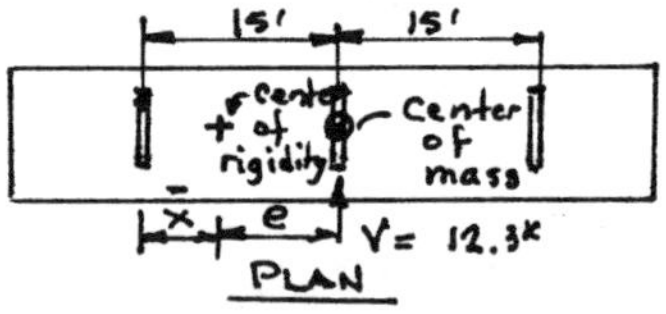

$K_i = 3EI / L^3$; EI = constant

$L_1 = 14 - 1 = 13'$; $K_1 = 3EI / 13^3 = K$

$L_2 = 12 - 1 = 11'$; $K_2 = 3EI / 11^3 = 1.6K$

$L_3 = 10 - 1 = 9'$; $K_3 = 3EI / 9^3 = 3.0K$

$\Sigma K_i = K + 1.6K + 3.0K = 5.6K$

$\bar{x} = [1.6K(15) + K(30)] / 5.6K = 9.6'$ from shortest column

An accidental eccentricity equal to 0.05 times the total length of the bridge will be included per 1630.7, UBC-97

$$e + 0.05B = (15 - \bar{x}) + 0.05 \times 45 = 15 - 9.6 + 2.25 = 7.65'$$

$M_t = V(e + 0.05B) = 12.3 \times 7.65 = 94^{k\text{-}ft}$

$\Sigma K_i d_i^2 = [3(9.6)^2 + 1.6(5.4)^2 + 1(20.4)^2]K = 739K$

$\therefore \quad V_1 = (12.3K) / (5.6K) + 94 \times (20.4K) / (739K) = 4.8^k$

similarly,

$V_2 = 12.3(1.6 / 5.6) + 94(5.4)(1.6K) / 739 = 4.6^k$

and $V_3 = 12.3(3 / 5.6) - 94(9.6)(3.0K) / 739) = 3.0^k$

Check for torsional irregularity per Table 16-M (i.e, if $\delta_{max} < 1.2\delta_{ave}$)

$\delta_1 = 4.8 \times 13^3 / 3EI = \delta_{max}$; $\delta_3 = 3.0 \times 9^3 / 3EI = 0.21\delta_1$

$\delta_{ave} = (1 + 0.21)\delta_1 / 2 = 0.61\delta_1$ $\therefore$ torsional irregularity requires that the accidental torsion be amplified per Eqn 30-16:

$A_x = [\delta_{max} / 1.2\delta_{ave}]^2 = [1 / (1.2 \times 0.61)]^2 = 1.87$

$\therefore$ accidental eccentricity $= 1.87 \times 0.05 \times 45 = 4.2'$. thus,

$M_t = V(e + 4.2) = 12.3 \times 9.60 = 118^{k\text{-}ft}$

This gives a maximum shear in the critical column

$V_1 = 12.3(1 / 5.6) + 118(20.4)(1) / 739 = 5.5^k$

$V_2 = 12.3(1.6 / 5.6) + 118(5.4)(1.6) / 739 = 4.9^k$

The eccentricity cannot be used to reduce the design force in the columns $\therefore$ $V_3 = 12.3(3/5.6) = 6.6^k$. Column moments (two columns per bent) are

$M_1 = (5.5 / 2)13 = 35.8^{k\text{-}ft}$; $M_3 = (6.6 / 2)9 = 29.7^{k\text{-}ft}$; column 2 is not critical by inspection. Design the reinforcement for column 1 – neglect axial load per problem statement. The given lateral forces are working load $\therefore$ apply 1.4 factor for design values:

$M_u = 1.1(1.4 \times 35.8) = 55.1^{k\text{-}ft}$. Design for flexure as a singly reinforced beam and use the same reinforcement in opposite face for load reversals ($d \cong 13 - 2.5 = 10.5"$).

$$\phi M_n = \phi \rho b d^2 f_y \{1 - 0.59\rho(f_y / f'_c)\} \geq M_u$$

$0.9 \times 13 \times 10.5^2 \times 40\rho\{1 - 0.59\rho(40 / 3)\} = 55.1 \times 12$

$\therefore \quad \rho = 0.0142$; $A_s = 0.0142 \times 13 \times 10.5 = 1.94$ in^2, say 2 # 9

The assumption of fixed base; pinned top, was used to compute deflections, but for shear strength calculations, V_e should be computed using the shortest column with M_{pr} moments at each end computed in accordance with 1921.3.4 ($f_{yp} = 1.25 f_y = 50$ ksi). By proportioning previously computed values

$M_{pr} \cong 1.25 \times 55.1(2.00 / 1.72) = 80.1^{k\text{-}ft}$

$\therefore \quad V_e = 2 \times 80.1 / 9 = 17.8^k$

For # 4 stirrups / ties ($A_v = 0.40$ in^2); taking $V_c = 0$ per 1921.3.4.1, and using $\phi = 0.6$

$s = A_v f_y d / V_s = 0.40 \times 40 \times 10.5 / (17.8 / 0.6) = 5.7$, say 6"

use # 4 closed hoops at 6" o.c.

e) maximum column deflection occurs in the 13' long column resisting $V_e = 1.4 \times 5.5 / 2 = 3.85^k$. For the doubly reinforced section

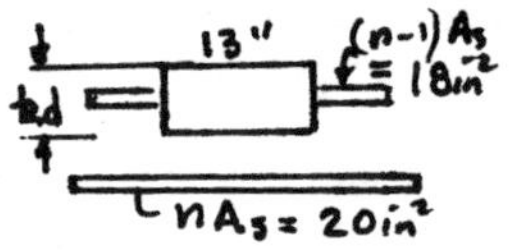

$k^2 + 0.56k - 0.36 = 0$ $\therefore$ $k = 0.382$; $kd = 4.01"$; $d - kd = 6.49"$

$$I_{cr} = b(kd)^3 / 3 + (n-1)A'_s(kd - d')^2 + nA_s(d - kd)^2$$

$$= 13 \times 4.01^3 / 3 + 9 \times 2.0(4.01 - 2.5)^2 + 10 \times 2.0 \times 6.49^2 = 1165\,\text{in}^4$$

$I_g = 13 \times 13^3 / 12 = 2380$ in^4; $M_a = 3.85 \times 13 = 50.1^{k\text{-}ft}$

$$M_{cr} = 7.5\sqrt{f'_c}\,I_g / (h/2) = [7.5\sqrt{3000}(2380) / (13/2)] / 12000 = 12.5^{k\text{-}ft}$$

$M_{cr} / M_a = 12.5 / 50.1 = 0.25$

$$I_e = (M_{cr} / M_a)^3 I_g + [1 - (M_{cr} / M_a)^3] I_{cr} \leq I_g$$

$$= 0.25^3 \times 2380 + [1 - 0.25^3] \times 1165 = 1184\,\text{in}^4$$

$$\Delta_S = VL_1^3 / (3EI) = 3.85 \times (13 \times 12)^3 / (3 \times 3600 \times 1184) = 1.14"$$

The deflection computed above is the elastic deflection corresponding to the base shear, which has been scaled down to account for inelastic response. The maximum probable deflection is estimated using Eqn. 30-17, UBC-97, which depends on the system response factor, R. The structural system in this problem is type 5 in Table 16-N, cantilevered column system, for which R = 2.2. Thus,

$\underline{\Delta_M = 0.7\,R\,\Delta_S = 0.7 \times 2.2 \times 1.14 = 1.75"}$

Concrete-48

Design the end walls and footing. The end walls span 60' simply supported by the caissons.

- For seismic loads: Given base shear coefficient = 0.186; let W = contributing weight for North-South ground motion = roof plus 6.5' of wall height.

  $W = 0.15 \times 60 \times 180 + 0.1 \times (2 \times 180) \times 6.5 = 1850^k$

Per problem statement, accidental torsion is to be neglected ∴ each wall resists one-half the lateral seismic force.

$V = 0.186 \times 1850 / 2 = 172^k$ per wall

Assume that the force transfers to the wall 1' from the top and is distributed uniformly along the wall length by a rigid diaphragm. Resistance at the bottom is at the bottom of the grade beams:

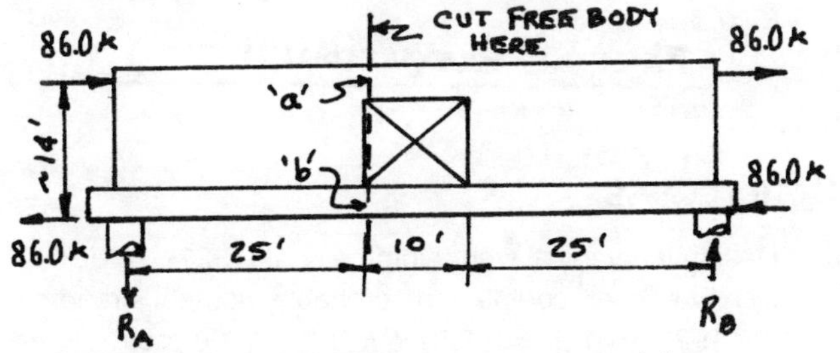

$\Sigma M_A = 0$: $172(14) - R_B(60) = 0$ ∴ $R_B = 40.1^k \uparrow$

$\Sigma F_y = 0$: $R_A = 40.1^k \downarrow$

Consider portion of wall as a free body diagram:

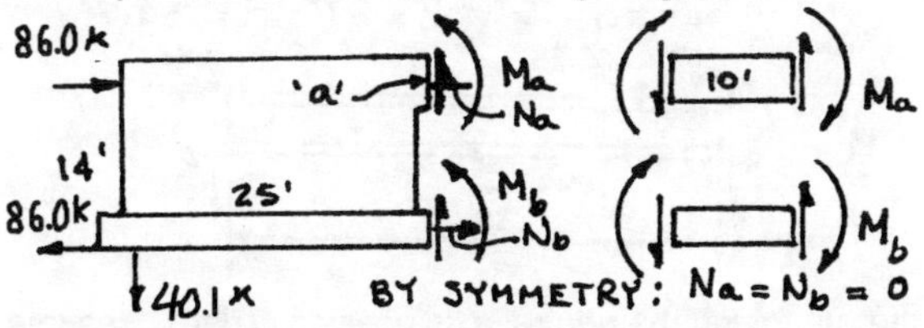

Since the lateral force is distributed uniformly along the 60' long length at top, then

$\Sigma F_x = -86 + (172 / 60)(25 + 5) + N_b = 0$ ∴ $N_b = 0$

$\Sigma M_a = 0$: $40.1(25) - (86.0)(14) + M_A + M_B = 0$

∴ $M_A + M_B = 202^{k\text{-}ft}$

Resistance in the coupling beams is in proportion to their flexural rigidities.

$I_a = 8 \times 36^3 / 12 = 31{,}104$ in^4; $I_b = 18 \times 24^3 / 12$
$= 20{,}736$ in^4

∴ $M_A = [31{,}104 / (31{,}104 + 20{,}736)]\,202 = 121^{k\text{-}ft}$;
$M_B = 81.0^{k\text{-}ft}$

For gravity loads:

w_d = roof + wall + grade beams

$w_d = 0.15 \times 30 / 2 + 0.1 \times 13 \times 1 + 0.15 \times 1.5 \times 2 = 4$ klf

$w_L = 0.012 \times 30 / 2 = 0.18$ klf

Governing load combination (1612.2.1, UBC-97):

$U = 1.1(1.2D + E + 0.2R)$

$w_{ug} = 1.1(1.2 \times 4 + 0.2 \times 0.18) = 5.34$ klf

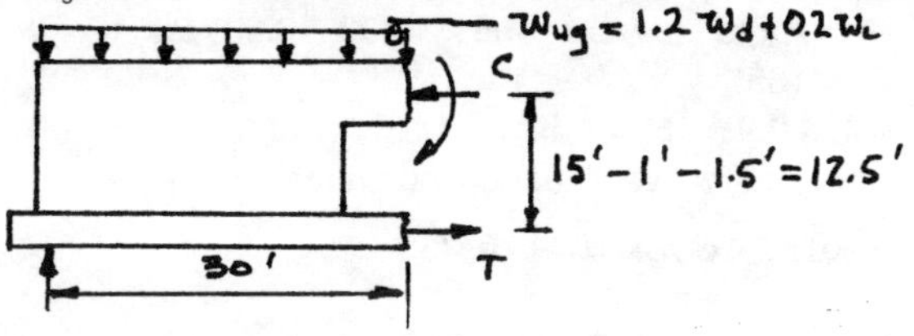

$M_{ug} = w_u L^2 / 8 = 5.34 \times 60^2 / 8 = 2403^{k\text{-}ft}$

∴ $C = T = M_{ug} / 12.5 = 192^k$

For gravity loads, the coupling beam is essentially fixed at both ends; therefore,

$M_g^- = w_u l'^2 / 12 = 5.34 \times 10^2 / 12 = 44.5^{k\text{-}ft}$; $V_g = w_u l' / 2$
$= 5.34 \times 10 / 2 = 26.7^k$

The critical combined axial plus bending on the coupling beam occurs at the right side where seismic compression force adds to gravity compression

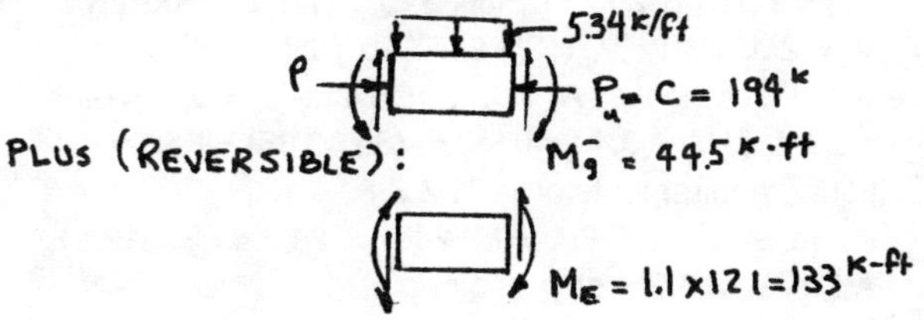

$P_u = 194$ kips

$M_u = 44.5 + 133 = 178^{k\text{-}ft}$

Check 1921.6.10, UBC-97: $l_n / d \cong 10 / 2.75 = 3.6 < 4$; however, $V_u < 4\sqrt{f'_c}b_w d = 58^k$ ∴ design per 1921.3 using straight reinforcement top and bottom for longitudinal steel (Note: the dimensional limits in 1921.3.1.3 and .4 require that the width be increased from 8" to > 10"; however, will proceed using the given 8" thick member). Try minimum steel $\rho_g = 0.01$; $A_{ST} = 0.01 \times 8 \times 36 = 2.88$ in^2, say 2 # 8 top and bottom ($A_{st} = 4 \times 0.79$ in^2).

Construct an approximate interaction diagram for the member

- For the concentric load case (M = 0):

  $\phi P_o = \phi[0.85f'_c(A_g - A_{st}) + f_y A_{st}]$
  $= 0.7[0.85 \times 3(288 - 3.16) + 40 \times 3.16] = 597^k$

- Balanced conditions: $\varepsilon_y = f_y / E_s = 40 / 29{,}000 = 0.00138$; By similar triangles

  $x = (0.003 / (0.003 + 0.00138)33 = 22.6"$

  $\varepsilon'_s = ((22.6 - 3) / 22.6)0.003 = 0.0026 > \varepsilon_y$ ∴ $f'_s = f_y$
  $= 40$ ksi

  $T = 40 \times 1.58 = 63.2^k$

  $C'_s = (40 - 0.85 \times 3) \times 1.58 = 59.1^k$

  $C_c = 0.85 \times 3 \times 8 \times 0.85 \times 22.6 = 392^k$

  ∴ $P_b = C_c + C'_s - T = 392 + 59.1 - 63.2 = 388^k$

  $(\phi P_b = 0.7 \times 388 = 271^k)$

  $M_b = P_b e_b = 392(18 - 0.85 \times 22.6 / 2) + 59.1(18 - 3)$
  $+ 63.2(18 - 3) = 5125^{k\text{-}in} = 427^{k\text{-}ft}$ ($\phi M_b = 0.7 \times 427$
  $= 298^{k\text{-}ft}$

- For pure bending (P = 0): For light, symmetrical reinforcement, in a relatively deep beam, the effect of compression steel on flexural strength is negligible

  $a = [(40 \times 1.58) / [0.85 \times 3 \times 8] = 3.08"$

  $M_n = A_s f_y(d - a / 2) = 1.58 \times 40(33 - 3.08 / 2) / 12$

  $= 165^{k\text{-}ft}$

  $\phi M_n > 0.9 \times 165 = 149^{k\text{-}ft}$

Thus, a conservative, linear bound on the strength is

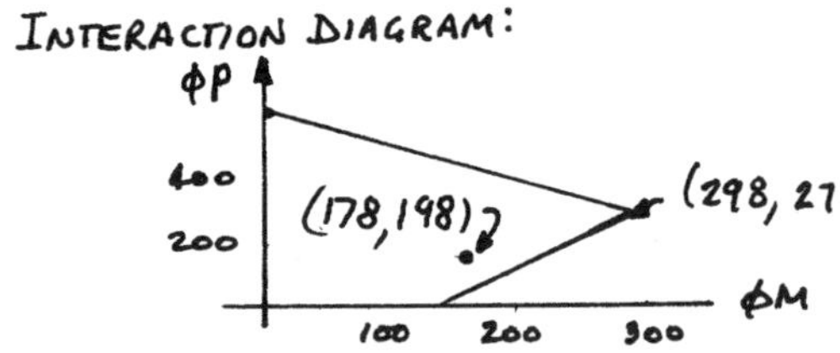

Thus, the 4 # 8 are adequate. For shear, compute the probable moment strength per 1921.3.4.1 ($f_{yp} = 1.25f_y$; $\phi = 1$)

$a_{pr} = [(1.25 \times 40 \times 1.58) / [0.85 \times 3 \times 8] = 3.87"$

$M_{pr} = A_s f_{yr}(d - a / 2)$

$= 1.58 \times 1.25 \times 40(33 - 3.87 / 2) / 12 = 205^{k\text{-}ft}$

$V_e = 2M_{pr} / 10 + V_g = 2 \times 205 / 10 + 24.2 = 65^k$

Use # 4 closed hoops ($A_v = 0.40$ in^2); take $V_c = 0$; $\phi = 0.85$

$s \le A_v f_y d / V_s = 0.4 \times 40 \times 33 / (65 / 0.85) = 6.9"$

Additional requirements of 1921.3.3

$s \le [d / 4 = 8.25"; 8d_{bl} = 8"; 24d_{bh} = 24 \times 4 / 8 = 12";$

$A_v f_y d / V_s = 6.9"]$; **say 7" o.c.**

For the grade beam, $d \cong 24 - 4 = 20"$; $T_u = 194^k$;

$M_u = 1.1M_B + 1.1 \times 1.2 \times 0.45 \times 10^2 / 12 = 94^{k\text{-}ft}$

Try 3 # 10 top and 3 # 10 bottom

$\phi T_n = 0.9 \times 6 \times 1.27 \times 40 = 275^k$

$a = 3 \times 1.27 \times 40 / (0.85 \times 3 \times 18) = 3.3"$

$\phi M_n = \phi A_s f_y(d - a / 2)$

$= 0.9 \times 3 \times 1.27 \times 40(20 - 3.3 / 2) / 12 = 210^{k\text{-}ft}$

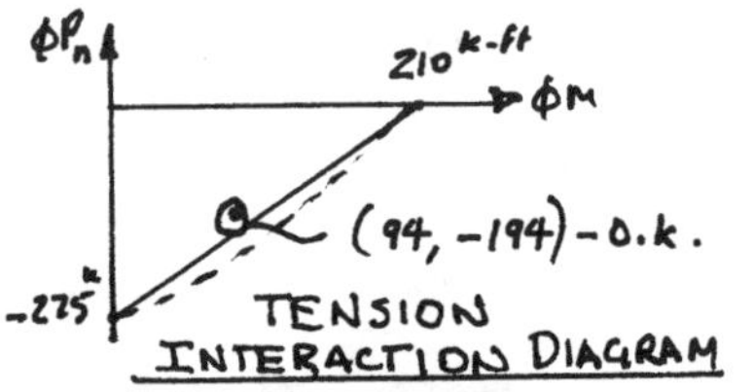

Thus, from the interaction diagram, 6 # 10 are adequate.

$a_{pr} = [(1.25 \times 40 \times 3 \times 1.27) / [0.85 \times 3 \times 18] = 4.1"$

$M_{pr} = A_s f_{yr}(d - a / 2) = 3.81 \times 1.25 \times 40(20 - 4.1 / 2) / 12$

$= 285^{k\text{-}ft}$

$V_e = 2M_{pr} / 10 + V_g = 2 \times 285 / 10 + 2.7 = 59.7^k$

Use # 4 closed hoops $A_v = 0.40$ in^2; take $V_c = 0$; $\phi = 0.85$

$s \le A_v f_y d / V_s = 0.4 \times 40 \times 20 / (59.7 / 0.85) = 4.6"$

$s \le [d / 4 = 5"; 8d_{bl} = 10"; 24d_{bh} = 12"; Avfyd / V_s = 4.6"]$;

say 4-1/2" o.c.

Extend the transverse reinforcement 2 times the effective depth on both sides of locations of plastic hinges (1921.3.3.1) $\therefore$ $l_o = 2 \times 33 = 66"$ top; $2 \times 20 = 40"$ bottom.

Check wall shear capacity per 1921.6.5.2: $A_{cv} = 8 \times 25 \times 12 = 2400$ in^2; $\rho_n = 0.0025$; $A_n = A_h = 0.0025\ 12 \times 8 = 0.24$ in^2/ft (say # 4 @ 10" o.c.).

$\phi V_n = \phi(A_{cv}(2\sqrt{f'_c} + \rho_n f_y)$

$= 0.6(2400(2(3000)^{1/2} + 0.0025 \times 40{,}000)$

$= 301{,}000^{\#} > V_u$

$\therefore$ wall with single curtain of horizontal and vertical steel # 4 @ 10" o.c. is adequate.

Tie to caisson – neglect gravity: $A_{st} = 1.1 \times 40.1 / (0.9 \times 40) = 1.22$ in^2; say 2 # 7.

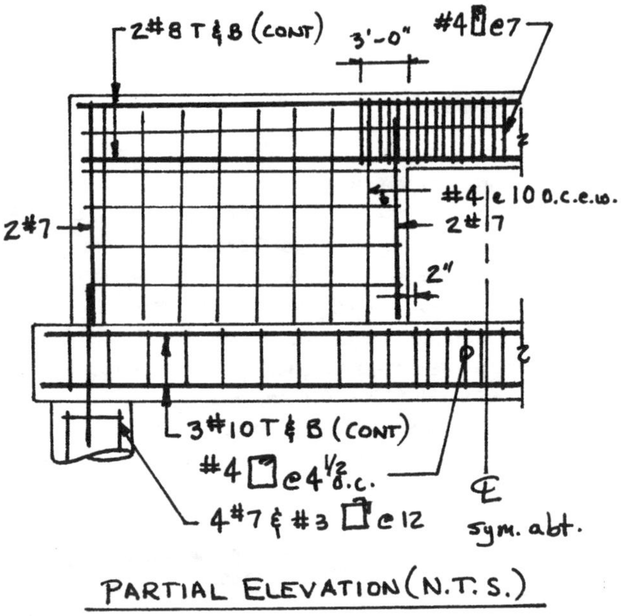

Concrete-49

a) Design the vertical web reinforcement (1921.3.4, UBC-97). For shear, compute the probable moment strength per 1921.3.4.1 ($f_{yp} = 1.25f_y$; $\phi = 1$; $f'_c = 3000$ psi; $f_y = 60{,}000$ psi; double reinforced with $A_s = A'_s = 4.00$ in^2).

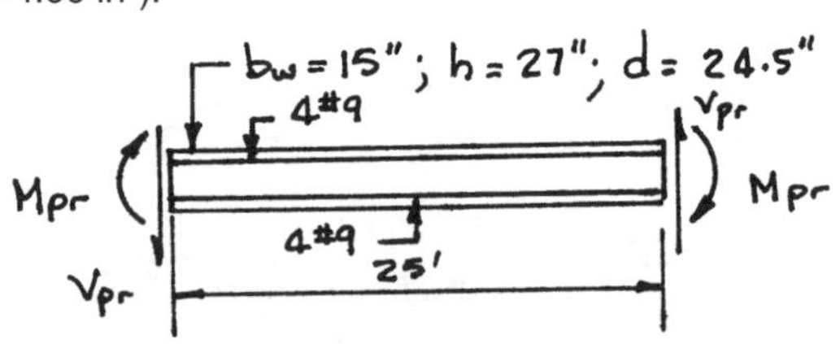

Iterate to find the neutral axis position: Trial 1. Assume $f'_s = 30$ ksi

$a_{pr} = [(1.25 \times 60 \times 4.00) - (30 \times 4.00)] / [0.85 \times 3 \times 15] = 4.97"$

$x = a_{pr} / \beta_1 = 4.97 / 0.85 = 5.85"$

$\varepsilon_s' = [(5.85 - 2.5) / 5.85]0.003 = 0.0017 < \varepsilon_y$

$f_s' = 29{,}000 \times 0.0017 = 50$ ksi – no good

Trial 2. Assume an intermediate value; say $f_s' = 40$ ksi

$a_{pr} = [(1.25 \times 60 \times 4.00) - (40 \times 4.00)] / [0.85 \times 3 \times 15] = 3.93"$

$x = a_{pr} / \beta_1 = 3.93 / 0.85 = 4.62"$

$\varepsilon_s' = [(4.62 - 2.5) / 4.62]0.003 = 0.00138 < \varepsilon_y$

$f_s' = 29{,}000 \times 0.00138 = 39.9$ ksi—close enough

$M_{pr} = [A'_s(f_s' - 0.85f'_c)(d - d') + 0.85f'_c ba(d - a/2)]$
$= [4.00(40 - 0.85 \times 3)(24.5 - 2.5) + 0.85 \times 3 \times 15$
$\times 3.93(24.5 - 3.93/2)]/12 = 557^{k\text{-}ft}$

Consider the tributary gravity load to act simultaneously with the shear associated with the plastic hinges:

$V_g = 50^k$

$\therefore \quad V_e = 2M_{pr}/L + V_g = 2 \times 557/25 + 50 = 94.6^k$

Since the earthquake-induced shear is less than one-half the design shear, the contribution of V_c can be included (1921.3.4.2). Use # 4 closed hoops ($A_v = 0.40$ in^2); $\phi = 0.85$:

$$\phi V_c = 2\phi\sqrt{f_c'}b_w d = 2 \times 0.85 \times \sqrt{3000} \times 15 \times 24.5$$
$$= 34{,}200^{\#} = 34.2^k$$

$V_s = V_e/\phi - V_c = (94.6 - 34.2)/0.85 = 71.0^k$

Need stirrups –

$s \le A_v f_y d/V_s = 0.4 \times 60 \times 24.5/(71.0) = 8.2$ in

Additional requirements of 1921.3.3

$s \le [d/4 = 6.1"; 8d_{bl} = 9"; 24d_{bh} = 24 \times 4/8 = 12";$
$A_v f_y d/V_s = 6.1"]$; **say 6" o.c.**

b) Reinforcement of the beam-column joint (1921.5, UBC-97)

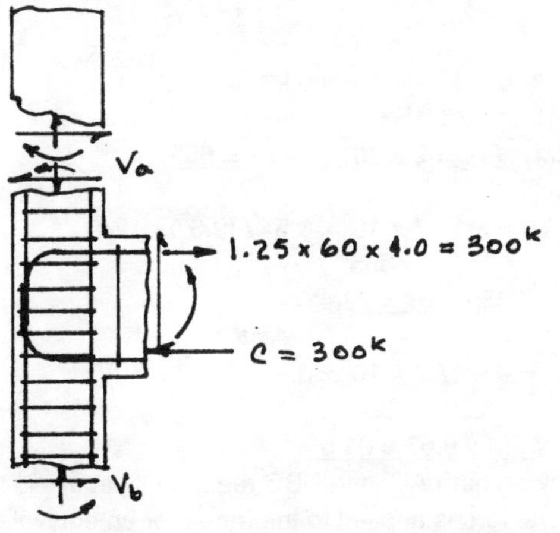

The shear transfer through the joint is 300^k, minus the shear in the column above. The shear above is not specified; therefore, disregard. Also, there is no information given regarding confinement from members framing perpendicular to the joint. Assume that the joint is part of an interior frame with confinement from spandrel beams on both sides. Thus,

$$\phi V_j = 15\phi\sqrt{f'_c}A_j = 15 \times 0.85(3000)^{1/2} \times 18 \times 24$$
$$= 301{,}700^{\#} = 302^k$$

The joint strength is adequate. For the confinement steel (1921.4.4), use # 4 hoops

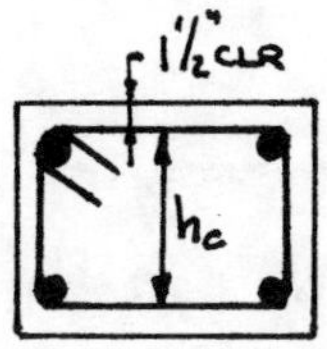

$h_c = 18 - 3 - 0.5 = 14.5"$; $A_{ch} = 15 \times 21 = 315$ in^2;
$A_g = 18 \times 24 = 432$ in^2

$$A_{sh} = \begin{cases} 0.3(sh_c f_c'/f_{yh})[(A_g/A_{ch}) - 1] \\ \quad = 0.3(s \times 15 \times 3/60)[(432/315) - 1] \\ \quad = 0.084s \\ 0.09(sh_c f_c'/f_{yh}) = 0.09(s \times 15 \times 3/60) \\ \quad = 0.0675s \end{cases}$$

Maximum spacing (1921.4.4.2)

$s \le [h_{min}/4 = 18/4 = 4.5"; 4"] = 4"$

$\therefore \quad A_{sh} = 0.084s = 0.084 \times 4 = 0.34$ in^2; use # 4 closed hoops; extend a distance l_o above and below joint where $l_o \ge (h_u/6; d; 18")$.

Thus, **provide # 4 closed hoops at 4" o.c. through the joint and d = 21" above and below joint.**

c) A ductile moment resisting space frame is designed and detailed to provide sufficient toughness to absorb the energy imparted by the design earthquake without collapse. To function properly, members and connections must possess ductility and strength to respond inelastically through many cycles of stress reversals. The base shear used in the design of the special moment resisting frame is significantly lower than the base shear that would be obtained from an elastic response spectrum for the design ground motion. Four basic principles in the design of ductile concrete members:

1. Confine the concrete – enclose the core with spirals or closely spaced hoops and ties.
2. Avoid brittle modes of failure – limit the ratio of longitudinal steel so that yielding will occur before shear or bond failure.
3. Recognize that material overstrength may be detrimental (e.g., yield strength in excess of specified minimum may result in shear or bond failure instead of the intended flexural yielding).
4. Design for plastic hinge formation in beams; not the columns.

Concrete-50

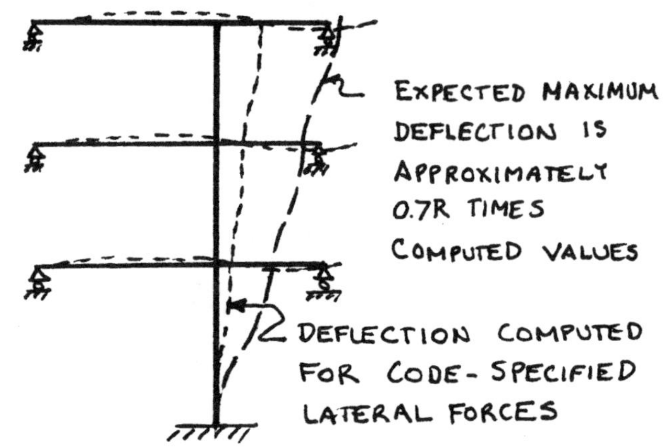

Design by UBC-97, section 1921.7, which requires that frame members that are not part of the lateral force-resisting system be detailed for the moments and shears induced by the maximum inelastic response displacements, Δ_M. The given displacement values were computed by elastic analysis using the design lateral forces, which were scaled down to account for inelastic behavior. To obtain the maximum values, computed displacements will be scaled up by the factor 0.7R. For this purpose, I will assume that the lateral force resisting system is such that R = 8.5 (Re: Table 16-N, UBC-97); thus, 0.7R = 6.0. The given drifts, Δ_s, and the scaled drifts, Δ_M, and computed interstory drifts for each level are tabulated below.

Level	Δ_S (in)	Δ_M (in)	Interstory Drift, δ_M (in)	Story Height (ft)	Clear Height, h_u (ft)
1-2	0.16	0.96	0.96	15	12.67
2-3	0.25	1.50	0.54	12	9.67
3-4	0.35	2.10	0.60	12	9.67

The given data are sufficient to permit an analysis of the frame using conventional elastic analysis (e.g., moment distribution). However, as noted in the SEAOC Commentary (1996, Appendix 1E), conventional analysis tends to underestimate the lateral drift under the design earthquake. The recommended approach is to consider the ends of the columns fixed and to use the clear heights of columns rather than centerline dimensions.

For purposes of assessing displacements under lateral loads, it is important that displacements not be underestimated. Therefore, SEACOC recommends that the stiffness of members that are part of the force-resisting system be computed as no more than one-half the gross section elastic stiffness. However, when calculating the moments and shears induced in members that are not part of the lateral force-resisting system, it is conservative to use a larger stiffness. Also, the factor β_d, which is used to reduce the rigidity of a column in code equations 10-12 and 10-13, accounts for the effect of creep in columns. It should not apply to the rigidity of a column under short duration lateral loads. A reasonable estimate of the average stiffness of the column depends on the magnitude of the axial force sustained during the displacement, which is not given. For purposes of this analysis, I will assume a sustained axial force of approximately $0.2f'_cA_g$, for which a reasonable estimate of I_e is one-half I_g (Paulay and Priestley, 1993, p. 163). Thus, for 18" x 18" columns with 4 # 9 longitudinal bars; f'_c = 4000 psi; normal weight concrete

$EI_{col} = 0.5E_cI_g = 0.5 \times 4030^{ksi} \times (18 \times 18^3 / 12)\ in^4$
$= 17.7 \times 10^6$ k-in^2

Thus, for the lowest level

$\delta_M = 0.96"$; $h_u = 12.67'$

$\Delta_4 = 6.0 \times 0.35 = 2.10"$

$\delta_3 = \Delta_4 - \Delta_3 = 2.10 - 1.50 = 0.60"$

$\Delta_3 = 6.0 \times 0.25 = 1.50"$

$\delta_2 = \Delta_3 - \Delta_2 = 1.50 - 0.96 = 0.54"$

$\Delta_2 = 6.0 \times 0.16 = 0.96"$

$\delta_1 = \Delta_2 - \Delta_1 = 0.96"$

$M_{1\text{-}2} = 6EI\ \delta_M / (h_u)^2$
$= 6 \times 17.7 \times 10^6 \times 0.96 / (12.67 \times 12)^2$
$4410^{k\text{-}in} = 367^{k\text{-}ft}$

<u>$V_{1\text{-}2} = 2M_{1\text{-}2} / h_u = 2 \times 367 / 12.67 = 58^k$</u>

Similarly,

$M_{2\text{-}3} = 6 \times 17.7 \times 10^6 \times 0.54 / (9.67 \times 12)^2$
$= 4258^{k\text{-}in} = 355^{\text{-}ft}$

<u>$V_{2\text{-}2} = 2 \times 355 / 9.67 = 73.4^k$</u>

and

$M_{3\text{-}4} = 6 \times 17.7 \times 10^6 \times 0.60 / (9.67 \times 12)^2$
$= 4732^{k\text{-}in} = 394^{\text{-}ft}$

<u>$V_{3\text{-}4} = 2 \times 394 / 9.67 = 81.5^k$</u>

b) Why do both ACI and UBC require this analysis? The lateral forces applied to the frame for an equivalent static force analysis are significantly smaller than the forces that the structure would experience if it were to respond elastically to the ground motion of a severe earthquake. To estimate the probable maximum drifts that would occur in an earthquake, it is necessary to scale the deflections that were computed from linear analysis using the equivalent static forces upward to reflect the larger values that are likely to occur. Thus, the UBC scales the computed Δ_S values by a factor 0.7R, which is greater than unity. The current ACI approach simply multiplies the computed static deflection by a factor of two, regardless of the structural system.

The columns in this example, though not part of the lateral force-resisting system are constrained by their attachment to the floor diaphragms to undergo similar drifts as frames in the lateral force-resisting systems (in fact, their drifts might actually be greater if deflection of the diaphragm is significant). In order for these members to continue to function (i.e., to support gravity loads) they must be detailed according to criteria in UBC Section 1921.7.3.

Concrete-51

a) Design the shear reinforcement based on the given 10" thick by 32' long wall: $A_{cv} = 10 \times 32 \times 12 = 3840$ in^2. Try minimum steel: $\rho_v = \rho_n = 0.0025$ (Re: 1921.6.2.1). From Eqn. 21-6 with $\phi = 0.6$ (Re: 1909.3.4.1). Given: service dead = 21 klf $\therefore$ $P_d = 21 \times 32 = 672^k$; Service live = 2.7 klf $\therefore$ $P_L = 2.7 \times 32 = 86.4^k$; Service moment = $\pm$ 20,264$^{k\text{-}ft}$; factored moment = $\pm$28,370$^{k\text{-}ft}$; service shear = 228^k; factored shear = 456^k; $f'_c = 3000$ psi; $f_y = 60{,}000$ psi. Design by UBC-97, Chapter 1921.

$$\phi V_n = \phi A_{cv}(2\sqrt{f'_c} + \rho_n f_y)$$
$$= 0.6 \times 3840[2(3000)^{1/2} + 0.0025 \times 60{,}000]$$
$$= 598{,}000^{lb}$$

$\phi V_n = 598^k > V_u = 456^k$ $\therefore$ shear strength is adequate with minimum reinforcement. Check 1921.6.2.2: $2A_{cv}\sqrt{f'_c} = 421^k < V_u$ $\therefore$ use two curtains of wall steel.

$A_v = A_n = 0.0025 \times 10 / 2 \times 12 = 0.15$ in^2/$_{ft}$ each face, **say # 4 @ 16" o.c.**

b) Calculate the service load stresses at the top of footing (note: this calculation is no longer relevant to the design procedure of the current UBC, but is included for completeness).

Stresses are computed using properties of the gross, uncracked section.

$$I_g = 10 \times (32 \times 12)^3 / 12 = 47.2 \times 10^6 \text{ in}^4$$
$$f_c = P_{d+L} / A_{cv} + M c / I_g$$
$$= 758 / 3840 + 20{,}264 \times 12 \times (16 \times 12) / (47.2 \times 10^6)$$

$f_c = 1.187$ ksi = 1187 psi

$$f_t = P_{d+L} / A_{cv} - M c / I_g$$
$$= 758 / 3840 - 20{,}264 \times 12 \times (16 \times 12) / (47.2 \times 10^6)$$

$f_t = -0.814$ ksi = 814 psi

$f_v = 3 / 2V/A_{cv} = (3 / 2)228 / 3840 = 0.089$ ksi = 89 psi

c) Boundary reinforcement (Re: 1921.6.6). The current UBC approach is to design the shear wall for combined axial force plus bending. Since failure will occur by yielding of the tensile steel, the controlling loading is that with maximum bending moment and minimum axial compression, i.e., 0.9D $\pm$ 1.1E.

$$P_u = 0.9\, P_d = 0.9 \times 672 = 605^k$$

The given data are insufficient to determine the inelastic deformation of the shear wall. Thus, the general method for determining the extent of confinement in the boundary zone (i.e., where compressive stresses exceed 0.003) cannot be used in this problem. Instead, use the alternative method of Section 1921.6.6.4.

Check $P_u = 605^k < 0.1A_g f'_c = 0.1 \times 3080$ in$^2 \times 3$ ksi $= 924^k$ $\therefore$ o.k.

$M_u / (V_u l_w) = 28{,}370 / (456 \times 32) = 1.94 > 1.0$

or $V_u < 3A_{cw}\sqrt{f'_c} = 3 \times 3840 \times (3000)^{1/2} = 631{,}000^{lb}$

$\therefore$ satisfied

$\therefore$ **Boundary zones are not required** at each end; however, to satisfy problem stipulations, minimum sized boundary zones will be designed and detailed per Section 1921.6.

For the 10" wall with minimum vertical steel:

$A_{st} = 0.0025 \times 10 \times 12 \times 32 = 9.6$ in^2

$$P_o = 0.85f'_c(A_g - A_{st}) + f_y A_{st}$$
$$= 0.85 \times 3(3840 - 9.6) + 60 \times 9.6 = 10{,}340^k$$

$P_u < 0.15P_o$ $\therefore$ minimum length boundary zone

$0.15l_w = 0.15 \times 32 = 4.8'$ controls; say 5'-0" long.

Thickness of boundary zone (1921.6.6.6): $b \geq l_u / 16$ = (13.7 × 12)" / 16 = 10.3; say 12" thick. For simplicity of formwork, use a 12" thick wall throughout with minimum steel $0.0025 \times 12 \times 12 = 0.36$ in^2/$_{ft}$ (say, **# 4 @ 12" o.c. each face**). The increased shear capacity is desirable and the increased longitudinal steel is beneficial in resisting the moment.

The design of wall reinforcement for the combined bending and axial force is relatively simple with design aids or computer, but it is tedious to do manually. An approximate solution can be found using procedures similar to those suggested by Paulay and Priestley (1993; p.392). Assume the center of compression lies at the centroid of the boundary zone at one end; tensile resistance from the stem lies at the center of wall; and the additional tensile force, T_2 acts at the centroid of the boundary zone at the opposite end. Then for equilibrium

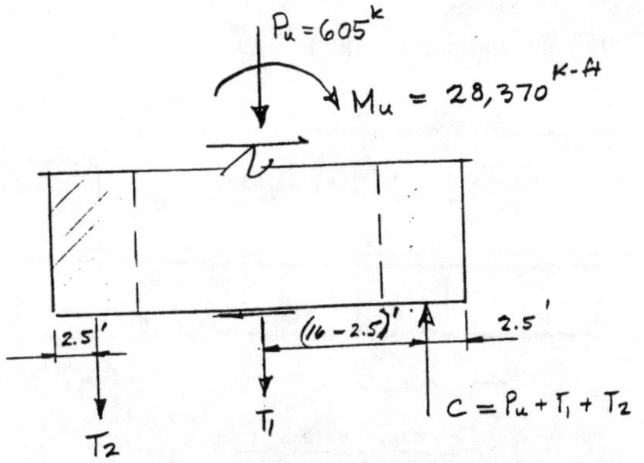

ELEVATION OF SHEARWALL AT BASE

$$A_{s2} \geq [M_u - T_1(16 - 2.5) - P_u(16 - 2.5)] / [\phi f_y(32 - 5)]$$
$$T_1 = (32 - 10) \times 12 \times 12 \times 0.0025 \times 60 = 475^k$$
$$A_{s2} = 28{,}370 - 475(13.5) - 605(13.5)] / [0.9 \times 60 \times 27] = 9.45 \text{ in}^2 \text{ (say 12 \# 8)}$$

The trial value can be used to refine the solution using the basic principles of 1921.10.2 & 3. A portion of the interaction diagram (the tension controls region) was constructed using a computer program and is drawn below. This solution confirms that the 12 # 8 computed by the approximate method is reasonably close to that of an "exact" solution. $\therefore$ **Use 12 # 8 distributed in two**

<u>faces uniformly over the 5' long boundary zone at each end</u>.

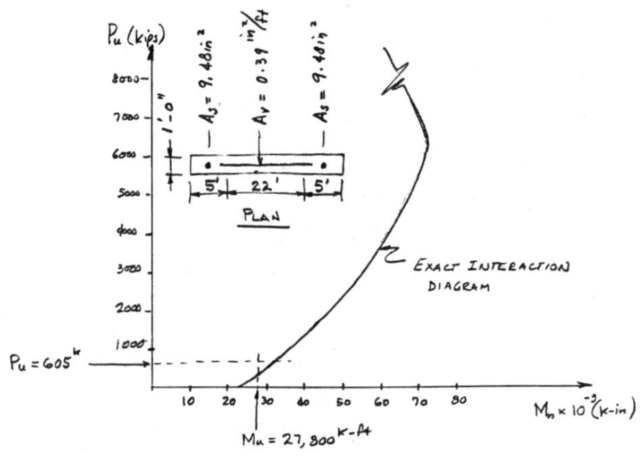

PARTIAL INTERACTION DIAGRAM FOR WALL

For confinement steel (1921.6.6.6 –2.):

$s \geq \{6d_b, 6"\} = 6"$

$h_c = 12 - 2(0.75 + 0.25) = 10"$

$A_{sh} \geq 0.09sh_cf'_c / f_y = 0.09 \times 6 \times 10 \times 3/60 = 0.27$ in²; say # 4 @ 6" o.c. (or # 3 at 4-1/2" o.c.).

Detail hoops and cross ties per 1921.6.6.6 (see figure below)

d) Anchorage of # 8 hooked tension bar in footing (1921.5.4.1):

<u>$l_{dh} \geq f_yd_b / 65\sqrt{f'_c} = 60{,}000 \times 1" / (65 \times 3000^{1/2}) = 17"$</u>

Since $V_u > A_{cv}\sqrt{f'_c} = 252^k$, 1921.6.2.2 requires that horizontal wall steel terminate in the boundary zone with a standard hook

e) Detail of wall reinforcement:

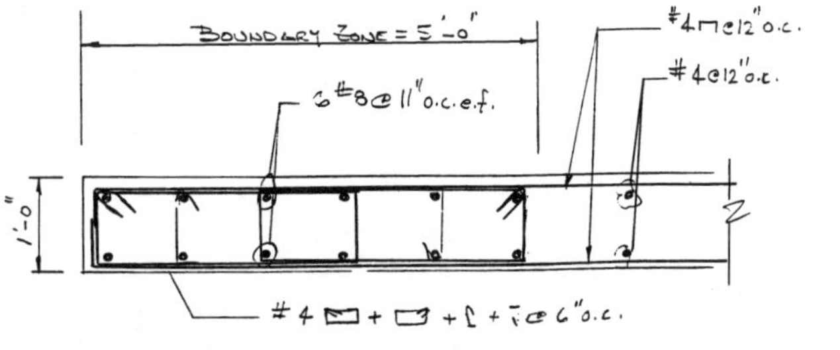

PARTIAL SECTION THROUGH WALL
(WALL SYMMETRICAL ABOUT CENTERLINE)

<u>Concrete-52</u>

a) Design the beam using an 18-in wide stem. Use the strength design method (ACI 318); $f'_c = 4$ ksi; $f_y = 60$ ksi; express the dead load as an equivalent uniform load per foot of horizontal projection.

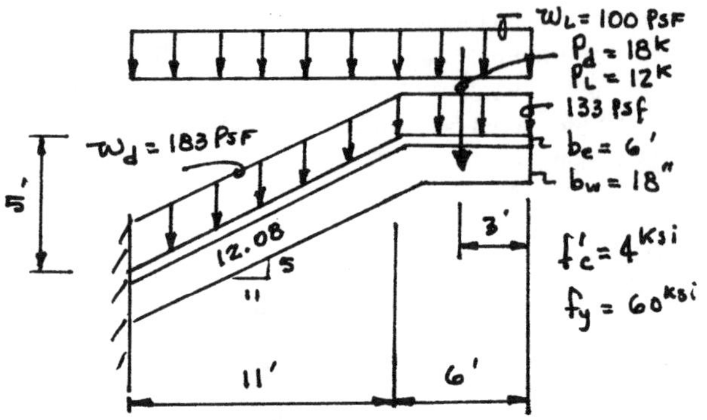

$w_u = 1.4w_d + 1.7w_l$

$w_{u1} = 1.4(183 \times (12.08 / 11) \times 6) + 1.7 \times 100 \times 6$
$= 2700 \text{ plf} = 2.7 \text{ klf}$

$w_{u2} = 1.4(133 \times 6) + 1.7 \times 100 \times 6 = 2130 \text{ plf}$
$= 2.13 \text{ klf}$

$P_u = 1.4P_d + 1.7P_l = 1.4 \times 18 + 1.7 \times 12 = 45.6^k$

$M_u^- = 45.6 \times 14 + 2.7 \times 11^2 / 2 + 2.13 \times 6 \times 14$
$= 981^{k\text{-}ft}$

The negative steel can easily be distributed over the 6' wide flange ∴ design for the maximum steel ratio, $\rho = \rho_{max} = 0.75\rho_b = 0.0214$; $b = b_w = 18"$; $d =$ overall depth – 2.5"

$$\phi M_n = \phi \rho b d^2 f_y \{1 - 0.59\rho(f_y / f'_c)\} \geq M_u$$

$f_y = 60$ ksi; $f'_c = 4$ ksi

$0.9 \times 18 \times d^2 \times 60 \times 0.0214\{1 - 0.59 \times 0.0214 \times (60 / 4)\} = 981 \times 12$

∴ $d^2 = 698$ in²; $d = 26.4"$, say overall depth = 30"

$A_s = \rho bd = 0.0214 \times 18 \times 26.4 = 10.17$ in²,

<u>Use 14 # 8; distributed uniformly over the 6' width</u>

Shear: the critical shear occurs at d-distance from face of support

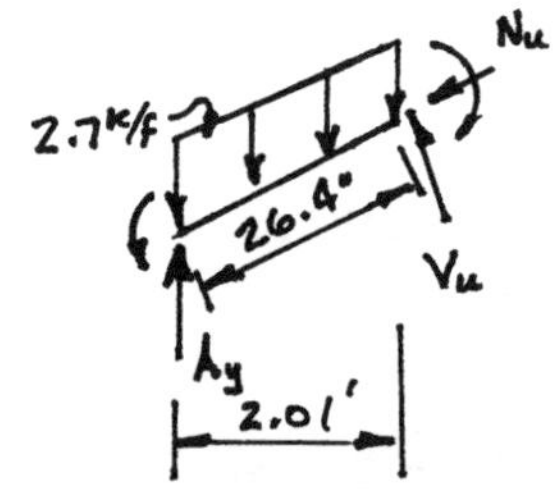

$\Sigma F_y = 0$: $A_y - 45.6 - 2.13 \times 6 - 2.7 \times 11 = 0$

∴ $A_y = 88.1^k$

$V_u = (88.1 - 2.7 \times 2.01)(11 / 12.08) = 75.3^k$

$N_u = (88.1 - 2.7 \times 2.01)(5 / 12.08) = 34.2^k$

The axial compressive force, N_u, has a beneficial effect on shear strength (11.3.2.2, ACI 318) and can be conservatively neglected. Thus,

$$\phi V_c = 2\phi\sqrt{f'_c}b_wd = 2 \times 0.85 \times \sqrt{4000} \times 18 \times 26.4$$
$$= 51{,}000^\# = 51^k$$

Need stirrups – try # 4 stirrup ($A_v = 2 \times 0.20 = 0.40\ in^2$). Per problem statement, the effect of torsional loading does not have to be explicitly considered in this problem, but recognizing its potential, use a closed stirrup configuration to resist accidental torsion which may occur. Also, to permit easier bending of stirrups, use grade 40 for stirrup steel.

$$V_s = V_u / 0.85 - V_c = (75.3 - 51) / 0.85 = 28.6^k$$

$$s \le \begin{cases} A_v f_y d / V_s = 0.40 \times 40 \times 26.4 / 28.6 \\ \quad = 14.7\text{" o.c.} \\ A_v f_y / 50 b_w = 0.40 \times 40{,}000 / (50 \times 18) \\ \quad = 18\text{" o.c.} \\ d/2 = 26.4/2 = 13\text{" o.c} \end{cases}$$

∴ **use # 4 U-stirrups @ 13" o.c.**

Transverse reinforcement (Re:8.10.5, ACI 318); use grade 40:

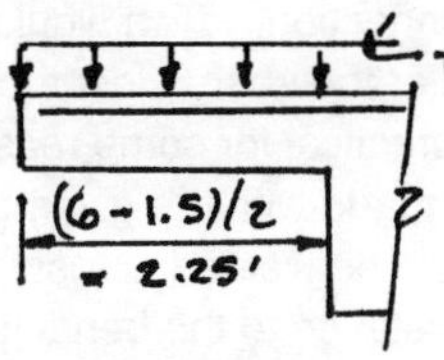

$$M_u^- = (2.7/6)^{klf}(2.25')^2/2 = 1.1^{k\text{-}ft}/_{ft}$$

Minimum Slab reinforcement:

$$A_{s,min} = 0.0018bh = 0.0018 \times 12 \times 6 = 0.13^{sq.in}/_{ft}$$

with minimum steel, the flexural strength is

$$a = 0.13 \times 60 / (0.85 \times 4 \times 12) = 0.19"$$

$$\begin{aligned} \phi M_n &= \phi A_s f_y (d - a/2) \\ &= 0.9 \times 0.13 \times 60(4.5 - 0.19/2)/12 \\ &= 2.6^{k-ft}/_{ft} \end{aligned}$$

Thus, minimum steel is more than adequate: use # 4 @ 18" o.c. in the form of closed stirrups (i.e., for torsional resistance); provide 2 # 4 bottom bars for stirrup support and add 2 # 4 at middepth of beam for crack control. Considering the short length involved and the embedment needed to develop the top flexural steel, there is little benefit to terminating the top bars in regions of lower moment. Therefore, use the 14 # 8 throughout. If possible, let the # 8 top bars extend straight into the wall a distance = 1.3 × 60,000 / (20 × $(4000)^{1/2}$) = 62" = l_d (Re: 12.2, ACI 318).

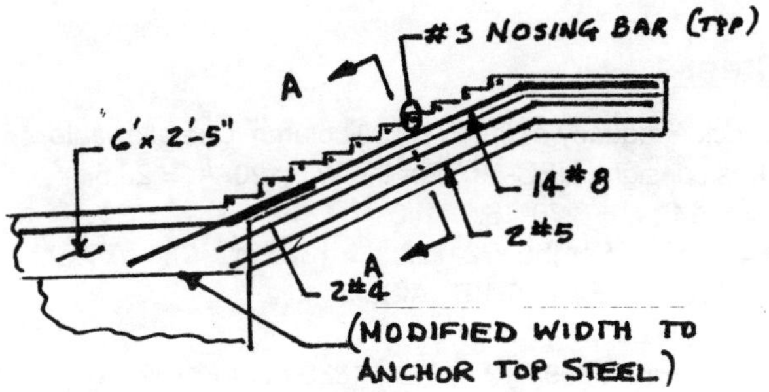

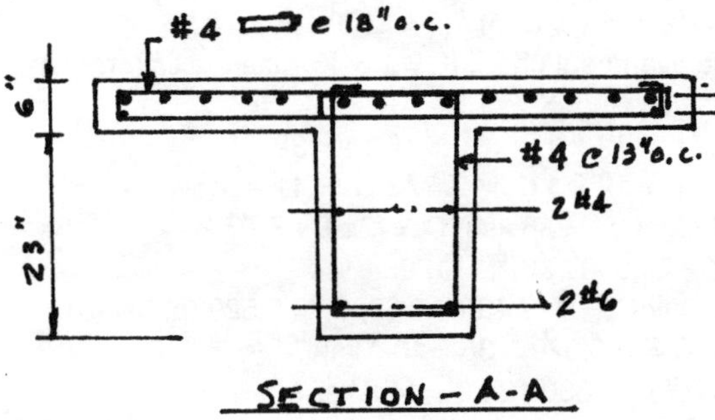

Steel-1

Check adequacy of the W14x90 column. Use allowable stress design (AISC-89). For the W14x90: $A_g = 26.5$ in^2; $r_x = 6.14$"; $r_y = 3.70$"; $S_x = 143$ in^3; $L_c = 15.3'$; $L_u = 34'$; $I_x = 999$ in^4;. For the W24x68: $I_x = 1830$ in^4; W24x76: $I_x = 2100$ in^4. Re p. 5-137, AISC

$$G_A = \Sigma(I_c / L_c) / \Sigma(I_g / L_g)$$
$$= (2 \times 999 / 12.5) / (2 \times 1830 / 25) = 1.09$$
$$G_B = \Sigma(I_c / L_c) / \Sigma(I_g / L_g)$$
$$= (2 \times 999 / 12.5) / (2 \times 2100 / 25) = 0.95$$

From the alignment chart, $K_x = 1.3$; For the y-axis (braced against sidesway), $K_y = 1.0$

$$KL_x / r_x = 12 \times 1.3 \times 12.5 / 6.14 = 32$$
$$KL_y / r_y = 12 \times 1.0 \times 12.5 / 3.70 = 41 \leftarrow \text{governs}$$

For A36 steel ($F_y = 36$ ksi), $C_c = [2\pi^2 E / F_y]^{1/2}$
$= [2 \times \pi^2 \times 29{,}000 / 36]^{1/2} = 126$

$$(KL / r) / C_c = 41 / 126 = 0.33;\ C_a = 0.529 \text{ (p. 5-119)}$$
$$F_a = C_a F_y = 0.529 \times 36 = 19.0 \text{ ksi}$$
$$f_a = P / A_g = 380 / 26.5 = 14.3 \text{ ksi}$$

Assume that load combinations are such that a one-third increase in allowable stress is permitted.

$$\therefore \quad f_a / F_a = 14.3 / (1.33 \times 19.0) = 0.57$$

For bending about the x-axis

$$L_b = 12.5' < L_c = 15.3' \ \therefore\ F_{bx} = 1.33 \times (0.66F_y)$$
$$= 1.33 \times 24 = 32 \text{ ksi}$$
$$f_{bx} = M_{x,max} / S_x = 130 \times 12 / 143 = 10.9 \text{ ksi}$$

For $KL_x / r_x = 32$: $F'_{ex} = (4 / 3)(146) = 195$ ksi

Unbraced frame $\therefore C_{mx} = 0.85$

$$f_a / F_a > 0.15 \ \therefore \text{ Check Eqns H 1-1 and H 1-2:}$$
$$f_a / F_a + C_{mx} f_{bx} / (1 - f_a / F'_{ex}) F_{bx} < 1.0$$
$$0.57 + 0.85 \times 10.9 / (1 - 14.3 / 195)32 = 0.88 < 1.0$$
$$f_a / 0.6F_y + f_{bx} / F_{bx} < 1.0$$
$$14.3 / [(4 / 3) \times 22] + 10.9 / 32 = 0.83 < 1.0 \ \therefore \text{ O.K.}$$

The **W14x90 is adequate in combined axial plus bending.**

Steel-2

Check the given E70xx fillet weld using the elastic method. Use Allowable Stress Design (AISC 1989).

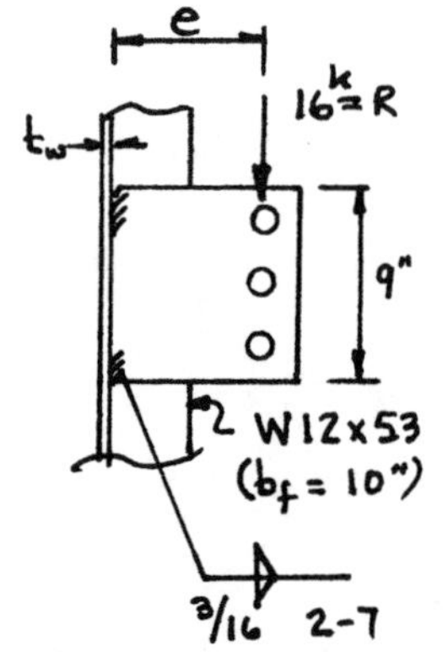

$$e = (b_f / 2) - (t_w / 2) + 1.5" = 5.0 - 0.17 + 1.5 = 6.3"$$
$$M = Re = 16 \times 6.3 = 101^{\text{k-in}}$$

$$L_w = 2 \times 2 \times 2 = 8"$$
$$I_w = 2[(2^3 / 12 + 2 \times 3.5^2) \times 2] = 100.7 \text{ in}^3$$
$$S_w = I_w / 4.5 = 22.4 \text{ in}^2$$
$$f_v = R / L_w = 16 / 8 = 2^{k}/_{in}$$
$$f_b = M / S_w = 101 / 22.4 = 4.5^{k}/_{in}$$
$$f_r = \sqrt{f_v^2 + f_b^2} = \sqrt{2^2 + 4.5^2} = 4.9^k /_{in}$$

For 3/16" fillet weld deposited on both sides of a 3/8" Plate (A36 steel):

$$q \le \begin{cases} 3 \times 0.928 = 2.8^k /_{in} \\ 0.4 \times 36 \times (3/8)/2 = 2.7^k /_{in} \leftarrow \text{governs} \end{cases}$$

Thus, the weld is overstressed (4.9 / 2.7)100% = 181%

$\therefore$ **The connection is no good**.

Note: even had the weld strength been adequate, the detail is poor. The Plate 3/8 x 9 would create high punching shear stress in the relatively thin web of the column. A better detail would be a seated connection, which would reduce the eccentricity and transfer the shear closer to the column flanges. If this is not practical for some reason, and the connection had to be made with the given eccentricity, horizontal plates should be welded across between the flanges to provide resistance to the bending in the plate.

Steel-3

Check the beam for the proposed loading. Use LRFD (AISC 1994); take U = 1.6 (conservative); A36 steel ($F_y = 36$ ksi). 4" thick concrete slab; normal weight; $f'_c = 3000$ psi; 18–3/4" diameter headed studs, uniformly spaced – Re: p. 5-9, AISCM, $Q_n = 21.0^{k}/_{stud}$.

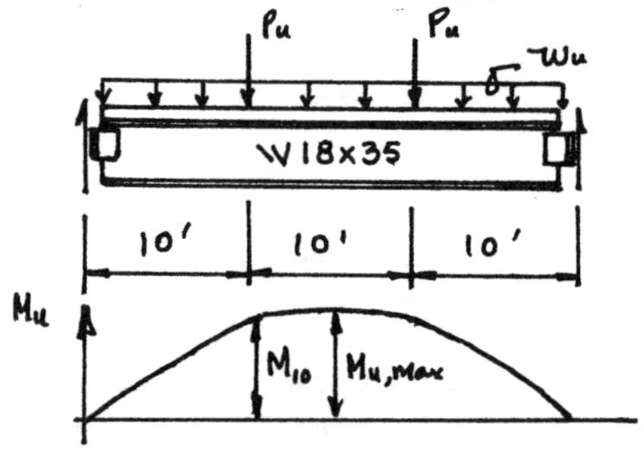

The effective flange width (Re: I-3) is

$$b_e \le \begin{cases} L / 4 = 30 \times 12/4 = 90" \leftarrow \text{governs} \\ \text{spacing} = 8 \times 12 = 96" \end{cases}$$

Per problem statement, shear and deflections are presumed adequate, therefore, need only check flexural strength.

$h_c / t_w = 15.5 / 0.3 = 52 < 640/\sqrt{F_y} = 107 \ \therefore$ use the plastic stress distribution with $\phi = 0.85$.

$$w_u = 1.6w = 1.6 \text{ klf}$$
$$P_u = 1.6P = 1.6 \times 3 = 4.8^k$$
$$M_{10'} = 4.8 \times 10 + 24 \times 10 - 1.6 \times 10^2 / 2 = 208^{\text{k-ft}}$$
$$M_{u,max} = 4.8 \times 10 + 1.6 \times 30^2 / 8 = 228^{\text{k-ft}}$$

Uniformly spaced studs are acceptable per AISC; however, I5.6 requires that the number of studs between a concentrated load and a point of zero moment must develop the moment at the load point. Therefore, check the design strength, ϕM_n, for the W18x35 required to develop the moment at the load point, $M_{10'} = 208^{k\text{-}ft}$.

Using 6–3/4" studs

$\Sigma Q_n = 6 \times 21.0 = 126^k$

$a = \Sigma Q_n / (0.85 f'_c b_e) = 126 / (0.85 \times 3 \times 90) = 0.55"$

$Y_2 = h_s - a/2 = 4 - 0.55/2 = 3.73"$

Re: page 5-28, AISCM, interpolating between $Y_2 = 3.5"$ and 4.0"; and $\Sigma Q_n = 92.7^k$ and 140^k:

$\phi M_n \cong 255^{k\text{-}ft} > 208^{k\text{-}ft}$

Since the capacity at the load point exceeds the $M_{u,max}$ there is no need to recompute the capacity at midspan. ∴ **The W18x35 composite beam is adequate in flexure.**

Steel-4

a) Determine the filllet weld size. Given: ASTM A36 steel (F_y = 36 ksi); E70xx electrodes.

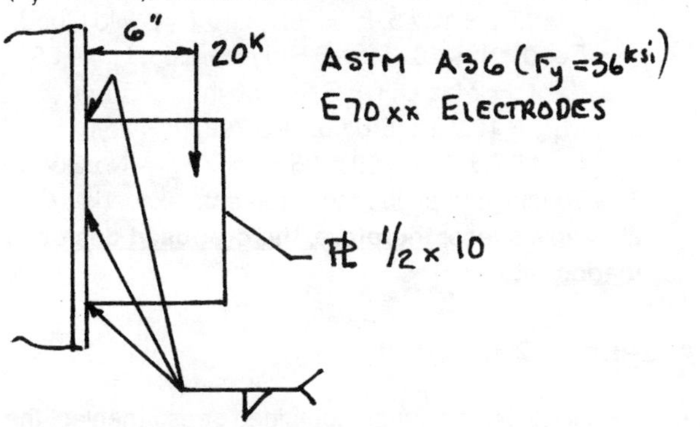

Note: the 1/2" wrap at the top of the weld is not a recommended weld detail (AISC 1989). The top and bottom welds cause notches (stress raisers). Nevertheless, calculations are made by the elastic method based on the given weld pattern.

For the 20^k as a static load, the minimum weld size (Table J.4) is ω_{min} = 1/4" :

$L_w = 10"$; $I_w = 10^3/12 + 2 \times 5^2 \times 0.5 = 108.3\ in^3$

$S_w = I_w / 5 = 21.7\ in^2$

$f_v = R / L_w = 20/10 = 2^k/_{in}$

$f_b = M / S_w = 20 \times 6 / 21.7 = 5.5^k/_{in}$

$f_r = \sqrt{f_v^2 + f_b^2} = \sqrt{2^2 + 5.5^2} = 5.9^k\ /_{in}$

For weld deposited against the 1/2" Plate (A36 steel):

$\omega \geq 5.9/0.928 = 6.3$, say 7/16

$q \leq 0.4 \times 36 \times 1/2 = 7.2^k\ /_{in}$

∴ base material is adequate

Use 7/16" fillet weld

b) For 2,000,000 cycles of fully reversible stress on the effective weld throat (stress ranges from $5.9^k/_{in}$ compression to $5.9^k/_{in}$ tension). Re: Appendix A; Loading Condition 4; Category F:

$F_{Sr4} = 8$ ksi

∴ $F_{max} = 8/2 = 4$ ksi for each cycle

Capacity of the weld (per 1/16" leg size)

$q = 0.707 \times 1/16 \times 1/4 \times 4 \times 1 = 0.177^{k/in}/_{16th}$

∴ $\omega = 5.9 / 0.177 = 33$ --- impossible!

Thus the detail cannot be developed for the specified fatigue loading.

Steel-5

a) Friction type, or slip critical, fasteners are used whenever slippage of the joint cannot be tolerated. Applications include joints subjected to stress reversals; where bolts share load with welds groups in a connection; where oversized holes are used to facilitate erection; and where slippage would impair serviceability (e.g., splices in longspan trusses),

b) For friction–type connections, the surface condition of the faying surfaces has an important influence on capacity in that it governs the coefficient of friction. For most applications, faying surfaces are cleaned of mill scale and left unpainted. Proper bolt tension is also important in a friction-type fastener; whereas, for bearing type fasteners, the bolts need only be tightened to a "snug tight" condition.

c)

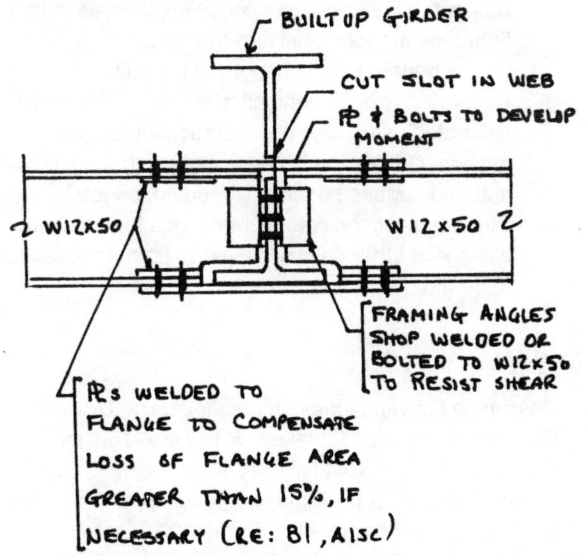

The reason I would choose to connect the flanges in the manner shown is that the alternatives to a plate through the web are either stiff tension hangers (e.g., structural tees) bolted through the web or welded splice plates on opposite sides of the web. The hanger type is inefficient and would interfere with the shear splice in such a shallow member. Welding splice plates to the relatively thick (1-1/2") web is bad because of the danger of lamellar tearing; viz.:

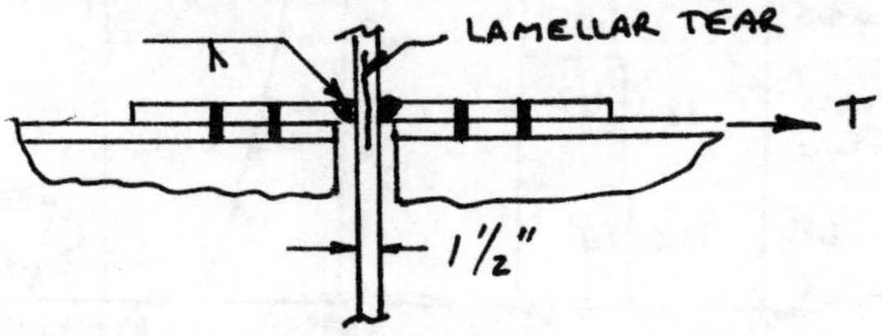

A slot large enough to pass the connection splice plate could probably be made without reinforcing the web opening (Darwin 1988). The proposed connection also has the advantage of field bolting rather than field welding.

Steel-6

a) Possible corrective measures for the oversize holes:
   1) Replace the A307 bolts with appropriate friction-type fasteners with hardened washers
   2) Ream the holes and use larger diameter A307 bolts
   3) Field weld the shear tab-to-web connection to carry the total reaction.

b) Assume $t_\omega \geq {}^3/_8$ (not specified). Use allowable stress design (AISC 1989). For 5 – 1" dia A325-N, single shear : $r_v = 16.5^{k}/_{bolt}$; $r_b = 26.1\ ^{k}/_{bolt}$

   $\therefore\quad R \leq nr_v = 5 \times 16.5 = 82.5^k$

   Given that the plate and weld are adequate, the capacity of the bolts is more than enough (not necessary to check the bolts for eccentric shear)

   $\therefore$ **Connection is adequate for 82^k**

c) Three methods of installing A325-F bolts:
   1) Turn-of-nut tightening: All bolts in the group are brought to the snug-tight condition, then each bolt is tightened an additional turn (ranging from ${}^1/_3$ to 1 full turn depending on the length of the bolt)
   2) Calibrated wrench tightening: a torque wrench is calibrated using a device that measures bolt tension. Calibration is performed using the specific bolt and surface conditions used on the job.
   3) Direct tension indicator tightening: a load indicator washer or other device is used to permit the tension to be gauged.

Steel-7

a) Determine the wind loads on the pipe using UBC-97, Chapter 16, Division III: take $I_w = 1.0$; $q_s = 16.4$ psf (Table 16-F); $C_q = 1.4$ (Table 16-H for flagpole); C_e varies (Table 16-G for Exposure condition C)

$p = C_eC_qq_sI_w = 1.4 \times 16.4 \times 1.0 \times q_s = 23.0C_e$

w = pD; where D = outside diameter of pipe

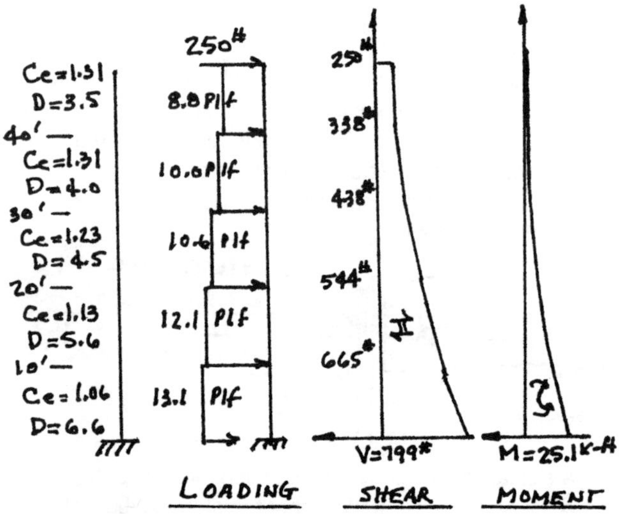

Check the stresses using Allowable Stress Design (ASIC 1989). For 40' $\leq h \leq$ 49':

Pipe 3 Std: D = 3.5"; $t_w = 0.216$"; $S_x = 1.72\ in^3$; Assume A53 pipe: $F_y = 36$ ksi

$M_{max} = 0.5(250 + 338)10 = 2940^{\#\text{-ft}} = 2.94^{k\text{-ft}}$

$f_b = M_{max} / S_x = 2.94 \times 12 / 1.72 = 20.5$ ksi

$D / t_w = 3.5 / 0.216 = 16 < 3300 / F_y = 92$

$\therefore\quad F_b = (4 / 3)(0.66F_y) = (4 / 3)(0.66 \times 36)$

$= 31.7$ ksi O.K.

- For 30' $\leq h \leq$ 40': Pipe 3-1/2 Std: D = 4.0; $t_w = 0.226$" ($D / t_w = 4.0 / 0.226 = 18$ is O.K.); $S_x = 2.39\ in^3$

  $M_{max} = 2.9 + 0.5(0.338 + 0.438)10 = 6.8^{k\text{-ft}}$

  $f_b = 6.8 \times 12 / 2.39 = 34.2\ \text{ksi} > F_b \leftarrow$ No Good

- For 20' $\leq h \leq$ 30': Pipe 4 Std: D = 4.5; $t_w = 0.237$" ($D / t_w = 19$ is O.K.); $S_x = 3.21\ in^3$

  $M_{max} = 6.8 + 0.5(0.438 + 0.544)10 = 11.7^{k\text{-ft}}$

  $f_b = 11.7 \times 12 / 3.21 = 43.9\ \text{ksi} > F_b \leftarrow$ No Good

- For 10' $\leq h \leq$ 20': Pipe 5 Std: D = 5.6; $t_w = 0.258$" ($D / t_w = 22$ is O.K.); $S_x = 5.45\ in^3$

  $M_{max} = 11.7 + 0.5(0.544 + 0.665)10 = 17.7^{k\text{-ft}}$

  $f_b = 17.7 \times 12 / 5.45 = 39.1\ \text{ksi} > F_b \leftarrow$ No Good

- For 0' $\leq h \leq$ 10': Pipe 6 Std: D = 6.63; $t_w = 0.28$" ($D / t_w = 24$ is O.K.); $S_x = 8.50\ in^3$

  $M_{max} = 17.7 + 0.5(0.665 + 0.799)10 = 25.0^{k\text{-ft}}$

  $f_b = 25.0 \times 12 / 8.50 = 35.3\ \text{ksi} > F_b \leftarrow$ No Good

The bending stress in most segments exceeds the allowable stress; **therefore, the proposed design is inadequate.**

Steel-8

a) Calculate the maximum combined stress (neglect the vertical shear stress). For the Pipe 12 Std: $I_x = I_y = 279\ in^4$; $S = 43.8\ in^3$; $d_o = 12.75$ in.

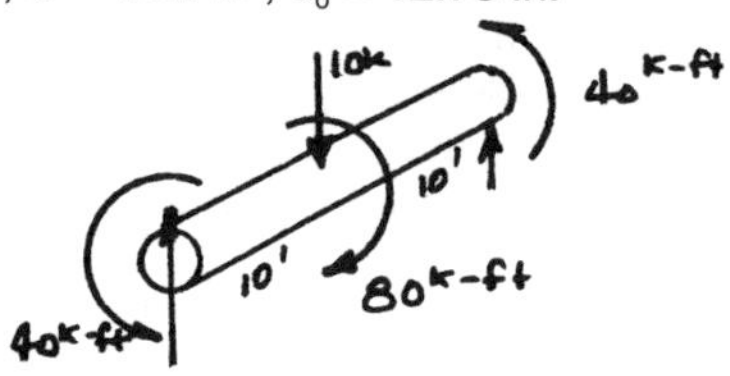

$J = I_x + I_y = 558\ in^4$

$\tau = Tr / J = 40 \times 12 \times (12.75 / 2) / 588 = 5.48$ ksi

$\sigma_b = M / S$; $M = PL / 4 = 10 \times 20 / 4 = 50^{k\text{-ft}}$

$\sigma_b = 50 \times 12 / 43.8 = 13.7$ ksi

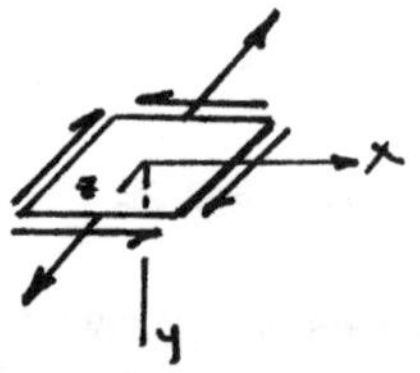

The critical stresses occur at top and bottom fibers where $\sigma_x = 0$; $\sigma_z = \pm 13.7$ ksi;

$\tau_{zx} = -\tau_{xz} - 5.5$ ksi; $\sigma_y = 0$. The principal stresses can be found using Mohr's Circle. For example at the bottom (tension) fiber:

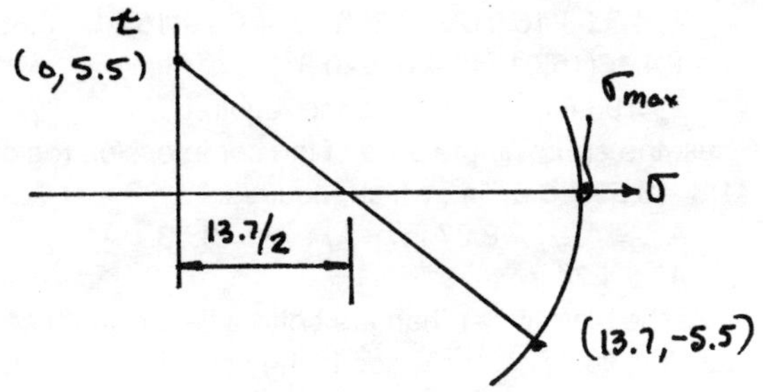

$$R = \sqrt{(13.7\ /2)^2 + 5.5^2} = 8.8 \text{ ksi}$$

$\therefore$ **σ_{max} = (13.7/2) + 8.8 = 15.7 ksi**

b) Compute the smallest diameter solid shaft that can resist the applied loading if $\tau \le 14.5$ ksi and $\sigma \le 22$ ksi:

$J = 2I = \pi D^4 / 32$

$\tau = Tr / J = [40 \times 12 \times (D / 2)] / (\pi D^4 / 32) \le 14.5$ ksi

$D^3 \ge 169$ in^3; $D \ge 5.52$"

Flexure: $\sigma \le 22$ ksi

$\sigma = Mc / I = 50 \times 12 \times (D / 2) / (\pi D^4 / 64) \le 22$ ksi

$D^3 \ge 278$ in^3; $D \ge 6.5$" ← governs.

Thus, flexure governs: **D = 6.5".**

Note: I assume that vertical shear stress is to be neglected in Part b, as was stipulated for Part a. If the vertical shear stress is included, the combined stress due to shear and torsion is

$\tau = VQ / Ib + Tr / J \le 14.5$ ksi

Steel-9

a) Calculate the girder and hanger reactions. The problem involves a rigid bar supported on elastic supports. Replace the cross beams by equivalent linear elastic springs.

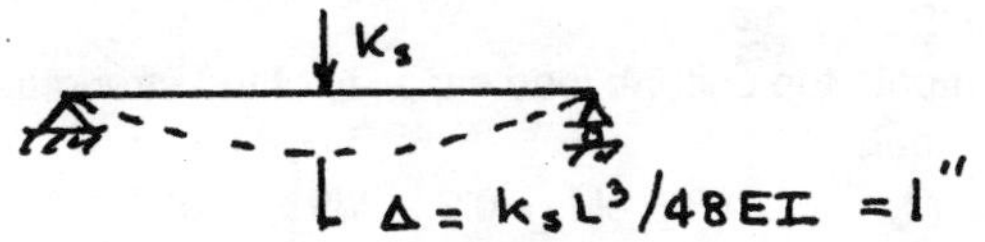

For girder 1: L = 14'; I = 1000 in^4

$k_{s1} = 48 \times 29{,}000 \times 1000 / (14 \times 12)^3 = 294^k/_{in}$

For girder 2: L = 12'; I = 500 in^4

$k_{s2} = 48 \times 29{,}000 \times 500 / (12 \times 12)^3 = 233^k/_{in}$

Let the tension in the rod be the redundant and release it:

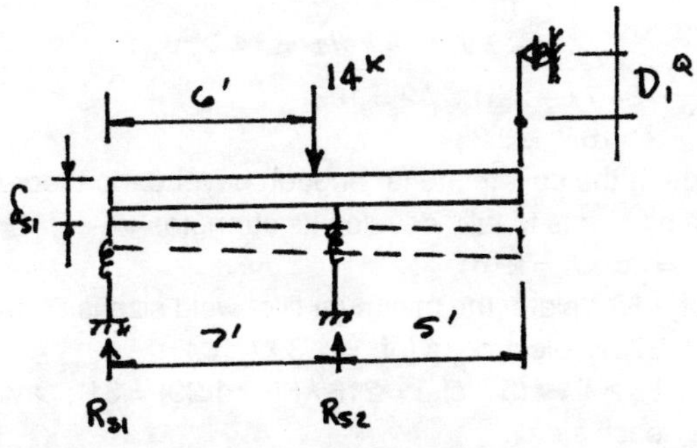

$\Sigma M_1 = 0$: $R_{s2} = 14 \times 6 / 7 = 12^k$

$\Sigma F_y = 0$: $R_{s1} = 2^k$

$\therefore$ $\delta_{s1} = R_{s1} / k_{s1} = 2 / 294 = 0.0068$" ↓

$\delta_{s2} = R_{s2} / k_{s2} = 12 / 233 = 0.0515$"↓

$D_1^Q = 0.0068 + [(0.0515 - 0.0068) / 7]12 = 0.0834$"↓

Apply a unit force at the top of the rod at the released point:

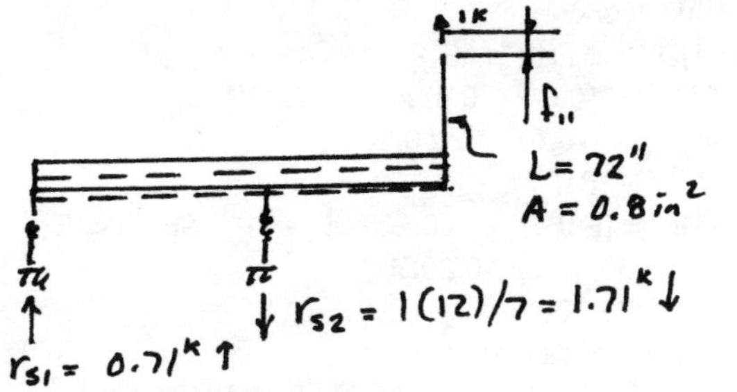

$\delta_{s11} = 0.71 / 294 = 0.0024"/_{kip}$ ↓

$\delta_{s21} = 1.71 / 233 = 0.00734"/_{kip}$ ↑

$\therefore$ $f_{11} = 1 \times 72 / (0.8 \times 29{,}000) - 0.0024$
$+ [(0.00734 + 0.0024) / 7] \times 12 = 0.0174"/_{kip}$ ↑

For consistent displacement the defection at the top must be zero; therefore:

$Tf_{11} = D_1^Q$: **T = 0.0834 / 0.0174 = 4.79^k (tension)**

From statics:

R_1 = (4.79 × 5 + 14 × 1) / 7 = 5.42^k ↑

R_2 = 14 − 5.42 − 4.79 = 3.79^k ↑

b) Calculate the displacements:

For the hanger:

δ = TL / AE = 4.79 × 72 / (0.8 × 29,000) = 0.0149" ↓

For girder 1:

$\delta_1 = R_1 / k_{s1}$ = 5.42 / 294 = 0.0184" ↓

For girder 2:

$\delta_2 = R_2 / k_{s2}$ = 3.79 / 233 = 0.0163" ↓

Steel-10

Calculate the unit stresses in each hanger.

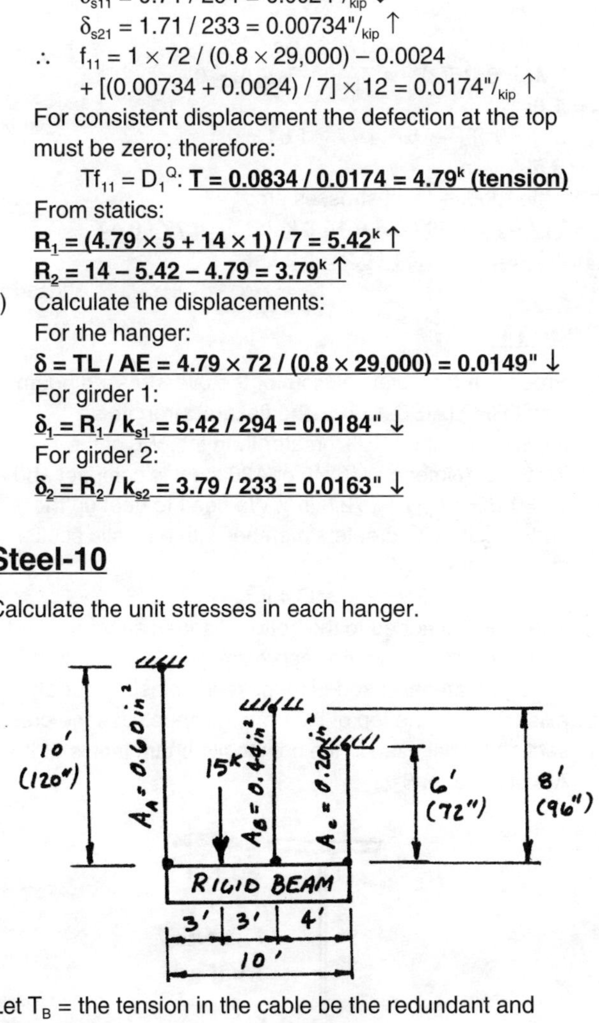

Let T_B = the tension in the cable be the redundant and release it:

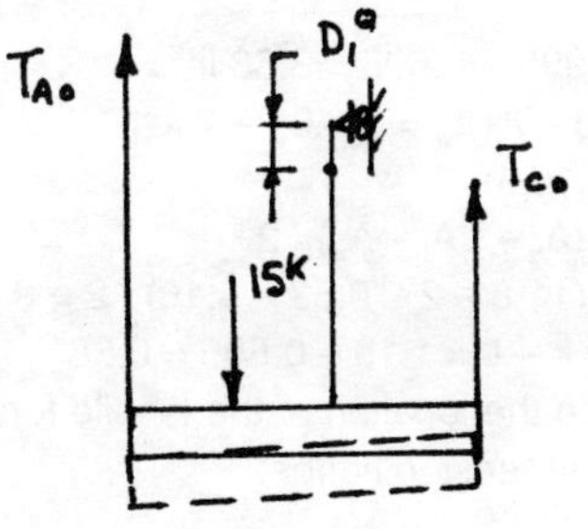

From statics:

$T_{CD} = 15(3) / 10 = 4.5^k$; $T_{Ao} = 15 - 4.5 = 10.5^k$

Apply a unit dummy force at the released point: $t_b = +1$

$t_c = -1 \times 6 / 10 = -0.6$; $t_a = -0.4$

By virtual work:

$(1)D_1^Q = \Sigma(TtL / (AE))_i$

$\therefore$ $ED_1^Q = [10.5 \times (-0.4) \times 120 / 0.60]$
$+ [4.5 \times (-0.6) \times 72 / 0.20]$

$D_1^Q = -1812 / E$

$(1)f_{11} = [(-0.4)^2 \times 120 / 0.60E] + [1^2 \times 96 / 0.44E]$
$+ (-0.60)^2 \times 72 / 0.20E]$

$f_{11} = 379.78 / E$

For consistent displacement:

$T_Bf_{11} + D_1^Q = 0 \therefore T_B(379.78 / E) - 1812 / E = 0$

$T_B = 4.77^k$

From statics:

$\Sigma M_A = 0$: $15(3) - 4.77(6) - T_C(10) = 0$

$T_C = 1.64^k$

$\Sigma F_y = 0$: $T_A - 15 + 4.77 + 1.64 = 0$

$T_A = 8.59^k$

Thus, the required unit stresses are:

$f_{tA} = T_A / A_A = 8.59 / 0.6 = 14.3$ ksi; $f_{tB} = 4.77 / 0.44 = 10.84$ ksi; $f_{tC} = 1.64 / 0.2 = 8.2$ ksi

Steel-11

a) Prepare a remedial solution for the overstressed beam. Use LRFD (AISC 1994). The flexural strength at midspan must be 35% greater than the ϕM_n of the W16x40. (Note: a W16x40 of A36 steel is compact and $L_B = 1.3' << L_p$; $Z_x = 72.9$ in^3). We need to beef up the bottom flange to create a member with a plastic section modulus $\geq 1.35(Z_x)_{W16x40}$.

$Z_{req} = 1.35 \times 72.9 = 98.4$ in^3

The material added to the bottom flange cannot exceed 1-1/2" in depth. As a first approximation, assume that enough material is added to make the plastic neutral axis (PNA) at the top of the bottom flange. Use the cross section model used in the composite beam tables of the AISCM (Re: p. 5-7)

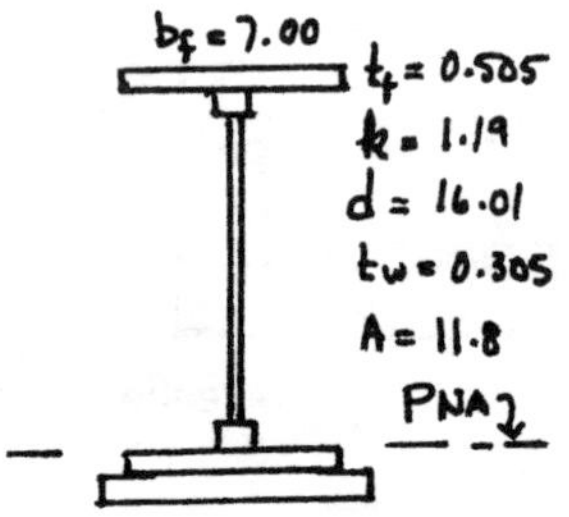

$A_f = 6.995 \times 0.505 = 3.53$ in^2

$A_w = (d - 2k)t_w = (16.01 - 2 \times 1.19) \times 0.305 = 4.16$ in^2

$k_{area} = (A_s - 2A_f - A_w) / 2 = (11.8 - 2 \times 3.53 - 4.16) / 2 = 0.29$ in^2

$k_{depth} = k - t_f = 1.19 - 0.505 = 0.68"$

Approximate the location of the tensile force as at the bottom of the tension flange:

$Z_x \cong 3.53(16.012 - 0.505 / 2) + 0.29(16.01 - 0.85)$
$+ 4.16(16.01 / 2) + 0.29(0.85)$

$Z_x = 93.6$ in^3 $\rightarrow$ close to the required Z_x

Thus, the approximate area of the compression region is $11.8 - 3.53 = 8.27$ in^2, which requires:

$A_{ten} = A_{comp} = 8.27$ in^2 $= A_f + A_{pl} = 3.53 + A_{pl}$

$\therefore$ $A_{pl} \geq 4.74$ in^2

Make the plate wider than the bottom flange width so that fillet welds can be made in the down hand position in the field: try a PL 8 x 0.75 ($A_{pl} = 6.0$ in^2).

Locate the plastic neutral axis:

$A_{ten} = A_{comp} = (11.8 + 6.0) / 2 = 8.9$ in^2

$6.995y + 6.0 = 8.9 \therefore y = 0.41"$

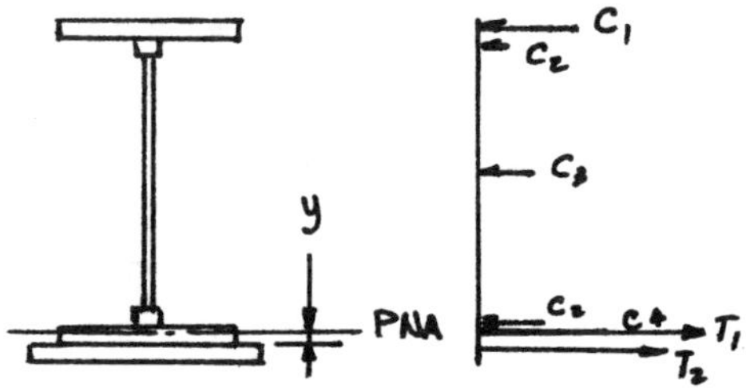

$C_1 = F_yA_f = 36 \times 3.58 = 129^k$

$C_2 = F_yA_k = 36 \times 0.29 = 10.4^k$

$C_3 = F_yA_w = 36 \times 4.16 = 150^k$

$T_1 = 36(b_fy) = 36 \times 6.995 \times 0.41 = 103^k$

$T_2 = 36 \times 6.0 = 216^k$

$C_4 = 216 + 103 - 129 - 2 \times 10.4 - 150 = 19.2^k$

Sum moments about the bottom of the W16x40:

$M_p = 129(16.01 - 0.505/2) + 10.4(16.01 - 0.85)$
$+ 150(16.01 / 2) + 10.4(0.85) + 19.2(0.46) \leftarrow$
$103(0.41 / 2) + 216(0.75 / 2) = 3469^{k\text{-}in} = 289^{k\text{-}ft}$

$(M_p)_{req} = F_yZ_{req} = 36 \times 98.4 / 12 = 295^{k\text{-}ft}$

$(M_p)_{req} / M_p = 295 / 289 = 1.02$ – close enough.

Use PL 8 x 3/4

Compute the uniform load capacity of the strengthened member:

$M_u = w_uL^2 / 8 = \phi M_p = 0.9 \times 289 = 260^{k\text{-}ft}$

$w_u \leq 260 \times 8 / 30^2 = 2.31$ klf

$w_u = 1.2w_d + 1.6w_L$

The coverplate can be terminated where the bending moment is less than $260 / 1.35 = 193^{k\text{-}ft}$.

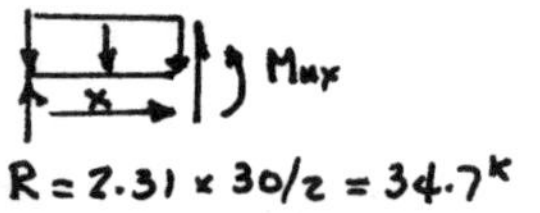

$M_{ux} = 34.7x - 2.31x^2 / 2 = 193$

$x = (6.7', 23.3')$

Extend the coverplate far enough beyond the theoretical cut off points to fully develop its strength:

$T_{PL} = 36 \times 6 = 216^k$

For a 3/4" plate, the minimum fillet weld size is 5/16"; use E70xx electrodes ($\phi f_w = 1.39^k/_{in/16th}$)

$L_w \geq T_{PL} / (5 \times \phi f_w) = 216 / (5 \times 1.39) = 31"$; say 16" each side

Actual cutoff points = (6.7 – 1.33 = 5.4'; 23.3 + 1.33 = 24.7')

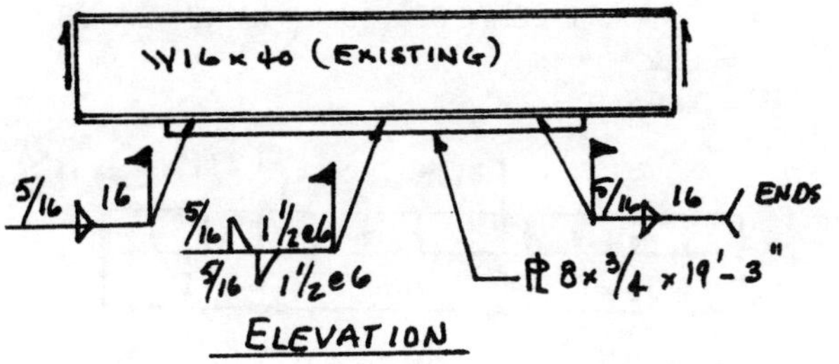

b) Construction sequence (note: the weld heat may be sufficient to significantly lower the yield point ∴ shore the floor during installation).
   1. Shore the joists on both sides adjacent to the W16x40.
   2. Clamp the PL 8 x 3/4 to the lower flange
   3. Weld the plate; remove clamps and shores.

Steel-12

a) Design the connection to transfer 50% of the shear and moment strength of the smaller (i.e., W14x176) section. The axial force is not specified and will not be included in the design (axial compression would have a beneficial effect on the resisting moment). Use the LRFD method (AISC 1994). For the W14x176: $Z_x = 320 \text{in}^3$; $d = 15.22"$; $b_f = 15.65"$; $t_w = 0.83"$; $t_f = 1.31"$ – A36 steel: $F_y = 36$ ksi.

$0.5M_p = 0.5 \times 36 \times 320 = 5760^{k\text{-}in}$

Re: Table J2.1: For a 45° bevel, the effective throat is 1/8" smaller than the weld size. The tension strength of a partial penetration weld is $0.9F_{yc}$.

∴ $0.5M_p \le \phi F_y b_f D(d - D)$; where D = required throat thickness.

$5760 \le 0.9 \times 36 \times 15.65D(15.22 - D)$

$D^2 - 15.22D + 11.36 = 0$

$D = 0.786"$

∴ $S = D + 1/8" = 0.786 + 0.125 = 0.91"$

<u>Use 15/16 Weld (BTC-P4)</u>. (Note: the size computed for strength exceeds the minimum per J2.1).

b) Determine the bolts required to transfer 50% of the shear capacity of the W14x176. Since bolts are slip critical, use the service load values from Table 8-16, AISCM-II, and the allowable shear capacity

$0.5V_{all} = 0.5(0.4F_y dt_w) = 0.5(0.4 \times 36 \times 15.22 \times 0.83)$
$= 91^k$

Use 1" diameter bolts arranged as close together as code permits to alleviate the stresses caused by eccentricity of the connection. Re: J3.9, bearing should not be critical on the $t_w = 0.83"$ material, therefore, space the bolts at (2-2/3)d, say 2.75" o.c. Use an 11" wide plate with 4 bolts across:

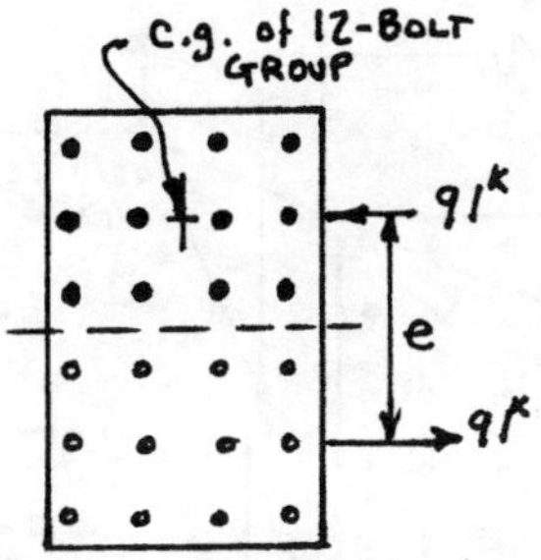

Edge Distance = (11 – 3 × 2.75) / 2 = 1-3/8"

The edge distance is no good for sheared plate (Table J3.7); therefore, specify that plates be gas cut. Try a 12-bolt group on each side of splice (note: 8 bolts were tried and found to be inadequate). Use the elastic method of analysis:

$e = 3 \times 2.75 = 8.25"$

$M = (V_{all})e = 91 \times 8.25 = 750^{k\text{-}in}$

$r_p = V_{all} / 12 = 91 / 12 = 7.6^k/_{bolt}$

$\Sigma(x_i^2 + y_i^2) = 8 \times 2.75^2 + 6 \times 4.125^2 + 6 \times 1.375^2$
$= 174 \text{ in}^2$

$r_{mx} = My / \Sigma(x_i^2 + y_i^2) = 750 \times 2.75 / 174 = 11.9^k/_{bolt}$

$r_{my} = Mx / \Sigma(x_i^2 + y_i^2) = 750 \times 4.125 / 174 = 17.8^k/_{bolt}$

$r = \sqrt{(r_p + r_{mx})^2 + r_{my}^2}$

$= [(7.6 + 11.9)^2 + (17.8)^2]^{1/2}$

$= 26.4$ k / bolt

From Table 8-16, AISCM-II, for a 1" diameter slip critical A325 bolt in double shear the tabulated capacity is $26.7^k/_{bolt}$, which is adequate.

∴ **<u>Use 12-1" diameter A325-SC each side.</u>**

Check the ultimate shear in the plates: $V_u = 1.5V_{all} = 1.5 \times 91 = 136^k$ ($68^k/_{plate}$).

$A_{ns} = (11 - 4(1 + 0.125))t = 6.5t$

Shear fracture (C-J4-2):

$V_u = 68 \le \phi 0.6F_u A_{ns} = 0.75 \times 0.6 \times 58 \times 6.5t$

∴ $t \ge 0.40"$

Shear yield by Eqn J5-3:

$V_u = 68 \le \phi 0.6F_y A_{ng} = 0.9 \times 0.6 \times 36 \times 11t$

∴ $t \ge 0.31"$

Fracture governs: **<u>Use 2Pls 11 x 7 / 16 x 1'-5".</u>**

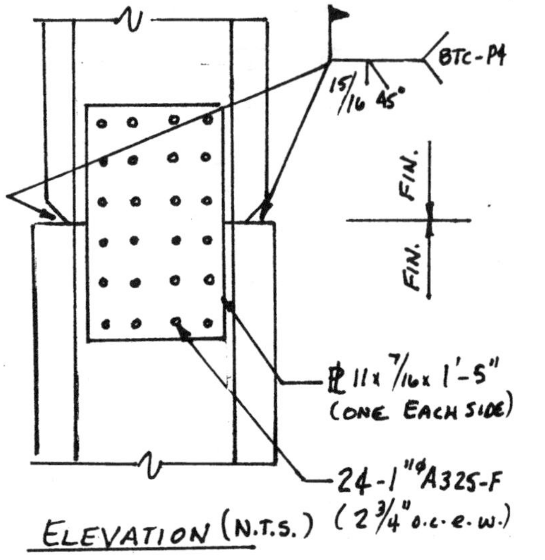

Steel-13

Design girder G1 using either standard or modified wide flange sections. Use LRFD (AISC 1994).

w_d = slab + partitions + ceiling / mechanical + beam

$w_d = 48 + 20 + 10 + 5 = 83$ psf

Assume that the wall weight around the elevator is included in the partition load.

$w_L = 80$ psf (reducible per UBC-97)

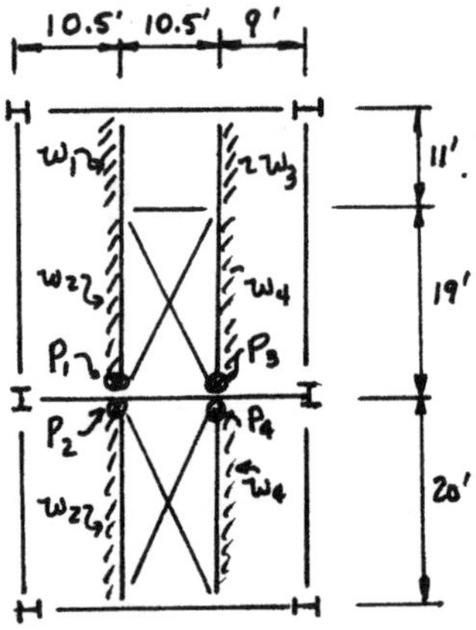

For the girder:

$A_{trib} \cong (15 + 10)(5.25 + 4.5) = 245\ \text{ft}^2$

$$R \le \begin{cases} r(A_{trib} - 150) = 0.08(245 - 150) = 7.5\% \leftarrow \text{Governs} \\ 23(1 + 83/80) = 47\% \\ 40\% \end{cases}$$

$\therefore\ w_L' = (1 - 0.075)80 = 74$ psf

$w_1 = 1.2w_d + 1.6w_L$

$= 1.2 \times 0.083 \times 10.5 + 1.6 \times 0.074 \times 10.5 = 2.3$ klf

$w_2 = 0.5w_1 = 1.15$ klf

$w_3 = (w_1 / 10.5)9.75 = 2.14$ klf

$w_4 = (w_1 / 10.5)4.5 = 0.99$ klf

$P_1 = [2.3 \times 11^2 / 2 + 1.15 \times 19 \times (11 + 19 / 2)] / 30$

$= 19.6^k$

$P_2 = 1.15 \times 10 = 11.5^k$

$P_3 = [2.14 \times 11^2 / 2 + 0.99 \times 19 \times (11 + 19 / 2)] / 30$

$= 17.2^k$

$P_4 = 0.99 \times 10 = 9.9^k$

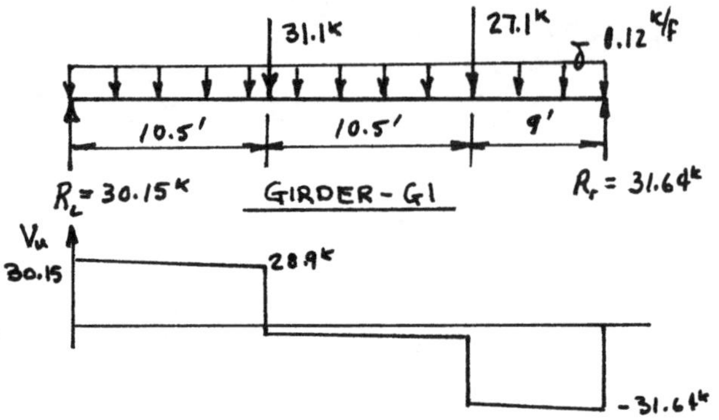

$M_u = 0.5(30.15 + 28.9)(10.5) = 310^{k\text{-ft}}$

$L_b = 19.5'$; $C_b = 1.0$; $F_y = 36$ ksi

Try a W24x62: $b_f = 7.0"$; Re: p. 4-18, AISCM, $L_p = 5.8'$; $\phi M_p = 413^{k\text{-ft}}$; $L_r = 17.2'$; $\phi M_r = 255^{k\text{-ft}}$. Since $L_r > L_b = 20.5' > L_p$, use Eqn F1-3:

$\phi M_n = C_b[\phi M_p - (\phi M_p - \phi M_r)(L_b - L_p) / (L_r - L_p)] \le \phi M_p$

$\phi M_n = [413 - (413 - 255)(10.5 - 5.8) / (17.2 - 5.8)]$

$= 348^{k\text{-ft}}$

$\phi M_n > M_u \therefore$ the section is O.K. in flexure

Check the live load deflection ($I_x = 1550\ \text{in}^4$):

$P_{L1} = [(1 - 0.075)0.08 \times 10.5 / 2.3]31.1 = 10.5^k$

$P_{L2} = [(1 - 0.075)0.08 \times 10.5 / 2.3]27.1 = 9.1^k$

Approximate as two equal, equally spaced loads of magnitude $P_{Lave} = 9.8^k$

$\delta = 0.036\, P_{Lave}L^3/EI$

$\delta = 0.036 \times 9.8 \times (30 \times 12)^3 / (29{,}000 \times 1550)$

$= 0.37" < L/360 = 1"$

$\therefore$ **use a W24x62 w/o modification.**

In this case, a standard shape with 7.0" flange width was available. If a standard section were not available, it would be necessary to select a heavier shape furnishing the required moment strength when it was trimmed to a 7" width. The flange tips could be gas cut the same amount each side to maintain a doubly symmetrical cross section. The reduced section would taper back to full width to alleviate stress concentrations as shown in the partial plan view below.

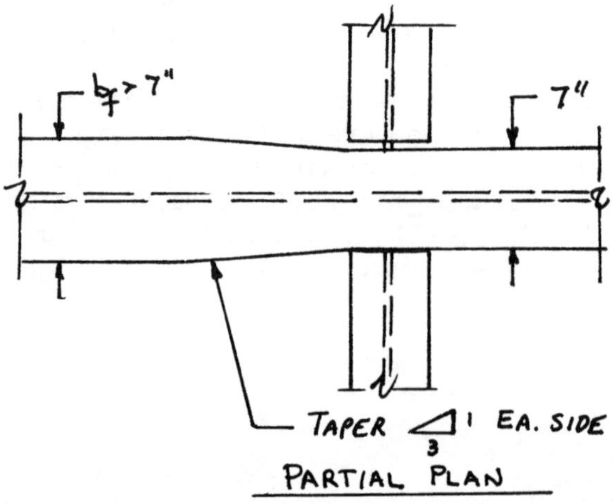

Steel-14

a) Determine the stiffener requirements to develop the M_p of the W21x44 shown below. Assume loads are non-seismic. Since column shears and axial forces are not shown, the shear will be neglected (conservative) and the axial load will be assumed to be less than $0.4P_y$. Use LRFD (AISC 1994) and A36 steel.

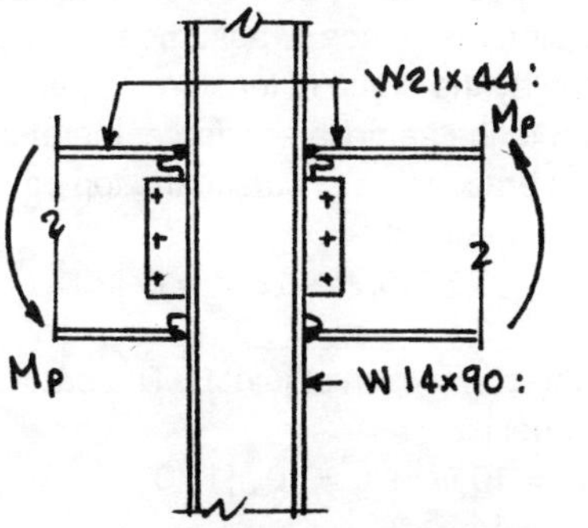

For the W21x44: $Z_x = 95.4$ in^3; d = 20.66"; $t_w = 0.35$"; $b_f = 6.5$"; $t_f = 0.45$". For the W14x90: d = 14.02"; $t_w = 0.44$"; $b_f = 14.52$"; $t_f = 0.71$"; k = 1.375".

$M_p = F_y Z_x = 36 \times 95.4 = 3434^{\text{k-in}}$

$P_f = M_p / (d - t_f) = 3434 / (20.66 - 0.45) = 170^k$

For bending of the tension flange (K1-1, AISCS):

$\phi R_n = \phi(6.25 t_f^2 F_y) = 0.9 \times 6.25 \times 0.71^2 \times 36 = 102^k$

Local web yielding (K1-2), N = beam flange thickness:

$\phi R_n = \phi(5K + t_{bf}) F_{yw} t_w$

$= 1.0 \times (5 \times 1.375 + 0.45) \times 36 \times 0.44 = 116^k$

Web crippling (K1-4):

$$\phi R_n = \phi 135 t_w^2 \left[1 + 3\left(\frac{N}{d}\right)\left(\frac{t_w}{t_f}\right)^{1.5}\right]\sqrt{F_{fw} t_f / t_w}$$

$$= 0.75 \times 135 \times 0.44^2 \left[1 + 3\left(\frac{0.45}{14.02}\right)\left(\frac{0.44}{0.71}\right)^{1.5}\right]$$

$$\times \sqrt{36 \times 0.71 / 0.44}$$

$$= 156^k$$

Compression web buckling (K1-8) does not apply to this problem. Since loads are reversible, bending of the tension flange is critical at the top and bottom ∴ design stiffeners for the difference between the design force, P_f, and the bending resistance, ϕR_n

$F_y A_{st} \geq 170 - 102 = 68^k$

$A_{st} \geq 68 / 36 = 1.89$ in^2

$t_{st} \geq t_{bf} / 2 = 0.45 / 2 = 0.22$" try 3/8" plate

$w_{st} \geq 1.89 / 0.375 = 5.0$", say 3" each side

The stiffeners sized for strength satisfy the limits of Section K1.9 ∴ **<u>Use 2PLs 3 x 3 / 8</u>**

($A_{st} = 2 \times 3 \times 0.375 = 2.25$ in^2)

b) Check for high web shear stresses (K1.7, $P_u < 0.4P_y$)

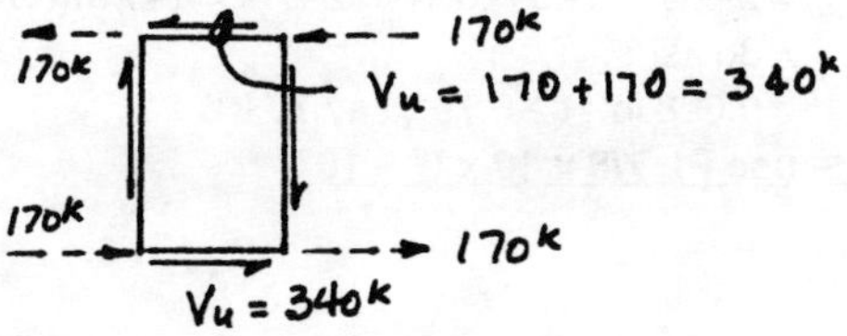

$\phi R_v = \phi 0.6 F_y d_c t_w = 0.9 \times 0.6 \times 36 \times 14.02 \times 0.44$

$= 120^k$

$V_u = 340^k > \phi R_n = 120^k$ ∴ need either a thicker web or to add web reinforcement.

If web reinforcement (i.e., doubler plate) is used:

$t_{req'd} \geq 340 / (0.9 \times 0.6 \times 36 \times 14.02) = 1.25$"

∴ $T_{PL} \geq t_{req'd} - t_w = 1.25 - 0.44 = 0.81$", **<u>say 7/8" Plate</u>**

Note: this problem was worked using LRFD criteria assuming non-seismic loading. If lateral forces were caused by seismic loading, additional and more restrictive criteria would apply.

Steel-15

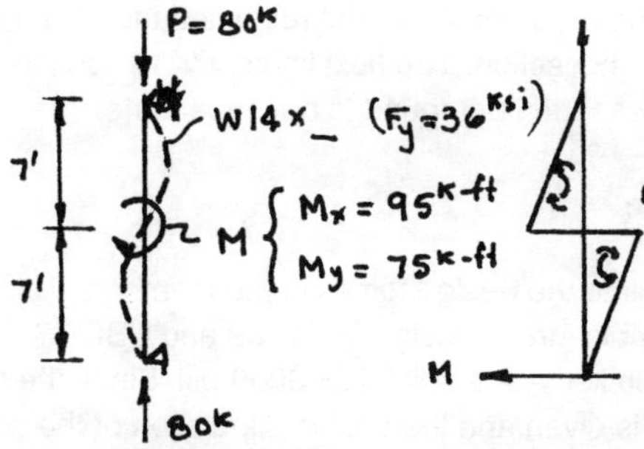

Determine the lightest column for the given loads using LRFD (AISC 1994); assume 50% live; 50% dead

U = 1.2D + 1.6L = 1.4; F_y = 36ksi.

$P_u = 1.4 \times 80 = 112^k$

$M_{nt,x} = 1.4 \times 95 / 2 = 66.5^{\text{k-ft}}$

$M_{nt,y} = 1.4 \times 75 / 2 = 52.5^{\text{k-ft}}$

Select a trial W14x section – Re: P.3-12, AISCM: m ≅ 2.0; U ≅ 2.0 for W14x

$P_{trial} = P_u + M_{ux}m + M_{uy}mu$

$= 112 + 66.5 \times 2 + 52.5 \times 2 \times 2 = 455^k$

Try a W14x68 for $P_{trial} = 455^k$ with KL_y = 14' (Re: P.3-21):

$A_g = 20.0$ in^2; $r_y = 2.46$"; $r_x = 6.00$"; $L_p = 10.3$'; $L_r = 37.3$'; $\phi M_p = 311^{\text{k-ft}}$; $\phi M_r = 201^{\text{k-ft}}$; d= 14.04"; $t_w = 0.415$"; k = 1.5"; $Z_y = 36.9$ in^3

Check local buckling (flange is O.K.):

$h_c / t_w = (d - 2k) / t_w = (14.04 - 2 \times 1.50) / 0.415 = 27$

(Re: Table B 5.1, AISCS)

$P_u / \phi_b P_y = 112 / (0.9 \times 36 \times 20.0) = 0.17 > 0.125$

$\lambda_p = 191 / \sqrt{F_y}\,(2.33 - P_u / \phi_b P_y) \geq 253 / \sqrt{F_y}$

$= 191 / \sqrt{36}\,(2.33 - 0.17) = 69 > h_c / t_w$ ∴ o.k.

Check combined axial plus bending

$KL / r)_x = 14 \times 12 / 6.0 = 28$

$KL / r)_y = 14 \times 12 / 2.46 = 68$ ← governs $\phi_c F_{cr} = 23.99$ ksi

$\phi P_n = \phi_c F_{cr} A_g = 23.99 \times 20.0 = 480^k$

$P_u / \phi P_n = 112 / 480 = 0.23$

$L_p = 10.3' < L_b = 14' < L_r = 37.3'$: Use Eqn F1-3:

$\phi M_n = C_b[\phi M_p - (\phi M_p - \phi M_r)(L_b - L_p) / (L_r - L_p)] \leq \phi M_p$

$\phi M_n = [311 - (311 - 201)(14 - 10.3) / (37.3 - 10.3)]$

$= 296^{\text{k-ft}}$

$\phi M_{ny} = \phi F_y Z_y = 0.9 \times 36 \times 36.9 / 12 = 99.6^{\text{k-ft}}$

(Re: C1, AISCS): For combined axial plus bending the concentrated moment at midheight produces reverse curvature bending that is equivalent to a column with

KL / r = one-half the actual KL / r; moment at one end and zero moment at the other $\therefore$ $C_m = 0.6$

$KL' / r)_x = 0.5 \times 28 = 14$: $P_{ex} = (\pi^2 E / 14^2) \times 20.0$

$= 29{,}000^k$

$KL' / r)_y = 0.5 \times 68 = 34$: $P_{ex} = (\pi^2 E / 34^2) \times 20.0 = 4950^k$

$B_1 = C_m / (1 - P_u / P_e) \geq 1.0 = 1.0$

$\therefore$ $M_{ux} = B_1 M_{nt,x} = 1.0 \times 66.5 = 66.5^{k\text{-}ft}$

$M_{uy} = B_1 M_{nt,y} = 1.0 \times 52.5 = 52.5^{k\text{-}ft}$

$P_u / \phi P_n = 0.23 > 0.2$ $\therefore$ check Eqn. H1-1a:

$P_u / \phi P_n + 8 / 9(M_{ux} / \phi M_{nx} + M_{uy} / \phi M_{ny})$

$= 0.23 + 8 / 9(66.5 / 296 + 52.5 / 99.6) = 0.90$

$\therefore$ **Use W14x68**

Note: an interaction value of 0.90 suggests that a lighter section might be available that would satisfy the design loading. The problem statement restricted the column to the lightest W14x section. The next lightest W14 section, a W14x61, was tried but found to be inadequate.

Steel-16

a) Calculate the design forces in the members. Use allowable stress design (AISC-89 and UBC-97); $F_y = 36$ ksi; A307 bolts; $f'_c = 3000$ psi. Since the seismic force is given and there is no stipulation of the seismic zone in the problem statement, I will assume that the structure is not in zone 3 or 4; therefore, the special load cases of 2213.5.1, UBC-97 are not applicable.

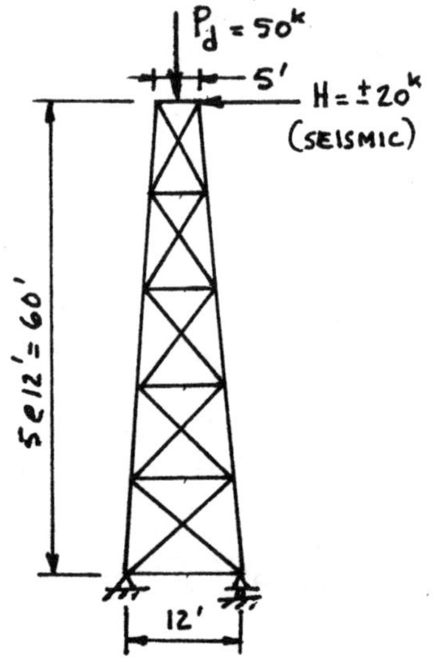

The tower legs batter (12 – 5) / 2 = 3.5' in 60' in two planes: $L = (2 \times 3.5^2 + 60^2)^{1/2} = 60.2'$. Consider the forces acting on one bent: Dead = 50 / 4 = $12.5^k/_{leg}$.

$\cos \alpha = 60 / 60.2 = 0.9966$

$\therefore$ $P_d = 12.5 / 0.9966 = 12.54^k$ (negligible increase)

Lateral force = 20 / 2 = $10^k/_{bent}$.

$P_e = (10 \times 60 / 12)/\cos \alpha = \pm 50.2^k$

For design of the tower leg, consider two load cases: Dead plus lateral causing axial compression, and 85% dead plus live causing tension in the leg.

$\therefore$ $P_{max} = 12.54 + 50.2 = 62.7^k$ (comp.)

$T_{max} = 0.85 \times 12.54 - 50.2 = -39.5 = 39.5^k$(ten.)

Take the 1 / 3 increase in allowable stresses:

$P_{equiv} = (0.75)(62.7) = 47^k$ with KL = 12.05'. Re: page 3-36, AISCM

Select Pipe 5 Std ($P_{all} = 68^k$)

Re: 2214.6.6, UBC-97, the trussed tower need only satisfy the requirements of 2213.9, which requires that connections be designed for $\Omega_0 = 2.0$ times the design force. The given brace, L3 x 3 x 1/4 has negligible compression resistance $\therefore$ assume that all panel shear is resisted by the brace in tension. All panels are subjected to the same lateral force; therefore, the maximum brace force occurs in the top panel where braces are steepest.

$\cos \alpha = 5.7 / (5.7^2 + 12^2)^{1/2} = 0.4291$

$(P_{brace})_{max} = 10 / \cos \alpha = 23.3^k$

b) Design the connection at joint B. The brace force in the lowest panel is

$P_{brace} = 10 / \cos \alpha = 10 / [11.3 / (11.3^2 + 12^2)^{1/2}]$

$= 14.6^k$

Connection must be designed to transfer $\Omega_o = 2.0$ times this force; i.e., $2 \times 14.6 = 29.2^k$

Try 3/4" diameter A307 bolts in single shear:

$r_v = 4 / 3(4.4) = 5.9^k/_{bolt}$

$n \geq 29.2 / 5.9 = 4.9$, say 5 bolts

Use PL 6 x 1/4, which is stronger than the L3 x 3 x 1/4.

Shop weld the gusset plate using 3/16 fillet welds:

$$q \leq \begin{cases} 4/3 \times 2 \times 3 \times 0.928 = 7.5^k/_{in} \\ 4/3 \times 0.4 \times 36 \times (1/4) = 4.8^k/_{in} \leftarrow \text{governs} \end{cases}$$

$P = qL_w > 4.8 \times 6 = 29^k \cong P_{brace}$ $\therefore$ o.k.

Design the base plate: try 4-1" diameter anchor bolts ($A_b = 0.79$ in^2). Re: Table J3.3

$f_v = 10 / (4 \times 0.79) = 3.1$ ksi

$F_t = (4 / 3)(26 - 1.8f_v) \leq (4 / 3)(20) = 26.7$ ksi

$T_{all} = 4 \times 26.7 \times 0.79 = 84^k > T_{max} = 2.0 \times 39.5^k$

$\therefore$ **Use 4-1" diameter anchor bolts**

Space bolts to allow 2" to the edge of Pipe 5 Std: $0.707(5 + 4) \cong 6"$, say a 10" x 10" plate with bolts centered 2" from edges

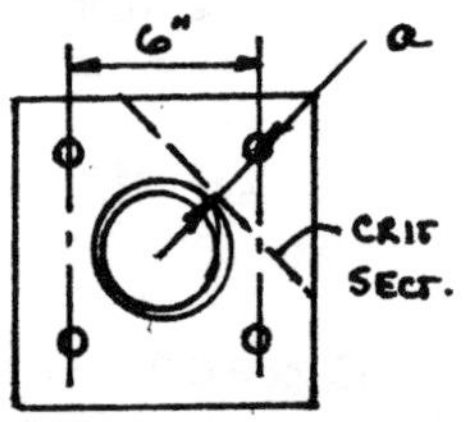

Diagonal = 10 / 0.707 = 14.14"

Width of critical section = 2(14.14 – 5) / 2 = 9.1"

a = ((6 / 0.707) – 5) / 2 = 1.74"

$\therefore$ $M = (\Omega_o T_{max} / 4)a = (2.0 \times 39.5 / 4)1.74 = 34.3^{k\text{-}in}$

$f_b = M / S = 34.3 / (9.1t^2 / 6) \leq F_b = (4 / 3)(0.75 \times 36)$

$= 36$ ksi

$t^2 \geq 0.628$ in^2; $t \geq 0.79"$, say 7 / 8"

– **use PL 7/8 x 10 x 0' – 10"**

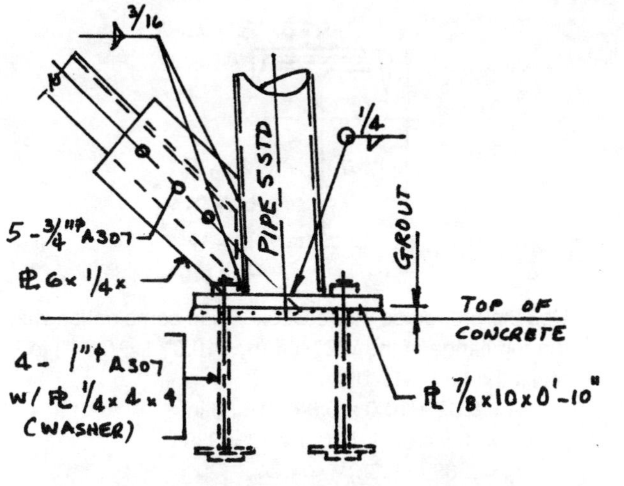

DETAIL – JOINT B (N.T.S.)

Steel-17

Find the load on the critical bolt. Re: A4.2 (AISC 1989). Assume a cab-operated trolley ∴ 25% impact factor applies to the wheel load:

$V = 1.25(50 + 4 + 2) = 70^k \downarrow$

A lateral force equal to 20% of the lifted load plus trolley weight is applied to the top of rails with one-half distributed to each side (A4.3):

$H = 0.2(50 + 4) / 2 = 5.4^k \leftarrow$ or $\rightarrow$

Use the elastic method of analysis (Salmon and Johnson, 1990)

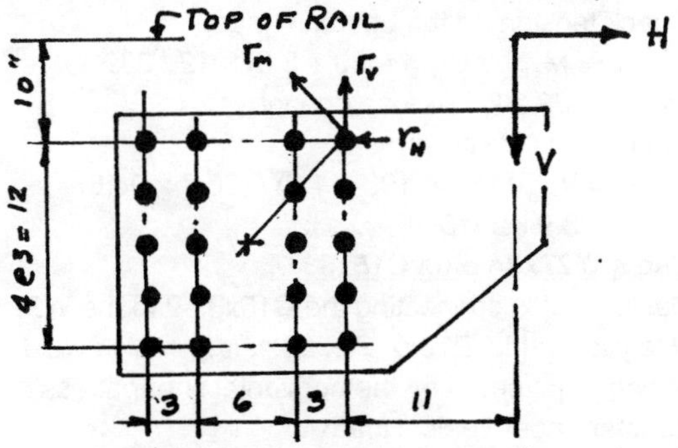

The centroid of the fastener group is found by inspection from symmetry. The critical loading occurs with the load acting → (moments additive). The critical fastener is at the top, where all components are additive:

$r_H = H / n = 5.4 / 20 = 0.27^k/_{bolt} \leftarrow$

$r_V = V / n = 70 / 20 = 3.50^k/_{bolt} \uparrow$

$M = He_y + Ve_x = 5.4 \times 16 + 70 \times 17 = 1276^{k\text{-}in}$

$\Sigma(x_i^2 + y_i^2) = 2 \times 4 \times (6^2 + 3^2) + 2 \times 5 \times (6^2 + 3^2) = 810\ in^2$

$r_{mx} = My / \Sigma(x_i^2 + y_i^2) = 1276 \times 6 / 810 = 9.5\ ^k/_{bolt} \leftarrow$

$r_{my} = Mx / \Sigma(x_i^2 + y_i^2) = 1276 \times 6 / 810 = 9.5\ ^k/_{bolt} \uparrow$

$$r = \sqrt{(r_H + r_{mx})^2 + (r_V + r_{my})^2}$$

$$= \sqrt{(0.27 + 9.5)^2 + (3.5 + 9.5)^2} = 16.3^k / \text{bolt}$$

Steel-18

Design the smallest size W14 section for the column. Given: columns spaced 20' o.c. each direction; frame unbraced (sidesway) in each direction; A36 steel; 1/3 increase in allowable stresses to be disregarded. Use Allowable Stress Design (AISC 1989). For a trial section, assume $K_x = K_y \cong 1.5$; for W14x sections, $B_x \cong 0.18$ and $B_y \cong 0.45$; thus,

$P_{trial} = P + 0.75B_xM_x + 0.6B_yM_y$

$P_{trial} = 500 + 0.75 \times 0.18 \times 230 \times 12 + 0.6 \times 0.45 \times 96 \times 12 = 1180^k$

For $KL_y = 1.5 \times 12 = 18'$, Re: p.3-21, select a W14x233 for a trial section: $A_g = 68.5\ in^2$; $r_x = 6.63"$; $r_y = 4.10"$; $S_x = 375\ in^3$; $Sy = 145\ in^3$; $L_c = 16.8'$; $I_x = 3010\ in^4$; $I_y = 1150\ in^4$. For the strong axis, the column frames to a W24x76 ($I_x = 2100\ in^4$) each side. Thus,

$G_A = G_B = \Sigma(I_c / L_c) / \Sigma(I_g / L_g) = (2 \times 3010 / 12) / (2 \times 2100 / 20)$

$= 2.4$

From P. 5-137, AISCM, $K_x = 1.7$. The weak axis is framed to a W18x60 ($Ix = 984\ in^4$) on one side only

$G_A = G_B = \Sigma(I_c / L_c) / \Sigma(I_g / L_g)$

$= (2 \times 1150 / 12) / (984 / 20) = 3.9$

From the alignment chart, $K_y = 2.0$.

$KL_x / r_x = 1.7 \times 12 \times 12 / 6.63 = 37$

$KL_y / r_y = 2.0 \times 12 \times 12 / 4.10 = 70 \leftarrow$ governs

For A36 steel ($F_y = 36$ ksi), $C_c = [2\pi^2E / F_y]^{1/2}$ $= [2 \times \pi^2 \times 29{,}000 / 36]^{1/2} = 126$

$(Kl/r) / C_c = 70 / 126 = 0.56$; $C_a = 0.455$ (p. 5-119)

$F_a = C_aF_y = 0.455 \times 36 = 16.4$ ksi

$f_a = P / A_g = 500 / 68.5 = 7.30$ ksi

∴ $f_a / F_a = 7.3 / 16.4 = 0.45$

For bending about the x-axis:

$L_b = 12.5' < L_c = 16.8'$ ∴ $F_{bx} = 0.66F_y = 24$ ksi

$f_{bx} = M_{x,max} / S_x = 230 \times 12 / 375 = 7.36$ ksi

$f_{by} = M_{y,max} / S_x = 96 \times 12 / 145 = 7.90$ ksi

Combined axial and bending (H1) Unbraced frame

∴ $Cmy = Cmx = 0.85$; $Kl / r)_x = 37$, $F'_{ex} = 109$ ksi; $KL / r)_y = 70$, $F'_{ey} = 30.5$ ksi

$f_a / F_a > 0.15$ ∴ Check Eqns H 1-1 and H 1-2:

$f_a / F_a + C_{mx}f_{bx} / (1 - f_a / F'_{ex})F_{bx} + C_{my}f_{by} / (1 - f_a / F'_{ey})F_{by} < 1.0$

$0.45 + 0.85 \times 7.36 / (1 - 7.30 / 109)24 + 0.85 \times 7.90 / (1 - 7.30 / 30.5)27 = 1.06$

$f_a / 0.6F_y + f_{bx} / F_{bx} + f_{by} / F_{by} < 1.0$

$7.30 / 22 + 7.36 / 24 + 7.90 / 27 = 0.93 < 1.0$

The trial section is overstressed about 6% (per Eqn. H1-1). Thus, we need to try the next heavier section, a W14x257. Repeating the above calculations with the properties of the W14x257 gives an interaction value by Eqn. H1-1 of 0.95, which is acceptable ∴ **use a W14x257**.

Steel-19

a) Design the crane runway girder using allowable stress design (AISC-89). Re: A4.2 (assume a cab-operated trolley) impact factor = 25%, which applies to the crane bridge wheel load.

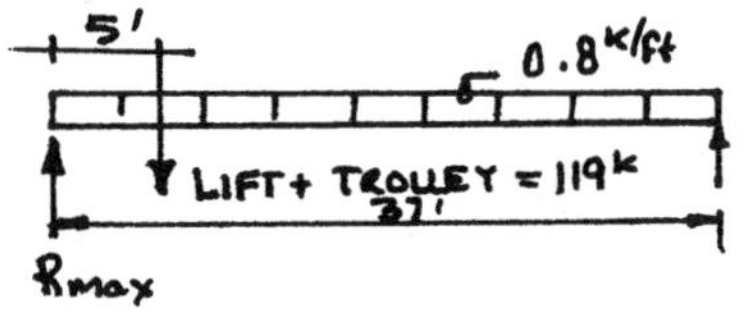

$R_{max} = [119(37 - 5) + 0.8 \times 37^2 / 2] / 37 = 118^k$

Increase R_{max} by 25% (impact) and distribute equally to two wheels spaced 12' on center.

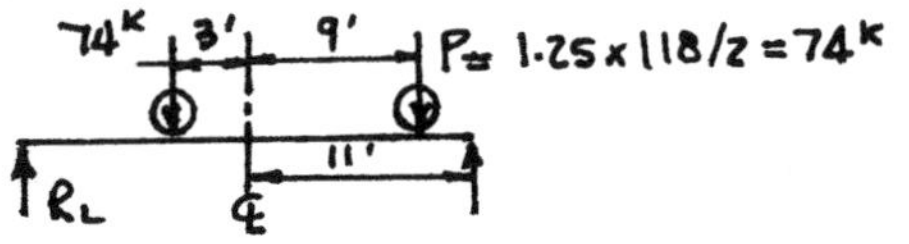

The crane girder spans 22' simply supported. The maximum bending moment occurs under the wheel closest to the centerline when positioned as shown above:

$R_L = 2 \times 74 \times (11 - 3) / 22 = 53.8^k$

$M_{max} = (53.8 \times (11 - 3) = 430^{k\text{-ft}}) + M_{dead}$

Maximum shear occurs when the wheel is adjacent to a support:

$V_{max} = (74 + 74 \times (22 - 12) / 22 = 108^k) + V_{dead}$

A lateral force equal to 20% of the trolley plus lift is applied at the top of rail with one-half to each side:

$H = 0.2 \times 119 / 2 = 11.9^k \; (5.95^k/_{wheel})$

Follow the usual procedures for design of crane runway girders (Fisher 1979). The class of service for the crane is not specified. Assume that maximum stresses occur infrequently so that no reduction in the stress range is required for fatigue. The recommended limits on live load deflections are $\Delta_y \le L / 1000$; $\Delta_x = L / 400$ (Fisher 1979). Thus, with one wheel at midspan and the other off the girder:

$\Delta_y = PL_3 / 48EI_x$

$= 74(22 \times 12)^3 / (48EI_x) \le 22 \times 12 / 1000$

$\therefore \quad I_x \ge 3705 \text{ in}^4$

$\Delta_y = 5.95(22 \times 12)^3 / (48EI_y) \le 22 \times 12 / 400$

$I_y \ge 119 \text{ in}^4$

Re: P. 1-85, AISC Manual, Try W27x84 in combination with a C15x33.9: $I_x = 4050 \text{ in}^4$; $I_y = 421 \text{ in}^4$; $S_{x1} = 237 \text{ in}^3$; $S_{x2} = 404 \text{ in}^3$; w = 118 + 28 = 146 plf;

$\therefore \quad V_{dead} = 0.146 \times 22 / 2 = 1.6^k$

$M_{dead} = 0.146 \times 22^2 / 8 = 8.8^{k\text{-ft}}$

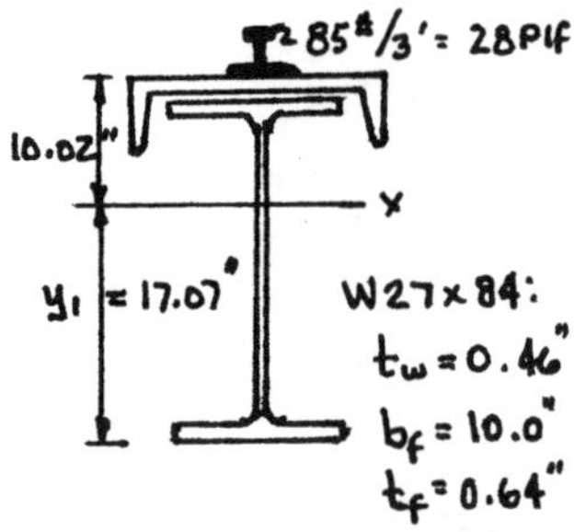

Use the properties of the C15x33.9 in combination with the top flange of the W27x84 (b_f = 10.0"; t_f = 0.64") to resist the lateral loads:

$A = 9.96 + 10.0 \times 0.64 = 16.36 \text{ in}^2$

$I_{yt} = 315 + 0.64 \times 10.0^3 / 12 = 368 \text{ in}^4$

$r_{yt} = (368 / 16.36)^{1/2} = 4.74"$

Re: F1.3, AISC, L = 22'; $L / r_T = 22 \times 12 / 4.74 = 56$

$F_{bx} = 170 \times 10^3 \times C_b / (L / r_T)^2 \le 0.6F_y$

$F_{bx} = 170{,}000 \times 1 / 56^2 = 54 \;\therefore$ use $F_{bx} = 0.6F_y$ = 22 ksi

$F_{by} = 0.6F_y = 22$ ksi

The moment about the y-axis is maximum in the same location as when the wheels produce $M_{x,max}$. Thus, by proportioning:

$M_{y,max} = [11.9 / (2 \times 74)]430 = 34.6^{k\text{-ft}}$

$f_{by} = 34.6 \times 12 \times 7.5 / 368 = 8.5$ ksi

Check combined compressive stress on the top flange (H1-3 with $f_a = 0$):

$f_{bx} = (430 + 8.8) \times 12 / 404 = 13.0$ ksi

$f_{bx} / F_{bx} + f_{by} / F_{by} = 13.0 / 22 + 8.5 / 22 = 0.98 \le 1.0$

$\therefore$ o.k.

Check tension on the bottom flange:

$f_{bx} = M_{x,max} / S_{x1} = (430 + 8.8) \times 12 / 237$

= 22.2 ksi – close enough

Check vertical shear:

$f_v = V_{max} / dt_w = (108 + 1.6) / (26.7 \times 0.46)$

$= 8.9 \text{ ksi} < 0.4F_y \;\therefore$ o.k.

$\therefore$ **Use a W27x84 plus C15x33.9.**

b) Design welds connecting the C15x33.9 to the W27x84. Use fillet welds; E70xx; Re: J2.4: ω_{min} = 1/4"; weld strength governed by the horizontal shear stress transfer under vertical load:

$Q = A_{C15x33.9}\bar{y} = 9.96(10.02 - 0.79) = 92 \text{in}^3$

$q = VQ / I = 109.6 \times 92 / 4050 = 2.3 \; ^k/_{in}$

The shear flow, $2.3^k/_{in}$, is shared by two welds. Thus, the minimum size, 1/4". weld is adequate

$f_w = 0.928 \times 4 = 3.71^k/_{in} > q$

If fatigue is not a consideration, intermittent welds can be used (note: intermittent welds are fatigue category A in the AISC Spec; continuous fillet welds are in the more favorable category B. I recommend using continuous welds for this application).

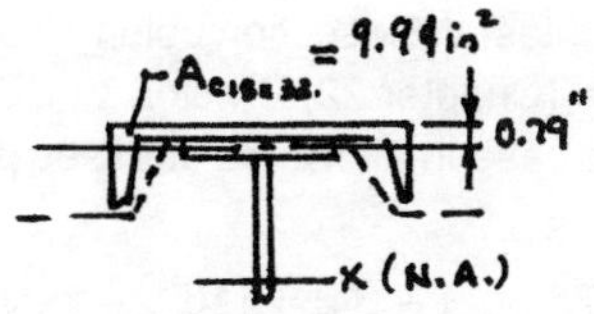

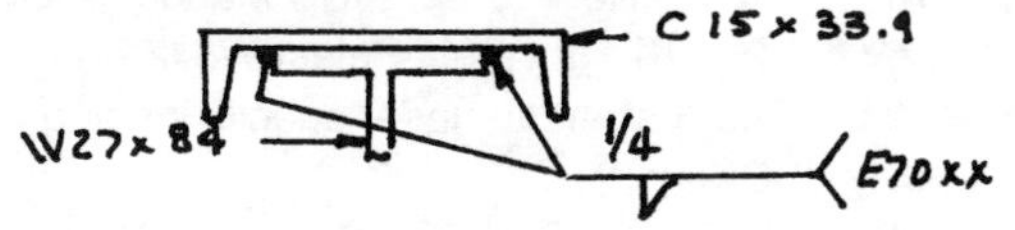

Steel-20

a) Design the lightest W30 member for the transfer girder. Use LRFD (AISC 1994). Given that lateral support is provided at supports and midspan: $L_b = 20'$; $C_b \cong 1.67$; ASTM A36 steel ($F_y = 36$ ksi); assume a dead-to-total load ratio of 0.5 ∴ U = 1.2D = 1.6L = 1.4; Approximate girder weight as 100 plf: $w_u = 0.12$ klf.

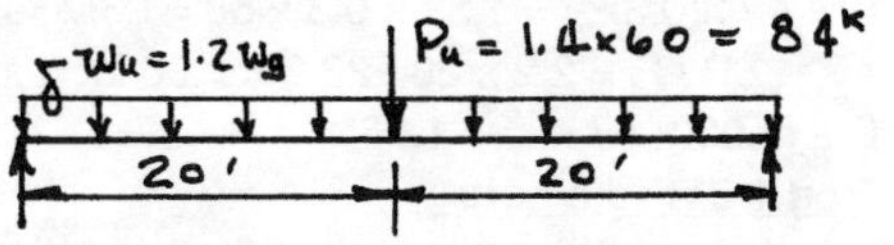

$M_u = P_uL / 4 + w_u L^2 / 8 = 84 \times 40 / 4 + 0.12 \times 40^2 / 8$
$= 864^{k\text{-}ft}$

Re: p. 4-17, AISC Manual, try a W30x108: $\phi M_p = 934^{k\text{-}ft}$; $L_p = 9.0'$; $\phi M_r = 583^{k\text{-}ft}$; $L_r = 26.3'$.

$L_p = 9.0' < L_b = 20' < L_r = 26.3'$: Use Eqn F1-3:

$\phi M_n = C_b[\phi M_p - (\phi M_p - \phi M_r)(L_b - L_p) / (L_r - L_p)] \leq \phi M_p$

$\phi M_n = 1.67[934 - (934 - 583)(20 - 9.0)$
$/ (26.3 - 9.0)]$
$= 1187^{k\text{-}ft}$

∴ $\phi M_n = \phi M_p = 934^{k\text{-}ft} > \phi M_u$ ∴ o.k.

Check live load deflection ($I_x = 4470$ in^4)

$\Delta_L = P_L L^3 / 48EI_x = 30(40 \times 12)^3 / (48E \times 4470)$
$= 0.53" < L / 360$

Check shear (d = 29.8"; $t_w = 0.55"$; $h / t_w < 260$):

$\phi V_n = \phi(0.6F_y)dt_w = 0.9 \times 0.6 \times 36 \times 29.8 \times 0.55$
$= 319^k$

$V_u = (84 / 2) + 1.2 \times 0.108 \times 20 = 44.6^k < \phi V_n$ ∴ o.k.

<u>**Use W30x108**</u>

b) Select a bolted seated connection. Re: Table 9-6, AISC Manual, Vol. II: For a 6"wide seat; 4" outstanding leg; $t_w = 0.55 \cong 9 / 16"$ a 5 / 8" thick A36 angle provides $V_u = 47.1^k$ capacity. A307 bolts in single shear; 3/4" diameter: $\phi R_v = 8.0$ $^k/_{bolt}$ ∴ $n \geq 44.6/8 = 6$ bolts. For 6" width with 6 bolts, use 9" angle leg.

Check the seat assuming critical bending occurs at the toe of a 3 / 8" fillet; 3/4" setback.; $N \geq k = 1\text{-}9/16"$

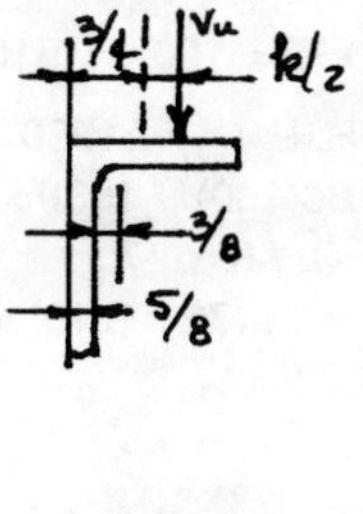

$e = 0.75 + k / 2 - 3 / 8 - t = 0.53"$

$M_u = V_ue = 44.6 \times 0.53 = 23.6^{k\text{-}in}$

$\phi M_n = \phi F_y(wt^2 / 4) = 0.9 \times 36 \times 6 \times (5 / 8)^2 / 4$
$= 19^{k\text{-}in} < M_u$ ∴ no good

Need (23.6 / 19)6 = 7.45, say 8" wide seat. Thus, for the connection: <u>**Use L9 x 4 x 5 / 8 x 0'-8" seat with 6-3/4" diameter A307 bolts.**</u>

c) Design the column: take $e_x \cong 6" = 0.5'$

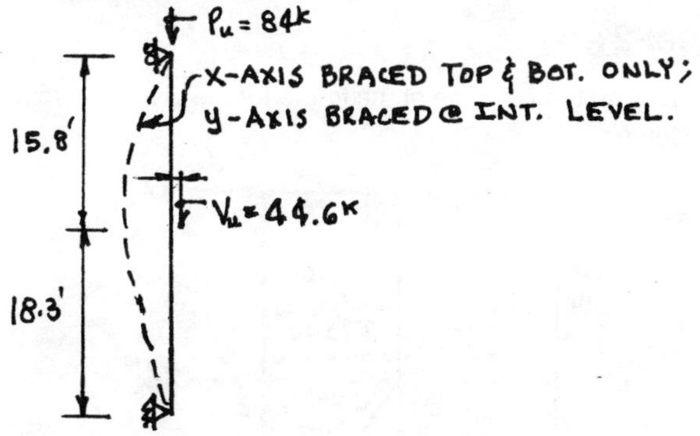

$M_x = V_ue_x = 44.6 \times 0.5 = 22.3^{k\text{-}ft}$

Distribute the moment to the top and bottom column segments:

$k_{bot} = 3EI / L = 3EI / 18.3$; $k_{top} = 3EI / 15.8$

∴ $M_{bot} = 1 / 18.8 / (1 / 18.3 + 1 / 15.8)22.3 = 10.3^{k\text{-}ft}$

$KL_x = 15.8 + 18.3 = 34.1'$; $KL_y = 18.3'$; $L_b = 18.3'$

For an efficient column section, $r_x / r_y \cong 1.75$; $m \cong 1.6$ (Table 3-2, AISC)

∴ $P_{trial} = P_u + mM_{nt} = 129 + 1.6 \times 10.3 = 146^k$

From page 3-28 AISC Manual, try W8x35 ($KL_{eq} = 34.1 / 1.75 = 19.5' > KL_y = 18.3$)': A = 10.3 in^2; $L_p = 8.5'$; $L_r = 35.1'$; $Z_x = 34.7$ in^3; $r_x = 3.51"$; $r_y = 2.03"$

Re: p. 4-20, AISC, $\phi M_p = 93.7^{k\text{-}ft}$; $\phi M_r = 60.8^{k\text{-}ft}$

$KL / r)_x = 34.1 \times 12 / 3.51 = 120 \leftarrow$ governs

$KL / r)_y = 18.3 \times 12 / 2.03 = 108$

P.6-147, $\phi F_{cr} = 14.34$ ksi ∴ $\phi P_n = 14.34 \times 10.3 = 148^k$

$P_u / \phi P_n = 129 / 148 = 0.87 > 0.2$ ∴ check Eqn. H1-1a.

$P_u / \phi P_n + 8 / 9(B_1M_{nt} / (I - P_u / P_{e1}) \leq 1.0$

From p 6-42, $C_m = 1.0$ (ends unrestrained and loaded between supports); $P_{ex} = 19.9 \times 10.3 = 205^k$ (p. 6-154).

$B_1 = C_{mx} / (1 - P_u / P_{e1}) = 1 / (1 - 129 / 205) = 2.7$

$L_p = 8.5' < L_b = 18.3' < L_r = 35.1'$: Use Eqn F1-3;

$C_b = 1.67$:

$\phi M_n = C_b[\phi M_p - (\phi M_p - \phi M_r)(L_b - L_p) / (L_r - L_p)] \leq \phi M_p$

$\phi M_n = 1.67[93.7 - (93.7 - 60.8)(18.3 - 8.5)$
$/ (35.1 - 8.5)] = 136^{k\text{-}ft}$

But ϕM_n cannot exceed $\phi M_p = 93.7^{k\text{-}ft}$. Check H1-1a:

$0.87 + 8 / 9(2.7 \times 10.3 / 93.7) = 1.13$ – no good.

Try a W8x40: A = 11.7 in^2; $L_p = 8.5'$; $L_r = 39.1'$; $Z_x = 39.8$ in^3; $r_x = 3.53"$; $r_y = 2.04"$

Re: p. 4-20, AISC, $\phi M_p = 107^{k\text{-}ft}$; $\phi M_r = 69^{k\text{-}ft}$

$KL / r)_x = 34.1 \times 12 / 3.53 = 116 \leftarrow$ governs

$KL / r)_y = 18.3 \times 12 / 2.04 = 108$

P.6-147, $\phi F_{cr} = 15.07$ ksi ∴ $\phi P_n = 15.07 \times 11.7 = 176^k$;

$P_u / \phi P_n = 129 / 176 = 0.74 > 0.2$ ∴ check Eqn. H1-1a; $P_{e1} = 21.3 \times 101.7 = 249^k$ (p. 6-154).

$B_1 = C_{mx} / (1 - P_u / P_{e1}) = 1 / (1 - 129 / 249) = 2.08$

$L_p = 8.5' < L_b = 18.3' < L_r = 35.1'$: Use Eqn F1-3;
$C_b = 1.67$:
$\phi M_n = 1.67[107 - (107 - 69)(18.3 - 8.5) / (39.1 - 8.5)] = 158^{k\text{-ft}}$

But ϕM_n cannot exceed $\phi M_p = 107^{k\text{-ft}}$.
$0.74 + 8/9(2.08 \times 10.3/107) = 0.92 \therefore$ o.k.
Thus, for the column: **Use W8x40.**

Steel-21

a) Design the columns of the frame for the given seismic forces.

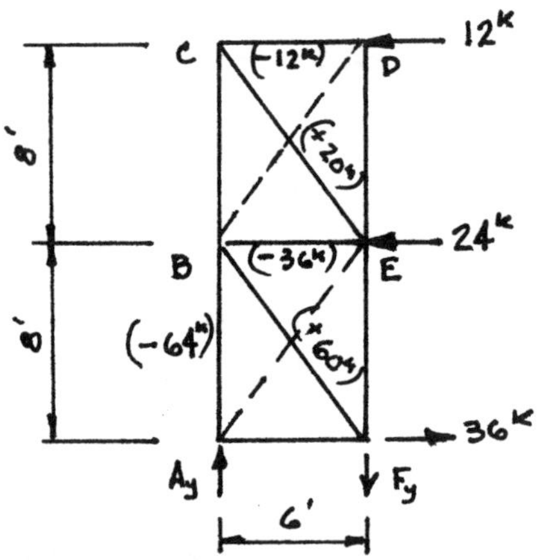

$\Sigma M_A = 0$: $12(16) + 24(8) - F_y(6) = 0$
$F_y = 64^k$
$\Sigma F_y = 0$: $A_y = -F_y = 64^k$
$P_{AB} = 64^k$ (Comp.)
$T_{BF} = (5/3)36 = 60^k$
$T_{CE} = (5/3)12 = 20^k$

Take a freebody diagram of ABC by cutting vertically:

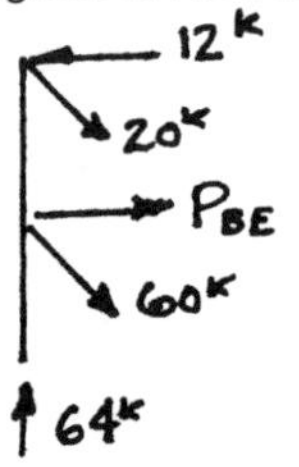

$\Sigma M_A = 0$: $(12 - (3/5)20)16 + (P_{BE} - (3/5)60)8 = 0$
$P_{BE} = -36^k = 36^k$ (comp.)

For gravity loads:

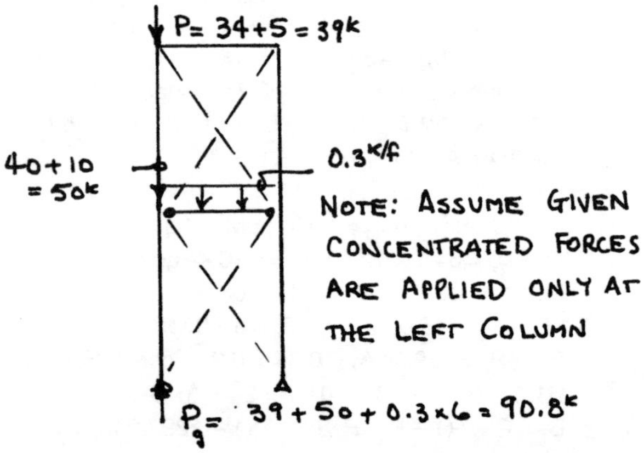

Select the lightest W8x section column. Use allowable stress design (Chapter 22, Division V, UBC-97); A36 steel; 1/3 increase in allowable stresses permitted.

$$P \geq \begin{cases} 90.8^k \\ 0.75(90.8 + 64) = 116^k \leftarrow \text{governs} \end{cases}$$

$KL_x = 16'$; $KL_y = 8'$ – Re: P.3-32, AISC Manual (AISC 1989): $r_x / r_y > 2 \therefore KL_y$ governs. Select W8x24 ($P_{all} = 124^k$). Check strength under special loading (2213.5.1, UBC-97):

$1.0P_{DL} + 0.7P_L + \Omega_o P_E = 75 + 0.7 \times 15 + 2.2 \times 64 = 226^k$

$P_{sc} = 1.7F_aA = 1.7P_{all} = 1.7 \times 124 = 211^k$ – no good – pick the next largest, a W8x28 ($P_{all} = 144^k$):

$\therefore$ $P_{sc} = 1.7 \times 144 = 245^k > 226^k$; tension strength is adequate by inspection.

$\therefore$ **Use W8x28.**

b) Select the lightest threaded rod for the upper tier, brace BD. Section 2213.8.6 applies and the tension member must resist Ω_o times the seismic force:

$F_t = 4/3(0.3F_u) = 1.33 \times 0.3 \times 58 = 23.2$ ksi
$A_g \geq \Omega_o P / F_t = 2.2 \times 20 / 23.2 = 1.90 \text{ in}^2 = \pi D^2 / 4$
$D \geq [1.90 \times 4 / \pi]^{1/2} = 1.56"$

$\therefore$ **Use Rod 1-5/8" diameter.**

Similarly, for the unthreaded rod in the lower tier:
$F_t = 4/3(0.6F_y) = 1.33 \times 0.6 \times 36 = 28.7 \text{ in}^2$
$A_g \geq 2.2 \times 60 / 28.7 = 4.60 \text{ in}^2$

Use Rod 2-1/2" diameter.

c) Check adequacy of the horizontal struts. Assume lateral support at the ends only. For CD: $P = 12^k$;
$KL_x = KL_y = 6'$. For an S3x5.7: $A_g = 1.67 \text{ in}^2$; $r_y = 0.52"$
$KL/r_y = 6 \times 12 / 0.52 = 138 > C_c = 126$

$\therefore$ $F_a = 4/3(12/23)\pi^2 E / (KL/r)^2 = 4/3(12/23)\pi^2 \times 29{,}000 / (138)^2 = 10.4$ ksi
$P_{all} = F_a A_g = 10.4 \times 1.67 = 17^k > P = 12^k$

$\therefore$ **S3x5.7 is O.K.**

For BE: $P = 36^k$; $M = 0.3 \times 6^2 / 8 = 1.35^{k\text{-ft}}$; combined axial and bending. For the S5x10: $A_g = 2.94 \text{ in}^2$; $r_x = 2.05"$;
$S_x = 4.92 \text{ in}^3$; $r_y = 0.643"$; $r_T = 0.72"$; $d/A_f = 5.10$.
$KL/r_x = 72/2.05 = 35$
$KL/r_y = 72/0.64 = 113 \leftarrow$ governs
$(KL/r)/C_c = 113/126 = 0.896$; Re: P.5-119, AISC
$C_a = 0.311 \therefore F_a = 4/3(0.311) \times 36 = 14.9$ ksi
$f_a = 36/2.94 = 12.2$ ksi
$f_a / F_a = 12.2/14.9 = 0.82 \therefore$ check H1-1
$f_{bx} = 1.35 \times 12 / 4.92 = 3.3$ ksi

$$L_c \leq \begin{cases} 76b_f / \sqrt{F_y} = 76 \times 3.0/6 = 38" \prec L_b = 72" \\ 20{,}000/(d/A_f)F_y \end{cases}$$

Check Eqns F1-6 and F1-8: $L_b = 72"$; $C_b = 1.0$:

$L_b / r_T = 72/0.72 = 100 < \sqrt{510 \times 10^3 C_b / F_y} = 119$

$\therefore$ $F_{b6} = [2/3 - F_y(L_b/r_T)^2 / 1530 \times 10^3 C_b]F_y = [0.67 - 36(100)^2 / 1530 \times 10^3 \times 1]36 = 15.6$ ksi
$F_{b8} = 12 \times 10^3 C_b / (L_b d / A_f) = 12 \times 10^3 \times 1 / (72 \times 5.10) = 32.6 \leq 0.6F_y$

$\therefore$ $F_b = 0.6F_y = 22$ ksi

$C_{mx} = 0.85$ (transverse load)

$F'_{ex} = 4 / 3(121.9) = 163$ ksi (Re: p. 5-122)

$f_a / F_a + C_{mx} f_{bx} / (F_{bx}(1 - f_a / F'_{ex}) \leq 4 / 3$

$0.82 + 0.85 \times 3.3 / (22 \times (1 - 12.2 / 163))$

$= 0.96$ – O.K.

$f_a / 0.6F_y + f_{bx} / F_{bx} = 12.2 / 22 + 3.3 / 22$

$= 0.70$ – O.K.

∴ **S5x10 is O.K.**

d) Design the base plate (Re: P. 3-106 – Note: the 1 / 3 increase in allowable stress has been included in the computation of $P_{max} = 0.75(P_g + P_e) = 116^k$)

$P_{min} = 0.75(0.3 \times 3 - 64) = -47.3^k$ (uplift)

$F_p = 1000$ psi (given); N = 12.5" (given)

$A_{PL} \geq P_{max} / F_p = 116 / 1 = 116$ in^2

$B \geq A_{PL} / N = 116 / 12.5 = 9.3$"; say 9.5"

For the W8x28: d = 8.06"; $b_f = 6.535$"

$n = (B - 0.8b_f) / 2 = (9.5 - 0.8 \times 6.535) / 2 = 2.14$"

$m = (N - 0.95d) / 2 = (12.5 - 0.95 \times 8.06) / 2$

$= 2.42$" ← critical

$f_p = P_{max} / BN = 116 / (9.5 \times 12.5) = 0.98$ ksi

Bending stress per unit width ($S = 1t^2 / 6$):

$f_b = M / S = (0.98 \times 2.42^2 / 2)/(t^2 / 6) \leq 0.75F_y$

$= 27$ ksi

$t^2 \geq 0.64$ in^2; $t \geq 0.80$"; say 7/8"

Use PL 9.5 x 7/8 x 1'-1/2"

The anchor bolts are located such that the bending moment in the base plate is negligable in the uplift condition. The anchor bolts must be sized and embedded to resist the combined axial tension of $47.3 / 4 = 11.8^k/_{bolt}$ and shear of $(36 / 4)0.75 = 6.8^k/_{bolt}$, in which the forces have been reduced to account for the 1 / 3 increase in allowable stresses.

e) Check the strut connection at D: $P = 12^k$; assume standard size holes. For 3/4" diameter A325-SC, single shear:

$r_v = 4 / 3 \times 7.5 = 10^k/_{bolt}$

$r_b = 4 / 3 \times 9.8$ – not critical

∴ $P \leq 2 \times 10 = 20^k$ – O.K.

Check the PL 1 1/2 x 1-1/2; KL = 4"; r = 0.3t

$KL / r = 4 / (0.3 \times 0.5) = 27 < C_c = 126$

Re: p. 5-119, $(KL / r) / C_c = 0.21$; $C_a = 0.56$

∴ $P \leq 4 / 3(0.56 \times 36)(1.5 \times 0.5) = 20^k > 6^k$

Check 2-5/16 E70xx welds each side:

$$q \leq \begin{cases} 2 \times 4/3 \times 5 \times 0.928 = 12.4\ ^k/_{in} \\ 4/3 \times 0.4 \times 36 \times 0.5 = 9.6^k\ /_{in} \leftarrow \text{governs} \end{cases}$$

∴ $P \leq 9.6 \times 1.5 = 14.4^k > 12^k$

∴ **The connection is adequate.**

Steel-22

Check whether the second tier columns are sized to ensure that yielding will occur first in the girders. The critical location is at the fourth floor where the W21x62 girders frame to the W14x120 column. For the W21x62: d = 20.99"; $Z_x = 144$ in^3; $S_x = 127$ in^3. For the W14x120: $A_g = 35.3$ in^2; d = 14.48"; $Z_x = 212$ in^3. Follow guidelines of FEMA 267A, Advisory No. 1 (SAC-96-03). For yielding to occur in the girders, the strength of the columns must satisfy Eqn 7.5.2.5-1-1:

$\Sigma Z_c(F_{yc} - f_a) / \Sigma M_c \geq 1.0$

Above the floor:

$f_a = 245 / 35.3 = 6.9$ ksi

Below the floor:

$f_a = [245 + 2 \times (30 + 15)] / 35.3 = 9.5$ ksi

∴ $\Sigma Z_c(F_{yc} - f_a) = [212(50 - 6.9) + 212(50 - 9.5)] / 12$

$= 1477^{k\text{-}ft}$

To determine ΣM_c, the probable moment strength of the girders at the reduced section must be computed:

$M_{pr} = Z_{RBS}\beta F_y$,

where β = factor to account for overstrength = 1.2 (Re: 7.5.2.2) and Z_{RBS} is the section modulus of the reduced section, given as $0.6Z_x$. Thus,

$M_{pr} = (0.6 \times 144) \times 1.2 \times 50 / 12 = 432^{k\text{-}ft}$

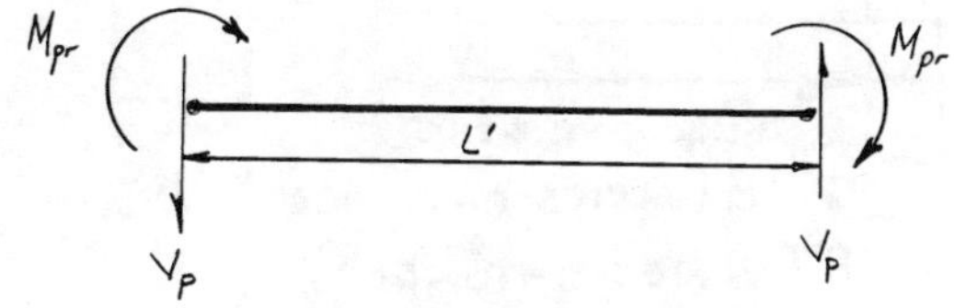

$L' = 30 - 14.48 / 12 - 2(5 + 7) / 12 = 26.8'$

$V_p = 2M_{pr} / L' = 2 \times 432 / 26.8 = 32.2^k$

Verify that stresses at the face of column are acceptable:

$f_b = (M_{pr} + V_p \times (5 + 7) / 12) / S_x$

$= (432 + 32.2 \times 1) \times 12 / 127 = 43.9$ ksi

The maximum bending stress, 43.9 ksi, satisfies the limit of $0.9F_y$ of the FEMA 267A guidelines.

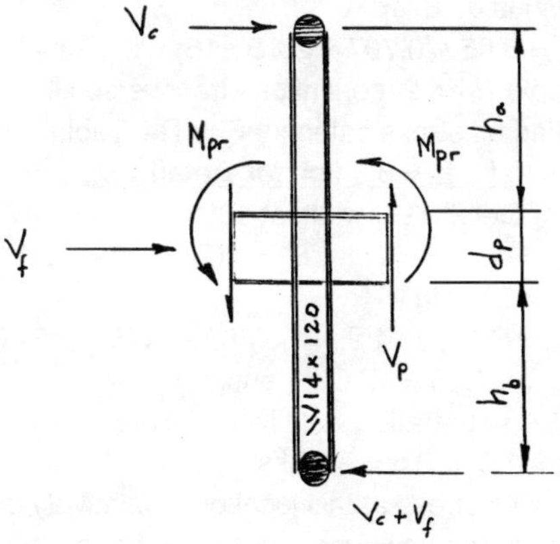

$h_a = [12.5 - (20.99 + 18) / (2 \times 12)] / 2 = 5.44'$

$d_p = 20.99 / 12 = 1.75'$

$h_b = [12.5 - (20.99 + 21) / (2 \times 12)] / 2 = 5.38'$

From statics:

$V_c = \{\Sigma[M_{pr} + V_p(L - L') / 2] - V_f(h_b + d_p / 2)\}$

$/ (h_b + d_p + h_a)$

$V_c = \{2[432 + 32.2(30 - 26.8) / 2] - 5(5.38 + 1.75 / 2)\}$

$/ (5.44 + 1.75 + 5.38)$

$V_c = 74.1^k$

$M_{ct} = V_c h_a = 74.1 \times 5.44 = 403^{k\text{-}ft}$

$M_{cb} = (V_c + V_f)h_b = (74.1 + 5) \times 5.38 = 425^{k\text{-}ft}$

∴ $\Sigma M_c = 403 + 425 = 828^{k\text{-}ft}$

$\Sigma Z_c(F_{yc} - f_a) / \Sigma M_c = 1477 / 828 = 1.78 > 1.0$

Therefore, the weak beam, strong column criterion is satisfied – yielding at the reduced beam section is reasonably assured.

Steel-23

a) Compute the stress category and fatigue loading conditions for the structure. Use Allowable Stress Design (AISC 1989). The number of loading cycles:

$N = (300 \text{ day/year}) \times 25 \text{ years} \times 300 \text{ cycles/day}$
$= 2.25 \times 10^6$ cycles

Re: Table A-K4.1, AISC Manual, Classifies as Loading Condition 4.

b) For Detail X-X:

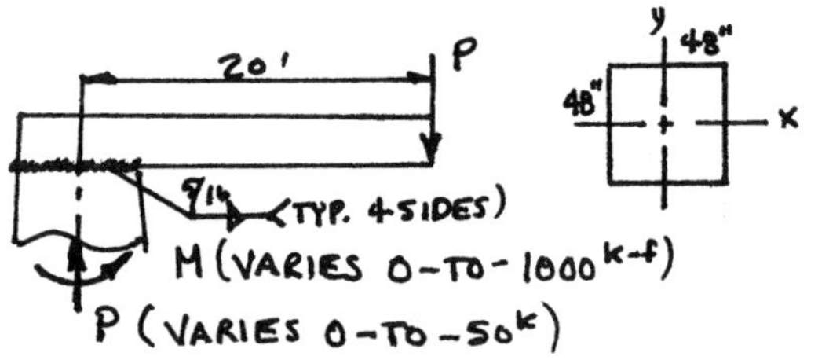

$L_w = 4 \times 48 = 192"$; $S_w = 2(48^3 / 12 + 48 \times 24^2) / 24$
$= 3072 \text{ in}^3$

$q_w = P / L_w \pm M / S_w = 50 / 192 \pm 1000 \times 12 / 3072$
$= 0.26 \pm 3.91$

For fatigue considerations, tensile stresses are critical $\therefore q_w = -3.65^k/_{in}$ (tension). For 2-5/16" fillet welds on 1/2" plate, the weld throat controls. The effective throat (SMAW) is 0.707ω:

$f_w = 3.65 / (0.707 \times 2 \times 5 / 16) = 8.3$ ksi

Re: Table A-K4.2, continuous transverse fillet welds are classified as stress category F $\therefore$ Re: Table F-K4.3:
$F_{sr4} = 8$ ksi $< f_w = 8.3$ ksi. for Detail X-X.

- For Detail Y-Y: width at 10' from top = 6' = 72"

  $L_w = 4 \times 72 = 288"$; $S_w = 2(72^3 / 12 + 72 \times 36^2) / 36$
  $= 6912 \text{ in}^3$

  $q_w = P / L_w \pm M/S_w = 50/288 - 1000 \times 12 / 6912$
  $= -1.56 = 1.56^k/_{in}$ (tension)

For a full penetration weld in 1/2" plate

$f_w = 1.56 / 0.5 = 3.12$ ksi

The tensile stress at this location is relatively low in the base material; however, the backup bar on the weld has apparently been left in place. This creates a bad stress raiser similar to the attachment in Case 24, Table A-K4.2. Thus, I would take this as stress category C (a < 2") $\therefore$ **$F_{sr4} = 10$ ksi $> f_w$ for Detail Y-Y.**

- For Detail Z-Z: $L_w = 4 \times 144 = 576"$;

  $S_w = 2(144^3 / 12 + 144 \times 72^2) / 72 = 27{,}648 \text{ in}^3$

  $q_w = P / L_w \pm M / S_w = 50 / 576 \pm 1000 \times 12 / 27{,}648$
  $= -0.35 = 0.35\ ^k/_{in}$ (tension)

  $f_w = 0.35 / 0.5 = 0.70$ ksi (tension)

It is not possible to grind the weld surface parallel to the direction of stress in this detail. Thus, this case corresponds to a groove weld with reinforcement not removed. Since non-destructive testing was specified, category C applies $\therefore$ **$F_{sr4} = 10$ ksi for Detail Z-Z.**

Steel-24

Determine the plastic moments in the beams and columns of the frame. Assume that the given loads are the design (i.e., factored) loads. Neglect axial and shear effects and consider joints as rigid.

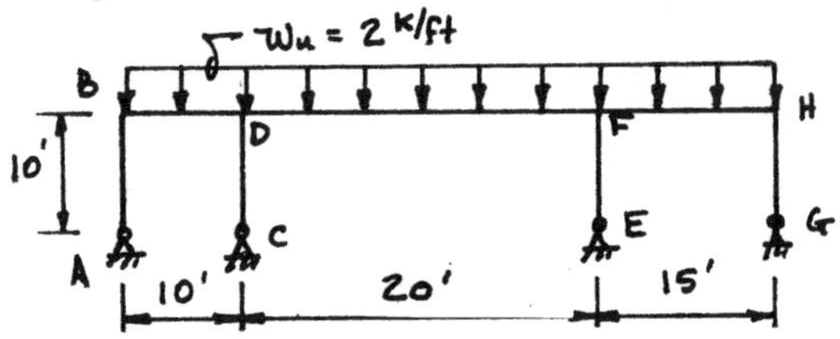

Use the statical approach of plastic analysis with the most favorable distribution of moments in the girders (i.e., $M_p = wl^2 / 16$)

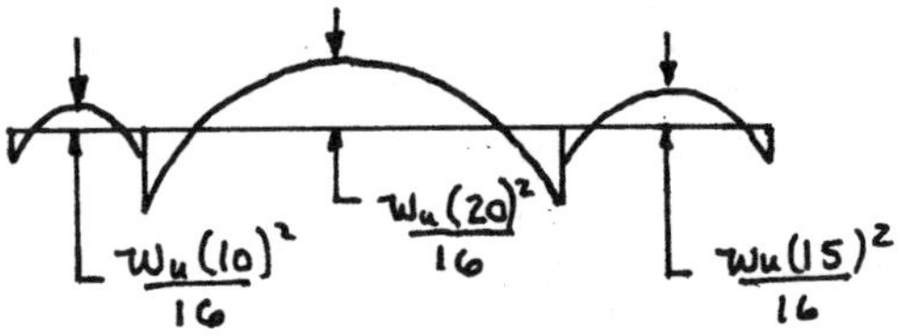

Thus,

$(M_p)_{BA} = (M_p)_{BD} = 2 \times 10^2 / 16 = 12.5^{k\text{-ft}}$
$(M_p)_{DF} = 2 \times 20^2 / 16 = 50^{k\text{-ft}}$
$(M)_{HF} = (M_p)_{HG} = 2 \times 15^2 / 16 = 28.1^{k\text{-ft}}$

For the interior columns (columns must have plastic moment capcity large enough that hinges will form first in the girders on either side):

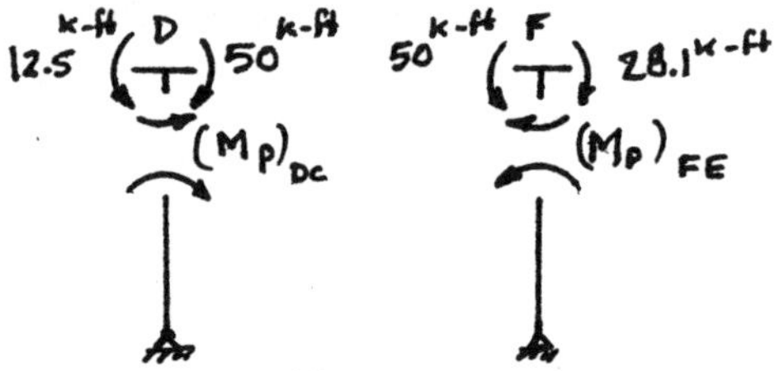

$(M_p)_{DC} \geq 50 - 12.5 = 37.5^{k\text{-ft}}$
$(M_p)_{FE} \geq 50 - 28.1 = 21.9^{k\text{-ft}}$

Steel-25

Check to see if the proposed 28" wide by 16" deep opening is feasible. Use LRFD (AISC 1994; Darwin 1988). Take the load factor, U = 1.2D + 1.6L = 1.6, conservative; $F_y = 36$ ksi.

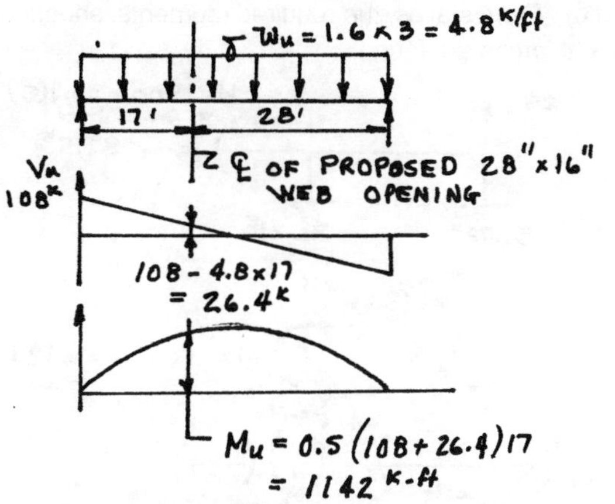

For the W33x130: b_f = 11.51"; t_f = 0.86"; d = 33.09"; t_w = 0.58"; S_t = (33.09 – 16) / 2 = 8.55; Z_x = 467 in^3.

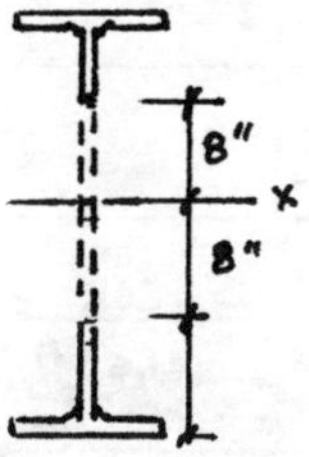

A section with unreinforced opening must satisfy the following interaction equation (Darwin 1988):

$$\left(\frac{M_u}{\phi_o M_m}\right)^3 + \left(\frac{V_u}{\phi_o V_m}\right)^3 \le 1.0$$

where ϕ_o = 0.9; $M_u = 1142^{k\text{-ft}}$; $V_u = 26.4^k$

M_m = nominal moment strength when shear is zero:

$M_m = F_y(Z_{x,W33x130} - t_w d_o^2 / 4) = 36(467 - 0.58 \times 16^2 / 4)$
$= 15{,}476^{k\text{-in}} = 1290^{k\text{-ft}}$

V_m = sum of the shear capacities of the tee-sections above and below the opening

$V_m = V_{mt} + V_{mb}$

$$V_{mt} = \frac{\sqrt{6} + \mu}{\nu + \sqrt{3}} V_{pt} \le V_{pt}$$

$V_{mt} = 0.6F_y s_t t_w = 0.6 \times 36 \times 8.55 \times 0.58 = 107^k$

$\nu = a_o / s_t$; a_o = opening length = 28"

∴ ν = 28 / 8.55 = 3.3

For a non-composite, unreinforced web opening, μ = 0

∴ $V_{pt} = (\sqrt{6} / (3.3 + \sqrt{3})) \times 107 = 52^k$

$V_m = 52 + 52 = 104^k$

Thus, the interaction equation gives:

$$\left(\frac{1142}{0.9 \times 1290}\right)^3 + \left(\frac{26.4}{0.9 \times 104}\right)^3 = 0.97 \therefore \text{o.k.}$$

The opening satisfies the shear-flexure interaction equation without reinforcement.

Check dimensional limits:

Opening depth ≤ 0.7d = 0.7 × 33.09 = 23" _

$s_t \ge 0.15d = 0.15 \times 33.09 = 5"$ _

corner radii ≥ $2t_w = 2 \times 0.58 = 1.16"$ _

Check local buckling:

$(d - 2t_f) / t_w = (33.09 - 2 \times 0.86) / 0.58 = 54 < 420 / \sqrt{F_y}$
$= 70$

$a_o / h_o = 28 / 16 = 1.75 < 3$

$V_m = 104^k < (2/3)(0.6F_y d t_w)$
$= (2/3)(0.6 \times 36 \times 33.09 \times 0.58) = 277^k$

$M_u / V_u d = 1142 / (26.4 \times 33.09 / 12) = 15.7 \le 20$

∴ Buckling of the tee section as a column does not have to be checked. Re: LRFD Spec (B5):

$s_t / t_w = 8.55 / 0.58 = 14.7 < 127 / \sqrt{F_y} = 21$

Therefore, the proposed opening, with corner radii ≥ 1.16, say 1-1/4", is adequate.

Steel-26

a) Determine the maximum moments in the frame below. Take advantage of structural symmetry and loading (axial deformation is neglected)

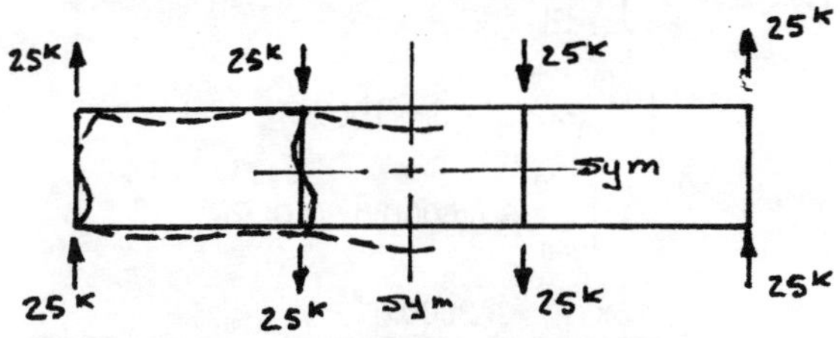

Doubly symmetrical – equivalent model:

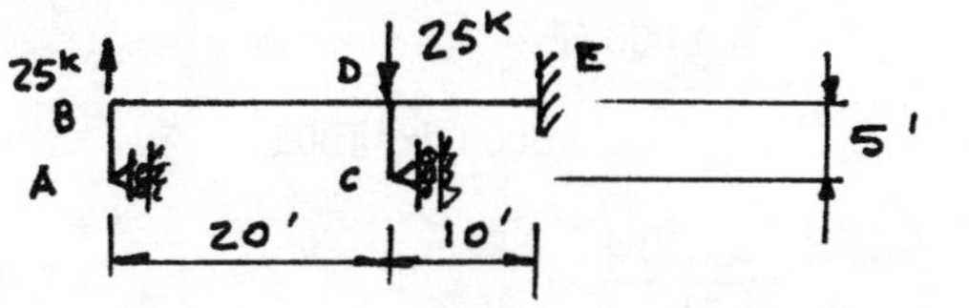

E = 29 × 10^3 ksi; I = 100 in^4; all except member CD:
I_{CD} = 50 in^4

Analyze by consistent displacement; take H_A and H_C as redundants; release:

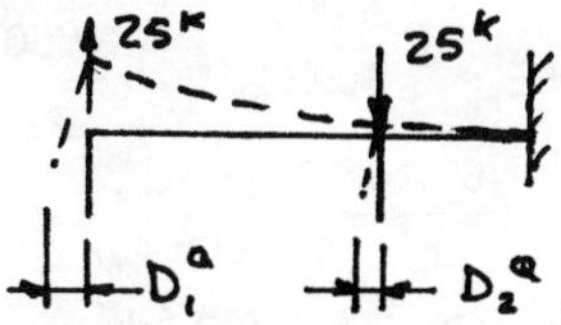

Members AB and CD:

M = 0

Member BDE:

M = 25x; 0 ≤ x ≤ 240

M = 6000; x > 240

Apply unit dummy force to the released structure horizontally at 'A'

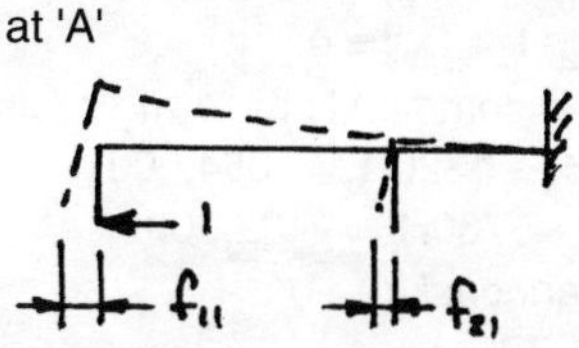

Member AB:

m = 1x; 0 ≤ x ≤ 60

Member BDE:

$m = 60;\ 0 \le x \le 360$

Member CD:

$m = 0$

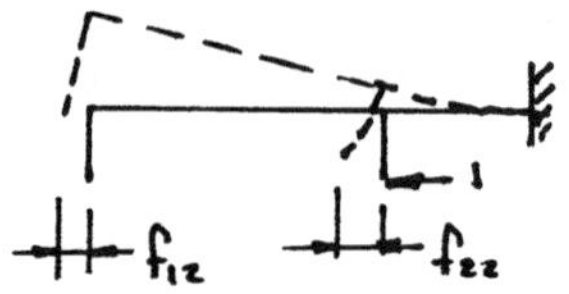

Members AB and BD:

$m = 0$

Member CD:

$m = 1x;\ 0 \le x \le 60$

Member DE:

$m = 60$

Using virtual work:

$$(1)\quad \Delta = \sum \int_L \frac{mMdx}{EI}$$

$$(1)\quad D_1^Q = \int_0^{240} (60)(25x)dx / 100E$$

$$+ \int_{240}^{360} (60)(6000)dx / 100E$$

$$= \frac{7.5x^2}{E}\Bigg|_0^{240} + \frac{3600x}{E}\Bigg|_{240}^{360}$$

$$= 864{,}000 / E \leftarrow$$

$$(1)\quad D_2^Q = \int_{240}^{360} (60)(6000)dx / 100E$$

$$= \frac{3600x}{E}\Bigg|_{240}^{360}$$

$$= 432{,}000 / E \leftarrow$$

$$(1)\quad f_{11} = \int_0^{60} \frac{x^2dx}{EI} + \int_0^{360} \frac{(60)^2dx}{EI}$$

$$= \frac{x^3}{300E}\Bigg|_0^{60} + \frac{3600x}{100E}\Bigg|_0^{360} = 13{,}680 / E \leftarrow$$

$$(1)\quad f_{12} = \int_{240}^{360} \frac{(60)^2dx}{EI} = \frac{3600x}{100E}\Bigg|_{240}^{360} = 4320 / E \leftarrow$$

$$(1)\quad f_{22} = \int_0^{60} \frac{x^2dx}{EI_2} + \int_{240}^{360} \frac{(60)^2dx}{EI_1}$$

$$= \frac{x^3}{3(50E)}\Bigg|_0^{60} + \frac{3600x}{100E}\Bigg|_{240}^{360} = 5760 / E$$

For consistent displacement:

$f_{11}H_a + f_{12}H_c + D_1^Q = 0$

$f_{21}H_a + f_{22}H_c + D_2^Q = 0$

Note that $^1/_E$ is common to all terms ∴ cancels. Thus,

$13{,}680H_a + 4320H_c = -864{,}000$

$4320H_a + 5760H_c = -432{,}000$

Solve simultaneously:

$H_a = -51.72^k = 51.72^k \rightarrow$

$H_c = -36.21^k = 36.21^k \rightarrow$

b) and c) From statics, the required moments, shears and axial forces are found:

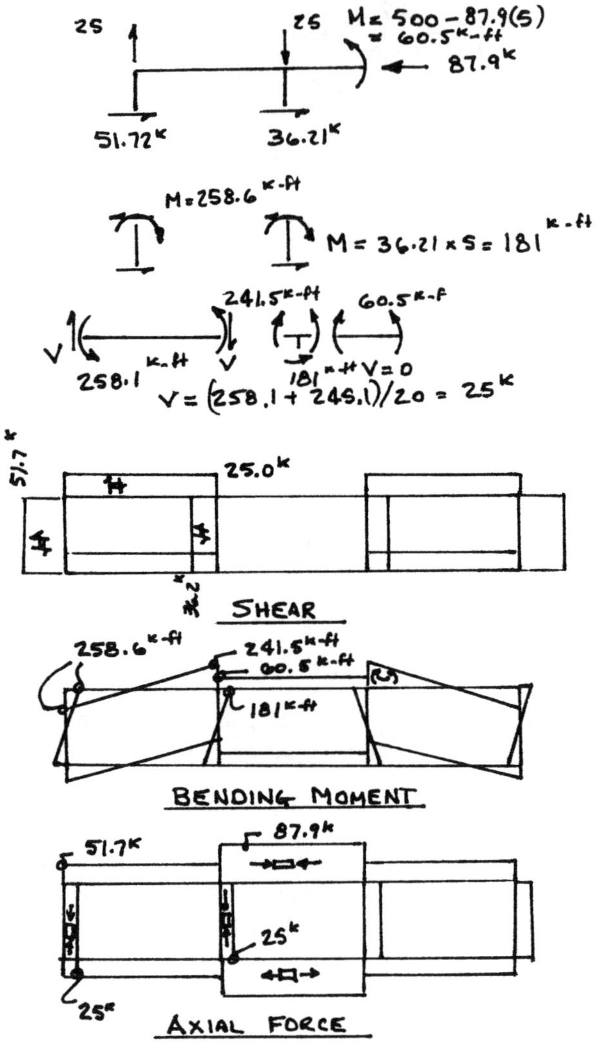

Steel-27

Since the problem statement asks for stresses at various locations, use Allowable Stress Design Specification (AISC 1989).

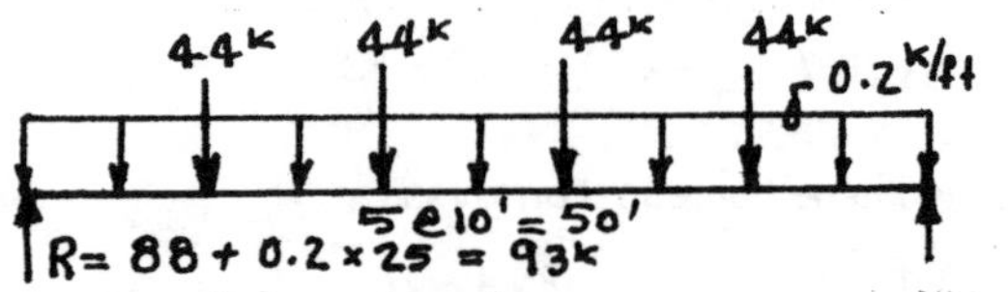

$M_{max} = 0.2 \times 50^2 / 8 + 88 \times 25 - 44(5 + 15) = 1383^{k\text{-ft}}$

$M_D = 0.55 \times 1383 = 761^{k\text{-ft}}$

$M_L = 0.45 \times 1383 = 622^{k\text{-ft}}$

Per AISC Specification, stress computations are based on the modular ratio of normal weight concrete having the same specified compressive strength (i.e., f'_c = 4000 psi).

Therefore (Re: I2.2, AISC):

$n = E_s / E_c = 8$

Properties of the steel section:

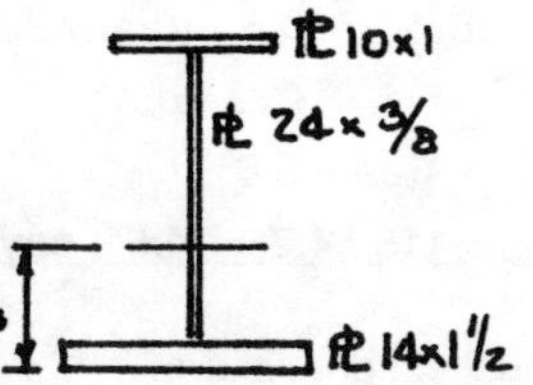

$A = 10 \times 1 + 24 \times 3/8 + 14 \times 1.5 = 40 \text{ in}^2$

$$\bar{y}_s = \frac{(21 \times 0.75 + 9 \times 13.5 + 10 \times 26)}{40} = 9.93"$$

$I_s = 14 \times 1.5^3 / 12 + 21(9.93 - 0.75)^2 + 0.375 \times 24^3 / 12$
$+ 9(13.5 - 9.93)^2 + 10 \times 1^3 / 12 + 10(26 - 9.93)^2$
$= 4904 \text{ in}^4$

$S_s = I_s / \bar{y}_s = 4904 / 9.93" = 494 \text{ in}^3$

For the transformed section:

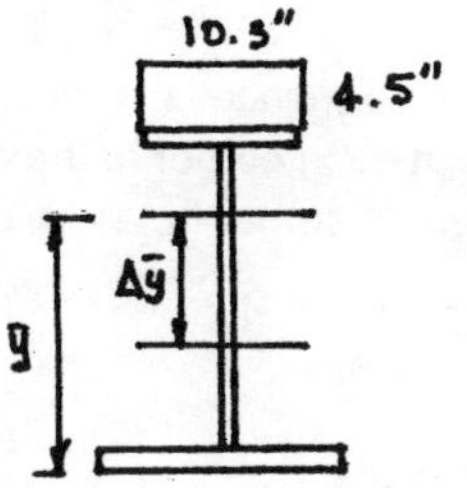

$b / n = 82 / 8 = 10.3"$

$$\Delta\bar{y}_s = \frac{(10.3 \times 4.5 \times (26.5 + 2.25 - 9.93)}{86.35} = 10.1"$$

$A_{tr} = 40 + 10.3 \times 4.5 = 86.35 \text{ in}^2$

$\bar{y} = \bar{y}_s + \Delta\bar{y}_s = 10.1 + 9.93 = 20.0"$

$I_{tr} = 4904 + 40(10.1)^2 + 4.5^3 \times 10.3 / 12$
$+ 46.35(26.5 + 2.25 - 20.0)^2 = 12{,}600 \text{ in}^4$

a) Flexural stresses in the concrete:

- If shored:
  $\underline{f_c = (M_{max} c_t / I_{tr}) / n}$
  $\underline{= 1383 \times 12 \times (26.5 + 4.5 - 20.0) / (8 \times 12{,}600)}$
  $\underline{= 1.81 \text{ ksi}}$
- If unshored:
  $\underline{f_c = (M_L c_t / I_{tr}) / n}$
  $\underline{= 622 \times 12 \times (26.5 + 4.5 - 20.0) / (8 \times 12{,}600)}$
  $\underline{= 0.81 \text{ ksi}}$

b) Flexural stress in bottom plate (note: theoretically, the stress in the bottom is different in an unshored vs. shored condition. However, the AISC Specification recognizes that the flexural strength is practically the same for both cases and, therefore, the stress is computed using M_{max} and the properties of the transformed section).

$S_{tr} = 12{,}600 / 20.0 = 630 \text{ in}^3$
$f_b = M_{max} / S_{tr} = 1383 \times 12 / 630 = 26.3 \text{ ksi}$

c) Flexural stresses in the web plate

$f_{bbw} = M_{max} y_b / I_{tr} = 1383 \times 12 \times 18.5 / 12{,}600$
$= 24.4 \text{ ksi}$

d) Check computed stresses against allowable (RE; I.2, AISC):

- Concrete: $f_c \le 1.81 \text{ ksi} < 0.45 f'_c = 0.45 \times 4000$
  $= 1.8 \text{ ksi} \therefore$ o.k.
- Bottom flange (Re: P. 1-7 and –8, $F_y = 46$ ksi):
  $F_b = 0.66 F_y = 0.66 \times 46 = 30.4 \text{ ksi} > f_b \therefore$ o.k.
- Steel web ($F_y = 36$ ksi):
  $F_b = 0.66 F_y = 0.66 \times 36 = 24 \text{ ksi} \cong f_{bbw}$

This slightly exceeds the allowable but would be acceptable in that local yielding of the web would be limited. Assuming adequate lateral bracing during all stages of construction, **<u>the flexural stresses are acceptable</u>**.

e) Shear connectors required for fully composite design (Re: I4, AISC):

$$V_h \le \begin{cases} \sum A_s F_y / 2 = (21.0 \times 46 + 19.0 \times 36) / 2 = 825^k \\ 0.85 f'_c A_c / 2 = 0.85 \times 4 \times 4.5 \times 82 / 2 = 627^k \\ \quad \leftarrow \text{controls} \end{cases}$$

For 3/4" diameter stud in light weight concrete (110 pcf); $f'_c = 4000$ psi:

$q_n = 13.3 \times 0.83 = 11.0^k/_{stud\text{-}}$

$\therefore$ **<u>$N \ge 627 / 11.0 = 57$ ea side of M_{max}; 114 total</u>**

f) Check to see if shoring is required

$f_D = M_D / S_s = 761 \times 12 / 494 = 18.5 \text{ ksi}$
$f_L = M_L / S_{tr} = 622 \times 12 / 630 = 11.9 \text{ ksi}$
$\therefore \quad f_D + f_L = 18.5 + 11.9 = 30.4^{ksi} < 0.9 F_y = 0.9 \times 46$
$= 41.4 \text{ ksi}$

$\therefore$ **<u>Shoring is not required.</u>**

<u>Steel-28</u>

a) Verify adequacy of the reinforcement across the 48" wide by 10" deep web opening. Use LRFD (AISC 1994; Darwin 1988). Assume compression flange fully braced and beam weight is negligible (given); 50% dead load; 50% live load, $U = 1.2D + 1.6L = 1.4$; $F_y = 36$ ksi.

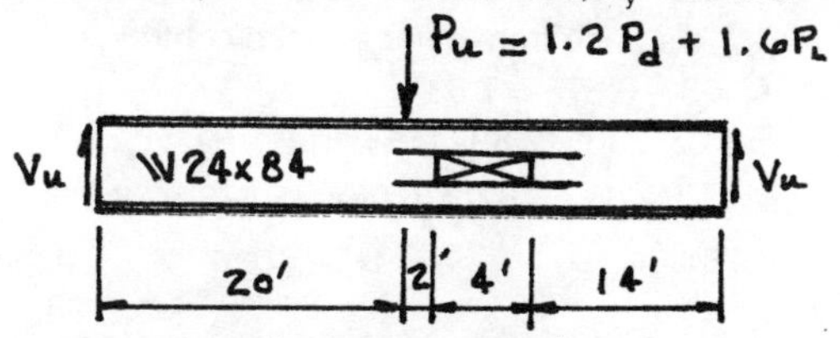

$P_u = 1.4 \times 30 = 42^k$; $V_u = P_u / 2 = 21^k$; at centerline of web opening $M_u = 21 \times 16 = 336^{k\text{-ft}}$.

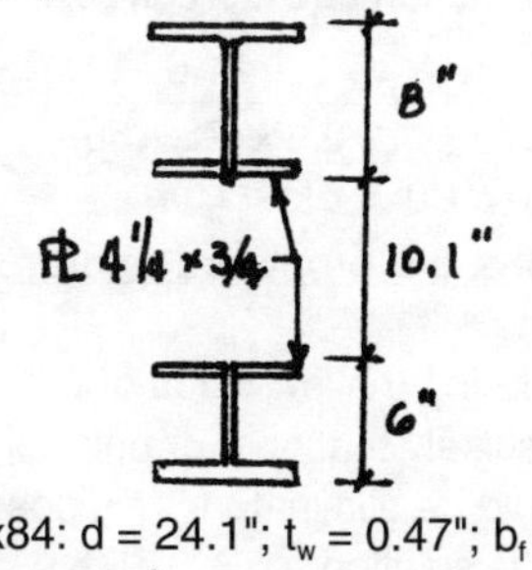

For the W24x84: $d = 24.1"$; $t_w = 0.47"$; $b_f = 9.0"$; $t_f = 0.77"$; $Z_x = 467 \text{ in}^3$.

A section with reinforced opening must satisfy the following interaction equation (Darwin 1988):

$$\left(\frac{M_u}{\phi_o M_m}\right)^3 + \left(\frac{V_u}{\phi_o V_m}\right)^3 \le 1.0$$

where $\phi_o = 0.9$; $M_u = 336^{k\text{-ft}}$; $V_u = 21.0^k$

M_m = the nominal moment strength when the shear is zero (must be less than or equal to the plastic moment capacity of the section without an opening: $M_p = F_yZ_x = 672^{k\text{-ft}}$). Locate the plastic neutral axis of the reinforced section:

$A_{ten} = A_{comp} = (24.7 - 10.1 \times 0.47 + 4 \times 4.25 \times 0.75) / 2$
$= 16.35 \text{ in}^2$

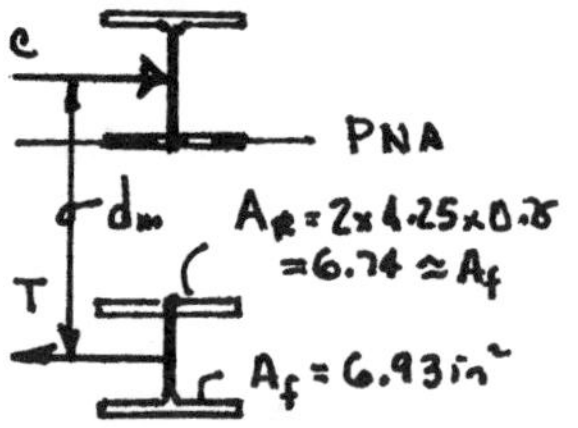

$M_m = Td_m \cong 16.35 \times 36 \times 17.1 = 10{,}100^{k\text{-in}}$
$= 839^{k\text{-ft}} > M_p$

$\therefore$ $M_m = M_p = 672^{k\text{-ft}}$

V_m = sum of the shear capacities of the tee-sections above and below the opening in the absence of bending moment

$V_m = V_{mt} + V_{mb}$

$$V_{mt} = \frac{\sqrt{6} + \mu}{\nu + \sqrt{3}} V_{pt} \le V_{pt}$$

$\nu = a_o / \bar{s}_t$; a_o = opening width = 48";

$\bar{s}_t = s_t - A_r / (2b_f)$;

$V_{pt} = F_yt_ws_t / \sqrt{3} = 36 \times 0.47 \times 8 / \sqrt{3} = 78^k$

A_r = area of reinforcement

$\nu = 48 / [8 - 2 \times 4.25 \times 0.75 / (2 \times 9.0)] = 6.3$

$\mu = 2P_rd_r / V_{pt}s_t$; $P_r = A_rF_y = 2 \times 4.25 \times 0.75 \times 36$
$= 230^k$

d_r = distance from outer edge to centroid of $A_r = 8 - 0.375 = 7.63"$

$\therefore$ $\mu = 2 \times 230 \times 7.63 / (78 \times 8) = 5.62$

$\therefore$ $V_{mt} = [(\sqrt{6} + 5.62) / (6.3 + \sqrt{3})]78 = 78.4^k \le V_{pt}$
$= 78^k$

Similarly,

$V_{mb} = 36 \times 0.47 \times 6 / \sqrt{3} = 58^k = V_{pt}$

Thus, the interaction equation gives:

$$\left(\frac{336}{0.9 \times 672}\right)^3 + \left(\frac{21.0}{0.9 \times (78 + 58)}\right)^3$$
$$= 0.18 \prec\prec 1.0 \therefore \text{o.k.}$$

Therefore, the proposed reinforced opening is adequate.

Note: Considering the low value of the interaction equation, it is likely that one, or both, of the reinforcing stiffeners could be eliminated. This possibility will not be pursued in this solution.

b) Design fillet welds to develop the full strength of the plates beyond the end of the opening.

$P_{ru} = A_rF_y / 2 = 115^k$

Welds are to 3/4" thick plate $\therefore$ ω_{min} = 1/4"; welds both sides $\therefore$ one-half of t_w is the available base material thickness:

$$\phi f_w \le \begin{cases} 0.8 \times 0.7 \times 36 \times (0.47 / 2) \\ \quad = 4.7^k / \text{in} \leftarrow \text{governs} \\ 4 \times 1.39 = 5.6^k / \text{in} \end{cases}$$

$L_w \ge P_{ru} / \phi f_w = 115 / 4.7 = 24.5"$; say 13" top and bottom

c) The required weld details are:

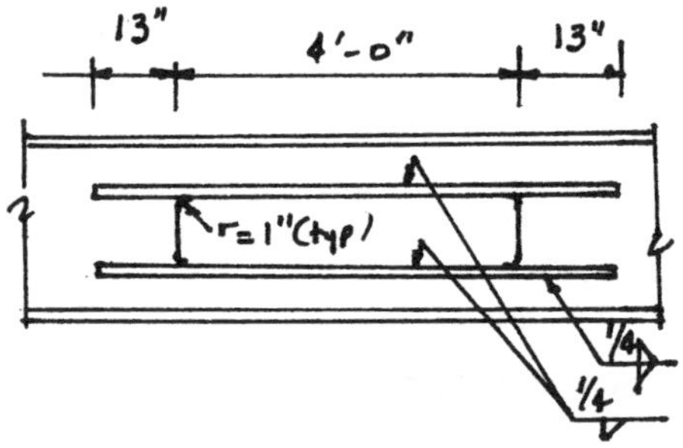

d) The construction sequence is: 1) shore; 2) drill 2" diameter holes at each corner of the opening; 3) weld stiffeners tangent to corner holes; 4) cut and remove the web portion.

Steel-29

Design a constant section pipe column to support the sign. Use LRFD (AISC 1994); ASTM A501 pipe (F_y = 36 ksi).

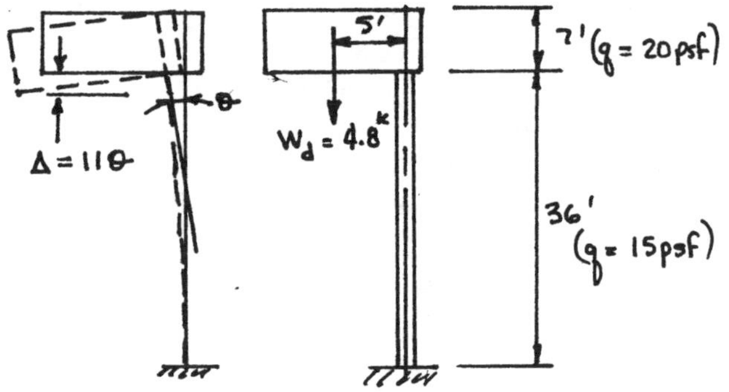

The tip of the sign must not deflect more than 1" vertically. The sign frame may be assumed rigid (given); therefore, the vertical deflection is the rotation at the top of the pole (radians) times the length of the sign (11'):

$\delta = \theta L = (MH / EI)L$
$= [(4.8 \times 5 \times 12)(36 \times 12) / (29{,}000I)](11 \times 12) \le 1"$

$\therefore$ $I \ge 566 \text{ in}^4$

Try a Pipe 16 Std ($I = 562 \text{ in}^4$ is close enough); D = 16.0"; $A_g = 18.4 \text{ in}^2$; w = 63 plf; r_x = 5.53"

Compute the lateral forces:

- Wind (pressures given) – take drag coefficient of 1.4 for a solid sign and 2/3(1.4) = 0.93 for a smooth pipe (Table 16-H, UBC-97):

  $V = 1.4 \times 0.02 \times 7 \times 12 + 0.93 \times 0.015 \times (16 / 12)$
  $\times 36 = 3.02^k$

- Seismic (Base Coefficient = 0.3 given as working load value)

  $V = 0.3 \times 4.8 + 0.3 \times 0.063(36 + 7) = 2.3^k$

Thus wind governs. Wind causes both bending and torsional moments about the base:

$M_w = 1.4 \times 0.02 \times 7 \times 12 \times (36 + 3.5) + 0.93 \times 0.015 \times (16 / 12) \times 36^2 / 2$
$= 105^{k\text{-}ft}$

$M_t = 1.4 \times 0.02 \times 7 \times 12 \times 5 = 11.8^{k\text{-}ft}$

The moment due to wind is maximum at 90° to the direction of the moment caused by the weight of the sign ($4.8^k \times 5'$). Factor the two moments per 1612.2, UBC-97 and combine them vectorially:

$\therefore \quad M_{lt} = [(1.2\ M_d)^2 + (1.3M_w)^2]^{1/2}$
$= [(1.2 \times 4.8 \times 5)^2 + (1.3 \times 105)^2]^{1/2} = 140^{k\text{-}ft}$

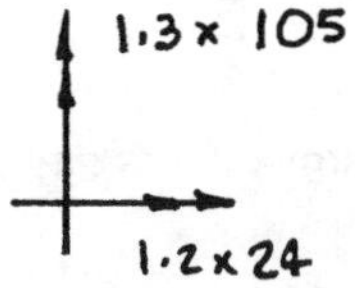

Check combined axial plus bending at the base of the pipe (Chapter H, AISC Spec.):

$kL / r = 2.1 \times 36 \times 12 / 5.53 = 164$

Re: P. 6-147: $\phi P_n = 7.93 \times 18.4 = 146^k$; $P_e = 10.64 \times 18.4 = 195^k$

$P_u = 1.2 \times (4.8 + 0.063 \times (36 + 7)) = 9^k$

Consider the total P_u to act at the top of the pipe (conservative)

$B_2 = 1 / (1 - P_u / P_e) = 1 / (1 - 9 / 195) = 1.05$

$M_u = B_2 M_{lt} = 1.05 \times 140 = 147^{k\text{-}ft}$

Check local buckling (Table B5): $D/t = 16.0 / 0.375 = 43 < \lambda_p = 2070 / F_y = 58$

$\therefore \quad \phi M_n = \phi M_p = \phi F_y Z_x$

$Z_x = (D^3 - (D - 2t)^3) / 6 = (16.0^3 - (16.0 - 2 \times 0.375)^3) / 6$
$= 91.5\ in^3$

$\phi M_n = 0.9 \times 36 \times 91.5 / 12 = 247^{k\text{-}ft}$

$P_u / \phi P_n = 9 / 146 = 0.06 < 0.2$ $\therefore$ check Eqn. H1-1b:

$P_u / 2\phi P_n + M_u / \phi M_n = 0.06 + 147 / 247 = 0.66 < 1.0$

$\therefore$ o.k.

Check shear (direct plus torsion):

$V_u = 1.3 \times 3.1^k = 4.03^k$; $M_t = 1.3 \times 11.8 = 15.3^{k\text{-}ft}$

$f_v = V_u Q / Ib + M_t r / J$

$Q = (A_g / 2)(0.58D / 2) = (18.4 / 2)(0.58 \times 16.0 / 2)$
$= 43\ in^3$

$f_v = 4.03 \times 43 / (562 \times 2 \times 0.375) + 15.3 \times 12 \times 8.0 / (2 \times 562) = 1.7$ ksi

$q_u = f_v t = 1.7 \times 0.375 = 0.64^k/_{in} << \phi(0.6F_y)t$ $\therefore$ o.k.

The design is governed by the deflection limitation

$\therefore$ **Use Pipe 16 Std.**

Steel-30

a) Design the notched portion of the beam below. Use LRFD (AISC 1994); assume a dead-to-total load ratio = 0.5 $\therefore$ U = 1.2D + 1.6L = 1.4; F_y = 36 ksi; E70xx electrodes; top flange continuously braced and beam weight neglected (given).

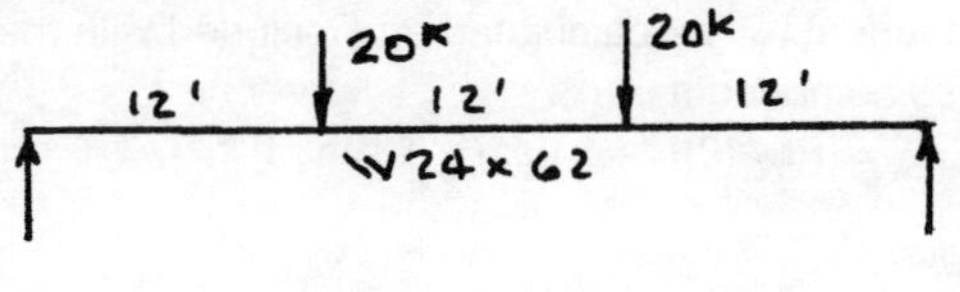

$P_u = 1.2P_d + 1.6P_L = 1.4 \times 20 = 28^k$

Check shear strength of the dapped segment:

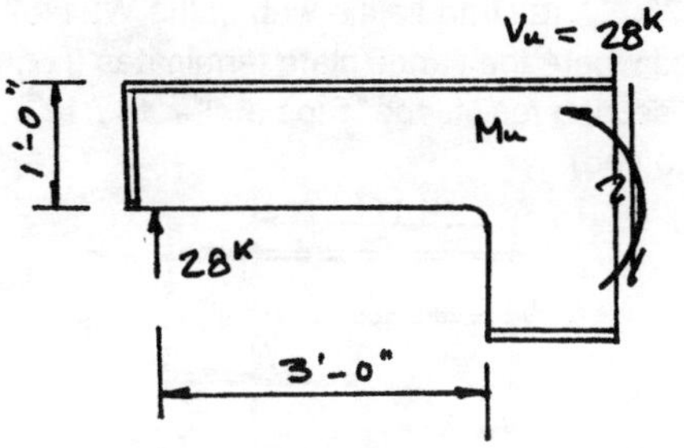

For the W24x62: $A_g = 18.2\ in^2$; d = 23.7"; t_w = 0.43"; b_f = 7.04"; t_f = 0.59"; $Z_x = 153\ in^3$; $S_x = 131\ in^3$.

$\phi V_n = \phi 0.6 F_y d t_w$ (d = 12" at the dap)

$\phi V_n = 0.9 \times 0.6 \times 36 \times 12 \times 0.43 = 100^k > V_u = 28^k$

$\therefore$ adequate in shear.

We do not need to reinforce the web for shear. The end plate, 3/16" thick, is adequate to prevent web crippling. Since the top flange is laterally braced, bolting to a cap plate on top of the TS 4 x 4 x 0.25 is sufficient to maintain stability about the longitudinal axis. Check the bending resistance of the 3' long dap:

$M_u = 28 \times 3 = 84^{k\text{-}ft}$

Re p. 1-72, AISC Manual, the Z_x for a WT12x31 is 28.4 in^3 (the stem is in tension and the top flange is braced):

$\therefore \quad \phi M_p = 0.9 \times 36 \times 28.4 / 12 = 77^{k\text{-}ft} < M_u$ $\therefore$ need to reinforce the bottom to provide flexural resistance. Try adding a PL7 x 5/8 to roughly match the size of the top flange of the W24x62.

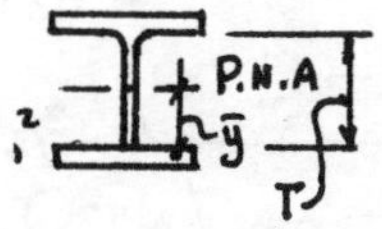

Neglecting fillets, T = 12 – 5 / 8 – 0.59 = 10.78"

$A_{tf} = b_f t_f = 7.04 \times 0.59 = 4.15\ in^2$

$A_{bf} = 7 \times 5 / 8 = 4.38\ in^2$

$A_{ten} = A_{comp}$: $4.38 + 0.43y_p = 4.15 + (10.78 - y_p)0.43$

$\therefore \quad y_p = 5.12"$

$C_{tf} = F_y A_{tf} = 36 \times 4.15 = 149.4^k$

$C_w = F_y t_w (T - y_p) = 36 \times 0.43 \times (10.68 - 5.12)$
$= 87.5^k$

$T_w = F_y t_w y_p = 36 \times 0.43 \times 5.12 = 79.3^k$

$T_{bf} = F_y A_{bf} = 36 \times 7.0 \times 5 / 8 = 157.7^k$

$\therefore$ Taking moments about the force T_{bf}

$M_p = 149.4(12 - (0.59 + 5 / 8) / 2) + 87.5(5.12 + 3.45) - 79.3 \times (5/8 + 5.12) / 2$

$M_p = 2225^{k\text{-}in} = 185^{k\text{-}ft} >> M_u$ $\therefore$ o.k.

Use PL 7 x 5/8

b) Develop the full strength of the flange on both sides of the point of maximum moment, which is conservative. Use E70xx fillet welds on both sides. Welds are to a 5/8" thick plate $\therefore \omega_{min}$ = 1/4"; welds both sides of A36 material (F_u = 58 ksi):

$$\phi f_w \le \begin{cases} 0.75 \times 0.6 \times 58 \times 0.43 = 11.2^k /_{in} \\ 2 \times 4 \times 1.39 = 11.1^k /_{in} \leftarrow \text{governs} \end{cases}$$

$L_w \ge F_y A_{PL} / \phi f_w = (36 \times 7 \times 5/8)/11.1 = 14"$

For simplicity, weld the top side of the plate only. Split the PL7 x 5/8 to straddle the web of the W24x62. Check stresses where the flange plate terminates (i.e., where the full section resists the loads at 3' + 14 / 12 = 4.17' from the end).

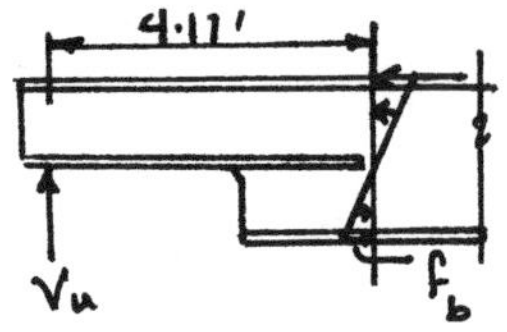

$M = 28 \times 4.17 = 117^{k\text{-}ft}$

$f_b = 117 \times 12 / 131 = 10.7$ ksi $< F_y$ ∴ still elastic

Check to ensure that the bottom flange can resist its share of the load at and beyond the point of termination:

$F_{flange} \cong f_b A_f = 10.7 \times 7.04 \times 0.59 = 44^k$

$f_{vh} = F_{flange} / (14t_w) = 44 / (14 \times 0.43) = 7.4$ ksi $< 0.6F_y$

∴ O.K.

The required details are:

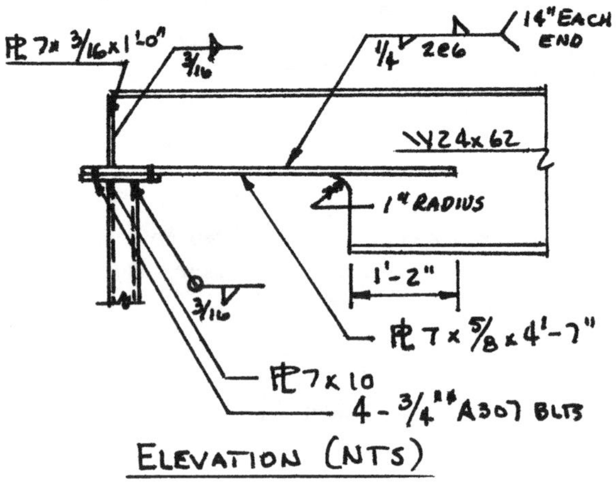

Steel-31

a) Determine the maximum flexural stress in the bottom flange of the crane girder. The problem requires only a computation of maximum stresses in the top and bottom flanges of the crane girder; not a check on the adequacy of the design. Follow AISC provisions for impact and lateral forces (AISC 1989). Re: A4.2 (assume a cab-operated trolley) impact factor = 25%, which applies to the lift plus trolley plus crane bridge weight.

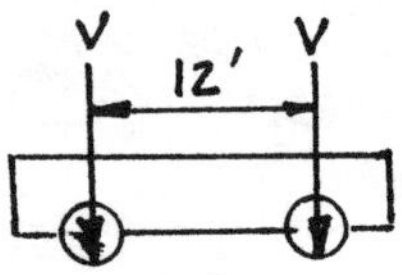

$V = 1.25 \times 53 = 66.3^k$

A lateral force equal to 20% of the trolley plus lift is applied at the top of rail with one-half to each side:

$H = 0.2 \times (80 + 20) / 2 = 10^k$ ($5.0\ ^k/_{wheel}$) → or ←

The lateral force is resisted by the top flange of the girder, which is a C15x33.9 in combination with a W27x94. Additional dead load consists of the girder plus rail:

$w = 94 + 34 + 85 / 3 = 156$ plf

The maximum moment occurs under the wheel closest to midspan with the crane positioned equidistance from midspan between wheel and resultant of the moving load:

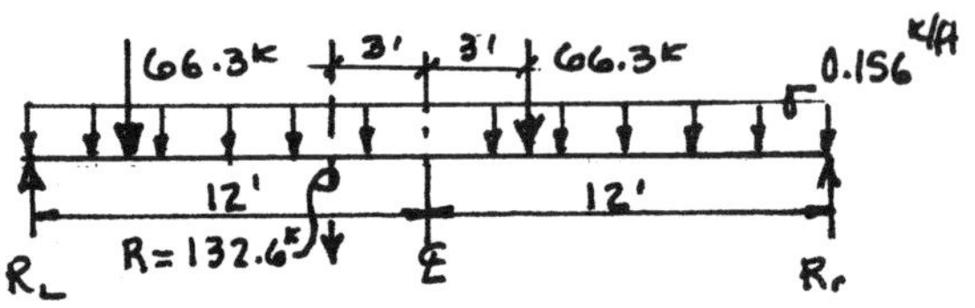

Thus, for $M_{x,max}$:

$\Sigma M_L = 0$: $132.6(9) + 0.156(24)^2 / 2 - R_r(24) = 0$

∴ $R_r = 51.6^k$

$M_{x,max} = 51.6 \times 9 - 0.156 \times 9^2 / 2 = 458^{k\text{-}ft}$

Maximum bending about the weak axis occurs with the wheels in the same position as above

$H_r = 10 \times 9 / 24 = 3.75^k$

$M_{y,max} = 3.75 \times 9 = 33.8^{k\text{-}ft}$

For the W27x94 in combination with a C15x33.9: $I_x = 4530$ in^4; $I_y = 421$ in^4; $S_{x1} = 268$ in^3 (bottom flange); $S_{x2} = 436$ in^3 (top flange).

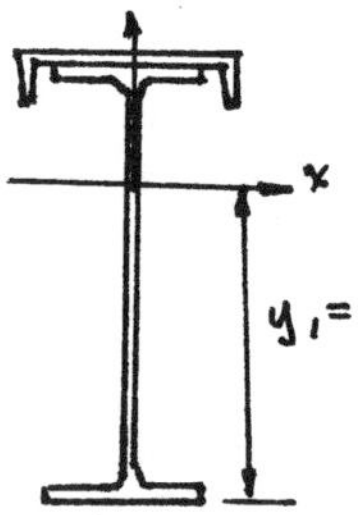

For the C15x33.9: $I_x = 315$ in^4; $I_y = 8.13$ in^4

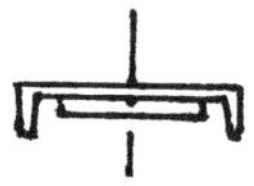

Compute properties of the C15x33.9 in combination with the top flange of the W27x94 (b_f = 9.99"; t_f = 0.745") to resist the lateral loads:

$I_{yt} = 315 + 0.745 \times 9.99^3 / 12 = 377$ in^4

$S_{yt} = 377 / 7.5 = 50.3$ in^3

Therefore, the bottom fiber flexural stress is

$f_{b,bot} = M_{x,max} / S_{x1} = 458 \times 12 / 268 = 20.5$ ksi

b) The top fiber stress is found by superposition of the stresses caused by vertical and lateral stresses:

$f_{b,t1} = M_{x,max} / S_{x2} = 458 \times 12 / 436 = 12.6$ ksi

$f_{b,t2} = M_{y,max} / S_{yt} = 33.8 \times 12 / 50.3 = 8.1$ ksi

There is also a local normal stress under the wheel that is generally negligible for girders of the proportion of this problem, but which can be large in deep girders with thin webs (AISC 1984, p. 1-11). For completeness, this stress component will be calculated and included with the other stress components:

$f_{bw} = [Pt_f / (8(I_R + I_F))][2(I_R + I_F)h / t]^{1/4}$

where I_R = moment of inertia of rail (in^4); I_F = moment of inertia of flange (in^4); and P= wheel load (kips). In this case, the centroid of the flange practically coincides with that of the C15x33.9:

$I_F = I_{yC} + b_f t_f^3 / 12 = 8.13 + 9.99 \times 0.745^3 / 12$
$= 8.5$ in^4

$h = d - 2t_f = 26.92 - 2 \times 0.745 = 25.43"$

$t = t_w = 0.49"$

$P = 66.3^k$

$f_{bw} = [66.3 \times 0.745 / (8(30.1 + 8.5))][2(30.1 + 8.5)25.43 / 0.49]^{1/4} = 1.3$ ksi

$\therefore \quad f_{b,top} = f_{b,t1} + f_{b,t2} + f_{bw} = 12.6 + 8.1 + 1.3 = 22.0$ ksi

Steel-32

a) Sketch the deflected shape of the laterally loaded frame:

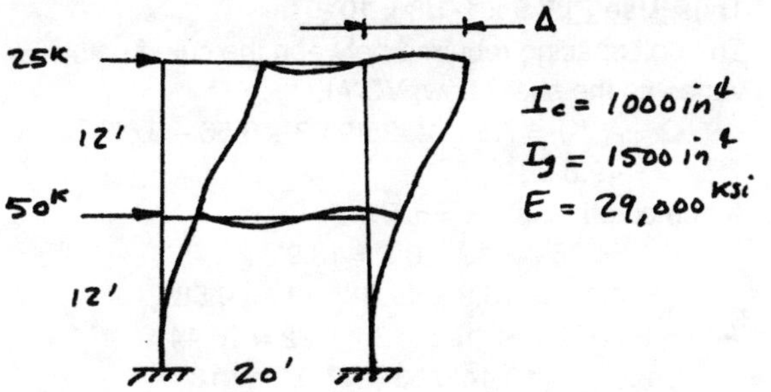

b) Moment diagrams: use the portal method, which assumes that points of inflection occur at midheight of columns and midspan of girders.

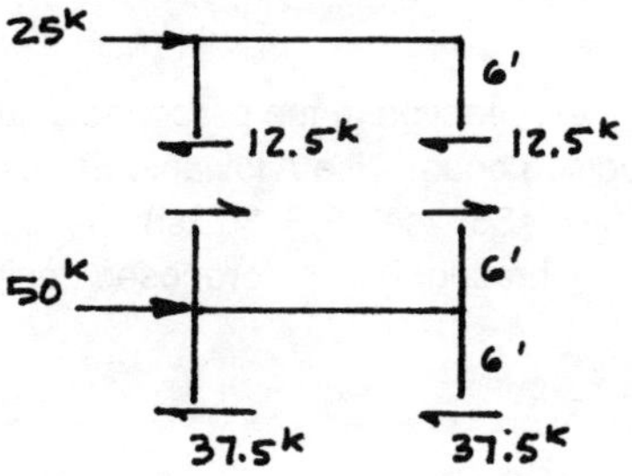

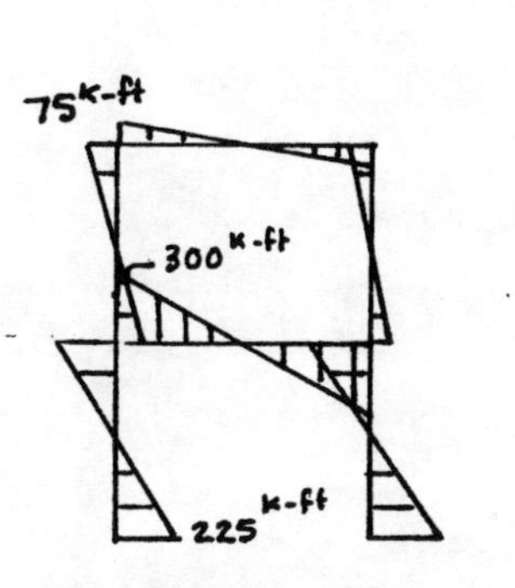

BENDING MOMENT

c) Approximate the lateral drift of the top story. For this approximation, I will assume that panel zone deformations are negligible and that only flexural deformations are significant. To analyze, apply a unit virtual force horizontally at the top and analyze for this force using the same approach as in Part B:

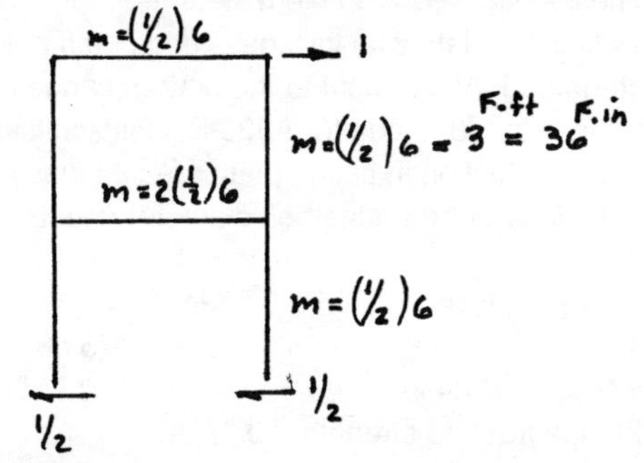

By virtual work:

$$(1) \quad \Delta = \Sigma \int m \frac{Mdx}{EI}$$

all m-diagrams and M-diagrams are triangles

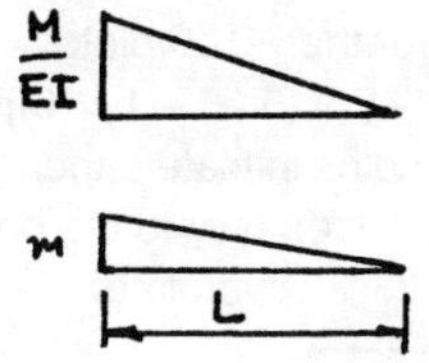

$$m \frac{Mdx}{EI} = \bar{m}A$$

where A = area of the $\frac{M}{EI}$ diagram and $\bar{m}$ is the ordinate of the m – diagram at the centroid of the $\frac{M}{EI}$ diagram

$$\therefore \quad \int m \frac{Mdx}{EI} = \frac{L}{2}\left(\frac{M}{EI}\right)\frac{2}{3}m = \frac{MmL}{3EI}$$

LOCATION	m (force-in)	M (kip-in)	number of segments, n	nmM/3E
Roof Girder	36	900	2	0.0596
Top Columns	36	900	4	0.1073
1st Floor Girders	72	3600	2	0.4777
Bottom Columns	36	2700	4	0.3217
Σ =				0.9663

$\therefore$ <u>Δ = 0.9963 ≅ 1" →</u>

Steel-33

a) Design a tension-side coverplate for the existing beam.

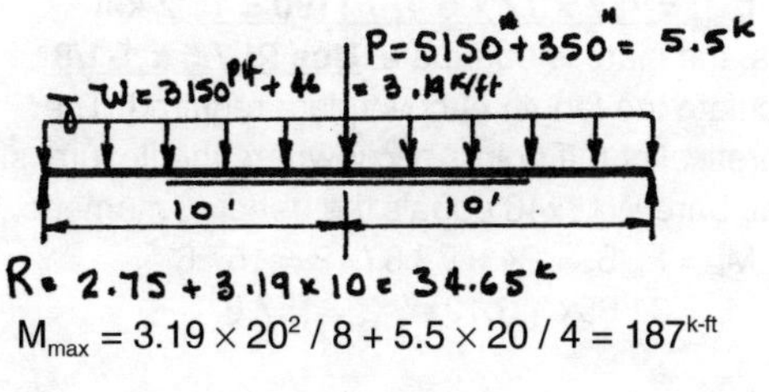

$M_{max} = 3.19 \times 20^2 / 8 + 5.5 \times 20 / 4 = 187^{k\text{-}ft}$

For the W18x46 of A36 steel: A_g = 13.5 in^2; d = 18.06"; b_f = 6.06"; t_f = 0.605"; I_x= 712 in^4; S_x = 78.8 in^3. Check the stresses without a coverplate: the section is compact; L_b = 0; Use Allowable Stress Design (AISC 1989).

$F_{bx} = 0.66F_y = 24$ ksi

$f_{bx} = M_{max} / S_x = 187 \times 12 / 78.8 = 28.5 \text{ ksi} > F_{bx}$

Bending is critical. The problem statement requires a 7-1/2" wide coverplate with thickness and cutoff points to be determined. Attachment to the bottom flange is to be by bolting with 5/8" diameter A325-SC (slip critical). Check the reduction in flange plate area if holes are drilled 1/16" larger than the bolt diameter (Re: B10-1, AISC):

$A_{fg} = t_f b_f = 6.06 \times 0.605 = 3.67$ in^2

$A_{nf} = A_{gf} - 2(5/8 + 1/16)t_f = 3.67 - 2(5/8 + 1/16) \times 0.605 = 2.84$ in^2

Flange fracture strength = $0.5F_uA_{nf}$
$= 0.5 \times 58 \times 2.84 = 82.4^k$

Flange yield strength = $0.6F_yA_{gf} = 0.6 \times 36 \times 3.67 = 79^k$

Thus, the tensile strength of the flange at its net area exceeds the yield strength ∴ do not have to deduct for holes. Similarly, for the 7-1/2" plate:

$0.5 \times 58[7.5 - 2 \times (5/8 + 1/16)]t > 0.6F_y 7.5t$

∴ No reduction for holes in plate either.

The plate can be added only to the bottom flange. As a first trial, try a 1" thick coverplate:

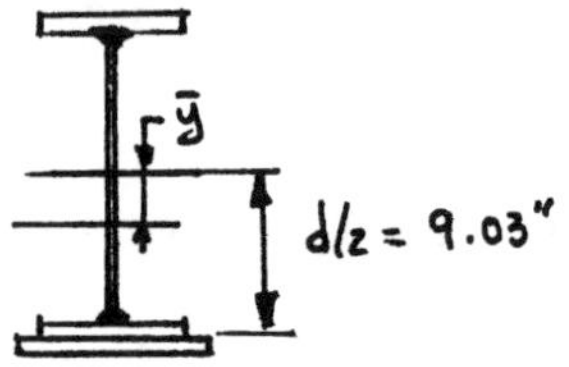

$A = 13.5 + 7.5 = 21.0$ in^2

$\bar{y} = 7.5(9.03 + 0.5)/21.0 = 3.40"$

$$I_x = \Sigma(I_{xc} + Ad^2) = 712 + 13.5(3.40)^2 + 7.5(1)^3/12 + 7.5(9.53 - 3.40)^2 = 1150 \text{ in}^4$$

$S_{xb} = I_x / c_b = 1150 / (10.03 - 3.40) = 173$ in^3

$S_{xt} = I_x / c_t = 1150 / (9.03 + 3.40) = 92.5$ in^3

$f_{b,max} = M_{max} / S_{xt} = 187 \times 12 / 92.5 = 24.3 \text{ ksi} > F_{bx}$

∴ Slightly overstressed – try again with a thicker plate, say, 1-1/8" thick:

$A = 13.5 + 8.44 = 21.94$ in^2

$S_{xt} = I_x / c_t = 1150 / (9.03 + 3.69) = 93.5$ in^3

$f_{b,max} = M_{max} / S_{xt} = 187 \times 12 / 93.5 = 24.0 \text{ ksi} = F_{bx}$

b) The flexural stresses at midspan are:

$f_{b,top} = 187 \times 12 / 93.5 = 24.0$ ksi

$f_{b,bot} = 187 \times 12 \times 6.47 / 1190 = 12.2$ ksi

Thus, the plate is adequate: **Use PL7.5 x 1-1/8**

c) Calculate the length of coverplate required. The theoretical cutoff points occur where the flexural strength of the bare W18x46 equals the bending moment:

$M_R = F_{bx}S_x = 24 \times 78.8 / 12 = 157.6^{k\text{-}ft}$

$M_x = 34.65x - 3.19x^2 / 2 = 157.6$

$x^2 - 21.72x + 98.9 = 0$

∴ x = 6.5' (only one valid root); the other point is at x = 20 – 6.5 = 13.5' (by symmetry).

Re: B-10, the coverplate must extend past its theoretical cutoff point far enough to develop its share of the flexural stress at the cutoff point:

$f_{bPL} = 157.6 \times 12 \times (9.03 + 0.56 - 3.69) / 1190 = 9.38$ ksi

$F_{PL} = f_{bPL}A_{PL} = 9.38 \times 7.5 \times 1.125 = 79^K$

For 5/8" diameter slip critical bolts in single shear: $r_v = 5.22$ $^k/_{bolt}$; bearing is not critical.

∴ $n \ge F_{PL} / r_v = 79.0 / 5.22 = 15.1$, say 16 bolts each end (2 rows, 8 ea)

Extension = $7 \times 3 + 1.5 = 22.5"$

Total plate length = $(20 - 2 \times 6.5) + 2 \times 22.5 / 12 = 10.75'$

Thus, **Use PL7.5 x 1-1/8 x 10'-9"**

The bolt spacing required between the cutoff points varies as the shear flow, VQ / I:

$Q = A_{PL}\bar{y} = 7.5 \times 1.125(9.03 + 0.56 - 3.69) = 49.8$ in^3

At the cutoff point , x = 6.5'

$V = 34.65 - 3.19 \times 6.5 = 13.9^k$

$q = VQ / I = 13.9 \times 49.8 / 1190 = 0.58$ $^k/_{in}$

Two-5/8", A325-SC resist $2 \times 5.22 = 10.44^k$

∴ Spacing ≤ 10.44/0.58 = 17.9, say 18"

Re: D2, maximum spacing ≤ (24t, 12") = 12" governs

Use 7 rows of 5/8" bolts at 3" o.c. at ends; 6 rows of two-5/8" bolts at 12" o.c. over the middle 7' of plate.

Steel-34

a) Determine the reaction at the proposed column location of the modified beam. Use Allowable Stress Design (AISC 1989); A36 steel (F_y = 36 ksi); the top flange is continuously braced. For the proposed design:

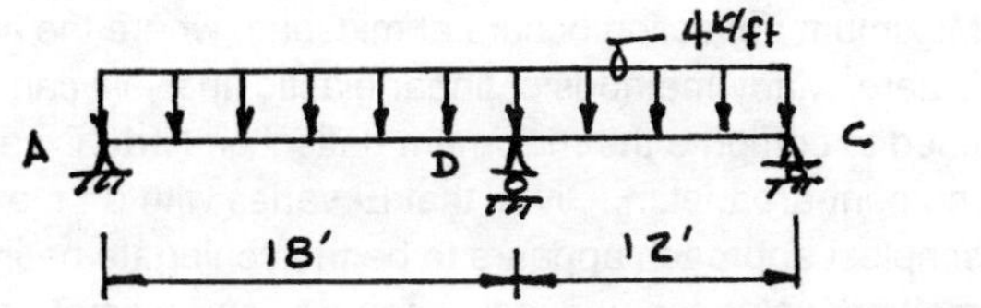

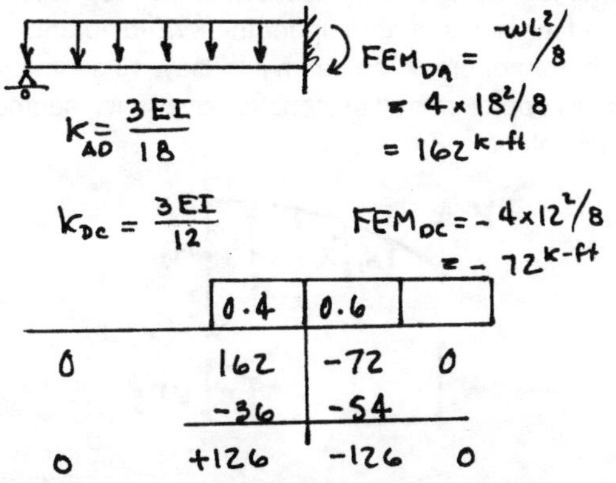

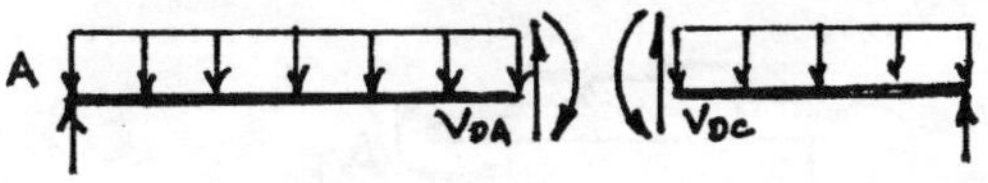

$\Sigma M_A = 0$: $V_{DA}(18) - 4 \times 18^2 / 2 - 126 = 0$

$V_{DA} = 43^k \uparrow$

$\Sigma M_C = 0$: $V_{DC} = 34.5^k \uparrow$

Thus, the reaction at D is

$\underline{R_D = V_{DA} + V_{DC} = 43 + 34.5 = 77.5^k}$

b) Verify the W18x46: d = 18.06"; t_w = 0.36"; S_x = 78.8 in^3; L_c = 6.4'; L_u = 9.4'. (Re: p.2-11).

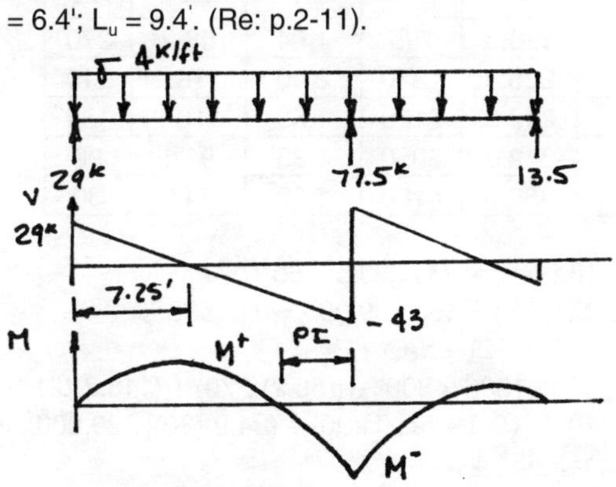

$M^+ = 0.5 \times 29 \times 7.25 = 105^{k\text{-ft}}$

$M^- = 105 - 0.5 \times 43 \times (18 - 7.25) = -126^{k\text{-ft}}$

Check flexure given that the top flange is braced continuously; the bottom flange is braced at the supports only. The point of inflection is not a braced point; however, (Re: p. 5-47), for a cantilever braced against twist only at the support, the AISC Spec. (p. 5-47) says that the unbraced length may conservatively be taken as the actual length with a C_b = 1.0. The segments of the beam from support to inflection points are no more severely stressed than would be a cantilever with the same load and length ∴ take the unsupported length as $L_b = 18 - 2 \times 7.25 = 3.5'$ with C_b = 1.0. Since the section is compact with L_c = 6.4' > L_b, the resisting moment is

$M_R = 0.66F_yS_x = 24 \times 78.8 / 12 = 158^{k\text{-ft}} > M^-$ ∴ O.K.

Check shear:

$f_v = V_{max} / dt_w = 43 / (18.06 \times 0.36) = 6.6$ ksi $< F_v = 0.4F_y$ ∴ O.K.

∴ **The existing W18x46 is O.K. for the proposed modification.**

c) Design a new column at D: P = 77.5^k with $KL_x = KL_y$ = 10'. Re: page 3-32, ASIC Manual: **Select a W6x20** (P_{allow} = 90^k).

d) The column-to-girder connection is to be made by field welding. The top flange is braced; in order for the W18x46 to be laterally braced on the bottom flange at the support, provide fitted stiffeners to the lower flange and web directly over the column. For simplicity, use two sets of stiffeners directly in line with the flanges of the W6x20 and furnishing roughly the same area.

$A_{st} = 6.02 \times 0.365 / 2 = 1.1$ in^2; say PL3 x 3/8. Local web crippling will not be a problem. Stop stiffeners short of the top flange (L < T = 15-1/2"; say 15"); use minimum size fillet welds: ω_{min} = 1/4"; use a plate to cap the W6x20 that is wider than the flange of the W18x46 (b_f = 6.06"). Thus, the proposed connection is:

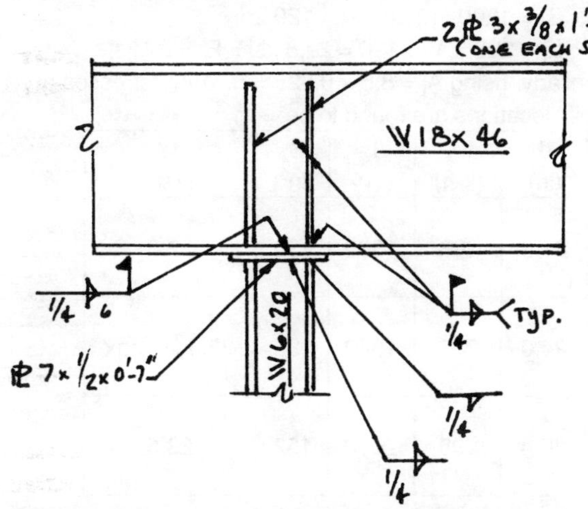

CONNECTION AT D – ELEVATION
(N.T.S.)

Steel-35

a) Compute the minimum flange width at the critical section.

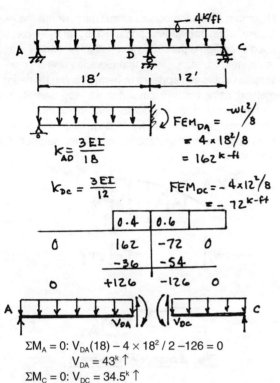

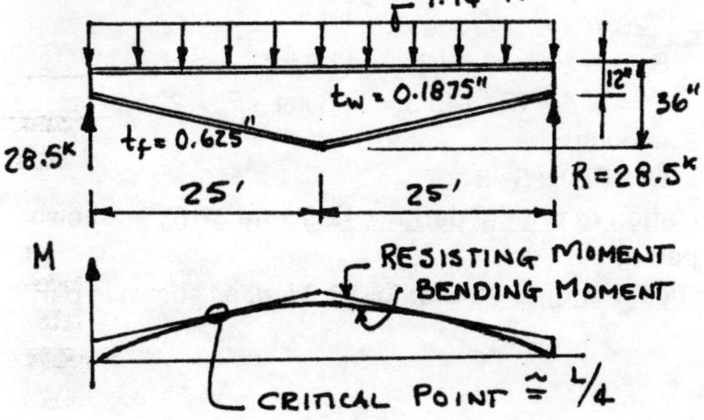

The depth varies linearly from d = 12" at ends to d = 36" at midspan; $t_f = 5/8"$; $t_w = 3/16"$; $F_b = 22$ ksi. As an approximation:

$$I_x \cong 2A_f(d/2)^2 + t_w d^3/12$$
$$S_x = I_x/(d/2) \cong A_f d + t_w d^2/6$$
$$M_x = 28.5x - 1.14x^2/2$$

To get a trial value of A_f assume that the critical location for bending stress is at L / 4 = 12.5' from supports where d = 24":

$$M_{12.5} = 28.5 \times 12.5 - 1.14 \times 12.5^2/2 = 267^{\text{k-ft}}$$
$$M_R = F_b S_x = M_{12.5}$$

Given that $F_b = 22$ ksi,:

$$22(A_f(24) + 0.1875(24)^2/6) = 267 \times 12$$
$$\therefore \quad A_f = 5.31 \text{ in}^2$$

For a 5/8" thick plate (given): $b_f = 5.31/0.625 = 8.5"$

Actual $I_x = 2A_f[(d - t_f)/2]^2 + t_w(d - 2t_f)^3/12$. Check the actual bending stress in the vicinity of L / 4; say at x = 10'; 12.5'; 15'; etc.

At x = 10': d = 21.6"; M = $228^{\text{k-ft}}$

$$I_x = 2 \times 5.31(20.98/2)^2 + 0.1875(20.35)^3/12$$
$$= 1300 \text{ in}^4$$
$$S_x = 1300/(21.6/2) = 120 \text{ in}^3$$
$$f_{bx} = 228 \times 12/120 = 22.8 \text{ ksi} > F_b = 22 \text{ ksi}$$

Similarly, using $A_f = 8.5 \times 0.625 = 5.31$ in^2, stresses at other locations are found to be:

x (ft)	d (in)	M_x (k-ft)	I_x (in^4)	S_x (in^3)	f_b (ksi)
10	21.6	228	1300	120	22.8
12.5	24.0	267	1640	136	23.6 ← max
15.0	26.4	299	2010	152	23.5

This confirms that the critical point for bending stress is near the quarter point (for all practical purposes). The actual bending stress exceeds that allowed by roughly 7%.; therefore, increase the flange width proportionally:

$$b_f = (23.6/22) \times 8.5 \cong 9" \text{ (to nearest 1/4")}$$

Check: $A_f = 9 \times 0.625 = 5.63$ in^2

$$I_x = 2 \times 5.63[(24.0 - 0.625)/2]^2 + 0.1875(22.75)^3/12$$
$$= 1720 \text{ in}^4$$
$$S_x = 1720/(24.0/2) = 143.3 \text{ in}^3$$

$f_{bx} = 267 \times 12/143.3 = 22.3$ ksi $\cong F_b = 22$ ksi – close enough

Use PL9 x 5 / 8

b) **Location of critical depth ≅ 12.5' and 37.5',** as shown in part a.

c) **Critical bending stress: f_b= 22.3 ksi,** as shown in part a.

d) Maximum deflection occurs at midspan, where the slope is zero. Many methods of linear elastic analysis can be used to compute the maximum deflection (virtual work, moment-area, etc.). Given that EI varies with d^3, the simplest approach appears to be the conjugate beam method, using a numerical scheme to approximate areas and moments of area under the M/EI-diagrams. Therefore, break the beam into, say, 10 segments with M/EI approximated as constant over each segment (take advantage of symmetry):

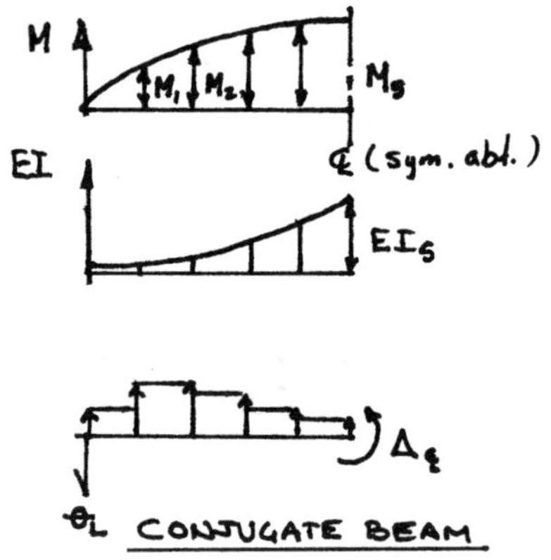

Location (ft)	M (k-ft)	I (in^4)	M/I (k/in^3)	Δ_i (k/in^2)	a_i (in)
0	0	383	0	---	---
1	128.3	795	1.94	58.2	270
2	228.0	1370	2.00	118.2	210
3	299.0	2120	1.69	110.7	150
4	342.0	3050	1.35	91.1	90
5	356.0	4180	1.02	84.9	30

$$A_i = [(M/I)_{i-1} + (M/I)_i]0.5 \times 60"$$
$$\theta_L = \Sigma(\Delta_i/E) = 463/29{,}000 = 0.01596^{\text{rad}}$$
$$\Delta_{max} = \theta_L(L/2) - \Sigma A_i a_i/E$$
$$\Delta_{max} = 0.01596 \times 300" - [(58.2)(270) + (118.2)(210) + (110.7)(150) + (91.1)(90) + (84.9)(30)]/29{,}000$$

Δ_{max} = 2.45" ↓

Steel-36

a) Calculate the maximum shear stress in the member. Assume that the end connections restrain rotation and translation, but not warping deformation. Replace the given loading by an equivalent system with the transverse loads acting through the shear center (centroid in this case) of the structural tube.

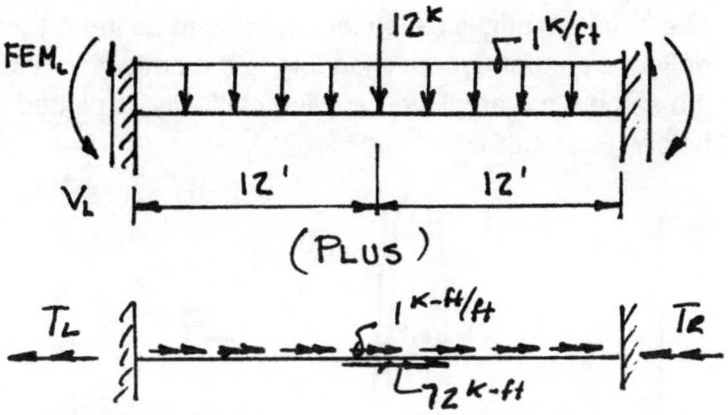

$FEM_L = wL^2 / 12 + PL / 8 = 1 \times 24^2 / 12 + 12 \times 24 / 8$
$= 84^{k\text{-}ft}$

$V_L = wL / 2 + P / 2 = 1 \times 24 / 2 + 12 / 2 = 18^k$

$T_L = T_R = m_tL / 2 + M_t / 2 = 1 \times 24 / 2 + 72 / 2 = 48^{k\text{-}ft}$

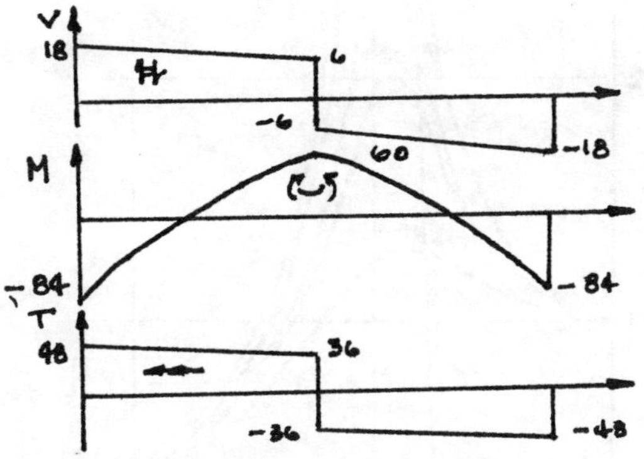

The shear stress due to transverse forces (maximum stress occurs at the middepth of vertical walls) is:

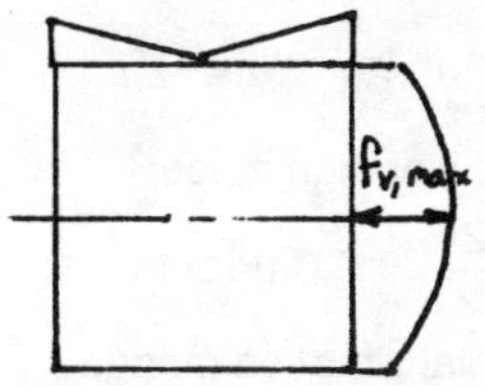

For the TS10x10x0.375: $A_g = 14.1$ in^2; $t = 0.375$";
$I_x = 214$ in^4; $S_x = 42.9$ in^3

$f_{v,max} = V_{max}Q_{max} / I_xb$

$Q_{max} = (14.1 / 4)[(5 - 0.375 / 2) + 2.5] = 25.8$ in^3

$f_{v,max} = (18 \times 25.8) / (214 \times 2 \times 0.375) = 2.9$ ksi

The maximum normal stress due to transverse loads:

$f_{b,max} = M_{max} / S_x = 84 \times 12 / 42.9 = 23.5$ ksi

The maximum shear stress due to torque occurs at midwall (Re: Roark, 1975, Table IX):

$S_{max} = T / [2t(a - t)(b - t)]$

$= 48 \times 12 / [2 \times 0.375(10 - 0.375)^2] = 8.3$ ksi

The maximum shear stress is obtained by superposition

$\tau_{max} = f_{v,max} + S_{max} = 2.9 + 8.3 = 11.2$ ksi

b) The intent in part b is to calculate the principal stress, which occurs at both top and bottom mid points. Using Mohr's circle:

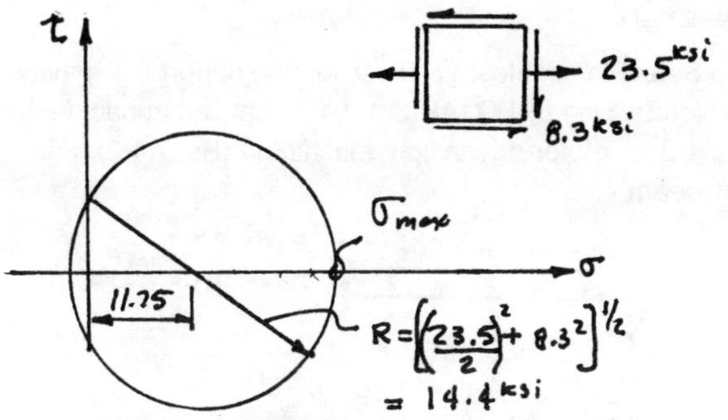

$\sigma_{max} = 11.75 + 14.4 = 26.2$ ksi

To design welds, we need the resultant shear flow ($^k/_{in}$) rather than the principal stress:

$q_v = \tau_{max}t = 11.2 \times 0.375 = 3.1^k/_{in}$

$q_b = \sigma_{max}t = 26.2 \times 0.375 = 8.9^k/_{in}$

$q_r = \sqrt{q_v^2 + q_b^2} = \sqrt{3.1^2 + 8.9^2} = 9.4^k /_{in}$

For E70xx fillet weld, $q = 0.928^k/_{in}$ per 16th

$\therefore \quad \omega \geq q_r / 0.928 = 9.4 / 0.928 = 10.1$, say, 10, which implies a 5/8" weld. However, for a 0.375" wall thickness of material with 46 ksi yield stress, the base material would be overstressed. **Therefore, use a full penetration groove weld.**

c) Discuss the relative merits of the given cross sections in resisting torsion: The closed sections are the most efficient. Of the given sections, the pipe is best; square structural tube is next best; rectangular tube is next. The closed sections offer relatively large resistance to torsional deformations. For the pipe, an applied torque induces only shear stress; the tubes experience some warping stresses, generally, but these are usually not significant. Open cross sections (e.g., the W, C, L-sections) are relatively flexible and weak in torsional resistance. Warping stresses may become significant for these sections and are additive to the normal stresses caused by the bending moment. Of the open sections listed in part d, the W-shape offers the advantage that its shear center coincides with its geometric centroid, which means that most applications involve loading that does not induce torsion. The channel, on the other hand, has its shear center outside the section, which makes it difficult to apply transverse loading without causing torsion (i.e., twisting) of the section. The shear center of the angle is at the centerline junction of the two legs, and it is also difficult to load without some twisting.

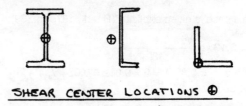

SHEAR CENTER LOCATIONS ⊕

Steel-37

a) Select the lightest W10 section to support the gravity loads. Use LRFD (AISC 1994). Apply an impact factor of 1.2 for motor-driven light machinery (Re: A4.2). For beam 1:

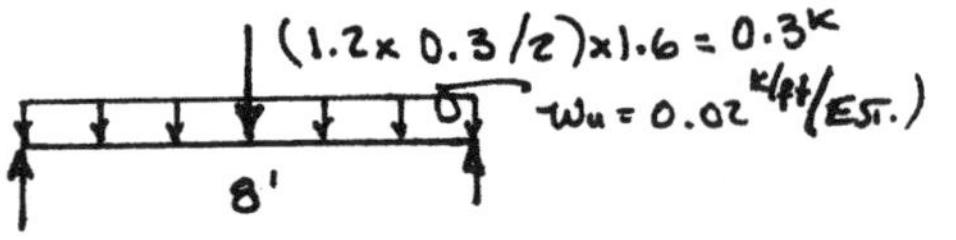

$M_u = P_uL / 4 + w_uL^2 / 8 = 0.75^{k\text{-}ft}$

$L_b = 8'$; $C_b = 1.0$

Per instructions, select the lightest W10x of A36 steel for this member ∴ try a W10x12 (Re: p. 4-21, AISC): $\phi M_p = 34^{k\text{-}ft}$; $L_p = 3.3'$; $L_r = 9.5'$; $\phi M_r = 21^{k\text{-}ft}$. Since $L_b = 8' < L_r$, $\phi M_n > 21^{k\text{-}ft}$; W10x12 is adequate in flexure. Use W10x12 for all beams (lightly stressed in flexure).

Check the vibration potential with the machine operating at 1725 rpm:

$\bar{\omega} = 1725 \text{ rpm} = 29 \text{ cps}$

Check the natural frequency of the beam-girder-motor system. Idealize it as a single degree of freedom system. The weight of the beams is significant relative to the motor weight. Approximate the equivalent mass for vertical vibration as that of the machine plus two beams-1, plus the central half of beam-2:

$\therefore \quad M = W / g = (300^{\#} + 2 \times 12 \times 8^{\#} + 2 \times 12 \times 4^{\#}) / g = 0.59 \text{ k} / g$

For the equivalent stiffness, treat the loads as concentrated forces at midspan:

$$\Delta = 1" = KL_b^3 / 48EI_b + (K/2)L_g^3 / 48EI_g = [K(8 \times 12)^3 / 48E][1 / (2 \times 53.8) + 0.5 / 53.8]$$

$\therefore \quad K = 84.6 \ ^k/_{in}$

For damping on the order of that in a steel frame (say, 5% of critical), the natural frequency is practically unaffected by damping

$$f_1 = \omega = \frac{1}{2\pi}\sqrt{K/M} = \frac{1}{2\pi}\sqrt{\frac{84.6 \times 32.2 \times 12}{0.59}} = 37.4 \text{ cps}$$

Since there is no lateral bracing in the system, vibration can occur in the principal horizontal directions. In either direction, the stiffness is furnished by 2 W10x12 bending about their weak axis ($I_y = 2.18 \text{ in}^4$).

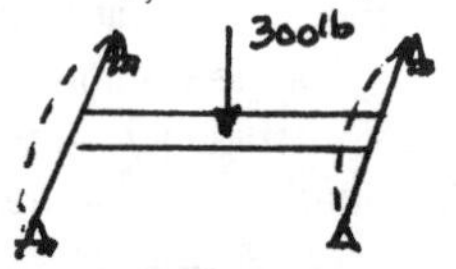

This case is given explicitly (Roark, 1975, p.369):

$$f = 3.13 / \sqrt{\frac{(W + 0.486wL)L^3}{48EI}}$$

where $W = 0.3^k$; $w = 2 \times 0.012 = 0.024 \ ^k/_{ft}$; $I = 2I_y = 4.36 \text{ in}^4$; $L = 8'$

$$f = 3.13 / \sqrt{\frac{(0.3 + 0.486 \times 0.024 \times 8)(8 \times 12)^3}{48 \times 29{,}000 \times 4.36}} = 11.7 \text{ cps}$$

Thus, for vertical motion, ω = 37.4 cps; for horizontal motion, ω = 11.7 cps.

The dynamic amplification factor, defined as the ratio of deflection under dynamic loading to the deflection under the same force amplitude applied statically, is plotted below (Paz 1985):

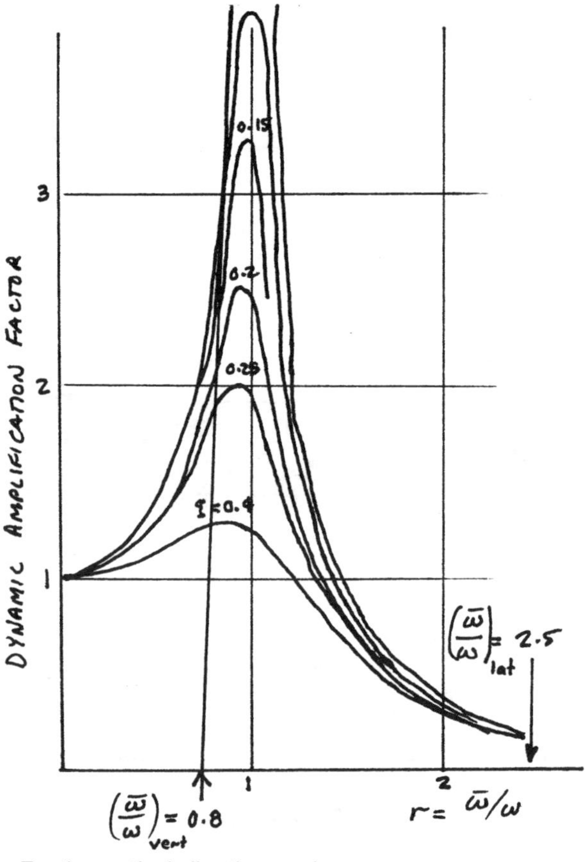

For the vertical vibration mode:

$$r = \frac{\bar{\omega}}{\omega} = \frac{29}{37.4} = 0.8 \text{ (too close to unity)}$$

For the horizontal vibration mode:

$$r = \frac{\bar{\omega}}{\omega} = \frac{29}{11.7} = 2.5 \text{ (no problem)}$$

For this system, the dynamic magnification of vertical deflection will be large and likely objectionable. The sensible solution is to use a lighter, more flexible beam that still has adequate strength and stiffness, but results in a lower natural frequency. Since the maximum moment under design loads is only $0.75^{k\text{-}ft}$, a W6x9 furnishes adequate strength ($\phi M_n = 14^{k\text{-}ft}$ with $L_b = 8'$), but $I_x = 16.4 \text{ in}^4$ (roughly one-third the stiffness of the W10x12) with $I_y = 2.2 \text{ in}^4$ practically the same as the W10x12. For this member:

$$K = [16.4 / 53.8]84.6 \ ^k/_{in} = 25.7 \ ^k/_{in}$$

$$\omega = \frac{1}{2\pi}\sqrt{\frac{25.7 \times 32.2 \times 12}{0.59}} = 20.6 \text{ cps}$$

$$r = \frac{\bar{\omega}}{\omega} = \frac{29}{20.6} = 1.4$$

For which the dynamic amplification is at a reasonable level.

Steel-38

(a) The P-Δ effect is the bending moment and deformation (so-called secondary bending) induced in a frame by the gravity forces (P-force) acting through the joint translations (Δ) caused by the externally applied forces.

(b) Prequalified welds are commonly used joint details that are exempt from testing for qualification provided that prescribed form, geometry and procedures are followed to deposit sound weld metal (AISC 1989).

(c) Delamination (i.e., lamellar tear) is a separation in the through-thickness direction of thick steel plates caused by stresses induced by restraint to weld shrinkage. For the detail shown, the potential for lamellar tearing exists in both plates, in the positions shown.

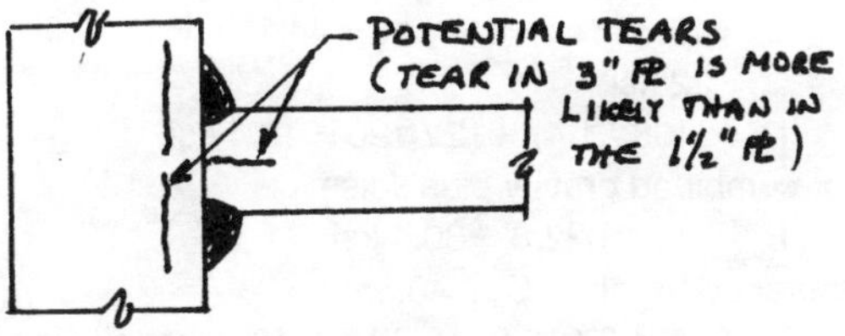

(d) Plastic hinges should form first in the girders, rather than the columns of a rigid frame. This permits energy imparted by ground motion to be absorbed by inelastic deformation without collapse of the structure.

(e) For a $^5/_{16}$" E80XX fillet weld, the strength based on the effective weld throat is:

- By LRFD (AISC 1986; Table J2.3)

  $\phi f_w = \phi(0.6F_{EXX})1"(0.707\omega)$

  $= 0.75 \times 0.6 \times 80 \times 1 \times 0.707(^5/_{16}) = 7.95\ ^k/_{in}$

- By ASD (AISC 1989)

  $q = 0.3F_u\ (1")(0.707\omega)$

  $= 0.3 \times 80 \times 0.707 \times (^5/_{16}) = \mathbf{5.3\ ^k/_{in}}$

(f) No. The hexagonal head of an A325 is marked:

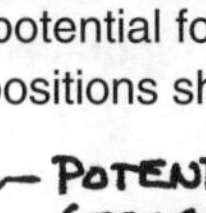

g) Calculate the ultimate capacity of details shown. Use LRFD (AISC 1986). The gusset plate is not specified ∴ assume it to be adequate.

- For $2L3\times2\times^5/_{16}$ with single gage of 3 – 1" dia A325-SC (std size punched holes)

  $A_e = UA_n = 0.85(A_g - 2t(1+0.125))$

  $= 0.85(2.92 - 2 \times (^5/_{16})(1.125)) = 1.88\ in^2$

$$\phi P_n \le \begin{cases} \phi F_y A_g = 0.9 \times 36 \times 2.92 = 95^k \\ \phi F_u A_e = 0.75 \times 58 \times 1.88 = 82^k \end{cases}$$

For the bolts, use the design strength from table 8-17, AISC Manual-II: $\phi r_v = 38^k/_{bolt}$

$$\phi P_n \le \begin{cases} 3\phi r_v = 3 \times 33 = 114^k \\ 3\phi r_b = 3 \times 65.2 = 196^k \end{cases}$$

∴ Tensile fracture governs: $\boldsymbol{\phi P_n = 82^k}$

- For $2L3\times2\times^5/_{16}$ attached with a single leg, 1/4" both sides, E70XX (note: capacity of the erection bolt cannot be included – Re: J1.8, AISC Spec)

Assume long legs back-to-back; no end returns

$L_w = 2 \times 6 = 12"$

$S_w = 2 \times 6^2 / 6 = 12$

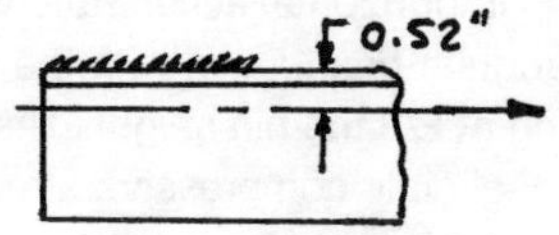

$f_w' = P / L_w = P / 12 = 0.083P$

$f_b' = Pe / S_w = P(0.52) / 12 = 0.043P$

$f' = P\sqrt{0.083^2 + 0.043^2} = 0.0936P$

$\phi f_w = 4 \times 1.39 = 5.56\ ^k/_{in}$ (each side)

∴ $0.0936P \le 5.56$

$P \le 59.4^k < \phi P_n = 95^k$ (yield; from above)

∴ $P_u < \phi P_n = 59.4^k$

I would expect the weld to fail, followed by shearing of the erection bolt. The capacity of the eccentrically loaded weld is significantly lower than the fracture strength, which governs the bolted connection.

Steel-39

a) Calculate the seismic forces per UBC-97. For bins or hoppers on braced or unbraced legs, Table 16-P permits R = 2.9 with $\Omega_o = 2.0$. For Zone 4 and near-fault factors of 1.0, Tables16-Q and 16-R, $C_a = 0.44$ and $C_v = 0.64$ for structures on soil profile S_D. Compute the period in the unbraced direction (1634.1.4, UBC-97):

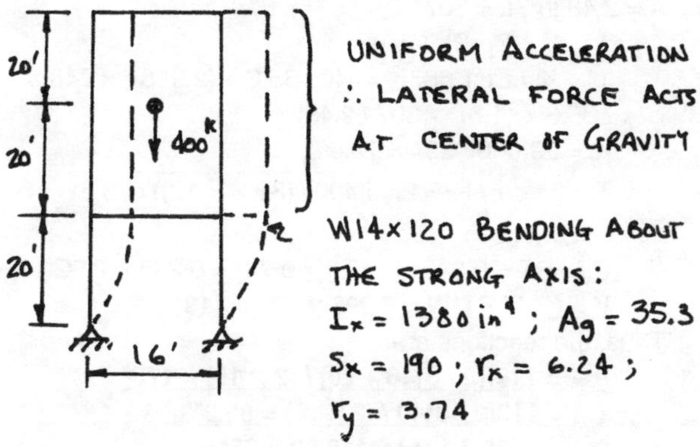

Assume $(I / L)_{girder} >> (I / L)_{column}$; compute the equivalent stiffness of four columns:

$K = 3EI_{total} / H^3 = 3 \times 29{,}000 \times 4 \times 1380 / (20 \times 12)^3$

$= 34.8\ ^k/_{in}$

$T = 2\pi\sqrt{M/K} = 2\pi\sqrt{(400 / (32.2 \times 12)) / 34.8}$

$= 1.08$ sec

Base shear coefficient for zone 4:

Limit $\ge \{0.11C_aI = 0.048;\ 0.8ZN_vI / R$

$= 0.8 \times 0.4 \times 1.0 / 2.9 = 0.11\} = 0.11$

$C_vI / RT = 0.64 \times 1 / (2.9 \times 1.08)$

$= 0.204 < 2.5C_aI / R$

∴ $\underline{\mathbf{V = (C_vI / RT)W = 0.204 \times 400 = 81.6^k}}$

Thus, reactions are:

$\underline{\mathbf{V_{col} = V / 4 = 81.6 / 4 = 20.4^k}}$

$\underline{\mathbf{M_{col} = V_{col}H = 20.4 \times 20 = 408^{k\text{-}ft}}}$

$\underline{\mathbf{P_{col} = (V(40) / 16) / 2 = (81.6 \times 40 / 16) / 2 = \pm 102^k}}$

For the braced direction, the rocking motion caused by axial deformation in the supporting members causes a linear variation in horizontal accelerations. Therefore, assume a triangular distribution with the lateral force resultant acting at 2/3 the bin height. The diagonal braces have negligible compressive strength.

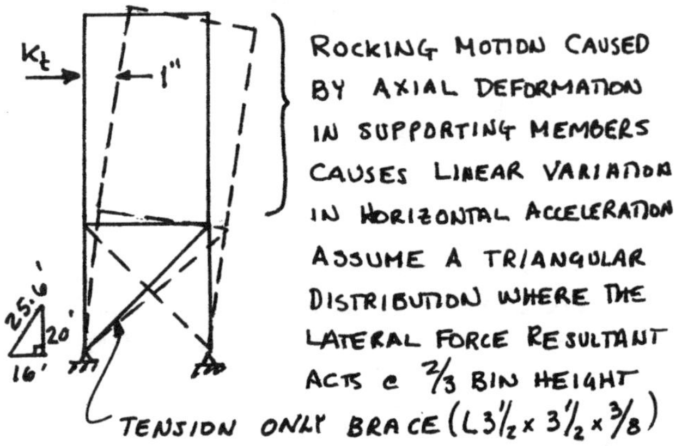

Compute the equivalent stiffness (use an energy method):

Force in the tension leg $= K_t\,(2/3)(40)/16$

Force in the compression leg $= K_t\,((2/3)40 + 20)/16$

Force in a diagonal $= K_t\,(25.6/16)$

For the legs: $A = 35.3\ \text{in}^2$; $L = 240"$. For the diagonal: $A = 2.48\ \text{in}^2$; $L = 307"$.

$\Delta = 1" = \Sigma p(PL/AE)$

$1" = K_t/E\{(1.667)^2 \times 240/35.3 + (2.916)^2 \times 240/35.3 + (1.6)^2 \times 307/2.48\}$

$\therefore\ K_t = 29{,}000/394 = 73.6^{k}/_{in}$

$T = 2\pi\sqrt{M/K} = 2\pi\sqrt{(400/(32.2 \times 12))/73.6} = 0.74\ \text{sec}$

$C_v I/RT = 0.64 \times 1/(2.9 \times 0.74) = 0.298 < 2.5C_a I/R$

$\therefore$ **$V = (C_v I/RT)W = 0.298 \times 400 = 119^k$**

Thus, the reactions are:

$P_{col} = 119[(2/3)40 + 20]/(2 \times 16) = 173^k$

$T_{col} = 119(2/3)40/(2 \times 16) = 99.2^k$

$P_{brace} = (25.6/16) \times 119/2 = 95^k$

Assume that the gravity load distributes equally to four columns and does not cause bending $\therefore P_D = 100^k$ each.

Check drift in the unbraced direction (1630.9, UBC):

$\Delta_s = V/K = 81.6/34.8 = 2.34"$

Check the P-Δ effect:

$\theta = \Delta_S W/VH = 2.34" \times 400^k/(81.6^k \times (20 \times 12)") = 0.047 < 0.1$

$\therefore$ P-Δ effect is negligible.

b) Verify member sizes using allowable stress design (AISC 1989). Use the load factors of UBC 1612.3.2 with a 1/3 increase in allowable stresses and divide the seismic forces by 1.4 to obtain working stress values.

For the W14x120 column: $A_g = 35.3\ \text{in}^2$; $d = 14.48"$; $t_w = 0.59"$; $b_f = 14.67"$; $t_f = 0.94"$; $r_x = 6.24"$; $r_y = 3.74"$; $S_x = 67.5\ \text{in}^3$; $K_x = 2.1$; $K_y = 1.0$.

$KL/r_x = 2.1 \times 20 \times 12/6.24 = 81$

$KL/r_y = 20 \times 12/3.74 = 64$

A36 Steel: $C_c = 126$; $(KL/r)/C_c = 81/126 = 0.64$; (Re: p 5-119, AISC) $C_a = 0.424 \therefore F_a = C_a F_y = 0.424 \times 36 = 15.3$ ksi

Check the special requirements of 2213.5, UBC

$P = 1.0P_D + 0.7P_L + \Omega_o P_E = 100 + 2.0 \times (102/1.4) = 245^k$

$P_{sc} = 1.7F_a A = 1.7 \times 15.3 \times 35.3 = 918^k > P \therefore$ O.K.

The column is part of two intersecting lateral-force resisting systems, but since $P_{max} = 118^k < 0.2P_{sc}$, orthogonal effects do not have to be considered, per exception of UBC 1633.1. Check combined axial and bending in the critical (unbraced) direction:

$f_a = P/A_g = (100 + 102/1.4)/35.3 = 4.9$ ksi

$L_c = 15.5' < L_b = 20' < L_u = 44.1' \therefore F_{bx} = 0.6F_y = 22$ ksi

$f_{bx} = (408/1.4) \times 12/190 = 18.4$ ksi

For combined gravity plus seismic:

$F'_{ex} = (4/3)22.8 = 30.4$ ksi

Unbraced frame $\therefore C_{mx} = 0.85$

$f_a/F_a = 4.9/15.3 = 0.32 > 0.15 \therefore$ check Eqn H1-1:

$f_a/F_a + C_{mx}f_{bx}/(1 - f_a/F'_{ex})F_{bx} < 1.0$

$0.32 + 0.85 \times 18.4/[(1 - 4.9/30.4)22] = 1.17 < 1.33$

$f_a/0.6F_y + f_{bx}/F_{bx} < 1.0$

$4.9/22 + 18.4/22 = 1.06 < 1.33 \therefore$ O.K.

The W14x120 is adequate in combined axial plus bending.

For the tension diagonals – provide connection at point of intersection $\therefore L = 25.6/2 = 12.8'$

$L/r_z = 12.8 \times 12/0.687 = 223$

The slenderness ratio exceeds the limit $720/\sqrt{F_y} = 120$, but this is permitted provided the brace can resist Ω_o times the seismic force: $T = 2.0(95.0/1.4) = 136^k$. Assume welded connections detailed such that full area is effective ($A = 2.48\ \text{in}^2$):

$T \le (4/3)F_t A_g = 1.33 \times 22 \times 2.48 = 73^k < 136^k$

$\therefore$ **L3-1/2 x 3-1/2 x 3/8 is inadequate.**

Steel-40

a) Calculate the steel and concrete stresses in the composite member. Use Allowable Stress Design (AISC 1989). Reduce live loads per 1607.5 , UBC-97:

Trib. Area $= 8 \times 24 = 192\ \text{ft}^2 > 150\ \text{ft}^2$

$$R \le \begin{cases} r(A - 150) = 0.08(192 - 150) = 3\% \leftarrow \text{Governs} \\ 23(1 + D/L) \\ 40\% \end{cases}$$

$\therefore\ w_L = (1 - 0.03) \times 0.075 \times 8 = 0.582$ klf

$w_D = 0.05 \times 8 + 0.026 = 0.426$ klf

$M_D = 0.426 \times 24^2/8 = 30.7^{\text{fk-ft}}$

$M_L = 0.582 \times 24^2/8 = 41.9^{\text{fk-ft}}$

For the W14x26: $A_g = 7.69\ \text{in}^2$; $d = 13.91"$; $I_x = 245\ \text{in}^4$; $S_x = 35.3\ \text{in}^3$. Assume full composite action (Re: I1, AISC):

$b_e \le \{l/4 = 24 \times 12/4 = 72"; \text{spacing} = 96"\} = 72"$
$b_e/n = 72/9 = 8$ (n = 9 is given)

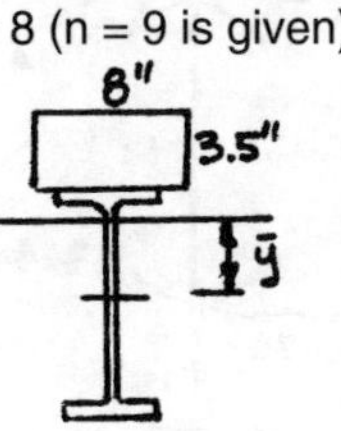

$A_t = 7.69 + 8 \times 3.5 = 35.7 \text{ in}^2$
$\bar{y} = 8 \times 3.5(6.96 + 1.75)/35.7 = 6.83"$
$I_{tr} = 245 + 7.69(6.83)^2 + 8(3.5)^3/12 + 8 \times 3.5(6.96 + 1.75 - 6.83)^2 = 731 \text{ in}^4$
$S_{tr} = I_{tr}/c_b = 731/(6.96 + 6.83) = 53 \text{ in}^3$
$S_t = I_{tr}/c_t = 731/(6.96 + 3.5 - 6.83) = 201 \text{ in}^3$

Construction is unshored ∴ dead loads are resisted by the W14x26 alone; live loads by the composite section:

$f_{bb} = M_D/S_S + M_L/S_{tr}$
$= 30.7 \times 12/35.3 + 41.9 \times 12/53 = 19.9$ ksi
$f_c = (M_L/S_b)/n = [(41.9 \times 12)/201]/9 = 0.28$ ksi

b) Compute the required number of studs for composite action (Re: I4, AISC):

$$V_h \le \begin{cases} 0.85f_c'A_c/2 = 0.85 \times 3 \times 72 \times 3.5/2 = 321^k \\ A_sF_y/2 = 7.69 \times 36/2 = 138^k \leftarrow \text{governs} \end{cases}$$

Given: q = 8 k / stud
$N \ge V_h/q = 138/8 = 17.3$, say 18 each side of M_{max}; 36 total

Thus, for full composite action, use 36-5/8" diameter studs at 8" o.c.

The stresses computed above indicate that there is no need for full composite action. From a strength standpoint, there is no difference between a shored and an unshored beam (ϕM_n is the same for both). For this reason, AISC limits the stress in steel to $(M_D + M_L)/S_{tr} \le 0.66F_y$. The computation of $M_D/S_S + M_L/S_{tr}$ is made only to assure that yielding in the bottom flange does not occur under service loads. Therefore, (Eqn I2-1):

$(M_D + M_L)/S_{req} \le 0.66F_y = 24$ ksi
$S_{req} \ge (30.7 + 41.9)12/24 = 36.3 \text{ in}^3$
$V_h' = V_h[(S_{req} - S_s)/(S_{tr} - S_s)^2]$
$= 138[(36.3 - 35.3)/(53 - 35.3)]^2 = 0.4^k$,

which is virtually nil; however, I4 requires that $V_h' \ge V_h/4 = 34.5^k$. Therefore,

$N' \ge V_h'/q = 35.4/8 = 4.3$, say 5 each side
stud spacing $\le \{24 \times 12/10 = 28"; 48"; 8h_s = 8 \times 3.5 = 28"\} = 28"$

Thus, for partial composite action, use 10-5/8" diameter studs @ 28" o.c.

c) Check the web shear connection. Assume bolts are 3/4" diameter A307 (not specified) acting in double shear through a pair of framing angles:

$r_v = 8.8^k/_{bolt}$; $r_b = 13.1^k/_{bolt}$ (bearing on $t_w = 0.26"$)
∴ $V \le nr_v = 3 \times 8.8 = 26.4^k > 12.1^k$ ∴ O.K.

Check block shear rupture of the coped web (J4, AISC):

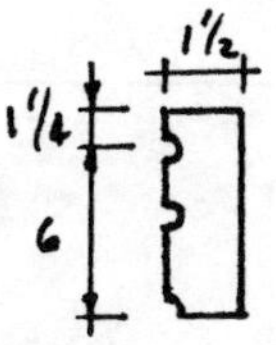

$V \le 0.3F_uA_{ns} + 0.5F_uA_{nt}$
$A_{ns} = (7.25 - 2.5(3/4 + 1/8))\,0.255 = 1.29 \text{ in}^2$
$A_{nt} = (1.5 - 0.5(3/4 + 1/8))0.255 = 0.27 \text{ in}^2$
∴ $V \le 0.3 \times 58 \times 1.29 + 0.5 \times 58 \times 0.27 = 30^k > 12.1^k$ ∴ O.K.

the capacity of 2L5x3-1/2x3/8 is adequate by inspection
∴ **the connection is adequate.**

Steel-41

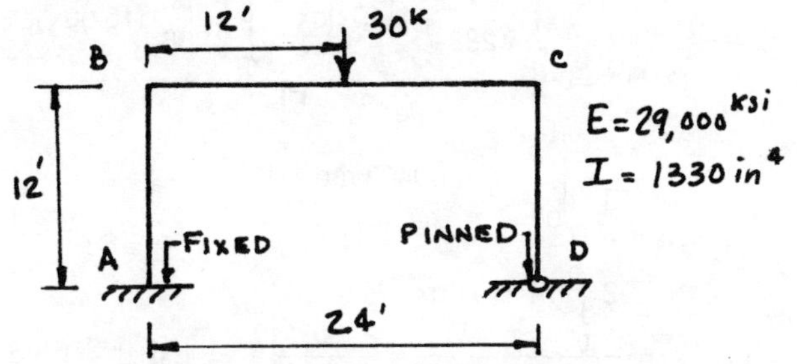

a) Compute the moment at "B" in the as-built condition (before settlement of 0.5" at support "D"). Use the method of consistent displacements with the moment at "A" and the horizontal reaction at "D" as the redundants. Release the structure:

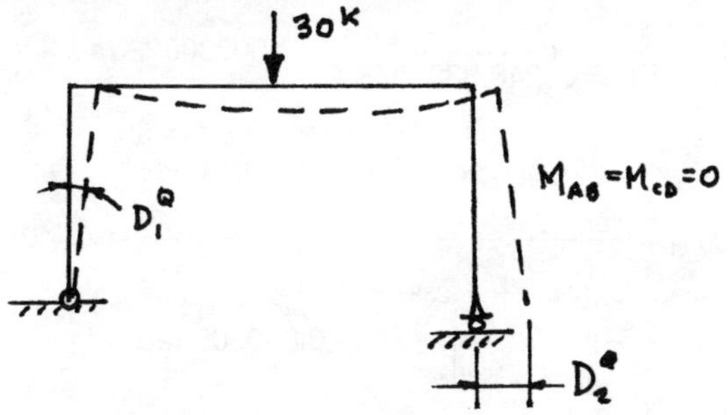

For member BC:
$M_x = 15x$; $0 \le x \le 144"$
$M_x = 4320 - 15x$; $144" \le x \le 288"$

Apply a unit dummy couple to the released structure at "A":

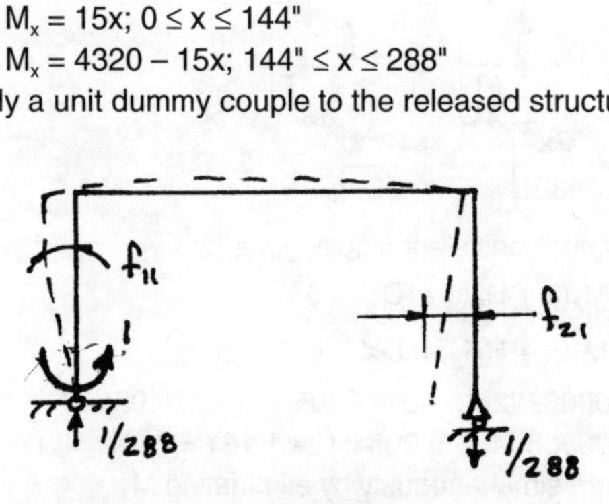

For member AB: m = –1; $0 \le x \le 144$
For member BC: m = –1 + x / 288; $0 \le x \le 288"$
Apply a unit horizontal force at D:

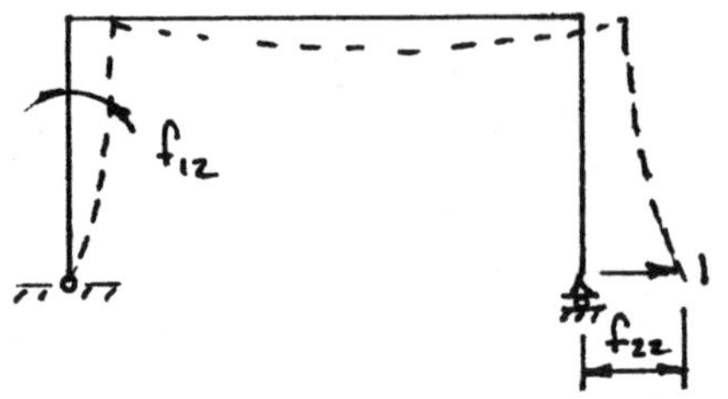

For member AB: $m = x$; $0 \le x \le 144"$
For member BC: $m = 144$; $0 \le x \le 288"$
For member DC: $m = -x$; $0 \le x \le 144"$

Use virtual work: $$1\Delta = \sum_{i=1}^{\#\,\text{mem}} \int_L \frac{m_i M_i dx}{EI_i}$$

$$\therefore \quad (1)D_1^Q = \frac{\int_0^{144}\left(\frac{x}{288}-1\right)(15x)dx + \int_0^{144}\left(\frac{-x}{288}\right)(15x)dx}{EI}$$

$$D_1^Q = \left.\frac{-7.5x^2}{EI}\right|_0^{144} = -0.00403 \text{ rad}$$

$$(1)D_2^Q = \frac{2\int_0^{144}144(15x)dx}{EI} = \left.\frac{2160x^2}{EI}\right|_0^{144} = 1.161" \rightarrow$$

$$(1)f_{11} = \frac{\int_0^{144}(-1)(-1)dx}{EI} + \frac{\int_0^{288}(x/288)^2dx}{EI}$$

$$f_{11} = \left.\frac{x}{EI}\right|_0^{144} + \left.\frac{x^3}{248{,}832EI}\right|_0^{288} = +0.00000622 \text{ rad}$$

$$(1)f_{21} = \frac{\int_0^{144}(-1)(x)dx}{EI} + \frac{\int_0^{288}(-x/288)(144)dx}{EI}$$

$$f_{21} = \left.\frac{-x^2}{2EI}\right|_0^{144} - \left.\frac{x^2}{4EI}\right|_0^{288} = -0.000805 \text{ rad} = f_{12}$$

$$(1)f_{22} = \frac{2\int_0^{144}x^2dx}{EI} + \frac{\int_0^{288}(144)^2dx}{EI}$$

$$f_{22} = \left.\frac{2x^3}{3EI}\right|_0^{144} \quad \left.\frac{144^2x}{EI}\right|_0^{288} = 0.2064"$$

Thus, for consistent displacements:

$$M_Af_{11} + H_Df_{12} + D_1^Q = 0$$
$$M_Af_{21} + H_Df_{22} + D_2^Q = 0$$

$0.000006222M_A - 0.000805H_D - 0.00403 = 0$
$-0.000805\ M_A + 0.2064H_D + 1.161 = 0$
Solving simultaneously by eliminating M_A:
$H_D = -6.25^k = 6.25^k \leftarrow$
$M_A = -160.9^{k\text{-in}} = 13.4^{k\text{-ft}}$ (clockwise)
From statics:

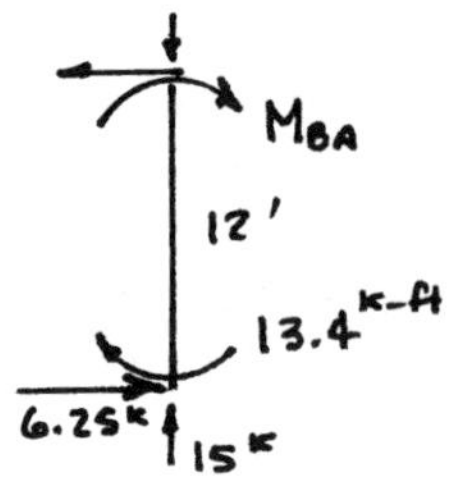

$\Sigma F_h = 0$: **$H_A = -H_D = 6.25^k \rightarrow$**
$\Sigma M_B = 0$: $M_{BA} + 13.4 - 6.25(12) = 0$ $\therefore$ **$M_{BA} = 61.6^{k\text{-ft}}$ (counterclockwise)**

b) Compute the moment at "B" if the support "D" settles 0.5" $\downarrow$:

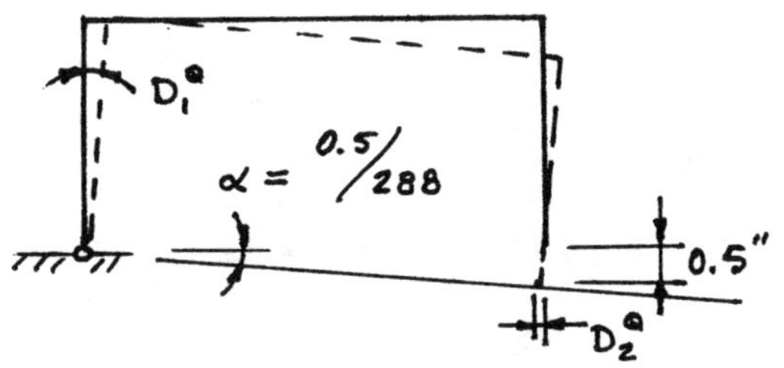

$$D_1^Q = -\alpha = -0.001736 \text{ rad}$$
$$D_2^Q = 0.5\sin\alpha \cong -0.5\alpha = -0.000868"$$

For consistent displacements:

$$M_Af_{11} + H_Df_{12} + D_1^Q = 0$$
$$M_Af_{21} + H_Df_{22} + D_2^Q = 0$$

$0.000006222M_A - 0.000805H_D - 0.001736 = 0$
$-0.000805\ M_A + 0.2064H_D - 0.000868 = 0$
$\therefore$ **$H_D = 2.20^k \rightarrow$**
$M_A = 564^{k\text{-in}} = 47.0^{k\text{-ft}}$ (counterclockwise)
From statics:

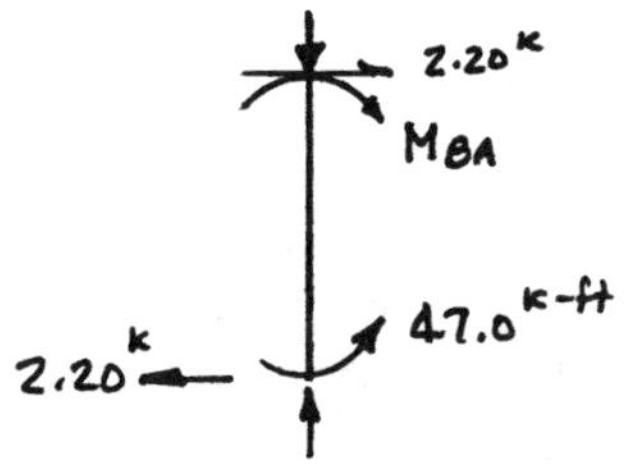

$\Sigma M_B = 0$: $M_{BA} - 47.0 + 2.20(12) = 0$ $\therefore$ **$M_{BA} = 20.6^{k\text{-ft}}$ (counterclockwise)**
Superimposing the moment due to 0.5" settlement on the moment due to gravity load:
$M_{BA} = 61.6 + 20.6 = 82.2^{k\text{-ft}}$ (counterclockwise)

Steel-42

Design the steel-to-concrete column connection using LRFD (UBC-97, Chapter 22, Division IV; AISC 1994). Given; $f'_c = 3000$ psi; ASTM A36 steel; gravity loads: $P_D = 500^k$, $P_L = 400^k$; seismic loads: $M_{xE} = 660^{k\text{-ft}}$, $V_{xE} = 110^k$.

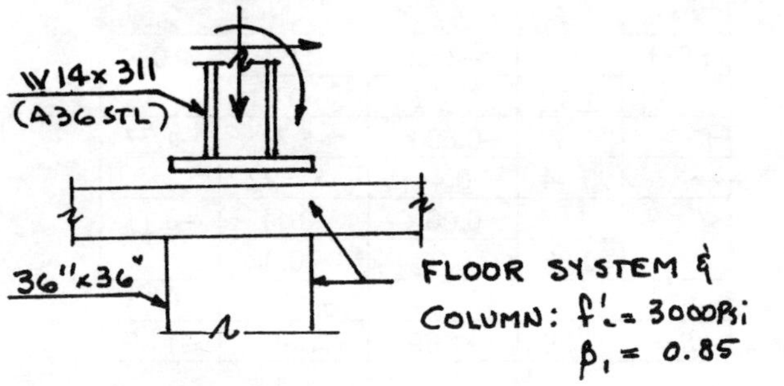

Design load combinations (Re: 1612.2.1, UBC):

$U_1 = 1.2D + 1.6\ L$

$U_2 = 1.2D + 0.5L + 1.0E$

$U_3 = 0.9D \pm 1.0E$

$\therefore\quad P_{u1} = 1.2 \times 500 + 1.6 \times 400 = 1240^k;\ M_{u1} = 0$

$P_{u2} = 1.2 \times 500 + 0.5 \times 400 = 800^k;\ M_{u2} = 660^{k\text{-}ft}$

$P_{u3} = 0.9 \times 500 = 450^k;\ M_{u3} = 660^{k\text{-}ft}$

Cases 1 and 3 are the critical conditions. Check bearing on the concrete (J9, AISC):

$P_p = 0.85f'_c A_1 \sqrt{A_2 / A_1}$; let $\sqrt{A_2 / A_1} = 2.0$ (limiting value) and $\phi = 0.6$

$P_u \le \phi P_p = 0.6 \times 0.85 \times 3 \times A_1 \times 2.0 = 1240^k$

$\therefore$ $A_1 \ge 405\ in^2$ for loading condition 1 (uniform bearing). For loading condition 3, the moment is best resisted by a plate with as large a length as practicable. The anchor bolts will extend through the floor and embed within the confined core of the concrete column below. Thus, to maintain clearance within the column core, the maximum distance from column centerline to anchor bolt is:

$c = h / 2 - \text{cover} - d_s - d_l - d_a$; where d_s is the tie bar diameter, d_l is longitudinal bar diameter and d_a is the anchor bolt diameter.

$\therefore\quad c \cong 18 - 1.5 - 0.5 - 14 / 8 - 1.5 = 12.8"$; say 12"

Thus, try a 20" x 28" base plate with the longer dimension resisting the moment.

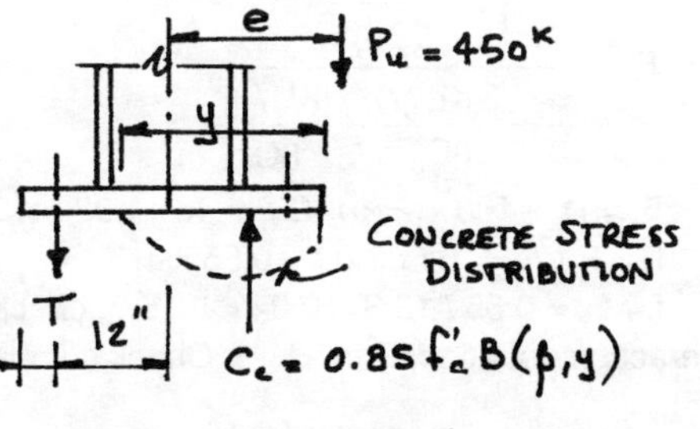

$e = M_u / P_u = 660 / 450 = 1.47' = 17.6"$

Let y = distance from the compression edge of base plate to the neutral axis under the design loading.

$\Sigma F_y = 0$: $T + P_u - C_c = 0$

$\therefore\quad T = 0.85 \times 3 \times 0.85y \times 20 - 450 = 43.35y - 450$

Take moments about the line of action of P_u:

$\Sigma M = 0$: $T(12 + 17.6) - C_c(3.6 + \beta_1 y / 2) = 0$

$(43.35y - 450)(29.6) - 43.35y(3.6 + 0.85y / 2) = 0$

$1283.16y - 13320 - 156.06y - 18.42y^2 = 0$

$\therefore\quad y = 16.0"$

$\therefore\quad T = 43.35 \times 16.0 - 450 = 244^k$

Try four anchor bolts each side (eight total). Assume all eight bolts resist the seismic shear:

$V = 110 / 8 = 13.7\ ^k/_{bolt}$

Re: Table J3.5, AISC – try 1-1/2" diameter A449 threaded rods ($A_b = 1.77\ in^2$; $F_u = 105$ ksi):

$\phi F_t = \phi\{0.98F_u - 1.9f_v\} \le \phi(0.75F_u)$; $\phi = 0.75$

$\phi F_t = 0.75\{0.98 \times 105 - 1.9 \times 13.7 / 1.77\}$

$= 66\ ksi \le 0.75 \times 0.75 \times 105 = 59\ ksi$

$\therefore\quad \phi T_n = \phi F_t A_b = 59 \times 1.77 = 105^k/_{bolt}$

$N \ge T / \phi T_n = 244 / 105$; **use 3, 1-1/2" A449 anchor bolts each side**

For the 28" x 20" plate, the actual bearing pressure under gravity loading is:

$f_p = 1240 / (28 \times 20) = 2.2\ ksi < 0.85f'_c$ $\therefore$ bearing pressure under case 3 loading is more critical.

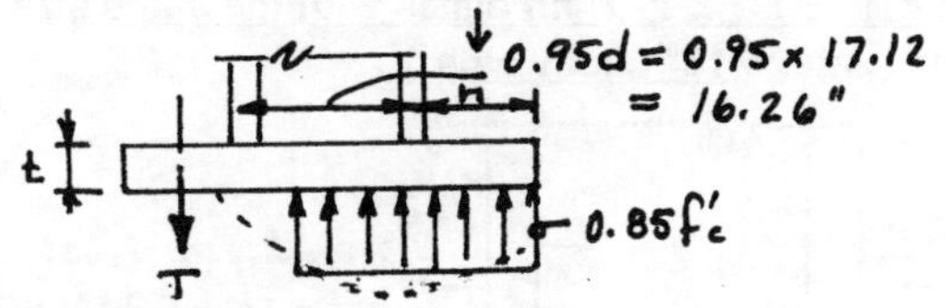

Determine the required plate thickness (A36 steel) based on the compression side bending moment:

$n = (28 - 0.95d) / 2 = (28 - 0.95 \times 17.12) / 2 = 5.87"$

$m_u = 0.85f'_c n^2 / 2 = 0.85 \times 3 \times 5.87^2 / 2 = 44^{k\text{-}in}/_{in}$

On the tension side:

$m_u = 224(12 - 0.95 \times 17.12 / 2)/20 = 43.3^{k\text{-}in}/_{in}$

Compression side governs

$m_u \le \phi m_n = \phi F_y Z = 0.9 \times 36 \times t^2 / 4$

$t^2 \ge 4 \times 44 / (0.9 \times 36) = 5.43\ in^2$ $\therefore$ $t \ge 2.33"$

Use PL 20 x 2-3/8 x 2'-4"

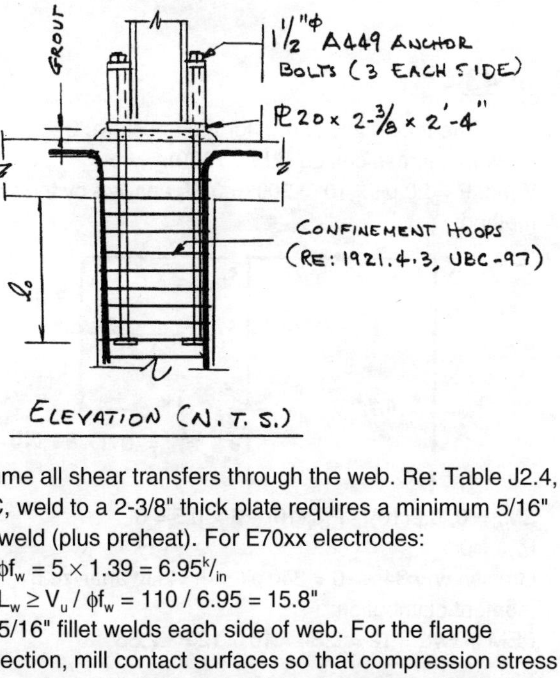

Assume all shear transfers through the web. Re: Table J2.4, AISC, weld to a 2-3/8" thick plate requires a minimum 5/16" fillet weld (plus preheat). For E70xx electrodes:

$\phi f_w = 5 \times 1.39 = 6.95^k/_{in}$

$L_w \ge V_u / \phi f_w = 110 / 6.95 = 15.8"$

Use 5/16" fillet welds each side of web. For the flange connection, mill contact surfaces so that compression stress is not critical. Follow current FEMA 267A recommendation and avoid partial penetration weld of flange-to-plate connection for seismic loading. Transfer the tension force, 244^k, through side plates welded to the column flange. Assuming three plates are fully effective:

$V = 244 / 3 = 81.3^{k}/_{plate}$

$M = 81.3 \times (12 - 17.12 / 2) = 280^{k\text{-}in}$

Use plates thick enough to develop the strength of 5/6" fillet welds to both sides:

$t \geq 2 \times 5 \times 1.39 / (0.9 \times (0.6 \times 36) = 0.71$"; say 3/4" plate

Determine the length required to resist the combined shear and bending (elastic method):

$L_w >> V / \phi f_w = 81.3 / (2 \times 5 \times 1.39) = 5.8$"; try 12" long plates

$q_v = 81.3 / (2 \times 12) = 3.4^{k}/_{in}$

$q_m = M / S_w = 280 / (2 \times 12^2 / 6) = 5.83^{k}/_{in}$

$q = \sqrt{q_v^2 + q_m^2} = \sqrt{3.4^2 + 5.83^2} = 6.7^{k}/_{in} < 5 \times 1.39$

$= 6.95\ ^{k}/_{in}$

∴ use 4, PL5 x x 1'-0" fin plates with the same size bearing plate; 5/16" fillet welds.

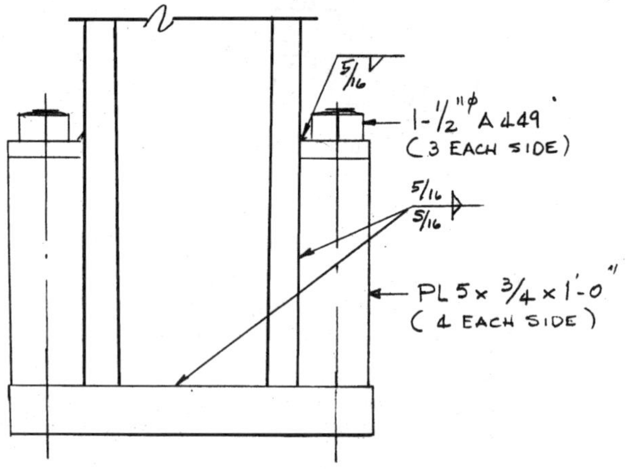

Steel-43

a) Verify the tube sizes shown for the structure. Use allowable stress design (AISC 1989).

Wind: $P = 20$ plf $\times$ 10' $= 200^{\#} = 0.2^{k}$; analyze by the portal method:

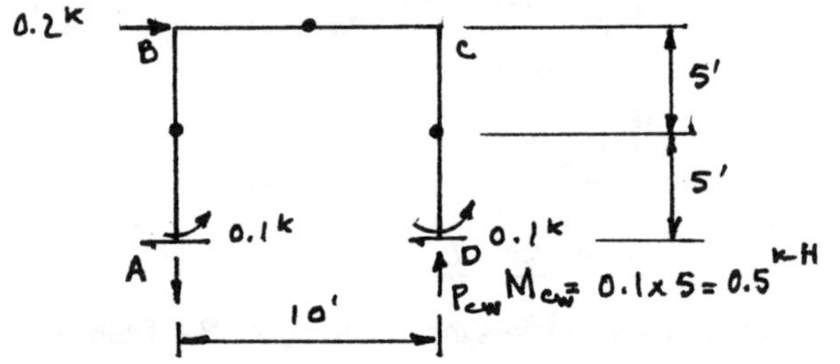

$\Sigma M_A = 0$: $0.2(10) - P_{CW}(10) - 2 \times 0.5 = 0$

$P_{CW} = 0.1^{k}$

Gravity: $w = 34 \times 10 = 340$ plf $= 0.34$ klf; analyze by moment distribution:

$FEM = \pm wL^2 / 12 = 0.34 \times 10^2 / 12 = \pm 2.83^{k\text{-}ft}$

DF	0.5	0.5	0.5	0.5
FEM:	0	−2.83	2.83	0
	1.42	1.42	→ 0.71	
		−0.88 ←	−1.77	−1.77
	0.44	0.44	→ 0.22	
		−0.06 ←	−0.11	−0.11
	0.03	0.03	→ 0.02	
			−0.01	−0.01
Σ =	1.89	−1.89	1.89	−1.89

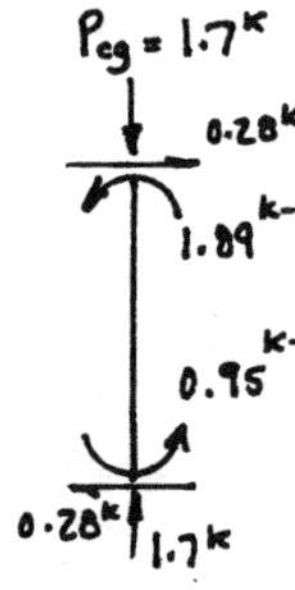

Loading cases:

Gravity: $P = 1.7^{k}$; $M = 1.89^{k\text{-}ft}$

Gravity plus wind: $P = 0.75(1.7 + 0.1) = 1.4^{k}$;

$M = 0.75(1.89 + 0.5) = 1.8^{k\text{-}ft}$

Thus, gravity controls. Check the TS 4 x 3 x 0.250:

$A_g = 3.09$ in^2; $r_x = 1.45$"; $r_y = 1.15$"; $S_x = 3.23$ in^3;

$F_y = 46$ ksi.

For the x-axis: $G_A = 1.0$; $G_B = 1.0$ (re: Fig. C-C2.2, AISC Manual): $K_x = 1.3$. The y-axis is braced out of plane

∴ $K_y = 1.0$

$KL / r_y = 1.0 \times 120 / 1.15 = 104$

$KL / r_x = 1.3 \times 120 / 1.45 = 108 \leftarrow$ governs

$KL / r = 108 < C_c = \sqrt{2\pi^2 E / F_y} = 112$ ∴ Inelastic buckling – use Eqn. E2-1:

$$F_a = \frac{\left[1 - \frac{(KL/r)^2}{2C_c^2}\right] F_y}{\frac{5}{3} + \frac{3(KL/r)}{8C_c} - \frac{(KL/r)^3}{8C_c^3}}$$

$F_a = [1 - 108 / (2 \times 112^2)] \times 46 / 1.92 = 12.8$ ksi

$f_a = P / A_g = 1.7 / 3.09 = 0.55$ ksi

$f_a / F_a = 0.55 / 12.8 = 0.04 < 0.15$ ∴ check the simple interaction equation, Eqn. H1-3. Check compactness (B5):

$b / t = (4 - 3 \times 0.25) / 0.25 = 13 < 190 / \sqrt{F_y} = 28$

Re: F3.1

$L_c = [1950 + 1200(M_1 / M_2)](b / F_y)$; $M_1 / M_2 = 0.5$

$L_c = [1950 + 1200 \times 0.5](3.0 / 46) = 166$"

$= 13.8' > L_b = 10'$

∴ $F_{bx} = 0.66F_y = 0.66 \times 46 = 30$ ksi

Combined axial plus bending (Eqn. H1-3):

$f_a / F_a + f_{bx} / F_{bx} = 0.04 + (1.89 \times 12 / 3.23)) = 0.27 <<$ 1.0 ∴ O.K.

∴ **TS 4 x 3 x 0.250 is O.K.**

b) Detail of the beam-to-column connection:

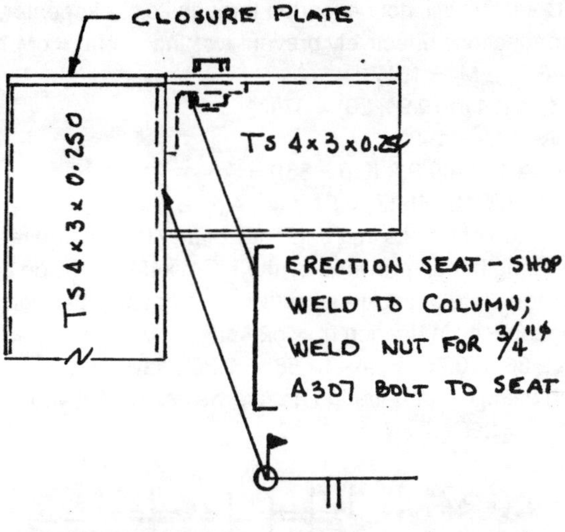

CONNECTION DETAIL (N.T.S.)

Steel-44

Design the lightest rolled section for the roof girder using plastic design. Use LRFD (AISC 1994; Salmon, et,al. 1989). Given: superimposed dead = 32 psf; allow 50 plf for beam weight; 12 psf roof live load.

$w_u = 1.2w_D + 1.6w_L = 1.2(32 \times 30 + 50) + 1.6 \times 12 \times 30$

$= 1788$ plf; say 1.8 klf

$P_u = 1.6P_L = 1.6 \times 6 = 9.6^k$

Analyze using the statical approach to plastic analysis:

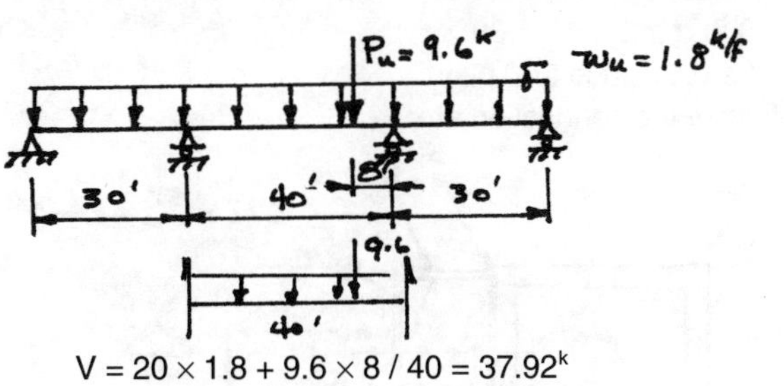

$V = 20 \times 1.8 + 9.6 \times 8 / 40 = 37.92^k$

$V = 0$ at $x = 37.92 / 1.8 = 21.07'$

$\therefore \quad M_{o,max} = 0.5 \times 37.92 \times 21.07 = 399.4^{k\text{-}ft}$

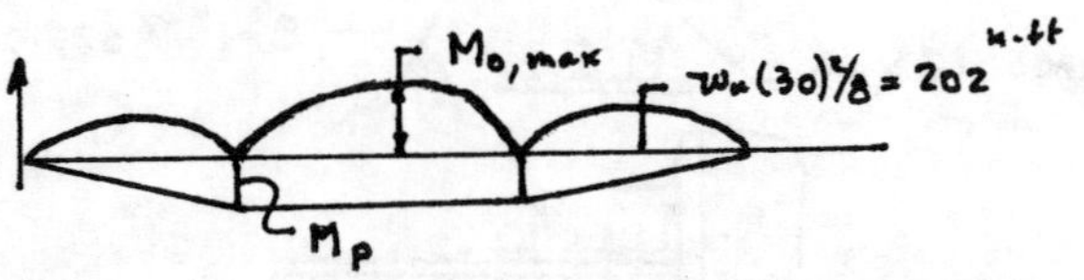

Assign $M_p = M_{o,max} / 2 = 399.4^{k\text{-}ft} / 2 = 199.7^{k\text{-}ft}$. The moment diagram at collapse will be:

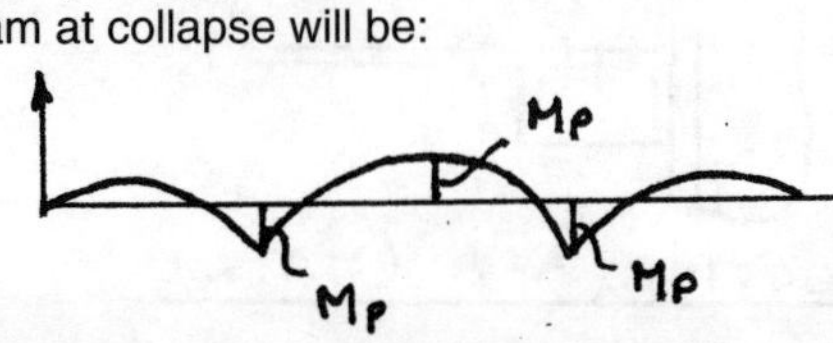

Sufficient hinges (3) form and the moment does not exceed M_p at any point; thus, the failure mechanism has been determined. As a check, use the mechanism method of analysis:

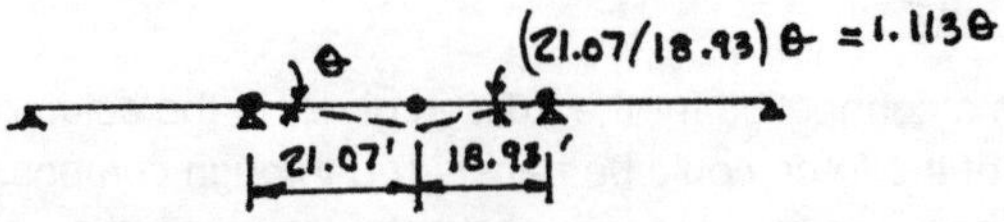

At collapse: $W_{ext} = U_{int}$

$M_p(\theta + 1.113\theta)2 = 1.8(21.07)^2\theta / 2 + [1.8(18.93)^2\theta / 2 + 9.6 \times 8\theta] \times 1.113$

$4.226M_p = 843.8 \therefore M_p = 199.6^{k\text{-}ft}$ checks!

For A36 steel ($F_y = 36$ ksi); $\phi = 0.9$

$\phi M_p = \phi F_y Z_x \geq 199.6^{k\text{-}ft}$

$\therefore \quad Z_x \geq 199.6 \times 12 / (0.9 \times 36) = 74 \text{ in}^3$

Re P.4-19, AISC Manual try a W18x40 ($Z_x = 78.4 \text{ in}^3$). The plastic hinges form first over the supports where the bottom flange is in compression. Thus, the maximum unbraced length of bottom flange in the vicinity of supports cannot exceed L_{pd} per Eqn F1-1, AISC-S. The last hinge to form is in the positive moment region where joists brace the top flange. Here, the maximum unbraced length cannot exceed L_p per Eqn. F1-4. Check the W18x40: $r_y = 1.27"$

$L_p = 300r_y / \sqrt{F_{yf}} = 300 \times 1.27 / \sqrt{36} = 63.5" = 5.3'$

The problem sketch shows joists spaced at 8' o.c., which exceeds the required distance between bracing for plastic design. We could either select a stockier beam (e.g., W14x48 has a $\phi M_p = 78.4^{k\text{-}ft}$ with $L_p = 8.0'$) or select lighter joists spaced no more than 5.3' o.c. Assuming the latter option is chosen, check the bracing requirement near support; Try $L_b = 5'$:

$M_1 = -199.6 + 0.5(37.92 + 37.92 - 1.8 \times 5) = -32.5^{k\text{-}ft}$

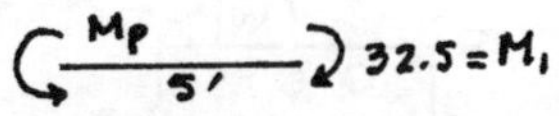

$L_{pd} = [3600 + 2200(M_1 / M_p)]\, r_y / F_y$

$= [3600 + 2200(-32.5 / 199.6)](1.27 / 36) = 114.6"$

$= 9.5' > 5' \therefore$ O.K.

Therefore, **<u>use W18x40 with top flange braced at 5 ' o.c. and bottom flange braced at 5' to either side of supports.</u>**

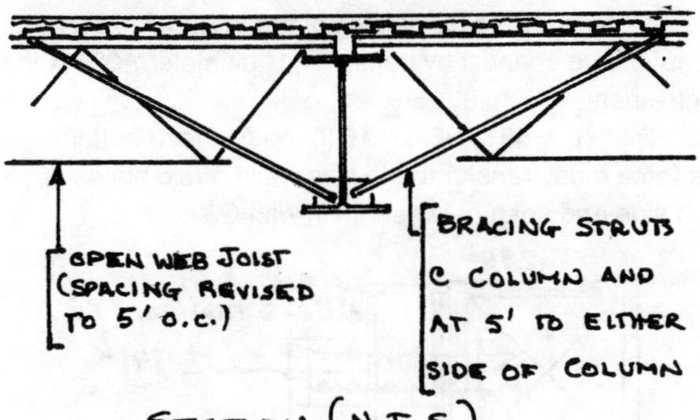

SECTION (N.T.S.)

Steel-45

Design a connection for the trussed girder to the column. Some of the force could be transferred through composite action between the steel and concrete. However, the problem statement requires current AISC Specifications, which do not cover composite connections. Therefore, I will assume that all forces must transfer through steel elements. Use the LRFD specification (UBC-97, Chapter 22, Division IV; AISC 1994) with an assumed dead-to-total load ratio of 0.6; A36 steel.

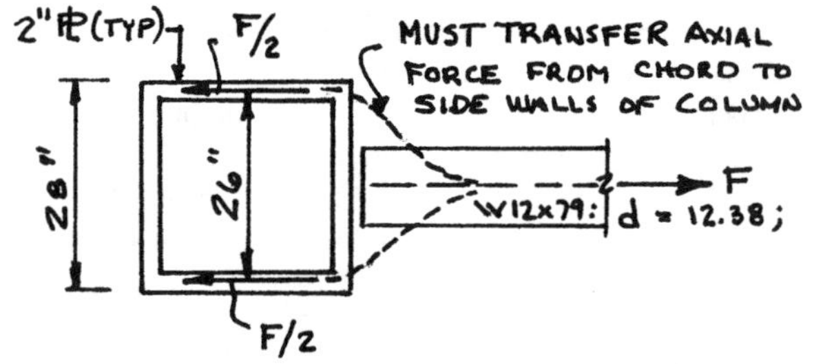

$P_u = 1.2P_D + 0.5P_L + P_E = 1.2 \times (0.6 \times 72) + 0.5 \times (0.4 \times 72) + 525$

$P_u = 591^k$ (compression)

$T_u = 0.9P_D - P_E = 0.9 \times (0.6 \times 72) - 525 = -486^k = 486^k$ (tension)

To simplify, design for $\pm 591^k$ (conservative). Try connecting with 1" diameter A325-X bolts through the flanges of the W12x79: $A_g = 23.2$ in^2; d = 12.38"; $t_w = 0.47$"; $b_f = 12.08$"; $t_f = 0.74$".

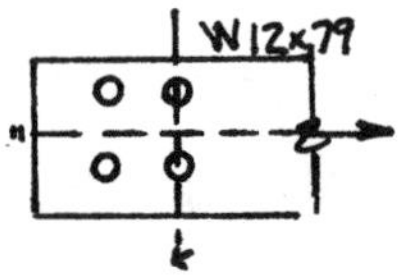

$A_n = A_g - 4\phi_h t_f = 23.2 - 4 \times 1.06 \times 0.74 = 20.1$ in^2

$$\phi T_n = \begin{cases} \phi F_y A_g = 0.9 \times 36 \times 23.2 = 752^k \leftarrow \text{governs} \\ \phi F_u A_e = 0.75 \times 58 \times (0.9 \times 20.1) = 787^k \end{cases}$$

Strength is governed by yielding of the W12x79 ∴ acceptable to connect by bolts. For 1" diameter A325-X in single shear, $r_v = 36.8^k/_{bolt}$

$n \geq T_u / r_v = 591 / 36.8 = 16$ (8 top flange; 8 bottom)

The force must transfer to the side walls. Weld plates on each side and span between them with C-sections.

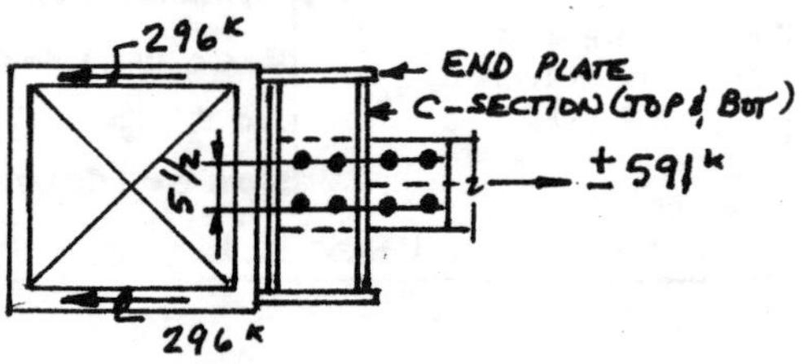

Each channel must transfer 591 / 2 = 296^k over a span 28 – 2 = 26". Using standard (5-1/2") gage between bolts in the W12x79 flange:

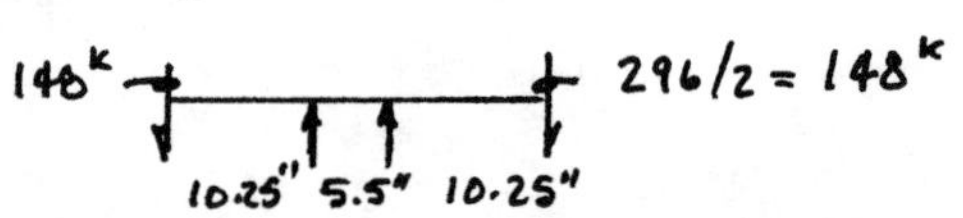

$V_u = 148^k$; $M_u = 148 \times 10.25 = 1517^{k\text{-in}}$

The C-section will not be loaded through its shear center, but the connections effectively prevent twisting ∴ $\phi M_n = \phi M_p$.

$\phi F_y Z_x \geq M_u = 1517^{k\text{-in}}$

$Z_x \geq 1517 / (0.9 \times 36) = 47$ in^3

$\phi V_n \geq V_u = 148^k$

$A_w \geq 148 / (0.9 \times (0.6 \times 36)) = 7.6$ in^2

∴ Use MC12 x45 ($Z_x = 51.7$ in^3; $A_w = 8.5$ in^2)

Connection of 8 bolts in the flanges of a W12x79 requires longer length than provided by the MC12x45 ∴ provide an extension plate of thickness sufficient to develop the bolt's shear capacity in bearing ($r_b = \phi 2.4dtF_u$)

t ≥ 36.8/(0.75 × 2.4 × 1× 58) = 0.35"; say 3/8"

Attach using full penetration groove welds. Try 12" wide plate: $T_u = 296 / 4 = 74^k$

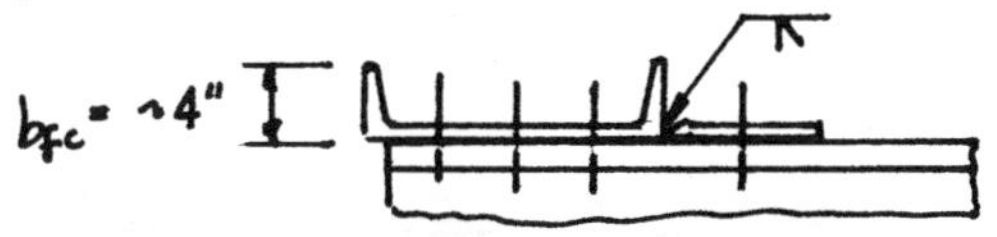

$\omega_n \leq \{12 - 2(1 + 1 / 8) = 9.88"; 0.85 \times 12 = 10.2"\} = 9.88"$

$$\phi T_n = \begin{cases} \phi F_y A_g = 0.9 \times 36 \times (12 \times 3 / 8) = 146^k \leftarrow \text{governs} \\ \phi F_u A_e = 0.75 \times 58 \times (1.0 \times 9.88 \times 3 / 8) = 161^k \end{cases}$$

∴ Use PL6 x 3 / 8 x 1'-0"

Side wall plates transfer 148^k over an effective length ≥ b_{fc} + d / 2 ≅ 10". Use E70xx fillet welds: $\omega_{min} = 5 / 16$

$\phi f_w = 2 \times 5 \times 1.39 = 13.9^k/_{in}$

$V_w = \phi f_w L_w = 13.9 \times 10 = 139^k < 148^k$

Use 3/8" fillet welds to get the required strength. Make the end plates thick enough to prevent failure of the base material:

$0.9F_y t \geq 2 \times 6 \times 1.39$

t ≥ 0.51"; use 5/8" plate

Details of the connection are:

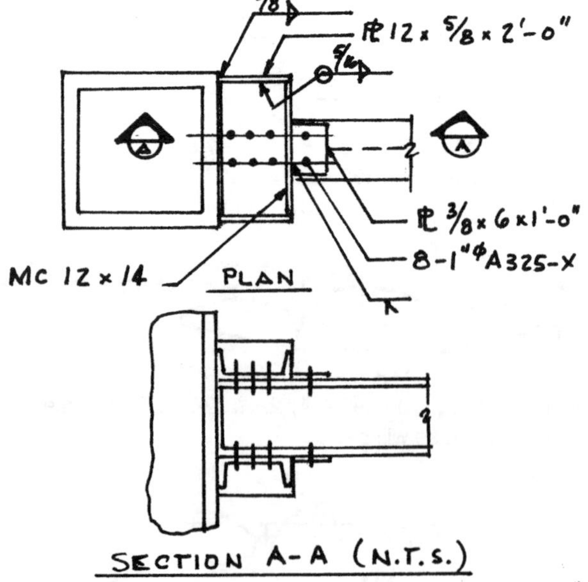

Steel-46

Complete the details for the rigid connection. Use LRFD (AISC 1994) assuming that the given loads are due to gravity only and that the dead-to-total load ratio is 0.5. Thus,

$$U = 1.2D + 1.6\,L = 1.2 \times 0.5 + 1.6 \times 0.5 = 1.4$$

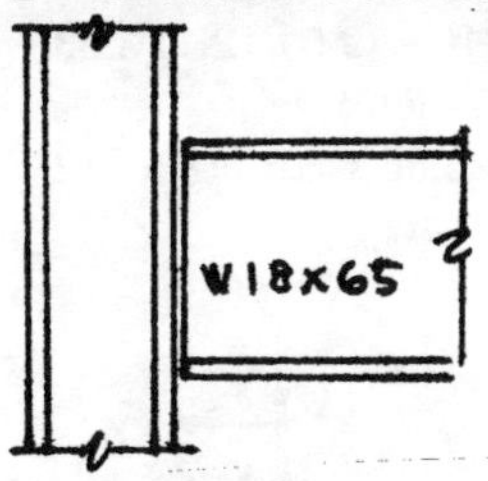

For the W18x65: $d = 18.35''$; $t_w = 0.45''$; $b_f = 7.59''$; $t_f = 0.75''$

For the W12x53: $d = 12.06''$; $t_w = 0.35''$; $b_f = 10.00''$; $t_f = 0.58''$

Steel is A36 (F_y = 36 ksi). Use E70xx welds and 3/4" diameter A325 bolts. Compute the forces and moments at the centerline:

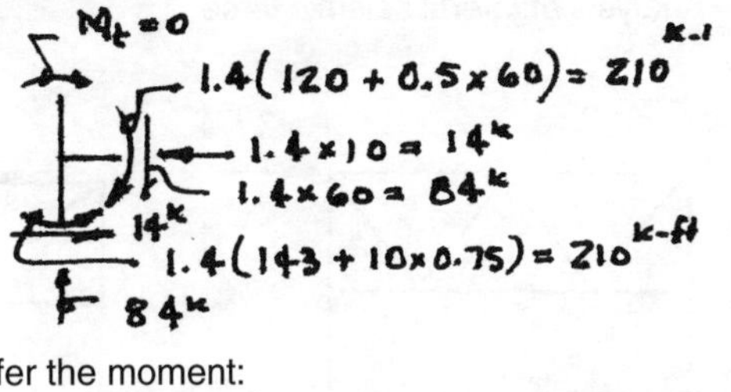

To transfer the moment:

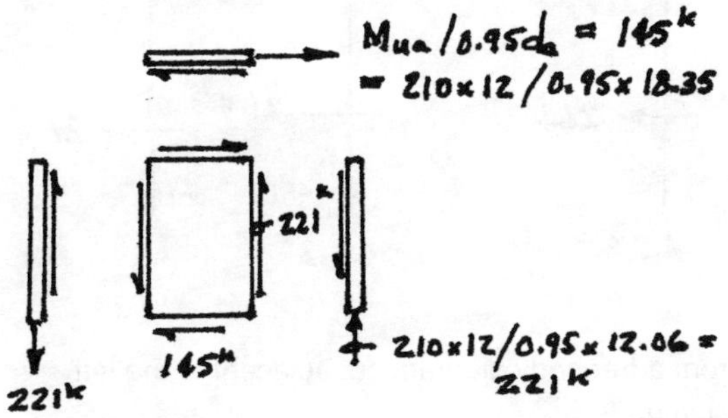

For the force transfer:

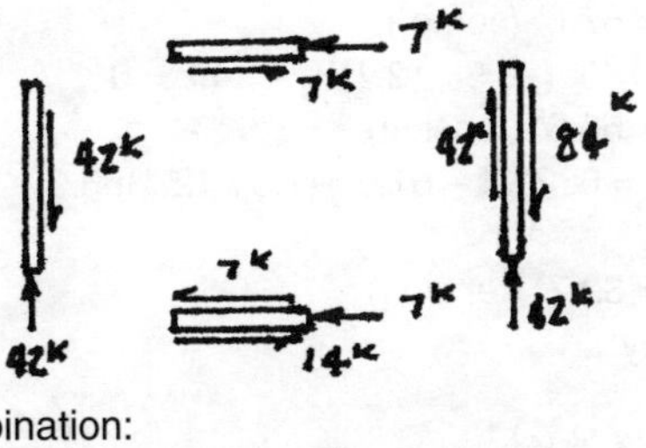

For the combination:

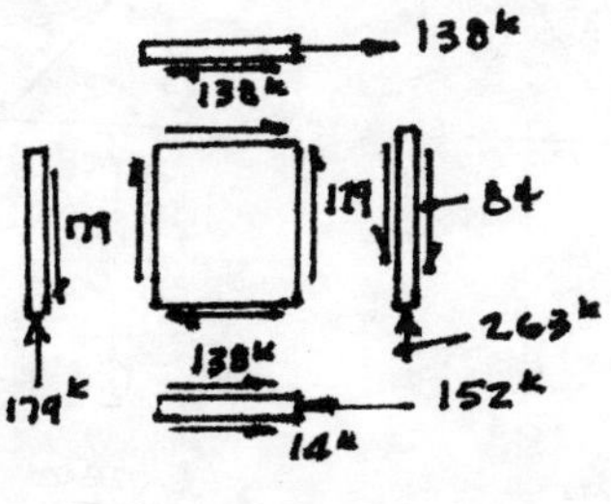

Check web shear under the assumption that $P_u < 0.4P_y$ and that panel zone deformations were not considered (Eqn. K1-9):

$$\phi R_v = \phi(0.6F_{yc}d_c t_w) = 0.9 \times 0.6 \times 36 \times 12.06 \times 0.35$$

$$\phi R_v = 82^k < V_{uc} = 138^k$$

∴ need to either provide a web doubler plate or a diagonal stiffener. Use a doubler plate:

$t_{req'd} \geq (138 - 82) / (0.9 \times 0.6 \times 36 \times 12.06) = 0.24''$ ∴ use 1/4" Plate

The doubler plate must clear the root of the fillets:

$$w = d_c - 2k_c = 12.06 - 2 \times 1.25 = 9.56''$$

∴ **<u>use PL 9 x 1/4 x 1'-9"</u>**

Check flange bending (Eqn.K1-1):

$$\phi R_n = \phi(6.25)\,t_f^2 F_{yf} = 0.9 \times 6.25 \times 0.58^2 \times 36 = 68^k < 138^k$$

∴ need a stiffener at the top. Check compression buckling of the web (Eqn. K1-8):

$$\phi R_n = \phi 4100 t^3 \sqrt{F_{yw}} / d_c$$

$$= 0.9 \times 4100 \times 0.35^3 \times (36)^{1/2} / 12.06 = 78^k < 152^k$$

∴ need a stiffener at the bottom as well. Follow the AISC guidelines for stiffeners (P 10-41, AISC Manual-II):

$$R_{u\ st} = P_{uf} - \phi R_{n\ min} = 152 - 78 = 74^k$$

$$A_{st} = R_{u\ st} / \phi F_y = 74 / (0.9 \times 36) = 2.28\ \text{in}^2$$

From section K1.9:

$$w_{st} + t_w / 2 \geq (1/3)b_f = 0.33 \times 7.6 = 2.5''$$

$$t_{st} \geq t_b / 2 = 0.75 / 2 = 0.375''$$

Beam is on one side of column only ∴ $L \geq d_c / 2 = 12.06 / 2 = 6.03''$

Try $w_s = 4''$: $t_{st} \geq (2.28 / 2) / 4 = 0.28''$; say 3 / 8"

Welds (Table J2) $\omega_{min} = 1/4''$

$$\phi f_w = 4 \times 1.39 = 5.56^k/_{in}$$

web thickness of the W12x53 is adequate to develop the 1/4" welds on both sides

$$L_w \geq (74 / 2) / 5.56 = 6.6''; \text{ say } 7''$$

∴ **<u>use 2PL 3 x 3 / 8 x 0'-7" for transverse stiffeners.</u>**

For the shear connection, use a single plate web connection to transfer $V_u = 84^k$. Try 3/4" diameter A325-N: from Table 8-17 AISC Manual-II, $\phi r_v = 15.9^k/_{bolt}$

$N \geq 84 / 15.9 = 6$ (too many); try 7/8" diameter A325-X:

$\phi r_v = 20.7^k/_{bolt}$

$N \geq 84 / 20.7 = 4.06$; say 4 bolts

let $t \cong t_w = 7/16''$; $\omega_{min} = 5/16''$ both sides of A36 material:

$$\phi f_w \leq \begin{cases} 0.9 \times 0.6 \times 36 \times 0.45 = 8.8^k\ /_{in} \leftarrow \text{governs} \\ 2 \times 5 \times 1.39 = 13.9^k\ /_{in} \end{cases}$$

$L_w \geq 84 / 8.8 = 9.5$; use 12"

Check shear rupture (re: J4):

$$w_{ns} = 12 - 4(0.75 + 1/16) = 8.75''$$

$\phi V_{ns} = 0.75 \times 0.6 \times 58 \times 8.75 \times 7/16 = 100^k > V_u$ ∴ O.K.

Ductility requires that the distance to the edge of the shear tab on either side be at least two bolt diameters ∴ **<u>use PL 4 x 7/16 x 1'-0"</u>**

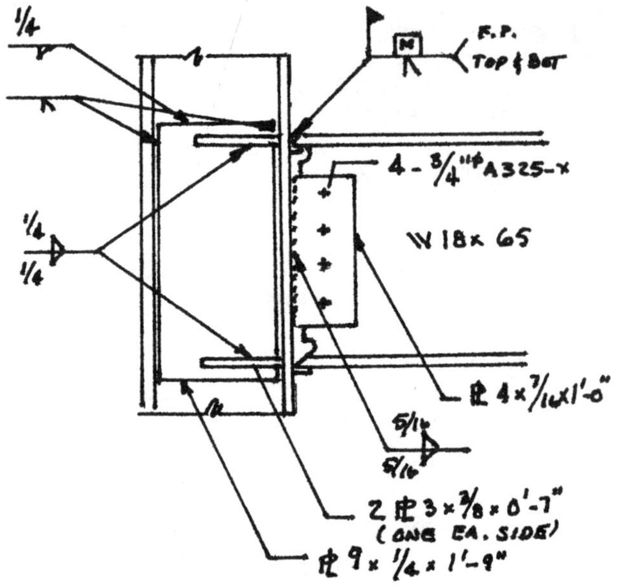

CONNECTION DETAIL (N.T.S.)

Steel-47

a) Analyze the modified frame for dead and lateral forces. I assume that the lateral force (given) was determined for a cantilevered column system (1629.6.6, UBC) and that the special detailing requirements of 2213.11, UBC do not have to be satisfied.

- Analysis of dead loads (Note: the flexibility of the W10x49 about its weak axis is such that the column shear is practically zero under the symmetrical dead loading).

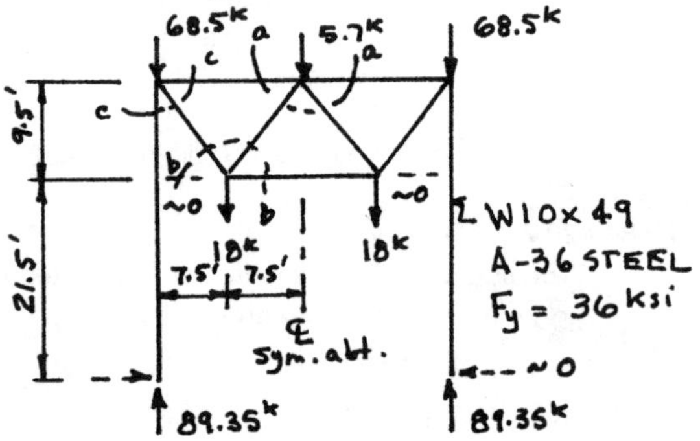

Freebody "a-a":

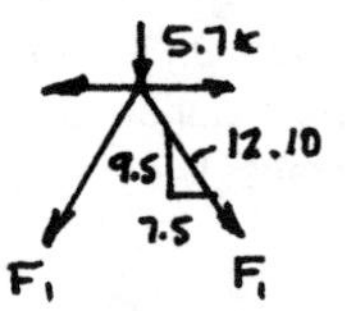

$\Sigma F_y = 0$: $2F_1(9.5 / 12.1) = 5.7$
$F_1 = 3.63^k$ (comp.)
Freebody "b-b":

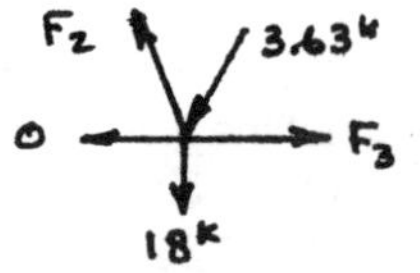

$\Sigma F_y = 0$: $-18 - (9.5 / 12.1)(3.63 - F_2) = 0$
$F_2 = 26.56^k$ (ten.)
$\Sigma F_x = 0$: $(26.56 + 3.63)(7.5 / 12.1) = F_3$
$F_3 = 18.7^k$ (ten.)
Freebody "c-c":

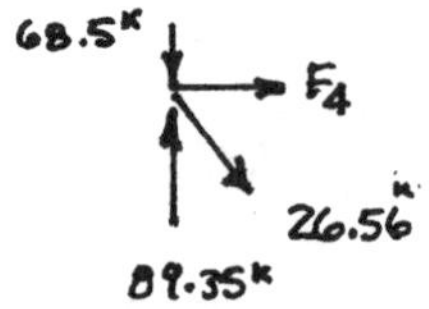

$\Sigma F_x = 0$: $F_4 + (26.56)(7.5 / 12.1) = 0$
$F_4 = 16.5^k$ (comp.)

- Analysis of lateral seismic force

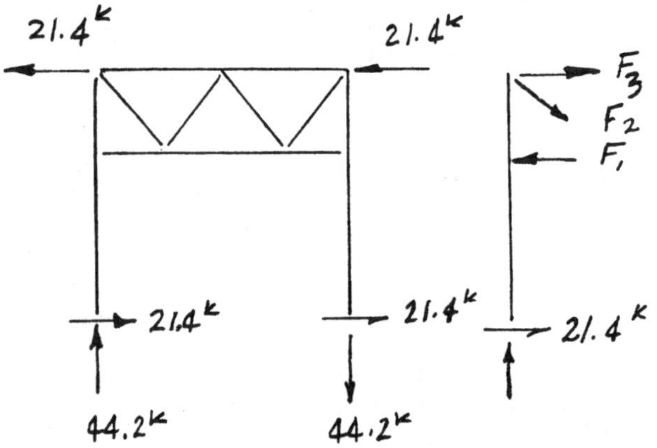

From a freebody diagram cut adjacent to the left column:

$\Sigma M_{y=30'} = 0$: $21.4(30) - 9.5F_1 = 0$
$F_1 = 67.6^k$ (comp.)
$\Sigma F_y = 0$: $(-9.5 / 12.1)F_2 + 44.2 = 0$
$F_2 = 56.3^k$ (tension)
$\Sigma F_x = 0$: $21.4 - 67.6 + (7.5 / 12.1)56.3 + F_3 - 21.4 = 0$
$F_3 = 32.7^k$ (tension)

Summary:

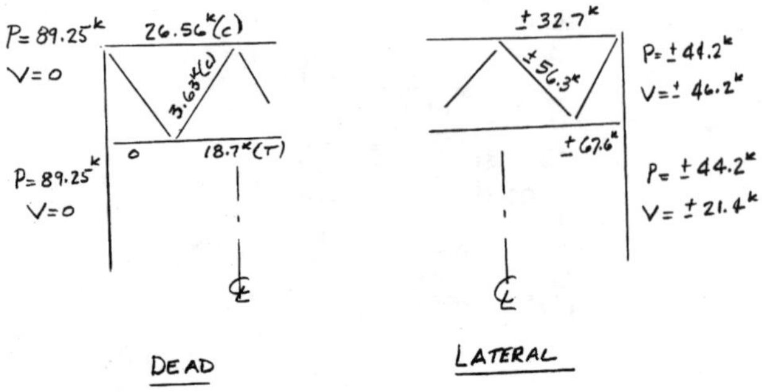

b) Reinforce the column. Use LRFD (UBC-97, Chapter 22, Division IV; AISC 1994). The critical condition is axial compression plus bending:

$P_u = 1.2D + E = 1.2 \times 89.35 + 44.2 = 151^k$

$M_u = M_E = 21.4 \times 21.5 = 460^{k\text{-}ft}$

Check bending in the weak axis at elevation 21.5': (W10x49: $Z_y = 28.3$ in^3)

$M_{py} = F_y Z_y = 36 \times 28.3 / 12 = 84.9^{k\text{-}ft} << M_u$

The modified frame requires a substantial increase in bending resistance at elevation 21.5'. The additional strength needed is

$\phi M_n = \phi_b(M_{py} + \Delta M_p) > 460^{k\text{-}ft}$

$\therefore \quad \Delta M_p = 460 / 0.9 - 84.9 = 426^{k\text{-}ft}$

Re: P. 4-18, AISC Manual, we would need the equivalent of a W21x68 bending about its strong axis to furnish this strength. Try 2 WT8x50 welded to the web of the W10x49:

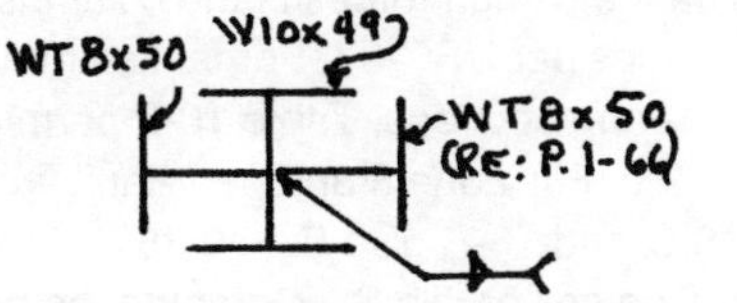

Properties of the builtup section (use data for a W16x100 as being approximately equivalent to the 2 WT8x50s): $A_g = 29.4 + 14.4 = 43.8$ in^2; $r_x > (r_x)_{W16x100} = 7.10"$; $r_y > (r_y)_{W16x100} = 2.51"$; For a W16x100: $L_b = 21.5'$; $L_p = 10.5'$; $L_r = 42.1'$

$M_r = 341 / 0.9 = 379^{k\text{-}ft}$

$M_p = F_y[Z_{y,W10} + 2A_{WT}(d_{WT} - y_p + t_w / 2)]$

$= 36[28.3 + 2 \times 14.7(8.49 - 0.71 + 0.34 / 2)] / 12$

$= 786^{k\text{-}ft}$

$L_p < L_b < L_r \therefore$ Use Eqn F1-3:

$\phi M_n = C_b[\phi M_p - (\phi M_p - \phi M_r)(L_b - L_p) / (L_r - L_p)] \leq \phi M_p$

$M_n = 1.67[786 - (786 - 379)(21.5 - 10.5) / (42.1 - 10.5)] = 1075^{k\text{-}ft}$

$\therefore \quad M_n = M_p = 786^{k\text{-}ft}$; $M_{nt} = 0$; $M_{lt} = 460^{k\text{-}ft}$

Check the combined stresses (H1.2):

$B_2 = 1 / (1 - \Sigma P_u / \Sigma P_e)$

$KL_x / r_x = 1.2 \times 21.5 \times 12 / 7.1 = 44$

$KL_y / r_y = 31 \times 12 / 2.51 = 149 \leftarrow$ governs

$\therefore \phi F_c = 9.6$ ksi

Re: P.6-154, AISC, $P_{ex} = 148 \times 43.8 = 6480^k$

$\therefore \quad B_2 = 1 / (1 - 151 / 6480) = 1.02$

$M_u = B_2 M_{lt} = 1.02 \times 460 = 469^{k\text{-}ft}$

$\phi P_n = 9.6 \times 43.8 = 420^k$

$P_u / \phi P_n = 151 / 420 = 0.36 > 0.2 \therefore$ check eqn. H1-1a

$P_u / \phi P_n + 8 / 9(M_{ux} / \phi M_{nx} + M_{uy} / \phi M_{ny})$

$= 0.36 + 8 / 9(469 / (0.9 \times 786)) = 0.95$

$\therefore$ **Use 2WT8x50 combined with the W10x49**

Size welds to attach the WT8x50 to the web of the W10x49. Use 1/4" E70xx (Re: J5.3); welds both sides of A36 material ($F_y = 36$ ksi):

$$\phi f_w \leq \begin{cases} 0.9 \times 0.6 \times 36 \times (0.34 / 2) \\ = 3.3^k / _{in} \leftarrow \text{governs} \\ 4 \times 1.39 = 5.6^k / _{in} \end{cases}$$

$L_w \geq F_y A_{WT} / \phi f_w = (36 \times 14.7) / 3.3 = 160"$ (total; 80" each side)

Connection of the W12x40 ($t_w = 5 / 16"$); $P_u = \pm 67.6^k$

$\phi f_w = 0.9 \times 0.6 \times 36 \times 5 / 16 = 6.1^k/_{in}$

$L_w = 67.6 / 6.1 = 11.1"$; say full 12" depth

Use A325 bolts to the flange of the WT8x50: $\phi R_n = 29.8^k/_{bolt}$ (tension) $\therefore$ $n \geq 67.6 / 29.8 = 2.3$; say 4 bolts. Use 5-1/2" gage: $M_u = 35 \times 2.75 = 96^{k\text{-}in}$. For 12" long connection plate:

$m_u = 96 / 12 = 8^{k\text{-}in}/_{in}$

$\phi M_p = 0.9 \times 36 \times t^2 / 4 \geq 8$

$t \geq 0.99"$; say 1" Plate

c) The required details are:

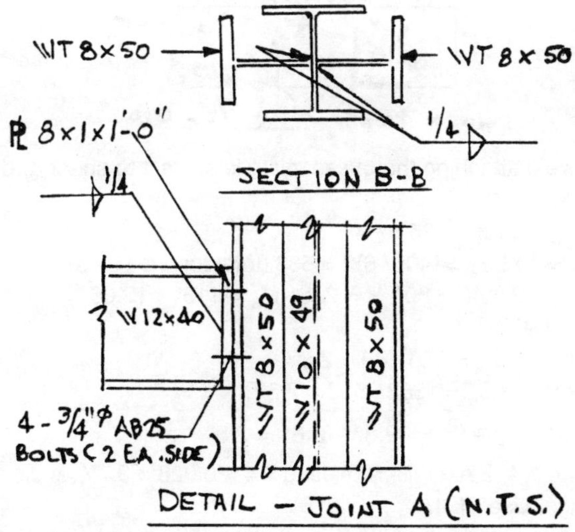

Steel-48

Design the truss connection using Allowable Stress Design (AISC 1989). Use 3/4" diameter A325-SC field bolts and E70xx shop welds.

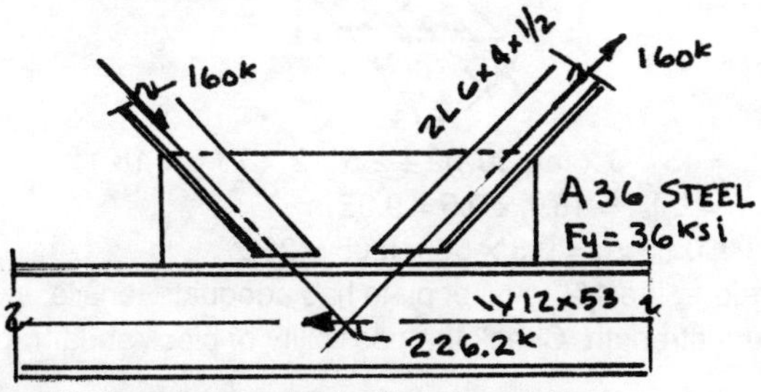

Assumed holes are punched rather than drilled: $\phi_h = \phi_b + 1 / 8"$. Calculate the number of bolts required to transfer $P = T = 160^k$ (Re p. 4-5, AISC):

$r_v = 15.0^k/_{bolts}$ (standard holes; double shear)

$r_b > 15.0^k/_{bolt}$ for bearing on plates 5/16" or thicker

$\therefore \quad n \geq 160 / 15 = 10.6$; say 11

Check the strength of the 2L6x4x1/2 with double gage of 3/4" diameter bolts (without staggering the holes):

$A_n = 2[6 + 4 - 0.5 - 2(3/4 + 1/8)](0.5) = 7.75\ \text{in}^2$

$$T \le \begin{cases} 0.6F_yA_g = 0.6 \times 36 \times 2(9.5 \times 0.5) = 209^k \\ 0.5F_uUA_n = 0.5 \times 58 \times 0.85 \times 7.75 \\ \quad = 191^k \leftarrow \text{governs} \end{cases}$$

$T = 191^k > P = 160^k$ ∴ holes without stagger are O.K.

Design welds to connect the gusset plate to the flange of the W12x53 (Re: Table J2.4):

$\omega_{min} = 1/4"$ (weld is to both sides)

$q = 2 \times 4 \times 0.928 = 7.4^k/_{in}$

Make the gusset plate thick enough so that shear of the base material is not the limiting criterion:

$0.4F_yt_{PL} = 0.4 \times 36t_{PL} > 7.4^k/_{in}$

∴ $t_{PL} \ge 0.51$; say 1/2"

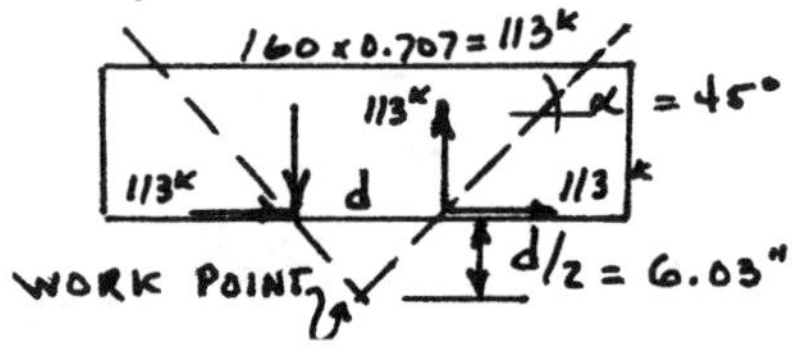

The weld attaching the gusset plate is subject to shear and bending:

$L_w > V/q = 226/7.4 = 31"$

Try L = 40": $S_w = (40^2/6)2 = 533$ (total for two sides)

$L_w = 2 \times 40 = 80"$; $V = 226^k$; $M_w = 113^k \times 12.06"$

$= 1363^{k\text{-}in}$

$f_v = V/L_w = 226/80 = 2.8^k/_{in}$

$f_b = M/S_w = 1363/533 = 2.6^k/_{in}$

$f_r = \sqrt{f_v^2 + f_b^2} = \sqrt{2.8^2 + 2.6^2} = 3.8^k/_{in}$

For the 1/4" E70xx fillet welds: $q = 4 \times 0.928 = 3.7^k/_{in}$ is close enough. Use PL 1/2" x _ x 3'-6".

Check the strength of the gusset plate at the bolt holes. Bolt pitch = 3"; gage = 2-1/2"

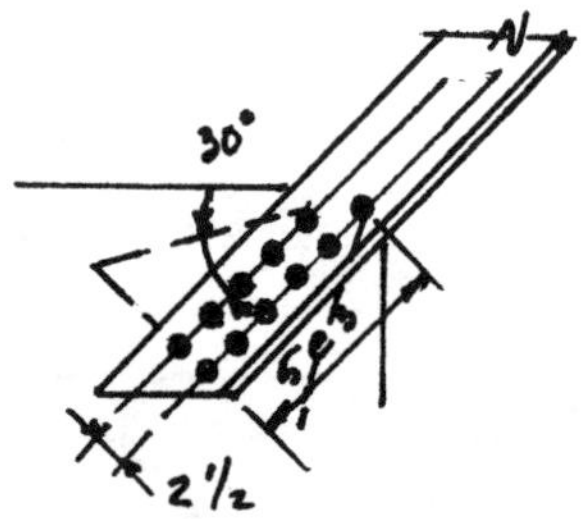

$L_e = (5 \times 3 \times \tan(30^\circ))2 + 2.5 - 2 \times 7/8 = 18.1"$

$A_e = L_et_{PL} = 18.1 \times 0.5 = 9.05\ \text{in}^2$

$T \le 0.5F_uA_e = 0.5 \times 58 \times 9.05 = 262^k$

Therefore, the 1/2" gusset plate has adequate tensile fracture strength. Check the possibility of block shear rupture (J4):

$A_{ns} \ge (13.5 - (4 \times 7/8) + 16.5 - (5 \times 7/8)]0.5$

$= 11.06\ \text{in}^2$

$A_{nt} = (2.5 - 7/8)0.5 = 0.813\ \text{in}^2$

$T \le 0.3F_uA_{ns} + 0.5F_uA_{nt}$

$= 0.3 \times 58 \times 11.06 + 0.5 \times 58 \times 0.813 = 216^k$

∴ the plate is adequate. Dimension the plate to accommodate five spaces at 3-1/2' o.c. plus 1-1/2". For edge distance: $w \ge 2 \times 1.5 + 15 \times 0.707 = 13.6"$.

<u>Use PL 14 x 1/2 x 3'-6"</u>

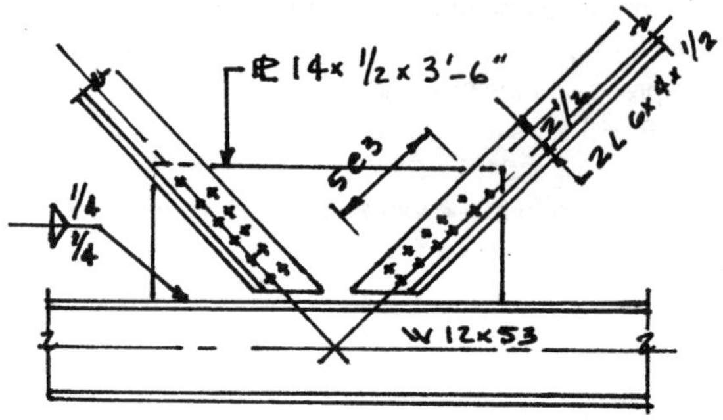

CONNECTION – ELEVATION (N.T.S.)

Steel-49

a) Determine the design forces in the columns. Calculate seismic forces per UBC-97. For tanks and vessels on braced or unbraced legs, Table 16-P permits R = 2.2 with $\Omega_o = 2.0$. For Zone 3 and near-fault factors of 1.0, Tables16-Q and 16-R give $C_a = 0.36$ and $C_v = 0.54$ for structures on soil profile S_D. Compute the period in a principal direction (1634.1.4, UBC-97):

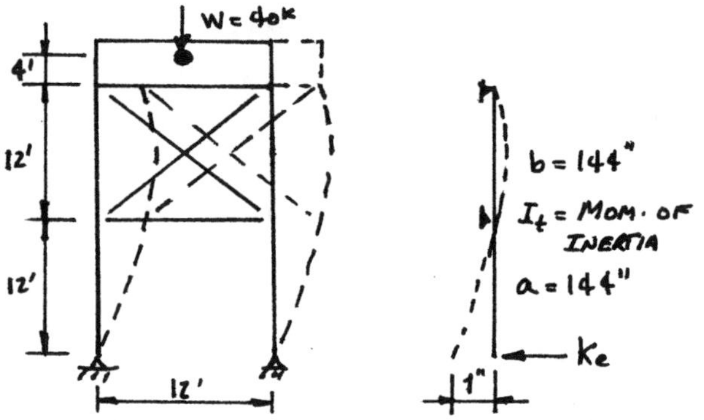

Determine the period by treating the structure as a single degree of freedom system. The equivalent stiffness of four legs is:

$\Delta = 1" = (K_e/3EI_t)(a^2b + a^3)$

$K_e = 3 \times 29{,}000 \times I_t/(2 \times 144^3) = 0.0146I_t$

where I_t is the moment of inertia of four legs. As a first approximation, choose a standard wall pipe with $KL/r > 100$; estimate $K \cong 2.0$

∴ $r > 2 \times 12 \times 12/100 = 2.9"$; Try Pipe 8 Std

($r = 2.94"$; $I = 73\ \text{in}^3$)

$K_e = 0.0146 \times 4 \times 73 = 4.26^k/_{in}$

$T = 2\pi\sqrt{M/K} = 2\pi\sqrt{(40/(32.2 \times 12))/4.26}$

$= 0.97$ sec

Compute the base shear coefficient for zone 3:

Limit $\ge \{0.11C_aI = 0.040;\ 0.8ZN_vI/R = 0.8 \times 0.3 \times 1.0/2.2 = 0.11\} = 0.11$

$C_vI/RT = 0.54 \times 1/(2.2 \times 0.97) = 0.25 < 2.5C_aI/R$

∴ $V = (C_vI/RT)W = 0.25 \times 40 = 10^k$

The total base shear is resisted equally by two parallel bents (neglecting any accidental torsion); thus, column reactions due to seismic loads are:

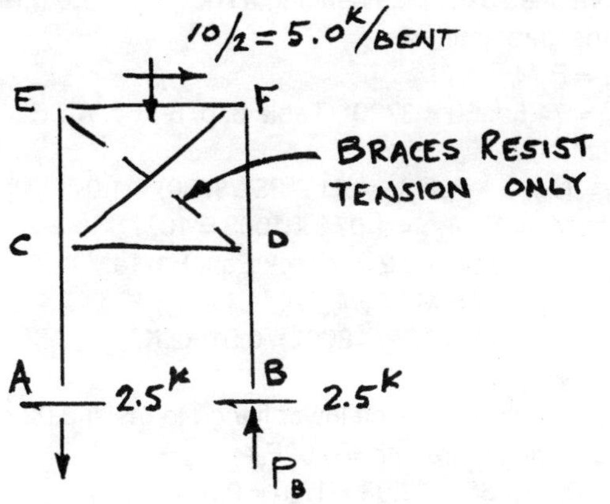

$V_{col} = V / 4 = 10 / 4 = 2.5^k$

$M_{col} = V_{col}H = 2.5 \times 12 = 30^{k\text{-ft}}$

$P_{col} = (V(24 + 4) / 12) / 2 = \pm 11.7^k$

b) The maximum column moment occurs at elevation 12':

$M_{max} = M_e = 30^{k\text{-ft}}$

c) For the braces and struts:

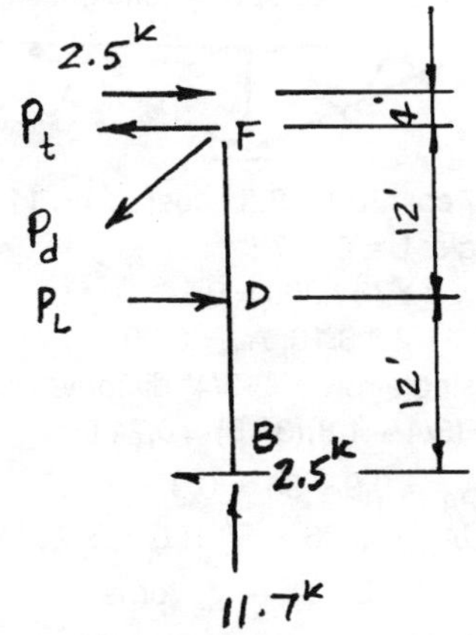

$\Sigma F_v = 0$: $0.707P_d = 11.7$

$\underline{P_d = 16.5^k \text{ (ten.)}}$

$\Sigma M_F = 0$: $P_L(12) = 2.5(24 - 4)$

$\underline{P_L = 4.17^k \text{ (comp.)}}$

$\Sigma F_x = 0$: $-P_t - 0.707 \times 16.5 + 4.17 = 0$

$\underline{P_t = 7.5\text{k (comp.)}}$

Symmetrical gravity load distributes equally to four legs

$\therefore P_g = 10^k$

d) Check the trial Pipe 8 Std. Use LRFD (UBC-97, Chapter 22, Division IV; AISC 1994); steel is A501 ($F_y = 36$ ksi). Treating the given gravity load as dead (U = 1.2, the design forces for the column are:

$P_u = 1.2P_g + P_e = 1.2 \times 10 + 11.7 = 23.7^k$

$P_u = 0.9P_g - P_e = 9 - 11.7 = -2.7^k$ (tension)

$M_{lt} = M_e = 30^{k\text{-ft}}$

$M_u = B_1M_{nt} + B_2M_{lt}$ (Re: H1-2)

$B_2 = 1 / (1 - \Sigma P_u / \Sigma P_e)$

$KL / r = 2.0 \times 12 \times 12 / 2.94 = 98$

$\therefore$ $\phi F_c = 18.46$ ksi; $P_e = 29.8 \times 8.40 = 250^k$

$B_2 = 1 / (1 - 23.7 / 250) = 1.10$

$M_u = B_2M_{lt} = 1.10 \times 30 = 33^{k\text{-ft}}$

$\phi M_n = \phi F_yZ_x = 0.9 \times 36 \times 22.2 / 12 = 60^{k\text{-ft}}$

$\phi P_n = 18.46 \times 8.40 = 155^k$

$P_u / \phi P_n = 23.7 / 155 = 0.15 < 0.2$ $\therefore$ check eqn. H1-1b

$P_u / 2\phi P_n + (M_{ux} / \phi M_{nx} + M_{uy} / \phi M_{ny})$

$= (0.15 / 2) + (33 / 60) = 0.63 < 1.0$ $\therefore$ O.K.

Note: the next smallest standard pipe is Pipe 6 Std, which is obviously no good because its Z_x is only one half that of the Pipe 8 Std. Since $P_u / \phi P_n < 0.5$, the special loading condition $1.2P_d + \Omega_o P_e$ does not govern.

$\therefore$ **Use Pipe 8 Std**

e) Discuss the effect of base fixity condition: The effect of base fixity versus hinged bases is beneficial in most respects. However, the increased stiffness would reduce the period, leading to an increase in the base shear for this structure. But this would be offset by the reduced moments in the columns and the reduced drift (lower moment amplification) and the effective length factor for the column would be roughly halved. Axial forces in the columns would also be reduced due to the resistance to overturning provided by the moments at the base.

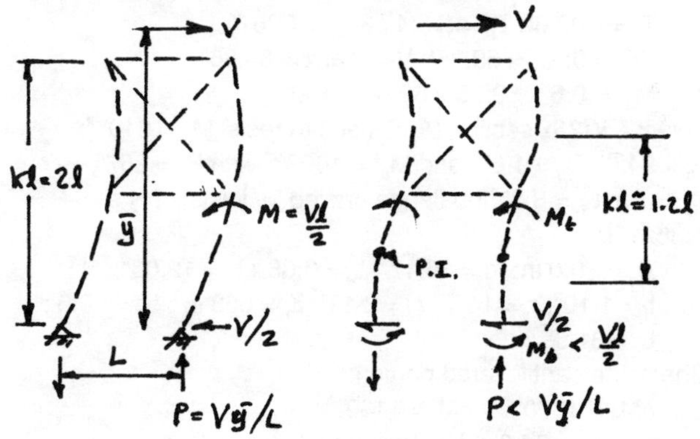

Steel-50

Use LRFD (AISC 1994). Consider strength only (i.e., deflections and ponding are not checked). Consider the effects of unbalanced roof live loads (1607.4.3, UBC-97). Provide lateral braces at ends and at 25' intervals.

$w_u = 1.2w_D + 1.6w_L = 1.2 \times 0.62 + 1.6 \times 0.63 = 1.75$ klf

$P_u = 1.2\ P_D = 1.2 \times 9.5 = 11.4^k$

For loading applied over the entire length:

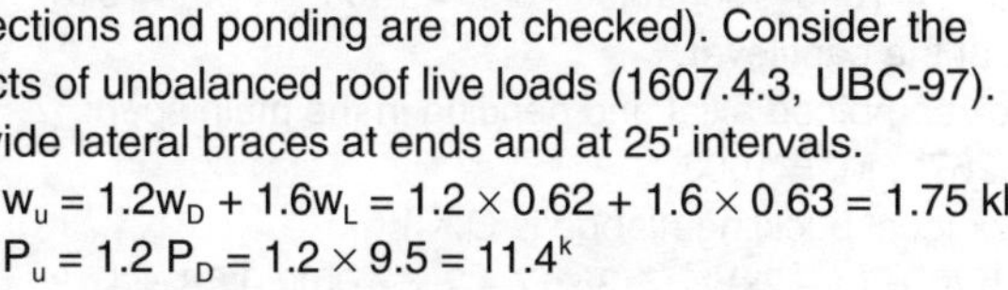

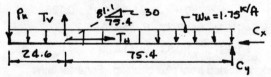

$\Sigma M_C = 0$: $11.4(100) + 1.75(100)^2 / 2 - T_V(75.4) = 0$

$T_V = 131.2^k$; $T_H = (75.4 / 30) \times 131.2 = 330^k$

$\Sigma F_y = 0$: $C_y = 55^k$; $\Sigma F_x = 0$: $C_x = 330^k$

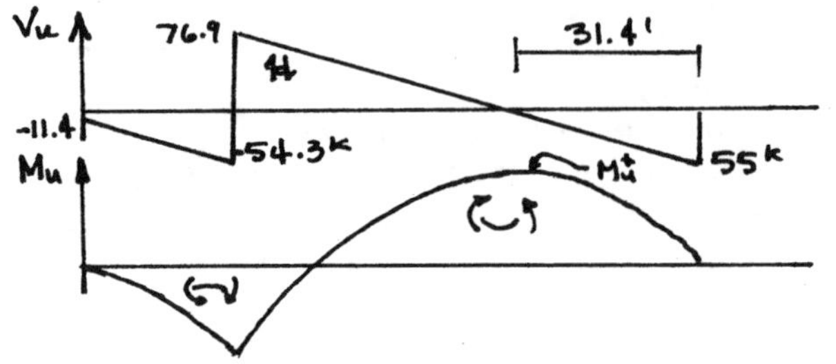

$M_u^- = (11.4 \times 24.6 + 1.75(24.6)^2 / 2) = -810^{k\text{-ft}}$

$M_u^+ = 0.5 \times 31.4 \times 55 = 864^{k\text{-ft}}$

For pattern live load with live on cantilever omitted:

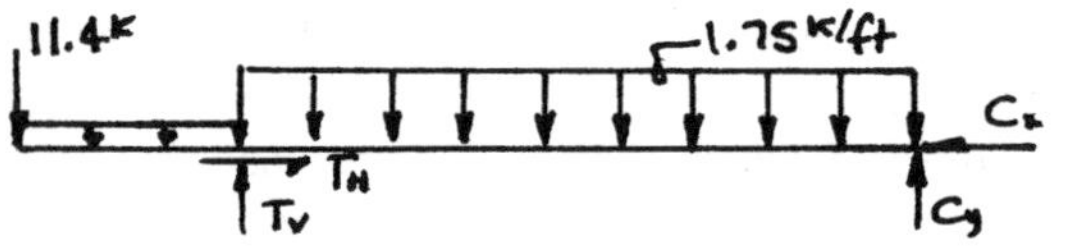

$w_{ud} = 1.2 \times 0.62 = 0.74$ klf

$\Sigma M_C = 0$: $0.74 \times 24.6(75.4 + 24.6/2) + 11.4(100) + 1.75 \times 75.4^2 / 2 - T_V(75.4) = 0$

$T_V = 102.3^k$; $T_H = (75.4 / 30)T_V = 257^k$

$\Sigma F_y = 0$: $C_y = 59.25^k$; $V = 0$ at $x = 33.86'$

$\therefore \quad M_u^+ = 0.5 \times 59.25 \times 33.86 = 1003^{k\text{-ft}}$

Select a W36x section (A36 steel) to resist $M_u^- = 810^{k\text{-ft}}$; $L_b = 24.6'$; $C_b = 1.67$; and $M_u^+ = 1003^{k\text{-ft}}$ with $P_u = 257^k$; $C_b = 1.0$; $L_b = 8.3'$ (top flange braced by joists). Try a W36x170:

$A_g = 50.0 \text{ in}^2$; $d = 36.17"$; $t_w = 0.68"$; $b_f = 12.03"$; $t_f = 1.10"$; $r_x = 14.5"$; $r_y = 2.43"$; $Z_x = 668 \text{ in}^3$; $L_p = 10.5'$; $L_r = 31.9'$.

Check the cantilevered portion:

$\phi M_p = 1800^{k\text{-ft}}$; $\phi M_r = 1130^{k\text{-ft}}$

$L_p < L_b = 24.6' < L_r$: Use Eqn F1-3:

$\phi M_n = C_b[\phi M_p - (\phi M_p - \phi M_r)(L_b - L_p)/(L_r - L_p)] \le \phi M_p$

$\phi M_n = 1.67[1800 - (1800 - 1130)(24.6 - 10.5) / (31.9 - 10.5)] = 2280^{k\text{-ft}}$

But $\phi M_p = 1800^{k\text{-ft}}$ controls $\therefore \phi M_n = 1800^{k\text{-ft}} > M_u^- = 810^{k\text{-ft}}$ is O.K. at the cantilever.

Check combined axial and bending in the main span:

$P_u = 257^k$; $M_u = 1003^{k\text{-ft}}$

Check local buckling (flange is O.K.):

$h_c/t_w = (d - 2k)/t_w = (36.17 - 2 \times 2.00) / 0.68 = 47.3$

(Re: Table B 5.1, AISCS)

$P_u/\phi_b P_y = 257/(0.9 \times 36 \times 50) = 0.15 > 0.125$

$$\lambda_p = (191/\sqrt{F_y})(2.33 - P_u / \phi_b P_y) \ge 253\sqrt{F_y}$$

$$= (191/\sqrt{F_y})(2.33 - 0.15) = 69 > h_c / t_w \quad \therefore \text{ o.k.}$$

Check combined axial plus bending. For axial only:

$KL / r)_x = 75.4 \times 12 / 14.5 = 62$

$KL / r)_y = 25.1 \times 12 / 2.43 = 124 \leftarrow$ governs

$\phi_c F_{cr} = 13.6$ ksi

$\phi P_n = \phi_c F_{cr} A_g = 13.6 \times 50.0 = 680^k$

$P_u / \phi P_n = 257 / 680 = 0.38$

For bending only:

$$L_b = 8.3' < L_p = 10.5' \therefore \phi M_n = \phi M_p = 1800^{k\text{-ft}}$$

Combined axial plus bending (H1-2): $C_m = 1.0$ (braced; ends unrestrained)

$M_u = B_1 M_{nt}$

$P_{ex} = 74.5 \times 50 = 3730^k$ (Table 8, p. 6-154, AISC Manual)

$B_1 = C_m/(1 - P_u/P_e) = 1/(1 - 257/3730) = 1.074$

$\therefore \quad M_{ux} = B_1 M_{nt,x} = 1.074 \times 1003 = 1077^{k\text{-ft}}$

$P_u / \phi P_n = 0.38 > 0.2 \therefore$ check Eqn. H1-1a:

$P_u / \phi P_n + 8 / 9(M_{ux} / \phi M_{nx} + M_{uy} / \phi M_{ny})$

$= 0.38 + 8 / 9(1077 / 1800) = 0.91$ O.K.

$\therefore$ **Use W36x170**

b) Provide a brace to the lower flange to resist 0.02 of the maximum flange strength $= 0.02 F_y A_f$

$P = 0.02 \times 36 \times 12.03 \times 1.10 = 9.5^k \leftarrow \text{or} \rightarrow$

We could use the lower chord of the joists provided that the joists are designed to resist the compression that would develop in the lower chord. An alternative is to provide a brace from the upper chord where the decking will furnish the necessary strength and stiffness. Place the braces on both sides of the W36x170 inclined at, say, 30° to the horizontal and design to resist only tension:

$P_{brace} = 9.5 / \cos(30°) = 9.5 / \cos(30°) = 11^k$

Use a single angle: $L \cong 6' = 72"$

$L / r \le 300 \therefore r \ge 73 / 300 = 0.24"$

Try an L2-1/2 x 2-1/2 x 3/16: $A_g = 0.90 \text{ in}^2$

Connect with a single row of 2-3/4" diameter A307 bolts

$A_n = 0.90 - (3/4 + 1/8)(3/16) = 0.74 \text{ in}^2$

$$\phi T \le \begin{cases} \phi F_y A_g = 0.9 \times 36 \times 0.9 = 29.1^k \\ \phi F_u U A_n = 0.75 \times 58 \times 0.75 \times 0.74 \\ \quad = 24.0^k \leftarrow \text{governs} \end{cases}$$

$\therefore$ **Use L 2-1/2 x 2-1/2 x 3/16.** Attach the brace at upper point using 3/16 Plate welded to resist $T_u = 11.0^k$.

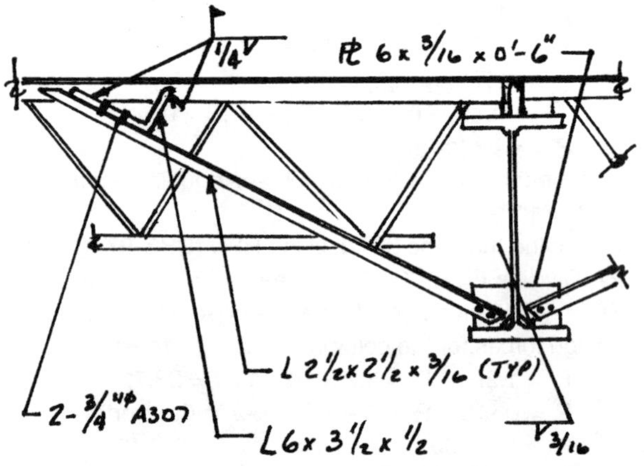

BRACE DETAIL (N.T.S.)

(BRACES TO BE LOCATED AT END OF CANTILEVER; AT POINT OF SUSPENSION; & AT QUARTER POINTS OF 75.4' SPAN)

Steel-51

Design the rigid connection following the guidelines of FEMA 267A, Advisory No. 1 (SAC-96-03). Of the non-proprietary connections that have been tested, the reduced beam section offers one of the more reliable approaches to assuring that plastic hinges will form away from the column faces, and is the choice for this solution. This approach typically increases drift on the order of 5% (Iwankiw 1997), which is tolerable. For the non-reduced W21x62 girder: d = 20.99"; t_w = 0.40";
b_f = 8.24"; t_f = 0.615"; Z_x = 144 in^3; S_x = 127 in^3. For the W14x120 column: A_g = 35.3 in^2; d = 14.48"; t_w = 0.59";
b_f = 14.67"; t_f = 0.940"; Z_x = 212 in^3.
Try reducing the flexural strength 40% using a 14" long "dogbone," starting 5" from the face of column (so that the probable plastic hinge centers at 12" = 1' from the column face). The required reduction in flange width is:

$$Z_{RBS} = 0.6Z_x = Z_x - b_R t_f(d - t_f)$$

$\therefore$ $b_R = 0.4 \times 144 / (0.615 \times (20.99 - 0.615)) = 4.60$", which results in a flange width at the plastic hinge location = 8.4 – 4.6 = 3.8".

Calculate the probable moment strength of the reduced section (Eqn. 7.5.3.2-2, FEMA 267A):

$$M_{pr} = Z_{RBS}\beta F_y = (0.6 \times 144) \times 1.2 \times 50 / 12 = 432^{k\text{-}ft}$$

Using the LRFD approach (AISC 1994), the design moments at column centerline are:

$$M_u \geq 1.2M_D + 1.6M_L = 1.2 \times (0.6 \times 187) + 1.6 \times (0.4 \times 187) = 254^{k\text{-}ft}$$

$$M_u \geq 1.2M_D + 0.5M_L + M_E = 1.2\ (0.6 \times 187) + 0.5 \times (0.4 \times 187) + 139^{k\text{-}ft} = 311^{k\text{-}ft} \leftarrow \text{governs}$$

The design moment at the reduced section will be less than $311^{k\text{-}ft}$, therefore the flexural strength is adequate.

For yielding to occur in the girders, the strength of the columns must satisfy Eqn 7.5.2.5-1-1 of FEMA 267A:

$$\Sigma Z_c(F_{yc} - f_a) / \Sigma M_c \geq 1.0$$

where f_a is given as 15 ksi for the column above and 18 ksi in the column below.

$$\therefore\ \Sigma Z_c(F_{yc} - f_a) = [212(50 - 15) + 212(50 - 18)] / 12 = 1184^{k\text{-}ft}$$

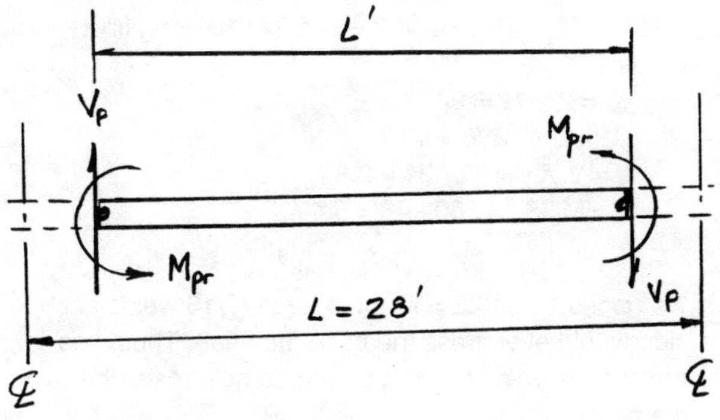

$$L' = 28 - 14.48 / 12 - 2(5 + 7) / 12 = 24.8'$$

$$V_p = 2M_{pr} / L' = 2 \times 432 / 24.8 = 34.8^k$$

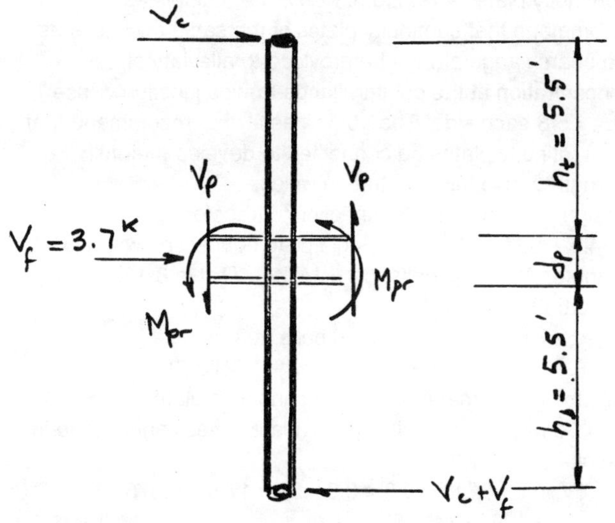

From statics:

$$V_c = \{\Sigma[M_{pr} + V_p(L - L') / 2] - V_f(h_b + d_p / 2)\} / (h_b + d_p + h_a)$$

$$V_c = \{2[432 + 34.8(28 - 24.8) / 2] - 3.7(5.5 + 1.75 / 2)\} / (5.5 + 1.75 + 5.5)$$

$$V_c = 74.6^k$$

$$M_{ct} = V_c h_a = 74.6 \times 5.5 = 410^{k\text{-}ft}$$

$$M_{cb} = (V_c + V_f)h_b = (74.6 + 3.7) \times 5.5 = 430^{k\text{-}ft}$$

$$\therefore\ \Sigma M_c = 410 + 430 = 840^{k\text{-}ft}$$

$$\Sigma Z_c(F_{yc} - f_a) / \Sigma M_c = 1184 / 840 = 1.40 > 1.0$$

$\therefore$ Yielding should occur in the girders. Verify that stresses at the face of column are acceptable:

$$M_f = M_{pr} + V_p x = 432 + 34.8 \times 1 = 467^{k\text{-}ft}$$

$$f_b = M_f / S_x = 467 \times 12 / 127 = 44.1 \text{ ksi}$$

Thus, the stress imposed on the through-thickness of the column flange satisfies the $0.9F_y$ limit recommended in the FEMA 267A guidelines.

Check panel zone strength (7.5.2.6, FEMA 267A). Since gravity moments are balanced, the panel zone shear is:

$$V \geq \Sigma M_E / (0.95d_b) - V_{c1} = 2 \times 139 / (0.95 \times 20.99 / 12) - 20.3 = 147^k$$

$$V \leq 0.8\Sigma M_f / (0.95d_b) = 0.8 \times 2 \times 467 / (0.95 \times 20.99 / 12) = 450^k$$

Thus, required strength is 147^k. Check Eqn. 8-1, AISC

$$\phi_v V_n = \phi_v 0.6F_y d_c t_p[1 + 3b_{cf}t^2_{cf} / (d_b d_c t_p)]$$

$$\phi_v V_n = 0.75 \times 0.6 \times 50 \times 14.48 \times 0.59[1 + 3 \times 14.67 \times 0.94^2 / (20.99 \times 14.48 \times 0.59)] = 234^k > V = 147^k$$

Check eqn. 8-2, AISC:

$$t_z \geq (d_z + w_z) / 90 = [(20.99 - 2 \times 0.615) + (14.48 - 2 \times 0.94)] / 90 = 0.36"$$

t_z = 0.59" > 0.36" – O.K.

$\therefore$ Column web is adequate for panel zone shear.

Continuity plates – re: 7.8.3, FEM267A, guidelines recommend that continuity plates of the same thickness as the beam flange always be provided to alleviate stress concentration at the column flange-to-web junction ∴ use PL4 x 5/8 each side. The guidelines further recommend that the continuity plates be connected to develop their full strength ∴ use full penetration welds.

Design the web shear connection to transfer $V_p + V_g = 34.8 + 39.2 = 74^k$. Use single plate connection with 7/8" diameter A325-SC bolts: $\phi r_v = 14.5^k/_{bolt}$ (Table 8-17, AISC Manual-II)

$n \geq 74 / 14.5 = 5.1$, say 6 bolts at 3" o.c.

(L = 17-1/2" < T = 18-1/4" of the W21x62)

Assuming A36 material for the connection plate (F_y = 36 ksi; F_u = 58 ksi), try 3/8" plate. Check shear yield in the plate:

$$\phi V_n = \phi 0.6 F_y L t = 0.9 \times 0.6 \times 36 \times 17.5 \times 0.375 = 128^k > 74^k$$

Check shear rupture:

$$A_{ns} = (17.5 - 6 \times (7/8 + 1/8)) \times 3/8 = 4.31 \text{ in}^2$$

$$\phi V_n = \phi 0.6 F_u A_{ns} = 0.75 \times 0.6 \times 58 \times 4.31 = 112^k > 74^k$$ ∴ 3 / 8" connection plate is O.K.

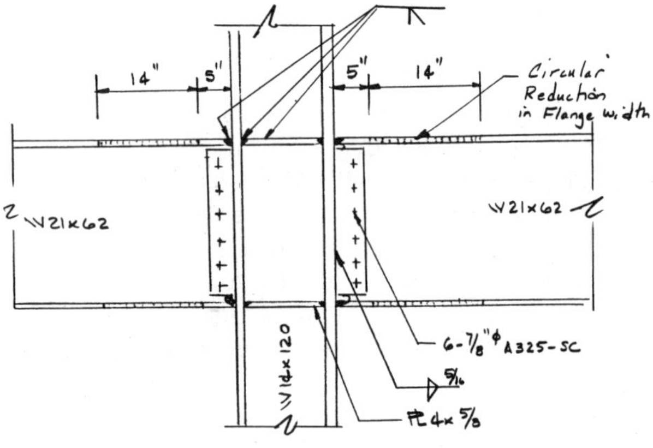

CONNECTION DETAIL (N.T.S.)

Steel-52

a) Design the joint to transfer forces into the W10x45. Use allowable stress design (AISC 1989). Given: steel is ASTM A36 (F_y = 36 ksi); E70xx electrodes. Design fillet welds to connect the angles to 1/2" thick gusset plates trying ω = 1/4" (larger than ω_{min}).

$q = 4 \times 0.928 = 3.7^k/_{in}$

Distribute the weld across the back leg and along edges to balance about the centroidal axis.

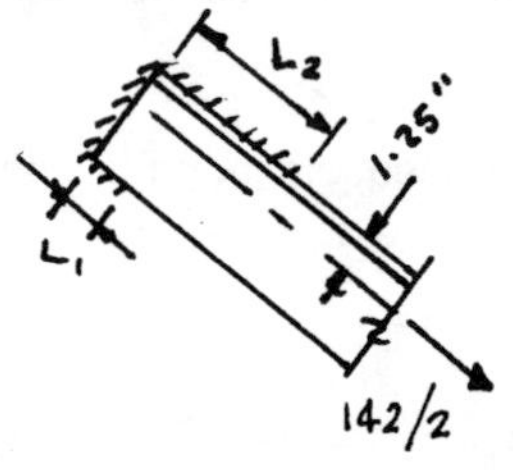

For the 2L4 x 3-1/2 x 1/2:

$\Sigma M_{L2} = 0$: $(142/2)1.25 = 3.7(4 \times 2 + 4L_1)$

∴ $L_1 = 4.0"$

$\Sigma F_{long} = 0$: $(142/2) = 3.7(4.0 + 4 + L_2)$

∴ $L_2 = 11.2"$

For the 2L3 1/2 x 2 1/2 x 5/16 (lightly loaded):

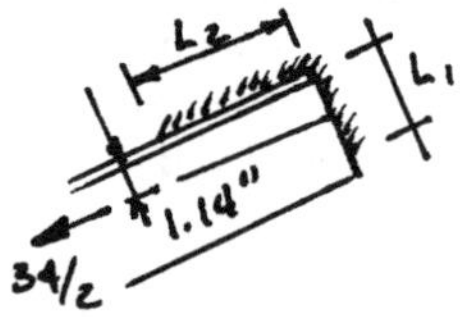

$(34/2)(1.14) = 3.7L_1(L_1/2)$

∴ $L_1 = 3.2"$ (< 3.5" ∴ O.K.)

$34/2 = 3.7(3.2 + L_2)$

∴ $L_2 = 1.4"$

The connection to the W10x45 is simple; that is, no moment transfer. Check the equilibrium of the joint:

$\Sigma F_y = 0$: $P - 24 - 36 - 60 + 14 - 28 = 0$

$P = 134^k$ (comp.)

$\Sigma F_x = 0$: $V = 124 + 24 - 129 - 37 + 18 = 0$

Transfer $P = 134^k$ through paired stiffeners as shown. Assume plates are 5 in wide to match the W10x45 depth: $r = 0.3h = 0.3 \times 10 = 3"$

$Kl/r \cong 12/3 = 4$ ∴ $F_a \cong 0.6F_y = 21.6$ ksi

Neglect the contribution of the 1/2" gusset plate to the stiffener strength (conservative):

$A_{st} \geq (P/F_a)/2 = (134/21.6)/2 = 3.1 \text{ in}^2$

$w_{st} \geq A_{st}/5 = 3.1/5 = 0.62"$; **try PL5 x 5 / 8**

Check local buckling (Re: Table B5.1):

$b/t = 5/0.625 = 8 < 95/\sqrt{F_y} = 95/6 = 15.8$

∴ O.K.

Design welds to attach the stiffeners to the gusset plate. For a WT5x22.5 (half the W10x45): y = 0.907"; d = 5.05"

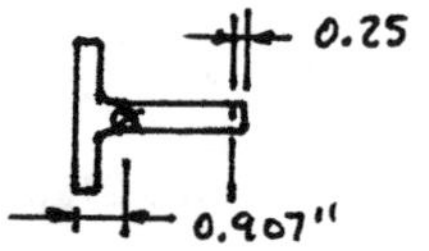

$M_w = F(d - y - 0.25) = (134/2)(5.05 - 0.907 - 0.25) = 260^{k\text{-}in}$; $V_w = P/2 = 134/2 = 67^k$. Try 12" long fillet welds:

$L_w = 2 \times 12 = 24"$

$S_w = 2 \times 12^2/6 = 48 \text{ in}^2$

$f_v = V_w/L_w = 67/24 = 2.8^k/_{in}$

$f_b = M_w/S_w = 260/48 = 5.4^k/_{in}$

$$f_r = \sqrt{f_v^2 + f_b^2} = \sqrt{2.8^2 + 5.4^2} = 6.1^k/_{in}$$

∴ ω > 6.1 / 0.928 = 7 ← too large (7/16 welds each side would overstress the base material). Thus, use transverse welds top and bottom to help resist the moment.

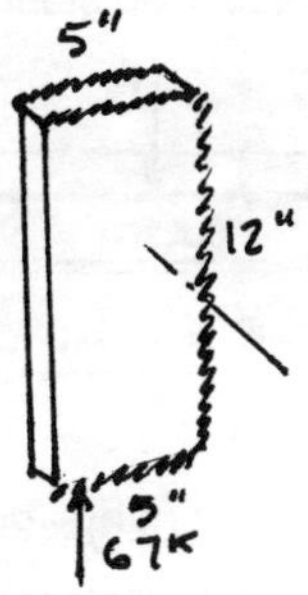

$I_w = 2(12^3/12 + 2 \times 5 \times 6^2) = 1008\ \text{in}^3$

$S_w = I_w/6 = 168\ \text{in}^2$

$f_b = M/S_w = 260 / 168 = 1.5^{k}/_{in}$

$f_r = \sqrt{f_v^2 + f_b^2} = \sqrt{2.8^2 + 1.5^2} = 3.2^{k}\ /_{in}$

$\therefore\quad \omega > 3.2 / 0.928 = 3.4$; **use 1/4" E70xx fillet welds**

Design a cap plate for the W10x45. Use limit analysis (similar to AISC Eqn. K1-1 for local flange bending). Assume a yield line on a 45° diagonal across the width and apply a load factor of 5/3 to the service load stresses:

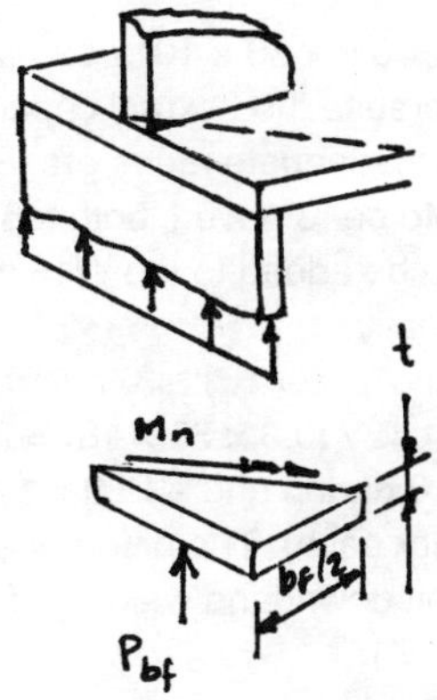

$P_{bf} = 5 / 3f_a(b_f t_f / 2) = 5 / 3(134 / 13.3)(8.02 \times 0.62/2)$
$= 42^k$

$M_u = P_{bf}\ (b_f / 4) \times 0.707 \le \phi M_n$
$= \phi F_y(b_f / (2 \times 0.707))t_f^2 / 4$

The term $(b_f / 4)$ cancels; therefore,

$t^2 \ge 1.1P_{bf} / F_y = 1.1 \times 42 / 36$

$t \ge 1.13"$; **say PL 8 x 1-1/8 x 0'-11"**

(Note: revise the fillet weld size to 5/16" based on the thickness of the cap plate (Re: J2.1, AISC). Try trimming the plate so that it is symmetrical about the vertical stiffener (it does not have to be as long as shown because the welds to angles can be made within the web plate.

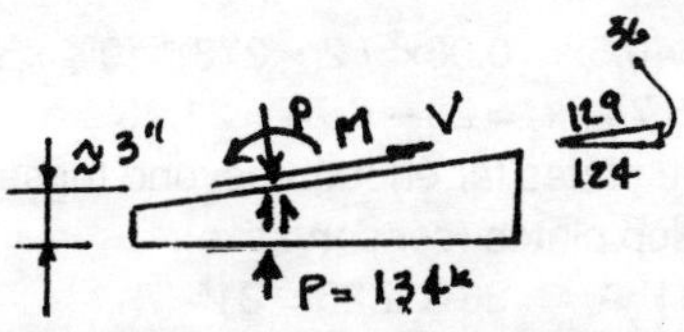

Axial stresses are transferred through the stiffeners. Use flush groove welds to attach gusset plate to the stem of the chord WT (the stresses are O.K. by inspection). Thus, the required details are:

b)

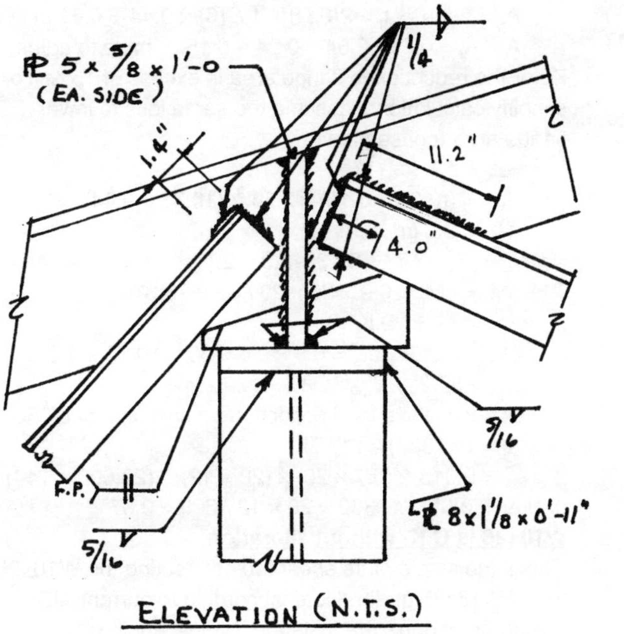

Steel-53

a) Review the existing floor framing. Use the live loads per UBC-97: for area devoted to composing and linotype in a printing plant, W_L = 100 psf (2500# concentrated load might be critical on the deck slab, but not the beams and girders).

Typical floor beam:

$A_{trib} = 7 \times 20 = 140\ \text{ft}^2 < 150\ \text{ft}^2\ \therefore$ no live load reduction permitted

w_L = 100 psf × 7' = 700 plf = 0.7 klf

The girder supports three beams per span (one beam spans directly to column); thus,

$A_{trib} = 3 \times 7 \times 20 = 420\ \text{ft}^2 > 150\ \text{ft}^2\ \therefore$ reducible

$$R \le \begin{cases} 0.08(420 - 150) = 22\% \leftarrow \text{governs} \\ 40\% \\ 23(1 + D/L) = 23(1 + 50/100) = 35\% \end{cases}$$

$\therefore\quad P_L = (1 - 0.22) \times 0.1^{ksf} \times 7' \times 20' = 10.9^k$

Use LRFD (AISC 1994). For the W10x26 beam:

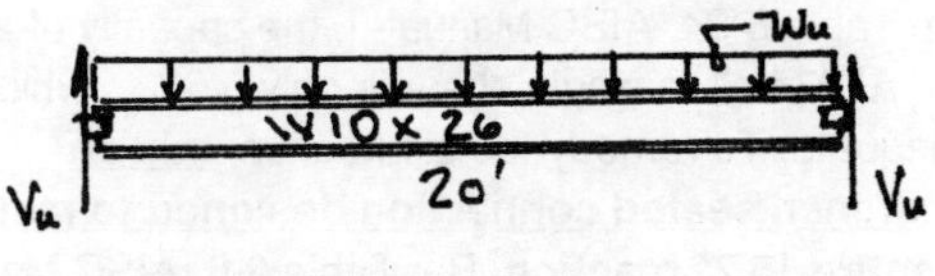

$w_u = 1.2w_D + 1.6w_L$
$= 1.2(0.05 \times 7 + 0.026) + 1.6(0.1 \times 7) = 1.57$ klf

$V_u = 1.57 \times 20 / 2 = 15.7^k$

$M_u = 1.57 \times 20^2 / 8 = 78.6^{k\text{-}ft}$

Properly installed metal deck will brace the compression flange $\therefore L_b = 0$. For the W10x26 (compact in A36 steel): d = 10.33"; t_w = 0.26"; b_f = 5.77"; t_f = 0.44"; Z_x = 31.3 in³; I_x = 144 in⁴.

Assume that the top flange is drilled for two 5 / 8" diameter bolts to attach the top clip angles – check the possible reduction in flexural strength (B1, AISC):

$A_f = b_f t_f = 5.77 \times 0.44 = 2.54\ \text{in}^2$

$A_h = 2 \times \phi_h \times t_f = 2(5/8 + 1/16) \times 0.44 = 0.61\ \text{in}^2$

$A_h / A_f = 0.61 / 2.54 = 0.24 > 0.15\ \therefore$ have to adjust Z_x for the reduction in flange area in excess of 15%. To simplify calculations, assume the same loss to lower flange area (conservative):

$\Delta Z_x = \Delta A_f\,(d - t_f)$

$= ((0.24 - 0.15) \times 2.54 \times (10.33 - 0.44)$

$= 2.3\ \text{in}^3$

$Z'_x = Z_x - \Delta Z_x = 31.3 - 2.3 = 29\ \text{in}^3$

$\phi M_n = \phi M_p = 0.9 \times 36 \times 29 / 12 = 78.3^{\text{k-ft}}$

$\cong M_u\ \therefore$ O.K.

$\phi V_n = \phi(0.6F_y)dt_w = 0.9 \times 0.6 \times 36 \times 10.33 \times 0.26$

$= 52.2^k > V_u\ \therefore$ O.K.

Check service live load deflection against limit of L/360

$\Delta_L = (5/384)w_L L^4 / EI$

$= 0.013 \times (0.7 \times 20) \times (20 \times 12)^3 / (29{,}000 \times 144)$

$\Delta_L = 0.6" < L/360 = 20 \times 12 / 360 = 0.67"$

$\therefore$ **W10x26 is O.K. without alteration.**

Check the single plate shear tab connecting the W10x26 to the W18x35 girder. Note: according to current AISC design procedure, the fasteners for the single plate connection must be designed for direct shear and eccentricity, e ≅ 2-1/4" and the shear tab should have fillet welds on both sides. Checking the bolts (assume standard 3" gage):

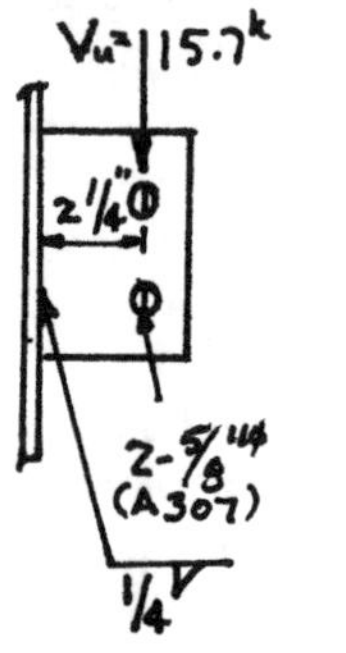

$\Sigma d_i^2 = 2 \times 1.5^2 = 4.5\ \text{in}^2$

$R_v = V/n = 15.7/2 = 7.9^k/_{\text{bolt}}$

$R_m = Ved / \Sigma d_i^2 = 15.7 \times 2.25 \times 1.5 / 4.5 = 11.8^k/_{\text{bolt}}$

$R = \sqrt{R_v^2 + R_m^2} = \sqrt{7.9^2 + 11.8^2} = 14.2^k/_{\text{bolt}}$

From Table 8-24, AISC Manual-II, the capacity of a 5 / 8" A307 bolt in single shear is only $5.5^k/_{\text{bolt}}$, which is inadequate. To remedy the connection, **add an unstiffened seated connection designed to resist the entire 15.7^k reaction**. Re: Table 9-6, AISC Manual-II, an L4 x 4 x 3 / 8 x 0'-6" with 2-3/4" diameter bolts through the web of the W18x50 girder will provide the 15.7^k capacity. The existing shear tab will prevent rotation of the beam about its longitudinal axis and all vertical force will be assumed to transfer through the seat (conservative).

For the girder (W18x50; A36 steel):

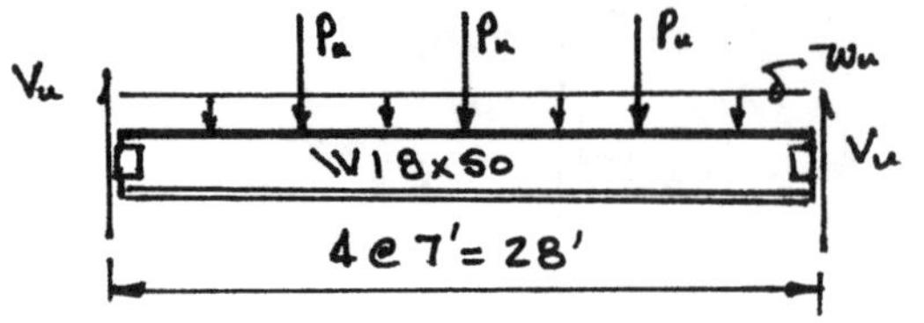

$P_u = 1.2P_D + 1.6P_L$

$= 1.2(0.05 \times 7 + 0.026) \times 20 + 1.6 \times 10.9 = 26.5^k$

$w_u = 1.2 \times 0.05 = 0.07$ klf

$V_u = 1.5P_u + w_u L / 2 = 1.5 \times 26.5 + 0.07 \times 20 / 2$

$= 40.5^k$

$M_u = w_u L^2 / 8 + 0.5 P_u L = 377^{\text{k-ft}}$

For the W18x50 (compact in A36 steel): d = 18.0"; $t_w = 0.36"$; $b_f = 7.50"$; $t_f = 0.57"$; $Z_x = 101\ \text{in}^3$; $I_x = 800\ \text{in}^4$.

Check live load deflection: $P_L = 10.9^k$

$\Delta_L = 0.0495 P_L L^3 / EI = 0.0495 \times 10.9(28 \times 12)^3$

$/ (29{,}000 \times 800)$

$\Delta_L = 0.88" \leq L/360 = 28 \times 12 / 360 = 0.93"\ \therefore$ O.K.

Check the flexural strength. Fully braced compression flange: $L_b = 0$:

$\therefore\quad \phi M_n = \phi M_p = 0.9 \times 36 \times 101 / 12 = 273^{\text{k-ft}} < M_u$

$\therefore$ need to increase the flexural capacity of the girder.

b) Strengthen the inadequate members. Since the existing floor system is to be removed, both top and bottom coverplates can be added to the W18x50 to strengthen the beam efficiently.

$\Delta(\phi M_n) = \phi F_y A_{PL}(d + t_{PL}) \geq 377 - 273 = 104^{\text{k-ft}}$

$A_{PL} \cong 104 \times 12 / (0.9 \times 36 \times 18) = 2.14\ \text{in}^2$

Try a PL6 x 3 / 8 on top and a PL9 x 1/4 bottom ($A_{PL} = 2.25\ \text{in}^2$ for each). The different sizes were chosen to permit downhand welding of the plates to the W18x50 ($b_f = 7.5"$).

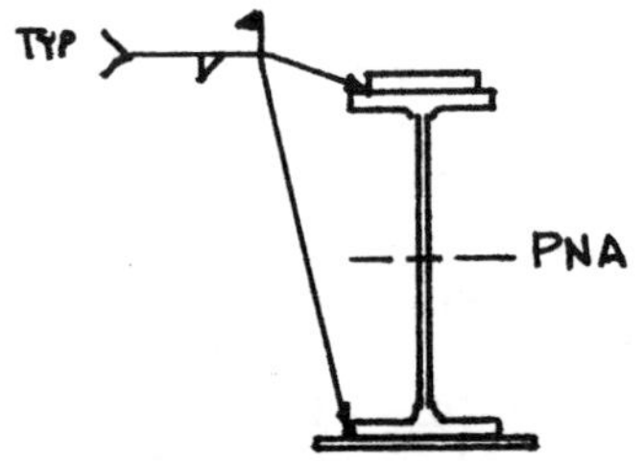

$A_{comp} = A_{ten}$

$\Delta(\phi M_n) = 0.9 \times 36 \times 2.25(18.0 + 0.31) = 1334^{\text{k-in}}$

$= 111^{\text{k-ft}}$

$\therefore\quad \phi M_n = 273 + 111 = 384^{\text{k-ft}} > M_u\ \therefore$ O.K.

The coverplates can terminate beyond the points where $M_u = 273^{\text{k-ft}}$.

$M_u = 40.5x - 0.06x^2 / 2 = 273^{\text{k-ft}}\ (0 \leq x \leq 7')$

$x_1 = 6.74'$; $x_2 = 28 - 6.74 = 21.26'$

Extend the plates far enough beyond these points to fully develop plates (conservative).

$F_{PL} = F_y A_{PL} = 36 \times 2.25 = 81^k$

Use ω = 1/4", which is greater than minimum per J2, but recommended for field welds.

$$q \leq \begin{cases} 4 \times 1.39 = 5.6^k/_{in} \\ 0.9 \times (0.6 \times 36) \times 0.25 = 4.9^k/_{in} \leftarrow \text{governs} \end{cases}$$

$L_w \geq F_{PL} / q = (81 / 2) / 4.9 = 8.2"$

∴ Total length of cover plates = $28 - 2 \times 6.74 + 2 \times 8.2 / 12 = 15.8'$; say 15'-9"

Use intermittent welds between ends:

$s \leq 24 \times 0.25 = 6"$ (D2, AISC); $L \geq 4 \times (0.25) = 1"$; use 1-1/2"

Thus, the details for strengthening the girders are:

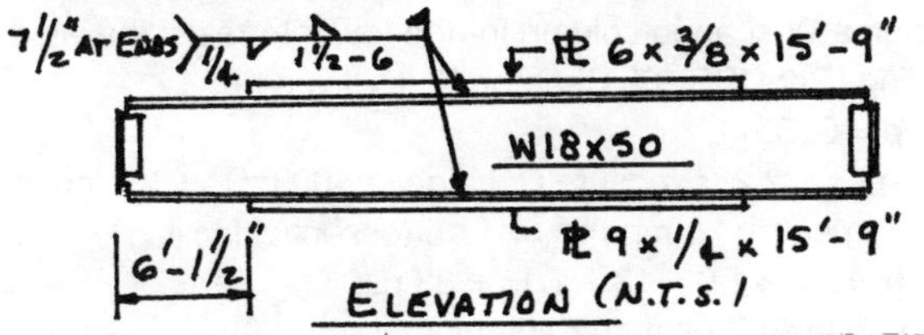

Check framed connections at ends of girder: three, 5 / 8" diameter A307 bolts in double shear:

$R = \min\{R_v = 19.5^k/_{bolt};\ R_b = 34.3^k/_{bolt}\} = 19.5^k/_{bolt}$

$V_u = 40.7^k < nR = 3 \times 19.5 = 58.5^k$

Check strength of the 2L3 x 3 x 1/4 x 0'-9"

$\phi R_n = 0.9 \times (0.6 \times 36) \times 2 \times 9 \times 0.25 = 87^k$

$V_u = 40.7^k < \phi R_n$ ∴ O.K.

∴ **the existing girder connection is O.K.**

Steel-54

a) Check the crane column with the boom rotated to 60°. Use Allowable Stress Design (AISC 1989).

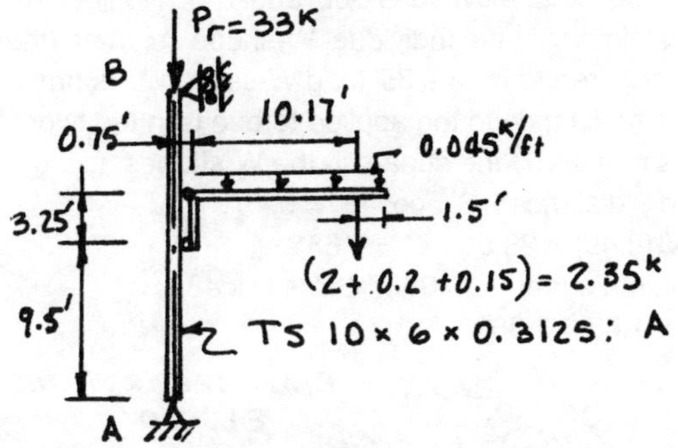

For the TS 10 x 6 x 0.3125: $A_g = 9.36\ in^2$; $S_x = 25\ in^3$; $S_y = 18.8\ in^3$; $r_x = 3.65"$; $r_y = 2.46"$; $F_y = 46$ ksi.

Apply an impact factor, 1.25, to the hoist, trolley and lift load:

$P = 1.25 \times 2.35 = 2.94^k$

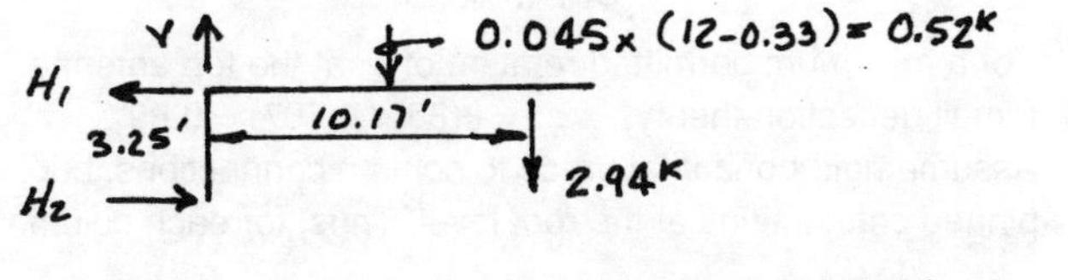

$\Sigma M_z = 0$: $3.25H_1 - 0.52(11.67) / 2 - 2.94(10.17) = 0$

$H_1 = 10.1^k \leftarrow$; $H_2 = 10.1^k \rightarrow$; $V = 3.46^k$

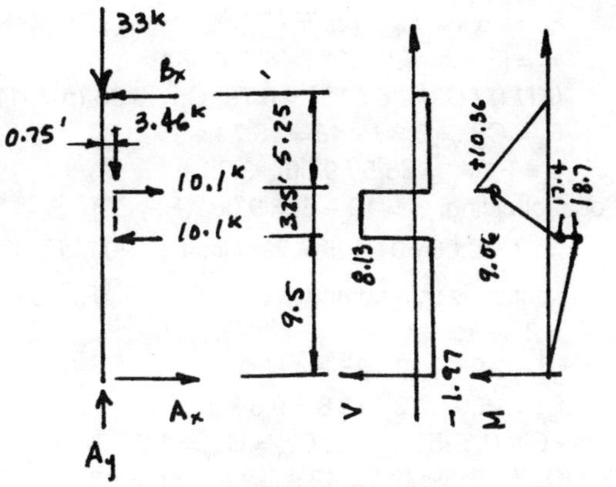

$\Sigma M_A = 0$: $18B_x - 3.46(0.75) - 10.1(3.25) = 0$

$B_x = 1.97^k \leftarrow$; $A_x = 1.97k \rightarrow$

$\Sigma F_y = 0$: $A_y = 36.5^k$

Assume that the $3.46^k \downarrow$ is shared equally by the hinged supports. Each point will experience a concentrated moment of $(3.46 / 2)0.75 = 1.3^{k\text{-}ft}$ (see V- and M-diagrams) $M_{max} = 18.7^{k\text{-}ft}$ occurs where $P = 36.5^k$. When the boom is at maximum swing: 60° with respect to the y-axis:

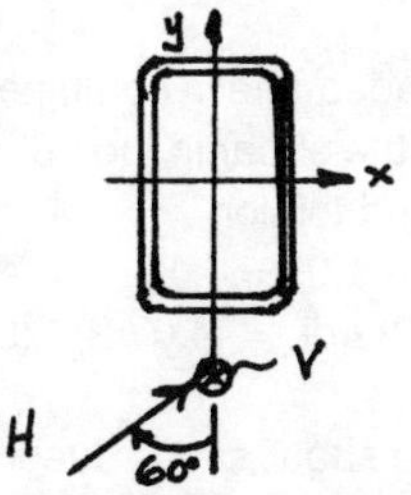

vertical load still causes x-axis bending; the horizontal reactions cause biaxial bending and torsion. Resolve the horizontal reactions:

$H_x = H\sin 60^\circ = 10.1\sin 60^\circ = 8.75^k \rightarrow$

$H_y = H\cos 60^\circ = 10.1\cos 60^\circ = 5.05^k$

$M_t = H_x e = 8.75(9) = 79^{k\text{-}in}$ (counterclockwise)

For bending about the x-axis:

$A_x = -B_x = (2 \times 1.3 + 5.05 \times 3.25) / 18 = 1.05^k$

$M_x = 1.05 \times 9.5 = 9.95^{k\text{-}ft}$

For y-axis bending:

$A_y = -B_x = (8.75 \times 3.25) / 18 = 1.58^k$

$M_x = 1.58 \times 9.5 = 15.0^{k\text{-}ft}$

The torsional moment acting between the hoist arm support causes warping stresses in the non-circular closed section. These warping stresses are additive to the normal stresses caused by bending and axial stresses. However, for rectangular tube sections, the warping stresses are small, localized and usually negligible (Bresler, et.al. 1971). Furthermore, the warping stresses are maximum midway between the hoist arm support points and diminish to zero at the support points, where maximum bending stresses occur. Therefore, we need to check only the combined axial and bending stresses per Chapter H, AISC

$KL_x / r_x = 18 \times 12 / 2.46 = 88 \leftarrow$ governs

$KL_y / r_y = 18 \times 12 / 3.65 = 59$

$F_y = 46$ ksi, $C_c = [2\pi^2 E / F_y]^{1/2}$
$= [2 \times \pi^2 \times 29{,}000 / 46]^{1/2} = 111$
$(Kl / r) / C_c = 88 / 111 = 0.79$; $C_a = 0.36$ (p. 5-119)
$F_a = C_a F_y = 0.36 \times 46 = 16.7$ ksi
$f_a = P / A_g = 36.5 / 9.36 = 3.9$ ksi

Local buckling: $w = 10 - 3t = 9.06"$ (Re: B5)
$w / t = 9.06 / 0.3125 = 29 >$ limit $= 190 / \sqrt{F_y} = 28$

$\therefore$ section is non compact: $F_{bx} = F_{by} = 0.6F_y = 0.6 \times 46 = 27.6$ ksi
$f_{bx} = 9.95 \times 12 / 25 = 4.8$ksi
$f_{by} = 15.0 \times 12 / 18.8 = 9.6$ ksi

Case Cii, P. 5-55, ASIC: $C_{mx} = C_{my} = 1.0$
For $KL_x / r_x = 59$: $F'_{ex} = 42.9$ ksi; $KL_y / r_y = 88$: $F'_{ey} = 19.3$ ksi

$f_a / F_a = 3.90 / 16.7 = 0.23 > 0.15$ $\therefore$ Check Eqns H 1-1 and H 1-2:
$f_a / F_a + C_{mx} f_{bx} / (1 - f_a / F'_{ex}) F_{bx} + C_{my} f_{by} / (1 - f_a / F'_{ey}) F_{by} < 1.0$
$0.23 + 4.8 / [27.6(1 - 3.9 / 42.9)] + 9.6 / [27.6(1 - 3.9 / 19.3)] = 0.86 < 1.0$
$f_a / 0.6F_y + f_{bx} / F_{bx} + f_{by} / F_{bxy} < 1.0$
$3.9 / 27.6 + 4.8 / 27.6 + 9.6 / 27.6 = 0.68 < 1.0$
$\therefore$ O.K.

The member is adequate in combined axial plus bending. Check the shear in the 10" wall caused by external shear and torsion

$V_{max} = 5.05 - 1.05 = 4.0^k$
$f_{v1} = (V_{max} / 2) / dt_w = (4.0 / 2) / (9.06 \times 0.3125)$
$= 0.71$ ksi

Superimpose the shear stress due to the torsional moment $= 79^{k\text{-}in}$ (Re: Roark 1975, Table IX)

$S = T / 2t(a - t)(b - t)$
$= 79 / [2 \times 0.3125 \times (10 - 0.3125) \times (6 - 0.3125)]$
$= 2.3$ ksi
$f_{v1} + S = 3.0$ ksi $< 0.4F_y$ $\therefore$ O.K.

$\therefore$ **TS 10 x 6 x 0.3125 is O.K.**

b) Substituting a W-shape for the structural tube would result in a section that is relatively weak and flexible in resisting the torsional moment. Whereas the torsional warping stress is insignificant for the structural tube (note: these stresses can be calculated using formulas in Roark and result in a maximum normal stress of 1.6 ksi), the stresses are generally significant in an open section such as the W-shape.

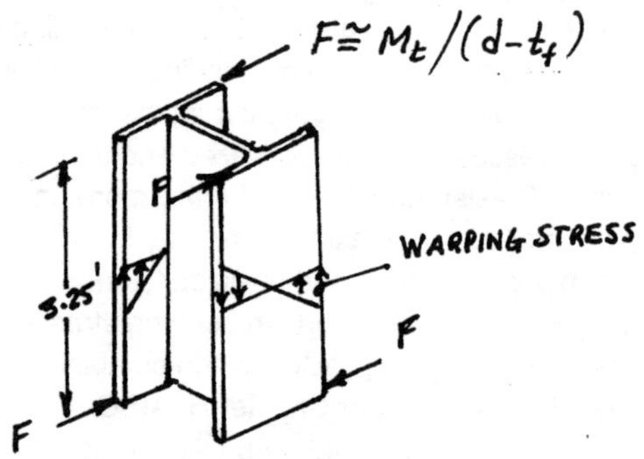

If the W-shape were used, it would be necessary to add torsional strength between the hoist arm support points (eg, by adding plates between the flange tips to produce a closed cross section in this region).

Steel-55

Design a W shaped column for the tower to resist the wind loading (Re: UBC-97, Chapter 16, Division III).

$p = C_e C_q q_s I_w$,

where $C_e = 2.2$ (exposure C; height $\geq 400'$); $C_q = 1.0$ (given; $q_s = 13$ psf (for 70 mph); $I_w = 1.15$ (essential facility).

$\therefore$ $p = 2.2 \times 1.0 \times 13 \times 1.15 = 33$ psf

Drag on the 10' diameter antenna:

$F = pA = 33 \times \pi \times 10^2 / 4 = 2590^{\#} = 2.59^k$

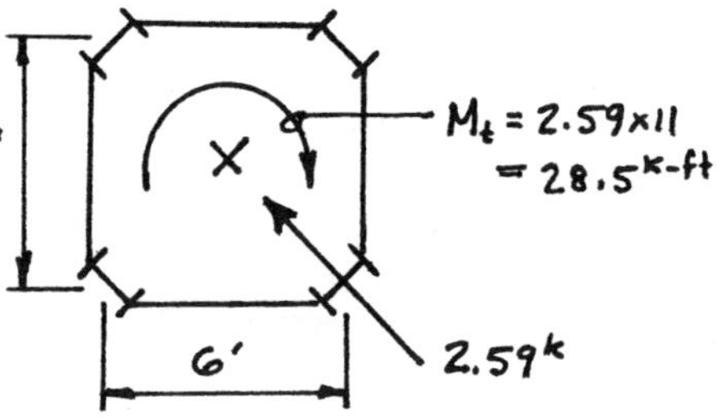

Braces located at levels +15'; +30'; +60' are assumed to be rigid with respect to both bending and torsional deformations (given). Consider the action of the torsional moments, $28.5^{k\text{-}ft}$, applied at elevations +30' and +60'. Neglect the torsional rigidity of the individual W-shape columns (the torque required to twist a 65' long W-section 1° is practically nil). The resistance to the applied torque is in the two couples formed by the shears in the W-shapes acting through a distance $= 6' / \cos 45° = 8.49'$.

$2V_1(8.49) = 28.5$ $\therefore$ $V_1 = 1.68^k$

The column shear from 0-to-+30' is twice V_1:
$V_2 = 2 \times 1.68 = 3.36^k$

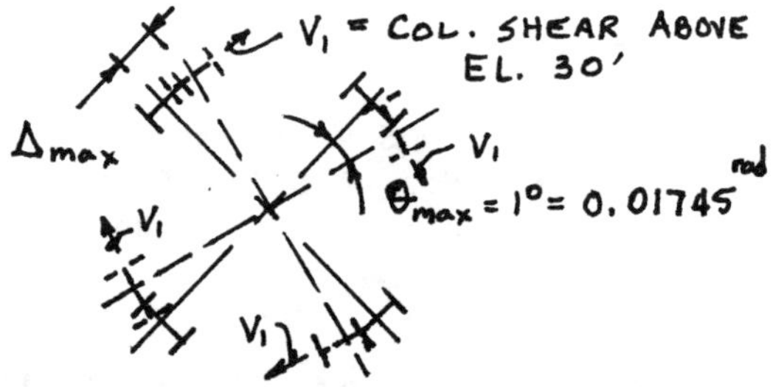

For a maximum permitted rotation of 1° at the top antenna (small deflection theory): $\Delta_{max} = \theta(36" / 0.707) = 0.89"$. Assume rigid horizontal brace to column connections, but pinned connections at the roof level. Thus, for each column:

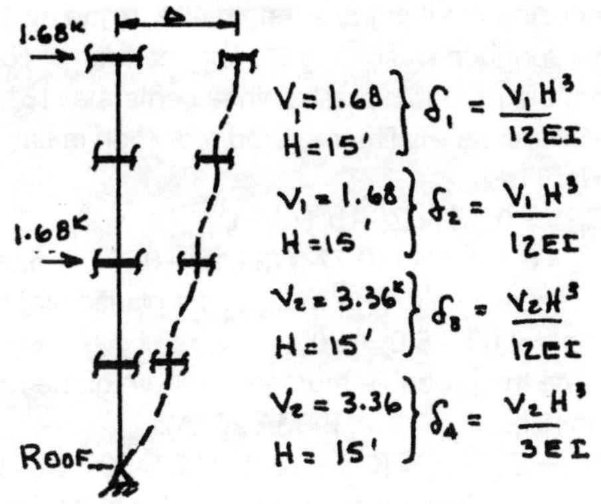

$\Delta = \delta_1 + \delta_2 + \delta_3 + \delta_4 = 0.89"$

$\therefore$ $[(15 \times 12)^3 / EI][2(1.68 / 12) + 3.36 / 12 + 3.36 / 3] \le 0.89$

$I \ge 380$ in^4

Try a W12x section with $I_x \ge 380$ in^4 and strength sufficient to resist the wind load. The applied torque causes maximum x-axis bending equal to V_2H in the lowest tier of the column: $M_x = 50.4^{k\text{-}ft}$. The total wind shear, which includes additional drag on columns, will distribute on the basis of I / H of the columns for wind blowing diagonally. Try a W12x65: $I_x = 533$ in^4; $I_y = 174$ in^4. Take the drag coefficients on the columns as $C_q = 2.0$; neglect shielding:

$V_w = 2 \times 1.68^k + 4 \times 1' \times 60' \times 2.0 \times 0.033^{ksf} = 19.2^k$

$V_x = [\, I_x / (I_x + I_y)]V / 2 = [533 / (533 + 174)](19.2 / 2) = 7.23^k$

$V_y = [174 / (533 + 174)](19.2 / 2) = 2.36^k$

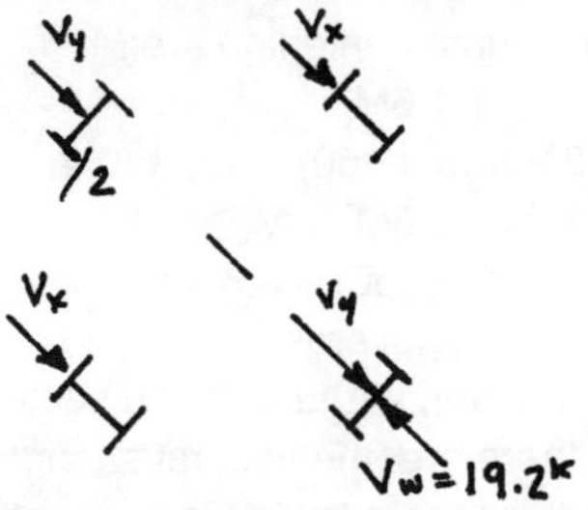

Additional bending is induced in the columns. For the the columns resisting V_x:

$M_x = 50.4 + 7.23 \times 15 = 159^{k\text{-}ft}$; $M_y = 0$; $P = 0$.

For columns resisting V_y:

$M_y = 2.36 \times 15 = 35.4^{k\text{-}ft}$

$P = M_o / (6 / 0.707)$; M_o = the overturning moment

$P = [1.68(30 + 60) + 15.8 \times 30] / 8.49 = 73.6^k$

Check the W12x65 under combined axial plus bending. Use LRFD (AISC 1994); take $P_d \cong 5^k$ in the critical column

$P_u = 1.2P_D + 1.3P_W = 1.2 \times 5 + 1.3 \times 73.6 = 102^k$

W12x65: $A_g = 19.1$ in^2; $r_x = 5.28$ "; $r_y = 3.02"$;
$Z_x = 96.8$ in^3; $Z_y = 44.1$ in^3; $L_p = 12.6'$; $L_r = 44.7'$;
$\phi M_p = 261^{k\text{-}ft}$; $\phi M_r = 171^{k\text{-}ft}$.

$KL / r)_x = 2.1 \times 15 \times 12 / 5.28 = 72$

$KL / r)_y = 2.1 \times 15 \times 12 / 3.02 = 125 \leftarrow$ governs

$\phi_c F_{cr} = 13.44$ ksi

$\phi P_n = \phi_c F_{cr} A_g = 13.44 \times 19.1 = 257^k$

$P_u / \phi P_n = 102 / 257 = 0.40$

$L_p = 12.6' < L_b = 15' < L_r = 44.7'$ $\therefore$ Use Eqn F1-3:

$\phi M_n = C_b[\phi M_p - (\phi M_p - \phi M_r)(L_b - L_p) / (L_r - L_p)] \le \phi M_p$

$\phi M_n = 1.67[261 - (261 - 171)(15 - 12.6) / (44.7 - 12.6)] = 425^{k\text{-}ft}$; use $\phi M_p = 261^{k\text{-}ft}$

$\phi M_{ny} = \phi F_y Z_y = 0.9 \times 36 \times 44.1 / 12 = 119^{k\text{-}ft}$

Combined bending:

$KL' / r)_x = 72$: $P_{ex} = (\pi^2 E / 72^2) \times 19.1 = 1054^k$

$KL' / r)_y = 125$: $P_{ey} = (\pi^2 E / 125^2) \times 19.1 = 350^k$

$\Sigma P_e = 2 \times (1054 + 350) = 2808^k$

$\Sigma P_u = \Sigma 1.2P_d = 1.2 \times 20 = 24^k$; $B_2 \cong 1.0$

$\therefore$ sidesway buckling is not likely

$M_{ux} = 50.4^{k\text{-}ft}$; $M_{uy} = 35.4^{k\text{-}ft}$

$P_u / \phi P_n = 0.4 > 0.2$ $\therefore$ check equation H1-1a:

$P_u / \phi P_n + 8/9(M_{ux} / \phi M_{nx} + M_{uy} / \phi M_{ny})$

$= 0.40 + 8 / 9(50.4 / 261 + 35.4 / 119) = 0.84$

W12x65 satisfies both serviceability and strength $\therefore$ **Use W12x65 columns**.

Steel-56

a) Compute preliminary moments and shears at the tenth floor using the portal method (Note: forces are the working stress values given in the problem statement. These values must be increased by 1.4 factor to agree with the design level forces of the current UBC-97).

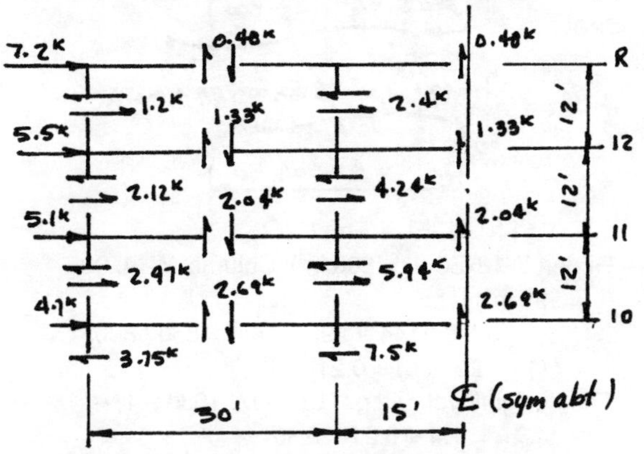

Distribution of shears shown above is based on the assumption that exterior columns resist one half as much shear as interior columns and that points of inflection occur at midspan of girders. Thus, at the tenth floor:

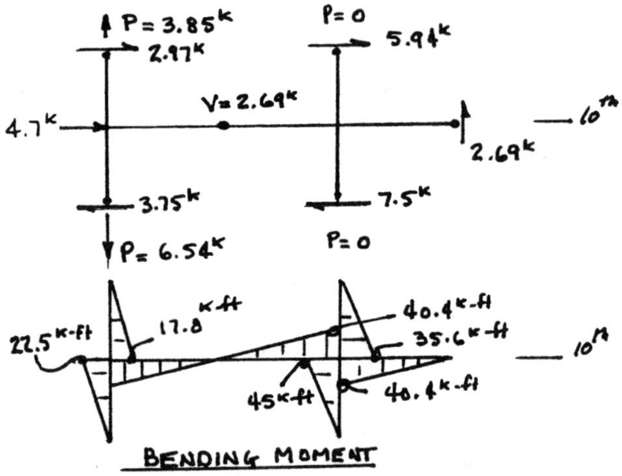

b) Estimate the 10th story drift. For a frame of this proportion, axial, shear and panel zone deformations are relatively small. The drift can be estimated based on member centerline dimensions and flexural deformations in the columns and girders (Becker, et.al. 1988). Again, these values would have to be increased by 1.4 to conform to design level of UBC-97.

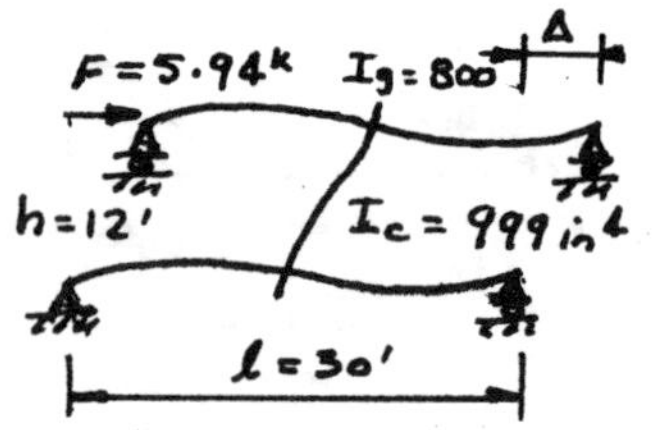

$\Delta \cong Fh^3 / 12EI_c + Fh^2l / 12EI_g$

Girder: W18x50 ($I_x = 800\ in^4$); Column: W14x90 ($I_x = 999\ in^4$)

$\therefore \quad \Delta_{10\text{-}11} = 5.94[144^3 / 999 + 144^2 \times 360 / 800] / (12 \times 29{,}000) = 0.21"$

The design level drift per UBC-97, 1630.9.1, is:

$\Delta_{S,10\text{-}11} = 1.4 \times 0.21 = 0.30"$

The corresponding maximum inelastic response is estimated using Eqn. 30-17, UBC-97, with R = 8.5 from Table 16-N:

$\Delta_{M,10\text{-}11} = 0.7\ R\ \Delta_{S,10\text{-}11} = 0.7 \times 8.5 \times 0.30 = 1.79"$

Finally, if the girder cross sections are reduced to assure that plastic hinges occur away from the face of columns (see Part c), the reduction in girder rigidity increases the drift on the order of 5% (Iwankiw 1997)

$\therefore$ **$\Delta_{M,10\text{-}11} = 1.05 \times 1.79 = 1.9"$**

c) Connection design -- follow the guidelines of FEMA 267A, Advisory No. 1 (SAC-96-03). Of the non-proprietary connections that have been tested, the reduced beam section offers one of the more reliable approaches to assuring that plastic hinges will form away from the column faces, and is the choice for this solution. For the non-reduced W24x84 girder: d = 24.10"; t_w = 0.47"; b_f = 9.02"; t_f = 0.77"; $Z_x = 224\ in^3$; $S_x = 196\ in^3$. For the W14x176 column: $A_g = 51.8\ in^2$; d = 15.22"; t_w = 0.83"; b_f = 15.65"; t_f = 1.31"; k = 2.0"; $Z_x = 320\ in^3$.

Try reducing the flexural strength 40% using an 18" long circular "dogbone," starting 6" from the face of column (so that the probable plastic hinge centers at 15" = 1.25' from column face). The required reduction in flange width is:

$Z_{RBS} = 0.6Z_x = Z_x - b_R t_f(d - t_f)$

$\therefore \quad b_R = 0.4 \times 224 / (0.77 \times (24.10 - 0.77)) = 5.00"$, which results in a flange width at the plastic hinge location = 9.02 – 5.0 = 4.0".

Calculate the probable moment strength of the reduced section (Eqn. 7.5.3.2-2, FEMA 267A):

$M_{pr} = Z_{RBS}\beta F_y = (0.6 \times 224) \times 1.2 \times 50 / 12 = 672^{k\text{-}ft}$

Increase the given seismic loads by 1.4 to conform to design level of UBC-97:

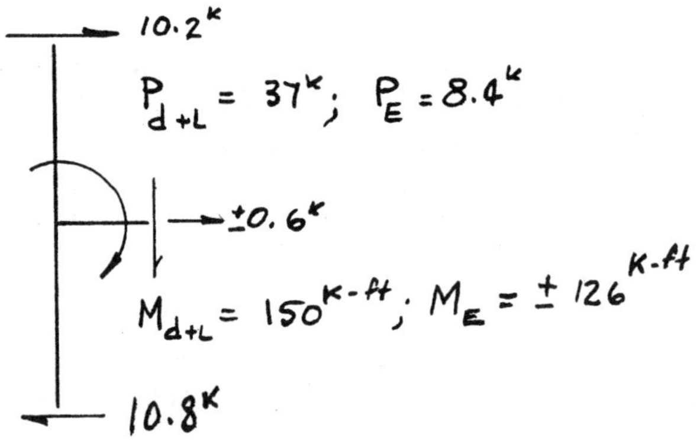

Using the LRFD approach (AISC 1994), the design moments at column centerline are:

$M_u \geq 1.2M_D + 1.6M_L$

$= 1.2 \times (0.6 \times 150) + 1.6 \times (0.4 \times 150) = 204^{k\text{-}ft}$

$M_u \geq 1.2M_D + 0.5M_L + M_E$

$= 1.2\ (0.6 \times 150) + 0.5 \times (0.4 \times 150) + 126^{k\text{-}ft}$

$= 264^{k\text{-}ft} \leftarrow$ governs

The design moment at the reduced section will be less than $264^{k\text{-}ft}$; therefore, the flexural strength is adequate. For yielding to occur in the girders, the strength of the columns must satisfy Eqn 7.5.2.5-1-4 of FEMA 267A:

$\Sigma Z_c(F_{yc} - f_a) / \Sigma M_c \geq 1.0$

The third floor column supports 9 floors plus roof and the additional 37^k gravity load at the third floor; thus,

$f_{a,above} \cong 10 \times 37 / 51.8 = 7.1$ ksi

$f_{a,below} \cong 11 \times 37 / 51.8 = 7.8$ ksi

$\therefore \quad \Sigma Z_c(F_{yc} - f_a) = [320(50 - 7.1) + 320(50 - 7.8)] / 12 = 2269^{k\text{-}ft}$

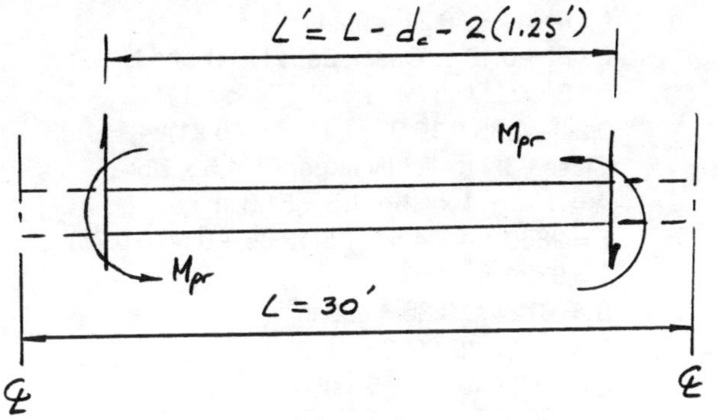

$L' = 30 - 15.22 / 12 - 2(6 + 9) / 12 = 26.2'$

$V_p = 2M_{pr} / L' = 2 \times 672 / 26.2 = 51.3^k$

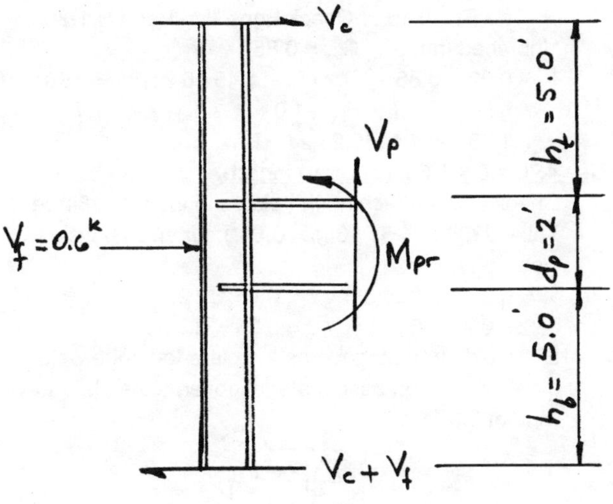

From statics:

$V_c = \{\Sigma[M_{pr} + V_p(L - L') / 2] - V_f(h_b + d_p / 2)\}$
$/ (h_b + d_p + h_a)$

$V_c = \{2[672 + 51.3(30 - 26.2) / 2]$
$- 0.6(5.0 + 2.0 / 2)\} / (5.0 + 2.0 + 5.0)$

$V_c = 128^k$

$M_{ct} = V_c h_a = 128 \times 5.0 = 640^{k\text{-ft}}$

$M_{cb} = (V_c + V_f)h_b = (128 + 0.6) \times 5.0 = 643^{k\text{-ft}}$

$\Sigma M_c = 640 + 643 = 1283^{k\text{-ft}}$

$\Sigma Z_c(F_{yc} - f_a) / \Sigma M_c = 2269 / 1283 = 1.76 > 1.0$

∴ Yielding should occur in the girders. Verify that stresses at the face of column are acceptable:

$M_f = M_{pr} + V_p x = 672 + 51.3 \times 1.25 = 736^{k\text{-ft}}$

$f_b = M_f / Sx = 736 \times 12 / 196 = 45.0 \text{ ksi} = 0.9F_y$

Thus, the stress imposed on the through-thickness of the column flange satisfies the $0.9F_y$ limit recommended in the FEMA 267A guidelines.

Check panel zone strength (7.5.2.6, FEMA 267A).

$V \geq (M_g + M_E) / (0.95d_b) - V_{c1}$
$= (150 + 126) / (0.95 \times 24.1 / 12) - 10.2 = 134^k$

$V \leq 0.8\Sigma M_f / (0.95d_b) = 0.8 \times 736 / (0.95 \times 24.1 / 12)$
$= 308^k$

Thus, required strength is 134^k. Check Eqn. 8-1, AISC

$\phi_v V_n = \phi_v 0.6F_y d_c t_p[1 + 3b_{cf}t^2_{cf} / (d_b d_c t_p)]$

$\phi_v V_n = 0.75 \times 0.6 \times 50 \times 15.22 \times 0.83[1 + 3 \times 15.65$
$\times 1.31^2 / (24.1 \times 15.22 \times 0.83)] = 360^k > V = 134^k$

Check Eqn. 8-2, AISC:

$t_z \geq (d_z + w_z) / 90 = [(24.1 - 2 \times 0.77)$
$+ (15.22 - 2 \times 1.31)] / 90 = 0.39"$

$t_z = 0.83" > 0.39"$ – O.K.

∴ Column web is adequate for panel zone shear.

Continuity plates – re: 7.8.3, FEM267A, guidelines recommend that continuity plates of the same thickness as the beam flange always be provided to alleviate stress concentration at the column flange-to-web junction ∴ use PL4 x 3/4 each side. The guidelines further recommend that the continuity plates be connected to develop their full strength ∴ use full penetration welds.

Design the web shear connection to transfer $V_p + V_g = 51.3 + 37 = 88.3^k$. Use a single plate connection with 7/8" diameter A325-SC bolts: $\phi r_v = 14.5^k/_{bolt}$ (Table 8-17, AISC Manual-II)

$n \geq 88.3 / 14.5 = 6.08$, say 6 bolts at 3" o.c.

(L = 18" < T = 22" of the W22x84)

Assuming A36 material for the connection plate (F_y = 36 ksi; F_u = 58 ksi), try 3 / 8" plate. Check shear yield in the plate:

$\phi V_n = \phi 0.6F_y Lt = 0.9 \times 0.6 \times 36 \times 0.375 \times 18$
$= 131^k > 88.3^k$

Check shear rupture:

$A_{ns} = (18 - 6 \times (7 / 8 + 1 / 8)) \times 3 / 8 = 4.5 \text{ in}^2$

$\phi V_n = \phi 0.6F_u A_{ns} = 0.75 \times 0.6 \times 58 \times 4.5$
$= 117^k > 88.3^k$ ∴ 3 / 8" connection plate is O.K.

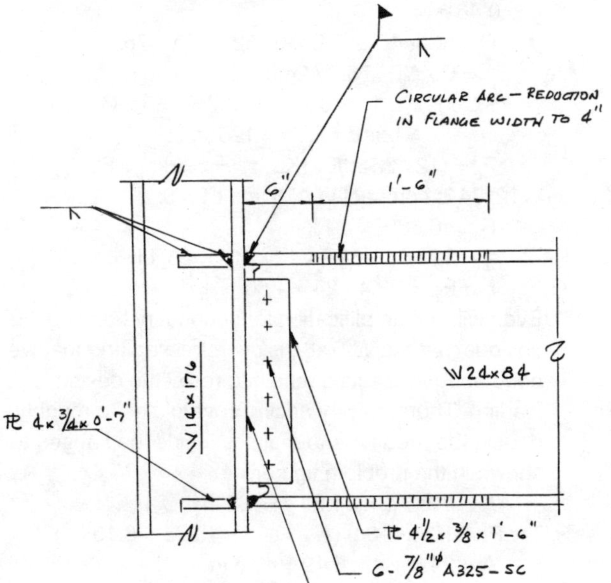

CONNECTION DETAIL (N.T.S.)

Steel-57

a) Determine the cutoff lengths of the added plates assuming that wind loading governs. Use Allowable Stress Design (AISC 1989) and UBC-97 to determine the design forces.

Wind loads (Re: UBC-97, Chapter 16, Division III).

$p = C_e C_q q_s I_w$,

where $C_q = 1.4$ (signs, flagpoles, minor structures); $q_s = 17$ psf (for 80 mph); $I_w = 1.0$; exposure condition C – consider the forces on one W27x84 (i.e., over an 8' tributary width).

$\therefore \quad p = C_e \times 1.4 \times 17 \times 1.0$

For h ≤ 20', $C_e = 1.2$; 20-to-40', $C_e = 1.3$; 40-to-60', $C_e = 1.5$; greater than 60', $C_e = 1.6$. Per problem statement, allowable stresses are not to be increased by 1/3 for wind loading.

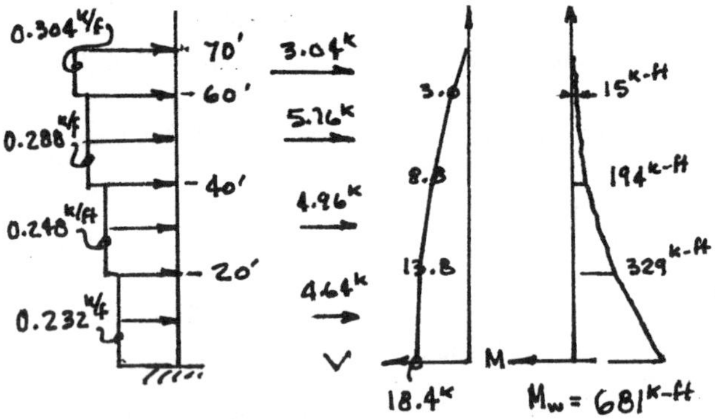

Check the adequacy of the W27x84: $A_g = 24.8$ in²; $I_x = 2850$ in⁴; $S_x = 213$ in³; $r_x = 10.7$ "; r_y = n.a. (fully braced about y-axis); $b_f = 9.96$"; $t_f = 0.64$"; d = 26.71"; $t_w = 0.46$"

$P = (0.084 + 2 \times 8 \times 0.062) \times 70 = 75.3^k$

$f_a = P / A_g = 75.3 / 24.8 = 3.0$ ksi

$KL_x / r_x = 2.1 \times 70 \times 12 / 10.7 = 165 > C_c$

$= [2\pi^2 E / F_y]^{1/2} = 126$

$\therefore \quad F_a = (12 / 23)\pi^2 E / (KL / r)^2 = 5.5 \text{ ksi} = F'_{ex}$

W12x84 is compact in A36 and $L_b = 0$:

$F_{bx} = 0.66F_y = 24$ ksi

$f_{bx} = M / S_x = 681 \times 12 / 213 = 38.4$ ksi

$f_{bx} / F_{bx} = 38.4 / 24 = 1.60$

Even without amplification for secondary bending it is obvious that the W27x84 is overstressed and that we need about twice its strength to resist the design loading. Therefore, try adding coverplates to roughly double the area (use four plates inside the flanges as shown in the problem figure).

$A_{PL} \cong A_g / 4 = 24.8 / 4 = 6.2 \text{ in}^2$

Try PL4-1/2 x 1-3/8 ($A_{PL} = 4.5 \times 1.375 = 6.19$ in²)

$A = 24.8 + 4 \times 6.19 = 49.6 \text{ in}^2$

$I_x = 2850 + 4 \times 6.19(13.36 - 0.64 - 0.69)^2$

$= 6433 \text{ in}^4$

$S_x = 6433 / 13.36 = 482 \text{ in}^3$

$f_{bx} = 681 \times 12 / 482 = 16.95$ ksi

$f_a = 75.3 / 49.6 = 1.5$ ksi

$r_x = \sqrt{I_x / A} = \sqrt{6433 / 49.6} = 11.39$"

$KL_x / r_x = 2.1 \times 70 \times 12/ 11.39 = 155 > C_c = 126$

$\therefore \quad F_a = (12 / 23)\pi^2 E / (KL / r)^2 = 6.21 \text{ ksi} = F'_{ex}$

$f_a / F_a = 1.5 / 6.21 = 0.24$

$f_a / F_a > 0.15 \therefore$ Check Eqns H 1-1 and H 1-2:

$f_a / F_a + C_{mx} f_{bx} / (1 - f_a / F'_{ex}) F_{bx} < 1.0$

$0.24 + 0.85 \times 16.95 / [(1 - 1.5 / 6.21)24] = 1.03 \rightarrow$

Close – try a slightly larger PL 4.5 x 1.5:

$A = 24.8 + 4 \times 4.5 \times 1.5 = 51.8 \text{ in}^2$

$I_x = 2850 + 4 \times 4.5 \times 1.5(13.36 - 0.64 - 0.75)^2$

$= 6724 \text{ in}^4$

$S_x = 6724 / 13.36 = 503 \text{ in}^3$

$f_{bx} = 681 \times 12 / 503 = 16.2$ ksi

$f_a = 75.3 / 49.6 = 1.5$ ksi

$r_x = \sqrt{I_x / A} = \sqrt{6724 / 51.8} = 11.4$"

$KL_x / r_x = 2.1 \times 70 \times 12 / 11.4 = 155 > C_c$

$\therefore \quad F_a = (12 / 23)\pi^2 E / (KL / r)^2 = 6.21 \text{ ksi} = F'_{ex}$

$f_a / F_a = 1.45 / 6.21 = 0.23$

$f_a / F_a > 0.15 \therefore$ Check Eqns H 1-1 and H 1-2:

Unbraced frame $\therefore C_{mx} = 0.85$

$0.23 + 0.85 \times 16.2 / (1 - 1.45 / 6.21)24 = 0.98$

$f_a / 0.6F_y + f_{bx} / F_{bx} < 1.0$

$1.45 / 21.6 + 16.2 / 24 = 0.74 \therefore$ O.K.

Use 4PL 4.5 x 1.5 x required length.

Check welds attaching the plates to the W27x84: re: Table J2, $\omega_{min} = 5 / 16$ ($t_f = 0.64$"); for an E70 weld to the web:

$q \le \{ 5 \times 0.928 = 4.64; \ 0.4F_y t_w / 2$

$= 0.4 \times 36 \times 0.46 / 2 = 3.3 \} = 3.3 \ ^k/_{in}$

E70 weld strength governs the plate-to-flange weld: $q = 4.64 \ ^k/_{in}$. The base material governs the plate-to-web weld at $3.3^k/_{in}$.

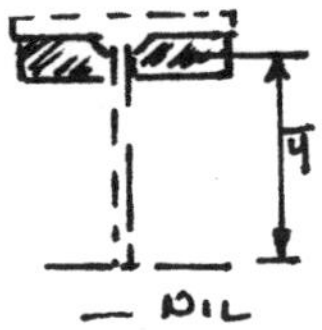

$Q = A_{PL} \bar{y} = 4.5 \times 1.5(13.36 - 0.64 - 0.75) = 81 \text{ in}^3$

b = 4.5 + 1.5 = 6"

$q_v = VQ / I = 18.4 \times 81 / 6724 = 0.22 \ ^k/_{in}$ -- Nil (use continuous welds, though, so that the member qualifies as compact). Terminate the coverplates where the S_x of the W27x84 is adequate. Use the f_a / F_a and $C_{mx} / (1 - f_a / F'_{ex})$ from above (conservative).

$0.23 + [0.85 / (1 - 1.5 / 6.21)](M / 213) / 24 \le 1.0$

$M \le [((1 - 0.23) \times 213 \times 24) / 1.12] / 12 = 292^{k\text{-ft}}$

The theoretical cutoff point (see M-diagram) occurs at:

$329 - 13.8x = 292$

x = 2.7' (20 – 2.7 = 17.3' above base)

The plate must extend far enough beyond the theoretical cutoff point to develop the plate's share of stresses:

$f_{bPL} = My / I$

$= 292 \times 12 \times (13.36 - 0.64 - 0.75) / 6724$

$= 6.2$ ksi

$F_{PL} \cong f_{bPL} A_{PL} = 6.2 \times 4.5 \times 1.5 = 42^k$

$L_P \ge 42 / (4.64 + 3.3) = 5.3" > a' = 4.5"$

∴ Use PL 4-12/x 1-1/2 x 17'-9" w/ 5/16" continuous fillet welds.

Note: since fillet welds are lightly stressed, a better solution might be to lower the allowable bending stress to $0.6F_y$ and use intermittent welds. This alternative is not explored in this solution.

b) Compute the period of the sign support (Re: 1634.1.4, UBC-97). There is continuous distribution of mass and stiffness. Idealize the structure as a lumped mass model (Paz 1985) and use Rayleigh's method to approximate the fundamental period. In this solution, the mass will be lumped at three levels:

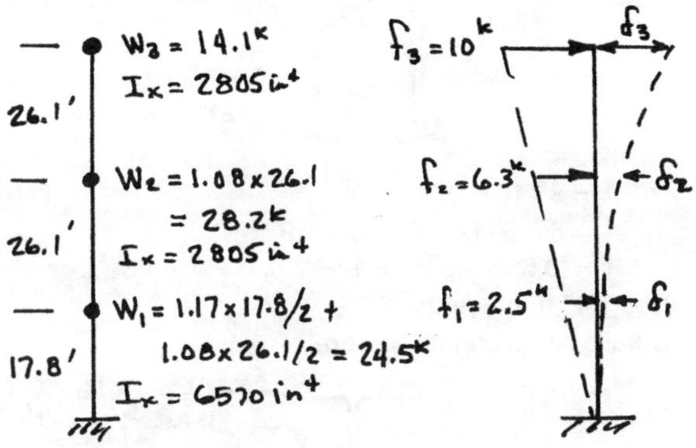

Apply an arbitrary force system that varies linearly to 10^k at the top:

$f_1 = (17.8 / 70)10 = 2.5^k$

$f_2 = (43.9 / 70)10 = 6.3^k$

The deflections, δ_i, at node points can be obtained by many methods of elastic analysis (e.g., conjugate beam) and found to be: $\delta_1 = 1.96"$; $\delta_2 = 9.03"$; $\delta_3 = 19.35"'$

$$T = 2\pi\sqrt{\Sigma w_i\delta_i^2 / (g\Sigma f_i\delta_i)}$$

$$T = 2\pi\sqrt{\frac{24.5(1.96)^2 + 28.2(9.03)^2 + 14.1(19.35)^2}{32.2\times 12[2.5(1.96) + 6.3(9.03) + 10(19.3)]}}$$

$= 1.8$ sec

Re: 1634, UBC-97, use the static force procedure of 1630.2 and divide by 1.4 to obtain a working load value comparable to the wind loads of part a:

$$(0.11C_aI)W \le V = (C_vI / RT)W \le (2.5C_aI / R)W,$$

where R = 2.9 (treating as distributed mass cantilevered structure rather than a sign or billboard; Table 16-P, UBC); I = 1; $C_v = 0.64$; $C_a = 0.44$.

$0.11C_aI = 0.11 \times 0.44 \times 1 = 0.048$

$C_vI / RT = 0.64 \times 1 / (2.9 \times 1.8) = 0.122$ ← controls

$2.5C_aI / R = 2.5 \times 0.44 / 2.9 = 0.380$

$V_w = V / 1.4 = 0.122 \times 75.4 / 1.4 = 6.6^k$

Assuming a triangular distribution of lateral force (uniformly distributed mass vibrating in the fundamental mode), the resultant lateral force will act at roughly 2/3 the height above the base. Thus,

$M_e = V(2/3)H = 6.6(2/3)70 = 310^{k\text{-}ft} < M_w$ ∴ wind governs.

Steel-58

a) Design a new steel beam to support the roof when the existing column is removed. Assume that the specified 20 psf roof live load is non reducible. The new steel beam must support a concentrated load equal to the column reaction on the existing column at line 2. The spacings of the 10" x 24" and 6" x 19" beams are not specified. The 4" roof slab would require support (Re: Table 9-5a, ACI 318-95):

$l \le 24h = 24 \times 4 / 12 = 8'$

∴ Assume beams to be spaced at 8' on centers.

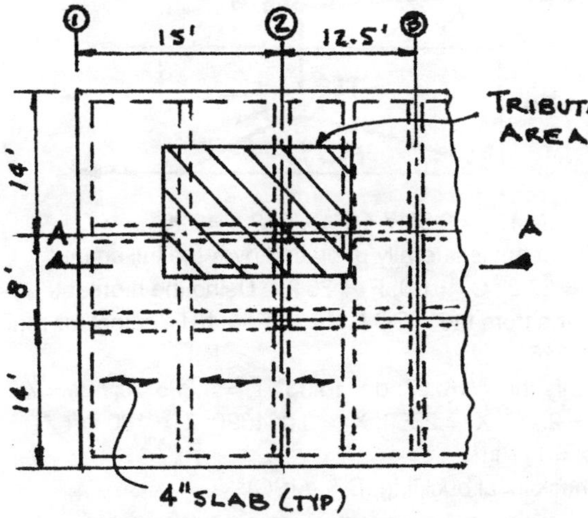

PARTIAL ROOF FRAMING PLAN

Service loads:

Dead = 4" slab + roofing + ceiling + beams

For 10" x 24" beams at 8' o.c.:

$w_{bm} = (10(24 - 4) / 144)150 / 8 = 26$ psf

For 8" x 19" beam at (14 + 8) / 2 = 11' o.c.:

$w_{bm} = (8(19 - 4) / 144)150 / 11 = 11$ psf

∴ $w_D = 48 + 10 + 12 + 26 + 11 = 107$ psf

$w_L = 20$ psf

Compute the column reaction based on its tributary area. The first interior support supports approximately 5/8 of the load on the exterior span plus one half the load on the adjacent interior span

∴ $A_{trib} = [(5/8) \times 15 + (1/2) \times 12.5][(5/8) \times 14 + (0.5) \times 8] = 200\ \text{ft}^2$

$P_{2D} = 200 \times 107 = 21{,}400^\# = 21.4^k$

$P_{2L} = 200 \times 20 = 4000^\# = 4^k$

Design a new steel beam to span from line "1" to line "3":

$L = 14.67 + 12.5 = 27.2'$

Use LRFD (AISC 1994):

$$P_u \ge \begin{cases} 1.4P_{2D} = 1.4\times 21.4 = 30^k \\ 1.2P_{2D} + 1.6P_{2L} = 1.2\times 21.4 + 1.6\times 4 \\ \qquad = 32^k \leftarrow \text{governs} \end{cases}$$

Estimate the factored beam weight as 100 plf (w_u = 0.1 klf):

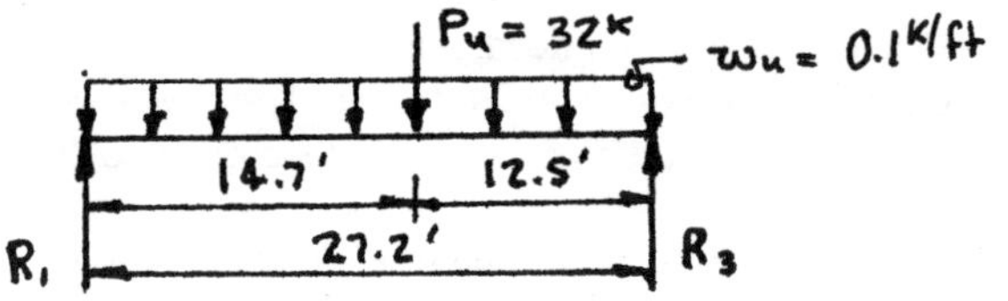

$\Sigma M_1 = 0 = [32(14.7) + 0.1(27.2)^2 / 2] - 27.2R_3 = 0$

$\therefore$ $R_3 = 18.7^k \uparrow$

$\Sigma F_y = 0$: $R_1 = 16.0^k \uparrow$

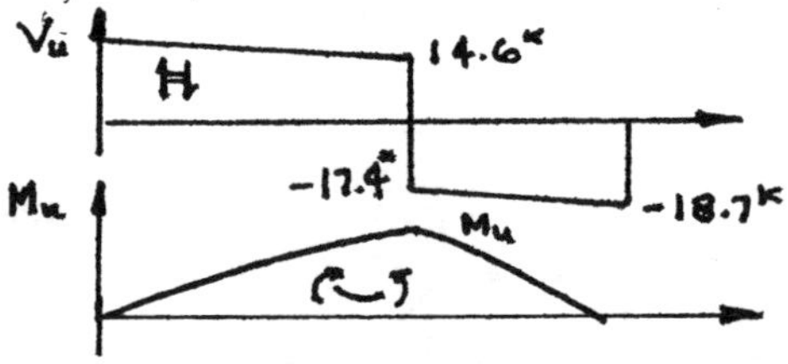

$(M_u)_{max} = 0.5(17.4 + 18.7)12.5 = 226^{k\text{-}ft}$

The beam is laterally unbraced over the full span, L_b = 27.2'; C_b = 1.0; F_y = 36 ksi. Using the moment charts from the AISC Manual (P. 4-131), select for trial a W16x67.

Verify the W16x67: d = 16.33"; t_w = 0.395"; $b_f / 2t_f$ = 7.7"; r_y = 2.46"; X_1 = 2350; X_2 = 0.004690; Z_x = 130 in^3; Sx = 117 in^3

Check local buckling (B-5, AISC):

$b_f / t_f = 7.7 < 65 / \sqrt{F_y} = 10.8 = \lambda_p$ $\therefore$ section is compact.

Flexural strength:

$L_p = 300r_y / \sqrt{F_{yf}} = (300 \times 2.46 / 6) / 12 = 10.3'$

$$L_r = \frac{r_y X_1}{(F_{yw} - F_r)} \sqrt{1 + \sqrt{1 + X_2(F_{yw} - F_r)^2}}$$

$$L_r = \frac{2.46 \times 2350}{(36 - 10)} \sqrt{1 + \sqrt{1 + 0.00469(36 - 10)^2}}$$

= 32.3'

$\therefore$ $L_p < L_b < L_r$, check Eqn F1-3

$\phi M_n = C_b[\phi M_p - (\phi M_p - \phi M_r)(L_b - L_p) / (L_r - L_p)] \le \phi M_p$

$M_p = F_y Z_x = 36 \times 130/12 = 390^{k\text{-}ft}$

$M_r = (F_{yw} - F_r)S_x = (36 - 10)117/12 = 253^{k\text{-}ft}$

$\phi M_n = 0.9[390 - (390 - 253)(27.2 - 10.3) / (32.3 - 10.3)] = 255^{k\text{-}ft}$

$\phi M_n = 255 > M_u = 226^{k\text{-}ft}$ $\therefore$ O.K. in flexure.

Check shear (section F2.2):

$\phi V_n = \phi(0.6F_y)dt_w = 0.9(0.6 \times 36) \times 16.33 \times 0.395 = 125^k > V_u = 18.7^k$

$\therefore$ **Use W16x67.**

b) Design a suitable connection. Use four hanger rods to support the beam from below.

$T_u = 32/4 = 8^k$/rod

Re: P. 5-3, AISCM, select 4-5/8" dia. rods

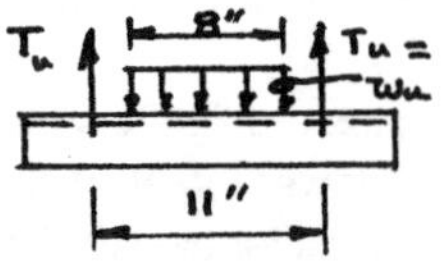

$w_u = (32/2)/8 = 2^k/_{in}$

$M_u = 8(5.5) - 8(2) = 28^{k\text{-}in} \le \phi_b M_p = 0.9 \times 36Z_y$

$\therefore$ $Z_y = 0.9$ in^3 – **use C6 x 8.2**

Connect the rods to the W16x67 web with vertical plates using 1/4" E70 welds at 5" eccentricity:

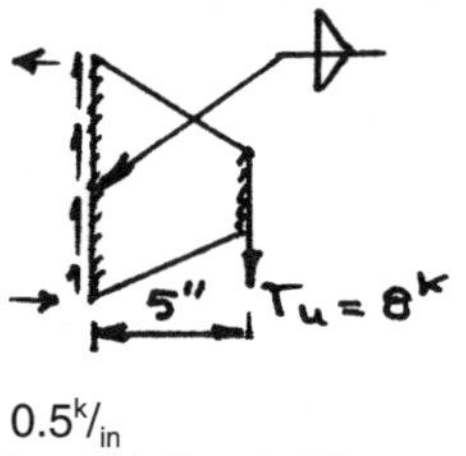

$f_v = 8 / 16 = 0.5^k/_{in}$

$f_b = 8 \times 5 / (2 \times 8^2 / 6) = 1.9^k/_{in}$

$f_r = (0.5^2 + 1.9^2)^{1/2} = 1.96^k/_{in} < q = 4 \times 1.39^k/_{in}$

$\therefore$ **use 1/4" fillet welds**

Details of the connection are:

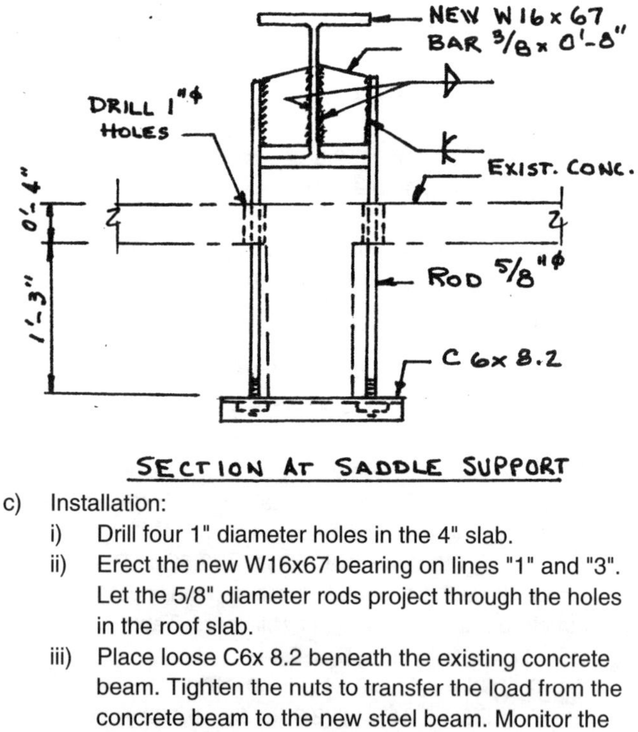

SECTION AT SADDLE SUPPORT

c) Installation:

i) Drill four 1" diameter holes in the 4" slab.

ii) Erect the new W16x67 bearing on lines "1" and "3". Let the 5/8" diameter rods project through the holes in the roof slab.

iii) Place loose C6x 8.2 beneath the existing concrete beam. Tighten the nuts to transfer the load from the concrete beam to the new steel beam. Monitor the deflection in the steel beam.

iv) Remove the column (the dead load has been transferred $\therefore$ no deflection occurs in the beam).

Steel-59

a) Calculate the upward component of the concrete compression stress at the ridge. Given:

$M_{CL} = 2200^{k\text{-}ft}$; f'_c = 3 ksi; w = 110 pcf;

I_{tr} = 63,700 in^4; n=14; b_e = 80"

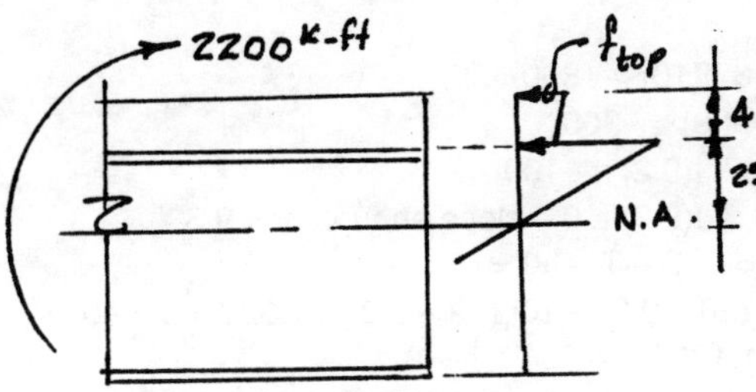

$f_{top} = (Mc_{top} / I_{tr}) / n$
$= (2200 \times 12 \times 29 / 63{,}700) / (14 \times 63{,}700)$
$= 0.85$ ksi
$f_{y=25} = (25 / 29)0.85 = 0.74$ ksi

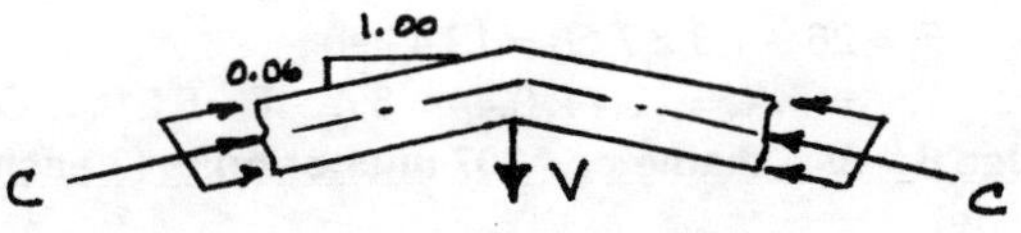

$C = 0.5(f_{top} + f_{y=25})(4)b_e = 0.5(0.85 + 0.74)(4)(80)$
$= 254^k$
$\Sigma F_y = 0$: $2C\sin\alpha - V = 0$
$\alpha \cong \sin\alpha \cong \tan\alpha = 0.06 / 1.0 = 0.06$
$\therefore$ $V = 2 \times 254 \times 0.06 = 30.5^k$

b) Develop a plan to strengthen the ridge beam. Use allowable stress design (AISC 1989). Distribute the 30.5^k over the effective width, b_e = 80" = 6.67'.
$w = 30.5/6.67 = 4.58$ klf ↑
Neglect gravity loading (per problem statement).

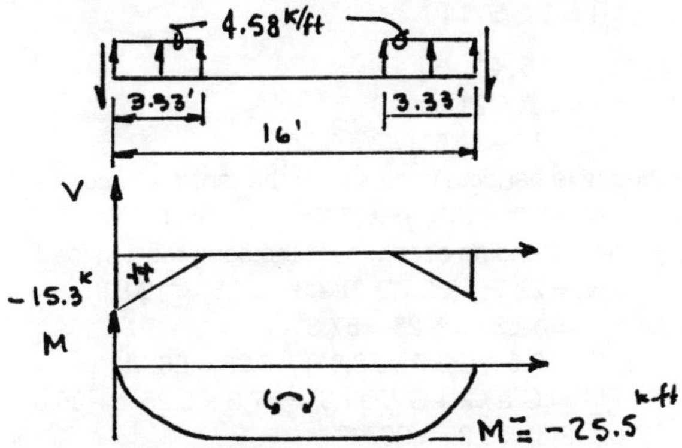

For the existing W8x10: $A_g = 2.96$ in²; d = 7.89"; t_w = 0.170"; b_f = 3.94"; t_f = 0.205"; $I_x = 30.8$ in⁴; $S_x = 7.81$ in³; r_T = 0.99"; 2-3/4" A307 bolts in single shear; $R_v = 4.4^k/_{bolt}$

$V = 15.3^k > 2 \times 4.4 = 8.8^k$ $\therefore$ The bolts are overloaded

$f_b = 25.5 \times 12/7.81 = 39$ ksi $> F_b$ $\therefore$ the beam is overstressed in bending.

To strengthen the beam, add another W8x10 and provide additional 3/4" bolts in single shear to plates in the tapered girder.

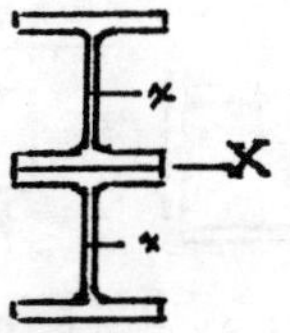

For the builtup section:
$I_x = 2(I_{xo} + Ad^2) = 2(30.8 + 2.96 \times (7.89/2)^2) = 154$ in⁴

$S_x = I_x/d = 154/7.89 = 19.5$ in³

The bottom flange is in compression. Brace the bottom flange using counters to the top flange of the adjacent roof beams.

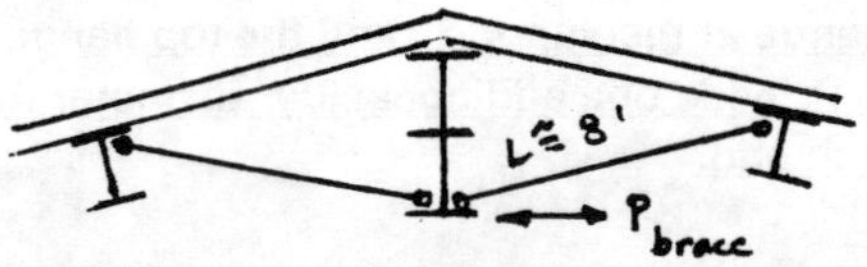

$P_{brace} = 0.02F_yA_f = 0.02 \times 36 \times 4.0 \times 0.205 = 0.6^k$

Slenderness ratio governs: $r \geq L/300 = 8 \times 12/300 = 0.32$"

Select L 1-3/4 x 1-3/4 x 3/16 (r_z = 0.34")

Try bracing the midspan: $L_b = 16/2 = 8'$; $C_b = 1.0$
$L_b/r_T = 8 \times 12 / 0.99 = 97 < 119$
$\therefore$ $F_b \geq [0.67 - F_y(L_b/r_T)^2/1{,}530{,}000C_b]F_y \leq 0.6F_y$
$F_b \geq [0.67 - 36 \times 97^2/1{,}530{,}000]36 = 16.0$ ksi
$f_b = 25.5 \times 12 / 19.5 = 15.7$ ksi $< F_b$ $\therefore$ O.K.

Check the welds to connect the 2 W8x10:
$q = VQ/I$; $Q = A(d/2) = 2.96(7.89/2) = 11.7$ in³
$q = 15.3 \times 11.7/ 154 = 1.1^k/_{in}$

Minimum size E70 fillet welds are adequate; use continuous welds at ends and intermittent in the midportion.

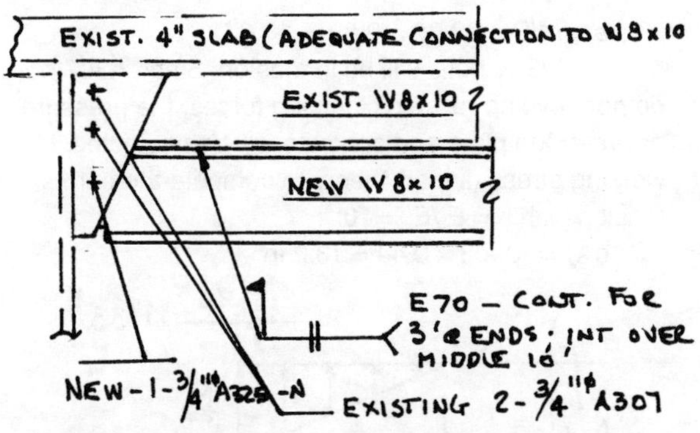

Check the adequacy of the existing stiffener plate to transfer the 15.3^k reaction:
2-1/4" E70 welds; 15" long: $q = 2 \times 8 \times 0.928 \times 15 = 222^k$ $\therefore$ O.K.

Clearance problems prevent two additional bolts $\therefore$ try only one additional:
$V = 15.3 - 2 \times 4.4 = 6.5^k$

One 3/4" A307 bolt is not adequate; need to attach the new member with an A325-N bolt, which has $R_v = 9.3^k/_{bolt}$. Cope the new W8x10 to clear the stiffener. Web tearout is no problem with the A307 bolts, but may be with the stronger A325-N (Re: J4, AISC).

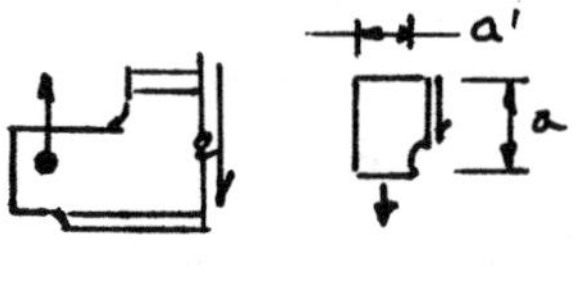

For t_w = 0.17"; k = 5/8"; a' = 2 – 0.5 – 0.375 = 1.125";
a = (7.89/2) – 5/8 – 3/8 = 2.9".

$V \leq 0.3A_vF_u + 0.5F_uA_t = 0.3 \times 2.9 \times 0.17 \times 58 + 0.5 \times 1.1 \times 0.17 \times 58 = 14^k$

The web has capacity to resist 14^k and only 6.5^k is needed ∴ O.K. Add one bolt in the new W8x10.

∴ **Strengthen the ridge beam by adding a W8x10 welded flange-to-flange to the existing W8x10. Brace the bottom flange at the supports and the top flange at midspan. Provide one additional 3/4" diameter A325-N bolt at each end.**

Steel-60

a) Determine the anchor bolt size. Use allowable stress design (AISC 1989); F_y = 36 ksi; F_p = 1.125 ksi; A307 anchor bolts.

Assume the 20^k shear is shared equally, V = 20/6 = 3.3^k.
Try 3/4" diameter bolts ($R_v = 4.4^k/_{bolt}$ in single shear).

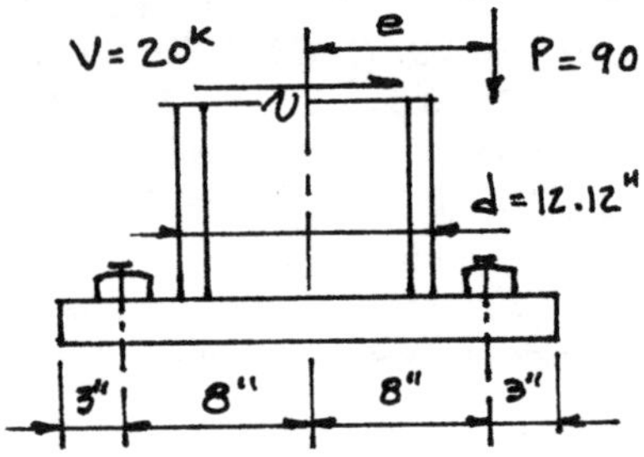

e = M/P = 80/90 = 0.89' = 10.67"

e > L/3 ∴ bolts will be in tension. Since A307 bolts do not develop reliable clamping forces, the pressure between the plate and concrete will be neglected. Use working stress (linear) theory to compute the bolt tension with $n = E_s/E_c = 10$:

$nA_s = 10 \times 3 \times 0.44 = 13.2\ \text{in}^2$

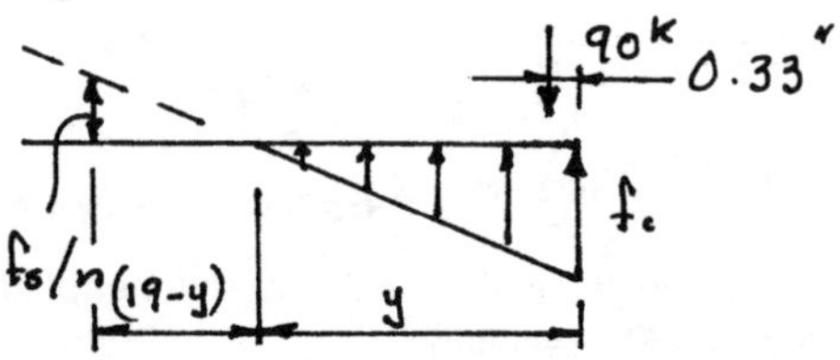

Let y = the distance from the compression edge to the neutral axis. By similar triangles:

$$\frac{f_s/n}{(19-y)} = \frac{f_c}{y} \therefore f_s = \frac{nf_c(19-y)}{y}$$

Take moments about the line of action of the force P:

$\Sigma M_{P\text{-line}} = 0$: T(19 – 0.33) – C(y/3 – 0.33) = 0

∴ $[nf_c(19 - y)/y] \times 3 \times 0.44 \times (19 - 0.33) = 0.5f_c \times 20y(y/3 - 0.33)$

Canceling f_c and simplifying:

$3.33y^3 - 3.3y^2 + 246.4y - 4682 = 0$

Solve the cubic equation by trial-and-error (or any other method):

Try y = 10: f(10) ≅ +800
Try y = 9: f(9) ≅ –300
Try y = 9.2: f(9.2) ≅ –100
Try y = 9.3; f(9.3) ≅ 0 – close enough: y = 9.3"

$\Sigma F_y = 0$: C – T – 90 = 0

$0.5f_c(20 \times 9.3) - 10f_c(19 - 9.3) \times 1.32 / 9.3 = 90$

∴ $f_c = 1.135\ \text{ksi} \cong F_p$

$f_s = 10 \times 1.135(19 - 9.3)/9.3 = 11.84$ ksi

∴ $T = f_sA_b = 11.84 \times 0.44 = 5.21^k/_{bolt}$

Check the anchor bolt under combined shear and tension (Table J3.3, AISC):

$f_v = 3.33/0.44 = 7.56$ ksi

$F_t = 26 - 1.8f_v \leq 20$

$F_t = 26 - 1.8 \times 7.56 = 12.4$ ksi

∴ $T_{max} = F_tA_b = 12.4 \times 0.44 = 5.4 > T = 5.21^k$ ∴ O.K.

Use 6 – 3/4" diameter A307 anchor bolts (3 each side)

b) The concrete bearing stress was found in part a to be **1.135 ksi**.

c) Determine the plate thickness required. The critical location for bending is at (22 – 0.95d)/2 from the edges.

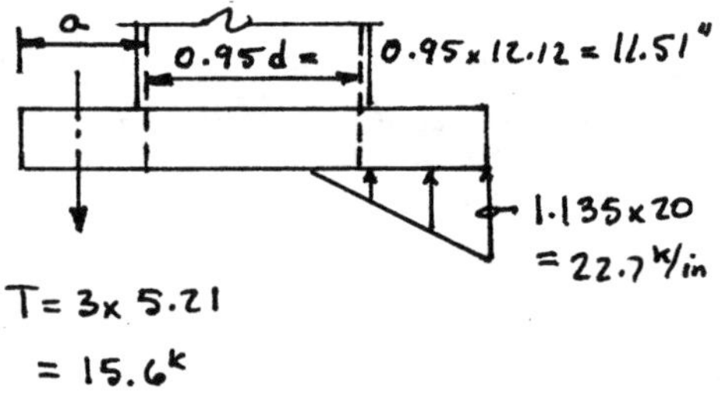

$T = 3 \times 5.21 = 15.6^k$

a = (22 – 11.51) / 2 = 5.25"

Negative bending occurs near the tension force:

$M^- = T(a - 3) = 15.6(5.25 - 3) = 35.1^{k\text{-in}}$

Positive bending occurs near the compression loading:

$w_2 = 22.7 - (22.7/9.3)(9.3 - 5.25) = 12.81^k/_{in}$

∴ $C_1 = 12.81 \times 5.25 = 67.3^k$

$C_2 = 0.5 \times (22.7 - 12.81) \times 5.25 = 26.0^k$

$M^+ = C_1a/2 + C_2(2a/3) = 67.3 \times 5.25/2 + 26.0 \times (2/3) \times 5.25 = 267^{k\text{-in}}$

The flexure strength is governed by positive bending, $267^{k\text{-in}}$.

For the 20" wide plate: $f_b \leq F_b = 0.75F_y = 27$ ksi:

$f_b = M_{max}/S = 267/(20t^2/6) \leq 27$

$t^2 \geq 2.97\ \text{in}^2$; ∴ $t \geq 1.72$"

Use 1-3/4" PL of the dimensions given.

Seismic-1

a) Calculate the story shear at each level (UBC 1997, Section 1630.2). For a regular building less than 160' in height the static force procedure is applicable. For Seismic Zone 3 and soil profile type S_D: $C_a = 0.36$, $C_v = 0.54$; importance factor: $I = 1.0$; assume the frame will be detailed to qualify as a special moment-resisting space frame with $R = 8.5$. Approximate the period using Eqn. 30-8 assuming steel moment-resisting frames:

$$T = C_t(h_n)^{3/4} = 0.035(156)^{3/4} = 1.5 \text{ sec}$$

The design-level base shear coefficient is:

Lower Limit $\geq 0.11C_aI = 0.11 \times 0.36 \times 1 = 0.04$

Upper Limit $\leq 2.5C_aI / R = 2.5 \times 0.36 \times 1.0 / 8.5 = 0.106$

$$C_vI / RT = 0.54 \times 1 / (8.5 \times 1.5) = 0.0423$$

$\therefore \quad V = (C_vI / RT)W = 0.0423W$

From the given dead loads:

$$W = \sum_{i=1}^{n} W_i = 18{,}500 \text{ kips}$$

UBC-1630.1.1 requires that the base shear be increased by a reliability/redundancy factor,

$\rho = 2 - \dfrac{20}{r_{max}\sqrt{A_B}}$. The value of r_{max} must give a calculated $\rho \leq 1.25$ for special moment-resisting frames. Given that only the exterior frames are special moment-resisting frames, and assuming that the shear in an interior column is twice that of an exterior column (i.e., using the portal analysis assumption), and noting that the element shear is defined as 70% of the sum of the shears in the two interior columns, the value of r_{max} is

$$r_{max} = 0.7[(V/6)+(V/6)]/V = 0.233$$

$$\rho = 2 - \frac{20}{(0.233)\left(\sqrt{90 \times 90}\right)} = 1.05$$

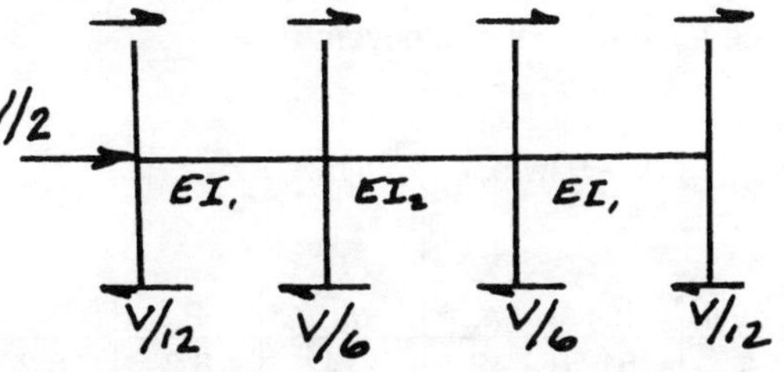

Thus, the minimum design-level base shear for the building is:

$$V = 1.05 \times 0.0423 \times 18{,}500 = 821 \text{ kips}$$

Calculate the story shear at each level:

$$F_t = 0.07TV = 0.07 \times 1.5 \times 821 = 86 \text{ kips}$$

$$V - F_t = 821 - 86 = 735 \text{ kips}$$

From Equation 30-13, UBC-97:

$$F_x = \frac{(V - F_t)w_xh_x}{\sum_{i=1}^{n} w_ih_i}$$

Level	w_x	h_x	w_xh_x	F_x
2	1500	13	19,500	9.0
3	1500	26	39,000	18.0
4	1500	39	58,500	27.0
5	1500	52	78,000	36.0
6	1500	65	97,500	45.0
7	1500	78	117,000	54.0
8	1500	91	136,500	63.0
9	1500	104	156,000	72.0
10	1500	117	175,500	81.0
11	1500	130	195,000	90.0
12	2000	143	286,000	132.0
13	1500	156	234,000	108.0
		$\Sigma =$	1,592,500	735.0 kips

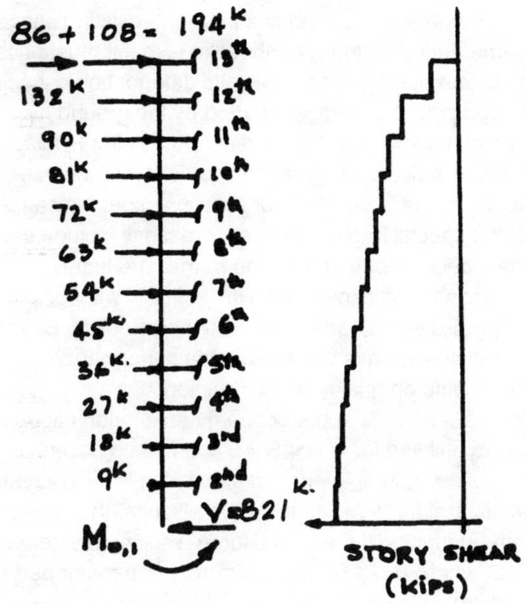

b) Calculate the overturning moment about the base:

$$M_{01} = F_th_n + \sum_{i=1}^{n} F_ih_i$$

$= (86 + 108)156 + 132(143) + 90(130)$
$+ 81(117) + 72(104) + 63(91) + 54(78)$
$+ 45(65) + 36(52) + 27(39) + 18(26)$
$+ 9(13)$

M_{01} = 94,200 k-ft

c) Calculate the overturning moment about level 4. Use the value from part b, M_{01}, and take the area under the shear diagram to get the change in moment from the base to level 4:

$$M_{04} = M_{01} - \sum_{i=1}^{4} V_i\Delta h_i$$

$= 94{,}200 - 821(39) + 9(26) + 18(13)$

M_{04} = 62,650 k-ft

Seismic-2

a) Elastic response spectra are plots of maximum response (e.g., displacement, velocity or acceleration) of a single degree of freedom (SDF) linear oscillator subjected to a specific load function (e.g., base motion) over a practical range of periods of the SDF oscillator.

b) The most important factor that reconciles the discrepancy between the earthquake and the design level of the response spectra is the inelastic behavior, which results in hysteretic energy dissipation by the structure.

c) Other factors that might modify the response of a structure to an earthquake are the changes in stiffness and damping characteristics as the structure degrades

d) In addition to the obvious requirements of strength, stability and continuous load path, a fundamental requirement for survival of a moment-resisting space frame during a strong earthquake is that it must possess sufficient ductility -- i.e., inelastic deformation capacity -- to dissipate the energy imparted by the ground motion.

e) The primary reason for the difference in the R for a bearing wall as compared to a special moment resisting space frame is the superior energy absorption capability of the special frame. Secondary reasons include the fact that special moment resisting frames are highly redundant and can respond inelastically without jeopardizing the vertical capacity and stability. Bearing wall systems have shown variable performance depending on quality of construction.

f) In addition to the superior energy absorption capability of the building for which R = 5.5, it has an essentially complete system to support gravity loads independent of the lateral force resisting system. Hence, the difference is redundancy. If the lateral force resisting elements fail (i.e., where R = 4.5), the structure – or a major part of it – might collapse.

g) The primary purpose of the UBC earthquake regulations is to prevent loss of life by setting minimum standards to prevent structural collapse. The performance to be expected (SEAOC 1996):
   1. Minor earthquake = no damage
   2. Moderate earthquake = no structural damage; possibly damage to non structural elements
   3. Major earthquake = possibly damage to structural and non structural elements, but avoids collapse

h) Influence of soil conditions:
   1. Increasing soil thickness generally increases predominate period of ground motion and amplifies the structural response
   2. Structures most likely to be subjected to greatest seismic force when underlain by rock are the relatively stiff structures with periods in the range 0.1 to 0.5 sec.
   3. Conversely, relatively flexible structures, those with periods greater than 0.5 sec, are most likely to experience greatest seismic force when underlain by soft soils. The worst situation occurs when the building's period and the predominate period of the ground motion nearly coincide.

Seismic-3

a) Compute the base shear and overturning moment in the north-south direction (UBC 1997). Check both wind and seismic loadings.

- Wind (UBC 1997, Section 1621):
- C_eq_s = 20 psf; C_p = 1.3 for the building; C_p = 1.4 for the penthouse

∴ $V_w = C_pC_eq_sA_{trib}$
$= 1.3 \times 20 \times 90 \times 39 + 1.4 \times 20 \times 20 \times 12$
$= 97{,}980 \text{ lb} = 98 \text{ kips}$

- Seismic (UBC 1997, Section 1630). For the combination braced and moment resisting frame along the NS axis, UBC 1630.4.4 requires that the R-factor for the braced bents be used ∴ use R = 5.6 and Ω_o = 2.2 for an ordinary steel braced frame (Table 16-N). Take I = 1.0 and the soil profile factor S_C, for which C_a = 0.33 and C_v =0.45. Approximate the period using Equation 30-8:
  $T = C_th^{3/4} = 0.02 \times (39)^{3/4} = 0.3$ sec
  The design-level base shear coefficient is:
  Lower Limit $\geq 0.11C_aI = 0.11 \times 0.33 \times 1 = 0.036$
  Upper Limit $\leq 2.5C_aI / R = 2.5 \times 0.33 \times 1.0 / 5.6 = 0.147$
  $C_vI / RT = 0.45 \times 1 / (5.6 \times 0.3) = 0.268$

∴ The upper limit governs: V = 0.147W.
From the given dead loads:
W = 923 + 909 + 772 + 60 + 42 = 2706 kips

Converting to working stress (i.e., dividing the design-level seismic force by 1.4):
$V = 0.147 \times 2706 / 1.4 = 284$ kips

Since $V_e >> V_w$, seismic governs
$T < 0.7$ sec ∴ $F_t = 0$

$$F_x = (V - F_t)w_hh_x / \sum_{i=1}^{n} w_ih_i$$

Level	h_x	w_x	w_xh_x	F_x	V_x	M_x
4	51	42	2142	8.3	8.3	99.6
3	39	832	32,448	126.3	134.6	1714
2	27	909	24,543	95.5	230.1	4476
1	15	923	13,845	53.9	284	8736
	Σ =	2706	72,980			

b) Compute the torsional moment at the 2nd floor including accidental eccentricity (1630.7, UBC-97): $\Delta e = 0.05L = 0.05 \times 90 = 4.5'$. Let $\bar{x}$ = the distance from line 'A' at the center of rigidity $= \Sigma R_ix_i / \Sigma R_i = (6 + 6)60 / 15 = 48'$

∴ $e_x = (48' - 38') + \Delta e = 14.5'$
$M_t = \Sigma F_xe_x = (8.3 + 126.3 + 95.5)$ kips × 14.5 feet
M_t = 3336 k-ft

c) Design braces B1 and B2 neglecting gravity load and torsion per the problem statement. Use allowable stress design (AISC 1989). With torsion neglected, the proportion of story shear and overturning moment resisted by the braced bents is:

$\% = [(6 + 6) / (6 + 6 + 3)] / 2 = 0.4$

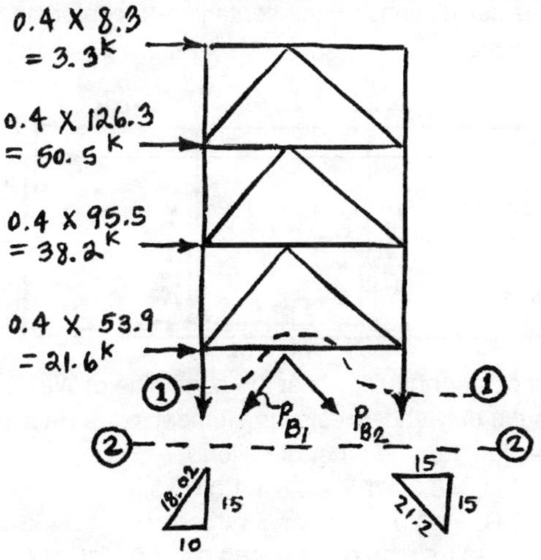

Design the braces to resist both tension and compression. From the free body denoted as 1-1:

$\Sigma M_{LT} = 3.3 \times 36 + 50.5 \times 24 + 38.2 \times 12 + P_c(25)$
$= 0$

$\therefore$ $P_c = 71.6$ kips $\uparrow$

From the free body 2-2:

$\Sigma F_y = 0$: $(15 / 18.02)P_{B1} + 0.707P_{B2} = 0$

$P_{B1} = -0.85P_{B2}$

$\Sigma F_x = 0$: $3.3 + 50.5 + 38.2 + 21.6 + 0.707P_{B2}$
$+ (10 / 18.02)(+0.85P_{B2}) = 0$

$P_{B2} = -96.4$ kips; $P_{B1} = 81.9$ kips (reversible)

Check the reliability/redundancy factor of 1630.1.1, UBC. The maximum horizontal force component in a brace is:

$P_{hor} = 96.4(0.707) = 68.2$ kips

$r_{max} = P_{hor} / V = 68.2/284 = 0.24$

$A_B = 90 \times 60 + 30 \times 40 = 6600\ \text{ft}^2$

$\therefore$ $\rho = 2.0 - 20 / [r_{max}\sqrt{A_B}] = 2.0 - 20 / [0.24 \times 6600^{1/2}]$
$= 0.97$; use 1.0

Checking Section 2213.8.2.3, UBC, the maximum horizontal component of brace force does not exceed 70% of the in-line force in the braced bent.

Per 2213.8.4.1, the chevron brace must be designed to resist 1.5 times the calculated force:

$\therefore$ $P_{max} = 1.5 \times 96.4 = 145$ kips w / $Kl = 21.2'$

Try 2Ls separated by 1/2" gusset plate:

$KL / r)_{max} = 21.2 \times 12 / 120 = 2.1"$

Try 2L $8 \times 8 \times 1/2$ ($r_{min} = 2.5"$; $A_g = 15.5\ \text{in}^2$)

$KL / r = 21.2 \times 12 / 2.5 = 102$; $F_a = 12.7$ ksi

$B = 1 / [1 + (KL / r) / 2 C_c]$; $C_c = 126$
$= 1 / [1 + 102 / (2 \times 126)] = 0.71$

$F_{as} = {}^4/_3 BF_a = ({}^4/_3)(0.71)(12.7) = 12.0$ ksi

$f_a = 145 / 15.5 = 9.35$ ksi $< F_{as}$ $\therefore$ OK

Check tension assuming a double gage of 3/4" diameter bolts in one leg $\therefore$ $U = 0.85$.

$A_n = A_g - A_h = 15.5 - 4({}^3/_4 + {}^1/_8){}^1/_2 = 13.8\ \text{in}^2$

$f_t = P_{max} / UA_n = 145 / (0.85 \times 13.8) = 12.4$ ksi $<< F_t$
$= {}^4/_3(0.5F_u)$ $\therefore$ OK

<u>Use 2L8×8×1/2 w/ 1/2" gusset plate</u>

d) Design the connection of braces to the W16× member (Re 2213.8.3.1, UBC-97):

$$P \le \begin{cases} 0.5A_eF_u = 0.5 \times 13.8 \times 58 = 400 \text{ kips} \\ 0.6A_gF_y = 0.6 \times 15.5 \times 36 = 335 \text{ kips} \\ \Omega_0 96.4 = 2.2 \times 96.4 = 212 \text{ kips} \leftarrow \end{cases}$$

Use ${}^7/_8$" dia A325-x; double shear bearing on plate ${}^5/_8$" thick; A36 steel:

$r \le {}^4/_3(36.1, 38.1) = 48$ kips / bolt

$\therefore$ $n \ge P / r = 212$ kips / (48 kips / bolt) = 4.4, say 5

<u>Use 5 – ${}^7/_8$" dia A325-x @ 3" pitch</u>

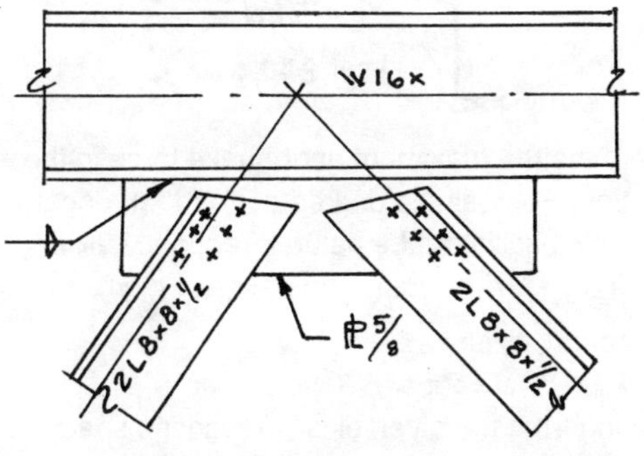

BRACE-TO-GIRDER CONNECTION

e) Design column D4 for the given gravity loads plus computed seismic load of 71.6 kips. Section 2213.5.1, UBC-97, applies with $\Omega_0 = 2.2$.

$$P \ge \begin{cases} 0.75(P_{DL} + 0.7P_{LL} + \Omega_0 P_E) = 0.75(240 + 0.7(35) + 2.2(71.6)) \\ \qquad = 317 \text{ kips} \\ P_{DL} + P_{LL} = 240 + 35 = 275 \text{ kips} \end{cases}$$

For $Kl_y = 15'$; $P = 317$ kips; $F_y = 36$ ksi (p. 3-28, AISCM)

<u>Select W 12×65 ($P_{all} = 334$ kips)</u>

f) Verify a W10×100 for the given loads ($A = 29.4$; $r_x = 4.6$; $r_y = 2.65$; $S_x = 112$; $P = 140$ kips; $M_x = 165$ k-ft):

$Kl / r_x = 2.0 \times 15 \times 12 / 4.6 = 78 \leftarrow$ governs

$Kl / r_y = 1.0 \times 15 \times 12 / 2.65 = 68$

$Kl / r_x = 78$ governs $\therefore$ $F_a = 15.6$ ksi

$f_a = P / A = 140 / 29.4 = 4.8$ ksi

$f_a / F_a = 4.8 / 15.6 = 0.31 > 0.15$

$L_b = 15' > L_c = 10.9'$; $< L_u = 48.2'$

$\therefore$ $F_{bx} = 0.6F_y = 22$ ksi

$f_{bx} = M_x / S_x = 165 \times 12 / 112 = 17.7$ ksi

$f_{bx} / F_{bx} = 17.7 / 22 = 0.80$

Combined stress check ($f_a / F_a > 0.15$):

$C_{mx} = 0.85$ (sidesway)

$F_{ex}' = {}^4/_3\ (24.5) = 32.7$ ksi (p. 5-122: $Kl / r_x = 78$)

$f_a / F_a + [C_{mx} / (1 - f_a / F_{ex}')]f_{bx} / F_{bx}$
$= 0.31 + [0.85 / (1 - 4.8 / 32.7)](0.80) = 1.1 < {}^4/_3$

$\therefore$ **<u>The W10×100 is OK</u>**

<u>Seismic-4</u>

a) Develop a lumped mass mathematical model of the structure. The structure is regular and symmetrical $\therefore$ a

2-dimensional single degree of freedom model is sufficient. Lumped masses should include the contribution from the roof plus one-half the mass of the exterior walls (use the total 20'×20' roof):

$M = W / g$

$W = 40\text{ psf} \times 20' \times 20' + 15\text{ psf} \times 4 \times 20' \times 6'$

$= 23{,}200^{\#}$

$\therefore \quad M = 23{,}200 / 32.2 = 720\ \# \cdot \text{sec}^2/\text{ft}$

Column stiffness is given as 5.0 kips/in per column ∴ the total for 4 columns is:

$K = 4k_c = 4 \times (5000^{\#}/_{\text{in}})\ 12^{\text{in}}/_{\text{ft}} = 240{,}000\ ^{\#}/_{\text{ft}}$

Thus, the equivalent lumped mass model is:

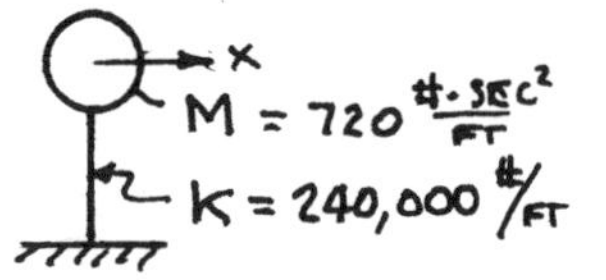

b) Determine the maximum earthquake force in the X-X direction. For a single degree of freedom, undamped harmonic oscillator, the natural frequency, ω, is:

$\omega = \sqrt{K/M} = \sqrt{240{,}000/720} = 18.3\ ^{\text{rad}}/_{\text{sec}}$

The period, T, is:

$T = 2\pi / \omega = 2\pi / 18.3 = 0.34\text{ sec}$

Reading from the given elastic response spectrum:

$S_a = 0.45g = 0.45 \times 32.2\ \text{ft/sec}^2 = 14.5\ \text{ft/sec}^2$

Thus, the base shear is:

$V = MS_a = 720\ \# \cdot \text{sec}^2/\text{ft} \times 14.5\ \text{ft/sec}^2$

$V = 10{,}400^{\#} = 10.4$ kips

c) Determine the maximum column shear corresponding to the force in part b. Assume a 5% minimum eccentricity to be consistent with building codes:

$M_t = Ve = 10.4 \times 0.05 \times 20' = 10.4$ k-ft

$\Sigma(x_i^2 + y_i^2) = 4(10^2 + 10^2) = 800\text{ ft}^2$

$\therefore \quad V_{col} = [V/2 + M_t x / \Sigma(x_i^2 + y_i^2)] / 2$

$V_{col} = [10.4/2 + 10.4 \times 10/800]/2 = 2.7$ kips

d) The spectral displacement is related to the spectral acceleration by the equation $S_D = S_a / \omega^2$. Thus, maximum displacement of the roof in the x-x direction is:

$S_D = S_a / \omega^2 = 14.5 / 18.3^2 = 0.043\text{ ft} = 0.52\text{ in}$

Seismic-5

a) Calculate the forces in the walls for North-South ground motion. Include one-half the weight of the shearwalls in the dead load for computation of seismic base shear (conservative in both directions):

$W_{wall} = 0.15\text{ kcf} \times (\text{wall area}) \times {}^1/_2\text{ height}$

$= 0.15\,[0.83(30 + 20 + 10 + 8) + 0.67 \times 40] \times 8$

$= 100$ kips

$\therefore \quad W = W_{roof} + W_{wall} = 0.100 \times 30 \times 90 + 100 = 370$ kips

Use the static force procedure (UBC 1997; section 1630.2). For a concrete bearing wall system, R = 4.5; zone 3 with soil profile type S_D gives $C_a = 0.36$ and $C_v = 0.54$. Approximate the period using Equation 30-8:

$T = C_t(h)^{3/4} = 0.02 \times 16^{0.75} = 0.16$ sec

The design-level base shear coefficient is:

Lower Limit $\geq 0.11C_aI = 0.11 \times 0.36 \times 1 = 0.04$

Upper Limit $\leq 2.5C_aI / R = 2.5 \times 0.36 \times 1.0 / 4.5$
$= 0.20$

$C_vI / RT = 0.54 \times 1 / (4.5 \times 0.16) = 0.75$

$\therefore \quad V = 0.20W = 0.20 \times 370 = 74$ kips

Base calculations on the centerline-to-centerline dimensions.

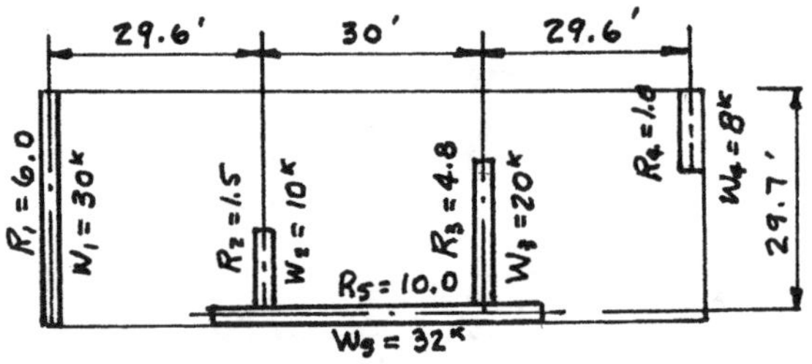

The center of rigidity is at the centerline of Wall 5, since it is the only shear-resisting element in the East-West direction. For the other direction:

$\Sigma R_y = 6.0 + 1.5 + 4.8 + 1.0 = 13.3$

$\Sigma R_x = 10.0$

$\bar{x} = [1.5 \times 29.6 + 4.8 \times 59.6 + 1.0 \times 89.2] / 13.3$

$= 31.6'$ (right of centerline of Wall 1)

The center of mass (including one-half the mass of walls) is located:

$x_m = [270 \times 44.6 + 10 \times 29.6 + 20 \times 59.6 + 8 \times 89.2 + 32 \times 44.6] / 370 = 42.4'$

$y_m = [270 \times 14.7 + 30 \times 14.7 + 10 \times 5.3 + 20 \times 10.3 + 8 \times 25.7] / 370 = 13.2'$

Wall	R_x	R_y	dx	dy	Rd^2
1	-	6.0	–31.6	-	5990
2	-	1.5	–2.0	-	6
3	-	4.8	28.0	-	3760
4	-	1.0	57.6	-	3320
5	10	-	-	0	0
Σ =	10	13.3			13,080

Torsional moment (1630.7, UBC-97):

$\Delta e = 0.05L = 0.05 \times 90 = 4.5'$

$\therefore \quad e_x = (42.4 - 31.6) + \Delta e = 15.3'$

$M_t = Ve_x = 74 \times 15.3 = 1132$ kip-ft

Note: the wall shear due to M_t adds algebraically to the shear due to the base shear force. However, building codes (e.g., UBC-97) do not allow the shear due to M_t to be used to reduce the design shear.

$V_i = VR_{xi} / \Sigma R_{xi} \pm M_t R_{xi} d_{xi} / \Sigma R_i d_i^2$

$V_1 = 74 \times 6.0 / 13.3 - 1132 \times 6.0 \times 31.6 / 13{,}080$

$V_1 = 17.0$ kips; use 33.4 kips for design

$V_2 = 74 \times 1.5 / 13.3 - 1132 \times 1.5 \times 2.0 / 13{,}080$

$V_2 = 8.1$ kips; use 8.3 kips for design

$V_3 = 74.0 \times 4.8 / 13.3 + 1132 \times 4.8 \times 28.0 / 13{,}080$

$V_3 = 38.3$ kips

$V_4 = 74.0 \times 1.0 / 13.3 + 1132 \times 1.0 \times 57.6 / 13{,}080$

$V_4 = 10.5$ kips; $V_5 = 0$

b) Calculate the distribution of wall shear for East-West ground motion:

$\Delta e = 0.05L = 0.05 \times 30 = 1.5'$

$e_y = (y_m - \bar{y}) + \Delta e = 13.2 + 1.5 = 14.7'$

$M_t = Ve_y = 74 \times 14.7 = 1088$ kip-ft

$V_1 = 1088 \times 6.0 \times 31.6 / 13{,}080 = 15.8$ kips

$V_2 = 1088 \times 1.5 \times 2.0 / 13{,}080 = 0.2$ kips

$V_3 = -1088 \times 4.8 \times 28.0 / 13{,}080 = -11.2$ kips

$V_4 = -1088 \times 1.0 \times 57.6 / 13{,}080 = -4.8$ kips

$V_5 = 74.0 \times 10 / 10 = 74$ kips

Seismic-6

a) Calculate the lateral force acting on the rigid frame at line 'J'. The resultant lateral forces at the center of mass are:

$F_2 = 2.5$ kips / ft $\times$ 160' = 400 kips

$F_R = 2.0$ kips / ft $\times$ 160' = 320 kips

Locate the center of rigidity (use given rigidities and overall dimensions):

$\Sigma R_x = 30$; $e_y = 32'$ (symmetry)

$\Sigma R_y = 20 + 0.7 = 20.7$

$\bar{x} = 0.7 \times 160 / 20.7 = 5.4'$

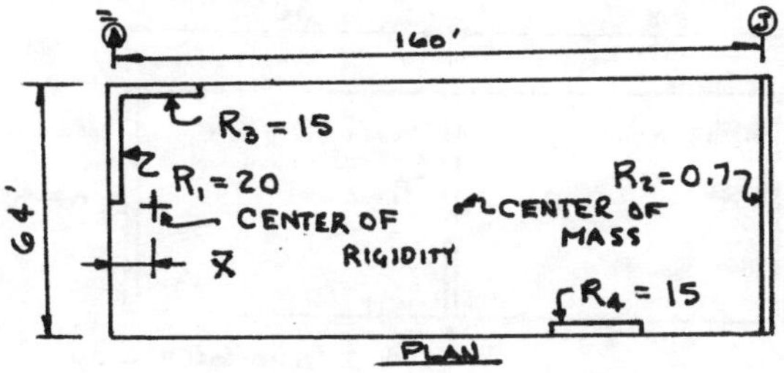

Element	R_x	R_y	dx	dy	Rd^2
1	-	20	−5.4	-	583
2	-	0.7	154.6	-	16,731
3	15	-	-	32	15,360
4	15	-	-	32	15,360
Σ =	30	20.7			48,000

Eccentricity: $e_x = x_m - \bar{x} = 80 - 5.4 = 74.6'$

$\therefore$ $M_{tR} = 320 \times 74.6 = 23{,}870$ kip-ft

$V_R = F_R R_{y2} / \Sigma R_y + M_{tR} R_{y2} d_{x2} / \Sigma Rd^2$

$= 320 \times 0.7 / 20.7 + 23{,}870 \times 0.7 \times 154.6 / 48{,}000$

$V_R = 64.6$ kips

Similarly, $M_{t2} = F_2 e_x = 400 \times 74.6 = 29{,}840$ kip-ft

$V_2 = 400 \times 0.7 / 20.7 + 29{,}840 \times 0.7 \times 154.6 / 48{,}000$

$V_2 = 80.8$ kips

b) Analyze Column J3 for the lateral forces using the portal method (first story only).

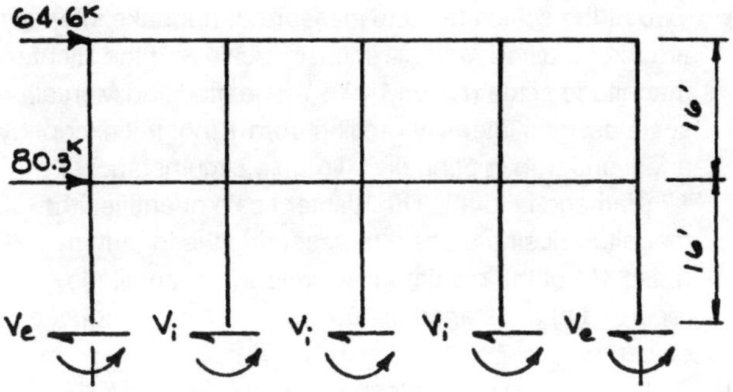

Portal analysis: $V_i = 2V_e$

$\therefore$ $V_e = (64.6 + 80.3) / 8 = 18.2$ kips

Points of inflection assumed to occur at the midheight of columns. Note that the portal analysis yields zero axial force in an interior column. Thus,

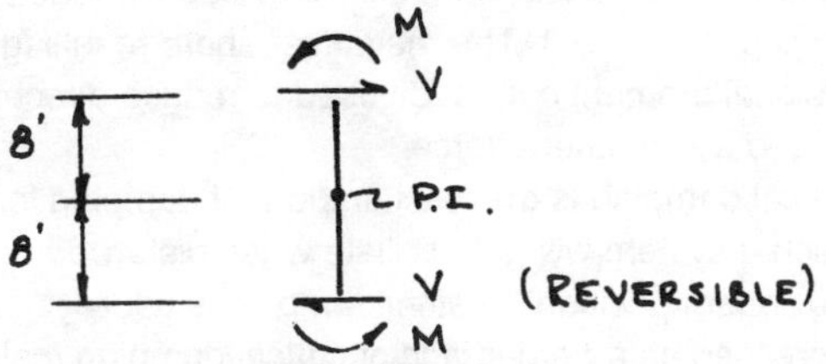

$V = V_i = 2 \times 18.2 = 36.4$ kips

$M = Vh / 2 = 36.4 \times 16 / 2 = 291$ kip-ft

Note: accidental eccentricity neglected per problem statement.

Seismic-7

a) The bearing wall system (SEAOC 1996) is characterized by shearwalls or braced frames that carry substantial vertical loads. The concern is that failure of the wall system could lead to loss of vertical support. The appropriate R-value depends on the specific material and system, and ranges from 2.8-to-5.5 (UBC 1997; Table 16-N).

b) The seismic zone referred to in the Uniform Building Code represents the maximum effective peak ground acceleration having a 90% chance of being exceeded in 50 years. Five zones are recognized (UBC 1997; Table 16-I):

1 = No damage

2A = Minor damage (EPA = 0.15g)

2B = Moderate damage (EPA = 0.2g)

3 = Major Damage (EPA = 0.3g)

4 = Major Damage (EPA = 0.4g)

For structures in zones 1, no special seismic requirements are required. A structure located in zone 2 would likely be governed by wind rather than seismic, but might be subject to some special detailing to ensure toughness under cyclic loading. Structures in zones 3 & 4 are likely to experience ground motions producing large inelastic response demands. A structure situated in zone 4, properly designed to current codes, would likely experience both structural and nonstructural damage in the event of maximum ground motion.

c) Two of the scales used to measure earthquake intensity are the modified Mercalli intensity scale and the Richter magnitude scale (NAEIM 1989). The modified Mercalli scale assigns intensity ranging from 1 (not felt except by a few under exceptionally favorable circumstances) to 12 (damage is total). The Richter scale quantifies the magnitude using a scale representing the logarithm (base 10) of the maximum seismic wave amplitude recorded at a distance of 100 km from the earthquake epicenter.

d) With respect to the Uniform Building Code (UBC 1997; Section 1630.7), the minimum accidental eccentricity for torsional moment is 5% of the building dimension perpendicular to the force under consideration. For a building with torsional irregularity (Table 16-M), the minimum eccentricity is further amplified by a factor ranging from 1 to 3. The "negative" shear resulting from torsional moment cannot be used to reduce direct shear caused by the lateral force.

e) Critical damping is the least amount of damping for which a system will not oscillate when disturbed from its equilibrium position. System damping is usually expressed as a percentage of critical damping (called the damping ratio). The damping ratio for real structures varies from virtually nil up to a maximum of about 20%. For most practical building structures, damping in the range of 2% to 7% is more typical of steel structures (2% for rigid, welded structures; 7% for flexible riveted or bolted with A307 connectors).

f) Confined concrete is enclosed by hoops or spirals such that lateral strains are resisted when an element is subjected to uniaxial compression. The compressive strength and ductility are increased as compared to an unconfined concrete element.

g) A fundamental period of 1.5 sec corresponds to a relatively flexible structure (e.g., a 15-story frame). For such structures, a reasonable approximation to the periods of the higher modes is $T_2 = T_1 / 3$; $T_3 = T_1 / 5$; ... (Taranath 1988, p. 135).

Thus,

$\underline{T_1 = 1.5 \text{ sec;}}$

$\underline{T_2 \cong T_1 / 3 = 1.5 / 3 = 0.5 \text{ sec;}}$

$\underline{T_3 \cong T_1 / 5 = 1.5 / 5 = 0.3 \text{ sec.}}$

Seismic-8

Data are given only for the first mode. Therefore, the modal analysis will be based on generalized coordinates for a single degree of freedom system (NAEIM, et al, 1989, Chapter 3):

$m^* = \Sigma m_i\phi_i^2 = [100 \times 1.0^2 + 200 \times 0.8^2 + 200 \times 0.6^2 + 200 \times 0.3^2] / g = 318/g$

$L = \Sigma m_i\phi_i = [100 \times 1.0 + 200 \times 0.8 + 200 \times 0.6 + 200 \times 0.3] / g = 440 / g$

Level	ϕ_i	$m_i\phi_i / L$
4	1.0	0.227
3	0.8	0.364
2	0.6	0.272
1	0.3	0.136
		$\Sigma = 1.000$

$q^{(x)}_{max} = \phi(x)m(x)\, L\, S_{pa} / m^*$

For a fundamental period = 0.35 sec, the response spectrum yields $S_{pa} = 0.3g$. Thus,

$\underline{q^{(4)}_{max} = 1.0 \times (100 / g)(440 / g)(0.3g) / (318 / g)}$
$\underline{= 41.5 \text{ kips}}$

$\underline{q^{(3)}_{max} = 0.8 \times (200 / g)(440 / g)(0.3g) / (318 / g)}$
$\underline{= 66.4 \text{ kips}}$

$\underline{q^{(2)}_{max} = 0.6 \times (200 / g)(440 / g)(0.3g) / (318 / g)}$
$\underline{= 49.8 \text{ kips}}$

$\underline{q^{(1)}_{max} = 0.3(200 / g)(440 / g)(0.3g) / (318 / g)}$
$\underline{= 24.9 \text{ kips}}$

Seismic-9

a) Calculate the shear resisted by each wall for the two types of diaphragms described in the problem statement.

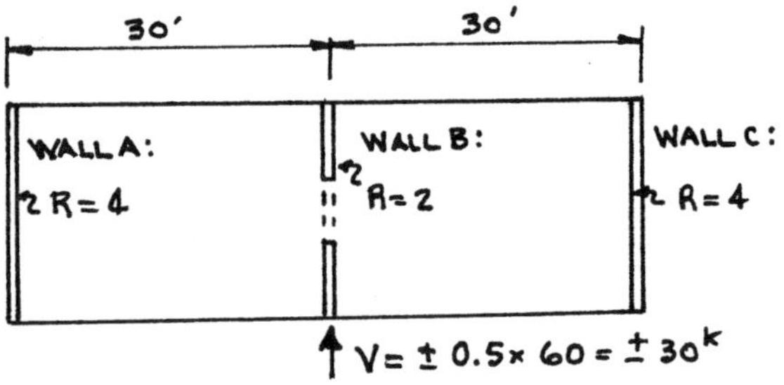

1. The roof diaphragm is a 6-in thick reinforced concrete slab, which qualifies as a rigid diaphragm. The base shear distributes to the walls on the basis of their relative rigidities, and the effect of an accidental eccentricity of $0.05 \times 60 = 3$ ft must be included. Thus,

   $\Sigma R_i = 4 + 2 + 4 = 10$

   $\Sigma R_i d_i^2 = 2 \times 4 \times 30^2 = 7200$

   $M_t = Ve = 30 \times 3 = 90$ k-ft

   $V_A = VR_x / \Sigma R_i + M_t(R_A d_x) / \Sigma R_i d_i^2$

   $\underline{V_A = 30 \times 4 / 10 + 90 \times 4 \times 30 / 7200 = 13.5 \text{ kips}}$

   By symmetry,

   $\underline{V_C = V_A = 13.5 \text{ kips}}$

   $\underline{V_B = VR_B / \Sigma R_i = 30 \times 2 / 10 = 6 \text{ kips}}$

2. The roof is 1/2" plywood nailed to wood joists. This is a flexible diaphragm and shear distributes to the resisting elements on the basis of the tributary width of loading to each wall. In this case, the interior wall resists twice as much force as each exterior wall:

   $\underline{V_A = V_C = 0.5 \times 30 \times 0.5 = 7.5 \text{ kips}}$

   $\underline{V_B = 30 \times 0.5 = 15 \text{ kips}}$

b) Discuss qualitatively how each of the following affects the walls:

1. Walls are founded on compressible soil rather than rock – This has no effect if the diaphragm is flexible; however, for the rigid diaphragm, the compressibility of the soil affects the relative rigidity of the walls. Assuming identical footings under the walls, the rigidity of walls A and C would be lowered relative to wall B; thus, the shear in wall B would be increased and shears in walls A and C would be lower than when the footings were founded on solid rock.
2. The roof diaphragm consists of metal deck over open-web steel joists with fiberglass insulation and roof covering. Depending on the method of deck attachment, side-lap connection and so forth, the rigidity of the diaphragm would be somewhere between that of the rigid and the flexible diaphragm. Thus, the shear in walls A and C would be between 7.5-to-13.5 kips and the shear in wall B would be between 6-to-15 kips.

Seismic-10

a) Calculate the period of each building (UBC 1997; Eqn 30-8):

$T = C_t(h)^{3/4}$

Building A: h = 30'; shearwalls:

$\underline{\mathbf{T_A = 0.02(30)^{3/4} = 0.25\ sec}}$

Building B: h = 100'; steel moment resisting space frame:

$\underline{\mathbf{T_B = 0.035(100)^{3/4} = 1.1\ sec}}$

b) Compare the damage potential of each site. (Note: I assume that the 40' layer of sand is dense, or densified prior to construction, so that failure by liquefaction is not a concern). The soil profile for site II would likely correspond to the soil type S_C of UBC-97, Table 16-J. Soils at site I are softer and the soil profile type is closer to that of soil profile type S_D. Qualitatively, the response spectra for the two sites would appear (smoothed design spectra):

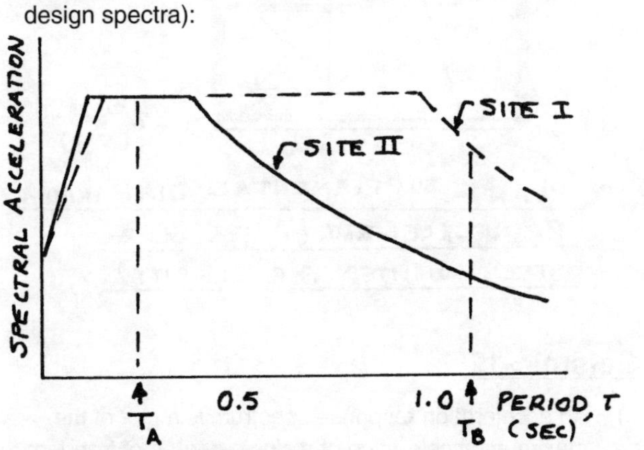

For the stiff building (T_A = 0.25 sec), the response is the same at either site. For the more flexible building (T_B = 1.1 sec), the response is significantly worst at site I than if the building were to be constructed at site II.

c) Compare responses at the two sites if the lateral force resisting systems are interchanged. The periods of the two buildings would change

$T_A = 0.035(30)^{3/4} = 0.44$ sec

$T_B = 0.02(100)^{3/4} = 0.63$ sec

In this case, the response of the 30' high building would be the same at both sites. The response of the 100' high building would be greater if constructed at site I than if constructed at site II.

Assuming building 'B' were properly designed and constructed, it would not likely have experienced structural damage during a moderate (magnitude 5.2) earthquake. Thus, there should be no appreciable effect on response during subsequent earthquakes.

Seismic-11

a) Use an approximate analysis to compute the maximum diaphragm shear stress. Use strength design (UBC-97).

$V_u = 1.1V = 1.1 \times 52 = 57.2$ kips

The endwalls have equal rigidities. Allow for an accidental eccentricity of 5% of the building length (UBC 1997, Section 1630.7):

$M_t = Ve = 57.2 \times 0.05 \times 40 = 114$ kip-ft

$\Sigma Rd^2 = 2 \times 20^2 R = 800R$

$V_{max} = (V / 2) + M_t(Rd) / \Sigma Rd^2$

$= 57.2 / 2 + 114 \times 20R / 800R = 31.5$ kips

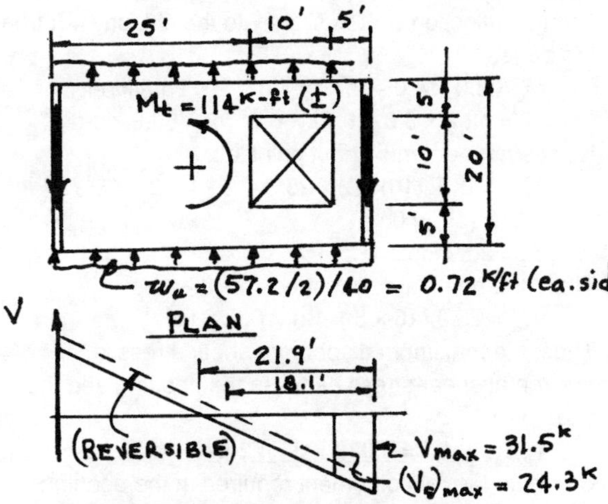

The maximum chord force in the diaphragm is:

$M_{max} = 0.5 \times 31.5^k \times 21.9' = 345$ kip-ft

$\therefore \quad T_{max} = C_{max} \cong M_{max} / 20' = 17.3$ kips

The maximum shear stress is:

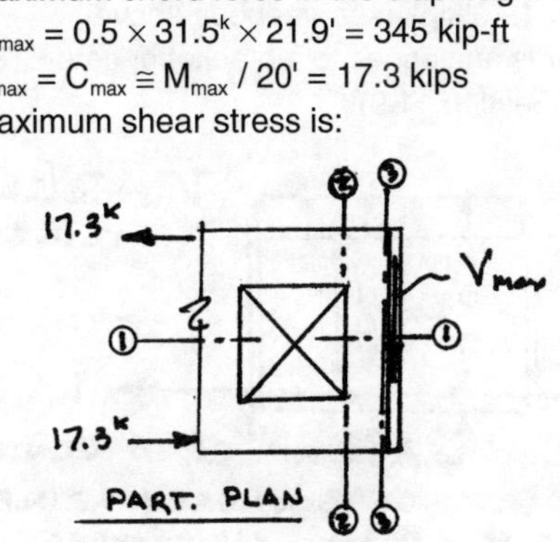

Calculate the nominal shear stress in the 6" thick diaphragm:

1. Horizontal shear stress on section 1-1:
   $v_{u1} = 17.3^k / ((6")(18.1 - 10)' \times 12\ ^{in}/_{ft}) = 0.03$ ksi
2. Vertical shear stress on section 2-2:
   $V = 31.5 - 1.44 \times 5 = 24.3$ kips
   $v_{u2} = 24.3^k / (6 \times 10 \times 12) = 0.034$ ksi
3. Vertical shear stress on section 3-3:
   $v_{u3} = 31.5 / (6 \times 20 \times 12) = 0.022$ ksi
   Now, consider a free body of the right hand side cut 15 ft from the east wall:

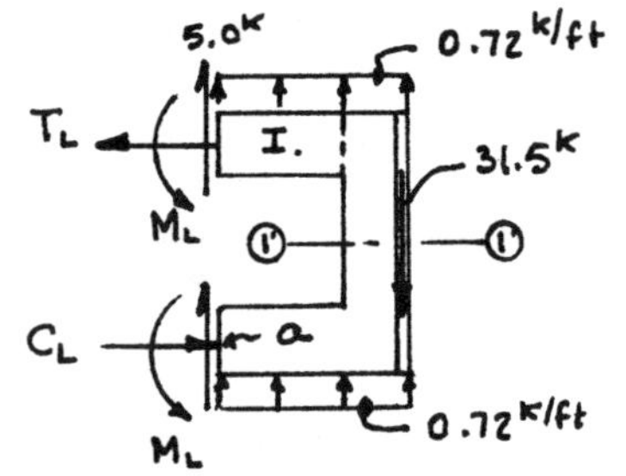

Consider the deformation of panel 'I':

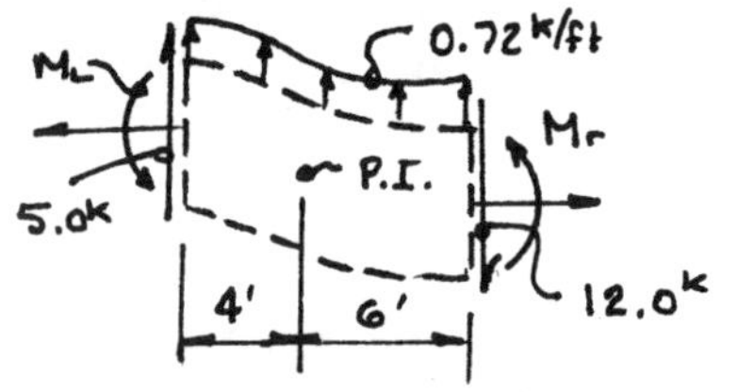

The rotation will be larger at the left end ∴ assume a point of inflection occurs slightly to the left, say 4 ft from the edge:

∴ $M_L = 5.0 \times 4.0 + 0.72 \times 4.0^2 / 2 = 25.8$ kip-ft
$M_r = 12.2 \times 6.0 - 0.72 \times 6.0^2 / 2 = 60.2$ kip-ft

Now, sum moments about point 'a':

$\Sigma M_a = 0$: $T_L(10) + 2 \times 25.8 + 2 \times 0.72 \times 15^2 / 2$
$- 31.5(15) = 0$

∴ $T_L = 25.9$ kips $= C_L$

Thus, the nominal shear stress on 1' – 1' is:

$v_{u1}' = 25.9 / (6 \times 5 \times 12) = 0.072$ ksi

Thus, the maximum diaphragm shear stress is the horizontal shear stress between the opening and the shearwall:

$v_{u,\,max} = v_{u1}' = 0.072$ ksi = 72 psi

b) Compute the reinforcement required at the opening:

$M_u = M_r = 60.2$ kip-ft; $T_u = T_L = 25.9$ kips
$V_u = 12.0$ kips (see freebody above)

Behavior is analogous to a bracket or corbel (Re: ACI 318-95, Section 11.9):

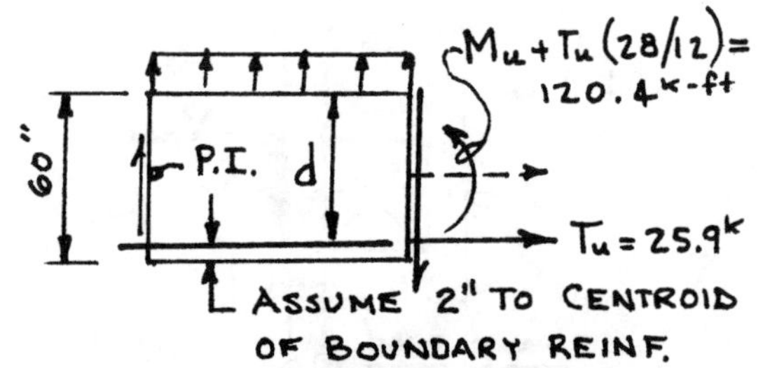

PARTIAL PLAN – SUBDIAPHRAGM I

Tension reinforcement ($f_y = 40$ ksi):

$A_f \geq (M_u + T_u(28 / 12)) / (\phi_s f_y(0.95d))$
$= (120.4 \times 12) / (0.85 \times 40 \times 0.95 \times 58) = 0.77\ in^2$
$A_n \geq T_u / (\phi_s f_y) = 25.9 / (0.85 \times 40) = 0.77\ in^2$

Shear friction ($a / d \cong 1.0$; $\mu = 1.4$):

$A_{vf} \geq V_u / (\phi_s \mu f_y) = 12 / (0.85 \times 1.4 \times 40) = 0.25\ in^2$

Check the shear strength using Eqn. 21-6, UBC 1921.6.5.2 ($\phi = 0.6$):

$V_u \leq \phi V_n = \phi[A_{cv}(2\sqrt{f'_c} + \rho_n f_y)]$
$A_{cv} = 6 \times 60 = 360\ in^2$

Take ρ_n as the minimum steel percentage required for temperature and shrinkage in the slab; i.e., 0.002 for grade 40 rebars.

∴ $\phi V_{nl} = 0.6[360(2\sqrt{3000} + 0.002 \times 40{,}000)]$
$= 40{,}900$ lbs $= 40.9$ kips $> V_u$

Thus, no special shear reinforcement is required.
Calculate the stress index for section 1921.6.2.3, UBC, to see if confinement steel is required in the boundary members.

$A = 360\ in^2$; $S = 6 \times 60^2 / 6 = 3600\ in^3$
$f_c = P / A + M / S$
$= 25{,}900 / 360 + 120{,}400 \times 12 / 3600$
$= 474$ psi $< 0.2f'_c = 600$ psi

Thus, confinement steel is not required around the opening. Since the temperature and shrinkage stresses are high at the corners, provide 2 #5 diagonally at each corner of openings. Thus, reinforcement required at the north-south edges is:

-- for the main diaphragm:
$A_s \geq (M_{max} / 20) / (\phi f_y) = 17.3 / (0.9 \times 40) = 0.48\ in^2$
Use 1 # 7 continuous at edges

-- at the opening:
$A_s \geq A_f + A_n = 0.77 + 0.77 = 1.54\ in^2$
Use 2 # 8 at edges of opening

Detail reinforcement as follows:

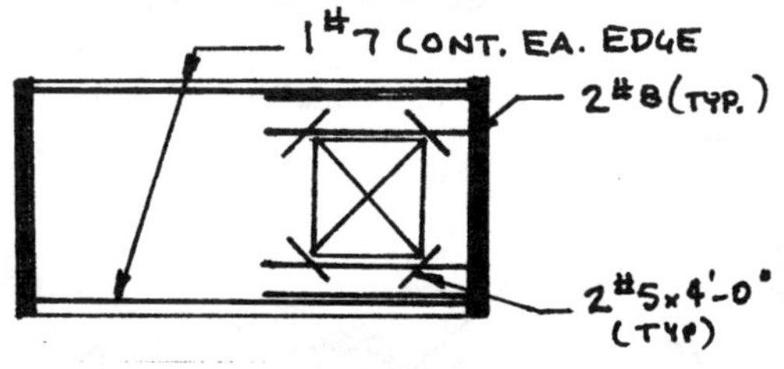

PLAN – SUPPLEMENTAL DIAPHRAGM REINFORCEMENT (OTHER SLAB STEEL OMITTED FOR CLARITY)

Seismic-12

a) An acceleration response spectrum is a plot of the maximum acceleration of a single degree of freedom, linear, harmonic oscillator when subjected to a specific forcing function or base excitation. The dependent variable for the plot is either the natural frequency or the period of the harmonic oscillator.

b) Soil liquefaction is the loss of strength and stiffness that occurs in a loose, saturated soil – usually sand – when the soil particles are "jarred" and tend to compact into a denser configuration. When this occurs, the soil loses its shear strength (i.e., it behaves as a liquid); hence, it loses its bearing capacity. Buildings founded over the soil can settle, tilt and sustain damage.

c) Tsunamis are long water waves caused by sudden faulting of the ocean floor.

d) The equation of motion is:

$$m\ddot{u}(t) + k(u(t) - u_g(t)) + c(\dot{u}(t) - \dot{u}_g(t)) = 0$$

(or)

$$m\ddot{u}(t) + ku(t) + c\dot{u}(t) = f(t)$$

where

$$f(t) = ku_g(t) + c\dot{u}(t)$$

e) By comparison to a frame, a shearwall structure will experience significantly less drift and interstory drift when subjected to lateral forces. As a consequence, there is generally less damage to nonstructural items in a shearwall structure than would occur to similar items in a frame structure.

f) An earthquake of magnitude 7.0 would likely cause several cycles of inelastic response in the structure. This would likely cause changes to the stiffness and damping characteristics of the building (especially if it were constructed of reinforced concrete). Three reasons that the building's response might not be acceptable in a subsequent earthquake are:

1. Duration of strong motion may be significantly longer and site-specific response more severe than in the first event.
2. Cumulative damage of inelastic response cycles might be too much and the building fails.
3. Strength and stiffness degradation might increase drift, causing instability due to P-Δ effects.

g) The choice of system factor for the cases described:

1. R = 4.5 (bearing wall system) – with columns embedded in the shearwalls, failure of the lateral force-resisting system would destroy the vertical support capability as well.
2. R_w = 4.5. Same reason as for the embedded steel column
3. R_w = 5.5 (building frame system). For this case, the frame provides nominal backup to the lateral force resisting system and failure of the lateral system would not lead to loss of vertical system.

h) The structure contains a vertical irregularity called a soft story, which places too large an inelastic rotation demand on the first story columns and beam-column joints. Furthermore, the design based on R = 8.5 yields a base shear that is too small for a structure with this irregularity.

Seismic-13

a) Determine an appropriate mathematical model for the structure. For the two-story, symmetrical, regular structure, an appropriate mathematical model is a planar structure undergoing horizontal displacements only (i.e., no torsion). Since girders are relatively rigid, a two-story shear building (Paz 1988) with mass lumped at the story levels is appropriate. Consider one-half of the building and take the weight of one-half the exterior walls as contributing mass at each level:

$m_i = w_i / g$; w_i = (floor + exterior wall + column)

$\therefore\ m_2 = [0.2 \times 20 \times 10 + 0.01 \times 6 \times 40 + 0.05 \times 2 \times 6] / g = 43.1 \text{ kips} / 32.2 = 1.34 \text{ kip-sec}^2/\text{ft}$

$m_1 = [0.2 \times 20 \times 10 + 0.01 \times 12 \times 40 + 0.05 \times 2 \times 12] / g = 46.0 \text{ kips} / 32.2 = 1.43 \text{ kip-sec}^2/\text{ft}$

Stiffness:

$k_1 = k_2 = 0.5(10^k / 0.1^{in}) = 50 \text{ kips/in}$

The equivalent model is:

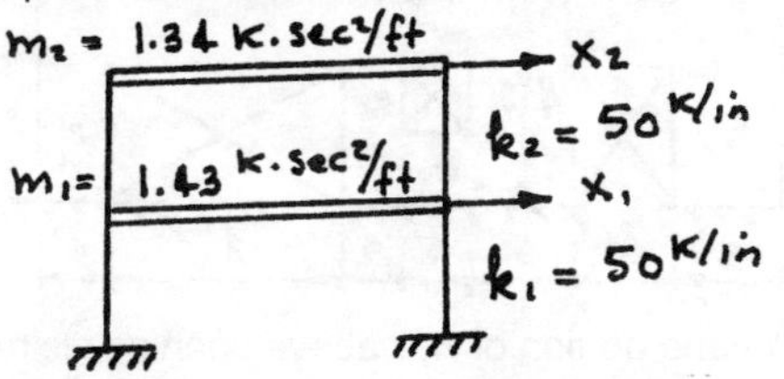

b) Approximate the first mode shape and fundamental period of the structure:

Fundamental period – use Rayleigh's Method. Apply an arbitrary system of lateral forces (e.g., 1 kip at the top and 0.5 kip at the first story to give a triangular distribution).

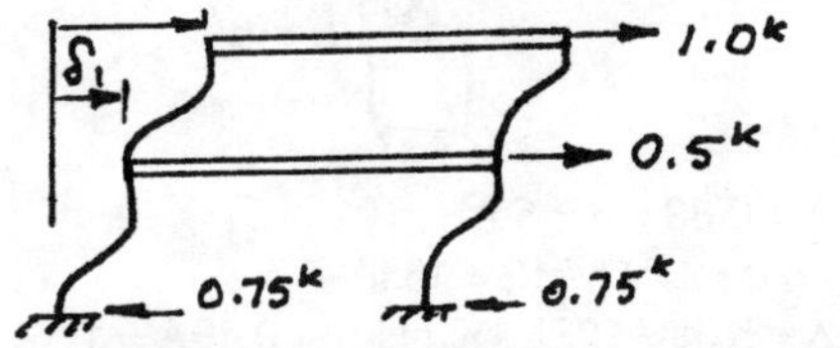

Each story deflects 0.1" per 10 kips ∴each column deflects 0.1" per 2.5 kips:

$\therefore\ \delta_1 = 0.75^k / 25^{k/in} = 0.03" = 0.0025'$

$\delta_2 = \delta_1 + 0.5^k / 25^{k/in} = 0.05" = 0.00417'$

Rayleigh's Equation is:

$$T_1 = 2\pi\sqrt{\sum w_i\delta_i^2 / \left(g\sum f_i\delta_i\right)}$$

$$= 2\pi\sqrt{\frac{\left(46.0 \times 0.0025^2 + 43.1 \times 0.00417^2\right)}{32.2(0.5 \times 0.0025 + 1.0 \times 0.00417)}}$$

T_1 = 0.48 sec

c) Assume (for part C) that T_1 = 0.5 sec, ϕ_1 = 0.67 and ϕ_2 = 1.0, calculate the first mode story shears. The lowest curve on the response spectra corresponds to 7% damping. Thus, $S_a = 0.2g = 0.2 \times 32.2 = 6.44 \text{ ft/sec}^2$.

$\Gamma_i = \sum m_i\phi_i / \sum m_i\phi_i^2 = (1.43 \times 0.67 + 1.34 \times 1.0) / (1.43 \times 0.67^2 + 1.34 \times 1.0^2) = 1.16$

$F_i = m_i\,\phi_i\,S_a\Gamma_1$

$\therefore$ **$F_1 = 1.43 \times 0.67 \times 6.44 \times 1.16 = 7.2$ kips**

$F_2 = 1.34 \times 1.0 \times 6.44 \times 1.16 = 10.0$ kips

d) Determine the effective peak ground acceleration. Since for a rigid structure ($T_1 \rightarrow 0$), the acceleration of the structure is the same as the ground, for the given response spectra, the curves approach 0.2g as $T \rightarrow 0$. Thus,

$S_{g,\text{-max}} = 0.2g = 6.44 \text{ ft / sec}^2$

Seismic-14

a) Determine the maximum ultimate shear stress in the wall. I will assume that the given lateral force is the design-level seismic force adjusted by the reliability/redundancy factor of section 1630.1.1, UBC-97. Distribute the base shear, $V = \pm 105$ kips, based on the relative rigidities of wall elements:

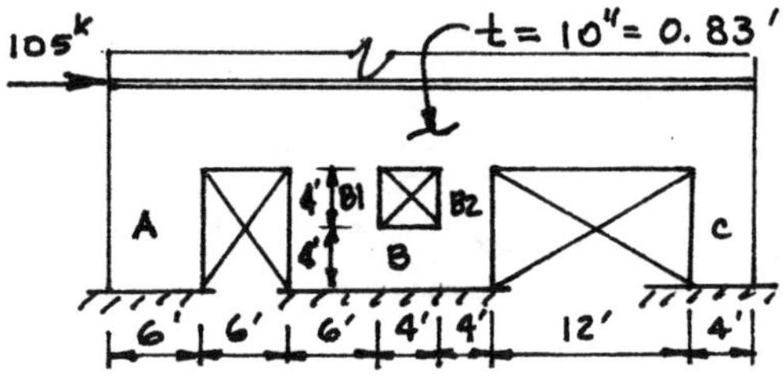

Assume the portion of wall above openings is relatively rigid. Consider both shear and flexural deformations; except that only shear deformation is significant in Element B. Take G = 0.4E. For Wall A:

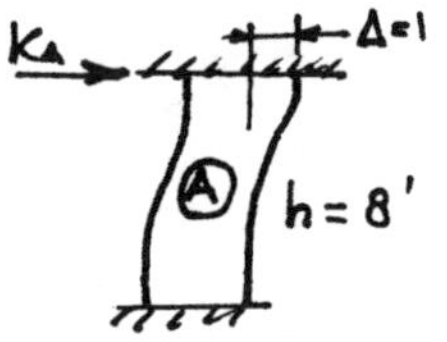

$A = 0.83 \times 6 = 5 \text{ ft}^2$

$I = 0.83 \times 6^3 / 12 = 15 \text{ ft}^4$

$\Delta = K_A h^3 / 12EI + K_A h(1.2) / 0.4EA = 1$

$K_A = E / [8^3 / (12 \times 15) + 1.2 \times 8 / 0.4 \times 5] = 0.131E$

For Wall B:

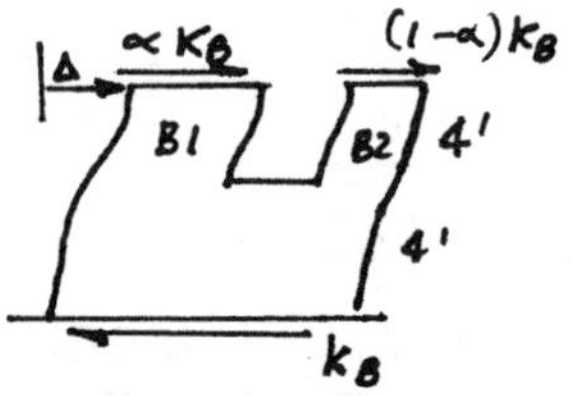

B1: $A = 5 \text{ ft}^2$; $I = 15 \text{ ft}^4$

B2: $A = 3.32 \text{ ft}^2$; $I = 4.42 \text{ ft}^4$

Let α = percentage of the force K_B resisted by element B1; then $(1 - \alpha)$ will be resisted by B2. For compatibility:

$\alpha K_B[4^3 / (12E \times 15) + 1.2 \times 4 / (0.4E \times 5)]$
$= (1 - \alpha)K_B[4^3 / (12E \times 4.42) + 1.2 \times 4$
$/ (0.4E \times 3.32)]$

$2.76\alpha = 4.82(1 - \alpha)$, $\therefore \alpha = 0.64$

For Wall B, take αK_B deflecting segment B1 and K_B deflection by shear only in segment B:

$\Delta = \alpha K_B[4^3 / (12E \times 15) + 1.2 \times 4 / (0.4E \times 5)]$
$+ K_B(1.2 \times 4 / (0.4E \times 0.83 \times 14)) = 1$

$\therefore$ $K_B = E / 2.79 = 0.36E$

For Wall C:

$A = 3.32 \text{ ft}^2$; $I = 4.42 \text{ ft}^4$; $h = 8$ ft

$\Delta = K_C[8^3 / (12E(4.42)) + 1.2 \times 8 / (0.4E \times 3.32)]$

$K_C = E / 16.9 = 0.059E$

Thus, the percentage of the base shear resisted by Wall B is:

$V_B = (K_B / \Sigma K)V = [0.36 / (0.131 + 0.36 + 0.059)]$
$\times 105 = 68.7$ kips

$V_{B1} = \alpha V_B = 0.64 \times 68.7 = 44$ kips

$V_{B2} = (1 - \alpha)V_B = 0.36 \times 68.7 = 24.7$ kips

$V_A = (0.131 / 0.55)105 = 25$ kips

$V_C = (0.059 / 0.55)105 = 11.3$ kips

Compute the nominal (ultimate) horizontal shear stresses in psi. The ultimate shear due to seismic loading is 1.1 times the design-level base shear (1612.2.1, UBC).

$v_u = V_u / A_{av}$

$v_{uA} = 1.1 \times 25{,}000 / (10 \times 72) = 38$ psi

$v_{uB1} = 1.1 \times 44{,}000 / (10 \times 72) = 67$ psi

$v_{uB} = 1.1 \times 68{,}700 / (10 \times 168) = 45$ psi

$v_{uB2} = 1.1 \times 24{,}700 / (10 \times 48) = 57$ psi

$v_{uC} = 1.1 \times 11{,}300 / (10 \times 48) = 26$ psi

The maximum shear stress occurs in pier B1 (6'×4'):

$v_{u,\text{-max}} = v_{uB1} = 67$ psi

b) Determine the reinforcement required in Element marked B1 due to the bending induced by the 44 kips shear. Given: $f'_c = 3000$ psi; $f_y = 40{,}000$ psi; neglect axial compression; use strength design per UBC-97.

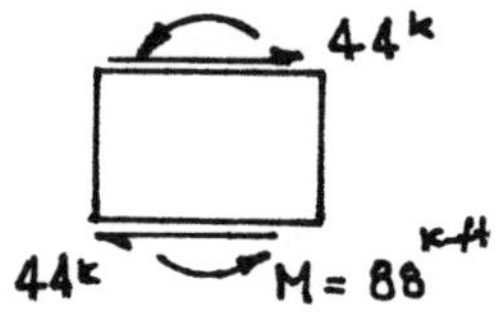

$V_u = 1.1 \times 44 = 48.4$ kips

$M_u = 1.1 \times 44 \times 2 = 96.8$ kip-ft

Check the stress index (1921.6.2.3, UBC):

$S_x = 10 \times 72^2 / 6 = 8640 \text{ in}^3$

$f_{bu} = M_u / S_x = 96.8 \times 12 / 8640 = 0.134 = 134$ psi

$f_{bu} = 134 \text{ psi} < 0.2 f'_c = 600$ psi $\therefore$ No special boundary member is required around the opening.

Check 1921.6.2.2:

$A_{CV} = 72 \times 10 = 720 \text{ in}^2$

$V_u = 48{,}400 \text{ lb} < 2A_{CV}(f'_c)^{1/2} = 78{,}800$ lb

$\therefore$ Single curtain of wall steel is OK.

Let $p_v = p_n = 0.0025$:

$A_{SV} = 0.0025 \times 10 \times 12 = 0.30 \text{ in}^2 / \text{ft}$

say #5 @ 12" o.c.e.w.

Check the shear strength using $\phi = 0.6$ (Equation 21-6):

$V_n = A_{CV}(2(f'_c)^{1/2} + p_n f_y)$

$\phi V_n = 0.6 \times 720(2(3000)^{1/2} + 0.0025 \times 40{,}000)$
$= 90{,}500 \text{ lb} = 90.5 \text{ kips} > V_u \therefore$ OK

$\therefore$ **Use #5 @ 12" o.c.e.w. – Provide 2#5 × 6'-0" diagonally at corners**

Seismic-15

a) Determine the fundamental period for vibration in the North-South direction. The horizontal truss spans 80' and is simply supported by the walls at lines 5 & 6. The tributary weight to each panel point is:

W = 60 psf × 20' × 30' = 36,000 lbs = 36 kips

Use Rayleigh's Method to compute the fundamental period. Apply (arbitrarily) equal concentrated forces of 10 kips at the interior panel points.

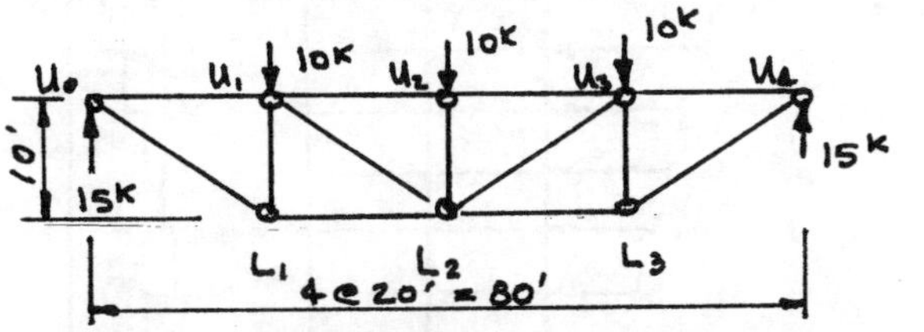

For 2L6 × 4 × $^3/_8$: A = 7.22 in^2; 2L5 × 3 1/2 × $^3/_8$: A = 6.09 in^2; W14×22: A = 6.49 in^2; W21×62: A = 18.20 in^2. Use the virtual work method to compute the panel point deflections. For a dummy unit force applied at Joint U_1:

$$(1)\ \delta_{u1} = \sum_{i=1}^{mem} \frac{p_i P_i L_i}{A_i E}$$

Values of p_i and P_i are tabulated below. from which:

$E\delta_{u1} = 10{,}376$; $\delta_{u1} = 10{,}376 / 29{,}000 = 0.358$ in

By symmetry, $\delta_{u3} = \delta_{u1} = 0.358$ in.

$\delta_{u2} = 14{,}422 / 29{,}000 = 0.497$ in

By Rayleigh's Method:

$$T_1 \cong 2\pi\sqrt{\Sigma w_i \delta_i^2 / (g(\Sigma f_i \delta_i))}$$

$$= 2\pi\sqrt{\frac{36(0.358^2 + 0.497^2 + 0.358^2)}{(32.2\times 12)\times 10(0.358 + 0.497 + 0.358)}}$$

$T_1 = 0.39$ sec

Mem	L_i / A_i	P_i (kips)	Dummy at U_1		Dummy at U_2	
			p_i	$(P_i p_i L_i) / A_i$	p_i	$(P_i p_i L_i) / A_i$
U_0U_1	37.0	–30	–1.5	1665	–1.0	1110
U_1U_2	37.0	–40	–1.0	1480	–2.0	2960
U_2U_3	37.0	–40	–1.0	1480	–2.0	2960
U_3U_4	37.0	–30	–0.5	555	–1.0	1110
L_1L_2	37.0	+30	1.5	1665	1.0	1110
L_2L_3	37.0	+30	0.5	555	1.0	1110
U_0L_1	37.2	+33.5	1.7	2119	1.12	1396
L_3U_4	37.2	+33.5	0.6	748	1.12	1396
U_1L_2	44.0	+11.2	–0.6	–296	1.12	552
L_2U_3	44.0	+11.2	0.6	+296	1.12	552
L_1U_1	6.6	–15.0	–0.8	79.2	–0.5	50
L_2U_2	6.6	–10.0	0	0	–1.0	66
L_3U_3	6.6	–15.0	–0.3	30	–0.5	50
		Σ =		**10,376**		**14,422**

b) Comment on the relative magnitudes of lateral force on the equipment for the specified periods. If the truss oscillates horizontally at its natural frequency, the equipment mounted on the mezzanine will be subjected to support motion given by the time-varying displacement of the truss (neglecting any coupling of motion between the truss and equipment). The dynamic response – in terms of lateral force on the equipment – can be plotted as a function of the frequency ratio $r = \overline{\omega} / \omega$, where $\overline{\omega}$ = frequency of support motion (i.e., the truss) and ω = frequency of the equipment. Typically, this plot appears (Paz, 1987):

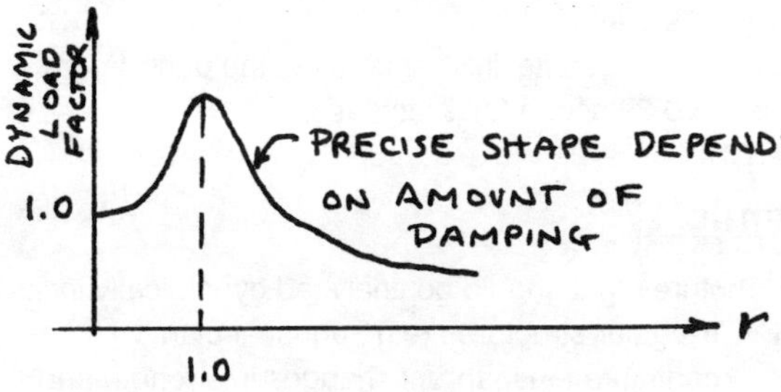

For $r = \overline{\omega} / \omega = 1.0$, resonance occurs and the force acting on the equipment is significantly magnified. Thus, for this example, equipment having the same period as the truss would experience significantly greater force than would the equipment having a period $^1/_{10}{}^{th}$ that of the truss.

Seismic-16

Determine the effect of the footing flexibility on the fundamental frequency. The problem statement requires that only the increased flexibility be considered (i.e., the rotational inertia of the footing and soil-structure is neglected).

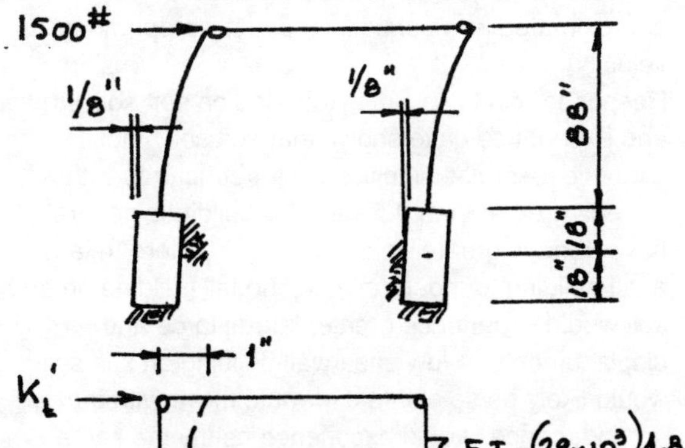

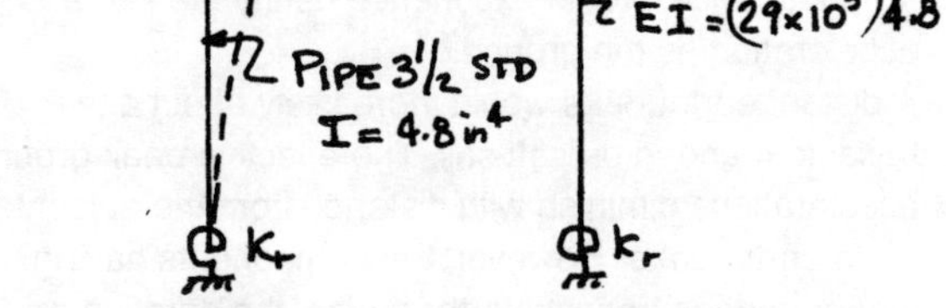

The soil is such that 1500 lbs applied horizontally at the top of the bent causes the top of the footing to displace $^1/_8$". Thus, the corresponding horizontal displacement at the top of the bent due only to the footing rotation is:

$\Delta_t / (88 + 18) = (^1/_8)'' / 18$

$\Delta_t = 0.736''$ (per 1500 lbs)

Thus, the modified translational stiffness of the bent is:

$$K_t'\left[\frac{88^3}{3\times 29\times 10^3\times 2\times 4.8}+\frac{0.736}{1.5}\right]=1''$$

$K_t' = 0.765\ ^k/_{in} = 765$ #/in

The tributary mass to each bent is:

$m = w / g = 1125^\# / (32.2 \text{ ft} / \text{sec}^2)$

The fundamental period is:

$T = 2\pi / \sqrt{k/m}$

$= 2\pi / [765 \text{ lbs} / \text{in} \times (32.2 \times 12) \text{ in} / \text{sec}^2 / 1125 \text{ lbs}]^{1/2}$

$\underline{T = 0.39 \text{ sec}}$

Thus, the increased flexibility increases the period from 0.31 sec to 0.39 sec, a 25% increase.

Seismic-17

a) Structures that should be analyzed dynamically include:
   1. Irregular structures (e.g., unusual plan configurations, abrupt changes in strength and stiffness from floor-to-floor, large variations in mass or other irregularities (see UBC 1997, Tables 16-L and 16-M).
   2. Tall, regular structures (i.e., those for which response is significantly affected by the higher modes of vibration).
   3. Structures founded on deep deposits of soft soil (e.g., where dynamic response may be amplified by soil-structure interaction)

b) Damping characteristics relate to the mechanisms that dissipate energy in an oscillating system (e.g., Coulomb friction; hysteretic material behavior). To simplify computations, structural damping is usually approximated as viscous (i.e., directly proportional to velocity).

c) Response of a tall building founded on soft soil: studies and experience have shown that soft soil amplifies the dynamic response significantly for structures with periods greater than 0.5 sec. Tall buildings generally have periods greater than 0.5 sec; therefore, relative to a tall building founded on rock, the tall building on soft soil would experience greater lateral force and displacements. A low shearwall building on stiff soil would likely be so stiff that it would move with the ground; hence, would experience nearly the same acceleration as the ground.

d) A distant earthquake would more likely affect a tall building founded on soft soil. The effective peak ground accelerations diminish with distance from the epicenter of an earthquake. However, the components having longer periods transmit farther. Also, the duration of strong ground motion is greater on soft soil. The periods of motion for soft soil and tall building are relatively close. All of these factors can contribute to create a measurable response in the tall building caused by an earthquake several hundreds of miles away.

Seismic-18

a1) Under extended high intensity ground motion the stiffness of the 10-story building would likely degrade, resulting in an increase in the period.

a2) The typical code specified seismic load distribution and shear diagram for a 10-story building is approximately as follows:

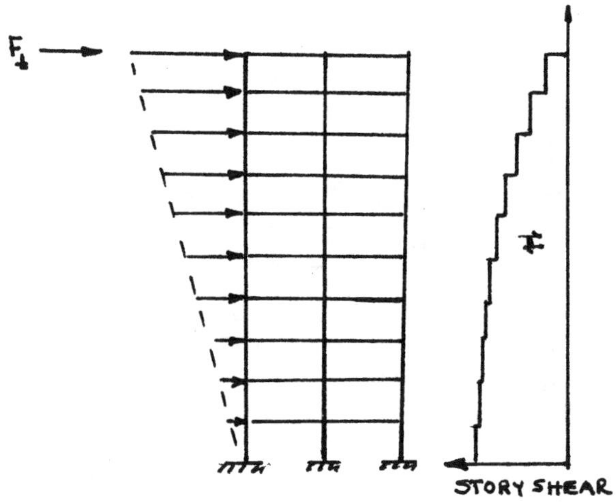

a3) Critical damping is that amount of damping for which a system will not oscillate when disturbed from its equilibrium position.

b1) The 20-story building might have a smaller base shear coefficient because its period of vibration is generally such that it is on the descending branch of the design response spectrum (i.e., a period greater than about 0.5 sec for a building founded on stiff soil).

b2) Concrete confinement increases the toughness; hence, the ability of a member to dissipate energy.

b3) An acceleration response spectrum is a plot of the maximum acceleration of a single degree of freedom, linear, harmonic oscillator of a varying frequency (or period) when subjected to the specified ground motion. The plot is usually made with the spectral acceleration as ordinate and the period as the abscissa.

c1) The strain corresponding to 20% elongation is 0.2. The yield strain for A-36 steel is $F_y / E_s = 36 / 29 \times 10^3 = 0.00124 = \varepsilon_y$. Thus, the proportion of inelastic to elastic strain is $\varepsilon_{inelastic} / \varepsilon_y = (0.2000 - 0.00124) / 0.00124 = 160$:

$\therefore \quad \varepsilon_{inelastic} = 160\ \varepsilon_{yield}$

c2) Bearing type bolts depend on their shear strength rather than the frictional resistance created by tightening the bolts. Thus, the bolts would develop their design strength. The only effect would be slippage in the joint (on the order of $^1/_{16}''$), which might impair serviceability. (Note: I assume that the connection is not slip critical; otherwise, bearing type bolts would not have been specified).

c3) Three factors that influence ductility in a reinforced concrete frame are:
   1. The longitudinal steel ratio, $\rho = A_s / bd$ (the higher the steel ratio, the lower the ductility).

2. Concrete confinement (i.e., hoops or spirals).
3. Combined axial force and bending moment.

d1) Three reasons that buildings subjected to a base shear higher than prescribed showed limited distress:
1. Resistance is furnished by non-structural elements (e.g., partitions and curtain walls).
2. Buildings were designed to earlier, more conservative codes with respect to member strength (e.g., using working stress design).
3. Buildings might have been proportioned for non-seismic loading conditions (e.g., low, light-frame structures that were governed by wind rather than seismic).

d2) Two effects of the infill wall:
1. Stiffens the frame significantly which causes a change in its dynamic response (e.g., shortening its fundamental period).
2. Creates an abrupt change in stiffness which produces high inelastic rotation demand on the columns above and below the in-filled story.

e1) The most influential factors affecting dynamic response are the magnitude and distribution of mass and stiffness.

e2) The deep, soft soil can amplify the input ground motion. Response is especially severe if the building's period is close to the period of the soil layer, in which case resonance may occur.

e3) The wall arrangement has poor torsional rigidity. Thus, slight eccentricity of an applied lateral force can cause large twisting deformation.

Seismic-19

a) Determine the force to the wall A resulting from North-South ground motion (UBC 97). The contributing weight for North-South ground motion is:

W = Roof + Walls in long direction

$W = 0.1 \times 80 \times 40 + 0.5 \times 2 \times 80 \times 0.02 \times 15 + 0.5 \times 2 \times 20 \times 0.15 \times 15 = 389$ kips

The base-shear coefficient is given as 0.1:

$\therefore \quad V = 0.1 \times 389 = 38.9$ kips

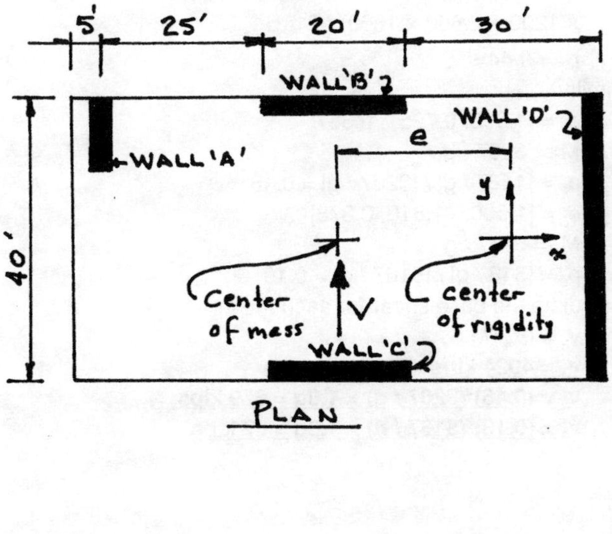

Since only shear deformation is to be considered, the rigidity of the walls is directly proportional to wall length (1.2H / (tG) = constant). Let:

e = Eccentricity of the lateral force with respect to the center of rigidity

x_i = Distance from the center of rigidity to the centerline of North-South wall

y_i = Distance from the center of rigidity to the centerline of East-West wall

Include accidental eccentricity (1630.7, UBC):

$\Delta e = 0.05L = 0.05 \times 80 = 4.0'$

$e = [40 \times 39.5 - 10 \times 35] / [40 + 10] = 24.6'$

$M_t = V(e + \Delta e) = 38.9(24.6 + 4.0) = 1113$ kip-ft

The center of rigidity in the North-South direction is located by symmetry. The polar moment of inertia of the wall group is:

$J = 10(35 + 24.6)^2 + 40(39.5 - 24.6)^2 + 2 \times 20 \times 19.5^2 = 59{,}600$

Thus, the force to Wall A is:

$V_A = Vh_a / \Sigma h_i + M_t(h_a x_a) / J = [38.9 \times 10 / (10 + 40)] + [10 \times (35 + 24.6) \times 1113 / 59{,}600]$

V_A = 18.9 kips

b) Discuss the effect on force distribution if wall A were masonry with one-half the stiffness of concrete.
Substituting masonry for concrete in Wall A lowers the rigidity of Wall A relative to the other walls. Thus, the shear resisted by Wall A is reduced.

c) The direction of force in Wall D caused by the torsional moment is opposite to the direction of force caused by the lateral force. Thus, the torsional moment has the beneficial effect of reducing the shear in Wall D. (Note: UBC-97 requires that negative shears due to torsional moment be disregarded when computing the design shear force for the wall).

Seismic-20

a) Qualitative plots of fundamental mode shapes

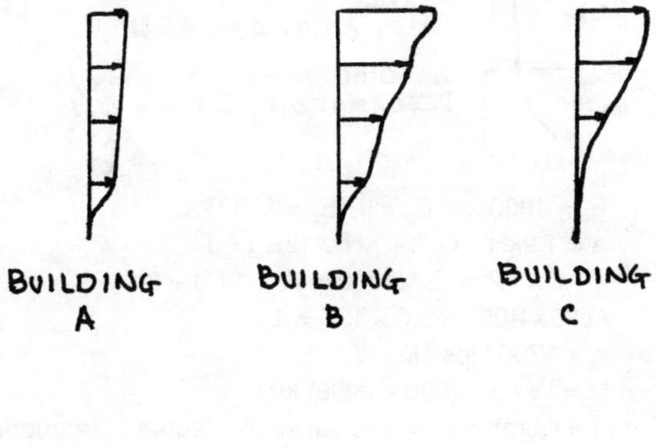

b) Comparative performance of the three buildings under strong ground motion:

Building A - The abrupt change in stiffness at the second floor level is such that practically all lateral deflection is due to bending of the first story columns. This places high rotation demand on the columns. If large inelastic rotations (i.e., plastic hinges) form in the first story columns, collapse would occur. Thus, the structural response would need to remain essentially elastic (creating relatively large base shear).

Buildings B and C should both respond well to strong ground motion, with inelastic rotation demand distributed throughout the buildings. Building B is more flexible and will exhibit larger drift and interstory drifts than Building C. Thus, the damage to nonstructural components would likely be greater in B than C.

c) Buildings B and C could both qualify for R = 8.5: Building B as a special moment resisting frame (SMRF) and Building C as a dual system (concrete shearwall with SMRF), in which the frame is designed to resist at least 25% of the base shear.

Building A represents a special case of vertical irregularity involving a building of more than two stories, over 30' height, with a weak story. This structure would be permitted under the current UBC only if the columns of the weak story are capable of reisting the lateral forces under an essentially elastic response (i.e., $\Omega_0 = 2.8$ times the seismic force computing using an R = 4.5).

d) Compute the period of Building A. For this purpose, idealize Building A as a single degree of freedom system with mass lumped at the midheight of the upper three stories.

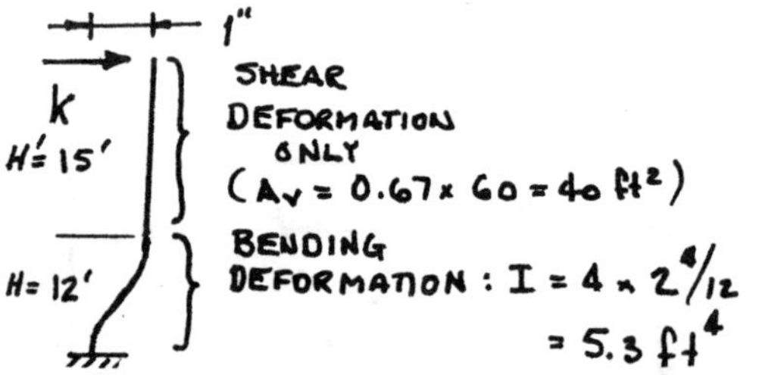

E_c = 4000 ksi; $G_c = 0.4E_c$ = 1600 ksi

$\Delta = 1.2kH' / G_cA + KH^3 / 12E_cI = 1''$

$k[1.2 \times 12 \times 15 / (1600 \times 40 \times 144) + (12 \times 12)^3 / (12 \times 4000 \times 5.3 \times 12^4) = 1$

$\therefore$ K = 1700 kips / in

M = W / g = (800 + 3000) kips / g

For free vibration, undamped, single degree of freedom system:

$T = 2\pi(M / k)^{1/2} = 2\pi[(3800 / (32.2 \times 12)) / 1600]^{1/2}$

T = 0.48 sec

Seismic-21

Mode shapes

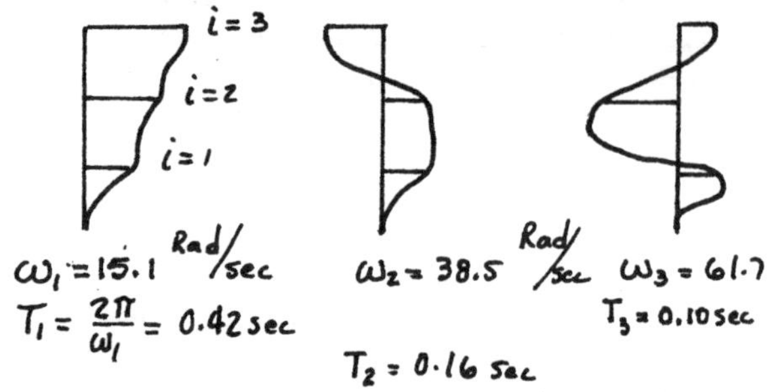

Given: Velocity response spectra for $\xi = 0$; $\xi = 0.02$; $\xi = 0.05$, etc. Use the spectrum for $\xi = 0.05$: For T_1 = 0.42 sec; S_v = 25 in / sec. The spectral acceleration is

$S_A = \omega S_v = 15.1 \times 25 = 378$ in / sec^2 = 0.98 g

For T_2 = 0.16 sec; S_v = 10 in / sec

$S_A = 38.5 \times 10 = 385$ in / sec^2 = g

For T_3 = 0.10 sec; S_v = 8 in / sec

$S_A = 61.7 \times 8 = 493$ in / sec^2 = 1.3g

For 100'×100' floor plan:

$M_1 = (100 \times 100 \times 0.18 = 1800 \text{ kips}) / g$

$M_2 = (100 \times 100 \times 0.12 = 1200 \text{ kips}) / g$

$M_3 = 1800 / g$

Use the response spectrum analysis outlined in Appendix IF, SEAOC Commentary (SEAOC 1988) for a two-dimensional building model: NP = N = 3 stories

$S_{a,1} = 0.98g$

$S_{a,2} = 1.00g$

$S_{a,3} = 1.30g$

$M = \Sigma M_i = (1800 + 1200 + 1800) / g = 4800 / g$

$\phi_1 = [1.000, 1.959, 2.860]^T$

$$M_1 = \sum_{i=1}^{3} \phi_{i1}^2 M_i = (1.000^2 \times 1800 + 1.959^2 \times 1200 + 2.860^2 \times 1800) / g$$

$M_1 = 21{,}211 / g$

Participation factor

$$p_1 = \left(\sum_{i=1}^{3} \phi_{i1} M_i\right) / \left(\sum_{i=1}^{3} \phi_{i1}^2 M_i\right) = (1.000 \times 1800 + 1.959 \times 1200 + 2.860 \times 1800) / 21{,}211$$

$p_1 = 0.44$

Similarly,

$\phi_2 = [1.000, 0.725, -0.657]^T$

$M_2 = 3207 / g$

$p_2 = [1487 / g] / [3207 / g] = 0.46$

$\phi_3 = [1.000, -1.610, 0.378]^T$

$M_3 = 5167 / g$

$p_3 = [548 / g] / [5167 / g] = 0.10$

Calculate the base shear for each mode:

$v_n = (P_n)^2 M_n S_{a,n}$

$\therefore$ **v_1 = 4024 kips**

$v_2 = (0.46)^2(3207 / g) \times 1.0g = 679$ kips

$v_3 = (0.10)^2(5167 / g) \times 1.3g = 67$ kips

b) Calculate the lateral force at each level:

For Mode 1: $F_{in} = M_i P_n \phi_{in} S_{a,n}$

$F_{11} = (1800 / g) \times 0.44 \times 1.0 \times 0.98g = 776$ kips

$F_{21} = (1200 / g) \times 0.44 \times 1.959 \times 0.98g = 1014$ kips

$F_{31} = (1800 / g) \times 0.44 \times 2.860 \times 0.98g = 2220$ kips

$V_1 = \Sigma = 4010$ kips

For Mode 2:

$F_{12} = (1800 / g) \times 0.46 \times 1.0 \times 1.0g = 828$ kips

$F_{22} = (1200 / g) \times 0.46 \times 0.725 \times 1.0g = 400$ kips

$F_{32} = (1800 / g) \times 0.46 \times (-0.657) \times 1.0g$
$= -544$ kips

$V_1 = \Sigma = 684$ kips

For Mode 3:

$F_{13} = (1800 / g) \times 0.10 \times 1.0 \times 1.3g = 234$ kips

$F_{23} = (1200 / g) \times 0.10 \times (-1.610) \times 1.3g$
$= -251$ kips

$F_{33} = (1800 / g) \times 0.10 \times 0.378 \times 1.3g = 88$ kips

$V_1 = \Sigma = 71$ kips

c) The most probable design base shear can be estimated busing the CQC Method (p. 88-C, SEAOC-88)

$\rho_{ij} = [\, 8\xi^2(1 + r)r^{3/2}] / [(1 - r^2)^2 + 4\xi^2 r(1 + r)^2]$

$\rho_{11} = \rho_{22} = \rho_{33} = 1;\ \xi = 0.05$

ρ_{21}: $r = T_1 / T_2 = 0.42 / 0.16 = 2.625$

$\rho_{21} = [8(0.05)^2(1 + 2.625)(2.625)^{3/2}] / [(1 - 2.625^2)^2 + 4(0.05)^2(2.625)(1 + 2.625)^2]$

$\rho_{21} = 0.1033 = \rho_{12}$

ρ_{31}: $r = T_1 / T_3 = 0.42 / 0.10 = 4.2$

$\rho_{31} = 0.0787 = \rho_{13}$

ρ_{32}: $r = T_2 / T_3 = 0.16 / 0.10 = 1.6$

$\rho_{32} = 0.2248 = \rho_{23}$

$$\rho = \begin{pmatrix} 1.0000 & 0.1033 & 0.0787 \\ 0.1033 & 1.0000 & 0.2248 \\ 0.0787 & 0.2248 & 1.0000 \end{pmatrix}$$

$$V = \left[\sum_{i=1}^{3} \sum_{j=1}^{3} V_i \rho_{ij} V_j \right]^{1/2}$$

$$= \{4010[1.0 \times 4010 + 0.1033 \times 684 + 0.0787 \times 71] + 684[0.1033 \times 4010 + 1.0 \times 684 + 0.2248 \times 71] + 71[0.0787 \times 4010 + 0.2248 \times 684 + 1.0 \times 71]\}^{1/2}$$

V = 4146 kips

Note: For this problem, the higher modes have little effect on base shear. The response can be obtained accurately enough using the simpler square root of the sum of the squares (SRSS) method of combining modes:

$$V = \left[\sum_{i=1}^{3} V_i^2 \right]^{1/2} = [4010^2 + 684^2 + 71^2]^{1/2}$$

V = 4069 kips

Seismic-22

a) Compute the seismic forces at the roof and second floor. Assume that the contributing weight of the columns and walls is included in the given dead loads.

Compute the base shear by the static force procedure of UBC-97. The period is calculated using Eqn. 30-8, UBC-97:

$T = C_t h^{3/4} = 0.02 \times 20^{3/4} = 0.2$ sec

$W = (0.025 + 0.120)40 \times 80 = 464$ kips

For a bearing wall system, R = 4.5. UBC Eqn. 30-5 will govern the base shear with $C_a = 0.24$:

$V = 2.5C_a I / R = (2.5 \times 0.24 \times 1.0/4.5) \times 464 = 62$ kips

$T < 0.7$ sec $\therefore F_t = 0$

$\Sigma w_i h_i = (0.12 \times 10 + 0.025 \times 20)40 \times 80 = 5440$

$\therefore$ **$F_1 = 62 \times 0.12 \times 40 \times 80 \times 10 / 5440 = 43.8$ kips**

$F_2 = 62 \times 0.025 \times 40 \times 80 \times 20 / 5440 = 18.2$ kips

V = 62.0 kips

b) Determine the distribution of shears on the second story walls. The roof diaphragm is flexible (plywood sheathing); therefore, distribute the lateral force on the basis of tributary areas:

$V_C = V_D = 18.2 / 2 = 9.1$ kips

$V_A = V_B = 0$

c) Determine the chord force in the roof diaphragm:

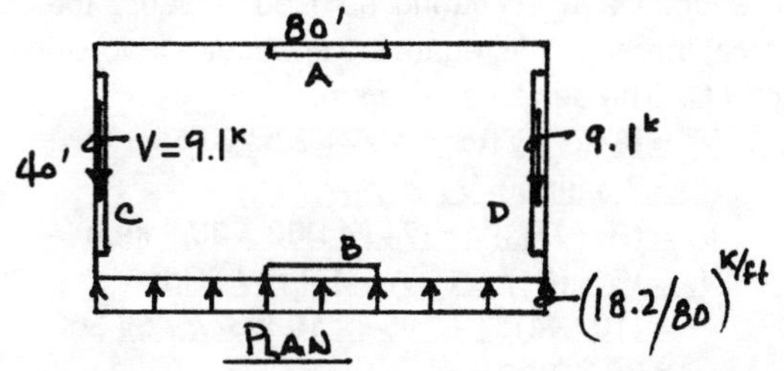

$M_{max} = wl^2 / 8 = (18.2 / 80)(80)^2 / 8 = 182$ kip-ft

Assume the resultant chord forces act 0.5 ft in from the edges: a = 40 – 2(0.5) = 39'

$\therefore$ **$C = T = M_{max} / a = 182 / 39 = 4.7$ kips**

d) Determine the distribution of shears on the first story walls. Neglect the rigidity of columns under Wall D. Consider only shear deformation in the walls. Walls are all 8 in thick concrete; therefore, rigidity is directly proportional to wall length.

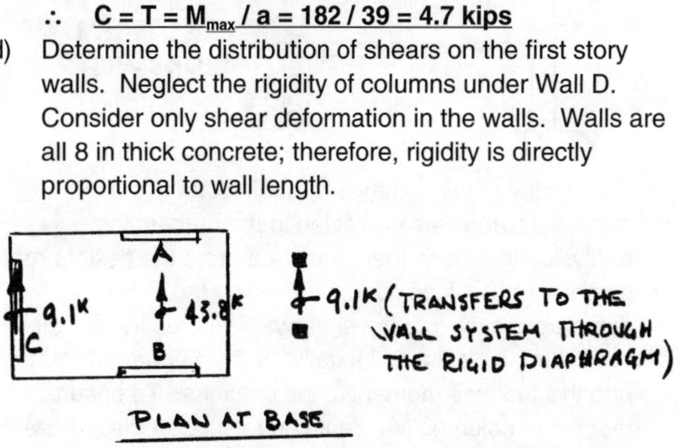

Locate the center of rigidity ($\bar{y}$ = 20 ft by symmetry):

$\bar{x} = 2 \times 10 \times 60 / (2 \times 10 + 24) = 27.3'$

Include the accidental eccentricity (1630.6, UBC-97):

$\Delta e = 0.05L = 0.05 \times 80 = 4.0$ ft

$M_t = \pm V(e + \Delta e) = 64(40 - 27.3 + 4.0) = 1069$ kip-ft

Wall	R_x	R_y	d_x	d_y	Rd^2
A	20			20	8000
B	20			20	8000
C		24	–27.3		17,887
E		10	32.7		10,693
F		10	32.7		10,693
Σ =					55,300

Due to the direct lateral force:

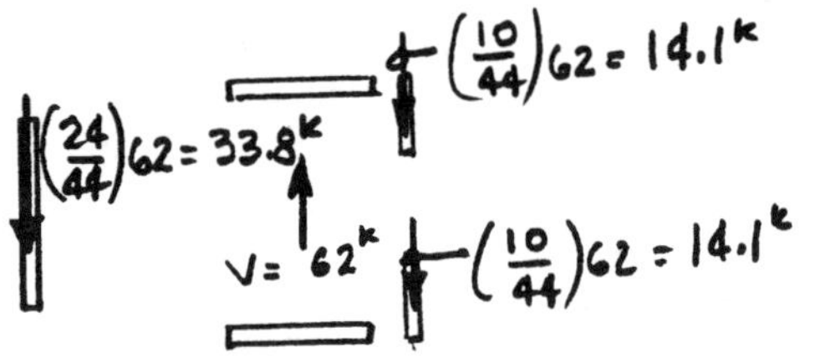

Due to the direct torsional moment:

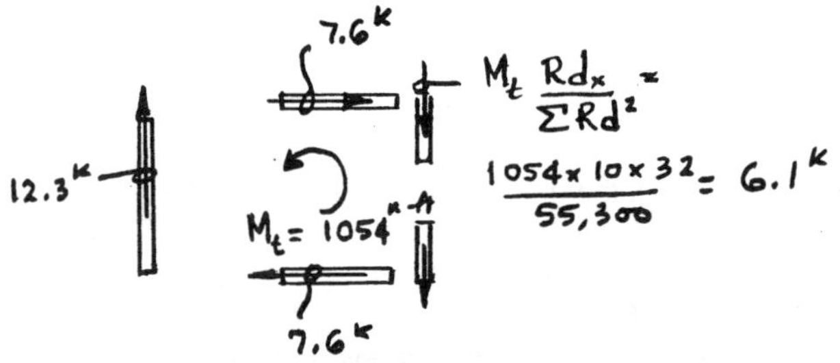

The torsional effect cannot be used to reduce the design force; therefore, disregard the shear due to torsion in Wall C. The design shears are:

$\underline{V_A = M_t R d_y / \Sigma R d^2 = 1054 \times 20 \times 20 / 55{,}300}$

$\underline{= 7.6 \text{ kips} = V_B}$

$\underline{V_C = (R_y / \Sigma R_{yi})V = (24 / 44)62 = 33.8 \text{ kips}}$

$\underline{V_E = V_F = (R_y / \Sigma R_{yi})V + M_t R_y d_x / \Sigma R d^2}$

$\underline{= (10 / 44)62 + 1054 \times 10 \times 32.7 / 55{,}300}$

$\underline{= 20.3 \text{ kips}}$

e) Discuss the design of columns supporting Wall D.

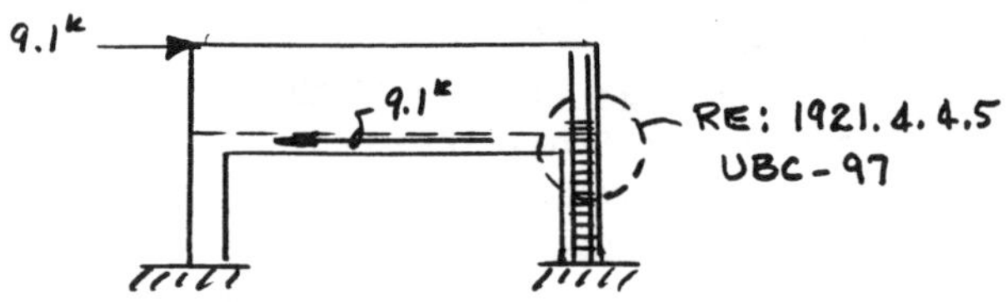

The rigidity of the columns is small relative to the shearwalls and was neglected in the preceeding analysis. In theory, then, the shear from the flexible roof diaphragm (9.1 kips) is transferred by the rigid diaphragm back into the shearwalls. In reality, the top of the columns displaces laterally as the diaphragm rotates and this induces moment in the columns. To ensure that these columns can continue to support their share of the gravity loads at this displacement, the special provisions of UBC 1921.7 must be satisfied. This requires that the induced moments be calculated under the maximum inelastic response displacement, Δ_m, as defined in 1630.9, UBC-97.

The abrupt change in stiffness between the shearwall above and the columns below creates stress concentrations that place considerable inelastic demand on these columns. For this reason, UBC 1921.4.4.5 requires that if the axial force due to gravity plus earthquake exceeds ($A_g f_c'$ / 10), then transverse confinement steel (hoops or spirals) must be provided over the full height of column and for a distance equal to the development length of the longitudinal bars into the wall.

Seismic-23

a) Determine the seismic forces in the connections marked 'A' and 'B'. Details of the roof construction are not given; however, the timber roof beams would likely support a plywood panel system. Therefore, assume a flexible roof diaphragm such that the seismic forces distribute on a tributary area basis. Calculate the base shear using the static force procedure of UBC-97. For a steel braced bearing wall system ($R = 4.4$; $\Omega_o = 2.2$); $C_a = 0.44$. Estimate the period using Eqn. 30-8, UBC-97:

$T = C_t h^{3/4} = 0.02 \times 18^{3/4} = 0.17$ sec

The base shear is governed by Eqn. 30-5:

$\therefore \quad V = (2.5C_a I/R)W = (2.5 \times 0.44 \times 1.0 / 4.4)250$

$= 62.5$ kips

Convert to working stress design by dividing the design-level base shear by 1.4:

$V = 62.5/1.4 = 44.6$ kips

To compute the seismic force in the glulam beams at Joints A & B there are two cases to consider:

1. As chord forces for the North-South ground motion:

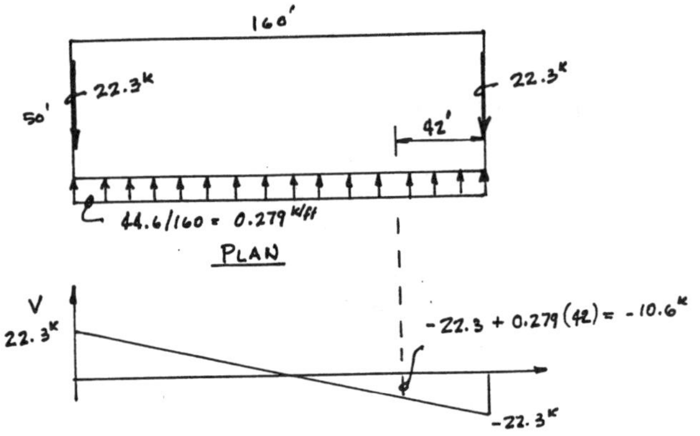

$M = 0.5(10.6 + 22.3)42 = 691$ kip-ft

$T = C = M / 50 = 691 / 50 = 13.8$ kips ($\pm$)

2. As collector forces for the East-West ground motion:

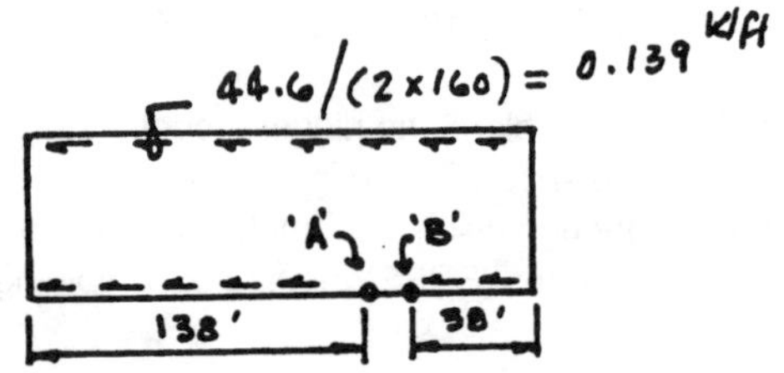

$T_A = 138 \times 0.139 = 19.2$ kips (reversible)

$C_B = 38 \times 0.139 = 5.3$ kips (reversible)

Thus, the horizontal reaction at Joint A is governed by case 2 and at Joint B is governed by case 1:

$\underline{H_A = \pm 19.2 \text{ kips}}$

$\underline{H_B = \pm 13.8 \text{ kips}}$

b) Design the connections at 'A' and 'C' using the given values: $V_A = \pm 14.5$ kips; $V_{total} = 19.3$ kips (gravity loads to be neglected per problem statement).

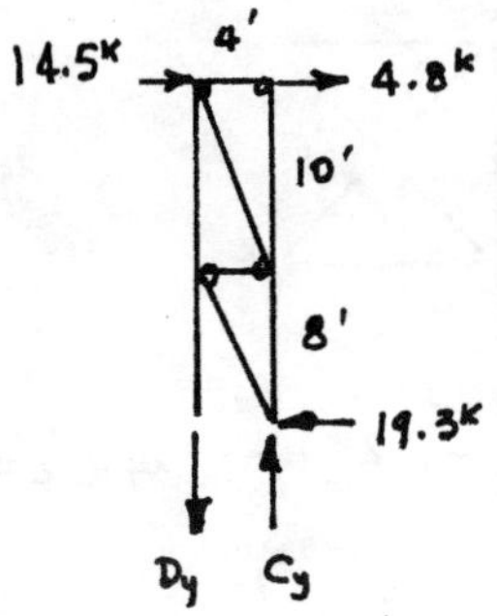

$\Sigma M_D = 0$: $19.3(18) - C_y(4) = 0$

$C_y = \pm 87$ kips

$\Sigma F_x = 0$: $C_x = \pm 19.3$ kips

- For the connection of the glulam at 'A'. Since only the horizontal force due to seismic loading is to be considered, the connection can be made by symmetrically arranged bolts through steel side plates. The bolt design value (Re: NDS-91, Table 8.3D) for a 3-1/8-in wide, 2400F combination Douglas Fir-Larch with $^3/_4$ in diameter bolts is:

$z_{//} = 3170$ lbs / bolt (normal load duration)

$z_{//}' = 1.33 \times 3170 = 4220$ lb / bolt (for seismic)

$n \geq 14{,}500 / 4220 = 3.4$ ∴ Use 4 bolts with an end distance (for tension) = 7d = 7 × 0.75 = 5.25 in. Steel side plates $^3/_{16}$ in are adequate; try 12 in length to accommodate bolts; use allowable stress design (AISC 1989):

$\omega_g = 12$ in; $\omega_n = 12 - 4(0.75 + 0.125) = 8.5$ in

$$T \leq \frac{4}{3}\begin{cases} 0.6F_yA_g = 0.6 \times 36 \times 12 \times 3/16 = 65 \text{ kips} \\ 0.5F_u(0.85A_g) = 0.5 \times 58 \times 0.85 \times 2.25 = 74 \text{ kips} \\ 0.5F_uA_n = 0.5 \times 58 \times 8.5 \times 3/16 = 62 \text{ kips} \end{cases}$$

The plates are more than adequate ∴ use 2 plates 7 × 3/16 × 1'-0".

Use E70xx electrodes; $\omega = {}^3/_{16}$ in: $q = (^4/_3)(3 \times 0.928) = 3.7$ kips / in

$L \geq 14.5 / 3.7 = 4"$ – no problem

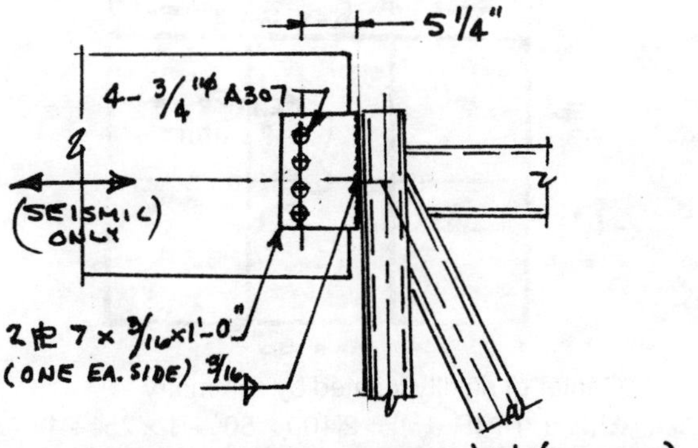

DETAIL – CONNECTION 'A' (N.T.S.)

For the anchor bolts in 3000 psi concrete (Table 19-D, UBC-97): Try 1 in diameter A307; 100% increase for installation with special inspection and $^4/_3$ increase for seismic loading:

- shear: $V_t = (4 / 3)4.5 = 6.0$ kips / bolt
- tension: $P_t = 2 \times (^4/_3)3.25 = 8.7$ kips / bolt

Try 12 – 1 in diameter (6 each side) spaced at minimum spacing S = 12d = 12 in. Check combined shear and tension:

$$(P_s / P_t)^{5/3} + (V_s / V_t)^{5/3} = [(87 / 12) / 8.7]^{5/3} + [(19.3 / 12) / 6.0]^{5/3} = 0.85 < 1.0$$

Since 12 bolts are not fully stressed, the minimum spacing can be proportionately reduced to 0.85 × 12 = 10.2; say 10.5 in on centers. The base connection requires a stiff member; therefore, try connecting to a W-section at base, say W8×28:

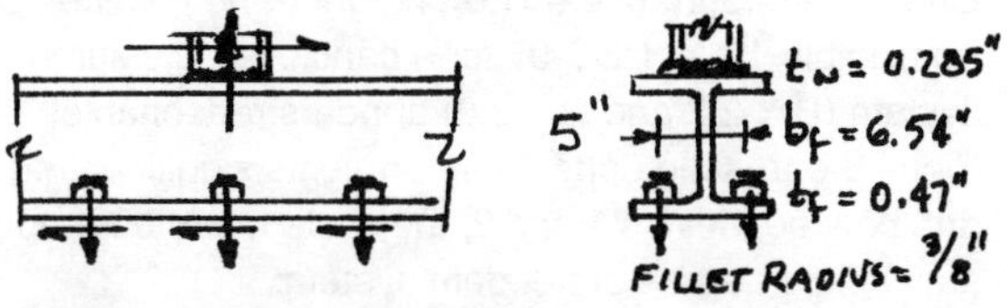

Check stresses in the flange:

$F_b = {}^4/_3(0.75F_y) = {}^4/_3(0.75 \times 36) = 36$ ksi

$S = 10.5 \times 0.47^2 / 6 = 0.39$ in^3

$M_{toe} = (87 / 12)(5.0 - 0.29 - 0.75) / 2 = 14.3$ kip-in

$f_b = 14.3 / 0.39 = 36$ ksi – OK

Attach the TS 4×4 using minimum size $^1/_4$" E70 fillet welds:

$f_v = 19.3 / (2{\times}4) = 2.4$ kips / in

$f_t = 87 / (4{\times}4) = 5.4$ kips / in

$f_r = (f_v^2 + f_t^2)^{1/2} = 5.9$ kips / in

$q = 4 \times 0.928 \times {}^4/_3 = 4.9$ kips / in – No good.

Need $^5/_{16}$ in welds: $q = 6.1$ kips / in > $f_r = 5.9$ kips / in

Web yield check: $N + 5k = 4 + 5(0.93) = 8.7"$

$P / (N + 5k)t_w = 87 / (8.7 \times 0.285) = 35$ ksi – Too high.

Add a web stiffener to be conservative and to add ductility.

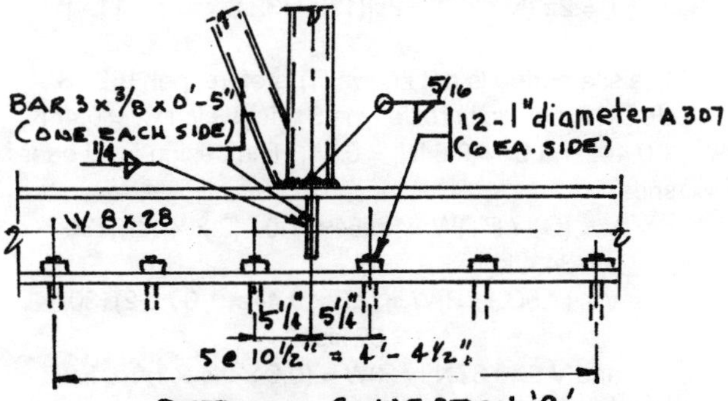

DETAIL – CONNECTION 'C'

d) Comments: even though the diaphragm was considered flexible, it is important to provide approximately the same rigidity on the line containing the plywood shearwalls as for the line with the steel shear frame. This is to alleviate torsional response, which would increase the shear on one line of the walls. Also, if possible, it would be desirable to use more than one shear frame, to provide redundancy. Increasing the spread of the shear frame would be beneficial in reducing forces throughout the frame (i.e., both verticals, diagonals and the base connection).

Seismic-24

a) Analyze the frame for seismic loads (UBC-97). The structure has vertical irregularity (soft story and weak story). Under seismic loading, maximum bending moments occur in the columns at Joints B and D. The vertical load capacity is destroyed if plastic hinges form at these joints. Thus, a conservative approach would be to design the structure to respond elastically to the ground motion (R = 1.0). However, some inelastic energy dissipation can occur (e.g., partial flexural yielding; secondary bending in braces and gussets). Given the importance (I=1.0), height (< 30'), values of R from Table 16-N, UBC-97 for a cantilevered column system (R = 2.2 and Ω_o= 2.0) appears reasonable. Since the trussed upper story prevents significant story drift between levels 1 and 2, the system can be modeled as a single degree of freedom system:

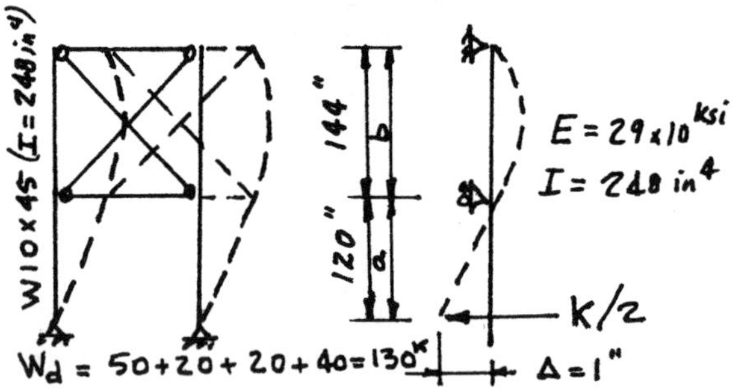

The equivalent stiffness is:

$\Delta = 1 = [(k / 2) / 3EI][a^2b + a^3]$

$\therefore$ $k = (6 \times 29 \times 10^3 \times 248) / (120^2 \times 144 + 120^3)$
$= 11.4$ kips / in

Lumped mass: M = 130 kips / g

$\therefore$ $T = 2\pi\, (M / k)^{1/2} = 2\pi[(130 / (32.2 \times 12)) / 11.4]^{1/2}$
$= 1.1$ sec

The site period is not known. Therefore, per 1629.3, UBC-97, use soil profile type S_D, for which $C_a = 0.44N_a = 0.44$, and $C_v = 0.64N_v = 0.64$. The design-level base shear is:

$V = (C_vI / RT)W = [(0.64 \times 1.0 / (2.2 \times 1.1)]130$
$= 34.4$ kips

$V \le (2.5C_aI / R)W = (2.5 \times 0.44 \times 1.0 / 2.2)130$
$= 65$ kips

and $V \ge (0.8ZN_vI / R)W = (0.8 \times 0.4 \times 1.0 \times 1.0 / 2.2)130 = 18.9$ kips

Thus, V = 34.4 kips. Distributing the design-level base shear in direct proportion to the structure's mass gives:

$F_1 = [(40 + 20) / 130]\, V = 0.46 \times 34.4 = 15.8$ kips

$F_2 = [(50 + 20) / 130]\, V = 0.54 \times 34.4 = 18.6$ kips

Analyzing for the seismic loads gives:

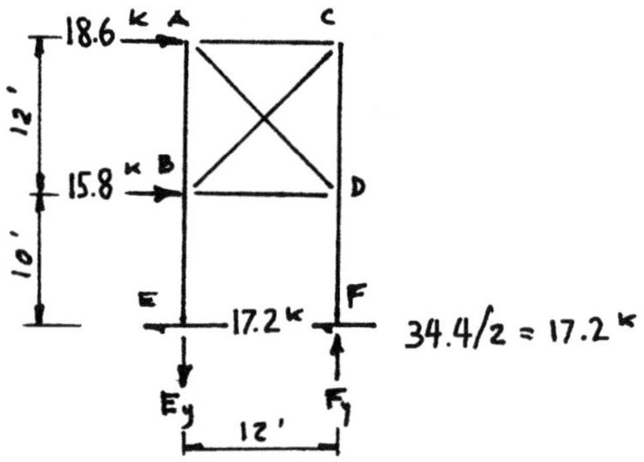

$\Sigma M_E = 0 = 18.6(22) + 15.8(10) - 12F_y$

$\therefore$ **$F_y = \pm 47.3$ kips**

$\Sigma F_v = 0$: **$E_y = \pm 47.3$ kips**

b) Check adequacy of the W10×45 (50 ksi) columns under dead plus seismic loading. Use LRFD (AISC LRFD 1995). Assuming the frame is part of a building system, UBC 1630.1.1 requires an increase in the seismic forces by a reliability/redundancy factor, ρ. For a two-column frame, ρ is limited by its maximum value, 1.5. Thus, for axial compression plus bending:

$P_u \ge 1.2P_d + \rho P_e = 1.2 \times 90 + 1.5 \times 47.3 = 179$ kips

$M_u = 1.5 \times (47.3/2) \times 10 = 355$ kip-ft

For A W10×45: $Z_x = 54.9$ in^3

$\phi_b M_p = \phi_b F_y Z_x = 0.9 \times 50 \times 54.9 / 12 = 206$ kip-ft

Thus, $M_u > \phi_b M_p$; the column is understrength under the most favorable condition $\therefore$ no need to check combined axial plus bending.

$\therefore$ **W10×45 columns are not adequate**

c) The frame is overloaded under dead plus seismic loading $\therefore$ the live load capacity is zero.

Seismic-25

a) If the roof diaphragm is rigid: distribute shear on the basis of relative rigidities and include the torsional effect of an assumed accidental eccentricity equal to 5% of building's length

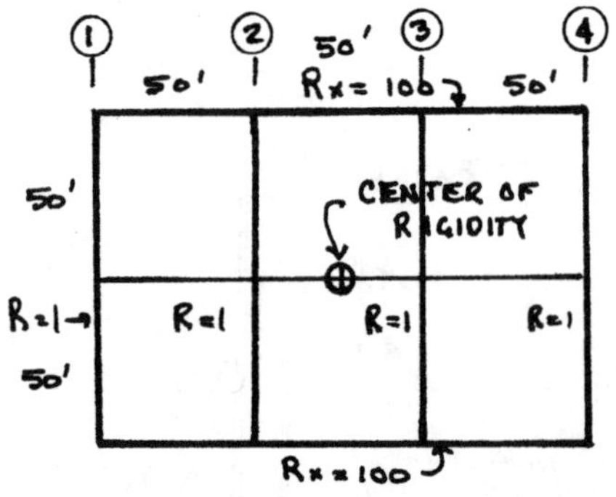

Center of rigidity located by symmetry

$\Sigma[(R_yd_x^2) + (R_xd_y^2)] = 2[100 \times 50^2 + 1 \times 75^2 + 1 \times 25^2] = 512{,}500R = \Sigma Rd^2$

$e_x = e_{min} = 0.05 \times 150 = 7.5'$

Let V = North-South base shear

$V_4 = V(R_4 / \Sigma R_y) + M_t(R_4d_x) / \Sigma Rd^2$

$\therefore$ **$V_4 = V(1 / 4) + 7.5V \times 1 \times 75 / 512{,}500 = 0.251V$**

b) If the roof diaphragm is flexible, the distribution is on the basis of tributary area. In this case, frame on Line 4 has tributary width = 25'; versus total length = 150'. Thus,

$V_4 = (25 / 150)V = (1 / 6)V = 0.167V$

c) If the frame on Line 4 is modified and other frames unchanged; rigid diaphragm; accidental eccentricity

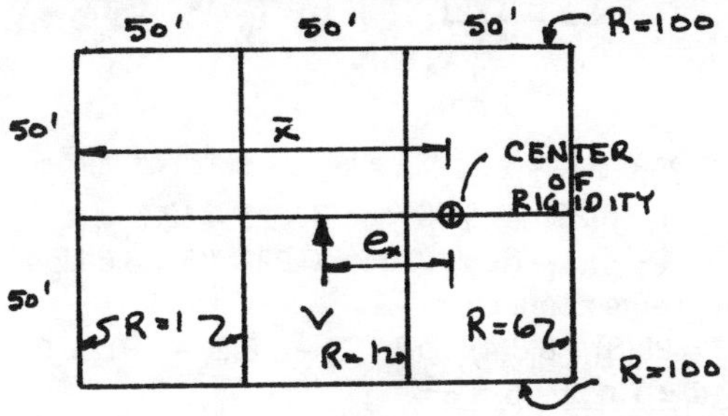

$\Sigma R_y = 1 + 1 + 1 + 6 = 9$

$\overline{x} = (1 \times 50 + 1 \times 100 + 6 \times 150) / 9 = 116.7$

$\Sigma Rd^2 = 1(116.7)^2 + 1(66.7)^2 + 1(16.7)^2 + 6(33.3)^2 + 2 \times 100 \times 50^2 = 525{,}000$

$e_x = 116.7 - 75 = 41.7$; $\Delta e = 0.05L = 7.5'$

$\therefore \quad M_t = V(e + \Delta e) = V(41.7 + 7.5) = 49.2V$

$V_4 = VR_4 / \Sigma R_y - M_t(33.3R_y) / \Sigma Rd^2$

$= V(6 / 9) - 49.2V(33.3 \times 6) / 525{,}000$

$V_4 = 0.667V - 0.0187V$

Note: UBC-97, Section 1630.7 specifies that only the increase in shear due to torsion be considered. Thus,

$V_4 = 0.667V$

d) The classification of a diaphragm as rigid, semirigid or flexible is based on the in-plane deflection of the diaphragm relative to the interstory drift of the lateral force resisting elements. The diaphragm deflection, which includes both shear and flexural deformation, is usually computed using empirical formulas developed for the specific type of diaphragm (e.g., plywood or metal deck). For a diaphragm to be considered rigid, its deflection under lateral load will be an order of magnitude smaller than the average interstory drift of adjacent vertical elements under the same loading. To qualify as flexible, 1630.7, UBC-97 stipulates that the diaphragm deflection be at least two times the deflection of the vertical resisting elements. In between these extremes, the diaphragm is considered semirigid.

Seismic-26

Determine the moments at the top and bottom of column D-2. The column is not part of the lateral force-resisting system. Column D-2 coincides with the center of rigidity, therefore, is unaffected by accidental torsion. Thus, the lateral force resisted by the end walls is:

$V = 4\ ^{kips}/_{ft} \times 150\ ft / 2 = 300\ kips$

The deflection at the top of the shearwall is:

$\Delta_1 = P / Et\,[4(h / d)^3 + 3(h / d)]$

Let E = modulus of elasticity of concrete, which is assumed to be the same for all concrete in the structure.

$\therefore \quad \Delta_1 = [300 / 0.67E]\,[4(10 / 15)^3 + 3(10 / 15)] = 1426 / E$

The maximum deflection of the diaphragm occurs at the midspan (Line D) and is given by:

$\Delta_2 = [WL / (6.4Et)]\,[(L / b)^3 + 2.13(L / b)]$

$= [4 \times 150 / (6.4E \times 0.75)]\,[(150 / 50)^3 + 2.13(150 / 50)]$

$= 4174 / E$

Thus, the top of the column on Line D at Line 2 will be displaced $\Delta_S = \Delta_1 + \Delta_2 = 5600 / E$. The deflection Δ_s is the calculated elastic deflection based on the code prescribed design-level seismic force. The actual response of the structure during an earthquake will be inelastic and the actual deflection will be significantly greater than Δ_S. Section 1633.2.4, UBC-97, requires that the column's vertical load-carrying capacity be investigated under an induced displacement, Δ_M, calculated in accordance with Section 1630.9, UBC. The strength of the concrete column must be investigated per UBC 1921.7. For this problem, only the magnitude of the induced moment is required, based on the given assumption of full fixity at top and bottom.

For a building frame system with concrete shearwalls, Table 16-N gives R = 5.5 and $\Omega_o = 2.8$. Thus, by Eqn. 30-17:

$\Delta_M = 0.7\,R\,\Delta_S = 0.7 \times 5.5 \times 5600 / E = 21{,}560/E$

Assuming the columns are fixed at the top and bottom (per problem statement), for a 14-in diameter circular column responding elastically:

$EI = E\,\pi\,d^4 / 64 = [\pi(14 / 12)^4 / 64]E = 0.091E$

Thus, for the 10-ft long column the induced elastic seismic moment is:

$M_e = 6EI\Delta_{col} / h^2$

$\therefore$ **$M_e = 6(0.091E)(21{,}560 / E) / 10^2 = 118$ kip-ft**

(Note that the moment calculated on the basis of the above assumptions is conservative. The induced moment corresponds to the peak inelastic response and the stiffness of the column is usually lower at that stage than computed using the elastic, uncracked section. Section 1633.2.4, UBC-97, permits consideration of inelastic deformation and restraint stiffness at the joint, which would result in a smaller induced moment.)

Seismic-27

Let α = percentage of lateral force resisted by Wall 1; $(1 - \alpha)$ resisted by Walls 2-to-6 – walls 8" concrete block; grouted solid

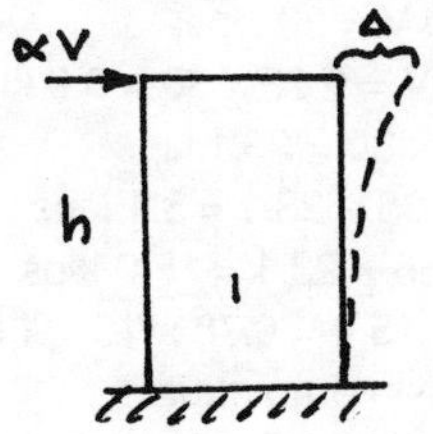

$\Delta = \alpha v\,[(h^3 / 3EI) + (1.2h / 0.4EA)]$

$A = (7.6 / 12)8 = 5.07\ ft^2$

$I = 5.07 \times 8^2 / 12 = 27\ ft^2$

$h = 16'$

$\therefore \quad E\Delta = \alpha v[(16^3 / (3 \times 27)) + (1.2 \times 16 / 0.4 \times 5.07)] = 60\alpha v$

For Walls 2-to-6:

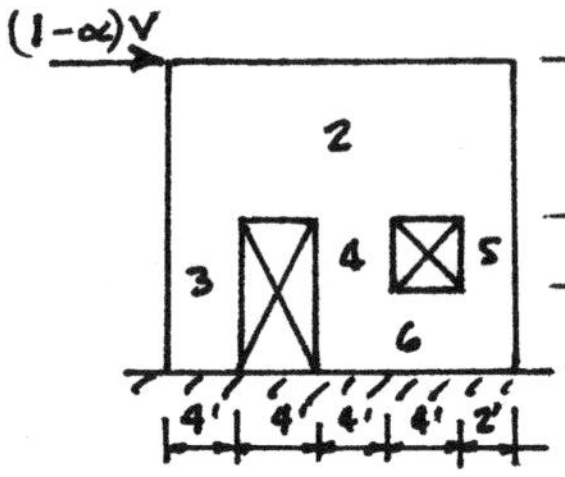

$A_2 = (7.6 / 12)18 = 11.4\ \text{ft}^2$

$A_3 = (7.6 / 12)4 = 2.53\ \text{ft}^2$

$I_3 = 2.5 \times 4^2 / 12 = 3.37\ \text{ft}^2$

$A_4 = A_3;\ I_4 = I_3$

$A_5 = (7.6 / 12)2 = 1.27\ \text{ft}^2$

$I_5 = 0.42\ \text{ft}^2$

$A_6 = (7.6 / 12)10 = 6.33\ \text{ft}^2$

Dimensions are such that only shear is significant in Segments 2 & 6; both shear and flexural deformations in other segments.

Let β = percentage of $(1 - \alpha)V$ resisted by Pier 3

$(1 - \beta)$ resisted by 4,5, and 6

r = percentage of $(1 - \alpha)(1 - \beta)$ resisted by Pier 4

$(1 - r)$ = percentage of $(1 - \alpha)(1 - \beta)$ resisted by Pier 5

Thus,

$$r\left[\frac{4^3}{12 \times 3.37} + \frac{1.2 \times 4}{0.4 \times 2.53}\right]$$

$$= (1 - r)\left[\frac{4^3}{12 \times 0.42} + \frac{1.2 \times 4}{0.4 \times 1.27}\right]$$

$6.32r = (1 - r)(22.15)$

$\therefore\quad r = 0.78;\ (1 - r) = 0.22$

$$\beta\left[\frac{8^3}{12 \times 3.37} + \frac{1.2 \times 8}{0.4 \times 2.53}\right]$$

$$= (1 - \beta)\left[0.78 \times 6.32 + 0.22 \times 22.15 + \frac{1.2 \times 4}{0.4 \times 6.33}\right]$$

$22.1\beta = (1 - \beta)(11.7)$

$\therefore\quad \beta = 0.35;\ 1 - \beta = 0.65$

$\therefore\quad 60\alpha = (1 - \alpha)[0.35 \times 22.10 + 0.65(11.70) + 1.2 \times 8 / (0.4 \times 11.4)]$

$\alpha = 0.23;\ (1 - \alpha) = 0.77$

Thus, since the drag strut is considered inextensible, the lateral force distributes on the basis of relative translational stiffness. For a given V = 30 kips:

$\alpha = 0.23;\ V_1 = \alpha V = 0.23 \times 30 = 6.9$ kips

$(1 - \alpha)V = 0.77 \times 30 = 23.1$ kips

$\beta = 0.35;\ V_3 = 0.35 \times 23.1 = 8.1$ kips

$(1 - \beta)23.1 = 0.65 \times 23.1 = 15.0$ kips

$r = 0.78;\ V_4 = r \times 15.0 = 0.78 \times 15 = 11.7$ kips

b) Anchorage force in drag strut:

<u>P = (1 − α)V = (1 − 0.23) × 30 = 23.1 kips</u>

c) Assume the force at the top of Wall B is 25 kips, determine the axial force in Pier 5. Cut a freebody through the midnight of Piers 3, 4, and 5 (points of inflection occur at these locations)

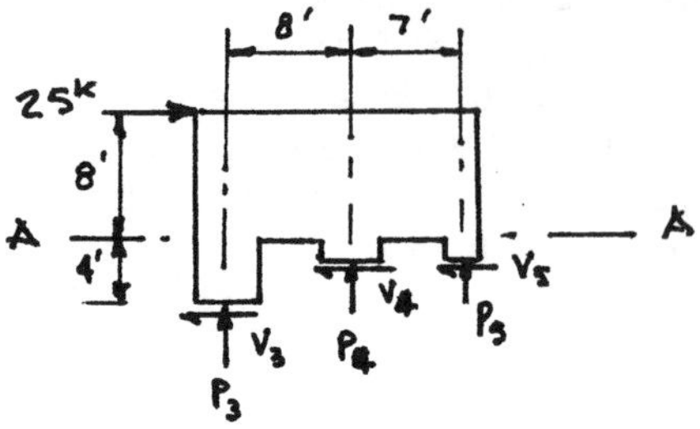

$V_3 = 25\beta = 0.35 \times 25 = 8.8$ kips

$V_4 = (1 - \beta)25(r) = 0.65 \times 0.78 \times 25 = 12.7$ kips

$V_5 = (1 - \beta)(1 - r)25 = 0.65 \times 0.22 \times 25 = 3.6$ kips

Take moments about Line A-A:

$\Sigma M = 25(8) - 8.8(4) - 12.7(2) - 3.6(2) - \Sigma P_i d_i = 0$

<u>$\Sigma P_i d_i = 132.2$ kip-ft = M</u>

Thus, 132.2 kip-ft of the overturning is resisted by linear elastic theory. Walls have same thickness = t

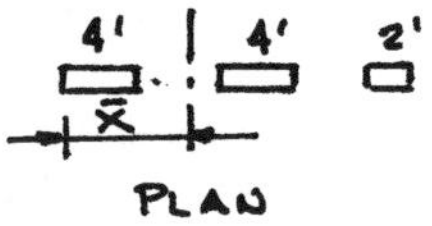

A = 10t

$\bar{x} = [4t(2) + 4t(10) + 2t(17)] / 10t = 8.2'$

$I = t[4^3 / 12 + 4(8.2 - 2)^2 + 4^3 / 12 + 4(10 - 8.2)^2 + 2^3 / 12 + 2(17 - 8.2)^2] = 333t$

Thus, for Pier 5:

$P_5 = M(ad_5) / I = 132.2 \times 2t(17 - 8.2) / 333t = 7.0$ kips ↑

<u>$P_5 = 7.0$ kips</u>

Foundations-1

a) Calculate the load resisted by the tension piles and design the piles.

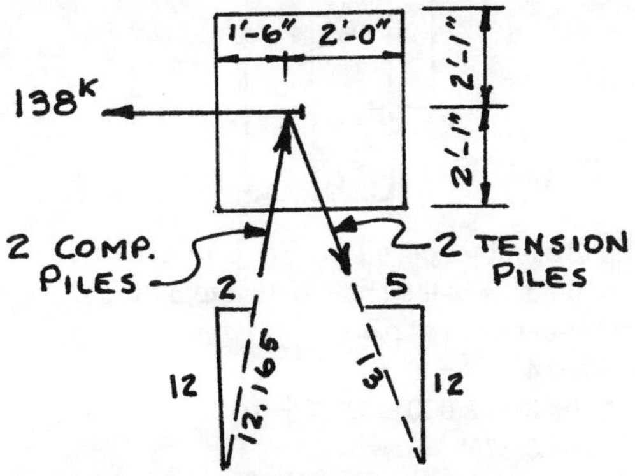

$\Sigma F_x = 0$: $-138 + 2(2 / 12.165)C + 2(5 / 13)T = 0$

$\Sigma F_y = 0$: $(12 / 12.165)C - (12 / 13)T = 0$

$\therefore$ $C = 0.936T$

$T = 128^k$; $C = 120^k$

Design the tension pile using Grade 60 reinforcement (UBC 1997, Section 1804.4). Assume that handling and driving stresses are not critical. Therefore, design for the tension working load of 128^k. To control crack width, limit the rebar stress to 24^{ksi}.

$\therefore$ $A_{st} \geq T / f_s = 128 / 24 = 5.33\ in^2$

$p_g = A_{st} / bh = 5.2 / (12 \times 12) = 0.036 < 0.04$

O.K. – **Use 8# 8 longitudinal**

Re: 1808.4.2, UBC-97: Use No. 5 gauge wire spiral; pitch = 3" o.c. for the first 2 ft; 8" o.c. elsewhere.

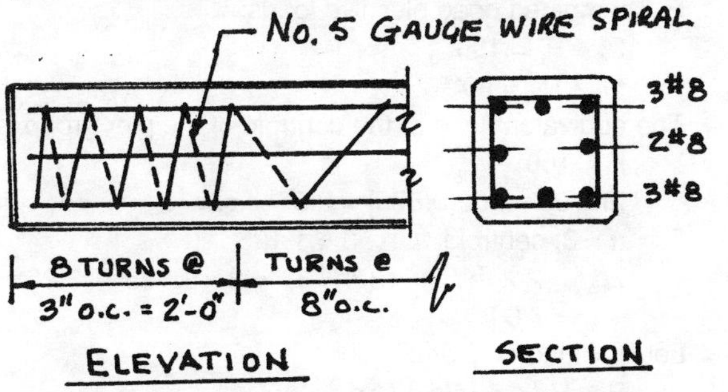

b) Calculate the required tension pile penetration. Neglect the group action reduction of the given skin friction values. Perimeter (12" × 12" pile):

$b_o = 4h = 4ft$

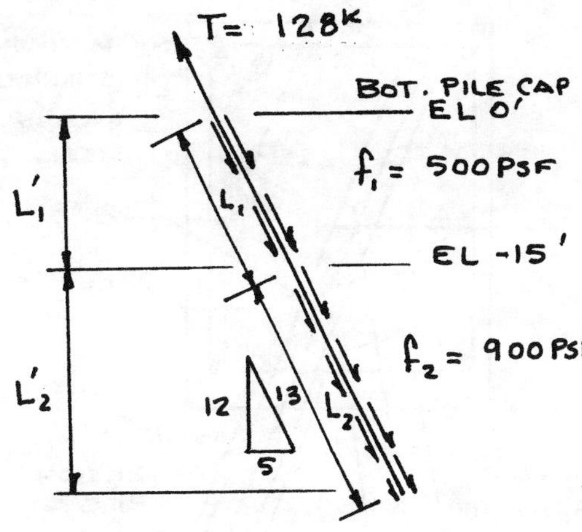

$L_1 = (13 / 12)15 = 16.25'$

$\Sigma F_{long} = 0$: $T - f_1L_1b_o - f_2L_2b_o = 0$

$\therefore$ $L_2 = (128 - 0.5 \times 16.25 \times 4) / (0.9 \times 4) = 26.5'$

$\therefore$ **Required penetration = 16.25 + 26.5 = 42.8'**

c) For a precast concrete element exposed to earth, the minimum concrete cover = 1 1/4" (ACI 318, Sec. 7.7). Thus, for 3#8 in a layer enclosed by No. 5 gauge wire (d_b = 1/4) and a 12" wide pile:

$S = [12 - 2(1.25 + 0.25) - 3(1)] / 2 = 3"$ $\therefore$ O.K.

The required section through the pile is:

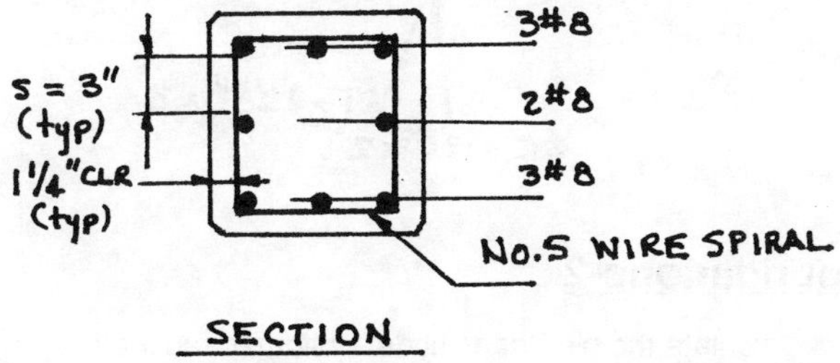

d) Detail the attachment of the piles to the pile cap. After driving, the longitudinal rebars must be exposed and extended past the point where lines of action intersect for a distance sufficient to develop the tension force in the #8. Use the strength design method with U = 1.7 (9.2.2, ACI 318):

$T_u = 1.7 \times 128 / 8 = 27.2^k$

$\phi T_n = \phi A_b f_y = 0.9 \times 0.79 \times 60 = 42.7^k$

Thus, the embedment length for a #8 "top" bar; $f_y = 60^{ksi}$; $f'_c = 3^{ksi}$ (ACI 318-95, Sec. 12-2) is:

$l_d = 1.3(T_u / \phi T_n)(f_y / \sqrt{f'_c})d_b / 20$

$= 1.3(27.2 / 42.7) \times (60{,}000 / (3000)^{1/2}) / 20 = 45"$

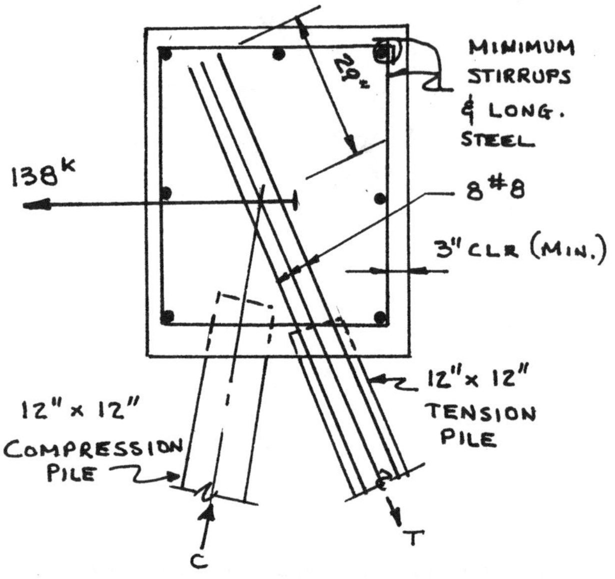

SECTION THROUGH PILE CAP

In the above section, the reinforcement for the tension piles must be developed above the line of action of the 138^k force; otherwise, the concurrent force system is not developed:

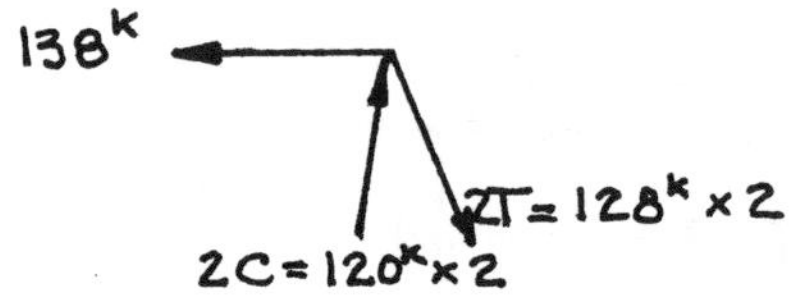

Foundations-2

a) Calculate the pier loads under uniform dead load neglecting the footing cap weight.

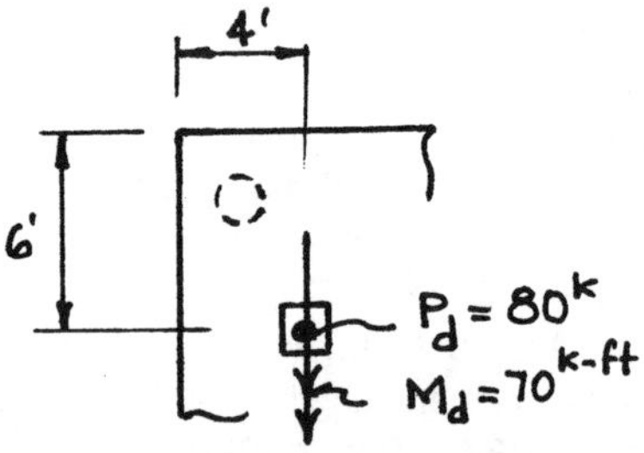

PARTIAL PLAN

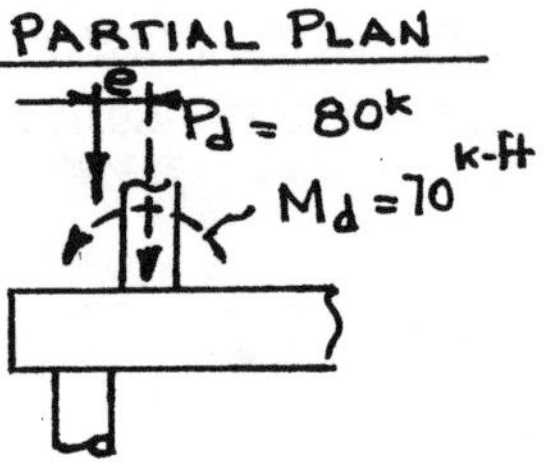

Position of resultant:

$e = M_d / P_d = 70 / 80 = 0.875$ ft

For uniform loading, the centroid of the pier group must coincide with the line of action of the resultant. Try 3 piers: $80 / 3 = 26.7^k < 35^k$ – OK.

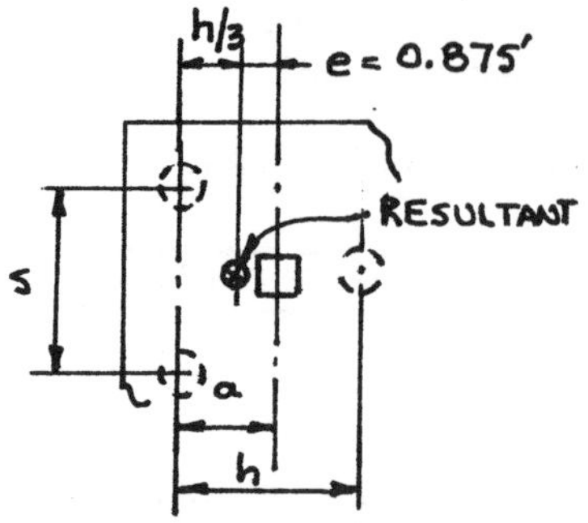

The 2' diameter drilled piers require a minimum of 1' from the side of the shaft to the property line and 5' from center-to-center of piers:

$\therefore$ $a = 4' - 2' = 2'$

$(h / 3) = 2.000 - 0.875 = 1.125'$

$h = 3.375'$

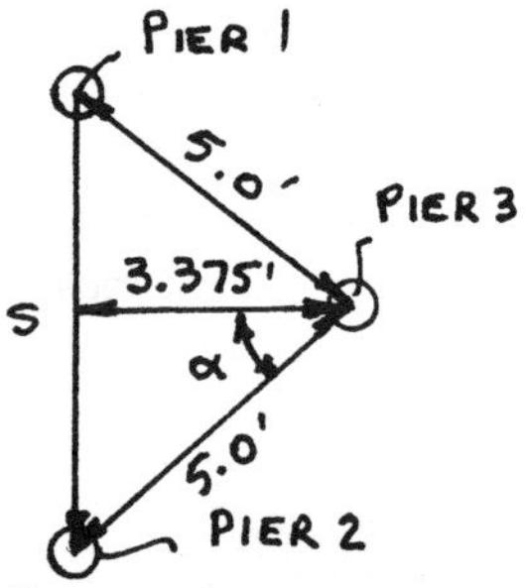

PIER LAYOUT

From the pier layout geometry:

$\alpha = \cos^{-1}(3.375 / 5) = 47.54°$

$\therefore$ $S = 2 \times 5 \sin \alpha = 7.37'$ – say 7' – 4 1/2"

For combined dead plus live loads:

$P_d + P_L = 100^k$

$M_d + M_L = 90^{k\text{-ft}}$

The equivalent load at the centroid of the pier group is:

$R = 100^k$

$M = 90 - 100(0.875) = 2.5^{k\text{-ft}}$ c.c.w.

$n = 3$; centroid @ $h / 3 = 1.125'$

$\Sigma d_x^2 = 2 \times (h / 3)^2 + (2h / 3)^2 = 2 \times 1.125^2 + 2.25^2$
$= 7.6 \text{ ft}^2$

Let P_i = load on pier i:

$P_i = R / n \pm M(d_{xi}) / \Sigma d_x^2$

$\therefore$ $\underline{P_1 = P_2 = 100 / 3 + 2.5 \times 1.125 / 7.6 = 33.7^k}$

$\underline{P_3 = 100 / 3 - 2.5 \times 2.25 / 7.6 = 32.5^k}$

b) The required pier layout for minimum cap dimensions is:

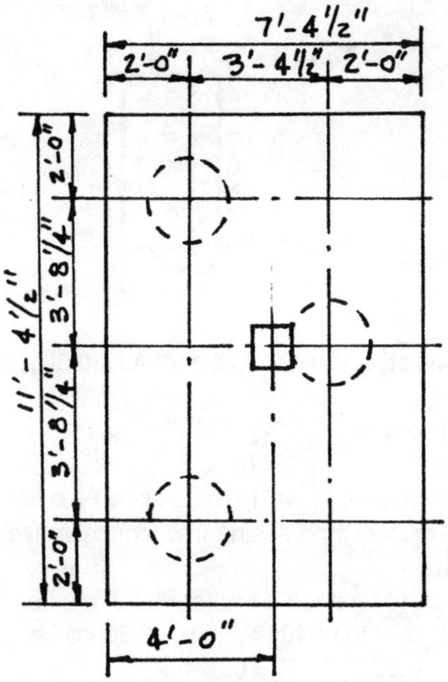

PLAN (N.T.S.)

Foundations-3

a) Calculate the shear in each wall as a function of F_Q.

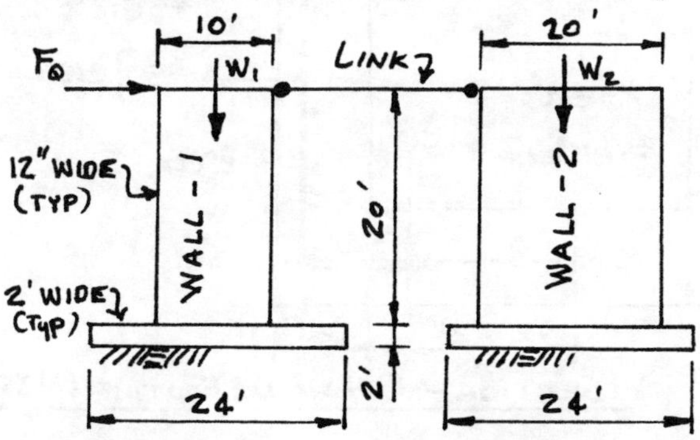

Given: the loads W_1 and W_2 are sufficient to maintain contact pressure beneath footings when lateral force acts; inextensible link connects walls; footings are 2' × 24'; E_c = 3000 ksi; G = 1200 ksi; the soil can compress 0.01" for each 1000 psf in soil pressure; both shear and flexural deformations in the wall are considered.

Let αF_Q = Shear in Wall 1, where α = a constant; $(1-\alpha)F_Q$ = Shear in Wall 2

$M_{O1} = \alpha F_Q H = \alpha F_Q(20 + 2) = 22\alpha F_Q$

$S = bh^2 / 6 = 2 \times 24^2 / 6 = 192 \text{ ft}^3$

$\therefore$ f_1 = soil pressure change due to $M_{O1} = M_{O1} / S$

$= 22\alpha F_Q / 192 = 0.1146\alpha F_Q$

$\Delta_{f1} = (1/100)" \, f_1 = 0.01(0.1146\alpha F_Q) = 0.001146\alpha F_Q$

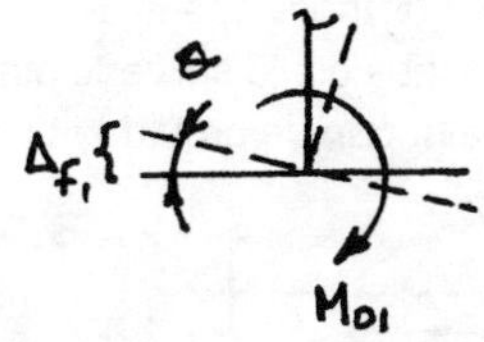

The deflection at the top of the wall due to support rotation is:

$\Delta_1 = \theta H = 0.001146\alpha F_Q(22 / 12) = 0.0021\alpha F_Q$

Similarly, for Wall-2 (same size footing):

$\Delta_2 = 0.0021(1-\alpha)F_Q$

For shear and flexural deflection in Wall 1 (H' = 20'):

$\Delta_1' = \alpha F_Q[H'^3 / 3EI + 1.2H' / AG]$

$A_1 = 12 \times 120 = 1440 \text{ in}^2$

$I_1 = 12 \times 120^3 / 12 = 1.728 \times 10^6 \text{ in}^4$

$\therefore$ $\Delta_1' = \alpha F_Q(0.001055)$

For Wall 2 (H' = 20' = 240"):

$A_2 = 12 \times 240 = 2880 \text{ in}^2$

$I_2 = 12 \times 240^3 / 12 = 13.824 \times 10^6 \text{ in}^4$

$\Delta_2' = (1-\alpha)F_Q[H'^3 / 3EI_2 + 1.2H' / A_2G]$

$= (1-\alpha)F_Q(0.0001943)$

Thus, the total deflections are:

$\Delta_{1T} = \Delta_1' + \Delta_1 = [0.0021 + 0.0001055]\alpha F_Q$

$= 0.003155\alpha F_Q$

$\Delta_{2T} = [0.0021 + 0.0001943](1-\alpha)F_Q$

$= 0.0022943(1-\alpha)F_Q$

For consistent displacements:

$\Delta_{1T} = \Delta_{2T}$

$\therefore$ $0.003155\alpha F_Q = 0.0022943(1-\alpha)F_Q$

Solve for α by canceling F_Q:

$\alpha = 0.42$; $\therefore$ $(1-\alpha) = 0.58$

That is, 42% of F_Q is resisted by Wall 1; the remaining 58% by Wall 2.

b) If shear and flexural wall deflections are neglected, the footings are of identical size. **Therefore, 50% is resisted by Wall 1 and 50% resisted by Wall 2.**

c) If soil compression is neglected, the compactibility equation is simply

$\Delta_1' = \Delta_2'$

$0.001055\alpha F_Q = 0.0001943(1-\alpha)F_Q$

$\therefore$ **$\alpha = 0.16$; $(1-\alpha) = 0.84$. Under this assumption, Wall 1 resists only 16% of F_Q with Wall 2 resisting 84%.**

Foundations-4

Design the continuous footing. The given data (i.e., passive pressure and friction factor) imply that mobilizing passive resistance on the outside face of the wall should stabilize the foundation. This is not a reliable way to stabilize the footing. The wall would have to tilt outward about 1/4" to 1/2" to mobilize resistance. Furthermore, if the adjacent site were to be excavated, the footing might fail. A more reliable approach is to tie the top of the foundation wall into the floor system. This might not be practical if the slab is closely jointed or if an isolation joint is required between the slab and foundation wall. For these cases, a system of grade beams beneath the slabs could serve to stabilize the footing. Assuming that restraint can be provided by the slab:

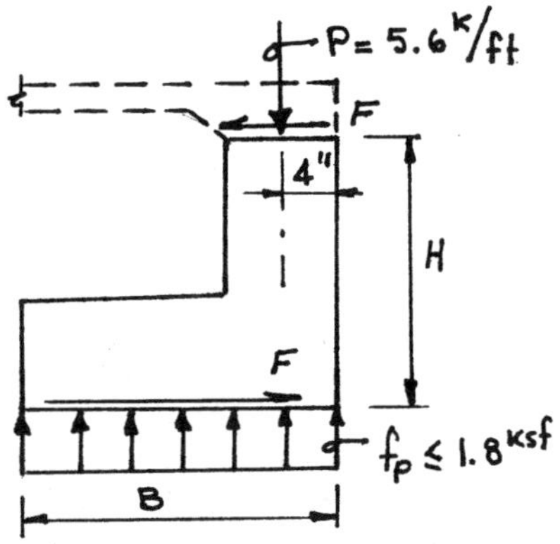

The maximum friction that can develop under the footing is $F = 0.40P$.

$F = 0.40 \times 5.6 = 2.24$ kips

If the footing is restrained against twisting, the bearing pressure beneath the footing will be uniform. Thus, the required footing width is:

$B = P / 1.8 = 5.6 / 1.8 = 3.11'$ (say 3'-3")

For stability:

$M_O = M_r = FH$

$\therefore \quad (2.24)H = 5.6((3.25 / 2) - 0.33)$

$H = 3.24'$ (say 3'-3")

Design the wall and footing reinforcement. Use the strength design method with $U = 1.7$; $f'_c = 3000$ psi; $f_y = 40{,}000$ psi.

$w_u = 1.7 \times 5.6 / 3.25 = 2.9$ klf

For a 12" thick footing; the minimum cover is 3" (ACI 318, Sec 7.7).

$\therefore \quad d = 12 - 3 - 1 = 8"$

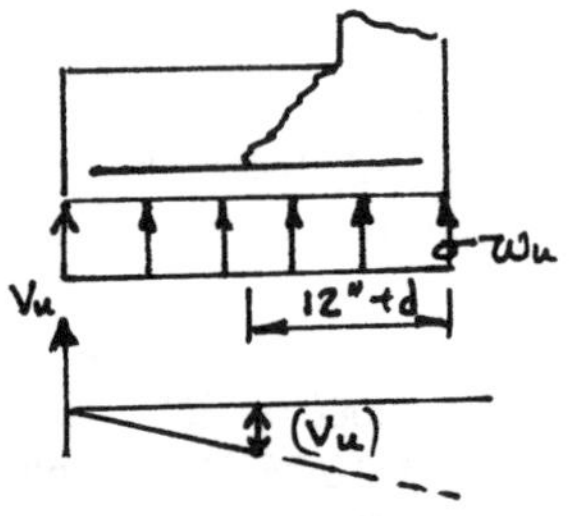

Check shear (Sec. 11, ACI 318-90):

$V_u = w_u\,(3.25' - 1' - 0.67') = 4.6$ kips

$\phi V_c = 2\phi\sqrt{f'_c}\,b_w d = 2 \times 0.85\sqrt{3000} \times 12 \times 8 = 8940$ lbs

$= 8.9$ kips $>> V_u \therefore$ OK

Flexural reinforcement (critical at the face of wall):

$M_u = w_u(3.25 - 1)^2 / 2 = 2.9 \times 2.25^2 / 2 = 7.3$ k-ft / ft

$$A_s^+ \geq \begin{cases} M_u / (\phi f_y 0.95d) = 7.3 \times 12 / (0.9 \times 40 \times 0.95 \times 8) \\ = 0.32 \text{ in}^2 / \text{ft} \\ 0.002bh = 0.002 \times 12 \times 12 = 0.29 \text{ in}^2 / \text{ft} \end{cases}$$

For wall reinforcement (under combined axial and bending):

$d \cong 12 - 2 = 10"$

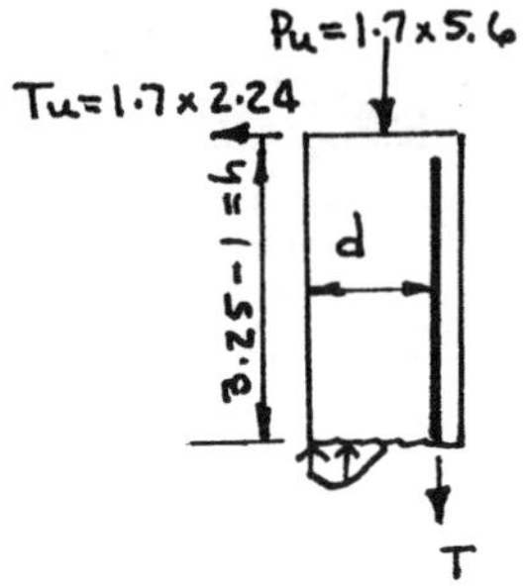

Minimum wall steel (14.3.2, ACI 318: $A_V = 0.0015bh$ $= 0.22$ in^2 / ft);

Say #4@11" o.c. ($A_V = 0.22$ in^2 / ft)

$\Sigma F_y = 0$: $P_u + T - 0.85f'_c ba = 0$

$\therefore \quad a = (1.7 \times 5.6 + 0.22 \times 40) / (0.85 \times 3 \times 12) = 0.60"$

Neglect the axial compression force (conservative): $M_u = T_u h$ $= 8.6$ k-ft / ft

$\phi M_n = \phi A_s f_y(d - a / 2) \geq M_u$; $(d - a / 2) \cong 0.9d$

$\therefore \quad A_s \geq 8.6 \times 12 / (0.9 \times 40 \times 0.9 \times 10) = 0.30$ in^2 / ft

Use #5@12" o.c.; $T_u \leq \phi T_n = 0.9 \times 40A_s$

$A_s \geq 0.11$ in^2 / ft (dowels to slab) $\therefore$ **#5 @ 12" o.c. is OK**.

Thus, the required footing is:

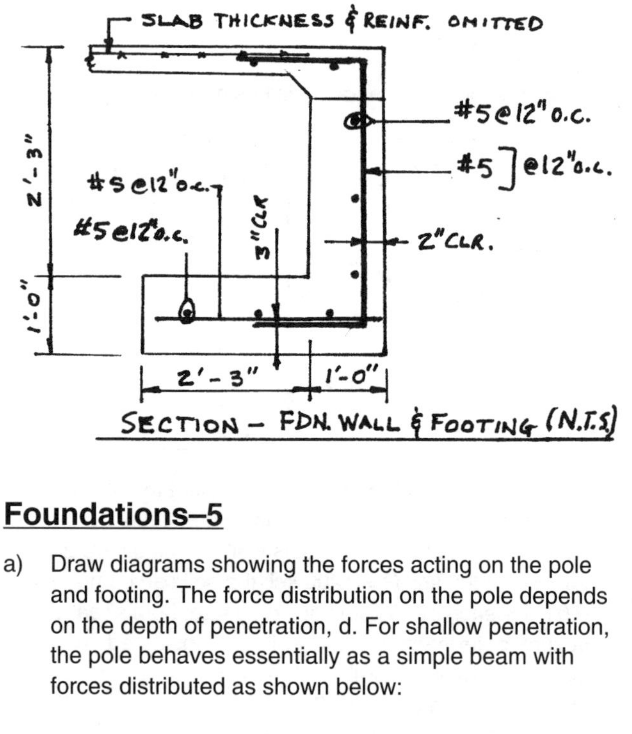

Foundations–5

a) Draw diagrams showing the forces acting on the pole and footing. The force distribution on the pole depends on the depth of penetration, d. For shallow penetration, the pole behaves essentially as a simple beam with forces distributed as shown below:

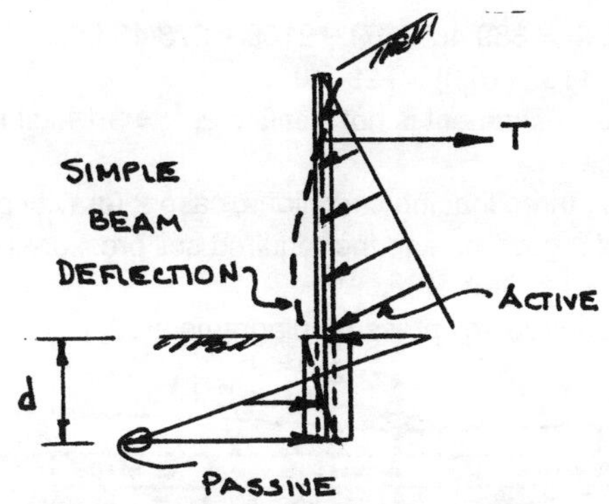

For deep penetration, the pole acts more as a propped cantilevered beam, with rotation fixed at some distance below ground. For this case, the soil pressure would appear (qualitatively) as shown below:

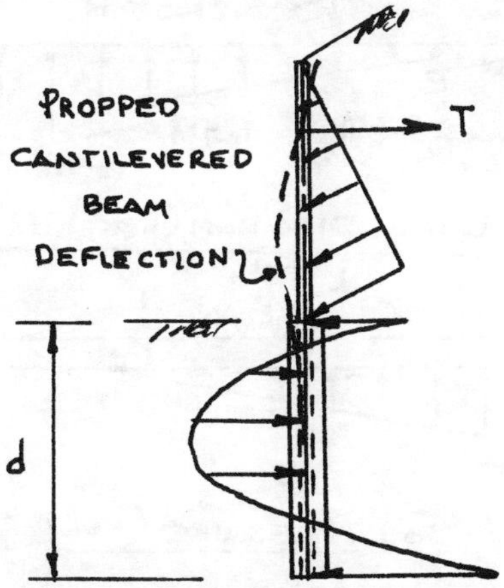

b) Determine the required penetration corresponding to backfill height = 9'. Assume the short embedment condition. Also, assume that the given values of 40 pcf for active pressure and 350 pcf for passive pressure correspond to the horizontal components of actual pressures.

With the poles spaced 10' o.c., the resultant lateral forces are

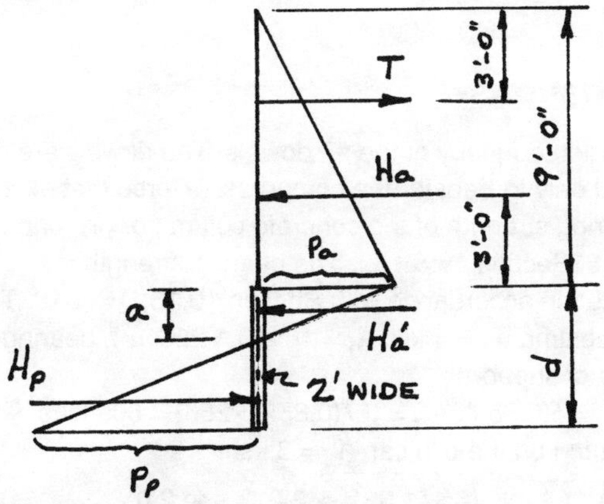

p_a = (40 pcf)9'(10') = 3600 plf

$H_a = 0.5p_ah$ = 0.5(3600)(9) = 16,200 lb

Try d = 5' (2' wide concrete encasement):

p_p = 2'(350 pcf × 5' – 40 pcf × 9') = 2780 plf

By similar triangles:

a / 720 = d / (720 + 2780)

∴ a = 5×720 / (720 + 2780) = 1.03'

H_a' = 0.5 × 1.03 × 720 = 370 lbs

H_p = 0.5 × (5 – 1.03) × 2780 = 5530 lbs

Check to see if the distribution satisfies equilibrium by summing the moments about the line of action of T.

$\Sigma M_{T\text{-force}} = 0$:

5530(6 + 1.03 + $^2/_3$(5 – 1.03)) – 370(6 + 1.03 / 3) – 16,200(3) = 2566

Equilibrium is not satisfied by d = 5'. Try d = 4.5':

p_p = 2'(350 pcf × 4.5' – 40 pcf × 9') = 2430 plf

a = 4.5 × 720 / (720 + 2430) = 1.02'

H_a' = 0.5 × 1.02 × 720 = 367 lbs

H_p = 0.5 × (4.5 – 1.02) × 2430 = 4228 lbs

$\Sigma M_{T\text{-force}} = 0$:

4228(6 + 1.02 + $^2/_3$(4.5 – 1.02)) – 367(6 + 1.02 / 3) – 16,200(3) = –11,380

Too small! To the nearest 6": **embed the pole a minimum of 5'**. The tension in the tie rod is then found to be:

$\Sigma F_x = 0$: $T - H_a - H_a' + H_p = 0$

∴ T = –5530 + 16,200 + 370

T = 11,000 lbs (tension)

Foundations-6

a) Find the maximum soil pressure beneath the wall footings. The seismic overturning moment is based on working load; the gravity loads have been multiplied by load factors. Convert the gravity loads to working loads by dividing the dead loads by 1.4 and the live loads by a factor of 1.7. Thus,

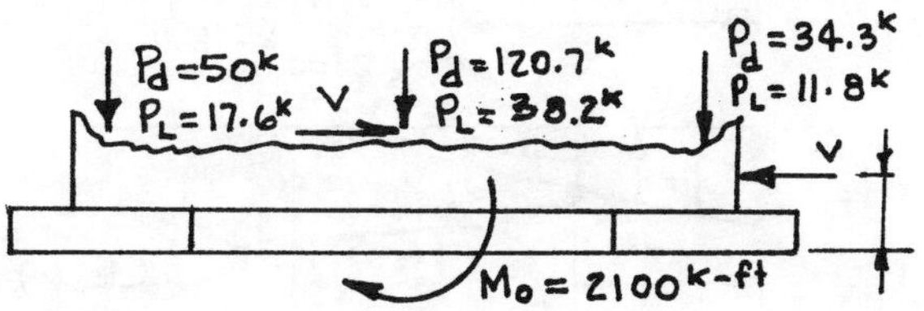

Neglect the slight reduction in overturning moment caused by the lateral force transfer to the slab 2' above the bottom of footing. Assume – initially – that gravity loads are sufficient to maintain compression over the entire area of foundation. Properties of the combined footing are:

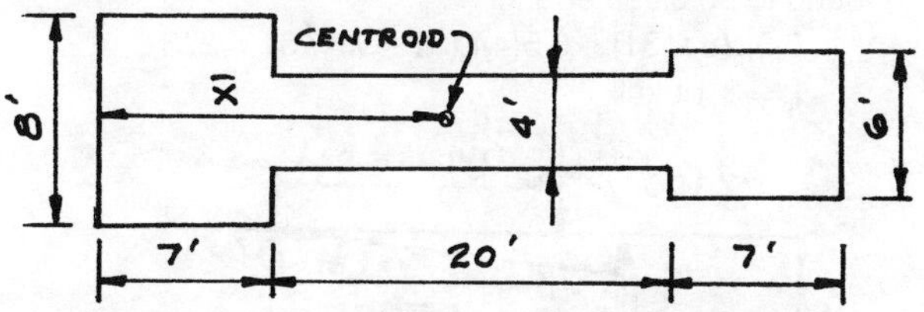

PLAN

$A = 4 \times 34 + 4 \times 7 + 2 \times 7 = 178 \text{ ft}^2$

$\bar{x} = [4 \times (34^2 + 7^2) / 2 + 14 \times 30.5] / 178 = 15.94'$

$I = 4 \times 34^3 / 12 + 4 \times 34(17 - 15.94)^2 + 4 \times 7^3 / 12 + 4 \times 7(15.94 - 3.5)^2 + 2 \times 7^3 / 12 + 2 \times 7(34 - 3.5 - 15.94)^2 = 20{,}727 \text{ ft}^4$

Calculate the equivalent gravity force system acting at the foundation centroid (neglect the foundation weight)

$P = 50 + 17.6 + 120.7 + 38.2 + 34.3 + 11.8$
$= 273 \text{ kips} \downarrow$

$M_g = (50 + 17.6)(15.94 - 4) - (34.3 + 11.8)(30 - 15.94) - (120.7 + 38.2)(17 - 15.94) = -9 \text{ kip-ft} \approx 0!$

Thus, the foundation pressure is uniform under gravity load condition. The corresponding soil pressure is:

$f_{pg} = P / A = 273 / 178 = 1.53$ ksf (comp)

For gravity plus seismic, $P = 273$ kips $\downarrow$;

$M = M_o = \pm 2100$ k-ft (reversible)

For $M = 2100$ k-ft c.w.:

$f_{pl} = P / A - M\bar{x} / I$
$= 273 / 178 - 2100(15.94) / 20{,}727 = -0.08 \text{ ksf}$
$\cong 0$

$f_{pr} = P / A + M(34 - \bar{x}) / I$
$= 273 / 178 + 2100(34 - 15.94) / 20{,}727$
$= 3.37 \text{ ksf}$

For this service load condition, the footing is, for all practical purposes, compressed throughout the 34' length.

For $M = 2100$ k-ft c.c.w.:

$f_{pr} = 273 / 178 - 2100(34 - 15.94) / 20{,}727$
$= -0.3 \text{ ksf} \rightarrow \text{tension!}$

Since soil cannot resist tension, we must recalculate the soil pressure based on a triangular distribution:

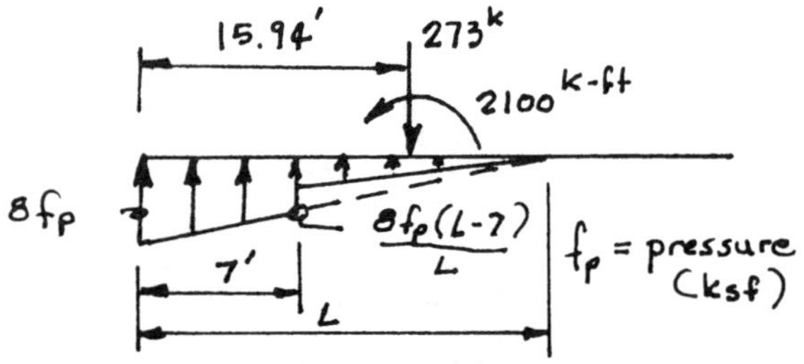

$\Sigma F_y = 0$: $0.5(8f_p)L - 0.5(4)(f_p / L)(L - 7)^2 = 273$

Equilibrium requires $\Sigma F_y = 0$ & $\Sigma M_{rt} = 0$, yielding two equations in the unknowns f_p and and L. By trial and error, neglecting the slight increase at the right end where the width changes from 4' to 6', a value L = 31' is found to be close enough:

$0.5f_p(8 \times 31) - 0.5(4)f_p(31 - 7)^2 / 31 = 273$

$f_p = 3.14$ ksf

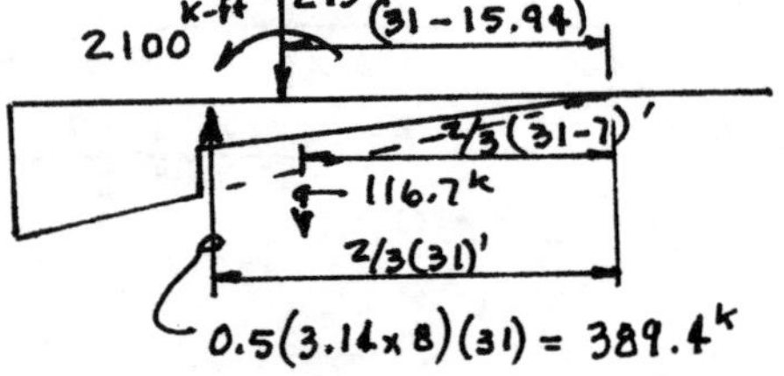

$\Sigma F_y = 389.4 - 273 - 116.7 \approx 0$

$\Sigma M_{rt} = 389.4(20.67) - 2100 - 273(15.06) - 116.7(16.0) = -29 \approx 0$

Further refinement is not justified $\therefore$ $f_p = 3.14$ ksf is close enough.

For the three loading conditions, case 2 (gravity plus seismic) governs and the required soil pressure is:

$f_{p,\,max} = 3.37$ ksf

b) The required soil pressure diagrams are:

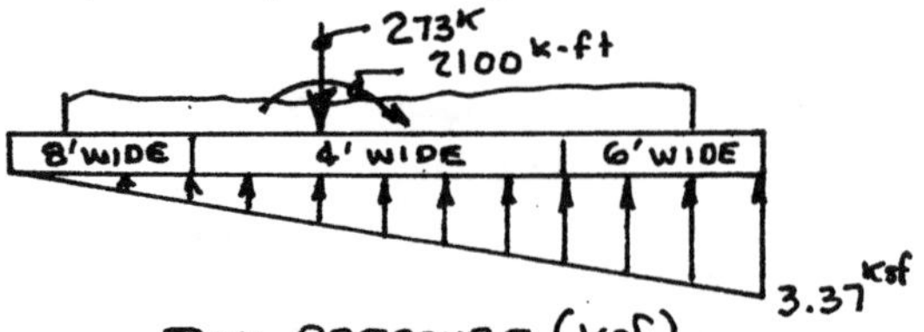

SOIL PRESSURE (ksf)

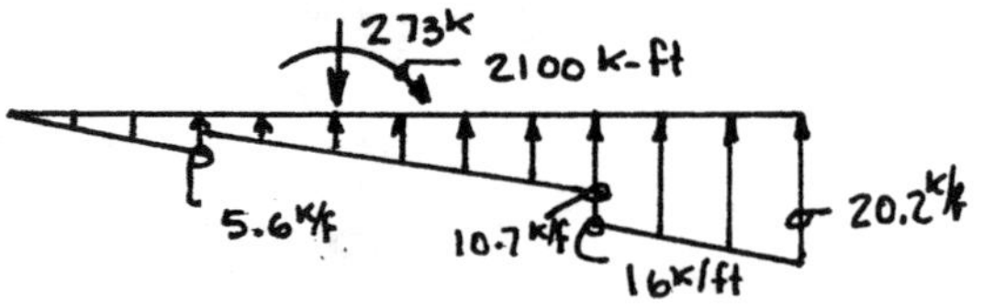

LOAD DIAGRAM (kips; klf)

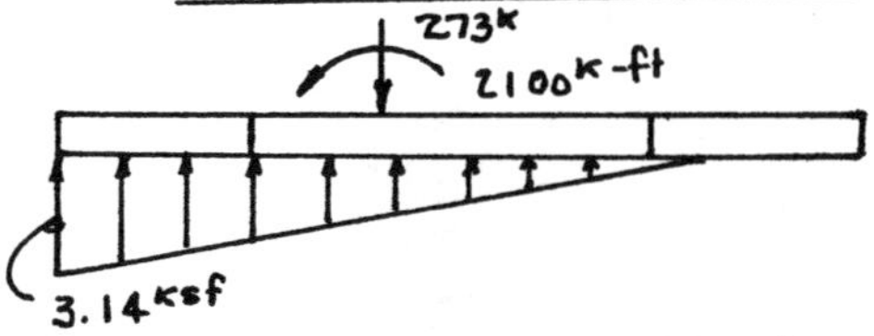

SOIL PRESSURE (ksf)

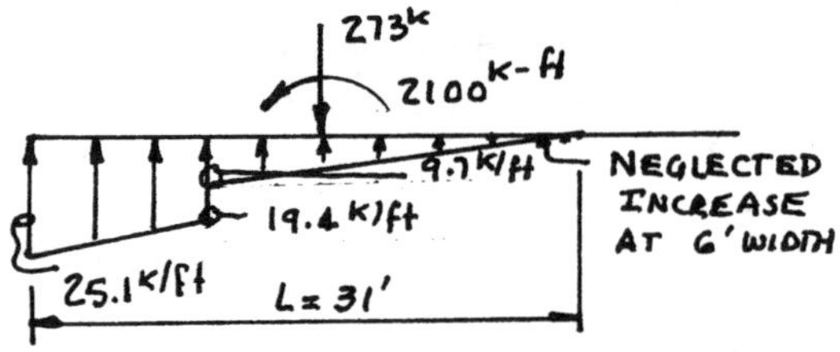

LOAD DIAGRAM (kips; klf)

Foundations-7

Determine adequacy of the #7 dowels. The dowels are required only to transfer the compressive force that exceeds the bearing strength of the concrete column or pile cap (ACI 318, Section 15.8.1.2). The bearing strength is calculated in accordance with Section 10.15, ACI 318. Thus, for the column ($f'_c = 4$ ksi; $A_1 = 16 \times 16 = 256 \text{ in}^2$), bearing on full area of support;

$\phi P_b = \phi(0.85 f'_c A_1) = 0.7(0.85 \times 4 \times 256) = 609 \text{ kips}$

For bearing on the pile cap ($f'_c = 3$ ksi):

$\sqrt{A_2 / A_1} = \sqrt{6^2 / 1.33^2} > 2.0 \therefore \text{ use } 2.0$

$\phi P_b = \phi(2)(0.85 f'c A1) = 0.7(2)(0.85 \times 3 \times 256)$
$= 914 \text{ kips}$

Thus, column bearing is critical. The dowels must transfer:

$P_u - \phi P_b = 800 - 609 = 191 \text{ kips}$

Required area of dowels is:

$A_{st} \geq (191) / \phi f_y = 191 / (0.7 \times 60) = 4.54 \text{ in}^2$

Section 15.8.2.1 also sets a minimum steel area of 0.5%

$\therefore \quad A_{st} \geq 0.005A_1 = 0.005 \times 256 = 1.28 \text{ in}^2$

Thus, the number of #7 dowels required is:

$n \geq A_{st} / A_b = 4.54 / 0.60 = 7.6$

$\therefore$ **Need 8 #7 dowels; only have 4; must provide 4 more #7.**

The compression embedment needed to develop a #7 grade 60 bar in 3000 psi concrete is (12.3, ACI 318):

$l_{db} = 22d_b = 22(^7/_8) = 19"$

To correct the field problem, drill 4 – 1 1/4" dia holes, 20" deep into the pile cap. Epoxy 4 #7 × 3'-4" grade 60 dowels to supplement the existing 4 #7 dowels. Use high modulus epoxy gel with installation per manufacturer's specifications.

Note: Since the #7's project into a column with $f'_c = 4000$ psi, the development length into the column is slightly less than required into the pile cap ($l_d = 19d_b = 16.6"$). Thus, 19 + 17 = 36" dowels are adequate.

Foundations-8

a) Calculate the axial loads on the piles:

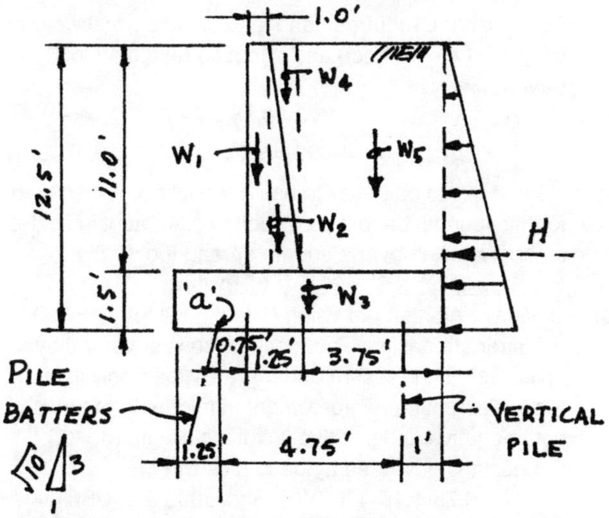

Consider a unit length of wall and footing:

Lateral force: $H = 0.5(36 \times 12.5) \times 12.5 = 2810$ lb ←

Stem weight: $W_1 = 150 \times 1 \times 11 = 1650$ lb ↓

$W_2 = 150 \times 0.5 \times 11 \times 0.25 = 200$ lb ↓

Footing weight: $W_3 = 150 \times 1.5 \times 7.0 = 1575$ lb ↓

Backfill weight: $W_4 = 100 \times 0.5 \times 11 \times 0.25 = 138$ lb ↓

$W_5 = 100 \times 11 \times 3.75 = 4125$ lb ↓

Take moments about the point where the batter pile centerline intersects the bottom of the footing. Let p_v = the force in the vertical pile per foot of wall length:

$\Sigma M_a = 0$:

$4.75\, p_v + 2810(12.5 / 3) - 1650(2.5 - 1.25)$
$- 200(3 + 0.25 / 3 - 1.25) - 1575(3.5 - 1.25)$
$- 138(3 + 2 \times 0.25 / 3 - 1.25) - 4125(7 - 3.75 / 2$
$-1.25) = 0$

$\therefore \quad p_v = 2210$ lb / ft ↑

Let p_b = the force in batter pile per foot:

$\Sigma F_y = 0$: $(3 / \sqrt{10})p_b + 2210 - 1650 - 200 - 1575$
$- 138 - 4125 = 0$

$\therefore \quad p_b = 5770$ lb / ft ↑

$\Sigma F_x = 0$: $(1 / \sqrt{10})5770 - 2810 - H_{key} = 0$

$\therefore$ key must resist $H_{key} = 985$ lb / ft →

Batter piles are spaced 6' on centers; vertical piles are spaced 12' on centers:

$\therefore \quad P_{batter} = 5770 \times 6 = 34{,}600$ lb / pile

$p_{vertical} = 2210 \times 12 = 26{,}500$ lb / pile

b) Calculate the maximum spacing of #5 vertical bars at the bottom of stem ($f_y = 40{,}000$ psi; $f'_c = 3000$ psi); use the strength design method; U = 1.7 (ACI 318, Section 9.2). Design on a per unit length of wall basis:

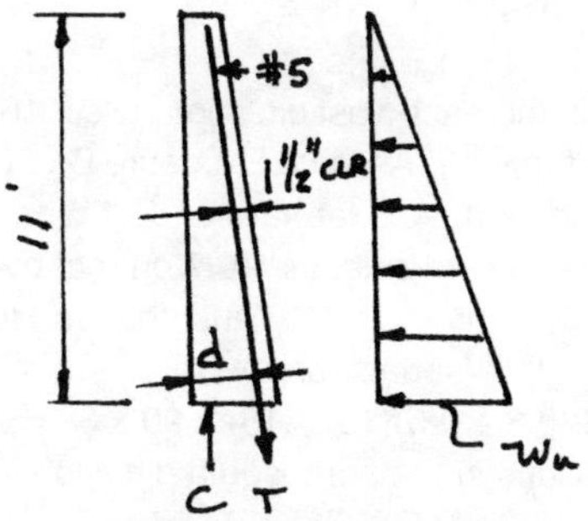

For #5 bars exposed to earth, minimum cover = 1 1/2" (Sec. 7.7, ACI 318)

$\therefore \quad d \cong 15" - 1.5" - 0.63 / 2 = 13.2"$

$w_u = 1.7 \times 36 \text{ psf} \times 11' = 675$ lb / ft

$M_u = (w_u h / 2)(h / 3) = 675 \times (11 / 2)(11 / 3)$
$= 13{,}500$ lb-ft / ft

Minimum steel (Sec. 14.3.2, ACI 318):

$A_s \geq 0.0015bt = 0.0015 \times 12 \times 15 = 0.27 \text{ in}^2 / \text{ft}$

Flexural strength: $\phi M_n \geq M_u$

$\phi A_s f_y(d - a / 2) \geq M_u = 13.5$ k-ft / ft

Approximate $(d - a / 2) \cong 0.95d = 12.5"$

$0.9A_s \times 40 \times 12.5 = 13.5 \times 12$

$A_s = 0.36 \text{ in}^2 / \text{ft}$

$a = A_s f_y / 0.85f'_c b = 0.36 \times 40 / (0.85 \times 3 \times 12)$
$= 0.47"$

$d - a / 2 = 12.26" = 0.98d \therefore$ OK

Section 14.1.2, ACI 318 requires that flexural requirements of Chapter 10 be satisfied, which includes the minimum steel requirement of Sec. 10.5.

$\therefore \quad A_{s,\,min} = (200 / f_y)bd$, or $(^4/_3)A_s = (200 / 40{,}000) \times 12 \times 12.5 = 0.75 \text{ in}^2 / \text{ft}$

The sensible approach would be to eliminate the batter; hence, lower the effective depth so that Sec. 10-5 does not govern. However, if we must stay with given dimensions, then increase calculated A_s by $^4/_3$ (Sec. 10.5.2):

$\therefore \quad A_s = (^4/_3)\, 0.36 = 0.48 \text{ in}^2 / \text{ft}$

Use #5 @ 7 1/2" o.c. (A_s = 0.50 in² / ft)

Foundations-9

Design a square spread footing for the rigid frame. Calculate the lateral forces in accordance with the Uniform Building Code (UBC 1997).

1. Seismic loading (Sec. 1630.2, UBC): zone 4; essential facility (I = 1.25); east-west ground motion resisted by a steel braced frame (R = 5.6); include all exterior walls as contributing weight to the base shear (conservative).

   Take $V = (2.5C_aI/R)W$, with $C_a = 0.44$.

   W = dead load

   $= 60 \times 38 \times 0.004 + 20 \times 2 \times 0.003(60 + 38)$

   $= 20.9$ kips

   $V = (2.5C_aI / R)W = (2.5 \times 0.44 \times 1.25 / 5.6)20.9$

   $= 5.13$ kips

   Divide the design level base shear by 1.4 to obtain the working stress level:

   $V_e = 5.13 / 1.4 = 3.7$ kips

2. For wind in the east-west direction, (Sec. 1615, UBC): $q_s = 15$ psf (given). Assume Exposure B $\therefore$ $C_e = 0.72$; $C_q = +0.8$ windward, -0.5 leeward and -1.6 on roof over the open windward side and -0.7 on roof over the enclosed portion; $I_w = 1.15$. Thus, the drag forces are:

   $V_w = C_eC_qq_sI_w \times$ (surface area)

   $V_{NS} = 0.72(0.8 + 0.5)(15)(1.15) \times 20 \times 38 = 12{,}300$ lb

   $= 12.3$ kips in the north-south direction

   $V_{EW} = 0.72(0.8 + 0.5)(15)(1.15) \times 60 \times 20 = 19{,}400$ lb

   $= 19.4$ kips in the east-west direction.

   Thus, lateral force is governed by wind in both primary directions.

   Compute the gravity load (dead only) on column B-1:

   $p_d = 0.004(10 + 8) \times 30 + 0.003(48) \times 20 = 5.0^k$

   For wind blowing east-to-west:

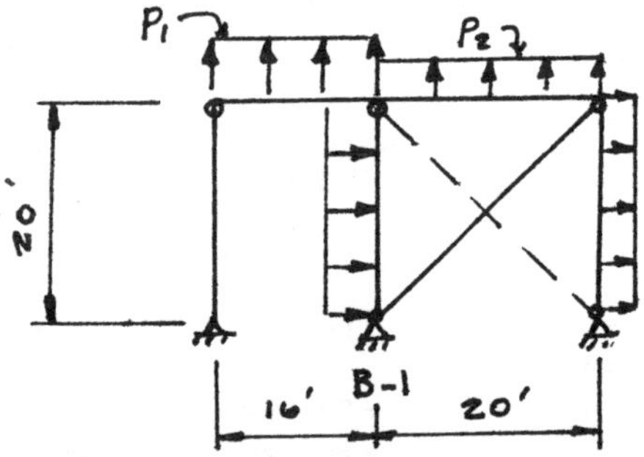

   Calculate the uplift on footing 'B-1' due to lift on the roof and the overturning effect of lateral forces:

   $p_1 = 0.72 \times 1.6 \times 15 \times 1.15 = 20$ psf $\uparrow$

   $p_2 = 0.72 \times 0.7 \times 15 \times 1.15 = 8.7$ psf $\uparrow$

   For the primary lateral force-resisting element:

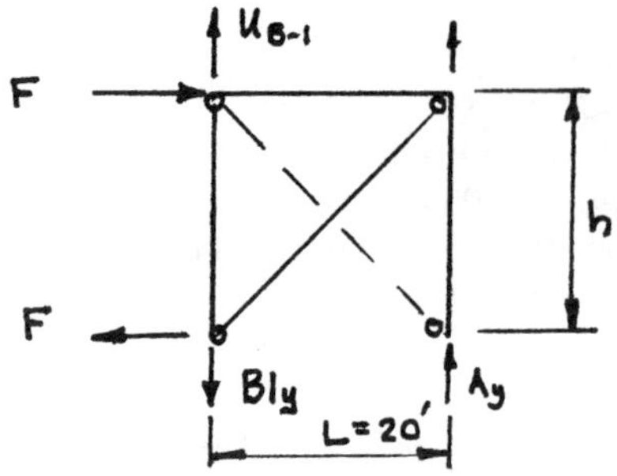

$u_{B\text{-}1} = p_1A_1 + p_2A_2 = 20 \times 8 \times 30 + 8.7 \times 10 \times 30$

$= 7410$ lb $= 7.4$ kips $\uparrow$

$F = 0.25V_{EW} = 19.4 / 4 = 4.9$ kips $\rightarrow$

UBC section 1621 requires that the base overturning moment for a structure with height-to-width > 0.5 shall be less than two-thirds the dead-load-resisting moment. With the base 1' below grade:

$M_o = F(h + 1) + u_{B\text{-}1}(L) = 4.9(20 + 1) + 7.4(20)$

$= 251$ k-ft c.w.

The resisting moment is furnished by the 5.0 kips $\downarrow$ dead load to the column plus the footing weight = W_f:

$M_r = (5.0 + W_f)20$

$M_o \leq {}^2/_3\, M_r \rightarrow M_r \geq 1.5M_o$

$\therefore$ $(5.0 + W_f)20 \geq 1.5 \times 251$

$W_f \geq 13.8$ kips

To furnish the required dead load using normal weight concrete (w = 150 pcf) and a footing restricted to 1' depth:

$A_{required} = W_f / w = 13.8 / 0.15 = 92\ \text{ft}^2$

$\therefore$ $B = \sqrt{A_{required}} = \sqrt{92} = 9.6'$, say 9'-9" x 9'-9"

Note: it would be better to use a monolithic ribbed slab for this foundation, but the problem statement indicates a separate slab over a square spread footing.

$\therefore$ **Use 9'- 9" × 9' – 9" × 1'-0" footing**

b) Design reinforcement using Grade 60 steel; $f'_c = 3000$ psi; strength design. The footing size is governed by uplift. Bearing pressure under gravity load only is nil. Place the footing reinforcement at middepth to furnish both negative and positive flexural strength (d = 6").

For designs governed by dead plus wind:

$U = 0.75(1.4D + 1.7W)$; (Sec. 1909.2.2, UBC)

For a unit width of footing:

$V_u = 0.75 \times 1.4 \times 0.15 \times 9.75 / 2 = 0.8$ k / ft

Shear capacity (Sec. 11, ACI 318)

$\phi V_n = 2\phi\sqrt{f'_c}\, b_wd = 2 \times 0.85\sqrt{3000} \times 12 \times 6$

$= 6700$ lb $= 6.7$ kips $>> V_u$ $\therefore$ OK

$M_u = 0.75 \times 1.4 \times 0.15 \times (9.75 / 2)^2 / 2 = 1.9$ k-ft / ft

Try minimum reinforcement:

$A_{s,\,min} = 0.0018bt = 0.0018 \times 12 \times 12 = 0.26\ \text{in}^2$ / ft (say #5 @ 14" o.c.)

$a = A_sf_y / 0.85f'_cb = 0.26 \times 60 / (0.85 \times 3 \times 12)$

$= 0.50$ in

$\phi M_n = \phi A_sf_y(d - a / 2)$

$= 0.9 \times 0.26 \times 60 \times (6 - 0.50 / 2) / 12$

$= 6.7$ k-ft / ft $>> M_u$

$\therefore$ **Use #5 @ 14" o.c.e.w.**

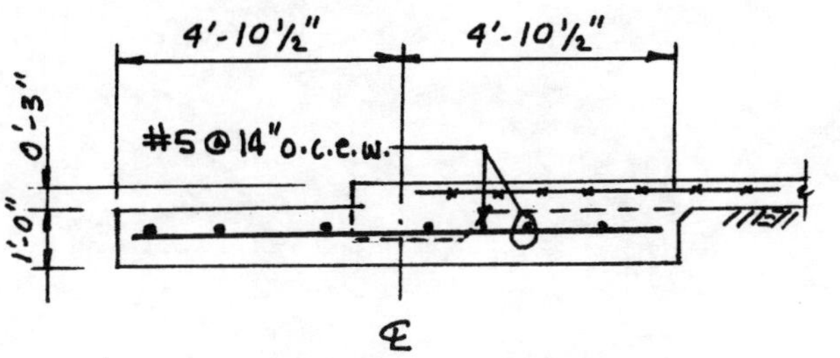

SECTION – FOOTING AT B-1
(N.T.S.)

Foundations-10

a) Determine the pad dimensions producing uniform bearing pressure under the given loads. Include the weight of the 3'×2' strap in the resultant. Use normal-weight (150 pcf = 0.15 kcf) concrete:

∴ $W = 0.15 \times 2' \times 3' \times 21.75' = 19.6$ kips

The strap is stiff enough to permit treating the combined footing as rigid. To achieve uniform bearing pressure, the centroid of the two footings must coincide with the center of gravity of the applied forces.

∴ $R = 150 + 19.6 + 225 = 394.6$ kips

$\bar{x} = [19.6((21.75 / 2) - 0.75) + 225(20)] / 394.6$
$= 11.91'$

Let A_e = area of exterior footing; A_i = area of interior footing

$\Sigma F_y = 0$: $A_e + A_i = 394.6$ kips / 3 ksf = 131.6 ft^2

There are infinite sets of A_e and A_i that produce a total area of 131.6 ft^2 with the centroid $\bar{x}$. One possibility is $A_i = A_e = 131.6 / 2 = 65.8$. For this case, the interior (square) footing would be $\sqrt{65.8} = 8.11$ square ft

$\bar{x} = 11.91' = 65.8(20 + e) / 131.6$,

where e = eccentricity of the exterior footing = 3.82'

Thus, the length of the exterior footing is $2 \times (e + 0.75) = 2 \times 4.57 = 9.14'$; from which, the required width of the exterior footing is 65.8 / 9.14 = 7.20'.

∴ **Use 7.20' × 9.14' for the exterior (practical size: say 7'-3" × 9'-0"). Use 8.11' × 8.11' for the interior.**

b) Determine the soil pressure profile when the forces in part 'a' are combined with 80 kips uplift on the interior. Use the theoretical footing dimensions rather than practical sizes

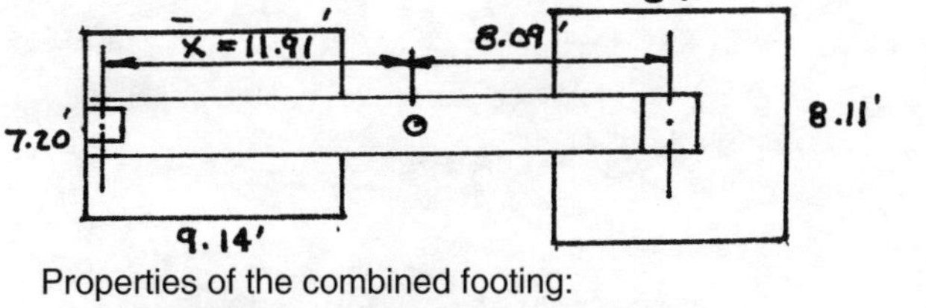

Properties of the combined footing:

$A = 2 \times 65.8 = 131.6$ ft^2

$I = \Sigma(I_{xo} + A_i d_i^2) = 7.20 \times 9.14^3 / 12$
$+ 65.8(11.91 - 3.82)^2 + 8.11 \times 8.11^3 / 12$
$+ 65.8(20 - 11.91)^2 = 9431$ ft^4

Pressures under the uplift force and moment combine algebraically with the uniform 3 ksf pressure of part 'a'.

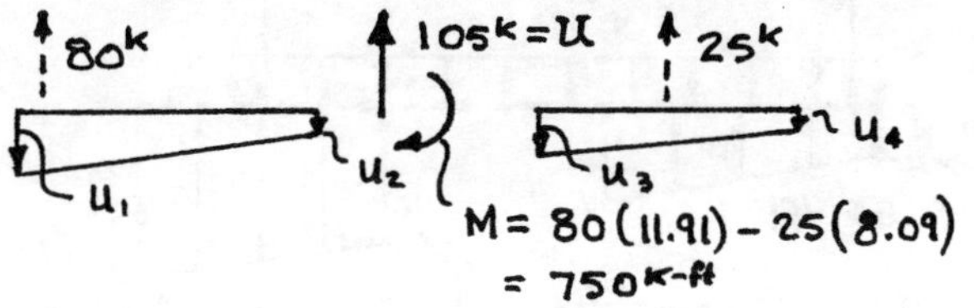

$u_i = U / A \pm Mx / I$

∴ $u_1 = -105 / 131.6 - 750(11.91 + 0.75) / 9431$
$= -1.80$ ksf

$u_2 = -105 / 131.6 - 750(11.91 + 0.75 - 9.14) / 9431$
$= -1.08$ ksf

$u_3 = -105 / 131.6 + 750(8.09 - 4.06) / 9431$
$= -0.48$ ksf

$u_4 = -105 / 131.6 + 750(8.09 + 4.06) / 9431$
$= +0.17$ ksf

Thus, combining forces and pressures, the required pressure diagrams are:

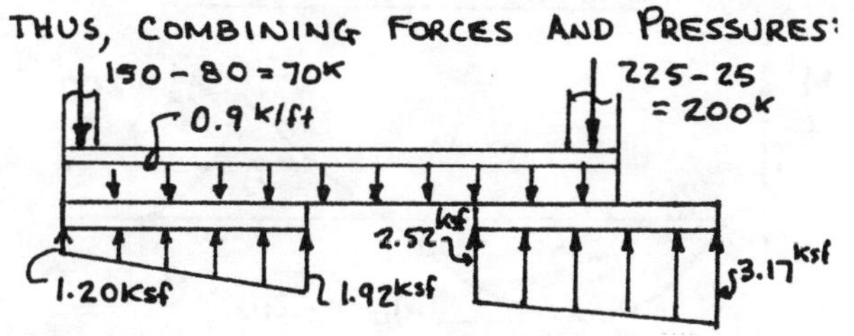

c) Calculate the maximum positive and negative bending moments for the loading in case 'b'.

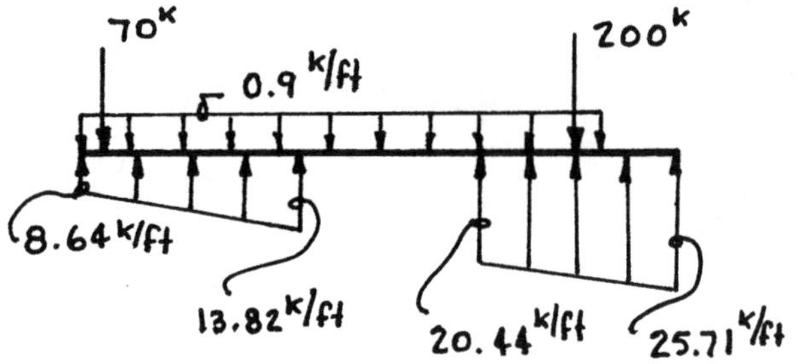

LOADING CASE 'b'

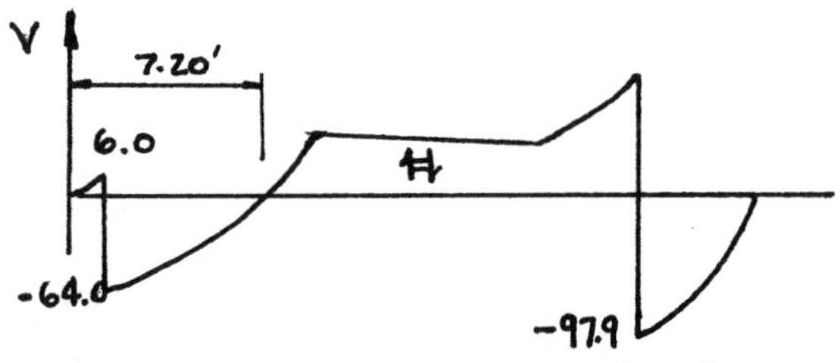

SHEAR DIAGRAM (KIPS)

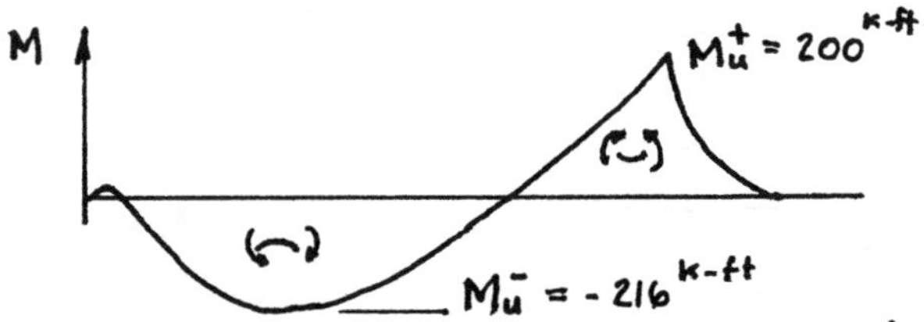

BENDING MOMENT DIAGRAM (k-ft)

$W_1 = 23 \times 4 = 92.4$ kips

$W_2 = 0.5 \times 2.7 \times 4 = 5.4$ kips

$\therefore\quad M_u^+ = (92.3 \times 2) + (^2/_3) \times 4 \times 5.4$

$\mathbf{M_u^+ = 200}$ **k-ft**

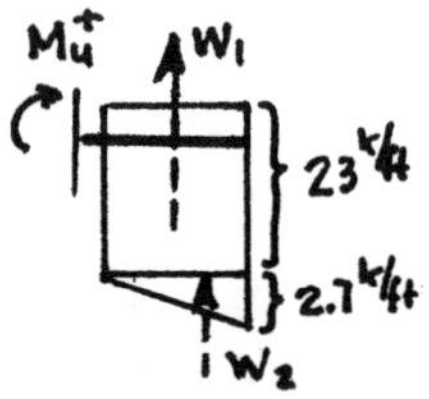

$W_1 = 0.9 \times 7.20 = 6.5$ kips

$W_2 = 8.6 \times 7.20 = 62.0$ kips

$W_3 = 0.5 \times 4.1 \times 7.20 = 14.9$ kips

$M_u^- = -70 \times 6.45 - 6.5 \times 3.6 + 62.0 \times 3.6 + 14.9 \times (^1/_3) \times 7.20$

$\mathbf{M_u^-} = -216$ k-ft

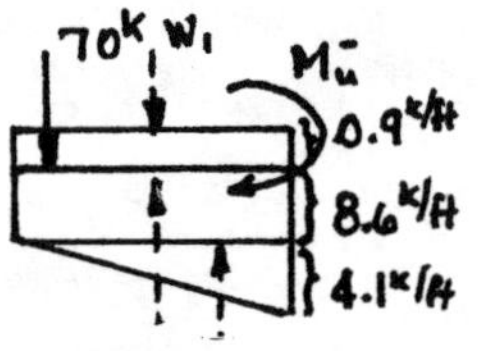

Note: the freebody diagrams shown above are cut at the points of maximum and minimum moments (i.e., where $V = 0$).

Timber-1

a) Design the floor joist. Take the allowable stresses for visually graded dimension lumber (Table 4A, NDS-91) for No.1 Douglas Fir-Larch (north): $F'_b = 1.15 \times 825 = 950$ psi (repetitive use and $C_F = 1.0$); $F'_v = 85$ psi; $F_{c\perp} = 625$ psi; E = 1,600,000 psi.

Estimate the dead loads:

w_d = 3/4" plywood + floor finish + ceiling + partitions + joists

= 3 + 1 + 1 + 8 + 2 = 15 psf to joists

Live load (residential): $w_L = 40$ psf

Try joists spaced at 12" o.c.:

w = 1.0'(15 + 40) psf = 55 plf

Dry use ($C_M = 1.0$); normal load duration ($C_D = 1.0$); assume $C_F \cong 1.0$

∴ **$F_b' = 950$ psi**

$M_{max} = wl^2 / 8 = 55 \times 17^2 / 8 = 1987$ #-ft

$S_{req'd} \geq M_{max} / F_b' = 1987 \times 12 / 950 = 25.1$ in^3

Limit the live load deflection to L / 360

$I_{req'd} \geq (5 / 384)360 (w_L L)(L^2) / E$

$= 0.013 \times 360(40 \times 1.0 \times 17)(17 \times 12)^2 / 1.6 \times 10.6$

= 83 in^4

∴ Try 2 × 12 (S = 31.6 in^3; I = 178 in^4)

Check shear:

$V = w(^L/_2 - d) = 55(17 / 2 - 11.25 / 12) = 416^\#$

$f_v = {^3/_2}V / bd = (1.5 \times 416) / (1.5 \times 11.25)$

$= 37$ psi $< F_v'$

Check bearing length = N:

$R = wL / 2 = 55 \times 17 / 2 = 467^\#$

$f_{c\perp} = R / (1.5N) \leq F_{c\perp} = 625$ psi

N ≥ 0.50" (any reasonably sized joist seat or ledger will provide this amount).

∴ **For the joists: Use 2×12 @ 12" o.c.**

b) Design the girder using Dense No. 1 Douglas Fir-Larch to obtain higher allowable flexural stress and stiffness: $F_b = 1550$ psi (single member use); $F_v = 85$ psi; E = 1,700,000 psi.

Tributary area = 17 × 20 = 340 ft^2 ≥ 150 ft^2

∴ Reduce live load 0.08(340 – 150) = 15%

The load to the girder is essentially uniformly distributed. Estimate the girder weight as 25 plf)

$w = (1 - 0.15)40 \times 17 + 15 \times 17 + 25 = 858$ plf

$M = 858 \times 20^2 / 8 = 42{,}900$ #-ft

$S_{req'd} \geq 42{,}900 \times 12 / 1550 = 332$in^3

$I_{req'd} \geq 0.013 \times 360(1 - 0.15)(40 \times 17 \times 20)(20 \times 12)^2 / (1.7 \times 10^6) = 1833$ in^4

Try 4 – 3 × 16 (S = 4 × 97 = 388 in^3; I = 4 × 740 = 2960 in^4)

Check shear:

$V = 858(20 / 2 - 15.25 / 12) = 7490$ #

$f_v = 1.5 \times 7490 / (4 \times 2.5 \times 15.25) = 74$ psi $< F_v = 85$ psi

∴ **For the girder: Use 4 – 3 × 16**

c) Design a wall ledger to support the joists. Try a 3×12 No.1 Douglas Fir-Larch and 1" diameter anchor bolts:

For gravity loads, R = 467 Plf

Provide joist hangers to support the 467# reaction from joists. For 1" dia. bolt bearing perpendicular to grain (re: 8.2.3, NDS-91 and interpolating table 8-2A for a 5" main member and 2.5" side member):

$Z' = 680 + 0.5(1290 - 680) = 985^\#$ / bolt

Spacing of bolts:

Spacing ≤ Z' / R = 985 / 467 = 2.1'

Check the shear stress near the end of member for a 3×12 ledger with the bolt supporting its maximum capacity of 985#:

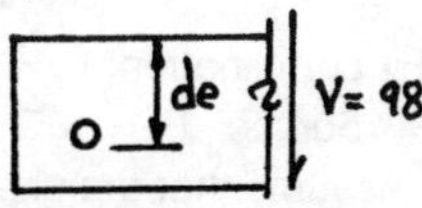

$V \leq {^2/_3} F_v' b d_e^2 / d$

$d_e \geq ([1.5 \times 985 / (95 \times 2.5)]11.25)^{1/2} = 8.8$" is O.K. for nominal 12" deep member

No information is given on the lateral loads. The diaphragm shear transfers into the top of the ledger by nails through plywood and the 1" dia. bolts would resist the shear by parallel to grain bearing. Lateral forces in the other direction would be resisted by ties anchored into the reinforced masonry wall designed to resist the larger of the code specified forces or a minimum design-level force of 420 plf (re: 1633.2.8.1, UBC-97). An acceptable detail is:

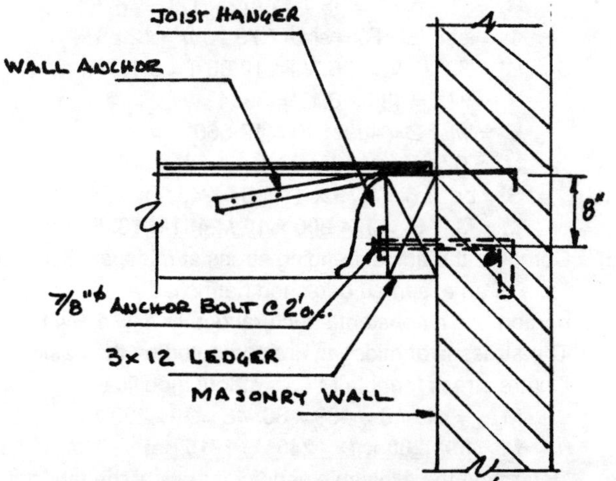

SECTION AT EXTERIOR WALL (N.T.S.)

Timber-2

a) Calculate the shear stress at the support. Analyze the beam as a double-tapered pitched beam with detached haunch (Re: AITC Manual, p. 5-243).

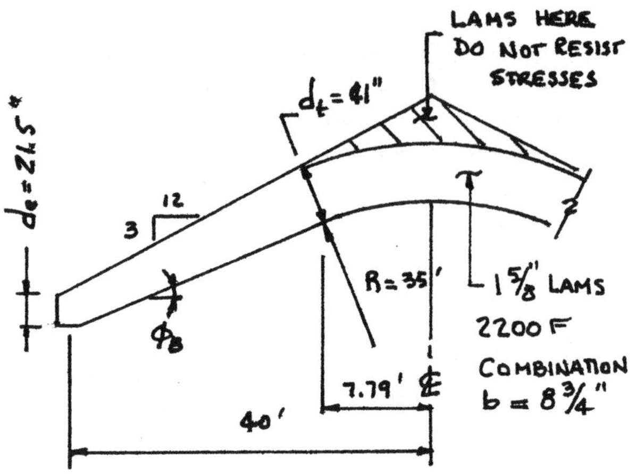

For the Douglas Fir combination: $F_b = 2200$ psi; $F_t = 1600$ psi; $F_c = 1500$ psi; $F_v = 165$ psi; $F_{c\perp} = 450$ psi; $E = 1.8 \times 10^6$ psi. Assume that the given dead load, 18 psf, is per foot of length along the top slope; live load is per foot of horizontal projection $\therefore$ w = ((12.36 / 12)18 + 12)16 = 489 plf. I assume also that the given 9" width is the nominal value $\therefore$ actual width = 8.75".

$V = wL_e / 2 = w(L - 2d_e) / 2$
$= 489\,(80 - 2 \times 21.5 / 12) / 2 = 18{,}680^{\#}$

$\therefore$ **$f_v = {}^3/_2 V / bd_e = {}^3/_2 \times 18{,}680 / (8.75 \times 21.5)$ $= 150$ psi**

b) Determine the allowable shear stress. The roof live load permits $C_D = 1.25$:

$\therefore$ **$F_v' = C_D F_V = 1.25 \times 165 = 206$ psi**

c) Calculate the actual bending stress at the tangent point:

$R_m = R + d_t / 2 = 35 + 41 / (2 \times 12) = 36.70'$
$\Phi_B = \sin^{-1}(L_t / R) = \sin^{-1}(7.79 / 35) = 12.86°$
$L_t' = R_m \sin \Phi_B = 36.7(\sin 12.86°) = 8.17'$
$V_t' = wL_t' = 489 \times 8.17 = 3994^{\#}$
$V_0 = wL / 2 = 489 \times 40 = 19{,}560^{\#}$
$\therefore$ $M_t' = 0.5(V_0 + V_t')(L / 2 - L_t') = 374{,}800^{\#\text{-ft}}$
$S_t = bd_t^2 / 6 = 8.75 \times 41^2 / 6 = 2451\,\text{in}^3$
$\therefore$ **$f_{bt} = M_t' / S_t = 374{,}800 \times 12 / 2451 = 1835$ psi**

d) Compute the actual bending stress at midspan. For the case of an assumed detached haunch, the curved portion has a constant structural depth: $d_e = d_t = 41"$. The stresses at midspan are computed by the basic flexure stress formula M / S, without modification.

$M_{max} = \omega L^2 / 8 = 489 \times 80^2 / 8 = 391{,}200^{\#\text{-ft}}$
$f_{bc} = 391{,}200 \times 12 / 2451 = 1915$ psi

e) Determine the allowable bending stress at the tangent point. For d = 41" > 12": $C_F = 0.87$ (Re: Table 3.5, AITC); $C_D = 1.25$

$F_b' = C_F C_D F_b = 0.87 \times 1.25 \times 2200 = 2390$ psi

f) Determine the allowable bending stress at midspan. The same factors apply as at the tangent point except an additional reduction must be taken for stresses induced by curving the lams (Re: p. 5 – 217, AITCM):

$C_F = 0.87$
$C_D = 1.25$
$C_C = 1 - 2000\,(t / R)^2$
$= 1 - 2000\,(1.625 / (35 \times 12))^2 = 0.97$

$F_b' = C_F C_D C_C F_b = 0.87 \times 1.25 \times 0.97 \times 2200 = 2320$ psi

g) Compute the radial tension stress (Re: p. 5 – 218, AITCM):

$f_r = 3M / 2R_m bd$
$= (3 \times 391{,}200 \times 12) / (2 \times 36.7 \times 12 \times 8.75 \times 41)$
$f_r = 45$ psi (tension)

h) Determine the allowable radial tension stress for Douglas Fir, without reinforcement, for gravity roof load conditions:

$F_r = C_D(15\text{ psi}) = 1.25 \times 15 = 19$ psi

i) Design a repair for the cracked central region. The analysis in parts 'g' and 'h' indicates that the member is overstressed by radial tension stresses in the central curved region. By reinforcing this region (e.g., adding radial dowels or lag bolts) the allowable radial tensile stress would increase to $F_r' = 1.25 \times 55$ psi = 69 psi, which is acceptable. Try adding 1" dia lag bolts centered on the member (Re: Table 5.16, AITC)

$$T \le \begin{cases} 9555\ \#\ \text{on the bolt} \\ 697\ \#/\text{in}(20.5 - 2)" = 12{,}900^{\#}\ (\text{withdrawal}) \end{cases}$$

Therefore, the tensile strength of the lag bolt governs. Space the bolts to resist the entire radial tensile stress resultant:

$f_r bs \le T = 9555^{\#}$
$s \le 9555 / (8.75 \times 45) = 24.2"$ say 2' o.c.

Bolts should be aligned radially in predrilled holes and extend to within 2" of the soffit. The holes for the lag bolts will reduce the flexural resistance in the curved region. Calculate the new section modulus by deducting the area of hole below the original neutral axis.

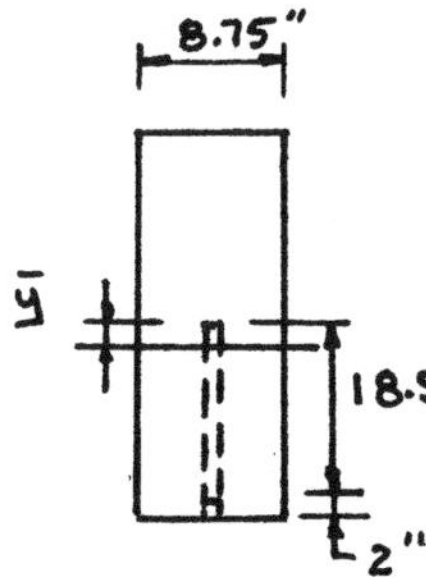

$A_n = 8.75 \times 41 - 1 \times 18.5 = 340\ \text{in}^2$
$\bar{y} = (-18.5^2 / 2) / 340 = 0.50"$
$I = [8.75 \times 41^3 / 12] + [359(0.5)^2] - [1(18.5)^3 / 12]$
$- [18.5(9.25 - 0.5)^2] = 48{,}400\ \text{in}^4$
$S_t = 48{,}400 / (20.5 + 0.5) = 2304\ \text{in}^3$
$\therefore$ $f_{bx} = 391{,}200 \times 12 / 2304 = 2037$ psi < F_b'
$\therefore$ Use 1" dia lag bolts at 2' o.c.

Arc length of central portion = $2\Phi_B R_m$
$A = 2(12.86° / 57.3\ {}^\circ/_{\text{rad}})\ 36.7 = 16.5'$

$\therefore$ **Need 8 lag bolts: use 8-1" dia at 2' o.c.**

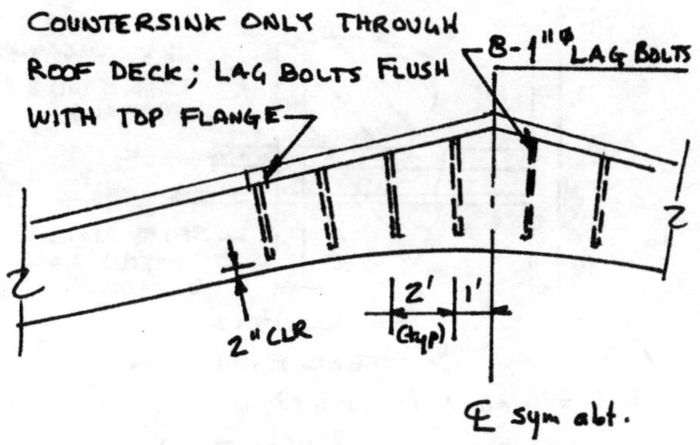

ELEVATION – PROPOSED REINFORCEMENT (N.T.S.)

Timber-3

a) Compute the lateral force resisted by the center wall using the Uniform Building Code (UBC-97).

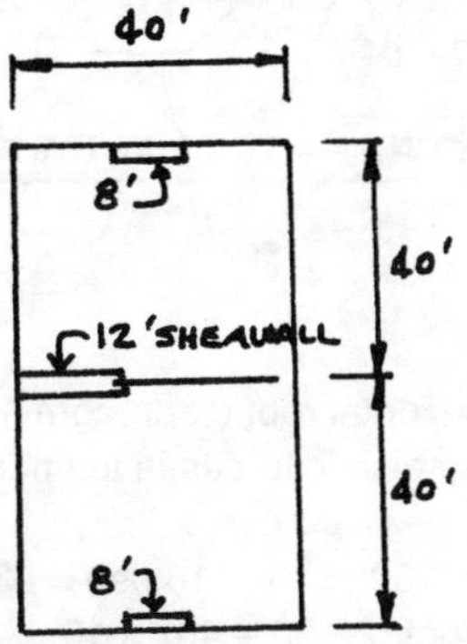

The exterior walls weigh 20 psf. Estimate the roof dead load as 15 psf.

$\therefore$ W = (15 psf × 40' × 80') + (20 psf × (11 / 2)' × 2 × 80') = 65,600#

For the seismic loading: C_a = 0.36; importance factor I = 1.0; R = 5.5:

$V = (2.5C_aI / R)W = (2.5 \times 0.36 \times 1.0 / 5.5)65,600 = 10,700^{\#}$

Convert the design-level force to working stress level by dividing by 1.4:

$V_e = 10,700 / 1.4 = 7640^{\#}$

For wind loading:

q_s = 16.4, say 17 psf (V_{30} = 80 mph)

C_e = 1.2 (exp. C; H ≤ 25')

C_q = 1.3; I_w = 1.0

$V_w = C_eC_q I_wq_s$

$= 1.2 \times 1.3 \times 17 \text{ psf} \times 1.0 \times 80' \times (11 / 2)$

$= 11,700^{\#}$

Thus, wind governs: $v = V_w / 80$ = 146 plf

For the flexible diaphragm the force is proportional to the tributary width. Therefore, the shear on the center wall is:

$\therefore$ <u>V_{center} = v(40') = 146 × 40 = 5840#</u>

b) Determine the nailing required to transfer the shear from the roof diaphragm to the center shearwall. The force in the top plate is transferred in part through the connection of the collector (see part c) and from the plywood-to-blocking-to-plate by toe nails, which must transfer 5840 / 40 = 146 plf. Use 10d nails with joists spaced at 2' o.c.

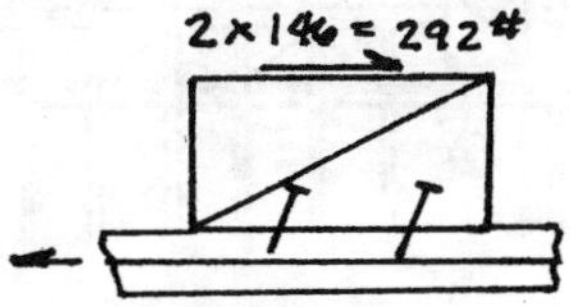

For 10-d toenails (NDS-91)

$Z' = ZC_{tn}C_D = (73)(0.83)(4 / 3) = 81\ ^{\#}/_{nail}$

$\therefore$ n ≥ 2 × 146 / 81 → 4 nails each block

<u>Use 4 × 10d (toe nails) for each 2 × 10 blocking over the center wall.</u>

c) Design the connection of the collector at point A:

$P = (146^{\#}/_{ft})(12' + 2') = 2044^{\#}$

Try $^7/_{16}$" diameter lag screws into Douglas Fir-Larch lumber. Assume a group factor, C_g, of 0.95 (conservative for 5 or less screws); C_D = 4/3:

$Z' = C_gC_DZ_{//} = 0.95 \times 1.3 \times 540 = 667^{\#}/_{bolt}$

n ≥ 2044 / 667 = 3.1; say 4 screws

Provide minimum 7d = 7($^7/_{16}$) = 3.0" end distance and a minimum spacing of 3" o.c.

<u>Use 4 - $^7/_{16}$" diameter × 6" lag screws</u>

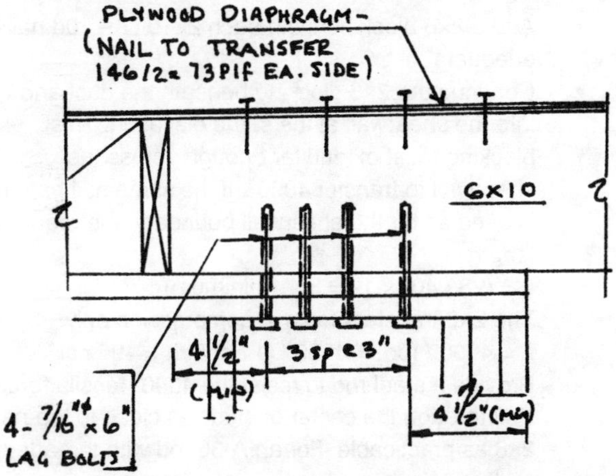

DETAIL – CONNECTION AT 'A' (N.T.S.)

d) Design to transfer the force at the cut top plate (point 'x'):

$P = 146 \text{ plf } (40 - 12)' = 4090^{\#}$

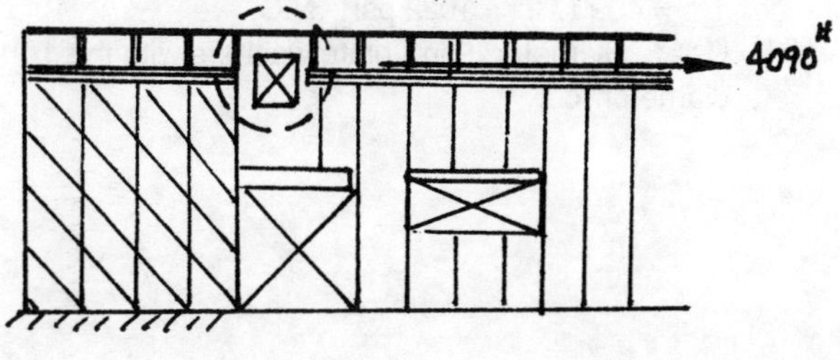

The more direct path for force transfer is above the duct. However, I assume that the stage of construction is such that framing over the duct is difficult; therefore, transfer the 4090# (reversible) force into the shear wall beneath the duct. The proposed scheme is as follows:

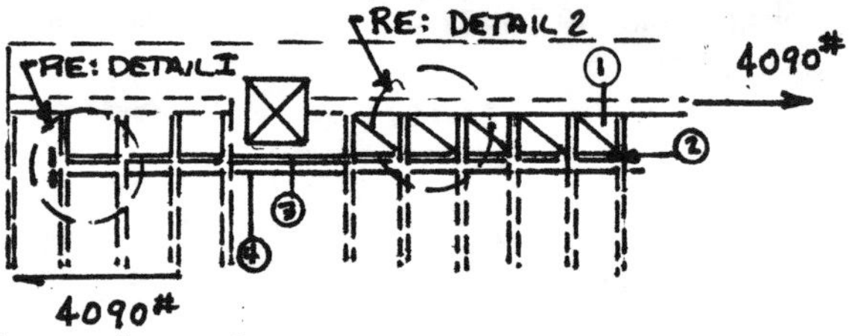

- Install 2×10 blocking beneath the top plate for the 2×6 stud wall 16' long (effectively 14' long with cut out for the duct). Thus, each 2' long blocking must resist:

  F = (4090# / 14') × 2' = 584#

  f_v = F / bl = 584 / (1.5 × (24 – 1.5)) = 17 psi

Toenail with 10d nails: Z' = 81 #/nail

V(22.5) = 584(10)

V = 260#

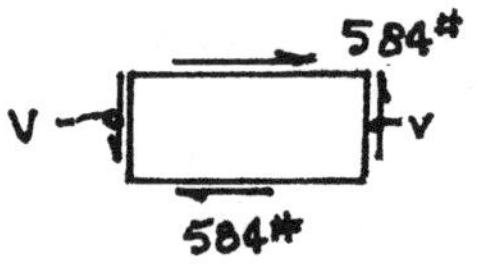

∴ Nails required: 584 / 81 = 7 at top; 260 / 81 = 3 at sides.

- Add a 2×6 block beneath each 2×10 (7 – 10d nails adequate).
- Continue the 2×6 blocking beneath the duct and into the shearwall at the same elevation. This blocking must extend far enough across the shearwall to transfer 4090#. If the same nail pattern is used as for the shearwall boundary, the required length is:

  L = (28 / 40) × 12 = 8.4' (minimum)

  The 2×6 transfers force in compression only:

  f_c = 4090 / (bd) = 4090 / (1.5 × 5.5) = 496 psi
- Provide a steel rod to resist the 4090# tensile force. Drill through the center of studs as close to the new 2×6 as practicable. For an A-36 rod with threaded ends:

  $A_{reg} \geq T / (^4/_3 \times 0.3F_u)$

  $= 4090$ #$/ (^4/_3 \times 0.3 \times 58{,}000) = 0.18$ in^2

Use 1 / 2" diameter rod and try a 2" diameter plate washer:

$f_{c\perp} = 4090 / 2^2 = 1022$ psi – N.G.

Use 3" diameter (9 in^2) plate washers with the 1/2" diameter rod.

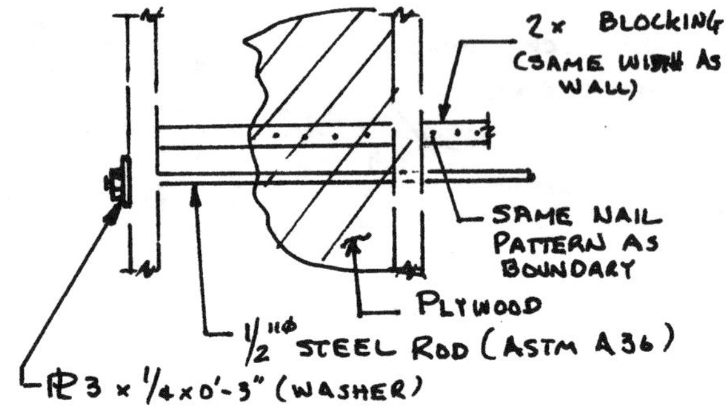

DETAIL-I (N.T.S.)

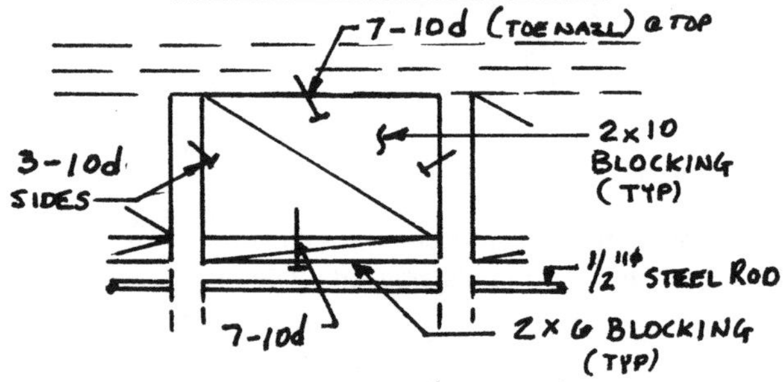

DETAIL-II (N.T.S.)

Timber–4

a) Design the horizontal roof diaphragm using UBC-97. For dead loads for seismic forces in the north-south direction:

- Main roof (130' × 40' × 16 psf) = 83.2^k
- Corridor roof (2(150 + 40)' × 10' × 12 psf) = 45.6^k
- Parapets (2(130 + 40)' × 5' × 10 psf) = 17.0^k
- Exterior Walls (2×130' × (9 / 2)' × 12 psf) = 14.0^k
- Interior Walls (130' × (9 / 2)' × 12 psf = 7.0^k

  Total Dead ≅ 167^k

Lateral seismic force in the north-south direction (1630.2, UBC):

$V = (2.5C_aI / R)W = (2.5 \times 0.36 \times 1.0 / 5.5)167$
$= 27.3^k$

Divide by 1.4 to convert the design-level base shear to working stress level:

$V_e = V / 1.4 = 27.3 / 1 / 4 = 19.5^k$

Wind loads:

$p = C_eC_qq_sI_w$

Where q_s = 16.4, say 17 psf; C_e = 1.2 (exp. C, H ≤ 25'); I_w = 1.0; C_q = 1.3 on building and parapets.

V_w = (1.3 × 1.2 × 17 × 1.0)(150' × 4.5' + 2 × 130' × 5') = 52,400# = 52.4^k

∴ **<u>Wind governs: V = 52.4^k</u>**

Diaphragm shear:

$\nu = V / L$

ν = 52,400 / 150 = 350 plf

For a flexible diaphragm the load to the shearwall is based on the tributary area.

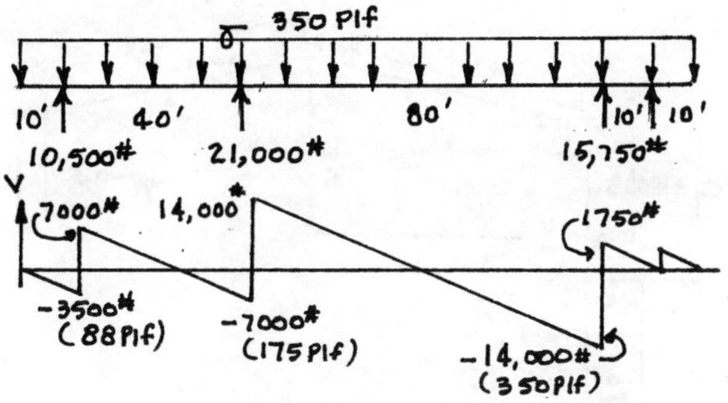

DIAPHRAGM SHEAR (#; Plf OVER 40')

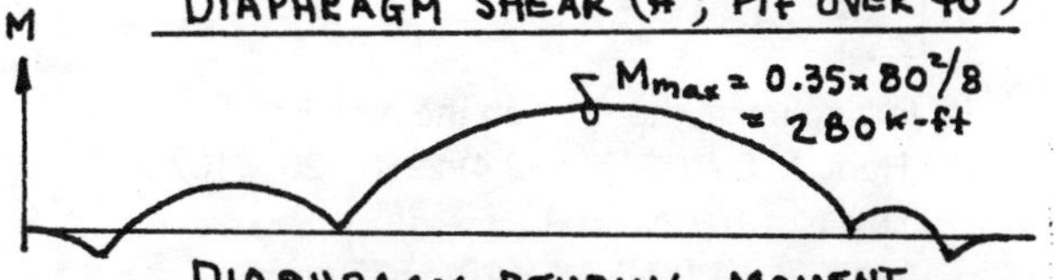

DIAPHRAGM BENDING MOMENT

Maximum diaphragm shear of 350 plf occurs along lines 'B' and 'C'. For a 1/2" Douglas Fir-Larch plywood (Re: Table No. 23-II-H, UBC-97): assume Grade C-C or equivalent; blocked; 8d nails at 4" o.c. along the boundary and continuous panel edges (case 4); 6" o.c. at the other edges; eliminate blocking where $q \leq 240$ plf.

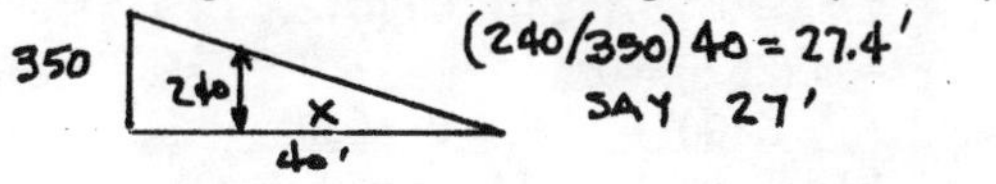

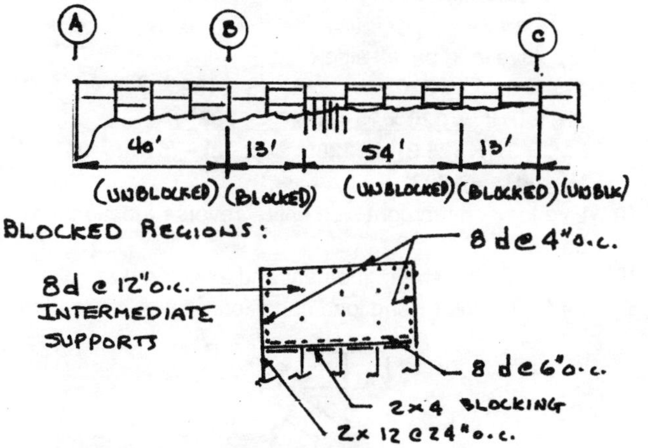

The maximum chord force is:

$M_{max} = 0.350\ (80)^2 / 8 = 280\ ^{k\text{-}ft}$

$T = C = M_{max} / 40' = 280 / 40 = 7.0^k$

Chords: 4×12 ($A = 3.5 \times 11.25 = 39.4\ in^2$)

Assume 3/4" diameter bolts with steel side plates are used to splice the chords.

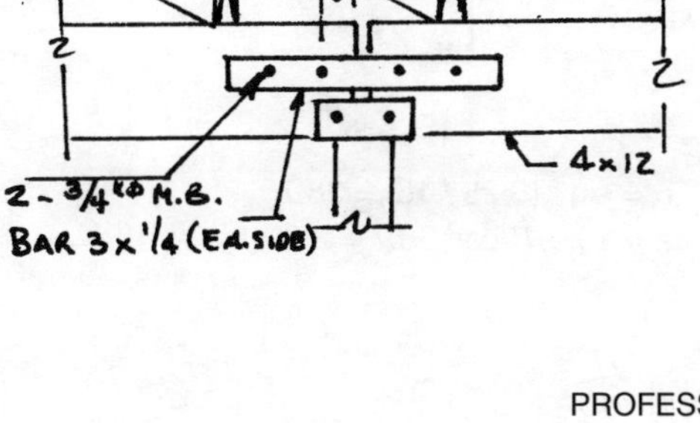

Try BAR 3 × 1/4 each side (ASTM A36 steel) for tension:

$A_n = (3 - {}^7/_8)\ 0.25 \leq 0.85A_g = 0.53\ in^2$ (Re: AISC ASD, 1989)

$$T \leq \begin{cases} (4/3)(0.6F_y)A_g = (4/3)(22)(3)(.25) = 22^k \\ (4/3)(0.5F_u)A_e = (4/3)(0.5 \times 58)(0.53) = 20.5^k \end{cases}$$

For compression: $kl = (14d_b + 0.5)$; $r = 0.3t$

$kl / r = (14 \times 0.75 \times 0.5) / (0.3 \times 0.25) = 147 > C_c$

$\therefore \quad F_a = ({}^4/_3)({}^{12}/_{23})\pi^2 E / (kl / r)^2 = 9.2\ ^{ksi}$

$P_a = 9.2 \times 3 \times 0.25 = 6.9\ ^k/_{bar}$ (13.8^k total)

For 3/4" bolts; 4" main member; steel side plates (Table 8.3B, NDS-91): $Z' = ({}^4/_3)(4290\#/_{bolt}) = 5600\ ^{\#}/_{bolt}$

$\therefore \quad n \geq 7.0 / 5.6\ ^k/_{bolt}$ Use 2 – 3/4" diameter machine bolts with 3 x 1/4" steel side plates.

b) Design the shearwall at line B:

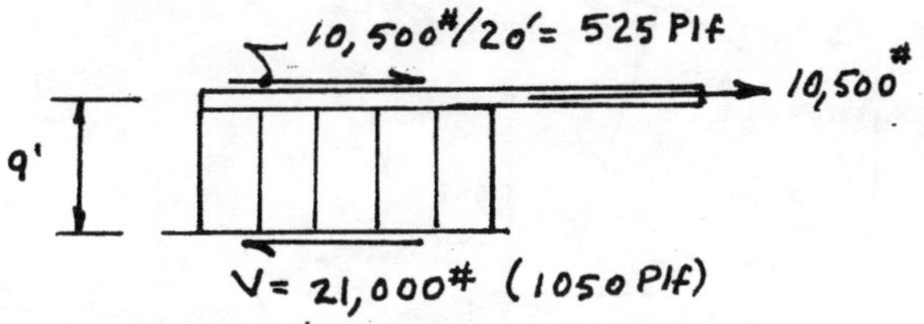

Provide plywood sheathing on both faces, thus, each face must transfer 525 plf (Re: Table No. 23-II-1-1, UBC-97, select 1/2" structural I plywood with 8d at 3" o.c. into nominal 2× sills and plates (furnishes 550 plf); vertical panels with 8d at 6" o.c. on intermediate studs at 24" o.c.

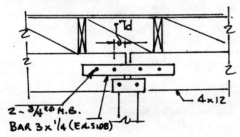

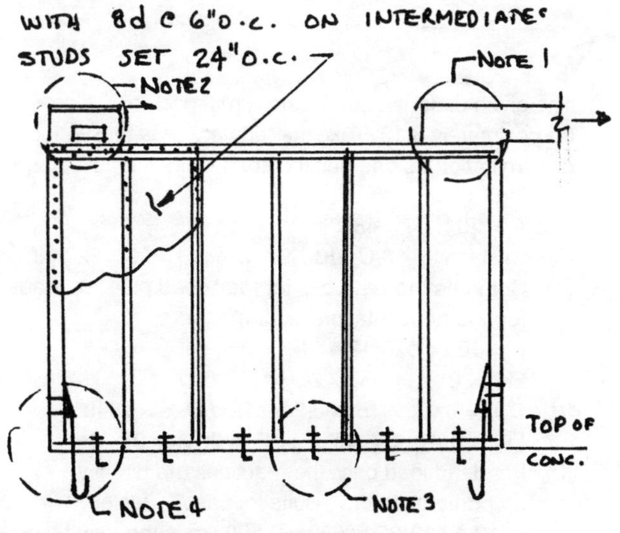

ELEVATION (N.T.S.)

Note 1: Connect collector to the top plate to transfer 10,500#.

Note 2: Provide connection from horizontal diaphragm to top plates to transfer 525 plf.

Note 3: Provide anchor bolts to transfer 21,000# (e.g., for 3/4" diameter bolts in 2" sill, Table 8.2A, NDS-91, single shear parallel to grain for $t_m = 2 \times 1.5 = 3$":

$Z = 4 / 3(1190) = 1590\ ^{\#}/_{bolt}$ (not critical in concrete)
$\therefore$ $n \geq 21{,}000 / 1590 = 13$ bolts).

Note 4: End posts and tiedown anchors to resist the overturning of $M_0 = 21{,}000(9) = 189{,}000^{\#\text{-ft}}$. If the dead load is neglected, $T = C = M_0 / 20 = 189{,}000 / 20 = 9450^{\#}$. Therefore, must provide anchorage and end post designed for $9450^{\#}$.

Timber-5

a) Compute the frame reactions for the given loading conditions.

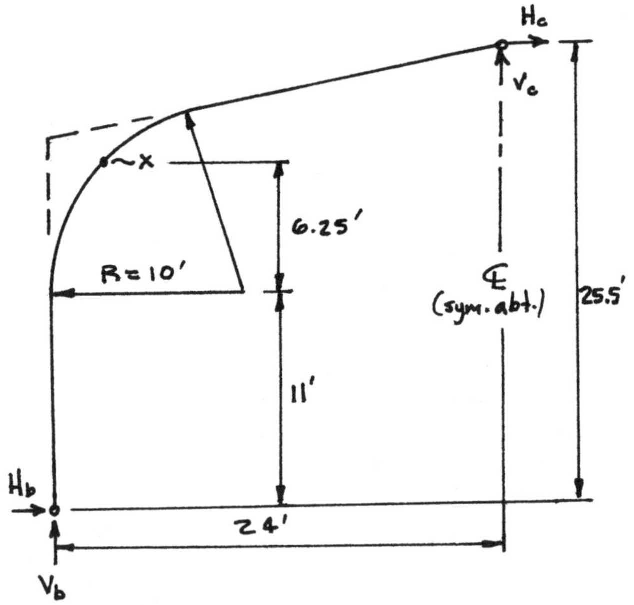

The given dimensions indicate a roof pitch of approximately 3:12. Take the tributary area for live load determination as one-half the span times the 20' spacing:

$A_t = 20 \times 24 = 480\ \text{ft}^2$

From Table No. 16-C, UBC-97, $w_L = 16 \times 20 = 320$ plf.

(1) Compute the reactions for total dead plus live loads (on the horizontal projection):

$w = 320 + 320(12.4 / 12) = 650$ plf

$\Sigma M_L = 0$: $V_b = 650 \times 24 = 15{,}600$ lb

$\Sigma M_C = 0$: $H_b = 15{,}600(24 - 12) / 25.5 = 7340^{\#} \rightarrow$

$\underline{\Sigma F_x = 0: H_c = 7340^{\#} \leftarrow; \Sigma F_y = 0: V_c = 0}$

(2) For dead load only, the reactions are directly proportional to the results for case 1 (factor $= (12.4 / 12)320 / 650 = 0.509$ times the results for part 1):

$\therefore$ $\underline{V_b = 7940^{\#}; V_c = 0; H_b = -H_c = 3740^{\#} \rightarrow}$

(3) Wind from the left for a 80 mph basic wind speed (neglect uplift):

$q_s = 16.4$, say 17 psf

Take $C_e = 1.13$ (mean roof height $\cong$ 20')

Assume an importance factor $I_w = 1.15$

$p = C_e C_q q_s I_w = 1.13 \times 17 \times 1.15 = 22.1 C_q$

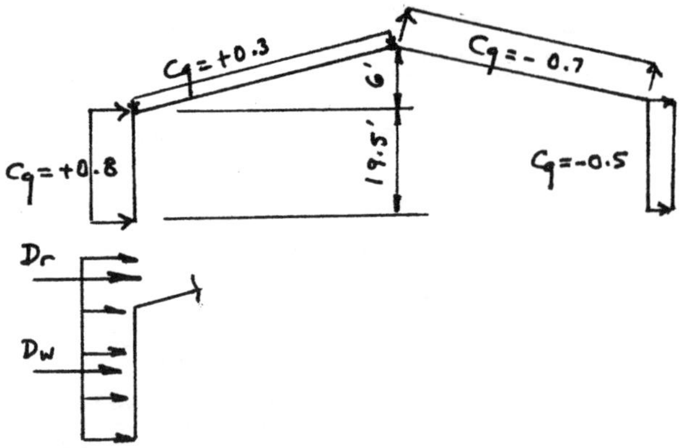

Neglecting uplift, the drag on the arch is:

Roof: $(0.3 + 0.7)(3 / 12.4)\ 22.1 \times 20 = 107$ plf

$D_r = 6' \times 107\ \text{plf} = 642^{\#}$

Wall: $(0.8 + 0.5)22.1 \times 20 = 575$ plf

$D_w = 575 \times 19.5 = 11{,}213^{\#}$

Take $(0.8 / 1.3) \cong 0.6$ of the drag on the windward side and the remaining 0.4 on the leeward side:

$\Sigma M_L = 11{,}213(19.5 / 2) + 642(22.5) - V_b(48) = 0$

$\underline{V_b = 2578^{\#}}$ (↑ on leeward; ↓ windward)

Take left side as a freebody:

$\Sigma M_C = 0$: $2578(24) + 0.6(642 \times 3 + 11{,}213 \times 15.75) - H_b(25.5) = 0$

$\therefore$ **$\underline{H_b = 6630^{\#} \leftarrow; \Sigma F_x = 0: 5230^{\#} \leftarrow}$**

(4) Wind from the right: simply reverse the results for part 3.

(5) Live load on left side only:

$\Sigma M_L = 0$: **$\underline{V_{br} = 320 \times 24 \times 12 / 48 = 1920^{\#} \uparrow}$**

$\Sigma F_y = 0$: **$\underline{V_b = 320 \times 24 - 1920 = 5760^{\#} \uparrow}$**

$\Sigma M_c = 0$: **$\underline{[H_b(25.5)] + (320 \times 24 \times 12) - (5760 \times 24) = 0}$**

$\underline{H_b = -H_{br} = 1807^{\#} \rightarrow}$

f) Live load on the right side only – reverse reactions for part 5.

b) Calculate the shear, moment and axial force at point X for full live plus dead load condition:

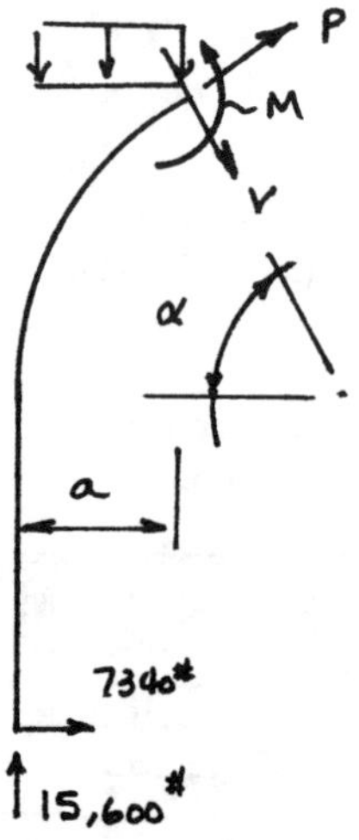

$\alpha = \sin^{-1}(6.25 / 10) = 38.7°$

$a = 10 - 10\cos 38.7 = 2.2'$

Let F_y = sum of applied forces in the y-direction

F_x = sum of the applied forces in the x-direction

$F_y = 15{,}600 - 650 \times 2.2 = 14{,}170^{\#}\uparrow$

$F_x = 7340 \rightarrow$

$\therefore$ $P(\sin 38.7°) - V(\cos 38.7°) + 14{,}170 = 0$

$P(\cos 38.7°) + V(\sin 38.7°) + 7340 = 0$

$P = -14{,}588^{\#} = 14{,}588^{\#}$(c)

$V = 6467^{\#}$

$M_X = 15{,}600(2.2) - 650(2.2)^2 / 2 - 7340(17.25)$

$= -93{,}900^{\#\text{-ft}}$

c) Determine the required depth at point X using the given values: $M = 100^{\text{k-ft}}$; axial force $= 18.4^{\text{k}}$; shear $= 2^{\text{k}}$. Laminations are 3/4" thick with an inside radius of 10' (Re: p. 5 – 127, AITCM, 1985):

$C_c = 1 - 2000\,(t / R)^2 = 1 - 2000\,((3/4)/120)^2$

$= 0.92$

$C_D = 1.25$ (7 days; roof live load)

Specify a 24F combination with $F_b = 2400$ psi; $F_v = 145$ psi; $F_c = 1600$ psi; $E = 1{,}600{,}000$ psi. Check the interaction equation: $f_c / F_c' + f_b / F_b' \le 1.0$.

Assume that lateral bracing is such that le / d < 11 at point X:

$\therefore$ $F_c' = C_D F_C = 1.25 \times 1600 = 2000$ psi

For bending, try $C_F = 0.85$

$F_b' = C_C C_D C_F F_b = 0.92 \times 1.25 \times 0.85 \times 2400$

$= 2345$ psi

Take b = 7" (given): $A = 7d$; $S = 7d^2 / 6$

$f_c = P / A = 18{,}400 / 7d = 2629 / d$

$f_b = M / S = 100{,}000 \times 12 / (7d^2 / 6) = 1{,}028{,}571 / d^2$

$f_c / F_c' + f_b / F_b' \le 1.0$

$(2629 / d) / 2000 + (1{,}028{,}571 / d^2) / 2345 \le 1.0$

$d^2 - 1.13d - 438.6 \ge 0$

$d \ge 21.6$" (say 29-3/4" lams, which gives an actual depth $= 29 \times 0.75 = 21.75$")

$C_F = (12 / d)^{1/9} = (12 / 21.75)^{1/9} = 0.94$

Thus, the calculated C_F is significantly larger than the assumed value, 0.85. Recalculate using a shallower section with the higher allowable bending stress.

Try 28 lams: d = 21"; A = 147 in^2; S = 515 in^3

$f_c = 18{,}400 / 147 = 125$ psi

$f_b = 100{,}000 \times 12 / 515 = 2330$ psi

$C_F = (12 / d)^{1/9} = (12 / 21.0)^{1/9} = 0.94$

$F_b' = 0.92 \times 1.25 \times 0.94 \times 2400 = 2590$ psi

$f_c / F_c' + f_b / F_b' = 125 / 2000 + 2330 / 2590$

$= 0.96$ – O.K.

Horizontal shear stress is nil at point X.

$\therefore$ **Use 28 – 3/4" lams (d = 21.0")**

d) The thickness at the crown is governed by horizontal shear stress. Check dead plus live load on the half span:

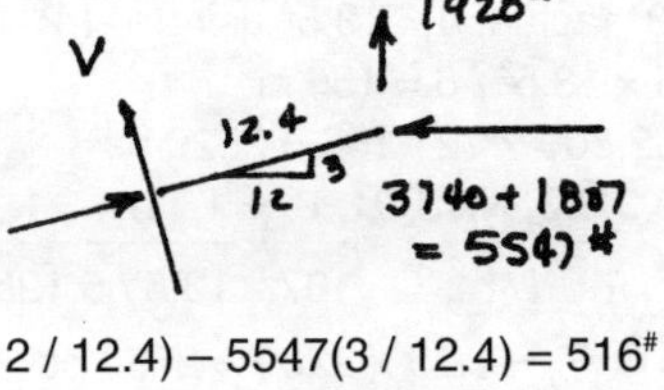

$V = 1920(12 / 12.4) - 5547(3 / 12.4) = 516^{\#}$

Check dead plus wind load:

$V = 2578(12 / 12.4) - (3740 + 483)(3 / 12.4) = 1473^{\#}$

$\therefore$ **Dead plus wind governs strength at the crown (actually, practical considerations usually dictate that d ≥ b, which governs)**

Timber-6

a) Check adequacy of the given connection details. I will assume that the given lateral forces satisfy the special loading required in 1633.2.6, UBC-97.

For the existing connection A under North-South lateral forces:

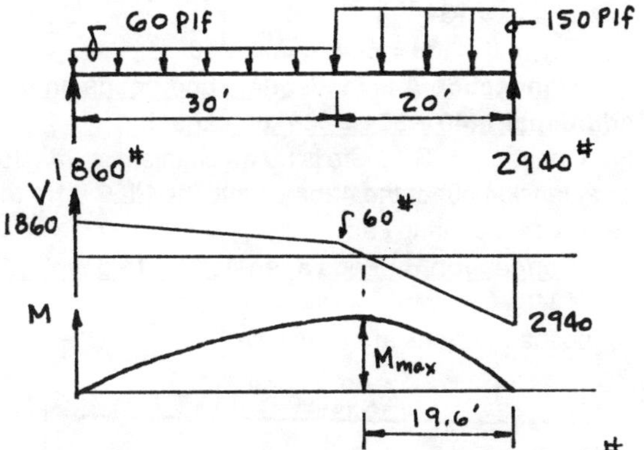

$M_{max} = 0.5 \times 2940 \times 19.6 = 28{,}812^{\#}$

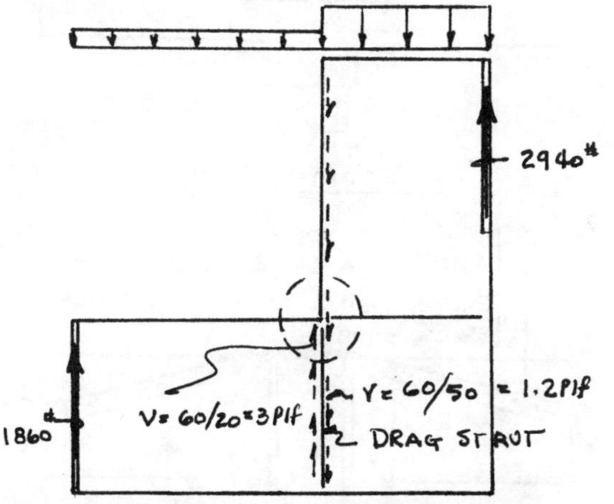

PLAN — N.S. LATERAL FORCES

The force transfer across the connection shown in Section A-A must tie the drag strut for resistance to North-South lateral forces and tie the chord for resistance to East-West lateral forces. Since the structure has plan irregularity type 2, Table 16-M, UBC, section 1633.2.9.7 applies and the seismic forces must be considered to act in the same and in opposite directions on the projecting wings. With forces acting in opposite directions, the tie force that must transfer across connection A-A is:

$(F_A)_{DRAG} = \pm\{[150 \text{ plf} \times 20' \times 10' / 50'](3/5) + [60 \text{ plf} \times 30' \times 15' / 50'](2/5)\} = 576\#$

The force transfer across the connection shown in Section B-B is to tie the diaphragm chord under the North-South lateral forces:

$(F_B)_{CHORD} = M_{max} / 20' = 28{,}812^{\#\text{-ft}} / 20' = \pm\ 1440^{\#}$

The dimensions and applied forces are such that the results for East-West lateral forces are obtained by simply interchanging the above results. Thus, the controlling force is:

$(F_A)_{CHORD} = (F_B)_{CHORD} = \pm 1440^{\#}$

Therefore, both connections must provide a tie for $\pm$ $1400^{\#}$.

Check the splice in Section A-A. Strength is furnished by 6 – 16d nails (3 each side) through 18 gage steel side plates. Since 1633.2.9.6 applies, a 4/3 increase in allowable stress is not permitted. Re: Table No. 12.3F, NDS-91:

$Z' = Z_{//} = 134\ ^{\#}/_{nail}$

$\therefore\ F_{max} = 6 \times 134 = 804^{\#} < F = 1440^{\#}$

$\therefore$ Connection A is inadequate and needs an additional 1440 – 804 = 636# tie capacity.

For Connection "B", there is no mechanism to transfer axial tension out of the 4" beam into the PL 6 × 1/4 that serves as a bearing cap.

$\therefore$ Connection B needs a tension tie to develop 1440#.

b) Proposed revisions are as follows:

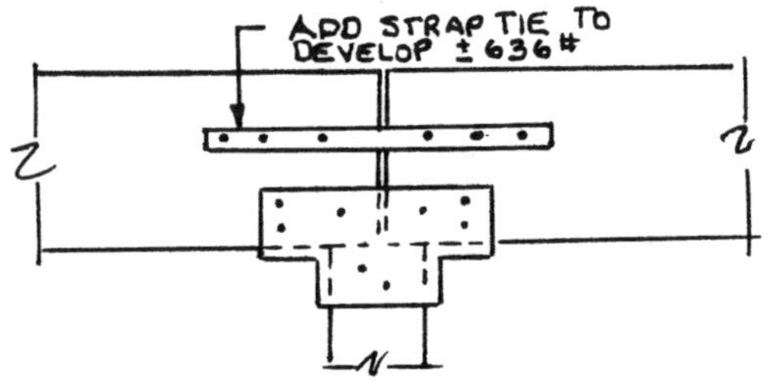

SECTION A-A

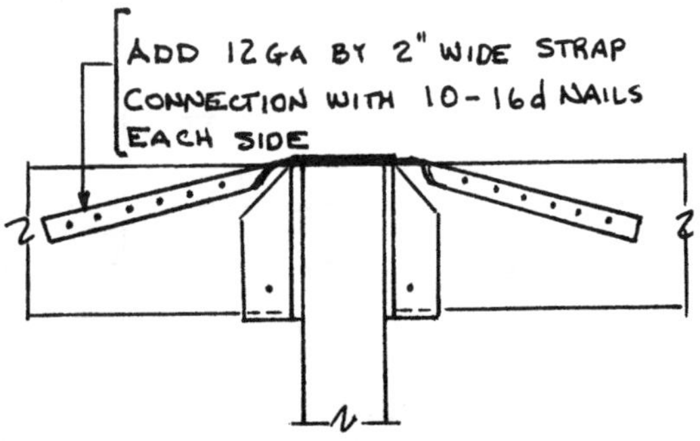

SECTION B-B

Timber-7

a) Calculate the maximum bending stress in the 5-1/8 x 30 glulam beam. The tributary area = 40' × 40' = 1600 ft^2; roof slopes 0.5" in 12"; Re: Table No. 16-C, UBC-97, Take w_L = 12 psf. Given that the loads are uniformly distributed, and assume that the given 10 psf includes the beam weight:

w = (10 psf + 12 psf) 20' = 440 plf

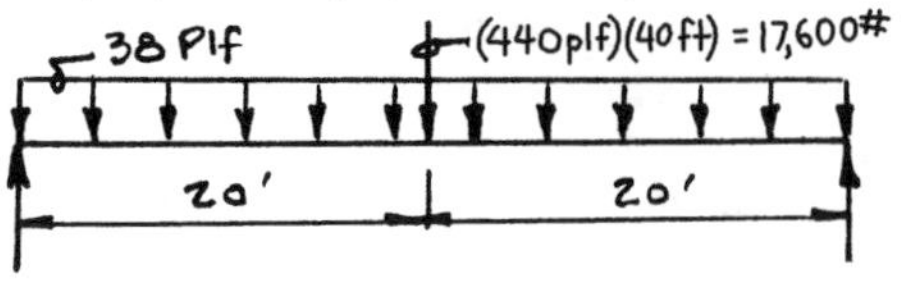

For the $5^1/_8 \times 30$ girder:

$A = 154$ in^2; $S = 769$ in^3; $I = 11{,}530$ in^4;

$w_g = (5.125 \times 30 / 144)\ 36 = 38$ plf

$M_{max} = w_g L^2 / 8 + PL / 4$

$= [38(40)^2 / 8] + [17{,}600(40) / 4] = 183{,}600^{\#\text{-ft}}$

$f_b = M_{max} / S = 183{,}600 \times 12 / 769$

$\therefore$ **$f_b = 2865$ psi**

b) Check adequacy of the beam as modified by the skylight. The skylight eliminates the lateral support to the top (compression) flange. Lateral support is now only at the ends and midspan. Check the slenderness effect, 3.3.3, NDS-91, with $l_u = 20'$; $C_D = 1.25$:

$l_e = 1.11 l_u = 1.11 \times 20 \times 12 = 266"$

$R_B = \sqrt{l_e d / b^2} = (266 \times 30 / 5.125^2)^{1/2} = 17.4$

$F_b^* = C_D F_b = 1.25 \times 2400 = 3000$ psi

$F_{bE} = 0.438E' / R_B^2 = 0.438 \times 1.6 \times 10^6 / 17.4^2$

$= 2315$ psi

$F_{bE} / F_b^* = 2315/3000 = 0.77$

$$C_L = \frac{1+F_{bE}/F_b^*}{1.9} - \sqrt{\left[\frac{1+F_{bE}/F_b^*}{1.9}\right]^2 - \frac{F_{bE}/F_b^*}{0.95}}$$

$$= \frac{1+0.77}{1.9} - \sqrt{\left[\frac{1+0.77}{1.9}\right]^2 - \frac{0.77}{0.95}}$$

$C_L = 0.69$

$\therefore\ F_b' = C_D C_L F_b = 1.25 \times 0.69 \times 2400 = 2070$ psi

$f_b = 2865 > F_b'$ (38% overstress)

$\therefore$ **The proposed skylight modification is bad**

Timber-8

a) Design the lintel using 2200F Glulam and UBC-97. Take the span as 16' (conservative) and provide lateral support at ends and at midspan.

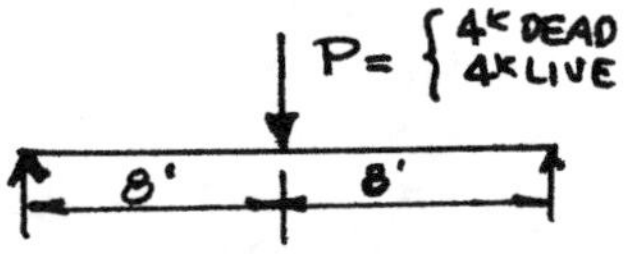

Estimate the member weight as 25 plf:

$M_{max} = 8 \times 16 / 4 + 0.025 \times 16^2 / 8 = 32.8^{\text{k-ft}}$

Compute a trial depth using $b = 5^1/_8$":

$S = 5.125 d^2 / 6$; Take $F_b' \cong C_D F_b$; $C_D = 1.25$

$f_b = M_{max} / S \le F_b' = 1.25 \times 2200 = 2750$ psi

$d^2 = 32{,}800 \times 12 \times 6 / (5.125 \times 2750) = 168$ in^2

Try d = 13" (actual d = 13.5" using 1-1/2" lams);

$S = 5.125 \times 13.5^2 / 6 = 156$ in^3

$f_b = 32{,}800 \times 12 / 156 = 2520$ psi

Re: Table 3.3.3, NDS-91, $l_e = 1.11 l_u = 1.11 \times 96 = 107"$.

$R_B = \sqrt{l_e d / b^2} = \sqrt{107 \times 13.5 / 5.125^2} = 7.4$

$F_b^* = C_D F_b = 1.25 \times 2200 = 2750$ psi

Specify a glulam combination for which $E' = E_{yy} = 1.6 \times 10^6$ psi:

$$F_{bE} = 0.438E' / R_B^2 = 0.438 \times 1.6 \times 10^6 / 7.4^2 = 12{,}500 \text{ psi}$$

$$F_{bE} / F_b^* = 12{,}500 / 2750 = 4.54$$

$$C_L = \frac{1+F_{bE}/F_b^*}{1.9} - \sqrt{\left[\frac{1+F_{bE}/F_b^*}{1.9}\right]^2 - \frac{F_{bE}/F_b^*}{0.95}}$$

$$= \frac{1+4.54}{1.9} - \sqrt{\left[\frac{1+4.54}{1.9}\right]^2 - \frac{4.54}{0.95}}$$

$$C_L = 0.99$$

∴ $F_b' = C_D C_L F_b = 1.25 \times 0.99 \times 2200 = 2720 \text{ psi} > f_b = 2520 \text{ psi}$

∴ $5^1/_8 \times 13.5$ in is OK in bending

Check shear: $V = 4.2^K$; $A = 69.2 \text{ in}^2$

$$f_v = {}^3/_2\, V / A = 1.5 \times 4200 / 69.2 = 91 \text{ psi}$$

∴ Check deflection: limit $\Delta_{LL} < L / 240$

$$\Delta_L = P_L L^3 / 48EI \le L / 240$$

$$I = 5.125 \times 13.5^3 / 12 = 1050 \text{ in}^4$$

$$E \ge 240 \times 4 \times (16 \times 12)^2 / (48 \times 1050) = 700^{ksi}$$

Deflection is no problem because $E > 700^{ksi}$ for all 2200F combinations.

Check bearing assuming support from the 2×6s at the ends:

$$f_{c\perp} = R / bn = 4200 / (5.125 \times 1.5) = 546 \text{ psi}$$

∴ **Use $5^1/_8$×13.5 in. (2200F) with $F_{c\perp}' > 550$ psi (e.g., 22F-V3, Table 5A, NDS-91).**

b) Design member B using No. 1 Douglas Fir-Larch. For nominal 6" dimension lumber, $F_b = C_F \times 1000 \text{ psi} = 1300$ psi; $F_v = 95$ psi; $E = 1.7 \times 10^6$. Assume 70 mph basic wind speed; Exposure C (UBC-97, Sec. 1620)

$p = C_e C_q C_s I_w$ (Take $I_w = 1.0$; $C_q = 1.2$)

$= 1.2 \times 1.2 \times 13 \times 1 = 19$ psf

Member B spans horizontally 16' to transfer wind load from the window above and the 4' high cripple wall below. Estimate the tributary width as 7.5' ∴ w = 19 psf × 7.5' = 143 plf

$$M_{max} = 143 \times 16^2 / 8 = 4580^{\#\text{-ft}}$$

$$F_b' = C_D F_b = {}^4/_3 \times 1300 = 1733 \text{ psi}$$

∴ $S \ge M_{max} / F_b' = 4580 \times 12 / 1733 = 31.7 \text{ in}^3$

$$S = b_e (5.5)^2 / 6 \ge 31.7$$

$$b_e \ge 6.3"$$

Rather than a solid sawn member that fits within a 6" stud wall to resist this lateral loading, try a flitch beam with steel plates between 2 - 2×6: $n = E_s / E_w = 29 / 1.7 = 17$

∴ $b_s = (6.5 - 2 \times 1.5) / 17 = 0.20"$; try a 1/4" steel plate.

Check deflection:

$$I_e = 2 \times (1.5 \times 5.5^3 / 12) + (17 \times 0.25 \times 5.5^3 / 12) = 100 \text{ in}^4$$

$$\Delta_{LL} = 5wl^4 / 384EI_e$$

$$= 0.013 \times (0.143 \times 16)(16 \times 12)^3 / (1700 \times 100) = 1.2"$$

$$= l / 155 \text{ (excessive)}$$

Say $\Delta_{max} = l / 240$ ∴ $I_e \ge (240 / 155) \times 100 = 155 \text{ in}^4$

∴ $b_e > 12 \times 155 / 5.5^3 = 11"$ ∴ $b_s \ge 0.48"$ use 1/2" plate.

Check shear (assume steel plate resists all shear):

$$V = 0.143 \times 8 = 1.14^k$$

$$f_v = V / td = 1.14 / (5.5 \times 0.5) = 0.4^{ksi} \ll F_v = 0.4F_y$$

∴ **Use 2 - 2×6 w/ PL 5 × 1/2**

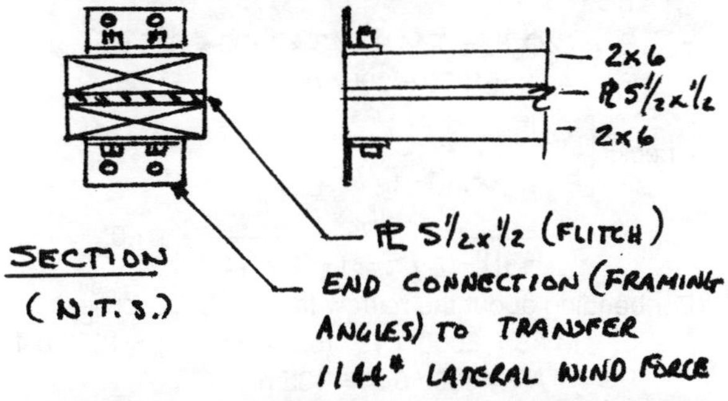

c) Design member C. Assume that bending induced by the reaction from member B is resisted by member D (in combination with the plywood sheathing). Design member C to carry the vertical reaction from A ($R = 4.2^k$) braced in plane:

∴ $l_e / d = (12 - 1.5) \times 12 / 5.5 = 23$; $F_c = 1300$ psi; $E = 1.7 \times 10^6$ psi

$$F_{cE} = \frac{0.3E'}{(l_e/d)^2} = \frac{0.3 \times 1{,}700{,}000}{23^2} = 964 \text{ psi}$$

$$F_c^* = C_D C_F F_c = 1.25 \times 1.1 \times 1300 = 1787 \text{ psi}$$

$$F_{cE} / F_c^* = 964 / 1787 = 0.54$$

$$C_p = \frac{1+F_{cE}/F_c^*}{2c} - \sqrt{\left[\frac{1+F_{cE}/F_c^*}{2c}\right]^2 - \frac{F_{cE}/F_c^*}{c}}$$

$$= \frac{1+0.54}{1.6} - \sqrt{\left[\frac{1+0.54}{1.6}\right]^2 - \frac{0.54}{0.8}}$$

$$C_p = 0.46$$

∴ $F_c' = C_p C_D C_F F_c = 0.46 \times 1.25 \times 1.1 \times 1300 = 822$ psi

$$A \ge R / F_c' = 4200 / 822 = 5.10 \text{ in}^2$$

∴ **Use 2x6 No. 1 Douglas Fir-Larch**

d) Design member D. There are two cases to consider: 1) dead plus the wind reaction from member B, and 2) dead plus or minus the seismic load. For case 1, check combined axial and bending stresses per 3.9.3, NDS-91. Use a trial and error approach using design values from Table 4D, NDS (timbers 5" x 5" or larger). Try a 6 x 6 Dense No.1 DF-L: $F_b = 1550$ psi; $F_c = 1100$; $E = 1.7 \times 10^6$ psi.

$P_d = 4^K$ (member C supports R = 4.2 kips)

$V = ({}^2/_3)Hw = 0.76^K$; $M_{max} = 0.76 \times 4 = 3^{k\text{-ft}}$

$$l_e / d = 12 \times 12 / 5.5 = 26$$

$$F_{cE} = \frac{0.3E'}{(l_e/d)^2} = \frac{0.3 \times 1{,}700{,}000}{26^2} = 754 \text{ psi}$$

$$F_c^* = C_D F_c = 4/3 \times 1100 = 1467 \text{ psi}$$

$$F_{cE} / F_c^* = 754 / 1467 = 0.51$$

$$C_p = \frac{1+F_{cE}/F_c^*}{2c} - \sqrt{\left[\frac{1+F_{cE}/F_c^*}{2c}\right]^2 - \frac{F_{cE}/F_c^*}{c}}$$

$$= \frac{1+0.51}{1.6} - \sqrt{\left[\frac{1+0.51}{1.6}\right]^2 - \frac{0.51}{0.8}}$$

$C_p = 0.44$

$\therefore \quad F_c' = C_pC_DF_c = 0.44 \times 4/3 \times 1100 = 645$ psi

$F_b' = 4/3 \times 1550 = 2067$ psi

$$\text{Check: } \left[\frac{f_c}{F_c'}\right]^2 + \frac{f_{b1}}{F_{b1}'[1-(f_c/F_{cE1})]} + \frac{f_{b2}}{F_{b2}'[1-(f_c/F_{cE2})-(f_b/F_{bE})^2]} \le 1.0$$

For bending about the narrow face, only f_{b1} applies:

$f_{b1} = M/S = 3000 \times 12/(5.5 \times 5.5^2/6) = 1298$ psi

$f_c = P/A = 4000/5.5^2 = 132$ psi

$$\left[\frac{132}{645}\right]^2 + \frac{1298}{2067[1-(132/754)]} = 0.80 \le 1.0 \therefore \text{O.K.}$$

For dead plus seismic, take $^2/_3$ of dead load resisting overturning:

$\therefore \quad P_{DT} = ((^2/_3)4 \times 8 - 3.36 \times 12)/8 = 2.4^K$ (ten)

$P_{DC} = 4 + 3.36 \times 12/8 = 9^k$ (comp)

$\therefore$ 6x6 Dense No.1 DF-L is adequate for member D; need $A \ge 9/0.755 = 11.9$ in^2 (say 4x6 is adequate for member E).

e) Design the plywood shearwall panel: $v = 3360/8 = 420$ plf (Note: the 2 x 6 sill plate shown in the problem statement should be increased to a 3 x 6 to provide adequate edge distance for nailing). Re: Table No. 23-II-I-1, UBC-97; For 1/2" plywood; 8d nails @ 3" o.c. (q = 490 plf) use 3 - 4×8 sheets horizontally. Anchor bolts (3×6 sill) Re: 8.2.3, NDS-91: $t_m = 2t_s = 5$"; $Z_{//} = 880\ ^\#/_{bolt}$ for $^5/_8$" diameter bolts.

$\therefore \quad n \ge 3360/(^4/_3 \times 880) = 2.8$ (use 3-5/8" bolts @ 32" o.c.)

Tie-down anchors must resist $T = 2.4^k$ tension at the base of members D and E. Details of the wall are:

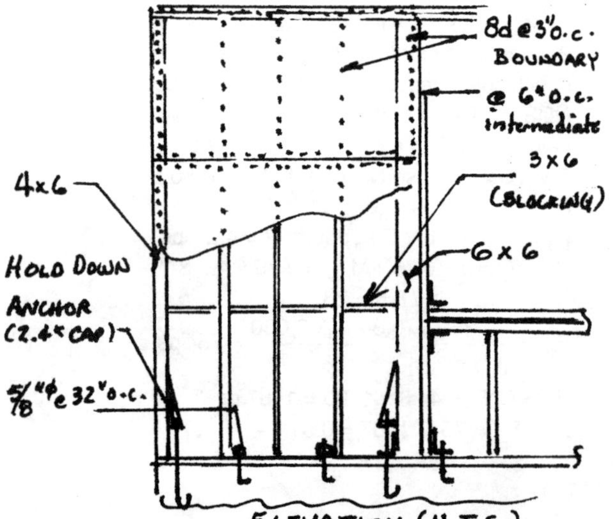

ELEVATION (N.T.S.)

Timber-9

Determine whether 2-3x4s have the equivalent axial load strength as the original 4 x 6. Check the capacity of the 4×6 as originally designed (note: the current NDS-91 does not provide allowable stresses for standard grade dimension lumber 6" or wider. I will assume that the material for the 4×6 and the proposed 2 - 3×4s substitution is the same and use $F_C = 1.1 \times 750 = 825$ psi; E = 1,400,000 psi with $C_D = 1.0$). For the 4×6: A = 19.25 in^2; d = 3.5".

$l_e/d = 12 \times 9/3.5 = 31$

$$F_{cE} = \frac{0.3E'}{(l_e/d)^2} = \frac{0.3 \times 1{,}400{,}000}{31^2} = 437\text{ psi}$$

$F_c^* = C_DC_FF_c = 1.0 \times 1.1 \times 750 = 825$ psi

$F_{cE}/F_c^* = 437/825 = 0.53$

$$C_p = \frac{1+F_{cE}/F_c^*}{2c} - \sqrt{\left[\frac{1+F_{cE}/F_c^*}{2c}\right]^2 - \frac{F_{cE}/F_c^*}{c}}$$

$$= \frac{1+0.53}{1.6} - \sqrt{\left[\frac{1+0.53}{1.6}\right]^2 - \frac{0.53}{0.8}}$$

$C_p = 0.45$

$\therefore \quad F_c' = C_pC_DC_FF_c = 0.45 \times 1.0 \times 1.1 \times 750 = 371$ psi

$P_a = F_c' A = 371 \times 19.25 = 7140^\#$

$P_{d+L} = 3600 + 2400 = 6000^\# < P_a$

$\therefore$ **The 4x6 is adequate**

Check the proposed 2 - 3×4 substitution.

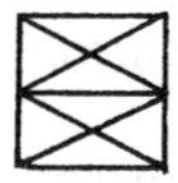

$A = 2 \times 2.5 \times 3.5 = 17.5$ in^2

Re: p. 5-124 AITCM: The strength of a column builtup of 2 or more pieces fastened with nails, bolts or other mechanical fasteners is not equivalent to the strength of a solid sawn section of the same cross section. The difference is caused by shear distortion, which is especially important for long columns. The AITC recommendation is that the strength be computed as the sum of the strengths of the two pieces computed separately. Thus,

$l_e/d = 12 \times 9/2.5 = 43$

$$F_{cE} = \frac{0.3E'}{(l_e/d)^2} = \frac{0.3 \times 1{,}400{,}000}{43^2} = 227\text{ psi}$$

$F_c^* = C_DC_FF_c = 1.0 \times 1.1 \times 750 = 825$ psi

$F_{cE}/F_c^* = 227/825 = 0.28$

$$C_p = \frac{1+F_{cE}/F_c^*}{2c} - \sqrt{\left[\frac{1+F_{cE}/F_c^*}{2c}\right]^2 - \frac{F_{cE}/F_c^*}{c}}$$

$$= \frac{1+0.28}{1.6} - \sqrt{\left[\frac{1+0.28}{1.6}\right]^2 - \frac{0.28}{0.8}}$$

$C_p = 0.26$

$\therefore \quad F_c' = C_pC_DC_FF_c = 0.26 \times 1.0 \times 1.1 \times 750 = 215$ psi

$P_a = 215$ psi $\times$ 17.5 in^2 = $3760^\# < P_{d+L}$

$\therefore$ **The proposed substitution is not adequate**

Timber-10

a) Design the joists spaced 16" o.c. Assume normal load duration ($C_D = 1.0$); exposed to weather: $C_M = 0.85$ (flexure); 0.97 (horizontal shear); 0.67 ($F_{c\perp}$); 0.9 (modulus of elasticity). From Table 4A, NDS-91, for No. 1 or better Douglas Fir-Larch: $F_b = 1.15 \times 1150 = 1320$ psi; $F_v = 95$ psi; $E = 1.8 \times 10^6$ psi; and $F_{c\perp} = 625$ psi.

The tributary area is small ∴ no reduction in live loads.

$w = (100 \text{ psf} + 20 \text{ psf})\ 1.33' = 160$ plf

$M_{max} = wL^2 / 8 = 160 \times 10^2 / 8 = 2000^{\#\text{-ft}}$

$S \geq M_{max} / F_b' = 2000 \times 12 / (0.85 \times 1320) = 21.4 \text{ in}^3$

Try 2 x 10: $C_F = 1.1$ ∴ $F_b' = 1.1 \times 1320 = 1450$ psi is O.K.

Check shear:

$L_e = L - 2d \cong 10 - 2(0.83) = 8.34'$

$V = wL_e / 2 = 160 \times 8.34 / 2 = 667^{\#}$

$f_v = {}^3\!/_2 V / bd = 1.5 \times 667 / (1.5 \times 9.25)$

$= 72 \text{ psi} \leq F_v' = 0.97 \times 95$ psi

∴ O.K. in shear.

Limit live load deflection to ${}^L\!/_{360}$:

$w_L = 100 \text{ psf} \times 1.33' = 133$ plf

$\Delta_L = ({}^5\!/_{384})(w_L L)L^3 / EI \leq {}^L\!/_{360}$

$I \geq 0.013 \times 360 \times (133 \times 10)(120)^2 / (0.9 \times 1.8 \times 10^6)$

$= 55.3 \text{ in}^4$

∴ **Use 2×10 (A = 13.9 in^2; S = 21.4 in^3; I = 99 in^4)**

b) Compute the shear that must be transferred between the two cantilevered beams at their intersection.

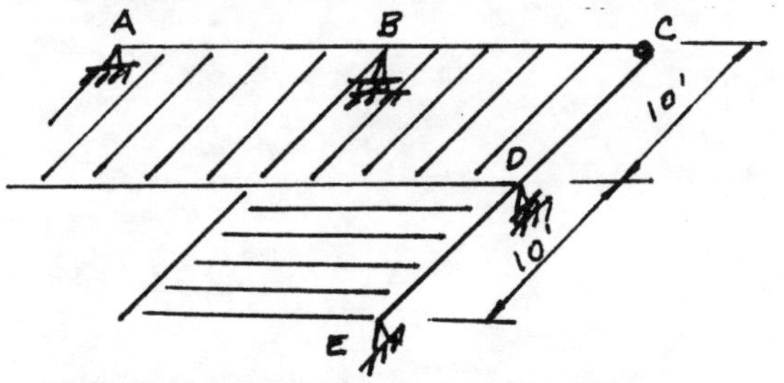

To simplify the representation, develop the two cantilevered beams into a planar idealization:

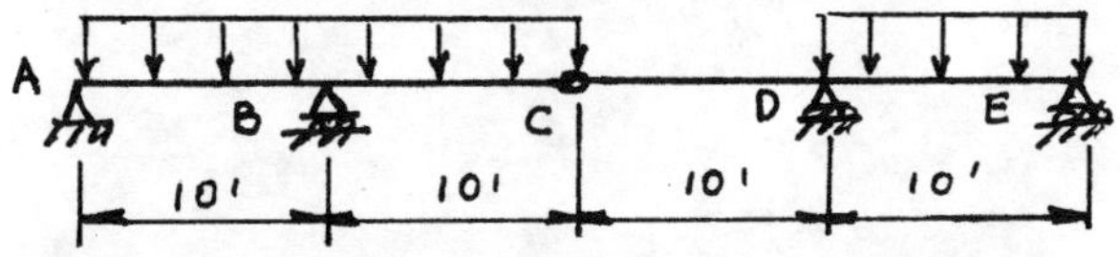

The influence line (qualitative sketch) for maximum shear at C is:

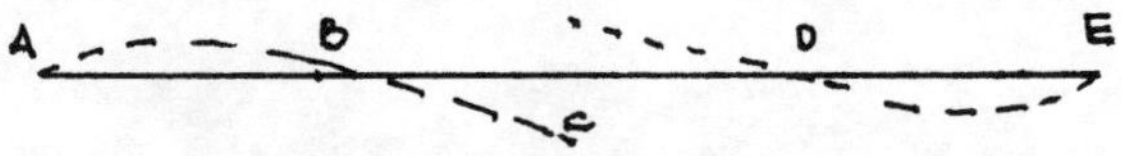

Thus, apply live and dead load on spans $\overline{BC}$ and $\overline{DE}$; dead load only on span $\overline{AB}$.

Tributary area = 10' × 10' = 100 ft^2 < 150 ft^2 ∴ No live load reduction.

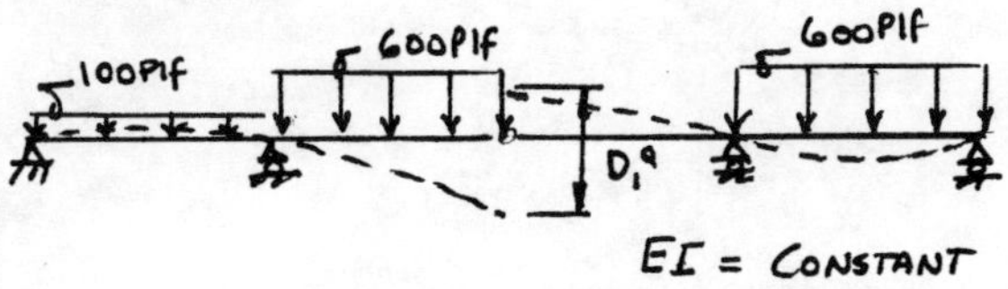

Release the connection at point C; Let D_1^Q = the relative deflection between tips of the cantilevers caused by the applied forces. The deflection at the tip is:

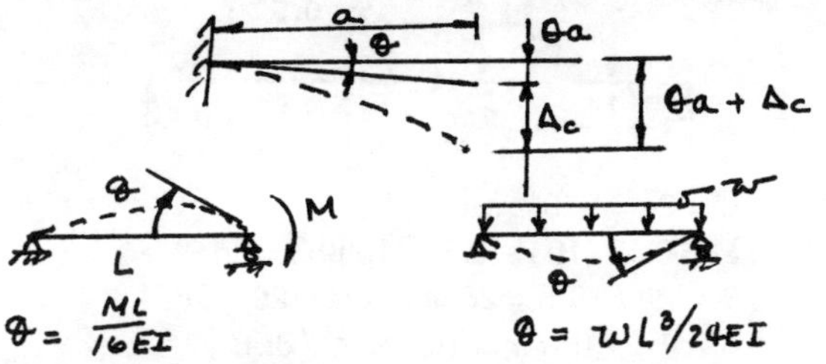

Substituting into the expressions above and simplifying (omitted for brevity):

$D_1^Q = 3.384 \times 10^9 / EI$

Apply a unit shear at 'C' in the released structure:

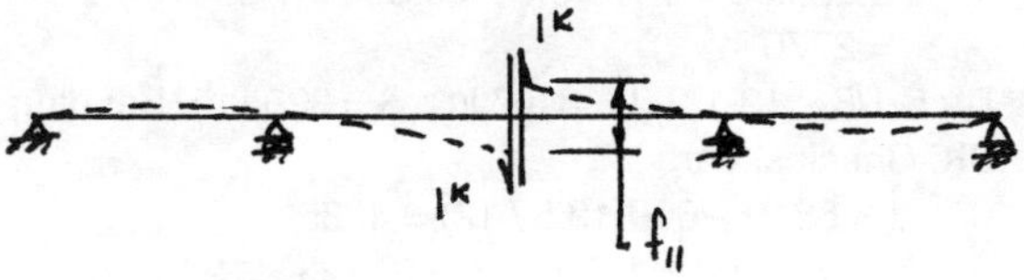

Solve as above:

$f_{11} = 2.302 \times 10^9 / EI$

For consistent displacement:

$D_1^Q + V_c f_{11} = 0$

$V_C = (-3.384 \times 10^9 / EI) / (2.302 \times 10^9 / EI)$

Therefore the required shear is:

$V_C = 1.47^k$

c) Design the glulam beams for appropriate strength and deflection limits. I assume that adequate bracing for bottom flange means that slenderness effects do not have to be considered. We have to repeat the analysis of the indeterminate beams for two additional live load patterns (steps omitted for brevity). Loading 2:

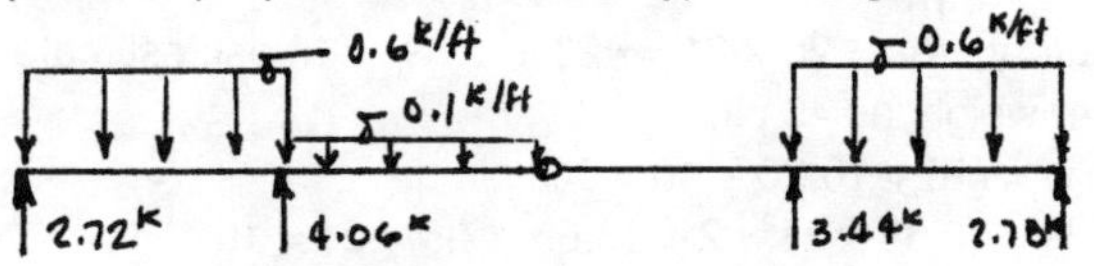

Maximum positive bending occurs in span DE:

$V = 0 \text{ @ } X = 2.78^k / 0.6^{k/ft} = 4.63'$

$M^+ = 0.5 \times 2.78^k \times 4.63' = 6.4^{k\text{-ft}}$

Loading 3:

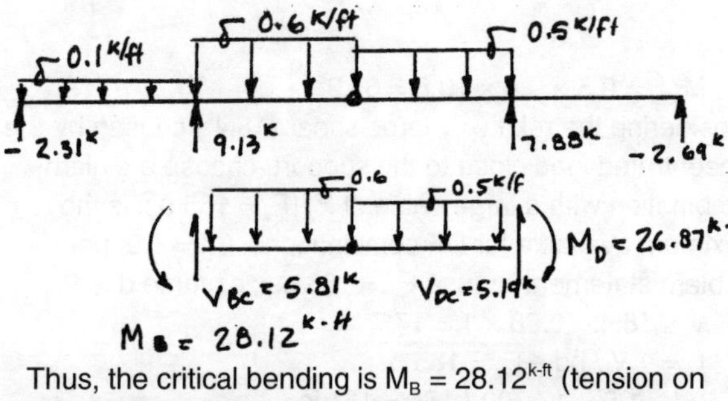

Thus, the critical bending is $M_B = 28.12^{k\text{-ft}}$ (tension on top) and the design shear is $V = V_{BC} - 0.6d$. Try a section assuming $C_F = 0.95$; $C_M = 0.8$: $F_b' = C_F C_M F_b$ = 1824 psi

$S \geq 28{,}120 \times 12 / 1824 = 185 \text{ in}^3$

Limit live load deflection to L' / 360. For loading condition 3:

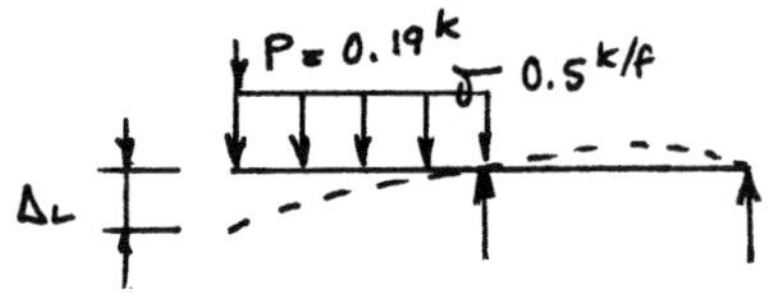

$L' = 2a = 240" \therefore \Delta_L < {}^{L'}/_{360} = 0.67"$

$M = 0.19^k(10') + 0.5^{k/ft}(10')(5') = 26.9^{k\text{-}ft}$

$\theta = ML / 16EI = 26{,}900 \times 12(120) / 16EI$

$\Delta_L = \theta a + (Pa^3 / 3EI) + (w_L a^4 / 8EI)$

$= 2.905 \times 10^8 / EI + 190(120)^3 / 3EI$

$+ (500 \times 10)(120)^3 / 8EI$

$= (2.7388 / EI)\, 10^9$

$\Delta_L < 0.66"$

$\therefore \quad I \geq 2.7388 \times 10^9 / (0.66 \times 0.833 \times 1.8 \times 10^6)$

$= 2770 \text{ in}^4$

Try 5-1/8 × 19-1/2 ($I = 3167 \text{ in}^4$; $S = 325 \text{ in}^3$). Bending is OK. Check shear:

$V = 5810 - 600(19.5 / 12) = 4835^\#$

$f_v = {}^3/_2 V / bd = 1.5 \times 4835 / (5.125 \times 19.5)$

$= 73 \text{ psi} < F_v' \therefore \text{OK.}$

$\therefore$ **Use 5-1/8 × 19-1/2 in.**

Timber-11

Design girder G1. Assume 6" from the outside wall face to the centerline of bearing

$\therefore \quad l = 22 - 1 = 21'$

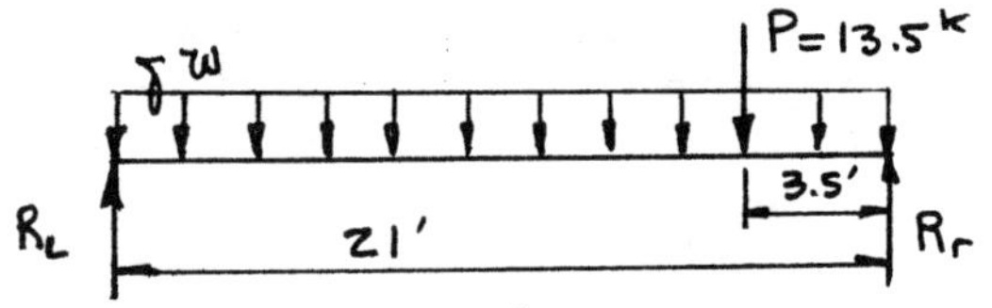

Tributary area = 20 × 21 = 420 ft^2 $\therefore$ w_L = 16 psf; estimate the girder weight as 40 plf.

$w = (16 + 16)20 + 40 = 680 \text{ plf}$

$R_r = (0.68 \times 21^2 / 2 + 13.5 \times 17.5) / 21 = 18.4^k$

$R_L = 13.5 + 0.68 \times 21 - 18.4 = 9.4^k$

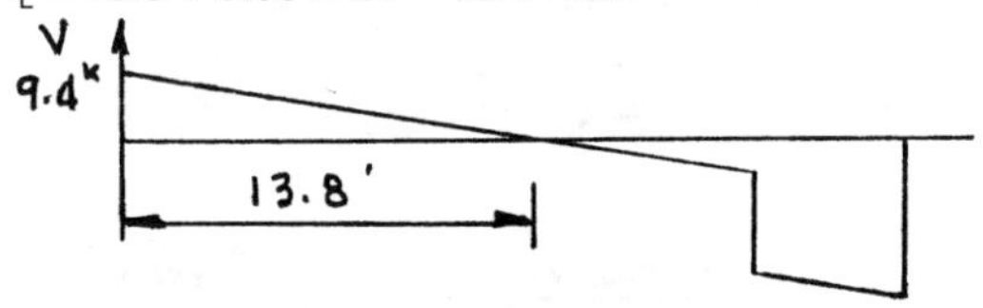

$M_{max} = 0.5 \times 9.4 \times 13.8 = 64.9^{k\text{-}ft}$

Considering the relatively large shear (18.4^k) caused by the concentrated load close to the support, choose a glulam combination with a large value of F_v (F_v = 165 psi is the maximum available for DF combinations; $C_D = 1.0$, per problem statement) $\therefore$ use F_v' = 165 psi; assume d > 1':

$\therefore \quad V \leq 18.4 - 0.68 \times 1 = 17.7^k$

$f_v = {}^3/_2 V / bd \leq F_v = 165 \text{ psi}$

$\therefore \quad bd \geq 1.5 \times 17{,}700 / 165 = 161 \text{ in}^2$

Take b = 6.75":

$d \geq 161 / 6.75 = 23.8$ (say 24 in)

Try 6.75" × 24": $I = 7780 \text{ in}^4$; $S = 648 \text{ in}^3$

$f_b = M_{max} / S = 64{,}900 \times 12 / 648 = 1200 \text{ psi}$

Need $F_b' > 1200$ psi – no problem

Check live load deflection by limiting the computed midspan deflection to L / 360 (plastered ceiling):

$$\Delta_L = \frac{P_L b}{48EI}\left(3L^2 - 4b^2\right) + \frac{5w_L L^4}{384EI} \leq L / 360$$

$P_L = 12.5^k$; $w_L = 0.016 \times 20 = 0.32^{\ k/ft}$

$$\therefore \quad E = \left[\frac{12.5 \times 42\left[3 \times 252^2 - 4 \times 42^2\right] \times 360}{252 \times 48 \times 7780}\right] + \left[\frac{5 \times (0.32 \times 21)(252)^2 \times 360}{384 \times 7780}\right]$$

$= 625^{ksi}$

$E \geq 625{,}000$ psi – no problem

Assume 6" bearing length:

$f_{c\perp} = R / bN \leq F_{c\perp}'$

$F_{c\perp}' > 18{,}400 / (6 \times 6.75) = 450$ psi – OK

$\therefore$ Strength is governed by horizontal shear.

Use $6^3/_4 \times 24$ in DF-Combination with $F_v \geq 165$ psi (e.g., 16F-V6 or equal)

Timber-12

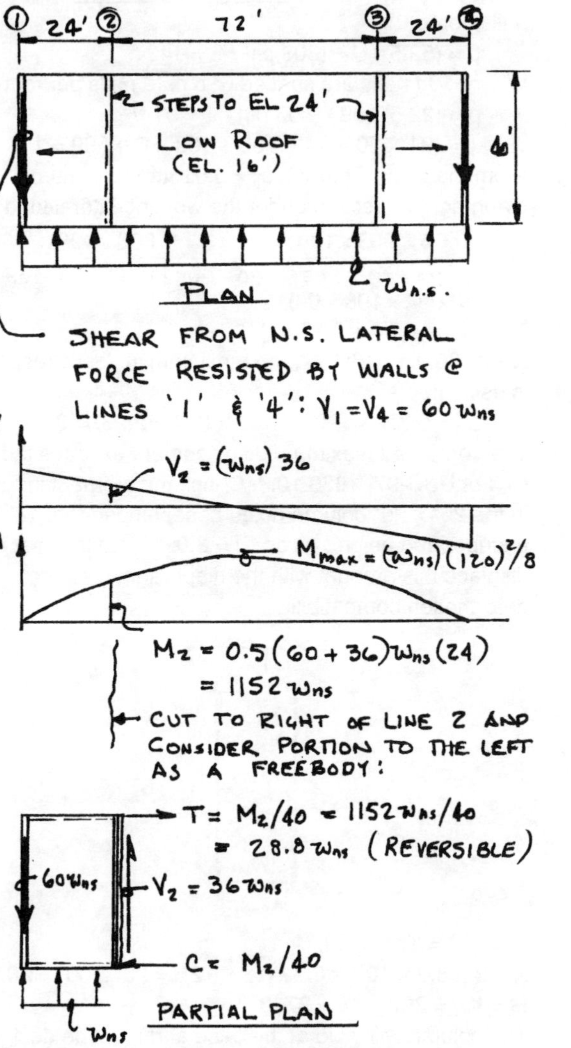

a) List two structural elements needed to maintain diaphragm integrity at the step (lines 2 and 3):
   1. Need a vertical diaphragm to transfer the shear from the low roof diaphragm, $V_2 = 36w_{ns}$, up to the sloped roof diaphragm at elevation 24'.
   2. Must transfer the chord force, $T = C = 28.8w_{ns}$, from the low roof at elevation 16' up to the sloped boundary member 8' above.

b) For a given lateral load w_{ns} = 190 lb / ft, determine the forces to transfer at lines 2 and 3:

   $V_1 = V_4 = 60w_{ns} = 11{,}400^{\#}$

   $V_2 = V_3 = 36w_{ns} = 6840^{\#}$

   $M_{max} = w_{ns}(120)^2 / 8 = 342{,}00^{\#\text{-ft}}$

   $M_2 = M_3 = 1152w_{ns} = 218{,}900^{\#\text{-ft}}$

   The required force transfer for elements of the lateral force-resisting system are as follows:

   1) Low roof diaphragm at lines '2' and '3': $v = V_2 / 40' = 6840^{\#} / 40' = 171$ plf.
   2) Beams at lines '2' and '3' serve as collectors for the 171 plf low roof diaphragm shear and they support the shearwalls that transfer the shear up to elevation 28'.
   3) Shearwalls (2, 8' long) between elevations 16' and 24' on lines '2' and '3'

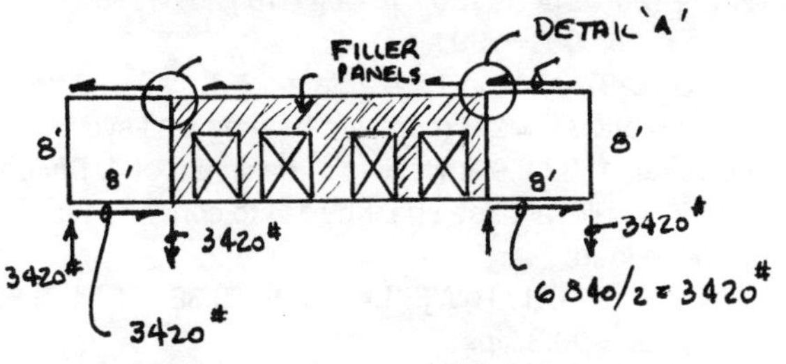

   Shearwall: $v = 3420 / 8 = 428$ plf

   The top plate over the full 40' width will transfer the diaphragm shear, 171 plf, into the sloped diaphragm. Connection of the top plate to the shearwall (detail 'A') must transfer $171^{\text{plf}}(20 - 8)' = 2052^{\#}$.

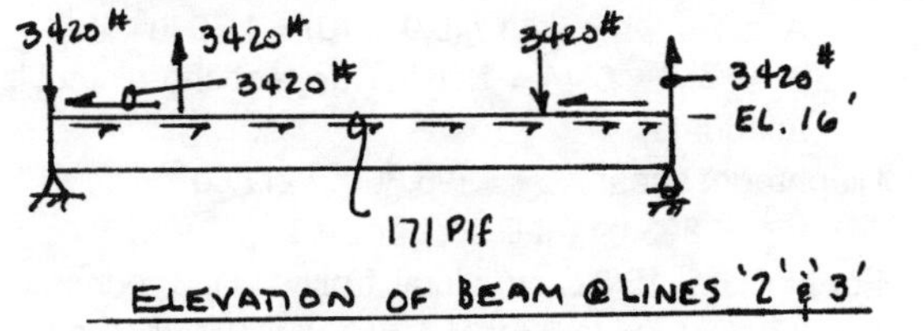

   4) Endwalls (north and south between lines '1' and '2'):

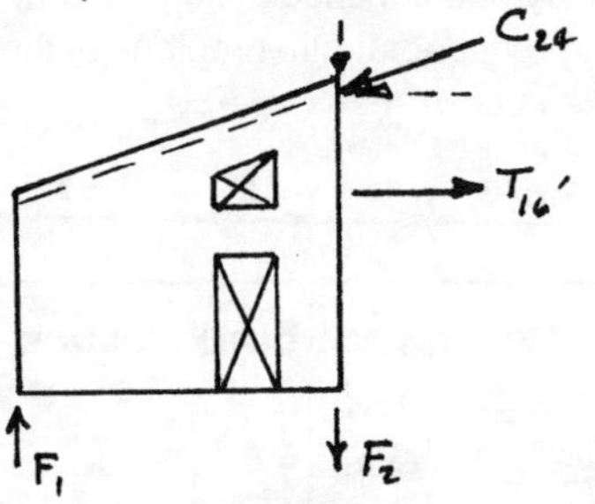

   The chord force, $T_{16} = M_2 / 40 = 218{,}900^{\#\text{-ft}} / 40' = 5480^{\#}$, must be transferred up to elevation 24'. This may be difficult given the location of the window opening. The transfer results in an overturning moment, $8'(T_{16})$, which must be resisted (e.g., by a couple F_1 & F_2 as indicated).

   5) The sloped diaphragm between lines '1' & '2' and '3' & '4' must resist $v = 11{,}400 / 40 = 285$ plf.
   6) The end shearwall at lines '1' and '4' must transfer the $11{,}400^{\#}$ shear from elevation 16' to the base.

c) The stepped diaphragm increases and complicates the force transfers required to bring the lateral force from the roof to the ground. Hence, the cost and failure risks are higher in the stepped than in a horizontal diaphragm at one level. The ideal case would be to maintain the diaphragm at the level of the low roof. If this cannot be done (as is the case for the problem), a better structural solution is to provide shear transfer to the ground at lines '2' and '3'.

Timber-13

a) Determine the diaphragm nailing and chord force for seismic forces in the North-South direction:

W = roof + (contributing weight of wall and columns)
$= 20 \text{ psf} \times 60' \times 198' + 2 \times 0.5\, [(6 \times 12.5) \text{ psf}$
$+ (20" \times 14" \times 150 / 144)\text{plf} / 18']\, 20' \times 198'$
$= 598{,}800^{\#} = 599$ kips

Re: UBC-97, section 1630.2, take $I = 1.0$; $C_a = 0.36$; $R = 4.5$ and $\Omega_o = 2.8$ (concrete bearing wall system); however, 1633.2.9 limits R to a maximum of 4. Divide the design-level base shear by 1.4 to convert to a working load:

$V = [(2.5C_aI / R)W] / 1.4 = [2.5 \times 0.36 \times 1 / 4.0)\, 599]$
$/ 1.4 = 96.3$ kips

$w_{ns} = 96{,}300^{\#} / 198' = 486 \text{ plf} = 0.486 \text{ klf}$

Check the given chord (4 #6; grade 40: $f_y = 40$ ksi)

$M = w_{ns}L^2 / 8 = 0.486 \times 198^2 / 8 = 2381^{\text{k-ft}}$

$C = T = M / b = 2381^{\text{k-ft}} / 60' = 39.7^{\text{k}}$

Use strength design (ACI 318; UBC-97,1612.2):

$T_u = 1.1 \times 1.4 \times 39.7 = 61.1^{\text{k}}$

$A_s \geq T_u / \Phi f_y = 61.1 / (0.9 \times 40) = 1.70 \text{ in}^2$

$A_{s,prov} = 4 \times 0.44 = 1.76 \text{ in}^2$ (area of chord steel is adequate)

Diaphragm shear: $v_{max} = (96{,}300^{\#} / 2) / 60'$

$v_{max} = 803$ plf (at lines 1 and 12)

For blocked, 15/32" structural 1 plywood; 3" nominal framing; Re: Table 23-II-H, UBC: use 10d @ 2" o.c. at the boundary and continuous edges (furnishes 820 plf). The nailing required at other regions of the diaphragm is summarized below:

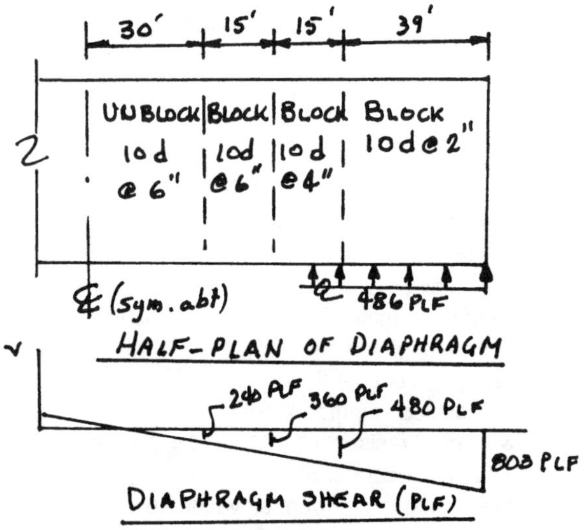

b) Compute the diaphragm deflection using the given equation:

$\Delta = 5vL^3 / 8EAb + vL / 4Gt + 0.094Le_n$

(See problem statement for definition of terms in the expression for Δ). The nail deformation depends on the load per nail, which varies over the length of the diaphragm. Use a weighted average value of the nail force. For the 10d @ 2" o.c. with 6 nails per foot:

$v_{ave} = [(99' - 39' / 2) / 99']\, 803 \text{ plf} = 645 \text{ plf}$

$\therefore \quad p = v_{ave} / 6 = 645 / 6 = 108\ {}^{\#}/_{nail}$

For the region where <u>10d nails are spaced 4</u>" o.c. (3 nails per foot):

$p = (52.5' / 99')(803 \text{ plf}) / 3 = 142\ {}^{\#}/_{nail}$

Where 10d nails are spaced 6" o.c. (2 nails per foot):

$p = (22.5' / 99')\, (803 \text{ plf}) / 2 = 91\ {}^{\#}/_{nail}$

$\therefore \quad \bar{p} = (91 \times 30 + 142 \times 30 + 108 \times 39) / 99 = 113\ {}^{\#}/_{nail}$

From the given Chart G: $e_n = 0.01$, the maximum diaphragm deflection under the working lateral load is:

$$\Delta = \frac{5 \times 803 \times 198^3}{8 \times 29 \times 10^6 \times 1.76 \times 60} + \frac{803 \times 198}{4 \times 110{,}000 \times 0.469} + 0.094 \times 198 \times 0.01$$

$\Delta = 1.27 + 0.77 + 0.19 = 2.23" \rightarrow$

Re: 1630.9.1, UBC-97, the maximum inelastic response is estimated as:

$\Delta_M = 0.7R\, \Delta_s = 0.7 \times 4.0 \times (1.4 \times 2.23) = 8.7" \rightarrow$

The computed maximum displacement exceeds the drift limit of UBC-97, 1630.10.2. Check the moment induced in the 20" × 20" column (fixed base; hinged top; 18' unsupported height; given $EI = 8.0 \times 10^{10}$ lb-in^2) that must displace this amount with the diaphragm (i.e., for deformation compatibility):

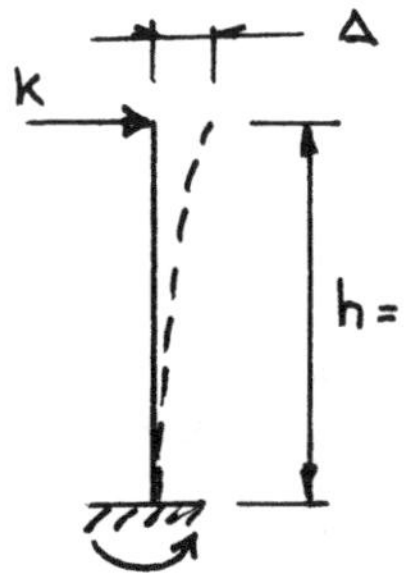

$\Delta = 8.7" = Kh^3 / (3EI)$

$K = 3 \times 8.0 \times 10^{10} \times 8.7 / (18 \times 12)^3 = 207{,}000^{\#} = 207^{\text{k}}$

<u>**$M = Kh = 207 \times 18 = 3730^{\text{k-ft}} >> M_n$**</u>

The column will yield at the base and must be designed to accommodate the inelastic rotation.

c) Check the proposed section C-C for North-South motion. The 3/4" diameter bolts @ 4' o.c. must transfer (East-West ground motion):

$V = [96{,}300^{\#} / (2 \times 198')] \times 4' = 973\ {}^{\#}/_{bolt}$

Re: 8.2.3, NDS-91, the equivalent length of bolt in the main member $= t_m = 2 \times 2.5 = 5"$

$\therefore \quad Z_{//}' = {}^4/_3 \times 1610 = 2150\ {}^{\#}/_{bolt} > V$ ∴ O.K. for transfer of diaphragm shear. However, for North-South ground motion, the proposed detail is not adequate because the transfer of lateral force from the precast wall subjects the ledger to cross-grain tension due to bending (Re: 1633.2.8, UBC-97). For the 6" precast wall (75 psf); $R_p = 3.0$; $a_p = 1.5$:

$F_p = (a_pC_aI_p / R_p)[1 + 3h_x / h_r]W_p$

$F_p = (1.5 \times 0.36 \times 1.0 / 3.0)[1 + 3 \times (18 / 18)]75\text{psf}$
$= 54$ psf

Reaction at the top of the precast wall is:

$V = (54 \times 20^2 / 2) / 18 = 600$ plf

This exceeds the limiting value, 420 plf, of UBC 1633.2.8.1 ∴ $V = 600$ plf.

Need a positive tie to resist the force:

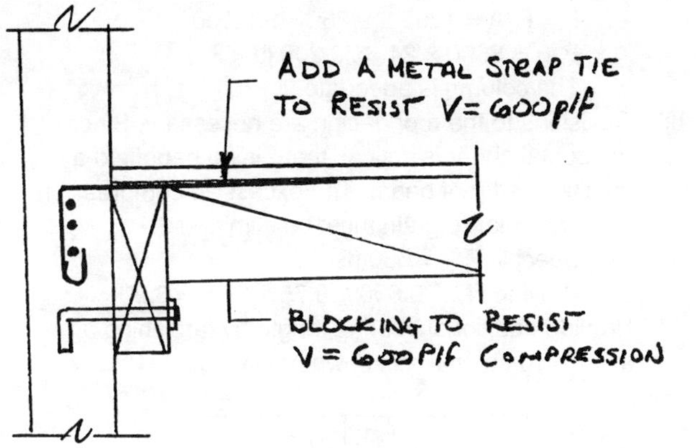

SECTION – 'C' (N.T.S.)

Timber-14

a) Calculate the capacity of the designated 1" diameter bolt. Re: 8.5.4, NDS-91, when the end distance for parallel to grain loading is less than required for full load (i.e., 7d), the design value for all bolts in the group must be taken as the lowest bolt design value in the group. Thus, for the designated bolt (assume dry use; normal load duration; single shear with 3-1/2" main and secondary member thickness; no group action reduction). Re: Table 8.2-A, NDS-91:

$$\underline{Z'_{//} = C_{\Delta}Z_{//} = (6.0 / 7.0)(2260) = 1940^{\#}/_{bolt}}$$

b) Determine the action to be taken on the use of "CDX" plywood for the diaphragm. Section 2315.5.3, UBC-97 requires that plywood used in a roof diaphragm be manufactured using exterior glue. Quoting from the American Plywood Association Construction Guide (APA 1989):

> "All-veneer APA rated sheathing exposure 1, commonly called "CDX" in the trade, is frequently mistaken as an exterior panel and erroneously used in applications for which it does not possess the required resistance to weather." (APA 1989, p.5)

∴ **The diaphragm must be replaced or strengthened.**

c) Illustrate by sketches the terms "toenail" and "slant nail" and give the reduction factor for each. For a toenail:

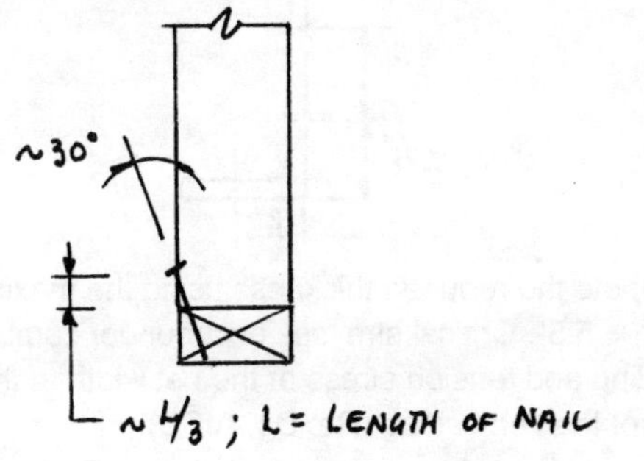

Reduction factor = $^5/_6$ for lateral strength (i.e., lateral strength is tabulated for nails driven perpendicular to grain).

A slant nail is face nailed but penetrates at an angle, θ, to the grain (0° < θ < 90°) as shown:

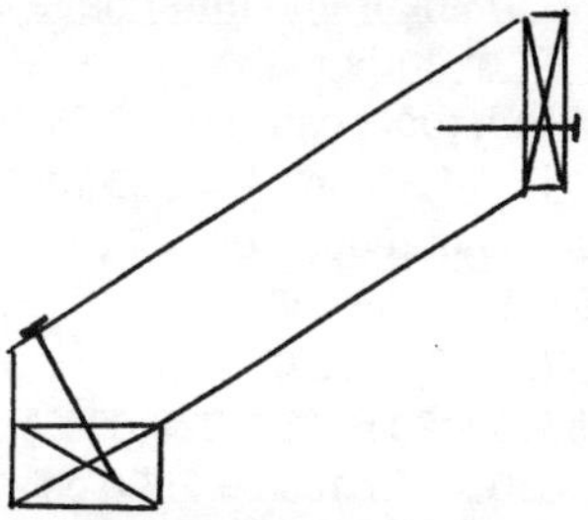

Current design specifications (e.g., NDS-91) provide lateral strength for nails driven perpendicular to grain and require that the strength when driven parallel to grain not exceed $^2/_3$ of the perpendicular to grain value. It seems logical to provide a transition formula (e.g., Hankison's Formula) to compute the strength for θ varying from 0-to-90°. However, the specifications do not provide for the transition for nailed joints. Thus, the conservative approach is to use the parallel to grain value for the slant nail.

d) The minimum dimensions for a clinched nailed are as follows (Re: p. 6 – 493, AITCM):

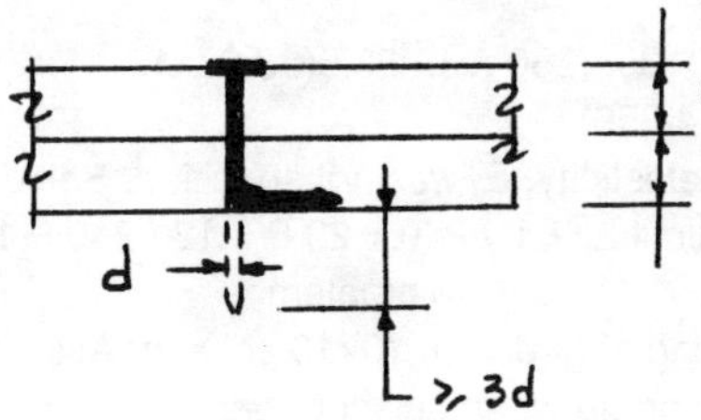

Timber-15

a) Verify adequacy of the design using the given allowable stresses. For the joists: F_b = 1500 psi (assume that the 15% increase permitted for repetitive use applies); snow load (2 month duration): C_D = 1.15; span = 12' – (9 / 12)' = 11.3'.

$M_{max} = wL^2 / 8 = 275 \times 11.3^2 / 8 = 4390^{\#\text{-ft}}$

For 2×12s @ 12" o.c.: A = 16.9 in^2; S = 31.6 in^4; I = 178 in^4.

Take F_v = 95 psi; $F_{c\perp}$ = 625 psi; E = 1,800,000 psi.

$f_b = M_{max} / S = 4390 \times 12 / 31.6 = 1670$ psi

$F_b' = 1.15(1.15 \times 1500) = 1980 \text{ psi} > f_b$ ∴ OK

Shear: $L_e = L - 2d = 11.3 - 2 \times 11.25 / 12 = 9.4'$

$V = wL_e / 2 = 275 \times 9.4 / 2 = 1296^{\#}$

$f_v = {}^3/_2 V / bd = 1.5 \times 1296 / (1.5 \times 11.25) = 115$ psi

$F_v' = 1.15 \times 95 = 109 \text{ psi} < f_v$

Joists are slightly (~5%) over stressed in shear.

Check live load deflection against the limit of L / 240:

$\Delta_L = {}^5/_{384}(w_L L)L^3 / EI \leq {}^L/_{240} = 0.57"$

$\Delta_L = 0.013 \times (250 \times 11.3)(11.3 \times 12)^3$
$/ (1.8 \times 10^6 \times 178)$
$= 0.29" < {}^L/_{240}$ ∴ OK

Support conditions for the joist are not specified. The required bearing length is:

$f_{c\perp} = R / bN \le F_{c\perp} = 625$ psi

$N \ge (275 \times 11.3 / 2) / (1.5 \times 625) = 1.7"$

Thus, 2×12 DF-Larch joists @ 12" o.c. are slightly overstressed in shear and must have bearing length of at least 1.7" at ends.

Check the glulam roof beams: b = 8.75"; d = 27"; S = 1063 in^3; I = 14,350 in^4; A = 236 in^2. The roof live load (250 psf) is not reducible:

$w = 275 \text{ psf} \times 12' = 3300$ plf

$L = 24 - (5 / 12) = 23.6'$

$M_{max} = 3300 \times 23.6^2 / 8 = 229{,}800^{\#\text{-ft}}$

$f_b = 229{,}800 \times 12 / 1063 = 2590$ psi

The roof beam is laterally braced by joists $\therefore$ $C_L = 1.0$

$C_V = K_L(21 / L)^{1/10}(12 / d)^{1/10}(5.125 / b)^{1/10}$

$= 1.0 \times (21 / 23.6)^{1/10} \times (12 / 27)^{1/10} \times (5.125 / 8.75)^{1/10}$

$= 0.86$

$F_b' = C_D C_V F_b = 1.15 \times 0.86 \times 2600 = 2570 \text{ psi} \cong f_b$

$= 2590$ psi

$\therefore$ OK in bending; Check shear:

$L_e = 23.6 - 2 \times 27 / 12 = 19.1'$

$V = 3300 \times 19.1 / 2 = 31{,}500^{\#}$

$f_V = {}^3/_2\,(31{,}500) / 236 = 199$ psi

$F_v' = 1.15 \times 165 = 190$ psi

Slightly overstressed (5%) in shear; Check the live load deflection:

$\Delta_L = 0.013 \times (250 \times 12)23.6(23.6 \times 12)^3 / (E \times 14{,}350)$

Modulus of elasticity, E, was not specified. Solve for E required to limit $\Delta_L < L / 240 = 23.6 \times 12 / 240 = 1.18"$

$\therefore$ $E \ge 1.2 \times 10^6$ psi – no problem

Check bearing of girder on 10×10 column. Assume a 1/2" separation between adjacent beams:

$N = (9.25 - 0.5) / 2$

$R = 3300 \times 24 / 2 = 39{,}600^{\#}$

$f_{c\perp} = R / bn = 39{,}600 / (8.75 \times 4.38) = 1033$ psi

Bearing perpendicular to grain is about twice the allowable!

Check the 10 x 10 column (Note: the column length was not given. I will assume $L_u = 10'$; $F_c = 1250$ psi; E = 1,800,000 psi. Also, I assume the column is braced by struts from the roof at its top).

$L_e = 10 \times 12 = 120"$

$L_e / d = 120 / 9.25 = 13$

$$F_{cE} = \frac{0.3E'}{(l_e / d)^2} = \frac{0.3 \times 1{,}800{,}000}{13^2} = 3190 \text{ psi}$$

$F_c^* = C_D C_F F_c = 1.15 \times 1.0 \times 1250 = 1468$ psi

$F_{cE} / F_c^* = 3190 / 1468 = 2.22$

$$C_p = \frac{1 + F_{cE} / F_c^*}{2c} - \sqrt{\left[\frac{1 + F_{cE} / F_c^*}{2c}\right]^2 - \frac{F_{cE} / F_c^*}{c}}$$

$$= \frac{1 + 2.22}{1.6} - \sqrt{\left[\frac{1 + 2.22}{1.6}\right]^2 - \frac{2.22}{0.8}}$$

$C_p = 0.88$

$\therefore$ $F_c' = C_p C_D C_F F_c = 0.88 \times 1.15 \times 1.0 \times 1250$

$= 1265$ psi

$P = F_c'A = 1265 \times 9.25^2 = 108{,}000^{\#}$

$P_{max} = 3300 \times 24 = 79{,}200 \text{ lb} < P$

$\therefore$ The column is adequate.

b) Revisions to the roof design are necessary. Since horizontal shear is critical, there is no benefit to a cantilevered roof beam. The excessive compression perpendicular to grain must be eliminated:

1) Specify $F_{c\perp} = 650$ psi

2) Provide $N \ge 39{,}600 / (8.75 \times 650) = 6.9"$

Provide additional bearing length by furnishing a cap on the 10×10 column as shown below:

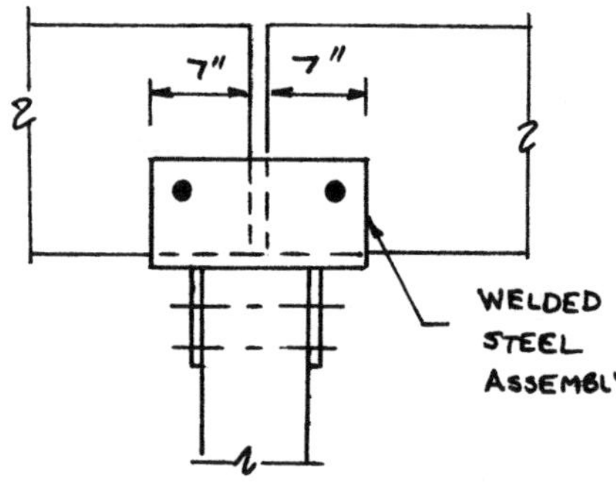

BEAM-TO-POST CONNECTION (N.T.S.)

Timber-16

a) Design the hold-down anchor to resist the $4000^{\#}$ uplift force. Assume the given force is due to wind or seismic and that the allowable stresses may be increased by ${}^1/_3$. For the bolts connecting the anchor to Douglas Fir-Larch timber: Re: Table 8.2B, NDS-91, try 3/4" diameter bolts: $Z'_{//} = {}^4/_3(1580) = 2100\ {}^{\#}/_{bolt}$ $\therefore$ $n \ge 4000 / 2100$ – **Use 2 – 3/4" dia bolts.**

For the anchor bolt: Re: Table No. 19-D, UBC-97; assume no special inspection: for 1" diameter bolt with a minimum 6" embedment: $T \le {}^4/_3 \times 3250 = 4333^{\#}$ is OK.

$\therefore$ **Use 1" diameter A307 anchor bolt**

For the angle, use an unstiffened angle positioned as shown in the problem sketch. Use allowable stress design (AISC 1989) and ASTM A36 Steel. Based on the required gages for the bolts, use an L7×4×t, where t is to be determined.

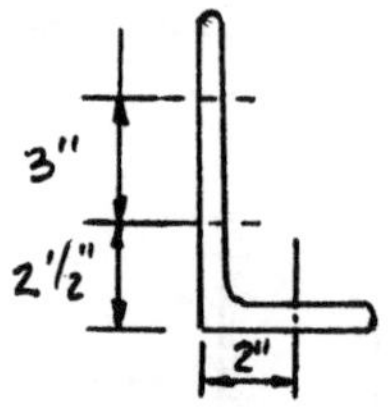

Compute the required thickness using the maximum width = 5.5". Critical stresses occur under combined bending and tension stress at the net width at the lower hole of the vertical leg (Re: D1, AISC).

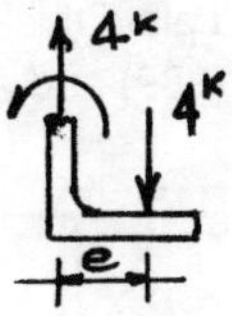

$e = 2 - {}^{t}/_{2}$

$w_n = 5.5 - ({}^{3}/_{4} + {}^{1}/_{8}) = 4.63"$

$S = 4.63t^2 / 6 = 0.77t^2$

$A_n = 4.63t$

Check the controlling tensile failure:

$$T \le \begin{cases} \left(0.6F_yA_g\right)\dfrac{4}{3} = \dfrac{4}{3}(22)(5.5t) = 161t \\ \left(0.5F_uA_c\right)\dfrac{4}{3} = \dfrac{4}{3}(0.5 \times 58)(4.63t) = 179t \end{cases}$$

Yield governs $\therefore$ $f_t = 4 / A_g = 4 / (5.5t)$

Combined stresses (H1, ASIC):

$f_t / F_t + f_b / F_b \le {}^{4}/_{3}$

Solve by trial: try t = 0.5"; e = 1.75"; $F_t = 0.6F_y = 22^{ksi}$; $F_b = 0.75F_y = 27^{ksi}$:

$4 / [(5.5 \times 0.5) \times 22] + 4 \times 1.75 / (0.77 \times 0.5^2 \times 27) = 1.4$

No good – try t = ${}^{5}/_{8}$": e = 1.69":

$4 / [5.5 \times 0.625 \times 22] + 4 \times 1.69 / (0.77 \times 0.625^2 \times 27)$ = 0.89

O.K. **Use L7 × 4 × ${}^{5}/_{8}$ × 0' – 5 "**

The required details are:

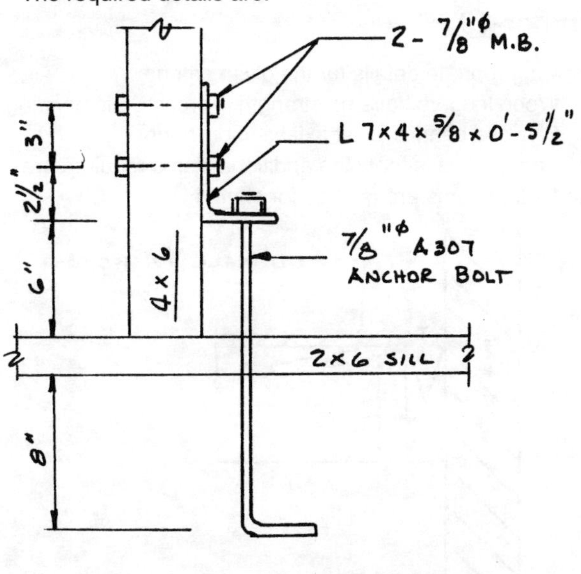

HOLD-DOWN ANCHOR DETAIL

Timber-17

Show the required force transfers for all horizontal and vertical ties in the lateral-force resisting system.

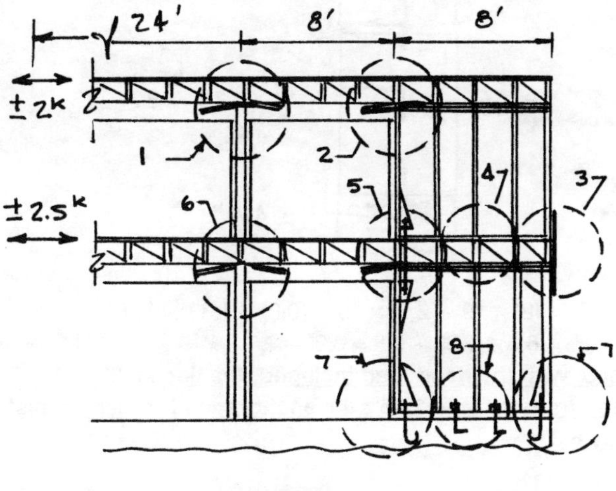

ELEVATION

- The 2^k in the roof diaphragm transfers to the shearwalls and collectors at the top (v = 2000 / 40 = 50 plf):

1. Tie collectors for 50 plf×24' = $\pm 1.2^k$
2. Collector to top plate of roof shearwall must transfer $50^{plf} \times 32' = \pm 1.6^k$
3. Tie across the 2nd floor to connect the outside post of shearwall: $T = 2^k \times 9' / 8' = 2.3^k$
4. Transfer 2^k shear from the upper wall into blocking plus an additional (8' / 40') × 2.5 = 0.5^k from the second floor diaphragm into blocking (total $\pm 2.5^k \rightarrow$ 2,500 / 8 = 313 plf into the top plate of the lower wall)
5. Tie the outside post of the upper and lower walls to transfer $T = 2.3^k$; connect the second floor collector to the top plate of lower wall to transfer ± (32 / 40) 2.5 = $\pm 2^k$
6. Tie collectors to transfer $\pm 1.5^k$
7. Tie the outside posts of lower shearwall to the foundation to transfer $(2 \times 19 + 2.5 \times 10) / 8 = 7.9^k$
8. Transfer $\pm(2 + 2.5) = \pm 4.5^k$ from the lower shearwall to the foundation (shear)

Note: the contribution of dead load to resisting overturning moment was neglected in the analysis above. If dead load is to be included, only ${}^{2}/_{3}$ of the given nominal values should be used (1621.1 UBC-97):

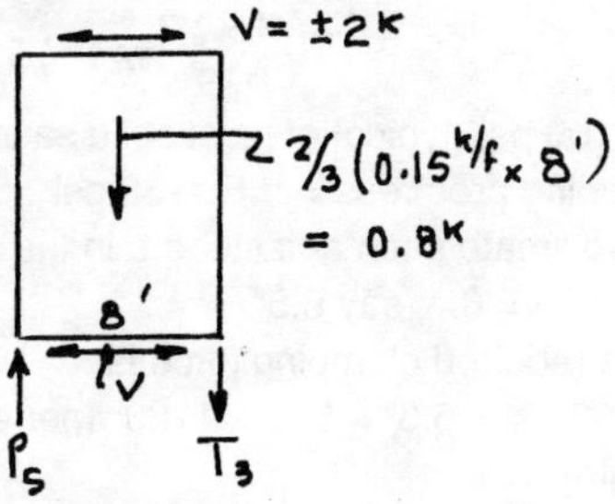

$T_3 = (2^k \times 9' - 0.8^k \times 4') / 8' = 1.85^k$

$P_5 = 1.85 + 0.8 = 2.65^k$

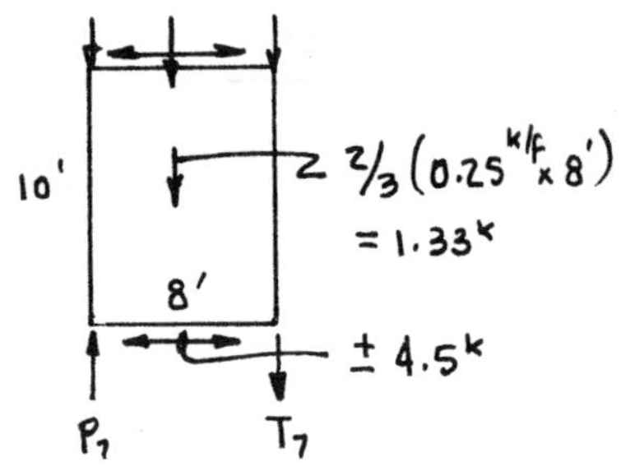

$T_7 = [2^k \times 19' + 2.5^k \times 10' - (0.8^k + 1.33^k)4'] / 8'$

$P_7 = 6.8 + 1.33 + 0.8 = 8.9^k$

Thus, with the dead load included, the ties at '3' and '5' need transfer only 1.85^k; the anchorage at '7' need resist only 6.8^k.

Timber-18

a) Describe a repair method for the damaged truss.
Proposed repair: Clamp the split region to induce a prestress across the grain of approximately $^1/_3\ F_v$ (the usual magnitude of radial tension stress permitted in curved timber members). The given 2" clearance between the bottom chord and the ceiling will permit installation of plates and bolts as indicated:

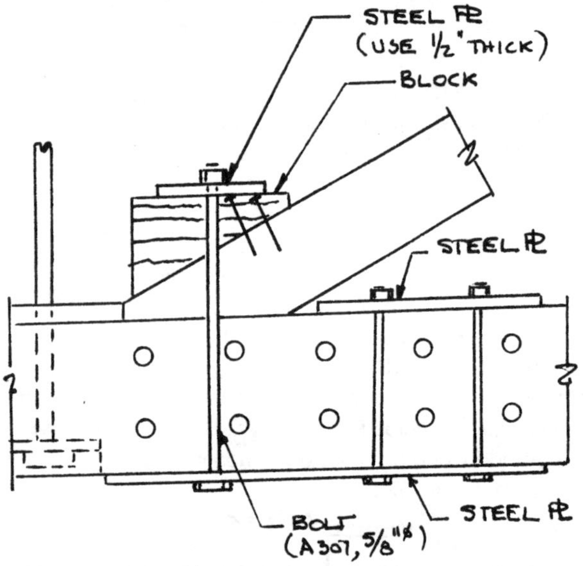

For $F_v \cong 100$ psi (typical of most solid sawn timber)

∴ clamping prestress = $^1/_3 F_v \cong 33$ psi

The approximate pitch of fasteners in the splice $\cong (64" / 2) / 5 = 6.4$, say 6.5"

Thus, the required clamping force is:

$T = 33 \text{ psi} \times 6.5" \times 5.5" = 1180^\#$ (per each 6.5" length)

Use 3 sets of bolts arranged as shown above. The force per bolt pair is:

$T \cong 2 \times 1180 = 2360^\#$

For tension on A307 bolts (AISC 1989), a $^5/_8$" diameter A307 furnishes $6.1^k/_{bolt}$, which is well above the required tension capacity.

∴ **Use 6 - $^5/_8$" diameter A307 bolts (3 each side)**

Compute the required plate thickness. Use allowable stress design (AISC 1989) and ASTM A36 steel ($F_y = 36^{ksi}$):

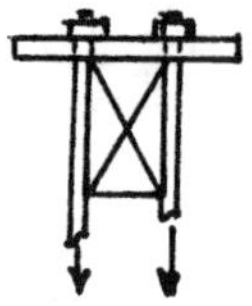

The critical plate is ~ 6" long and distributes its clamping force over the width of the lower chord (5-1/2"). Thus,

$M = (2.36^k / 5.5")(5.5")^2 / 8 = 1.63^{k\text{-}in}$

$F_b = 0.75\ F_y = 0.75 \times 36 = 27^{ksi}$

$S \geq M / F_b = bt^2 / 6;\ b = 6"$

∴ $t^2 \geq 1.63 / 27 = 0.06 \text{ in}^2$

t = 0.25"; 1/4" PL is adequate for strength, but use 1/2" PL to provide stiffness; hence more uniform bearing pressure (additional cost is nil for such a small quantity).

Use PL 10 × 1/2 × (length varies)

b) Describe the construction sequence:

1. Remove ceiling tiles in the vicinity of split chord
2. Shore the ceiling joists
3. Position wood block, steel plates and bolts
4. Tighten bolts sufficiently to close split, but avoid crushing wood fibers
5. Remove the shoring; reinstall the ceiling tiles

Timber-19

Show appropriate details for the given criteria:

- Wood ledgers have no strength in cross-grain bending
- Plywood cannot resist axial compression

Appropriate details (only the additional items required are listed; other items are omitted for clarity):

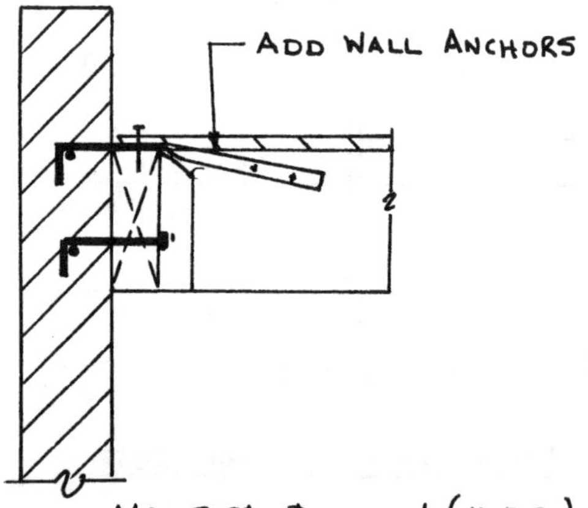

MODIFIED DETAIL 1 (N.T.S.)

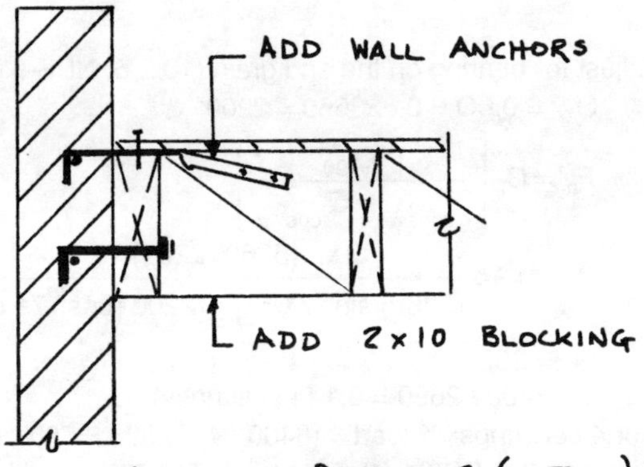

MODIFIED DETAIL 2 (N.T.S.)

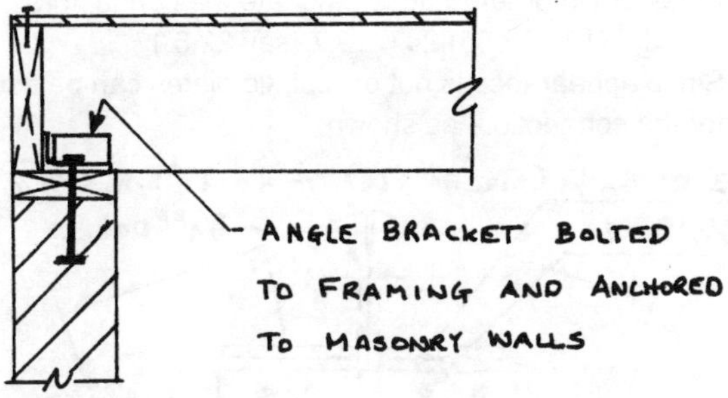

MODIFIED DETAIL 3 (N.T.S.)

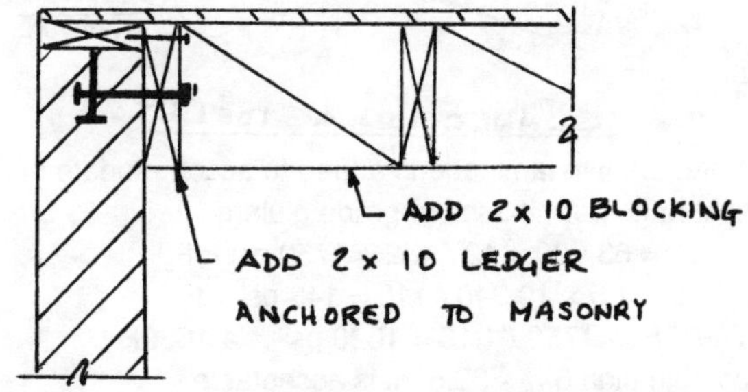

MODIFIED DETAIL 4 (N.T.S.)

Timber-20

a) Design beams A and B using combination C glulam.

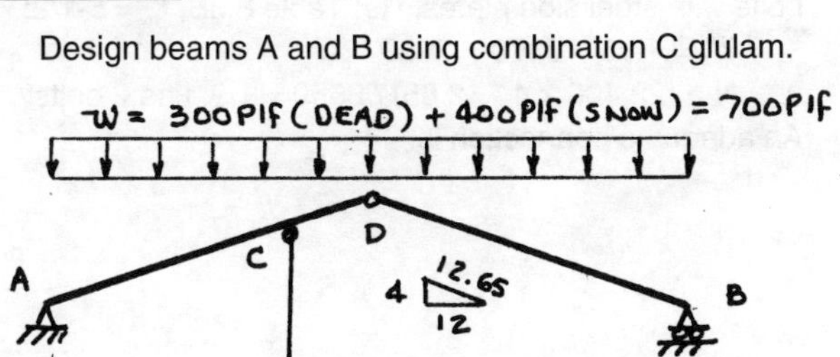

$\Sigma M_D = 0$: $B_y(24) - 700(24)^2 / 2 = 0$

$B_y = 8400^{\#}\uparrow$

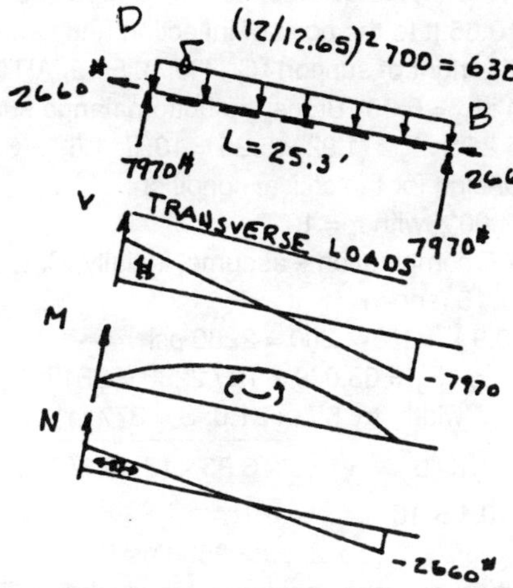

$M_{max} = 630 \times 25.3^2 / 8 = 50{,}400^{\#\text{-ft}}$ (comp. top)

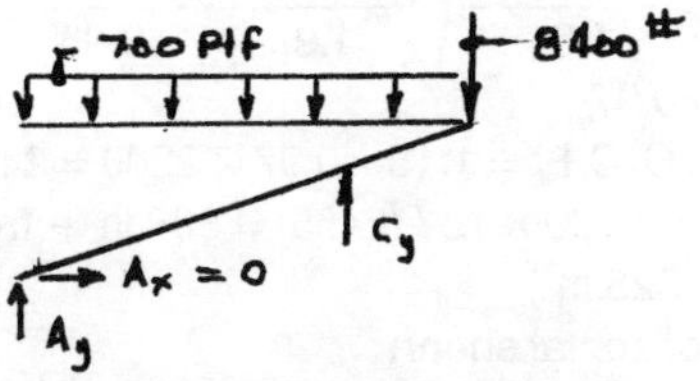

$\Sigma M_A = 0$: $8400(24) + 700(24)^2 / 2 - 18C_y = 0$

$C_y = 22{,}400^{\#}\uparrow$

$\Sigma F_y = 0$: $A_y = 700\times24 + 8400 - 22{,}400 = 2800^{\#}\uparrow$

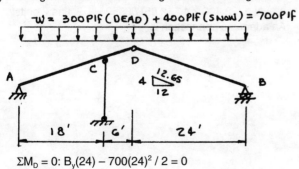

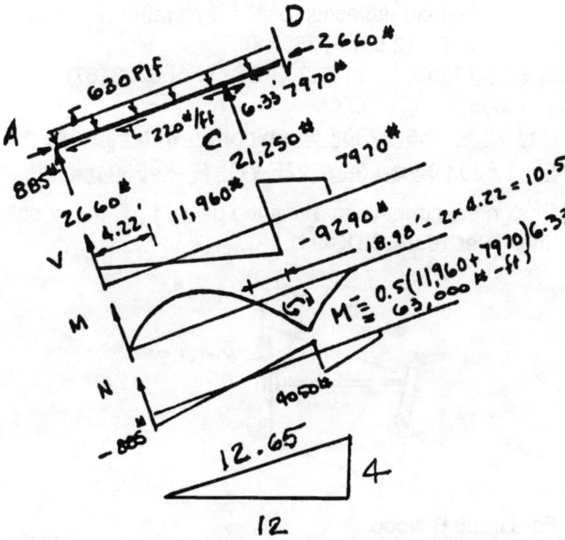

The compression flange extends from support "C" a distance 10.65 ft to the point of inflection and cantilevers 6.33 ft to the right of support "C". Re: p. 5.15, AITCM – l_{u1} = 10.54'; l_{u2} = 6.33'. Brace the bottom flange such that $l_{u1} \cong 5.25'$ $\therefore$ $l_e \geq (1.92 \times 5.25 = 10.1'; 1.69 \times 6.33 = 10.7')$. Design for the critical condition, $M_{max} = 63{,}000^{\#\text{-ft}}$ with $l_e = 10.7'$.

Try a 2200F combination – assume, initially, $C_L \cong 0.9$ and $C_D = 1.15$ (snow).

$\therefore$ $F_b' = 0.9 \times 1.15 \times 2200 = 2280$ psi

$S \geq M_{max} / F_b' = 63{,}000 \times 12 / 2280 = 331$ in^3

For $5^1/_8$" width, try $5^1/_8 \times 21.0$: $S = 377$ in^3

$$R_B = \sqrt{l_e d / b^2} = \sqrt{1.69 \times 6.33 \times 12 \times 21 / 5.125^2}$$

$= 10.1 > 10$

$F_b^* = C_D F_b = 1.15 \times 2200 = 2530$ psi

Specify a glulam combination for which $E' = E_{yy} = 1.6 \times 10^6$ psi:

$F_{bE} = 0.438E' / R_B^2 = 0.438 \times 1.6 \times 10^6 / 10.1^2 = 6870$ psi

$F_{bE} / F_b^* = 6870 / 2530 = 2.71$

$$C_L = \frac{1 + F_{bE}/F_b^*}{1.9} - \sqrt{\left[\frac{1 + F_{bE}/F_b^*}{1.9}\right]^2 - \frac{F_{bE}/F_b^*}{0.95}}$$

$$= \frac{1 + 2.71}{1.9} - \sqrt{\left[\frac{1 + 2.71}{1.9}\right]^2 - \frac{2.71}{0.95}}$$

$C_L = 0.97$

$\therefore$ $F_b' = C_D C_L F_b = 1.15 \times 0.97 \times 2200 = 2455$ psi

$\therefore$ $S_x \geq 63{,}000 \times 12 / (2455) = 308$ in^3 – try $5^1/_8 \times 19.5$ ($S = 325$ in^3).

Check horizontal shear:

$V = 11{,}960 - 630 \times 19.5 / 12 = 10{,}940^\#$

$f_v = {}^3/_2 V / bd = 1.5 \times 10{,}940 / (5.125 \times 19.5) = 164$ psi

Select glulam with $F_v' = F_v(1.15) > 164$ psi

$\therefore$ $F_v \geq 143$ psi

Check live load deflection on the right rafter ($I = 3167$; $w_L = 400(12 / 12.65)^2 = 360$ plf):

$\Delta_L = {}^5/_{384}(360 \times 25.3)(25.3 \times 12)^3 / (1.6\text{E}6 \times 3167)$

$= 0.65" = {}^L/_{465}$ $\therefore$ OK

Use $5^1/_8 \times 19.5$, 2200F combination with $F_V \geq 145$ psi; E = 1,600,000 psi (e.g., 22F-V7); $F_{c\perp} = 560$ psi.

b) Design the connection at the top using 4" diameter split rings (Re: 10.2A, NDS-91).

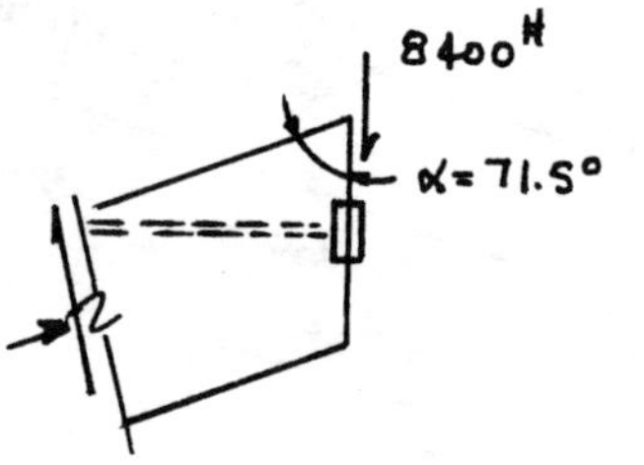

For Group B wood:

$$\left.\begin{aligned} P &= 5260^\# \\ Q &= 3660^\# \end{aligned}\right\} \text{Table 10.2A, NDS-91}$$

Adjust for bearing on the end grain (10.2-6, NDS-91):

$Q'_{90} = 0.6Q = 0.6 \times 3660 = 2200^\#$

$$P_a' = C_D \left[\frac{P' Q'_{90}}{P'_{\sin^2 \alpha} + Q'_{\cos^2 \alpha}} \right]$$

$$= 1.15 \frac{(5260)(2200)}{(5260)(\sin^2 71.5^\circ) + (2200)(\cos^2 71.5^\circ)}$$

$= 2690^\#$

$n \geq 8400 / 2690 = 3.1$ (4 split rings)

For 4 split rings: % load = (8400 / 4) / 2690 = 78%. From Table 10.3, NDS-91, 4" rings loaded to 78% require center-to-center spacing $\cong$ 6" o.c. $\therefore$ need to increase the depth in order to accommodate the connectors:

$d \geq (12 / 12.65)\ 24 = 22.7$ (say 22.5")

Since appearance is not critical, tie plates can be used for the connection, as shown:

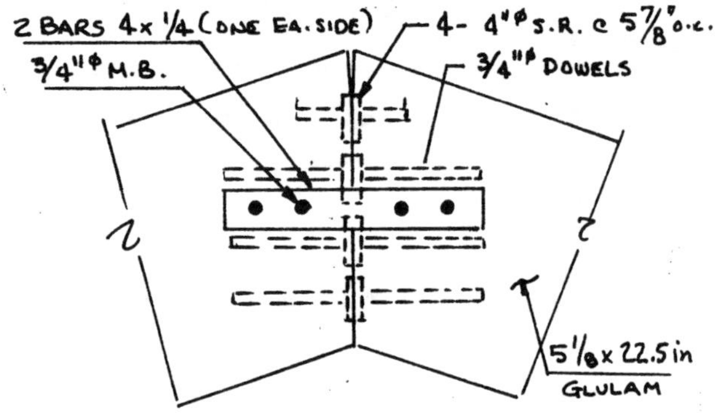

CONNECTION AT TOP (N.T.S.)

Note: the two laminations added to accommodate the split rings permit a lower grade glulam:

$f_b = 63{,}000 \times 12 / 432 = 1750$ psi $< F_b'$

$f_v = 1.5 \times 10{,}940 / 115 = 143$ psi $< F_v'$

Thus, $F_b > 1750 / 1.15 = 1520$ psi $\therefore$ a 1600F combination $5^1/_8 \times 22.5$ in. is acceptable.

For the connection to the pipe column, the normal component of the reaction requires a bearing length ($F_{c\perp} = 560$ psi):

$N \geq R_\perp / F_{c\perp} = 21{,}250 / (5.125 \times 560) = 7.4"$

For the parallel to grain component, try 3/4" diameter bolts with steel side plates. Re: Table 8.3B, t_m = 5-1/2:

$Z'_{//} = 1.15 \times 3170 = 3650\ ^\#/_{bolt}$

$\therefore$ $N \geq (22{,}400 \times 4 / 12.65) / 3650 = 1.9$; use 2 bolts

An adequate connection is:

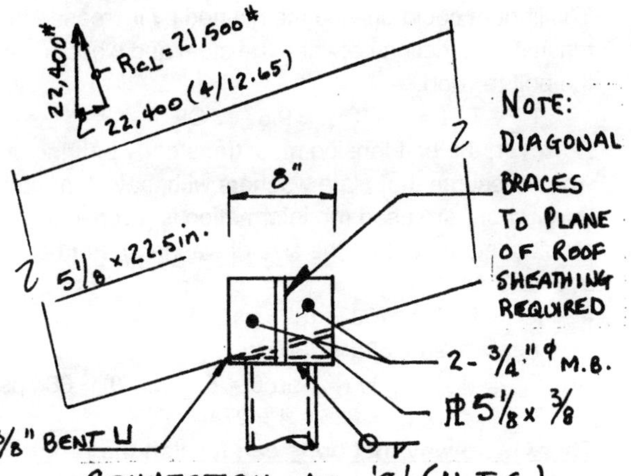

CONNECTION AT 'C' (N.T.S.)

Timber-21

a) Check adequacy of the beams using 22F combination glulam. The typical header is laterally supported on the top flange; $L \leq 20' - (5 + 0.5 + 0.5) / 12 = 19.5'$; use 2200-F combination, Douglas Fir; dry use; normal load duration.

$M = wL^2 / 8 = 700 \times 19.5^2 / 8 = 33{,}270^{\#\text{-ft}}$

For the given $5^1/_8$"×15" member: $A = 76.9$ in^2; $S = 192$ in^3; $I = 1441$ in^4

$f_b = M / S = 33{,}270 \times 12 / 192 = 2080$ psi

For d = 15": The roof beam is laterally braced by joists

$\therefore C_L = 1.0$

$C_V = K_L(21 / L)^{1/10}(12 / d)^{1/10}(5.125 / b)^{1/10}$

$= 1.0 \times (21 / 19.5)^{1/10} \times (12 / 15)^{1/10} \times (5.125 / 5.125)^{1/10} = 0.985$

$F_b' = C_D C_V F_b = 1.0 \times 0.985 \times 2200 = 2170 \text{ psi} > f_b$

$\therefore$ OK in bending – check shear:

$L_e = L - 2d = 19.5 - 2 \times 15 / 12 = 17'$

$V = wL_e / 2 = 700 \times 17 / 2 = 5950^{\#}$

$f_v = {}^3/_2 V / bd = 1.5 \times 5950 / 76.9 = 116$ psi

Shear requires $F_v > 116$ psi (e.g., a 22F – VI (DF / WW) is adequate: $F_v = 140$ psi; $E = 1.6 \times 10^6$ psi; $F_{c\perp} = 650$ psi (bottom)). Check live load deflection – limit to $^L/_{360}$:

$\Delta_L = (^5/_{384})(\omega_L L)L^3 / EI \leq {}^L/_{360} = 0.66"$

$\Delta_L = 0.013(400 \times 19.5)(19.5 \times 12)^3 / (1.6 \times 10^6 \times 1441)$

$= 0.56" < {}^L/_{360} \therefore$ OK

$\therefore$ **$5^1/_8$"×15.0", 22F Glulam is adequate**

b) Design a connection to the Pipe 5 Std. Use a detail similar to the partially concealed type shown in Figure 4.6, AITCM (the purpose is to facilitate connection to the pipe, not to conceal the connection hardware). Assume a 1/2" saw kerf in the center of the beam:

$b_e = b - 0.5 = 5.125 - 0.5 = 4.63"$

Take $F_{c\perp} = 650$ psi:

$N \geq R / b_e F_{c\perp} = (700 \times 19.5 / 2) / (4.63 \times 650)$

$= 2.26"$, say 2-1/2"

Allow 1/2" setback from the beam end to the edge of pipe; use ASTM A36 steel; E70xx Electrodes

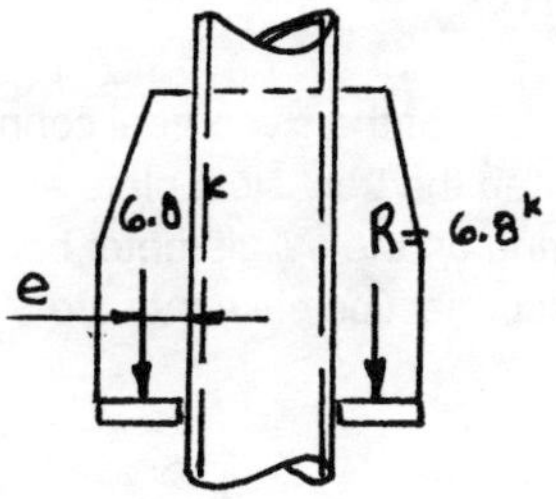

$e = 0.5 + 2.5 / 2 = 1.75"$

Try a 12" long web plate:

$f_v = R / Lt_{PL}$

Use ASIC-ASD (1989):

$F_v = 0.4F_y = 0.4 \times 36 = 14.5^{\text{ksi}}$

$\therefore \quad t_{PL} \geq 6.8 / (12 \times 14.5) = 0.04"$

use PL 1/4" × 1' – 0"

Check bending in the vertical plate:

$S = 0.25 \times 12^2 / 6 = 6$ in^3

$f_b = M / S = 6.8 \times 1.75 / 6 = 2^{\text{ksi}} << F_b$

Seat plate (say 3" long):

$e \cong 5.125 / 4 = 1.28"$

$S = bt^2 / 6 = 3t^2 / 6$

$F_b = 0.75F_y = 27^{\text{ksi}}$

$f_b = 3.4 \times 1.28(t^2 / 2) \leq 27$

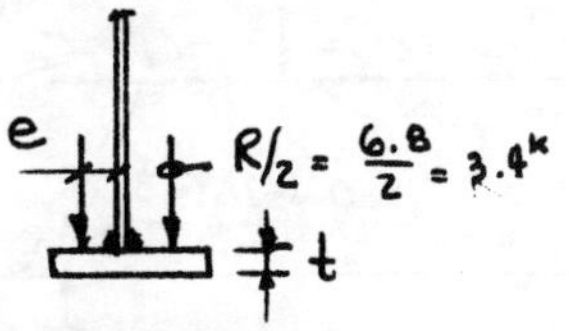

$t^2 \geq 0.32$; $t \geq 0.57$ say PL$^5/_8$

Weld to PL 1/4" using 1/4" (minimum size) fillet welds (stresses in the base material govern):

$q \leq 0.4F_y t_{PL} = 14.5 \times 0.25 = 3.6^{\text{ k / in}}$

$q_t = R / L_w = 6.8 / 3 = 2.3^{\text{ k / in}} < 3.6^{\text{ k / in}} \therefore$ OK

The connection detail is:

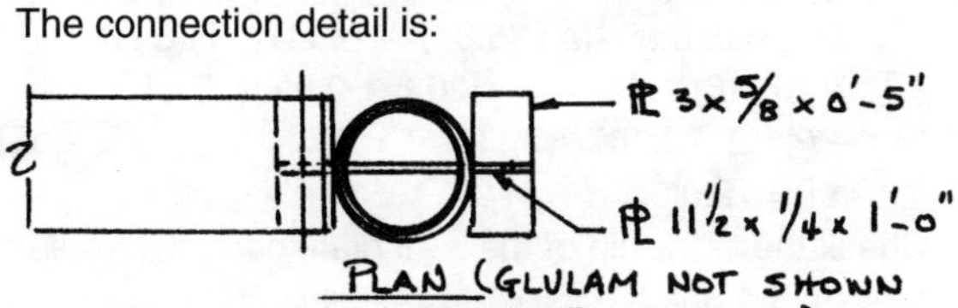

PLAN (GLULAM NOT SHOWN ON RIGHT SIDE)

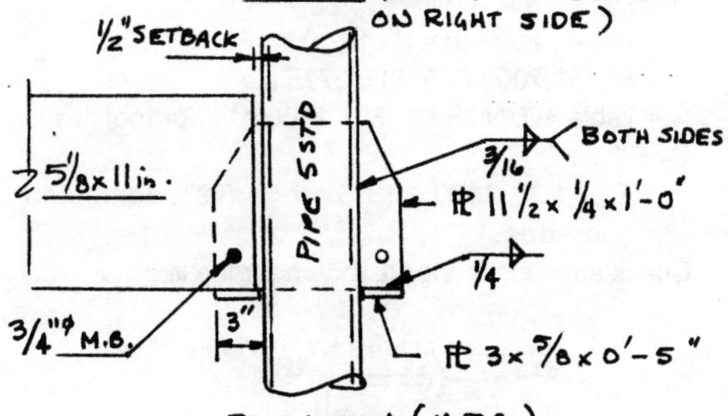

ELEVATION (N.T.S.)

Timber-22

a) Check adequacy of the given heel connection. Neglect friction between the bent steel plate and wood and neglect bearing on the $^7/_8$" diameter bolts. The freebody diagrams of connection elements are:

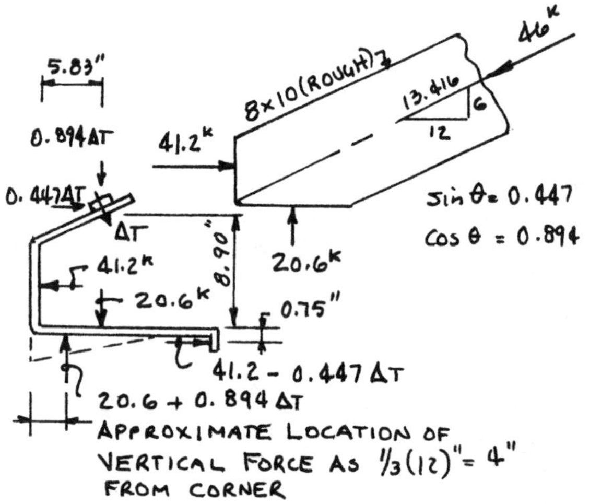

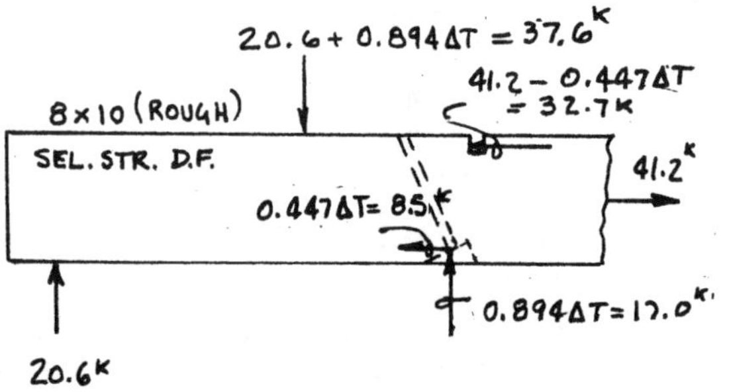

For equilibrium of the bent plate under dead plus live loading under these assumptions (let ΔT = tension in the 2 – $^7/_8$" diameter bolts induced by the applied forces):

$\Sigma M_a = 0$: $0.894(\Delta T)(5.83)" + 0.447(\Delta T)(8.90)"$
$+ 20.6(6) - 41.2(3) - (20.6 + 0.894(\Delta T))4$
$- (41.2 - 0.447(\Delta T))\ 0.75" = 0$
$\Delta T = 19.0^k$

Check the 1-1/2" lip of the bent plate bearing parallel to grain in the notch:

$V = 41{,}200 - 0.447 \times 19{,}000 = 32{,}700^\#$
$f_c = 32{,}700 / (1.5 \times 8) = 2725$ psi

Re: Table 4D, NDS-91: $F_c = 1100$ psi. For roof live load, $C_D = 1.25$:

$F_c' = 1.25 \times 1100 = 1375$ psi $< f_c$ **(98% computed overstress)**

Check stresses in the PL $8 \times ^5/_8$ near the lip:

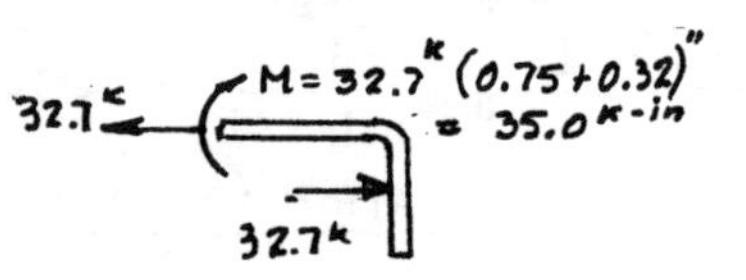

$S = 8 \times (^5/_8)^2 / 6 = 0.52$ in^3
$A = 8 \times (^5/_8) = 5$ in^2
$f_t = 32.7 / 5 = 6.5$ ^{ksi}
$f_b = 35.0 / 0.52 = 67.3$ ^{ksi}
$f_t + f_b > F_y$ ∴ the plate yields!

Equilibrium could only be maintained by increased bolt tension and friction between the steel and wood. Check the bolt tension:

$T = \Delta T / 2 = 19.0 / 2 = 9.5^{\ k}$ ∴ OK

However, the bolt tension must transfer by bearing on wood. Assume that plate washers with, say, 4 in^2 at the lower chord are used (no information is given in the problem statement on the size of washers). For bearing at $\theta = 63°$:

$F_C' = 1.25 \times 1100 = 1375$ psi
$F_{C\perp} = 1.25 \times 625 = 780$ psi
$F_{C\theta}' = (F_C'\ F_{C\perp}') / (F_C' \sin^2\theta + F_{C\perp} \cos^2\theta) = 686$ psi
∴ $T \le F_{C\theta}'\ A_{washer} = 686 \times 4 = 2744^\#$

<u>There is no way that bolts can furnish the required resistance!</u>

Check combined axial tension and bending in the lower chord at the notch:

$A_n = (10 - 1.5)\ 8 = 68$ in^2
$S_n = 8\ (10 - 1.5)^2 / 6 = 96.3$ in^3
$M = 20.6(34) - 37.6(34 - 20) - 32.7(5 - 0.75)$
$+ 8.5(5 - 0.75) + 17.0(2) = 105^{k\text{-}in}$ (tension bottom)

Compute the reference stress without an increase for stress concentration due to the notch:

$f_t = 41{,}200 / 68 = 606$ psi
$f_b = \pm 105{,}000 / 96.3 = \pm 1090$ psi
$f_{top} = 606 - 1090 = -484$ psi $= 484$ psi (c)
$f_{bot} = 606 + 1090 = 1696$ psi (ten)

The combined stresses are OK. Check the horizontal stress at the depth of the notch:

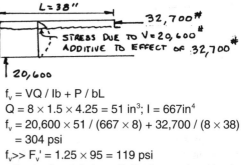

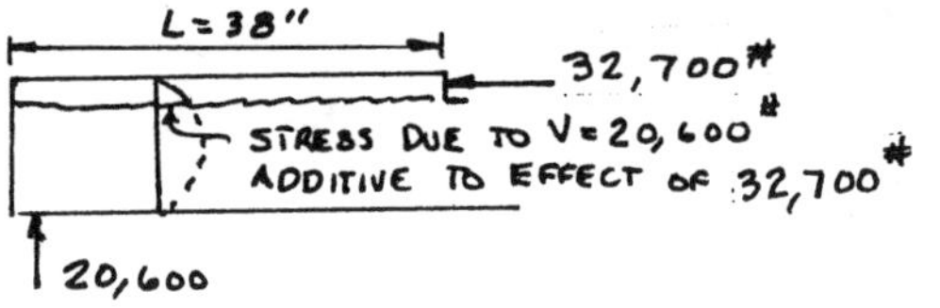

$f_v = VQ / Ib + P / bL$
$Q = 8 \times 1.5 \times 4.25 = 51$ in^3; $I = 667$in^4
$f_v = 20{,}600 \times 51 / (667 \times 8) + 32{,}700 / (8 \times 38)$
$= 304$ psi
$f_v >> F_v' = 1.25 \times 95 = 119$ psi

Thus, the horizontal shear in the lower chord is about 2.5 times the code allowable. Check the maximum bending stress in the lower chord:

$M_{max} = 20{,}600 \times (4 + 12 + 4) = 412{,}000^{\#\text{-}in}$
$f_b = 412{,}000 / (8 \times 10^2 / 6) = 3090$ psi
$F_b' = 1.25 \times 1600 = 2000$ psi

Bending stress is about 50% over the code allowable.

b) **<u>Conclusion: The connection is bad!</u>** Several elements are grossly overstressed under the design loading. Since the dead load represents $^2/_3$ of the design forces, the elements exceed safe levels under dead load alone. Given that timber weakens under sustained loading, it is possible that a failure could be initiated without an application of live load. The risk of failure is too great to permit the structure to be used in its present condition.

Timber-23

a) Check the given plywood nailing in the horizontal diaphragm. To simplify analysis, split the lateral force equally between the windward and leeward sides, 100 plf each side. The diaphragm is flexible ∴ the shearwalls at the east and west ends resist an equal share of the total lateral force:

$V_W = V_E = (0.5)(200 \text{ plf} \times 100') = 10{,}000^{\#}$

Assuming a collector between walls at the west end, the diaphragm shear distributes over the full 40'

$v_w = 10{,}000^{\#} / 40' = 250 \text{ plf}$

Re: Table No. 23-II-H, UBC-97, for a 1/2" structural I plywood; blocked with 8d @ 6" o.c. at edges; load case 3; 2" framing:

$v = 270 \text{ plf} > v_w$

∴ **<u>the proposed nailing is adequate at the west end</u>**

The skylight opening near the east end causes force concentrations near the opening.

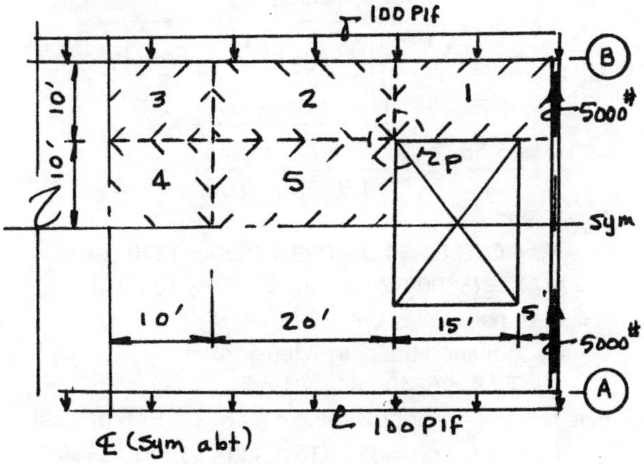

Divide the upper half of the east side into 5 subdiaphragms as shown above. The chord force at the midspan of the full diaphragm is:

$M = 200 \times 100^2 / 8 = 250{,}000^{\text{ \#-ft}}$

∴ $C = T = M / 40 = 250{,}000^{\text{ \#-ft}} / 40' = 6250^{\#}$

Thus, from a freebody of the subdiaphragms:

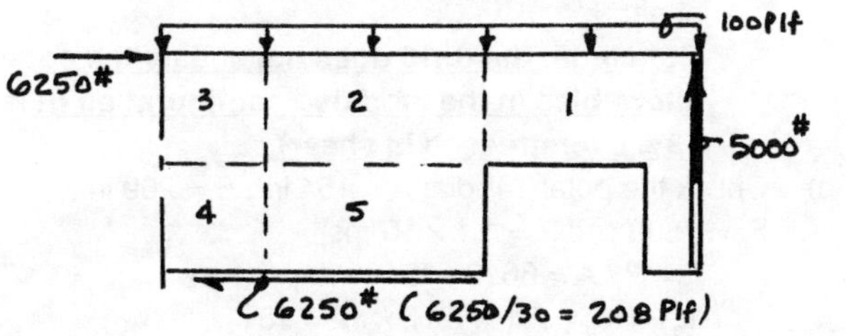

TAKE FREE BODY OF PANELS 4 & 5:

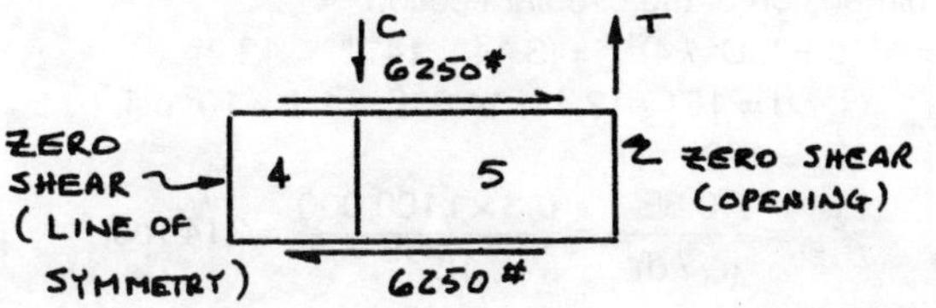

To maintain equilibrium, we need to develop a couple by forces acting vertically on the top edge (e.g., at the corners of panel 5):

∴ $C = T = 6250^{\#} \times 10' / 20' = 3125^{\#}$

For panels 1, 2, and 3:

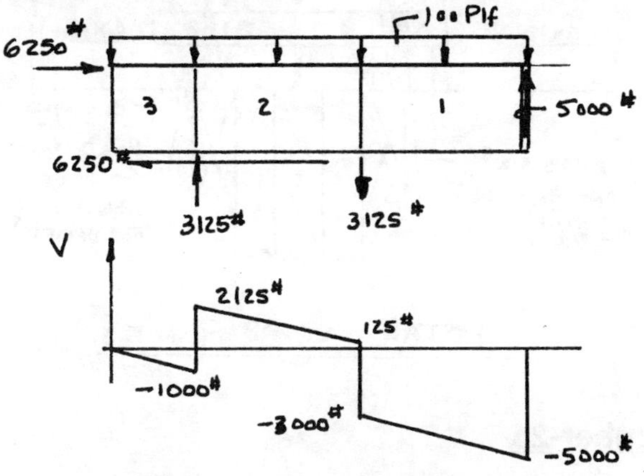

SHEAR IN SUBDIAPHRAGMS 1, 2 & 3

In subdiaphragm 1, $v = 5000 / 10 = 500$ plf,

∴ **<u>the proposed nailing near the opening is inadequate. We need to provide 8d @ 2 1/2" o.c. (with nominal width ≥ 3") at the boundary; 4" o.c. at other edges.</u>** This will provide $v \geq 500$ plf. For all other regions 8d @ 6" o.c. is adequate.

b) The maximum chord force at line A (and B) occurs at the boundary member for subdiaphragm 1:

$M = 0.5\,(3000 + 5000)\,20 = 80{,}000^{\text{ \#-ft}}$

<u>$C = T = M / 10 = 8000^{\#}$</u>

c) At the connection marked 'P', we need continuity ties to transfer the ±8000# from the 4×12 boundary member. If 3/4" diameter bolts with 1/4" steel side plates are used (Re: Table 8.3B, NDS-91):

$Z'_{/\!/} = 4/3 \times 3170 = 4230^{\#}/_{\text{bolt}}$

∴ Use 2 – 3/4" dia A307 bolts each side. Provide 4×12 along the edges of subdiaphragms 2 & 5. Space the bolts 7d from end of member

∴ $Kl = (2 \times 7 \times 0.75 + 0.5) = 11"$. Limit $Kl / 0.3t \leq 100$

∴ $t \geq 0.36"$

Use $^3/_8$" plate, say 3" wide:

$F_a = {}^4/_3(13.0)^{\text{ksi}}$ (AISC-ASD, 1989)

∴ $P \leq 2 \times 3 \times {}^3/_8 \times {}^4/_3 \times 13 = 39^k >> 8^k$

The connection at "P" is detailed as follows:

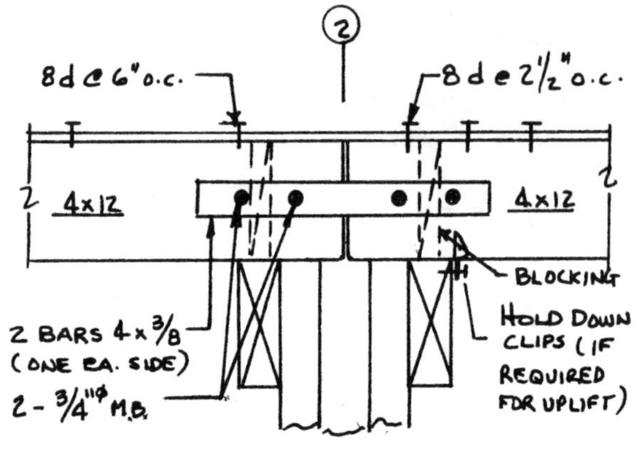

DETAIL AT 'P' (N.T.S.)

Timber-24

a) Check the 6x16 beam using the given allowable stresses (F_b = 1350 psi; $F_{c\perp}$ = 385 psi; F_a = 925 psi; F_v = 85 psi; E = 1.6×10^6 psi). The maximum reaction from the monorail with 2 – 2750# wheel loads @ 4' o.c. is:

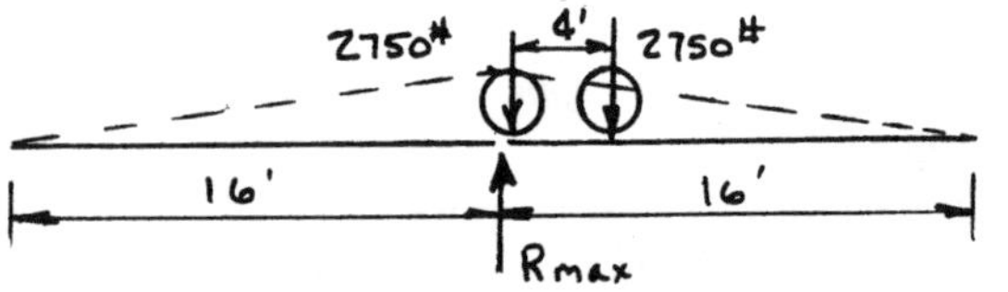

For the 6" × 18" beam + rail: w ≅ 50 plf.

R_{max} = 2750(1 + 0.75) + 50 × 16 = 5600#

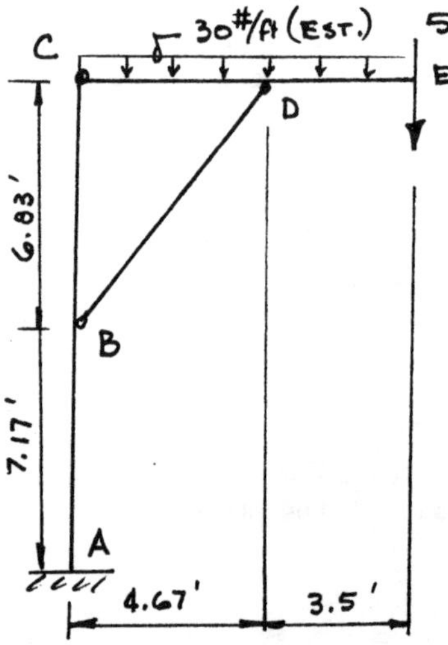

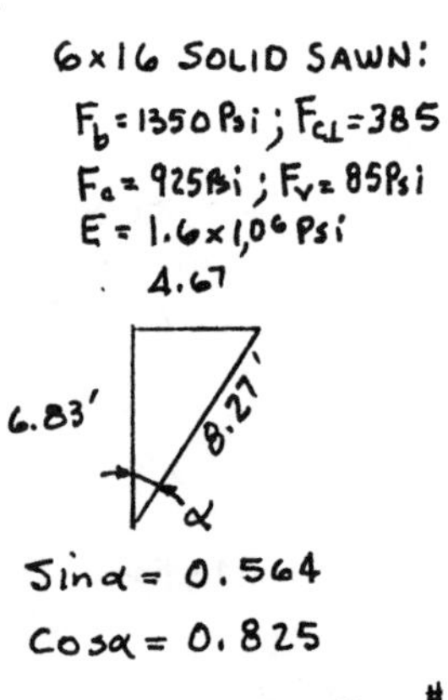

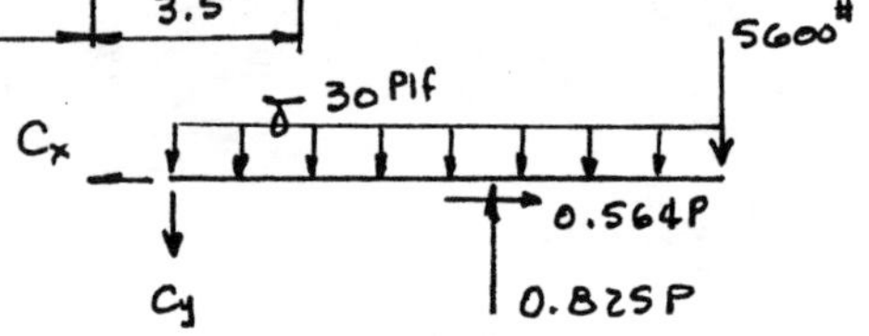

$\Sigma M_c = 0 = (0.825P)4.67 - 30(8.17)^2/2 - 5600(8.17)$ + (0.825P) = 10,000# ↑ (P = 12,140)

$\Sigma F_x = 0$: $C_x = 0.564 \times 12{,}140 = 6840^{\#}$ ←

$\Sigma F_y = 0$: $C_y = 30 \times 8.17 + 5600 - 10{,}000 = 4160^{\#}$ ↓

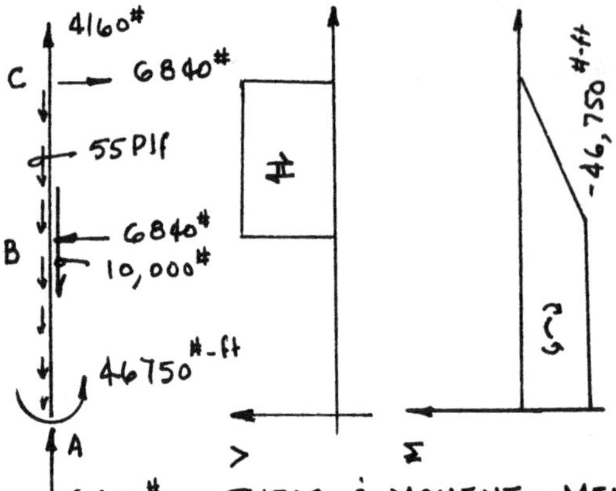

SHEAR & MOMENT — MEMBER ABC

Check the 6×16: b = 5.5"; d = 15.25"; A = 84 in²; S = 213 in⁴; neglect impact ∴ C_D = 1.0; assume beam is laterally braced at ends only: $l_e = 1.61 l_u$ (analogous to case b, p. 5-160, AITCM)

$$R_B = \sqrt{l_e d / b^2} = \sqrt{1.61 \times 8.17 \times 12 \times 15.25 / 5.5^2} = 8.9$$

$$F_{bE} = 0.438E' / R_B^2 = 0.438 \times 1.6 \times 10^6 / 8.9^2 = 8847 \text{ psi}$$

$$F_{bE} / F_b^* = 8847 / 1350 = 6.55$$

$$C_L = \frac{1 + F_{bE}/F_b^*}{1.9} - \sqrt{\left[\frac{1 + F_{bE}/F_b^*}{1.9}\right]^2 - \frac{F_{bE}/F_b^*}{0.95}}$$

$$= \frac{1 + 6.55}{1.9} - \sqrt{\left[\frac{1 + 6.55}{1.9}\right]^2 - \frac{6.55}{0.95}}$$

$C_L = 0.99$

∴ $F_b' = C_D C_L F_b = 1.0 \times 099 \times 1350 = 1340$ psi

$f_b = M/S = (5600(3.5) + 30(3.5^2)/2) \times 12/213$
= 1114 psi < F_b' ∴ OK

Check combined bending + tension:

$f_t = T/A = 6840/84 = 81$ psi

F_t is not given – approximate it as $F_t = {}^2/_3 F_c = 610$ psi.

$f_t / F_t' + f_b / F_b' = 81/610 + 1114/1340 = 0.96$

∴ OK

Check shear:

$V_{max} = 5600 + 30(3.5 - 15.25/12) = 5667^{\#}$

$f_v = {}^3/_2 V_{max} / bd = 1.5 \times 5667 / (5.5 \times 15.25)$
= 101 psi

F_v' = 85 psi < f_v ∴ 19% overstress in horizontal shear

<u>**Conclude: the 6×16 does not satisfy the code allowables in the modified configuration (it is 19% overstressed in shear).**</u>

b) Check the pole: 14" dia; A = 154 in²; S = 269 in³; F_b = 1800 psi; E = 1.1×10^6 psi

$f_c = P/A = 6610/154 = 43$ psi

$1_e = K_e l_u = 2.1 \times 7.17 \times 12 = 181$

Re: 3.7.3, NDS-91, compute the equivalent square dimension of the circular section:

$d = (\pi D^2/4)^{1/2} = (3.14 \times 14^2)^{1/2} = 12.4"$

$l_e / d = 181/12.4 = 14.6$; E = 1.1×10^6 psi;

F_c = 1200 psi:

$$F_{cE} = \frac{0.3E'}{(l_e/d)^2} = \frac{0.3 \times 1{,}100{,}000}{12.4^2} = 2146 \text{ psi}$$

$F_c^* = C_D F_c = 1.0 \times 1200 = 1200$ psi

$F_{cE} / F_c^* = 2146 / 1200 = 1.79$

$$C_p = \frac{1+F_{cE}/F_c^*}{2c} - \sqrt{\left[\frac{1+F_{cE}/F_c^*}{2c}\right]^2 - \frac{F_{cE}/F_c^*}{c}}$$

$$= \frac{1+1.79}{1.6} - \sqrt{\left[\frac{1+1.79}{1.6}\right]^2 - \frac{1.79}{0.8}}$$

$C_p = 0.94$

$\therefore$ $F_c' = C_pC_DF_c = 0.94 \times 1200 = 1130$ psi

For a round member (2.3-8. NDS-91):

$C_fC_F = 1.18(12 / D)^{1/9} = 1.16$

$F_b' = C_fC_FF_b = 1.16 \times 1800 = 2088$ psi

$f_b = M / S = 46{,}750 \times 12 / 269 = 2086$ psi

For combined axial and bending (Eqn. 3-9-3, NDS):

$$\text{Check: } \left[\frac{f_c}{F_c'}\right]^2 + \frac{f_{b1}}{F_{b1}'[1-(f_c/F_{cE1})]} + \frac{f_{b2}}{F_{b2}'[1-(f_c/F_{cE2})-(f_b/F_{bE})^2]} \le 1.0$$

For bending about one axis, only $f_{b1} = 2086$ psi applies:

$$\left[\frac{43}{1130}\right]^2 + \frac{2086}{2088[1-(43/2146)]} = 1.02 > 1.0$$

Close enough (2% overstress). Check shear stress:

$Q = 0.424\pi R^3 = 454$ in^3; $I = 1885$ in^4

$f_v = VQ / Ib = 6840 \times 457 / (1885 \times 14) = 118$ psi

F_v is not given; typical value for poles is 110-to-120 psi

$\therefore$ likely OK.

$\therefore$ **<u>The 14" dia pole is slightly overstressed as modified (2% overstress).</u>**

c) Determine the required size of 4× brace: $P = 12.1^k$; $l_e = 8.27 \times 12 = 99$"; assume 4" (nominal) is the smaller dimension:

$\therefore$ $l_e / d = 99 / 3.5 = 28.3$

$$F_{cE} = \frac{0.3E'}{(l_e/d)^2} = \frac{0.3 \times 1{,}600{,}000}{28.3^2} = 599\,\text{psi}$$

$F_c^* = C_DC_FF_c = 1.0 \times 1.15 \times 925 = 1063$ psi

$F_{cE} / F_c^* = 599 / 1063 = 0.56$

$$C_p = \frac{1+F_{cE}/F_c^*}{2c} - \sqrt{\left[\frac{1+F_{cE}/F_c^*}{2c}\right]^2 - \frac{F_{cE}/F_c^*}{c}}$$

$$= \frac{1+0.56}{1.6} - \sqrt{\left[\frac{1+0.56}{1.6}\right]^2 - \frac{0.56}{0.8}}$$

$C_p = 0.47$

$\therefore$ $F_c' = C_pC_DC_FF_c = 0.47 \times 1.0 \times 1.15 \times 925 = 500$ psi

$P \le F_c'A = 500(3.5)d = 12{,}100$

$\therefore$ $d \ge 6.90"$ $\therefore$ **<u>Use 4×8</u>**

d) Design the connection at point "1". Force transfer at top at the centerline of the 14" diameter pole:

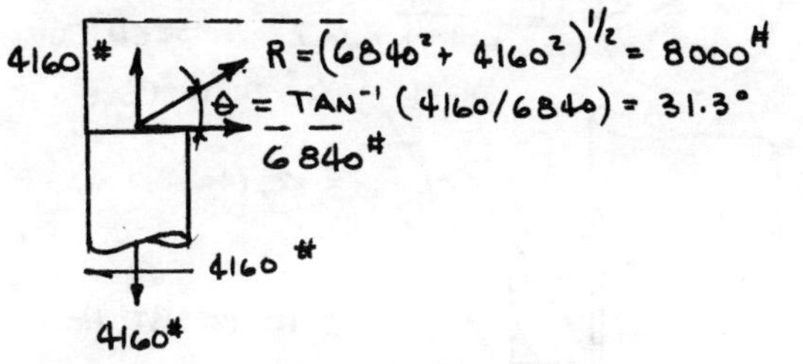

Use steel straps (A36):

$A_g \ge 8^k / (0.6 \times 36^{ksi}) = 0.40$ in^2

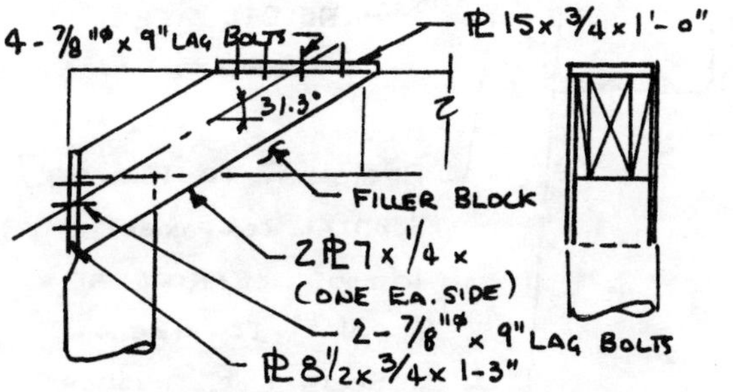

Re: Table 9.3B, NDS (Assume wood with G = 0.5) try 7/8" diameter lag bolts with 1/4" steel side plates: $Z'_{//} = 2130$ $^{\#}/_{bolt}$. For transfer to the the 6×16:

$V = 6840^{\#}$ $\therefore$ $n \ge 6840 / 2130 = 3.2$, say 4 bolts

For connection to into the dapped pole: $V = 4160^{\#}$: $n \ge 1.9$; use 2 bolts

Design the plate across the steel straps (6.84^k is critical); use LRFD with plastic design:

need $3 \times 7 / 8 = 2.625$" spacing $\therefore$ use PL 1/4 for straps: $M_p = F_yZ = 36 \times 7 \times 0.25^2 / 4 = 3.9^{k\text{-}in}$ each side; 14" width across the diameter.

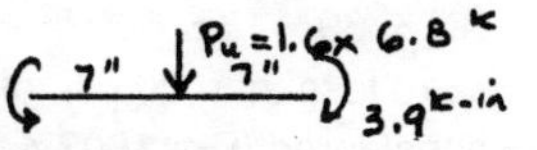

$(M_p)_{mid} = P_uL / 4 - M_p = 10.9 \times 14 / 4 - 3.9 = 34.2^{k\text{-}in}$

$\phi M_n = 0.9F_ybt^2 / 4 \ge 34.2$

$t^2 \ge 34.2 / (0.9 \times 36 \times 7 / 4) = 0.60$ $\therefore$ $t \ge 0.77"$

Use PL 3/4" thick (top and sides). Minimum size (1/4") welds are adequate. Check bearing perpendicular to grain:

$f_{c\perp} = 4160 / (5.5 \times 12) = 63$ psi (nil)

e) Sketch the connection at point '2'. Create a seat attached to the shallow plane cut into the 14" diameter pole. Orient the 4×8 brace to bear with the wide face vertical, as shown below:

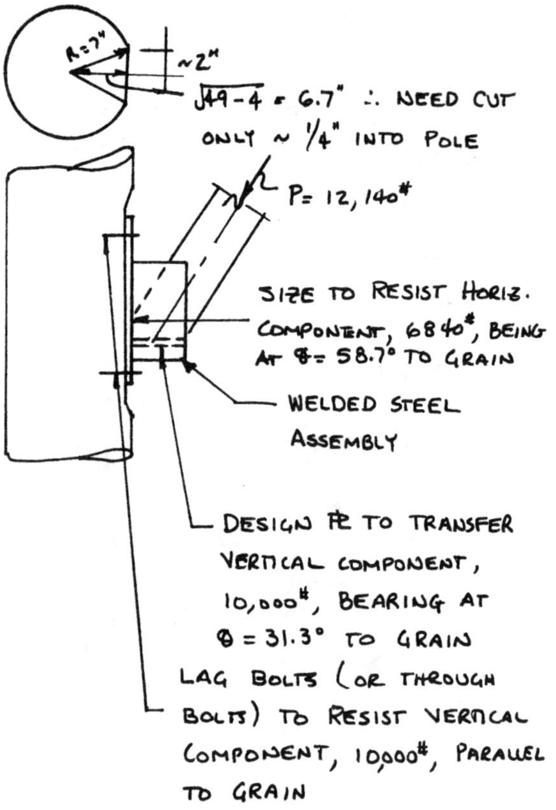

DETAIL – JOINT '2' (N.T.S.)

Timber-25

a) Design beam B2. Re: Table No. 16-C, UBC-97, for member CD, $A_{trib} = 36 \times 16 = 576\ \text{ft}^2$ ∴ $w_L = 16$ psf. For member ABC, $A_{trib} = 60 \times 16 > 600\ \text{ft}^2$ ∴ $w_L = 12$ psf. Since live loads are less than 20 psf, alternate span live loads must be considered (Re: 1607.4.2, UBC). Estimate the member weight as 50 plf for beam B2 and 100 plf for beam B1.

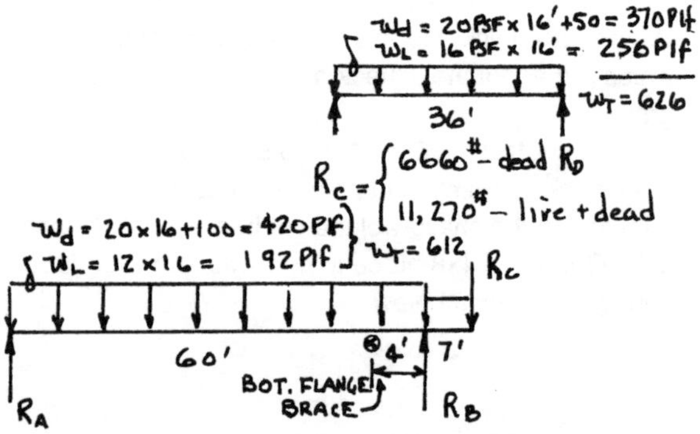

The maximum positive bending occurs with the dead load only on span B-to-D:

$\Sigma M_A = 0$: $R_B(60) - 612 \times 60^2/2 - 6660(67) - 370(7)(63.5) = 0$ ∴ $R_B = 28{,}540^{\#}$

$\Sigma F_y = 0$: $R_A = 612 \times 60 + 370 \times 7 + 6660 - 28{,}540 = 17{,}430^{\#} \uparrow$

$V = 0$: $x = 17{,}430 / 612 = 28.48'$

∴ $M^+_{max} = 0.5 \times 17{,}430 \times 28.48 = 248{,}200^{\#\text{-ft}}$

The critical negative moment and maximum shear occurs with the live load on both spans:

$R_b = [612 \times 60^2/2 + 626(7)(63.5) + 11{,}270 \times 67] / 60 = 35{,}580^{\#}$

$V_{max} = 35{,}580 - 11{,}270 - 626(7) = 19{,}930^{\#}$

$M^- = 11{,}270 \times 7 + 626(7^2)/2 = 94{,}200^{\#\text{-ft}}$

Beam B2 has full lateral support; take $C_D = 1.25$ (roof live); $C_v \cong 0.85$:

$F_b' = C_D C_v F_b = (1.25)(0.85)(2200) = 2340$ psi

$M_{max} = wL^2/8 = 626 \times 36^2/8 = 101{,}400^{\#\text{-ft}}$

$S \geq M_{max}/F_b' = 101{,}400 \times 12/2340 = 520\ \text{in}^3$

Try $5^1/_8 \times 25.5$ in ($S = 555\ \text{in}^3$; $I = 7082\ \text{in}^4$). Check the asumed F'_b:

$C_V = K_L(21/L)^{1/10}(12/d)^{1/10}(5.125/b)^{1/10}$

$= 1.0 \times (21/36)^{1/10} \times (12/25.5)^{1/10} \times (5.125/5.125)^{1/10}$

$= 0.88$

$F_b' = C_D C_V F_b = 1.25 \times 0.88 \times 2200 = 2420$ psi OK

Check the deflection between the hinge points (Re: Table 5.7, AITCM) Limit $\Delta_L = {}^L/_{360}$; $\Delta_{TL} \leq {}^L/_{240}$ (plastered ceilings):

$\Delta_L = {}^5/_{384}(w_L L)L^3/EI \leq {}^L/_{360} = 1.2"$

$\Delta_L = 0.013(256 \times 36)(36 \times 12)^3/(1.8 \times 10^6 \times 7082) = 0.76" < {}^L/_{360}$

$\Delta_{TL} = (626/256)\,0.76 = 1.86" > {}^L/_{240} = 1.8"$

Need to stiffen the beam:

$I \geq (1.86/1.8)\,7082 = 7320\ \text{in}^4$

Select $5^1/_8 \times 27$ in ($I = 8406\ \text{in}^4$)

Check the assumed weight (36 pcf):

$w = (5.125 \times 27/144) \times 36 = 35$ plf < 50 plf OK

Check shear:

$L_e = L - 2d = 36 - 2 \times 27/12$

$V = wL_e/2 = 626(31.5/2) = 9860^{\#}$

$f_v = {}^3/_2 V/bd = 1.5 \times 9860/(5.125 \times 27) = 107$ psi

$f_v < F_v' = 1.25F_v = 1.25 \times 165 = 206$ psi

∴ **Use $5^1/_8 \times 27$ in. for B2**

b) Design beam B1. Two flexural checks:

$M^+ = 248{,}200^{\#\text{-ft}}$, $w/l_u = 0$; $M^- = 94{,}200^{\#\text{-ft}}$ $w/l_u = 7'$ ($l_e = 1.87 \times 7 = 13.09'$). Select a trial section based on the positive moment and $F'_b \cong 1.25 \times 0.8 \times 2200 = 2200$ psi:

$S \geq M^+/F_b' \cong 248{,}200 \times 12/2200 = 1353\ \text{in}^3$

Try $6^3/_4 \times 36$ in ($S = 1458\ \text{in}^3$; $I = 26{,}244\ \text{in}^4$)

$C_V = K_L(21/L)^{1/10}(12/d)^{1/10}(5.125/b)^{1/10}$

$= 1.0 \times (21/60)^{1/10} \times (12/36)^{1/10} \times (5.125/6.75)^{1/10}$

$= 0.78$

$F_b' = C_D C_V F_b = 1.25 \times 0.78 \times 2200 = 2145$ psi $> f_b = 2042$ psi

Check the overhang:

$R_B = \sqrt{l_e d/b^2} = \sqrt{13.09 \times 12 \times 36/6.75^2} = 11.1$

$F_{bE} = 0.438E'/R_B^2 = 0.438 \times 1.8 \times 10^6/11.1^2 = 6400$ psi

$F_{bE}/F_b^* = 6400/(1.25 \times 2200) = 2.32$

$$C_L = \frac{1+F_{bE}/F_b^*}{1.9} - \sqrt{\left[\frac{1+F_{bE}/F_b^*}{1.9}\right]^2 - \frac{F_{bE}/F_b^*}{0.95}}$$

$$= \frac{1+2.32}{1.9} - \sqrt{\left[\frac{1+2.32}{1.9}\right]^2 - \frac{2.32}{0.95}}$$

$C_L = 0.97 < C_v = 0.78$

$\therefore$ $F_b' = C_D C_v F_b = 1.25 \times 0.78 \times 2200 = 2145$ psi

$f_b = M^- / S = 94{,}200 \times 12 / 1225 = 925$ psi

$f_b < F_b'$ $\therefore$ OK at the overhang

Check deflection with live load only on span AB:

$\Delta_L = {}^5/_{384}(w_L L)L^3 / EI$

$= 0.013(192 \times 60)(60 \times 12)^3 / (1.8E6 \times 26{,}244)$

$= 1.18"$

${}^L/_{360} = 60 \times 12 / 360 = 2" > \Delta_L$ $\therefore$ OK

Check total load deflection $< {}^L/_{240}$:

$M_d^- = (370 / 626) \times 94{,}200 = 55{,}700^{\#\text{-ft}}$

$\Delta_d = {}^5/_{384}(w_d L)L^3 / EI - M_d L^2 / 16EI$

$= 0.013(420 \times 60)(60 \times 12)^3 / (1.8E6 \times 26{,}244)$

$- 55{,}700 \times 12(60 \times 12)^2 / (16 \times 1.8E6 \times 26{,}244)$

$= 2.57 - 0.46 = 2.11"$

${}^L/_{240} = 60 \times 12 / 240 = 3"$

$\Delta_{TL} = \Delta_L + \Delta_d = 1.18 + 2.11 = 3.29" > {}^L/_{240}$

$\therefore$ Need more stiffness

$I \geq (3.29 / 3.00)\ 26{,}244 = 28{,}780$ in^4

Select 6.75 × 37.5 in. ($I = 29{,}660$ in^4). Check shear:

$V = V_{max} - wd = 19{,}930 - 612(37.5 / 12) = 18{,}000^{\#}$

$f_v = 1.5 \times 18{,}000 / (6.75 \times 37.5) = 107$ psi $< F_v'$

$\therefore$ **Use 6.75×37.5 in. for B1**

c) Compute the required camber for B1. Re: Table 5.9, AITCM, the recommended camber is $1.5\Delta_d$.

$\therefore$ **Camber = 1.5(370 / 256) 0.91 = 1.97", say 2"**

Timber-26

Check the maximum stresses in the pile. Assume a single, isolated pile for which a reduction in allowable stresses applies (sec. 1808.1.2 UBC-97) and that the live load is for normal (10 yr) duration. Therefore, $C_D = 1.0$ for gravity loads and $C_D = 1.33$ for dead plus seismic. Re: Table 6A, NDS-91, with adjustment per 6.3-11:

$F_C = (1250 / 1.25)(1 + 0.002(12)) = 1025$ psi

$F_b = 2450 / 1.3 = 1885$ psi

$E = 1{,}500{,}000$ psi

The tabulated F_b includes the form factor of 1.18 for a round section in flexure. Check axial compression under live plus dead load (3.7-2 and 3.7-3, NDS). The equivalent square tapered column is:

$d_{min} = (\pi D^2 / 4)^{1/2} = (\pi \times 10^2 / 4)^{1/2} = 8.86"$;

$d_{max} = (\pi D^2 / 4)^{1/2} = (\pi \times 12^2 / 4)^{1/2} = 10.6"$;

$d = d_{min} + [a - 0.15(1 - d_{min} / d_{max})](d_{max} - d_{min})$

$d = 8.86 + [0.3 - 0.15(1 - 8.86 / 10.6)](10.6 - 8.86)$

$= 9.33$

$K_e = 2.1$ (fixed-free column)

$l_e = K_e l_u = 2.1 \times 16 \times 12 = 403"$

$l_e / d = 403 / 9.33 = 43.2$

Use the current NDS-91 for F_c':

$E' = 1.5 \times 10^6$ psi

$$F_{cE} = \frac{0.3E'}{(l_e/d)^2} = \frac{0.3 \times 1{,}500{,}000}{43.2^2} = 241 \text{ psi}$$

$F_c^* = C_D C_F F_c = 1025$ psi

$F_{cE} / F_c^* = 241 / 1025 = 0.24$

$$C_p = \frac{1+F_{cE}/F_c^*}{2c} - \sqrt{\left[\frac{1+F_{cE}/F_c^*}{2c}\right]^2 - \frac{F_{cE}/F_c^*}{c}}$$

$$= \frac{1+0.24}{1.6} - \sqrt{\left[\frac{1+0.24}{1.6}\right]^2 - \frac{0.24}{0.8}}$$

$C_p = 0.23$

$\therefore$ $F_c' = C_p C_D C_F F_c = 0.23 \times 1.0 \times 1.0 \times 1025 = 236$ psi

$f_c = P_{ltd} / A = 15{,}000 / 9.33^2 = 172$ psi $< F_c'$ $\therefore$ OK

Check combined axial and bending under seismic plus dead. Critical stresses occur at the point of fixity:

$A = 78.5$ in^2; $S = \pi d^3 / 32 = \pi(10)^3 / 32 = 98.1$ in^3.

$M_e = \pm 1^k(16') = \pm 16{,}000^{\#\text{-ft}}$

$$\text{Check: } \left[\frac{f_c}{F_c'}\right]^2 + \frac{f_{b1}}{F_{b1}'[1-(f_c/F_{cE1})]} + \frac{f_{b2}}{F_{b2}'[1-(f_c/F_{cE2})-(f_b/F_{bE})^2]} \leq 1.0$$

For bending about one axis, only f_{b1} applies:

$f_{b1} = M / S = 16{,}000 \times 12 / 98.1 = 1957$ psi

$F_b' = C_D F_b = 1.33 \times 1885 = 2500$ psi

$f_c = P_d / A = 10{,}000 / 78.5 = 127$ psi

$F_c' = C_p C_D C_F F_c = 0.23 \times 4 / 3 \times 1.0 \times 1025 = 314$ psi

$$\left[\frac{127}{314}\right]^2 + \frac{1957}{2500[1-(127/241)]} = 1.82 \succ 1.0$$

Excessive! The pile is not adequate under dead plus seismic loading.

Timber-27

a) Determine the nailing required for the plywood wall sheathing.

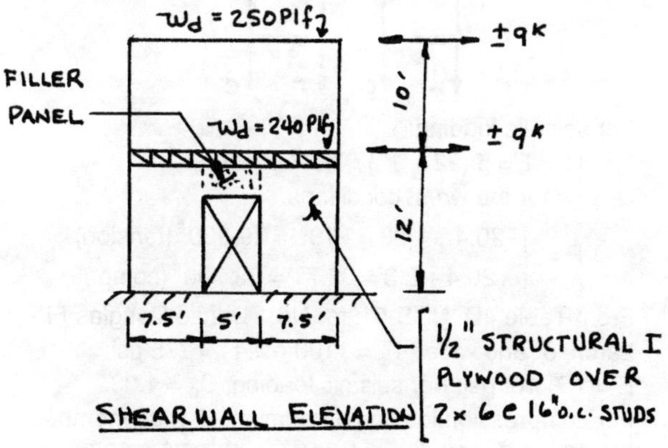

(Re: Table No. 23-II-I-1, UBC-87)

For the top story: $v = 9000 / 20 = 450$ plf

Use **8d @ 3" o.c. along panel edges**
8d @ 12" o.c. at interior supports

For the bottom story: $v = 18{,}000 / (2 \times 7.5) = 1200$ plf. The shear is too large to resist with sheathing on only one side; therefore, sheath both sides: $v = 1200 / 2 = 600$ plf:

Use 8d @ 2-1/2" o.c. along panel edges

By interpolation between values for 3" o.c. and 2" o.c., the allowable shear for 8d @ 2-1/2" o.c. is:

$v = 550 + 0.5(730 - 550) = 640$ plf – OK

∴ Use **8d @ 2-1/2" o.c. along panel edges / 8d @ 12" o.c. at interior supports** } on both sides

Panel joints must be offset to fall on different framing members and the nails staggered along each edge of the sheathing.

For the filler panel over the door, sheath one side only and nail with 6d @ 6" o.c. (no structural coupling exists between the two lower walls).

b) Design boundary members for the lower walls. Per problem statement, include only 75% of the dead load as resisting overturning:

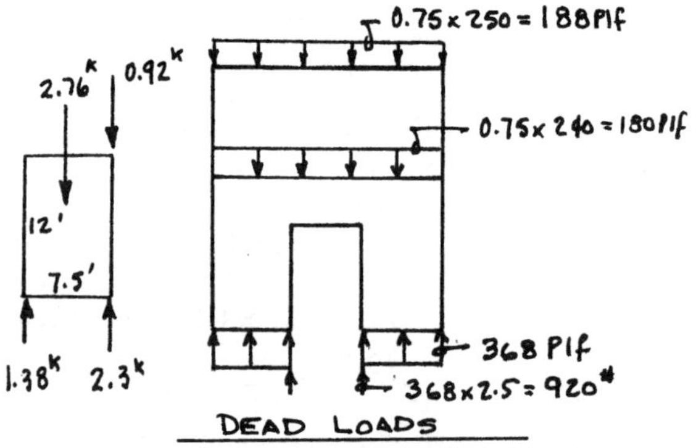

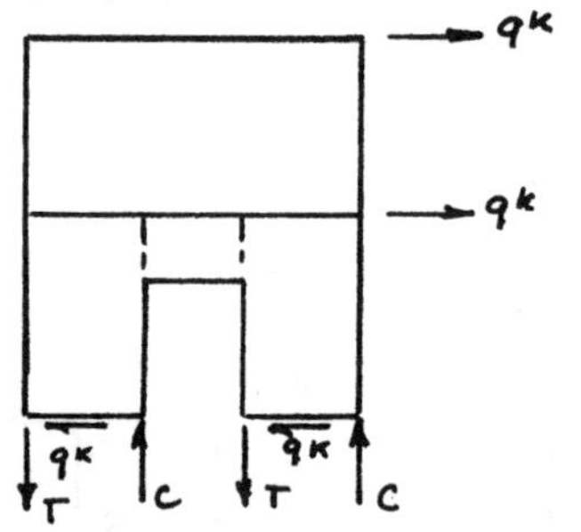

For seismic loading:

$C = T = [9(22 + 12) / 7.5] / 2 = 20.4^k$

Design for the worst conditions:

$$P = \begin{cases} -20.4 + 1.38 = -19^k = 19{,}000^{\#}\text{(tension)} \\ +20.4 + 2.3 = 22.7^k = 22{,}700^{\#}\text{(comp.)} \end{cases}$$

From Table 4D, NDS-91, for No.1 Dense Douglas Fir-Larch, 6" and wider, $F_c = 1100$ psi; $F_t = 775$ psi; $E = 1.7 \times 10^6$ psi. For seismic loading, $C_D = 1.33$.

For compression loading on nominal 6" wide member:

$l_u = 12 - 1.5 = 10.5'$; $l_e / d = 10.5 \times 12 / 5.5 = 22.9$:

$$F_{cE} = \frac{0.3E'}{(l_e/d)^2} = \frac{0.3 \times 1{,}700{,}000}{22.9^2} = 972 \text{ psi}$$

$F_c^* = C_D C_F F_c = 1.33 \times 1.0 \times 1100 = 1463$ psi

$F_{cE} / F_c^* = 972 / 1463 = 0.66$

$$C_p = \frac{1 + F_{cE}/F_c^*}{2c} - \sqrt{\left[\frac{1 + F_{cE}/F_c^*}{2c}\right]^2 - \frac{F_{cE}/F_c^*}{c}}$$

$$= \frac{1 + 0.66}{1.6} - \sqrt{\left[\frac{1 + 0.66}{1.6}\right]^2 - \frac{0.66}{0.8}}$$

$C_p = 0.54$

∴ $F_c' = C_p C_D C_F F_c = 0.54 \times 4/3 \times 1.0 \times 1100 = 790$ psi

$A \geq P / F_c' = 22{,}700 / 790 = 28.7 \text{ in}^2$

$h \geq A / b = 28.7 / 5.5 = 5.2"$

Try 6×6: assume 3/4" dia bolts through the member for holddown anchors:

$A_n = 5.5(5.5 - {}^{13}/_{16}) = 25.78 \text{ in}^2$

$f_t = P_t / A_n = 19{,}000 / 25.78 = 737$ psi

$F_t' = C_D F_t = 1.33 \times 775 = 1030 \text{ psi} > f_t$ ∴ OK

Check bearing against the 2×6 sill:

$f_{c\perp} = P / A = 22{,}700 / (5.5 \times 5.5) = 750$ psi

$F_{c\perp} = 625 \text{ psi} < f_{c\perp}$ (Note: $F_{c\perp}$ cannot be increased by C_D) $< f_{c\perp}$ **∴ need a larger post:**

$A \geq P / F_{c\perp} = 22{,}700 / 625$

$A \geq 36.3 \text{ in}^2$ ∴ $h > A / b$

$h \geq 36.3 / 5.5 = 6.6"$

∴ **Use 6×8 boundary members**

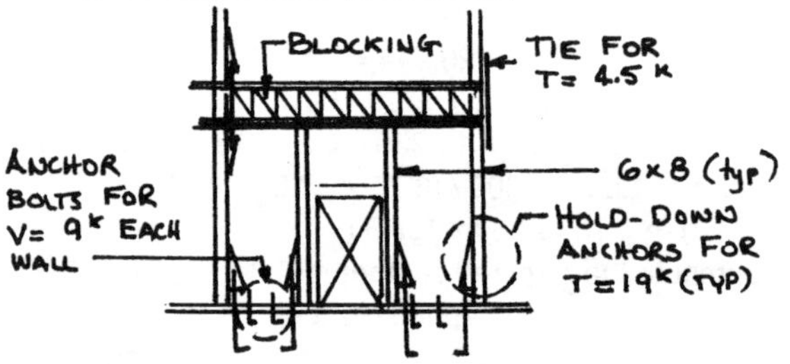

Timber-28

Comment on the given construction details (Re: AITC 104-84 "Typical Construction Details," (AITCM, 1985)).

a) Primary problem is that the 8 1/2" gage in fasteners in the 6×14 will create restraint to cross grain shrinkage, which will likely cause splitting of the 6×14. Also, high support of the 6×14 may cause loss of bearing; hence, overload of fasteners as shrinkage occurs. A better detail – assuming load transfer is primarily bearing rather than uplift – is the T-plate shown in Figure 5.3, AITCM:

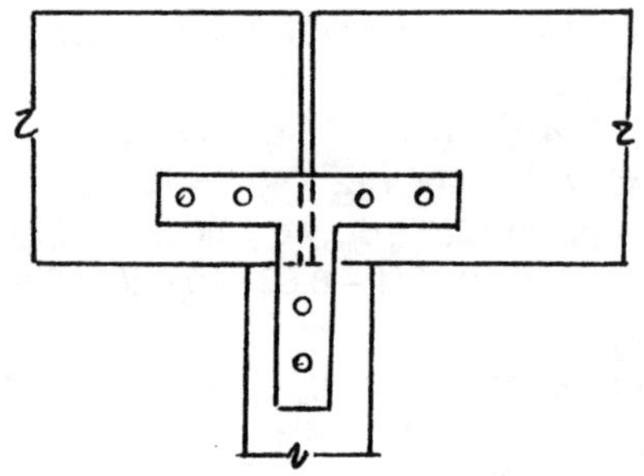

b) The location of bolts high in the beam will likely result in loss of bearing in the beam seat. The effective depth resisting shear would become $d_e = 6"$. The end joint shear stress, $f_v = (^3/_2V / bd_e)(d / d_e)$ would likely be too large, resulting in horizontal shear failure at the level of the lower bolt. Better details are shown in Figures 4.1, 4.2, and 4.3, ATICM, in which the fastener is located near the bearing seat:

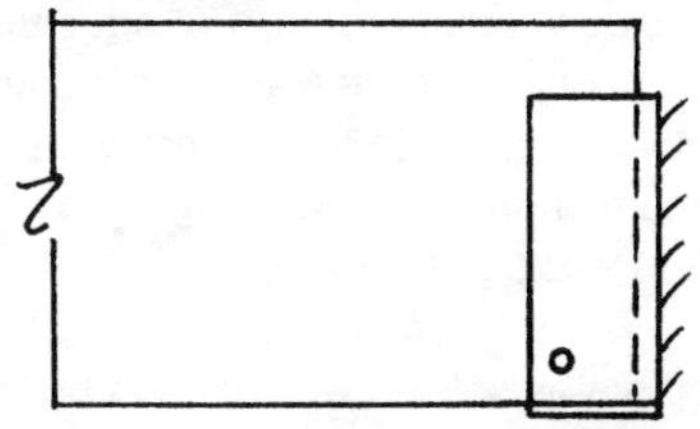

c) Lack of support of the outer lams will create large shear stresses, which will likely lead to splitting in the glulam. Support should be provided over the full width of the tapered cut, as shown in Figure 2.6, AITCM:

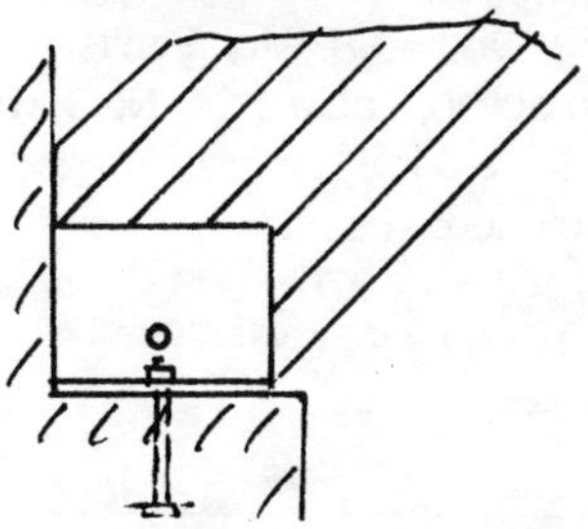

d) The connection induces high cross grain tension; hence, likely splitting failure at the level of the bolts. The best connection would be a hanger that crosses the top of the beam and transfers force by bearing perpendicular to grain. If a bolted connection must be used, the connection should be made well above the neutral axis. Preferred types are shown in Figure 10.1, AITCM:

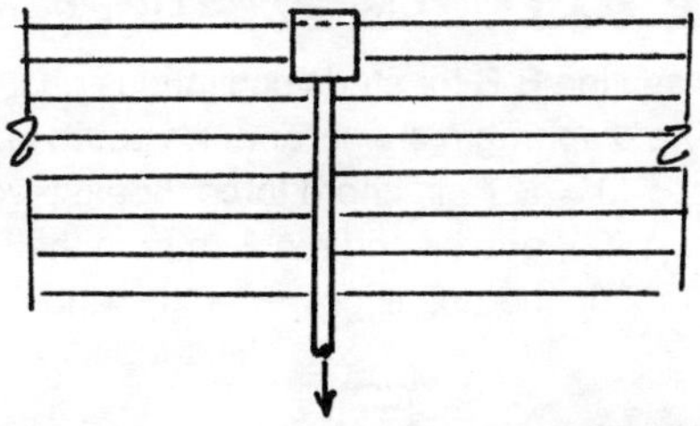

e) The new connection has the same potential problem with bolts located too high that was discussed in part 'b'. In addition, a potential problem is differential vertical deflection of the new floor where it abuts the existing floor. This can be caused by shrinkage of the new 4×14 if, as is generally the case, the moisture content of the 4×14 is higher at time of erection that it will be during service. Either the beam should be allowed to reach its final moisture content prior to attaching the flooring, or provision made to compensate for the shrinkage.

Timber-29

a) Design the typical roof joists that are spaced 24" o.c. and span 20'. Tributary area = $20 \times 2 = 140$ ft^2; Re: Table No. 16-C, UBC-97: $w_L = 20$ psf.

w_d = builtup roof + plywood + ceiling + elec / mech + joist = 7 + 1 + 3 + 2 + 2 = 15 psf

$\therefore \quad w_{TL} = w_L + w_d = 35$ psf

$w = 35 \text{ psf} \times 2' = 70$ plf

$M = wL^2 / 8 = 70 \times 20^2 / 8 = 3500^{\#\text{-ft}}$

$R = wL / 2 = 70 \times 20 / 2 = 700^{\#}$

Douglas Fir-Larch No. 1: $F_b = 1.15 \times 1150 = 1320$ psi (repetitive use); $F_V = 95$ psi; $F_{c\perp} = 625$ psi; $E = 1.8 \times 10^6$ psi; $C_D = 1.25$ (roof live).

$F_b' = C_D C_F F_b = 1.25 \times 1.0 \times 1320 = 1650$ psi

$S \geq M / F_b' = 3500 \times 12 / 1650 = 25.5$ in^3

Neglect ponding; however, limit the deflection: $\Delta_L \leq {}^L/_{240}$; $\Delta_{TL} \leq {}^L/_{180}$.

Total load deflection governs:

$\Delta_{TL} = (^5/_{384})(wL)L^3 / EI \leq {}^L/_{180}$

$\therefore \quad I \geq 180 \times 0.013(70 \times 20)(20 \times 12)^2 / 1.8E6 = 105$ in^4

Approximate d = 12": $L_e = L - 2d = 18'$

$\therefore \quad V = wL_e / 2 = 70 \times 18 / 2 = 630^{\#}$

$f_v = {}^3/_2V / bd \leq F_v' = 1.25 \times 95 = 119$ psi

$\therefore \quad bd \geq 1.5 \times 630 / 119 = 7.9$ in^2

Select a 2×12; stiffness governs: I = 178 in^4; A = 16.9 in^2; S = 31.6 in^3.

$\therefore$ **<u>Use 2×12 @ 24" o.c. for joists.</u>**

For 2×12: nominal depth-to-width = 6 requires full depth blocking at the ends with the top edge restrained. Provide two lines of bridging or blocking near the third points.

b) Design the glulam roof beams, which are laterally braced by the joists and span 40' simply supported. Tributary area = $40 \times 20 = 800$ ft^2 $\therefore$ $w_L = 12$ psf. Treat loads as uniformly distributed and estimate the beam weight as 50 plf.

$\therefore \quad w = (12 + 15) \text{ psf} \times 20' + 50 \text{ plf} = 590$ plf

$M_{max} = 590 \times 40^2 / 8 = 118{,}000^{\#\text{-ft}}$

For 22F glulam, estimate $C_V = 0.9$

$F_b' = 1.25 \times 0.9 \times 2200 = 2475$ psi

$\therefore \quad S \geq M_{max} / F_b' = 118{,}000 \times 12 / 2475 = 572$ in^3

Take $E = 1.6 \times 10^6$ psi; limit $\Delta_{TL} \leq L / 180$.

$\Delta_{TL} = (^5/_{384})(590 \times 40)(40 \times 12)^3 / ((1.6 \times 10^6)I) \leq 40 \times 12 / 180$

$I \geq 180 \times 0.013 \times 590 \times 40 \times (40 \times 12)^2 / 1.6 \times 10^6 = 7954$ in^4

Try $5^1/_8 \times 27.0$ in: S = 623 in^3; I = 8406 in^4:

$f_b = M_{max} / S = 118{,}000 \times 12 / 623 = 2273$ psi

$C_V = K_L(21 / L)^{1/10}(12 / d)^{1/10}(5.125 / b)^{1/10}$

$= 1.0 \times (21 / 40)^{1/10} \times (12 / 27)^{1/10} \times (5.125 / 8.75)^{1/10} = 0.86$

$F_b' = C_D C_V F_b = 1.25 \times 0.86 \times 2200 = 2365 \text{ psi} > f_b$ $\therefore$ OK

Check shear: $L_e = L - 2d = 40 - 2 \times 27 / 12 = 35.5'$

$V = 590 \times 35.5 / 2 = 10{,}500^{\#}$

$f_v = {}^3/_2V / bd = 1.5 \times 10{,}500 / (5.125 \times 27.0)$

$= 113 < F_v'$ all combinations

∴ **Use $5^1/_8 \times 27.0$ in. glulam (22F-V6) for roof beams.**

c) Compute lateral loads to the roof diaphragm per UBC-97. For wind (Re: 1620):

$p = C_e C_q q_s I_w$; $I_w = 1.0$; $C_e = 1.2$; $q_s = 17$ psf

$p = C_q(1.2)(17) = (20.4 \text{ psf})C_q$

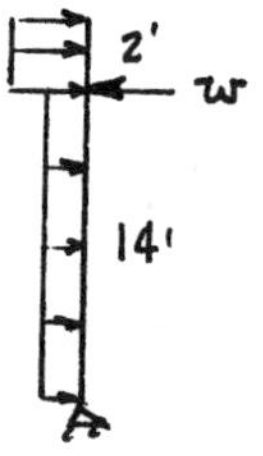

Take $C_q = 1.3$ on building; $C_q = 1.3$ on each parallel parapet. The wind load at the roof level is the same in all directions:

$w = [2 \times 1.3 \times 20.4 \times 2 \times 15 + 1.3 \times 20.4 \times 14^2 / 2] / 14 = 300$ plf

Seismic (Re: 1630, UBC-97):

$V = (2.5C_aI / R)W$

Most of the gravity load is supported by the masonry bearing walls ∴ take $R = 4.5$ and $\Omega_o = 2.8$; $C_a = 0.44$; divide the design level base shear by 1.4 to convert to working stress level:

$V = (2.5 \times 0.44 \times 1 / 4.5)\ W / 1.4 = 0.175W$

For North-South ground motion:

$w_1 = 0.175(15 \text{ psf} \times 80' + (2 \times 15 \text{ psf} \times 16^2 / 2) / 14)$

w_1 = 258 plf (wind governs between 1 & 2)

$w_2 = 0.175(15 \times (80 + 40) + (2 \times 15 \times 16^2 / 2) / 14)$

w_2 = 363 plf (seismic governs) between 2 & 3

For East-West ground motion:

$w_3 = 0.175(15 \times 200 + ((15 + 80)16^2 / 2) / 14)$

w_3 = 677 plf

$w_4 = w_1$ = 258 plf

d) Calculate the plywood nailing and chord forces for north-south loads on the flexible diaphragm.

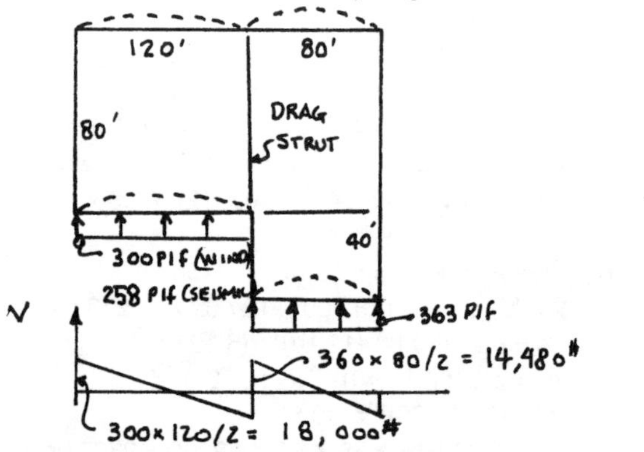

DIAPHRAGM SHEAR : NORTH-SOUTH LOADS

Wind (300 plf) governs the diaphragm and the west wall; seismic (258 plf) acts in combination with 363 plf between lines 2 and 3 governs the collector and wall on line 2.

Diaphragm nailing: Line 1: $v = 18{,}000 / 80 = 225$ plf; at Line 3: $v = 14{,}480 / 120 = 121$ plf. Thus, the maximum diaphragm shear is $v = 225$ plf. Re: Table No. 23-II-H, UBC-97, use a blocked diaphragm with 6d @ 4" o.c., or, 8d @ 6" o.c. on the boundary and continuous edges. Eliminate the blocking where $v < 180$ plf. Thus, for the maximum condition (225 plf):

Use a blocked diaphragm, 1/2" structural I, with 8d @ 6" o.c. on boundaries and continuous edges; 6" o.c. other edges; 12" o.c. on intermediate supports.

The maximum chord force is:

$M = 300 \times 120^2 / 8 = 540{,}000^\#$

$T = C = M / 80 = 540{,}000 / 80 = 6750^\#$

Try 2×6 top plates spliced with 3/4" dia bolts:

$F_t = {}^4/_3(1000) = 1333$ psi

$A_n = (5.5 - {}^{13}/_{16})1.5 = 7.03 \text{ in}^2$

$f_t = 6750 / 7.03 = 960 \text{ psi} < F_t$ ∴ OK

Bolts (Re: Table 8.2A, NDS-91):

$Z'_{//} = (1.33)(720) = 960\ ^\#/_{\text{bolt}}$

$n \geq 6750 / 960 = 7.1$ – try 8 bolts

Check group action, Table 3.6A, NDS, with

$A_s = 5.5 \times 1.5 = 8.3 \text{ in}^2$:

$C_g = 0.7$ – no good – use a larger bolt size, say 1" dia:

$Z'_{//} = 0.7 \times 1.33 \times 970 = 903\ ^\#/_{\text{bolt}}$

$n \geq 6750 / 903 \cong 8$ –OK

Detail as follows:

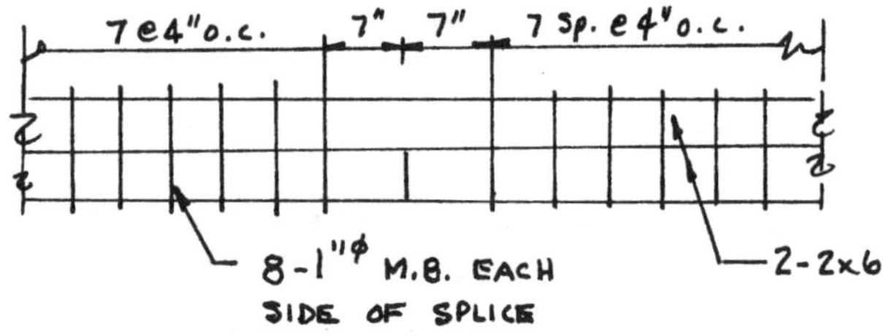

TOP PLATES (CHORD) SPLICE DETAIL

e) Design section B-B for studs continuous past the roof level. The diaphragm shear for north-south loads = 15,200 / 120 = 127 plf; chord force for east-west loads is given as 1500# gravity and reaction from the joist = 700#; wall-to-roof anchorage is governed by wind.

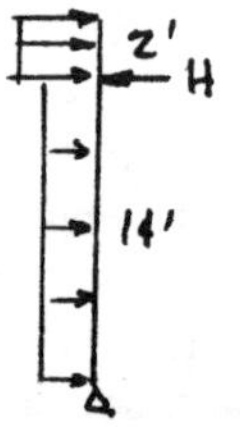

$C_q = 1.1$ outward; $= 1.3$ on parapet

$H = (1.3 \times 20.4 \times 2 \times 15 + 1.1 \times 20.4 \times 14^2 / 2) / 14 = 214 \text{ plf} \rightarrow$

Bearing length for $R = 700^\#$:

$F_{c\perp} = 1.0 \times 625 = 625$ psi

Assume studs spaced @ 16" o.c.: $R_s = 700(16 / 24) = 470^\#$

$N \geq R_s / bF_{c\perp} = 470 / (1.5 \times 625) = 0.5"$

Use a 2×12 ledger let-in 1/2" into 2×6 studs; provide strap ties to anchor wall to roof at 4' o.c. for P = 4 × 214 = 856 $^{\#}/_{anchor}$; use metal joist hangers nailed to ledgers to support the 700# joist reaction; splice the ledger (i.e., chord) using 2 – 3/4" diameter bolts with PL 4×1/4 each side. Detail as follows:

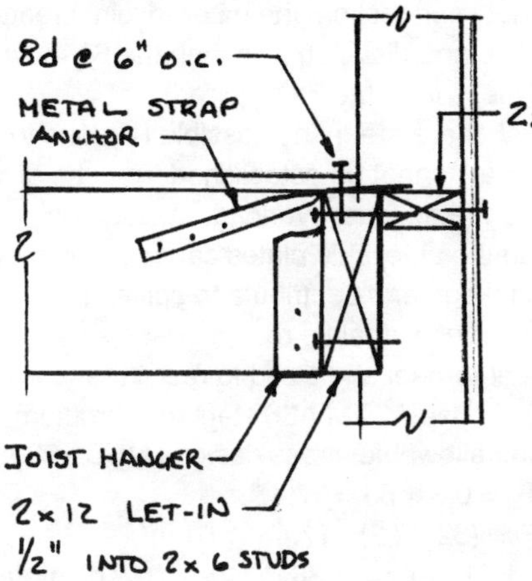

DETAIL - CONNECTION B-B. (N.T.S.)

Timber-30

a) Design the beam marked No.3. Given: F_b = 2000 psi; F_V = 165 psi; E = 1.8×10^6 psi; C_D = 1.0; beam weight to be neglected; do not check deflection; b = 3$^1/_8$"; bearing is sufficient.

Assume that the beam is laterally braced at ends only: l_u = 20'. The 3-1/8" width specified in the problem statement is inefficient for this condition and better approaches are either to use a wider section or to brace the compression flange at midspan. However, I assume that conditions will not permit either of these options and that a deeper, less efficient member is required.

Solve by trial and error: Assume initially that $F_b' = C_L F_b$ = 0.75 × 2000 psi = 1500 psi

$M = 3100 \times 20 / 4 = 15{,}500^{\#\text{-ft}}$

$S \geq M / F_b' = 15{,}500 \times 12 / 1500 = 124 \text{ in}^3$

Try 3$^1/_8$×16.5" (S = 141.2 in^3). Re: Table 3.3.3, NDS-91:

$l_e = 1.37 l_u + 3d = 1.37 \times 20 \times 12 + 3 \times 16.5 = 378"$

$R_B = (l_e d / b^2)^{1/2} = (378 \times 16.5 / 3.125^2)^{1/2} = 25.3$

$F_b^* = C_D F_b = 1.0 \times 2000 = 2000 \text{ psi}$

$F_{bE} = 0.438E' / R_B^2 = 0.438 \times 1.8 \times 10^6 / 25.3^2$
$= 1231 \text{ psi}$

$F_{bE} / F_b^* = 1231 / 2000 = 0.62$

$$C_L = \frac{1 + F_{bE}/F_b^*}{1.9} - \sqrt{\left[\frac{1 + F_{bE}/F_b^*}{1.9}\right]^2 - \frac{F_{bE}/F_b^*}{0.95}}$$

$$= \frac{1+0.62}{1.9} - \sqrt{\left[\frac{1+0.62}{1.9}\right]^2 - \frac{0.62}{0.95}}$$

$C_L = 0.58$

∴ $F_b' = C_D C_L F_b = 1.0 \times 0.58 \times 2000 = 1160 \text{ psi}$

$f_b = M / S = 15{,}500 \times 12 / 141.2 = 1320 \text{ psi}$

∴ No good. Try 3-1/8 × 18: S = 169 in^3:

$l_e = 1.37 l_u + 3d = 1.37 \times 20 \times 12 + 3 \times 18 = 383"$

$R_B = (l_e d / b^2)^{1/2} = (383 \times 18 / 3.125^2)^{1/2} = 26.6$

$F_{bE} = 0.438E' / R_B^2 = 0.438 \times 1.8 \times 10^6 / 26.6^2$
$= 1115 \text{ psi}$

$F_{bE} / F_b^* = 1090 / 2000 = 0.56$

$$C_L = \frac{1 + F_{bE}/F_b^*}{1.9} - \sqrt{\left[\frac{1 + F_{bE}/F_b^*}{1.9}\right]^2 - \frac{F_{bE}/F_b^*}{0.95}}$$

$$= \frac{1+0.56}{1.9} - \sqrt{\left[\frac{1+0.56}{1.9}\right]^2 - \frac{0.56}{0.95}}$$

$C_L = 0.53$

∴ $F_b' = C_D C_L F_b = 1.0 \times 0.53 \times 2000 = 1060 \text{ psi}$

$f_b = M / S = 15{,}500 \times 12 / 169 = 1100$ psi – no good.

Increase to 19.5" (S = 198 in^3):

$l_e = 1.37 \times 20 \times 12 + 3 \times 19.5 = 387"$

$R_B = (l_e d / b^2)^{1/2} = (387 \times 19.5 / 3.125^2)^{1/2} = 27.8$

$F_{bE} = 0.438 \times 1.8 \times 10^6 / 27.8^2 = 1020 \text{ psi}$

$F_{bE} / F_b^* = 1020 / 2000 = 0.51$

$$C_L = \frac{1 + F_{bE}/F_b^*}{1.9} - \sqrt{\left[\frac{1 + F_{bE}/F_b^*}{1.9}\right]^2 - \frac{F_{bE}/F_b^*}{0.95}}$$

$$= \frac{1+0.51}{1.9} - \sqrt{\left[\frac{1+0.51}{1.9}\right]^2 - \frac{0.51}{0.95}}$$

$C_L = 0.49$

∴ $F_b' = 1.0 \times 0.49 \times 2000 = 980 \text{ psi}$

$f_b = M / S = 15{,}500 \times 12 / 198 = 940 \text{ psi} < F'_b$ ∴ Ok

Check shear:

$f_v = {}^3/_2 V / bd = 1.5 \times 1550 / (3.125 \times 19.5)$
$= 38 \text{ psi} << F_v$

∴ **Use 3$^1/_8$ × 19.5 in glulam**

b) Design the post-to-beam connection. The post bears directly on the top flange and the problem statement indicates that the bearing area is sufficient. The beam was designed to span 20' with lateral support at the ends only. Theoretically, there should be no tendency for lateral displacement at the midspan for the design stress level. However, the condition at midspan is one of unstable equilibrium, which requires a positive connection to ensure against lateral displacement. Accordingly, design a connection to resist 2% of the vertical force applied laterally at the connection:

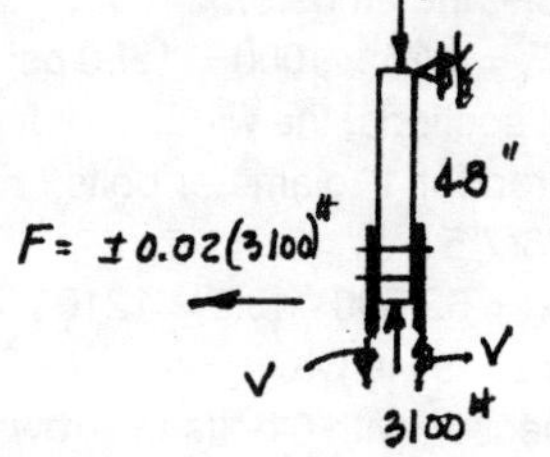

$F = 0.02 \times 3100 = 62^{\#}$

$M = 62^{\#}(48)" = 2976^{\#\text{-in}}$

Resist M by bolts through steel side plates:

$V = M / b = 2976 / 3.5 = 850^{\#}$

Re: Table No. 8.3B, NDS, For 3" main member; 1/4" steel side plates; $\perp$ to grain loading in the beam; single shear; 1/2" diameter bolts:

$Z'_{\perp} = 450\ ^{\#}/_{bolt}$

$\therefore$ use 2 bolts; space 4d = 2"; detail as follows:

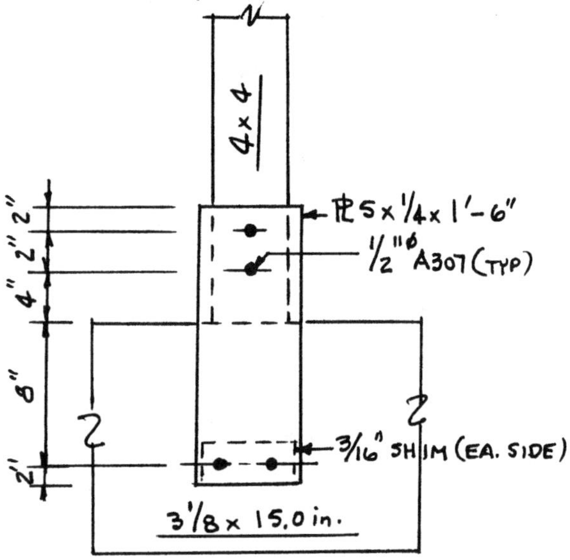

DETAIL. POST-TO-BEAM CONNECTION
(N.T.S.)

Timber-31

a) Check adequacy of the bottom chord.

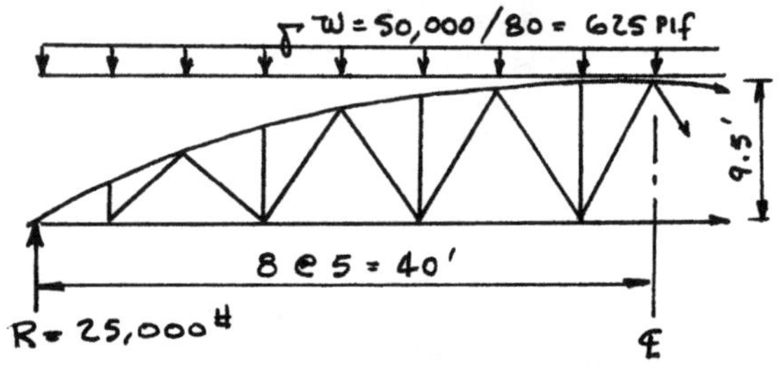

For a bowstring truss, the tension in the lower chord is constant throughout.

$T = M_{max} / \text{rise} = (625 \times 80^2 / 8) / 9.5 = 52{,}600^{\#}$

For 8×8 Douglas Fir-Larch Select, Re: Table 4D, NDS-91 (posts and timbers):

$F_t' = C_D F_t = 1.25 \times 1000 = 1250$ psi

At the critical sections, the width is reduced to 6.75" and holes are bored for 1" diameter bolts:

$\therefore$ $A_n = 6.75(7.5 - {}^{17}/_{16}) = 43.5\ \text{in}^2$

$f_t = T / A_n = 52{,}600 / 43.5 = 1210\ \text{psi} < F_t'$

$\therefore$ Tension stress is OK

For the connection with 6 bolts in a row; 1" diameter; 6.75" wide main member with steel side plates, Re: Table 7.3.6C:

$A_m = 6.75 \times 7.5 = 50.6\ \text{in}^2$; $A_s = 2 \times 5 \times 0.375 = 3.8\ \text{in}^2$

$A_m / A_s = 13.3$ $\therefore$ $C_g = 0.92$

Re: Table 8.3B, NDS:

$Z' = C_D C_g (5520)^{\#}/_{bolt} = 1.25 \times 0.92 \times 5520 = 6350\ ^{\#}/_{bolt}$

$\therefore$ $T \le nr = 6 \times 6{,}350 = 38{,}100^{\#}$ - No Good

The bolts are overloaded! (38% overstress)

Possible causes of splitting in the heel connection are:

- High stresses near the outside bolt initiates splitting.
- For the heel connection, the mitered joint creates an effective end distance less than the 8" shown (possibly less than the 7d required).
- In general, there are many possible reasons for splitting that cannot be established from the given information (e.g., local overload; overtightening the bolts; and the steel side plates can restrain cross-grain shrinkage and contribute to splitting.

b) Proposed repair: use steel tie rods on each side connected heel-to-heel and sized to resist the tension force: $T = 52.6^k$. Use ASTM A36 steel (per problem statement) and allowable stress design (AISC 1989):

$F_t = 0.3F_u = 0.3 \times 58 = 17.4^{ksi}$.

$A \ge T / F_t = (52.6 / 2) / 17.4 = 1.51\ \text{in}^2$

$D \ge (4 \times 1.51 / \pi)^{1/2} = 1.38$ in – use $1^3/_8$" diameter threaded rods

Check capacity of 1" diameter A307 bolts in double shear: $r_v = 15.7\ ^k/_{bolt}$

$n \ge T / r_v = 52.6 / 15.7 = 3.3$ say 4 bolts

Connect the rod $1^3/_8$ into a new plate using the arrangement below:

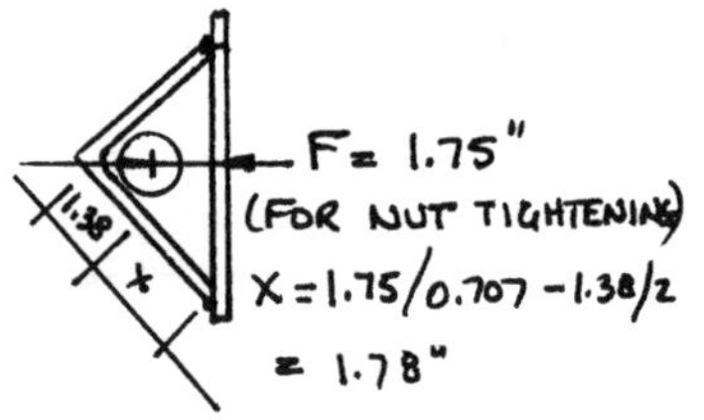

Leg size required:

$l = 1.38 + 1.78 + t = 3.16 + t$

Use L 3 1/2 × 3 1/2 × t

Plate width required:

$\omega > 3.5 / 0.707 = 4.9$ – say 6" to allow for welds

Try 2 – 1/4" E70 welds to connect the angles to the plate; say 2 , 6" long welds:

$f_v = (52.6 / 2) / (2 \times 6) = 2.19\ ^{k/in}$

$f_b = M_w / S_w = (52.6 / 2)1.75 / (2 \times 6^2 / 6) = 3.8\ ^{k/in}$

No good . Try 2 , 8" long welds:

$$\left.\begin{aligned} f_v &= 1.6^{k/in} \\ f_b &= 46 / \left(2 \times 8^2 / 6\right) = 2.1^{k/in} \end{aligned}\right\} f_r = 2.6^{k/in}$$

$f_r = 2.6^{k/in} < q = 4 \times 0.928 = 3.7\ ^{k/in}$ $\therefore$ OK

Detail the connection as follows:

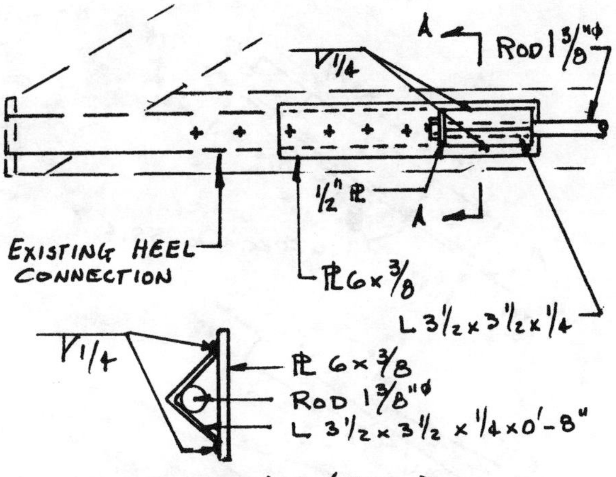

SECTION A-A (N.T.S.)

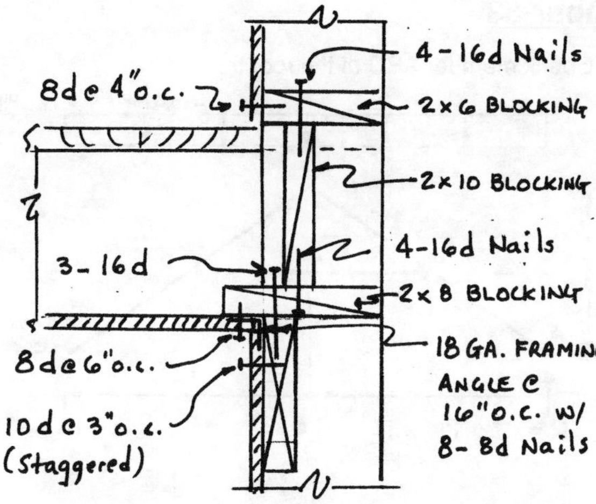

DETAIL - 2nd FLR TO WALL (N.T.S.)

Timber-32

a) Complete the details for section A. Assume the plywood is structural I ; Re: Table 23-II-H, UBC-97: 8d @ 6" o.c. is adequate for the 200 plf ceiling diaphragm whether blocked or unblocked. Re: Table No. 23-II-I-1, UBC-97, use 8d @ 4" o.c. along panel edges for the wall above the second floor (v = 430 plf) and use 10d @ 3" o.c. for the lower wall (v = 665 plf ; nailing is into the existing 2×8 ribbon – stagger nails).

Use No. 1 Douglas Fir-Larch blocking to transfer the shear to the lower story shearwall. Try 2×'s nailed with 16d nails (Re: Table No. 12.3B, NDS-91) ; lateral load parallel to grain loading:

$$Z' = C_D(105\ ^{\#}/_{nail}) = {}^4/_3 \times 105 = 140\ ^{\#}/_{nail}$$

Provide a blocking assembly between the joists and studs, which are spaced 16" o.c. Thus, the force from the second floor wall (400 plf) requires:

$$n \geq (400 \text{ plf} \times 1.33') / 140\ ^{\#}/_{nail} = 3.8 \rightarrow 4 \text{ nails}$$

From the ceiling diaphragm (200 plf):

$$n \geq 600 \times 1.33 / 140 = 6 \text{ ; too many!}$$

Too many nails (potential splitting) use an 18 gage framing anchor with 8 – 8d nails (provides 500# capacity) and additional nails to resist the difference:

$n \geq (600 \times 1.33 - 500) / 140 = 2.1$ say 3 additional nails. Detail the assembly as follows:

b) Detail a splice for the existing 2×8s to serve as a chord resisting a tension force of 1000#. The splice point occurs midway between the studs spaced 16" o.c. Use a 2×8 (16 – 1.5 = 14.5, say 14" long). Position bolts to provide 7d end distance in all locations:

$\therefore$ $2(7d) \leq 14 / 2 \rightarrow d \leq 0.5"$

Try 1/2" diameter bolt; Re: Table No. 8.2A, NDS-91:

$$Z' = C_d Z = ({}^4/_3)(480) = 640\ ^{\#}/_{bolt}$$

$$n \geq T / r' = 1000 / 640 = 1.6, \text{ say 2 bolts each side;}$$

thus:

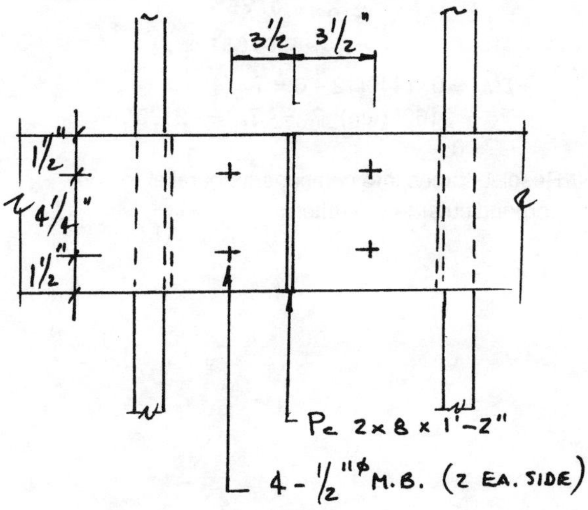

CHORD SPLICE DETAIL (N.T.S.)

Timber-33

a) Check member ABC of the roof truss.

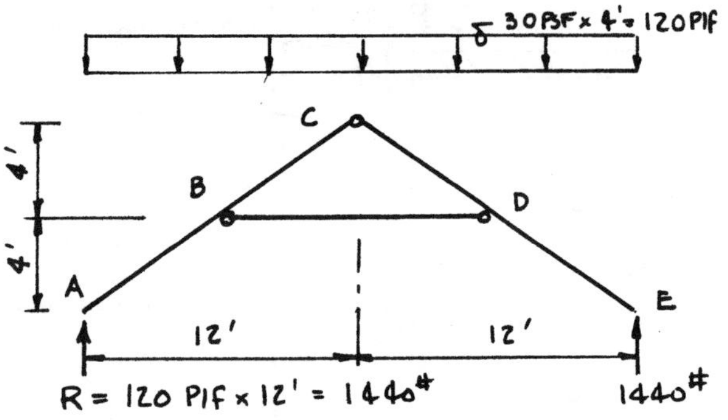

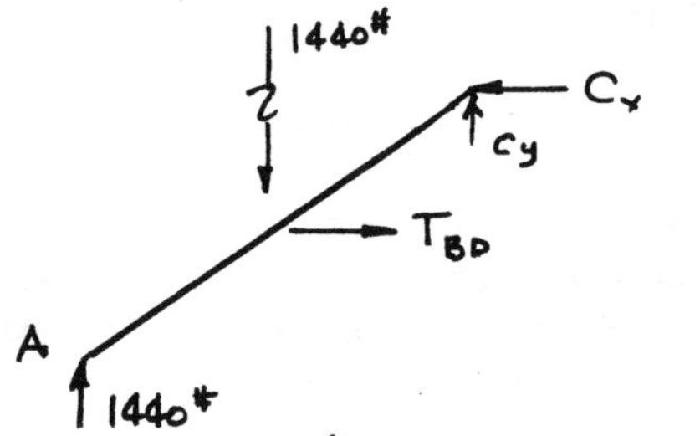

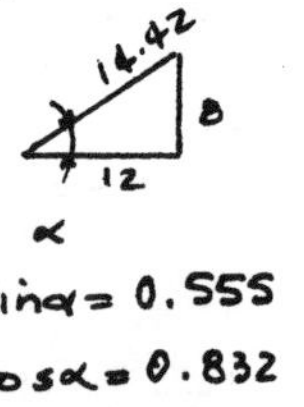

$\Sigma M_C = 0$: $1440(12 - 6) = T_{BD}(4)$

$T_{BD} = 2160^{\#}$ (ten); $C_X = -T_{BD} = -2160^{\#}$

$C_y = 0$

Resolve forces into components parallel to and perpendicular to the rafter:

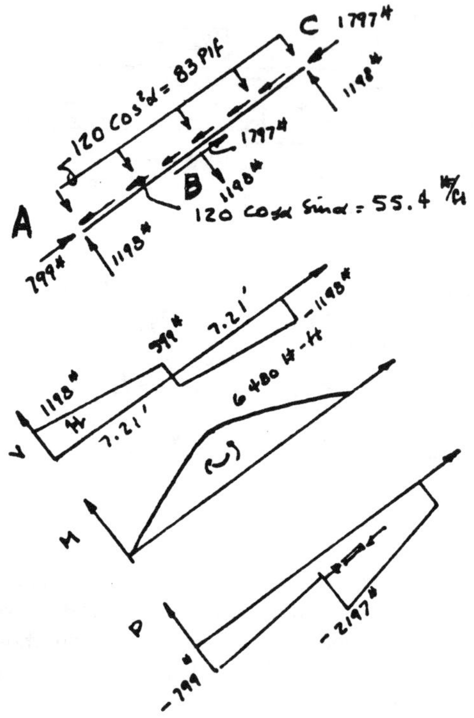

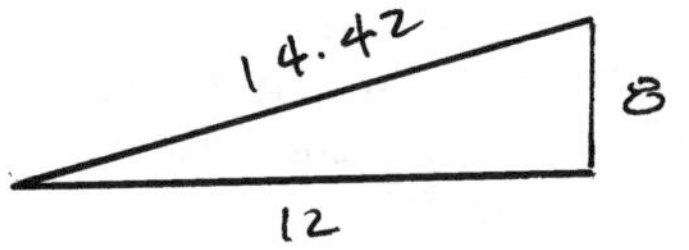

Check the rafter : Select Structural Douglas Fir-Larch. Re: Table 4A, NDS-91. The trussed rafters are spaced 4' o.c. ∴ non repetitive for establishing bending stress: $F_b = 1450$ psi; $F_v = 95$ psi; $F_t = 1000$ psi; $F_{c\perp} = 625$ psi; $F_c = 1700$ psi; $E = 1.9 \times 10^6$ psi; $C_D = 1.25$; $C_F = 1.0$. For the 3×12 rafter: b = 2.6"; d = 11.25"; S = 52.7 in^3; I = 297 in^4.

Check shear:

$V = 1198 - 83 \times 11.25 / 12 = 1120^{\#}$

$f_v = {}^3/_2 V / bd = 1.5 \times 1120 / (2.5 \times 11.25) = 60$ psi

$f_v < F_v'$ ∴ OK

Check combined axial and bending stress at the critical point (point B):

$l_e / d = (14.42 / 2) \times 12 / 11.25 = 7.7$

$$F_{cE} = \frac{0.3E'}{(l_e / d)^2} = \frac{0.3 \times 1{,}900{,}000}{7.7^2} = 9613 \text{psi}$$

$F_c^* = C_D F_c = 1.25 \times 1700 = 2125$ psi

$F_{cE} / F_c^* = 9613 / 2125 = 4.52$

$$C_p = \frac{1 + F_{cE} / F_c^*}{2c} - \sqrt{\left[\frac{1 + F_{cE} / F_c^*}{2c}\right]^2 - \frac{F_{cE} / F_c^*}{c}}$$

$$= \frac{1 + 4.52}{1.6} - \sqrt{\left[\frac{1 + 4.52}{1.6}\right]^2 - \frac{4.52}{0.8}}$$

$C_p = 0.95$

∴ $F_c' = C_p C_D F_c = 0.95 \times 1.25 \times 1700 = 2020$ psi

$F_b' = 1.25 \times 1450 = 1810$ psi

$$\text{Check: } \left[\frac{f_c}{F_c'}\right]^2 + \frac{f_{b1}}{F_{b1}'[1-(f_c/F_{cE1})]} + \frac{f_{b2}}{F_{b2}'[1-(f_c/F_{cE2})-(f_b/F_{bE})^2]} \le 1.0$$

For bending about the narrow face, only f_{b1} applies:

$f_{b1} = M/S = 6480 \times 12 / 52.7 = 1475$ psi

$f_c = P/A = 2197 / (2.5 \times 11.25) = 78$ psi

$$\left[\frac{78}{2020}\right]^2 + \frac{1475}{1810[1-(78/9613)]} = 0.82 \le 1.0 \therefore \text{O.K.}$$

Check transverse deflection under the total load: limit $= {}^L/_{180} = 14.42 \times 12 / 180$

$\Delta = ({}^5/_{384}(wL)L^3 + {}^1/_{48}PL^3) / EI$

$= [0.013 \times 83 \times 14.42(14.42 \times 12)^3 + 0.0208 \times 1198 \times (14.42 \times 12)^3] / (1.9 \times 10^6 \times 297)$

$= 0.37" < {}^L/_{180} = 0.96"$

$\therefore$ **the 3×12 rafter is OK**

b) Design the connection of the 2×6 to the 3×12 rafter. Try 2-1/2" split rings:

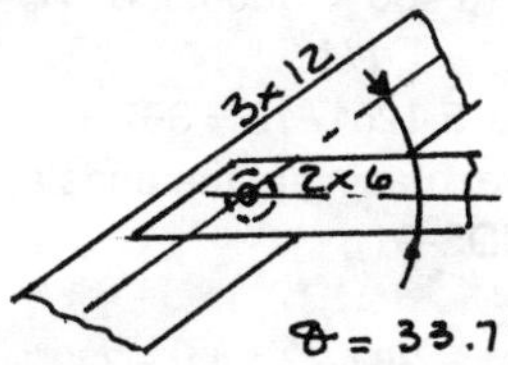

For connection group B; bearing 33.7° to grain; Re: Table 10.2A, NDS-91:

$Z_q = 2730^\#$

$Z_\perp = 1620^\#$

$Z_\theta = (Z_q Z_\perp) / (Z_q \sin^2\theta + Z_\perp \cos^2\theta) = 2255^\#$

$Z_\theta' = C_D Z_\theta = 1.25 \times 2255 = 2820^\# > T = 2160^\#$

% loaded = 2160 / 2820 = 0.77 ($\therefore$ 4" end distance is adequate)

Use 1 – 2-1/2" diameter split ring

Check end joint shear in rafter (Re: 3.4.5, NDS-91):

$d_e = (11.25 + 2.5) / 2 = 6.9"$

$f_v = ({}^3/_2 V / bd_e)(d / d_e)$

$= 1.5 \times [(599 - 80) / (2.5 \times 6.9)][11.25 / 6.9]$

$= 74$ psi

$f_v < F_v' \therefore$ OK

Check tension in the collar :

$A_n = 1.5 \times 5.5 - 1.73 = 6.5 \text{ in}^2$

$f_t = T / A_n = 2160 / 6.5 = 333 \text{ psi} << F_t' \therefore$ OK

The connection is adequate.

Timber-34

a) Find the forces at joint A of the roof under a symmetrical gravity load of 32 psf on the horizontal projection. Consider the hip rafter AE and assume that roof joists are spaced close enough so that the load on the rafter is uniformly distributed. The load varies linearly from its maximum at the top to zero at the bottom.

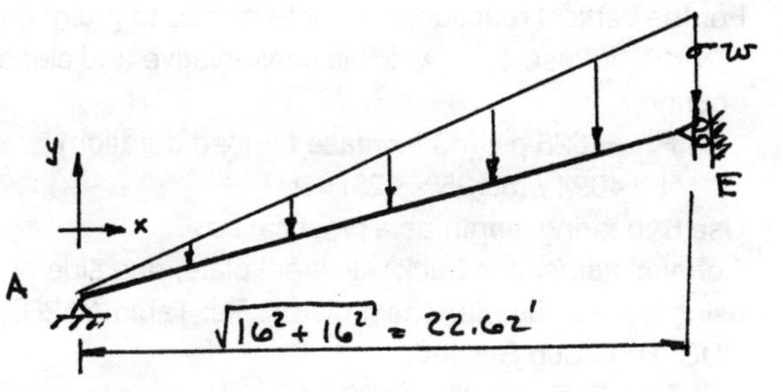

Distribute the reactions from the roof joists over the horizontal projection of the rafter:

$w = (2 \times 32 \text{ psf} \times 16' / 2)(16 / 22.62) = 362$ plf

$\Sigma M_A = 0: 6E_x - 0.5 \times 362 \times (22.62)^2 \times {}^2/_3 = 0$

$E_x = 10{,}290^\# \leftarrow$

$\Sigma F_x = 0$: **$A_x = 10{,}290^\# \rightarrow$**

$\Sigma F_y = 0$: **$A_y = 0.5 \times 362 \times 22.62 = 4094^\# \uparrow$**

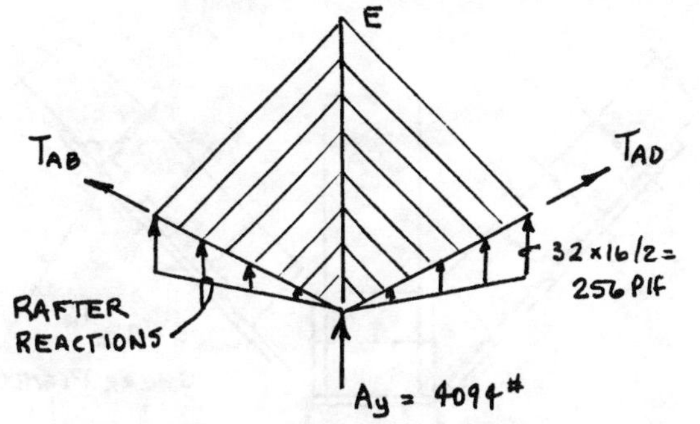

Check vertical equilibrium in one quadrant:

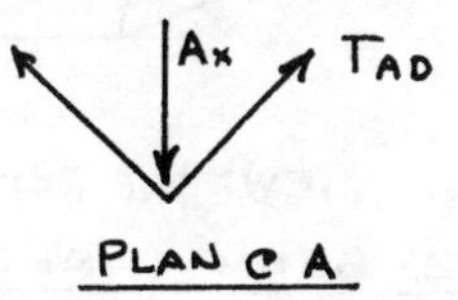

$\Sigma F_y = 32 \text{ psf} \times 16' \times 16' - 4094 - 2(256)\, 0.5 \times 16' = 0$

$\Sigma F_x = 0$ at joint A:

$T = T_{AB} = T_{AD}$

$\therefore \quad 2T(0.707) = 10{,}290^\#$

$T = 7277^\#$

b) Design the connection at A. Use a welded steel assembly with the rafter cut to transfer 4094# vertically and 10,290# horizontally. Weld steel side plates and transfer 7277# into the eave beam. For the rafter bearing:

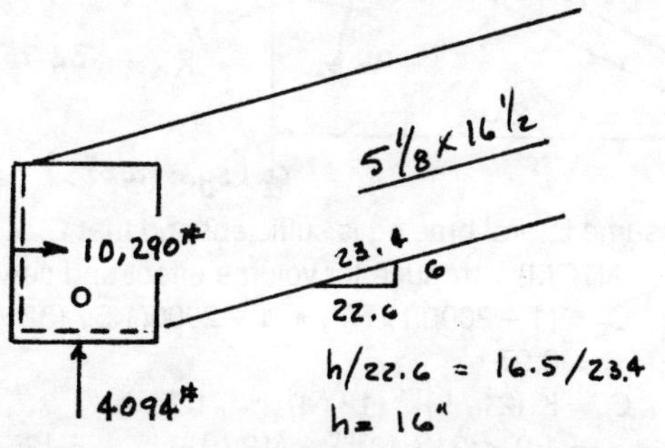

$f_c = 10{,}290 / (5.125 \times 16) = 125$ psi – nil

For the vertical reaction, the angle of load to grain = 75.1°, but use $F_{c\perp}$ (which is conservative and close enough):

$F_{c\perp}$ = 625 psi (no increase for load duration)

$N \geq 4094 / (5.125 \times 625) = 1.3"$

Use 6" bearing length as a practical size.

Connect across the back with steel plate, one side only, using 2-1/2" diameter shear plates. Re: Table 10.2B, NDS-91; Group B wood:

$Z_q' = C_D Z_q = 1.25 \times 2670 = 3340\ ^{\#}/_{\text{shear PL}}$

$n \geq 7277 / 3340 = 2.2$ use 3

Try PL $5 \times {}^3/_8$:

$A_n = (5 - {}^{13}/_{16})0.375 = 1.57\ \text{in}^2$

$$T \leq \begin{cases} 5 \times 0.375 \times 22 = 41^k \leftarrow \text{governs} \\ 1.57 \times 0.5 \times 58 = 45.5^k \end{cases}$$

Strength is adequate. Detail the connection as follows:

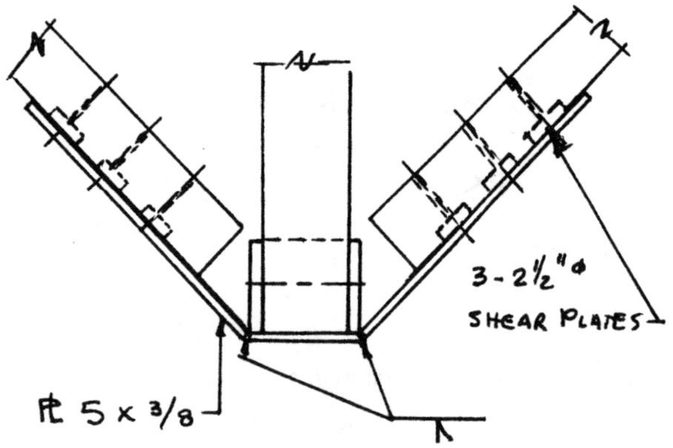

PLAN VIEW: HIP RAFTER TO EAVE BEAM CONNECTION

Timber-35

a) Compute the allowable bending stress. Given: 2400F glulam; 1.5" lams ; $E = 1.8 \times 10^6$ psi; no increase for load duration.

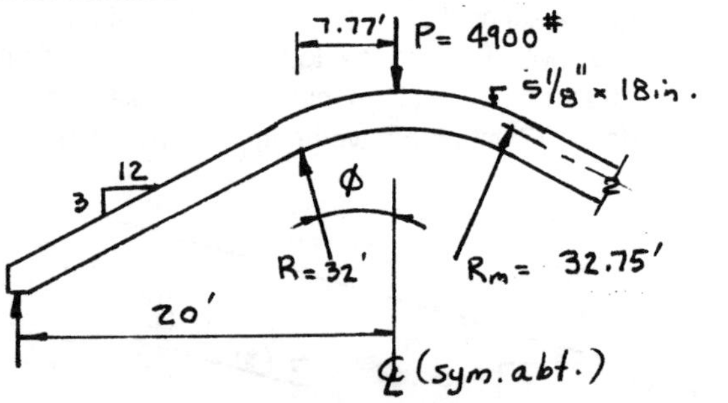

Assume lateral bracing is sufficient and that $C_V < C_L$ (Re: AITCM) ∴ reduce for volume effect and curvature:

$C_C = (1 - 2000(t / R)^2) = (1 - 2000(1.5 / (32 \times 12))^2)$
$= 0.97$

$C_V = K_L(21 / L)^{1/10}(12 / d)^{1/10}(5.125 / b)^{1/10}$
$= 1.0 \times (21 / 40)^{1/10} \times (12 / 18)^{1/10} \times (5.125 / 5.125)^{1/10} = 0.90$

∴ **$F_b' = C_C C_F F_b = 0.97 \times 0.90 \times 2400 = 2100$ psi**

b) Check the bending stress including the member weight based on γ = 36 pcf:

$w_d = (12.36 / 12)(5.125 \times 18 / 144)\ 36 = 24$ plf

$M_{max} = w_d L^2 / 8 + PL / 4$
$= 24 \times 40^2 / 8 + 4900 \times 40 / 4 = 53{,}800^{\#\text{-ft}}$

For a pitched and curved beam of constant depth:

$f_b = M_{max} / S = 53{,}800 \times 12 / 277 = 2330$ psi

Note: $f_b > F_b'$ (11% overstress) so in a strict sense, there is no value of l_u for which the given loading is acceptable (F_b' = 2100 psi is an upper bound to the allowable stress). However, I will interpret part b to mean the maximum unsupported length for which $C_L \geq C_V$. Equating the expression for C_L to 0.9:

$$C_L = \frac{1 + F_{bE}/F_b^*}{1.9} - \sqrt{\left[\frac{1 + F_{bE}/F_b^*}{1.9}\right]^2 - \frac{F_{bE}/F_b^*}{0.95}}$$
$$= 0.90$$

By transposing the radical and 0.9, then squaring both sides of the equation, the appropriate value of F_{bE}/F^*_b is found to be 1.305. Thus,

$F_{bE} / F_b^* = (0.438 \times 1{,}800{,}000 / R_B^2) / 2400 = 1.305$

$R_B^2 = 251.7 = l_e d / b^2$

∴ $l_e = 251.7 \times 5.125^2 / 18 = 367.3"$

Assuming lateral bracing at the ends and load point (Table 3.3.3, NDS-91):

$l_e = 1.11 l_u$

$l_u = l_e / 1.11 = 367.3 / 1.11 = 330" = 27.6'$

∴ **bracing at the ends and at midspan is adequate.**

c) Compute the radial tension stress (Re: p. 5-243, AITCM):

$f_r = 3M / (2R_m bd)$
$= 3 \times 53{,}800 \times 12 / (2 \times 32.75 \times 12 \times 5.125 \times 18)$

$f_r = 26.7$ psi > $F_r' = 15$ psi

d) Radial tension stress is excessive. Use lag screws to reinforce the curved central region. Orient the lag bolts radially and install from the top flange into prebored holes to within 2" of the soffit. In this way, the computed radial tension stress is permitted as $F_v' / 3$, which is acceptable. The capacity of the lag screw reinforcement is (p. 5-129):

$$T \leq \begin{cases} 5240^{\#}\ \text{(bolt strength)} \leftarrow \\ 528\ ^{\#}/_{\text{in}}\ (18 - 2)" = 8450^{\#} \end{cases}$$

Let S = spacing measured along the arc of radius R_m. Provide bolts to resist the total 26.7 psi computed stress:

$\Sigma F_{radial} = 0 = T - (f_{rt})bs$

∴ $S \leq 5240 / (5.125 \times 26.7) = 38.3"$

For the curved central region:

$\phi = \tan^{-1}(7.77 / 32) = 13.6° = 0.237$ rad

$C_C = 2R_m\phi = 2 \times 32.75 \times 12 \times 0.237 = 186"$

∴ $n \geq 186 / 38.3 = 4.9$; say 5

$S = 186 / 5 = 37.2"$ (38.4" o.c. outside)

Use 5, 3/4" diameter lag screws in the central region spaced symmetrically at 38.4" o.c. along the top flange.

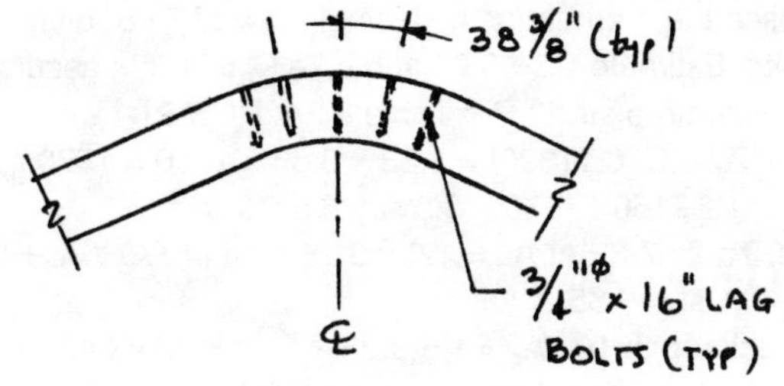

PARTIAL ELEVATION

Timber-36

Compute the maximum load permitted on the bolted connection.

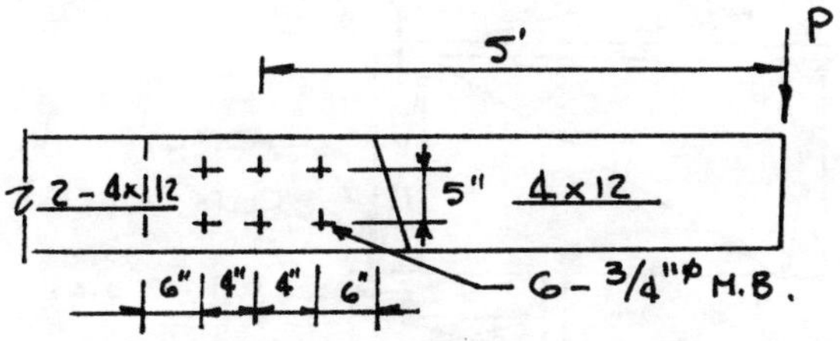

For Douglas Fir (assume No. 1 Douglas Fir-Larch), neglect the duration of load factor and check the force on the critical bolt.

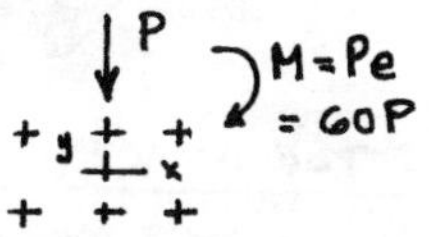

$r_{py} = P / 6 = 0.17P$

$\Sigma(x_i^2 + y_i^2) = 4(2.5^2 + 4^2) + 2(2.5^2) = 101.5$

$r_{my} = Mx / \Sigma(x_i^2 + y_i^2) = 60P(4) / 101.5 = 2.36P$

$r_{mx} = My / \Sigma(x_i^2 + y_i^2) = 60P(2.5) / 101.5 = 1.48P$

$r = P[(2.366 + 0.17)^2 + 1.48^2]^{1/2} = 2.93P$

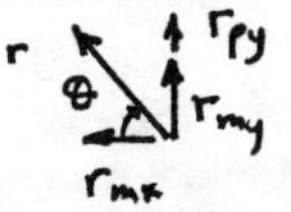

$\theta = \tan^{-1}(2.36 + 0.17) / 1.48 = 60°$

Re: Table No. 8.3A, NDS-91 for 3.5" main and side members:

$Z_q = 3220\ ^{\#}/_{bolt}$

$Z_{\perp} = 1750\ ^{\#}/_{bolt}$

For parallel to grain loading and end distance = 6" > 7d, which is adequate; 4" o.c. spacing is adequate. For 3 fasteners in a row, Re: Table 7.3.6A: $A_m = 3.5 \times 11.25 = 39\ in^2$, use $A_m / A_s = 0.55$: $C_g = 0.99$.

$Z_{\perp}' = 0.99 \times 3220 = 3190^{\#}/_{bolt}$

For perpendicular to grain loading, the edge distance = 3.13" > 4d is adequate and no group action reduction is needed

$\therefore \quad Z_{\perp}' = 1750\ ^{\#}/_{bolt}$

On the critical bolt, the force acts at an angle θ = 60° to grain ∴ use Hankinson's Formula;

$Z_\theta = Z_q' Z_{\perp}' / [Z_q' \sin^2\theta + Z_{\perp}' \cos^2\theta]$

$= 3190 \times 1750 / [3190 \sin^2 60° + 1750 \cos^2 60°]$

$= 1970\ ^{\#}/_{bolt}$

Thus,

$2.93P \le 1970;\ P \le 672^{\#}$

Check shear at the joint:

$f_v = (^3/_2 V / bd_e)d / d_e$

$d_e = 5 + 3.13 = 8.13"$

Take $F_v' = 95$ psi:

$[(^3/_2)P / (3.5 \times 8.13)](11.255 / 8.13) \le 95$

$P \le 1300^{\#}$

Check bending at the net section:

$S = I / c = [3.5(11.25)^3 / 12 - 2(3.5)(^{13}/_{16})2.5^2] / 5.63$

$= 67.5\ in^3$

$F_b' = 1150$ psi ; $M = (5 - 0.33)P$

$4.67 \times 12P / 67.5 \le 1150$

$P \le 1390^{\#}$

Thus, bolt strength governs and the required capacity is:

$\underline{\mathbf{P_{max} = 672^{\#}}}$

Timber-37

a) A 24F combination of Hem Fir rather than Douglas Fir will have a lower modulus of elasticity (typ, $E = 1.4 \times 10^6$ psi vs. 1.6×10^6 psi). The deflection in the Hem Fir member will be proportionately increased (i.e., 1.6 / 1.4 = 1.14 times the intended deflection). The lateral buckling tendency will also be increased by the lower E. The allowable horizontal shear stress, F_V, may also be lower, so horizontal shear must be checked. The species groups (Re: Table 6.2, AITCM) differ so connections involving mechanical fasteners must be checked. Essentially, a complete check of the beam is necessary.

b) Penetration of the nail head past the outer ply reduces the lateral resistance of the nail to about one-half its intended value. I assume that the same problem exists at the top plate and sills so that providing companion studs and renailing will not solve the problem. Since the nailing shown is 4" o.c., additional nails driven into the nominal 2" studs might cause splitting (need 3" or wider for spacings ≤ 3"). The only obvious solution, if the sheathing is on only one side, is to sheath the other side to resist one half or more of the wall shear. If this is not possible, alternative force paths or replacement of the wall would need to be considered.

c) If the wall is long relative to 3", it may be practical to shift the 4×4 post 3" to the right to connect to the anchor bolt. This will shorten the wall; hence, increase the unit wall shear (plf) and axial force in the boundary member. If the increase is small, and there was some extra margin of safety in the original design (as is frequently the case), the movement would not overstress the components. If this is the case, 2-2×4 studs can be used to fill the 3" gap caused by shifting the 4×4. If it is not possible to move the post, a stiffened connection angle could be designed to handle the force transfer with the additional 3" eccentricity.

Timber-38

a) Determine the loads caused by seismic forces on wall line 3.

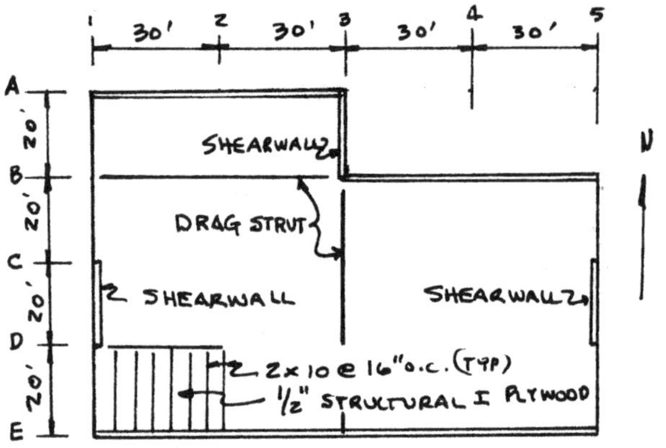

PLAN

Calculate the forces at the working stress level by dividing the design level force by 1.4. Re: 1630.2, UBC-97, given that $C_a = 0.33$; take I = 1.0 and R = 5.5:

$V = [(2.5C_aI / R)W] / 1.4 = [(2.5 \times 0.33 \times 1.0 / 5.5)$ $W] / 1.4 = 0.107W$

Between lines 1 and 3:

$W_1 = (\text{roof} + 2 \text{ side walls})0.107 = 0.107(80' \times 20 \text{ psf}$ $+ 2 \times 8' \times 16 \text{ psf}) = 199 \text{ plf}$

From lines 3-to-5:

$W_2 = 0.107(60' \times 20 \text{ psf} + 2 \times 16 \text{ psf} \times 8') = 156 \text{ plf}$

With 30 ft tributary from the roof diaphragm on either side and including the wall weight in the base shear, the load on the shearwall is:

$V_3 = (199 \text{ plf} + 156 \text{ plf})\ 30' + 0.107 \times 16 \text{ psf} \times 16'$ $\times 20')$

$V_3 = 11{,}200$ lb

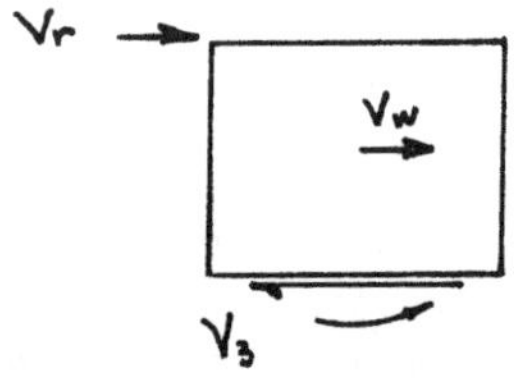

b) Specify the nailing and connection details for the wall on line 3. The shearwall is 20' long and resists 11,200#:

$v = 11{,}200 / 20 = 560 \text{ plf}$

Re: Table No. 23-II-I-1, UBC-97, the required nailing for 15/32" structural 1 plywood wall is:

8d @ 3" o.c. panel edges; 8d @ 12" o.c. for intermediate members (provides 550 plf, which is close enough).

For the collector and boundary members, use No. 1 Douglas-Fir Larch framing:

$F_t = C_D(1.1 \times 675 \text{ psi})$ for 2×10 joists

The collector element (drag strut) on line 3 must transfer:

$T_{BC} = (199 \text{ plf} \times 30' / 80' + 156 \text{ plf} \times 30' / 60')\ 60'$ $= 9160^{\#}$

Splice the member using a single row of 7 / 8" diameter bolts. Estimate $C_g = 0.85$ and take $C_D = 1.33$; assume a 3" wide drag strut ; Re: Table 8.2A, NDS-91:

$Z_q' = C_DC_g(1390) = 1.33 \times 0.85 \times 1390 = 1570^{\#}/_{\text{bolt}}$

$\therefore\ n \geq 9160 / 1570 = 5.8$ say 6 bolts

Re: Table 7.3.6A: $A_m = 1.5 \times 9.25 = 14 \text{ in}^2$; $A_m / A_s = 0.5$: $C_g = 0.81 \approx 0.85\ \therefore$ OK

Check the 2×10 : $A_n = 1.5(9.25 - {}^{15}/_{16}) = 12.47 \text{ in}^2$

$f_t = T / A_n = 9160 / 12.47 = 735 \text{ psi} < F_t'$

Transfer $V_r = (199 + 156)30' = 10{,}650^{\#}$ from the 2×10 joist to the top plate of the shearwall:

$v = 10{,}650 / 20 = 533 \text{ plf}$

Use lightgage metal clips to transfer 533 plf (e.g., Simpson strong-tie, A35F, or equivalent @ 12" o.c.).

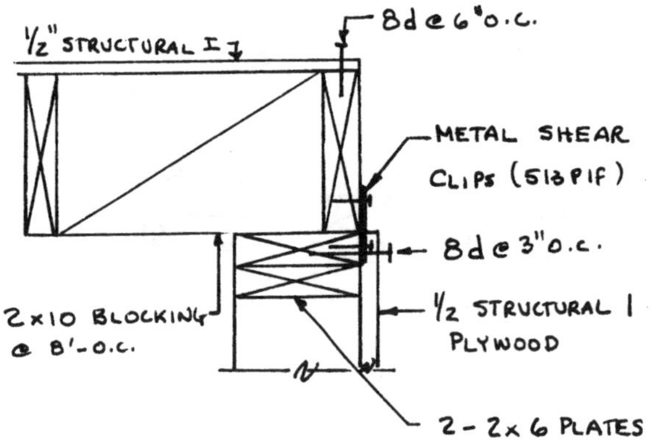

SECTION – TOP OF WALL ON LINE 3

c) Detail the wall-to-foundation connection. Dead load is negligible $\therefore$ from part a:

$M_0 = 10{,}650^{\#}\ (16') + 560^{\#}\ (8') = 174{,}900^{\#\text{-ft}}$

$T = C = M_0 / 20 = 174{,}900 / 20 = 8750^{\#}$

Provide hold-down anchors both sides for 8750# ↑

Connect the sill plate for 11,200# shear. Try 3/4" diameter anchor bolts – wood governs:

$Z_q' = 1.33(1190) = 1580^{\#}$

$n \geq 11{,}200 / 1580 = 6.8$ say 7, 3/4" anchor bolts equally spaced.

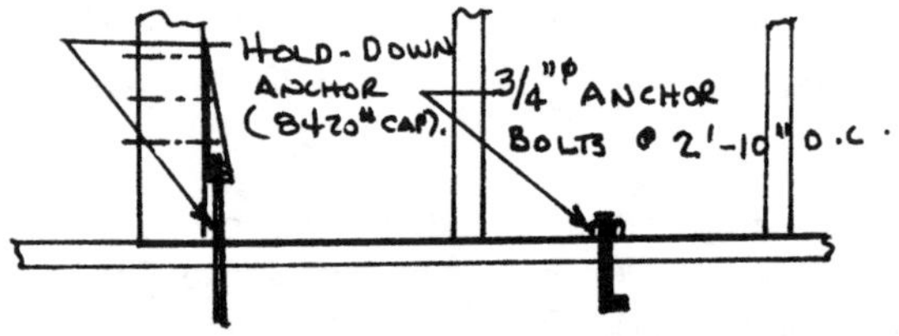

DETAIL AT WALL BASE

Timber-39

a) Compute the load, shear and moment diagrams for the low roof due to seismic forces. Calculate the forces at the working stress level by dividing the design level force by 1.4. Re: 1630.2, UBC-97, given that $C_a = 0.33$; take I = 1.0 and R = 5.5:

$V = [(2.5C_aI / R)W] / 1.4$

$= [(2.5 \times 0.33 \times 1.0 / 5.5)\ W] / 1.4 = 0.107W$

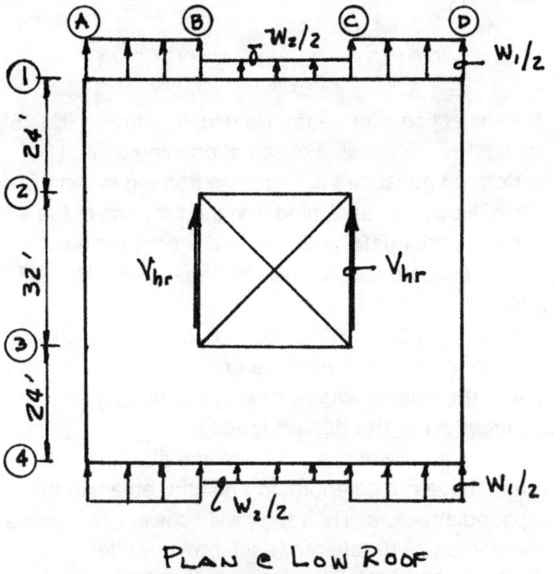

PLAN @ LOW ROOF

The seismic load from the high roof is:

$$V_{hr} = 0.107(W_{hr} + W_{walls}) / 2$$
$$= 0.107(20\ \text{psf} \times 32' \times 32' + 4 \times 20\ \text{psf} \times 4' \times 32') / 2$$
$$= 1644^{\#}$$

Load on the low roof from the roof and exterior walls (take tributary wall height as 6.5'):

$$W_1 = 0.107(20\ \text{psf} \times 80' + 2 \times 20\ \text{psf} \times 6.5')$$
$$= 199\ \text{plf}$$
$$W_2 = 0.107(20\ \text{psf} \times 48' + 2 \times 20\ \text{psf} \times 6.5')$$
$$= 130\ \text{plf}$$

The required values are as follows:

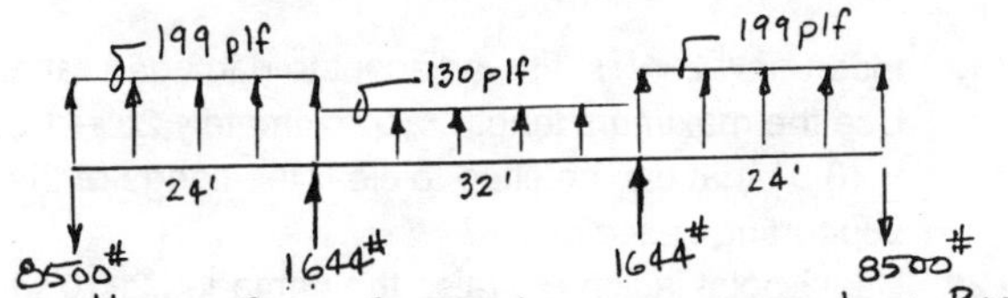

NORTH - SOUTH SEISMIC LOADS ON THE LOW ROOF

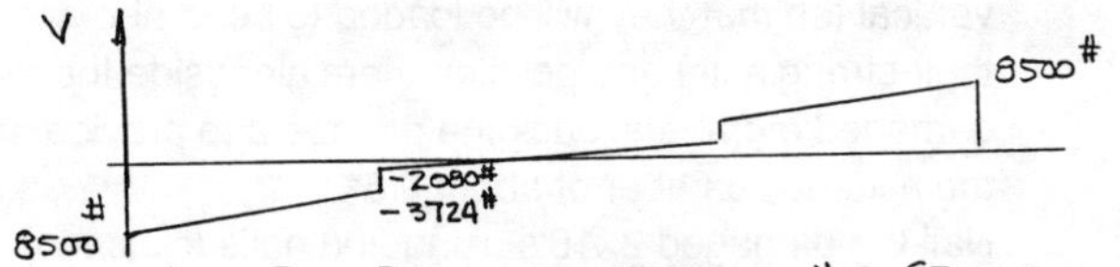

LOW-ROOF DIAPHRAGM SHEAR — N.S. SEISMIC

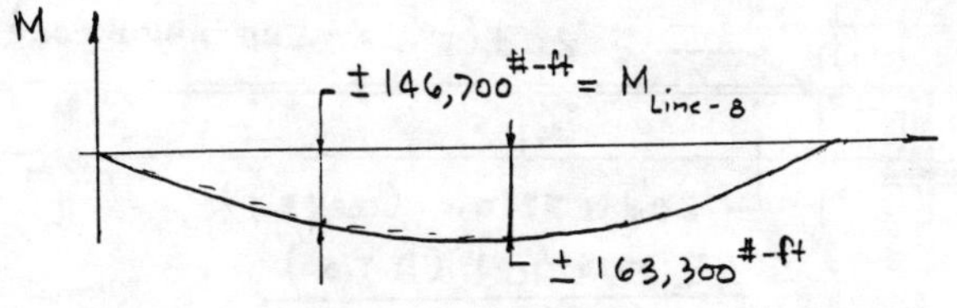

LOW-ROOF DIAPHRAGM MOMENT — N.S. SEISMIC

b) Compute the chord forces at lines 'B' and 'C' (and '2' and '3' by symmetry):

$$T = C = M / 80 = 146{,}700 / 80 = 1834^{\#}$$

Take a portion of the roof between lines 'B' and 'C' as a freebody:

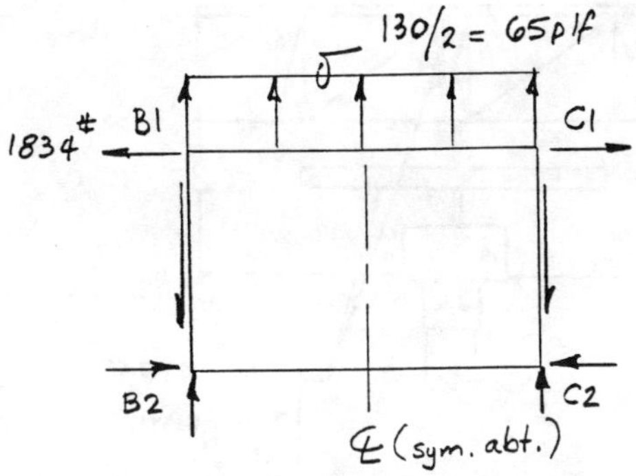

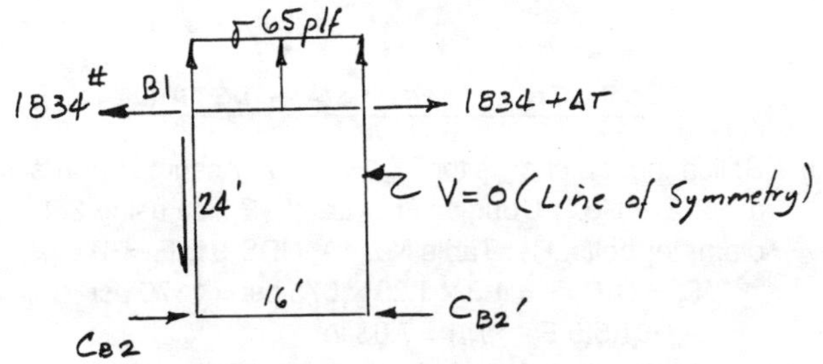

$$\Sigma M_{B2} = 0\text{: } 24\Delta T - 65(16)8 = 0$$
$$\Delta T = 347^{\#}$$

From a freebody through the middle of the building (between lines B and C), sum moments about a point on line A:

$$\Sigma M_A = 0\text{: } 199 \times 24 \times 12 + 1644 \times 24 + 130 \times 16(24 + 8) - (1834 + 347)80 + C_{B2'}(32) = 0$$
$$C_{B2'} = 349^{\#}\ (\text{Thus, } C_{B2} = 0)$$

Thus, ties across lines B and C on lines 2 and 3 are not necessary for North-South forces.

The maximum chord force is:

$$\underline{\mathbf{T = C = 1834 + 349 = 2183^{\#}}}$$

c) Calculate the force at the corner of openings. Take a freebody by slicing adjacent to line "B":

$$v_1 = 65 \times 16 / 24 = 43\ \text{plf}$$
$$v_2 = (65 \times 16 + 822) / 40 = 47\ \text{plf}$$

Thus, the corner tie force is:

$$\underline{\mathbf{T = (47 - 43)24 = 96^{\#}}}$$

As shown in part b, there is no tie force in the orthogonal direction due to north-south lateral forces. However, by symmetry, a force of 96# develops under east-west lateral forces.

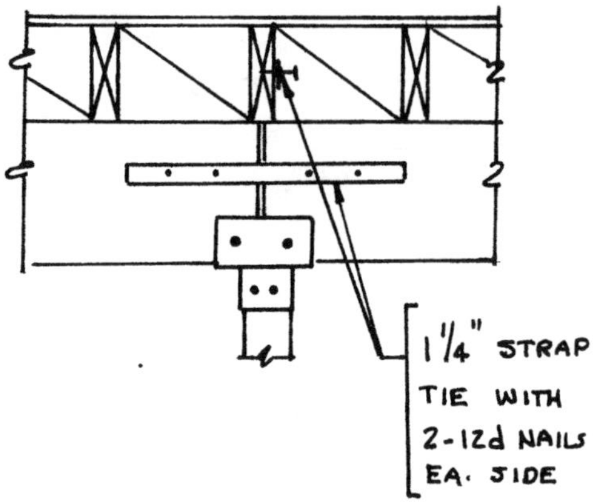

CORNER @ OPENING — DETAIL

d) Splice the top chords for $T_{max} = 2183^{\#}$. Assume chords are 2-2×6 No. 1 Douglas Fir – Larch. Splice using 3/4" diameter bolts. Re: Table No. 4A, NDS-91: $F_t = 675$ psi

$F_t' = C_F C_D F_t = 1.3 \times 1.33 \times 675$ psi = 1170 psi

$A_n = 1.5(5.5 - {}^{13}/_{16}) = 7.03$ in^2

$f_t = 2183 / 7.03 = 310$ psi << F_t ∴ OK

Re: Table No 8.2A, NDS-91: 1.5" main member; estimate $C_g = 0.85$ (group action); $C_D = {}^4/_3$

$Z_q' = C_g C_D(720) = 815\ {}^{\#}/_{bolt}$

$n \geq 2183 / 815 = 2.7 \rightarrow 3$ bolts (∴ $C_g > 0.85$)

∴ **Use 3 – 3/4" diameter bolts each side**

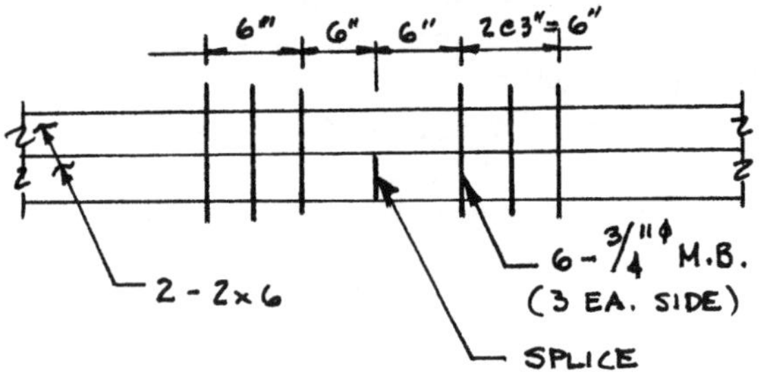

CHORD SPLICE DETAIL

Timber-40

a) Check the bending stress in the rafters using the given allowable stresses. The scab splice provides negligible restraint to rotation; therefore, assume that rafters are simply supported. For the given allowable stresses: $F_b = 1500$ psi (increase to 1750 psi for repetitive members); $F_v = 95$ psi; $E = 1.8 \times 10^6$ psi. For the 2×10 rafters: b = 1.5"; d = 9.25 in; S = 21.4 in^3; I = 99 in^4. The given information suggests a flexural failure. Check flexural stresses under the design load of 39 psf. For a 20 psf live load, a load duration factor $C_D = 1.25$ should apply:

∴ $F_b' = C_D F_b = 1.25 \times 1750 = 2190$ psi

$w = 39$ psf × 2' = 78 plf

$M_{max} = wL^2 / 8 = 78 \times 20^2 / 8 = 3900$ $^{\#\text{-ft}}$

$f_b = M_{max} / S = 3900 \times 12 / 21.4 = 2186$ psi

$f_b \approx F_b'$ ∴ OK in flexure

Check Δ_{TL}:

$\Delta_{TL} = {}^5/_{384}(wL)L^3 / EI$

$= 0.013(78 \times 20)(20 \times 12)^3 / (1.8 \times 10^6 \times 99)$

= 1.57" = L / 152

The deflection is more than is usually allowed (L / 152 vs. L / 180). A detailed check of ponding cannot be performed because the complete framing system is not given. However, assuming that the stiffness of the steel beams is adequate, a simple check on ponding is that the deflection under a 5 psf loading < 0.25" (Re: AITCM, p. 4-89):

$\Delta_S = (5 / 39)\ 1.57 = 0.20" < 0.25"$

∴ Ponding should not be a problem.

Thus, the failure was not likely caused by application of the design loads.

b) Describe a repair scheme. I assume that the broken rafters have not deteriorated and can remain in the repaired structure. The repair will consist of providing a companion 2×10 rafter for each broken rafter.

Check the shear and bearing stresses in the 2×10's:

$L_e = L - 2d = 20 - 2 \times 9.25 / 12 = 18.5'$

$V = wL_e / 2 = 78 \times 18.5 / 2 = 722^{\#}$

$f_v = {}^3/_2 V / bd = 1.5 \times 722 / (1.5 \times 9.25) = 78$ psi < F_v' ∴ OK

$R = 78 \times 10 = 780^{\#}$

Take $F_{c\perp} = 650$ psi:

$N \geq R / b F_{c\perp} = 780 / (1.5 \times 650) = 0.8"$ nil

For properly installed rafters, lateral support for h / d = 10 / 2 = 5 (Re: NDS-91) requires that only one edge be held in line. Thus, it may not be necessary to remove the intermediate bridging to install the new 2×10's from below.

1. Insert new 2×10's, flat, adjacent to damaged rafters. Use the maximum length (approximately 20' – 1.5' = 18.5') that can be tilted to clear the flange of the supporting w-section.
2. Use 2-point jacking to raise the damaged 2×10's to the horizontal. Force tilt the new 2×10's to the vertical (so that they will be loaded to bend about their strong axis) and position them alongside the damaged members. Jack the members to provide the midspan camber of about 0.5".
3. Nail the damaged 2×10's using 20d nails (clinched) in the pattern below :

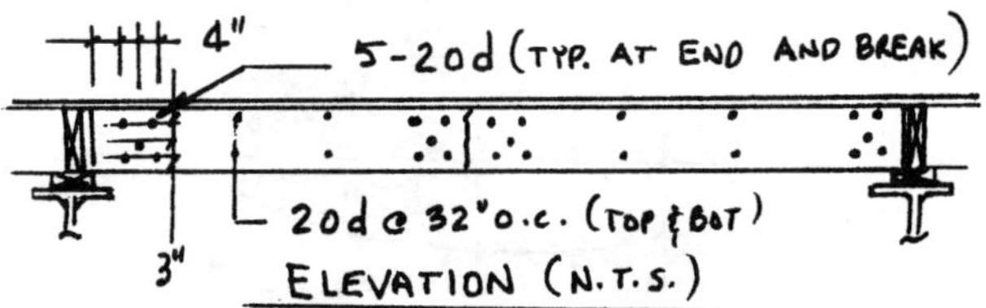

ELEVATION (N.T.S.)

4. Remove the jacks.

c) Several likely causes of the localized failure:

- Local overload
- Cuts or holes (e.g., for electrical mechanical installations) that weakened a series of rafters and was not detected until after failure.

- Water damage caused by roof or drain pipe leaks that significantly weakened several rafters.

Timber-41

Modify the existing ceiling framing to support the new equipment loads.

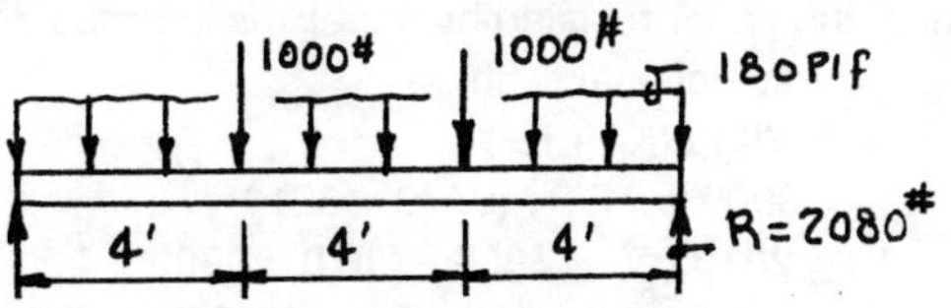

Allow 5 psf attic load in addition to the equipment

$\therefore$ w = (5 + 10) × 12 = 180 plf

$M_{max} = 180 \times 12^2 / 8 + 1000(4) = 7240^{\#\text{-ft}}$

For a 4x8, No.1 and better Douglas Fir-Larch:

$F_b = 1.3 \times 1150 = 1500$ psi; $F_v = 95$ psi; $F_{c\perp} = 625$ psi; $E = 1.8 \times 10^6$ psi. For the existing 4×8 beam: b = 3.5"; d = 7.25"; I = 111 in^4; S = 30.7 in^3. Check the existing 4×8 under the proposed loading:

$f_b = M_{max} / S = 7240 \times 12 / 30.7 = 2830$ psi

$V = R - wd = 2080 - 180(7.25 / 12) = 1970^{\#}$

$f_v = {}^3/_2 V / bd = 1.5 \times 1970 / (3.5 \times 7.25) = 117$ psi

$\Delta_{TL} = ({}^5/_{384})(wL)L^3 / EI + Pa(3L^2 - 4a^2) / 24\,EI$

$= [0.013(180 \times 12)(144)^3 + 1000 \times 48(3 \times 144^2 - 4 \times 48^2) / 24] / [1.8 \times 10^6 \times 111]$

$= 0.42 + 0.53 = 0.95$ in

The deflection limit is ${}^L/_{240} = 12 \times 12 / 240 = 0.6$". Thus, the 4×8 is overstressed and the deflection is excessive. Since the beam is accessible from above, try to reinforce the beam by adding a 4" wide piece at top. Assume that the 4×8 is deflected by the uniformly distributed load:

$\Delta_{unif} = 0.42$"

The beam stiffness must be sufficient to limit the deflection due to the 1000$^{\#}$ equipment loads:

$\Delta_{conc} = {}^L/_{240} - \Delta_{unif} = 0.60 - 0.42 = 0.18$"

$\therefore$ $1000 \times 48(3 \times 144^2 - 4 \times 48^2) / (24 \times 1.8 \times 10^6 (I)) \le 0.18$"

$\therefore$ $I \ge 327$ in^4

Let h = overall depth: $I = bh^3 / 12$

$bh^3 / 12 \ge 327$

$h \ge (327 \times 12 / 3.5)^{1/3} = 10.4$"

$\therefore$ Try adding a 4×4 above the existing 4×8: h = 7.25 + 3.5 = 10.75"

$I' = 3.5 \times 10.75^3 / 12 = 362$ in^4

$S' = 67.4$ in^3

Superimpose the stresses due to 180 plf acting on the 4×8 with the stresses due to the 1000$^{\#}$ equipment load on the built up section:

$f_b = M_1 / S + M_2 / S'$

$M_1 = 180 \times 12^2 / 8 = 3240^{\#\text{-ft}}$; $M_2 = 4000^{\#\text{-ft}}$

$f_b = 3240 \times 12 / 30.7 + 4000 \times 12 / 67.4 = 1266 + 712 = 1980 \text{ psi} > F_b'$

No Good! Stresses due to the ceiling load are too high. Shore up the 4×8 prior to installing the top piece and then release the builtup section. The deflection will obviously be acceptable. The built up section will resist stresses:

$f_b = (M_1 + M_2) / S' = 7240 \times 12 / 67.4 = 1290 \text{ psi} < F_b'$

$V = 180(12 - 2) / 2 + 1000 = 1900^{\#}$

$f_v = 1.5 \times 1900 / (3.5 \times 10.75) = 76 \text{ psi} < F_v'$

Try connecting with 16d toenails as shown:

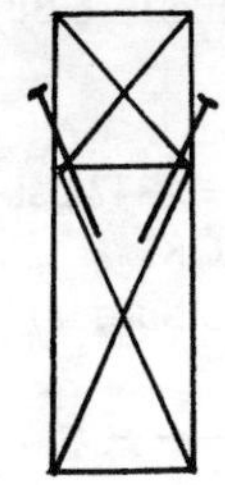

$p = ({}^5/_6)(108) = 90\ {}^{\#}/_{nail}$

$q = VQ / I$

$Q = A\bar{y} = 3.5^2(10.75 - 3.5) / 2 = 44.4$ in^3

$V_{max} = 2080^{\#}$

$q = 2080 \times 44.4 / 362 = 255\ {}^{\#}/_{in}$

Cannot make the connection using nails. Try lag screws – Re: Table 9.3A ; say 3/4" diameter × 9" bolt: P = 1570 $^{\#}/_{bolt}$

For V = 2080$^{\#}$; S ≥ 1570 / 255 = 6.2"

For V = 1900$^{\#}$; S ≥ 1570 / 233 = 6.7"

For V = 1720$^{\#}$; S ≥ 1570 / 211 = 7.5"

For V = 1540$^{\#}$; S ≥ 1570 / 188 = 8.4"

For V = 360$^{\#}$; S ≥ 1570 / 44 = 36"

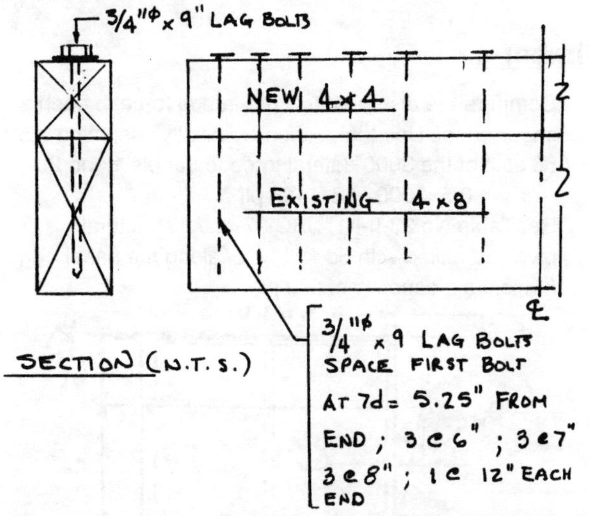

Recheck flexural stresses assuming that the portion of the lag bolt in the tension zone resists no tension (Re: p. 5-238, AITCM):

$c = (bd^2 - b'c^2 + b'a^2) / [2(bd - b'c + b'a)]$

b' = 0.75"

a = 1.75"

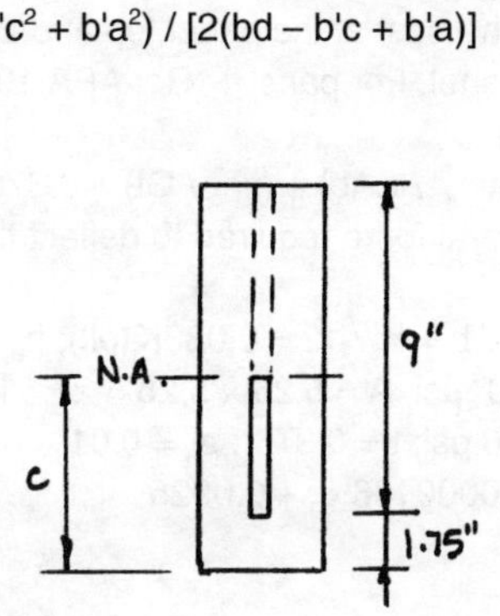

$c = [3.5 \times 10.75^2 - 0.75c^2 + 0.75(1.75)^2] / [2(10.75 \times 3.5 - 0.75c + 0.75(1.75))]$

$c^2 - 105.588C + 542.0 = 0$

$c = 5.42"$

$I = 362 + 3.5(10.75)(5.42 - 5.38)^2 - 0.75(5.42 - 1.75)^3 / 3 = 349.7 \text{ in}^4$

$S = 349.7 / 5.42 = 64.5 \text{ in}^3$

$f_b = 7240 \times 12 / 64.5 = 1347 \text{ psi} < F_b'$

∴ The built up beam is OK.

Note: an alternative way to build up the beam is illustrated below.

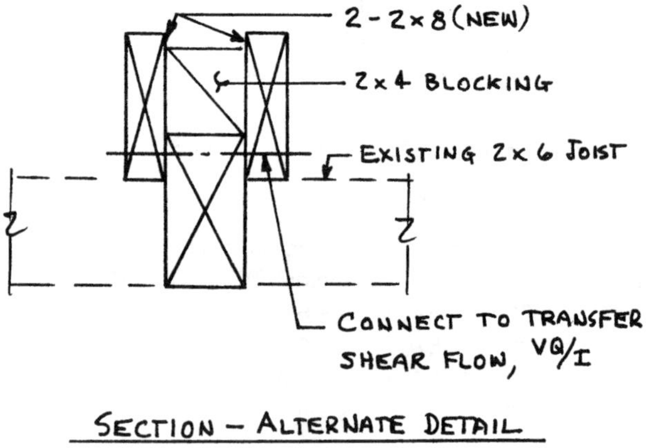

SECTION – ALTERNATE DETAIL

Timber-42

a) Compute the distribution of the lateral force to each shear panel. Use trial and error, initially assuming a 50-50 split of the 3000# lateral force to panels 1 and 2:

$v = 0.5 \times 3000 / 12 = 125 \text{ plf}$

Re: Table No.23-II-I-1, UBC-97, a $^3/_8$" structural I plywood nailed with 6d @ 6" o.c. along the panel edges furnishes a capacity of 200 plf.

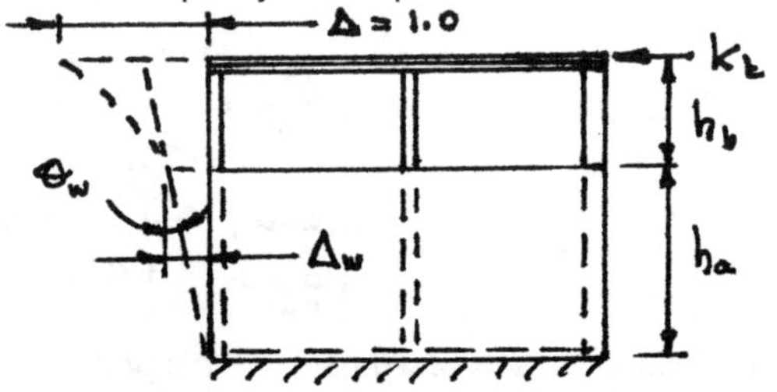

The deflection at the top is the sum of the bending deformation in the 6×6 posts; the deflection due to rotation of the post at the panel top and the deflection of the shear panel. For panel 1 (Re: APA 1988), the deflection is:

$\Delta_w = (8vh_a^3 / EAb) + (vh_a / Gt) + 0.376h_a e_n + {}^1/_{16}"$

Let K_{t1} = lateral force required to deflect the top of panel 1, 1".

Take $v = K_t / b = K_t / 12 = 0.083K_t(\text{plf})$; $h_a = 8'$; $E = 1.8 \times 10^6$ psi; $A = 5.25 \times 5.25 = 27.6 \text{ in}^2$; $G = 110{,}000$ psi; $t = 0.375"$; $e_n = 0.01"$

∴ $\Delta_w = 0.0000166K_{t1} + 0.0925$

$\theta_w = \Delta_w / h_a$

$\Delta = \Delta_w + (\Delta_w / h_a)h_b + K_{t1}h_b^3 / 3EI = 1.0$

$= 0.0000249K_{t1} + 0.139 + 0.000108K_{t1} = 1.0$

$K_{t1} = 6479 \ ^\#/_{in}$

Similarly, for panel 2: $v = K_t / 24$; $h_a = 6'$; $h_b = 6'$; other parameters are the same as panel 1. Substituting into the expression for Δ yields $K_{t2} = 3210 \ ^\#/_{in}$.

Thus, the shear resisted by panel 1 is in proportion to its relative translational stiffness:

$V_1 = VK_{t1} / (K_{t1} + K_{t2})$

$= 3000 \times 6479 / (6479 + 3210)$

$V_1 = 2010^\#$; $V_2 = 3000 - 2010 = 990^\#$

$v_1 = 2010^\# / 12' = 168 \text{ plf} < v_2 < 200 \text{ plf}$

Use $^3/_8$" structural I plywood with 6d @ 6" o.c. along panel edges and boundaries; 6d @ 12" along intermediate studs.

b) Check the 6×6 post (re: NDS-91). Maximum gravity load ($C_D = 1.0$; $F_c = 1000$ psi, $E = 1.8 \times 10^6$ psi).

$P_{L+D} = (200 \text{ plf} + 200 \text{ plf})6' = 2400^\#$

$l_e = Kl_u = 2.1 \times 72 = 151$

$l_e / d = 151 / 5.25 = 28.8$

$$F_{cE} = \frac{0.3E'}{(l_e/d)^2} = \frac{0.3 \times 1{,}800{,}000}{28.8^2} = 651 \text{ psi}$$

$F_c^* = 1000$ psi

$F_{cE} / F_c^* = 651 / 1000 = 0.65$

$$C_p = \frac{1 + F_{cE}/F_c^*}{2c} - \sqrt{\left[\frac{1 + F_{cE}/F_c^*}{2c}\right]^2 - \frac{F_{cE}/F_c^*}{c}}$$

$$= \frac{1 + 0.65}{1.6} - \sqrt{\left[\frac{1 + 0.65}{1.6}\right]^2 - \frac{0.65}{0.8}}$$

$C_p = 0.53$

∴ $F_c' = C_p C_D C_F F_c = 0.53 \times 1.0 \times 1.0 \times 1000 = 530$ psi

$A \geq R / F_c' = 4200 / 530 = 7.92 \text{ in}^2$ -- OK

Check earthquake plus dead load – critical condition is in panel 1:

$P_D = 1200^\#$

$M_E = (2010 / 3)^\# (4') = 2680^{\#\text{-ft}}$

$F_b' = C_D F_b = (^4/_3)1200 = 1600$ psi

Check combined axial and bending:

$F_c' = 530$ psi

$F_b' = 1.33 \times 1200 = 1600$ psi

$$\text{Check: } \left[\frac{f_c}{F_c'}\right]^2 + \frac{f_{b1}}{F_{b1}'[1 - (f_c / F_{cE1})]} + \frac{f_{b2}}{F_{b2}'[1 - (f_c / F_{cE2}) - (f_b / F_{bE})^2]} \leq 1.0$$

For bending about the narrow face, only f_{b1} applies:

$f_{b1} = M_E / S = 2680 \times 12 / 24.2 = 1329$ psi

$f_c = P_d / A = 1200 / 5.25^2 = 44$ psi

$$\left[\frac{44}{530}\right]^2 + \frac{1329}{1600[1 - (44/651)]} = 0.90 \leq 1.0 \therefore \text{O.K.}$$

∴ **The 6×6 posts are OK**

c) Detail connection A, which is governed by gravity loads (P_{L+D})

$f_{c\perp} = P_{L+D} / A = 2400 / 5.25^2 = 87$ psi << $F_{c\perp}$

For lateral loading, the maximum shear transfer = 2010 / 3 = 670#. Use an 18 gauge post cap with allowable lateral capacity ≥ 670# (e.g., Simpson strong-tie BC 6):

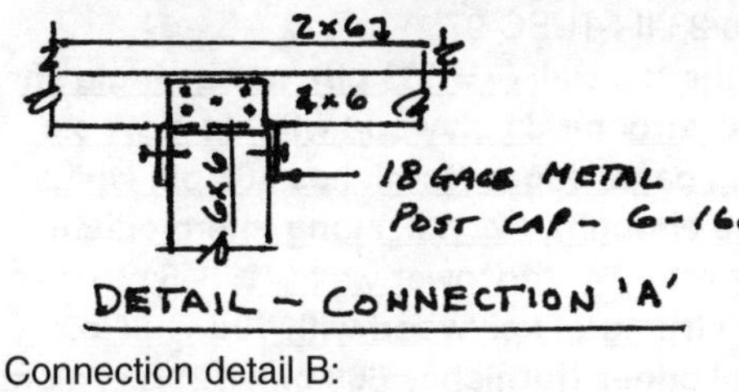

Connection detail B:

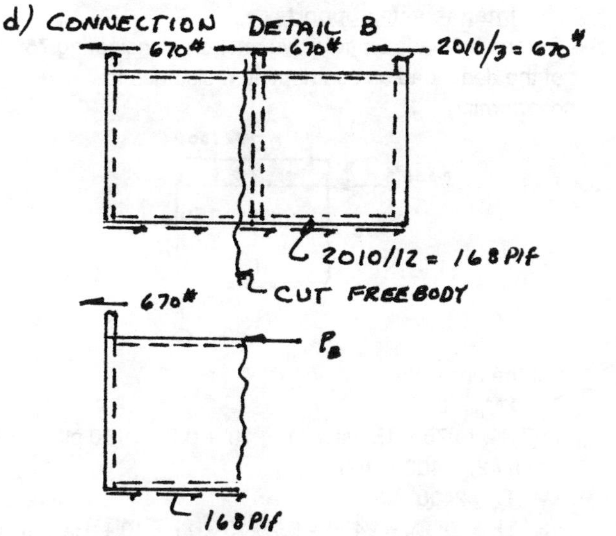

$\Sigma F_x = 0$: (168 plf)6' – 670 – P_B = 0

$P_B = 338^{\#}$ (reversible)

∴ Provide a strap tie to develop 338# tension across the post at B.

Try 16d nail (Re: Table No. 23-III-C-2, UBC-97)

$Z' = C_{st}C_D(108) = 1.25 \times {}^4/_3 \times 105 = 175\ {}^{\#}/_{nail}$

∴ n ≥ 338 / 175 = 1.9 → 2 each side

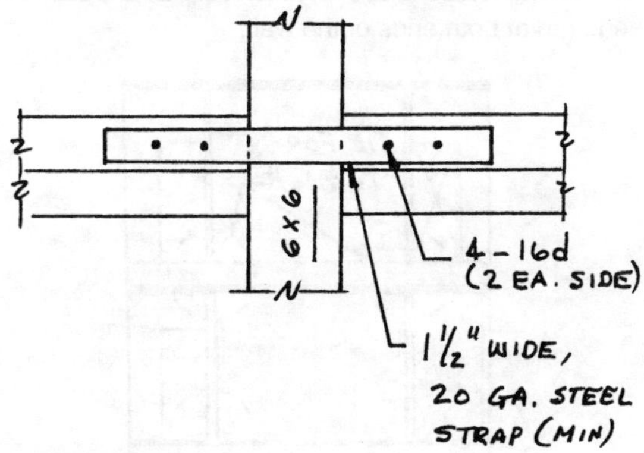

Timber-43

a) Determine the required number of bolts for the beam-to-post connection.

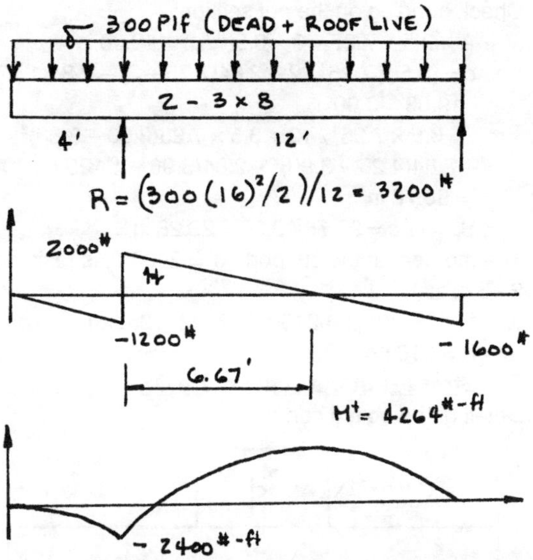

For the 2 - 3×8 Douglas Fir-Larch members connected to a 4×4 post: Re: Table 7.3.3 , NDS-91, for bolts exposed to weather, C_M= 0.75. Assume the roof live is not a snow loading: C_D = 1.25. Re: Table 8.3A, NDS-91: try 3/4" diameter bolts; 3.5" main member length; perpendicular to grain loading

$Z_\perp' = C_DC_mQ = 1.25 \times 0.75 \times 1370 = 1285\ {}^{\#}/_{bolt}$

∴ n ≥ 3200 / 1285 = 2.5→ use 3

% loaded = (3200 / 3) / 1285 = 0.83

Spacing limits for full bolt capacity are:

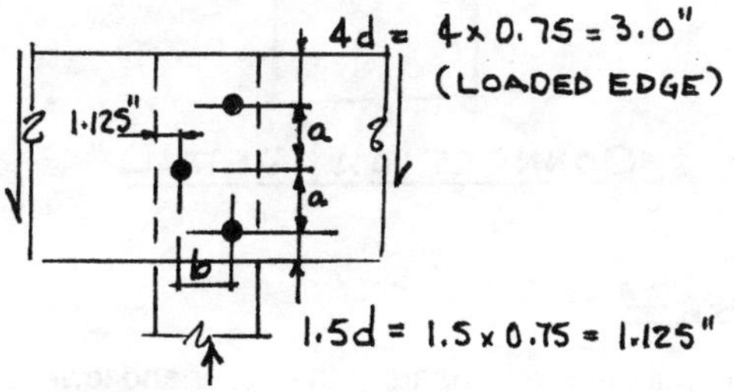

a = (7.25 – 3.0 – 1.125) / 2 = 1.56"

b = 3.5 – 2 × 1.125 = 1.25"

a > b / 4 ∴ can compute as 1 row of 1 plus 1 row of 2 bolts: no reduction for group action; spacing = 2a = 3.12" = 4.16D.

Check the required spacing (Table 8.5.3, NDS-91):

l / D = 3.5 / 0.75 = 4.67:

For l / D = 2; spacing = 2.5D

For l / D ≥ 6; spacing = 5.0D

Interpolate to find spacing for l / D = 4.67

Spacing = D[2.5 + 2.5((4.67 – 2) / (6 – 2))]

= 4.16D = 3.12" = 2a ∴ OK

∴ No reduction for edge or spacing

Use 3 – 3/4" diameter bolts arranged as shown in the figure above.

b) Check stresses at the connection (re: NDS-91). Shear:

$V_{max} = 2000^{\#}$; d_e= 7.25 – 1.125 = 6.12"

$f_v = ({}^3/_2\ V / bd_e)(d / d_e)$

= [1.5 × 2000 / (3.5 × 2 × 6.13)](7.25 / 6.13)

f_v = 83 psi < F_v'

Check bending on the net section:

$A = 3.5 \times 7.25 - 2 \times 3.5(^{13}/_{16}) = 19.68 \text{ in}^2$

$c = [3.5 \times 7.25^2 / 2 - 2.84(7.25 - 3) - 2.84(1.125)] / 19.68 = 3.90"$

$I = 3.5 \times 7.25^3 / 12 + 3.5 \times 7.25(3.90 - 3.63)^2 - 2.84(4.25 - 3.90)^2 - 2.84(3.90 - 1.125)^2 = 90.78 \text{ in}^4$

$S_b = I / c = 90.78 / 3.90 = 23.28 \text{ in}^3$

The moment at the support, $M^- = 2400^{\#\text{-ft}}$, is resisted by 2 - 3×8's:

$\therefore \quad f_b = M / (2S_b) = 2400 \times 12 / (2 \times 23.28)$

$f_b = 616 \text{ psi} < F_b'$

$\therefore$ **<u>Stresses at the support are OK</u>**

c) Detail of the bolted connection:

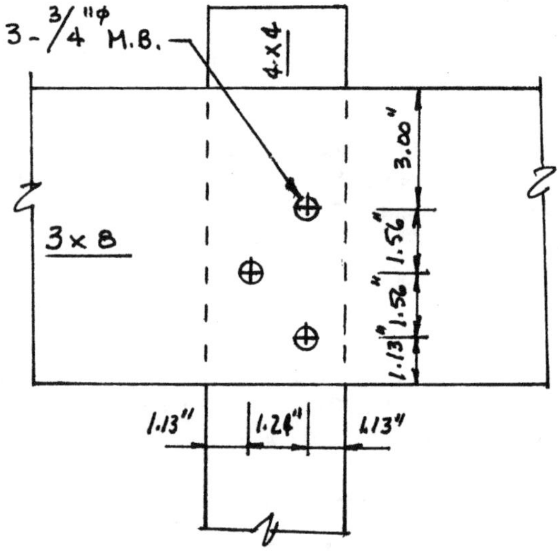

CONNECTION DETAIL

Timber-44

a) Compute the wall shears in the upper and lower stories and determine the required nailing.

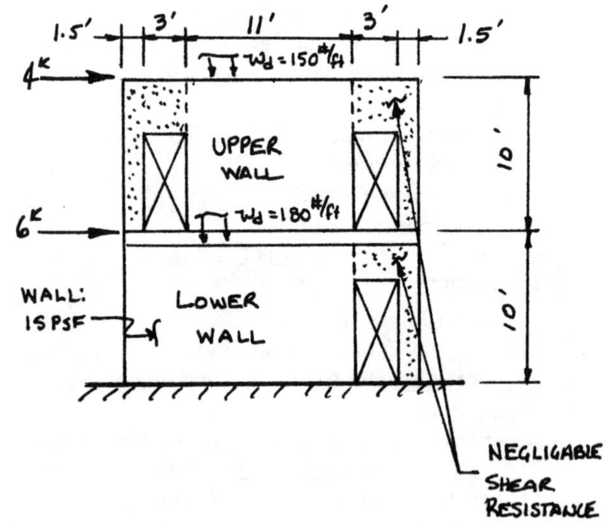

For the upper wall:

$V = 4000^\#$

$\nu_2 = 4000^\# / 11' = 364 \text{ plf}$

For the lower wall:

$V = 4000 + 6000 = 10{,}000^\#$

$\nu_1 = 10{,}000^\# / 15.5' = 645 \text{ plf}$

Re: Table 23-II-I-1UBC-97:

<u>For the top wall (ν = 363 plf), use a single sheath of $^3/_8$" structural I plywood with 8d @ 4" o.c. along panel edges (furnishes 360 plf, which is close enough); 12" o.c. along intermediate supports. For the lower wall (ν = 645 plf), use 1/2" structural I plywood with 10d @ 3" o.c. on panel edges (furnishes 660 plf); 12" o.c. along intermediate supports.</u>

b) Determine loads on tie-down anchors considering 75% of the dead load as contributing resistance to overturning.

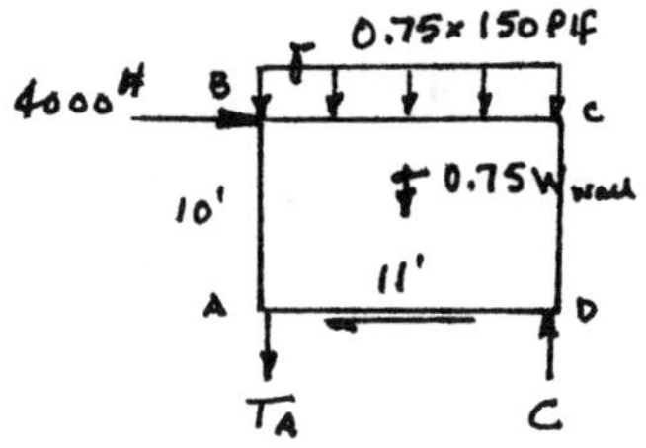

For the upper wall:

$\Sigma M_D = 0$:

$11(T_A) + [0.75 \times 15 \text{ psf} \times 11' \times 10' + 0.75 \times 150 \text{ plf} \times 11'] \times (11' / 2) - 4000^\#(10') = 0$

<u>$T_A = 2400^\#$</u>

$\Sigma F_y = 0$: $C_D = 2400 + 0.75(15 \times 11 \times 10 + 150 \times 11)$

<u>$C_D = 4875^\#$</u>

For the lower wall:

$\Sigma M_{D'} = 0$:

$15.5(T_{A'}) + 0.75[(150 + 180)\text{plf} \times 15.5' + 15 \text{ psf} \times 15.5' \times 20'](15.5 / 2) - 4000(20) - 6000(10) = 0$

<u>$T_A' = 5370^\#$</u>

The lateral forces are reversible $\therefore$ tie-downs are required at both ends of the wall.

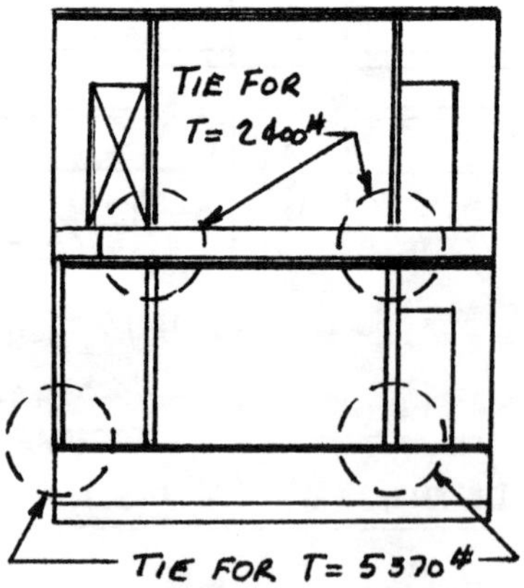

c) Several types of tie-down anchors are illustrated below. Across the second floor ($T = 2400^\#$):

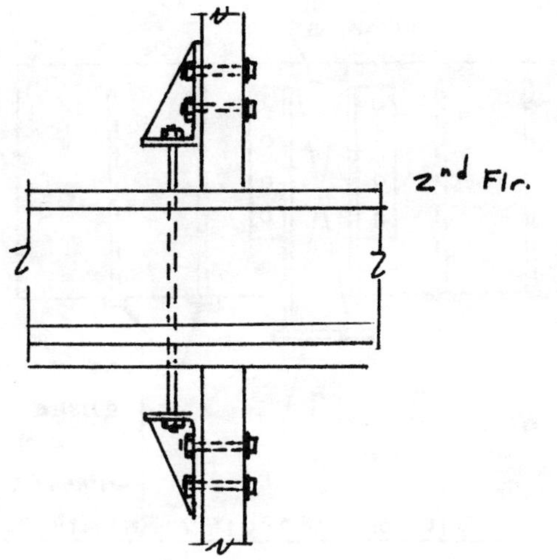

(or)

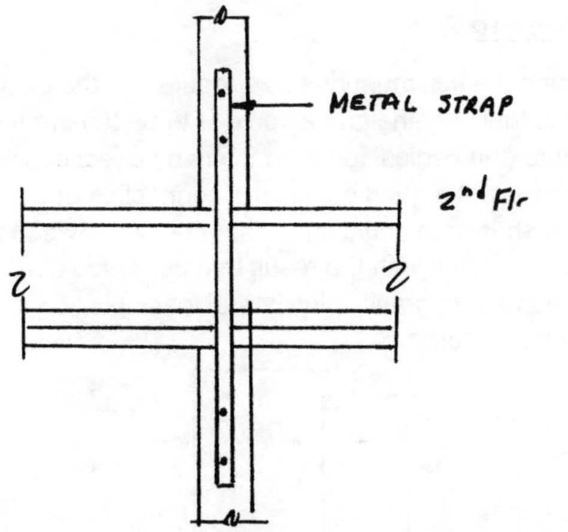

At the first floor (T = 5370#):

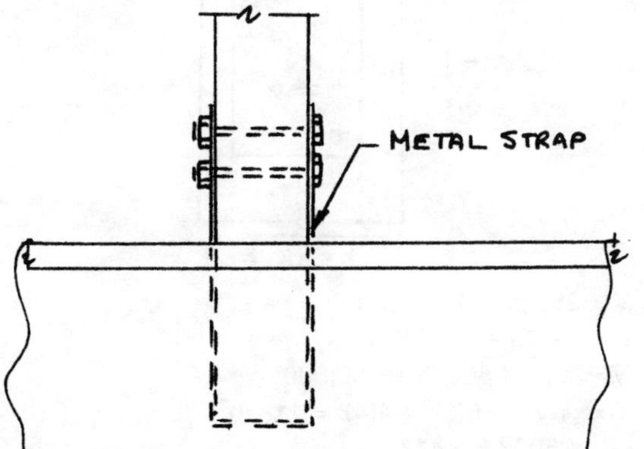

(or)

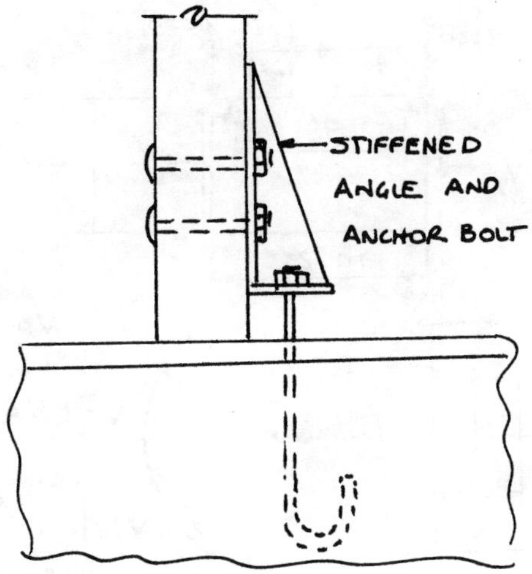

Timber-45

a) Describe the probable mode of failure in beams A and B. Check the shear stress in beam B:

$V = 16{,}000^{\#}$

$d_e = 16"$

$d = 24"$

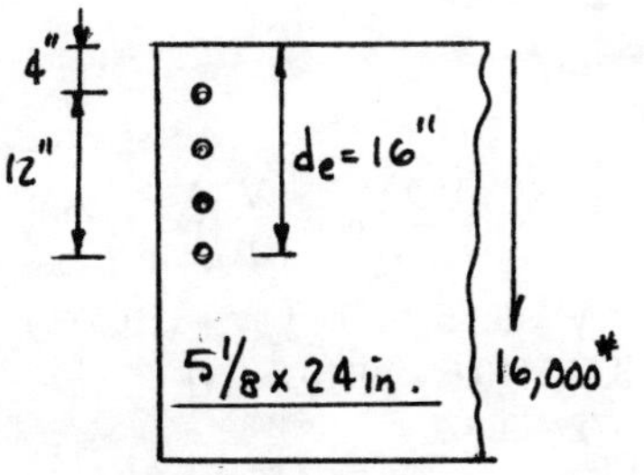

$f_v = (^3/_2 V / bd_e)(d / d_e)$

$f_v = [1.5 \times 16{,}000 / (5.125 \times 16)](24 / 16)$

$= 439 \text{ psi} >> F_v'$

Thus, splitting of beam B was likely caused by excessive horizontal shear at the end joint (i.e., a poor connection detail). Restraint to cross grain shrinkage as the wood dried from its as-built to final moisture content would be an additional contributing factor.

For splitting in beam A:

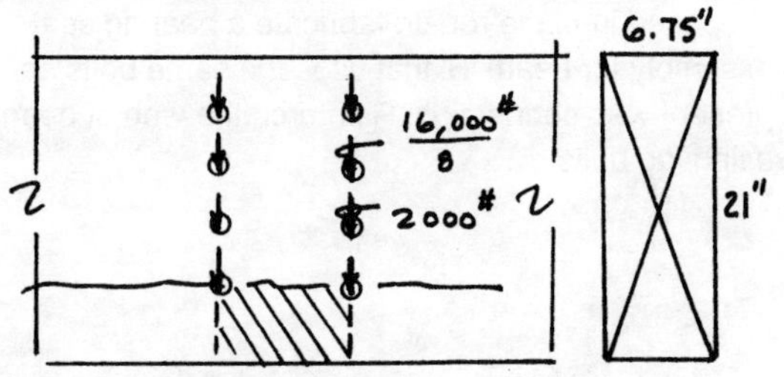

Consider each bolt to resist an equal share of the reaction from beam B:

$r = R / 8 = 16{,}000 / 8 = 2000 \ ^{\#}/_{bolt}$

Take a freebody of the crosshatched region:

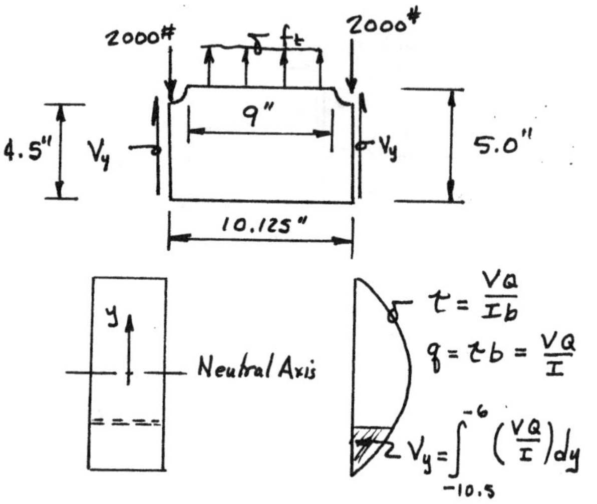

$I = bd^3 / 12 = 6.75(21.0)^3 / 12 = 5209 \text{ in}^4$

For the 16,000# reaction from beam 'B' at the midspan of beam 'A', the shear to either side of the connection is 8000#. The portion of the resisting shears in beam 'A' below the lowest bolt is obtained by integrating the shear flow, $w = \tau b = VQ / I$, from the lowest fiber, $y = -10.5"$, to the level at the bottom of the bolt, $y = -6.0"$

Thus,

$$V_y = \int_{-10.5}^{-6.0} \left(\frac{VQ}{I}\right) dy = \left(\frac{V}{I}\right) \int_{-10.5}^{-6.0} Q dy$$

$$Q = A\bar{y} = 6.75(10.5 - y)[y + (10.5 - y) / 2]$$
$$= 3.375(10.5 - y)(10.5 + y)$$
$$= 372 - 3.375y^2$$

$$\therefore \quad V_y = (V / I)\left[372y - \frac{3.375y^3}{3}\right]_{-10.5}^{-6.0}$$
$$= (8000 / 5209)(615) = 944^{\#}$$

Take a summation of the vertical forces:

$$\Sigma F_y = 0{:}\ 2 \times 944 + f_t(6.75 \times 9) - 2 \times 2000 = 0$$
$$f_t = 35 \text{ psi}$$

Thus, the cross-grain tensile stress is about twice the allowable for Douglas Fir (Re: p. 5-229, AITC). This is the likely cause of the splitting in beam A.

c) Proposed in-place repair: fabricate a bearing seat assembly for beam 'B' that uses the same bolts as the present web connection. Reinforce the web of beam A using lag bolts.

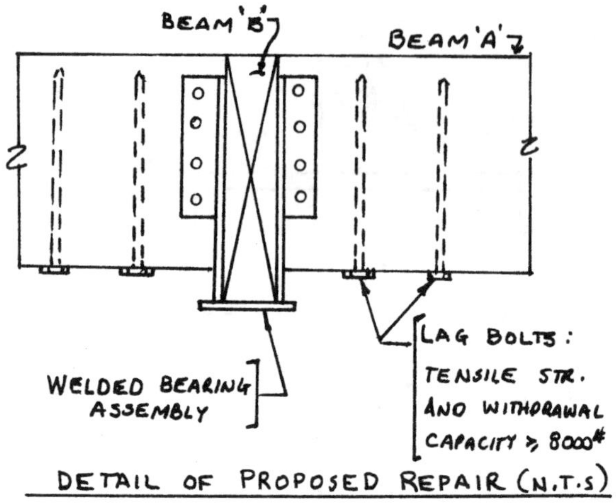

Timber-46

Determine the minimum bolt size required in the connection. I assume that only the lateral force is to be transferred by the bolt group (the vertical force transfers by direct bearing). Note: this is not a good detail for a connection of this type because shrinkage could cause separation between the beam and column with the result that bolts would be subjected to both gravity plus lateral loading.

For the lateral force:

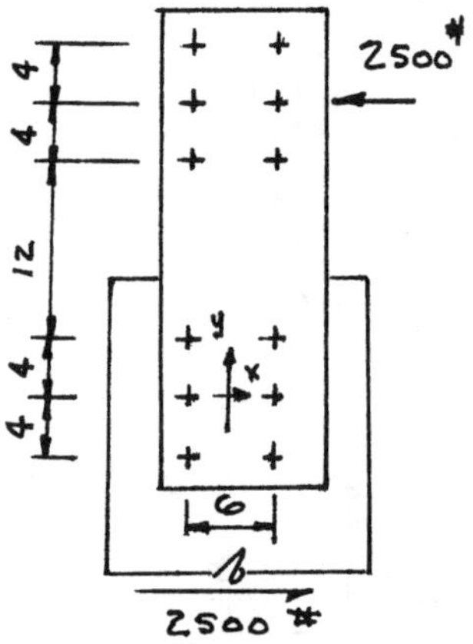

$V = 250 \text{ plf} \times 10' = 2500^{\#}$

$e = 4 + 12 + 4 = 20"$

$M_t = Ve = 2500(20) = 50{,}000^{\#\text{-in}}$

$\Sigma(x_i^2 + y_i^2) = 6(3)^2 + 4(4)^2 = 118 \text{ in}^2$

$r_{px} = 2500 / 6 = 417^{\#} \rightarrow$

$r_{mx} = M_t y / \Sigma(x_i^2 + y_i^2) = 50{,}000 \times 4 / 118 = 1695^{\#} \rightarrow$

$r_{my} = Mx / \Sigma(x_i^2 + y_i^2) = 50{,}000(\pm 3) / 118 = 1270^{\#} \uparrow$

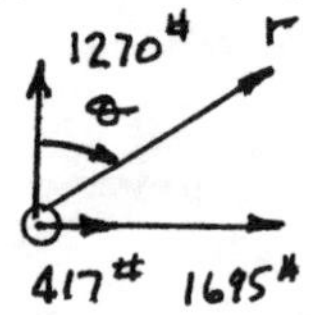

$\theta = \tan^{-1} [(417 + 1695) / 1270] = 59°$

$r = [(417 + 1695)^2 + 1270^2]^{1/2} = 2465^{\#}$

The critical loading occurs in the two top bolts of the six bolts in the column. End and edge distances are adequate for

$D \le {}^7/_8$, $\therefore$ try ${}^5/_8$" diameter bolts (Re: Table 8.3C, NDS-91) for 5-1/2" main member with steel side plates, Douglas Fir-Larch timber:

$Z_q = 2250^\#$

$Z_\perp = 1330^\#$

Duration of load (wind): $C_D = 1.33$; $C_g \cong 1.0$; Assume dry use: $C_m = 1.0$

$Z_q' = 1.33 \times 2250 = 2992^\#$

$Z_\perp' = 1.33 \times 1330 = 1769^\#$

Use the Hankison Formula:

$Z_\theta' = (Z_q'\ Z_\perp') / (Z_q'\sin^2\theta + Z_\perp'\cos^2\theta)$

$Z_\theta' = 2992(1769) / [2992\sin^2 59° + 1769 \cos^2 59°]$

$= 1985^\# < r = 2465^\#$

$\therefore$ ${}^5/_8$" diameter bolts are inadequate; try 3/4" diameter:

$Z_q = 3170^\#$; $Z_\perp = 1800^\#$

$Z_\theta' = 1.33[3170(1800)] / [3170\sin^2 59° + 1800 \cos^2 59°]$

$= 2700^\# > r = 2465^\#$

$\therefore$ **Use 3/4" diameter bolts.**

Timber-47

a) Investigate compliance of beams A and B with the UBC-97. Include a 20 psf moveable partition load as part of the total live load : $w_L = 50 + 20 = 70$ psf; $w_D = 30$ psf. Reduce live loads in accordance with Sec. 1807, UBC – 97:

$$R \le \begin{cases} r(A-150) = 0.0008(18 \times 40 - 150) = 0.46 \\ 0.40 \\ 0.23(1 + D/L) = 0.23(1 + 30/70) = 0.33 \leftarrow \text{governs} \end{cases}$$

$\therefore$ $P = 8' \times 18'$ (30 psf + (1 – 0.33) 70 psf)

$= 4320^\# + 6750^\# = 11{,}070^\#$ say 11.1^k

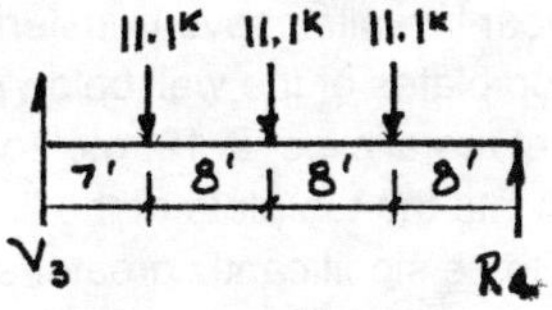

$\Sigma M_4 = 0$: $11.1(8 + 16 + 24) = V_3(31)$

$V_3 = 17.19^k$; $\Sigma F_y = 0$: $R_4 = 16.11^k$

$M^+_{max} = 17.19 \times 7 + (17.19 - 11.1)8 = 169.05^{k\text{-}f}$

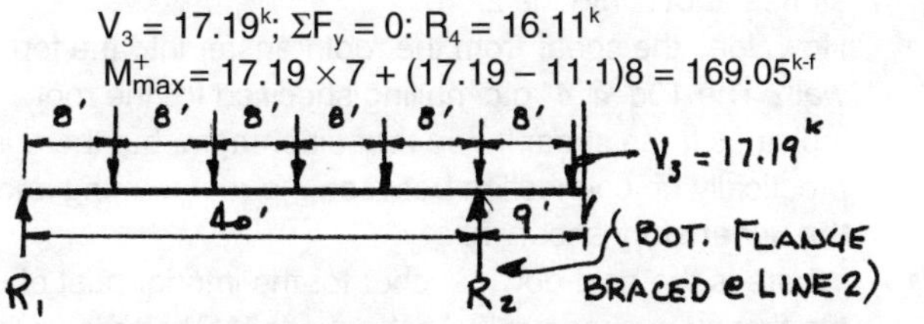

$\Sigma M_1 = 0$: $11.1(8 + 16 + 24 + 32 + 40 + 48)$

$+ 17.19(49) - 40R_2 = 0$

$R_2 = 67.67^k \uparrow$; $\Sigma F_y = 0$: $R_1 = 16.12^k$

$M^+ = 16.12 \times 8 + (16.12 - 11.1)8 = 169.1^{k\text{-ft}}$

$M^- = 11.1 \times 8 + 17.19 \times 9 = 243.5^{k\text{-ft}}$

$$V_{max} \le \begin{cases} 67.67 - 2 \times 11.1 - 17.19 = 28.28^k \\ 17.19 + 11.1 = 28.29^k \end{cases}$$

For the existing 6.75×33: S = 1225 in^3; $F_b = 2400$ psi; $E = 1.7 \times 10^6$ psi (assumed). The critical condition is at the overhang:

$l_e = 1.87 l_u = 1.87 \times 9 \times 12 = 202$

$R_B = \sqrt{l_e d / b^2} = (202 \times 33 / 6.75^2)^{1/2} = 12.1$

$F_b^* = C_D F_b = 1.0 \times 2400 = 2400$ psi

$F_{bE} = 0.438E' / R_B^2 = 0.438 \times 1.7 \times 10^6 / 12.1^2$

$= 5090$ psi

$F_{bE} / F_b^* = 5090 / 2400 = 2.12$

$$C_L = \frac{1 + F_{bE}/F_b^*}{1.9} - \sqrt{\left[\frac{1 + F_{bE}/F_b^*}{1.9}\right]^2 - \frac{F_{bE}/F_b^*}{0.95}}$$

$$= \frac{1 + 2.12}{1.9} - \sqrt{\left[\frac{1 + 2.12}{1.9}\right]^2 - \frac{2.12}{0.95}}$$

$C_L = 0.96$

Re: 5.3.2, NDS-91, check the volume factor:

$C_V = K_L(21/L)^{1/10}(12/d)^{1/10}(5.125/b)^{1/10}$

$= 1.0 \times (21/49)^{1/10} \times (12/33)^{1/10} \times (5.125/6.75)^{1/10}$

$= 0.81$

$F_b' = C_D C_V F_b = 1.0 \times 0.81 \times 2400 = 1945$ psi

$f_b = M^- / S = 243{,}500 \times 12 / 1225 = 2385$ psi

$f_b > F_b'$ (22% overstress)

Check shear: $V = 28{,}290^\#$

$f_v = {}^3/_2 V / bd = 1.5 \times 28{,}290 / (6.75 \times 33) = 191$ psi

Highest available F_v for Douglas Fir ($C_D = 1.0$) is 165 psi $\therefore$ 15% overstress in horizontal shear. Check bearing perpendicular to grain:

$(f_{c\perp})_2 = R / bN = 67{,}670 / (6.75 \times 20) = 501$ psi OK – can specify a combination with $F_{c\perp} > 500$ psi. Similar calculations for beam B yields:

$f_b = 2220$ psi $\approx F_b' = 0.91 \times 2400$

$f_v = 1.5 \times 17{,}190 / (6.75 \times 28.5) = 134$ psi

$\therefore$ **Beam B is OK; however, beam A needs to be resized.**

b) Revisions: since horizontal shear controls beam A, it is desirable to reduce the amount of overhang. By trial and error, an overhang of 4.5' appears to be most economical for a combination with $F_b = 2400$ psi; $F_v = 165$ psi; $E = 1.7 \times 10^6$

$V_3 = 11.1(8 + 16 + 24 + 32) / (40 - 4.5) = 25.01^k$

$M^- = 25.01 \times 4.5 = 112.56^{k\text{-ft}}$

$R_2 = [11.1(8 + 16 + 24 + 32 + 40) + 25.01 \times 44.5] / 40$

$= 61.1^k$; $R_1 = 19.4^k$

$M^+ = 19.4 \times 8 + (19.4 - 11.1)8 = 221.7^{k\text{-ft}}$

Try 6.75×34.5:

$C_V = K_L(21/L)^{1/10}(12/d)^{1/10}(5.125/b)^{1/10}$

$= 1.0 \times (21/44)^{1/10} \times (12/34.5)^{1/10}$

$\times (5.125/6.75)^{1/10} = 0.81$

$F_b' = C_D C_V F_b = 1.0 \times 0.81 \times 2400 = 1945$ psi

$f_b = M^- / S = 221{,}700 \times 12 / 1339 = 1986$ psi $\cong F_b'$

$f_v = 1.5 \times 25{,}010 / (6.75 \times 34.5) = 161$ psi $\cong F_v'$

Check the total load midspan deflection assuming ${}^L/_{240}$ limit is acceptable:

$\Delta_{TL} \cong ({}^5/_{384})\, WL^3 / EI + M^- L^2 / 16EI$

Take $W = (11{,}100 / 8)40 = 55{,}500^\#$

$\Delta_{TL} = [0.013 \times 55{,}500(40 \times 12)^3 - 112{,}560 \times 12(40 \times 12)^2 / 16] / (1.7 \times 10^6 \times 23{,}100) = 2.03" \downarrow -0.50" \uparrow = 1.53" \downarrow$

$L / 240 = 2" > 1.53"$ $\therefore$ OK

$\therefore$ **Use 6.75 × 34.5 in. with 4.5' overhang**

For beam B: $V_{max} = V_3 = 25.01^k$

$\Sigma F_y = 0$: $R_4 = 4 \times 11.1 - 25.01 = 19.4^k$

$\therefore$ $M^+ = 221.7^{k\text{-}ft}$ w / $l_u = 8'$

Same V, M^+ and deflection as beam A

$\therefore$ **Use 6.75 × 34.5 in. for beam B**

Note: an alternative – perhaps a better approach -- is to increase the depth of beam 'A' to 39.0" and leave the shear splice at its given location. Beam 'B' would remain 6.75×28.5 in. The relative board footage required of the two options is:

$B_1 / B_2 = (6.75 \times 33 \times 80) / [6.75(39 \times 49 + 28.5 \times 31] = 0.95$

Thus, option 1 saves about 5% of materials compared to option 2. Also, beams of constant depth may be more desirable from a construction standpoint.

c) Determine the camber for beam B. Use 1.5 times the dead load deflection:

$\Delta_{DL} \cong (5 / 384)W_D L^3 / EI$; $W_D = (4.3 / 8)35.5$

$= 0.013 \times 19{,}081(35.5 \times 12)^3 / (1.7 \times 10^6 \times 23{,}100)$

$\Delta_{DL} = 0.49"$

$\therefore$ **Camber = 1.5 × 0.49 = 0.74, say 3/4"**

d) The feasible camber diagram for beam A is simply a mirror image of the deflected shape under the dead load. Thus,

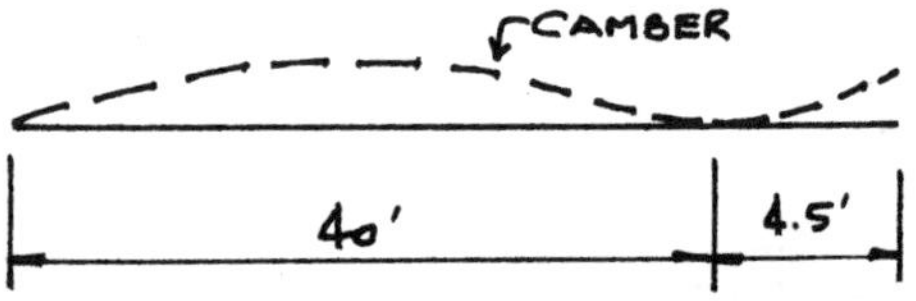

e) Design the connection at point 3. Compute the bearing length required for V = 25.01^k using $F_{c\perp}$ = 650 psi:

$N \geq V / (b F_{c\perp}) = 25{,}010 / (6.75 \times 650) = 5.7"$, say 6"

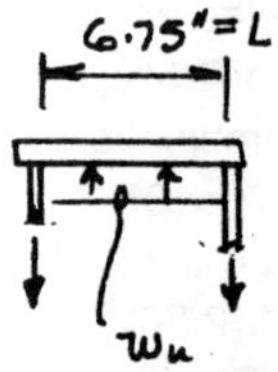

Use $^3/_8$" side plates; ASTM A36 steel; LRFD (AISC 1994); Re: A4.1

$w_u = 1.6 \times 25.01 / 6.75 = 6\ ^k/_{in}$

$M_u = w_u L^2 / 8 = 6 \times 6.75^2 / 8 = 34.2^{k\text{-}in}$

$\phi M_n = \phi_b F_y Z \geq M_u = 34.2$

$Z = bt^2 / 4 = 6t^2 / 4$; $\phi_b = 0.9$

$\therefore$ $t^2 \geq 34.2 \times 4 / (0.9 \times 36 \times 6) = 0.70$ in^2

$t \geq 0.84"$ – use $^7/_8$" PL

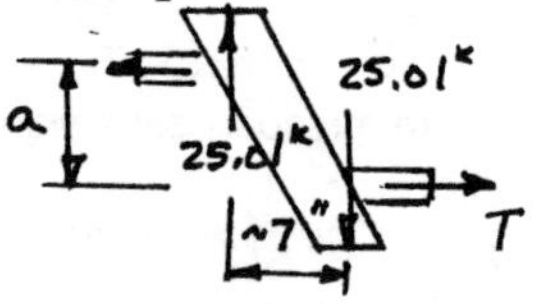

$a \cong 34.5 - 2 \times 4 = 26.5"$

$T = 25.01 \times 7 / 26.5 = 6.6^k = 6600^\#$

Re: Table 8.3, NDS-91; 7/8" diameter {$Z_{//} = 4260\ ^\#/_{bolt}$

Use 2 –7/8" diameter bolts each side

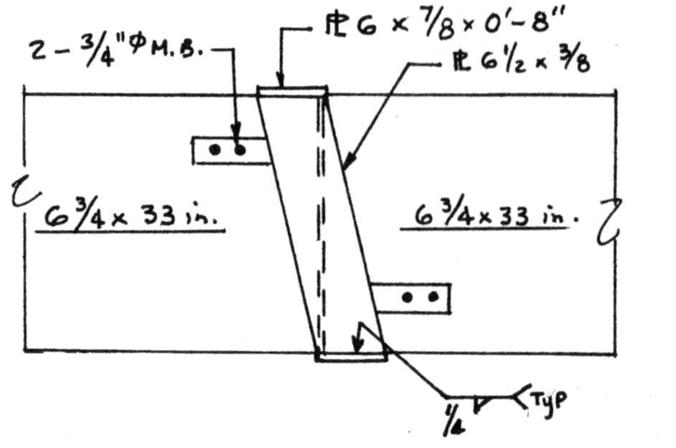

DETAIL – CONNECTION AT '3' (N.T.S.)

Timber-48

List any deficiencies in the construction details.

- Details 'B' and 'C' are suspects. The tie-down anchors for detail 'B' are connected with 3 – $^7/_8$" machine bolts; the connection for detail 'C' is by 2 – 3/4" diameter bolts. Thus, detail 'B' is significantly stronger than detail 'C'; however, detail 'B' applies to the story above where one would expect lower tension to exist. Also, the 4" end distance is less than required for full bolt capacity (7d minimum) and the 3" pitch for the $^7/_8$" diameter bolt is less than the 4d = 3.5" required for full strength.
- For detail 'C', the plywood edge is nailed close to the outer edge of the 2×6 stud. Is the nailing between the stud and 4×4 post (i.e., 16d @ 16" o.c.) adequate to transfer the shear from stud to post?
- How is the shear in walls above transferred into the floor and into the top plates of the wall below? Section 'A' shows 9" sheet metal clips @ 12" o.c. to transfer shear from the 2×10 into the top plate of the first story wall. This appears to be significantly greater shear transfer capability than is available to transfer the shear from two stories above into the 2×10.
- How does the shear from the roof transfer into the top wall? The 10d @ 4" o.c. nailing specified for the roof indicates large shear in the roof diaphragm, but there is practically no connection between the roof framing and the upper story shearwall.
- Where is the hold-down anchor for the interior post of the first story shearwall? Anchor bolts, 1/2" diameter @ 48" o.c., seem light for a three-story wall. Nailing, 10d @ 2-1/2" o.c., in the first story wall is nailed to nominal 2" members on top and side and along adjoining panel edges. This requires 3" or wider framing to accommodate this nail spacing. Also, the pattern must be staggered, not as shown in detail 'C'.
- For detail 'D': 2×10 joists serve as collectors for the diaphragm shear. Can 3 – 16d nails in a scab splice transfer the axial force in these members?

Thus, there are too many suspected deficiencies to approve the construction. A complete review of the

lateral load analysis should be conducted, and change orders issued to correct the deficiencies.

Timber-49

a) Verify the adequacy of the 3x10 floor joists under gravity loads (per UBC-97). Assume the given 12 psf roofing and ceiling dead load is an equivalent load over the horizontal projection. The tributary area to the roof truss is less than 200 ft^2 $\therefore w_L = 20$ psf. Assume no moveable partitions and a floor live load = 50 psf (office occupancy). The ceiling live load (10 psf) does not act concurrently with other live loads. For the exterior wall weight:

1/2" gypsum wallboard	2
2×6 @ 16" o.c.	2
$^5/_8$" gypsum wallboard	2.5
1" siding + paper	3
Insulation / Electrical / Etc	0.5
$\therefore$	$w_d = 10$ psf

Compute the bearing wall reaction on top of the overhang (joist @ 16" = 1.33' o.c.):

P_d = [12 psf × 12.5' + 14 psf × 15.75' / 2 + 10 psf (8' + 1' + 8')] 1.33' = 572$^\#$

P_L = [20 psf × 12.5' + 50 psf × 15.75' / 2] 1.33' = 856$^\#$

Loading condition 1: live + dead on overhang; dead only on the back span.

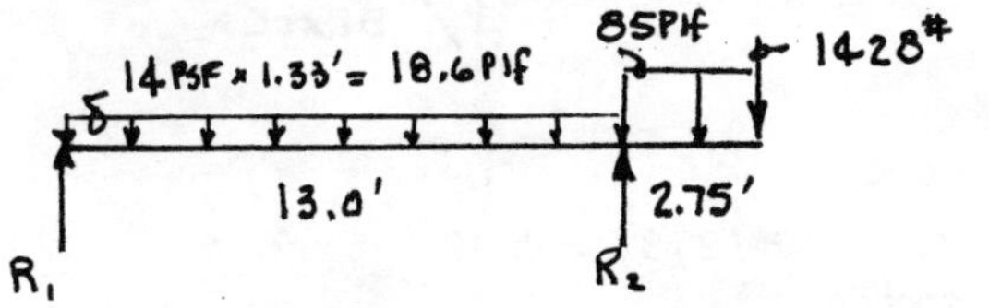

$\Sigma M_1 = 0$: $13R_2 - 18.6(13)^2/2 - 85(2.75)(14.38) - 1428(15.75) = 0$

$R_2 = 2109^\# \uparrow$

$\Sigma F_y = 0$: $R_1 = 18.6(12) + 85(2.75) + 1428 - 2109$

$R_1 = -205^\# = 205^\# \downarrow$ (uplift)

$M^- = 1428(2.75) + 85(2.75)^2/2 = 4248^{\#\text{-ft}}$

Loading condition 2: dead on overhang; live plus dead on backspan.

$\Sigma M_2 = 0$: $13R_1 + 572(2.75) + 18.6(2.75)^2/2 - 85(13)^2/2 = 0$

$R_1 = 426^\# \uparrow$

V = 0 at x = 426 / 85 = 5.01'

$\therefore \quad M^+ = 0.5 \times 426 \times 5.01 = 1067^{\#\text{-ft}}$

The critical condition is loading condition 1. Check the existing 3×10 for this loading and no increase in allowable stress for load duration. For nominal 10" No. 1 and better Douglas Fir-Larch: $F_b = 1.1 \times 1.15 \times 1150$ = 1455 psi (repetitive use); F_v = 95 psi; $F_{c\perp}$ = 625 psi; E = 1.8 × 10^6 psi. Check bending (3×10's):

$S = 2.5 \times 9.25^2 / 6 = 35.7$ in^3

$f_b = M^- / S = 4248 \times 12 / 35.7 = 1428 \text{ psi} < F_b$

OK in bending – check shear stress near the support:

$V = 1428 + 85(2.75 - 1) = 1577^\#$

$f_v = {}^3/_2 V / bd = 1.5 \times 1577 / (2.5 \times 9.25) = 102$ psi (8% overstress)

Check bearing: $R_{max} = 2540^\#$ (85 plf on the entire beam):

$f_{c\perp} = R / bn = 2540 / (2.5 \times 5.25) = 194 \text{ psi} < F_{c\perp}$

Bearing at the new stud wall is OK – check the end joint shear at detail 'A' (note: the effective depth at the connection is not specified. Approximate d_e as ($^2/_3$)d at the connection:

$d_e = 0.67 \times 9.25" = 6.2"$

$V_e = 1428^\#$

$f_v = ({}^3/_2 V_e / bd_e)(d / d_e)$

$= (1.5 \times 1428 / 2.5 \times 6.2)(9.25 / 6.2)$

$= 206 \text{ psi} >> F_V$

Thus, the 3x10 is overstressed in shear.

b) Verify detail 'A'. This detail is inadequate due to excessive endjoint shear (calculated above). Furthermore, the maximum force in the wall, 1428$^\#$, cannot be safely transferred by the 650$^\#$ capacity clip angle and 3 – 16d nails. The total capacity of connection at 'A' is:

$V = 650 + 3 \times 105 = 965^\#$

Additional shear transfer capacity could be furnished; e.g., a 1×6 ribbon let into the studs above the joists would add:

$\Delta V = F_{c\perp} bt = (625 \text{ psi}) 2.5" \times 0.75" = 1170^\#$

The combined resistance then becomes 1170 + 965 = 2135$^\#$ would be more than adequate. However, the framing scheme would not satisfy sec. 2320.8.5, UBC-97, which requires that bearing partitions (wall, in this case) shall not be offset from the supporting wall more than the joist depth. An alternative scheme needs to be devised. One possibility, which involves losing additional space, is to frame the space above the overhang to transfer load into the new wall as shown below:

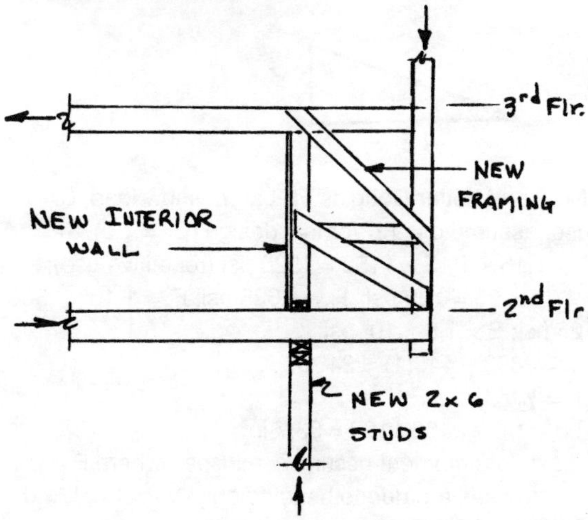

c) Check detail 'B'. This connnection is adequate. Bearing on the top plate is less under loading condition 2 than existed prior to modification. The uplift of 205$^\#$ that occurs under loading condition 1 is counterbalanced by the dead load from the wall directly above:

P_{above} > [14 psf(9.25 + 15.75 / 2) + 10 psf × 8'] 1.33'
= $425^{\#}$

This is roughly double the uplift.

Timber-50

Design the roof system for the given conditions.

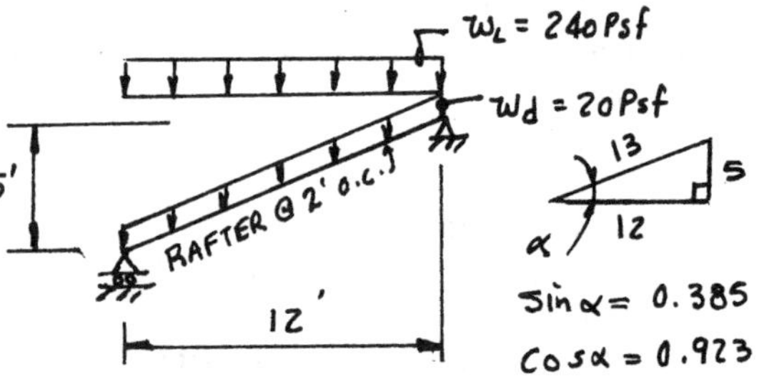

Express the load as an equivalent force per lineal foot of length (rafters 2' o.c.):

w = 20 psf × 2' + 240(12 / 13)psf × 2' = 483 $^{\#}/_{ft}$

R = wL / 2 = 483 × 13 / 2 = $3140^{\#}$

Resolve forces into components parallel to and perpendicular to the rafter:

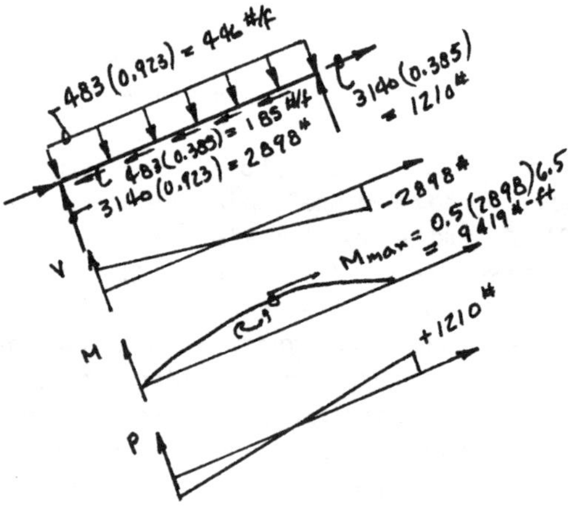

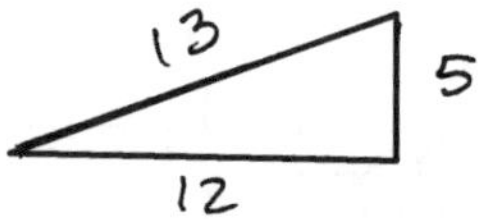

For No.1 and better Douglas Fir-Larch; snow load, C_D = 1.15; dry use; assume d ≅ 12" for trial design (C_F = 1.0)

∴ F'_b = 1.15 × 1.15 × 1150 = 1520 psi (repetitive use);

F'_v = 1.15 × 95 = 109 psi; $F_{c\perp}$ = 625 psi; F_c = 1.15 × 1500 = 1725 psi; E = 1.8 × 10^6 psi:

V = 2898 – 446(1) = $2452^{\#}$

$f_v = {}^3/_2 V / bd \le F_v'$

∴ bd ≥ 1.5 × 2452 / 109 = 33.7 in^2

The maximum moment occurs @ midspan where P = 0 ∴ normal stresses are due to bending only: V_D ≅ 1.0; l_e = 0

$f_b = M_{max} / S \le F_b'$

S ≥ 9419 × 12 / 1520 = 74.4 in^3

Deflection limit: $\Delta_L \le {}^L/_{240}$

$\Delta_L = ({}^5/_{385})(w_L L)L^3 / EI \le {}^L/_{240}$

w_L = 2 × 240(12 /13) = 443 plf

I ≥ 240(0.013)(443 × 13)(13 × 12)^2 / 1.8 × 10^6

I ≥ 243 in^4

∴ **<u>Use 4×12 (A = 39.4 in^2; S = 73.8 in^3 (close enough); I = 415 in^4)</u>**

Check bearing of the rafter on the top plates:

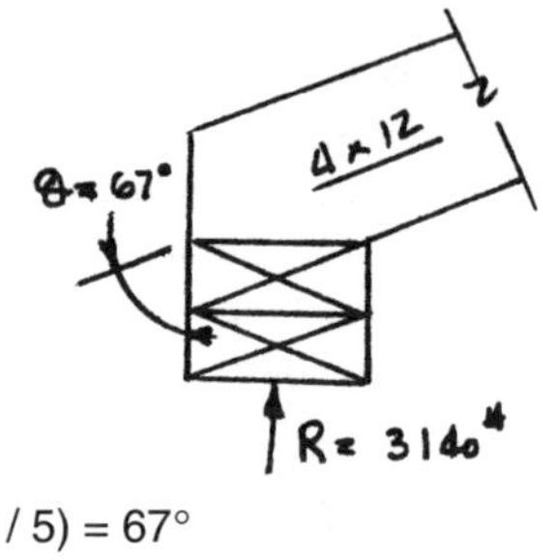

θ = tan^{-1}(12 / 5) = 67°

$F_{c\perp}'$ = 625 psi

Use Hankison Formula:

$F_{c\theta} = (F_{c\perp}' F_c') / (F_c' \sin^2\theta + F_{c\perp}' \cos^2\theta)$

= 625 × 1725 / (1725 sin^{2}67° + 625 cos^{2}67°)

= 692 psi

N ≥ R / b$F_{c\theta}$ = 3140 / (3.5 × 692) = 1.3"

OK – any reasonable width top plate will furnish adequate support to the rafter.

Design the center post. Assume that bracing from the centerline of ridge beam to centerline of girder can be provided: $l_e = K\, l_u$ = 5'. Posts and girders are spaced at 8' o.c..

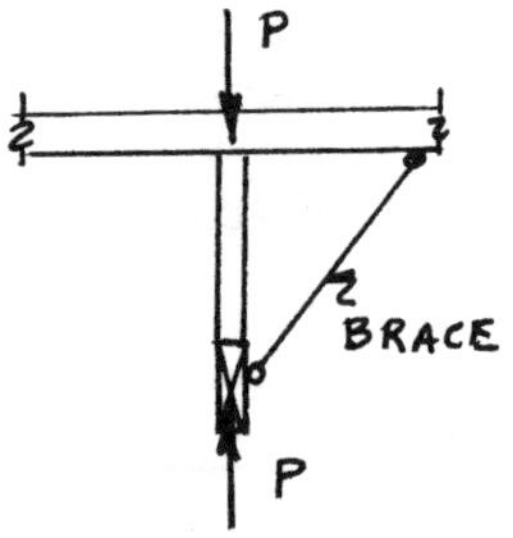

P = [(3140$^{\#}$ / 2') 8']2 = 25,120$^{\#}$

Trial section:

∴ A ≥ P / F_c' = 25,120 / 1725 = 14.6 in^2

b ≥ 3.8" (for square column)

Try 6×6 (A = 5.25^2 = 27.6 in^2)

l_e / d = 60 / 5.25 = 11.4

$$F_{cE} = \frac{0.3E'}{(l_e/d)^2} = \frac{0.3 \times 1{,}800{,}000}{11.4^2} = 4155\,\text{psi}$$

F_c^* = $C_D C_F F_c$ = 1.15 × 1.1 × 1500 = 1900 psi

F_{cE} / F_c^* = 4155 / 1900 = 2.19

$$C_p = \frac{1+F_{cE}/F_c^*}{2c} - \sqrt{\left[\frac{1+F_{cE}/F_c^*}{2c}\right]^2 - \frac{F_{cE}/F_c^*}{c}}$$

$$= \frac{1+2.19}{1.6} - \sqrt{\left[\frac{1+2.19}{1.6}\right]^2 - \frac{2.19}{0.8}}$$

C_p = 0.88

∴ F_c' = $C_p C_D C_F F_c$ = 0.88 × 1.15 × 1.1 × 1500 = 1670 psi

f_c = P / A = 25,120 / 27.6 = 910 psi < F_c'

∴ **<u>Use 6×6 center post</u>**

Design the glulam girder. Assume lateral support at ends and midspan and estimate the girder weight as 50 plf:

$M = PL/4 + w_g L^2/8 = 25{,}120 \times 24/4 + 50 \times 24^2/8$

$= 154{,}300^{\#\text{-ft}}$

$S \approx M/F_b' = 154{,}300 \times 12/2400 = 772 \text{ in}^3$

Try 6.75×27 in ($S = 820 \text{ in}^3$; $I = 11{,}070 \text{ in}^4$)

$l_e = 1.11 l_u = 1.11 \times 144 = 160"$; Re: 3.3.3, NDS-91:

$R_B = (l_e d/b^2)^{1/2} = (160 \times 27/6.75^2)^{1/2} = 9.7$

$F_b^* = C_D F_b = 1.15 \times 2400 = 2760 \text{ psi}$

Specify a glulam combination for which $E' = E_{yy} = 1.6 \times 10^6$ psi:

$F_{bE} = 0.438E'/R_B^2 = 0.438 \times 1.6 \times 10^6/9.7^2 = 7450 \text{ psi}$

$F_{bE}/F_b^* = 7450/2760 = 2.70$

$$C_L = \frac{1+F_{bE}/F_b^*}{1.9} - \sqrt{\left[\frac{1+F_{bE}/F_b^*}{1.9}\right]^2 - \frac{F_{bE}/F_b^*}{0.95}}$$

$$= \frac{1+2.70}{1.9} - \sqrt{\left[\frac{1+2.70}{1.9}\right]^2 - \frac{2.70}{0.95}}$$

$C_L = 0.97$

Check the volume factor:

$C_V = K_L(21/L)^{1/10}(12/d)^{1/10}(5.125/b)^{1/10}$

$= 1.0 \times (21/24)^{1/10} \times (12/27)^{1/10} \times (5.125/6.75)^{1/10}$

$= 0.89 \leftarrow$ governs

$F_b' = C_D C_V F_b = 1.15 \times 0.89 \times 2400 = 2450 \text{ psi}$

$S \geq 154{,}300 \times 12/2450 = 755 \text{ in}^3$ (6.75×27 in is OK in bending)

Check shear:

$L_e = 24 - 2 \times 27/12 = 19.5'$

$V = P/2 + w_g L_e/2 = 25{,}120/2 + 50 \times 19.5/2 = 13{,}050^\#$

$f_v = {}^3/_2 V/bd = 1.5 \times 13{,}050/(6.75 \times 27) = 107 \text{ psi}$

OK – requires combination w/ $F_v' > 107$ psi

Check live load deflection: $I = 11{,}070 \text{ in}^4$; $P_L = 240 \text{ psf} \times 12' \times 8' = 23{,}040^\#$

$\Delta_L = P_L L^3/48EI \leq L/240$

$\therefore \quad E \geq 23{,}040 \times (24 \times 12)^2 \times 240/(48 \times 11{,}070)$

$= 863{,}000$ psi – OK

Use 6.75 × 27.0 in with 24F-E1 combination

Check bearing of 6×6 post on the top flange: $F_{c\perp} = 650$ psi

$f_{c\perp} = P/A = 25{,}120/(5.25^2) = 911 \text{ psi} > F_{c\perp}$

Need to provide a metal bearing plate – say PL 6.75 ×

$N \geq P/F_{c\perp} b = 25{,}120/(6.75 \times 650) = 5.7"$

Say PL 6 × $^3/_8$ × 0' – 6.75"; use $^5/_8$" diameter machine bolt attached as shown below:

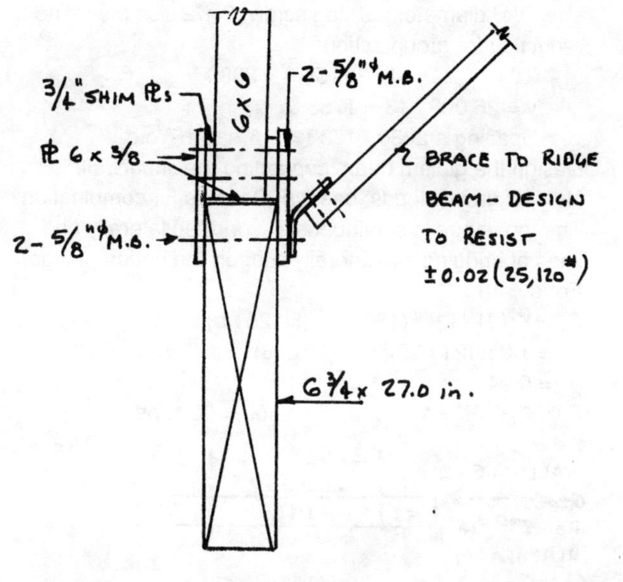

POST – TO – GIRDER CONNECTION (N.T.S.)

Timber-51

a) Design the shearwall from the second floor to roof:

$\nu = 26{,}000/15 = 1733 \text{ plf}$

The shear flow is too large to resist by sheathing only one side ∴ use both sides:

$\nu = 26{,}000/(2 \times 15) = 867 \text{ plf}$

Re: Table No. 23-II-I-1, UBC-97, use 1/2" structural plywood with 10d @ 2" o.c. (staggered at panel edges). Studs, plates and blocking to be 3" nominal width. The design furnishes a shear capacity of 870 plf.

b) Complete the details in section A-A. Use the full 25' width to transfer shear from the ceiling diaphragm:

$\nu = 26{,}000/25 = 1040 \text{ plf}$

Provide 2 lines of nailing along the top plate (I assume here that the interior wall collects an equal shear from each side. Re: Table No. 23-II-H, UBC-97, assume a blocked diaphragm; 8d @ 2.5" o.c. along both top edges of top plates will furnish $\nu \geq 600$ plf each edge; 1200 plf total.

Use 3×6 collectors nailed from the underside with 2 rows of 8d @ 2.5" o.c.. The maximum collector force over the 6' width is:

$V = \pm 1040 \text{ plf} \times 6' = \pm 6240^\#$

Connect to the top plates through bolts – try 3/4" diameter; $l = 2.5"$; single shear:

$Z_q' = ({}^4/_3)(1020) = 1360\ {}^\#/_{\text{bolt}}$

$n \geq 6240/1360 = 4.6$, say 5 bolts

Over the 4' opening, $V = \pm 4160^\#$ ∴ 3 bolts are OK

Check stresses:

$A_n = 2.5(5.25 - {}^{13}/_{16}) = 11.1 \text{ in}^2$

$f_t = 6240/11.1 = 562 \text{ psi} \ll F_t'$

Transfer the total shear into the glulam beam below. Use the maximum depth 18" beam. Re: Table No. 9.3A

– try 3/4" diameter × 9" lag screw; 2-1/2 side piece; no reduction for group action:

$Z_q' = C_D Z_q = (4/3)(1560) = 2080\ ^{\#}/_{bolt}$

$v = 26{,}000 / 15 = 1733\ ^{\#}/_{ft}$

∴ Spacing ≤ 2080 / 1733 = 1.2' say 15" o.c.

c) Design the glulam beam supporting the shearwall. Neglect gravity loads; use 24F Douglas Fir combination. The top flange is continuously supported laterally and lines of bridging can laterally support the bottom flange.

For d = 18":

$C_V = K_L(21/L)^{1/10}(12/d)^{1/10}(5.125/b)^{1/10}$

$= 1.0 \times (21/25)^{1/10} \times (12/18)^{1/10} \times (5.125/5.125)^{1/10}$

$= 0.94$

$F_b' \le C_D C_V F_b = 1.33 \times 0.94 \times 2400 = 3000$ psi

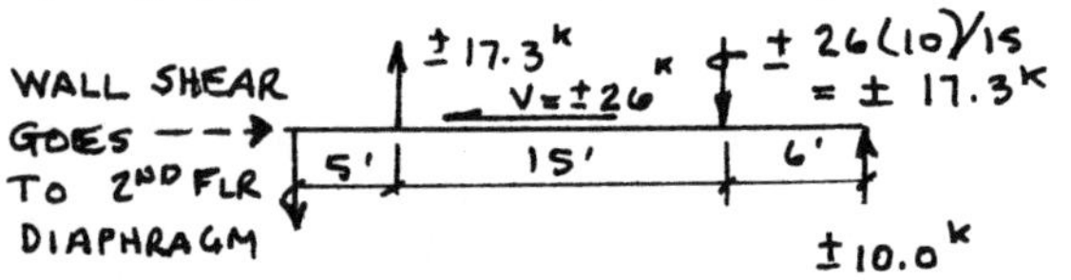

$M_{max} = \pm 10.0(6) = \pm 60^{k\text{-}ft}$

$S \ge M_{max} / F_b' = 60{,}000 \times 12 / 3000 = 240\ in^3$

Try $5^1/_8 \times 18$ in member; effective section includes reduction for 3/4" lag bolts in tension zone (6" deep):

$\Delta \bar{y} = 6 \times 0.75 \times 6 / (5.125 \times 18 - 0.75 \times 6) = 0.31"$

$I = 5.125 \times 18^3 / 12 + 5.125 \times 18 \times 0.31^2 - 0.75 \times 6^3 / 12 - 0.75(6)(9 - 3 - 0.31)^2 = 2340\ in^4$

$S = 2340 / (9 + 0.31) = 251\ in^3 > 240\ in^3$

Check shear: $V = 10{,}000^{\#}$

$f_v = {}^3/_2 V / bd = 1.5 \times 10{,}000 / (5.125 \times 18) = 162$ psi

$< F_v' = {}^4/_3 F_v$

∴ Specify a combination with $F_v > 122$ psi.

Try bracing the bottom flange at the quarter points:

$l_u = 25/4 = 6.25'$. Re: Table 3.3.3, NDS-91, $l_e \le 2.06 l_u = 2.06 \times 6.25 \times 12 = 155"$:

$$R_B = \sqrt{l_e d / b^2} = \sqrt{155 \times 18 / 5.125^2} = 10.3$$

$F_b^* = C_D F_b = 1.33 \times 2400 = 3190$ psi

Specify a glulam combination for which $E' = E_{yy} = 1.6 \times 10^6$ psi:

$F_{bE} = 0.438E' / R_B^2 = 0.438 \times 1.6 \times 10^6 / 10.3^2 = 6600$ psi

$F_{bE} / F_b^* = 6600 / 3190 = 2.06$

$$C_L = \frac{1 + F_{bE}/F_b^*}{1.9} - \sqrt{\left[\frac{1 + F_{bE}/F_b^*}{1.9}\right]^2 - \frac{F_{bE}/F_b^*}{0.95}}$$

$$= \frac{1 + 2.06}{1.9} - \sqrt{\left[\frac{1 + 2.06}{1.9}\right]^2 - \frac{2.06}{0.95}}$$

$C_L = 0.96 < C_V$ ∴ bending is OK

∴ provide 3 lines of bracing to bottom flange; use $5^1/_8 \times 18$ in. glulam with 24F / DF combination; $F_v \ge 140$ psi; $F_{c\perp} \ge 650$ psi

d) Design a hold-down anchor for the shearwall to beam.

$T = (26^k)(10') / 15' = 17.3^k$

Try 3/4" diameter bolts; 4×6 post; No.1 Douglas Fir-Larch: $F_t = 775$ psi

$A_n = 3.5(5.25 - {}^{13}/_{16}) = 15.5\ in^2$

$f_t = T / A_n = 17{,}300 / 15.5 = 1160$ psi $< F'_t = C_D C_F F_t = 1.33 \times 1.3 \times 675 = 1170$ psi

Check the 4×6 in compression ($F_c = 1450$ psi, $E = 1{,}700{,}000$ psi):

$l_e / d = 120 / 5.25 = 23$

$$F_{cE} = \frac{0.3E'}{(l_e/d)^2} = \frac{0.3 \times 1{,}700{,}000}{23^2} = 964\ \text{psi}$$

$F_c^* = C_D C_F F_c = 1.33 \times 1.1 \times 1450 = 2120$ psi

$F_{cE} / F_c^* = 964 / 2120 = 0.45$

$$C_p = \frac{1 + F_{cE}/F_c^*}{2c} - \sqrt{\left[\frac{1 + F_{cE}/F_c^*}{2c}\right]^2 - \frac{F_{cE}/F_c^*}{c}}$$

$$= \frac{1 + 0.45}{1.6} - \sqrt{\left[\frac{1 + 0.45}{1.6}\right]^2 - \frac{0.45}{0.8}}$$

$C_p = 0.40$

∴ $F_c' = C_p C_D C_F F_c = 0.40 \times 4/3 \times 1.1 \times 1450 = 850$ psi

$A \ge P / F_c' = 17{,}300 / 850 = 20.4\ in^2$

Thus, the 4x6 is inadequate; use a 6×6 post (A = 27.6 in²); connect to the hold-down anchor with steel side plates (Re: Table 8.2B, NDS-91); $l = 5.5"$; 3/4" diameter bolts:

$Z_q' = C_D Z_q = ({}^4/_3)(1580) = 2100\ ^{\#}/_{bolt}$; low – try 1" diameter bolts

$Z_q' = ({}^4/_3)(2760) = 3680\ ^{\#}/_{bolt}$

$n \ge 17{,}300 / 3680 = 4.7$; try 5 bolts

Check group action with 5 bolts; Re: Table 7.3.6C, NDS, $C_g \cong 0.9$ ∴ use 6 bolts.

Transfer $T = 17.3^k$ to the underside of beam by direct bearing:

$N \ge T / b F_{c\perp} = 17{,}300 / (650 \times 5.125) = 5.19"$; C5×9 is close enough

Attach the C5×9 to the PL 5×1/4 side plates using 2 – 3/4" diameter A307 bolts ($r_v = {}^4/_3 \times 4.4^k$) each side. Weld the side PLs to pieces of C5×9 using 2 – ${}^3/_{16}$ E70 welds. Thus, the connection details are as follows:

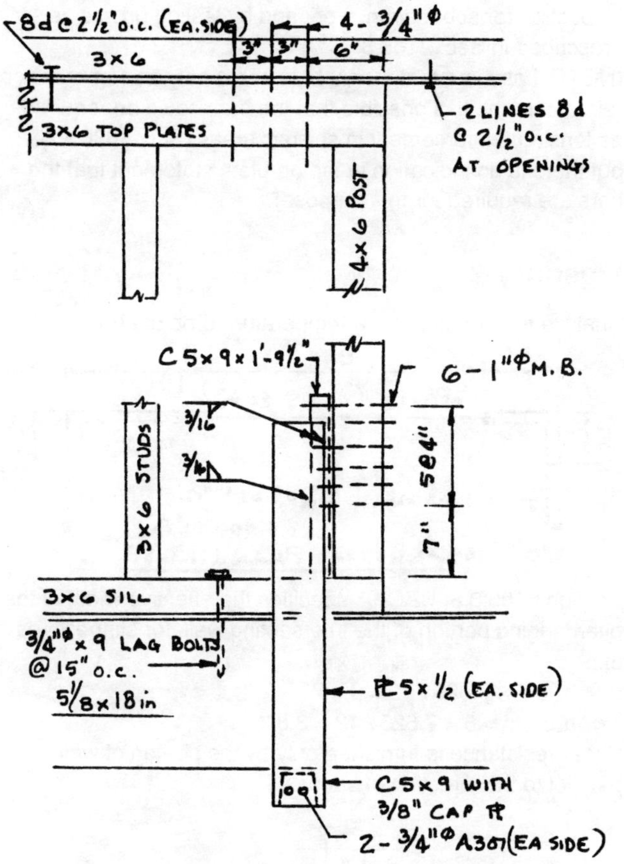

PARTIAL ELEVATION (N.T.S.)

Masonry-1

Suggest corrective measures for the missing bottom reinforcement. The wall and cantilever are complete to within 10" of the top and bars 'A' were omitted.

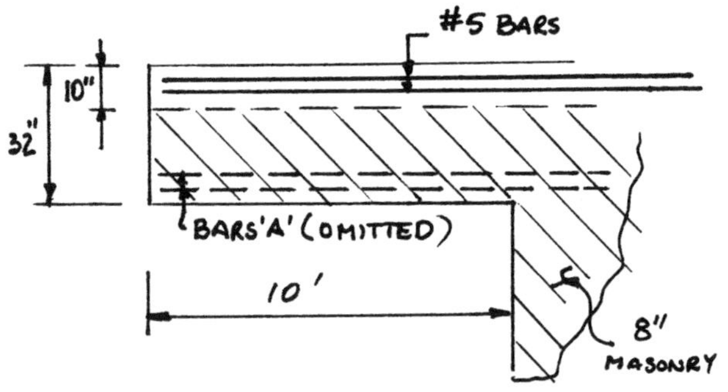

Design load for the cantilever = 0.5 k/ft. Use the Uniform Building Code (UBC–97), Chapter 21.

$n = E_s / E_m = 44$

$\therefore \quad E_m = E_s / 44 = 29 \times 10^6 / 44 = 6.6 \times 10^5$ psi

$E_m = 750f'_m$

$\therefore \quad f'_m \cong E_m / 750 = 6.6 \times 10^5 / 750 = 880$ psi

The allowable flexural compressive stress (wall was constructed without special inspection) is:

$F_c = 0.5\ (0.33\ f'm \le 2000)$

$= 0.5\ (0.33 \times 880) = 145$ psi

Check flexural stresses in the cantilever assuming it is singly reinforced: d = 32 – 4 = 28"; b = 7.625 in; A_s^- (2 # 5) $= 2 \times 0.31 = 0.62\ in^2$

$M^- = wa^2 / 2 = 0.5 \times 10^2 / 2 = 25^{\text{ k-ft}}$

$p = A_s / bd = 0.62 / (7.625 \times 28) = 0.0029$

$pn = 0.0029 \times 44 = 0.128$

$k = \sqrt{(pn)^2 + 2pn} - pn = 0.39; \; j = 1 - {}^k/_3 = 0.87$

$f_s = M / A_s jd = 25 \times 12 / (0.62 \times 0.87 \times 28)$

$= 19.9$ ksi $< F_s = 24$ ksi (Grade 60)

$f_c = (M / bd^2)(2 / jk)$

$= (25 \times 12) / (7.625 \times 28^2)(2 / (0.87 \times 0.39))$

$= 0.29$ ksi $= 290$ psi $> F_c = 145$ psi

Thus, the beam does not satisfy UBC-97 allowable compressive stress as a singly reinforced member; the 'A' bars were needed as compression reinforcement. There is no way to add reinforcement within the top 10" to compensate for the high compressive stress (i.e., if d' > 22", the stresses in the bottom fiber would exceed F_c well before the compressive stress developed in the steel). The only practical way to salvage the completed work is to justify a higher allowable F_c, which seems reasonable to do.

If special inspections had been made, the allowable F_c would have been 2×145=290 psi; which is acceptable. The critical inspection items for the cantilever are placement and grouting of the top steel, which can still be inspected. Since f'_m < 1500 psi, Section 7.1, UBC-97 permits periodic inspections and it is likely that no inspection would have been performed on the cantilever on work done to this point. Thus, there is a rationale for increasing F_c by providing inspection for subsequent work and by testing prisms as prescribed in Sec. 2105.3.

(NOTE: I assume that the 'A' bars serve only as compression reinforcement. It is possible that the bars could be required as tension reinforcement in seismic areas (Sec. 1630.11), but there is no indication in the problem statement that the bars are required for this purpose).

Masonry-2

Analyze the structure for a temperature drop of 80°F.

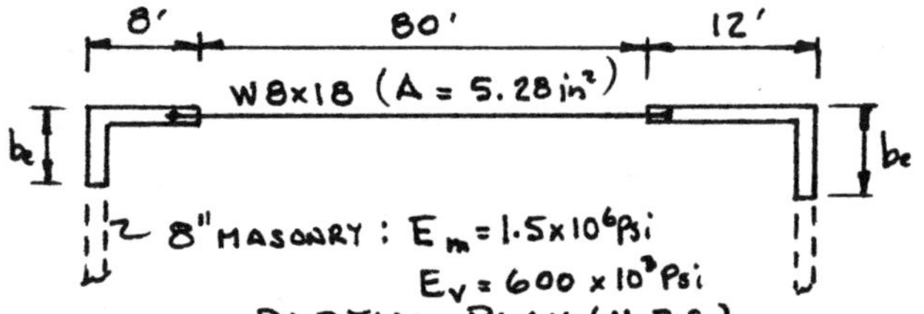

PARTIAL PLAN (N.T.S.)

Section 2106.3.8, UBC-97 specifies the effective width of the overhanging portion of the intersecting wall, for stiffness, as 6t.

$\therefore \quad b_e = 7 \times 7.625 / 12 = 4.44'$

$b_{eo} = 6t = 6 \times 7.625 / 12 = 3.81'$

Shear resistance is furnished only by the portion of wall parallel to the direction of shear.

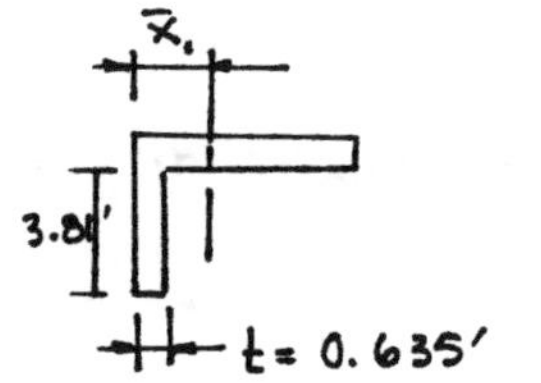

$A = 0.635(8 + 3.81) = 7.5\ ft^2$

$$\bar{x}_1 = \frac{0.635(8 \times 4 + 3.81 \times 0.318)}{7.5} = 2.81'$$

$I_1 = 0.635(8)^3 / 12 + 0.635(8)(4 - 2.81)^2 + 3.81 \times 0.635^3 / 12 + 0.635 \times 3.81(2.81 - 0.318)^2$

$= 49.4\ ft^4$

$A_{v1} = 0.635(8) = 5.1\ ft^2$

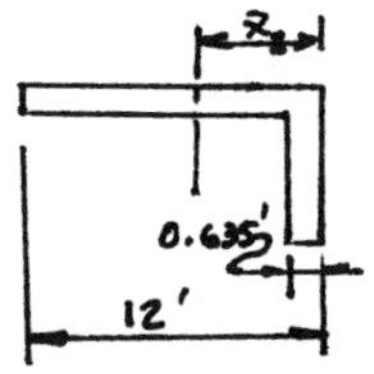

$A = 0.635(12 + 3.81) = 10.0\ ft^2$

$$\bar{x}_2 = \frac{0.635(12 \times 6 + 3.81 \times 0.318)}{10.0} = 4.64'$$

$I_2 = 0.635(12)^3 / 12 + 0.635 \times 12 \times (6 - 4.64)^2 + 3.81 \times 0.635^3 / 12 + 0.635 \times 3.81(4.64 - 0.318)^2 = 150.9\ ft^4$

$A_{v2} = 0.635(12) = 7.62\ ft^2$

Compute the deflection at the top of walls caused by a temperature drop of 80°F. Consider the base of walls fixed; no slip in the wall-to-eave strut connection. The system is one degree statically indeterminate. Analyze by consistent displacement. Release the connection between the strut and right wall:

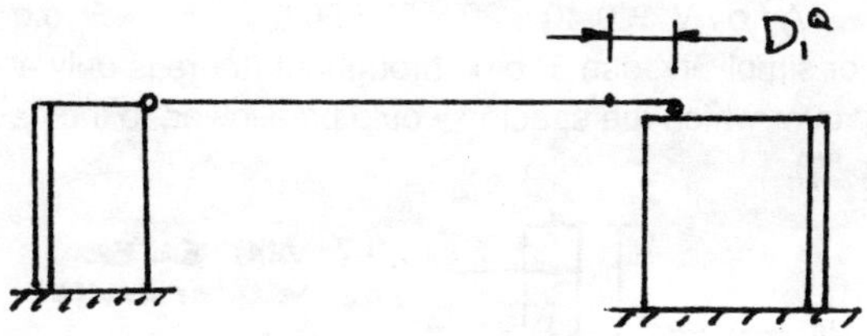

D_1^Q = movement between the release point and the end of eave strut when temperature drops 80°F

$D_1^Q = C_{TE}\,\Delta_T L = (6.5 \times 10^{-6}/_{°F})(-80°F)(80') = -0.0416'$

Apply equal and opposite unit forces at the release points:

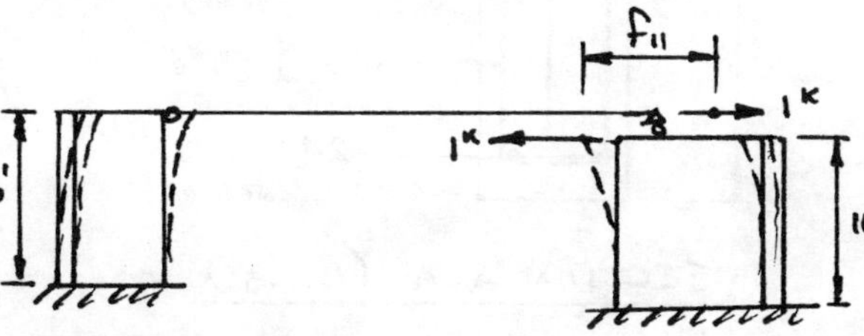

f_{11} = bending + shear deformations in walls + elongation of 80' strut under a unit load

$$f_{11} = \Sigma\left[\frac{(1)h^3}{3E_mI} + \frac{(1)h}{E_vA}\right] + \frac{(1)L}{E_sA_s}$$

In the above expression, the shear deflection is computed as $(1)h / E_vA$, rather than $1.2(1)h / E_vA$, which is used for rectangular walls. This is approximately correct for the flanged walls in which only the parallel wall element resists shear (Roark, 1975).

$$E_mI_1 = 1500 \text{ ksi} \times 144 \ \frac{\text{in}^2}{\text{ft}^2} \times 49.5 \text{ ft}^4 = 1.0692 \times 10^7 \text{ k-ft}^2$$

$$E_vA_1 = 600 \text{ ksi} \times 144 \ \frac{\text{in}^2}{\text{ft}^2} \times 5.1 \text{ ft}^2 = 4.4064 \times 10^5 \text{ k}$$

$$E_sA_s = 29 \times 10^3 \times 5.28 = 1.5312 \times 10^5 \text{ k}$$

Similarly,

$E_mI_2 = 3.2594 \times 10^7$ k-ft²;

$E_vA_2 = 6.5836 \times 10^5$ k

Thus,

$f_{11} = [16^3 / (3 \times 1.0692 \times 10^7) + (16 / (4.4064 \times 10^5)]$
$+ [16^3 / (3 \times 3.2594 \times 10^7) + 16 / (6.5836 \times 10^5)] + 80 / 1.5312 \times 10^5$

$f_{11} = 7.53 \times 10^{-4}$ ft/kip

For consistent displacements:

$T_sf_{11} - D_1^Q = 0$

∴ $T_s = D_1^Q / f_{11} = 0.0416 / 7.53 \times 10^{-4} = 55.2^k$

From the flexibility coefficients above:

$f_{wall\text{-}1} = [16^3 / (3 \times 1.0692 \times 10^7) + 16 / (4.4064 \times 10^5)]$
$= 0.000164$ ft/kip

Thus, the required displacements are:

$\Delta_{wall\text{-}1} = T_sf_{wall\text{-}1} = 0.00906' = 0.11"\rightarrow$

$f_{wall\text{-}2} = 0.0000662$ ft/kip

$\Delta_{wall\text{-}2} = 55.2 \times 0.000662 = 0.00365' = 0.044"\leftarrow$

Masonry-3

a) Design a typical grouted masonry shearwall for the given seismic forces. Neglect the gravity loads (per problem statement); f'_m = 1500 psi; (special inspection); Grade 60 reinforcement. For a typical wall:

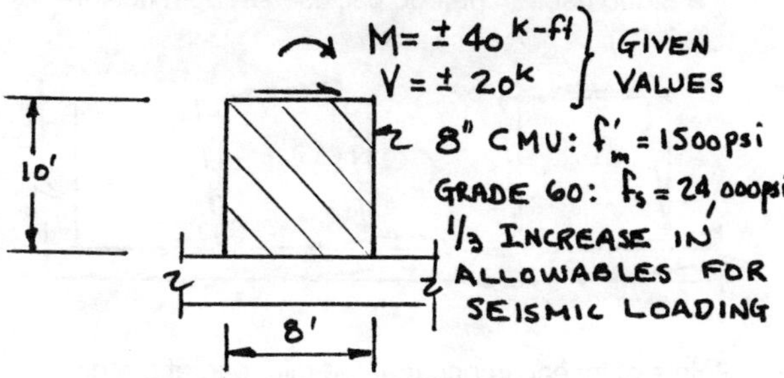

At the base:

$M_o = (40 + 20 \times 10) = \pm 240$ k-ft

Try $d \cong 0.9h = 0.9 \times 8 \times 12 = 86.4"$

Follow the Uniform Building Code (UBC-97). Check shear (eqn 7-37):

$f_v = V / bjd \cong 20{,}000 / (7.625 \times 7 / 8 \times 86.4)$
$= 35$ psi

$M / Vd = 240 \times 12 / (20 \times 86.4) = 1.67 > 1.0$

Check eqn. 7-20 with a 1/3 stress increase for seismic loading:

$F_v \le 4 / 3(1.0(f'_m)^{1/2}; 35 \text{ psi}) = 4 / 3((1500)^{1/2}; 35)$
$= 47 \text{ psi} > f_v$

∴ Shear reinforcement is not required.

$A_s \ge M / f_sjd = 240 \times 12 / (4 / 3 \times 24 \times 7 / 8 \times 86.4)$
$= 1.19$ in² (say 2 #7 at each end)

Check the trial design:

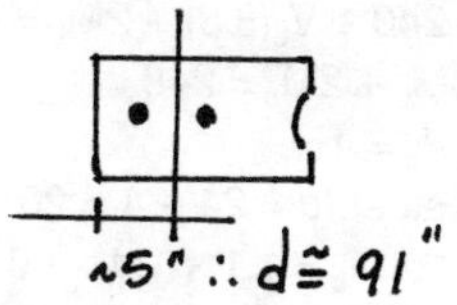

$\rho = A_s / bd = 1.20 / (7.625 \times 91) = 0.00173$

$$n = \frac{E_s}{E_m} = \frac{29 \times 10^6}{750f'_m} = 26$$

$\rho n = 0.045$

$k = \sqrt{(\rho n)^2 + 2\rho n} - \rho n = (0.045^2 + 0.045)^{1/2} - 0.045$
$= 0.26$

$j = 1 - k / 3 = 1 - 0.26 / 3 = 0.91$

$f_s = M / A_sjd = 240{,}000 \times 12 / (1.20 \times 0.91 \times 91)$
$= 28{,}900 \text{ psi} < 4 / 3(24{,}000)$ ∴ OK

$f_c = (M / bd^2)(2 / jk)$
$= [240{,}000 \times 12 / (7.625 \times 91^2)][2 / (0.91 \times 0.26)]$
$= 386 \text{ psi} < {}^4/_3(0.33 f'_m) = 667$ psi ∴ OK

Check the flexural bond stress:

$\Sigma_o = 2 \times 2.75 = 5.5"$

$u = V / \Sigma_ojd = 20{,}000 / (5.5 \times 0.91 \times 91)$
$= 44 \text{ psi} \ll U = {}^4/_3(160)$ ∴ OK

∴ **Use 2 #7 each side.**

Re: 2106.1.12.3, UBC-97, 1#4 is required at top and bottom and at a maximum spacing of 4 ft o.c. horizontally across the wall. The total wall steel is:

$A_{s,ver} + A_{s,hor} \geq 0.002A_g = 0.002 \times 7.625 \times 12$
$= 0.183\ in^2 / ft$

For the horizontal reinforcement, use #4 @ 32" o.c.

b) Design the typical interior portion of the 16×24 reinforced concrete beam supporting the wall. Given: $f'_c = 3{,}000$ psi; $f_y = 60{,}000$ psi; use strength design (UBC-97).

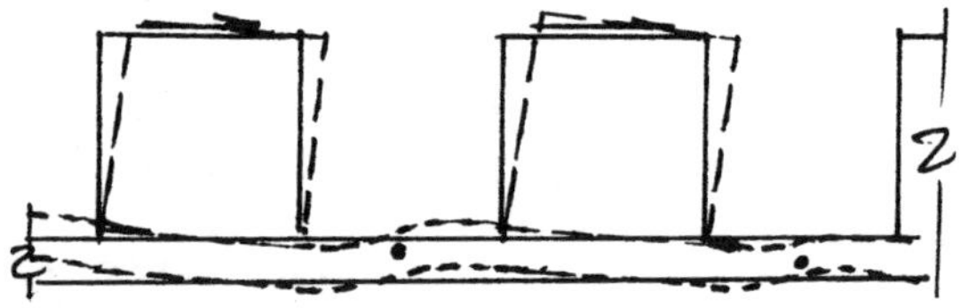

Points of inflection occur at the midspan of the girder:

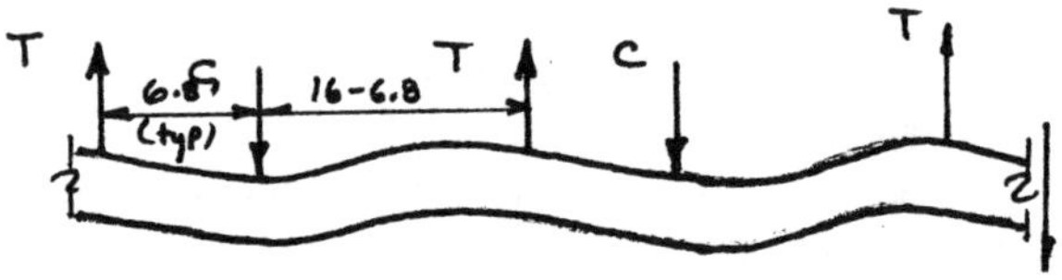

$T = C = {}^{M}/_{jd} = 240 / (0.9 \times 91 / 12) = 35.1^k$

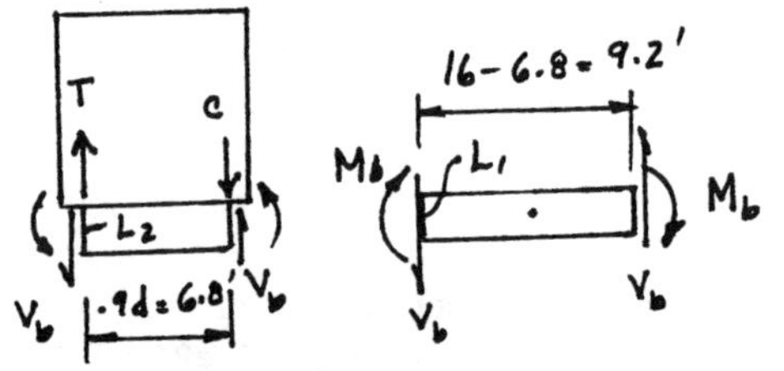

$\Sigma M_{L1} = 0$: $2M_b - 9.2V_b = 0$

$V_b = (2 / 9.2)M_b$

$\Sigma M_{L2} = 0$: $-240 + V_b(6.8) + 2M_b = 0$

$(2 / 9.2)6.8M_b + 2M_b = 240$

$M_b = 69^{k\text{-}ft}$; $V_b = 15^k$

Design flexural steel ($d = 24 - 4 = 20$"):

$M_u = 1.1 \times 1.4M_b = 1.1 \times 1.4 \times 69 = 106^{k\text{-}ft}$

$M_u \leq \phi M_n = \phi bd^2 \rho f_y(1 - 0.59\rho f_y / f'c)$

$106 \times 12 = 0.9 \times 16 \times 20 \times 60\rho(1 - 0.59\rho(60 / 3))$

$\rho = 0.0035 > \rho_{min}$

$\therefore\ A^-_s = A^+_s = 0.0035 \times 16 \times 20 = 1.12\ in^2$

Use 2 #7 continuous top and bottom.

Design for shear based on the probable flexural strength (1921.3.4, UBC-97):

$a_{pr} = 1.25f_yA_s / 0.85f'_cb$

$= 1.25 \times 60 \times 1.20 / (0.85 \times 3 \times 16) = 2.20"$

$M_{pr} = 1.25f_yA_s(d - a_{pr} / 2)$

$= 1.25 \times 60 \times 1.20(20 - 1.1) / 12 = 142^{k\text{-}ft}$

Add the gravity load shear of the grade beam:

$w_{bm} = 16 \times 24 \times 150 / 144 = 400$ plf.

$V_{ug} = w_{bm}L / 2 = 0.4 \times 9.2 / 2 = 1.8^k$

$V_e = V_{ug} + 2M_{pr} / L$

$1.8 + 2 \times 142 / 9.2 = 32.7^{\,k}$

Neglect V_c and take $\phi = 0.6$:

$V_s = V_e / \phi = 32.7 / 0.6 = 54.5^k$

Use #4 hoops ($A_v = 0.40\ in^2$):

$s \leq \{d / 4 = 5"; 8d_b = 8 \times {}^7/_8 = 7"; 24d_n = 24 \times {}^4/_8$
$= 12"$;

$A_vf_yd / V_s = 0.40 \times 20 \times 60 / 54.5 = 8.8"\} = 5"$ o.c.

For simplicity, use 5" o.c. throughout (there is only about 1' over which the spacing would be allowed to increase to ${}^d/_2$).

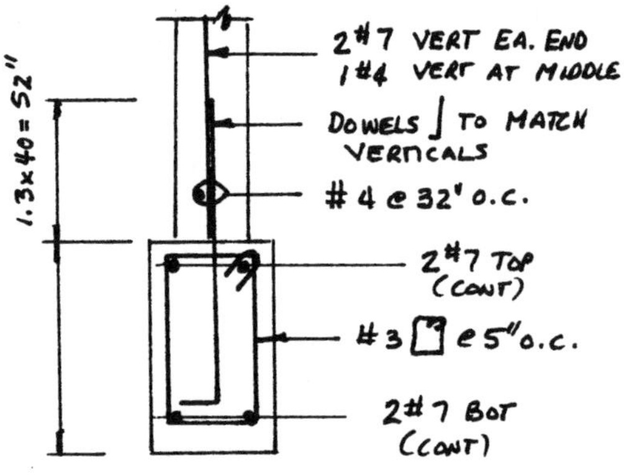

SECTION A-A (N.T.S.)

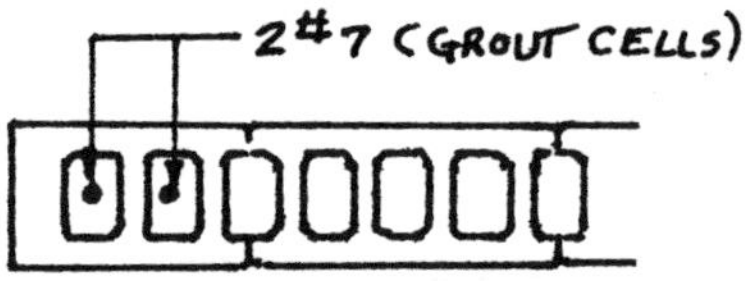

SECTION B-B (N.T.S)

Masonry-4

a) Design the masonry retaining wall and size the footing. Given: $\gamma_s = 90$ pcf; $K_a\gamma_s = 30$ pcf; $F_p = 2500$ psf; $f'_m = 1500$ psi; $f_y = 60{,}000$ psi.

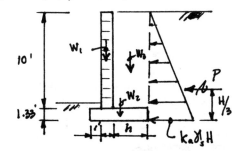

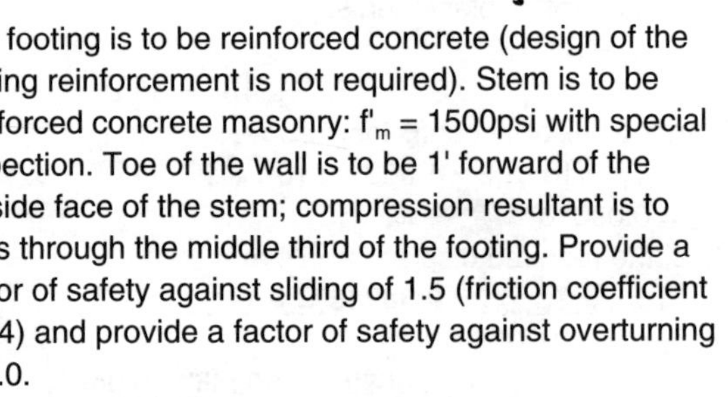

The footing is to be reinforced concrete (design of the footing reinforcement is not required). Stem is to be reinforced concrete masonry: $f'_m = 1500$psi with special inspection. Toe of the wall is to be 1' forward of the outside face of the stem; compression resultant is to pass through the middle third of the footing. Provide a factor of safety against sliding of 1.5 (friction coefficient = 0.4) and provide a factor of safety against overturning of 2.0.

Try a 12" wall (uniform thickness, but vary the vertical reinforcement). The unit weight of reinforced masonry $\cong$ 100 pcf. Design on a unit length basis:

$P = 0.5(K_a\gamma_s H)H = 0.5 \times 30 \times 11.33^2 = 1926^{\#}$

$M_o = PH / 3 = 1926(11.33 / 3) = 7274^{\#\text{-ft}}$

Neglect the passive resistance to sliding:

$\therefore$ $0.4W \geq 1.5P$

$W \geq 1.5 \times 1926 / 0.4 = 7200^{\#}$

$W_1 = 100 \times 10 = 1000^{\#}$

$w = 10 \times 90 + 1.33 \times 150 = 1100\text{plf}$

$W_1 + wh \geq 7200$

$h \geq (7200 - 1000) / 1100 = 5.6'$

Try 7'-6" total width of footing:

$M_r = 1000(1.5) + 1.33 \times 150(7.5)^2 / 2 + 10 \times 90(7.5 - 2)(7.5 - 5.5 / 2) = 30{,}620^{\#\text{-ft}}$

$\text{F.S.} = M_r / M_o = 30{,}620 / 7274 = 4.2$

$\bar{x} = (M_r - M_d) / W$

$W = 1000 + 1.33 \times 150 \times 7.5 + 900 \times 5.5 = 7450^{\#}$

$\bar{x} = (30{,}620 - 7274) / 7450 = 3.1'$

$B / 3 = 7.5 / 3 = 2.5'$

Thus, the resultant falls within the middle third of the footing. Check the bearing pressure:

$W = 1000 + 4950 + 1500 = 7450^{\#}$

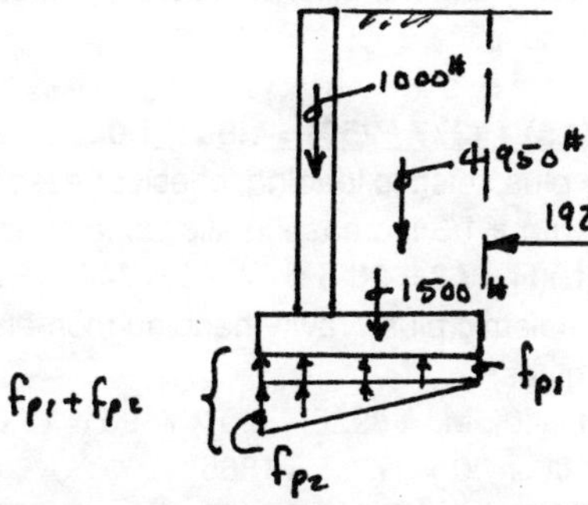

With the resultant within the middle third, the footing exerts compression throughout the width. Properties of the footing are:

$A = 7.5(1); S = (1)7.5^2 / 6 = 9.38\ \text{ft}^2$

Take moments about the midwidth (7.5'/ 2 from the toe):

$M = 1926(11.33 / 3) + 1000(3.75 - 1.5) - 4950(3.75 - 5.5 / 2) = 4573^{\#\text{-ft}}$

$f_{p1} = W / A - M / S = 7450 / 7.5 - 4573 / 9.38 = 506\ \text{psf} \ll F_p = 2500\ \text{psf}$

$f_p = W / A + M / S = 7450 / 7.5 + 4573 / 9.38 = 1480\ \text{psf} < F_p = 2500\ \text{psf}$

$\therefore$ **Use 7'-6" footing width.**

b) Design the wall reinforcement at the base of stem (H = 10'):

$V = 0.5(30 \times 10)10 = 1500^{\#}$

$M = 1500(10 / 3) = 5000^{\#\text{-ft}}$

For a 12" (nominal) block, allow 3" from the back face to the centroid of steel. Use grade 60 rebar (f_s = 24,000 psi); d = 9.63"

$A_s \geq M / f_s jd \cong 5000 \times 12 / (24{,}000 \times 7 / 8 \times 9.63) = 0.30\ \text{in}^2 / \text{ft}$ (say #5 @ 12" o.c.)

Check stresses based on the trial A_s (Re: UBC-97, chapter 21).

$\rho = A_s / bd = 0.31 / (12 \times 9.63) = 0.0027$

$n = E_s / E_m = 29 \times 10^6 / 750 f'_m = 26$

$\rho n = 0.0702$

$k = \sqrt{(\rho n)^2 + 2\rho n} - \rho n = 0.311;\ j = 1 - k / 3 = 0.9$

$f_s = M / jd A_s = 5000 \times 12 / (0.31 \times 0.9 \times 9.63) = 22{,}300\ \text{psi} < F_s = 24{,}000\ \text{psi}$

$f'_c = (M / bd^2)(2 / jk) = [5000 \times 12 \times / (12 \times 9.63^2)][2 / (0.9 \times 0.311] = 385\ \text{psi} < 0.33f'_m = 0.33 \times 1500 = 500\ \text{psi}$

$f'_v = V / bjd = 1500 / (12 \times 0.9 \times 9.63) = 14\ \text{psi}$

$f_v < F_v = (f'm)^{1/2} \leq 50 = (1500)^{1/2} = 39\ \text{psi}$

Check flexural bond with #5 @ 12:

$\Sigma_o = 1.96"$

$u = V / \Sigma_o jd = 1500 / (1.96 \times 0.9 \times 9.63) = 88\ \text{psi}$

$u = 88\ \text{psi} < U = 140\ \text{psi}$

Check minimum steel required:

$A_{s,min} = A_{s,vert} + A_{s,hor} \geq 0.002bt = 0.002 \times 12 \times 11.625 = 0.28\ \text{in}^2 / \text{ft}$

Computed $A_{s,vert} = 0.31\ \text{in}^2 / \text{ft}$ exceeds $A_{s,min}$; however, must have at $A_{s,min} / 3$ in the horizontal steel:

$\therefore$ $A_{s,hor} \geq 1 / 3(0.28) = 0.093\ \text{in}^2 / \text{ft}$

Use #4 @ 24"o.c. horizontal and #5 @ 12" o.c. vert.

Note: need #5 @ 12"o.c. full height to satisfy the minimum steel requirement.

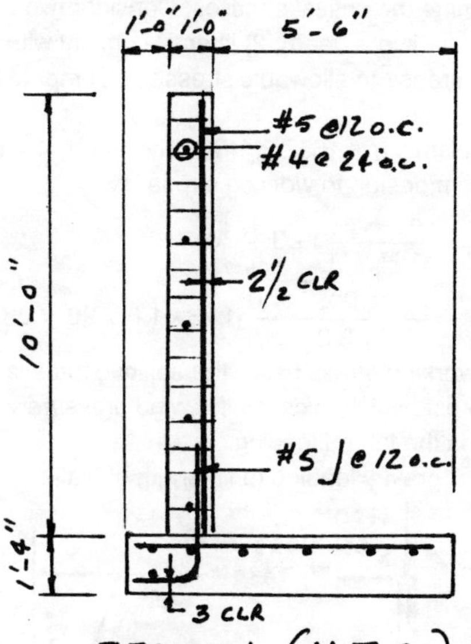

SECTION (N.T.S.)

Masonry-5

a) Compute stresses in the steel and masonry under the specified loading. Given: P = 600 $^{\#}/_{ft}$ (Dead) and 500 $^{\#}/_{ft}$ (Live); 8" (nominal) CMU reinforced with #4 @ 32" o.c. vertical, w = 90 psf; f'_m =1500 psi (no special inspection); f_y = 40,000 psi; wind = 15 psf; seismic zone 3; masonry design per chapter 21, UBC-97 by working stress design.

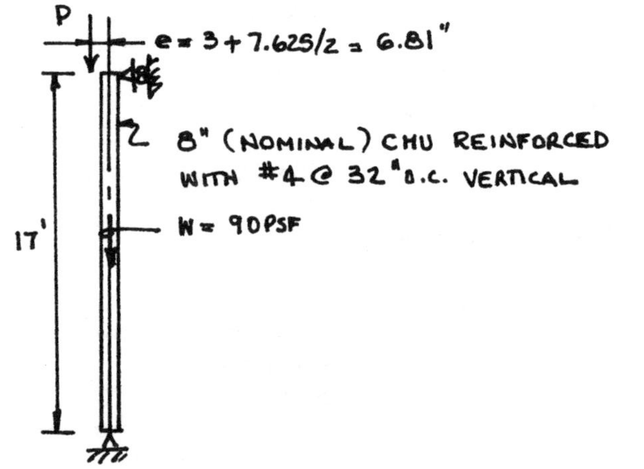

Determine the critical stresses. Consider two cases: 1) gravity (live + dead); 2) lateral + gravity with a one-third increase in allowable stresses. Compute the lateral forces:

- Seismic (re: 1632.2, UBC, dividing by 1.4 to convert from design to working stress level):

$$F_p = \frac{a_p C_a I_p}{R_p}\left(1+3\frac{h_x}{h_r}\right)W_p$$

$$= \frac{1.0 \times 0.24 \times 1.0}{3.0}(1+3\times 1.0)\times 90 = 28.8 \text{ psf}$$

On a working stress basis, the seismic force is 28.8/1.4 = 20.6 psf, which exceeds the wind pressure ∴ seismic governs the lateral loading.

- For gravity loading (unit length of wall):

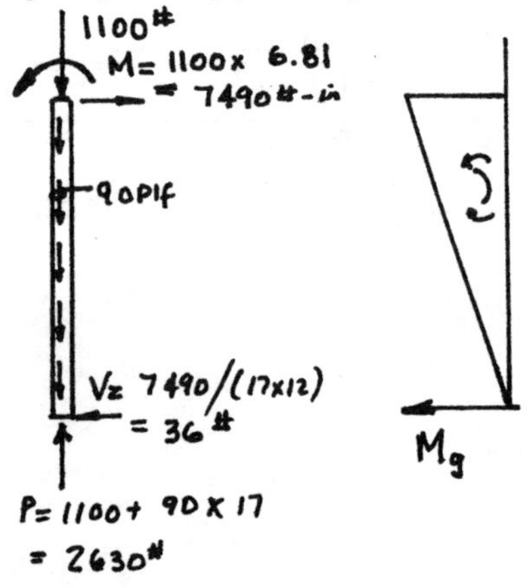

The wall bending moment varies linearly from a maximum at the top to zero at the bottom; axial force varies linearly from 1100$^{\#}$ at the top to a maximum of 2630$^{\#}$ at the base. Thus, we have to check the combined axial and bending stresses at several locations. Check, initially, the stresses at the top:

$d = 7.625 / 2 = 3.81$"; $b = 12$"; $A_s = (0.20 / 32)12 = 0.075$ in^2/ft; $n = E_s / E_m = 26$

$\rho = A_s / bd = 0.075 / (12 \times 3.81) = 0.00164$

$\rho n = 0.0427$

$k = \sqrt{(\rho n)^2 + 2\rho n} - \rho n = 0.252; \quad j = 1 - k/3 = 0.91$

Steel stress (F_s= 20,000 psi):

$f_s = M / A_s jd = 7490 / (0.075 \times 0.91 \times 3.81)$

$= 28{,}800 \text{ psi} > F_s$

Concrete compressive flexural stress:

$f_c = (M / bd^2)(2 / jk)$

$= [7490 / (12 \times 3.81^2)](2 / (0.91 \times 0.251) = 377$ psi

$F_c = \_(0.33f'_m) = 0.5(0.33 \times 1500) = 250$ psi

∴ $f_c > F_c$

Axial compression at the top:

$f_a = P / A_e = 1100 / (12 \times 7.63) = 12$ psi

Re: 2107.2.5, UBC, Table 21-H-1, for cells grouted, r = 2.53 in.

$h' / r = 17 \times 12 / 2.53 = 80.6 < 99$ ∴ use Eqn. 7-11

$F_a = 1/2 \times 0.25f'_m[1 - (h' / 140r)^2]$

$= 0.5 \times 0.25 \times 1500[1 - (17 \times 12 / (140 \times 2.53))^2]$

$= 125$ psi

Combined axial and bending at the top is critical for the gravity only case:

$f_a / F_a + f_c / F_c \le 1.0$

$(12 / 125) + (377 / 250) = 1.60 > 1.0$

For gravity plus seismic loading, check stresses at the midheight with a 1/3 increase in allowable stresses:

$M_e = (f_p)(H')^2 / 8 = 20.6 \times 17^2 / 8 = 744^{\#\text{-ft}} = 8928^{\#\text{-in}}$

Combined seismic plus gravity bending moment at midheight gives:

$M = M_e + 0.5M_g = 8928 + 0.5 \times 7490 = 12{,}675^{\#\text{-in}}$

$P = 1100 + 90 \times 17 / 2 = 1865^{\#}$

$f_s = 12{,}675 / (0.075 \times 0.91 \times 3.81) = 48{,}700$ psi

$f_s \gg 4 / 3(20{,}000 \text{ psi}) = 26{,}700$ psi

$f_c = [12{,}675 / (12 \times 3.81^2)][2 / 0.91 \times 0.252]$

$= 635$ psi

$4 / 3F_c = (4 / 3)250 = 333\text{psi} \ll f_c$

$f_a = 1865 / (12 \times 7625) = 20$ psi

$4 / 3F_a = (4 / 3)125 = 167$ psi

$f_a / (4 / 3\ F_a) + f_c / (4 / 3\ F_c)$

$= 20 / (4 / 3 \times 125) + 675 / (4 / 3 \times 125) = 2.03$

Thus, the critical stresses occur near the midheight of wall under gravity plus seismic loading. Stresses in concrete and steel are about twice the code-specified allowables.

b) There are several ways to bring the wall into compliance with the allowable stresses. One approach is to specify special inspection, which will result in the full allowable masonry stresses, and to increase the amount of vertical steel. Try this approach using a proportional increase in the area of vertical steel:

$A_s \approx (48{,}700 / 26{,}700)0.075 = 0.137$ in^2/ft

Try #5 @ 24"o.c.: $A_s = 0.155$ in^2/ft

$\rho n = 26 \times 0.155 / (12 \times 3.81) = 0.088$

$k = 0.340; j = 0.886$

$f_s = 12{,}675 / (0.155 \times 0.886 \times 3.81)$
$= 24{,}225$ psi $< 4/3(20{,}000)$ psi
$f_c = [12{,}675 / (12 \times 3.81^2)][2 / 0.886 \times 0.340)]$
$= 483$ psi
$(4/3)F_c = 4/3(500) = 666$ psi
$f_a = 1865 / (12 \times 7.625) = 20$ psi
$(4/3)F_a = 4/3(2 \times 125) = 333$ psi
$f_a / (4/3F_a) + f_c / (4/3F_c) = 20/333 + 483/666$
$= 0.79 < 1.0$

Check shear and bond stress:
$V = 36 + 20.6 \times 17 / 2 = 211^{\#}$
$f_v = V / bjd = 211 / (12 \times 0.886 \times 3.81) = 5$ psi – NIL
$\Sigma_o = 0.98^{in}/_{ft}$
$\mu = V / \Sigma_o jd = 211 / (0.98 \times 0.886 \times 3.81)$
$= 64$ psi $\ll U = 4/3(160$ psi$)$

Therefore, with f'$_m$ = 1500 psi and special inspection, the 8" wall with #5 @ 24" o.c. vertical is adequate.

Masonry-6

a) Calculate the reinforcement in the 4'-8" wide pier under the dead plus seismic loading. Assume that the given dead and live loads are applied concentrically. Check two load cases: 1) Dead plus Live, and 2) Dead plus Live plus Seismic. For seismic loading (UBC-97, use Eqn. 32-2 and divide by 1.4 to convert from design to working stress level). From Table 16-O, $a_p = 1.0$ on the main wall and 2.5 on the unbraced cantilever; $R_p = 3.0$ for both elements. For the main wall with the given 135 psf dead weight, $C_a = 0.32$; and $I_p = 1.0$:

$$F_p = \frac{a_p C_a I_p}{R_p}\left(1 + 3\frac{h_x}{h_r}\right)W_p$$

$$= \frac{1.0 \times 0.32 \times 1.0}{3.0}(1 + 3 \times 1.0) \times 135 = 57.6\,\text{psf}$$

On a working stress basis, the seismic force is 57.6 / 1.4 = 41.1 psf, which exceeds the lower limit imposed by Eqn. 32-3. The force on the cantilevered paraphet is 2.5 times this level and, for vibration in the fundamental mode, will act in the opposite sense to the lateral force on the wall. Note that the tributary wall width is 4.67 ft in the lower 16 ft and is 4.67 + 16 = 20.67 ft wide in the upper region and paraphet. Thus, the lateral loading is:

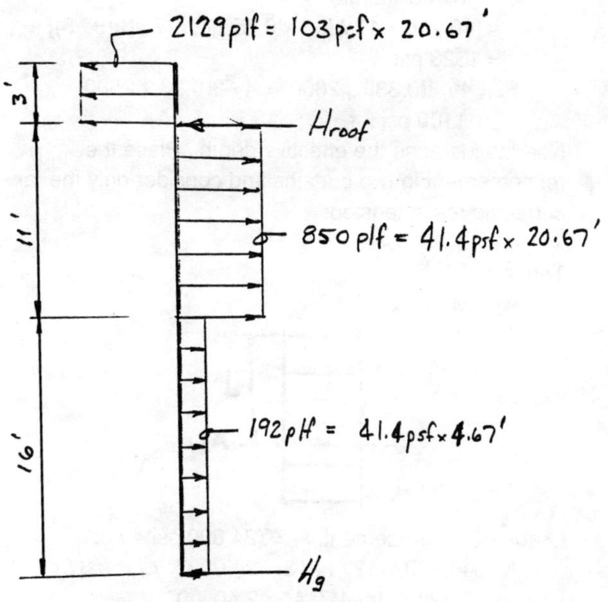

$\Sigma M_{roof} = 0$: $H_g(27) - 192(16)(11 + 8) - 850(11)(5.5) - 2129(3)(1.5) = 0$
$H_g = 4421^{\#}$

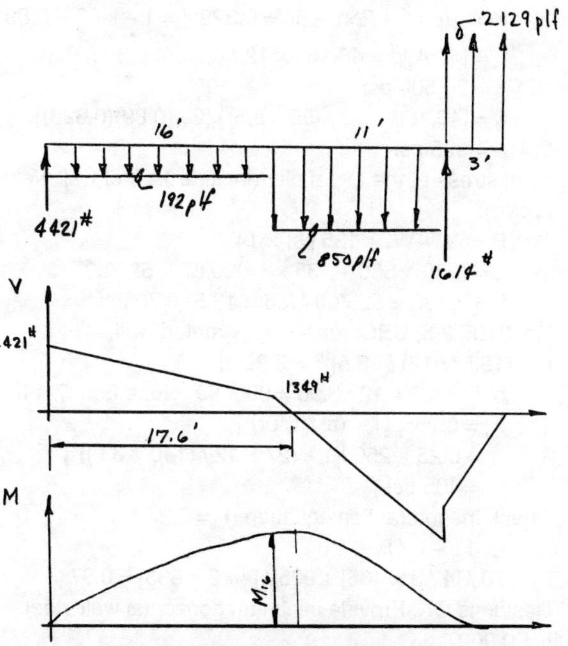

The maximum bending moment occurs 17.6 ft above the ground; however, the stresses are critical at the top of pier where the moment is practically equal to M_{max} and only 4.67' = 56" width is available to resist stresses:

$M_{16'} = 0.5(4421 + 1349)16 = 46{,}160^{\#\text{-ft}} = M$

Design per chapter 21, UBC-97 using working stress theory with $f'_m = 2500$psi; $f_s = 24^{ksi}$. Approximate the compressive stress assuming the reinforcement is centered in the wall:

$d = 13.5 / 2 = 6.75"$; $j = 7/8$

$f_c = (M / bd^2)(2 / jk)$

$\cong [46{,}160 \times 12 / (56 \times 6.75^2)][2 / (7 / 8)(3 / 8)]$

$= 1323$ psi

$F_c = 4 / 3(0.33f'_m; 2000) = 4 / 3(0.33 \times 2500)$

$= 1100 \text{ psi} > f_c$

∴ Need to increase the effective depth. Place the reinforcement in two curtains and consider only the far curtain to resist tension:

$d = 13.5 - 3 - 2 = 8.5"$

Trial A_s:

$A_s = M / F_s jd$

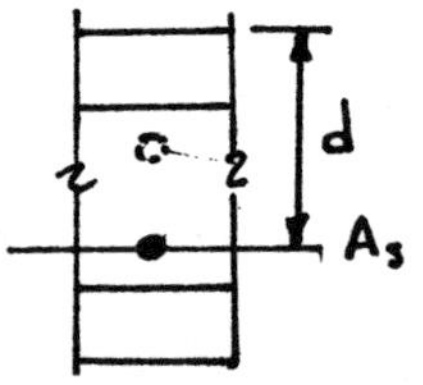

Grade 60 reinforcement: F_s = (24,000 psi)4 / 3:

$A_s \approx 46{,}160 \times 12 / (4 / 3 \times 24{,}000 \times 7 / 8 \times 8.5)$

$= 2.32 \text{ in}^2$ (try 4#7: $A_s = 2.40 \text{ in}^2$)

$\rho = A_s / bd = 2.40 / (56 \times 8.5) = 0.00504$

$n = E_s / E_m = 29 \times 10^6 / (750 \times 2500) = 16$

$\rho n = 0.0807$

$k = \sqrt{(\rho n)^2 + 2\rho n} - \rho n = 0.329; j = 1 - k / 3 = 0.890$

$f_s = M / A_s jd = 46{,}160 \times 12 / (2.40 \times 0.89 \times 8.5)$

$= 30{,}500$ psi

$f_c = [46{,}160 \times 12 / (56 \times 8.5^2)][2 / (0.89)(0.329)]$

$= 935$ psi

Axial stress at the top of pier (include 14 ft of wall weight above):

$P = W_l + W_d + 135 \text{ psf} \times 14'$

$= (300 + 360 + 135 \times 14)20.67 = 52{,}700^{\#}$

$f_a = P / A_e = 52{,}700 / (56 \times 13.5) = 70$ psi

Re: 2107.2.5, UBC, for 13-1/2" grouted wall,

$r = [(13.5^3 / 12) / 13.5]^{1/2} = 3.90$ in.

$h' / r = 27 \times 12 / 3.90 = 83 < 99$ ∴ use Eqn. 7-11

$F_a = 0.25f'_m[1 - (h' / 140r)^2]$

$= 0.25 \times 2500[1 - (27 \times 12 / (140 \times 3.9))^2]$

$= 405$ psi

Check the interaction equation ($F_b = 0.33 f'_m$)

$f_a / F_a + f_b / F_b \le 1.0$

$70 / [4 / 3 \times 405] + 935 / [4 / 3 \times 833] = 0.97$

∴ Design is OK. Provide minimum horizontal wall steel:

$\rho > 0.0007$

$A_{s,hor} \ge 0.0007 \times 12 \times 13.5 = 0.11 \text{ in}^2/\text{ft}$

Provide #3 ties @ 8"o.c. (more than adequate for horizontal steel); use 4 #7 vertical each face.

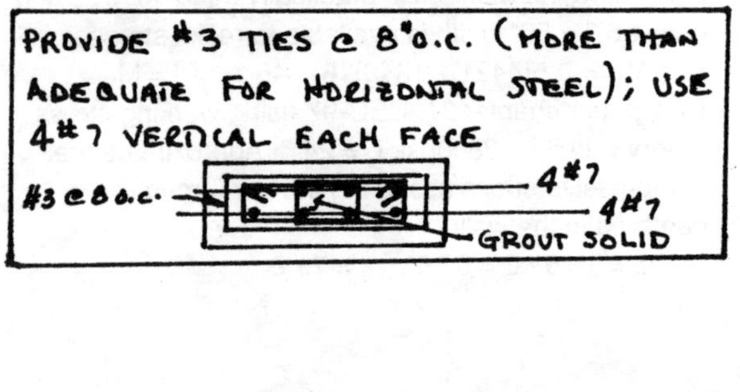

Masonry-7

a) Calculate the allowable vertical load on the connection. This connection is not covered explicitly in a design specification. The force transfer will be investigated using an allowable stress approach with criteria for reinforced masonry (chapter 21, UBC-97), structural steel (AISC 1989) and reinforced concrete (ACI 318-95). Assume the force, P, is due to gravity only (the wall anchorage requirement of 1633.2.8, UBC-97 is to be furnished by a separate connection). Assume ASTM A36 steel; E70xx electrodes.

a) Compute capacity of steel components:

Side plates (1/4" thick) weld to the back plate (1/2" thick):

$$q \le \begin{cases} 2 \times 4 \times 0.928 = 7.4 \ \text{k/in} \\ 0.4F_y t_p = 0.4 \times 36 \times 0.25 = 3.6 \ \text{k/in} \end{cases}$$

Base material governs. Each side plate resists shear of P / 2 and moment = 3P / 2. Weld properties are:

$L_w = 21"; S_w = L_w^2 / 6 = 73.5$

$f_v = P / 2L_w = P / (2 \times 21) = 0.024P$

$f_b = M / S_w = 3P / (2 \times 73.5) = 0.020P$

$f_r = (f_v^2 + f_b^2)^{1/2} = P(0.024^2 + 0.020^2)^{1/2}$

$= 0.0312P \le q = 3.6$

∴ $P \le 3.6 / 0.0312 = 115$ kips

Shear transfer from the back plate into the wall. The direct bearing of the plate and the L1 $^1/_2$×1 $^1/_2$× $^1/_4$ on the grout will be disregarded due to the low bearing strength of the grout near an edge and the difficulty of placing grout reliably under the angle. Shear transfer will be assumed to occur by shear friction (11.7, ACI 318) with 4-1"$^\phi$ anchors of A-36 steel ($F_y = 36^{ksi}$) assumed to develop fully. Use a factor of safety of 2 to convert the nominal shear-friction strength to the working load value:

∴ $P \le V_n / 2 = (A_{vf} f_y \mu); \mu = 0.7$

$P \le 4 \times 0.79 \times 36 \times 0.7 / 2 = 40$ kips

Moment transfer to the wall from the back plate:

$M = P(3 + 0.5) = 3.5$ P

$E_m = 750 f'_m = 750 \times 1500 = 1.125 \times 10^6$ psi

$n = E_s / E_m = 29 / 1.125 = 26$

Transform the top 2-1" diameter anchors:

$nA_s = 26 \times 2 \times 0.79 = 41.1 \text{ in}^2$

Sum moments of area about the lower bars: 41.1(12) > 8 × 7 × 3.5 ∴ the lower bars are in the compression zone. Let y = distance from the bottom of plate to the centroid of transformed section:

$nA_s(19 - y) = (n - 1)A_s(y - 7) + 8y(y / 2)$

$41.1(19 - y) = 39.5(y - 7) + 4y^2$

∴ $y = 9.03"$

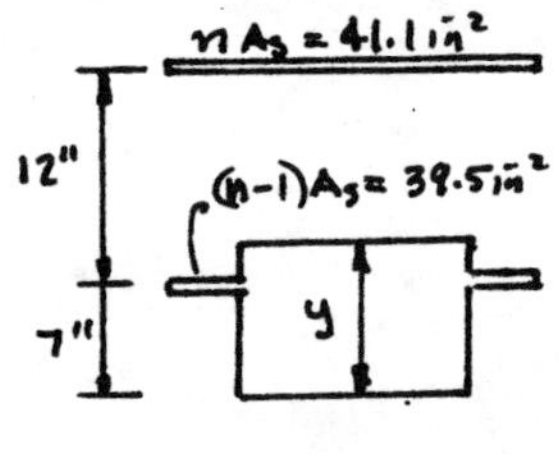

Moment of inertia of the transformed section:

$I = nA_s(19 - y)^2 + (n - 1)A_s(y - 7)^2 + by^3 / 3$
$= 41.1(19 - 9.03)^2 + 39.5(9.03 - 7)^2 + 8 \times 9.03^3 / 3$
$= 6212 \text{ in}^4$

Limit the stress in steel to $0.6F_y = 22$ ksi:

$f_s = n(M(19 - y) / I) \leq 22$ ksi
$26(3.5P(19 - 9.03) / 6212 \leq 22$
$P \leq 150$ kips

Limit stress in grout to $0.33\ f'_m = 500$ psi:

$f_c = M_y / I \leq 0.5$ ksi
$3.5P \times 9.03 / 6212 \leq 0.5$
$P \leq 98$ kips

Thus, the connection strength is limited by shear-friction to 40 kips.

b) Check the wall strength near the connection. The wall terminates above the connection at elevation 12 ft. The 4 #6 grade 40 rebars are located at the middle of the 8" masonry wall:

$d = 7.6 / 2 = 3.8$ in
$e = 3 + 3.8 = 6.8$ in

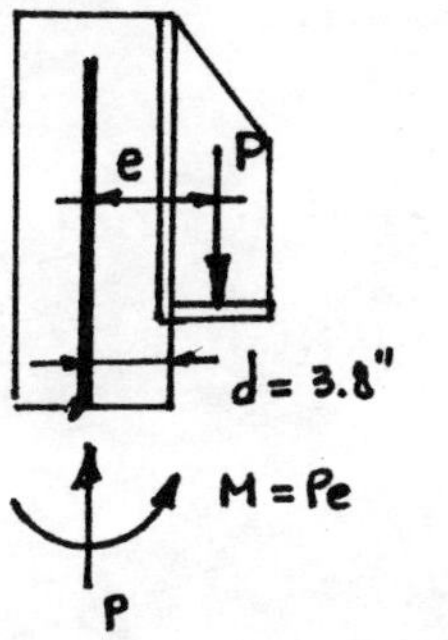

Consider the moment transfer to be analogous to the column-to-slab transfer at the edge of a flat plate (re: Chapter 13, ACI 318). A portion of the transfer will be through eccentricity of shear (Eqn. 11-41, ACI 318-95):

$\gamma_v = (1 - \gamma_f) = 1 - [1 + (2 / 3)\sqrt{b_1 / b_2}\ ]^{-1}$

The critical perimeter for shear transfer is shown below:

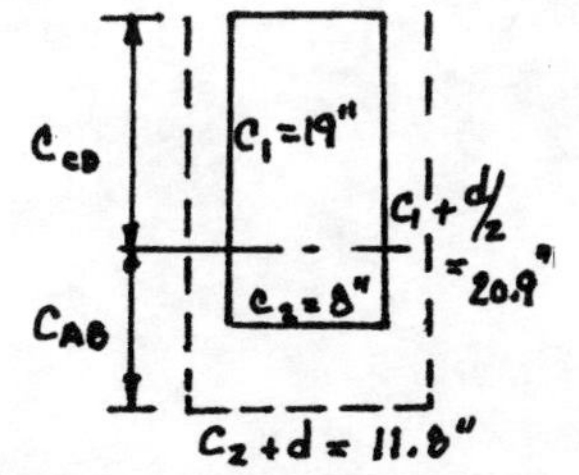

Equations for the section properties can be found in several texts (e.g. Rice and Hoffman 1975). For the critical section:

$c_{AB} = 8.2$ in; $c_{CD} = 12.7$ in
$J_c = 9792 \text{ in}^4$
$\gamma_v = 1 - [1 + 2 / 3\sqrt{20.9 / 11.8}\ ]^{-1} = 0.47$
$f_v = \gamma_v Mc_{CD} / J_c = 0.47 \times 6.8P \times 12.7 / 9792$
$= 0.0041P$

The allowable shear stress for reinforced masonry is given by:

$F_v = 1.0(f'_m)^{1/2} \leq 50$ psi
$= (1500)^{1/2} = 39$ psi

$\therefore\ 0.0041P \leq 39$ psi
$\therefore\ P \leq 9500$ lbs = 9.5 kips

For bending in the absence of axial load:

$b = b_e = 6t = 6 \times 7.6 = 45.6$ in
$\rho = A_s / bd = 4 \times 0.44 / (45.6 \times 3.8) = 0.010$
$\rho n = 0.26$; $k = \sqrt{\rho n^2 + 2\rho n} - \rho n = 0.51$;
$j = 1 - k / 3 = 0.83$
$F_s = 0.5F_y = 20$ ksi
$f_s = M / A_s jd \leq F_s = 20$ ksi
$6.8P \leq 20 \times 4 \times 0.44 \times 0.83 \times 3.8$
$P \leq 16.3$ kips

Check combined axial load plus bending. For a concentrated load, the effective width of wall is (re: 2106.2.7, UBC-97):

b_e = bearing width + 4t = $8 + 4 \times 7.6 = 38.4$ in
$f_a = P / A_e = P / [7.6 \times 38.4] = 0.0034P$

Re: Table 21-H-1, r = 2.19 in for a solid grouted 8-in wall. For $h' / r = 123 / 2.19 < 99$, Eqn. 7-11 gives:

$F_a = 0.25f'_m[1 - (h' / 140r)^2]$
$= 0.25 \times 1500[1 - (123 / (140 \times 2.19)^2] = 315$ psi
$f_b = (M / bd^2)(2 / jk)$
$= [6.8P / (6 \times 7.6 \times 3.8^2)][2 / (0.51 \times 0.83)]$
$= 0.049P$
$F_b = 0.33f'_m = 0.33 \times 1500 = 500$ psi

Combined stresses are computed by Eqn. 7-16, UBC:

$f_a / F_a + f_b / F_b = 0.0034P / 315 + 0.049P / 500 \leq 1.0$
$\therefore\ P \leq 9360$ lbs = 9.4 kips

Thus, the vertical load capacity is governed by the combined axial load and bending stresses in the reinforced wall: P ≤ 9.4 kips.

Masonry-8

a) Design the wall reinforcement for the given loading. Given: $f'_m = 1400$ psi; grade 60 reinforcement; no special inspection; end connections are adequate. Design according to Chapter 21, UBC-97 by working stress design theory.

$$\text{Span} \leq \begin{cases} \text{clearheight} + 2t = 9 + 2 \times 7.63 / 12 = 10.3\text{ ft} \\ \text{center-to-center of supports} \\ \quad \cong 9.5\text{ ft} \leftarrow \text{governs} \end{cases}$$

Allowable stresses must be reduced by 50% because special inspection is not specified, but a one-third increase can be taken for the wind loading. For flexural compression:

$F_c = (4 / 3)(1 / 2)(0.33f'_m)$
$= (4 / 3)(1 / 2)(0.33 \times 1400) = 308$ psi

Allowable shear stress:

$F_v = (4 / 3)(1 / 2)(f'_m)^{1/2} = (4 / 3)(1 / 2)(1400)^{1/2}$
$= 25$ psi

Allowable tensile stress in the reinforcement:

$F_s = (4 / 3)(1 / 2)40{,}000 = 26{,}600$ psi

Allowable flexural bond stress:

$U = (4 / 3)(100) = 133$ psi

Design the flexural reinforcement neglecting the beneficial effect of the dead load. Work on a per unit width of wall basis and compute a trial area of reinforcement assuming that steel governs.

$M = wH^2 / 8 = 50 \times 9.5^2 / 8 = 564^{k\text{-}ft}$

$d = t / 2 = 7.63 / 2 = 3.8''$; $j \cong 7 / 8$

$\therefore$ $A_s = M / f_s jd \cong 564 \times 12 / (26{,}600 \times 7 / 8 \times 3.8)$
$= 0.077\ in^2/ft$

Try #4 @ 24" o.c., $A_s = 0.10\ in^2/ft$

$E_m = 750f'_m = 750 \times 1400 = 1.05 \times 10^6$ psi

$n = E_s / E_m = 29 / 1.05 = 28$

Re: 2106.3.8, UBC-97, $b \le [6t = 6 \times 7.63 = 46''; 24'']$ $= 24''$.

$\therefore$ $\rho n = A_s / bd = 0.20 / (24 \times 3.8) \times 28 = 0.061$

$k = \sqrt{\rho n^2 + 2\rho n} - \rho n = \sqrt{0.061^2 + 2 \times 0.061} - 0.061$
$= 0.294$

$j = 1 - k / 3 = 1 - 0.294 / 3 = 0.902$

For the design width, b = 12":

$f_c = (M / bd^2)(2 / jk)$
$= (564 \times 12 / (12 \times 3.8^2))[2 / (0.902 \times 0.294)]$
$= 295$ psi

$f_c \cong F_c \therefore$ OK

$f_s = M / (A_s jd) = 564 \times 12 / (0.10 \times 0.902 \times 3.8)$
$= 19{,}700$ psi $\le F_s \therefore$ OK

Check shear:

$V = 50 \times 9.5 / 2 = 238^{\#}/ft$

$f_v = V / bjd = 238 / (12 \times 0.902 \times 3.8) = 6$ psi $< F_v$

$\therefore$ OK

Flexural bond:

$\Sigma_o = (1 / 2) \times \pi \times 0.5 = 0.79$ in

$u = V / \Sigma_o jd = 238 / (0.79 \times 0.902 \times 3.8) = 88$ psi $< U$

$\therefore$ OK

b) Check the combined axial plus bending stresses (re: 2107.2.7, UBC). Check at the midheight where the moment is maximum and include the wall weight at 85 psf:

$P = 1000 + 85 \times 9.5 / 2 = 1400\ ^{\#}/ft$

$A_e = [2(24 - 8.75) \times 1.5 + 7.73 \times 8.75] / 2 = 57\ in^2/ft$

$f_a = P / A_e = 1400 / 57 = 25$ psi

Re: 2107.2.5, UBC, Table 21-H-1, for cells grouted 24 in o.c., r = 2.53 in.

$h' / r = 9 \times 12 / 2.53 = 43 < 99 \therefore$ use Eqn. 7-11

$F_a = 0.25f'_m[1 - (h' / 140r)^2]$
$= 0.25 \times 1400[1 - (9 \times 12 / (140 \times 2.53))^2]$
$= 317$ psi

Considering wind loading and the reduction for no special inspection:

$F_a = (4 / 3)(1 / 2)317 = 211$ psi

Check the interaction equation:

$f_a / F_a + f_b / F_b = (25 / 211) + (295 / 308) = 1.08 > 1.0$

$\therefore$ No Good

Revise the reinforcement to #4 @ 16 in o.c., which gives $A_s = 0.15\ in^2/ft$; $\rho n = 0.092$; $k = 0.346$; $j = 0.884$. Thus,

$f_c = (M / bd^2)(2 / jk)$
$= (564 \times 12 / (12 \times 3.8^2))[2 / (0.884 \times 0.346)]$
$= 255$ psi

$A_e = 7.63 \times 12 = 91.6\ in^2/ft$ (grout solid)

$f_a = P / A_e = 1400 / 91.6 = 15$ psi

$h' / r = 9 \times 12 / 2.19 = 49 < 99$

$F_a = (4 / 3)(1 / 2) \times 0.25 \times 1400[1 - (9 \times 12 / (140 \times 2.19))^2] = 204$ psi

$f_a / F_a + f_b / F_b = (15 / 204) + (255 / 308) = 0.91$

$\therefore$ OK

Check the minimum wall steel requirements, $\rho \ge 0.002$:

$A_{st} \ge 0.002 \times 12 \times 7.63 = 0.18\ in^2/ft$

$A_{st,\,hor} \ge 0.0007 \times 12 \times 7.63 = 0.064\ in^2/ft$

Thus, $A_{st} = 0.20 / (16 / 12) + 0.064 = 0.21 > A_{st}$

$\therefore$ **Use #4 @ 16" o.c. with solid grout.**

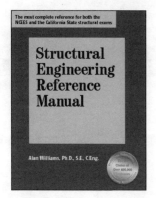

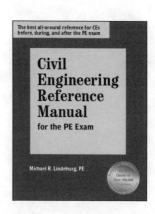

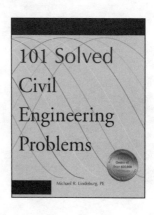

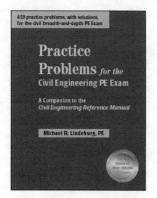